Biotechnology

Second Edition

Volume 7

Products of Secondary Metabolism

Biotechnology

Second Edition

Fundamentals

Volume 1
Biological Fundamentals

Volume 2
Genetic Fundamentals and
Genetic Engineering

Volume 3
Bioprocessing

Volume 4
Measuring, Modelling, and Control

Products

Volume 5
Recombinant Proteins,
Monoclonal Antibodies, and
Therapeutic Genes

Volume 6
Products of Primary Metabolism

Volume 7
Products of Secondary Metabolism

Volume 8
Biotransformations

Special Topics

Volume 9
Enzymes, Biomass, Food and Feed

Volume 10
Special Processes

Volumes 11a and b
Environmental Processes

Volume 12
Legal, Economic and
Ethical Dimensions

A Multi-Volume Comprehensive Treatise

Biotechnology

Second, Completely Revised Edition

Edited by
H.-J. Rehm and G. Reed
in cooperation with
A. Pühler and P. Stadler

Volume 7

Products of Secondary Metabolism

Edited by
H. Kleinkauf and H. von Döhren

Series Editors:
Prof. Dr. H.-J. Rehm
Institut für Mikrobiologie
Universität Münster
Corrensstraße 3
D-48149 Münster
FRG

Dr. G. Reed
1914 N. Prospect Ave. #61
Milwaukee, WI 53202-1401
USA

Prof. Dr. A. Pühler
Biologie VI (Genetik)
Universität Bielefeld
P.O. Box 100131
D-33501 Bielefeld
FRG

Prof. Dr. P. J. W. Stadler
Bayer AG
Verfahrensentwicklung Biochemie
Leitung
Friedrich-Ebert-Straße 217
D-42096 Wuppertal
FRG

Volume Editors:
Prof. Dr. H. Kleinkauf
Dr. H. von Döhren
Institut für Biochemie
Technische Universität
Franklin-Straße 29
A-10587 Berlin
Germany

Executive Editor: Dr. Hans-Joachim Kraus
Editorial Director: Karin Dembowsky
Production Manager: Hans-Jochen Schmitt

Library of Congress Card No.: applied for

British Library Cataloguing-in-Publication Data:
A catalogue record for this book is available from the British Library

Die Deutsche Bibliothek – CIP-Einheitsaufnahme
Biotechnology : a multi volume comprehensive treatise / ed. by H.-J. Rehm and G. Reed. In cooperation with A. Pühler and P. Stadler. – 2., completely rev. ed. – VCH.
ISBN 3-527-28310-2 (Weinheim …)

NE: Rehm, Hans J. [Hrsg.]

Vol. 7. Products of secondary metabolism / ed. by H. Kleinkauf and H. von Döhren – 1997
ISBN 3-527-28317-X

Printed on acid-free and chlorine-free paper.

Composition and Printing: Zechnersche Buchdruckerei, D-67330 Speyer.
Bookbinding: J. Schäffer, D-67269 Grünstadt.
Printed in the Federal Republic of Germany

Preface

In recognition of the enormous advances in biotechnology in recent years, we are pleased to present this Second Edition of "Biotechnology" relatively soon after the introduction of the First Edition of this multi-volume comprehensive treatise. Since this series was extremely well accepted by the scientific community, we have maintained the overall goal of creating a number of volumes, each devoted to a certain topic, which provide scientists in academia, industry, and public institutions with a well-balanced and comprehensive overview of this growing field. We have fully revised the Second Edition and expanded it from ten to twelve volumes in order to take all recent developments into account.

These twelve volumes are organized into three sections. The first four volumes consider the fundamentals of biotechnology from biological, biochemical, molecular biological, and chemical engineering perspectives. The next four volumes are devoted to products of industrial relevance. Special attention is given here to products derived from genetically engineered microorganisms and mammalian cells. The last four volumes are dedicated to the description of special topics.

The new "Biotechnology" is a reference work, a comprehensive description of the state-of-the-art, and a guide to the original literature. It is specifically directed to microbiologists, biochemists, molecular biologists, bioengineers, chemical engineers, and food and pharmaceutical chemists working in industry, at universities or at public institutions.

A carefully selected and distinguished Scientific Advisory Board stands behind the series. Its members come from key institutions representing scientific input from about twenty countries.

The volume editors and the authors of the individual chapters have been chosen for their recognized expertise and their contributions to the various fields of biotechnology. Their willingness to impart this knowledge to their colleagues forms the basis of "Biotechnology" and is gratefully acknowledged. Moreover, this work could not have been brought to fruition without the foresight and the constant and diligent support of the publisher. We are grateful to VCH for publishing "Biotechnology" with their customary excellence. Special thanks are due to Dr. Hans-Joachim Kraus and Karin Dembowsky, without whose constant efforts the series could not be published. Finally, the editors wish to thank the members of the Scientific Advisory Board for their encouragement, their helpful suggestions, and their constructive criticism.

H.-J. Rehm
G. Reed
A. Pühler
P. Stadler

Scientific Advisory Board

Contents

Contributors

Prof. Dr. Timm Anke
Lehrbereich Biotechnologie
Universität Kaiserslautern
Postfach 3049
D-67618 Kaiserslautern
Germany
Chapter 11

Dr. Jochen Berlin
Gesellschaft für Biotechnologische Forschung
Mascheroder Weg 1
D-38124 Braunschweig
Germany
Chapter 13

Dr. Mervin Bibb
John Innes Centre
Norwich Research Park
Colney Lane
Colney, Norwich NR4 7UH
UK
Chapter 2

Dr. Bruno Cavalleri
MMDRI – Lepetit Research Center
Via R. Lepetit, 34
I-21040 Gerenzano (Varese)
Italy
Chapter 9

Prof. Keith Chater
John Innes Centre
Norwich Research Park
Colney Lane
Colney, Norwich NR4 7UH
UK
Chapter 2

Jürgen Distler
Bergische Universität GH
Mikrobiologie – FB 19
Gauss-Straße 20
D-42097 Wuppertal
Germany
Chapter 10

Dr. Hans von Döhren
Institut für Biochemie
Technische Universität
Franklin-Str. 29
D–10587 Berlin
Germany
Chapters 1, 7

Dr. Klausjürgen Dornberger
Hans-Knöll-Institut für Naturstoff-Forschung
Bereich Naturstoffchemie
Beutenbergstraße 11
D–07745 Jena
Germany
Chapter 14

Dr. Hartmut Drechsel
Mikrobiologie und Biotechnologie
Universität Tübingen
Auf der Morgenstelle 1
D–72076 Tübingen
Germany
Chapter 5

Dr. Gerhard Erkel
Lehrbereich Biotechnologie
Universität Kaiserslautern
Postfach 3049
D–67618 Kaiserslautern
Germany
Chapter 11

Dr. Hans Fliri
Rhône Poulonc Rorer S.A.
Centre de Recherche de Vitry-Alfortville
13, quai Jules Guesde
F–94403 Vitry-sur-Seine Cedex
France
Chapter 12

Dr. Friedrich Götz
Lehrstuhl für Mikrobielle Genetik
Universität Tübingen
Auf der Morgenstelle 18
D–72076 Tübingen
Germany
Chapter 8

Prof. Dr. Udo Gräfe
Hans-Knöll-Institut für Naturstoff-Forschung
Bereich Naturstoffchemie
Beutenbergstraße 11
D–07745 Jena
Germany
Chapters 1, 14

Dr. Ralph Jack
Universität Tübingen
Institut für Organische Chemie
Auf der Morgenstelle 18
D-72076 Tübingen
Germany
Chapter 8

Prof. Dr. Günter Jung
Universität Tübingen
Institut für Organische Chemie
Auf der Morgenstelle 18
D–72076 Tübingen
Germany
Chapter 8

Dr. Jörg Kallen
Sandoz Pharma Ltd.
Preclinical Research
CH–4002 Basel
Switzerland
Chapter 12

Prof. Dr. Horst Kleinkauf
Institut für Biochemie
Technische Universität
Franklin-Str. 29
D–10587 Berlin
Germany
Chapter 7

Dr. Giancarlo Lancini
MMDRI - Lepetit Research Center
Via R. Lepetit, 34
I–21040 Gerenzano (Varese)
Italy
Chapter 9

Dr. Henk van Liempt
DRL, BT-FDG
Südstraße 125
D–53175 Bonn
Germany
Chapter 6

Dr. Vincent Mikol
Sandoz Pharma Ltd.
Preclinical Research
CH–4002 Basel
Switzerland
Chapter 12

Prof. Dr. Satoshi Ōmura
School of Pharmaceutical Sciences
Kitasato University
The Kitasato Institute
9-1, Shirokane 5-chome
Minato-ku, Tokyo 108
Japan
Chapter 3

Prof. Dr. Wolfgang Piepersberg
Bergische Universität GH
Mikrobiologie - FB 19
Gauss-Straße 20
D–42097 Wuppertal
Germany
Chapter 10

Dr. Valérie F.J. Quesniaux
Sandoz Pharma Ltd.
Preclinical Research
CH–4002 Basel
Switzerland
Chapter 12

Prof. Colin Ratledge
The University of Hull
Department of Applied Biology
Hull HU6 7RX
UK
Chapter 4

Prof. Dr. habil. Hans-Peter Saluz
Hans-Knöll-Institut für Naturstoff-Forschung
Bereich Naturstoffchemie
Beutenbergstraße 11
D–07745 Jena
Germany
Chapter 14

Dr. Elisabeth Schneider-Scherzer
Biochemie GmbH
A-6330 Kufstein-Schaftenau
Austria
Chapter 12

Dr. Kurt Schörgendorfer
Biochemie GmbH
A-6330 Kufstein-Schaftenau
Austria
Chapter 12

Dr. Torsten Schwecke
Institute of Biochemistry
University of Cambridge
Tennis Court Road
Cambridge, CB2 1QW
UK
Chapter 6

Dr. Paul L. Skatrud
Infectious Diseases Research
Eli Lilly and Company
Lilly Corporate Center
Indianapolis, IN 46285
USA
Chapter 6

Dr. Haruo Tanaka
School of Pharmaceutical Sciences
Kitasato University
The Kitasato Institute
Minato-ku, Tokyo 108
Japan
Chapter 3

Dr. Matthew B. Tobin
Infectious Diseases Research
Eli Lilly and Company
Lilly Corporate Center
Indianapolis, IN 46285
USA
Chapter 6

Dr. Malcolm D. Walkinshaw
Sandoz Pharma Ltd.
Preclinical Research
CH-4002 Basel
Switzerland
Chapter 12

Dr. Gerhard Weber
Biochemie GmbH
A–6330 Kufstein-Schaftenau
Austria
Chapter 12

Prof. Dr. Günther Winkelmann
Mikrobiologie und Biotechnologie
Universität Tübingen
Auf der Morgenstelle 1
D–72076 Tübingen
Germany
Chapter 5

Introduction

This volumes provides an overview of secondary metabolites illustrating most aspects of their discovery, formation, exploitation, and production. Compared to the first edition the focus when has clearly shifted towards the molecular genetic background of the producing organisms. These efforts serve not only our understanding of the production processes to permit improvements by genetic manipulations, but also promote our appreciation of the environmental significance of secondary metabolites.

The term "secondary metabolite" has been discussed widely, and a shift in perception took place in the last years. From a playground of nature leading to mostly disparable products ideas focus now on special purpose products promoting evolutionary advantages. This shift is connected to the impressive elucidation of the genetics of multistep synthetic processes of secondary metabolite formation. Genes encoding biosynthetic reaction sequences have been found clustered together with resistance or export genes and are under the control of specific signals. Biosynthetic functions or unit operations reside on modules, and these modules in their functional protein state interact to assure the fidelity of the multistep processes. The genetic burden for many of these processes seems remarkable, and genes assembled from modules often display sizes of 10 to more than 45 kilobases. Since some of the now established microbial genomes are devoid of such multistep pathways, their unique placement in other genomes indicates important functions for their producers.

Still largely unconnected to the background of their producers secondary metabolites generally are high-value compounds established mainly in pharmacology, veterinary medicine, agriculture, and biochemical and medical research. The introductory chapter points to product fields and to the genetic investigation of biosynthetic unit operations. Regulatory mechanisms are then considered in the most advanced fields of the prokaryotes. As the central field of present drug discovery approaches target-based screenings are discussed. Compound groups considered are lipids siderophores, aminoglycosides, and peptides (β-lactams, dalbaheptides, cyclosporins, lantibiotics). Producer groups presented are basidiomycetes and plant cells. As a target group antitumor drugs are evaluated.

An updated chapter on macrolides as secondary metabolites including reprogramming strategies will be included in Volume 10 of the Second Edition of *Biotechnology* (see also Volume 4 of the First Edition).

Further chapters to be consulted are especially on biopolymers and surfactants (Volume 6), on the overproduction of metabolites and the treatment of producer organisms like bacilli, streptomycetes and filamentous fungi (Volume 1) as well as on reactor modeling (Volume 3). We thank our colleagues for their valuable contributions, the publisher for their patience and cooperativity, and the series editors for many helpful suggestions.

Berlin, March 1997

Hans von Döhren
Horst Kleinkauf

1 General Aspects of Secondary Metabolism

Hans von Döhren

Berlin, Germany

Udo Gräfe

Jena, Germany

1 Introduction: The Importance of Secondary Metabolites as Drugs

Today, bioactive secondary metabolites of microorganisms and of plants, and their synthetic derivatives as well, are among the most frequently used therapeutics in human and veterinary medicine (Scrip, 1993). The invention of antibiotic therapy contributed greatly to the successful control of most of the epidemic infectious diseases and even promoted their disappearance. Moreover, it contributed to the general increase in the lifespan of man, not only in industrialized countries. New applications for bioactive biotechnical products in medical care like their use as immunosuppressants or antiatherosclerotics, and as animal growth promoters and pesticides in agriculture rendered research on new secondary metabolites an apparently endless story (SANGLIER and LARPENT, 1989; Comité Editorial, 1992; LANCINI and LORENZETTI, 1993; VINING and STUTTARD, 1995).

In the past, natural products supplied 5–8% of the annual increase in the world's total pharmaceutical market. The list of the 25 worldwide best-selling drugs for application in humans in 1992 includes a series of drugs of microbial origin which are used either in their native structures or as chemical derivatives (see, e.g., Mevacor, Cefaclor or other cephalosporins, Augmentin, Sandimmun) (Scrip, 1993).

Many plant products, from digitalis glycosides and neuroactive alkaloids to the pyrethrines, serve as therapeutics for human diseases and as agricultural agents (Comité Editorial, 1992). Sometimes, the experiences made in folk medicine initiated the discovery of new plant-derived antitumor drugs, antineuralgic, antihypertonic, antidepressant, insecticidal, nematicidal, and other bioactive compounds.

Antiinfective chemotherapy once was the classical domain of biotechnical drug production due to the discovery of β-lactam antibiotics, such as penicillins, cephalosporins, clavulanates, and carbapenems. Even today, the increase in resistant nosocomial and opportunistic pathogens (particularly dangerous to immunosuppressed AIDS and tumor patients) requires both improvement of known drugs and search for new drugs (GRÄFE, 1992; LANCINI and LORENZETTI, 1993; HUTCHINSON, 1994).

Microbial products such as doxorubicin, bleomycin, and mitomycin C are indispensable as cancerostatics (FOX, 1991). The same is true for plant metabolites such as the vinca alkaloids, taxol, and their chemical derivatives which exert excellent antitumor activity by interaction with the cellular mitotic system (NOBLE, 1990; FOX, 1991; HEINSTEIN and CHANG, 1994; POTIER et al., 1994).

However, even the non-therapeutic fields of application, such as in animal husbandry and plant protection, contributed to a high degree to the continuing interest in secondary metabolite production. Last but not least, natural products of biotechnical and agricultural origin play an important role as "biochemical tools" in molecular biology and in the investigation of cellular functions.

More than 10000 antibiotics and similar bioactive secondary metabolites have been isolated so far from microbes, and a comparably higher number of drugs was derived from plants and even from animals (see, e.g., marine tunicates, molluscs, toxic insects, snakes, and toads) (BERDY et al., 1980; LAATSCH, 1994). Approximately 500 new representatives of low-molecular weight compounds are published every year.

In addition to this huge and still growing number of bioactive molecules, more than 100000 derivatives as representatives of some few basic structures (e.g., β-lactams, macrolides, aminoglycosides, tetracyclines, anthracyclines) were obtained by means of synthetic derivatizations (LAATSCH, 1994). Irrespective of this plethora of drug molecules a little more than a hundred basic structures gained practical importance.

We owe much progress in the detection of new drug structures to modern physicochemical approaches such as mass-spectrometry, high-field nuclear magnetic resonance spectroscopy and X-ray diffractometry. Compilations of the numerous structural data (BERDY et al., 1980; BYCROFT, 1988;

LAATSCH, 1994) provide indispensable assistance in the identification of new drug molecules. Thus, the enormous number of already known metabolites from microbes and plants increased the detection and isolation of already known structures dramatically.

A compilation of about 200 recently described products illustrates the current trends in screening efforts (Tab. 1). These have been published during the last two years. It is evident from these data that highly selective screens prevail and yet the majority of compounds originate from the classical Actinomycete pool. Rare bacteria and fungi, marine microorganisms and plants now have a significant share. It is obvious that well-known organisms again contribute with newly isolated substances to new, e.g., receptor targeted screens. Strategies of such screens are discussed in this volume in Chapter 3 by TANAKA and ŌMURA.

The development of new drugs from natural sources is common practice of the pharmaceutical industry. 6000 to 10000 chemicals have to be tested in a given assay system to obtain one single compound suitable as a therapeutical agent (ŌMURA, 1992; KROHN et al., 1993). No wonder that research and development for a new approved drug may cost up to one billion US$. In most cases, a new natural "leading structure" is intensively modified by chemical means to improve its activity and to reduce side effects. Chemistry is also extremely helpful if rather rare natural products occurring in low amounts or in organisms from sensitive ecological areas have been proposed as drugs. For example, 40000 yew trees, i.e., the whole population of Northern America, would be required to produce 25 kg of taxol, a new promising cancerostatic drug, and even this amount would not be sufficient to treat every cancer patient. Fortunately, taxol derivatives of similar activity (taxotere) can be obtained by chemical derivatization of taxoid metabolites which are obtainable in large quantities from the dried leaves of European yews (HEINSTEIN and CHANG, 1994). Alternatively, cell cultures (ELLIS et al., 1996) or endophytic fungi such as *Pestalotiopsis microspora* (STIERLE et al., 1994, 1995; STROBEL et al., 1996) of *Taxus* species could be exploited for production.

From the recently completed chemical synthesis of taxol it is evident that, as in bicyclic β-lactams, classical approaches cannot compete with natural producers. Instead, increasing attention is given to the recruitment of biocatalysts for certain key reactions in metabolite production. In addition, directed biosynthesis in microbial cultures (THIERICKE and ROHR, 1993), production of plant products in cell cultures (BERLIN, Chapter 14, this volume), and cell free *in vitro* systems of enzymatic synthesis and peptide and protein producing translation systems are considered as complementary methods in structure–function studies (ALAKHOV and VON DÖHREN, unpublished data).

Only 30% of the total developmental efforts have been spent to the search for a new drug. However, for the estimation of its efficacy and evaluation of safety often more than 50% are needed. Taking into account a quota of approximately 1:15000 for a hit structure, the challenges of modern pharmaceutical development become visible. In general, natural products seem to offer greater chances than synthetically derived agents. Hence, a great research potential is still dedicated to the discovery of new natural drugs and their biotechnical production. Classical strategies of drug development are being more and more supplemented by new biomedical approaches and ideas and by the use of genetically engineered microbes and cells as screening organisms (TOMODA and ŌMURA, 1990; ELDER et al., 1993). These tools initiated a "renaissance" in the search for new leading structures. New sources of bioactive material, such as marine organisms, and new microbes from ecological "niches" promoted the recent advances in the discovery of drugs (WILLIAMS and VICKERS 1986; RINEHART and SHIELD 1988; MONAGHAN and TKACZ, 1990; JACOB and ZASLOFF, 1994; JENSEN and FENICAL, 1994) (Tab. 1).

Present research activities were also stimulated by the discovery of block busters (Scrip, 1993) such as cyclosporin A (KAHAN, 1987), avermectins (CAMPBELL, 1989), acarbose (MÜLLER, 1989), and monacolin (ENDO, 1979) in microbial cultures. A series of very promising new screening drugs (zaragozic acid, squalestatins) (HASUMI, 1993), erbstatin

Tab. 1. Selected Natural Products Detected by Screening Efforts Published in 1995/96

Compound Reference	Producing Organisms	Structural Type[1]	Selected Properties	Research Group Involved
Antimicrobial Drugs:				
Griseusin derivatives	Actinomycete (unidentified)	PK	antibacterial	Institute of Microbial Chemistry
BE-24566B	*Streptomyces violaceus-niger*	PK	antibacterial	Banyu Pharm. Co.
Amicenomycin	*Streptomyces* sp.	PK-GLYC	antibacterial	Institute of Microbial Chemistry
Kalimantacins	*Alcaligenes* sp.	PK, mod.	antibacterial, MDR strains	Yamanouchi Pharm. Co. and PT Kalbe Pharma
A21459	*Actinoplanes* sp.	PEP	antibacterial	Lepetit
Epoxyquinomycins	*Amycolatopsis*	acyl AA	antibacterial	Institute of Microbial Chemistry
GE 37468	*Streptomyces* sp.	PEP	antibacterial	Lepetit
Phencomycin	*Streptomyces* sp.	PK	antibacterial	Hoechst
Chrysoapermin	*Apiocrea chrysosperma*	PEP	antibacterial antifungal	Hans Knöll Institute and Univ. Tübingen
Bacillaene	*Bacillus subtilis*	PK	antibacterial	Bristol Myers Squibb
GE2270	*Planobispora rosea*	PEP	antibacterial	Lepetit
AL072	*Streptomyces* sp.	PK, mod.	antilegionella	Cheil Foods & Chem. Inc and NIH Korea
Ripostatin	*Sorangium cellulosum*	PK	antibacterial	GBF
Sorangiolid	*Sorangium cellulosum*	PK	antibacterial	GBF
Thiomarinol	*Alteromonas rava* (marine)		antibacterial	Sankyo
Echinoserine	*Streptomyces tendae*	PEP+PK	antibacterial	Univ. Tübingen and Hans Knöll Institute
O7F275	unidentified fungus	PK	antibacterial	Lederle
Pyralomycins	*Actinomadura spiralis*	PK, mod.	antibacterial	Institute of Microbial Chemistry
RS-22	*Streptomyces violaceusniger*	PK	antibacterial	RIKEN
Ochracenomycins	*Amycolatopsis* sp.	PK	antibacterial	Institute of Microbial Chemistry
Azicemycins	*Amycolatopsis sulphurea*	PK	antibacterial	Institute of Microbial Chemistry
Amythiamycin	*Amycolaotopsis* sp.	PEP	antibacterial	Institute of Microbial Chemistry
APHE 3/4	*Streptoverticillium griseocarnum*	ALK	antibacterial	Univ. Alcala
Aurantimycin	*Streptomyces aurantiacus*	PEP	antibacterial, cytotoxic	Hans Knöll Institute
Cineromycins	*Streptomyces griseoviridus*		antibacterial	Univ. Tübingen, Univ. Göttingen, Hans Knöll Institute
Papyracon	*Lachnum payraceum*	TERP	antibacterial	Univ. Lund and Univ. Kaiserslautern
Cephem derivatives	*Penicillium chrysogenum*	PEP, mod.	antibacterial	Panlabs
Sorrentanone	*Penicillium chrysogenum*	PK	antibacterial	Bristol Myers Squibb

Tab. 1. (Continued)

Compound Reference	Producing Organisms	Structural Type[1]	Selected Properties	Research Group Involved
Benzastatin	*Streptomyces nitrosporeus*	ALK	antifungal, antiviral free radical scavenger	KRIBB
Jerangolides	*Sorangium cellulosum*	PK	antifungal	GBF
BE29602	*Fusarium* sp.	PK-GLYC	antifungal	Banyu Pharm. Co.
Dibefurin	unidentified mushroom	PK	antifungal	Abbott
Darlucins	*Sphaerellopsis filu*	PK-mod.	antibacterial, antifungal	Univ. Kaiserslautern and Univ. Munich
Fusaricidin	*Bacillus polymyxa*	PEP-PK	antifungal, antibacterial	Wakunaga Pharm. Co. and PT Kalbe Pharma
Helioferin	*Mycogone rosea*	PEP-PK	antifungal, antibacterial	Hans Knöll Institute
Azalomycin	Actinomycete	PK	antifungal	Hoechst, AgrEvo
Liposidolide	*Streptomyces* sp.	PK	antifungal	RIKEN
Chirosazol	*Sorangium cellulosum*	PK	antifungal cytotoxic	GBF
Ratjadon	*Sorangium cellulosum*	PK	antifungal	GBF
Chrysospermin	*Apiocrea chrysosperma*	PEP	antifungal	Hans Knöll Institute
Hydroxystrobilurin	*Pterula* sp.	PK	antifungal	Univ. Kaiserslauern and Univ. Munich
Fusacandin	*Fusarium sambucinum*	PK-GLYC	antifungal	Abbott
Favolon	*Favolaschia*	TERP	antifungal	Univ. Kaiserslautern and Univ. Munich
Aureobasidins	*Aureobasidium pullulans*	PEP	antifungal	Takara Shuzo Co.
Phosmidosine	*Streptomyces* sp.	NUC	antifungal	RIKEN and SynPhar Lab. Inc.
NP-101A	*Streptomyces aurantiogriseus*	A-mod.	antifungal	Hokkaido Univ.
YM-47522	*Bacillus* sp.	PK	antifungal	Yamanouchi Pharm. Co
Australifungin	*Sporomiella australis*	PK	antifungal	Merck Sharp & Dohme
AKD-2	*Streptomyces* sp.	PK	antibacterial antifungal	Univ. Osaka City
UK-2A/B/C/D	*Streptomyces* sp.	PEP	antifungal	Osaka Univ. and Suntory Ltd.
Prodimicin	*Actinomadura spinosa*	PK	antifungal	Meijo U, Toyama Pref. Univ.
Pradimicin	*Actinomadura spinosa*	PK	antifungal	Toyama Pref. Univ. and Bristol Myers Squibb
Ascosteroside	*Ascotricha amphitricha*	TERP-GLYC	antifungal	Bristol Myers Squibb
Epothilone	*Mucor hiemalis*	PK-AA	antifungal, cytotoxic	GBF
Fusarielin	*Fusarium* sp.	PK	antifungal	Univ. Tokyo
Saricandin	*Fusarium* sp.	PK	antifungal	Abbott
Furanocandin	*Tricothecium* sp.	PK-GLYC	antifungal	Meiji Seika Kaishi Ltd and Mitsubishi Chem. Corp.
Siamycin	*Streptomyces* sp.	PEP	antiviral, HIV	Bristol Myers Squibb
L-671, 776, derivatives	*Stachybotrys* sp.	PK-AA	HIV protease inhib., endothelin antag.	Ciba-Geigy

Tab. 1. (Continued)

Compound Reference	Producing Organisms	Structural Type[1]	Selected Properties	Research Group Involved
Benzastatins	*Streptomyces nitrosporeus*	ALK	free radical scavenger anti-fungal, antiviral	KRIBB
Triterpene-sulfates	*Fusarium compactum*	TERP	rhinovirus protease inhib.	Abbott
Quinoxa-peptides	*Betula papyrifera*	PEP-PK	antiviral: HIV1,2,RT	Merck Sharp & Dohme
Karalicin	*Pseudomonas fluorescens*	PK	antiviral: HSV	Univ. Cagliari and Univ. Cattolin (Rome)
AH-758	*Streptomyces* sp.	PK	antiviral: HSV	Kumamoto Univ.
Eulicin	*Streptomyces* sp.	PEP-mod.	antiviral: HIV1	Jikei Univ., Institute of Microbial Chemistry
Sattabacin	*Bacillus* sp.	A-mod.	antiviral: HSV	Univ. Cagliari and Univ. Rome
Sattazolin	*Bacillus* sp.	AA-mod.	antiviral: HSV	Univ. Cagliari and Univ. Rome
GE20372	*Streptomyces* sp.	PEP	antiviral: HIV	Lepetit
Isochromo-philones	*Penicillium* sp.	PK	antiviral: HIV DGAT, ACTAT inhib.	Kitasato
Antitumor Drugs				
Sch52900/1	*Gliocladium* sp.	PEP	antitumor	Schering-Plough
Rakicidins	*Micromonospora* sp.	PEP-PK	cytotoxic	Bristol Myers Squibb
Esperamicin	*Actinomadura verrucosospora*	PK-GLYC	antitumor	Bristol Myers Squibb
Ossamycin	*Streptomyces hygroscopicus*	PK	cytotoxic	Lilly
Acetophthalidin	*Penicillium* sp. (marine)	PK	cell cycle inhibitor	RIKEN
Tryprostatins	*Aspergillus fumigatus*	PEP-TERP	cell cycle inhib.	RIKEN
Sparoxomycin	*Streptomyces sparsogenus*	NUC-mod.	proliferation mod.	Toyama Pref. Univ.
Cochleamycins	*Streptomyces* sp.	PK	antitumor	Kirin Brewery Co.
Himastatin	*Streptomyces hygroscopicus*	PEP	antitumor	Bristol Myers Squibb
Chondramide	*Chondromyces crocatus*	PEP	cytotoxic	GBF
Anguinomycin	*Streptomyces* sp.	PK	cytotoxic	Univ. Tokyo
Clovalicin	*Sporothrix* sp.	PK	cytocidal	Kitasato
Clecarmycin	*Streptomyces* sp.	PK	antitumor	Kyowa Hakko Kogyo
Piericidin derivatives	*Streptomyces* sp.	PK-GLYC	antitumor	Snow Brand Milk Co. and Kamagawa Univ.
Hydroxymyco-trienine	*Bacillus* sp.	PK	cancerostatic	Institute of Microbial Chemistry and Showa College
FR901537	*Bacillus* sp.	PEP-PK	aromatase inhib.	Fujisawa
Medelamine	*Streptomyces* sp.	PK, mod.	anticancer	Nippon Kayaku
Naphthablin	*Streptomyces* sp.	PK	oncogen function inhib.	Keio Univ. and Institute of Microbial Chemistry
Macquarimicin	*Micromonospora* sp.	PK	cytotoxic	Abbott

Tab. 1. (Continued)

Compound Reference	Producing Organisms	Structural Type[1]	Selected Properties	Research Group Involved
Thiazinotrienomycin	*Streptomyces* sp.	PK-PEP	cytostatic (cancer)	Institute of Microbial Chemistry and Showa College
Cremeduycin	*Streptomyces cremeus*	A, mod.	cytotoxic	Univ. Illinois
Tryprostatin	*Aspergillus fumigatus*	PEP, mod.	cell cycle inhib.	RIKEN
Sch50673,6	*Nattrassia mangiferae*	PK	antitumor	Schering-Plough
Terpentecin	*Streptomyces* sp.	PK	antitumor: topoisomerase inhib.	Kyowa Hakko Kogyo
FD-211	*Myceliophthora lutea*	PK	cytotoxic: MDR	Taisho Pharm. Co.
Cytogenin	*Streptoverticillium eurocidium*	PK	antitumor	Institute of Chemotherapy (Shizuoka) and Institue of Microbial Chemistry
Enaminedonin	*Streptomyces* sp.	PEP-PK	detransforming tumor cells	RIKEN
Dihydroepiepoformin	*Penicillium patulum*	PK-mod.	IL-1 receptor antag.	Upjohn
EI-1507-1/2	*Streptomyces* sp.	PK	IL-1-converting enz. inhib.	Kyowa Hakko Kogyo
TAN-1511	*Streptosporangium amethystogenes*	PEP-PK	induces cytokines	Takeda
CJ-12,371,2	unidentified fungus	PK	DNA gyrase inh.	Pfizer
Pharmacological Activities				
FR901,483	*Cladybotryum* sp.	ALK-P-ester	immunosuppr.	Fujisawa Pharm. Co.
PA-48,153	*Streptomyces prunicolor*	PK	immunosuppr.	Shionogi
27-O-demethyl-Rapamycin	*Steptomyces hygroscopicus*	PK-AA	immunosuppr.	Smith Kline Beecham
NFAT 68,133	*Streptomyces* sp.	PK	immnuosuppr.	Abbott
Stevastatin	*Penicillium* sp.	PEP-PK	immunosuppr.	Nippon Kayaku
Trichstatin	*Streptomyces* sp.	PK	immunosuppr., histidine decarboxylase inhib.	Kyowa Hakko Kogyo
Cytosporin	*Cytospora* sp.	PK	angiotensin bdg. inhib.	Merck Sharp & Dohme
Leustroducsin	*Streptomyces platensis*	PK-P-ester	thrombocytosis inhib.	Sankyo Co.
Plactins	Agonomycetales	PEP	stimulates fibrinolytic activity	Tokyo Noko Univ.
TAN1323C/D	*Streptomyces purpurescens*	PK	angiogenesis inhib.	Takeda
Monamidocin	*Streptomyces* sp.	PEP	fibrinogen rec. antag.	Nippon Roche
A-72363	*Streptomyces nobilis*	GLYC	heparanase inhib.	Sankyo Co.
Trachyspic acid	*Talaromyces trachyspermus*	PK	heparanase inhib.	Sankyo and Univ. Tokyo
Carbazoquinocins	*Streptomyces violaceus*	AA-PK	antioxidant	Univ. Tokyo

Tab. 1. (Continued)

Compound Reference	Producing Organisms	Structural Type[1]	Selected Properties	Research Group Involved
Phenopyrazin	*Penicillium* sp.	PK-AA	radical scavenger	Kitasato
Balmoralmycin	*Streptomyces* sp.	PK	protein kinase inhib.	Ciba-Geigy
Staurosporine analogs	*Streptomyces longisporoflavus*	ALK	protein kinase C inhib.	Ciba Geigy
Paeciloquinones	*Paecilomyces carneus*	PK	protein tyrosine kinase inhib.	Ciba Geigy and Panlabs
Factor A/C	unidentified fungus	PK, mod.	myoinositol Pase inhib.	Lepetit
MS-444	*Micromonospora* sp.	PK	myosin light chain kinase inhib.	Kyowa Hakko Kogyo
WS79089B	*Streptosporangium roseum*	PK	endothelin converting enzyme inhib.	Fujisawa
Stachybocin	*Stachybotrys* sp.	PK-AA	endothelin rec. antag.	Asahi
RES-1149	*Aspergillus* sp.	PK	endothelin rec. antag.	Kyowa Hakko Kogyo
RES-701	*Streptomyces* sp.	PEP	endothelin rec. antag.	Kyowa Hakko Kogyo
L-671,776 derivatives	*Stachybotrys* sp.	PK-AA	endothelin rec. antag.	Ciba-Geigy
Mer-A2026	*Streptomyces pactum*	PK	vasodilatory	Mercian Corp.
ET	*Penicillium sclerotium*	PK	endothelin rec. antag.	Xenova and Parke Davis
Drimane-sesquiterpenes	*Aspergillus ustus*	TERP-PK	entothelin rec. bdg.	Xenova
Bassiatin	*Beauveria bassina*	PEP	platelet aggr. inhib.	Taisho Pharm.
Herquline	*Penicillium herquei*	ALK	platelet aggr. inhib.	Kitasato
Schizostatin	*Schizophyllum commune*	TERP	squalene synth. inhib.	Sankyo
Macrosphelide	*Microsphaeropsis* sp.	PK	cell adhesion inhib.	Kitasato
Sulfobacins	*Chryseobacter* sp.	PK-S	Willebrand factor rec. antag.	Nippon Roche
Lateritin	*Gibberella lateritium*	PEP	ACAT inhib.	Tokyo Noko Univ.
Isohalobacillin	*Bacillus* sp.	PEP-PK	ACAT inhib.	Tokyo Noko Univ.
GERI-BP002-A	*Aspergillus fumigatus*	PK	ACAT inhib.	KRIBB
Pyripyropenes	*Aspergillus fumigatus*	PK	ACAT inhib.	Kitasato and Pfizer
Amidepsine	*Humicola* sp.		DGAT inhib.	Kitasato
Terpendole	*Albophoma yamanashiensis*	TERP-mod.	ACAT inhib.	Kitasato
Epi-cochlio-quinone	*Stachybotrys bisbyi*	TERP-PK	ACAT inhib.	Sankyo
F1839	*Stachybotrys*	TERP-PK	cholesterol esterase inhib.	Tokyo Tanabe Co. and Univ. Tokyo
CETPI	*Cytospora* (insect associated)	PK	cholesteryl ester transfer protein inhib.	Cornell Univ. and Schering-Plough

Tab. 1. (Continued)

Compound Reference	Producing Organisms	Structural Type[1]	Selected Properties	Research Group Involved
Fluvirucin	*Streptomcyces* sp.	PK-GLYC	phospholipase inhib.	Univ. Keio
Thermorubin	*Thermoactinomyces* sp.	PK	aldose reduct-ase inhib.	UNITIKA Co. and Univ. Osaka
Salfredins	*Crucibulum* sp.	PK-mod.	aldose reduct-ase inhib.	Shionogi
Panosialins	*Streptomyces* sp.	PK-mod.	glycosidase inhib.	Kitasato
Xenovulene	*Acremonium strictum*	TERP	GABA-benzo-diazepine re-ceptor binding	Xenova
Arisugacin	*Penicillium* sp.	TERP	AChE-inhib.	Kitasato
Nerfilin I	*Streptomyces halstedii*	PEP-PK	neurite out-growth ind.	Somtech and Univ. Tokyo
Michigazones	*Streptomyces halstedii*	PEP	neuronal cell protecting	Univ. Tokyo
Aestivophoerin	*Streptomyces purpeofuscus*	PK-mod.	neuronal cell protecting	Univ. Tokyo
Lavandu-quinocin	*Streptomyces virdochromogenes*		neuronal cell protecting	Univ. Tokyo
Epolactaene	*Penicillium sp.* (marine)	PK	neuritogenic	RIKEN and Kaken Pharm. Co.
MQ-387	*Streptomyces nayagawaensis*	PEP	aPase N inhib.	KRIBB
YL-01869P	*Actinomadura ultramentaria*	PEP-mod.	matrix metallo-proteinase inhib.	Sankyo
YM 47141/2	*Flexibacter* sp.	PEP	elastase inhib.	Yamanouchi Pharm. Co.
Poststatin	*Streptomyces virdochromogenes*	PEP	Pro-endopeptid-ase inhib.	Institute of Microbial Chemistry
Cathstatins	*Microascus longirostris*	PEP-mod.	proteinase inhib.	SynPhar Lab Inc. and Institute of Marine Bioscience (Halifax)
BE-40644	*Actinoplanes* sp.	PK	thioredoxin inhib.	Tsukuba Res. and Banyo Pharm. Co.
RPR113228	*Chrysosporium lobatum*	TERP	farnesyl protein transferase inhib.	Rhône Poulenc Rorer
Andrastin	*Penicillium* sp.	TERP-PK	farnesyl protein transferase inhib.	Kitasato and Keio Univ.
Saquayamycins	Actinomycetes	PK	farnesyl protein transferase inhib.	Keio Univ. and Institute of Microbial Chemistry
Agricultural Uses				
Rotihibin	*Streptomyces graminofaciens*	PEP-PK	plant growth regulator	Univ. Tokyo and Ajinimoto
Pironetin	*Streptomyces* sp.	PK	plant growth regulator	Nippon Kayaku
Phthoxazolin	*Streptomyces griseoaurantiacus*	PK	herbicidal	Univ. Paul Sabatier (Toulouse)

Tab. 1. (Continued)

Compound Reference	Producing Organisms	Structural Type[1]	Selected Properties	Research Group Involved
Methylstrept-imidon-derivatives	*Streptomyces* sp.	PK-mod.	herbicidal	Hoechst India
Fudecalone	*Penicillium* sp.	PK	anticoccidial	Kitasato
Arohynapene	*Penicillium* sp.	PK	anticoccidial	Kitasato
Xanthoquinodin	*Humicola* sp.	PK	anticoccidial	Kitasato
Hydrantomycin	*Streptomyces* sp.	PK	herbicidal antibiotic	Kitasato
Iturins	*Bacillus subtilis*	PEP-PK	phytopathogens	USDA, Univ. Texas and Univ. Purdue
Trichorzins	*Trichoderma harzianum*	PEP	antifungal	CNRS (Paris)
Azalomycin	Actinomycete *Streptomyces hygroscopicus*	PK	antifungal	Hoechst and AgrEvo Merck Sharp and Dohme
Phthoxazolines	*Streptomyces* sp.	PK-mod.	antifungal	Kitasato
Phenamide	*Streptomyces albospinus*	AA-mod.	antifungal	Monsanto
Patulodin	*Penicillium urticae*	PK	antifungal	Osaka Univ.
Gualamycin	*Streptomyces* sp.	GLYC	acaricidal	Nippon Kayaku Co.
NK-374200	*Taralomyces* sp.	NUC-PEP	insecticidal	Nippon Kayaku Co.
Melanoxadin	*Trichoderma* sp.		melanine bios. inhib.	Teikyo Univ. and Tokyo Univ.
Albocycline	*Streptomyces* sp.	PK	melanogenesis inhib.	Kitasato
CI-4	*Pseudomonas* sp. (marine)	PEP	chitinase inhib.	Shimizu Labs.
Oligosperons	*Arthrobotyrys oligospora*	TERP	nematocidal	Australian National Univ.
Isocoumarins	*Lachnum* sp. (Ascomycete)	PK	nematocidal	Univ. Kaiserslautern (FRG) and Univ. Lund (Sweden)
Milbemycins	*Streptomyces* sp.	PK	antihelminthic	Smith Kline Beecham
Sulfinemycin	*Streptomyces albus*	PK-mod.	antihelminthic	Lederle
Musacins	*Streptomyces griseoviridis*		antihelminthic	Univ. Göttingen, Univ. Tübingen, Hans Knöll Institute
Lachnum-lactone	*Lachnum papyraceum*	PK	nematocidal, cytotoxic	Univ. Lund and Univ. Kaiserslautern

[1] Structural type: PEP – peptide, PK – polyketide, TERP – terpenoid, GLYC – glycoside, AA – amino acid, NUC – nucleoside, mod. – modified.
[2] Property: antag. – antagonist; bios. – biosynthesis; ind. – inducer; inhib. – inhibitor; rec. – receptor.
[3] Group identification: Univ. – University of.

(Azuma, 1987), bestatin (Ochiai, 1987), topostins (Suzuki et al., 1990), etc., are to be introduced into future therapy.

The large-scale biotechnical production of bioactive compounds has been developed in a highly effective manner. Fermentations of high-producing microorganisms are carried out up to a volume of more than 300 m^3. The yield is sometimes more than 40 $g\ L^{-1}$ (Vandamme, 1984), and up to 100 $g\ L^{-1}$ in penicillin fermentations. This demonstrates the efficiency of strain selection which supported knowledge of biosynthesis and strain genetics. Optimum bioprocess control and suitable fermentation equipment were developed as further prerequisites of a highly efficienct production of biotechnical drugs.

As an introduction to this volume, this chapter summarizes some of the general aspects of secondary metabolism in microorganisms such as:

- the biological role of bioactive compounds in the producer strains,
- the biosynthetic pathways and their organization,
- natural and induced variations of secondary metabolite structures and problems of their structural classification.

Finally, future perspectives of drug screening from microbial sources are discussed.

2 Secondary Metabolism, an Expression of Cellular and Organismic Individuality

2.1 Roles of Secondary Metabolites in Producing Organisms

The majority of bioactive products of microorganisms and plants is generated by secondary metabolism. This part of the metabolic machinery of microbes, plants, and animals may play no essential role in the vegetative development of the producing organisms, but seems to convey advantages to the pertinent species concerning its long-term survival in the biological community and environment (LUCKNER et al., 1977; KLEINKAUF and VON DÖHREN, 1986; WILLIAMS et al., 1989; LUCKNER, 1989; VINING, 1992; WILLIAMS et al., 1992; CAVALIER-SMITH, 1992; OLESKIN, 1994; VINING and STUTTARD, 1995) (Tab. 2).

Further interpretations imply the formation of certain secondary metabolites by relatively small, but systematically defined groups of organisms (e.g., special species and genera of microbes, plants, animals) and point to the enormous variability of chemical structures (Comité Editorial, 1992). In microbes, the capacity to generate secondary metabolites is frequently lost by genomic mutations, but this feature misses any concomitant effect on the vegetative development of the pertinent strains (SHAPIRO, 1989; OLESKIN, 1994). An inverse correlation is usually observed between specific growth rate and the formation of secondary metabolites such as antibiotics. Particular features of morphological differentiation in surface or submerged cultures, such as the formation of spores and conidia, seem to be related to the production capacity of secondary metabolism. Moreover, a maximum production rate of antibiotics and other secondary metabolites (pigments, alkaloids, mycotoxins, enzyme inhibitors, etc.) has frequently been observed when growth-promoting substrates were depleted from the medium (DEMAIN, 1992). This phenomenon was called "catabolite regulation" (DEMAIN, 1974). This may be one of the reasons for the phase-dependency of biosynthesis of many microbial drugs.

Thus, during the microbial growth phase (trophophase) secondary metabolism is often suppressed, but increased later during the "idiophase" (VINING, 1986). Sometimes this feature is not present and depends on the particular strains and growth conditions. For instance, the formation of phytotoxins by some phytopathogenic microbes such as *Alternaria* and *Fusarium* strains is not a subject of catabolite regulation and even occurs in a growth-associated manner (REUTER, 1989). On the other hand, the production of antifungal effectors including peptaibol trichorzianine may be induced, as shown in *Trichoderma harzianum* by cell walls of the plant pathogen *Botrytis cinerea* (SCHIRMBOCK et al., 1994). Likewise, certain plant metabolites may induce the synthesis of peptide antibiotics in the respective pathogenic *Pseudomonas* strains (MAZZOLA and WHITE, 1994; MO et al., 1995). In general, the phase-dependency or specific inducibility indicates that the secondary metabolism is strictly governed by inherent regulatory systems (see Sect. 2.2).

Tab. 2. Presumed "Roles" of Secondary Metabolites in Their Producer Organism

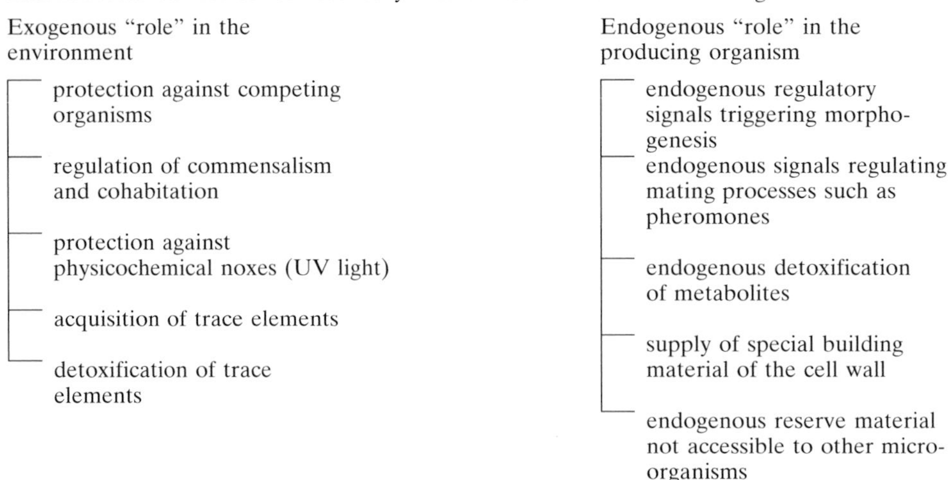

Most of the secondary metabolites are biosynthesized in microbes and plants via complex multistep pathways involving many enzymatic and even non-enzymatic events. These appear to be integrated in a coordinated manner into the global microbial processes of cytodifferentiation such as formation of spores, conidia, and aerial mycelia (LUCKNER, 1989), or in the processes of invasion or defense. The same is true for plants in which secondary metabolite formation occurs in different tissues, e.g., roots, leaves, flowers, and seeds. Hence, it seems obvious that secondary metabolism does not reflect an occasional feature but is the result of a very long evolutionary development. As was shown for the tetracycline antibiotics from *Streptomyces* spp. more than 200 genes may affect the biosynthetic pathway (VANEK and HOSTALEK, 1985). No wonder that speculation about the endogenous "function" and "roles" of secondary metabolites in the producing organisms themselves never came to an end (VANEK et al., 1981; VINING, 1992; OLESKIN, 1994; VINING and STUTTARD, 1995).

To maintain such a great number of genes, generally linked into clusters, during evolution should be of advantage to the pertinent organism. Obviously, in plants many secondary metabolites are involved in the protection against microorganims and animals (CUNDLIFFE, 1992; JOHNSON and ADAMS, 1992). Others act as chemoattractants or as repellents towards insects fructifying flowers or damaging plant tissues. A series of plant hormones (cytokinins, gibberelic acid, jasmonic acid, etc.) are similar in structure but *per definitionem* are not secondary metabolites. Another function of secondary metabolites in plants is the detoxification of poisonous metabolites via an endogenous compartmentized storage (LUCKNER, 1989). The role of secondary metabolism in microbes is even more difficult to understand. Cellular efforts needed for secondary pathways are rather low in the wild-type strains (only a small amount of the overall substrate intake is converted to bioactive secondary metabolites). This part of metabolism would possibly have been eliminated during phylogenesis without any selective advantage of secondary metabolite production. It appears to be a generally accepted view that microbial secondary metabolites play an important but not generalizable rote, at least in special situations, e.g., in warranting the survival in particular environmental systems, during limitation of nutrient supply or even in the course of morphological development (LUCKNER et al., 1977; KLEINKAUF and VON DÖHREN, 1986; VINING, 1992; KELL et al., 1995; VINING and STUTTARD, 1995). From this point of view, the formation of large amounts of antibiotics by high-producing strains (substrate conversion rates $Y_{glucose/drug} > 0.1$) would be considered as a "pathophysiological" problem (VANEK et al., 1981). In order to better understand the general roles of secondary metabolites in microbes one could refer to the color of hairs and feathers in animals, their odorous pheromones, and other metabolic products which do not contribute *per se* to the vegetative life of the pertinent species. But they could have outstanding importance during the adaptation to changing media, in the protection against competing organisms, and in the regulation of sexual and asexual processes of genetic exchange. General discussions of secondary metabolite formation in microbes consider four major fields of importance (LUCKNER et al., 1977; KLEINKAUF and VON DÖHREN, 1986; LUCKNER, 1989; WILLIAMS et al., 1989, 1992; VINING, 1992; CAVALIER-SMITH, 1992; OLESKIN, 1994; VINING and STUTTARD, 1995) (Tab. 2):

(1) The formation of secondary metabolites facilitates the adaptation to metabolic imbalances as a kind of a "metabolic valve", which is needed to remove an excess of toxic, endogenous metabolites that otherwise are accumulated during a partial limitation of substrates.
(2) Secondary metabolism could be a source of individual building blocks of cells or of metabolic reserves which warrant the individuality and particular functionality of the given strain.
(3) Secondary metabolites could be regarded as endogenous signals triggering particular stages of morphogenesis and the exchange of genetic material (see Fig. 1). This hypothesis was particularly supported by the observation that the majority of the "good" producers (e.g., actinomycetes, fungi, bacteria) display a life cycle involving several stages of morphological differentiation.
(4) Secondary metabolite formation is particularly important in biosystems as a signal of interspecific "communication" between microbes and other microbes, plants, and animals. Symbiosis, commensalism, and antagonism could be regulated by secondary metabolites in heterologous populations.

The self-protecting mechanism in antibiotic-producing microbes should be mentioned as a further evidence of an ecological function of antibiotics, as a "weapon" against competitors (ZÄHNER et al., 1983; BRÜCKNER et al., 1990; CUNDLIFFE, 1989, 1992; WILLIAMS and MAPLESTONE, 1992). By this means the microbe prevents suicide due to its own secondary metabolite either by enzymatic modifications of the drug, by alteration of its biological target, or by an active transport-directed export (see, e.g., the tetracycline efflux) (JOHNSON and ADAMS, 1992; NIKAIDO, 1994). Usually, resistance mechanisms of the antibiotic-producing microorganisms are the same as in antibiotic-resistant bacteria. The analysis of the gene sequences encoding resistance determinants support the idea that the transfer of resistance occurs from the antibiotic producers to the non-producing microbes (JOHNSON and ADAMS, 1992; SALYERS et al., 1995; HIRAMATSU, 1995; DAVIES, 1994). In addition, the emergence of new types of resistance factors by the formation of mosaic genes has been analyzed in β-lactam-resistant pneumococci (SPRATT, 1994; COFFEY et al., 1995).

The great variation of both active and inactive secondary metabolites, that were observed in microorganims and plants supplied the main arguments against their determined function. Obviously, the formation of a bioactive secondary metabolite, such as an antibiotic, rather appears as an exception than as a rule. Frequently, many inactive "shunt-metabolites" and congenors are produced in addition to the few active metabolites. It is not reasonable to believe that all these metabolites are needed in a single organism. It might be that a "function" of a secondary metabolite could become apparent only in a particular, exceptional situation or in special stages of development. Hence, the selection pressure on structures and secondary pathways is necessarily low (ZÄHNER et al., 1983). As a consequence, spontaneously evolving variants and mutants could survive with the same probability as their parents, and modifications of secondary pathways and structures would

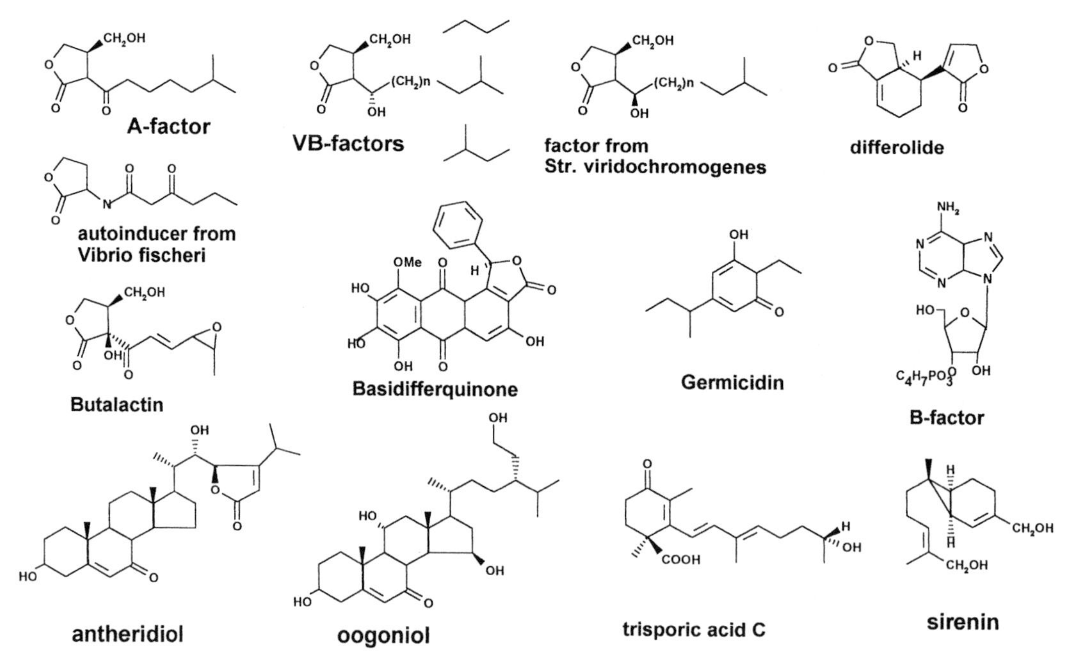

Fig. 1. Structures of some representatives of signaling molecules from bacteria (streptomycetes) and fungi (for references, see text).

be preserved (secondary metabolism as a "playground of evolution") (ZÄHNER et al., 1983). This might explain the existence of the numerous similar structures. According to this hypothesis, the limited substrate specificity of some enzymes of secondary metabolism has to be mentioned (LUCKNER, 1989). However, it should be noted that in many multistep processes this limited specificity is restricted to certain steps and thus less restricted structural regions of the compounds (KLEINKAUF and VON DÖHREN, in press). A few secondary metabolites, out the pool of the many non-functional metabolites, have apparently acquired an essential role in growth and differentiation. The siderophores, e.g., are microbial vehicles of iron transport formed in variable structures as constitutive parts of the iron uptake system (VON DER HELM and NEILANDS, 1987; WINKELMANN, 1991; WINKELMANN and DRECHSEL, Chapter 5, this volume). *Per definitionem,* they should not be regarded as secondary metabolites. Highly specialized biomolecules such as cytochromes, chlorophylls, sexual pheromones of fungi and bacteria, etc. might have been evolved similarly. Some of them may be attested to defined "functions" of microbial secondary metabolites (Tab. 2, Fig. 1).

A role of secondary metabolism in the adaptation to changing nutrient conditions is a realistic position since an excessive supply of metabolic intermediates (precursors) usually induces or stimulates drug production (DEMAIN 1974, 1984, 1992). Growth may become imbalanced and precursors are accumulated during the limitation of a given substrate in the medium, while others are still available in excess. Excessive precursors could be released into the medium or converted to hardly metabolizable products which would not support the growth of competitors. Moreover, colored secondary metabolites, such as pigments, could protect cells and spores from damage by ultraviolet radiation or also could promote the acquisition of rare elements via complex formation as, e.g., siderophores. Complex formation could also protect the cells from high concentrations of toxic heavy metals.

The incorporation of secondary metabolites into cellular structures has been suggested to contribute to their individual characteristics. Thus, streptomycin and its building moiety, streptidine, were established as a constitutent of the cell wall of the producing *Streptomyces griseus* (DEMAIN, 1984; DISTLER et al., 1992). Otherwise, the production of secondary metabolites (so-called "idiolites") (DEMAIN, 1992), could serve as a kind of a metabolic reserve which cannot be metabolized by other microbes. Some antibiotics (anthracyclines, tetracyclines, cyclosporins, etc.), e.g., are stored within the mycelium and their complete degradation requires a series of specialized enzymatic steps. Otherwise, bioconversions of antibiotics are a constitutive part of the self-protecting mechanisms of the producer strain.

Moreover, concentrations of several antibiotics were shown to decrease in the course of prolonged cultivation, thus indicating the onset of degradative processes. Some fungi are well-known to degradate their own polyketides such as, e.g., citrinin (BARBER et al., 1988) and zearalenon and even to use them for additional syntheses. Active antibiotics were usually not detected in soil samples, although recently sensitive procedures have permitted the detection of phenazines (COOK et al., 1995). Their complete degradation under natural conditions seems very likely.

Most likely, a series of signaling molecules is supplied by the secondary metabolism that possess interspecific (ecological) or species-dependent functions, e.g., as signals triggering morphogenesis and the exchange of genetic material (Fig. 1). By growth inhibition of competing microbes a producer strain could attain an advantage (c.f. the production of herbicidal antibiotics by phytopathogenic bacteria which damage plant tissues and facilitate nutrient acquisition from the host) (KOHMOTO and YODER, 1994; MAZZOLA and WHITE, 1994; MO et al., 1995). *Vice versa*, secondary metabolism could confer a particular advantage in symbiotic systems, such as *Pseudomonas*/plant roots, to both the producing strain and the symbiont. An example is the control of phytopathogenic *Fusarium* or *Rhizoctonia* fungi on plant roots by products of cohabiting streptomycetes and bacteria. Interspecific effects have also been postulated for volatile compounds which are formed, e.g., by streptomycetes and cyanobacteria. Geosmin, isoborneol, and mucidon are the constituents of the typical earthy odor. It has been shown that sclerin and scleroid from the fungus *Sclerotinia libertiana* stimulate the biosynthesis of aminoglycosides by streptomycetes, but also the growth of some plants (KUBOTA et al., 1966; OXFORD et al., 1986). The formation of phytotoxins by phytopathogenic microbes is mentioned as another interspecific communication system (KOHMOTO and YODER, 1994). Constituents of the microbial cell wall (elicitors such as β-1,3-1,6-glucans from *Phytophtora megasperma*) are recognized by specific plant cell membrane receptors. Subsequently, a series of protective mechanisms is induced in the plant (e.g., hypersensitivity reactions, *de novo* synthesis of tissues, secretion of enzymes lysing microorganisms, and formation of antimicrobial phytoalexins). On the other hand, some of the phytoalexins are inactivated by enzymes of phytopathogenic microbes.

In the natural habitat genetic information can be transferred from one microbe to another interspecifically. Both biosynthetic procedures and resistance mechanisms thus can be spread among various heterologous species and genera. Apparently this is also true for genetic exchanges between plants and microbes. A recent intriguing example is the discovery of a taxol producing fungus living in taxol producing yew trees (STIERLE et al., 1994). Typical plant hormones such as gibberellins and jasmonic acid are also produced by some microorganisms. Aflatoxins formed via complicated biosynthetic pathways in fungi, such as *Aspergillus*, have been established in actinomycetes. Sequence analyses of the genes encoding penicillin and cephalosporin biosynthetic clusters (ACV synthase, isopenicillin N-synthase, acyltransferase, deacetoxycephalosporin C-synthase, and deacetoxycephalosporin C-hydroxylase) in *Penicillium chrysogenum, Acremonium chrysogenum,* and *Streptomyces* spp. strongly suggested that fungi received the pertinent genes from the prokaryotic actinomycetes during evolution

(LANDAN et al. 1990; MILLER and INGOLIA, 1993; BUADES and MOYA, 1996). The production of cephabacins, chitinovorins, clavulanates, olivanic acids, carbapenems, and thiopeptides by unicellular bacteria and streptomycetes may indicate that an original biosynthetic pathway was spread horizontally among different microbes, thus giving rise to evolutionary variations of structures and pathways.

The evolution of secondary metabolism even appears to create hybrid structures by the combination of genetic material originating from heterologous hosts. Recently, thiomarinol (SHIOZAWA et al., 1993) was isolated from the marine bacterium *Alteromonas rava* as a composite compound formed by the esterification of pseudomonic acid (found in *Pseudomonas fluorescens*) and holomycin (a pyrrothine antibiotic, found in *Streptomyces* sp.).

The involvement of secondary metabolism in the regulation of microbial cytodifferentiation seems to be important, at least in some cases. The morphogenesis of antibiotic-producing microorganisms (streptomycetes, fungi, *Mycobacteria,* etc.) is obviously mediated by a plethora of biochemical steps, which display a high specificity for the given organism. The pathways are regulated by individual signals in a highly coordinated manner (Fig. 1) (LUCKNER, 1989). During morphogenesis, silent genes are activated that have not been expressed during the growth phase. Accordingly, several endogenous non-antibiotic regulators of the cell cycle were discovered in *Streptomyces* cultures, and their structure was elucidated (see below) (KHOKHLOV, 1982; GRÄFE, 1989; HORINOUCHI and BEPPU, 1990, 1992a, b, 1995; BEPPU, 1992, 1995). Correlations between the biogenesis of some peptidic antibiotics and morphogenesis were also described for synchronously growing *Bacillus* cultures (MARAHIEL et al., 1979). Tyrocidin, gramicidin, and bacitracin are produced during the onset of sporulation, suggesting that their function concerns the control of transcription, spore permeability, dormancy of spores, and their temperature stability (MARAHIEL et al., 1979, 1993).

The γ-butyrolactones represent a particularly important group of endogenous regulators of *Streptomyces* differentiation (Fig. 2) (KHOKHLOV, 1982; GRÄFE 1989; HORINOUCHI and BEPPU, 1990, 1992a, b, 1995; BEPPU, 1995). They are required as microbial "hormone-like" substances in few species such as streptomycin, virginiamycin or anthracycline producing strains. These effectors permit the formation of antibiotics and aerial mycelium by some blocked, asporogenous, antibiotic-negative mutants even in very low concentrations. Several other autoregulators of morphogenesis have been investigated (see, e.g., factor C) (SZESZAK et al., 1991). Otherwise, germicidin B (PETERSEN et al., 1993) from *Streptomyces violaceusniger* inhibits germination of its own spores by interference with endogenous ATPase. Antibiotics such as hormaomycin (RÖSSNER et al., 1990) and pamamycin (KONDO et al., 1986) were shown to have autoregulatory functions. Moreover, streptomycetes can produce interspecific inducers such as anthranilic acid and basidifferquinone (Fig. 1) which affect basidiomycetes and the formation of fruiting bodies (AZUMA et al., 1980; MURAO et al., 1984).

Moreover, regulatory molecules inducing cytodifferentiation were isolated from fungi and molds confirming that morphogenesis can be mediated by the aid of an agency of specialized endogeneous factors (HAYASHI et al., 1985). They can be regarded as secondary metabolites since they do not possess any function in vegetative development.

In addition, sexual factors from fungi and yeasts can be considered as functionalized secondary metabolites. They trigger zygospore formation by haploid cells belonging to different mating types (GOODAY, 1974). During the evolution of signal systems, from the simple pro- and eukaryotes up to the hormonal control in mammalians, some structures and activities have been conserved. The alpha-factor of the yeast *Saccharomyces cerevisiae* as one of its sexual pheromones, e.g., appears to be partially homologous to the human gonadotropin releasing hormone (LOUMAYE et al., 1982). Moreover, inducers of differentiation of Friend leukemia cells were isolated from soil organism such as *Chaetomium* sp. These chlorine containing substituted diphenols (Fig. 1) also induce morphogenesis (stalk cell differentiation) of *Dictyostelium*

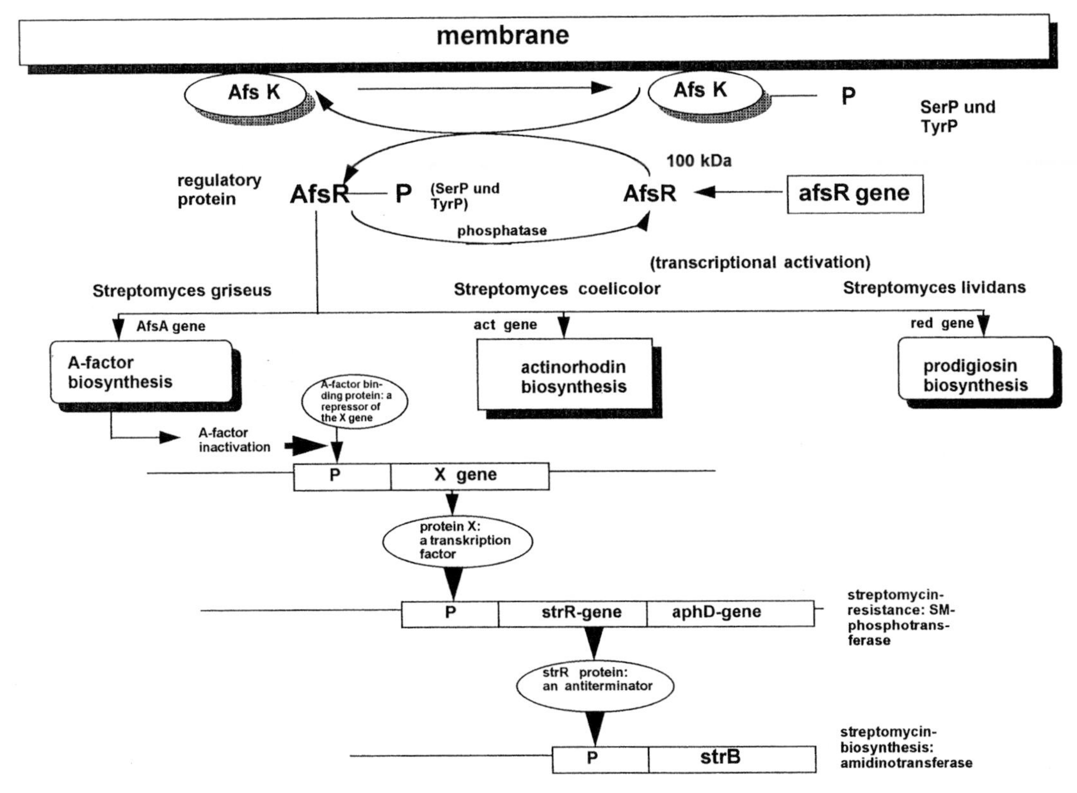

Fig. 2. Regulatory events suggested to be involved in morphogenesis and secondary metabolism of *Streptomyces griseus* (P: promotor) (HORINOUCHI and BEPPU, 1992a).

discoideum, suggesting the similarity of mammalian and fungal control of the cell cycle (KUBOHARA et al., 1993). Recently, the occurrence of sexual pheromones was even established for the prokaryote *Streptococcus faecalis*. Its pheromones stimulate or inhibit the transfer of conjugative plasmids from donor to recipient strains (WIRTH et al., 1990). Peptides triggering competence in *Bacillus subtilis* have been characterized and were termed pheromones (D'SOUZA et al., 1994; SOLOMON et al., 1995; HAMOEN et al., 1995).

2.2 Regulation of Microbial Secondary Metabolism

2.2.1 Genetic Organization of Product Formation

A large number of biosynthetic genes were isolated and characterized and, in general, they have been found assembled in clusters (Tab. 3). Such clusters may contain information for the biosynthesis of the basic structure of the metabolite, its modification, resistance determinants, e.g., promoting modification of products, targets, altered targets, or export systems, as well as regulatory elements; individual gene products which might as well exert regulatory functions.

Tab. 3. Biosynthetic Clusters Identified

Compound	Type	Organism	Selected References[1]
A54145	acylpeptidolactone	*Streptomyces fradiae*	BALTZ et al., 1996[1]
Aflatoxins	polyketide	*Aspergillus parasiticus, Aspergillus flavus*	BROWN et al. 1996; MAHANTI et al., 1996
Actinomycin	chromopeptidolactone	*Streptomyces chrysomallus*	KELLER et al., 1996[2]
Anguibactin	modified peptide	*Vibrio anguillarum*	CHEN et al., 1996
Astaxanthin	carotinoid	*Agrobacterium aurantiacum*	MISAWA et al., 1995
Avermectin	polyketide	*Streptomyces avermitilis*	MACNEIL, 1995
Avilamycin	polyketide	*Streptomyces viridochromogenes*	BECHTHOLD et al., 1996[2]
Bacitracin	branched cyclopeptide	*Bacillus licheniformis*	HERZOG-VELIKONJA et al., 1994
Bialaphos	peptide	*Streptomyces viridochromogenes*	SCHWARTZ et al., 1996
Carbomycin	polyketide	*Streptomyces thermotolerans*	ARISAWA et al., 1995
Carotinoids	terpenoids	*Rhodobacter capsulatus*	ARMSTRONG, 1994
Carotinoids	terpenoids	*Myxococcus xanthus*	ARMSTRONG, 1994
Carotinoids	terpenoids	*Synecococcus* PCC7942	ARMSTRONG, 1994
Clavulanic acid	modified peptide	*Streptomyces clavuligerus*	HODGSON et al., 1995
Cephalosporin	modified peptide	*Acremonium chrysogenum*	MARTÍN and GUTIERREZ, 1995
Cephamycin	modified peptide	*Nocardia lactamdurans*	COQUE et al., 1993, 1995a, b; PETRICH et al., 1994
Coronatin	modified polyketide	*Pseudomonas syringae*	BENDER et al., 1996
Cyclosporin	cyclopeptide	*Tolypocladium niveum*	WEBER et al., 1994
Daptomycin	acylpeptidolactone	*Streptomyces roseosporus*	BALTZ et al., 1996[1]
Daunomycin, Daunorubicin, Doxorubicin	polyketide	*Streptomyces C5/peucetius*	YE et al., 1994; GRIMM et al., 1994; FILIPPINI et al., 1995; MADDURI and HUTCHINSON, 1995a, b; DICKENS et al., 1996
Destruxin	peptidolactone	*Metarhizium anisopliae*	BAILEY et al., 1996
Elloramycin	polyketide	*Streptomyces olivaceus*	DECKER et al., 1995
Fatty acids	polyketide	*Streptomyces glaucescens*	SUMMERS et al., 1995
Fatty acids	polyketide	*Escherichia coli*	ROCK and CRONAN, 1996
Fengymycin	peptide	*Bacillus subtilis*	LIU et al., 1996[2]
Ferrichrome	cyclopeptide	*Ustilago maydis*	LEONG et al., 1996[1]
Frenolicin	polyketide	*Streptomyces roseofulvus*	BIBB et al., 1994
Geldanamycin	polyketide	*Streptomyces hygroscopicus*	ALLEN and RITCHIE, 1994
Gramicidin S	cyclopeptide	*Bacillus brevis* ATCC9999	TURGAY and MARAHIEL, 1995
Granaticin	polyketide	*Streptomyces violaceoruber*	SHERMAN et al. 1989; BECHTHOLD et al., 1995
Griseusin	polyketide	*Streptomyces griseus*	YU et al., 1994
HC-toxin	cyclopeptide	*Helminthosporium carbonum*	PITKIN et al., 1996
HET?	polyketide?	*Anabaena* sp.	BLACK and WOLK, 1994
Immunomycin	modified polyketide	*Streptomyces* sp.	MOTAMEDI et al., 1996[2]
Jadomycin B	polyketide	*Streptomyces venezuelae*	YANG et al., 1995b, 1996b

Tab. 3. (Continued)

Compound	Type	Organism	Selected References[1]
Melanin		*Aspergillus nidulans* *Colletotrichum lagenarium*	TAKANO et al., 1995
Landomycin	polyketide, glycosylated	*Streptomyces* sp.	BECHTHOLD et al., 1996[2]
6-Methylsalicylic acid	polyketide	*Penicillium patulum*	BECK et al., 1990
Microcystin	cyclopeptide	*Microcystis aeruginosa*	MEISSNER et al., 1996
Mithramycin	polyketide	*Streptomyces argillaceus*	LOMBÓ et al., 1996
unknown	polyketide	*Streptomyces cinnamonensis*	ARROWSMITH et al., 1992
Nikkomycin	modified peptide	*Streptomyces tendae*	BORMANN et al., 1996
Nodusmicin	polyketide	*Saccharopolyspora hirsuta*	LE GOUILL et al., 1993
Nogalamycin	polyketide	*Streptomyces nogalater*	YLIHONKO et al., 1996
Oleandomycin	polyketide	*Streptomyces antibioticus*	QUIRÓS and SALAS, 1995
Oxytetracyclin	polyketide	*Streptomyces rimosus*	KIM et al., 1994
Penicillin	modified peptide	*Aspergillus nidulans,* *Penicillium chrysogenum*	SMITH et al., 1990; MACCABE et al., 1990; DÍEZ et al., 1990
Phenazin	heterocycle	*Pseudomonas aureofaciens*	PIERSON et al., 1995
Pristinamycin A	acylpeptidolactone	*Streptomyces pristinaespiralis*	DE CRÉCY-LAGARD, personal Communication
Pristinamycin M	polyketide/peptide	*Streptomyces* sp.	BECK et al., 1990
Puromycin	modified aminoglucoside	*Streptomyces alboniger*	TERCERO et al., 1996
Pyoverdin	branched cycloacylpeptide	*Pseudomonas fluorescens*	STINTZI et al., 1996
Rapamycin	modified polyketide	*Streptomyces hygroscopicus*	SCHWECKE et al., 1995
Saframycin	modified peptide	*Myxococcus xanthus*	POSPIECH et al., 1996
Soraphen A	modified polyketide	*Sorangium cellulosum*	SCHUPP et al., 1995
Sterigmatocystin	polyketide	*Aspergillus nidulans*	BROWN et al., 1996
Streptomycin	aminoglycoside	*Streptomyces glaucescens* *Streptomyces griseus*	BEYER et al., 1996
Streptothricin	modified aminoglucoside	*Streptomyces rochei*	FERNÁNDEZ-MORENO et al., 1996
Surfactin	peptidolactone	*Bacillus subtilis*	COSMINA et al., 1993
Tetracenomycin	polyketide	*Streptomyces glaucescens*	SHEN and HUTCHINSON, 1994
Tylosin	polyketide	*Streptomyces fradiae*	MERSON-DAVIES and CUNDLIFFE, 1994
Urdamycin	polyketide	*Streptomyces fradiae*	DECKER et al., 1995
Whi, spore pigment	polyketide	*Streptomyces coelicolor*	DAVIS and CHATER, 1990
Zeaxanthin	terpenoid (carotinoid)	*Erwinia herbicola,* *Erwinia uredovora*	ARMSTRONG, 1994, HUNDLE et al., 1994

[1] Presented at the conference *Genetics and Molecular Biology of Industrial Microorganisms.* Bloomington 1996.
[2] Presented at the symposium *Enzymology of Biosynthesis of Natural Products.* Berlin 1996.
Abstracts available from the authors on request.

The techniques employed include reverse genetics if sequence data of relevant enzymes is available, the use of homologous gene probes or probes constructed from key sequences, the generation by PCR of specific probes flanked by conserved key motifs, complementation of idiotrophic mutants, expression of pathways or single step enzymes in heterologous hosts, cloning of resistance determinants followed by isolation of flanking sequences, identification and cloning of regulatory genes or sequences (promoters, regulatory protein binding sites, pleiotropic genes, "master" genes, etc.).

To improve product levels, the addition of extra copies of positive regulators (CHATER, 1992; HOPWOOD et al., 1995; CHATER and BIBB, Chapter 2, this volume), extra copies of biosynthetic genes possibly representing bottlenecks (SKATRUD et al., Chapter 6, this volume), or the alteration of promoters of key enzymes are under investigation.

The analysis of clusters has revealed a wealth of information including biosynthetic unit operations and their surprisingly complex organization. The majority of large proteins now known are multifunctional enzymes involved in peptide and polyketide formation, with sizes ranging from 165 kDa to 1.7 MDa. Other systems also forming polyketides, peptides, aminoglycosides, etc., are comprised of non-integrated enzyme activities, still performing the synthesis of highly complex structures. The details of various biosynthetic clusters are described in the respective chapters on regulatory mechanisms (CHATER and BIBB, Chapter 2, this volume), peptides (VON DÖHREN and KLEINKAUF, Chapter 7, this volume), β-lactams (SKATRUD et al., Chapter 6, this volume), lantibiotics (JACK et al., Chapter 8, this volume), and aminoglycosides (PIEPERSBERG and DISTLER, Chapter 10, this volume). Recent highlights of the elucidation of such data have been the rapamycin and immunomycin clusters in *Streptomyces*, the erythromycin cluster in *Saccharopolyspora*, the surfactin and gramicidin S clusters in *Bacillus*, various β-lactam clusters, and the sterigmatocystin cluster in *Aspergillus nidulans*. An overview of examples is presented in Tab. 3.

The amplification of biosynthetic clusters in highly selected strains has been a fascinating key result, as shown for the industrial penicillin producer (FIERRO et al., 1995; MARTÍN and GUTIERREZ, 1995). The main findings with regard to sequencing of complete genomic fragments are as follows:

- The identification of biosynthetic genes follows by the detection of core sequences. Such sequences permit the recognition of types of biosynthetic unit operations like polyketide condensation reactions, the specificities of the respective transferase sites (HAYDOCK et al., 1995), the number of elongation steps, amino acid activation sites; in the case of repetitive cycles where certain sites are reused, as in type II polyketide forming systems or, e.g., cyclodepsipeptide synthetases, where the number of steps remains uncertain.
- Additional genes for modification reactions like oxygenases and transferases are readily identifiable by standard structural alignments as well as possible regulatory proteins.

At present, however, the unambiguous correlation of product and biosynthetic machinery is not possible without the support of various genetic techniques or, if not available due to the lack of transformation systems, structural details from protein chemistry of isolated enzymes or multienzymes.

To illustrate a few concepts, we will point to some recent examples of cluster analysis: β-Lactam antibiotics as classical examples of modified peptides are still leading antibacterial drugs. Some efforts have been directed to understand at the molecular level the performance of industrial overproducers selected for decades (SKATRUD et al., Chapter 6, this volume). Following the reverse genetics approach in isolation of the isopenicillin N synthase gene (SAMSON et al., 1985), which catalyzes the formation of the penem bicycle from the tripeptide precursor ACV, the clustering of biosynthetic genes was demonstrated in both pro- and eukaryotic producers (BARTON et al., 1990). The two key enzymes, ACV synthetase and isopenicillin N synthase showed extensive similarities in both bacteria and fungi, and a horizontal intergenic transfer has been suggested (LANDAN et al., 1990; MILL-

ER and INGOLIA, 1993; BUADES and MOYA, 1996). The linkage of these adjacent genes illustrates well basic principles of cluster organization (Fig. 3) (AHARONOWITZ et al., 1992). In bacteria both genes are transcribed unidirectionally within an operon linked to sets of other genes the products of which are required for the modifying reactions of the cephem nucleus to cephamycin, and the formation of the β-lactamase inhibitor clavulanic acid (WARD and HODGSON, 1993). Such extensive linkages have been termed superclusters. In fungi the encoding genes for ACVS and isopenicillin N synthase are bidirectionally transcribed, separated by intergenic regions of about 1 kbp. A variety of environmental conditions are known to affect fungal β-lactam production at the transcriptional level (ESPESO and PEÑALVA, 1996; SUAREZ and PEÑALVA, 1996; BRAKHAGE and TURNER, 1995). The bidirectionally oriented promoters between *acvA* (*pcbAB*) and *inpA* (*pcbC*) may permit the asymmetrical expression of both genes, and indeed different levels of expression have been obtained in constructs employing different reporter genes which allowed to measure the expression of both genes simultaneously (BRAKHAGE et al., 1992; BRAKHAGE and TURNER, 1995; BRAGKHAGE and VAN DEN BRULLE, 1995; THEN BERG et al., 1996). Such results suggest possible additional functions for the penicillin tripeptide precursor, besides its role in the formation and the still unclear excretion of penicillins. The 872 bp intergenic region between the *A. nidulans acvA* (*pcbAB*) and *ipnA* (*pcbC*) permits the complex and sensitive regulation involving several protein factors (for *P. chrysogenum*, see FENG et al., 1995; CHU et al., 1995). The current knowledge of regulatory factors and putative factors implied by the identification and characterization of *trans*-acting mutations specifically involved in the regulation of the *A. nidulans* biosynthetic genes is summarized in Fig. 3b. One of these factors, designated PACC, was shown to activate at least the *ipnA* gene transcription in response to shifts to alkaline pH values (SHAH et al., 1991; ESPESO et al., 1993; TILLBURN et al., 1995; ARST, 1996). For

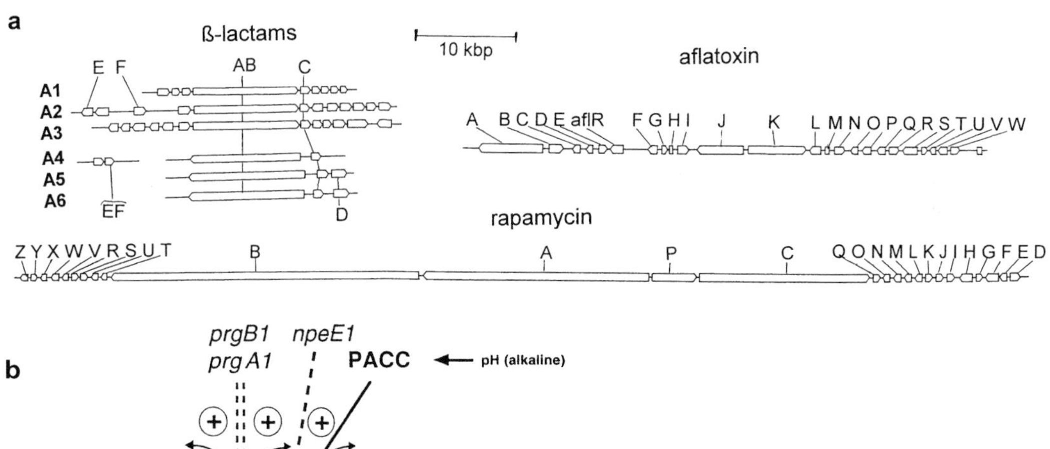

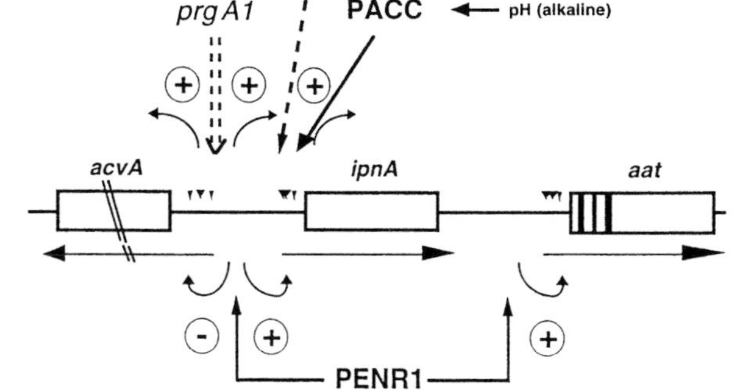

Fig. 3a. Organization of the biosynthetic clusters of β-lactams, rapamycin, and sterigmatocystin, **b** regulatory sites identified in the penicillin biosynthetic cluster in *Aspergillus nidulans*.

PACC seven binding sites with different affinities have been mapped in this intergenic region (SUAREZ and PEÑALVA, 1996). Another binding site containing a CCAAT motif was detected, bound by a protein complex designated PENR1 (THEN BERG et al., 1996). PENR1 also binds to a CCAAT-containing DNA region in the promoter of the *aat* gene encoding acyl-CoA:isopenicillin N acyltransferase which is located 3′ of the *ipnA* gene (LITZKA et al., 1996). Deletion analysis and mutagenesis experiments indicated that the binding of PENR1 represses the expression of *acvA* and increases that of both *ipnA* and *aat* (THEN BERG et al., 1996; LITZKA et al., 1996). PENR1 thus represents the first example of a regulatory protein controlling the regulation of the whole β-lactam biosynthesis gene cluster in fungi. However, many promoters of eukaryotic genes are known to contain CCAAT motifs which are bound by distinct gene regulatory proteins (JOHNSON and MCKNIGHT, 1989). At the time being, it is unknown what kind of CCAAT binding protein PENR1 represents and whether it is a global acting factor specific for the regulation of β-lactam biosynthesis genes.

Using a genetic approach which is feasible for the ascomycete *A. nidulans,* three recessive *trans*-acting mutations were identified designated *prgA1/prgB1* for penicillin regulation (BRAKHAGE and VAN DEN BRULLE, 1995) and *npeE1* (PÉREZ-ESTEBAN et al., 1995). These mutations formally correspond to positively acting regulatory genes. Mutants carrying one of the mutations mentioned produced reduced amounts of penicillin. For *prgA1* and *prgB1* it was shown that the expression of both genes *acvA* and *ipnA* was affected (BRAKHAGE and VAN DEN BRULLE, 1995), whereas *npeE1* controls at least *ipnA* expression (PÉREZ-ESTEBAN et al., 1995). The major nitrogen regulatory protein NRE of *Penicillium chrysogenum* has also been found to specifically attach to three GATA/GATT pairs within this intergenic region (HAAS and MARZLUFF, 1995). The pairwise attachment sites indicate a possible dimeric state of this GATA family transcription factor and as well connect this regulatory site with nitrogen assimilation. This example illustrates that similar biosynthetic genes are under the regime of organizationally specific mechanisms of regulation. The respective regulatory mechanisms will be evaluated comparatively in a variety of pro- and eukaryotic hosts.

Regulation of the formation of secondary metabolites in eukaryotes, however, does not need to be this complex, as will be discussed below in the case of sterigmacystin/aflatoxin biosynthesis. As a second example for the organization of biosynthetic information the polyketide immunosuppressant rapamycin has been selected (SCHWECKE et al., 1995). This polyketide with an iminoacyl residue is of interest as an immunosuppressor in autoimmune disease and transplantation. Its biosynthesis proceeds by 16 successive condensation and 21 modification reactions of 7 acetyl and propionyl residues, respectively, followed by pipecolate onto the cyclohexane carboxylic acid starter unit. The respective cluster has been identified in *Streptomyces hygroscopicus* by LEADLAY et al. (SCHWECKE et al., 1995) using polyketide synthase gene probes of erythromycin synthase from *Saccharopolyspora erythrea*). The sequence of 107.3 kbp has been determined as well as the boundary sequences, to assure the completeness of the effort. The key part of the cluster is represented by four genes encoding multifunctional enzymes with sizes of 900 (A), 1070 (B), 660 (C), and 154.1 kDa (P) responsible for the formation of the macrolactam ring. These four genes of 25.7, 30.7, 18.8, and 4.6 kb unambiguously correlate with the structural features of the product, however, module 3 and 6 contain catalytic sites for the reduction of the polyketide intermediates, which actually are not found in rapamycin. The solution of this problem remains to be found and plausible explanations are either non-functionality due to, e.g., point mutations, or a possible transient reduction of the intermediates to facilitate folding, which is reversed later.

These key genes are flanked by additional 24 open reading frames, most of which have been assigned tentative functions including modification of the macrolactam, export, and regulation. Standard identification procedures are hampered by the non-availability of genetic operations for this strain.

The essential data in this case are the presence of large polyfunctional genes in prokaryotic clusters and the surprising lack of strict correlation of expected biosynthetic unit operations within the predicted modules with the actual gene structures found. A similar observation has also been made in the case of the avermectin biosynthetic cluster (McNeil et al., 1995).

As a recent eukaryotic example the sterigmatocystin biosynthetic cluster in *A. nidulans* is considered (Brown et al., 1996). Sterigmatocystin is the penultimate intermediate in the biosynthesis of aflatoxins. Both polyketides are highly mutagenic and thus carcinogenic. They spoil food upon fungal colonization, especially by *A. flavus* and *A. parasiticus*. These losses may be reduced by a detailed understanding of the regulation of the biosynthetic events. So, e.g., the induction of aflatoxin formation has been shown to be strongly suppressed by jasmonate, a phytohormone (Goodrichtanrikulu et al., 1995). Detailed genetic studies have confirmed the linkage and coregulation of sterigmatocystin and aflatoxin biosynthesis (Trail et al., 1995a, b; Keller and Adams, 1995; Brown et al., 1996). The recent sequencing of the sterigmatocystin biosynthetic cluster in *A. nidulans* revealed within a 60 kb region 25 transcripts, the expression of which is coordinated under conditions of toxin production. The cluster is flanked by genes also expressed under non-production conditions. The regulatory gene *aflR* and its *A. flavus* homolog both specifically induce gene expression within the cluster. Among the identified genes are a fatty acid synthase, five monoxoygenases, four dehydrogenases, an esterase, an O-methyltransferase, a reductase, and an oxidase, all functionally implied in the proposed reaction sequence. Comparative evaluation of the respective cluster in *A. parasiticus* shows conservation of clustering, but no strict conservation of the gene order (Trail et al., 1995a, b; Yu and Leonard, 1995). Conservation of clustering has been suggested to serve both purposes of global regulation and horizontal movement of biosynthetic activities among species. The striking features of the tremendous efforts so far show the integration of a specific fatty acid synthase into a secondary product cluster. These types of genes have been commonly referred to as primary pathway enzymes. The respective hexanoyl structure serves as a starter and is elongated by a type II system forming an aromatic polyketide. So far, such systems have been found only in prokaryotes. Gene characteristics, however, do not suggest a horizontal transfer as in the β-lactam case (Brown et al., 1996). Finally, a specific transcription factor is a key element in the expression of the enzyme system, and no evidence has yet been obtained for complex timing and differential gene expression as in the penicillin pathway in *A. nidulans*.

Inspection of other clusters included in Tab. 3 suggests extensive similarities of certain groups which, at first sight, look like structurally unrelated compounds. Certain types of regulatory genes are implied in the formation of various metabolites. There seems to be a non-species-related separation of type I and type II systems, e.g., in polyketide formation, but the various degrees of integration of biosynthetic modules catalyzing unit operations may be dictated by the chemistry of their products. Finally, the clustering of pathways also suggests their genetic transfer between various hosts. Within the evolutionary frame, adaptation of pathways to various targets has been proposed, e.g., for Aspergilli adapting to insect colonization and perhaps moving to other target organisms (Wicklow et al., 1994). The structures of metabolites with key roles in invasive processes would then adapt to new targets by evolutionary processes.

2.2.2 Regulatory Mechanisms

Mechanisms involved in the regulation of secondary metabolite expression have been reviewed recently, focussing on global control in bacterial systems (Doull and Vining, 1995), bacterial mechanisms in detail (Chater and Bibb, Chapter 2, this volume), antibiotic formation in *Streptomyces coelicolor* (Hopwood et al., 1995), and autoregulators (Horinouchi and Beppu, 1995; Beppu, 1995). Eukaryotic systems except for β-lactams have not been in focus regarding special metabolites. Recent reviews cover β-lactams (BRAKHAGE and TURNER, 1995; SKATRUD et al., Chapter 6, this volume; JENSEN and DEMAIN, 1995).

A variety of stress conditions have been documented to lead to secondary metabolite production (DEMAIN, 1984; DOULL and VINING, 1995; VINING and STUTTARD, 1995). Besides physical parameters (temperature shock, radiation) chemical signals will trigger the formation of various small response molecules, which are the subject of this volume. Such signals include both high and low concentrations of oxygen (oxidative stress, lack of oxygen, or shift to anaerobic growth), acidity (pH shift), but generally the response to nutrient alterations. Phase-dependency of secondary metabolite formation in microbial cultures and its correlation to morphological changes suggest that secondary metabolism is subject to general regulatory mechanisms governing cellular development (BARABAS et al., 1994). Only some of the regulatory features have been elucidated in the past and many are still to be unraveled. Nutrient shift regulation of growth is closely coupled to differention through a series of common metabolic signals and regulations such as mediated by sigma factors and transcriptional enhancers. In this context, two major questions are addressed:

(1) Why are microbial secondary metabolism and morphogenesis suppressed during growth on media which are rich in carbon, nitrogen, or phosphorus and what is the cause of catabolite regulation?
(2) What is the nature of the general signals governing a plethora of metabolic events and how do they cooperate within the cellular frame of developmental programs?

There are indeed drastic variations in the extent of responses upon nutritional stress. Obvious morphological changes like sporulation or formation of aerial mycelia are caused by an undetermined number of respective genes, reading to sets of proteins and mediators promoting alterations in the cellular composition. Such changes include altered cell wall composition and changes in the metabolic spectrum. The changes may not be obvious and some work has been conducted on model systems such as *Escherichia coli, Bacillus subtilis,* and *Aspergillus nidulans.* Besides nutrient depletion as envisioned and studied in chemostate-like environments employed in fermentation, a generally neglected field is the response to environmental factors indicating the presence of alike or competing organisms. According to our understanding of the basic role of many of the metabolites employed in the control of invasive processes. Such approaches seem obvious. It has been shown that cell density critically affects antibiotic production (WILLIAMS et al., 1992; FUCQUA et al., 1994; SANCHEZ and BRANA, 1996). The induction of nisin formation by nisin itself, as mediated by its cluster-inherited signal system, is another intriguing example (RA et al., 1996; DERUYTER et al., 1996). Likewise the presence of phytopathogenic fungi induces responses, e.g., in rhizosphere colonizing bacteria including the production of antifungals (KAJIMURA et al., 1995; PIERSON and PIERSON, 1996). While the presence of resistant microorganisms has been applied in selection processes for antimicrobial agents the identification of response signals is still an open field.

Stress Conditions Related to Nutrient Limitations

In connection with nutrient depletion carbon, nitrogen, and phosphate starvation are considered in general. The differential induction of metabolite forming processes has been excellently demonstrated by BUSHELL and FRYDAY (1983). Extensive studies of this aspect have also been conducted in the antibiotic fermentation of gramicidin S in *Bacillus brevis.* Formation of this cyclopeptide has been found in a variety of stress conditions, including sporulation and non-sporulation conditions and, surprisingly, two phosphate concentration ranges (KLEINKAUF and VON DÖHREN, 1986) differing from other phosphate-effected systems (LIRAS et al., 1990). Thus, in many cases less specific induction and maintenance by interacting regulatory devices are implied and manipulation may be exerted by growth rate control.

Nutritional downshift in the media caused by limitation of particular metabolites (amino acids, ATP, sugars, etc.) promotes excessive formation of some metabolites due to an imbalanced metabolism (supra) (MARTÍN et al., 1986; LIRAS et al., 1990). Accumulation of these "precursors" is known to induce secondary pathways (see, e.g., the induction of ergotamin alkaloid formation by tryptophane in *Claviceps* strains) (HÜTTER, 1986). On the other hand, limitation of some endogenous metabolites could be important which inhibit global regulatory mechanisms governing aerial mycelium and spore formation. In this repressing or inhibitory effects on the secondary pathways and on morphogenesis could be diminished. Both features, accumulation of precursors and limitation of repressing metabolites seem to be involved (DEMAIN, 1974, 1992; MARTÍN et al., 1986; HORINOUCHI et al., 1990; LIRAS et al., 1990). The pertinent regulatory mechanism may be similar to those shown for other global microbial regulations.

Metabolite formation has been studied in detail in model cases of surfactin (*B. subtilis*), streptomycin (*Streptomyces griseus*), or penicillin (*A. nidulans* and *P. chrysogenum*). It is controlled by superimposed regulatory cascades or networks. Such networks include intracellular and extracellular components and might include regulators, transducers, signaling systems, interacting repressors, and activators, as well as modification and expression systems. In the case of streptomycin the term "decision phase" has been coined as a model for a variety of production processes (PIEPERSBERG, 1995; PIEPERSBERG and DISTLER, Chapter 10, this volume). Despite this complexity, manipulations of single genes may have substantial effects on production levels.

Most information on the respective depletion events have come from model organisms, but they proved to be useful in a variety of cases. Carbon sources are known but poorly understood tools in natural product processes. Readily assimilated compounds, e.g., glucose repress production while other carbon sources causing slow growth promote production. This has been demonstrated nicely in the case of bacitracin formation in *Bacillus licheniformis* (HANLON and HODGES, 1981). Glucose-6-phosphate suppresses the synthetase enzymes in penicillin biosynthesis (JENSEN and DEMAIN, 1995). However, as was shown for ACV-synthase, IPN-synthase, and expandase in penicillin and cephalosporin producing fungal and *Streptomyces* strains inhibition or repression by glucose-6-phosphate, ammonium, and phosphate ions depend on the given strain (AHARONOWITZ et al., 1992).

Carbon uptake systems have been studied in several organisms including enteric bacteria in which the phosphoenolpyruvate–carbohydrate phosphotransferase system controls uptake and transport (POSTMA et al., 1993). This phosphorylation-controlled multistep process involves adenylate cyclase and cAMP-mediated gene regulation. Other mechanisms operate in gram-positive bacteria (STEWART, 1993) and streptomycetes (CHATER and BIBB, Chapter 2, this volume).

Nitrogen depletion again is a determining factor in many antibiotic fermentations (SHAPIRO, 1989). These effects are attributed to nitrogen catabolite repression. A two-component system sensing the glutamine and α-ketoglutarate levels activates transcription of catabolic enzymes releasing ammonia or other nitrogen sources by autophosphorylation of a His protein kinase (DOULL and VINING, 1995). The activation of glutamine synthetase is included in this process, the activity of which is as well controlled by several factors including the glutamine level. Actinomycetes contain two types of glutamine synthetases. In process analysis ammonia has been found to repress secondary metabolite formation. Roles of various nitrogen sources have not been evaluated in detail, but are discussed in the case of β-lactams (SKATRUD et al., Chapter 6, this volume). Ammonium ions are also catabolite repressors of β-lactam biosynthesis (cephalosporin C, cephamycin C) in *Acremonium* and some *Streptomyces* spp. (JENSEN and DEMAIN, 1995; DEMAIN, 1989). Deamination of L-valine in the biosynthesis of tylosin is subject to catabolite regulation by ammonium ions (TANAKA, 1986).

In bacteria, a stringent response is caused by nitrogen limitation (CASHEL, 1975) and the appearance of non-acylated tRNAs. A concomitant increase of guanosine-3′,5′-tetraphosphate (ppGpp) concentration switches off unfavorable biosynthetic processes. Ribosomal protein synthesis is reduced, but the degradation of amino acids continues. This fact is due to the binding of ppGpp to RNA polymerase and the alteration of its promoter recognition. Thus, transcription of many genes might be stimulated while the expression of others declines in a coordinated manner. The molecule of guanosine-3′,5′-tetraphosphate might be involved in the regulation of the secondary metabolism and also in sporulation of streptomycetes (OCHI, 1990).

The heterogeneity of promotor structures and the complementation of bacterial RNA polymerases by sigma factors could provide another rational basis for the understanding of the developmental regulation of gene expression (CHATER and BIBB, Chapter 2, this volume). RNA polymerase consists of a core enzyme composed of each of two α- and two β-subunits. Bacterial promotor recognition is regulated by sigma factors (σ^{37}, σ^{43}, etc.) attached to the core enzyme. Depending on the type of the individual sigma factor, either general (e.g., the factors needed for vegetative growth) or specialized genes (e.g., those responsible for secondary metabolism and cytodifferentiation) can be transcribed. In *Streptomyces griseus* MARCOS et al. (1995) identified three sigma factors differentially expressed under specific nutritional conditions. The sigma factors *whiG* and *sigF,* each controlling certain events in the development of spore chains in *Streptomyces coelicolor,* are controlled by transcriptional and posttranscriptional events involving additional proteins (KELEMEN et al., 1996).

Recent approaches of molecular genetics showed that DNA-binding protein factors are crucial for the transcription of both eukaryotic and prokaryotic genes (HORINOUCHI and BEPPU, 1992b; CHATER, 1992; THEN BERG et al., 1996). They often occur as dimers and stimulate activity by binding to particular promotor regions. An example is the regulatory system of the γ-butyrolactones (A-factor) involving proteinaceous transcriptional activators (AfsR protein) (HORINOUCHI and BEPPU, 1992b; BEPPU, 1995).

The response to exogenous phosphate has been studied in *E. coli* and the involvement of more than 30 genes in the PHO regulon has been established (WANNER, 1993). Respective efforts in antibiotic production have been reviewed (LIRAS et al., 1990). So *p*-aminobenzoate synthetase by *S. griseus* as a key enzyme of candicidin synthesis is negatively regulated by inorganic phosphate (MARTÍN, 1989). An upstream promotor region of 113 bp length and rich in AT was identified as a binding site of a general phosphate-dependent repressor protein. If phosphate-insensitive genes such as the β-galactosidase gene were coupled to this fragment and transferred in other *Streptomyces* hosts (such as *S. lividans*) they became subject to phosphate control.

2.2.3 Genetic Instability

The formation of secondary metabolites often is genetically instable and many explanations for this phenomenon have been given (DYSON and SCHREMPF, 1987; ALTENBUCHNER, 1994). The occurrence of extracellular plasmids containing transposon structures and IS elements was discussed initially. These could be integrated into the genome and induce genomic rearrangements and gene disruptions (HORNEMANN et al., 1993).

Streptomycetes contain only one single linear chromosome (8 Mb) (ALTENBUCHNER, 1994; CHEN et al., 1994; REDENBACH et al., 1996). Gene mapping experiments, complementation of blocked mutants, and heterologous expression of genes in different *Streptomyces* hosts have shown that the genes of secondary metabolite production are localized on chromosomal gene clusters (HOPWOOD et al., 1983; LIU et al., 1992; STUTTARD and VINING, 1995). Clusters which are responsible for the polyketide and aminoglycoside syntheses contain the genes of self-resistance protecting against the toxicity of the own secondary metabolite (SENO and BALTZ, 1989). Moreover, regulatory gene products are involved which integrate secondary metabolism into cellular developmental regulations (DISTLER et al., 1988; HORINOUCHI and BEPPU, 1990; BEPPU, 1992).

An organizational principle is the formation of large "amplified" genomic structures (DYSON and SCHREMPF, 1987). Such sequences could be used as amplifying tools for biosynthetic pathways in the future. A few examples demonstrate that the enlarged "gene dosage" contributes to improved drug production in high-yielding strains (TURNER, 1992; MARTÍN and GUTIERREZ, 1995). Commercial penicillin producing strains derived in decades of random selections may contain more than 20 copies of the biosynthetic cluster linked by conserved hexanucleotide spacer elements (FIERRO et al., 1995). The frequent deletion of the cluster has been related to these hot spots of recombination, since non-producer mutants have lost the entire region within these boundaries (FIERRO et al., 1996).

In the event of transpositions, frameshift mutations may lead to disruptions of both structural and regulatory genes (gene deletions) (HORNEMANN et al., 1993). Activation of "silent" gene sequences may occur in mutant strains because of the same reason. Thus, regeneration of protoplasts or curing of plasmids may yield mutants of streptomycetes experiencing a completely altered pattern of secondary metabolism (see, e.g., the formation of curromycin, indolizomycin, and iremycin) (OGARA et al., 1985; OKAMI et al., 1988).

2.2.4 Developmental Processes

Another feature of regulation of gene expression in the development of streptomycetes implies regulatory genes, such as *whi, bld, afs,* and *abs* in *S. lividans* and *S. coelicolor* (HOPWOOD, 1988; DAVIES and CHATER, 1992). Thus, deletion of the *whiB* gene causes the concomitant loss of aerial mycelium formation. BldA was shown to specify the leu-tRNA (UUA codon). Further evidence suggested that TTA codons in the DNA are absent from all genes involved in vegetative growth, but are present in the regulatory or resistance gene clusters of antibiotic biosynthesis. Accordingly, leucyl-tRNA could signify a marker governing gene transcription of differentiation-dependent pathways, at least in streptomycetes (DAVIES and CHATER, 1992; CHATER, 1992).

As a summary, initial factors suppressing or stimulating cytodifferentiation and secondary metabolism of the microbes are excessive nutrients converted to regulatory metabolites, nutrient downshifts, the accompanying changes of general regulatory metabolites (such as ppGpp), and accumulations of precursors due to metabolic imbalances. Low and high molecular-weight mediators are needed to trigger the coordination of numerous pathways and cellular events in cytodifferentiation. The A-factor and similar γ-butyrolactones are mentioned here as a particularly intriguing example of how the complex developmental programs are organized in streptomycetes (HORINOUCHI and BEPPU, 1990, 1992a, b; BEPPU, 1995).

2.2.5 The A-Factor and the Signal Cascade of Cytodifferentiation in *Streptomyces*

The A-factor and its dihydro derivatives (Fig. 1) are formed by numerous streptomycetes. In some *Streptomyces* strains such as *S. griseus* and *S. virginiae* γ-butyrolactone autoregulators are required as a kind of a microbial hormone for antibiotic production and even for sporulation (PLINER et al., 1976; HORINOUCHI and BEPPU, 1990; ISHIZUKA et al., 1992).

In 1975, KHOKHLOV and coworkers (PLINER et al., 1976) found an idiotrophic mutant strain of *S. griseus* which neither produced streptomycin (an aminoglycoside antibiotic) nor formed aerial mycelia and spores. It regained normal cytodifferentiation in the presence of the culture liquid of the parental strain. Later, the structure of the "autoregulatory" A-factor and a series of 2′-dihydro derivatives was elucidated (PLINER et al., 1976; ISHIZUKA et al., 1992). The A-factor (2*R*-hydroxymethyl - 3 - oxocaproyl) - γ - butyrolactone, its homologs and analogs are capable not only to induce streptomycin biosynthesis and aerial mycelium formation but also daunorubicin, virginiamycin, and carbapenem production in other *Streptomyces* strains, bioluminescence in *Vibrio fischeri,* nodulation of plant associated bacteria, and toxin production in *Pseudomonas aeruginosa* (BEPPU, 1995).

For the dihydro derivatives of the A-factor isolated from *Streptomyces viridochromogenes* and *S. bikiniensis,* the 2*R*,3*R*- and the 2*S*,3*R*-configuration was initially proposed (SAKUDA and YAMADA, 1991) but later the absolute stereochemistry was established as 2*S*,3*R*,2′*R* and 2*S*,3*R*,2′*S*, respectively (YAMADA et al., 1987; SAKUDA et al., 1992; LI et al., 1992). The latter structure, but not the A-factor-type 2′-oxo-butyrolactones, induce the production of the peptide antibiotic virginiamycin by *S. virginae* (virginiae butanolides) (YAMADA et al., 1987). Recently, 2′-deoxy derivatives (NFX factors) were even shown to stimulate virginiamycin production in the same manner, and NFX-2 ((2*R*,3*R*,4*S*)-2-hexyl-3-hydroxy-4-pentanolide) proved to be identical with blastomycinol lactole (a component of antimycin A1) (YAMADA et al., 1987; KIM et al., 1990; OKAMOTO et al., 1992).

Thus γ-butyrolactones play an outstanding role as regulatory signals inducing cytodifferentiation and formation of quite different secondary metabolite structures such as aminoglycosides, polyketides, and peptides (HORINOUCHI and BEPPU, 1990, 1992a, b; SAKUDA et al., 1992; BEPPU, 1995).

Genetical and biochemical experiments contributed much to the present knowledge of the regulatory cascade of cytodifferentiation of *S. griseus* which involves the A-factor and its congeners as signal transmitters (Fig. 2). AfsR (a 100 kDa protein encoded by the *afsR* gene containing ATP and DNA-binding domains) represents an early event in the cytodifferentiation of *S. griseus.* It is active in its phosphorylated form AfsR-P as a transcriptional activator of several other genes and it can be phosphorylated by *afsK,* a respective kinase. The N-terminal region of this kinase shows significant similarity to other Ser/Thr kinases including the β-adrenergic receptor kinase, the Rous sarcoma oncogene product, and a *Myxococcus* enzyme (BEPPU, 1995). Gene disruption of *afsK* in *S. coelicolor* caused the reduction of actinorhodin formation without effecting growth. Residual biosynthetic activities may be regulated by other kinases and the two-component system *afsQ1/afsQ2* controlling actinorhodin production in *S. coelicolor.*

So the phosphorylated *afsR* protein seems to bind to regulatory DNA sequences near the *afsA* gene (in *S. griseus*), *act* genes (in *S. coelicolor*), and *red* genes (in *S. lividans*) and to enhance their transcription. The *afsA* gene encodes for the biosynthesis of A-factor-like molecules, which is accomplished by the fusion of phosphorylated glycerol and β-keto-fatty acids (SAKUDA et al., 1992). Intracellular recognition of the A-factor occurs via an A-factor binding protein acting as the repressor of the X-gene (HORINOUCHI and BEPPU, 1992a). Inactivation by A-factor thus permits formation of the X-protein acting as a transcriptional enhancer of the *strR* and *aphD* genes. While AphD is responsible for the self-resistance of the producer strain to streptomycin, *strR* appears as a transcriptional antiterminator of streptomycin biosynthesis. Although the scheme is still incomplete it suggests that numerous events of sporulation and secondary metabolism could be governed by *afsR, X-* and *strR* gene products in a concerted manner.

In analogy to *S. griseus,* a virginiae butanolide (VB-C) binding protein (Mr 36000 Da) was isolated from *S. virginiae* which is suggested to be involved in the mechanism of pleiotropic signal transduction. The binding activity of this protein towards VB-C decreased by 40% in presence of DNA in a similar manner as shown for other regulatory proteins and transcriptional factors (KIM et al., 1990; SAKUDA et al., 1992). The pertinent gene was sequenced and displayed considerable homology (62–64%) to the amino acid sequences of ribosomal protein L 11 of diverse origins (rpIK) and to the essential protein of *E. coli.* This suggested it to be a part of an essential gene cluster encoding general components of the transcriptional and translational systems. Exogenously added factors have been used here to improve metabolite production in the case of virginiamycin (YANG et al., 1995a, 1996a).

Possibly, the *afsR* gene of *S. griseus* is also controlled by other genes which have not been identified so far. The *whiB* gene of *S. coelicolor,* e.g., is responsible for early sporulation events due to the formation of a small transcription factor-like protein which is dispensable for growth, but essential for sporulation (DAVIES and CHATER, 1992; CHATER, 1992).

Moreover, *S. griseus* mutants which were recently investigated, produce the A-factor but nevertheless miss the normal sporulation behavior (MCCUE et al., 1992). An open coding gene sequence (ORF 1590) was identified which is possibly responsible for the synthesis of two polypeptidic transcription factors (P 56 and P 49.5). Dimerization of P 56 was suggested to induce the onset of sporulation, but P 49.5 prevents this event. In the above mutant imbalanced regulation of the syntheses of P 56 and P 49.5 have been proposed to cause lack of sporulation.

As another type of event, ADP-ribosylation of proteins catalyzed by NAD-glycohydrolase and ADP-ribosyltransferase seems to participate in cytodifferentiation of *S. griseus.* Failure to ADP-ribosylate certain cellular proteins in mutant strains was thought to cause impaired differentiation (SZESZAK et al., 1991; OCHI et al., 1992).

A γ-butyrolactone derivative, β-hydroxybutyryl-homoserine lactone, is the autoinducer of light emission by *Vibrio harveyi* (MEIGHEN, 1991; WILLIAMS et al., 1992; FUQUA et al., 1994; GEIGER, 1994). Similar to other photobacteria, luminescence is strongly influenced by the density of the cell culture. *V. harveyi* synthesizes the above small extracellular molecule, which accumulates in the growth medium and induces luminescence by luciferase and $FMNH_2$-coupled oxidation of a long-chain fatty aldehyde. *Vibrio fischeri* forms a similar autoinducer, β-ketocaproyl homoserine lactone (FUQUA et al., 1994). Previously, similar molecules have been reported to regulate carbapenem biosynthesis by *Erwinia carotovora* (BAINTON et al., 1992).

The signaling pathway of light emission which is induced in presence of the above mentioned butyrolactones seems to involve transmembrane signaling proteins as receptors. They possess enzymic domains at the inner site of the membrane (MEIGHEN, 1991; FUQUA et al., 1994; GEIGER, 1994).

Early evidence for autoregulatory functions of special metabolites in the differentiation and diploidization was presented for a series of fungi (GOODAY, 1974; ZAKELJMAVRIC et al., 1995). In some molds there are sex hormones like antheridiol, sirenin, oogoniol, and trisporic acids (Fig. 1), which trigger zygospore formation and the subsequent exchange of genetic material. In the aquatic fungus *Achlya* the signaling chain of the fungal sterol antheridiol displays similarity to mammalian cells. Here the response to steroidal sex hormones is also mediated by membrane receptors (ZAKELJMAVRIC et al., 1995).

2.2.6 Overproduction of Microbial Secondary Metabolites and Precursor Pools

Much experience has been obtained in the past with empirical selections of high-yielding strains of antibiotic producing microorganisms. Comparison of the high-producing mutants of streptomycetes and fungi with the low-yielding wild-type strains suggested that a series of heritable metabolic changes had been introduced (OCHI et al., 1988; VANEK and HOSTALEK, 1988), for instance:

- the elimination of "bottle-necks" in the production of biosynthetic precursors,
- the suppression of negative catabolite regulations concomitant with increased production of synthetases,
- improved resistance of the producer strain against its own toxic product, and
- the absence of negative feedback regulation of the formed secondary metabolite on its biogenesis.

To realize these prerequisites of high productivity, the natural regulatory mechanism of the wild-type strains, permitting only little product formation had to be altered in a step-by-step selection procedure. The alterations concern both the genetic and the physiological system of the pertinent strain, the secondary pathways, and the cellular morphology (VANEK and MIKULIK, 1978).

Many of the high-producing strains overproduce the pertinent precursors. An excessive precursor supply thus appears to determine high secondary metabolite production. Moreover, when several alternative precursors can be used by the same biosynthetic pathway the availability of the individual precursors governs the quality of formed products. Wild-type strains often produce a series of homologous structures due to the usage of several intracellularly supplied precursors (SANGLIER and LARPENT, 1989). During strain improvement by mutagenesis and selection empirical pathway engineering was done. Sometimes, the selection promoted excessive formation of a single precursor and, consequently, a single product was formed instead of a series of homologous structures (CLARIDGE, 1983; THIERICKE and ROHR, 1993).

"Precursor-directed biosynthesis", "mutational" and "hybrid" biosyntheses signify microbiological techniques (CLARIDGE, 1983; THIERICKE and ROHR, 1993), which have successfully been used in the past to alter product formation by excessive feeding of precursors or biosynthetic intermediates to parental strains and their mutants. Even when structural analogs of the special precursor were fed to the medium they could be used as a substitute of the natural structure. In this manner, the formation of many new and unusual secondary metabolites was demonstrated (SHIER et al., 1969).

During the rapid (balanced) growth of microbial cultures no excess of intermediary metabolites is available, but when some substrates become rate-limiting while others are still available a metabolic imbalance arises which promotes the accumulation of precursors (imbalanced growth) (DEMAIN, 1974, 1992). Apparently, the size of precursor pools is of regulatory importance in secondary metabolite formation and determines the production rate.

Investigations of the β-lactam biosynthesis illustrate well that penicillin formation by *P. chrysogenum* is subject to negative feedback control by L-lysine, and to a lesser extent by L-valine (MARTÍN et al., 1986). The former inhibits and suppresses homocitrate synthetase in the low-producing strains as a negative feedback regulator. The high-producing strains display greatly reduced sensitivity to lysine (MARTÍN and DEMAIN, 1980). This "branched-pathway" model of regulation was also reported for the biogenesis of candicidin by *S. griseus*. It is reduced by excessive tryptophan in the medium due to the feedback inhibition of the *p*-aminobenzoic acid synthetase (MARTÍN, 1978).

L-cysteine needed for β-lactam production can be produced either from sulfide and O-acetylserine or by reverse transsulfuration of O-acetylhomoserine using L-methionine as a donor of sulfur. In *P. chrysogenum* (forming penicillin G) cysteine is produced mainly by the sulfate reduction pathway, in *Acremonium chrysogenum* (producing cephalosporin C) via transsulfuration (MARTÍN, 1978). In the latter strain, feeding of L-methionine highly stimulates cephalosporin biosynthesis concomitant with the formation of arthrospores in submerged fermentations (MARTÍN et al., 1986).

High-producing strains were shown to synthesize precursors by particular metabolic sequences. Carboxylation of acetyl coenzyme A by oxaloacetate to yield malonyl coenzyme A and the activation of D-glucose by polyphosphate glucokinase are characteristics of some streptomycetes (QUEENER et al., 1986; VANEK et al., 1978). These peculiar pathways enhance precursor supply in the biosyntheses of tetracyclines, erythromycin, and macrolide polyenes.

Compartmentation of the precursor- and energy-generating metabolism plays an important but yet incompletely understood role in eukaryotic microorganisms. The biosynthesis of benzodiazepines by *Penicillium cyclopium* depends on precursor pools stored within vacuoles. Their membranes become permeable during the production phase due to the appearance of a particular permeabilizing factor (ROOS and LUCKNER, 1986).

2.2.7 Biotechnical Production of Secondary Metabolites

For more than 50 years semi-empirical rules determined the scaling-up of microbial procedures for the production of secondary metabolites.

Maximum production rate of a given secondary metabolite usually is attained below the maximum growth rate. Consequently, fermentations are carried out under partial substrate limitation. Mostly, complex nutrient sources are employed or slow feeding of substrates such as glucose which cause a vigorous development of biomass concomitant with catabolite repression of secondary pathways. But, an optimized fermentation process is characterized by moderate development of biomass. Secondary metabolism thus occurs parallel to submaximal but continuous growth. A major goal of the bioengineer is to grow high concentrations of producing biomass in the fermenter. Finally, the available oxygen concentration in the fermenter is the critical value for high productivity (CALAM, 1987; FIECHTER, 1988). Oxygen intake is dependent on fermenter geometry and impeller performance.

Promotion of impeller speed increases shear stress of the mycelia and causes fragmentations and reduction of the mycelial production rate. More than other microbial processes fermentations of secondary metabolites require producer strains displaying an optimal morphological behavior under the given technical conditions. Changes of the mycelial morphology not only cause alterations in the rheological behavior of the fermentation broth, but also affect the intake of oxygen into the culture. Moreover, nutrient penetration into the cells is affected by the formation of pellets and mycelial aggregations (STEELE and STOWERS, 1991).

Another serious problem in large-scale biotechnical production of secondary metabolites is heat formation. Maintenance metabolism of high biomass concentrations burns a great part of substrate without product formation. Hence, strains selected for low heat production appear particularly promising.

3 The Biosynthetic Pathways

3.1 Precursors and the Main Biosynthetic Pathways

Secondary metabolites are formed from few starter molecules acting as precursors. They will either be modified to yield new chemical derivatives of the initial molecule or they will be coupled to oligomeric material which is subsequently modified. An outstanding variability of structures arises from the latter biosynthetic principle which combines homologous and even heterologous building moieties in a polycondensation process. Moreover, oligomeric structures once formed, such as the aglycones of macrolides, angucyclines, and anthracyclines, can be linked to other moieties like biosynthetically modified sugars.

Only a few precursor structures are used in secondary metabolite formation: coenzyme A derivatives of lower fatty acids (acetyl-, propionyl-, *n*- and isobutyryl-CoA, etc.), mevalonate (also derived from acetyl-CoA), amino acids and shikimate, sugars (preferably glucose), and nucleosides (purines and pyrimidines). These precursors are also needed in primary metabolism to form cellular materials such as proteins, nucleic acids, cell wall constituents, and membrane lipids (ZÄHNER and ZEECK, 1987). Numerous biotransformations of single molecules are known, but oligomerizations of the above mentioned basic structures only occur by three pathways: glycosylation of activated sugars and polycondensations involving either activated fatty acids or amino acids.

3.2 Secondary Metabolites Formed through Biosynthetic Modifications of a Single Precursor

Structurally modified monosaccharides, such as valienamin in acarbose (MÜLLER, 1989), are derived from glucose via a series of

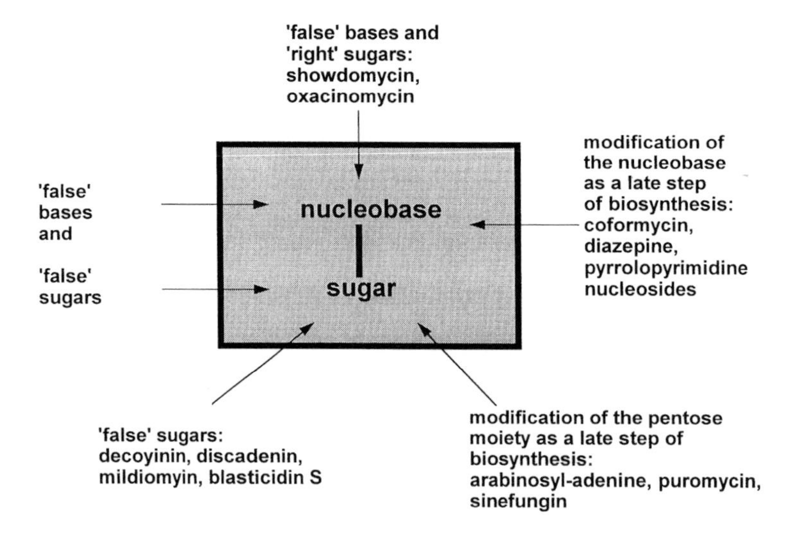

Fig. 4. Biosynthetic strategies in the formation of nucleoside antibiotics (ISONO, 1990).

individual biotransformations. Similarly, amino acids are used as precursors of numerous antimetabolites and enzyme inhibitors (VOGEL, 1984).

Shikimate and chorismate (as the intermediates of the aromatic amino acid pathway) provide the starting material for chloramphenicol and C_7N-derived metabolites. The ansamycins (VINING et al., 1986), e.g., are formed from aminoshikimate as a shunt metabolite of the deoxyarabinoheptulosonate phosphate pathway (CHIAO et al., 1995).

Various higher fatty acids are precursors in the pathway to secondary metabolites including polyketides and lipopeptides.

The variety of naturally occurring nucleoside antibiotics is formed from cellular nucleosides (c.f. tubercidin) through a series of transformations involving oxidations, dehydrations, and rearrangements of the carbon skeleton. Another biosynthetic strategy uses ribose or free nucleobases as starters which are subsequently coupled to separately formed analogs of the nucleobases and sugars, respectively (ISONO, 1990) (Fig. 4).

3.3 Polyketides

Microbial polyketide-type secondary metabolites are generated by head-to-tail decarboxylating condensation of an acyl coenzyme A unit (as a starter) with an additional malonyl-CoA moiety. This mechanism is similar to fatty acid biosynthesis (condensing enzyme). Thereafter, chain elongation occurs by subsequent condensation with additional substituted malonyl-CoA units (BIRCH and ROBINSON, 1995). As a major difference to fatty acid biosynthesis, the enzymic steps following the initial condensation (β-ketoacid reduction, dehydratation, hydrogenation) are not carried out regularly, but quite occassionally depending on the given polyketide synthase. The final structures are compatible with the assumption of intermediate formation of poly-β-ketones (polyketides). These highly unstable enzyme-bound intermediates are reduced, dehydrated, hydrogenated, and sometimes condensed intramolecularly to yield numerous cyclic and polycyclic structures (ROHR et al., 1993). In contrast to fatty acid synthesis of mammalian cells and plants, various acyl-coenzyme A building moieties are used as starters and as chain extenders (Tab. 4).

The polyketide synthases (PKS) of several microbes investigated so far by extensive gene sequencing experiments are comparable to the fatty acid synthases both in the mechanistic and the reaction kinetic manner (HOPWOOD and SHERMAN, 1990). In streptomycetes there are two different kinds of PKS: those similar to the fatty acid synthase II of bacteria and plants and polyketide synthases comparable to the type I fatty acid synthases (multienzyme synthases) of vertebrates and fungi. The former consist of at least eight monofunctional polypeptides, each responsible for one single reaction step. The latter are characterized by one or two polypeptides carrying sites for all catalytic activities and the attachment point of the acyl carrier protein prosthetic group (O'HAGAN, 1991; HOPWOOD and SHERMAN, 1990) (Fig. 5).

Type II polyketide synthases accomplish the formation of polycyclic or aromatic compounds such as actinorhodin, granaticin, tetracenomycin, oxytetracyclin, and frenolicin (VINING et al., 1986; BÊHAL and HUNTER, 1995; HUTCHINSON and FUJII, 1995). Otherwise, PKS from some macrolide producing *Actinomyces* strains resemble the eukaryotic type I enzyme (O'HAGAN, 1991; DONADIO et al., 1991). Erythronolide B synthase from *Saccharopolyspora erythraea* displays a modular organization which combines all activities of PKS I on three identical colinear polypeptides. These are encoded by three open reading frames (ORF) of each 5 kb (Fig. 5). Every ORF contains (in two modules) the double set of information coding for acyltransferase, acyl carrier protein and β-ketoacid ACP synthase (DONADIO et al., 1991). Due to the "modular" organization the polyketide synthesizing enzymes appear sixfold amplified. The identical polyketide I synthase subunits located on the three polypeptides cooperate closely during the elongation of the polyketide chain. The sixth modul is a terminating enzyme (TE) which is responsible for the cyclization to yield erythronolide B as the first

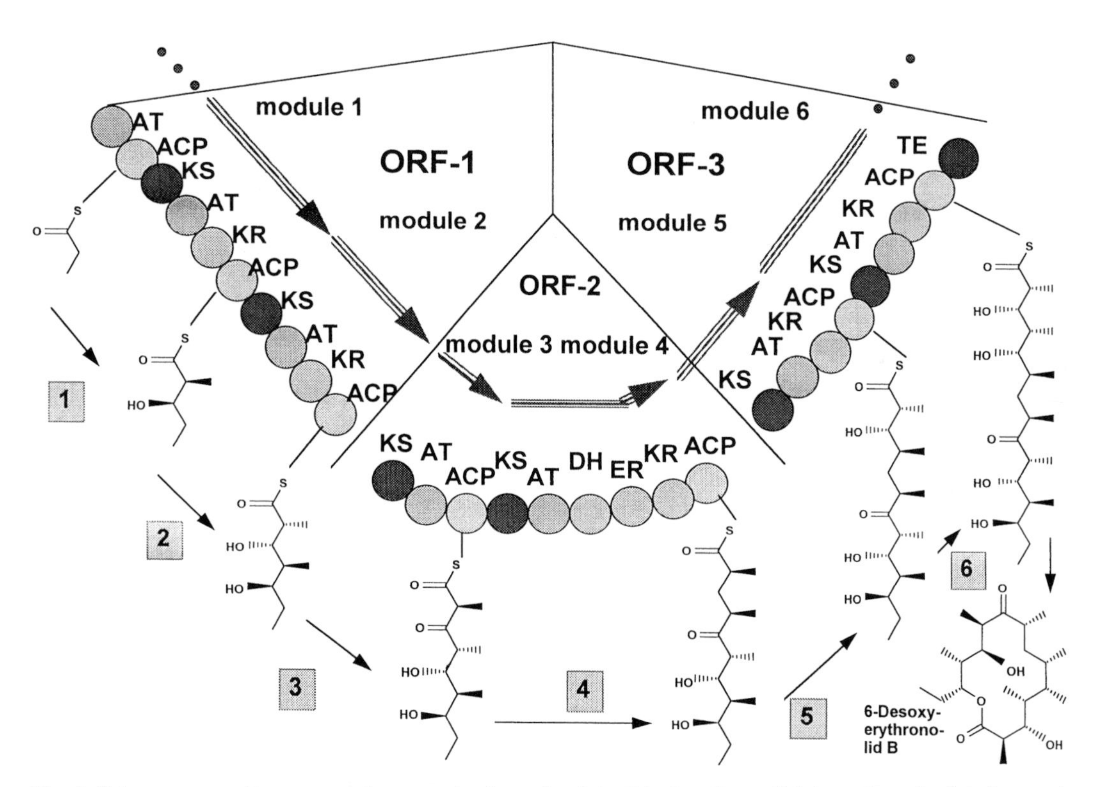

Fig. 5. Scheme suggesting a modular organization of polyketide (erythronolide) synthase in *Saccharopolyspora erythraea* (HOPWOOD and KHOSLA, 1992). Abbreviations. AT: Acyl-transferase; ACP: acyl-carrier proteins; KS: β-ketoacid-ACP-synthase; KR: β-ketoacid-ACP-reductase; DH: dehydrase; ER: enoyl reductase; TE: thioesterase (cyclase).

detectable intermediate of erythromycin biosynthesis. Altered structures of polyketides may be engineered by point mutations within functional domains (KATZ and DONADIO, 1995), by positional alterations of domains, e.g., the terminating thioesterase domain (WIESMANN et al., 1995), or by domain exchanges (BEDFORD et al., 1996; OLIYNYK et al., 1996).

The genes of the aromatic type II polyketide synthases from different streptomycetes display extensive sequence homology suggesting only minor differences in the substrate specificity and in the sequence of reactions (O'HAGAN, 1991; DONADIO et al., 1991; HOPWOOD and KHOSLA, 1992). But the individual manner of folding of the intermediate enzyme-bound polyketides determines in a large measure what kind of cyclic aromatic system is formed from the same intermediate polyketide (ROHR et al., 1993). Obviously, daunomycin, tetracyclines, tetracenomycines, and some angucyclines arise from nonaketide precursors which are cyclicized in a quite different manner in the course of polyketide processing (Fig. 6). Various successful attempts have been made to deduce the functions of proteins detected in type II polyketide biosynthetic clusters (KIM et al., 1995). This has led to the concept of a minimal polyketide forming system containing the condensing enzyme, the acyl carrier protein, and a malonyl-CoA transferase (MCDANIEL et al., 1994). Additional proteins may then function as chain length factors determining the number of elongation steps and as cyclases directing the mode of cyclization (HUTCHINSON and FUJII, 1995). A number of new polyke-

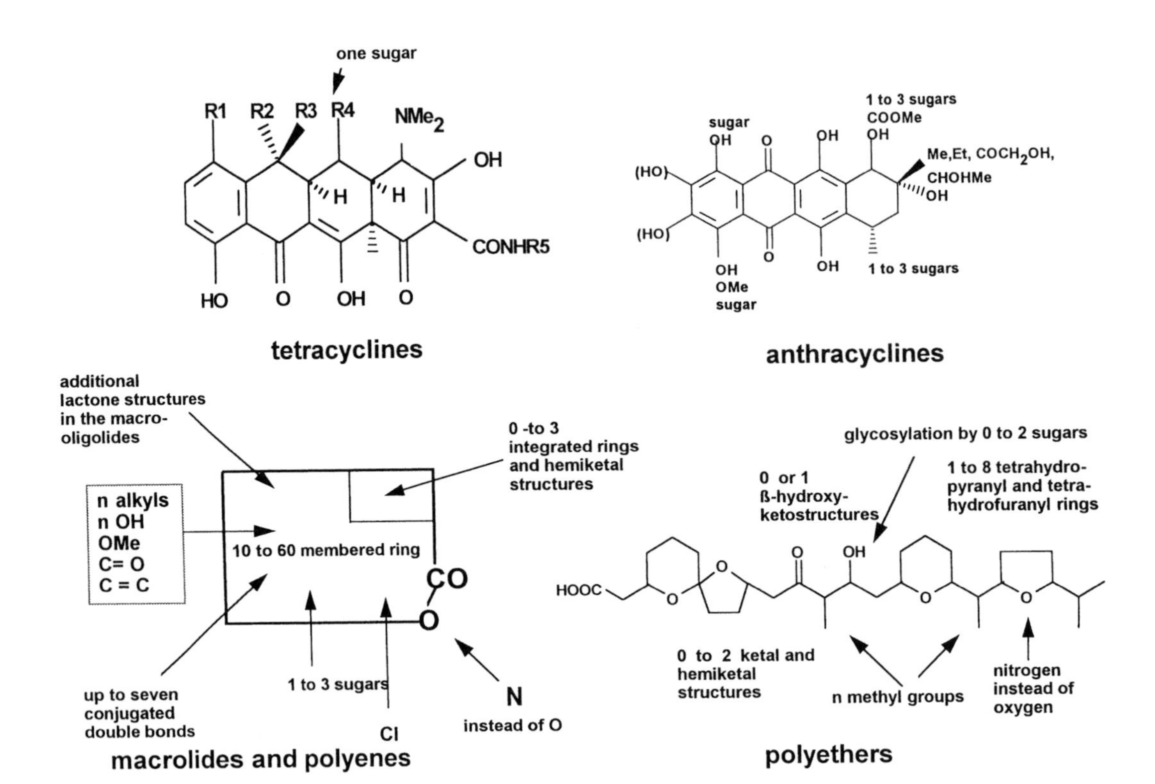

Fig. 6. Variations of polyketide structures (tetracyclines, anthracyclines, macrolides, polyethers) occurring in microorganisms. The substituents may vary in dependence of the given compound. The polyether structure shown above is highly variable with regard to the arrangement of the structural elements (rings, β-hydroxyketo structure, substituents).

tides have been formed by new strains with various combinations of minimal systems and factors leading to first combinatorial biosynthetic approaches (TSOI and KHOSLA, 1995; KAO et al., 1995). Without detailed structural knowledge of the proteins involved the results remain highly unpredictable (MEURER and HUTCHINSON, 1995), but exciting procedures for the generation of new compounds have been opened up (HUTCHINSON, 1994).

3.4 Terpenes

A plethora of mono-, sesqui-, di-, and triterpenoid structures of secondary metabolites are formed from acetyl coenzyme A via mevalonate and isopentenyl pyrophosphate. The initial steps of their biosynthesis (e.g., formation of β-hydroxy methylglutaryl coenzyme A, isopentenyl pyrophosphate, geranyl pyrophosphate, farnesyl pyrophosphate) are the same as in the formation of triterpenoid steroids and hopanoids as essential cellular constituents of fungi and bacteria (CANE et al. 1992; CANE, 1992, 1995). Terpenoid secondary metabolites frequently occur as secondary metabolites in plants and fungi, but they are rather unusual in bacteria (see, e.g., pentalenolacton, arenaemycin) (BERDY et al., 1980). Final steps of fungal terpenoid biosynthesis (e.g., trichothecens, germacrine, aristolochene, etc.) are carried out by specialized cyclases (CANE, 1992). Many cyclizations involve the protonation or alkylation of a double bond or an epoxide and the ionization of an allylic diphosphate ester. Thereafter, carbocationic intermediates are formed by the electrophilic attack of the resulting species to an olefinic bond followed by proton elimination and a reaction with water as a nucleophile. A series of terpenoid cyclases have been investigated recently by labeling and gene cloning experiments (CANE et al., 1992; CANE, 1992, 1995).

3.5 Sugar-Derived Oligomeric Structures

Biotransformations of simple monosaccharides, their activation as 1-O-nucleosides such as 1-O-dTDP and 1-O-dUDP derivatives and mutual coupling to other activated sugars generate more than 200 oligosaccharide structures in actinomycetes (BERDY et al., 1980; BYCROFT, 1988; LAATSCH, 1994). Aminocyclitols and other secondary metabolites thus originate from a few sugar moieties (HOTTA et al., 1995). The biosynthetic pathways leading to some therapeutically important representatives of sugar-derived structures such as streptomycin, kanamycin, and lincomycin have been investigated in detail (WRIGHT, 1983; PIEPERSBERG, 1994, 1995; PIEPERSBERG and DISTLER, Chapter 10, this volume). L-Glucosamine, streptidine, and L-streptose as constitutive parts of the streptomycin molecule are formed via three independent, multistep pathways. Thus, dTDP-L-dihydrostreptose formation is started from 1-O-dTDP-glucose followed by dehydratation, 3,5-epimerization, and reduction in an initial series of reactions (WRIGHT, 1983; PIEPERSBERG, 1994, 1995). Streptidine is synthesized by *S. griseus* from glucose via a series of at least twelve enzymic steps. By linkage of the three subunits hydrostreptomycin-6-O-phosphate is formed intracellularly which is inactive as an antibiotic (PIEPERSBERG, 1994). During its transport through the cytoplasmic membrane outside the cells oxidation occurs and phosphate is split off to yield the active streptomycin (WALKER and WALKER, 1978). Streptomycin biosynthesis was studied in more detail by the investigation of the pertinent genes and the corresponding enzymes (PIEPERSBERG, 1994, 1995; PIEPERSBERG and DISTLER, Chapter 10, this volume). It provides a nice example of the formation of sugar-derived secondary metabolites. Moreover, the same biosynthetic mechanism, steps of sugar activation and transformation, are suggested to be involved in the formation of mixed-type structures such as, e.g., macrolides (KATZ and DONADIO, 1995), anthracycline antibiotics (HUTCHINSON, 1995), and glycopeptides (ZMIJEWSKI and FAYERMAN, 1995; LANCINI and CAVALLERI, Chapter 9, this volume).

3.6 Oligo- and Polypeptides

Three ways of peptide bond formation are known in secondary metabolism (KLEINKAUF and VON DÖHREN, 1990, 1996):

- coupling of amino acids by single enzymes to form small peptides with up to five amino acids (e.g., glutathione, peptidoglycan),
- non-ribosomal biosynthesis of larger peptides (containing up to about 50 amino acids) by multienzyme complexes, and
- ribosomal mechanisms.

Oligopeptide biosynthesis on multienzyme complexes (as the most important mechanism) has been described for many bacterial products such as gramicidins, bacitracin, tyrocidin, and fungal secondary metabolites such as enniatins and cyclosporins (KLEINKAUF and VON DÖHREN, 1987, 1990, 1996).

The individual amino acids are first activated via adenylate formation and thereafter are bound as thioesters to the non-ribosomal synthase multienzyme complex. Subsequently, they are coupled in a step-by-step procedure to form large polypeptides which are sometimes composed of several subunits. The sequence of the amino acids in the peptide is exactly the same as that of the amino acids activated on the multienzyme complex ("thiotemplate-directed non-ribosomal peptide synthesis on a protein matrix"). Stepwise formation of the peptidic bonds occurs through translocation of the growing nascent peptide chain involving a phosphopantothenoyl carrier moiety (Fig. 7).

The terminating reactions are carried out by specified enzymic subunits of the same multienzyme complex. Cyclizations can occur to form cyclo- and depsipeptides as well as reductions, oxidations, and methylations which introduce, e.g., a disulfide bond (see, e.g., triostins) (VON DÖHREN, 1990; BERDY et al., 1980) or reduce a carboxylic acid to the pertinent aldehyde (see, e.g., pepstanone) (BERDY et al., 1980).

A major difference of template-directed mechanisms as compared to the ribosomal formation of peptidic bonds is the acceptance of non-proteinogenic amino acids and even of hydroxy acids and fatty acids either as building blocks of the oligomer formation or as carbon and nitrogen terminal substituents (KLEINKAUF and VON DÖHREN, 1987; VON DÖHREN, 1990). This peculiarity of the non-ribosomal mechanism contributes in a particular manner to the structural diversity of low-molecular weight peptides produced as secondary metabolites by so many microorganisms (BERDY et al., 1980; BYCROFT, 1988; LAATSCH, 1994).

Genetic analysis of peptide forming enzyme systems has revealed a modular structure of the enzymes involved. As in the case of polyketides various degrees of integration

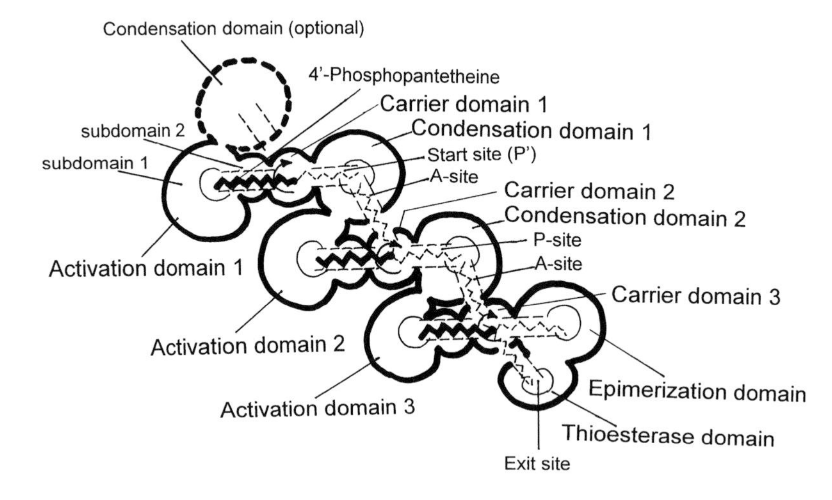

Fig. 7. Schematic view of the multiple carrier protein model of enzymatic peptide formation (thiotemplate model).

are found, with eukaryotic systems generally being fully integrated (KLEINKAUF and VON DÖHREN, 1996, and Chapter 7, this volume). Genetic exchange of modules specifying amino acids or related substrates in the protein code may lead to new peptides of altered composition (STACHELHAUS et al., 1995a, b).

Polypeptide-type secondary metabolites such as, e.g., microcins, tendamistat, subtilin, and lantibiotics (epidermin, gallidermin, nisin) are biosynthesized in microorganisms on the ribosomes as larger prepeptides. During their export into the medium, proteolytic processing occurs to yield the bioactive structures. A series of posttranslational alterations, such as the linkage to chromogenic and other groups, the formation of lanthionine, methyl lanthione, and disulphide units increases the number of possible homologs and creates the bioactive structures (SAHL et al., 1995; GASSON, 1995; JACK et al., Chapter 8, this volume; MORENO et al., 1995).

3.7 Biosynthetic Modifications of Structures and Precursor-Directed Biosyntheses

The secondary metabolism is carried out by specified enzymes acting within the frame of long biosynthetic chains. Modified structures can frequently be obtained due to the comparably low substrate specificity of some enzymes (LUCKNER, 1989). In many cases, feeding of a tentative precursor molecule or interruption of its biosynthesis, e.g., by the addition of metabolic inhibitors, has been used successfully to direct the secondary metabolism toward the formation of one single component of a mixture of naturally occurring metabolites (precursor-directed biosynthesis) (SADAKANE et al., 1983; DUTTON et al., 1991). Some of the producer strains even accept structural homologs of the natural precursor to form unusual derivatives of the original molecule(s) (SADAKANE et al., 1983; BALDWIN et al., 1991; MARTINEZ-BLANCO et al., 1991; LUENGO, 1995).

The term "mutational biosynthesis" signifies the use of blocked "idiotrophic" mutants which are unable to carry out the complete biosynthetic pathway (SHIER et al., 1969; CLARIDGE, 1983; THIERICKE and ROHR, 1993). Biosynthesis is initiated, again, when the missing intermediate is fed to the medium. Feeding of chemically derived analogs of the pertinent intermediate can yield new structural variants of the initial products. This technique was invented already in 1969 by SHIER (SHIER et al., 1969), and in a few cases (avermectins, cyclosporins) (see, e.g., DUTTON et al., 1991) more powerful compounds were obtained. Similar to the mutational biosynthesis the "hybrid biosynthesis" employs idiotrophic mutants which are blocked in a particular step of the secondary biosynthetic pathway (SADAKANE et al., 1983). Some of them accumulate intermediates of the interrupted biosynthetic chain due to the lack of a transforming enzymic step. Such kinds of intermediates (e.g., the protylonolide from *Streptomyces fradiae*) were fed to blocked mutants of another strain missing the formation of a similar intermediate (e.g,. spiramycino lactone in idiotrophs of *Streptomyces ambofaciens* forming spiramycin). Sometimes the fed heterologous metabolite (e.g., protylonolide) can be used in the same manner as the native metabolite (spiramycino lactone). In this way, chimeramycins were formed as hybrids of secondary metabolite structures from two different *Streptomyces* strains (SADAKANE et al., 1983).

As was demonstrated with biosynthetic enzymes, e.g., acyltransferase and isopenicillin-N-synthase of *Penicillium chrysogenum,* cyclosporin synthase of *Beauveria niveum,* enniatin synthase of *Fusarium oxysporum,* and gramicidin S synthases of *Bacillus brevis,* directed biosyntheses can also be carried out very efficiently by cell-free enzymes (BALDWIN et al., 1991; MARTINEZ-BLANCO, 1991; LAWEN and TRABER, 1993; KLEINKAUF and VON DÖHREN, 1996). The above biocatalysts convert a series of synthetic acyl coenzyme-A derivatives and homologs of the ACV-tripeptide to form novel penicillins which have not occurred as microbial products so far. Reference should also be given here to the use of enzymes in biotransformations of secondary metabolites (see, e.g., the enzymatic hydrolyses of the side chains of penicillins and cephalosporin C). The total synthesis of various cyclopeptides and depsipeptides has been carried out up to the milligram scale.

Moreover, growing evidence attests to the outstanding possibilities of molecular genetics in the modification of already known structures, and in the generation of new structures (HOPWOOD, 1989; HOPWOOD and SHERMAN, 1990; HUTCHINSON, 1994; HUTCHINSON and FUJII, 1995). Genes of *Streptomyces* type II polyketide synthases have recently been transferred to other *Streptomyces* hosts, and the biosynthesis of new and modified aromatic structures (hybrid antibiotics) is being exploited (HOPWOOD, 1989; HUTCHINSON and FUJII, 1995).

4 Variability of Structures of Secondary Metabolites

4.1 Secondary Metabolites as Products of Biological "Unit Operations"

Starting from a few molecular structures as precursors, the secondary pathways of the microbial kingdom produce much more than 10000, and the secondary pathways of plants produce more than 100000 different chemical individuals (VERALL, 1985; GROOMBRIDGE, 1992). At first glance, this huge number appears to be incredibly high, but the observer soon recognizes that the majority of structures are representatives of some few structural classes. Many homologs of a basic structure have been disclosed, not only in a given strain but also in different species and genera (BERDY et al., 1980; BYCROFT, 1988; LAATSCH, 1994). The detection of a novel structural class of natural drugs structurally unrelated to the already known compounds appears to be rather rare. Some organisms are characterized by the preferred production of a particular secondary class of metabolites (c.f., the frequent formation of polyene macrolides by streptomycetes or of sesqui- and diterpenes by some plants). This is the reason why the search for new structures turns to unusual sources such as plants, animals, and microorganisms from special ecosystems (e.g., marine animals and bacteria, special fungi, lichens, algae). Plants referred to in folk medicine and marine tunicates, toxic snakes, and toads offer an advantageous field of research on new "leading" structures. Moreover, the biosynthetically available modifications of basic structures such as macrolides, peptides, polyethers, etc. follow distinct rules: some derivatives occur frequently, but others are very rare. In general, the anthracyclines, e.g., occur as glycosylated derivatives, whereas the tetracyclines are usually non-glycosylated. But previously, the dactylocyclins were detected in cultures of *Dactylosporangium* sp. as the first glycosylated representatives of the tetracycline family (TYMIAK et al., 1993).

Otherwise, small structural changes of a given basic structure will often cause major changes in biological activities. The macrolide antibiotics from streptomycetes are an example which are similar in structure but possess antibacterial, antifungal, insecticidal, nematocidal, immunosuppressant, and cytotoxic properties. Traditional rescreening of compounds in newly established biological screens leads to the detection of unsuspected biological activities. In addition, chemical derivatization of side chains is an established and especially effective procedure to arrive at functionally improved structures.

4.2 Structural Classifications of Secondary Metabolites

The large number of known secondary metabolites needs classification. This could be achieved by considering their biosynthesis, the producing organisms (bacteria, fungi, plants, animals, etc.), their biological activities, and also their chemical structures. Few examples can be mentioned here to show how the structural variability of secondary metabolites is channeled by classifications according to biosynthetic origin and chemical nature (BERDY et al., 1980; LANCINI and LORENZETTI, 1993).

Peptides

Peptidic drugs are produced by numerous bacteria, fungi, plants and even animals (c.f. the magainins and other skin antibiotics of toxic toads) (JACOB and ZASLOFF, 1994). Peptide antibiotics from microbial sources occur as linear homopeptides so far composed of a maximum of 45 amino acids (KLEINKAUF and VON DÖHREN, 1996). Substitutions by fatty acids are a characteristic feature of the lipopeptides. Many cyclic peptides are known (c.f., the cyclosporins as undecapeptides) and the amino acids are often replaced by α- and β-hydroxy acids in an irregular or even a regular manner (peptidolactones and depsipeptides).

Even non-proteinaceous amino acids can be constitutive parts of peptidic drugs (KEINKAUF and VON DÖHREN, 1986, 1987, 1990). Small peptide chains can be linked to other unique structures such as fatty acids and chromophoric groups whereas combinations with sugars, macrolides, and anthracyclines seem to be unusual.

Variable structures have also been unraveled in the high-molecular weight peptide antibiotics from streptomycetes such as lantibiotics (SAHL et al., 1995; GASSON, 1995; JACK et al., Chapter 8, this volume) and enzyme inhibitors (subtilin, streptinoplasmin, etc.) (BERDY et al., 1980). The peptide chains are formed by ribosomal mechanisms and posttranslational modifications create the individual bioactive structures.

Polyketide Drugs

Actinomycetes are rich sources of polyketide metabolites like macrolides, polycyclic aromatic and semi-aromatic compounds like tetracyclines, anthracyclines and angucyclines, polyethers, and ansamycins.

The variability of macrolide structures involves ring sizes ranging from 10 to 60 (as recently found in quinolidomycin) (HAYAKAWA et al., 1993). Up to seven conjugated and additionally isolated double bonds can be present in the macrocycle (BERDY et al., 1980; BYCROFT, 1988; LAATSCH, 1994).

Moreover, one to three sugars are attached to the non-polyene macrolide aglycones which are excessively substituted by hydroxy, methoxy, methyl, and epoxy functions.

Even open-chain polyenic fatty acids (e.g., enacyloxin) are produced by strains which cannot carry out the final step of lactonization (WATANABE et al., 1992).

Structures of the antitumor anthracyclines also demonstrate the diversity which has been introduced by a few modifications into a basic structure of a tetrahydro naphthacenequinone backbone. Up to ten sugars are linked to several molecule positions. In addition, the number of hydroxy, carboxymethyl, and keto groups varies in the individual representatives to form approximately 300 different structures.

Many aromatic polycyclic compounds are also derived from the polyketide pathway. During their biosynthesis ring closures involving nitrogen and oxygen substituents are frequent features. In this way, even heterocyclic structures such as carbazols, phenoxazins, and phenazines are formed (BERDY et al., 1980; BYCROFT, 1988; LAATSCH 1994).

In fungi mycotoxins such as the aflatoxins and ochratoxins are likewise polyketide metabolites (TURNER and ALDRIDGE, 1983; BROWN et al., 1996).

Terpenoid Structures

Rich sources are plants and fungi, while terpenoid structures rarely occur in bacteria. Characteristic fungal terpenoids are mycotoxins such as trichothecens (BERDY et al., 1980; TURNER and ALRIDGE, 1983; LUCKNER, 1989). Important bioactive terpenoid structures from plants are the vinca alkaloids (NOBLE, 1990; FOX, 1991) and taxol (HEINSTEIN and CHANG, 1994). Moreover, the triterpenoid steran backbone is widely distributed in natural structures such as digitalis glycosides, saponins, etc. Fungal antibiotics such as fusidic acid, cephalosporin P, azasterols, and toad toxins (see, e.g., bufadienolides) are also representatives of the same structural class. Marine tunicates offer a rich source of unique steroid-type molecules (CAVALIER-SMITH, 1992).

Oligoglycosides

Up to several hundred sugars (see, e. g., shizophyllan, lentinan, avilamycins) can be linked in a linear or even a cyclic manner. Therapeutically useful oligomeric representatives of the oligosaccharides are the aminocyclitols which contain amino inositols such as streptidine (in streptomycin) and 2-deoxystreptamin (in neomycins, gentamicins, kanamycins, istamycins, etc.) (UMEZAWA and HOOPER, 1982; DIMITRIU, 1996).

Nucleoside Antibiotics

Approximately 150 nucleoside analogs are known from natural sources like actinomycetes and fungi. They are characterized by the presence of "false" nucleobases as aglycones and/or by "false" sugars (ISONO, 1990, 1995). The biosynthetic strategies imply their formation from a normal nucleoside (see, e. g., guanosine in the biogenesis of tomaymycin) and/or sugars such as ribose.

These few examples mentioned above demonstrate that the structural characteristics of secondary metabolites can serve as a tool for their classification despite their outstanding diversity. Many compounds combine structural elements of several basic classes as can be seen in Tab. 1. Thus, biosynthetic pathway enzymes of different types interact in a highly specific manner.

5 Future Perspectives: New Products of the Secondary Metabolism

As DAVID PERLMAN, one of the pioneers of modern industrial microbiology once stated microbial capacity is rather unlimited, and if mankind asks the microbes the proper questions they will truly answer. This idea applies to all of the living organisms when they are considered as a source of new bioactive structures. Screening for new structures is still a growing business stimulating other fields of biotechnology and biomedicine (VERALL, 1985; MONAGHAN and TKACZ, 1990) (Tab. 1).

As far as the sources of new drugs are concerned, only a small percentage of the presumed microbial world population has been explored so far, and only a minor part of the existing microbial strains has been deposited in strain collections (CHICARELLI-ROBINSON et al., 1994). The traditional sources of bioactive microbial metabolites, actinomycetes, bacilli, and sporulating fungi, still appear promising. Special genera such as the *Myxobacteria* (REICHENBACH and HÖFLE, 1993) have been demonstrated as particularly rich producers of unique structures. Future interest is focused on ecological "niches" and microhabitats which might harbor peculiar organisms. They look promising because they could miss special metabolic control due to their particular adaptation to the natural environment. In this context, microorganisms from extremly poor grounds, marine systems, plant rhizo- and endospheres became subject to detailed investigations (ŌMURA, 1992). Plants still provide an apparently inexhaustible reservoir of new bioactive structures. More than in the past, increasing knowledge on the molecular, cellular, and organismic causes of diseases promotes the search for new drugs possessing more specific activity (TOMODA and ŌMURA, 1990; ŌMURA, 1992; TANAKA and ŌMURA, Chapter 3, this volume). Today, the screening assay determines what kind of novel natural product will be detected. In the last decade, an increasing number of publications on enzyme inhibitors and receptor antagonists attests that classical screening for "simply" antibiotic molecules has been extended and rationalized on the basis of modern and biochemical pharmacy and molecular biology (see Tab. 1). Even viral targets such as, e.g., viral proteinase and adhesive proteins (GP 120) became amenable to the search for new inhibitors.

The following screening assays are now commonplace: Mammalian cell cultures to detect receptor agonists and antagonists, inducers and inhibitors of cytodifferentiation, and effectors of cell-to-cell and cell-to-virus interactions. Mammalian cell lines transformed by the expression of oncogenes are used in the search for new antitumor agents promoting redifferentiation of cancer cells. A particular advantage of these assays is that they reduce animal trials which otherwise would be necessary in the development of new pharmacological agents. In the same manner the use of plant and insect cell cultures permits more rational and time-saving drug discovery in screening for new phytoeffectors and insecticides.

Cloning of genes encoding for enzymes, receptors, protein factors, etc. and their expression in heterologous organisms supplies another promising approach to drug discovery. The insertion of regulatory and reporter gene sequences into microorganisms and cells may be useful for the detection of compounds which interfere with DNA-binding proteins and transcriptional regulators. In this way, new specific inhibitors of oncogenesis and viral replication may be uncovered.

As a conclusion, secondary metabolism in microbes, plants, and animals still promises new leading structures for future drug development. This promise is due to the apparently inexhaustible pool of organisms and structures and the rapid development in biomedical and biochemical disease research. Moreover, secondary metabolism supplies biochemical tools which allow deeper insights into cellular processes (see, e.g., the previous discovery of inhibitors of protein kinases and protein phosphatases as effectors of the mammalian cell cycle). It seems reasonable to believe that the discovery of new leading structures of antibiotics, anticancer and antiviral drugs, pharmacological agents, crop protecting and insecticidal compounds, etc. will continue to promote the future development of biotechnology and medicine. Every new structure provides a challenge to the biochemist exploring its mode of action, to the chemist wanting to disclose the structure–activity relationships, to the pharmacologist studying the activity in macroorganisms, to the molecular biologist investigating the genes of the biosynthetic pathway, to the fermentation engineer developing a new biotechnical procedure, and last but not least, to those who could benefit for their health from new and better drugs.

Acknowledgements

We are gratefully indebted to Dr. HANS-PETER SALUZ for helpful comments and critical manuscript revision and to Dr. AXEL BRAKHAGE for helpful comments and part of Fig. 3.

6 References

AHARONOWITZ, Y., COHEN, G., MARTÍN, J. F. (1992), Penicillin and cephalosporin biosynthetic genes: structure, organization, regulation, and evolution, *Annu. Rev. Microbiol.* **46**, 461–495.

ALLEN, I. W., RITCHIE, D. A. (1994), Cloning and analysis of DNA sequences from *Streptomyces hygroscopicus* encoding geldanamycin biosynthesis, *Mol. Gen. Genet.* **243**, 593–599.

ALTENBUCHNER, J. (1994), Hohe genetische Instabilität von Streptomyceten durch chromosomale Deletionen und DNA-Amplifikationen, *Bio/Engineering* **3**, 33–46.

ARISAWA, A., TSUNEKAWA, H., OKAMURA, K., OKAMOTO, R. (1995), Nucleotide sequence analysis of the carbomycin biosynthetic genes including the 3-O-acyltransferase gene from *Streptomyces thermotolerans, Biosci. Biotech. Biochem.* **59**, 582–588.

ARMSTRONG, G. A. (1994), Eubacteria show their true colors: genetics of carotinoid pigment biosynthesis from microbes to plants, *J. Bacteriol.* **176**, 4795–4802.

ARROWSMITH, T. J., MALPARTIDA, F., SHERMAN, D. H., BIRCH, A., HOPWOOD, D. A., ROBINSON, J. A. (1992), Characterization of *actI*-homologous DNA encoding polyketide synthase genes from the monensin producer *Streptomyces cinnamonensis, Mol. Gen. Genet.* **234**, 254–264.

ARST, H. N., JR. (1996), Regulation of gene expression by pH, in: *The Mycota III* (ESSER, K., LEMKE, P. A., Eds.), *Biochemistry and Molecular Biology* (BRAMBLE, R., MARZLUFF, G. A., Eds.), pp. 235–240. Berlin, Heidelberg: Springer-Verlag.

AZUMA, I. (1987), Developments of immunostimulants in Japan, in: *Immunostimulants Now and Tomorrow* (AZUMA, I., JOLLES, G., Eds.), pp. 41–56. Tokyo, Berlin: Japan Scientific Press/Springer.

AZUMA, M., HORI, K., HASHI, Y.-O., YOSHIDA, M., HORINOUCHI, S., BEPPU, T. (1980), Basidifferquinone, a new inducer of fruiting-body formation of a basidiomycete *Farvolus arcularius* from a *Streptomyces* strain. II. Structure of basidifferquinone, *Agric. Biol. Chem.* **54**, 1447–1452.

BAILEY, A. M., KERSHAW, M. J., HUNT, B. A., PATERSON, I. C., CHARNLEY, A. K., REYNOLDS, S. E., CLARKSON, J. M. (1996), Cloning and sequence analysis of an intron-containing domain from a peptide synthetase of the entomopathogenic fungus *Metarhizium anisopliae, Gene* **173**, 195–197.

BAINTON, N. J., STEAD, P., CHABRA, S. R., BYCROFT, B. W., SALMOND, G. P. C., STEWARD, G. S. A. B., WILLIAMS, P. (1992), N-(3-Oxohexanoyl)-L-homoserine lactone regulates carbapenem antibiotic production in *Erwinia carotovora, Biochem. J.* **288**, 977–1004.

BALDWIN, J. E., BRADLEY, M., ABBOTT, S. D., ADLINGTON, R. M. (1991), New penicillins from isopenicillin N-synthase, *Tetrahedron* **47**, 5309–5328.

BARABAS, G., PENYIGE, A., HIRANO, T. (1994), Hormone-like factors influencing differentiation of *Streptomyces* cultures, *FEMS Microb. Rev.* **14**, 75–82.

BARBER, J., CHAPMAN, A. C., HOWARD, T. D., TEBB, G. (1988), Recycling of polyketides by fungi: the degradation of citrinin by *Penicillium citrinum, Appl. Microb. Biotechnol.* **29**, 387–391.

BEDFORD, D., JACOBSEN, J. R., LUO, G., CANE, D. E., KHOSLA, C. (1996), A functional chimeric modular polyketide synthase generated via domain replacement, *Chemistry & Biology* **3**, 827–831.

BECHTHOLD, A., SOHNG, J. K., SMITH, T. M., CHU, X., FLOSS, H. G. (1995), Identification of *Streptomyces violaceoruber* Tü22 genes involved in the biosynthesis of granaticin, *Mol. Gen. Genet.* **248**, 610–620.

BECK, J., RIPKA, S., SIEGNER, A., SCHILTZ, E., SCHWEIZER, E. (1990), The multifunctional 6-methylsalicylic acid synthase gene of *Penicillium patulum.* Its gene structure relative to that of other polyketide synthases, *Eur. J. Biochem.* **192**, 487–498.

BÊHAL, V., HUNTER, I. S. (1995) Tetracyclines, in: *Genetics and Biochemistry of Antibiotic Production* (VINING, L. C., STUTTARD, C., Eds.), pp. 359–384. Boston: Butterworth-Heinemann.

BENDER, C., PALMER, D., PENALOZAVAZQUEZ, A., RANGASWAMY, V., ULLRICH, M. (1996), Biosynthesis of coronatine, a thermoregulated phytotoxin produced by the phytopathogen *Pseudomonas syringae, Arch. Microbiol.* **166**, 71–75.

BEPPU, T. (1992), Secondary metabolites as chemical signals for cellular differentiation, *Gene* **115**, 159–165.

BEPPU, T. (1995), Signal transduction and secondary metabolism: prospects for controlling productivity, *Trends Biotechnol.* **13**, 264–269.

BERDY, J., ASZALOS, A., BOSTIAN, M., MCNITT, K. L. (1980), CRC *Handbook of Antibiotic Compounds, 1980–1984.* Boca Raton, FL: CRC Press.

BEYER, S., DISTLER, J., PIEPERSBERG, W. (1996), The Str cluster for the biosynthesis of 5'-hydroxystreptomycin in *Streptomyces glaucescens* Gla.0(ETH 22794) – new operons and evidence for pathway specific regulation by StrR, *Mol. Gen. Genet.* **250**, 775–784.

BIBB, M. J., SHERMAN, D. H., ŌMURA, S., HOPWOOD, D. A. (1994), Cloning, sequencing and deduced functions of a cluster of *Streptomyces* genes probably encoding biosynthesis of the polyketide antibiotic frenolicin, *Gene* **142**, 31–39.

BIRCH, A. W., ROBINSON, J. A. (1995), Polyethers, in: *Genetics and Biochemistry of Antibiotic Production* (VINING, L. C., STUTTARD, C., Eds.), pp. 443–476. Boston, MA: Butterworth-Heinemann.

BIRCH, A. W., LEISER, A., ROBINSON, J. A. (1993), Cloning, sequencing, and expression of the gene encoding methylmalonyl coenzyme A mutase from *Streptomyces cinnamonensis, J. Bacteriol.* **175**, 3511–3519.

BLACK, T. A., WOLK, C. P. (1994), Analysis of a Het(−) mutation in *Anabaena* Sp strain PCC 7120 implicates a secondary metabolite in the regulation of heterocyst spacing, *J. Bacteriol.* **176**, 2282–2292.

BORMANN, C., MÖHRLE, V., BRUNTNER, C. (1996), Cloning and heterologous expression of the entire set of structural genes for nikkomycin synthesis from *Streptomyces tendae* Tü901 in *Streptomyces lividans, J. Bacteriol.* **178**, 1216–1218.

BRAKHAGE, A. A., TURNER, G. (1995), Biotechnical genetics of antibiotic biosynthesis, in: *The Mycota II* (ESSER, K., LEMKE, P. A., Eds.), *Genetics and Biotechnology* (KÜCK, U., Ed.), pp. 263–385. Berlin, Heidelberg: Springer-Verlag.

BRAKHAGE, A. A., VAN DEN BRULLE, J. (1995), Use of reporter genes to identify recessive *trans*-acting mutations specifically involved in the regulation of *Aspergillus nidulans* penicillin biosynthetic genes, *J. Bacteriol.* **177**, 2781–2788.

BRAKHAGE, A. A., BROWNE, P., TURNER, G. (1992), Regulation of *Aspergillus nidulans* penicillin biosynthesis and penicillin biosynthesis genes *acvA* and *ipnA* by glucose, *J. Bacteriol.* **174**, 3789–3799.

BROWN, D. W., YU, J.-H., KELKAR, H. S., FERNANDES, M., NESBITT, T. C., KELLER, N. P., ADAMS, T. H., LEONARD, T. J. (1996), Twenty-five coregulated transcripts define a sterigmatocystin gene cluster in *Aspergillus nidulans, Proc. Natl. Acad. Sci. USA* **93**, 1418–1422.

BRÜCKNER, B., BLECHSCHMIDT, D., SEMBDNER, G. SCHNEIDER, G. (1990), Fungal gibberellin production, in: *Biotechnology of Vitamins, Pigments, and Growth Factors* (VANDAMME, E., Ed.), pp. 383–429. London, New York: Elsevier.

BUADES, C., MOYA, A. (1996), Phylogenetic analysis of isopenicillin N synthetase horizontal gene transfer, *J. Mol. Evol.* **42**, 537–542.

BUSHELL, M. E., FRYDAY, A. (1983), The application of materials balancing to the characterization of sequential secondary metabolite formation in *Streptomyces cattleya* NRRL 8057, *J. Gen. Microbiol.* **129**, 1733–1741.

BYCROFT, B. J. (1988), *Dictionary of Antibiotics.* London, New York: Chapman and Hall.

CALAM, C. T. (1987), *Process Design and Control in Antibiotic Fermentations.* Cambridge: Cambridge University Press.

CAMPBELL, W. C. (Ed.) (1989), *Ivermectin and Avermectin.* New York, Heidelberg: Springer-Verlag.

CANE, D. E. (1992), Terpenoid cyclases: design and function of electrophilic catalysts, in: *Secondary Metabolites: Their Function and Evolution, Ciba Foundation Symposium 171* (CHADWICK, D. J., WHELAN, J., Eds.), pp. 163–168. Chichester, New York: J. Wiley & Sons.

CANE, D. E. (1995), Isoprenoid antibiotics, in: *Genetics and Biochemistry of Antibiotic Production* (VINING, L. C., STUTTARD, C., Eds.), pp. 633–656. Boston, MA: Butterworth-Heinemann.

CANE, D. E., YANG, G., COATES, R. M., PYNN, H., HOLM, T. M. (1992), Trichodiene synthase. Synergistic inhibition by inorganic phosphate and aza-analogs of the bisabolyl cation, *J. Org. Chem.* **57**, 3454–3462.

CASHEL, M. (1975), Regulation of bacterial ppGpp and pppGpp, *Ann. Rev. Microbiol.* **29**, 301–335.

CAVALIER-SMITH, T. (1992), Origins of secondary metabolism, in: *Secondary Metabolites: Their Function and Evolution, Ciba Foundation Symposium 171* (CHADWICK, D. J., WHELAN, J., Eds.), pp. 64–87. Chichester, New York: J. Wiley & Sons.

CHATER, K. F. (1992), Genetic regulation of secondary metabolic pathways in streptomyces, in: *Secondary Metabolites: Their Function and Evolution, Ciba Foundation Symposium 171* (CHADWICK, D. J., WHELAN, J., Eds.), pp. 144–162. Chichester, New York: J. Wiley & Sons.

CHEN, C. W., LIN, Y.-S., YANG, Y.-L., TSOU, M.-F., CHANG, H.-M., KIESER, H. M., HOPWOOD, D. A. (1994), The linear chromosomes of *Streptomyces:* structure and dynamics, *Actinomycetologica* **8**, 103–112.

CHEN, Q., WERTHEIMER, A. M., TOLMASKY, M. E., CROSA, J. H. (1996), The AngR protein and the siderophore anguibactin positively regulate the expression of iron-transport genes in *Vibrio anguillarum, Mol. Microbiol.* **22**, 127–134.

CHIAO, J.-S., XIA, T.-H., MEI, B.-G., JIN, Z. K., GU, W.-L. (1995), Ansamycins, in: *Genetics and Biochemistry of Antibiotic Production* (VINING, L. C., STUTTARD, C., Eds.), pp. 477–498. Boston, MA: Butterworth-Heinemann.

CHICARELLI-ROBINSON, M. I., FOX, F., NISBETH, C. (1994), The value of microbial screening in drug discovery (abstract), in: *New Strategies in Searching Drug Leads,* pp. 1–3. Airport Symposium, Frankfurt, 28 February.

CHU, Y. W., RENNO, D., SAUNDERS, G. (1995), Detection of a protein which binds specifically to the upstream region of the *pcbAB* gene in *Penicillium chrysogenum, Curr. Genet.* **28**, 184–189.

CLARIDGE, C. A. (1983), Mutasynthesis and directed biosynthesis for the production of new antibiotics, in: *Basic Biology and New Developments in Biotechnology* (HOLAENDER, A., LASKIN, A. I., ROGERS, P., Eds.), pp. 231–269. New York: Plenum Press.

COFFEY, T. J., DOWSON, C. G., DANIELS, M., SPRATT, B. G. (1995), Genetics and molecular biology of beta-lactam-resistant Pneumococci, *Microb. Drug Res. – Mechanisms, Epidemiology and Disease* **1**, 29–34.

COMITÉ EDITORIAL (1992), Médicaments antibiotiques. Traité de chimie thérapeutique, Vol. 2. Paris: Tec & Doc Lavosier.

COOK, R. J., THOASHOW, L. S., WELLER, D. M., FUJIMOTO, D., MAZZOLA, M., BANGERA, B., KIM, D. (1995) Molecular mechanisms of defense by rhizobacteria against root disease, *Proc. Natl. Acad. Sci. USA* **92**, 4197–4201.

COQUE, J. J. R., LIRAS, P., MARTÍN, J. F. (1993), Genes for a β-lactamase, a pencillin binding protein and a transmembrane protein are clustered with the cephamycin biosynthetic genes in *Nocardia lactamdurans, EMBO J.* **12**, 631–639.

COQUE, J. J. R., ENGUITA, F. J., MARTÍN, J. F., LIRAS, P. (1995a), A two protein component 7-cephem-methoxylase encoded by two genes of the cephamycin C cluster converts cephalosporin to 7-methoxycephalosporin C, *J. Bacteriol.* **177**, 2230–2235.

COQUE, J. J. R., PÉREZ-LLARENA, F. J., ENGUITA, F. J., FUENTE, J. L., MARTÍN, J. F., LIRAS, P. (1995b), Characterization of the *cmcH* genes of

Nocardia lactamdurans encoding a functional 3'-hydroxymethylcephem O-carbamoyltransferase for cephamycin biosynthesis, *Gene* **162**, 21–27.

Cosima, P., Rodriguez, F., de Ferra, F., Grandi, G., Perego, M., Venema, G., van Sinderen, D. (1993), Sequence and analysis of the genetic locus responsible for surfactin synthesis in *Bacillus subtilis, Mol. Microbiol.* **8**, 821–831.

Cundliffe, E. (1989), How do antibiotic-producing organisms avoid suicide, *Annu. Rev. Microbiol.* **43**, 207–233.

Cundliffe, E. (1992), Self-protection mechanisms in antibiotic producers, in: *Secondary Metabolites: Their Function and Evolution, Ciba Foundation Symposium 171* (Chadwick, D. J., Whelan, J., Eds.), pp. 199–214. Chichester, New York: J. Wiley & Sons.

Davies, J. (1994), Inactivation of antibiotics and the dissemination of resistance genes, *Science* **264**, 375–382.

Davis, N. K., Chater, K. F. (1990), Spore colour in *Streptomyces coelicolor* A3(2) involves the developmentally regulated synthesis of a compound biosynthetically related to polyketide antibiotics, *Mol. Microbiol.* **4**, 1679–1691.

Davis, N. K., Chater, K. F. (1992), The *Streptomyces coelicolor whiB* gene encodes a small transcription factor-like protein dispensable for growth but essential for sporulation, *Mol. Gen. Genet.* **232**, 351–358.

Decker, H., Rohr, J., Motamedi, H., Zähner, H., Hutchinson, C. R. (1995), Identification of *Streptomyces olivaceus* Tü2353 genes involved in the production of the polyketide elloramycin, *Gene* **166**, 121–126.

Demain, A. L. (1974), Nutrition and the function of secondary metabolites, *Adv. Appl. Microbiol.* **16**, 177–202.

Demain, A. L. (1984), Biology of antibiotic formation, in: *Biotechnology of Industrial Antibiotics* (Vandamme, E. J., Ed.), pp. 33–39. New York: Marcel Dekker.

Demain, A. L. (1989), Carbon source regulation of idiolite biosynthesis in actinomycetes, in: *Regulation of Secondary Metabolism in Actinomycetes* (Shapiro, S., Ed.), pp. 127–134. Boca Raton, FL: CRC Press.

Demain, A. L. (1992), Microbial secondary metabolism: a new theoretical frontier for academia, a new opportunity for industry, in: *Secondary Metabolites: Their Function and Evolution, Ciba Foundation Symposium 171* (Chadwick, D. J., Whelan, J., Eds.). Chichester, New York: J. Wiley & Sons.

De Ruyter, P. G. G. A., Kuipers, O. P., Beerthuyzen, M. M., van Alenboerrigter, I., Devos, W. M. (1996), Functional analysis of promoters in the nisin gene cluster of *Lactobacillus lactis, J. Bacteriol.* **178**, 3434–3439.

Dickens, M. L., Ye, J. S., Strohl, W. R. (1996), Cloning, sequencing, and analysis of aklaviketone reductase from *Streptomyces* sp. strain C5, *J. Bacteriol.* **178**, 3384–3388.

Díez, B., Gutiérrez, S., Barredo, J. L., van Solingen, P., van der Voort, L. H. M., Martín, J. F. (1990), The cluster of penicillin biosynthetic genes. Identification and characterization of the *pcbAB* gene encoding α-aminodipyl-cysteinyl-valine synthetase and linkage to the *pcbC* and *penDE* genes, *J. Biol. Chem.* **265**, 16358–16365.

Dimitriu, S. (Ed.) (1996), *Polysaccharides in Medical Applications*. New York, Basel, Hongkong: Marcel Dekker.

Distler, W., Ebert, A., Heinzel, P., Mansouri, K., Mayer, G., Pissowotzki, K. (1988), Expression of genes for streptomycin biosynthesis, in: *Biology of Actinomycetes '88* (Okami, Y., Beppu, T., Ogawara, H., Eds.), pp. 86–91. Toyko: Japan Scientific Society Press.

Distler, J., Mansouri, K., Stockmann, M., Piepersberg, W. (1992), Streptomycin biosynthesis and its regulation in Streptomycetes, *Gene* **115**, 105–111.

Donadio, S., Starer, M. J., McAlpine, J. B., Swanson, S. J., Katz, L. (1991), Modular organization of genes required for complex polyketide biosynthesis, *Science* **252**, 675–679.

Doull, J. L., Vining, L. C. (1995), Global physiological controls, in: *Genetics and Biochemistry of Antibiotic Production* (Vining, L. C., Stuttard, C., Eds.), pp. 9–64. Boston, MA: Butterworth-Heinemann.

D'Souza, C., Nakano, M. M., Zuber, P. (1994), Identification of *comS*, a gene of the *srfA* operon that regulates the establishment of genetic competence in *Bacillus subtilis, Proc. Natl. Acad. Sci. USA* **91**, 9397–9401.

Dutton, C. J., Gibson, S. P., Goodie, A. C., Holdom, K. S., Pacey, M. S., Ruddock, J., Bu'Lock, J. D., Richards, M. K. (1991), Novel avermectins produced by mutational biosynthesis, *J. Antibiot.* **44**, 357–365.

Dyson, P., Schrempf, H. (1987), Genetic instability and DNA-amplification in *Streptomyces lividans* 66, *J. Bacteriol.* **169**, 4769–4803.

Elder, J. H., Schnölzer, M., Hasselkus-Light, C. S., Henson, M., Lerner, D. A., Phillips, T. R., Wagaman, P. C., Kent, S. B. H. (1993), Identification of proteolytic processing sites within the gag and pol polyproteins of feline immunodeficiency virus, *J. Virol.* **67**, 1869–1876.

ELLIS, D. D., ZELDIN, E. L., BRODHAGEN, M., RUSSIN, W. A., MCCOWN, B. H. (1996), Taxol production in nodule cultures of *Taxus*, *J. Nat. Prod.* **59**, 246–250.

ENDO, A. (1979), Monacolin K, a new hypercholesterolemic agent produced by *Monascus species, J. Antibiot.* **32**, 852–854.

ESPESO, E. A., PEÑALVA, M. A. (1996), Three binding sites for the *Aspergillus nidulans* PacC zinc-finger transcription factor are necessary and sufficient for regulation by ambient pH of the isopenicillin N synthase gene promoter, *J. Biol. Chem.* **271**, 28825–28830.

ESPESO, E. A., TILBURN, J., ARST, H. N., JR., PEÑALVA, M. A. (1993), pH regulation is a major determinant in expression of a fungal biosynthetic gene, *EMBO J.* **12**, 3947–3956.

FENG, B., FRIEDLIN, E., MARZLUFF, G. A. (1995), Nuclear DNA-binding proteins which recognize the intergenic control region of penicillin biosynthetic genes, *Curr. Genet.* **27**, 351–358.

FERNÁNDEZ-MORENO, M. A., VALLÍN, C., MALPARTIDA, F. (1996), Isolation and characterization of streptothricin biosynthetic genes from a newly *Streptomyces* isolated strain, *Proc. Conf. Biol. Streptomycetes,* Ohrbeck, Abstract V51.

FIECHTER, A. (1988), Bioprocess development, in: *Overproduction of Macrolide Metabolites* (VANEK, Z., HOSTALEK, Z., Eds.), pp. 231–259. Boston, London: Butterworth.

FIERRO, F., BARREDO, J. L., DIEZ, B., GUTIERREZ, S., FERNANDEZ, F. J., MARTÍN, J. F. (1995), The penicillin gene cluster is amplified in tandem repeats linked by conserved hexanucleotide sequences, *Proc. Natl. Acad. Sci. USA* **92**, 6200–6204.

FIERRO, F., MONTENEGRO, E., GUTIERREZ, S., MARTÍN, J. F. (1996), Mutants blocked in penicillin biosynthesis show a deletion of the entire penicillin gene cluster at a specific site within a conserved hexanucleotide sequence, *Appl. Microbiol. Biotechnol.* **44**, 597–604.

FILIPPINI, S., SOLINAS, M. M., BREME, U., SCHLÜTER, M. B., GABELLINI, D., BIAMONTI, G., COLOMBO, A. L., GAROFANO, L. (1995), *Streptomyces peucetius* daunorubicin biosynthesis gene, *dnrF*: sequence and heterologous expression, *Microbiology* **141**, 1007–1016.

FOX, B. W. (1991), Natural products in cancer treatment from bench to the clinic, *Trans. R. Soc. Trop. Med. Hyg.* **85**, 22–85.

FUQUA, W. C., WINANS, S. C., GREENBERG, E. P. (1994), Quorum sensing in bacteria: the LuxR–Lux I family of cell-density-responsive transcriptional regulators, *J. Bacteriol.* **176**, 269–275.

GASSON, M. J. (1995), Lantibiotics, in: *Genetics and Biochemistry of Antibiotic Production* (VINING, L. C., STUTTARD, C., Eds.), pp. 283–306. Boston, MA: Butterworth-Heinemann.

GEIGER, O. (1994), N-Acyl-homoserinlacton-Autoinduktor-Signalmoleküle bei gramnegativen Bakterien, *BioEngineering* **5**, 40–46.

GOODAY, G. W. (1974), Fungal sex hormones, *Ann. Rev. Biochem.* **1974**, 35–49.

GOODRICHTANRIKULU, M., MAHONEY, M. E., RODRIGUEZ, S. B. (1995), The plant growth regulator methyl jasmonate inhibits aflatoxin production by *Aspergillus flavus, Microbiology* **141**, 2831–2837.

GRÄFE, U. (1989), Autoregulatory secondary metabolites from actinomycetes, in: *Regulation of Secondary Metabolism in Actinomycetes* (SHAPIRO, S., Ed.), pp. 75–126. Boca Raton, FL: CRC Press.

GRÄFE, U. (1992), *Biochemie der Antibiotika.* Heidelberg: Spektrum Verlag.

GRIMM, A., MADDURI, K., ALI, A., HUTCHINSON, C. R. (1994), Characterization of the *Streptomyces peucetius* ATCC 29050 genes encoding doxorubicin polyketide synthase, *Gene* **151**, 1–10.

GROOMBRIDGE, B. (Ed.) (1992), *Global Biodiversity. Status of the Earth's Living Resources.* London: Chapman and Hall.

HAAS, H., MARZLUFF, G. A. (1995), NRE, the major nitrogen regulatory protein of *Penicillium chrysogenum,* binds specifically to elements in the intergenic promoter regions of nitrate assimilation and penicillin biosynthetic gene clusters, *Curr. Genet.* **28**, 177–183.

HAMOEN, L. W., ESHUIS, H., JONGBLOED, J., VENEMA, G., VAN SINDEREN, D. (1995), A small gene, designated *comS,* located within the coding region of the fourth amino acid-activation domain of *srfA,* is required for competence development in *Bacillus subtilis, Mol. Microbiol.* **15**, 55–63.

HANLON, G. W., HODGES, N. A. (1981), Bacitracin and protease production in relation to sporulation during exponential growth of *Bacillus licheniformis* on poorly utilized carbon and nitrogen source, *J. Bacteriol.* **147**, 427–431.

HASUMI, K. (1993), Competitive inhibition of squalene synthetase by squalestatin-I, *J. Antibiot.* **46**, 689–691.

HAYAKAWA, Y., MATSUOKA, M., SHIN-YA, K., SETO, H. (1993), Quinolidomycins, A_1, A_2, and B_1, novel 60-membered macrolide antibiotics. I. Taxonomy, fermentation, isolation, physicochemical properties and biological activity, *J. Antibiot.* **46**, 1557–1562.

HAYASHI, H., TARUI, N., MURAO, S. (1985), Isolation and identification of cyclooctasulfur, a fruiting-body-inducing substance, produced by *Strep-*

tomyces albulus TO447, *Agric. Biol. Chem.* **49**, 101.

HAYDOCK, S. F., APARICIO, J. F., MOLNAR, I., SCHWECKE, T., KHAW, L. E., KÖNIG, A., MARSDEN, A. F. A., GALLOWAY, I. S., STAUNTON, J., LEADLAY, P. F. (1995), Divergent sequence motifs correlated with the substrate specificity of (methyl)malonyl-CoA-acyl carrier protein transacylase domains in modular polyketide syntheses, *FEBS Lett.* **374**, 246–248.

HEINSTEIN, P. F., CHANG, C. J. (1994), Taxol, *Ann. Rev. Plant. Physiol. Plant Mol. Biol.* **45**, 663–674.

HERZOG-VELIKONJA, B., PODLESEK, Z., GRABNAR, M. (1994), Isolation and characterization of Tn917-generated bacitracin deficient mutants of *Bacillus licheniformis, FEMS Microbiol. Lett.* **121**, 147–152.

HIRAMATSU, K. (1995), Molecular evolution of MRSA, *Microbiol. Immunol.* **39**, 531–543.

HODGSON, J. E., FOSBERRY, A. P., RAWLINSON, N. S., ROSS, H. N. M., NEAL, R. J., ARNELL, J. C., EARL, A. J., LAWLOR, E. J. (1995), Clavulanic acid biosynthesis in *Streptomyces clavuligerus*: gene cloning and characterization, *Gene* **166**, 49–55.

HOPWOOD, D. A. (1988), Towards an understanding of gene switching in *Streptomyces,* the basis of sporulation and antibiotic production, *Proc. R. Soc. London B: Biological Sciences* **235**, 121–138.

HOPWOOD, D. A. (1989), Antibiotics: opportunities for genetic manipulation, *Philos. Trans. R. Soc.* (London) **324**, 549–562.

HOPWOOD, D. A., KHOSLA, C. (1992), Genes for polyketide secondary metabolic pathways in microorganisms and plants, in: *Secondary Metabolites: Their Function and Evolution, Ciba Foundation Symposium 171* (CHADWICK, D. J., WHELAN, J., Eds.), pp. 88–112. Chichester, New York: J. Wiley & Sons.

HOPWOOD, D. A., SHERMAN, D. A. (1990), Molecular genetics of polyketides and its comparison to fatty acid biosynthesis, *Annu. Rev. Genet.* **24**, 37–66.

HOPWOOD, D. A., KIESER, T., WREIGHT, E. M., BIBB, M. J. (1983), Plasmids, recombination and chromosome mapping in *Streptomyces lividans* 66, *J. Gen. Microb.* **129**, 2257–2269.

HOPWOOD, D. A., CHATER, K. F., BIBB, M. J. (1995), Genetics and antibiotic production in *Streptomyces coelicolor* A3(2), a model Streptomycete, in: *Genetics and Biochemistry of Antibiotic Production* (VINING, L. C., STUTTARD, C., Eds.), pp. 65–102. Boston, MA: Butterworth-Heinemann.

HORINOUCHI, S., BEPPU, T. (1990), Autoregulatory factors of secondary metabolism and morphogenesis in actinomycetes, *CRC Crit. Rev. Biotechnol.* **10**, 191–204.

HORINOUCHI, S., BEPPU, T. (1992a), Regulation of secondary metabolism and cell differentiation in streptomycetes: A-factor as a microbial hormone and the AfsR protein as a component of a two-component regulatory system, *Gene* **115**, 167–172.

HORINOUCHI, S., BEPPU, T. (1992b), Autoregulatory factors and communication in actinomycetes, *Annu. Rev. Microbiol.* **46**, 377–393.

HORINOUCHI, S., BEPPU, T. (1995), Autoregulators, in: *Genetics and Biochemistry of Antibiotic Production* (VINING, L. C., STUTTARD, C., Eds.), pp. 103–120. Boston, MA: Butterworth-Heinemann.

HORINOUCHI, S., KITO, M., NISHIYAMA, M., FURUYA, K., HONG, S. K., MIYAKE, K., BEPPU, T. (1990), Primary structure of AfsR, a global regulatory protein for secondary metabolite formation in *Streptomyces coelicolor* A3(2), *Gene* **95**, 49–56.

HORNEMANN, U., ZHANG, X. Y., OTTO, J. C. (1993), Transferable *Streptomyces* DNA amplification and coamplification of foreign DNA sequences, *J. Bacteriol.* **175**, 1126–1133.

HOTTA, K., DAVIES, J., YAGISAWA, M. (1995), Aminoglycosides and aminocyclitols, in: *Genetics and Biochemistry of Antibiotic Production* (VINING, L. C., STUTTARD, C., Eds.), pp. 571–596. Boston, MA: Butterworth-Heinemann.

HUNDLE, B., ALBERTI, M., NIEVELSTEIN, V., BEYER, P., KLEINIG, H., ARMSTRONG, G. A., BURKE, D. H., HEARST, J. E. (1994), Functional assignment of *Erwinia herbicola* Eho10 carotinoid genes expressed in *Escherichia coli, Mol. Gen. Genet.* **245**, 406–416.

HÜTTER, R. A. (1986), Overproduction of microbial metabolites, in: *Biotechnology,* 1st Edn., Vol. 4 (REHM, H. J., REED, G., Eds.), pp. 3–14. Weinheim: VCH.

HUTCHINSON, C. R. (1994), Drug synthesis by genetically engineered microorganisms, *Bio/Technology* **12**, 375–380.

HUTCHINSON, C. R. (1995), Anthracyclines, in: *Genetics and Biochemistry of Antibiotic Production* (VINING, L. C., STUTTARD, C., Eds.), pp. 331–358. Boston, MA: Butterworth-Heinemann.

HUTCHINSON, C. R., FUJII, I. (1995), Polyketide synthase gene manipulation: a structure–function approach in engineering novel antibiotics, *Annu. Rev. Microbiol.* **49**, 201–238.

ISHIZUKA, H., HORINOUCHI, S., KIESER, H. M., HOPWOOD, D., BEPPU, T. (1992), A putative two component regulatory system involved in secondary metabolism in *Streptomyces* species, *J. Bacteriol.* **174**, 7585–7594.

ISONO, K. (1990), Nucleoside antibiotics: Structure, biological activity, and biosynthesis, *J. Antibiot.* **41**, 1711–1752.

ISONO, K. (1995), Polyoxins and related nucleotides, in: *Genetics and Biochemistry of Antibiotic Production* (VINING, L. C., STUTTARD, C., Eds.), pp. 597–618. Boston, MA: Butterworth-Heinemann.

JACOB, L., ZASLOFF, M. (1994), Potential therapeutic applications of magainins and other antimicrobial agents of animal origin, in: *Antimicrobial Peptides, Ciba Foundation Symposium 186* (BOMAN, H. G., MARSH, J., GOODE, J. A., Eds.), Chichester: J. Wiley & Sons.

JENSEN, S. E., DEMAIN, A. L. (1995), Beta-lactams, in: *Genetics and Biochemistry of Antibiotic Production* (VINING, L. C., STUTTARD, C., Eds.), pp. 239–268. Boston, MA: Butterworth-Heinemann.

JENSEN, P. R., FENICAL, W. (1994), Strategies for the discovery of secondary metabolites from bacteria: ecological perspectives, *Annu. Rev. Microbiol.* **48**, 559–584.

JOHNSON, R., ADAMS, J. (1992), The ecology and evolution of tetracycline resistance, *TREE* **7**, 295–299.

JOHNSON, P. F., MCKNIGHT, S. L. (1989), Eukaryotic transcriptional regulatory proteins, *Annu. Rev. Biochem.* **58**, 799–839.

KAHAN, B. D. (Ed.) (1987), *Cyclosporins.* London: Grune & Stratton.

KAJIMURA, Y., SUGUYAMA, M., KANDEA, M. (1995), Bacillopeptins, new cyclic lipopeptide antibiotics from *Bacillus subtilis* FR-2, *J. Antibiot.* **48**, 1095–1103.

KAO, C. M., LUO, G., KATZ, L., CANE, D. E., KHOSLA, C. (1995), Manipulation of macrolide ring size by directed mutagenesis of a modular polyketide synthase, *J. Am. Chem. Soc.* **117**, 9105–9106.

KATZ, L., DONADIO, S. (1995), Macrolides, in: *Genetics and Biochemistry of Antibiotic Production* (VINING, L. C., STUTTARD, C., Eds.), pp. 385–420. Boston, MA: Butterworth-Heinemann.

KELEMEN, G. H., BROWN, G. L., KORMANEC, J., POTUCKOVA, L., CHATER, K. F., BUTTNER, M. J. (1996), The positions of the sigma factor genes, *whiG* and *sigF*, in the hierarchy controlling the development of spore chains in the aerial hyphae of *Streptomyces coelicolor* A3(2), *Mol. Microbiol.* **21**, 593–603.

KELL, D. B., KAPRELYANTS, A. S., GRAFEN, A. (1995), Pheromones, social behaviour and the functions of secondary metabolism in bacteria, *Trends Ecol. Evol.* **10**, 126–129.

KELLER, N. P., ADAMS, T. H. (1995), Analysis of a mycotoxin gene cluster in *Aspergillus nidulans, SAAS Bull. Biochem. Biotechnol.* **8**, 14–21.

KHOKHLOV, A. S. (1982), Low molecular weight microbial bioregulators of secondary metabolism, in: *Overproduction of Microbial Products* (KRUMPHANZ, V., SIKYTA, B., VANEK, Z., Eds.), pp. 97–109. London: Academic Press.

KIM, H. S., TADA, H., NIHIRA, T., YAMADA, Y. (1990), Purification and characterization of virginiae butanolide C-binding protein, a possible pleotropic signal-transducer in *Streptomyces virginiae, J. Antibiot.* **43**, 692–706.

KIM, E. S., BIBB, M. J., BUTLER, M. J., HOPWOOD, D. A., SHERMAN, D. H. (1994), Sequences of the oxytetracycline polyketide synthase-encoding *otc* genes from *Streptomyces rimosus, Gene* **141**, 141–142.

KIM, E.-S., CRAMER, K. D., SHREVE, A. L., SHERMAN, D. H. (1995), Heterologous expression of an engineered biosynthetic pathway: functional dissection of type II polyketide synthase components in *Streptomyces* species, *J. Bacteriol.* **177**, 1202–1207.

KLEINKAUF, H., VON DÖHREN, H. (1986), Peptide Antibiotics, in: *Regulation of Secondary Metabolite Formation* (KLEINKAUF, H., VON DÖHREN, H., DORNAUER, H., NESEMANN, G., Eds.), pp. 173–207. Weinheim: VCH.

KLEINKAUF, H., VON DÖHREN, H. (1987), Biosynthesis of peptide antibiotics, *Annu Rev. Microbiol.* **41**, 259–289.

KLEINKAUF, H., VON DÖHREN, H. (Eds.) (1990), *Biochemistry of Peptide Antibiotics. Recent Advances in the Biochemistry of β-Lactams and Microbial Bioactive Peptides.* Berlin: DeGruyter.

KLEINKAUF, H., VON DÖHREN, H. (1996), A nonribosomal system of peptide biosynthesis, *Eur. J. Biochem.* **236**, 335–51.

KLEINKAUF, H., VON DÖHREN, H. (in press), Biosynthesis of cyclosporins and related peptides, in: *Fungal Biotechnology* (ANKE, T., Ed.). London: Chapman and Hall.

KOHMOTO, K., YODER, O. C. (Eds.) (1994), *Host-Specific Toxin. Biosynthesis, Receptor and Molecular Biology.* Tottori University Press.

KONDO, S., YASUO, K., MATSUMA, M., KATAYAMA, M., MARUMO, S. (1986), Pamamycin-607, a new macrodiolide antibiotic from *Streptomyces alboniger, J. Antibiot.* **41**, 1196–1204.

KROHN, K., KIRST, H. A., MAAG, H. (Eds.) (1993), *Antibiotics and Antiviral Compounds.* Weinheim: VCH.

KUBOHARA, Y., OKAMOTO, K., TANAKA, Y., ASALIN, K., SAKURAI, A., TAKAHASHI, N. (1993), Differanisol A, an inducer of the differentiation of Friend leukemic cells induces stalk cell differentiation in *Dictyostelium discoideum, FEBS Lett.* **322**, 73–75.

KUBOTA, T., TOKOROYAMA, KAMIKAWA, T., SATAMURA, Y. (1966), The structures of sclerin and scleroide, metabolites of *Sclerotinia libertiana, Tetrahedron Lett.* **1966**, 5205.

LAATSCH, H. (1994), *Antibase, Database.* Heidelberg: Chemical Concepts.

LANCINI, G., LORENZETTI, R. (1993), *Biotechnology of Antibiotics and Other Bioactive Microbial Metabolites.* New York, London: Plenum Press.

LANDAN, G., COHEN, G., AHARONOWITZ, Y., SHUALI, Y., GRAUR, D., SHIFFMAN, D. (1990), Evolution of isopenicillin N synthase may have involved horizontal gene transfer, *Mol. Biol. Evol.* **7**, 399–406.

LAWEN, A., TRABER, R. (1993), Substrate specificities of cyclosporin synthetase and peptolide SDZ 214-103 synthetase, *J. Biol. Chem.* **268**, 20452–20465.

LE GOUILL, C., DESMARAIS, D., DERY, C. V. (1993), *Saccharopolyspora hirsuta* 367 encodes clustered genes similar to ketoacyl synthase, ketoacyl reductase, acyl carrier protein, and biotin carboxyl carrier protein, *Mol. Gen. Genet.* **240**, 146–150.

LI, W., NIKIRA, T., SAKUDA, S., HISHIDA, T. YAMADA, Y. (1992), New inducing factors for virginiamycin production from *Streptomyces antibioticus, J. Ferment. Biotechnol.* **74**, 214–217.

LIRAS, P., ASTURIAS, J. A., MARTÍN, J. F. (1990), Phosphate sequences involved in transcriptional regulation of antibiotic biosynthesis, *Trends Biotechnol.* **8**, 184–189.

LITZKA, O., THEN BERG, K., BRAKHAGE, A. A. (1996), The *Aspergillus nidulans* penicillin biosynthesis gene *aat* (penDE) is controlled by a CCAAT containing DNA element, *Eur. J. Biochem.* **238**, 675–682.

LIU, Y. S., KIESER, H. M., HOPWOOD, D. A., CHEN, C. W. (1992), The chromosomal DNA of *Streptomyces lividans* 66 is linear, *Mol. Microbiol.* **10**, 923–933.

LOMBÓ, F., BLANCO, G., FERNÁNDEZ, E., MÉNDEZ, C., SALAS, J. A. (1996), Characterization of *Streptomyces argillaceus* genes encoding a polyketide synthase involved in the biosynthesis of the antitumor mithramycin, *Gene* **172**, 87–91.

LOUMAYE, E. THORNER, J., CATT, K. J. (1982), Yeast mating pheromone activates mammalian gonadotrophs: evolutionary conservation of a reproductive hormone, *Science* **218**, 1323–1325.

LUCKNER, M. (1989), *Secondary Metabolism in Microbes, Plants and Animals.* Jena: G. Fischer.

LUCKNER, M., NOVER, L., BÖHM, H. (1977), *Secondary Metabolism and Cell Differentiation.* New York, Heidelberg, Berlin, Tokyo: Springer-Verlag.

LUENGO, J. M. (1995), Enzymatic synthesis of hydrophobic penicillins, *J. Antibiot.* **48**, 1195–1212.

MACCABE, A. P., RIACH, M. B. R., UNKLES, S. E., KINGHORN, J. R. (1990), The *Aspergillus nidulans npeA* locus consists of three contiguous genes required for penicillin biosynthesis, *EMBO J.* **9**, 279–287.

MADDURI, K., HUTCHINSON, C. R. (1995a), Functional characterization and transcriptional analysis of the *dnrR*$_1$ locus, which controls daunorubicin biosynthesis in *Streptomyces peucetius, J. Bacteriol.* **177**, 1208–1215.

MADDURI, K., HUTCHINSON, C. R. (1995b), Functional characterization and transcriptional analysis of a gene cluster governing early and late steps in daunorubicin biosynthesis in *Streptomyces peucetius, J. Bacteriol.* **177**, 3879–3884.

MAHANTI, N., BHATNAGAR, D., CARY, J. W., JOUBRAN, J., LINZ, J. E. (1996), Structure and function of FAS-1A, a gene encoding a putative fatty acid synthetase directly involved in aflatoxin biosynthesis in *Aspergillus parasiticus, Appl. Environ. Microbiol.* **62**, 191–195.

MARAHIEL, M. A., DANDERS, W., KRAUSE, M., KLEINKAUF, H. (1979), Biological role of gramicidin S in spore function, *Eur. J. Biochem.* **99**, 49–58.

MARAHIEL, M. A., NAKANO, M. M., ZUBER, P. (1993), Regulation of peptide antibiotic production in *Bacillus, Mol. Microbiol.* **7**, 631–636.

MARCOS, A. T., GUTIERREZ, S., DIEZ, B., FERNANDEZ, F. J., OGUIZA, J. A., MARTÍN, J. F. (1995), Three genes *hrdB, hrdD* and *hrdT* of *Streptomyces griseus* IMRU 3570, encoding sigma-factor-like proteins, are differentially expressed under specific nutritional conditions, *Gene* **153**, 41–48.

MARTÍN, J. F. (1978), Manipulation of gene expression in the development of antibiotic production, in: *Antibiotics and Other Secondary Metabolites* (HÜTTER, R., LEISINGER, J., NÜESCH, J., WEHRLI, W., Eds.), pp. 19–38. London, New York: Academic Press.

MARTÍN, J. F. (1989), Molecular mechanisms for the control by phosphate of the biosynthesis of antibiotics and other secondary metabolites, in: *Regulation of Secondary Metabolism in Actinomycetes* (SHAPIRO, S., Ed.), pp. 213–233. Boca-Raton, FL: CRC Press.

MARTÍN, J. F., DEMAIN, A. L. (1980), Control of antibiotic biosynthesis, *Microbiol. Rev.* **44**, 230–251.

MARTÍN, J. F., GUTIERREZ, S. (1995), Genes for beta-lactam antibiotic biosynthesis, *Antonie van*

Leeuwenhoek Int. J. Gen. Mol. Microbiol. **67**, 181–200.

MARTÍN, J. F., LOPEZ-NIETO, M. J., CASTRO, J. M., CORTES, J., RAMOS, F. R., CANTORAL, J. N., ALVAREZ, E., RAMIREZ, M. G., BARREDO, J. C., LIRAS, P. (1986), Enzymes involved in β-lactam biosynthesis controlled by carbon and nitrogen regulation, in: *Regulation of Secondary Metabolite Formation* (KLEINKAUF, H., VON DÖHREN, H., DORNAUER H., NESEMANN, G., Eds.), pp. 41–47. Weinheim: VCH.

MARTINEZ-BLANCO, H., REGLERS, A., LUMENGO, J. M. (1991), "*In vitro*"-syntheses of different naturally-occurring semisynthetic and synthetic penicillins using a new effective enzymatic coupled system, *J. Antibiot.* **44**, 1252–1254.

MAZZOLA, M., WHITE, F. F. (1994), A mutation in the indole-3-acetic acid biosynthesis pathway of *Pseudomonas syringae* PV *syringae* affects growth in *Phaseolus vulgaris* and syringomycin production, *J. Bacteriol.* **176**, 1374–1382.

MCCUE, L. A., KWAK, J., BABCOCK, M. J, KENDRICK, K. E. (1992), Molecular analysis of sporulation in *Streptomyces griseus, Gene* **115**, 173–179.

MCDANIEL, R., EBERT-KOHOSLA, S., FU, H., HOPWOOD, D. A., KHOSLA, C. (1994), Engineered biosynthesis of novel polyketides: influence of a downstream enzyme on the catalytic specificity of a minimal aromatic polyketide synthase, *Proc. Natl. Acad. Sci. USA* **91**, 11542–11346.

MCNEIL, D. J. (1995), Avermectins, in: *Genetics and Biochemistry of Antibiotic Production* (VINING, L. C., STUTTARD, C., Eds.), pp. 421–442. Boston, MA: Butterworth-Heinemann.

MEIGHEN, E. A. (1991), Molecular biology of bacterial luminescence, *Microbiol. Rev.* **55**, 123–142.

MEISSNER, K., DITTMANN, E., BÖRNER, T. (1996), Toxic and nontoxic strains of the cyanobacterium *Microcystis aeruginosa* contain sequences homologous to peptide synthetase genes, *FEMS Microbiol. Lett.* **135**, 295–303.

MERSON-DAVIES, L. A., CUNDLIFE, E. (1994), Analysis of five tylosin biosynthetic genes from the tyIIBA region of the *Streptomyces fradiae* genome, *Mol. Microbiol.* **13**, 349–355.

MEURER, G., HUTCHINSON, C. R. (1995), Daunorubicin type II polyketide synthase enzymes DpsA and DpsB determine neither the choice of starter unit nor the cyclization pattern of aromatic polyketides, *J. Am. Chem. Soc.* **117**, 5899–5900.

MILLER, J. R., INGOLIA, T. D. (1993), Cloning and characterization of β-lactam biosynthetic genes, *Mol. Gen. Genet.* **3,** 689–695.

MISAWA, N., SATOMI, Y., KONDO, K., YOKOYAMA, A., KAJIWARA, S., SAITO, T., OHTANI, T., MIKI, W. (1995), Structure and functional analysis of a amrine bacterial carotinoid biosynthesis gene cluster and astaxanthin biosynthetic pathway proposed at the gene level, *J. Bacteriol.* **177**, 6575–6584.

MO, Y. Y., GEIBEL, M., BONSALL, R. F., GROSS, D. C. (1995), Analysis of sweet cherry (*Prunus avium L.*) leaves for plant signal molecules that activate the *syrB* gene required for synthesis of the phytotoxin, syringomycin, by *Pseudomonas syringae* pv. syringae, *Plant Physiol.* **107**, 603–612.

MONAGHAN, R. L., TKACZ, J. S. (1990), Bioactive microbial products: focus upon mechanism of action, *Annu. Rev. Microbiol.* **44**, 271–301.

MORENO, F., MILLÁN, J. L. S., HERNÁNDEZ-CHICO, C., KOLTER, R. (1995) Microcins, in: *Genetics and Biochemistry of Antibiotic Production* (VINING, L. C., STUTTARD, C., Eds.), pp. 307–322. Boston, MA: Butterworth-Heinemann.

MÜLLER, L. (1989), Chemistry, biochemistry and therapeutic potential of microbial α-glucosidase inhibitors, in: *Novel Microbial Products for Medicine and Agriculture* (DEMAIN, A. L., SOMKUTI, G. A., HUNTER-CEVERA, J. C., ROSSMORE, H. W., Eds.), pp. 109–116. Amsterdam: Elsevier.

MURAO, S., HAYASHI, H., TARUI, H. (1984), Anthranilic acid, as a fruiting body inducing substance in *Favolus arcularius,* from a strain TA7 of actinomycetes, *Agric. Biol. Chem.* **48**, 1669–1677.

NIKAIDO, H. (1994), Prevention of drug access to bacterial targets: permeability barriers and active efflux, *Science* **264**, 382–388.

NOBLE, R. L. (1990), The discovery of the vinca alkaloids – chemotherapeutic agents against cancer, *Biochem. Cell. Biol.* **68**, 1344–1351.

OCHI, K. (1990), *Streptomyces relC* mutants with an altered ribosomal protein ST-U1 and genetic analysis of a *Streptomyces griseus relC* mutant, *J. Bacteriol.* **172**, 4008–4016.

OCHI, K., TSURUMI, Y., SHIGEMATSU, N., IWANI, M., UMEHARA, K., OKUHARA, M. (1988), Physiological analysis of bicozamycin high-producing *Streptomyces griseoflavus* used at industrial level, *J. Antibiot.* **41**, 1106–1115.

OCHI, K., PENYIGE, A., BARABAS, G. (1992), The possible role of ADP-ribosylation in sporulation and streptomycin production by *Streptomyces, J. Gen. Microbiol.* **138**, 1745–1750.

OCHIAI, T. (1987), Critical view on the use of immunostimulants in cancer treatment, in: *Immunostimulants: Now and Tomorrow* (AZUMA, I., JOLLES, G., Eds.), pp. 167–172. Tokyo/Berlin: Japan Scientific Societies Press/Springer.

OGARA, M., NAKAYAMA, H., FURIHATA, K., SHIMADZU, A., SETO, H., OTAKE, N. (1985), Structure of a new antibiotic curromycin A, produced by a genetically modified strain of *Streptomyces hygroscopicus,* a polyether antibiotic producing organism, *J. Antibiot.* **38**, 669–673.

O'HAGAN, D. (Ed.) (1991), *The Polyketide Metabolites.* Chichester: Ellis Horwood.

OKAMOTO, S., NIHIRA, T., KATAOKA, H., SUZUKI, A., YAMADA, Y. (1992), Purification and molecular cloning of a butyrolactone autoregulator receptor from *Streptomyces virginiae, J. Biol. Chem.* **267**, 1093–1098.

OLESKIN, A. V. (1994), Social behaviour of microbial populations, *J. Basic Microbiol.* **34**, 425–439.

OLIYNYK, M., BROWN, M. J. B., CORTÉS, J., STAUNTON, J., LEADLAY, P. F. (1996), A hybrid modular polyketide synthase obtained by domain swapping, *Chemistry & Biology* **3**, 833–839.

ŌMURA, S. (Ed.) (1992), *The Search of Bioactive Metabolites.* New York: Springer-Verlag.

OXFORD, J. S., FIELDS, H. J., REEVES, D. S. (Eds.) (1986), *Drug Resistance in Viruses, Other Microbes and Eukaryotes.* London, New York: Academic Press.

PÉREZ-ESTEBAN, B., GÓMEZ-PARDO, E., PEÑALVA, M. A. (1995), A *lacZ* reporter fusion method for the genetic analysis of regulatory mutations in pathways of fungal secondary metabolism and its application to the *Aspergillus nidulans* penicillin pathway, *J. Bacteriol.* **177**, 6069–6076.

PETERSEN, F., ZÄHNER, H., METZGER, J. W., FREUND, S., HUMMEL, R. P. (1993), Germicidin, an autoregulative germination inhibitor of *Streptomyces viridochromogenes* NRRL B-1551, *J. Antibiot.* **46**, 1126–1138.

PETRICH, A. K., LESKIW, B. K., PARADKAR, A. S., JENSEN, S. E. (1994), Transcriptional mapping of the genes encoding the early enzymes of the cephamycin biosynthestic pathway of *Streptomyces clavuligerus, Gene* **142**, 41–48.

PIEPERSBERG, W. (1994), Biosynthesis of secondary carbohydrate components in actinomycetes and other bacteria. Perspectives of a pathway engineering (in German), *BioEngineering* **6**, 27–34.

PIEPERSBERG, W. (1995), Streptomycin and related aminoglycosides, in: *Genetics and Biochemistry of Antibiotic Production* (VINING, L. C., STUTTARD, C., Eds.), pp. 531–570. Boston, MA: Butterworth-Heinemann.

PIERSON, L. S., PIERSON, E. A. (1996), Phenazine antibiotic production in *Pseudomonas aureofaciens* – role in rhizosphere ecology and pathogen suppression, *FEMS Microbiol. Lett.* **136**, 101–108.

PIERSON, L. S., GAFFNEY, T., LAM, S., GONG, F. C. (1995), Molecular analysis of genes encoding phenazine biosynthesis in the biological control bacterium *Pseudomonas aureofaciens* 30-84, *FEMS Microbiol. Lett.* **134**, 299–307.

PITKIN, J. W., PANACCIONE, D. G., WALTON, J. D. (1996), A putative cyclic peptide efflux pump encoded by the *toxA* gene of the plant pathogenic fungus *Cochliobolus carbonum, Microbiology* **142**, 1557–1565.

PLINER, S. A., KLEINER, E. M., KORNITSKAYA, E. Y., TOVAROVA, I. I., ROZYNOV, B. V., SMIRNOVA, G. M., KHOKKLOV, A. S. (1976), Isolation and primary characteristics of the A-factor, *Sov. J. Bioorg. Chem.* (English translation) **2**, 825–832.

POSPIECH, A., BIETENHADER, J., SCHUPP, T. (1996), Two multifunctional peptide synthetases and an O-methyltransferase are involved in the biosynthesis of the DNA-binding antibiotic and antitumor agent saframycin Mx1 from *Myxococcus xanthus, Microbiology* **142**, 741–746.

POSTMA, P. W., LENGELER, J. W., JACOBSON, G. R. (1993), Phosphoenolpyruvate:carbohydrate phosphotransferase systems of bacteria, *Microbiol. Rev.* **57**, 543–594.

POTIER, P., GUERITTE-VOEGELEIN, F., GUENARD, D. (1994), Taxoids, a new class of antitumor agents of plant origin: recent results, *Nouv. Rev. Fr. Hematol.* **36**, 521–523.

QUEENER, S. W., WILKERSON, S., TUNIN, D. R., MCDERMOTT, J. P., CHAPMAN, J. K., NASH, C., WESTPHELING, J. (1986), Cephalosporin C fermentation biochemical and regulatory aspects of sulfur metabolism, in: *Biotechnology of Industrial Antibiotics* (VANDAMME, E. J., Ed.), pp. 141–170. New York/Basel: Marcel Dekker.

QUIRÓS, L. M., SALAS, J. A. (1995), Biosynthesis of the macrolide oleandomycin by *Streptomyces antibioticus, J. Biol. Chem.* **270**, 18234–18239.

RA, S. R., QIAO, M. Q., IMMONEN, T. I., SARIS, P. E. J. (1996), Genes responsible for nisin synthesis, regulation and immunity from a regulon of two operons and are induced by nisin in *Lactococcus lactis* N8, *Microbiology* **142**, 1281–1288.

REDENBACH, M., KIESER, H. M., DENAPAITE, D., EICHNER, A., CULLUM, J., KINASHI, H., HOPWOOD, D. A. (1996), A set of ordered cosmids and a detailed genetic and physical map for the 8 Mb *Streptomyces coelicolor* A3(2) chromosome, *Mol. Microbiol.* **21**, 77–96.

REICHENBACH, H., HÖFLE, G. (1993), Biologically active secondary metabolites from Myxobacteria, *Biotechnol. Adv.* **11**, 219–277.

REUTER, G. (1989), Enzymatic regulation of microbial phytoeffector biosynthesis, *Progr. Ind. Microbiol.* **27**, 271–282.

RINEHART, K. L., SHIELD, L. S. (1988), Biological active marine natural products, in: *Horizons on Antibiotic Research* (DAVIES, B. D., ICHIKAWA, T., MAEDA, K., MITCHER, L. A., Eds.), pp. 194–227. Tokyo: Japan Antibiotics Association.

ROCK, C. O., CRONAN, J. E. (1996), *Escherichia coli* as a model for the regulation of dissociable (type II) fatty acid biosynthesis, *Biochim. Biophys. Acta* **1302**, 1–16.

ROHR, J., SCHÖNEWOLF, M., UDVARNOKI, G., ECKARDT, K. SCHUMANN, G., WAGNER, C., BEALE, J. M., SOREY, S. D. (1993), Investigations on the biosynthesis of the angucycline group antibiotics, *J. Org. Chem.* **58**, 2547–2551.

ROOS, W., LUCKNER, M. (1986), The spatial organization of secondary metabolism in microbial and plant cells, in: *Cell Growth and Management* (SUBRAMANIAN, A. V. (Ed.), pp. 45–73. Boca Raton, FL: CRC Press.

RÖSSNER, E., ZEECK, A., KÖNIG, W. A. (1990), Structure elucidation of hormaomycin (in German), *Angew. Chem.* **102**, 84–85.

SADAKANE, N., TANAKA, Y., ŌMURA, S. (1983), Hybrid biosynthesis of a new macrolide antibiotic by a daunomycin-producing microorganism, *J. Antibiot.* **36**, 921–922.

SAHL, H.-G., JACK, R. W., BIERBAUM, G. (1995), Biosynthesis and biological activities of lantibiotics with unique post-translational modifications, *Eur. J. Biochem.* **230**, 827–853.

SAKUDA, S., YAMADA, Y. (1991), Stereochemistry of butyrolactone autoregulators from streptomycetes, *Tetrahedron Lett.* **32**, 1817–1820.

SAKUDA, S., HIGASHI, A., TANAKA, S., NIHIRA, T., YAMADA, Y. (1992), Biosynthesis of virginiae butanolide A, a butyrolactone autoregulator from *Streptomyces, J. Am. Chem. Soc.* **114**, 663–668.

SALYERS, A. A., SHOEMAKER, N. A., STEVENS, A. M., LI, L.-Y. (1995), Conjugative transposons: an unusual and diverse set of integrated gene transfer elements, *Microbiol. Rev.* **59**, 579–590.

SAMSON, S. M., BELAGAJE, R., BLANKENSHIP, D. T., CHAMPMAN, J. L., PERRY, D., SKATRUD, P. L., VAN FRANK, R. M., ABRAHAM, E. P., BALDWIN, J. E., QUEENER, S. W., INGOLIA, T. D. (1985), Isolation, sequence determination and expression in *Escherichia coli* of the isopenicillin N synthetase gene from *Cephalosporium acremonium, Nature* **318**, 191–194.

SANCHEZ, L., BRANA, A. F. (1996), Cell density influences antibiotic biosynthesis in *Streptomyces clavuligerus, Microbiology* **142**, 1209–1220.

SANGLIER, J. J., LARPENT, J. P. (1989), *Biotechnologie des Antibiotiques.* Paris, Milano, Barcelona, Mexico: Masson.

SCHIRMBOCK, M., LORITO, M., WANG, Y. L., HAYES, C. K., ARISAN-ATAC, I., SCALA, F., HARMAN, G. E., KUBICEK, C. P. (1994), Parallel formation and synergism of hydrolytic enzymes and peptaibol antibiotics, molecular mechanisms involved in the antagonistic action of *Trichoderma harzianum* against phytopathogenic fungi, *Appl. Environ. Microbiol.* **60**, 4364–4370.

SCHUPP, T., TUOPET, C., CLUZEL, B., NEFF, S., HILL, S., BECK, J. J., LIGON, J. M. (1995), A *Sorangium cellulosum* (Myxobacterium) gene cluster for the biosynthesis of the macrolie antibiotic sorphen A-cloning, characterization, and homology to polyketide synthase genes from Actinomycetes, *J. Bacteriol.* **177**, 3637–3679.

SCHWARTZ, D., ALIJAH, R., NUSSBAUMER, B., PELZER, S., WOHLLEBEN, W. (1996), The peptide synthetase gene *phsA* from *Streptomyces viridochromogenes* is not juxtaposed with other genes involved in nonribosomal biosynthesis of peptides, *Appl. Environ. Microbiol.* **62**, 570–577.

SCHWECKE, T., APARICIO, J. F., MOLNAR, I., KÖNIG, A., KHAW, L. E., HAYDOCK, S. F., OLIYNYK, M., CAFFREY, P., CORTES, J., LESTER, J. B., BÖHM, G. A., STAUNTON, J., LEADLAY, P. F. (1995), The biosynthetic gene cluster for the polyketide immunosuppressant rapamycin, *Proc. Natl. Acad. Sci. USA* **91**, 7839–7843.

Scrip, World Pharmacentical News (London) (1993), **1825**, 23.

SENO, E. T., BALTZ, R. H. (1989), Structural organization and regulation of antibiotic biosynthesis and resistance genes in actinomycetes, in: *Regulation of Secondary Metabolism in Actinomycetes* (SHAPIRO, S., Ed.), pp. 2–48. Boca Raton, FL: CRC Press.

SHAH, A. J., TILBURN, J., ADLARD, M. W., ARST, H. N., JR. (1991), pH regulation of penicillin production in *Aspergillus nidulans, FEMS Microbiol. Lett.* **77**, 209–212.

SHAPIRO, S. (Ed.) (1989), *Regulation of Secondary Metabolism in Actinomycetes.* Boca Raton, FL: CRC Press.

SHEN, B., HUTCHINSON, C. R. (1994), Triple hydroxylation of tetracenomycin A2 to tetracenomycin C in *Streptomyces glaucescens, J. Biol. Chem.* **269**, 30726–30733.

SHERMAN, D. H., MALPARTIDA, F., BIBB, M. J., KIESER, H. M., HOPWOOD, D. A. (1989), Structure and deduced function of the granaticin producing polyketide synthase gene cluster of *Streptomyces violaceoruber* Tü22, *EMBO J.* **8**, 2717–2725.

SHIER, W. T., RINEHART, K. L., JR., GOTTLIEB, D. (1969), Preparation of four new antibiotics from a mutant of *Streptomyces fradiae, Proc. Natl. Acad. Sci USA* **63**, 198–204.

SHIOZAWA, H., KAGASAKI, T., KINOSHITA, T., HARUYAMA, H., DOMON, H., UTSUI, Y., KODAMA, K., TAKAHASHI, S. (1993), Thiomarinol, a new hybride antimicrobial antibiotic produced by a marine bacterium, *J. Antibiot.* **46**, 1834–1842.

SMITH, D. J., BURNHAM, M. K. R., BULL, J. H., HODGSON, J. E., WARD, J. M., BROWNE, P., BROWN, J., BARTON, B., EARL, A. J., TURNER, G. (1990), β-Lactam antibiotic biosynthetic genes have been conserved in clusters in prokaryotes and eukaryotes, *EMBO J.* **9**, 741–747.

SOLOMON, J. M., MAGNUSON, R., SRIVASTAVA, A., GROSSMAN, A. D. (1995), Convergent sensing pathways mediate response to two extracellular competence factors in *Bacillus subtilis, Genes Dev.* **9**, 547–558.

SPRATT, B. G. (1994), Resistance to antibiotics mediated by target alterations, *Science* **264**, 388–393.

STACHELHAUS, T., SCHNEIDER, R., MARAHIEL, M. A. (1995a), Rational design of peptide antibiotics by targeted replacement of bacterial and fungal domains, *Science* **269**, 69–72.

STACHELHAUS, T., SCHNEIDER, R., MARAHIEL, M. A. (1995b), Engineered biosynthesis of peptide antibiotics, *Biochem. Pharmacol.* **52**, 177–186.

STEELE, D. B., STOWERS, M. D. (1991), Techniques for selection of industrially important microorganisms, *Annu. Rev. Microbiol.* **45**, 89–106.

STEWART, G. C. (1993), Catabolite repression in the gram-positive bacteria: generation of negative regulators of transcription, *J. Cell Biochem.* **51**, 25–28.

STIERLE, H., STIERLE, D., STROBEL, G., BIGNANI, G., GROTHANS, P. (1994), Endophytic fungi of pacific yew (*Taxus brevifolia*) as a source of taxol, taxanes and other pharmacophores, *ACS Symp. Ser.* **557** (Bioregulators for crop protection and pest control), 64–77.

STIERLE, A., STROBEL, G., STIERLE, D., GROTHAUS, P., BIGNAMI, G. (1995), The search for a taxol-producing microorganism among the endophytic fungi of the pacific yew, *Taxus brevifolia, J. Nat. Prod.* **58**, 1315–1324.

STINTZI, A., CORNELIS, P., HOHNADEL, D., MEYER, J.-M., DEAN, C., POOLE, K., KOURAMBAS, S., KRISHNAPILLAI, V. (1996), Novel pyoverdine biosynthesis gene(s) of *Pseudomonas aeruginosa* PAO, *Microbiology* **142**, 1181–1190.

STROBEL, G., YANG, X. S., SEARS, J., KRAMER, R., SIDHU, R. S., HESS, W. M. (1996), Taxol from *Pestalotiopsis microspora,* an endophytic fungus of *Taxus wallachiana, Microbiology* **142**, 435–440.

SUAREZ, T., PEÑALVA, M. A. (1996), Characterization of a *Penicillium chrysogenum* gene encoding a *pacC* transcription factor and its binding sites in the divergent *pcbAB-pcbC* promoter of the penicillin biosynthetic cluster, *Mol. Microbiol.* **20**, 529–540.

SUMMERS, R. G., ALI, A., SHEN, B., WESSEL, W. A., HUTCHINSON, C. R. (1995), Malonyl-coenzyme A-acyl carrier protein acyltransferase of *Streptomyces glaucescens* – a possible link between fatty acid and polyketide biosynthesis, *Biochemistry* **34**, 9389–9402.

SUZUKI, K., YAMAGUCHI, H., MIYAZAKI, S., NAGORI, K. (1990), Topostin, a novel inhibitor of mammalian DNA topoisomerase I from *Flexibacter topostinus* sp. nov. I. Taxonomy, and fermentation of producing strain, *J. Antibiot.* **43**, 154–157.

SZESZAK, F., VITALIS, S., BEKESI, I., SZABO, G. (1991), Presence of factor C in streptomycetes and other bacteria, in: *Genetics and Product Formation in Streptomycetes* (BAUMBACH, S., KRÜGEL, H., NOACK, D., Eds.), pp. 11–18. New York: Plenum Press.

TAKANO, Y., KUBO, Y., SHIMIZU, K., MISE, K., OKUNO, T., FURUSAWA, I. (1995), Structural analysis of *PKS1*, a polyketide synthase gene involved in melanin biosynthesis in *Colletotrichum lagenarium, Mol. Gen. Genet.* **249**, 162–167.

TANAKA, Y. (1986), Biosynthesis of tylosin and its regulation by ammonium and phosphate, in: *Regulation of Secondary Metabolite Formation* (KLEINKAUF, H., VON DÖHREN, H., DORNAUER, H., NESEMANN, G., Eds.), pp. 305–337. Weinheim: VCH.

TERCERO, J. A., ESPINOSA, J. C., LACALLE, R. A., JIMÉNEZ, A. (1996), The biosynthetic pathway of the aminonucleoside antibiotic puromycin, as deduced from the molecular analysis of the *pur* cluster of *Streptomyces alboniger, J. Biol. Chem.* **271**, 1579–1590.

THEN BERG, K., LITZKA, O., BRAKHAGE, A. A. (1996), Identification of a major *cis*-acting element controlling bidirectionally transcribed penicillin biosynthesis genes *acvA* (*pcbAB*) and *ipnA* (*pcbC*) of *Aspergillus nidulans, J. Bacteriol.* **178**, 3908–3916.

THIERICKE, R., ROHR, J. (1993), Biological variation of microbial metabolites by precursor-directed biosynthesis, *Nat. Prod. Rep.* **12**, 265–289.

TILBURN, J., SARKAR, S., WIDDICK, D. A., ESPESO, E. A., OREJAS, M., MUNGROO, J., PEÑALVA, M. A., ARST, H. N., JR. (1995), The *Aspergillus* PacC zinc finger transcription factor mediates regulation of both acidic- and alkaline-expressed genes by ambient pH, *EMBO J.* **14**, 779–790.

TOMODA, H., ŌMURA, S. (1990), New strategies for discovery of enzyme inhibitors: screening with intact mammalian cells or intact microorganisms having special functions, *J. Antibiot.* **38**, 1207–1222.

TRAIL, F., MAHANTI, N., LINZ, J. E. (1995a), Molecular biology of aflytoxin biosynthesis, *Microbiology* **141**, 755–765.

TRAIL, F., MAHANTI, N., RARICK, M., MEHIGH, R., LIANG, S. H., ZHOU, R., LINZ, J. E. (1995b), Physical and transcriptional map of an aflatoxin gene cluster in *Aspergillus parasiticus* and functional disruption of a gene involved early in the aflatoxin pathway, *Appl. Environ. Microbiol.* **61**, 2665–2673.

TSOI, C. J., KHOSLA, C. (1995), Combinatorial biosynthesis of "unnatural" natural products: the polyketide example, *Chem. Biol.* **2**, 355–362.

TURGAY, K., MARAHIEL, M. A. (1995), The *gtcRS* operon coding for two-component system regulatory proteins is located adjacent to the *grs* operon of *Bacillus brevis, DNA Sequence* **5**, 283–290.

TURNER, G. (1992), Genes for the biosynthesis of β-lactam compounds in microorganisms, in: *Secondary Metabolites: Their Function and Evolution, Ciba Foundation Symposium 171* (CHADWICK, D. J., WHELAN, J., Eds.), pp. 113–128. Chichester, New York: J. Wiley & Sons.

TURNER, W. B., ALDRIDGE, D. C. (1983), *Fungal Metabolites II.* London: Academic Press.

TYMIAK, A. A., AX, H. A., BOLGAR, M. S., KAHLE, A. D., PORUBCAN, M. A., ANDERSEN, N. H. (1993), Dactylocyclines, novel tetracycline derivatives produced by *Dactylosprangium* sp. II. Structure elucidation, *J. Antibiot.* **45**, 1899–1906.

UMEZAWA, H., HOOPER, I. R. (Eds.) (1982), *Aminoglycoside Antibiotics.* New York, Heidelberg, Berlin, Tokyo: Springer-Verlag.

VANDAMME, E. J. (Ed.) (1984), *Biotechnology of Industrial Antibiotics.* New York, Basel: Marcel Dekker.

VANEK, Z., HOSTALEK, Z. (1985), Tetracycline biosynthesis, in: *The Tetracyclines* (HLAVKA, J. J., BROTHE, J. H., Eds.), pp. 137–138. New York, Heidelberg, Berlin, Tokyo: Springer-Verlag.

VANEK, Z., HOSTALEK, Z. (Eds.) (1988), *Overproduction of Microbial Metabolites: Strain Improvement and Process Control Strategies.* Boston/London: Butterworth.

VANEK, Z., MAJOR, J. (1967), Macrolide antibiotics, in: *Antibiotics,* Vol. 2 (GOTTLIEB, D., SHAW, P. D., Eds.), pp. 154–188. New York, Heidelberg, Berlin, Tokyo: Springer-Verlag.

VANEK, Z., MIKULIK, K. (1978), Microbial growth and production of antibiotics, *Folia Microbiol.* **23**, 309–328.

VANEK, Z., BEHAL, V., JECHOVA, V., CURDOVA, E., BLUMAUEROVA, M., HOSTALEK, Z. (1978), Formation of tetracycline antibiotics, in: *Antibiotics and Other Secondary Metabolites* (HÜTTER, R., LEISINGER, I., NÜESCH, J., WEHRLI, W., Eds.), pp. 101–112. London, New York, San Francisco: Academic Press.

VANEK, Z., CUDLIN, J., BLUMAUEROVA, H., HOSTALEK, Z. (1981), *Physiology and Pathophysiology of the Production of Excessive Metabolites.* Institute of Microbiology, CS Academy of Sciences, Prague.

VERALL, M. S. (Ed.) (1985), *Discovery, and Isolation of Microbial Products.* Chichester: Ellis Horwood.

VINING, L. C. (1986), Secondary metabolism, in: *Biotechnology,* 1st Edn., Vol. 4 (REHM, H.-J., REED, G., Eds.), pp. 19–38. Weinheim: VCH.

VINING, L. C. (1992), Role of secondary metabolites from microbes, in: *Secondary Metabolites: Their Function and Evolution, Ciba Foundation Symposium 171* (CHADWICK, D. J., WHELAN, J., Eds.), pp. 184–198. Chichester, New York: J. Wiley & Sons.

VINING, L. C., STUTTARD, C. (1995), Chloramphenicol, in: *Genetics and Biochemistry of Antibiotic Production* (VINING, L. C., STUTTARD, C., Eds.), pp. 505–530. Boston, MA: Butterworth-Heinemann.

VINING, L. C., SHAPIRO, S., AHMED, Z., VATS, S., DOULL, J. STUTTARD, C. (1986), Genetic and physiological control of chloramphenicol production, in: *Regulation of Secondary Metabolite Formation* (KLEINKAUF, H., VON DÖHREN, H., DORNAUER, H., NESEMANN, G., Eds.), pp. 209–224. Weinheim: VCH.

VOGEL, R. (1984), *Natürliche Enzyminhibitoren.* Stuttgart: Thieme.

VON DER HELM, P., NEILANDS, J. B. (Eds.) (1987), *Iron Transport in Microorganisms, Plants and Animals.* Weinheim: VCH.

VON DÖHREN, H. (1990), Compilation of peptide structures: a biogenetic approach, in: *Biochemistry of Peptide Antibiotics* (KLEINKAUF, H., VON DÖHREN, H., Eds.), pp. 411–507. Berlin, New York: de Gruyter.

VON DÖHREN, H. (1995), Peptides, in: *Genetics and Biochemistry of Antibiotic Production* (VINING,

L., STUTTARD, C., Eds.), pp. 129–172. Boston, MA: Butterworth-Heinemann.

WALKER, J. B., WALKER, M. S. (1978), Enzymatic synthesis of streptomycin as a model system for study of the regulation and evolution of antibiotic biosynthesis, in: *Overproduction of Microbial Products* (KRUMPHANZL, V., SIKYTA, B., VANEK, Z., Eds.), pp. 423–438. London, New York: Academic Press.

WANNER, B. L. (1993), Gene regulation by phosphate in enteric bacteria, *J. Cell. Biochem.* **51**, 47–54.

WARD, J. M., HODGSON, J. E. (1993), The biosynthetic genes for clavulanic acid and cephamycin production occur as "super-cluster" in three *Streptomyces, FEMS Microbiol.* **110**, 239–242.

WATANABE, T., SHIRMA, K., IZAKI, K., SUGIYAMA, T. (1992), New polyenic antibiotic active against gram-positive and gram-negative bacteria. VII. Isolation and structure of enacyloxin IVa, possible biosynthetic intermediate of enacyloxin IIa, *J. Antibiot.* **45**, 575–576.

WEBER, G., SCHÖRGENDORFER, K., SCHNEIDER-SCHERZER, E., LEITNER, E. (1994), The peptide synthetase catalyzing cyclosporine production in *Tolypocladium niveum* is encoded by a giant 45.8 kilobase open reading frame, *Curr. Genet.* **26**, 120–125.

WICKLOW, D. T., DOWD, P. F., GLOER, J. B. (1994), Antiinsectian effects of *Aspergillus* metabolites, in: *The Genus Aspergillus from Taxonomy and Genetics to Industrial Applications* (POWELL, K., RENWICK, A., PEBERDY, J., Eds.), pp. 93–114. New York: Plenum Press.

WIESMANN, K. E. H., CORTÉS, J., BROWN, M. J. B., CUTTER, A. L., STAUNTON, J., LEADLEY, P. F. (1995), Polyketide synthesis *in vitro* on a modular polyketide synthase, Chemistry & Biology **2**, 583–589.

WILLIAMS, D. H., MAPLESTONE, R. A. (1992), Why are secondary metabolites synthesized? Sophistication in the inhibition of cell wall biosynthesis by vancomycin group antibiotics, in: *Secondary Metabolites: Their Function and Evolution, Ciba Foundation Symposium 171* (CHADWICK, D. J., WHELAN, J., Eds.), pp. 45–63. Chichester, New York: J Wiley & Sons.

WILLIAMS, S. T., VICKERS, J. L. (1986), The ecology of antibiotic production, *Microb. Ecol.* **12**, 43–52.

WILLIAMS, D. H., STONE, M. J., HAUCK, P. R., RAHMAN, S. R. (1989), Why are secondary metabolites (natural products) biosynthesized? *J. Nat. Prod.* **52**, 1189–1208.

WILLIAMS, P., BAINTON, N. J., SWIFT, S., SCHABRA, C. R., WILSON, M. K., STEWARD, G., MICHAEL, K., SALMAND, G. P. C., BYROFT, B. W. (1992), Small-molecule dependent density-control of gene expression in prokaryotes. Bioluminescence and the biosynthesis of carbapenem antibiotics, *FEMS Microbiol. Lett.* **100**, 161–167.

WINKELMANN, G. (Ed.) (1991), *Handbook of Microbial Iron Chelates.* Boca Raton, FL: CRC Press.

WIRTH, R., WANNER, G., GALLI, D. (1990), The sex pheromone system of *Enterococcus faecalis:* a unique mechanism of plasmid uptake (in German), *Forum Mikrobiol.* **6**, 321–332.

WRIGHT, D. C. (1983), The leucomycin–celesticidin–anthramycin group, in: *Biochemistry and Genetic Regulation of Commercially Important Antibiotics* (VINING, C. C., Ed.), pp. 311–328. Reading: Addison-Wesley.

YAMADA, Y., SUGAMURA, K., KONDO, K., YANAGIMOTO, M., OKADA, H. (1987), The structures of inducing factors for virginiamycin production in *Streptomyces virginiae, J. Antibiot.* **40**, 496–504.

YANG, Y. K., SHIMIZU, H., SHIOYA, S., SUGA, K.-I., NIHIRA, T., YAMADA, Y. (1995a), Optimum autoregulator addition strategy for maximum virginiamycin production in batch cultivation of *Streptomyces virginiae, Biotechnol. Bioeng.* **46**, 437–442.

YANG, K. Q., HAN, L., VINING, L. C. (1995b), Regulation of jadomycin B production in *Streptomyces venezuelae* ISP5230 – involvement of a repressor gene, *jadR(2), J. Bacteriol.* **177**, 6111–6117.

YANG, Y. K., MORIKAWA, M., SHIMIZU, H., SHIOYA, S., SUGA, K.-I., NIHIRA, T., YAMADA, Y. (1996a), Maximum virginiamycin production by optimization of cultivation conditions in batch culture with autoregulator addition, *Biotechnol. Bioeng.* **49**, 437–444.

YANG, K. Q., HAN, L., AYER, S. W., VINING, L. C. (1996b), Accumulation of the angucycline antibiotic rabelomycin after disruption of an oxygenase gene in the jadomycin B biosynthetic gene cluster of *Streptomyces venezuelae, Microbiology* **142**, 123–132.

YE, J., DICKENS, M. L., PLATER, R., LI, Y., LAWRENCE, J., STROHL, W. R. (1994), Isolation and sequence analysis of polyketide synthase genes from the daunomycin-producing *Streptomyces* sp. strain C5, *J. Bacteriol.* **176**, 6270–6280.

YLIHONKO, K., TUIKKANEN, J., JUSSILA, S., CONG, L. N., MANTSALA, P. (1996), A gene cluster involved in nogalamycin biosynthesis from *Streptomyces nogalater* – sequence analysis and complementation of early block mutations in the anthracycline pathway, *Mol. Gen. Genet.* **251**, 113–120.

YU, J. H., LEONARD, T. J. (1995), Sterigmatocystin biosynthesis in *Aspergillus nidulans* requires a novel type I polyketide synthase, *J. Bacteriol.* **177**, 4792–4800.

YU, T. W., BIBB, M. J., REVILL, W. P., HOPWOOD, D. A. (1994), Cloning, sequencing and analysis of the griseusin polyketide synthase gene cluster from *Streptomyces griseus, J. Bacteriol.* **176**, 2627–2634.

ZÄHNER, H., ZEECK, A. (1987), Mikrobieller Sekundärstoffwechsel, in: *Jahrbuch der Biotechnologie* (PRÄVE, P., Ed.), pp. 93–113. München, Wien: Oldenbourg.

ZÄHNER, H., ANKE, H., ANKE, T. (1983), Evolution and secondary pathways, in: *Secondary Metabolism and Cell Differentiation in Fungi* (BENNETT, J. W., CIEGLER, A., Eds.), pp. 51–70. London, New York: Academic Press.

ZAKELJMAVRIC, M., KASTELICSUHADOLC, T., PLEMENITAS, A., RIZNER, T. L., BELIC, I. (1995), Steroid hormone signalling system and fungi, *Comp. Biochem. Physiol.-B. Comp. Biochem.* **112**, 637–642.

ZMIJEWSKI, M. J., JR., FAYERMAN, J. T. (1995), Glycopeptides, in: *Genetics and Biochemistry of Antibiotic Production* (VINING, L. C., STUTTARD, C., Eds.), pp. 269–282. Boston, MA: Butterworth-Heinemann.

2 Regulation of Bacterial Antibiotic Production

Keith F. Chater
Mervyn J. Bibb

Norwich, UK

1 Introduction

1.1 The Scope of this Chapter

Study of the molecular basis for the regulation of antibiotic biosynthesis has gradually expanded since production genes were first cloned from *Streptomyces* spp. more than 10 years ago (see HOPWOOD et al., 1983 and MARTÍN and GIL, 1984 for reviews of the early history of this subject). This has led to a plethora of recent reviews (e.g., MARTÍN and LIRAS, 1989; SENO and BALTZ, 1989; CHATER, 1990, 1992; CHAMPNESS and CHATER, 1994; HUTCHINSON et al., 1994) which have dealt almost exclusively with *Streptomyces* and related actinomycetes. Recently, knowledge of the regulation of antibiotic biosynthesis in non-actinomycete bacteria has emerged almost as a by-product of studies of the switch from rapid growth to stationary phase in model organisms such as *Escherichia coli* and *Bacillus subtilis* and of mechanisms of plant pathogenicity in *Erwinia carotovora.* This has given a new opportunity, in this review, to examine both general themes common to the regulation of production of diverse antibiotics by diverse bacteria, and the idiosyncrasies of different bacterial genera or different pathways within particular bacteria.

1.2 Cellular Efficiency Involves Extensive Regulation of Metabolism in Response to Growth Conditions

When nutrients are abundant and readily available, microorganisms grow fast. In mixed communities, rapid conversion of nutrients to biomass is the overriding theme of metabolism and its efficiency is maximized by the well-known regulatory systems that govern such assimilation (many of which are reviewed by NEIDHARDT et al., 1987). Some of these are pathway-specific. Thus, in the feedback loops of amino acid biosynthetic pathways of *E. coli*, the end products typically inhibit the activity of enzymes for the earliest steps in the pathways and also cause repression of synthesis of many of the biosynthetic enzymes, mostly by mechanisms involving repressor proteins or transcriptional or translational attenuation. (We are not aware of any such feedback regulatory loops in antibiotic synthesis.) Likewise, the assimilation of carbon usually involves specific induction of genes for the relevant enzymes, usually by interaction of a specific transcriptional repressor or activator protein with the substrate or a simple derivative of it generated in the cell by a constitutive low level of the pathway enzyme(s). More global regulation is also necessary. For example, most free-living microbes possess an integrated system of carbon catabolite repression which ensures that the most readily utilized and ergogenic substrate (often glucose) is used in preference to less favorable substrates. In *E. coli*, such repression operates in two ways: via interference with uptake of the less favorable compound, preventing it from participating in induction of the enzymes needed for its assimilation ("inducer exclusion") and through more direct influences on transcription of the genes for these enzymes (SAIER, 1989; POSTMA et al., 1993; Fig. 1). In the latter case, the pattern of available carbon sources determines the intracellular pool of cAMP, which in turn directly interacts with a cAMP-binding protein, and modulates its ability to interact with the transcriptional initiation complex at promoters of various gene sets for utilization of carbon sources (Fig. 1). Carbon catabolite repression may, however, be exerted by different mechanisms in different microbes (SAIER, 1991).

1.3 Antibiotic Production Does Not Usually Occur in Rapidly Growing Cultures

In bacteria, antibiotic production nearly always takes place only after rapid growth has ceased. To a large extent, this chapter reviews the molecular and physiological mechanisms that underpin this general observation, but in this section we briefly reflect on the possible adaptive significance of this regulatory pat-

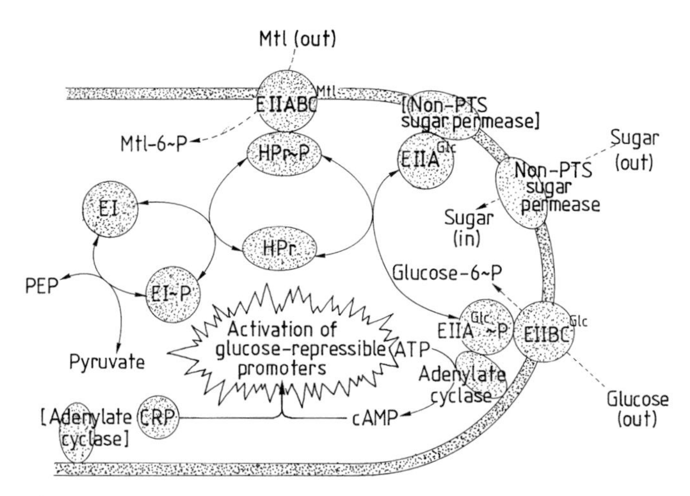

Fig. 1. Global regulation of carbon catabolite-repressible operons in *E. coli* involves multiple influences of membrane-bound and cytoplasmic components. During uptake of glucose by $EIIBC^{Glc}$ (components B and C of PTS^{Glc}, the glucose phosphotransferase system), the sugar is phosphorylated by the phosphorylated form of the cytoplasmic $EIIA^{Glc}$ which is thereby itself dephosphorylated. This has three important regulatory consequences. (1) The activity of adenylate cyclase is reduced, because it depends on interaction with $EII^{Glc} \sim P$. This causes a drop in the level of cAMP available to bind to and activate the catabolite repression protein (CRP). The cAMP–CRP complex is necessary for efficient transcription of many glucose-repressible promoters. (2) Unphosphorylated $EIIA^{Glc}$ directly inhibits non-PTS sugar permeases, excluding the relevant sugars from the cytoplasm and, therefore, preventing them from inducing the relevant genes for sugar catabolism. (3) The unphosphorylated $EIIA^{Glc}$ competes with other PTS systems such as $EIIABC^{Mfl}$, the uptake system for Mfl, for the phosphate-donating protein $HPr \sim P$, thereby indirectly inhibiting uptake of other PTS sugars EI, part of the enzyme cascade responsible for the transfer of phosphate from PEP to PTS sugars.

tern. We assume that antibiotic activities observed experimentally have actual evolutionary significance: i.e., that antibiotics help the producing organisms by inhibiting their competitors in natural environments. Why, then, do organisms not produce antibiotics throughout growth to maximize this competitive advantage? Perhaps the answer lies in the comparatively high diversion of resources away from biomass accumulation that might be required if the few cells present during early growth are to produce an inhibitory level of antibiotic. This might conflict with the need to grow as rapidly as possible in the competition for the nutritional resources of a new environment. On the other hand, the effective production of chemical weapons at relatively high population density (i.e., no earlier than the last few cell divisions before nutrient exhaustion) can be achieved with a much smaller proportion of each cell's metabolism being devoted to secondary metabolism. Even at this late stage in exploiting an environment there is still potential advantage to be gained from inhibiting competitors. This could take several forms: inhibiting the development of more persistent resting stages of competitors; greater competitiveness in the hidden population dynamics of stationary phase (during which minor subpopulations of cells grow at the expense of the majority of the population) (ZAMBRANO et al., 1993); or, in the case of developmentally complex organisms such as streptomycetes, to prevent invasion of colonies by competitors after the lysis of some of the cells within colonies, which may provide nutrition for spore development (CHATER and MERRICK, 1979; MÉNDEZ et al., 1985; but see also O'CONNOR and ZUSMAN, 1988). Viewed in this way, one important aspect of the regulation of antibiotic production should be the mechanisms by which information about population density or nutrient availability is perceived by cells. This information must then be interpreted by the cell and ultimately used to activate the specific relevant pathways of antibiotic biosynthesis.

2 Themes in the Regulation of Antibiotic Production Illustrated by Examples from Unicellular Bacteria

2.1 Intracellular Signals Associated with Starvation and Low Growth Rate Activate Microcin C7 Synthesis in *E. coli*

Bacterial antibiotic production is generally found in organisms such as *Streptomyces* spp. that undergo complex differentiation (CHATER and MERRICK, 1979), but many simple unicellular bacteria are also producers. Thus, some *E. coli* strains produce microcins, a heterogeneous collection of inhibitory compounds many (but not all) synthesized nonribosomally. The regulation of production of one such compound, microcin C7, provides a nice example of the activation of antibiotic synthesis in response to a shift-down in cellular metabolism.

Microcin C7 is a ca. 1000 Da oligopeptide antibiotic whose production, in post-exponential phase, is specified by genes (*mcc*) located on the plasmid pMccC7 (NOVOA et al., 1986). Only certain laboratory strains of *E. coli* K-12 support transcription of *mcc–lacZ* fusions (DÍAZ-GUERRA et al., 1989), other strains bearing mutations in a locus that turned out to coincide with *appR*, initially studied as a regulatory gene for acid phosphatase synthesis. The *appR* gene, formerly also referred to as *nur* and *katF*, is now known as *rpoS* and encodes an RNA polymerase sigma factor (σ^S) responsible for expression of many stationary-phase genes (MULVEY and LOEWEN, 1989; TANAKA et al., 1993; NGUYEN et al., 1993; see Fig. 2 for a summary of σ factor structure and function). (The variation in *mcc* transcription among *E. coli* strains is consistent with the finding that cultures left in stationary phase are often taken over by mutants in which the C-terminus of σ^S is deleted; ZAMBRANO et al., 1993.) Thus microcin C7 is produced only during stationary phase because, directly or indirectly, its production genes are activated via σ^S. The nature of *mcc* promoters and the conserved features of σ^S-dependent promoters in general remain to be fully elucidated. There is, however, some overlap between the promoter class recognized by σ^S and that recognized by the principal sigma factor, σ^{70} (TANAKA et al., 1993; NGUYEN et al., 1993). This is consistent with the close similarities between the regions of σ^S and σ^{70} expected to make sequence-specific DNA contacts (regions 2.4 and 4.2 in Fig. 2). Some promoters that are σ^{70}-dependent and require activators during rapid growth may perhaps be utilized during stationary phase in an activator-independent manner by RNA polymerase containing σ^S (KOLTER et al., 1993).

How is σ^S activity increased on entry into stationary phase? GENTRY et al. (1993) showed that σ^S levels respond to intracellular changes in the concentration of the important signaling molecule ppGpp, best known for its role in mediating the stringent response (Fig. 3). Levels of ppGpp increase in *E. coli* when cultures are limited for amino acids, inorganic nitrogen, or carbon (IRR, 1972; CASHEL and RUDD, 1987), and probably also for phosphate (GENTRY et al., 1993), so ppGpp is probably a regulator of entry into stationary phase. Consistent with this view, increasing the steady-state level of ppGpp either by the use of mutants deficient in ppGpp degradation or by manipulating the expression of a truncated ppGpp synthetase gene leads to a reduction in growth rate under conditions of nutritional sufficiency (SARUBBI et al., 1988; SCHREIBER et al., 1991). Interestingly an *E. coli* strain unable to make ppGpp has a phenotype somewhat like that of an *rpoS* mutant (GENTRY et al., 1993).

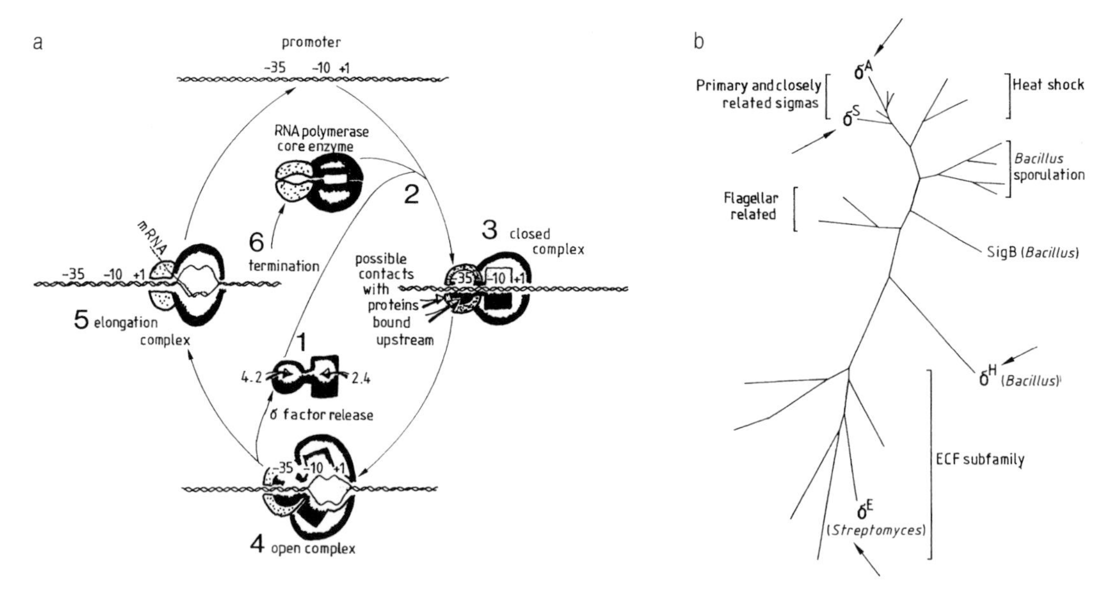

Fig. 2. (a) The role of σ factors in initiation of transcription. Nearly all σ factors are related to each other and share certain features. At (1) we emphasize two regions that interact with promoters; region 2.4 of any particular σ interacts about 10 bp upstream of the transcription start point (+1) at sequences characteristic of promoters dependent on that σ, and region 4.2 interacts with DNA about two helical turns further upstream (−35). (2) The σ–DNA interaction is largely dependent on association of σ with RNA polymerase core enzyme to give RNA polymerase holoenzyme which is thereby directed to appropriate promoters. (3) Initially a closed promoter complex is formed in which the DNA remains double-stranded. RNA polymerase–promoter interactions may be significantly affected at this and the next stage by contacts with regulatory proteins, especially transcriptional activators, bound a short distance upstream. (4) The σ factor plays an important role in melting the DNA around +1 to give an open complex. (5) RNA polymerase begins to transcribe, and the σ factor is ejected and may become associated with another core enzyme particle to reinitiate the cycle. (6) Eventually the completed mRNA and the RNA polymerase core enzyme are released. The core enzyme may now potentially associate with a different σ factor to initiate transcription from a promoter of a different class. **(b)** Phylogeny of σ factors. The familial relationships of most known σ factors from diverse bacteria are shown here. Arrows indicate σ factors referred to in this chapter. The diagram is basically that of LONETTO et al. (1994).

It is not clear how the increased ppGpp levels might cause increases in the level of σ^S (indeed, the mechanism of the stringent response is still elusive). HUISMAN and KOLTER (1994b) suggested a speculative model for the effect of ppGpp, contingent on the well-known association of increased ppGpp levels with increased expression of genes for amino acid biosynthesis (Fig. 4). It predicts that the resultant increase in the threonine pool would cause feedback inhibition of threonine biosynthesis, and so an increase in the pool of homoserine and homoserine phosphate (which normally feed into threonine biosynthesis). These intermediates could then be cyclized to homoserine lactone via a known interaction with tRNA synthetases. Homoserine lactone is proposed to be the critical intracellular regulator for induction of *rpoS*. The key pieces of evidence to support this model are: (1) the discovery of an *E. coli* gene, *rspA*, encoding a product resembling a known lactonizing enzyme, which switches off *rpoS* transcription when present at high copy number (RspA may degrade the homoserine lactone); and (2) the elimination of *rpoS* expression by mutations blocking homoserine synthesis. The model was proposed following

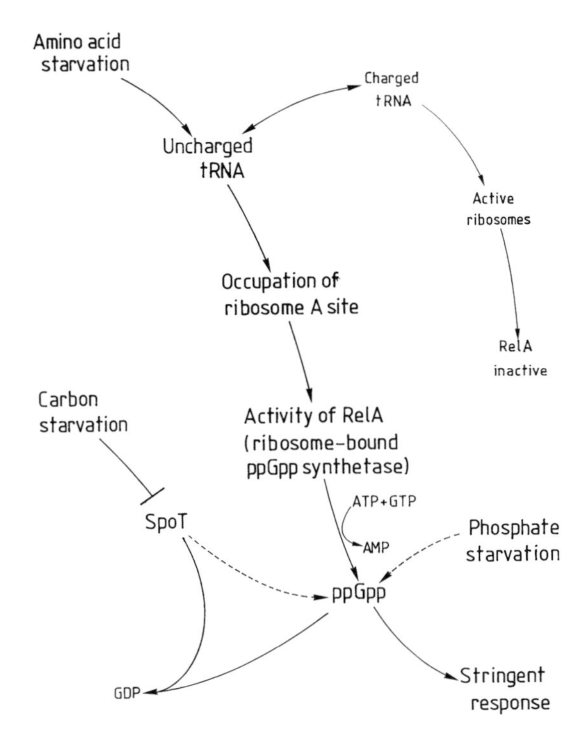

Fig. 3. Activating ppGpp synthesis in *E. coli*. Most is known about how amino acid starvation causes ppGpp synthesis. The effects of carbon and especially phosphate starvation are comparatively little studied. SpoT can apparently cause ppGpp synthesis in a *relA* null mutant (HERNANDEZ and BREMER, 1991; XIAO et al., 1991), but its major physiological role is its ppGpp-degrading activity which is inhibited under carbon-starved conditions (GENTRY et al., 1993).

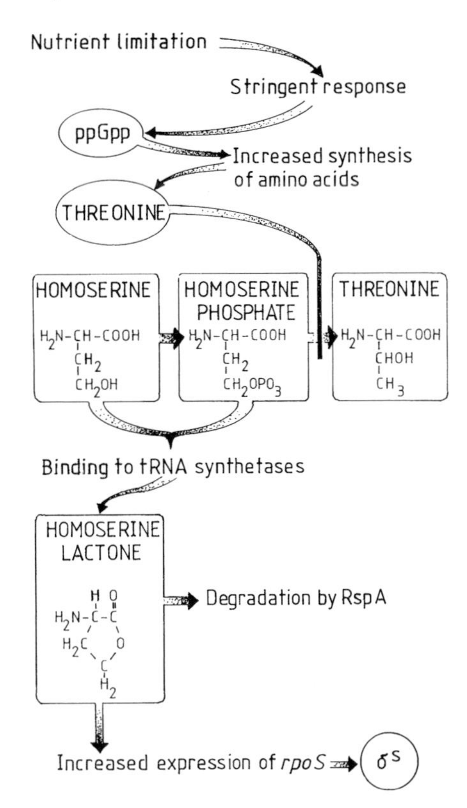

Fig. 4. Speculative model connecting increased ppGpp levels with increased production of σ^S in *E. coli*. The diagram represents the model formulated by HUISMAN and KOLTER (1994b).

the discovery that homoserine lactones are widespread as regulatory molecules (see Sect. 2.2).

The regulation of σ^S is proving to be very intricate (see HUISMAN and KOLTER, 1994a, for a review). Not only is its transcription apparently susceptible to subtle regulation by various metabolic influences, but there is also regulation at the levels of translation and protein stability (HECKER and SCHROETER, 1985; LOEWEN et al., 1993; MCCANN et al., 1993; TAKAYANAGI et al., 1994; LANGE and HENGGE-ARONIS, 1994): indeed, posttranscriptional regulation is the overriding influence on σ^S activity in minimal medium (Fig. 5). It has even turned out that a segment of the *rpoS* mRNA responds to osmotic stress, and a model has been proposed in which (by analogy with the situation for the mRNA for the heat shock sigma factor σ^{32} of *E. coli*) an internal segment of the mRNA binds to the translational initiation region, acting as an antisense inhibitor (LANGE and HENGGE-ARONIS, 1994). The stability of the sense–antisense interaction is postulated to depend on an osmotically sensitive protein. Also, σ^S half-life increases from about 2 min during rapid growth to 10–20 min during stationary phase (LANGE and HENGGE-ARONIS, 1994; TAKAYANAGI et al., 1994). Increased translation of *rpoS* mRNA at late exponential phase also implies that cell density information may be a component of σ^S regulation,

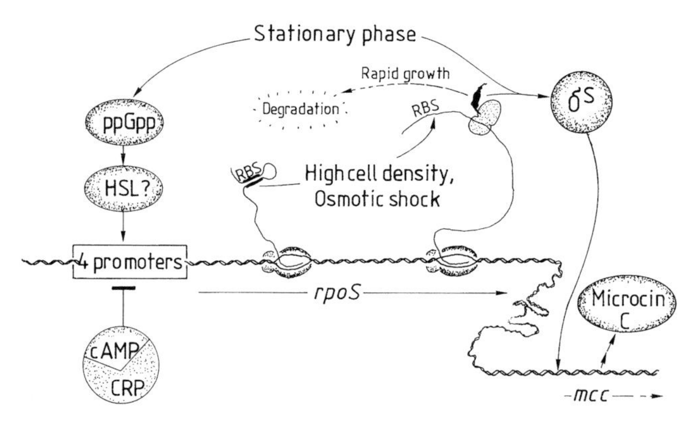

Fig. 5. Complex control of the σ^S subunit of *E. coli* RNA polymerase. The diagram is based on the data and models of LANGE and HENGGE-ARONIS (1994) and TAKAYANAGI et al. (1994). Transcription of *rpoS* is initiated from as many as four promoters. The cAMP–CRP complex (Fig. 1) inhibits use of one (or more) of these promoters, and ppGpp stimulates use of one or more of them, possibly by causing increased intracellular concentrations of homoserine lactone (HSL) (Fig. 4). Translation of the *rpoS* mRNA is thought to be limited by a (protein-stabilized?) secondary structure that sequesters the ribosome-binding site (RBS). This secondary structure is destabilized under some conditions, such as during osmotic shock, releasing the RBS for translation. The resulting σ^S protein is rapidly degraded during the growth phase, but is more stable in stationary phase, allowing it to direct RNA polymerase to transcribe stationary phase-associated genes such as those determining microcin C biosynthesis.

perhaps involving extracellular substances (LANGE and HENGGE-ARONIS, 1994). Examples of such situations in other bacteria are discussed in the following sections.

Not all stationary-phase activities in *E. coli* are regulated by σ^S (LANGE and HENGGE-ARONIS, 1991a; MCCANN et al., 1991). For example, induction of many proteins by glucose starvation is σ^S-independent and requires the cAMP/CRP system (Fig. 1). A case in point is provided by some of the *E. coli* genes (notably the *glgCAY* operon) for stationary phase-associated synthesis of glycogen under conditions of nitrogen limitation (ROMEO and PREISS, 1989). Transcription of *glgCAY* is unaffected by *rpoS* mutations (HENGGE-ARONIS and FISCHER, 1992), and it is not clear what form of RNA polymerase is involved in *glgCAY* promoter recognition (ROMEO and PREISS, 1989). Both ppGpp and cAMP/CRP have significant regulatory impact on *glgCAY*, particularly when the carbon source is not glucose. The *glgCAY* operon is also subject to repression during growth by a 6.8 kDa protein encoded by the *csrA* gene which may also be an important regulator of stationary phase genes (ROMEO et al., 1993). Interestingly, one of the genes needed for glycogen synthesis, *glgS*, maps away from the *csrA*/cAMP–CRP/ppGpp-regulated *glg* genes and shows a clear dependence on σ^S (HENGGE-ARONIS and FISCHER, 1992). The intricacy of stationary phase regulation is further illustrated by the finding that cAMP/CRP may also influence the expression of *rpoS* (LANGE and HENGGE-ARONIS, 1994).

Most of these stationary-phase regulatory devices are *trans*-acting, but at least one *cis*-acting mechanism is known: the "gearbox" promoter. Such promoters may be defined by their property of enhanced relative expression at low growth rates, coupled with resemblance to a particular unusual -10 consensus sequence (VICENTE et al., 1991). Gearbox promoters have been discussed mostly in the context of some cell cycle-related genes, the expression of which may have a specially important relationship to growth rate, as well as in relation to stationary phase. The promoter of the *mcbA-G* operon, which specifies the ribosomally synthesized antibiotic microcin B17, has gearbox kinetics and a gearbox-like

−10 region (Vicente et al., 1991). There is little information about what gives these promoters their property of gearbox kinetics, but it is not the result of recognition by a particular σ factor (Lange and Hengge-Aronis, 1991b).

2.2 A Critical Cell Population Density Signaled by an Autogenous Extracellular Signal Molecule Triggers Carbapenem Synthesis by *Erwinia carotovora*

β-Lactam antibiotics are produced by diverse bacteria including species of *Streptomyces*, *Nocardia*, *Flavobacterium*, and *Erwinia*, as well as by fungi. There is surprisingly little information about the genetic regulation of β-lactam biosynthesis. The extreme amenability of many purple gram-negative bacteria, including the carbapenem producer *Erwinia carotovora*, to rapid genetic manipulation has provided the most penetrating information so far. However, the different lifestyles and ecologies of the different producers may mean that their regulatory systems have diverged much more than the structural genes.

In apparent contrast to the situation described above for microcin C7 production by *E. coli*, production of carbapenem by *E. carotovora* is regulated by a specialized extracellular signal. Carbapenem non-producing (Car^-) mutants fall into two classes: group 1 mutants produce N-(3-oxohexanoyl)-L-homoserine lactone (OHHL) (Fig. 6; Eberhard et al., 1981) which restores the Car^+ phenotype to group 2 mutants which are themselves unable to make OHHL (Bainton et al., 1992a, b). Group 2 mutants are also pleiotropically defective in the production of various exoenzymes associated with the degradation of plant tissues during the disease process, and this entire phenotype can be reversed by OHHL.

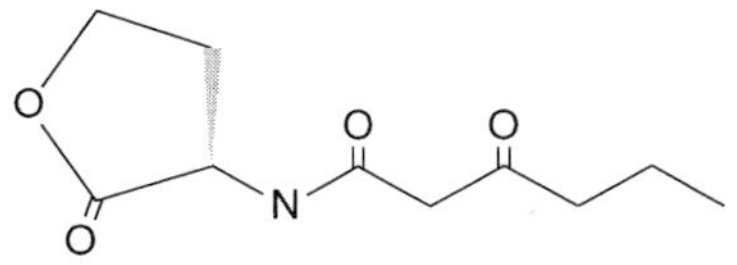

OHHL had been identified earlier as an extracellular signaling molecule in a different context: it is required to trigger light emission by the marine organism *Vibrio fischeri* (Meighen, 1991), as a very effective signal of increasing cell density (Williams, 1994; Fig. 7). OHHL synthesis requires the action of the *luxI* gene product in a single-step biosynthesis from intermediary metabolism (possibly from S-adenosyl methionine and 3-oxohexanoyl coenzyme A) (Eberhard et al., 1981). During growth at low cell densities, low-level expression of *luxI* results in a slow accumulation of OHHL in the environment. As cultures become denser, so the concentration of OHHL builds up. Since OHHL is predicted to be freely diffusible through membranes (because of its lipophilic side chain), the intracellular levels also increase until they are high enough (10^{-8} M) for effective binding to a specific cytoplasmic receptor protein (LuxR). The LuxR–OHHL complex can stimulate transcription of *luxI*, thereby causing increased OHHL synthesis and reinforcing the signal that activates luminescence. Since *luxI* is part of an operon that also encodes luciferase, light emission is strongly activated in response to a threshold level of OHHL. Regulatory systems that recognize critical levels of population density have been called "quorum sensors" (Fuqua et al., 1994). In the model described earlier implicating homoserine lactone as a hypothetical intra- (rather than inter-) cellular regulatory factor in *E. coli* it is supposed that the lactone has no lipophilic side chain, so it is not released from the cell (Huisman and Kolter 1994b).

OHHL and functional homologs of it have turned out to be widespread. This has been demonstrated by introducing a *V. fischeri lux* gene set deleted for *luxI* into *E. coli*, and exposing the transformed strain to culture supernatants of different bacteria. Luminescence was induced by samples from 18 strains

Fig. 6. Structure of the luminescence autoinducer N-(3-oxohexanoyl)-L-homoserine lactone (OHHL) of *Vibrio fischeri*.

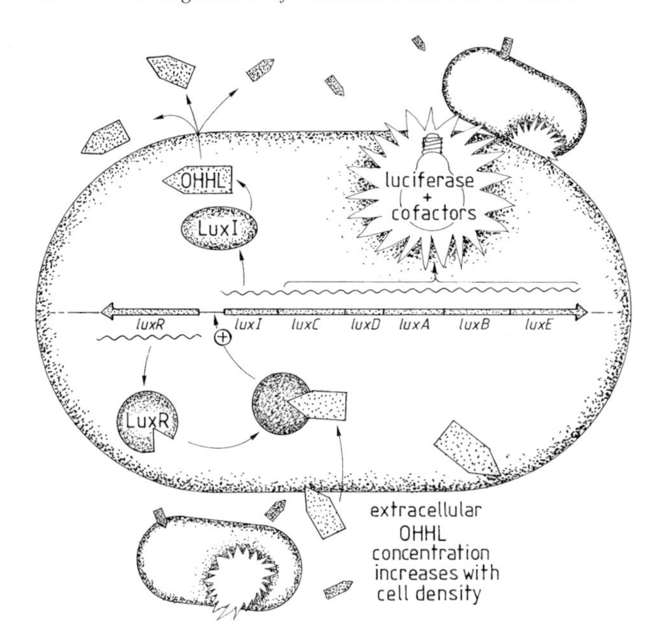

Fig. 7. Cell population-dependent expression of light emission mediated by the autoinducer OHHL of *Vibrio fischeri*. As the cell density increases, the concentration of freely permeating OHHL becomes high enough to bind to LuxR which then becomes an activator of the *luxICDABE* operon encoding a protein (LuxI) that causes OHHL synthesis (hence giving an autoinducing regulatory loop) and enzymes required for luminescence. The structure of OHHL is shown in Fig. 6.

of gram-negative bacteria of 13 genera and 6 strains of gram-positive bacteria of 5 genera (WILLIAMS, 1994). In several of the gram-negatives OHHL (or a related molecule) acts as an indicator of cell density, and each organism contains a gene that can substitute for *luxI* (SWIFT et al., 1993). The *luxI* homologs form a rather widely diverged family of genes, but many are accompanied by *luxR*-like genes encoding specific lactone-binding regulatory proteins (the LuxR family; FUQUA et al., 1994), each of which is presumably specific for a set of genes that determine a property responsive to cell population density. The emerging picture of these regulatory proteins is of an N-terminal domain for lactone binding and multimerization and a C-terminal domain with DNA-binding and transcription-activating regions. The C-terminal domain resembles those of some other transcriptional activators unconnected with quorum responsiveness, including the UhpA subfamily of response regulators (Sect. 2.3), and even of most σ factors (the LuxR superfamily; FUQUA et al., 1994). LuxR probably binds to a 20 bp inverted repeat centered at about -40 from the transcription start point, and similar sequences occupy equivalent positions in at least two other OHHL-regulated promoters, for *traA* in *Agrobacterium tumefaciens* and *lasB* of *Pseudomonas aeruginosa* (FUQUA et al., 1994).

Carbapenem production is the only example of antibiotic production known to be "quorum-regulated" by OHHL-like regulators in non-differentiating bacteria (though quorum regulation is a feature of production of some antibiotics by some differentiating bacteria including *Bacillus subtilis* and perhaps *Streptomyces griseus*; see Sects. 2.3, 3.2.4, and 3.3.3). However, existing screening systems may well miss some OHHL-related compounds because there is considerable specificity for binding. This has been revealed both by analyzing the efficacy of synthetic analogs of OHHL (WILLIAMS, 1994) and by the finding that the *lux* genes of another luminescent species, *Vibrio harveyi*, respond to a closely related molecule, N-(3-hydroxybutyryl) homoserine lactone (CAO and MEIGHEN, 1989), instead of OHHL.

2.3 The Non-Ribosomal Production of Peptide Antibiotics in Various *Bacillus* spp. Is One of a Number of Alternative Stationary-Phase Fates Determined by a Network of Transition State Regulators Involving Protein Phosphorylation

The dissection of stationary-phase functions has been most extensively analyzed in the gram-positive *Bacillus subtilis* 168. This reflects widespread interest in two of these functions, the formation of resistant endospores and the occurrence of natural competence for genetic transformation. *Bacillus* spp. often produce small peptide antibiotics during stationary phase, either through the action of large peptide synthetase enzymes or by the posttranslational modification of ribosomally synthesized propeptides. The non-ribosomally synthesized peptides include antibiotics such as bacitracin (*B. licheniformis*), gramicidin and gramicidin S (*B. brevis*), polymyxins (*B. polymyxa*), tyrocidine (*B. brevis* ATCC 8185), and surfactin (*B. subtilis* ATCC 21332). Understanding of the regulation of production of these antibiotics was reviewed by MARAHIEL et al. (1993) and has recently been extended in the case of surfactin which is the main subject of this section.

Surfactin production requires three large peptide synthetases encoded by three of the four consecutive genes in the 27 kb *srfA* operon (COSMINA et al., 1993). The classical genetic strain 168 of *B. subtilis* does not make surfactin, even though it contains an intact *srfA* operon. This deficiency is due to a frameshift mutation (the *sfp*° allele) of the adjacent *sfp* gene (NAKANO et al., 1992) which may be involved in the export of small peptides (GROSSMAN et al., 1993).

Transcription of the *srfA* operon is the major point of regulation of surfactin production, and it has turned out to involve a complex array of controlling elements including autoregulation, phosphorylation cascades, extracellular signaling, and interplay with the regulation of different developmental pathways. All these influences appear to be transmitted through a single transcriptional activator, ComA, which also plays a key role in the onset of competence for transformation (DUBNAU, 1993; Fig. 8). ComA belongs to the response regulator class of transcriptional activators (WEINRAUCH et al., 1989; Fig. 9), activity of which is generally controlled by phosphorylation of a conserved aspartate residue. This residue, and others involved in forming a phosphorylation pocket, are present in ComA, so it is not surprising that its ability to bind to the *srfA* promoter *in vitro* is greatly enhanced by phosphorylation (ROGGIANO and DUBNAU, 1993). In its active configuration, ComA binds to each of two similar dyad elements upstream of the *srfA* promoter, and the bound proteins are presumed to interact with each other bringing about DNA bending (ROGGIANO and DUBNAU, 1993; NAKANO and ZUBER, 1993) (Fig. 8). This multiple interaction stimulates *srfA* transcription by RNA polymerase bound to the promoter (probably via the major σ factor, σ^A; NAKANO et al., 1991). Phosphorylation of response regulators such as ComA typically occurs when appropriate environmental conditions are sensed by a membrane-located histidine protein kinase, and ComA is no exception (WEINRAUCH et al., 1990). Its partner kinase, ComP, appears to behave as a "quorum sensor", being activated by binding of an extracellular competence "pheromone" produced by the *B. subtilis* cells themselves. This pheromone, a 9–10 amino acid oligopeptide with a lipophilic modification, is formed from the C-terminus of the 55 amino acid product of the *comX* locus, in a process dependent on the product of the immediately adjacent upstream gene, *comQ* (*comQ* and *comX* are themselves immediately upstream of the genes encoding the ComP–ComA proteins involved in sensing and responding to the ComX pheromone; Fig. 8).

Remarkably, a segment of the *srfA* operon plays an additional role, as part of a regulatory cascade leading to competence (D'SOUZA et al., 1993) (Fig. 8). Surfactin itself is not involved in this cascade, since *srfA* mutations eliminating production do not all eliminate competence; and indeed, *B. subtilis* 168 which usually makes no surfactin is used as a labora-

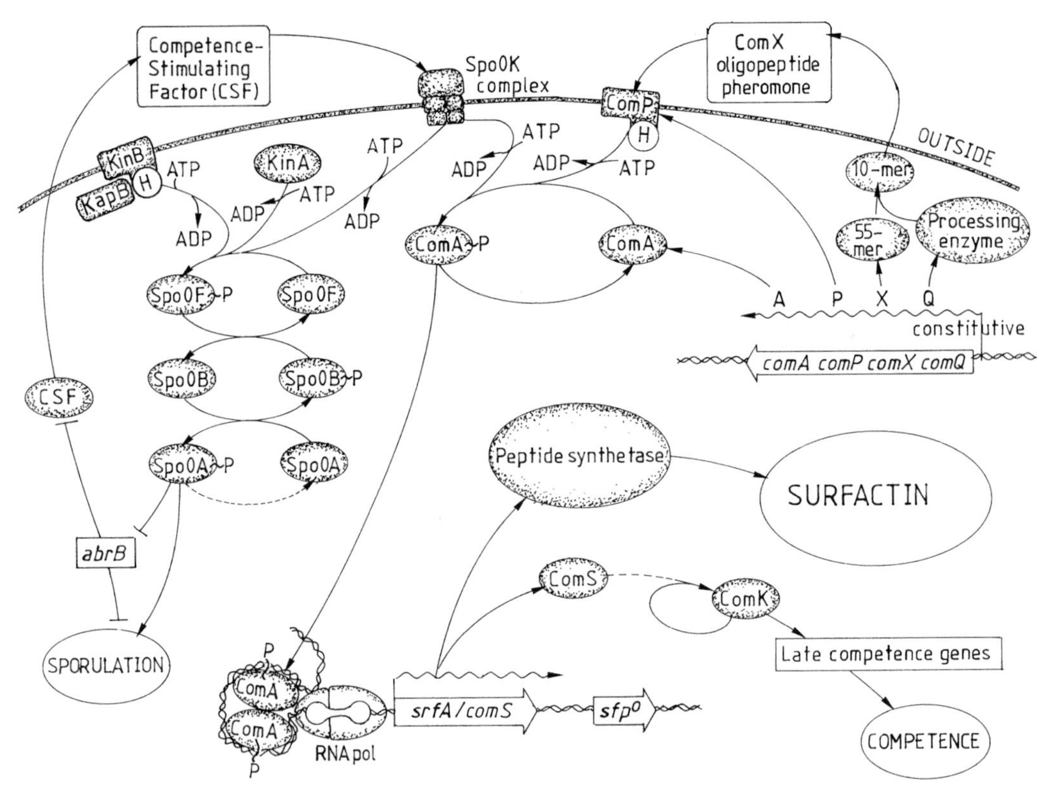

Fig. 8. Signal cascades and networks leading to surfactin synthesis and other stationary-phase processes in *Bacillus subtilis*. The onset of surfactin synthesis is transcriptionally dependent on the phosphorylated form of the ComA response regulator. ComA phosphorylation is carried out partly by a cognate histidine protein kinase, ComP, which autophosphorylates a conserved histidine residue (H) when it interacts with an extracellular oligopeptide pheromone, ComX, which is a processed and modified form of the *comX* gene product. This pheromone and a second one, CSF, accumulate to effective concentrations only under conditions of high cell density. CSF also stimulates ComA phosphorylation by an unknown route involving the Spo0K transporter complex. Production of CSF is autoregulated via the sporulation phosphorelay which is activated partially by Spo0K and partially by the histidine protein kinase KinB and another protein kinase, KinA, in response to unknown signals. The phosphorelay passes a phosphoryl group via Spo0F and Spo0B to Spo0A, a response regulator protein that is both an activator of sporulation genes and a repressor of *abrB*, itself encoding a repressor of some sporulation genes and of CSF synthesis. Thus, activation of the phosphorelay leads to relief of CSF synthesis from repression (the figure also shows that the activation of surfactin synthesis is accompanied by the activation of competence; see text for further details).

tory organism because of its competence. Studies of constructed *srfA* partial deletions had shown that regulation of competence requires only the DNA region encoding the valine-activating domain of one of the peptide synthetases (VAN SINDEREN et al., 1993), and that aminoacylation activity itself is not needed (D'SOUZA et al., 1993). This unusual situation has been clarified by the recent discovery that a distinct small open reading frame, *comS*, within *srfA* encodes a critical regulatory element for competence (D'SOUZA et al., 1994) (Fig. 8). *comS* is translated in a different reading frame from the sequence encoding the valine-activating domain of *srfA*, giving a protein which appears to activate competence by interfering with a protein-mediated inhibition of ComK, a critical activator of the late competence genes *comC*, *comG*, and *comDE* (MSADEK et al., 1994;

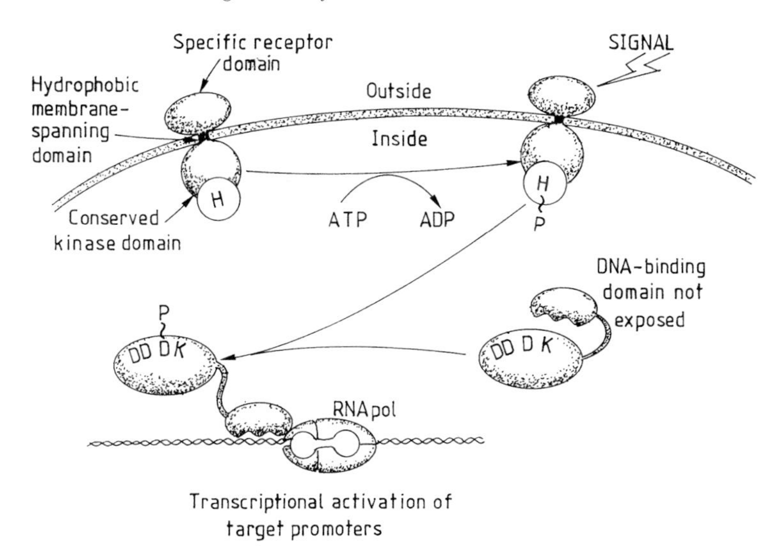

Fig. 9. Signal transduction in a bacterial two-component regulatory system. The conserved core elements of such systems are a protein kinase module capable of autophosphorylation at a conserved histidine residue (H), and a target response regulator module to which the activated phosphoryl group is transferred via an acidic pocket consisting of conserved aspartate residues, one of which (Asp^{57}) becomes phosphorylated. These modules are usually on separate proteins though they can also be found within single proteins (ALEX and SIMON, 1994). In the example shown, the histidine kinase module forms the C-terminal, cytoplasmic domain of a protein whose N-terminal domain is a surface receptor for an unspecified extracellular signal. Binding of the ligand activates the kinase which then phosphorylates the N-terminal domain of a cytoplasmic response regulator. This activates a C-terminal DNA-binding domain of the regulator. In this example, binding to a specific DNA target sequence leads to transcriptional activation by contact with RNA polymerase at an adjacent promoter. The DNA-binding/transcriptional activation modules of such response regulators themselves form subfamilies, some of which are found in regulatory proteins that are not part of two-component systems (e.g., LuxR; Sect. 2.2).

KONG and DUBNAU, 1994). It is not known why surfactin production has evolved this connection with competence.

Activation of ComA can be detected, albeit at a reduced level, in *comP* mutants that lack the cognate protein kinase. This is accounted for by an as yet incompletely defined pathway from a second membrane-bound signal receptor. This receptor, encoded by the complex *spo0K* locus, is an aggregate of several proteins and resembles various oligopeptide transport systems (hence the use of the acronym Opp to describe the Spo0K proteins). It is a member of the ATP-binding cassette (ABC) family of transporters (HIGGINS, 1992). The Opp complex responds to a second oligopeptide pheromone, CSF (competence stimulating factor), quite distinct from the ComX pheromone. The precise structure and biosynthetic origin of CSF are not known, but its production is regulated by a route different from that of ComX pheromone: it is repressed by the AbrB protein, a repressor of several stationary-phase pathways including sporulation. Relief of these pathways from AbrB-mediated repression at the onset of stationary phase is usually brought about by phosphorylation of Spo0A, the central regulator of transition stage genes, via a phosphorelay system that is best known for its role in initiating sporulation (HOCH, 1993). The Opp complex is also one of several routes by which largely undefined physiological signals lead to transmission of phosphoryl groups to Spo0A via the Spo0F and Spo0B proteins (RUDNER et al., 1991; HOCH, 1993). Spo0A has an NH_2-terminal portion homologous to the phosphorylated domain of response regulators (HOCH, 1993; Fig. 8). The Spo0A ~ P protein represses *abrB*, thereby derepressing CSF production in an autoregulatory loop that biases further development in the direction of sporulation, surfactin production, and competence development. Spo0A ~ P also directly activates some genes, notably some involved in sporulation. This constellation of Spo0A ~ P activities explains the pleiotropic sporulation, antibiotic production, and competence deficiencies of *spo0A* mutants, and of *spo0K*, *spo0B*, and *spo0F* mutants deficient in the phosphorelay. The full expression of *spo0A*, and hence the expression of *srfA*, also depends on a minor RNA polymerase σ factor, σ^H, which directs transcription of *spo0A* from an alternative, stationary phase-specific promoter. The regulation of σ^H is itself highly complex, with increased σ^H activity during entry into stationary phase involving both transcriptional and posttranscriptional control (HEALY et al., 1991).

The stimuli for activation of the Spo0 phosphorelay are not yet well understood. The multiple steps possibly provide the means to integrate sensory input from diverse sources (HOCH, 1993). For example, Spo0B is specified by the first gene in an operon that also encodes an essential GTP-binding protein, Obg, and Obg might cause phosphorylation of Spo0B in response to intracellular information about the cell cycle or the levels of GTP, which are critical in signaling the onset of sporulation (HOCH, 1993).

In a further twist of this increasingly complex system, *srfA* expression can be influenced by another regulatory protein, DegU, which belongs to the same subfamily of response regulators as ComA (HENNER et al., 1988; WEINRAUCH et al., 1989): hyperphosphorylated mutant forms of DegU repress *srfA* by an unknown mechanism (HAHN and DUBNAU, 1991).

2.4 What Has Been Learned from Studies of Antibiotic Production in Unicellular Bacteria?

The examples discussed so far have revealed that the association of antibiotic production with stationary phase is brought about by an almost bewildering variety of mechanisms. The initial switches typically result from a change in a constitutively synthesized regulatory protein, brought about by either intracellular or extracellular chemical signals (e.g., OHHL or oligopeptide pheromones). Extracellular signals may be membrane-diffusible and recognized by cytoplasmic binding proteins, or membrane-non-diffusible and recognized by membrane-bound proteins able to initiate an intracellular phosphorylation cascade. In all cases described so far, the end result of transmission of the initial signal is activation of transcription of structural genes for antibiotic biosynthesis. In the best characterized cases, this activation may arise either because of increased levels of a minor form of RNA polymerase containing a sigma factor such as σ^S of *E. coli* that can recognize the promoters of the structural genes, or by the posttranscriptional activation of transcription factors needed for the major form of RNA polymerase to initiate transcription at the relevant promoters (e.g., phosphorylation of ComA in *B. subtilis* or OHHL-mediated conformational change of a LuxR homolog in carbapenem-producing *E. carotovora*). It is abundantly clear from studies in *E. coli* and *B. subtilis* that antibiotic production requires the integration of diverse information through complex regulatory networks.

3 Regulation of Antibiotic Production in Streptomycetes and their Relatives

3.1 Introduction to the Organisms

Streptomyces spp. are the most versatile and commercially important producers of antibiotics and have traditionally been central to screening programs for new chemotherapeutic compounds. They are morphologically and phylogenetically distinct from the bacteria dealt with in the preceding sections. Their branching, mycelial growth habit is probably an adaptation that allows them to grow efficiently on the surface of insoluble organic debris in soil. Dispersal is by means of spores, typically borne on aerial hyphae, and in the laboratory mature *Streptomyces* colonies have a furry appearance because of this aerial mycelium. The *Actinomycetales* to which streptomycetes belong form a division of the gram-positive bacteria characterized by a high proportion of G+C in their DNA (on average 74 mol% G+C). *B. subtilis*, on the other hand, belongs to the division with low G+C content. These divisions resulted from a very ancient evolutionary separation. Striking progress has been made in the last decade in the molecular analysis of antibiotic production by streptomycetes, notably in two model species, *Streptomyces griseus*, the producer of streptomycin, and *Streptomyces coelicolor* A3(2), genetically the most-studied strain (reviewed by CHATER and HOPWOOD, 1993, and HOPWOOD et al., 1995). Studies in *S. coelicolor* have been helped by the extensive availability of natural and artificial genetic systems, the development of a combined physical and genetic linkage map of the chromosome (KIESER et al., 1992), and the fact that the strain produces at least four antibiotics, production of one of which (methylenomycin A) is plasmid-specified (Sect. 3.3.4.5); two others are conveniently pigmented (actinorhodin is blue at high pH and red at low pH while undecylprodigiosin is red). Moreover, diverse aspects of the physiology and developmental biology of *S. coelicolor* have been studied providing important information to relate to studies of secondary metabolism. Much of the information reviewed below is drawn from *S. griseus* and *S. coelicolor*.

3.2 General Physiological Aspects of the Regulation of Antibiotic Production in *Streptomyces*

Like the organisms already described, streptomycetes grown in liquid media generally produce antibiotics during stationary phase or at low growth rates. (In the latter case, this may reflect production by cells inside mycelial pellets that may be nutritionally limited and that have, therefore, entered sta-

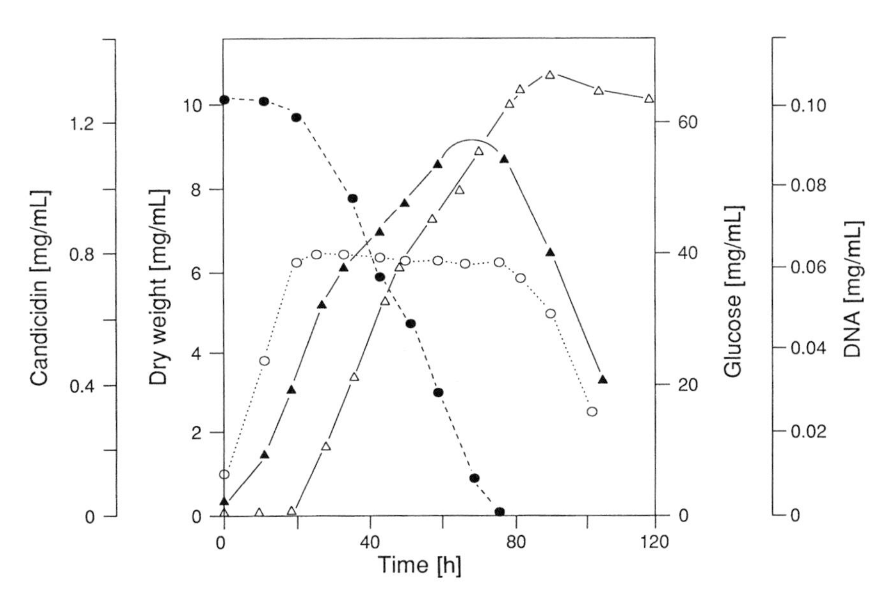

Fig. 10. Growth phase-dependent production of candicidin by *S. griseus* in liquid culture. △ Candicidin; ▲ dry weight; ● glucose; ○ DNA. Redrawn from MARTÍN and MCDANIEL (1975).

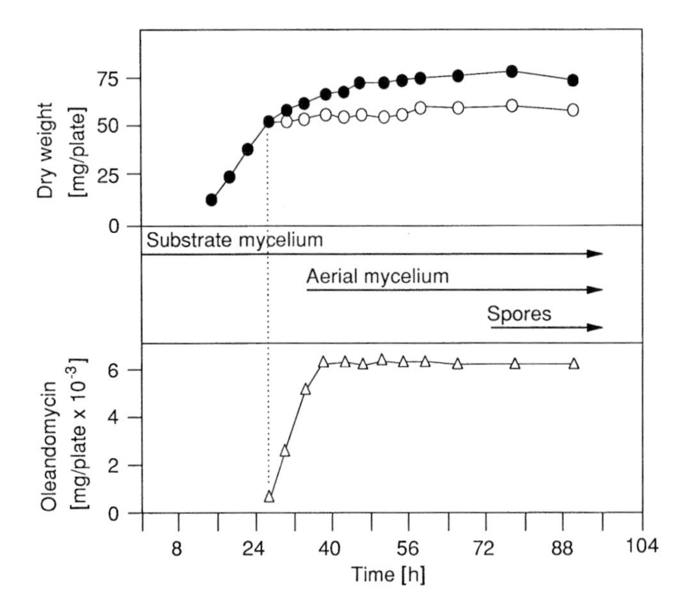

Fig. 11. Growth phase-dependent production of oleandomycin by surface-grown cultures of *S. antibioticus*. ● dry weight; ○ dry weight minus glycogen; △ oleandomycin. Redrawn from MÉNDEZ et al. (1985).

tionary phase.) For example, candicidin was produced by *S. griseus* in liquid culture only after net DNA synthesis had ceased (Fig. 10; MARTIN and MCDANIEL, 1975). Biomass continued to increase slowly, presumably through the accumulation of storage compounds such as glycogen (BRAÑA et al., 1986) or triacyl glycerol (OLUKOSHI and PACKTER, 1994).

Stationary phase antibiotic production could be viewed as a physiological abnormality that results from growing soil organisms in submerged culture, particularly since most streptomycetes do not differentiate normally in liquid culture. However, antibiotic production on solid media is also growth phase-dependent; thus, production of oleandomycin by *Streptomyces antibioticus* on agar began only after growth had ceased (Fig. 11; MÉNDEZ et al., 1985). Interestingly, oleandomycin production appeared to be confined largely to the substrate mycelium since synthesis of the antibiotic was apparently completed before aerial hyphae appeared. In other cases, antibiotic production coincides approximately with the onset of morphological differentiation, and the isolation from both *S. coelicolor* and *S. griseus* of *bld* mutants defective in both processes suggests at least some common elements of genetic control (Sect. 3.3.2.6).

In Sects. 3.2.1–3.2.5 we review current understanding of the physiological factors that might play a general role in triggering the onset of antibiotic synthesis, before moving to consider the genetics of antibiotic production in Sect. 3.3.

3.2.1 Metabolite Interference with Antibiotic Production in Streptomycetes

Growth is most rapid when readily utilized carbon, nitrogen, and phosphate sources are abundant, in part because of repression or inhibition by these metabolites of most inessential processes. In principle, antibiotic production could be subject to the same regulatory controls, and its occurrence during stationary phase might reflect relief from metabolite interference after nutrient depletion. While there are many examples of metabolite interference with antibiotic biosynthesis, especially by glucose, ammonium, and phosphate (Tab. 1), the underlying mechanisms are generally not known. In a few cases, repression of transcription appears to be involved, as in glucose repression of phenoxazinone synth-

Tab. 1. Metabolites that Interfere with Antibiotic Production in Streptomycetes

Interfering metabolite	Antibiotic
Carbon sources:	
Citrate	Novobiocin
Glucose	Actinomycin, chloramphenicol, chlortetracycline, kanamycin, mitomycin, neomycin, oleandomycin, puromycin, siomycin, streptomycin, tetracycline, tylosin
Glycerol	Actinomycin, cephamycin
Nitrogen sources:	
Ammonium ions	Actinorhodin, chloramphenicol, leucomycin, streptomycin, streptothricin, tetracycline, tylosin, undecylprodigiosin
L-Glu, L-Ala, L-Phe, D-Val	Actinomycin
L-Tyr, L-Phe, L-Trp, PABA	Candicidin
Inorganic phosphate:	
	Actinorhodin, candicidin, cephamycin, nanaomycin, nourseothricin, streptomycin, tetracycline, tylosin, undecylprodigiosin, vancomycin

Information from DEMAIN et al. (1983), DEMAIN (1992), and HOBBS et al. (1992).

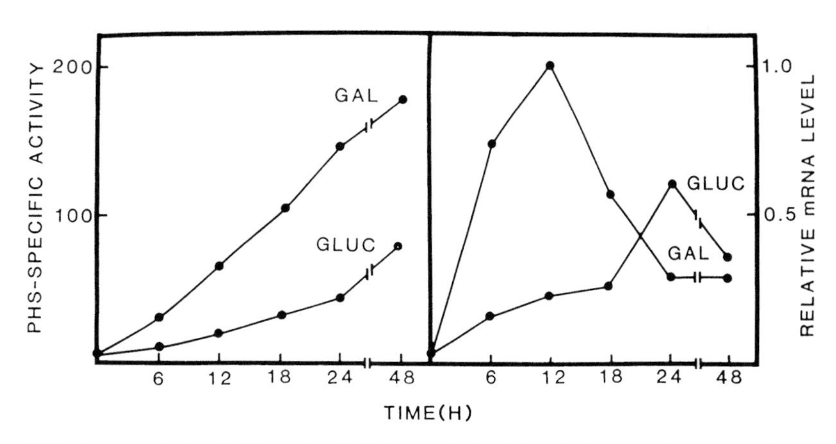

Fig. 12. Effect of glucose (Gluc) on activity (left) and mRNA levels (right) of phenoxazinone synthase (PHS) in *S. antibioticus* (Gal, galactose). Redrawn from JONES (1985).

ase, the final enzyme in actinomycin biosynthesis in *S. antibioticus* (JONES, 1985; Fig. 12). Phosphate also appears to repress transcription of genes required for candicidin biosynthesis by *S. griseus* (Fig. 13; ASTURIAS et al., 1990) and of those for actinorhodin production in *S. coelicolor* (HOBBS et al., 1992).

Unfortunately, the levels of nutrients and the identification of the growth-limiting component(s) during culture have rarely been reported, so the extent of the contribution of metabolite interference to growth phase-dependent antibiotic production is difficult to assess. Furthermore, although metabolite interference might account for some examples of growth phase dependence, there appears to be no general pattern. Even within one species, the production of different secondary

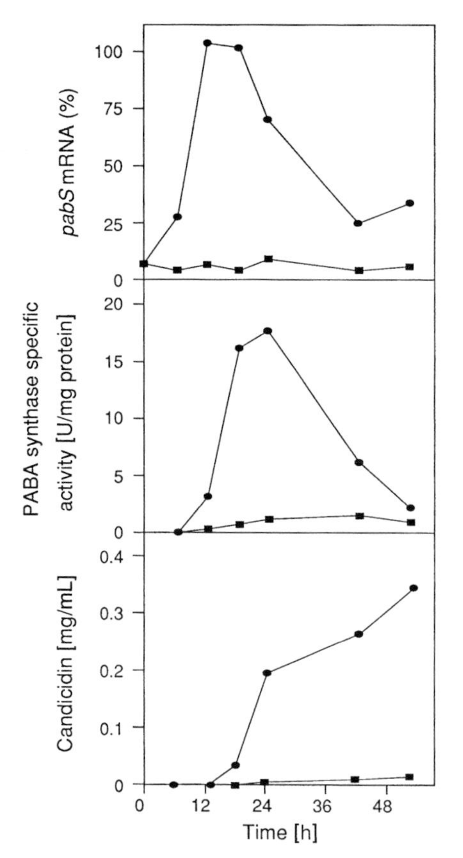

Fig. 13. Phosphate repression of candicidin production in *S. griseus*. *pabS* encodes PABA synthase, and plays an early role in candicidin production. SPG, soya peptone–glucose medium. ● SPG; ■ SPG+7.5 mM phosphate. Redrawn from ASTURIAS et al. (1990).

metabolites appears to be triggered by the depletion of different nutrients. In *Streptomyces cattleya*, the production of melanin, cephamycin C, and thienamycin in batch culture occurred on depletion of glucose, ammonia, and phosphate, respectively. In a chemostat, cephamycin C production occurred at low growth rates that could be brought about by limiting for carbon, nitrogen, or phosphate, whereas the production of thienamycin required a low growth rate specifically associated with phosphate deficiency, consistent with the observations made in batch culture (LILLEY et al., 1981).

3.2.2 Antibiotic Production and Imbalances in Metabolism

An alternative or additional possibility is that antibiotic production is triggered by an imbalance in metabolism. Undecylprodigiosin, the major component of the red antibiotic of *S. coelicolor* (TSAO et al., 1985), is derived partly from proline (WASSERMAN et al., 1974; GERBER et al., 1978). To determine whether the amino acid incorporated into the antibiotic was synthesized internally or taken up from outside, HOOD et al. (1992) isolated and characterized *put* mutants deficient in proline transport, which turned out to be defective also in proline catabolism. Since proline biosynthesis appears to be constitutive in *S. coelicolor*, such mutants might be expected to accumulate proline intracellularly. While this has not been determined experimentally, the mutants markedly overproduce the red antibiotic suggesting that undecylprodigiosin serves as a sink for excess proline. The need to remove surplus proline might reflect the role that it plays as an osmoregulant in other bacteria (KILLHAM and FIRESTONE, 1984a, b). It will be interesting to see whether the *put* mutants produce undecylprodigiosin earlier than the parental strain, as predicted by this hypothesis, and how this is mediated at the level of gene regulation.

An imbalance in carbon metabolism may be responsible for triggering the production of methylenomycin A, another *S. coelicolor* antibiotic. Methylenomycin biosynthesis began at the same time as a rapid drop in the pH of a *S. coelicolor* fermentation (Fig. 14) caused by the efflux of α-ketoglutarate and pyruvate from the mycelium (HOBBS et al., 1992). Indeed, acid shock alone caused transient methylenomycin biosynthesis, suggesting that this might be a stress response to the change in pH that presumably reflects the imbalance in carbon metabolism (G. HOBBS and S. G. OLIVER, personal communication).

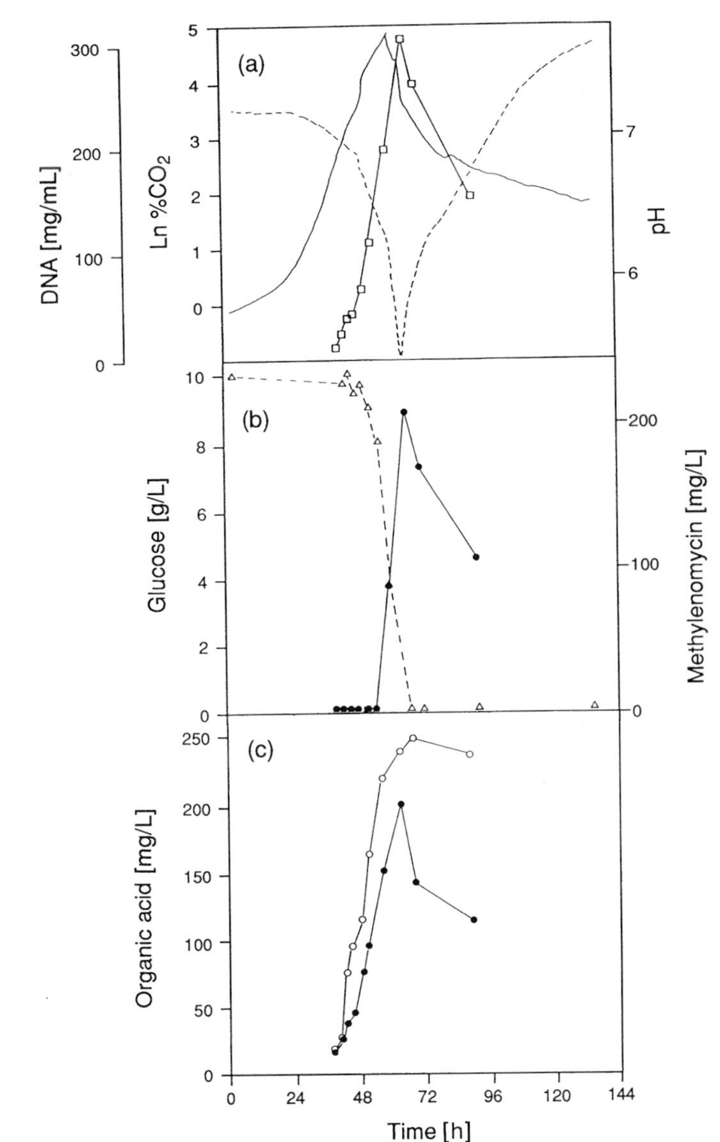

Fig. 14a–c. Methylenomycin production during growth of *S. coelicolor* A3(2) on minimal medium with alanine as nitrogen source. **a** Growth and change in extracellular pH. □ DNA; — Ln% CO_2 (logarithm of the percentage carbon dioxide concentration in the fermenter exhaust gases); – – – pH. **b** Glucose assimilation and methylenomycin production. △ glucose; ● methylenomycin. **c** Production of pyruvate and α-ketoglutarate. ○ α-ketoglutarate; ▲ pyruvate. Redrawn from HOBBS et al. (1992).

3.2.3 The Possible Role of Growth Rate and ppGpp in Antibiotic Production

Most of the published data are consistent with a role for growth rate, or the cessation of growth, in determining the onset of antibiotic production in streptomycetes. As discussed earlier (Sect. 2.1), ppGpp is believed by many to play a central role in the growth rate control of gene expression in *E. coli* (SARUBBI et al., 1988; HERNANDEZ and BREMER, 1990, 1993; SCHREIBER et al., 1991). A possible role for ppGpp in triggering the onset of antibiotic production was first addressed in *S. griseus* by AN and VINING (1978). They found that streptomycin production occurred only after

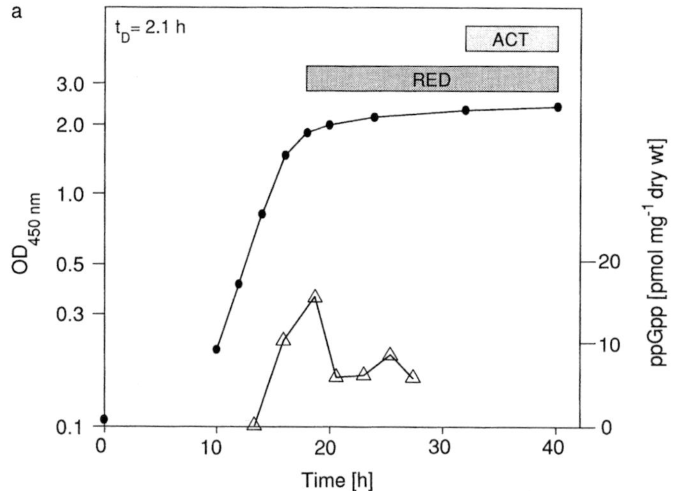

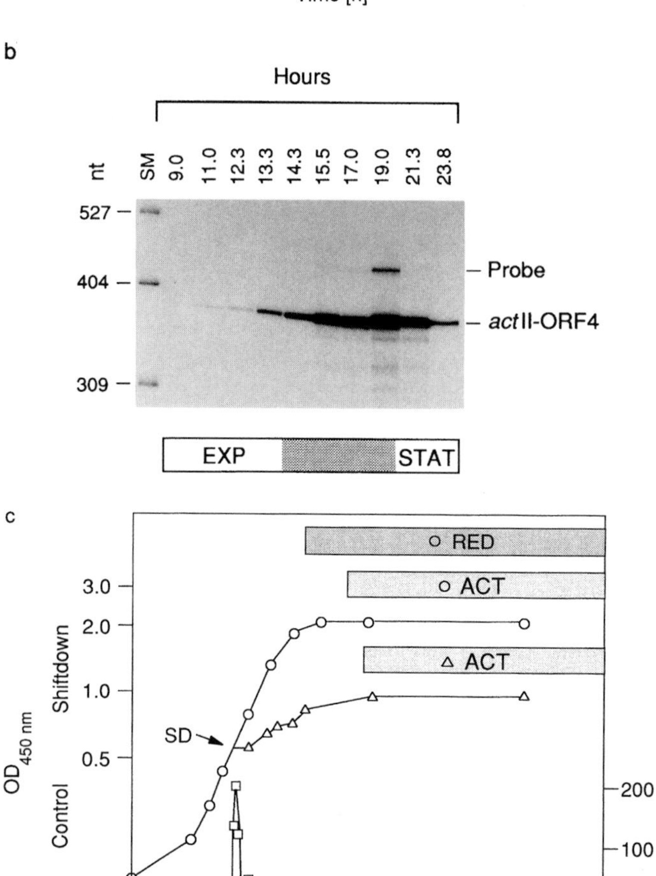

Fig. 15a–d. ppGpp and antibiotic production in *S. coelicolor* A3(2). **a** Growth (●), antibiotic, and ppGpp (△) production in *S. coelicolor* A3(2). The shaded boxes labeled ACT or RED denote the presence of actinorhodin or undecylprodigiosin, respectively. t_D, doubling time. **b** Transcription of *actII*-ORF4 was monitored by S1 nuclease protection studies using RNA isolated at the times indicated from the culture shown in (**a**). "Probe", the position of the full-length probe. EXP and STAT, exponential and stationary phases, respectively, and the shaded area between them indicates the transition phase. SM, size marker. Similar results were observed for *redD* transcription. **c** Growth curve of *S. coelicolor* A3(2) with (△) and without (○) nutritional shiftdown. The shaded boxes labeled (○ RED) or (○ ACT) denote the presence of undecylprodigiosin or actinorhodin, respectively in the control culture. SD indicates when the cultures were subjected to shiftdown, the shaded box labeled (△ ACT) denotes the presence of actinorhodin in the culture subjected to shiftdown, and (□) the ppGpp level in the shifted culture. **d** Transcription of *actII*-ORF4 after nutritional shiftdown. S1 nuclease protection analysis of *actII*-ORF4 transcripts in RNA isolated at the times indicated from the culture shown in (**c**).

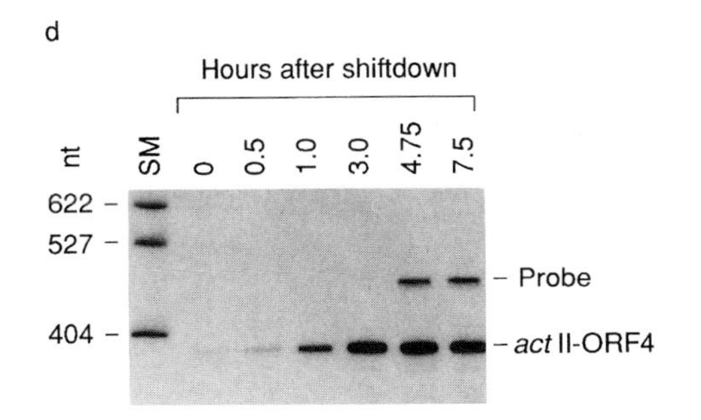

Fig. 15d

a drop in ppGpp levels, so it seemed that ppGpp did not stimulate the initiation of antibiotic synthesis. Subsequently, ppGpp was detected in several *Streptomyces* spp. (HAMAGISHI et al., 1980; SIMUTH et al., 1979; HAMAGISHI et al., 1981; NISHINO and MURAO, 1981; STASTNA and MIKULIK, 1981), and positive correlations were observed between ppGpp and antibiotic biosynthesis in *Streptomyces aureofaciens* (SIMUTH et al., 1979) and *Streptomyces galilaeus* (HAMAGISHI et al., 1981). OCHI (1986) found that an accumulation of ppGpp following nutritional shiftdown was accompanied by an 8-fold increase in formycin production by *Streptomyces lavendulae* MA406-A-1; moreover, a mutant deficient in ppGpp accumulation after amino acid depletion (the "relaxed" phenotype) was impaired in formycin production. Similar *relC* mutants were isolated from *S. antibioticus* 3720 (OCHI, 1987), *Streptomyces griseoflavus* 1805 (OCHI, 1988), *S. griseus* 13189 (OCHI, 1990a), and *S. coelicolor* (OCHI, 1990b). The mutants accumulated low levels of ppGpp after nutritional shiftdown (on average ca. 15% of wild-type levels) and were deficient in antibiotic production, leading to the idea that ppGpp plays a central role in triggering antibiotic production (OCHI, 1990b). A streptomycin-producing pseudorevertant of a *relC* mutant of *S. griseus* remained defective in (p)ppGpp accumulation. This suppressor mutation was proposed to identify a gene activated by ppGpp and involved in triggering streptomycin production. Consistent with OCHI's proposal, KELLY et al. (1991) found reduced levels of actinomycin biosynthetic enzymes and mRNA in a *relC* mutant of *S. antibioticus*, and HOLT et al. (1992) noted a peak of ppGpp accumulation that coincided with transcription of the pathway-specific activator gene, *brpA*, before production of bialaphos by *Streptomyces hygroscopicus*. However, BASCARAN et al. (1991) concluded that there was no obligate relationship between ppGpp and antibiotic production in *Streptomyces clavuligerus*. Production of cephalosporins (mainly cephamycin C) occurred during a phase of slow exponential growth (doubling times of ca. 7 h and 18 h in rich and minimal media, respectively) and increased in stationary phase, while ppGpp levels remained constant from the beginning of growth. *relC*-like mutants accumulated lower levels of ppGpp than the wild-type strain after nutritional shiftdown (varying between 8% and 85% of the wild-type levels), but there was no simple correlation between the levels of ppGpp and cephalosporin production (some produced more, and others less, cephalosporin than the wild-type strain).

Recently, the relationship between ppGpp synthesis and the production of undecylprodigiosin and actinorhodin has been examined in *S. coelicolor* (Fig. 15). Transcription of pathway-specific activator genes for each antibiotic (*redD* for undecylprodigiosin and *actII*-ORF4 for actinorhodin: see Sect. 3.3.4.1) conspicuously increased as the culture was in transition between growth and stationary phase, coinciding with the onset of ppGpp production; this was followed by transcription of representative *red* and *act* biosynthetic structural genes and production of the antibiotics (STRAUCH et al., 1991; TAKANO et al., 1992). In contrast, transcription of a typical rRNA gene set (*rrnD*) decreased markedly (E. TAKANO and M. J. BIBB, unpublished results). Thus a positive correlation was observed between ppGpp accumulation and transcription of *actII*-ORF4 and *redD* at the end of exponential growth. ppGpp synthesis could also be induced by a nutritional shiftdown during rapid growth, and again resulted in a rapid and marked decrease in *rrnD* transcription (STRAUCH et al., 1991). Transcription of *actII*-ORF4 was detected within 30 min of shiftdown (Fig. 15) and transcription of the *act* biosynthetic genes followed 30 min later (TAKANO and BIBB, 1994). However, there was no immediate stimulation of *redD* transcription (TAKANO and BIBB, 1994), suggesting that if ppGpp does play a role in triggering the onset of antibiotic production, it is not always sufficient: the activation of at least some biosynthetic pathways may depend on additional factors. Such variations in regulatory responses of different sets of pathway genes in *Streptomyces* spp. are perhaps not surprising in view of comparable variations observed in *E. coli* and *Bacillus* spp. (see Sects. 2.1 and 2.3).

Correlations between ppGpp synthesis and the onset of antibiotic production, where they occur, do not establish a causal relationship, and although the *relC* mutants isolated by OCHI (1986, 1987, 1988, 1990a, b) showed a marked reduction in antibiotic production, they also grow at about half the maximal rate of the parental strain, presumably reflecting impaired protein synthesis. Thus it is difficult to assess whether the effect on antibiotic production is a direct consequence of reduced levels of ppGpp or an indirect effect of the *relC* mutation on protein synthesis. It is interesting to note that mutants of *E. coli* unable to make ppGpp (ppGpp° mutants) do not show the marked reduction in growth rate characteristic of *relC* mutants (XIAO et al., 1991). Attempts to establish a role for ppGpp in triggering the onset of antibiotic production in *Streptomyces* spp. will require the isolation of the corresponding ppGpp° mutants and possibly the ability to regulate ppGpp levels in the absence of the physiological trauma associated with nutritional shiftdown, as was done by GENTRY et al. (1993) in their analysis of the effects of ppGpp on *rpoS* expression in *E. coli* (see Sect. 2.1).

3.2.4 Antibiotic Production and the Accumulation of Small Diffusible Signaling Compounds

The synthesis of threshold levels of small diffusible signaling molecules appears to play a central role in triggering production of at least some antibiotics in some streptomycetes (HORINOUCHI and BEPPU, 1992, 1995). Such molecules might act simply as indicators of cell population density or they might be synthesized in response to physiological conditions under which antibiotic production would be favorable to the organism.

γ-Butyrolactone compounds whose acylated lactone structures somewhat resemble the homoserine lactones found in gram-negative bacteria (see Fig. 6) have been detected in many streptomycetes, and have been implicated in antibiotic production and morphological differentiation in several species (Fig. 16). The most intensively studied example is A-factor (2-isocapryloyl-3*R*-hydroxymethyl-γ-butyrolactone) which is required for streptomycin production and morphological differentiation in *S. griseus* (Sect. 3.3.3). Five related compounds, virginiae butanolides A-E (Fig. 16a; VB-A-E), are inducers of virginiamycin production by *Streptomyces virginiae* (YAMADA et al., 1987), and compounds with the same biological activity are found in other streptomycetes (OHASHI et al., 1989). Such butyrolactones probably diffuse readily through membranes, so extracellular and intracellular levels are likely to be the same: hence, just as with LuxR and OHHL, monitoring of extracellular concentrations is done by cytoplasmically located binding proteins with very high affinity and specificity for their ligands (K_D ca. 10^{-9} M). To illustrate this specificity, virginiae butanolide C shows no biological activity *in vivo* with A-factor-deficient mutants of *S. griseus* and the A-factor receptor does not bind it *in vitro* (MIYAKE et al., 1989). However, ligand specificity can vary between strains, and A-factor analogs with acyl chains of different lengths, or with a hydroxyl group rather than a carbonyl group at position 6, had some activity in an anthracycline-producing *S. griseus* strain that requires A-factor for sporulation and antibiotic production (GRÄFE et al., 1982, 1983). The receptor proteins have not been characterized – an earlier report that a cytoplasmic binding protein for the *S. virginiae* factor VB-C showed significant sequence homology to NusG (OKAMOTO et al., 1992), an *E. coli* protein believed to play a role in transcriptional antitermination, appears to have been ill-founded (PUTTIKHUNT et al., 1993). How A-factor binding protein influences antibiotic production is addressed in Sect. 3.3.3.

S. coelicolor does not make A-factor. However, it does make a series of structurally very similar compounds (ANISOVA et al., 1984; EFREMENKOVA et al., 1985; Fig. 16b), and at least some of these can apparently stimulate streptomycin production by *S. griseus* mutants deficient in A-factor biosynthesis (HARA et al., 1983). *afsA* mutants of *S. coelicolor* that had lost the ability to cross-stimulate the *S. griseus* A-factor-deficient mutants were unaffected in antibiotic production and morphological differentiation. However, it may be that the *afsA* mutants lack only some of the A-factor-like compounds made by the parental strain, so γ-butyrolactones might conceivably play a role in triggering antibiotic production in *S. coelicolor*. Alternatively, antibiotic production in *S. coelicolor* might perhaps have lost a requirement for such A-factor-like compounds, a situation reminiscent of the A-factor receptor-deficient mutants of *S. griseus* (Sect. 3.3.3); the failure to detect any A-factor-binding protein in cell extracts of *S. coelicolor* is consistent with this hypothesis.

3.2.5 Summary

The various physiological factors that influence the onset of antibiotic production in streptomycetes do not fit into a simple unifying model. However, it seems reasonable to propose an overall regulatory influence of growth rate with superimposed pathway-specific regulatory effects influencing the production of individual antibiotics. Such effects could include responsiveness to catabolite repression or inhibition, imbalances in metabolism, and environmental signals and stresses. It is clear that low molecular-weight effectors, like A-factor, play essential roles for some antibiotics. Whether these are produced as a consequence of a reduction in growth rate, or in response to some extrinsic factor, or via some autoregulatory circuit such as that described for LuxR and OHHL remains to be determined, as does the potential role of ppGpp as an intracellular signaling molecule.

3.3 Genetics of Antibiotic Production

3.3.1 Organization of Antibiotic Biosynthetic Genes and Clusters

In order to clarify discussion of the genetic regulation of antibiotic biosynthesis, we first briefly summarize the organization of the biosynthetic genes themselves. Genes specifically involved in the production of a particular antibiotic are invariably found clustered together, and only one set, for methylenomycin production (Sect. 3.3.4.5), is known to be plasmid-located rather than chromosomal. Thus, all of the *act* and *red* genes of *S. coelicolor* occur in chromosomal segments of ca. 23 kb and 35 kb, respectively. Transfer of these segments to *Streptomyces parvulus*, a host not known to make any structurally similar compounds, caused synthesis of the antibiotics (MALPARTIDA and HOPWOOD, 1984; MALPARTIDA et al., 1990). Similarly, expression in surrogate hosts has led to the demonstration that the entire biosynthetic pathways for production of tetracenomycin by *Streptomyces glaucescens* and puromycin by *Streptomyces alboniger* are located in DNA segments of 12.6 kb and 15 kb, respectively (DECKER and HUTCHINSON, 1993; LACALLE et al., 1992). The biosynthetic genes generally seem to be organized into several transcription units of varying complexity. Pathway-specific regulatory genes have been identified in several of the clusters (Sect. 3.4), although there are exceptions that seem so far to lack such genes, e.g., the tetracenomycin pathway and the very large biosynthetic clusters (ca. 45 kb and 95 kb, respectively) for the macrolides erythromycin (made by *Saccharopolyspora erythraea*; DONADIO et al., 1993, and references therein) and avermectin (produced by *Streptomyces avermitilis*; MACNEIL et al., 1992). The biosynthetic clusters usually also contain one or more genes that confer immunity to the antibiotic, which vary considerably in their mode of action. For example, resistance to erythromycin in *S. erythraea* is accomplished by a 23S rRNA methylase that modifies the target of the antibiotic by N^6 dimethy-

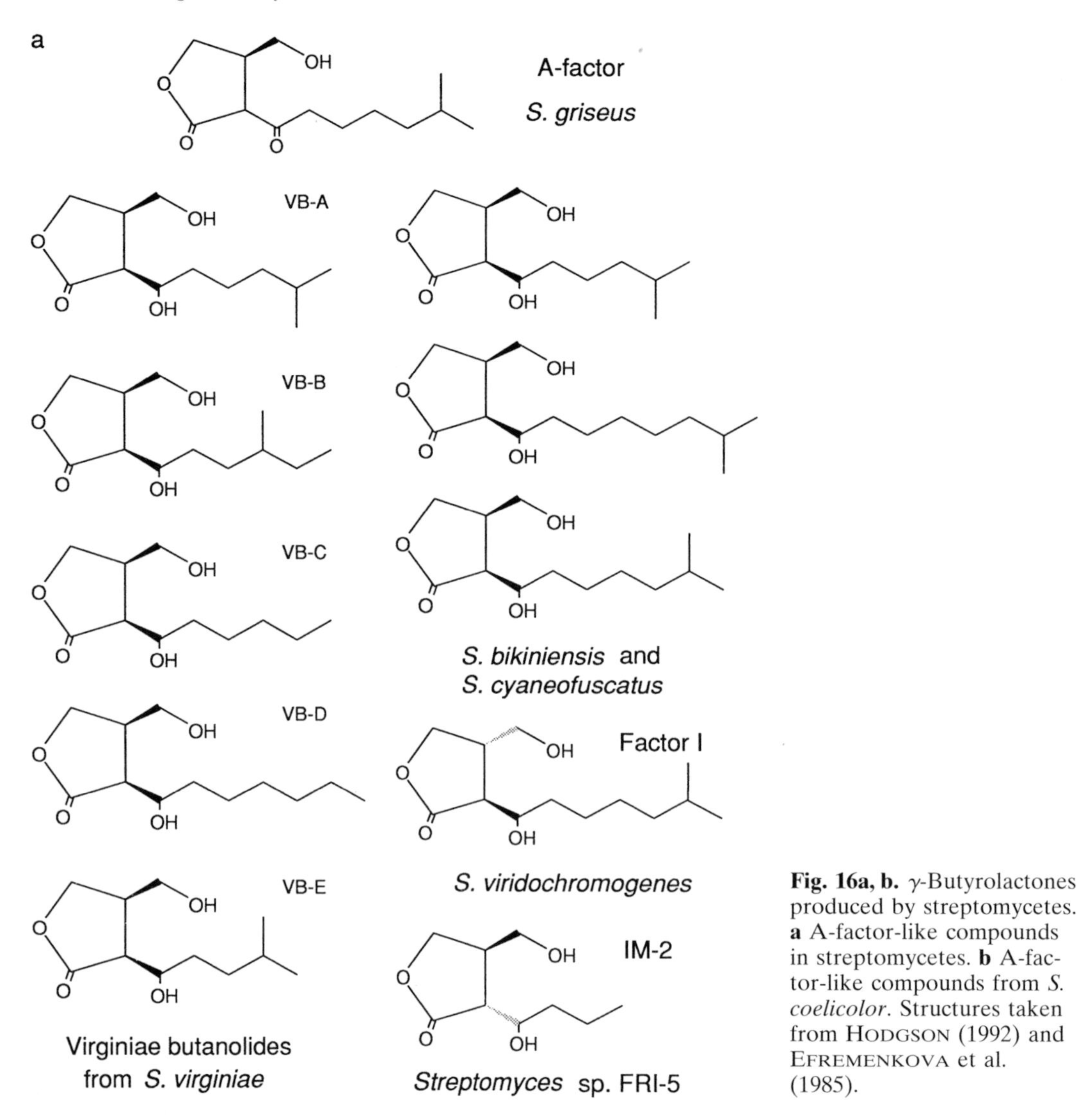

Fig. 16a, b. γ-Butyrolactones produced by streptomycetes. **a** A-factor-like compounds in streptomycetes. **b** A-factor-like compounds from *S. coelicolor*. Structures taken from HODGSON (1992) and EFREMENKOVA et al. (1985).

lation of a specific adenine residue (THOMPSON et al., 1982). Different types of efflux systems appear to confer resistance to different antibiotics; thus *srmB*, one of four spiramycin resistance genes in *Streptomyces ambofaciens*, appears to encode an ATP-dependent transport system for this macrolide antibiotic (SCHONER et al., 1992), whereas the efflux systems for tetracenomycin, methylenomycin, and probably actinorhodin encoded by the *tcmA* (GUILFOILE and HUTCHINSON, 1992a), mmr (NEAL and CHATER, 1987) and *actII*-ORF2 (FERNÁNDEZ-MORENO et al., 1991) genes of *S. glaucescens* and *S. coelicolor* are presumed to be driven by transmembrane electrochemical gradients. In some cases, expression of the resistance gene appears to be induced by the corresponding antibiotic (Sect. 3.3.5).

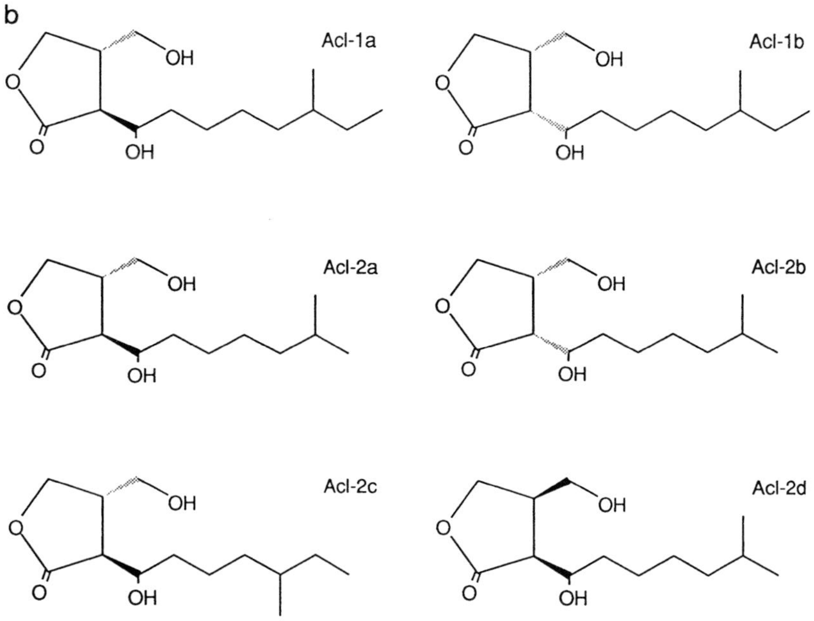

b

Acl-1a Acl-1b Acl-2a Acl-2b Acl-2c Acl-2d

Fig. 16b.

3.3.2 Genes that Pleiotropically Affect Antibiotic Production in *Streptomyces coelicolor*: Introduction and Overview

Many genes have been identified that pleiotropically affect antibiotic production in *S. coelicolor*, and several of these are likely to play a global role in regulating the onset (and perhaps maintenance) of antibiotic synthesis. Mutants in about half of these pleiotropic genes also show deficiencies in morphological differentiation. These "*bld*" mutants are discussed in Sect. 3.3.2.6. Some of the work on regulatory genes has involved use of a very close relative of *S. coelicolor*, *S. lividans* 66, which also contains *act* and *red* gene sets, but generally expresses them rather poorly. *S. lividans* is a slightly more convenient recipient strain for transformation than *S. coelicolor*.

3.3.2.1 *afsB*, *afsR*, and *afsK* – A Role for Protein Phosphorylation in Triggering the Onset of Antibiotic Production?

afsB mutants resemble *afsA* mutants (Sect. 3.2.4) in that they cannot induce streptomycin production and sporulation of A-factor-deficient mutants of *S. griseus* grown near them. However, unlike *afsA* mutants, they are defective in actinorhodin and undecylprodigiosin synthesis (HARA et al., 1983). (Production of the other two antibiotics known to be made by *S. coelicolor*, methylenomycin and a calcium-dependent antibiotic (CDA), appears to be normal; ADAMIDIS and CHAMPNESS, 1992.) Northern analysis failed to detect transcripts corresponding to *actI*, *actII* (including *actII*-ORF4, the pathway-specific activator gene), *actIII*, and *actVI* in the *afsB* mutant BH5 (HORINOUCHI et al., 1989a). Attempts to clone *afsB*, which was presumed to regulate production of the pigmented antibiotics and the A-factor-like compounds, have so far failed. However, they yielded *afsR*, which was obtained by screening of a library of *S. coeli-*

color DNA for overproduction of undecylprodigiosin and actinorhodin in an *afsB*-like mutant of *S. lividans* (HH21) that could not cross-feed an A-factor-deficient mutant of *S. griseus* (HORINOUCHI et al., 1983; STEIN and COHEN, 1989). At high copy number, a DNA fragment containing *afsR* and *afsR2* (see below) suppresses the *afsB* phenotype in *S. coelicolor* (HORINOUCHI et al., 1983) and stimulates transcription of *act* genes in *S. coelicolor* and *S. lividans* (HORINOUCHI et al., 1989a). *afsR* encodes a protein of 933 amino acids that contains putative DNA-binding helix-turn-helix motifs towards the C-terminus and potential ATP-binding sites towards the middle of the protein (HORINOUCHI et al., 1986, 1990); site-directed mutagenesis of the latter resulted in a 4-fold reduction (though not an elimination) of stimulatory activity in *S. lividans* (HORINOUCHI et al., 1990). A similar reduction was observed when fragments corresponding to the N-terminal 264 or 510 amino acids of AfsR, rather than the entire coding region, were expressed in *S. lividans*; even cloned fragments containing the C-terminal half of AfsR elicited a stimulatory effect in *S. lividans* and restored actinorhodin, undecylprodigiosin, and A-factor production to *afsB* mutants of *S. coelicolor* (HORINOUCHI et al., 1983, 1990), although this effect could also be due to *afsR2*, a recently identified small gene located immediately downstream of *afsR* and also capable of stimulating actinorhodin and undecylprodigiosin production (VÖGTLI et al., 1994; see below). Disruption of *afsR* in *S. coelicolor* using DNA fragments from the 5′, middle, and 3′ regions of the *afsR* coding sequence led to only a 4-fold reduction in actinorhodin production, which was also delayed (HORINOUCHI et al., 1990), suggesting that *afsR* is not essential for antibiotic biosynthesis. AfsR is phosphorylated *in vitro* by the membrane-bound product of *afsK* which lies downstream of, and in the opposite orientation to, *afsR* (HONG et al., 1991; HORINOUCHI, 1993). AfsK (799 amino acids) shows significant similarity to eukaryotic serine–threonine protein kinases, and phosphorylation of AfsR occurs at serine and threonine residues (MATSUMOTO et al., 1994). Moreover, phosphorylation of AfsR *in vitro* is severely reduced by K-252a and staurosporine, inhibitors of eukaryotic protein kinases (HONG et al., 1993). Disruption of *afsK*, like that of *afsR*, resulted in reduced levels of actinorhodin production, though the reason for this is not clear, since AfsR could still undergo phosphorylation at serine and threonine residues in the *afsK* mutant (MATSUMOTO et al., 1994). Clearly some other protein kinase can phosphorylate AfsR, a conclusion consistent with emerging evidence of multiple protein kinases in *S. coelicolor* (WATERS et al., 1994). Interestingly, the N-terminal region of AfsR resembles the pathway-specific regulatory proteins RedD, ActII-ORF4, and DnrI (Sect. 3.3.4.1). Unlike *redD* and *actII*-ORF4, whose transcription increases dramatically during transition phase, transcription of *afsR* occurs throughout exponential growth and declines slowly on entry into stationary phase (E. TAKANO and M. J. BIBB, unpublished). Thus, any major role of *afsR* in activating genes expressed in stationary phase may depend on posttranscriptional regulation of *afsR*, a deduction consistent with the modification of AfsR by phosphorylation.

A previously undetected gene, *afsR2*, has recently been identified in *S. lividans* (VÖGTLI et al., 1994). *afsR2*, which appears to occur in a similar location in *S. coelicolor* (MATSUMOTO et al., 1994; VÖGTLI et al., 1994), is transcribed in the same direction as *afsR* but from its own promoter and encodes a 63 amino acid protein. When cloned at high copy number, *afsR2* suppresses an *afsB* mutation in *S. lividans* and stimulates actinorhodin and undecylprodigiosin production in *S. coelicolor*. The stimulatory effect on actinorhodin production in both species appears to be mediated through transcription of *actII*-ORF4, and is retained after deletion of most of the C-terminal domain of the chromosomal copy of *afsR* in *S. lividans*. Fragments of *S. coelicolor* DNA containing the C-terminal portion of *afsR* stimulated actinorhodin production (HORINOUCHI et al., 1983, 1990; see above); since these fragments were likely to have contained *afsR2*, it is possible that the latter, rather than the C-terminus of AfsR, was responsible for the increase in antibiotic production. Since deletion of a chromosomal segment of *S. lividans* that included *afsR2* (and part of *afsR*) did not completely block blue pigment production, *afsR2* does not appear to play an essential role in actinorhodin production.

3.3.2.2 *afsQ1* and *afsQ2* – A Two-Component Regulatory System that Can Influence Antibiotic Production

afsQ1 and *afsQ2* were isolated in the same way as *afsR* (ISHIZUKA et al., 1992). Sequence analysis of a 1.3 kb *KpnI-PstI* fragment of *S. coelicolor* DNA that stimulated actinorhodin, undecylprodigiosin, and A-factor production in *S. lividans* HH21 identified *afsQ1* whose predicted product is homologous to bacterial response regulator genes (Fig. 9): AfsQ1 belongs to the OmpR subfamily (VOLZ, 1993). *afsQ2*, which was subsequently discovered downstream of *afsQ1*, appears to be translationally coupled to it. AfsQ2 belongs to the family of sensory histidine protein kinases; thus the genes appear to constitute a two-component regulatory system (Fig. 9; STOCK et al., 1990; ALEX and SIMON, 1994). AfsQ2 is presumed to be a membrane protein (it has putative membrane spanning domains towards its N-terminus) which is thought to be autophosphorylated at His^{294} in response to an unknown signal; the phosphate group may then be transferred to Asp^{52} of AfsQ1. AfsQ1 is, or interacts with, a transcriptional activator (ISHIZUKA et al., 1992). Evidence for this model of AfsQ2 action was obtained by changing His^{294} to Glu^{294}, which resulted in a loss of stimulatory activity in *S. lividans*. Cloned fragments containing only *afsQ1* gave the same level of stimulation as those containing both genes. However, disruption of both genes in *S. coelicolor* had no obvious phenotypic effect. Thus *afsQ1* and *afsQ2* either are inessential for antibiotic production or operate under as yet undefined physiological conditions. Alternatively, the stimulatory effects of *afsQ1* may reflect the ability of AfsQ1, when present at high levels, to substitute for AbsA, a response regulator that clearly does play a role in antibiotic production (Sect. 3.3.2.4).

3.3.2.3 *abaA* Influences the Production of Three of the Four Antibiotics Made by *Streptomyces coelicolor*

abaA of *S. coelicolor* was isolated by virtue of its ability to stimulate actinorhodin production in *S. lividans* when cloned on a high-copy number plasmid; the effect on undecylprodigiosin production was not reported (FERNÁNDEZ-MORENO et al., 1992). Sequencing of a 2 kb *PstI* fragment revealed five short ORFs, with ORFs A, B, and C transcribed divergently from ORFs D and E. ORFB and 137 nucleotides of downstream sequence were sufficient to give the same stimulatory phenotype in *S. lividans*, and disruption of the chromosomal copy of ORFB in *S. coelicolor* resulted in loss of actinorhodin production, almost complete loss of undecylprodigiosin synthesis, a reduction in CDA production, but no effect on methylenomycin. When cloned at high copy number, *abaA* was unable to confer actinorhodin production on a mutant deficient in the pathway-specific activator gene *actII*-ORF4, consistent with a location "higher up" in any putative regulatory cascade.

3.3.2.4 *absA* and *absB* – Mutants Isolated on the Basis of a Pleiotropic Defect in Antibiotic Production

An extensive screen for UV-induced mutants deficient in both actinorhodin and undecylprodigiosin production led to the identification of *absA* and *absB* (ADAMIDIS et al., 1990; ADAMIDIS and CHAMPNESS, 1992); mutants of both classes were also defective in CDA and methylenomycin synthesis (though they expressed methylenomycin resistance). The rarity of *absA* mutants ($5 \cdot 10^{-6}$ per survivor) suggested that they may represent a particular allelic form (CHAMPNESS et al., 1992). Pseudorevertants of an *absA* mutant were obtained that fell into two classes: *sab*(I) pseudorevertants which overproduced actinorhodin and undecylprodigiosin, and *sab*(II) which made wild-type levels of antibiotic. Both classes contain suppressor mutations mapping close to the starting *absA* mutation, but recombining with it. Sequencing of a DNA fragment that complements the *absA* mutation showed that its product is homologous to bacterial sensory histidine kinases, and preliminary data suggest that a homolog of response regulators is located immediately downstream (P. BRIAN and W. CHAMPNESS, personal communication). Interestingly, additional copies of *afsQ1* restored actinorhodin production to an *absA* (but not *absB*) mutant (ISHIZUKA et al., 1992), possibly indicating cross-talk between the two systems (Sect. 3.3.2.2).

absB mutants sporulate less well than their progenitor and produce low levels of actinorhodin, undecylprodigiosin, and methylenomycin on some media. Attempts to clone *absB* on a low copy number vector by screening for restoration of actinorhodin production led to the isolation, in addition to *actII*-ORF4 and *afsR*, of a cloned fragment which fully complements the *absB* mutation (T. ADAMIDIS and W. CHAMPNESS, personal communication).

3.3.2.5 *mia* – Multicopy Inhibition of Antibiotic Production

Attempts to clone *absA* on a high copy number plasmid led to the identification of *S. coelicolor* DNA that inhibited the production of all four antibiotics. The DNA fragment, which was isolated repeatedly, had no inhibitory effect at low copy number (CHAMPNESS et al., 1992). Transcription of *redD* and *actII*-ORF4 is undetectable in strains containing the fragment on a high copy number plasmid (W. CHAMPNESS, personal communication). Subcloning localized the inhibitory function, termed *mia*, to a 363 bp *Sau*3A1 fragment that does not appear to be protein-coding. It is not known whether the inhibitory effect results from the DNA itself or its transcript.

3.3.2.6 Genes that Affect Both Antibiotic Production and Morphological Differentiation

In surface-grown cultures of streptomycetes, antibiotic production generally coincides with the onset of morphological differentiation (Sect. 3.2). The isolation of *bld* mutants defective in both processes points to at least some common elements of genetic control. It is important to note that the morphological deficiencies of some classes of *bld* mutants can be suppressed nutritionally or in some cases by cross-stimulation by diffusible factors. Phenotypic suppression of the pleiotropic defect in antibiotic production has been observed in only a few cases and generally not under conditions that suppress the morphological deficiency (an exception to this is provided by *bldH* mutants; see below).

The most extensive studies of *bld* mutants have been made in *S. coelicolor* in which at least 10, and perhaps 11, different classes of *bld* mutants with deficiencies in antibiotic production – *bldA, B, D, E, F, G, H, I*, -17, -21, -830 – have been identified (reviewed by CHAMPNESS and CHATER, 1994). All mutant classes except *bldE* and *bldF* (which both produce abundant undecylprodigiosin) are deficient in both actinorhodin and undecylprodigiosin synthesis and most are also deficient in methylenomycin and CDA production. Antibiotic production is restored to *bldH* mutants grown on mannitol instead of glucose, and undecylprodigiosin is produced by *bldA* mutants grown at low phosphate concentrations. Four of the *bld* genes (*A*, *B*, *D*, and *G*) have been cloned, and detailed characterization has been reported for *bldA*. Remarkably, *bldA* encodes the only tRNA in *S. coelicolor* and *S. lividans* that can translate the rare leucine codon UUA efficiently (LAWLOR et al., 1987; LESKIW et al., 1991a). The lack of expression of the *xylE* reporter gene when fused to *act*, *red*, or *mmy* transcription units in *bldA* mutants suggests that the defect in antibiotic production reflects a failure to *transcribe* the biosynthetic structural genes even though *bldA* encodes a component of the *translational* apparatus (GUTHRIE and CHATER, 1990; BRUTON et al., 1991; A. WIETZORREK and K. F. CHATER, unpublished). An explanation for the failure to transcribe *act* genes is to be found in the presence of a TTA codon in the pathway-specific regulatory gene *actII*-ORF4. If this codon is changed to the synonymous codon TTG, actinorhodin production takes place even in a *bldA* mutant (FERNÁNDEZ-MORENO et al., 1991). Phenotypically similar *bldA* mutants have also been isolated in the phylogenetically more distant *S. griseus* (MCCUE et al., 1992), suggesting that the role of *bldA* in secondary metabolism and differentiation is widespread among streptomycetes. The unimpaired vegetative growth of *bldA* mutants indicates that TTA codons are absent from genes essential for primary metabolism and growth, but TTA codons have been found in several genes likely to be expressed late in growth. This, coupled with evidence that *bldA*-specific RNA is more abundant late in surface growth (LAWLOR et al., 1987), provided support for the idea that *bldA* regulates antibiotic production by allowing the translation of UUA codon-containing mRNA only under appropriate conditions (LESKIW et al., 1991b). However, caution is necessary in assuming an active *regulatory* role for *bldA*, since one series of detailed experiments on liquid-grown cultures of *S. coelicolor* failed to reveal any limitation of translation of the *actII*-ORF4 UUA codon during exponential growth (GRAMAJO et al., 1993). Together with the transition phase activation of *actII*-ORF4 transcription, these results were consistent with a more prosaic possibility that the absence of TTA codons from vegetatively expressed genes might reflect selection against codons that were inefficiently translated during growth, rather than a role for *bldA* in the temporal regulation of actinorhodin production. On the other hand, the observations of LESKIW et al. (1993) tend to support a regulatory role for *bldA*. Northern analysis of RNA from surface-grown cultures indicated that the amount of the *bldA* transcript increased with growth, and S1 nuclease protection assays revealed an increase in the level of the 5′ end of the mature *bldA* transcript late in growth, both in rich liquid media (this was not observed by GRAMAJO et al., 1993) and in surface-grown cultures; furthermore, the efficiency of translation of seven UUA codons of a heterologous reporter gene apparently increased in older cultures. The differences between the two sets of results may reflect the different growth conditions used; possibly the liquid culture conditions of GRAMAIO et al. (1993) overrode a regulatory role of *bldA* adapted for surface growth.

3.3.2.7 An Outline Scheme for the Interactions of Pleiotropic Antibiotic Regulatory Genes in *Streptomyces coelicolor*

No satisfactory integrated model has yet emerged for the roles of the various pleiotropic regulatory genes in antibiotic production (probably because not enough of the pieces of the jigsaw are yet available), but here we summarize some of the key features that must be taken into account. For actinorhodin and undecylprodigiosin, expression of the pathway-specific activator genes *actII*-ORF4 and *redD*, respectively, appears to play a major limiting role in determining the onset of antibiotic production (TAKANO et al., 1992; GRAMAJO et al., 1993), so it is attractive to propose that all the pleiotropic genes (*afs*, *aba*, *abs*, *mia*, *bld*) influence the synthesis of these two antibiotics via *actII*-ORF and *redD*. The generalized and simplified scheme in Fig. 17 is built from the following observations and deductions.

(1) Transcription of *actII*-ORF4 is virtually undetectable in an *afsB* mutant, at least under certain culture conditions, suggesting that the (still uncharacterized) AfsB gene product may be higher than ActII-ORF4 in a transcriptional cascade (HORINOUCHI et al., 1989a). (It should be noted that *afsB* mutants are noticeably leaky in their actinorhodin deficiency on a variety of different media.)

(2) *absA* and *absB*, whose mutant phenotype proves their importance for antibiotic biosynthesis, may perhaps play a role in maximizing expression of the pathway-specific activator genes, since the introduction of *actII*-ORF4 and *redD* on high copy number plasmids res-

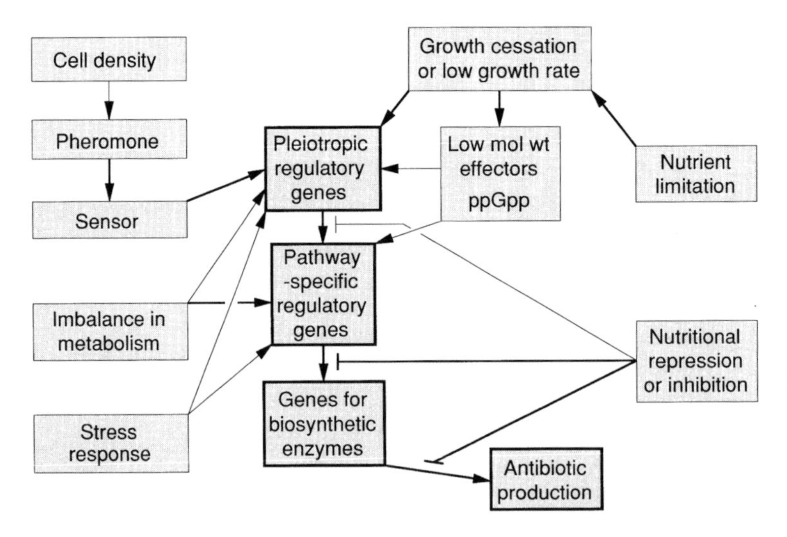

Fig. 17. Factors potentially determining the onset of antibiotic production in streptomycetes. Thinner lines present plausible interactions for which there is currently no direct evidence.

tores production of the relevant antibiotic in *absA* and *absB* mutants (T. ADAMIDIS and W. CHAMPNESS, personal communication). Alternatively, the *abs* genes may encode accessory elements normally required for actinorhodin and undecylprodigiosin synthesis which are rendered unnecessary by overproduction of ActII-ORF4 and RedD.

(3) Multiple copies of *afsR* and *afsR2* stimulate actinorhodin production, but appear to depend on *actII*-ORF4 for this effect (B. FLORIANO and M. J. BIBB, unpublished results; T. ADAMIDIS and W. M. CHAMPNESS, personal communication; VÖGTLI et al., 1994), while multiple copies of segments encoding the C-terminal portion of AfsR, and in retrospect containing *afsR2*, confer actinorhodin and undecylprodigiosin production on *absA* and *absB* mutants (CHAMPNESS et al., 1992).

(4) Taken together, observations (1)–(3) suggest a working model in which expression of *afsR* and *afsR2* or the activities of their products, depend on *absA* and *absB*, and in which AfsR2 and AfsR (perhaps in its phosphorylated form) stimulate expression of *actII*-ORF4 (and possibly *redD*).

(5) Extra copies of *afsQ1* restore actinorhodin and undecylprodigiosin production to *absA* (but not *absB*) mutants (ISHIZUKA et al., 1992), suggesting that *afsQ* may depend on *absA* for a role in enhancing *actII*-ORF4 and *redD* activity, unless a high copy number of *afsQ* takes the activity of its product above a threshold level. If *afsQ1* has such a role, it is redundant in the wild-type strain under laboratory conditions; perhaps this role can also be filled by *afsR*, a question that should be resolved by isolating *afsR* and *afsR afsQ* null-mutants. It remains possible that the high copy number effects of *afsQ* result from artificially induced cross-talk between normally separated regulatory elements.

(6) The recent sequence analysis of *absA*, together with the published data on *afsR*/*afsK* and *afsQ1*/*afsQ2*, strongly suggest that protein phosphorylation, and potentially phosphorylation cascades, play a role in triggering antibiotic production; presumably AfsK, AfsQ2, and AbsA sense external signals that cause phosphorylation of their regulatory counterparts (AfsR, AfsQ1, and the product of a gene located downstream of *absA*) which can then stimulate transcription of the antibiotic biosynthetic pathways, perhaps via the pathway-specific regulators.

(7) None of the mutants described above are unconditionally and completely defective in production of all four antibiotics; even *absA* produces actinorhodin on some media (W. CHAMPNESS, unpublished data), and antibiotic production in several of the others shows media dependence. This may indicate a complex regulatory network in which there are several different routes to activation of a par-

ticular pathway. (Alternatively, the available mutants may not be truly null.)
(8) Little is known of how *abaA* fits into this interactive scheme, but in affecting three of the four antibiotics it differs from both the *afs* loci (which appear to affect only actinorhodin and undecylprodigiosin) and the *abs* loci (which affect all four).
(9) *bldA* is the only *bld* gene whose mode of action is (partially) understood. *bldA* dependence of actinorhodin production appears to be exerted entirely at the level of translation of the unique UUA codon in the *actII*-ORF4 transcript. Undecylprodigiosin production, except at low phosphate levels, also requires *bldA*, though neither *redD* nor, apparently, any of the *red* biosynthetic structural genes contain TTA codons (NARVA and FEITELSON, 1990; GUTHRIE and CHATER, 1990). In contrast to *actII*-ORF4 whose *transcription* is not *bldA*-dependent, transcription of *redD* could not be detected in a *bldA* mutant (J. WHITE and M. J. BIBB, unpublished results). Presumably there is at least one other gene required for *redD* transcription whose transcript does contain a UUA codon. The *pwb* mutations which restore undecylprodigosin, but not actinorhodin or aerial mycelium formation, to *bldA* mutants and which may map within the *red* cluster (E. P. GUTHRIE and K. F. CHATER, unpublished) could be relevant here.
(10) Antibiotic production is associated with reduced growth rate. One of the signals implicated is an increased ppGpp level. No other candidate signal, intracellular or extracellular, has been described that might have pleiotropic activity, though extracellular acidification can activate methylenomycin production (see Sect. 3.2.2). Whatever the signals, evidence is growing that they may lead, directly or indirectly, to phosphorylation of several regulatory proteins such as AfsR, AfsQ1, and AbsA by specific kinases and thence, via pathway-specific regulatory genes and their products, to transcription of genes encoding antibiotic pathways.
(11) Model building is limited by major gaps in knowledge. It is not yet possible to deduce at what level (transcriptional, translational, or posttranslational) the phosphorylated pleiotropic regulators interact with the pathway-specific regulators, nor whether they act in a linear cascade or by convergence. Furthermore, there is little information about the regulation of expression of the characterized pleiotropic regulatory genes, and several relevant *bld* genes remain uncharacterized.

3.3.3 *Streptomyces griseus* – The A-Factor Cascade

Although diffusible factors have been implicated in the production of several antibiotics in streptomycetes (see Sect. 3.2.4), the role of A-factor in the production of streptomycin in *S. griseus* is by far the best characterized. A-factor was discovered by KHOKHLOV et al. (1967). It was found to be required for both streptomycin production and sporulation in *S. griseus* (KHOKHLOV, 1982) and also for streptomycin resistance (HARA and BEPPU, 1982a). Subsequent genetic and molecular analyses have provided considerable insights into its mode of action. A-factor accumulates to detectable levels in the culture medium just before the onset of streptomycin production (HARA and BEPPU, 1982b). It appears to be freely diffusible across the cytoplasmic membrane and binds to a cytoplasmic A-factor-binding protein of approximately 26 kDa with a stoichiometry of 1 : 1 and a dissociation constant of 0.7 nM, consistent with the similar low concentrations of A-factor required for biological activity. The receptor protein is present at about 30–40 copies per genome. It is thought that binding of A-factor prevents the receptor protein from acting as a repressor of a hypothetical gene (X) required for both sporulation and streptomycin production (Fig. 18). A negative regulatory role for the binding protein is indicated by the discovery that *S. griseus* mutants unable to make A-factor can undergo further mutations that restore streptomycin production and sporulation by eliminating A-factor-binding protein. Gene X is believed to activate transcription of a gene that in turn encodes an activator of *strR*, a streptomycin pathway-specific activator gene (Sect. 3.3.4.3). In support of this, a protein in extracts of A-factor-producing strains, but absent from A-factor-deficient

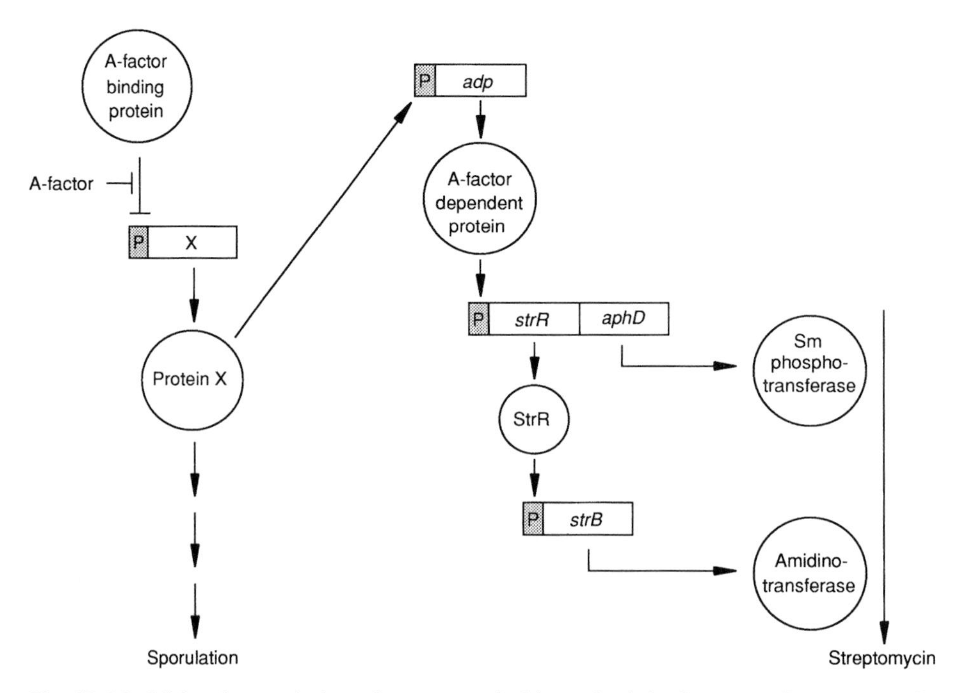

Fig. 18. Model for the regulation of streptomycin biosynthesis in *S. griseus*. P, promoter; Protein X, regulatory protein derived from unidentified gene X required for both sporulation and streptomycin production; *adp*, regulatory gene encoding the A-factor-dependent protein that binds to the promoter region of *strR*, the pathway-specific activator gene for streptomycin production; *aphD* and *strB* encode a resistance determinant and biosynthetic enzyme, respectively; Sm, streptomycin. Redrawn from HORINOUCHI (1993).

mutants, could bind to nucleotide sequences just upstream of *strR*. Little is known about how A-factor synthesis takes place or is controlled. A putative A-factor biosynthetic gene, *afsA*, was cloned from *S. griseus*, but its predicted translation product did not resemble any other known protein (HORINOUCHI et al., 1989b). Surprisingly, *afsA* did not hybridize to DNA from some streptomycetes that produce structurally similar γ-butyrolactones (HORINOUCHI et al., 1984). The earlier onset of streptomycin production and sporulation in mutants lacking the A-factor-binding protein (MIYAKE et al., 1990), and earlier production of streptomycin on elevation of A-factor levels, either by exogenous addition (BEPPU, 1992) or by cloning *afsA* on a multi copy plasmid (HORINOUCHI et al., 1984), are clearly consistent with a role for *afsA* and the γ-butyrolactone in determining the timing of both processes; they also suggest that the A-factor-binding protein is present during early growth. In view of the positive autoregulation of OHHL synthesis in *Vibrio fischeri* (see Sect. 2.2), one might anticipate that the A-factor-binding protein also represses – directly or indirectly – the expression of *afsA*.

3.3.4 Pathway-Specific Regulatory Genes

We have already made frequent references to the important role of pathway-specific regulatory genes. Here we review these genes and their (deduced) products in more detail.

3.3.4.1 The *actII*-ORF4, *redD*, and *dnrI* Family of Pathway-Specific Activator Genes

Early genetic analyses identified putative pathway-specific activator genes for the undecylprodigiosin (*redD*) and actinorhodin (*actII*-ORF4) biosynthetic pathways of *S. coelicolor* (reviewed by CHATER, 1992). The failure of *redD* and *actII*-ORF4 mutants to cosynthesize with representatives of any other *red* or *act* mutant class, the lack of expression of *red* and *act* biosynthetic structural genes in *redD* and *actII*-ORF4 mutants, and the ability of extra cloned copies of *redD* and *actII*-ORF4 to elicit overproduction of undecylprodigiosin and actinorhodin, respectively, all suggest that *redD* and *actII*-ORF4 are pathway-specific activator genes. Furthermore, the stationary-phase production of undecylprodigiosin and actinorhodin appears to result from transcriptional activation of *redD* (TAKANO et al., 1992) and *actII*-ORF4 (GRAMAJO et al., 1993), respectively. Production of the antibiotics in rapidly growing cultures appears to be limited only by the absence of the relevant pathway-specific activator protein, because overproduction of the putative activators causes increased transcription of the corresponding biosynthetic structural genes and can be used to cause antibiotic production prematurely during rapid growth.

The *redD* and *actII*-ORF4 genes are homologous to each other and to the positively-acting regulatory gene *dnrI* required for the production of daunorubicin in *Streptomyces peucetius* (STUTZMAN-ENGWALL et al., 1992). Insertional inactivation of *dnrI* blocks production of daunorubicin and all of its biosynthetic intermediates and prevents transcription of putative operons containing daunorubicin biosynthetic and resistance genes. The predicted *redD*, *actII*-ORF4, and *dnrI* gene products show 33–37% amino acid sequence identity in pairwise alignments (Fig. 19; STUTZMAN-ENGWALL et al., 1992). *dnrI* can complement mutations in *actII*-ORF4, and *actII*-ORF4 can stimulate daunorubicin production in *S. peucetius* (STUTZMAN-ENGWALL et al., 1992), but *redD* and *actII*-ORF4 do not show cross-complementation. Since computer analysis using the algorithm of DODD and EGAN (1990) failed to reveal likely helix-turn-helix DNA-binding motifs in these proteins, they may represent a novel family of DNA-binding regulatory proteins. Perhaps more likely, they may need to interact with other proteins

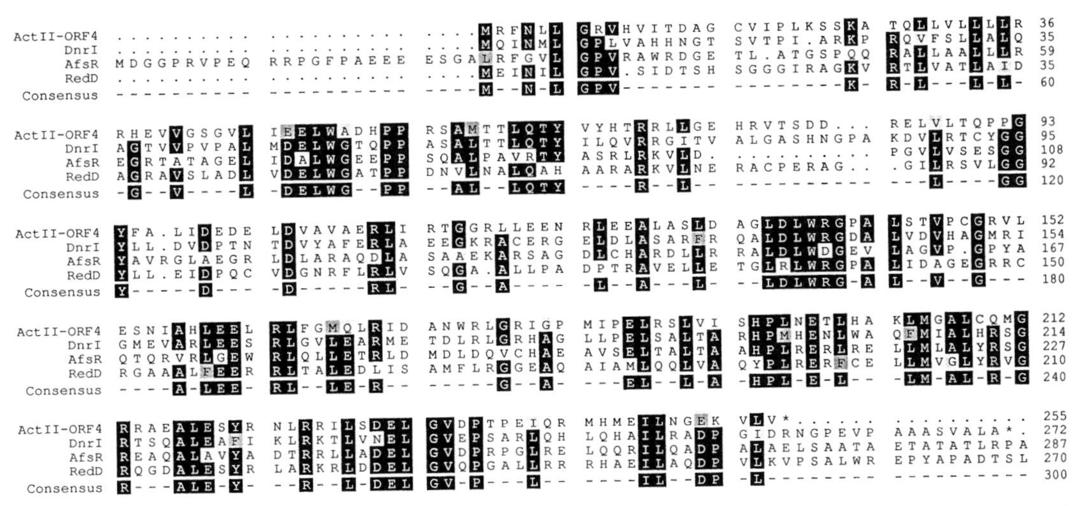

Fig. 19. Alignment of the amino acid sequences of RedD, ActII-ORF4, DnrI, and the N-terminal region of AfsR. The alignment was made using the PILEUP and PRETTYBOX programs contained in the UWG sequence analysis package (DEVEREUX et al., 1984).

to effect activation of biosynthetic structural gene promoters.

Intriguingly, the N-terminal region of AfsR, excluding the putative DNA-binding motifs in the C-terminal region (HORINOUCHI et al., 1990), also shows significant identity to the RedD–ActII–ORF4–DnrI family (Fig. 19), raising the possibility that the stimulation of actinorhodin and undecylprodigiosin production by multiple copies of *afsR* may reflect partial functional interchangeability of AfsR with ActII-ORF and RedD. However, since strong stimulatory effects were also observed with segments of AfsR (notably the C-terminal half) that are not homologous to RedD and ActII-ORF4, models that rely solely on functional substitution are at best an oversimplification.

3.3.4.2 *srmR* – A Regulatory Gene for Spiramycin Production in *Streptomyces ambofaciens*

Cloning and gene disruption revealed a putative regulatory gene, *srmR*, for spiramycin production in *Streptomyces ambofaciens*. *srmR* mutants fail to make spiramycin and do not cosynthesize the antibiotic with *srm* mutants that accumulate intermediates in the biosynthetic pathway. *srmR* was required not only for transcription of *srmG*, which encodes the polyketide synthase that produces the aglycone of spiramycin, but also for expression of the resistance gene *srmB* (GEISTLICH et al., 1992). Lack of expression of *srmB* in *srmR* mutants proved to be an indirect effect of the failure of *srmR* mutants to produce spiramycin, which is an inducer of its own resistance gene. *srmR* was also required for the transcription of another flanking gene, *srmX*, that is also likely to play a role in spiramycin production. Multicopy cloning of *srmR* in the wild-type strain led to a 4-fold increase in spiramycin production. The *srmG* and *srmX* promoters are strikingly similar to each other, with three blocks of conserved sequences, centered at about position −39 (CCNGNCGTTCCT), −27 (CCCGGC), and −10 (CTGTNN-GNT), one or more of which may be binding sites for the 65 kDa putative transcriptional activator (SrmR) encoded by *srmR*. SrmR shows no significant sequence similarity to any other known protein. *srmR* homologs have not been reported in gene clusters for the production of other macrolide antibiotics. The *srmR* gene contains a single TTA codon, so it may be a target for regulation by a *bldA* homolog (Sect. 3.3.2.6).

3.3.4.3 *strR* Encodes a DNA-Binding Protein that Regulates at Least One of the Streptomycin Biosynthetic Genes in *Streptomyces griseus*

Production of streptomycin and 5′-hydroxy-streptomycin has been studied in *S. griseus* and *S. glaucescens* GLA.0, respectively. In *S. griseus*, *strR* appears to be a positive regulator of at least one of the biosynthetic structural genes, *strB1*, which encodes amidinotransferase I, and the *strR* homolog of *S. glaucescens* GLA.0 is presumed to perform the same function. StrR contains a potential helix-turn-helix DNA-binding motif, and the protein binds to at least two specific sites inside the *str* clusters of both species. Analysis of the binding sites reveals 11 bp inverted repeats separated by 11 bp. These results make it more likely that StrR acts as a conventional transcriptional activator, rather than through transcriptional antitermination (RETZLAFF et al., 1993). The *strR* genes of both species contain single TTA codons and *bldA* mutants of *S. griseus* do not make streptomycin (MCCUE et al., 1992). Single TTA codons are also present in *strN*, encoding a biosynthetic enzyme of both species and in *strA*, the streptomycin resistance gene of *S. glaucescens* (DISTLER et al., 1992).

3.3.4.4 *brpA* and *dnrN* – Regulatory Genes for Bialaphos Production in *Streptomyces hygroscopicus* and for Daunorubicin Production in *Streptomyces peucetius* which Show Different Degrees of Similarity to Response Regulator Genes of Two-Component Systems

Bialaphos is made by *Streptomyces hygroscopicus* at the approach of stationary phase (HOLT et al., 1992). Early studies identified *brpA* as a likely pathway-specific activator gene for bialaphos production; *brpA* mutants were defective in at least 6 of the 13 steps leading to bialaphos production, lacked at least 7 of the bialaphos biosynthetic transcripts, and showed reduced levels of bialaphos resistance (ANZAI et al., 1987). Furthermore, a *brpA* mutant lacked 27 proteins implicated in bialaphos production (HOLT et al., 1992). *brpA* encodes a predicted product of 28 kDa whose C-terminal region resembles a region located towards the C-terminus of the response regulators of the UhpA subfamily of two-component regulatory systems (RAIBAUD et al., 1991; GROSS et al., 1989). This region includes a putative helix-turn-helix motif, but does not extend to the conserved region that includes the site of phosphorylation of the regulatory components. BrpA contains three hydrophobic regions towards its N-terminus, leading to suggestions that these might represent transmembrane domains or regions of hydrophobic interaction with other proteins (RAIBAUD et al., 1991). *brpA* is transcribed from three promoters expressed at a low level early in exponential growth but more strongly during a pause in growth shortly before stationary phase, and the activity of one of them (*brpA*p3) continued to increase on entry into stationary phase. *brpA* contains a single TTA codon located towards the C-terminus of the coding region making it potentially *bldA*-dependent (*bldA* mutants of *S. hygroscopicus* have not been described).

In addition to the *actII*-ORF4-like gene *dnrI*, the daunorubicin biosynthetic cluster of *S. peucetius* encodes at least one other putative regulatory gene, *dnrN*. DnrN shows significant sequence similarity throughout its length to the UhpA subfamily of two-component response regulator proteins, including a likely site for phosphorylation, although *dnrN* does not appear to be closely linked to a sensory histidine protein kinase gene. Transcription of *dnrI* is reduced in *dnrN* mutants (HUTCHINSON et al., 1994), and this may be the cause of their daunorubicin deficiency. Indeed, production is restored by adding extra copies of the cloned *dnrI* gene. (On the other hand, extra copies of *dnrN* do not restore production to a *dnrI* mutant.) A non-sporulating derivative (H6101) of *Streptomyces peuceticus* var. *caesius* that is deficient in daunorubicin production has been described. Cloned copies of either *dnrI* or *dnrN* restored daunorubicin production in H6101 suggesting that the mutant is defective in *dnrN* expression (presumably as a secondary result of the pleiotropic mutation) (HUTCHINSON et al., 1994; STUTZMAN-ENGWALL et al., 1992).

3.3.4.5 Negative Regulation of Methylenomycin Production in *Streptomyces coelicolor*

The biosynthetic genes for methylenomycin production in *S. coelicolor* reside on the 350 kb linear plasmid SCP1 (KINASHI et al., 1987). *mmy* genes were initially isolated by mutational cloning yielding over 20 kb of contiguous DNA (CHATER and BRUTON, 1983, 1985). Insert-directed prophage insertions into the leftmost 3 kb of the cluster caused overproduction of methylenomycin. Sequence analysis of this region revealed a gene (*mmyR*) whose predicted product resembles the TetR family of repressor proteins (C. J. BRUTON and K. F. CHATER, unpublished results; Sect. 3.3.5.1). Disruption or deletion of *mmyR* resulted in overproduction of methylenomycin, providing the only known example of pathway-specific negative regulation of antibiotic production. The absence of methylenomycin production from various pleiotropic mutants (see Sect. 3.3.2) suggests that there may also be a positively acting pathway-specific regulatory gene, and the absence of *mmy* gene transcription in a *bldA* mutant (A. Wietzorrek and K. F. Chater, unpublished results) leads to the prediction that this gene should contain a TTA codon.

3.3.5 Induction of Antibiotic Resistance in Antibiotic Producing Streptomycetes – Antibiotics as Inducers of Gene Expression

Resistance towards an antibiotic made by a streptomycete often develops only at the onset of antibiotic production. In some cases, resistance may be a consequence of export alone (and may therefore be regarded as a late step in antibiotic production). In others, specific resistance mechanisms operate, in addition to efflux, to ensure continued viability. Below, we consider examples of resistance mechanisms that appear to be induced by their antibiotic substrates or by intermediates in the pathway.

3.3.5.1 *actII*-ORF1/2 of *Streptomyces coelicolor* and *tcmR/A* of *Streptomyces glaucescens* GLA.0 – Regulatory Cassettes for Antibiotic Export

actII-ORF1/2 of *S. coelicolor* (Caballero et al., 1991) and *tcmR/A* of *S. glaucescens* GLA.0 (Guilfoile and Hutchinson, 1992a) are divergently transcribed gene pairs for export of actinorhodin and tetracenomycin C, respectively. Both ActII-ORF2 and TcmA are similar to tetracycline transport proteins from several other gram-positive bacteria (including Tet347 of *Streptomyces rimosus*) and gram-negative organisms, and ActII-ORF1 and TcmR are clearly members of the TetR family of repressor proteins. Inactivation of *actII*-ORF2 appeared to prevent the export of actinorhodin (Fernández-Moreno et al., 1991). Studies of *actII*-ORF1/2 expression in *E. coli* (Caballero et al., 1991) demonstrated that ActII-ORF1 repressed transcription of *actII*-ORF2 and of itself (and indicated that both genes could be expressed in the absence of the pathway-specific activator ActII-ORF4; Sect. 3.3.4.1). Both promoters were most active in *S. coelicolor* cultures that were making actinorhodin. Guilfoile and Hutchinson (1992a, b) showed that transcription of *tcmA* in *S. glaucescens* was induced by tetracenomycin C and that inactivation of *tcmR* resulted in constitutive *tcmA* expression; furthermore, *in vitro* binding of TcmR to the *tcmR/A* intergenic region was inhibited in the presence of tetracenomycin C. It thus seems likely that export of tetracenomycin C is induced by the antibiotic once production begins (Guilfoile and Hutchinson, 1992b).

Resistance of *S. coelicolor* to methylenomycin is conferred by *mmr* which encodes a protein with significant sequence similarity to the same family of transporter proteins as ActII-ORF2 and TcmA (Neal and Chater, 1987). Hobbs et al. (1992) found that transcripts corresponding to at least one of the methylenomycin biosynthetic genes appeared before that of *mmr*, suggesting that *mmr* expression might be induced by methylenomycin or by an intermediate in the pathway.

3.3.5.2 *srmB* of *Streptomyces ambofaciens* – A Probable ATP-Dependent Efflux System Induced by its Antibiotic Substrate

The *srmB*, *tlrC*, and *drrA* gene products respectively involved in the export of spiramycin from *S. ambofaciens*, tylosin from *S. fradiae*, and daunorubicin from *S. peucetius* show a high degree of amino acid sequence similarity (Schoner et al., 1992; Guilfoile and Hutchinson, 1991). The proteins each possess a putative ATP-binding motif, suggesting that they are components of ATP-dependent efflux systems. Although details of the regulation of *tlrC* and *drrA* have not been published, expression of *srmB* in *S. fradiae* is induced by spiramycin (Geistlich et al., 1992; Sect. 3.3.4.2); increased levels of *srmB* transcription occurred on addition of spiramycin to mutants blocked in production of the antibiotic (no increase was observed with the wild-type strain).

3.3.5.3 Induction of *tlrA* in *Streptomyces fradiae* – A Role for Transcriptional Attenuation?

tlrA and *tlrD* are two of at least four *S. fradiae* genes that confer tylosin resistance. They cause di- and monomethylation, respectively, of residue A-2058 of 23S rRNA. Unlike *tlrD*, which is expressed constitutively, expression of *tlrA* is induced by tylosin or its biosynthetic intermediates (KELEMEN et al., 1994). In the absence of inducer, transcription terminates at the beginning of the coding region of *tlrA* through the adoption of a particular secondary structure in the RNA; the presence of an inducer is thought to cause sensitive ribosomes to stall in the non-translated leader region, preventing transcriptional termination and allowing production of the methylase.

3.3.5.4 Induction of Resistance to Novobiocin in *Streptomyces sphaeroides* and *Streptomyces niveus* – Roles for DNA Supercoiling and a Diffusible Signaling Molecule

Novobiocin, produced by *Streptomyces sphaeroides* and *Streptomyces niveus*, is an inhibitor of bacterial DNA gyrase. Gyrase exists as a tetramer (A_2B_2) in which the two subunits have different functions that can be blocked by different groups of antibiotics. The B subunit, encoded by *gyrB*, is the target for novobiocin. *S. sphaeroides* has two *gyrB* genes (THIARA and CUNDLIFFE, 1989, 1993): one encoding a novobiocin-sensitive B subunit, $GyrB^S$, that is produced constitutively, and the other encoding a resistant B subunit, $GyrB^R$, that is produced in the presence of novobiocin. Transcriptional fusions showed that the $gyrB^R$ promoter responds to changes in DNA supercoiling. Transcription of $gyrB^R$ increased when DNA gyrase was inhibited by novobiocin or ciprofloxacin (an inhibitor of the A subunit), i.e., under conditions that reduce negative supercoiling, and decreased during growth in a medium of high osmotic strength that should increase negative supercoiling (HIGGINS et al., 1988). Thus resistance to novobiocin in *S. sphaeroides* probably occurs, at least in part, by production of the resistant gyrase following the reduction in negative supercoiling which results from inhibition of the sensitive enzyme by the antibiotic.

Two different, uncharacterized genes that confer novobiocin resistance on *S. lividans* have been isolated from *S. niveus* (HOGGARTH et al., 1994); one hybridizes with a second resistance determinant from *S. sphaeroides* (THIARA and CUNDLIFFE, 1988). The isolation of multiple resistance genes and a marked increase in the level of novobiocin resistance (from 25 to over 200 $\mu g\ mL^{-1}$) during growth of *S. niveus* suggest that resistance is determined by several mechanisms that may be subject to different regulatory controls. Recent studies on *S. niveus* (HOGGARTH et al., 1994) identified a diffusible γ-butyrolactone signaling molecule that induces high-level resistance to novobiocin well before the onset of production.

3.3.5.5 Regulation of Isoforms of the Target for Pentalenolactone Inhibition in the Producing Organism

Pentalenolactone (PL) is a potent inhibitor of glyceraldehyde-3-phosphate dehydrogenase (GAPDH). The producer, *Streptomyces arenae*, has two distinct isoforms of GAPDH: a PL-sensitive enzyme produced before antibiotic production and a PL-resistant form produced on induction of PL synthesis. The sensitive isoform rapidly disappears when PL is produced (FRÖHLICH et al., 1989). The two isoforms are encoded by two distinct genes, but the mechanisms responsible for their regulation are unknown.

4 Concluding Remarks

Studies of the regulation of antibiotic production in diverse prokaryotes are revealing several common themes. Typically, antibiotic biosynthetic genes appear to be regulated principally at the level of transcription, and this involves rather large numbers of apparent regulatory genes, most of them acting more or less globally on secondary metabolism. The regulatory genes might exert their effects by direct interaction with promoters of antibiotic production genes, though it seems more likely that they are mostly involved with producing, detecting, transmitting, and integrating information relevant to deciding the appropriateness of commitment to production. Different organisms have different ecologies, and because different antibiotics differ in their biosynthetic origins and modes of action, and thus in their potential effectiveness in different ecological situations, activation of any particular pathway of any particular organism might be expected to require its own combination of signals; it is, therefore, not surprising to find great diversity in the systems analyzed so far. Nevertheless, there are recurrent themes. One is the use of quorum sensing. Lipid-soluble lactones play such a role in widely different organisms, and in each case the receptor is cytoplasmic and capable of direct interaction with DNA to influence gene expression. However, in *B. subtilis* the most well-defined extracellular signaling is done by a small peptide probably modified with a hydrophilic side chain (MAGNUSON et al., 1994), and the receptor is a membrane-located protein kinase that does not interact with DNA directly but phosphorylates and thereby activates a transcription factor. This solution adopted by *B. subtilis* to the problem of quorum sensing resembles the mating pheromone systems of *Enterococcus faecalis* (CLEWELL, 1993). It is, therefore, interesting that the *B. subtilis* pheromone also controls the competence regulator, the function of which (like mating) is to permit the ingress of DNA.

Protein phosphorylation is widely implicated in activating antibiotic synthesis, notably involving members of the two component histidine kinase-response regulator family. A further example of this, beyond that described above for *B. subtilis*, regulates production of three different secondary metabolites (2,4-diacetyl phloroglucinol, HCN, and pyoluteorin) in *Pseudomonas aeruginosa* (LAVILLE et al., 1992). Such two-component systems are clearly important in *Streptomyces* spp., but it is interesting that serine–threonine kinases, previously described only in eukaryotes, are also involved (there is also evidence of multiple tyrosine kinases in streptomycetes, though their roles are unknown; WATERS et al., 1994).

While specialized minor σ factors are often required for transcription of genes for antibiotic production, there is no evidence that a particular sub-branch of σ factors is implicated (Fig. 2b). Thus, the stationary-phase factor σ^S in *E. coli* is sometimes involved, and in *Streptomyces* at least one antibiotic production gene is transcribed *in vitro* by RNA polymerase containing σ^E (G. H. JONES and M. J. BUTTNER, personal communication), a σ factor that was used as the paradigm of a new subfamily of σ factors (the ECF family; LONETTO et al., 1994). *Streptomyces* spp. appear to contain a rather large number of different σ factors (BUTTNER, 1989: LONETTO et al., 1994), and *Streptomyces* promoters are very diverse in their sequences and complexity (STROHL, 1992), so it is quite likely that other minor σ factors may prove to be involved in secondary metabolism.

As knowledge of the regulation of antibiotic synthesis increases, new genetic approaches to industrial overproduction will probably be devised which may be especially useful when a new product is being developed. A particularly attractive approach is to add extra copies of positive regulatory genes which are found in many pathways (CHATER, 1990). This could be done by self-cloning, but an alternative strategy is to collect a panel of cloned DNA fragments (isolated from various species) capable of heterologous stimulation of production of some antibiotics, such as are being discovered by F. MALPARTIDA et al. (personal communication) and to introduce these into any interesting new species. A similar approach might also lead to the discovery of new antibiotics in old strains (HOPWOOD et al., 1983).

Note added in proof

Recently, several papers have been published on the regulation of antibiotic production in streptomycetes, mostly in *S. coelicolor*, and are discussed briefly here. The *absA* locus of *S. coelicolor* (Sect. 3.3.2.4) has been shown to encode a two-component regulatory system, *absA1/A2*, which acts as a negative regulator of antibiotic production (BRIAN et al., 1996). Disruption of *absA* results in early hyperproduction of both actinorhodin (Act) and undecylprodigiosin (Red). All four previously isolated *absA* mutations lie in *absA1* encoding the predicted sensor histidine kinase; these mutations may lock the kinase in an active conformation preventing the relief of the negative influence of the phosphorylated form of AbsA2 on antibiotic synthesis. In a potentially similar fashion, the *cutRS* locus also acts to negatively regulate Act production in *S. lividans* and in *S. coelicolor* (CHANG et al., 1996). Thus, protein phosphorylation mediated by *absA1/A2* and *cutRS* acts to negatively regulate antibiotic production, in contrast to the positive effects of *afsK/afsR* (Sect. 3.3.2.1) and *afsQ1/Q2* (Sect. 3.3.2.2). Recent studies on *afsR* (FLORIANO and BIBB, 1996) revealed that while it is homologous to *act*II-ORF4 and *redD*, pathway-specific regulatory genes for Act and Red production, respectively, it cannot substitute for them. Moreover, an in-frame deletion that removed most of the *afsR* coding sequence resulted in loss of Act and Red production, and a marked reduction in the synthesis of the calcium-dependent antibiotic (CDA), but only under some (non-permissive) nutritional conditions. Although additional copies of *afsR* resulted in elevated levels of the *act*II-ORF4 and *redD* transcripts, transcription of the pathway-specific regulatory genes under non-permissive conditions was unaffected by deletion of *afsR*. While *afsR* may operate independently of the pathway-specific regulatory proteins to influence antibiotic production, the activity of ActII-ORF4 and of RedD under non-permissive conditions could depend on interaction with, or modification by, AfsR. To assess whether there might be a causal relationship between ppGpp synthesis and antibiotic production (Sect. 3.2.3), a PCR-based approach was used to clone the ppGpp synthetase gene (*relA*) of *S. coelicolor* (CHAKRABURTTY et al., 1996). The cloned gene was used to create a null-mutant that is totally deficient in ppGpp synthesis upon amino acid starvation (CHAKRABURTTY, 1996). The resulting mutant grows at the same rate as the $relA^+$ strain but fails to make Act or Red on some media, but does so on others; similar results were obtained by MARTÍNEZ-COSTA et al. (1996). This indicates an obligatory role for ppGpp in antibiotic biosynthesis, and together with the conditional phenotype of the *afsR* null-mutant (FLORIANO and BIBB, 1996), indicates the presence of multiple signal transduction pathways for the activation of antibiotic production. Further evidence for this stems from the isolation and characterization of an extracellular signaling molecule, a novel γ-butyrolactone, that elicits the precocious production of both Act and Red when added to the wild-type strain (E. TAKANO, T. NIHIRA, and M. J. BIBB, unpublished results). The *relA* and *afsR* mutants produce, but do not respond to, this factor. Genes encoding the binding proteins for the γ-butyrolactones made by *S. griseus* (A-factor) and *S. virginiae* (the virginae butanolides VB-A-E) (Sect. 3.2.4) have been cloned and sequenced (ONAKA et al., 1995; OKAMOTO et al., 1995). Another pleiotropic regulatory gene for antibiotic production in *S. coelicolor* is *afsB* (Sect. 3.3.2.1). Attempts to complement *afsB* using a genomic library made in a low copy-number plasmid led to the discovery that additional copies of *hrdB*, which encodes the major σ factor of *S. coelicolor* (BROWN et al., 1992), restored Red and Act production in *afsB* mutants (WIETZORREK, 1996). The effect of *hrdB* resembles a recent report in *Pseudomonas fluorescens* (SCHNIDER et al., 1995), in which production of the antibiotics pyoluteorin and 2,4-diacetylphloroglucinol, which are made during stationary phase, was stimulated by the presence of additional copies of *rpoD*, which encodes the major and essential σ factor of that organism. This may reflect a role for the major σ factor of both organisms in the transcription of antibiotic biosynthetic genes, or may result from an indirect effect (e.g., provision of precursors). Consistent with the former notion, *in vitro* transcription of the pathway-specific regulatory gene *redD* was observed upon addition of a protein corresponding in size to σ^{hrdB} to core RNA polymerase (FUJII et al., 1996). Recent studies have further elucidated the way in which *bldA* (Sect. 3.3.2.6) influences the activation of individual biosynthetic pathways. Analysis of the Pwb mutations (Sect. 3.3.2.7) has identified an additional regulatory gene, *redZ* (GUTHRIE, E. P., FLAXMAN, C. S., WHITE, J., HODGSON, D. A., BIBB, M. J., and CHATER, K. F., manuscript in preparation), and revealed a pathway-specific regulatory cascade. *redZ* is located approximately 4 kb downstream of *redD* and contains a single UUA codon. Disruption of *redZ* results in loss of Red production and loss of *redD* transcription, suggesting that RedZ is a transcriptional activator of *redD* (WHITE and BIBB, in press). RedZ shows end-to-end similarity to members of the response regulator family of proteins, and possesses a putative DNA-binding α-helix-turn-α-helix motif towards its C-terminus but lacks the charged amino acids normally essential for phosphorylation of a response regulator by its cognate sensory histidine protein kinase. The existence of two pathway-specific regulatory genes for Red production in *S. coelicolor* parallels the situation for daunorubicin synthesis in *S. peuceticus*, in which *dnrI* and *dnrN* are homologues of *redD* and *redZ*, respectively. In *S. peuceticus*, transcription of *dnrI* depends on *dnrN* (Sect. 3.3.4.4), and DnrN has been shown recently to bind to the *dnrI* promoter region (FURUYA and HUTCHINSON, 1996). Moreover, DnrI has also been shown to bind to the promoters of daunorubicin biosynthetic structural genes (TANG et al., in press), providing the elusive evidence that this family of pathway-specific regulatory genes are indeed likely to act directly as transcriptional activators (Sect. 3.3.4.1).

Acknowledgements

We are grateful to WENDY CHAMPNESS, ALAN GROSSMAN, DAVID HOPWOOD, MICHIKO NAKANO, GEORGE SALMOND, and PETER ZUBER for comments on parts of the manuscript, and to all those who have allowed us to cite their unpublished results. We also thank MEREDYTH LIMBERG and ANNE WILLIAMS for their patience in typing successive versions of the manuscript. Our laboratories' work in this area was funded by the Biotechnology and Biological Research Council and the European Community.

References

ADAMIDIS, T., CHAMPNESS, W. (1992), Genetic analysis of *absB*, a *Streptomyces coelicolor* locus involved in global antibiotic regulation, *J. Bacteriol.* **174**, 4622–4628.

ADAMIDIS, T., RIGGLE, P., CHAMPNESS, W. C. (1990), Mutations in a new *Streptomyces coelicolor* locus which globally block antibiotic biosynthesis but not sporulation, *J. Bacteriol.* **172**, 2962–2969.

ALEX, L. A., SIMON, M. I. (1994), Protein histidine kinases and signal transduction in prokaryotes and eukaryotes, *Trends Genet.* **10**, 133–138.

AN, G., VINING, L. C. (1978), Intracellular levels of guanosine 5′-diphosphate 3′-diphosphate (ppGpp) and guanosine 5′-triphosphate 3′-diphosphate (pppGpp) in cultures of *Streptomyces griseus* producing streptomycin, *Can. J. Microbiol.* **24**, 502–511.

ANISOVA, L. N., BLINOVA, I. N., EFREMENKOVA, O. V., KOZ'MIN, YU. P., ONOPRIENKO, V. V., SMIRNOVA, G. M., KHOKHLOV, A. S. (1984), Regulators of the development of *Streptomyces coelicolor* A3(2), *Izv. Akad. Nauk SSSR, Ser. Biol.* **1**, 98–108.

ANZAI, H., MURAKAMI, T., IMAI, S., SATOH, A., NAGAOKA, K., THOMPSON, C. J. (1987), Transcriptional regulation of bialaphos biosynthesis in *Streptomyces hygroscopicus*, *J. Bacteriol.* **169**, 3482–3488.

ASTURIAS, J. A., LIRAS, P., MARTÍN, J. F. (1990), Phosphate control of *pabS* gene transcription during candicidin biosynthesis, *Gene* **93**, 79–84.

BAINTON, N. J., BYCROFT, B. W., CHHABRA, S. R., STEAD, P., GLEDHILL, L., HILL, P. J., REES, C. E. D., WINSON, M. K., SALMOND, G. P. C., STEWART, G. S. A. B., WILLIAMS, P. (1992a), A general role for the *lux* autoinducer in bacterial cell signalling: control of antibiotic biosynthesis in *Erwinia*, *Gene* **116**, 87–91.

BAINTON, N. J., STEAD, P., CHHABRA, S. R., BYCROFT, B. W., SALMOND, G. P. C., STEWART, G. S. A. B., WILLIAMS, P. (1992b), N-(3-Oxohexanoyl)-L-homoserine lactone regulates carbapenem antibiotic production in *Erwinia carotovora*, *Biochem. J.* **288**, 997–1004.

Bascaran, V., Sanchez, L., Hardisson, C., Braña, A. F. (1991), Stringent response and initiation of secondary metabolism in *Streptomyces clavuligerus*, *J. Gen. Microbiol.* **137**, 1625–1634.

Beppu, T. (1992), Secondary metabolites as chemical signals for cellular differentiation, *Gene* **115**, 159–165.

Braña, A. F., Méndez, C., Díaz, L. A, Manzanal, M. B., Hardisson, C. (1986), Glycogen and trehalose accumulation during colony development in *Streptomyces antibioticus*, *J. Gen. Microbiol.* **132**, 1319–1326.

Brian, P., Riggle, P. J., Santos, R. A., Champness, W. C. (1996), Global negative regulation of *Streptomyces coelicolor* by an *absA*-encoded putative signal transduction system, *J. Bacteriol.* **178**, 3221–3231.

Brown, K. L., Wood, S., Buttner, M. J. (1992), Isolation and characterization of the major vegetative RNA polymerase of *Streptomyces coelicolor* A3(2)-renaturation of a sigma-subunit using GroEL, *Mol. Microbiol.* **6**, 1133–1139.

Bruton, C. J., Guthrie, E. P., Chater, K. F. (1991), Phage vectors that allow monitoring of transcription of secondary metabolism genes in *Streptomyces*, *Bio/Technology* **9**, 652–656.

Buttner, M. J. (1989), RNA polymerase heterogeneity in *Streptomyces coelicolor* A3(2), *Mol. Microbiol.* **3**, 1653–1659.

Caballero, J. L., Malpartida, F., Hopwood, D. A. (1991), Transcriptional organization and regulation of an antibiotic export complex in the producing *Streptomyces* culture, *Mol. Gen. Genet.* **228**, 372–380.

Cao, J. G., Meighen, E. A. (1989), Purification and structural identification of an autoinducer for the luminescence system of *Vibrio harveyi*, *J. Biol. Chem.* **264**, 21670–21676.

Cashel, M., Rudd, K. E. (1987), The stringent response, in: *Escherichia coli and Salmonella typhimurium: Cellular and Molecular Biology* (Neidhardt, F., Ingraham, J. L., Low, K. B., Magasanik, B., Schaechter, M., Umbarger, H. E., Eds.), pp. 1410–1438. Washington, DC: American Society for Microbiology.

Chakraburtty, R. N. (1996), The (p)ppGpp synthetase gene (*relA*) of *Streptomyces coelicolor*, A3(2). *Ph D Thesis,* University of East Anglia, Norwich, UK.

Chakraburtty, R., White, J., Takano, E., Bibb, M. J. (1996), Cloning, characterization and disruption of a (p)ppGpp synthetase gene (*relA*) from *Streptomyces coelicolor* A3(2), *Mol. Microbiol.* **19**, 357–368.

Champness, W. C., Chater, K. F. (1994), Regulation and integration of antibiotic production and morphological differentiation in *Streptomyces* spp., in: *Regulation of Bacterial Differentiation* (Piggot, P., Moran, C. P., Youngman, P., Eds.), pp. 61–93. Washington, DC: American Society for Microbiology.

Champness, W. C., Riggle, P., Adamidis, T., Vandervere, P. (1992), Identification of *Streptomyces coelicolor* genes involved in regulation of antibiotic synthesis, *Gene* **115**, 55–60.

Chang, H.-M., Shieh, Y.-T., Bibb, M. J., Chen, C. W. (1996), The *cutRS* signal transduction system of *Streptomyces lividans* represses the biosynthesis of the polyketide antibiotic actinorhodin, *Mol. Microbiol.* **21**, 1075–1085.

Chater, K. F. (1990), The improving prospects for yield increase by genetic engineering in antibiotic-producing streptomycetes, *Bio/Technology* **8**, 115–121.

Chater, K. F. (1992), Genetic regulation of secondary metabolic pathways in *Streptomyces*, in: *Secondary Metabolites. Their Function and Evolution* (Chadwick, D. J., Whelan, J., Eds.), pp. 144–162, Ciba Foundation Symposium 171. Chichester: Wiley.

Chater, K. F., Bruton, C. J. (1983), Mutational cloning in *Streptomyces* and the isolation of antibiotic production genes, *Gene* **26**, 67–78.

Chater, K. F., Bruton, C. J. (1985), Resistance, regulatory and production genes for the antibiotic methylenomycin are clustered, *EMBO J.* **4**, 1892–1893.

Chater, K. F., Hopwood, D. A (1993), *Streptomyces*, in: *Bacillus subtilis and Other Gram-Positive Bacteria: Biochemistry, Physiology, and Molecular Genetics* (Sonenshein, A. L., Hoch, J. A., Losick, R., Eds.), pp. 83–89. Washington, DC: American Society for Microbiology.

Chater, K. F., Merrick, M. J. (1979), *Streptomyces*, in: *Developmental Biology of Prokaryotes* (Parish, J. H., Ed.), pp. 93–114. Oxford: Blackwell.

Clewell, D. B. (1993), Bacterial sex pheromone-induced plasmid transfer, *Cell* **73**, 9–12.

Cosmina, P., Rodriguez, F., de Ferra, F., Grandi, G., Perego, M., Venema, G., van Sinderen, D. (1993), Sequence and analysis of the genetic locus responsible for surfactin synthesis in *Bacillus subtilis*, *Mol. Microbiol.* **8**, 821–831.

Decker, H., Hutchinson, C. R. (1993), Transcriptional analysis of the *Streptomyces glaucescens* tetracenomycin C biosynthesis gene cluster, *J. Bacteriol.* **175**, 3887–3892.

Demain, A. L. (1992), Microbial secondary metabolism: a new theoretical frontier for academia, a new opportunity for industry, in: *Secondary Metabolites. Their Function and Evolution* (Chad-

WICK, D. J., WHELAN, J., Eds.), pp. 3–23, Ciba Foundation Symposium 171. Chichester: Wiley.

DEMAIN, A. L., AHARONOWITZ, Y., MARTÍN, J.-F. (1983), Metabolic control of secondary biosynthetic pathways, in: *Biochemistry and Genetic Regulation of Commercially Important Antibiotics* (VINING, L. C., Ed.), pp. 49–72. London: Addison-Wesley.

DEVEREUX, J., HAEBERLI, P., SMITHIES, O. (1984), A comprehensive set of sequence analysis programs for the VAX, *Nucleic Acids Res.* **12**, 387–395.

DÍAZ-GUERRA, L., MORENO, F., SAN MILLÁN, J. L. (1989), *appR* gene product activates transcription of microcin C7 plasmid genes, *J. Bacteriol.* **171**, 2906–2908.

DISTLER, J., MANSOURI, K., MAYER, G., STOCKMANN, M., PIEPERSBERG, W. (1992), Streptomycin biosynthesis and its regulation in streptomycetes, *Gene* **115**, 105–111.

DODD, I. B., EGAN, J. B. (1990), Improved detection of helix-turn-helix DNA-binding motifs in protein sequences, *Nucleic Acids Res.* **18**, 5019–5026.

DONADIO, S., STASSI, D., MCALPINE, J. B., STAVER, M. J., SHELDON, P. J., JACKSON, M., SWANSON, S. J., WENDT-PIENKOWSKI, E., YI-GUANG, W., JARVIS, B., HUTCHINSON, C. R., KATZ, L. (1993), Recent developments in the genetics of erythromycin formation, in: *Industrial Microorganisms: Basic and Applied Molecular Genetics* (BALTZ, R. H., HEGEMAN, G. D., SKATRUD, P. L., Eds.), pp. 257–265. Washington, DC: American Society for Microbiology.

D'SOUZA, C., NAKANO, M. M., CORBELL, N., ZUBER, P. (1993), Amino-acylation site mutations in amino acid-activating domains of surfactin synthetase: effects on surfactin production and competence development in *Bacillus subtilis*, *J. Bacteriol.* **175**, 3502–3510.

D'SOUZA, C., NAKANO, M. M., ZUBER, P. (1994), Identification of *comS*, a gene of the *srfA* operon that regulates the establishment of genetic competence in *Bacillus subtilis*, *Proc. Natl. Acad. Sci. USA* **91**, 9397–9401.

DUBNAU, D. (1993), Genetic exchange and homologous recombination, in: *Bacillus subtilis and Other Gram-Positive Bacteria: Biochemistry and Molecular Genetics* (SONENSHEIN, A. L., HOCH, J. A., LOSICK, R., Eds.), pp. 555–584. Washington, DC: American Society for Microbiology.

EBERHARD, A., BURLINGAME, A. L., KENYON, G. L., NEALSON, K. H., OPPENHEIMER, N. J. (1981), Structural identification of autoinducer of *Photobacterium fischeri* luciferase, *Biochemistry* **20**, 2444–2449.

EFREMENKOVA, O. V., ANISOVA, L. N., BARTOSHEVICH, Y. E. (1985), Regulators of differentiation in actinomycetes, *Antibiotiki i Meditsinskaya Biotekhnologiya* **9**, 687–707.

FERNÁNDEZ-MORENO, M. A, CABALLERO, J. L., HOPWOOD, D. A., MALPARTIDA, F. (1991), The *act* cluster contains regulatory and antibiotic export genes, direct targets for translational control by the *bldA* transfer RNA gene of *Streptomyces*, *Cell* **66**, 769–780.

FERNÁNDEZ-MORENO, M. A., MARTÍN-TRIANA, A. J., MARTINEZ, E., NIEMI, J., KIESER, H. M., HOPWOOD, D. A., MALPARTIDA, F. (1992), *aba*A, a new pleiotropic regulatory locus for antibiotic production in *Streptomyces coelicolor*, *J. Bacteriol.* **174**, 2958–2967.

FLORIANO, B., BIBB, M. J. (1996), *afsR* is a pleiotropic but conditionally required regulatory gene for antibiotic production in *Streptomyces coelicolor* A3(2), *Mol. Microbiol.* **21**, 385–396.

FRÖHLICH, K. U., WIEDMANN, M., LOTTSPEICH, F., MECKE, D. (1989), Substitution of a pentalenolactone-sensitive glyceraldehyde-3-phosphate dehydrogenase by a genetically distinct resistant isoform accompanies pentalenolactone production in *Streptomyces arenae*, *J. Bacteriol.* **171**, 6696–6702.

FUJII, T., GRAMAJO, H. C., TAKANO, E., BIBB, M. J. (1996), *redD* and *act*II-ORF4, pathway-specific regulatory genes for antibiotic production in *Streptomyces coelicolor* A3(2), are transcribed *in vitro* by an RNA polymerase holoenzyme containing the non-essential σ factor, σ^{hrdD}, *J. Bacteriol.* **178**, 3402–3405.

FUQUA, W. C., WINANS, S. C., GREENBERG, E. P. (1994), Quorum sensing in bacteria: the LuxR–LuxI family of cell density-responsive transcriptional regulators, *J. Bacteriol.* **176**, 269–275.

FURUYA, K., HUTCHINSON, C. R. (1996), The DnrN protein of *Streptomyces peuceticus*, a pseudo-response regulator, is a DNA-binding protein involved in the regulation of daunorubicin biosynthesis, *J. Bacteriol.* **178**, 6310–6318.

GEISTLICH, M., LOSICK, R., TURNER, J. R., RAO, R. N. (1992), Characterization of a novel regulatory gene governing the expression of a polyketide synthase gene in *Streptomyces ambofaciens*, *Mol. Microbiol.* **6**, 2019–2029.

GENTRY, D. R., HERNANDEZ, V. J., NGUYEN, L. H., JENSEN, D. B., CASHEL, M. (1993), Synthesis of the stationary-phase sigma factor σ^S is positively regulated by ppGpp, *J. Bacteriol.* **175**, 7982–7989.

GERBER, N. N., MCINNES, A. G., SMITH, D. G., WALTER, J. A., WRIGHT, J. L. C. (1978), Biosynthesis of prodiginines. ^{13}C resonance assignments and enrichment patterns in nonyl-, cyclononyl-, methylcyclodecyl-, and butylcycloheptylprodiginine produced by actinomycete cultures supplemented with ^{13}C-labeled acetate and ^{15}N-labeled nitrate, *Can. J. Chem.* **56**, 1155–1163.

GRÄFE, U., SCHADE, W., ERITT, I., FLECK, W. F. (1982), A new inducer of anthracycline biosynthesis from *Streptomyces viridochromogenes*, *J. Antibiot.* **35**, 1722–1723.

GRÄFE, U., REINHARDT, G., SCHADE, W., ERITT, I., FLECK, W. F., RADICS, L. (1983), Interspecific inducers of cytodifferentiation and anthracycline biosynthesis from *Streptomyces bikiniensis* and *Streptomyces cyaneofuscatus*, *Biotechnol. Lett.* **5**, 591–596.

GRAMAJO, H. C., TAKANO, E., BIBB, M. J. (1993), Stationary-phase production of the antibiotic actinorhodin in *Streptomyces coelicolor* A3(2) is transcriptionally regulated, *Mol. Microbiol.* **7**, 837–845.

GROSS, R., ARICO, B., RAPPUOLI, R. (1989), Families of bacterial signal-transducing proteins, *Mol. Microbiol.* **3**, 1661–1667.

GROSSMAN, T. H., TUCKMAN, M., ELLESTAD, S., OSBURNE, M. S. (1993), Isolation and characterization of *Bacillus subtilis* genes involved in siderophore biosynthesis: relationship between *B. subtilis sfp^o* and *Escherichia coli entD* genes, *J. Bacteriol.* **175**, 6203–6211.

GUILFOILE, P. G., HUTCHINSON, C. R. (1991), A bacterial analog of the *mdr* gene of mammalian tumor cells is present in *Streptomyces peucetius*, the producer of daunorubicin and doxorubicin, *Proc. Natl. Acad. Sci. USA* **88**, 8553–8557.

GUILFOILE, P. G., HUTCHINSON, C. R. (1992a), Sequence and transcriptional analysis of the *Streptomyces glaucescens tcmAR* tetracenomycin-C resistance and repressor gene loci, *J. Bacteriol.* **174**, 3651–3658.

GUILFOILE, P. G., HUTCHINSON, C. R. (1992b), The *Streptomyces glaucescens* TcmR protein represses transcription of the divergently oriented *tcmR* and *tcmA* genes by binding to an intergenic operator region, *J. Bacteriol.* **174**, 3659–3666.

GUTHRIE, E. P., CHATER, K. F. (1990), The level of a transcript required for production of a *Streptomyces coelicolor* antibiotic is conditionally dependent on a transfer RNA gene, *J. Bacteriol.* **172**, 6189–6193.

HAHN, J., DUBNAU, D. (1991), Growth stage signal transduction and the requirements for *srfA* induction in development of competence, *J. Bacteriol.* **173**, 7275–7282.

HAMAGISHI, Y., YOSHIMOTO, A., OKI, T., INUI, T. (1980), Occurrence of guanosine-5′-diphosphate-3′-diphosphate and adenosine-5′-triphosphate-3′-diphosphate in *Streptomyces galilaeus*, *Agric. Biol. Chem.* **44**, 1003–1007.

HAMAGISHI, Y., YOSHIMOTO, A., OKI, T. (1981), Determination of guanosine tetraphosphate (ppGpp) and adenosine pentaphosphate (pppApp) in various microorganisms by radioimmunoassay, *Arch. Microbiol.* **130**, 134–137.

HARA, O., BEPPU, T. (1982a), Induction of streptomycin-inactivating enzyme by A-factor in *Streptomyces griseus*, *J. Antibiot.* **35**, 1208–1215.

HARA, O., BEPPU, T. (1982b), Mutants blocked in streptomycin production in *Streptomyces griseus* – the role of A-factor, *J. Antibiot.* **35**, 349–358.

HARA, O., HORINOUCHI, S., UOZUMI, T., BEPPU, T. (1983), Genetic analysis of A-factor synthesis in *Streptomyces coelicolor* A3(2) and *Streptomyces griseus*, *J. Gen. Microbiol.* **129**, 2939–2944.

HEALY, J., WEIR, J., SMITH, I., LOSICK, R. (1991), Post-transcriptional control of a sporulation regulatory gene encoding transcription factor σ^H in *Bacillus subtilis*, *Mol. Microbiol.* **5**, 477–487.

HECKER, M., SCHROETER, A. (1985), Synthese der alkalischen Phosphatase in einem stringent und relaxed kontrollierten Stamm von *Escherichia coli* nach Aminosäuren- und Phosphatlimitation, *J. Basic Microbiol.* **25**, 341–347.

HENGGE-ARONIS, R., FISCHER, D. (1992), Identification and molecular analysis of *glgS*, a novel growth-phase regulated and *rpoS*-dependent gene involved in glycogen synthesis in *Escherichia coli*, *Mol. Microbiol.* **6**, 1877–1886.

HENNER, D. J., YANG, M., FERRARI, E. (1988), Localization of *Bacillus subtilis sacU*(hy) mutations to two linked genes with similarities to the conserved procaryotic family of two-component signalling systems, *J. Bacteriol.* **170**, 5102–5109.

HERNANDEZ, V. J., BREMER, H. (1990), Guanosine tetraphosphate (ppGpp) dependence of the growth rate control of *rrnA* P1 promoter activity in *Escherichia coli*, *J. Biol. Chem.* **265**, 11605–11614.

HERNANDEZ, V. J., BREMER, H. (1991), *Escherichia coli* ppGpp synthetase II activity requires *spoT*, *J. Biol. Chem.* **266**, 5991–5999.

HERNANDEZ, V. J., BREMER, H. (1993), Characterization of RNA and DNA synthesis in *Escherichia coli* strains devoid of ppGpp, *J. Biol. Chem.* **268**, 10851–10862.

HIGGINS, C. F. (1992), ABC transporters: from microorganisms to man, *Annu. Rev. Cell Biol.* **8**, 67–113.

HIGGINS, C. F., DORMAN, C. J., STIRLING, D. A, WADDELL, L., BOOTH, I. R., MAY, G., BREMER, E. (1988), A physiological role for DNA supercoiling in the osmotic regulation of gene expression in *S. typhimurium* and *E. coli*, *Cell* **52**, 569–584.

HOBBS, G., OBANYE, A. I. C., PETTY, J., MASON, J. C., BARRATT, E., GARDNER, D. C. J., FLETT, F., SMITH, C. P., BRODA, P., OLIVER, S. G. (1992), An integrated approach to studying regulation of production of the antibiotic methylenomycin by *Streptomyces coelicolor* A3(2), *J. Bacteriol.* **174**, 1487–1494.

HOCH, J. A. (1993), *spo0* genes, the phosphorelay, and the initiation of sporulation, in: *Bacillus subtilis and Other Gram-positive Bacteria: Biochemistry, Physiology, and Molecular Genetics* (SONENSHEIN, A. L., HOCH, J. A., LOSICK, R., Eds.), pp. 747–755. Washington, DC: American Society for Microbiology.

HODGSON, D. A. (1992), Differentiation in actinomycetes, in: *Prokaryotic Structure and Function. A New Perspective* (MOHAN, S., DOW, C., COLE, J. A., Eds.), *SGM Symposium* Vol. 47, pp. 407–440. Cambridge: Society for General Microbiology.

HOGGARTH, J. H., CUSHING, K. E., MITCHELL, J. I., RITCHIE, D. A. (1994), Induction of resistance to novobiocin in the novobiocin-producing organism *Streptomyces niveus*, *FEMS Microbiol. Lett.* **116**, 131–136.

HOLT, T. G., CHANG, C., LAURENTWINTER, C., MURAKAMI, T., GARRELS, J. I., DAVIES, J. E., THOMPSON, C. J. (1992), Global changes in gene expression related to antibiotic synthesis in *Streptomyces hygroscopicus*, *Mol. Microbiol.* **6**, 969–980.

HONG, S. K, KITO, M., BEPPU, T., HORINOUCHI, S. (1991), Phosphorylation of the AfsR product, a global regulatory protein for secondary-metabolite formation in *Streptomyces coelicolor* A3(2), *J. Bacteriol.* **173**, 2311–2318.

HONG, S. K., MATSUMOTO, A., HORINOUCHI, S., BEPPU, T. C. (1993), Effect of protein kinase inhibitors on *in vitro* protein phosphorylation and cellular differentiation of *Streptomyces griseus*, *Mol. Gen. Genet.* **236**, 347–354.

HOOD, D. W., HEIDSTRA, R., SWOBODA, U. K., HODGSON, D. A. (1992), Molecular genetic analysis of proline and tryptophan biosynthesis in *Streptomyces coelicolor* A3(2) – interaction between primary and secondary metabolism – a review, *Gene* **115**, 5–12.

HOPWOOD, D. A., BIBB, M. J., BRUTON, C. J., CHATER, K. F., FEITELSON, J. S., GIL, J. A. (1983), Cloning *Streptomyces* genes for antibiotic production, *Trends Biotechnol.* **1**, 42–48.

HOPWOOD, D. A., CHATER, K. F., BIBB, M. J. (1995), Genetics of antibiotic production in *Streptomyces coelicolor* A3(2), in: *Genetics and Biochemistry of Antibiotic Production* (VINING, L., STUTTARD, C., Eds.), pp. 65–102. Newton, MA: Butterworth-Heinemann.

HORINOUCHI, S. (1993), "Eukaryotic" signal transduction systems in the bacterial genus *Streptomyces*, *Actinomycetologica* **7**, 68–87.

HORINOUCHI, S., BEPPU, T. (1992), Autoregulatory factors and communication in actinomycetes. *Annu. Rev. Microbiol.* **46**, 377–398.

HORINOUCHI, S., BEPPU, T. (1995), Autoregulators, in: *Genetics and Biochemistry of Antibiotic Production* (VINING, L., STUTTARD, C., Eds.), pp. 103–119, Newton, MA: Butterworth-Heinemann.

HORINOUCHI, S., HARA, O., BEPPU, T. (1983), Cloning of a pleiotropic gene that positively controls biosynthesis of A-factor, actinorhodin, and prodigiosin in *Streptomyces coelicolor* A3(2) and *Streptomyces lividans*, *J. Bacteriol.* **155**, 1238–1248.

HORINOUCHI, S., KUMADA, Y., BEPPU, T. (1984), Unstable genetic determinant of A-factor biosynthesis in streptomycin-producing organisms: Cloning and characterization, *J. Bacteriol.* **158**, 481–487.

HORINOUCHI, S., SUZUKI, H., BEPPU, T. (1986), Nucleotide sequence of *afsB*, a pleiotropic gene involved in secondary metabolism in *Streptomyces coelicolor* A3(2) and "*Streptomyces lividans*", *J. Bacteriol.* **168**, 257–269.

HORINOUCHI, S., MALPARTIDA, F., HOPWOOD, D. A., BEPPU, T. (1989a), *afsB* stimulates transcription of the actinorhodin biosynthetic pathway in *Streptomyces coelicolor* A3(2) and *Streptomyces lividans*, *Mol. Gen. Genet.* **215**, 355–357.

HORINOUCHI, S., SUZUKI, H., NISHIYAMA, M., BEPPU, T. (1989b), Nucleotide sequence and transcriptional analysis of the *Streptomyces griseus* gene (*afsA*) responsible for A-factor biosynthesis, *J. Bacteriol.* **171**, 1206–1210.

HORINOUCHI, S., KITO, M., NISHIYAMA, M., FURUYA, K., HONG, S. K., MIYAKE, K., BEPPU, T. (1990), Primary structure of AfsR, a global regulatory protein for secondary metabolite formation in *Streptomyces coelicolor* A3(2), *Gene* **95**, 49–56.

HUISMAN, G. W., KOLTER, R. (1994a), Regulation of gene expression at the onset of stationary phase in *Escherichia coli*, in: *Regulation of Bacterial Differentiation* (PIGGOT, P., MORAN, C. P., YOUNGMAN, P., Eds.), pp. 21–40. Washington, DC: American Society for Microbiology.

HUISMAN, G. W., KOLTER, R. (1994b), Sensing starvation: a homoserine lactone-dependent signaling pathway in *Escherichia coli*, *Science* **265**, 537–539.

HUTCHINSON, C. R., DECKER, H., MADDURI, K., OTTEN, S. L., TANG, L. (1994), Genetic control of polyketide biosynthesis in the genus *Strepto-*

myces, *Anton. Leeuwenhoek Int. J. Gen. Microbiol.* **64**, 165–176.

IRR, J. D. (1972), Control of nucleotide metabolism and ribosomal ribonucleic acid synthesis during nutrient starvation of *Escherichia coli*, *J. Bacteriol.* **110**, 554–561.

ISHIZUKA, H., HORINOUCHI, S., KIESER, H. M., HOPWOOD, D. A., BEPPU, T. (1992), A putative two-component regulatory system involved in secondary metabolism in *Streptomyces* spp., *J. Bacteriol.* **174**, 7585–7594.

JONES, G. H. (1985), Regulation of phenoxazinone synthase expression in *Streptomyces antibioticus*, *J. Bacteriol.* **163**, 1215–1221.

KELEMEN, G. H., ZALACAIN, M., CULEBRAS, E., SENO, E. T., CUNDLIFFE, E. (1994), Transcriptional attenuation control of the tylosin-resistance gene *tlrA* in *Streptomyces fradiae*, *Mol. Microbiol.* **14**, 833–842.

KELLY, K. S., OCHI, K., JONES, G. H. (1991), Pleiotropic effects of a *relC* mutation in *Streptomyces antibioticus*, *J. Bacteriol.* **173**, 2297–2300.

KHOKHLOV, A. S. (1982), Low molecular weight microbial bioregulators of secondary metabolism, in: *Overproduction of Microbial Products* (KRUMPHANZL, V., SIKYTA, B., VANEK, Z., Eds.), pp. 97–109. London: Academic Press.

KHOKHLOV, A. S., TOVAROVA, I. I., BORISOVA, L. N., PLINER, S. A., SCHEVCHENKO, L. A., KORNITSKAYA, E. Y., IVKINA, N. S., RAPOPORT, I. A. (1967), A-factor responsible for the production of streptomycin by a mutant strain of *Actinomyces streptomycini*, *Dokl. Akad. Nauk SSSR* **177**, 232–235.

KIESER, H. M., KIESER, T., HOPWOOD, D. A. (1992), A combined genetic and physical map of the *Streptomyces coelicolor* A3(2) chromosome, *J. Bacteriol.* **174**, 5496–5507.

KILLHAM, K., FIRESTONE, M. K. (1984a), Salt stress control of intracellular solutes in streptomycetes indigenous to saline soils, *Appl. Environ. Microbiol.* **47**, 310–306.

KILLHAM, K., FIRESTONE, M. K. (1984b), Proline transport increases growth efficiency in salt-stressed *Streptomyces griseus*, *Appl. Environ. Microbiol.* **48**, 239–241.

KINASHI, H., SHIMAJI, M., SAKAI, A. (1987), Giant linear plasmids in *Streptomyces* which code for antibiotic biosynthesis genes, *Nature* **328**, 454–456.

KOLTER, R., SIEGELE, D. A., TORMO, A. (1993), The stationary phase of the bacterial life cycle, *Annu. Rev. Microbiol.* **47**, 855–874.

KONG, L., DUBNAU, D. (1994), Regulation of competence-specific gene expression by Mec-mediated protein–protein interaction in *Bacillus subtilis*, *Proc. Natl. Acad. Sci. USA* **91**, 5793–5797.

LACALLE, R. A, TERCERO, J. A., JIMENEZ, A. (1992), Cloning of the complete biosynthetic gene cluster for an aminonucleoside antibiotic, puromycin, and its regulated expression in heterologous hosts, *EMBO J.* **11**, 785–792.

LANGE, R., HENGGE-ARONIS, R. (1991a), Identification of a central regulator of stationary-phase gene expression in *Escherichia coli*, *Mol. Microbiol.* **5**, 49–59.

LANGE, R., HENGGE-ARONIS, R. (1991b), Growth phase-regulated expression of *bolA* and morphology of stationary-phase *Escherichia coli* cells are controlled by the novel sigma factor σ^S, *J. Bacteriol.* **173**, 4474–4481.

LANGE, R., HENGGE-ARONIS, R. (1994), The cellular concentration of the σ^S subunit of RNA polymerase in *Escherichia coli* is controlled at the levels of transcription, translation, and protein stability, *Genes Dev.* **8**, 1600–1612.

LAVILLE, J., VOISARD, C., KEEL, C., MAURHOFER, M., DÉFAGO, G., HAAS, D. (1992), Global control in *Pseudomonas fluorescens* mediating antibiotic synthesis and suppression of black root rot of tobacco, *Proc. Natl. Acad. Sci. USA* **89**, 1562–1566.

LAWLOR, E. J., BAYLIS, H. A., CHATER, K. F. (1987), Pleiotropic morphological and antibiotic deficiencies result from mutations in a gene encoding a tRNA-like product in *Streptomyces coelicolor* A3(2), *Genes Dev.* **1**, 1305–1310.

LESKIW, B. K., LAWLOR, E. J., FERNANDEZ-ABALOS, J. M., CHATER, K. F. (1991a), TTA codons in some genes prevent their expression in a class of developmental, antibiotic-negative, *Streptomyces* mutants, *Proc. Natl. Acad. Sci. USA* **88**, 2461–2465.

LESKIW, B. K., BIBB, M. J., CHATER, K. F. (1991b), The use of a rare codon specifically during development, *Mol. Microbiol.* **5**, 2861–2867.

LESKIW, B. K., MAH, R., LAWLOR, E. J., CHATER, K. F. (1993), Accumulation of *bldA*-specified transfer RNA is temporally regulated in *Streptomyces coelicolor* A3(2), *J. Bacteriol.* **175**, 1995–2005.

LILLEY, G., CLARK, A. E., LAWRENCE, G. C. (1981), Control of the production of cephamycin C and thienamycin by *Streptomyces cattleya* NRRL 8057, *J. Chem. Tech. Biotechnol.* **31**, 127–134.

LONETTO, M. A., BROWN, K. L., RUDD, K. E., BUTTNER, M. J. (1994), analysis of the *Streptomyces coelicolor* σ^E gene reveals the existence of a subfamily of eubacterial RNA polymerase σ factors involved in the regulation of extracytoplasmic functions, *Proc. Natl. Acad. Sci. USA* **91**, 7573–7577.

LOEWEN, P. C., VON OSSOWSKI, I., SWITALA, J., MULVEY, M. R. (1993), KatF (σ^S) synthesis in *Escherichia coli* is subject to posttranscriptional regulation, *J. Bacteriol.* **175**, 2150–2153.

MACNEIL, D. J., OCCI, J. L., GEWAIN, K. M., MACNEIL, T., GIBBONS, P. H., RUBY, C. L., DANIS, S. J. (1992), Complex organisation of the *Streptomyces avermitilis* genes encoding the avermectin polyketide synthase, *Gene* **115**, 119–125.

MAGNUSON, R., SOLOMON, J., GROSSMAN, A. D. (1994), Biochemical and genetic characterization of a competence pheromone from *B. subtilis*, *Cell* **77**, 207–216.

MALPARTIDA, F., HOPWOOD, D. A. (1984), Molecular cloning of the whole biosynthetic pathway of a *Streptomyces* antibiotic and its expression in a heterologous host, *Nature* **309**, 462–464.

MALPARTIDA, F., NIEMI, J., NAVARRETE, R., HOPWOOD, D. A. (1990), Cloning and expression in a heterologous host of the complete set of genes for biosynthesis of the *Streptomyces coelicolor* antibiotic undecylprodigiosin, *Gene* **93**, 91–99.

MARAHIEL, M. A., NAKANO, M. M., ZUBER, P. (1993), Regulation of peptide antibiotic production in *Bacillus*, *Mol. Microbiol.* **7**, 631–636.

MARTÍN, J. F., GIL, J. A. (1984), Cloning and expression of antibiotic production genes, *Bio/Technology* **2**, 63–72.

MARTÍN, J. F., LIRAS, P. (1989), Organization and expression of genes involved in the biosynthesis of antibiotics and other secondary metabolites, *Annu. Rev. Microbiol.* **43**, 173–206.

MARTÍN, J. F., MCDANIEL, L. E. (1975), Kinetics of biosynthesis of polyene macrolide antibiotics in batch cultures: cell maturation time, *Biotechnol. Bioeng.* **17**, 925–938.

MARTÍNEZ-COSTA, O. H., ARIAS, P., ROMERO, N. M., PARRO, V., MELLADO, R. P., MALPARTIDA, F. (1996), A *relA/spoT* homologous gene from *Streptomyces coelicolor* A3(2) controls antibiotic biosynthetic genes, *J. Biol. Chem.* **271**, 10627–10634.

MATSUMOTO, A., HONG, S.-I., ISHIZUKA, H., HORINOUCHI, S., BEPPU, T. (1994), Phosphorylation of the AfsR protein involved in secondary metabolism in *Streptomyces* species by a eukaryotic-type protein kinase, *Gene* **146**, 47–56.

MCCANN, M. P., YIDWELL, J. P., MATIN, A. (1991), The putative σ factor KatF has a central role in development of starvation-mediated general resistance in *Escherichia coli*, *J. Bacteriol.* **173**, 4188–4194.

MCCANN, M. P., FRALEY, C. D., MATIN, A. (1993), The putative σ factor KatF is regulated posttranscriptionally during carbon starvation, *J. Bacteriol.* **175**, 2143–2149.

MCCUE, L. A., KWAK, J., BABCOCK, M. J., KENDRICK, K. E. (1992), Molecular analysis of sporulation in *Streptomyces griseus*, *Gene* **115**, 173–179.

MEIGHEN, E. A. (1991), Molecular biology of bacterial bioluminescence, *Microbiol. Rev.* **55**, 123–142.

MÉNDEZ, C., BRAÑA, A. F., MANZANAL, M. B., HARDISSON, C. (1985), Role of substrate mycelium in colony development in *Streptomyces*, *Can. J. Microbiol.* **31**, 446–450.

MIYAKE, K., HORINOUCHI, S., YOSHIDA, M., CHIBA, N., MORI, K., NOGAWA, N., MORIKAWA, N., BEPPU, T. (1989), Detection and properties of A-factor-binding protein from *Streptomyces griseus*, *J. Bacteriol.* **171**, 4298–4302.

MIYAKE, K., KUZUYAMA, T., HORINOUCHI, S., BEPPU, T. (1990), The A-factor-binding protein of *Streptomyces griseus* negatively controls streptomycin production and sporulation, *J. Bacteriol.* **172**, 3003–3008.

MSADEK, T., KUNST, F., RAPOPORT, G. (1994), MecB of *Bacillus subtilis*, a member of the ClpC ATPase family, is a pleiotropic regulator controlling competence gene expression and growth at high temperature, *Proc. Natl. Acad. Sci. USA* **91**, 5788–5792.

MULVEY, M. R., LOEWEN, P. C. (1989), Nucleotide sequence of *katF* of *Escherichia coli* suggests KafF protein is a novel σ transcription factor, *Nucleic Acids Res.* **17**, 9979–9991.

NAKANO, M. M., ZUBER, P. (1993), Mutational analysis of the regulatory region of the *srfA* operon in *Bacillus subtilis*, *J. Bacteriol.* **175**, 3188–3191.

NAKANO, M. M., XIA, L., ZUBER, P. (1991), Transcription initiation region of the *srfA* operon, which is controlled by the *comP–comA* signal transduction system in *Bacillus subtilis*, *J. Bacteriol.* **173**, 5487–5493.

NAKANO, M. M., CORBELL, N., BESSON, J., ZUBER, P. (1992), Isolation and characterization of *sfp*: a gene that functions in the production of the lipopeptide biosurfactant, surfactin, in *Bacillus subtilis*, *Mol. Gen. Genet.* **232**, 313–321.

NARVA, K. E., FEITELSON, J. S. (1990), Nucleotide sequence and transcriptional analysis of the *redD* locus of *Streptomyces coelicolor* A3(2), *J. Bacteriol.* **172**, 326–333.

NEAL, R. J., CHATER, K. F. (1987), Nucleotide sequence analysis reveals similarities between proteins determining methylenomycin A resistance in *Streptomyces* and tetracycline resistance in eubacteria, *Gene* **58**, 229–241.

NEIDHARDT, F. C., INGRAHAM, J. L., LOW, K. B., MAGASANIK, B., SCHAECHTER, M., UMBARGER, H. E. (1987), *Escherichia coli* and *Salmonella typhimurium*: Cellular and Molecular Biology. Washington, DC: American Society for Microbiology.

NGUYEN, L. H., JENSEN, D. B., THOMPSON, N. E., GENTRY, D. L., BURGESS, R. R. (1993), *In vitro* functional characterization of overproduced *Escherichia coli katF/rpoS* gene product, *Biochemistry* **32**, 11112–11117.

NISHINO, T., MURAO, S. (1981), Possible involvement of plasmid in nucleotide pyrophosphokinase production and the relationship between this productivity and cellular accumulation of guanosine tetraphosphate (ppGpp) in *Streptomycetes*, *Agric. Biol. Chem.* **45**, 199–208.

NOVOA, M. A., DÍAZ-GUERRA, SAN MILLÁN, J. L., MORENO, F. (1986), Cloning and mapping of the genetic determinants for microcin C7 production and immunity, *J. Bacteriol.* **168**, 1384–1391.

OCHI, K. (1986), Occurrence of the stringent response in *Streptomyces* sp. and its significance for the inhibition of morphological and physiological differentiation, *J. Gen. Microbiol.* **132**, 2621–2631.

OCHI, K. (1987), A *rel* mutation abolishes the enzyme induction needed for actinomycin synthesis by *Streptomyces antibioticus*, *Agric. Biol. Chem.* **51**, 829–835.

OCHI, K. (1988), Nucleotide pools and stringent response in regulation of *Streptomyces* differentiation, in: *Biology of Actinomycetes '88* (OKAMI, Y., BEPPU, T., OGAWARA, H., Eds.), pp. 330–337. Tokyo: Scientific Societies Press.

OCHI, K. (1990a), *Streptomyces relC* mutants with an altered ribosomal protein ST-L11 and genetic analysis of a *Streptomyces griseus relC* mutant, *J. Bacteriol.* **172**, 4008–4016.

OCHI, K. (1990b), A relaxed (*rel*) mutant of *Streptomyces coelicolor* A3(2) with a missing ribosomal protein lacks the ability to accumulate ppGpp, A-factor and prodigiosin, *J. Gen. Microbiol.* **136**, 2405–2412.

O'CONNOR, K. A., ZUSMAN, D. R. (1988), Reexamination of the role of autolysis in the development of *Myxococcus xanthus*, *J. Bacteriol.* **170**, 4103–4112.

OHASHI, H., ZHENG, Y. H., NIHIRA, T., YAMADA, Y. (1989), Distribution of virginiae butanolides in antibiotic-producing actinomycetes, and identification of the inducing factor from *Streptomyces antibioticus* as virginiae butanolide A, *J. Antibiot.* **42**, 1191–1195.

OKAMOTO, S., NAKAMURA, K., NIHIRA, T., YAMADA, Y. (1995), Virginiae butanolide binding protein from *Streptomyces virginiae*, *J. Biol. Chem.* **270**, 12319–12326.

OKAMOTO, S., NIHIRA, T., KATAOKA, H., SUZUKI, A., YAMADA, Y. (1992), Purification and molecular cloning of a butyrolactone autoregulator receptor from *Streptomyces virginiae*, *J. Biol. Chem.* **267**, 1093–1098.

OLUKOSHI, E. R., PACKTER, N. M. (1994), Importance of stored triacylglycerols in *Streptomyces* – possible carbon source for antibiotics, *Microbiology* **140**, 931–943.

ONAKA, H., ANDO, N., NIHIRA, T., YAMADA, Y., BEPPU, T., HORINOUCHI, H. (1995), Cloning and characterization of the A-factor receptor gene from *Streptomyces griseus*, *J. Bacteriol.* **177**, 6083–6092.

POSTMA, P. W., LENGELER, J. W., JACOBSON, G. R. (1993), Phosphoenol pyruvate: carbohydrate phosphotransferase systems of bacteria, *Microbiol. Rev.* **57**, 543–594.

PUTTIKHUNT, C., OKAMOTO, S., NAKAMURA, T., NIHIRA, T., YAMADA, Y. (1993), Distribution in the genus *Streptomyces* of a homolog to *nusG*, a gene encoding a transcriptional anti-terminator, *FEMS Microbiol. Lett.* **110**, 243–248.

RAIBAUD, A., ZALACAIN, M., HOLT, T. G., TIZARD, R., THOMPSON, C. J. (1991), Nucleotide sequence analysis reveals linked N-acetyl hydrolase, thioesterase, transport, and regulatory genes encoded by the bialaphos biosynthetic gene cluster of *Streptomyces hygroscopicus, J. Bacteriol.* **173**, 4454–4463.

RETZLAFF, L., MAYER, G., BEYER, S., AHLERT, J., VERSECK, S., DISTLER, J., PIEPERSBERG, W. (1993), Streptomycin production in streptomycetes: a progress report, in: *Industrial Microrganisms: Basic and Applied Molecular Genetics* (BALTZ, R. H., HEGEMAN, G. D., SKATRUD, P. L., Eds.), pp. 183–194. Washington, DC: American Society for Microbiology.

ROGGIANO, M., DUBNAU, D. (1993), ComA, a phosphorylated response regulator protein of *Bacillus subtilis*, binds to the promoter region of *srfA*, *J. Bacteriol.* **175**, 3182–3187.

ROMEO, T., PREISS, J. (1989), Genetic regulation of glycogen biosynthesis in *Escherichia coli: in vitro* effects of cyclic AMP and guanosine 5′-diphosphate 3′-diphosphate and analysis of *in vivo* transcripts, *J. Bacteriol.* **171**, 2773–2782.

ROMEO, T., GONG, M., LIU, M. Y., BRUN-ZINKERNAGEL, A. M. (1993), Identification and molecular characterization of *csrA*, a pleiotropic gene from *Escherichia coli* that affects glycogen biosynthesis, gluconeogenesis, cell size, and surface properties, *J. Bacteriol.* **175**, 4744–4755.

RUDNER, D. Z., LE DEAUX, J. R., IRETON, K., GROSSMAN, A. D. (1991), The *spo0K* locus of

Bacillus subtilis is homologous to the oligopeptide permease locus and is required for sporulation and competence, *J. Bacteriol.* **173**, 1388–1398.

SAIER, M. H. (1989), Protein phosphorylation and allosteric control of inducer exclusion and catabolite repression by the bacterial phosphoenolpyruvate: sugar phosphotransferase system, *Microbiol. Rev.* **53**, 109–120.

SAIER, M. H. (1991), A multiplicity of potential carbon catabolite repression mechanisms in prokaryotic and eukaryotic microorganisms, *New Biol.* **3**, 1137–1147.

SARUBBI, E., RUDD, K. E., CASHEL, M. (1988), Basal ppGpp level adjustment shown by new *spoT* mutants affect steady state growth rates and *rrnA* ribosomal promoter regulation in *Escherichia coli*, *Mol. Gen. Genet.* **213**, 214–222.

SCHNIDER, U., KEEL, C., BLUMER, C., TROXLER, J., DEFAGO, G., HASS, D. (1995), Amplification of the housekeeping sigma factor in *Pseudomonas fluorescens* CHA0 enhances antibiotic production and improves biocontrol abilities, *J. Bacteriol.* **121**, 416–421.

SCHONER, B., GEISTLICH, M., ROSTECK, P., RAO, R. N., SENO, E., REYNOLDS, P., COX, K., BURGETT, S., HERSHBERGER, C. (1992), Sequence similarity between macrolide-resistance determinants and ATP-binding transport proteins, *Gene* **115**, 93–96.

SCHREIBER, G., METZGER, S., AIZENMAN, E., ROZA, S., CASHEL, M., GLASER, G. (1991), Overexpression of the *relA* gene in *Escherichia coli*, *J. Biol. Chem.* **266**, 3760–3767.

SENO, E. T., BALTZ, R. H. (1989), Structural organization and regulation of antibiotic biosynthesis and resistance genes in actinomycetes, in: *Regulation of Secondary Metabolism in Actinomycetes* (SHAPIRO, S., Ed.), pp. 2–48. Boca Raton, FL: CRC Press.

SIMUTH, J., HUDEC, J., CHAN, H. T., DANYI, O., ZELINKA, J. (1979), The synthesis of highly phosphorylated nucleotides, RNA and protein by *Streptomyces aureofaciens*, *J. Antibiot.* **32**, 53–58.

STASTNA, J., MIKULIK, K. (1981), Role of highly phosphorylated nucleotides and antibiotics in the development of streptomycetes, in: *Proc. 4th Int. Symp. Actinomycete Biology*, Cologne, pp. 481–486 (SCHAAL, K. P., PULVERER, G. Eds.). Stuttgart, New York: Gustav Fischer Verlag.

STEIN, D., COHEN, S. N. (1989), A cloned regulatory gene of *Streptomyces lividans* can suppress the pigment deficiency phenotype of different developmental mutants, *J. Bacteriol.* **171**, 2258–2261.

STOCK, J. B., STOCK, A. M., MOTTONEN, J. M. (1990), Signal transduction in bacteria, *Nature* **344**, 395–400.

STRAUCH, E., TAKANO, E., BAYLIS, H. A., BIBB, M. J. (1991), The stringent response in *Streptomyces coelicolor* A3(2), *Mol. Microbiol.* **5**, 289–298.

STROHL, W. R. (1992), Compilation and analysis of DNA sequences associated with apparent *Streptomyces* promoters, *Nucleic Acids Res.* **20**, 961–974.

STUTZMAN-ENGWALL, K. J., OTTEN, S., HUTCHINSON, C. R. (1992), Regulation of secondary metabolism in *Streptomyces* spp. and overproduction of daunorubicin in *Streptomyces peucetius*, *J. Bacteriol.* **174**, 144–154.

SWIFT, S., WINSON, M. K., CHAN, P. F., BAINTON, N. J., BIRDSALL, M., REEVES, P. J., REES, C. E. D., CHHABRA, S. R., HILL, P. J., THROUP, J. P., BYCROFT, B. W., SALMOND, G. P. C., WILLIAMS, P., STEWART, G. S. A. B. (1993), A novel strategy for the isolation of *luxI* homologues: evidence for the widespread distribution of a LuxR: LuxI superfamily in enteric bacteria, *Mol. Microbiol.* **10**, 511–520.

TAKANO, E., BIBB, M. J. (1994), The stringent response, ppGpp and antibiotic production in *Streptomyces coelicolor* A3(2). *Actinomycetologica* **8**, 1–10.

TAKANO, E., GRAMAJO, H. C., STRAUCH, E., ANDRES, N., WHITE, J., BIBB, M. J. (1992), Transcriptional regulation of the *redD* transcriptional activator gene accounts for growth-phase dependent production of the antibiotic undecylprodigiosin in *Streptomyces coelicolor* A3(2), *Mol. Microbiol.* **6**, 2797–2804.

TAKAYANAGI, Y., TAMAKA, K., TAKAHASHI, H. (1994), Structure of the 5′ upstream region and the regulation of the *rpoS* gene of *Escherichia coli*, *Mol. Gen. Genet.* **243**, 525–531.

TANAKA, K., TAKAYANAGI, Y., FUJITA, N., ISHIHAMA, A., TAKAHASHI, H. (1993), Heterogeneity of the principal sigma factor in *Escherichia coli*: The *rpoS* gene product, σ^{38}, is a second principal sigma factor of RNA polymerase in stationary phase *Escherichia coli*, *Proc. Natl. Acad. Sci. USA* **90**, 3511–3515.

TANG, L., GRIMM, A., ZHANG, Y.-X., HUTCHINSON, C. R. (in press), Purification of the DNA-binding protein DnrI, a transcriptional factor of daunorubicin biosynthesis in *Streptomyces peuceticus*, *Mol. Microbiol.*

THIARA, A. S., CUNDLIFFE, E. (1988), Cloning and characterization of a DNA gyrase B gene from *Streptomyces sphaeroides* that confers resistance to novobiocin, *EMBO J.* **7**, 2255–2259.

THIARA, A. S., CUNDLIFFE, E. (1989), Interplay of novobiocin-resistant and -sensitive DNA gyrase activities in self-protection of the novobiocin producer, *Streptomyces sphaeroides*, *Gene* **81**, 65–72.

THIARA, A. S., CUNDLIFFE, E. (1993), Expression and analysis of two *gyrB* genes from the novobiocin producer, *Streptomyces sphaeroides*, *Mol. Microbiol.* **8**, 495–506.

THOMPSON, C. J., SKINNER, R. H., THOMPSON, J., WARD, J. M., HOPWOOD, D. A., CUNDLIFFE, E. (1982), Biochemical characterisation of resistance determinants cloned from antibiotic-producing streptomycetes, *J. Bacteriol.* **151**, 678–685.

TSAO, S. W., RUDD, B. A. M., HE, X., CHANG, C., FLOSS, H. G. (1985), Identification of a red pigment from *Streptomyces coelicolor* A3(2) as a mixture of prodigiosin derivatives, *J. Antibiot.* **38**, 128–130.

VAN SINDEREN, D., GALLI, G., COSMINA, P., DE FERRA, F., WITHOFF, S., VENEMA, G., GRANDI, G. (1993), Characterization of the *srfA* locus of *Bacillus subtilis*: only the valine-activating domain of *srfA* is involved in the establishment of genetic competence, *Mol. Microbiol.* **8**, 833–841.

VICENTE, M., KUSHNER, S. R., GARRIDO, T., ALDEA, M. (1991), The role of the 'gearbox' in the transcription of essential genes, *Mol. Microbiol.* **5**, 2085–2091.

VÖGTLI, M., CHANG, P.-C., COHEN, S. N. (1994), *afsR2*: a previously undetected gene encoding a 63-amino-acid protein that stimulates antibiotic production in *Streptomyces lividans*, *Mol. Microbiol.* **14**, 643–653.

VOLZ, K. (1993), Structural conservation of the CheY superfamily, *Biochemistry* **32**, 11741–11753.

WASSERMAN, H. H., SHAW, C. K, SYKES, R. J. (1974), The biosynthesis of metacycloprodigiosin and undecylprodigiosin, *Tetrahedron Lett.* **33**, 2787–3091.

WATERS, B., VUJAKLIJA, D., GOLD, M. R., DAVIES, J. (1994), Protein tyrosine phosphorylation in streptomycetes, *FEMS Microbiol. Lett.* **120**, 187–190.

WEINRAUCH, Y., GUILLEN, N., DUBNAU, D. A. (1989), Sequence and transcription mapping of *Bacillus subtilis* competence genes *comB* and *comA*, one of which is related to a family of bacterial regulatory determinants, *J. Bacteriol.* **171**, 5362–5375.

WEINRAUCH, Y., PENCHEV, R., DUBNAU, E., SMITH, I., DUBNAU, D. (1990), A *Bacillus subtilis* regulatory gene product for genetic competence and sporulation resembles sensor protein members of the bacterial two-component signal-transduction systems, *Genes Dev.* **4**, 860-872.

WHITE, J., BIBB, M. J. (in press), The *bldA*-dependence of undecylprodigiosin production in *Streptomyces coelicolor* A3(2) involves a pathway-specific regulatory cascade, *J. Bacteriol.*

WIETZORREK, A. (1996), Regulation of antibiotic production in *Streptomyces coelicolor* A3(2)-analysis of the *afsB* mutant BH5. *Ph D Thesis*, University of East Anglia, Norwich, UK.

WILLIAMS, P. (1994), Compromising bacterial communication skills, *J. Pharmacol.* **46**, 1–10.

XIAO, H., KALMAN, M., IKEHARA, K., ZEMEL, S., GLASER, G., CASHEL, M. (1991), Residual guanosine 3′,5′-bispyrophosphate synthetic activity of *relA* null mutants can be eliminated by *spoT* null mutations, *J. Biol. Chem.* **266**, 5980–5990.

YAMADA, Y., SUGAMURA, K., KONDO, K, YANAGIMOTO, M., OKADA, H. (1987), The structure of inducing factors for virginiamycin production in *Streptomyces virginiae*, *J. Antibiot.* **40**, 496–504.

ZAMBRANO, M. M., SIEGELE, D. A., ALMIRÓN, M., TORMO, A., KOLTER, R. (1993), Microbial competition: *Escherichia coli* mutants that take over stationary phase cultures, *Science* **259**, 1757–1760.

3 Screening of Novel Receptor-Active Compounds of Microbial Origin

HARUO TANAKA

SATOSHI ŌMURA

Tokyo, Japan

1 Introduction

Development of receptor agonists and antagonists as drugs had already begun before their receptors were characterized as substances. These drugs account for a fairly high percentage of the medications currently used. Most of them are produced by chemical synthesis. In recent years, many receptor-active compounds of microbial origin have been discovered, stimulating efforts for developing new drugs from these lead compounds. These are now providing an important tool for clarifying the functions of receptors and the relationship between receptor function and disease.

Receptors are specific proteins, which recognize exogenous signaling molecules or physical stimuli, and induce cellular responses. Receptors located in the membrane or the cytoplasm function as the first window in the transmission of extracellular information into the cell. Some receptors, e.g., virus receptors and the LDL receptor, are involved in the uptake of components cells and not directly in signal transduction.

Early in the 1980s, the nicotinic acetylcholine receptor gene was cloned and its primary structure was determined (NODA et al., 1983). Since then, knowledge regarding the receptor-constituting proteins has been cultivated, and cloning of many receptor genes has been performed, thus stimulating discussions about the relationship between receptor structure and function (PEROUTKA, 1994). However, before their characterization as substances, receptors had been recognized as a vague concept and had contributed greatly to the analysis of drug effects and to the development of new drugs. A number of derivatives and analogs of low molecular weight ligands have been synthesized. Through the study of the effects and the pharmacological actions of these substances on receptors, many agonists and antagonists have been developed and are currently used as drugs.

In recent years, the development of radiolabeled ligands has simplified receptor binding experiments using tissue homogenates or cells. With such techniques, many synthetic compounds and microbial cultures have been screened, resulting in the discovery of new subtances which act on receptors. They provide not only lead compounds for the development of new drugs but also important tools for clarifying the receptor functions. For the study of the individual functions of receptor subtypes, which have been discovered in recent years, the identification of substances selectively acting on receptor subtypes is needed now.

This paper reviews the general considerations of receptor-active compounds of microbial origin, the screening methods, and the physiological and pharmacological activities of those discovered to date. In addition, future perspectives for this class of compounds are also discussed.

2 Assay Methods for Screening of Receptor-Active Compounds

Although animal experiments are the most reliable method of assessing pharmacological actions, they require much cost and labor and are not suitable for the examination of numerous samples. A common alternative is screening using radio-labeled ligands and receptor-containing cells or tissue homogenates. Various simple methods of screening have been devised, some of which are presented below.

2.1 Assays Based on Physiological Activities in Animal Tissues and Cells

The Magnus test, using blood vessels, has been employed for the screening of numerous samples, although the detection of ligands is not very easy with this technique. FUJII et al. (1991) assessed the contraction inhibiting activity with the following technique: First, an isolated tracheal specimen was suspended in a Magnus bath. Subsequently, bronchial contraction was induced by neurokinin A. Microbial cultures were then added to assess the contraction inhibitory activity. Using this technique, active substances from 10000 samples were selected. This led to the isolation of actinomycin D. This compound inhibited the contraction induced by neurokinin A ($IC_{50}=1.8\times10^{-6}$M), but did not inhibit that by substance P, acetylcholine, etc. (FUJII et al., 1991).

Methods based on cellular responses to ligands allow an easier examination of many samples than those using tissues. One example for this is the use of platelets. Platelet aggregation is known to be induced by collagen, ADP, arachidonic, acid, thrombin, PAF etc. Their antagonists can be isolated by selecting substances which inhibit platelet aggregation (NAKAGAWA, 1992). With such a method, OKAMOTO et al. (1986a, b) discovered two PAF receptor antagonists, i.e., FR-49175 produced by *Penicillium terlikowskii* and FR-900452 produced by *Streptomyces phaeofaciens.* LAUER et al. (1991) screened thromboxane A2 receptor antagonists taking inhibition of platelet aggregation as an indicator. Another cell culture method is based on macrophage chemotaxis. TSUJI et al. (1992a), e.g., identified the leukotriene B4 antagonist WF11605 influencing chemotaxis of polymorphonuclear leukocytes.

These methods examine the responses of tissues and cells and hence are advantageous in that they allow a distinction of agonists from antagonists. However, since all the various responses of tissues or cells are shown in which many receptors other than the target receptor are included, selectivity may not be very high with these methods. However, if the examiner is experienced and performs careful observation, the discovery of compounds with novel physiological actions can be expected.

2.2 Assays Using Radio-Labeled Ligands

In recent years, a number of radio-labeled ligands (such as hormones, autacoids, cytokines and neurotransmitters) have become commercially available. They can be used for experiments on receptor binding in cells, tissue homogenates and cell fractions (receptors for steroid hormones etc. are located in the cytoplasm). The amount of a radio-labeled ligand bound to the receptor is estimated from the amount of the ligand bound to the receptor in the presence of excess cold ligand to obtain the amount of specific receptor binding, which can be used for selecting substances which specifically inhibit receptor binding. The receptor binding inhibitors include both agonists and antagonists. Agonists can be distinguished from antagonists by examining the influence of receptor binding inhibitors on the physiological actions of ligands, e.g., using the Magnus method. These methods are estimated to have been employed in screening of ligands from microorganisms frequently. However, many publications lack descriptions of the screening methods. Following recent success in the cloning of many receptor genes, it is now possible to examine the binding of radio-labeled ligands to recombinant receptors by radioimmunoassay or ELISA.

2.3 Functional Assays Using Recombinant Cells Transformed with a Receptor Gene

A new type of assay for ligands using recombinant microorganisms has been reported. KING et al. (1990) constructed a recombinant yeast with which ligands of the β_2-adrenergic receptors can be assessed by transfecting genes of the human β_2-adrenergic receptor and a G protein α-subunit into the yeast. *Saccharomyces cerevisiae* possesses the G protein but lacks the α-subunit which is necessary for intracellular signal transduction. Therefore, the human β_2-adrenergic receptor gene (*hβ-AR*) and the mammalian α-subunit gene (rat *Gsα*) were transfected into *S. cerevisiae,* resulting in coupling of β- and γ-subunits. Furthermore, a system for the identification of β_2-adrenergic receptor agonists by colorimetry was established by linking the β_2-receptor G protein to the β-galactosidase gene (Fig. 1).

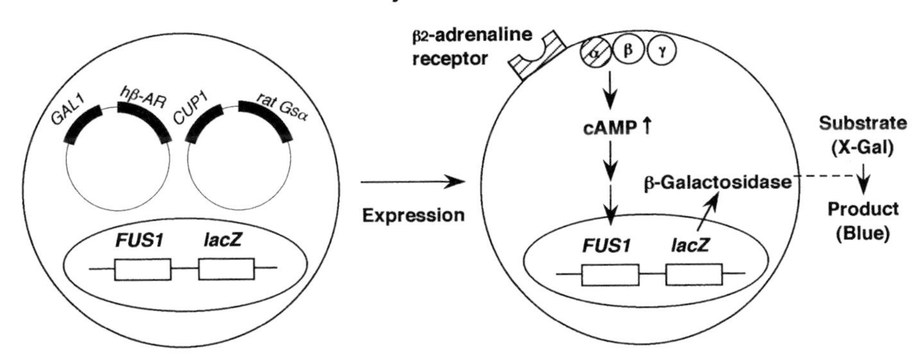

Fig. 1. A new screening system for agonists and antagonists of G protein coupled receptors.

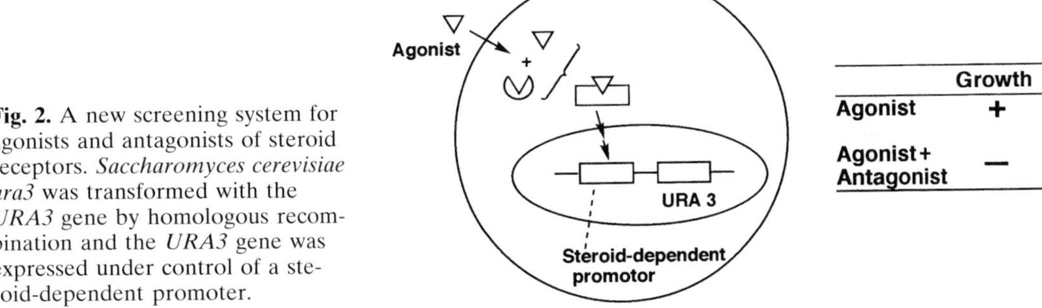

Fig. 2. A new screening system for agonists and antagonists of steroid receptors. *Saccharomyces cerevisiae ura3* was transformed with the *URA3* gene by homologous recombination and the *URA3* gene was expressed under control of a steroid-dependent promoter.

McDonell constructed another recombinant yeast for the assessment of ligands for steroid hormone receptors, utilizing the similarities of the transcription factors between yeasts and mammals (Abbott, 1991). As steroid hormone receptors are a family of transcription factor, an *in vivo* transcription system could be established using inducible expression vector system containing steroid hormone receptor genes. The linking of this system to the *URA3* gene in yeast resulted in the establishment of a system for measuring substances acting on the steroid hormone receptors in *S. cerevisiae.* As shown in Fig. 2, this system uses yeast growth as an indicator.

These approaches to assay systems using recombinant microorganisms can also be applied to establish assay systems for many other receptor ligands. Since the distinction between agonists and antagonists, which could not be made with binding experiments, is possible with such systems, they are expected to provide simpler assay systems if automation techniques are incorporated.

3 Receptor Antagonists

Ever since antibiotics were discovered from microorganisms, microorganisms have been regarded as a treasure-house of secondary metabolites. They serve as an important source of physiologically active substances, in addition to antibiotics. Plants and marine organisms have also been screened but the percentage of new substances discovered from microorganisms is much higher. Receptor-ac-

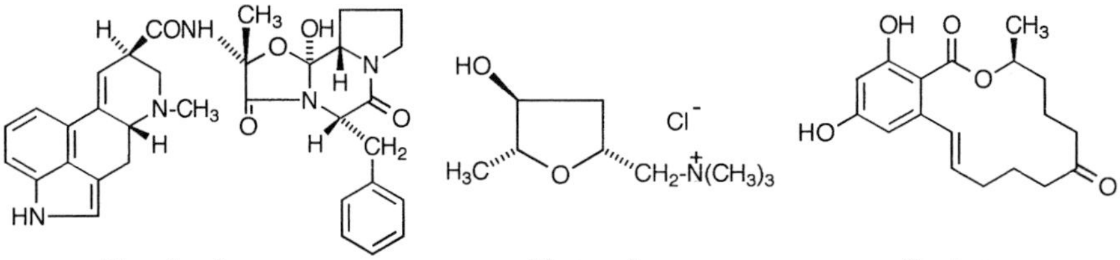

Ergotamine (*Claviceps purpurea*) α-Receptor antagosist

Muscarine (*Amanita muscaria*) Muscarinic acetylcholine receptor agonist

Zealarenone (*Gibberella zeae*) Estrogen receptor agonist

Fig. 3. Traditional receptor agonists and antagonists of microbial origin.

tive substances derived from microorganisms have a long history. It dates back to the discovery of ergot alkaloids (leading to the clinical use of ergotamine and ergometrine), muscarine (an acetylcholine receptor agonist) and zealarenone (an estrogen receptor agonist) (Fig. 3), although these substances were not found by screening for receptor-active compounds. There were more systematic approaches to discover receptor-active compounds of microbial origin after the identification of the cholecystokinin antagonist asperlicin in 1985. Receptor-active compounds discovered from microorganisms by such efforts are described below.

3.1 Antagonists of Low Molecular Weight Ligand Receptors

(Tab. 1, Fig. 4)

3.1.1 Muscarinic Acetylcholine Receptor Antagonists

TAKESAKO et al. (1988) screened microorganisms for muscarinic acetylcholine receptor antagonists in order to develop anticholinergic agents useful as anti-ulcer agents. Taking guinea pig brain homogenates as receptors and [^{3}H]quinuclidinyl benzilate as ligands,

Tab. 1. Antagonists of Microbial Origin of Low Molecular Weight Ligand Receptors

Ligand	Antagonist	Producer	Reference
Acetylcholine	IJ2702-I and -II	Soil isolate	TAKESAKO et al. (1988), UENO et al. (1988)
Dopamine	Sch42029	*Actinoplanes* sp.	HEDGE et al. (1991)
NMDA	ES-242-1 to -8	*Verticillium* sp.	TOKI et al. (1992a, b)
Leukotriene B_4	WF11605	Fungus	TSUJI et al. (1992a), SHIGEMATSU et al. (1992)
	Novobiocin	*Streptomyces* sp.	TSUJI et al. (1992b)
PAF	FR-49175	*Penicillium terlikowskii*	OKAMOTO et al. (1986a)
	FR-900452	*S. phaeofaciens*	OKAMOTO et al. (1986b)
	Phomatins (A, B, A_1, B_1)	*Phoma* sp.	SUGANO et al. (1991)
Fibrinogen	Tetrafibricin	*S. neyagawaensis*	KAMIYAMA et al. (1993a, b)
Estrogen	R1128 A, B, C, D	*Streptomyces* sp.	HORI et al. (1993a, b, c)
	Napiradiomycin A and B1	*Streptomyces* sp.	HORI et al. (1993e)
Androgen	3-Chloro-4-(2-amino-3-chlorophenyl)-pyrrole	*Pseudomonas* sp.	HORI et al. (1993d)
	WS9761 A and B	*Streptomyces* sp.	HORI et al. (1993f)

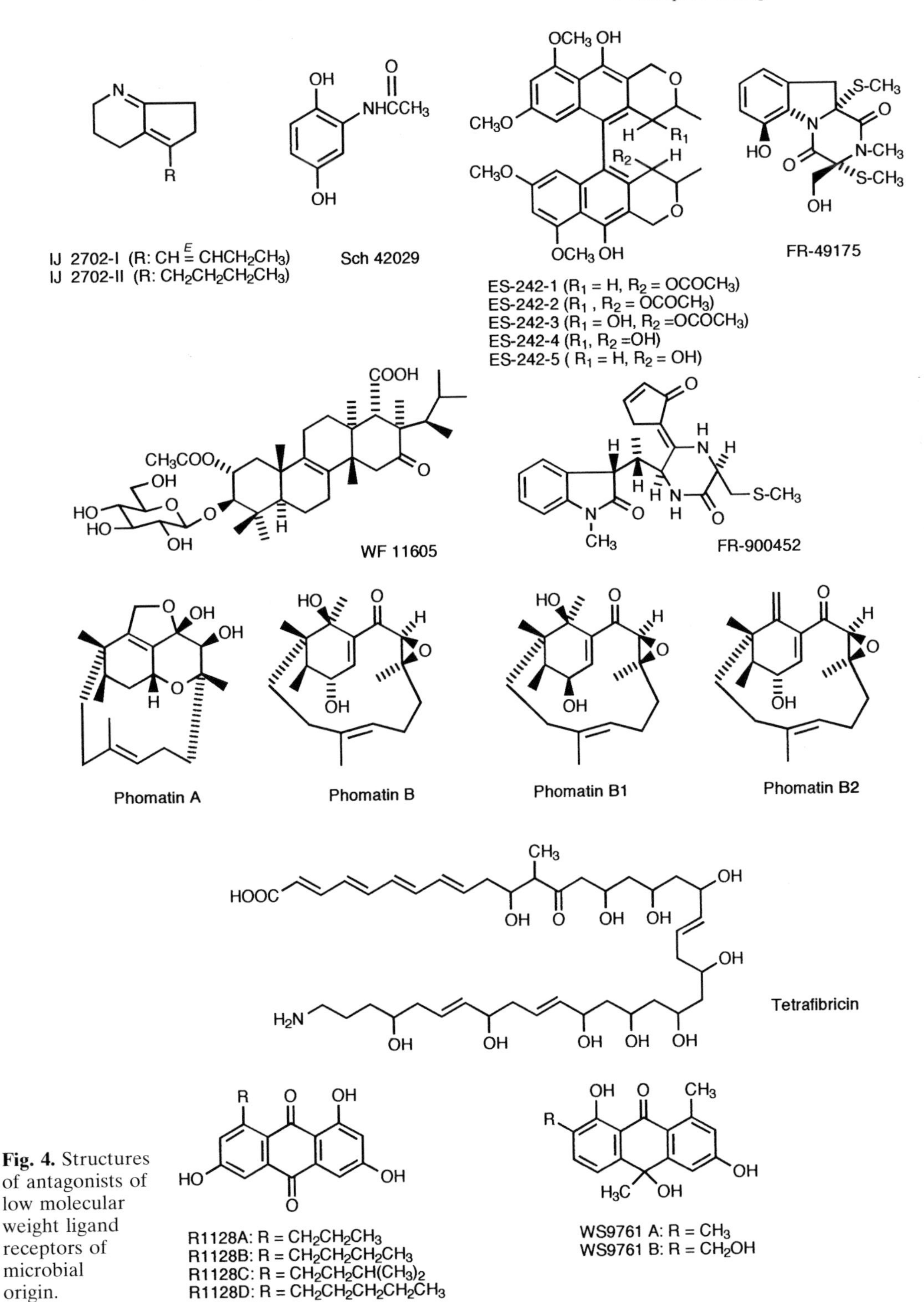

Fig. 4. Structures of antagonists of low molecular weight ligand receptors of microbial origin.

they assayed inhibitory effects on the receptor ligand binding of cultures of soil isolated actinomycetes, fungi and bacteria. The cultures showing inhibition were examined for anticholinergic activity, using isolated guinea pig ileum. In this way, two new compounds, IJ2702-I and IJ2702-II ($IC_{50}=0.3$ resp. 0.6 μg/mL), were isolated from the culture of an actinomycete strain, IJ2702 (TAKESAKO et al., 1988; UENO et al., 1988).

3.1.2 Dopamine Receptor Antagonists

HEDGE et al. (1991) searched for dopamine D_1 receptor-active compounds of microbial origin, using [^{3}H]Sch23390 and rat striatum. In their study, Sch42029 (2,5-dihydroxyacetoanilide) was found to be a D_1 receptor-specific ligand. Many of the drugs for the treatment of Parkinson's disease, which is related to abnormal dopamine metabolism in the brain, are D_2 receptor-specific. Sch42029 is the first natural substance specific to the D_1 receptor ($K_i=0.6$ μM).

3.1.3 NMDA Receptor Antagonists

TOKI et al. (1992a, b) carried out screening work with [^{3}H]TCP [1-(1-(2-thienyl)cyclohexylpiperidine] and a rat brain membrane fraction. They identified new compounds, ES-242-1 through ES-242-8, which are produced by *Verticillium* sp. and serve as NMDA receptor antagonists. These compounds inhibit the binding of [^{3}H]TCP to the synaptic membrane (IC_{50}: 0.1 μM for ES-242-1) but not the binding of [^{3}H]kainic acid. Although MK801 and ketamine are also known as synthetic compounds of this type, the ES-242 series are the first new compounds of microbial origin (TOKI et al., 1992d). They have recently been used in experiments to clarify the pharmacological actions at the molecular level of NMDA receptors.

3.1.4 Leukotriene B_4 Receptor Antagonists

Leukotriene B_4 (LTB4) is an autacoid which promotes aggregation, degranulation and chemotaxis of polymorphonuclear leukocytes. LTB4 is thought to be involved in inflammatory reactions. TSUJI et al. (1992a) examined microorganisms for substances that inhibit the LTB4-induced chemotaxis of rat polymorphonuclear leukocytes, leading to the discovery of WF11605 produced by a fungus. WF11605 is a new compound with a triterpene glucoside structure (SHIGEMATSU et al., 1992). WF11605 was found to inhibit not only chemotaxis ($IC_{50}=1.7\times10^{-7}$ M) but also the binding of [^{3}H]LTB4 to the membrane fraction of polymorphonuclear leukocytes ($IC_{50}=5.6\times10^{-6}$ M) and the LTB4-induced degranulation of polymorphonuclear leukocytes ($IC_{50}=3.0\times10^{-6}$ M). These results indicate that WF11605 is an LTB4 receptor antagonist. Its LD_{50} in mice was 1.0 g/kg or more (i.p.).

During screening for LTB4 antagonists, TSUJI et al. (1992b) recently found that novobiocin, an antibiotic in clinical use, acts as an antagonist of LTB4 receptors. This compound inhibited the binding of [^{3}H]LTB4 to the membrane fraction of polymorphonuclear leukocytes ($IC_{50}=1.0\times10^{-6}$ M). Since leukotrienes are known to be involved in the onset of ear edema in mice, the investigators examined the effect of novobiocin on ear edema in mice induced by arachidonic acid, and found that local treatment with this antibiotic (0.1 μg or more per ear) suppressed the formation of edema. The compound was effective even when administered orally ($ED_{50}=220$ μg/kg).

3.1.5 PAF Receptor Antagonists

Platelet activating factor (PAF) is produced by various cells and tissues. Even very small amounts of PAF exert various biological actions. It is a new physiologically active substance involved in different diseases such as bronchial asthma, inflammatory reactions, renal disease, collagen disease, and anaphylaxis. Therefore, PAF antagonists are expected not only to clarify the physiological actions and pathophysiological roles of PAF but also to provide an effective therapeutic agent for the treatment of these diseases.

OKAMOTO et al. (1986a, b) examined microorganisms for PAF antagonists using inhibition of PAF-induced platelet aggregation as an indicator. They found FR-49175 and FR-900452; FR-49175 was identified as bisdesthiobis(methylthio)gliotoxin, while FR-900452 was a new compound. The IC_{50} of the platelet aggregation inhibiting effect of FR-49175 was 8.5 μM. Intravenous injection (0.1 mg/kg) to guinea pigs inhibited PAF-induced bronchial stenosis (OKAMOTO et al., 1986a) FR-900452 is a compound with a unique structure, including piperidine and indolinone. It was able to inhibit PAF-induced rabbit platelet aggregation ($IC_{50}=3.7\times10^{-7}$ M), while its inhibitory effect on platelet aggregation induced by collagen, arachidonic acid or ADP was much weaker. The compound markedly suppressed PAF-induced bronchial stenosis, hypotension and elevation in vascular permeability in guinea pigs when it was administered intravenously, even in low doses below 10 μg/kg (OKAMOTO et al., 1986b).

SUGANO et al. (1991) examined marine microorganisms for secondary metabolites and isolated the phomatins A, B, B_1, and B_2 from *Phoma* sp., a species of *Fungi Imperfecti* living upon crabshells. These four compounds inhibited PAF-induced platelet aggregation, with an IC_{50} of 1.0×10^{-5} M, 1.7×10^{-5} M, 9.8×10^{-6} M, and 1.6×10^{-6} M, respectively.

3.1.6 Fibrinogen Receptor Antagonists

Platelet aggregation plays a key role in normal hemostasis and thrombosis. Platelets first adhere and spread onto the thrombogenic components of the vascular subendothelium at the sites of vascular lesions. When stimulated by an agonist, such as ADP, collagen or thrombin, the fibrinogen receptors acquire the ability to bind fibrinogen through some conformational changes within the molecule. Fibrinogen binding to the receptors on the surface of platelets is a prerequisite for platelet aggregation. Thus, fibrinogen receptor antagonism is a good target for a platelet aggregation inhibitor.

In the course of their screening program for fibrinogen binding antagonists, KAMIYAMA et al. (1993a, b) isolated a non-peptide antagonist, tetrafibricin, from the culture broth of an actinomycete. Tetrafibricin strongly inhibited the binding of fibrinogen to its receptors with an IC_{50} of 46 nM. It also inhibited ADP-, collagen-, and thrombin-induced aggregation of human platelets with an IC_{50} of 5.6 μM, 11.0 μM, and 7.6 μM, respectively. Tetrafibricin is a novel non-peptide antagonist of the fibrinogen receptor.

3.1.7 Estrogen Receptor Antagonists

Non-steroidal estrogen receptor antagonists, e.g., tamoxifen, have been used successfully in the therapy of advanced breast cancer, especially estrogen receptor positive breast cancer. Although this therapy results in remarkable improvements for breast cancer patients, the development of tamoxifen resistance frequently occurs and most patients eventually relapse. One potential method to overcome the resistance is the use of estrogen receptor antagonists with a new chemical structure different from tamoxifen and related compounds, containing the triphenyl ethylene moiety.

Based on such considerations HORI et al. screened microbial products for new non-steroidal estrogen receptor antagonists without the triphenyl ethylene moiety. They found new non-steroidal estrogen receptor antagonists – R1128 A, B, C, and D – from the culture broth of *Streptomyces* sp. No. 1128 (HORI et al., 1993a, b, c). These compounds inhibited estrogen binding to its receptor. The IC_{50} values of R1128 A, B, C, and D for partially purified rat uterine cytosol receptor were 1.1×10^{-7} M, 1.2×10^{-7} M, 2.6×10^{-7} M, and 2.7×10^{-7} M, respectively. R1128 B was a competitive inhibitor of estrogen receptor binding and inhibited the growth of estrogen-responsive human mammary adenocarcinoma MCF-7 cells in soft agar. This inhibition was reversed by addition of estradiol to the culture medium. R1128 B showed antitumor activities against MCF-7 when xenografted to nude mice by implantation into the subrenal capsule of mice (SRC assay). The potency of R1128 B was about 8-fold lower than that of tamoxifen both *in vitro* and *in vivo* (HORI et al., 1993c). A recent study by HORI et al. (1993d) revealed that napiradiomycin A and B1, which have been known to possess antimicrobial activities, are estrogen receptor antagonists ($IC_{50}=4.2\times10^{-6}$ M and 3.5×10^{-7} M, respectively.

3.1.8 Androgen Receptor Antagonists

Androgen plays an important role in the prostatic growth including benign prostatic hyperplasia and prostate cancer. Androgen actions are thought to be mediated through binding to its own receptor. Therefore, androgen receptor antagonists can be used in the treatment for androgen-responsive diseases.

During the course of search for non-steroidal androgen receptor binding inhibitors, HORI et al. (1993d) found that 3-chloro-4-(2-amino-3-chlorophenyl)-pyrrole (WB2838), a known antifungal antibiotic, is a non-steroidal androgen receptor antagonist. More recently, HORI et al. (1993f) discovered the novel androgen receptor antagonists WS9761 A and B. WS9761 A and B inhibited androgen receptor binding with IC_{50} values of 8.6×10^{-7} M and 4.5×10^{-7} M, respectively, and showed weak inhibitory activity against estrogen receptor binding.

3.2 Antagonists of Peptide Ligand Receptors

(Tab. 2, Fig. 5)

3.2.1 Cholecystokinin Receptor Antagonists

Cholecystokinin (CCK) is a digestive hormone which promotes lipid degradation and absorption by stimulating gallbladder contraction, pancreatic juice secretion and small bowel motility. Its involvement in the central regulation of appetite and pain has recently been noted. Known CCK receptors include CCK-A, primarily located in the periphery, and CCK-B, primarily located centrally. After the discovery of the CCK-A antagonists asperlicins (CHANG et al., 1985; GOETZ et al., 1985, 1988; LIESCH et al., 1985, 1988), the CCK-B antagonist tetronothiodin (IC_{50} against the CCK-B receptor = 3.6 nM) was found (OHTSUKA et al., 1992, 1993a, b; WATANABE et al., 1993). Both compounds are non-peptide antagonists. Using asperlicin as a lead compound, devazepide was synthesized (GOETZ et al., 1985; EVANS et al., 1986) and is under development now as an oral agent for the treatment of pancreatitis etc., as mentioned below (see Sect. 3.2.8). Recently, anthramycin (KUBOTA et al., 1989) and virginiamycin M1 (LAM et al., 1991) were found to be CCK-B antagonists. Anthramycin has a benzodiazepin moiety like asperlicin, but it binds to the CCK-B receptor unlike asperlicin.

3.2.2 Endothelin Receptor Antagonists

Endothelin (ET) was discovered in 1988 as a new peptide with potent activity to induce vascular contraction. During the subsequent five years, three isopeptides of ET (ET-1, -2, and -3) were found and there are at least two receptors (ET_A and ET_B) for ET. Studies of the agonists and antagonists of ET have also been carried out. Following the discovery of cyclic peptide antagonists of microbial origin (BE-18257A and B) (IHARA et al., 1991; KO-

Tab. 2. Antagonists of Microbial Origin of Peptide Ligand Receptors

Ligand	Subtype	Antagonist	Producer	Reference
Cholecystokinin	A	Asperlicin A, B, C, D, E[a]	*Aspergillus alliaceus*	CHANG et al. (1985), GOETZ et al. (1985, 1988), LIESCH et al. (1985, 1988)
	B	Anthramycin[a]	*S. spadicogriceus*	KUBOTA et al. (1989)
	B	Tetronothiodin[a]	*Streptomyces* sp.	OHTSUKA et al. (1992, 1993a, b), WATANABE (1993)
Gastrin/cholecystokinin	B	Verginiamycin M_1	*S. olivaceus*	LAM et al. (1991)
		L-156586, L-156587, L-156588, L-156906		
Endothelin	A	BE-18257 A and B	*S. misakiensis*	IHARA et al. (1991), KOJIRI et al. (1991), NAKAJIMA et al. (1991)
	A	WS-7338 C and D	*Streptomyces* sp.	MIYATA et al. (1992a, b, c)
	A	WS-0009 A and B[a]	*Streptomyces* sp.	MIYATA et al. (1992d, e)
	A and B	Cochinmicin I, II, III	*Microbispora* sp.	LAM et al. (1992), ZINK et al. (1992)
	A	Asterric acid[a]	*Aspergillus* sp.	OHASHI et al. (1992)
Substance P	NK-1 and 2	WS-9326 A	*S. violaceusniger*	HAYASHI et al. (1992)
	NK-1	Actinomycin D	*Streptomyces* sp.	FUJII et al. (1991)
	NK-1	Fiscalin A, B, C[a]	*Neosartorya fischeri*	WONG et al. (1993b)
	NK-1	Anthrotainin[a]	*Gliocladium catenulatum*	WONG et al. (1993a)
	NK-1 and 2	WIN 64821[a]	*Aspergillus* sp.	SEDLOCK et al. (1994)
ANP		Anantin	*S. coerulescens*	WEBER et al. (1991), WYSS et al. (1991)
		HS-142-1[a]	*Aureobasidium pullulans*	MORISHITA et al. (1991a, b)
Arginine-vasopressin		Hapalindolinone A and B[a]	*Fischerella* sp.	SCHWARTZ et al. (1987)
Oxytocin		L-156373	*S. silvensis*	PETTIBONE et al. (1989)
C5a		L-156602 = PD124966	*Streptomyces* sp.	HENSENS et al. (1991), HURLEY et al. (1986)

[a] Nonpeptide antagonist.

JIRI et al., 1991; NAKAJIMA et al., 1991) and WS7338C and D (MIYATA et al., 1992a, b, c), non-peptide antagonists such as WS009A and B (MIYATA et al., 1992e, d), cochinmicins (LAM et al., 1992; ZINK et al., 1992) and asterric acid (OHASHI et al., 1992) have been identified. The cyclic peptide BE-18257B exhibited IC_{50} values of 1.4 and 0.8 mM against [^{125}I]ET-1 binding to aortic smooth muscle tissue and to ventricle membranes from pig,

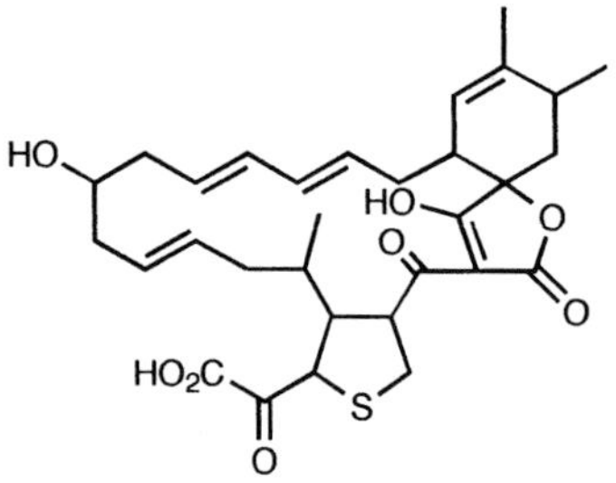

Tetronothiodin

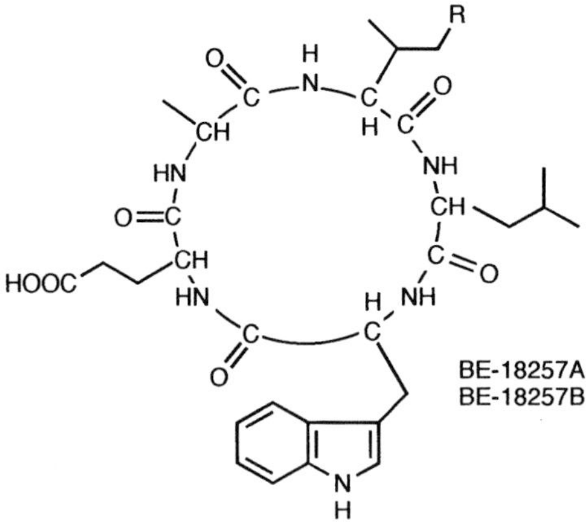

BE-18257A (R: H)
BE-18257B (R: CH_3)

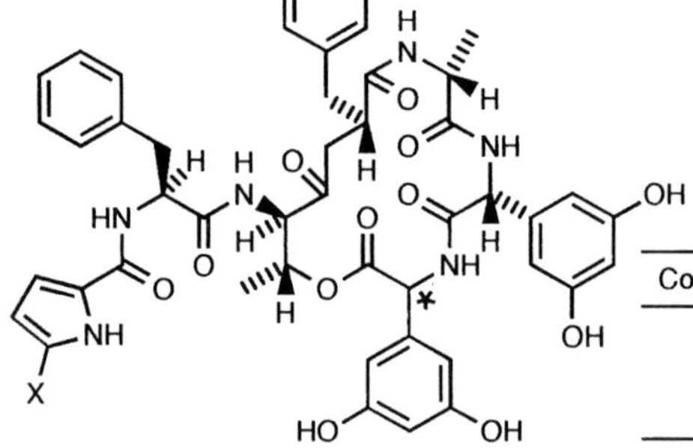

Cochinmicin	X	*
I	H	*R*
II	Cl	*S*
III	Cl	*R*

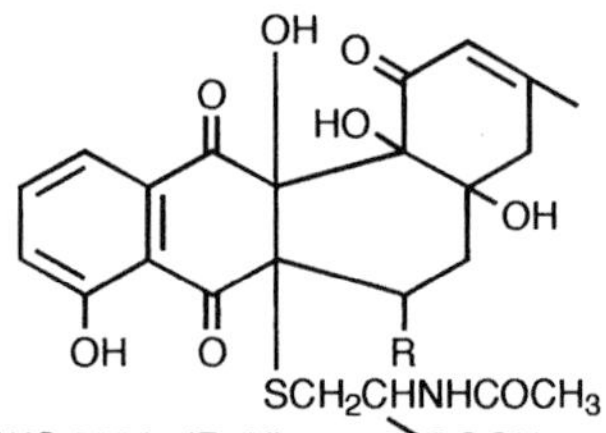

WS 009A (R: H)
WS 009B (R: OH)

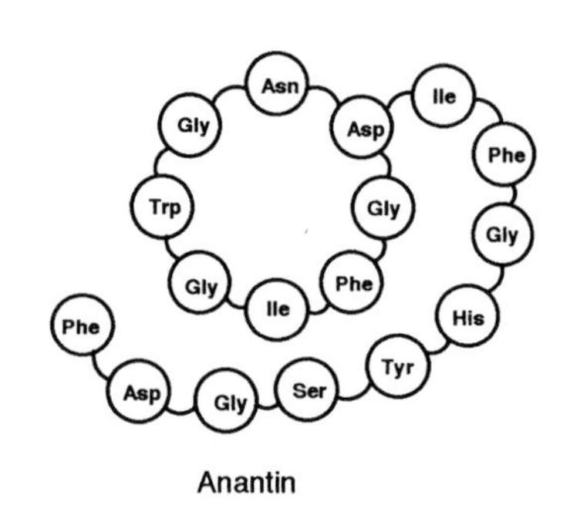

Anantin

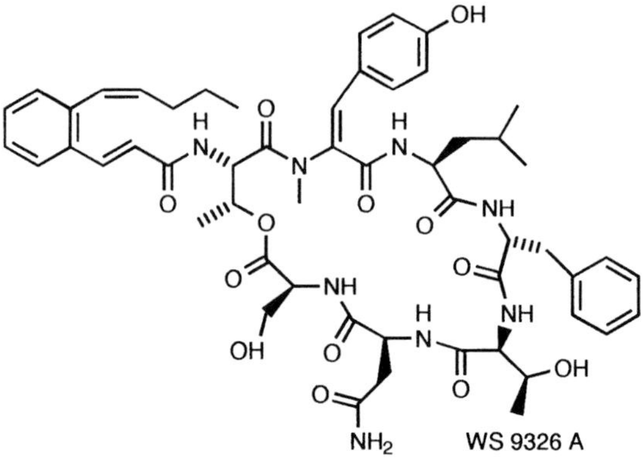

WS 9326 A

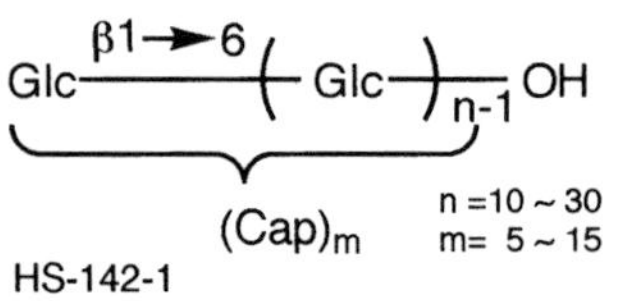

$Glc \xrightarrow{\beta 1 \rightarrow 6} (Glc)_{n-1} OH$ with $(Cap)_m$

n = 10 ~ 30
m = 5 ~ 15

HS-142-1
Glc; D-glucose, Cap; capronic acid

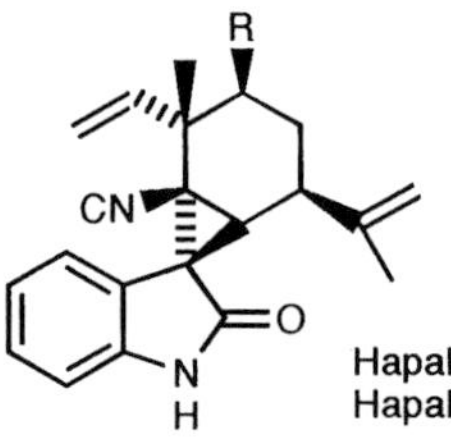

Hapalindolinone A (R: Cl)
Hapalindolinone B (R: H)

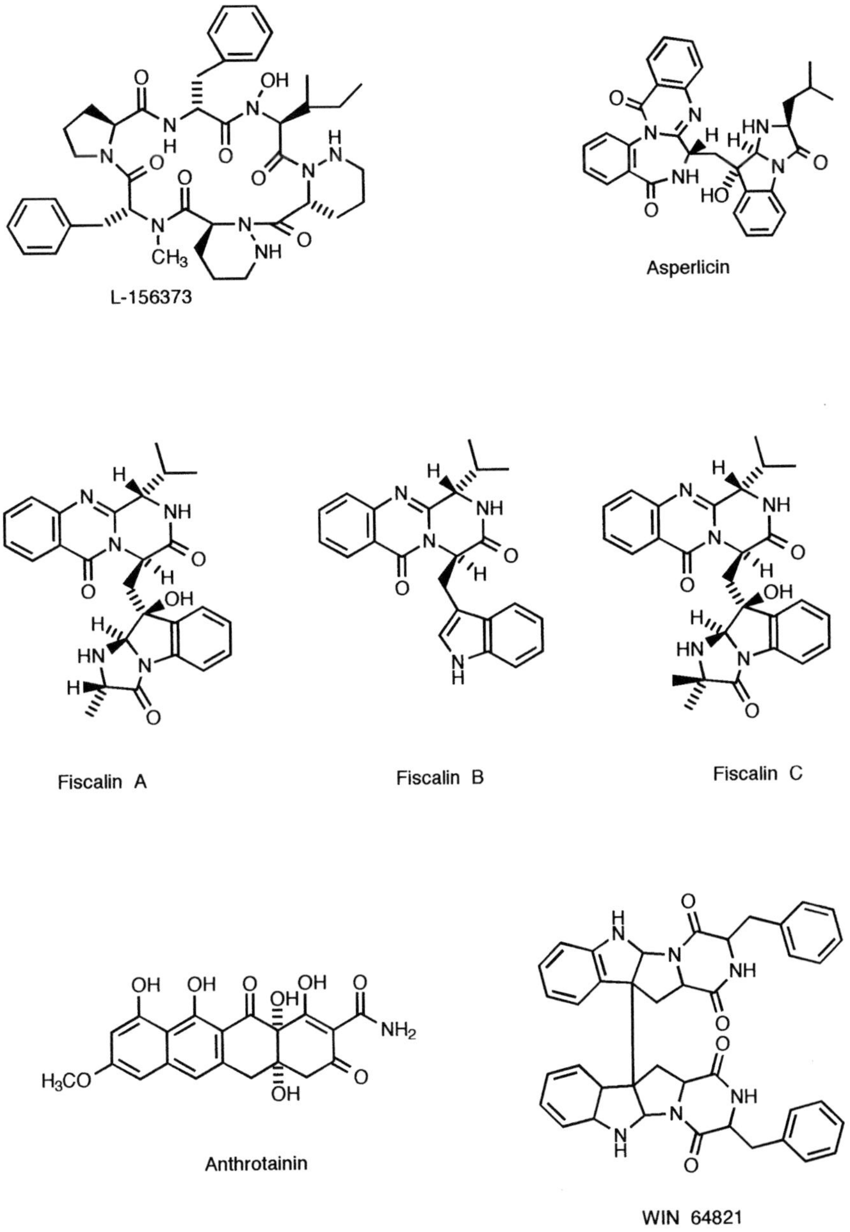

L-156373

Asperlicin

Fiscalin A

Fiscalin B

Fiscalin C

Anthrotainin

WIN 64821

Fig. 5. Structures of antagonists of peptide ligand receptors of microbial origin.

which are ET_A-rich tissues. It did not inhibit [^{125}I]ET-1 binding to ET_B-rich tissues. In isolated rabbit iliac arteries, BE-18257B antagonized ET-1-induced vasoconstriction. Thus, it was found that BE-18257B is an ET_A receptor antagonist (IHARA et al., 1991). On the other hand, the peptolide cochinmicin 1 is a nonselective antagonist for ET_A and ET_B sites (LAM et al., 1992). Studies using these antagonists are expected to clarify the physiological and pathophysiological significance of ET isopeptides and their receptors and to provide new therapeutic agents.

3.2.3 Substance P Receptor Antagonists

Neuropeptides such as substance P and neurokinin A, which markedly induce airway constriction and promote mucosal secretion, have been investigated because of their relationship to respiratory diseases such as asthma. HAYASHI et al. (1992) searched for inhibitors of [^{3}H] substance P binding to guinea pig lung membrane fractions and discovered WS9326A, a cyclic depsipeptide produced by an actinomycete. WS9326A exhibits an IC_{50} value of 3.6×10^{-6} M in the above assay and acts as a tachykinin antagonist in various functional assays. Its tetrahydro derivative, FK224, was more potent than WS9326A ($IC_{50} = 1.0 \times 10^{-7}$ M) and antagonizes both the neurokinin 1 receptor (involved in airway edema) and the neurokinin 2 receptor (involved in airway constriction) (HASHIMOTO et al., 1992). This compound is under development for clinical use. In addition, non-peptide inhibitors, termed fiscalins, with moderate neurokinin 1 binding activity have also been reported (WONG et al., 1993b). There has also been a report of a tetracyclic compound, anthrotainin, with neurokinin 1 activity (WONG et al., 1993a), but this compound was found to be a noncompetitive substance P antagonist.

More recently, WIN64821, a non-peptide secondary metabolite produced by *Aspergillus* sp. was found to inhibit radiolabeled substance P binding in a variety of tissues with K_i values ranging from 0.24 μM in human astrocytoma U-373 MG cells to 7.89 μM in submaxillary membranes. Additionally, WIN64821 was found to inhibit [^{125}I]-neurokinin A binding to the neurokinin 2 receptor in human tissue at a concentration equivalent to its neurokinin 1 activity (0.26 μM). WIN64821 was shown to be a functional antagonist of neurokinin 1 and neurokinin 2 receptors (OLEYNEK et al., 1994).

3.2.4 ANP Receptor Antagonists

Atrial natriuretic peptide (ANP) is a peptide hormone involved in the regulation of body water, electrolytes and blood pressure. It has potent diuretic and vasoconstrictive actions. ANP and two other peptides with different amino acid sequences (RNP and CNP) constitute a natriuretic peptide family. Two types of ANP receptors are known, one carrying a guanylate cyclase domain inside the cells and another without any guanylate cyclase domain. The receptors with a guanylate cyclase domain may be involved in the ANP-induced elevation of the intracellular cGMP level, while the receptor without this domain is thought to be involved in the ANP metabolism. However, the exact physiological roles of ANP and its pathophysiological significance have not yet been fully clarified. For this reason, the development of ANP receptor antagonists is needed.

WEBER et al. (1991) examined microbial metabolites for those substances which inhibit the binding of [^{125}I]-labeled rat ANP, i.e., [^{125}I]rANP to the bovine adrenocortical membrane. They discovered the peptide antagonist anantin produced by an actinomycete. This is a cyclic peptide composed of 17 amino acids (WYSS et al., 1991). The compound was suggested to be an ANP receptor antagonist because it inhibited both the binding of [^{125}I]rANP to receptors ($IC_{50} = 1.0$ μM) and the ANP-induced intracellular cGMP accumulation in bovine aortic smooth muscle cells, and it did not show any agonist effect (WEBER et al., 1991).

MORISHITA et al. (1991a, b) examined microbial cultures for substances inhibiting the binding of [^{125}I]rANP to the rabbit renal cortical membrane. They isolated HS-142-1 from the culture broth of a fungus belonging to the genus *Aureobasidium.* HS-142-1 is a new polysaccharide composed of a linear β-1,6-glucose chain conjugated to caproic acid. In the target tissues and cells of ANP, i.e., in the bovine adrenocortical membrane fraction (MORISHITA et al., 1992; ODA et al., 1992), bovine vascular smooth muscle cells (IMURA et al., 1992), rat renal glomerulus (SANO et al., 1992a, b, c), LLC-Pkl cells (TANAKA et al., 1992), and PC12 cells (TOKI et al., 1992c), this compound specifically antagonized the binding of [^{125}I]rANP to ANP receptors containing guanylate cyclase (IC_{50} for rabbit renal cortical membranes = 0.3 μg/mL), and inhibited the ANP-induced elevation in cGMP level. Although anesthetized rats treated intravenously with HS-142-1 alone showed no reaction, the diuretic response of the animals to exogenous or endogenous ANP was not seen after pretreatment with HS-142-1 (SANO et al., 1992b, c). HS-142-1 thus seems to be a new non-peptide ANP antagonists useful for the analysis of the physiological and pathophysiological role of ANP. In the future, this compound often will be used in cardiovascular studies.

3.2.5 Arginine–Vasopressin Receptor Antagonists

Arginine–vasopressin (AVP) is a peptide composed of 9 amino acids. It is a hormone secreted from the posterior lobe of the pituitary gland and possesses antidiuretic and hypertensive properties. The renal AVP-V_2 receptor is involved in the antidiuretic, the AVP-A_1 receptor of the cardiac smooth muscle in the hypertensive action.

SCHWARTZ et al. (1987) isolated substances from a microorganism of the genus *Fischerella* that inhibited the binding of [^{3}H]AVP to the renal tissue containing V_2 receptors. They called the substances hapalindolinone A and B. These compounds are non-peptide antagonists which carry cyclopropane and indolinone skeletons. They inhibit not only the binding of [^{3}H]AVP to renal tissue (IC_{50} = 37.5 μM) but also the AVP-induced activation of adenylate cyclase (IC_{50} = 44.6 μM).

3.2.6 Oxytocin Receptor Antagonists

Oxytocin is a peptide composed of 9 amino acids. It is a hormone secreted from the posterior lobe of the pituitary gland and induces uterine contraction and milk secretion.

PETTIBONE et al. (1989) isolated a substance from an actinomycete (L-156373) which inhibited the binding of [^{3}H]oxytocin to the rat uterine membrane fraction. The K_i value of this compound was 150 μM. Its affinity for oxytocin receptors was more than 20 times higher than that for the arginine–vasopressin receptors (AVP-V_1 and AVP-V_2). Its derivative L-365209, produced by dehydroxylation of the *N*-hydroxyleucine unit and oxidation of the piperazic acid residues of the L-156373, was 20 times as potent as L-156373 and had a K_i of 7.3 μM. This derivative was highly selective and antagonized the oxytocin action to the rat uterus (ID_{50}=460 μg/kg).

3.2.7 Complement C5a Receptor Antagonists

C5a is thought to be involved in the aggravation of various inflammatory allergic diseases. HENSENS et al. (1991) isolated a substance from actinomycete metabolites (L-156602) that inhibited the binding of human polymorphonuclear leukocytes. This compound was considered to be identical to PD-124966 which had been discovered as an antitumor antibiotic. The structure of this compound shown in Fig. 5, was proposed by HENSENS et al. (1991) and has never been determined.

3.2.8 Devazepide – A Non-Peptide Peptide Ligand Antagonist under Development for Medical Use

In 1985, CHANG et al. (1985) isolated the CCK receptor antagonist asperlicin from a microorganism (*Aspergillus alliaceus*), which is the first non-peptide antagonist of peptide ligand receptors. Asperlicin does not act on the CCK-B receptors primarily located in the center of the body, but selectively acts on the CCK-A receptors mainly located in the periphery. Its discovery confirmed the presence of CCK receptor subtypes and made it possible to distinguish receptor A from receptor B. CCK is known to be involved in diseases such as pancreatitis. Asperlicin was initially ex-

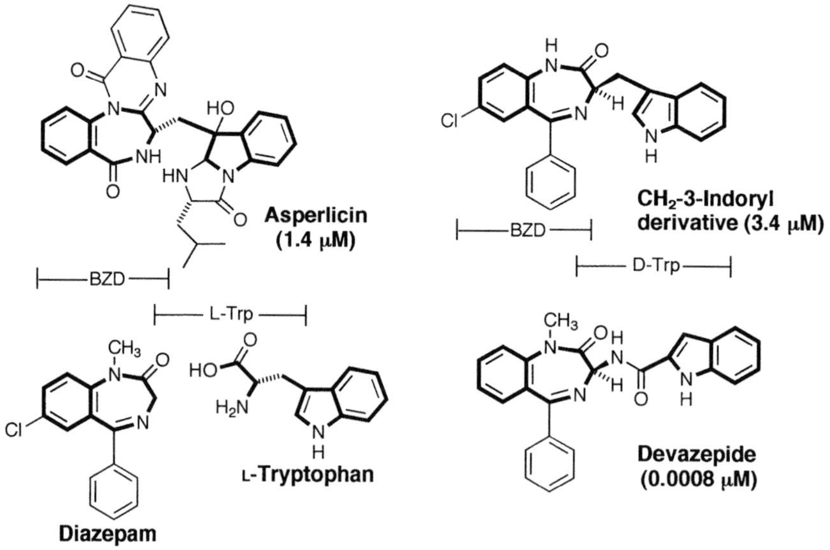

Fig. 6. Structures and activities of the CCK-A antagonist asperlicin and its analog derazepide.

pected to be useful as a therapeutic agent, but, because of low solubility, it is ineffective when administered orally.

In 1983, KUBOTA et al. (1983, 1985) found in an experiment with peripheral tissue and brain that diazepam, an anti-anxiety drug which can be administered orally, antagonized CCK. In *in vitro* binding experiments, however, it did not antagonize CCK (GOETZ et al., 1985). Since asperlicin has a benzodiazepine skeleton like diazepam, efforts have been started to synthesize highly soluble, orally applicable derivatives. EVANS et al. (1986) synthesized a number of derivatives and analogs containing benzodiazepine and indole because the asperlicin molecule contains benzodiazepine and L-tryptophan. Devazepide which contains D-tryptophan, carrying a 2-indolyl bond to benzodiazepine as shown in Fig. 6, is more than 1000 times more potent than asperlicin. It is highly soluble in water while retaining selective activity. Thus devazepide was selected as an excellent candidate for development (EVANS et al., 1986). At present, devazepide is under development for treatment of acute pancreatitis, biliary colic, abdominal pain, and anorexia (EVANS, 1989).

4 Receptor Agonists

4.1 Motilides (Macrolides with Motilin Activity)

ITOH et al. (1985) found that the side effect of erythromycin causing diarrhea etc., is similar to the effect of motilin which is a hormone promoting gastrointestinal motility. Subsequently, ŌMURA and coworkers (ŌMURA et al., 1987; TSUZUKI et al., 1989; SUNAZUKA et al., 1989) synthesized a number of erythromycin derivatives which exert only motilin activity and no antibacterial activity. EM536, one of these derivatives, showed 2890 times higher motilin activity than erythromycin A. Quaternary ammonium derivatives of the desosamine moiety of 6,9-hemiketal erythromycin A possess the most potent gastrointestinal motor stimulating activity. However, these ionized derivatives have low permeability. Consequently, EM523 (18 times as potent as erythromycin A) and EM574 (248 times as potent) were selected from the tertiary ammonium derivatives as candidates for medical use (Fig. 7, Tab. 3). At present, these two derivatives (EM523 and EM574) are under de-

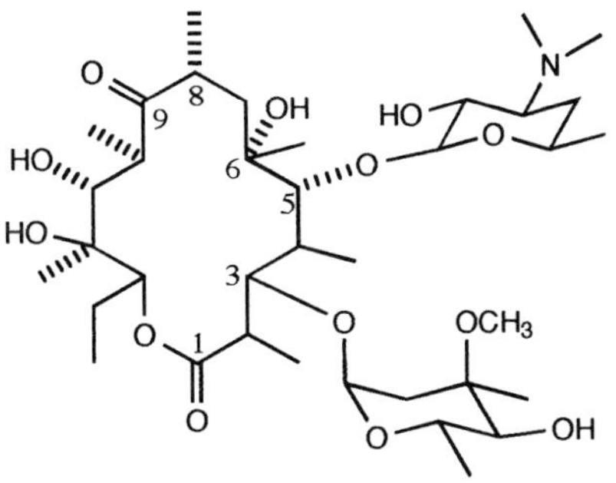

Erythromycin A (EMA)

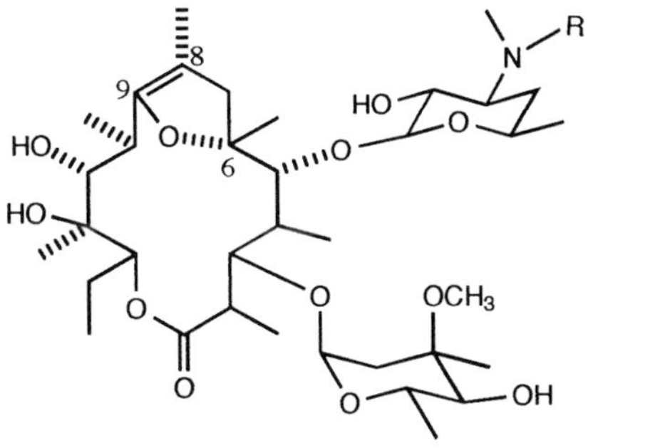

EM201 R = CH_3
EM523 R = CH_2CH_3
EM574 R = $CH(CH_3)_2$

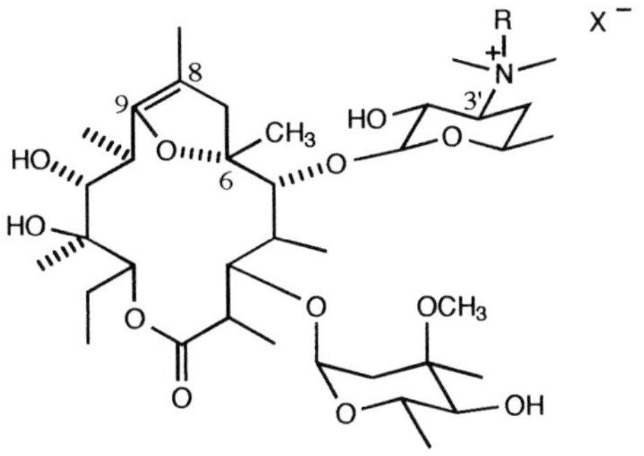

EM485 R = CH_3
EM491 R = CH_2CH_3
EM511 R = $CH_2CH{=}CH_2$
EM536 R = $CH_2C{\equiv}CH$

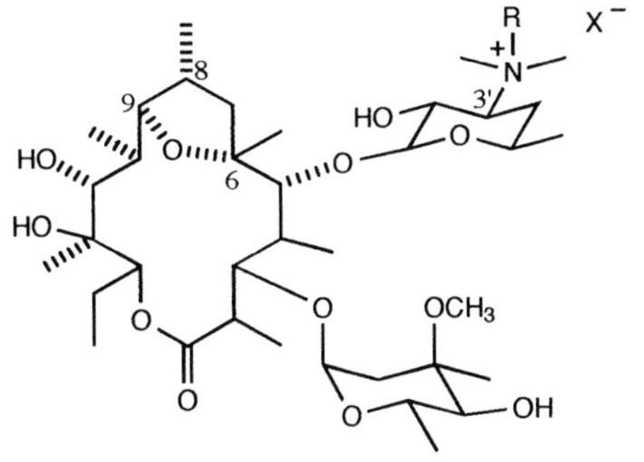

EM502 R = CH_3
EM506 R = $CH_2CH{=}CH_2$
EM507 R = $CH_2C{\equiv}CH$

Fig. 7. Structures of erythromycin A and its derivatives.

velopment to be used as gastrointestinal motility regulators when administered either by injection or orally. The macrolides exerting motilin activity were called "motilides" by TSUZUKI et al. (1989). KONDO et al. (1988) demonstrated that motilides are agonists of the motilin receptor. Because motilin is a peptide hormone composed of 22 amino acids, motilin itself is ineffective when administered orally while the abovementioned motilide EM574 (a non-peptide agonist) is effective. At present, morphine, an enkephalin agonist, is the only non-peptide agonist of peptide ligand receptors in clinical use, motilide EM574 is expected to be used as the second non-peptide agonist.

4.2 Other Agonists

The traditional receptor agonists muscarine and zearalenone (see Fig. 3) are well known to be isolated from microorganisms. Muscarine is an alkaloid from the red variety of *Amanita muscaria,* a poisonous mushroom (KUEHL et al., 1955; KÖGL et al., 1957). It is an acetylcholine receptor agonist and used as a important biochemical reagent.

Zearalenone is an estrogen receptor agonist isolated from the mycelia of the fungus *Gibberella zeae* (*Fusarium graminearum*) (STOB et al., 1962; URRY et al., 1966).

No agonists of peptide ligand receptors of microbial origin, except erythromycin A (see

Tab. 3. Antimicrobial Activities and Gastrointestinal Motor Stimulating Activities of Erythromycin A and its Derivatives

Compound	Antimicrobial Activity (MIC, μg/mL)					Gastrointestinal Motor Stimulating Activity (relative activity)
	SA	BS	BC	EC	KP	
EMA	0.2	0.1	0.1	12.5	6.25	1
EM201	50	25	25	>100	>100	10
EM523	>100	>100	>100	>100	>100	18
EM574	>100	>100	>100	>100	>100	248
EM485	>100	>100	>100	>100	>100	21
EM491	>100	>100	>100	>100	>100	111
EM511	100	>100	>100	>100	>100	256
EM536	100	100	100	>100	>100	2890
EM502	>100	>100	>100	>100	>100	65
EM506	100	100	100	>100	>100	115
EM507	>100	>100	>100	>100	>100	202

SA *Staphylococcus aureus* ATCC6358P; BS *Bacillus subtilis* ATCC6633; BC *Bacillus cereus* IFO3001; EC *Escherichia coli* NIHJ; KP *Klebsiella pneumoniae* ATCC10031.

Sect. 4.1), have been reported to date. Their discovery is desired for the development of new orally available drugs to replace peptide hormones and cytokines – in the same way as motilides are being developed as orally available gastrointestinal motor stimulating drugs.

5 Inhibitors of Virus Receptor Binding – gp120–CD4 Binding Inhibitors

The entry of viruses needs their specific binding to a receptor of the susceptible cell. Human immunodeficiency virus (HIV) entry begins with the highly specific binding of the HIV gp120 envelope glycoprotein with a CD4 molecule on the surface of most susceptible cells (McDougal et al., 1986; Sadroski et al., 1986; Lifson et al., 1986). Blocking of HIV entry is one of the most important targets for HIV therapy (Johnston and Hoth, 1993).

In the screening program for new inhibitors of gp120–CD4 binding from microorganisms, Ōmura et al. (1993) discovered the novel inhibitors isochromophilone I and II (Fig. 8) from the culture borth of *Penicillium* sp. FO-2338, and chloropeptin I and II (Fig. 9) from *Streptomyces* sp. WK-3419 (Ōmura et al., unpublished data). Chloropeptin II, however, was identified with complestatin (Kaneko et al., 1989).

The inhibitory activities against gp120–CD4 binding were determined by enzyme-linked immunosorbent assay (ELISA) using recombinant soluble CD4 and recombinant gp120 as described by Gilbert et al. (1991). Isochromophilone I and II inhibited gp120–CD4 binding with IC$_{50}$ values of 6.6 μM and 3.9 μM, respectively. The IC$_{50}$ values for chloropeptin I and II were 2.0 μM and 3.3 μM, respectively.

Anti-HIV activity was assayed as follows. Peripheral human lymphocytes were isolated by density gradient centrifugation. After stimulation by a mitogen, the cells were infected with a standardized preparation of HIV-1. Subsequently, the infected cells were cultured in the presence of the agent for 4 days. The amount of viral core protein p24 synthesized and released by the infected cells was determined by the capture-ELISA technique on

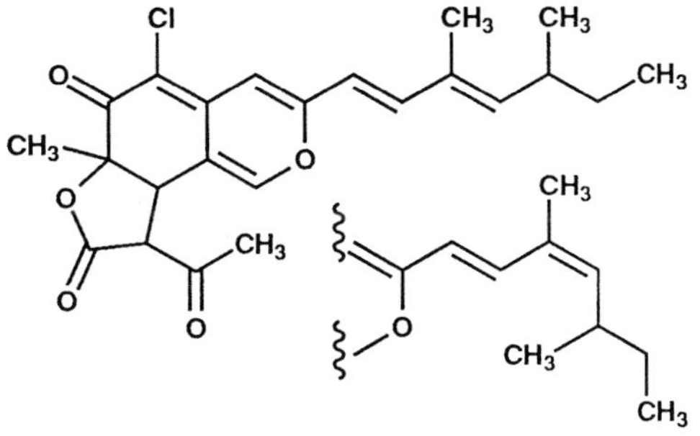

Cl CH3 CH3 CH3 Ia
CH3 O O O CH3 O O CH3 Ib CH3 CH3

Isochromophilone I

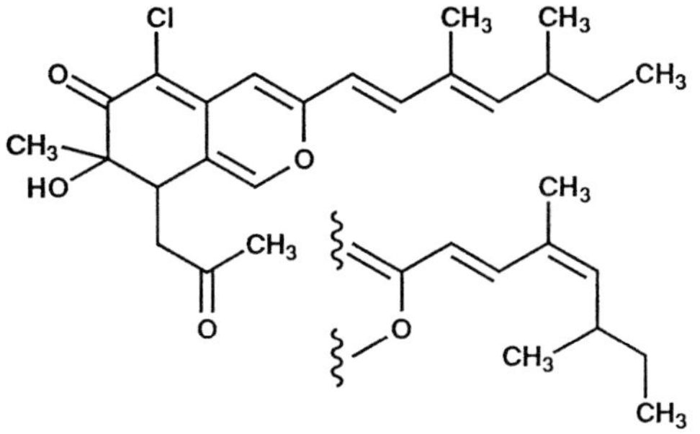

Cl CH3 CH3 CH3 IIa
CH3 HO O CH3 O CH3 IIb CH3 CH3

Isochromophilone II

Fig. 8. Structures of the gp120–CD4 binding inhibitors isochromophilone I and II.

Tab. 4. Inhibition of HIV Replication on the Viral Core Protein Level (for the assay method, see text)

Sample	Viral Core Protein p24 Synthesized (ng/mL)		
	Day 2	Day 3	Day 4
None	0	97.3	129.6
Isochromophilone II	0	0	13.5
Chloropeptin I	0	0	7.3

days 2, 3, and 4. By comparing with a standard preparation, the amount of protein (p24) produced by the virus infected cells was calculated. As shown in Tab. 4 isochromophilone II and chloropeptin I significantly inhibited HIV replication at 25 μM and 7.5 μM, respectively. The inhibition of HIV replication by isochromophilone II and chloropeptin I is considered to be due to blocking of HIV entry into the cells. Isochromophilone I and II are the first novel non-peptide compounds to inhibit gp120–CD4 binding. Isochromophilones and chloropeptins are expected to provide the lead compounds for development of HIV therapy.

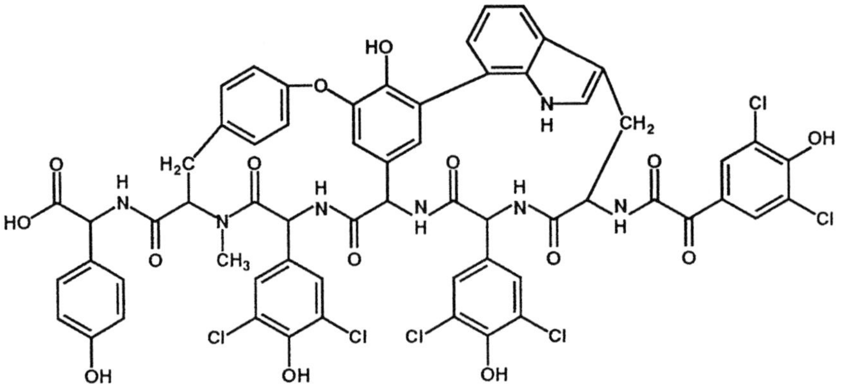

Chloropeptin I

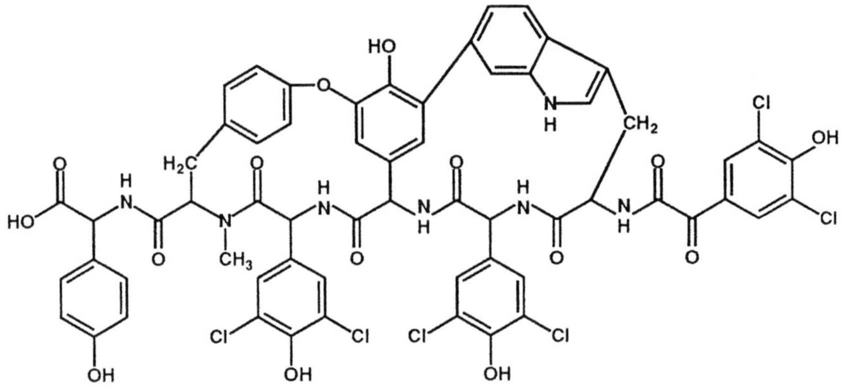

Chloropeptin II (complestatin)

Fig. 9. Structures of the gp120–CD4 binding inhibitors chloropeptin I and II.

6 Current State and Future Perspectives

As described above, substances acting on receptors (i.e., agonists and antagonists) have been synthesized before the exact nature of receptors was clarified. These substances have contributed greatly not only to treatment of diseases but also to advances in studies in the field of cellular biology and pharmacology.

Following recent commercialization of radio-labeled peptide ligands, screening of receptor-active compounds in various specimens such as microbial cultures has been performed in experiments involving the binding of these ligands to tissue or cells. In this way, new substances affecting receptors of peptide ligands have been discovered. At present, derivatives of these substances (i.e., devazepide, FK-224, and motilide) are under development for clinical use. This class of drugs will further increase in the future.

It is speculated that orally administered non-peptide agonists and antagonists will bring about an epochal reform of drug therapy. To date, however, no low molecular weight compound acting on the receptors of macromolecular peptide ligands (e.g., ligands with 100 or more amino acid residues) has been reported although many natural peptide ligands or their analogs have been clinically used by injection.

If the three-dimensional structure of ligands and their receptors is identified and the mode of the ligand–receptor binding is clarified in detail, the development of new drugs by computerized information processing will be possible. Although the primary structure of many receptors has been clarified to date, the three-dimensional structure of a receptor is not known until the crystalline structure of the growth hormone receptor complex (see below) has been determined.

The first analysis of the crystalline structure of a macromolecular peptide ligand receptor complex was reported in 1992 (DEVOS et al., 1992). The analysis of the three-dimensional structure of the complex formed between human growth hormone and the extracellular domain of its receptor revealed that this complex is composed of one hormone and two receptor molecules. The hormone has four helical structures with abnormal topology, and the receptor bound to it has two different binding domains. The two receptor molecules bind through the same amino acid residues in each domain to two structurally distinct sites of the hormone. At their C-terminal domains distant from the binding sites the two receptor molecules are in contact to each other. This contact may play a crucial role in intracellular signal transduction.

Since the three-dimensional features of the mode of binding between growth hormone and its receptor has been clarified, molecular designing of new agonists and antagonists using computer graphics technology will advance in the future. However, because crystallization of the ligand–receptor complex usually is not easy, search for new receptor-active compounds and subsequent chemical modification of the thus discovered compounds will, for the time being, continue to play a principal role in the development of new receptor-active compounds.

7 Concluding Remarks

This chapter provides a general review of receptor-active compounds. Studies of substances acting on receptors of peptide ligands still have only a very short history. We expect more simple assay techniques to be developed in the future, facilitating the discovery of many receptor-active compounds. These compounds will help to clarify the function of cells and elucidate the physiological and pathophysiological functions of receptors.

As studies on receptor-active compounds of microbial origin have been advancing, some known substances have been highlighted because of their additional action on receptors. Considerably small differences in the chemical structures of ligands or receptors often reflect quite different actions and it is quite likely that the same compound can have two or more target molecules. This indicates that it is not easy to discover a highly selective compound and that it is even possible to modify a compound so that its minor action replaces the major action – as seen in the case of motilide.

Acknowledgements

The authors are indebted to Dr. H. KLEINKAUF for providing the opportunity of this presentation, and also to Drs. S. TAKAMATSU and J. INOKOSHI and Mr. K. MATSUZAKI for their useful help in preparation of this manuscript.

8 References

ABBOTT, A. (1991), Receptor screens detect bioactivity, *Biotechnology* **9,** 694.

CHANG, R. S. L., LOTTI, V. J., MONAGHAM, R. L., BIRNBAUM, J., STAPLAY, E. O., GOETZ, M. A., ALBERS-SCHÖNBERG, G., PATCHETT, A. A., LIESCH, J. M., HENSENS, O. D., SPRINGER, J. P. (1985), A potent nonpeptide cholecystokinin antagonist selective for peripheral tissues isolated from *Aspergillus alliaceus, Science* **230,** 177–179.

DEVOS, A. M., ULTSCH, M., KOSSIAKOFF, A. A. (1992), Human growth hormone and extracellular domain of its receptor: crystal structure of the complex, *Science* **255,** 306–312.

EVANS, B. E. (1989), Recent developments in cholecystokinin antagonist research, Drugs of the Future, **14,** 971–979.

EVANS, B. E., BOCK, M. G., RITTLE, K. E., DIPARDO, R. M., WHITTER, W. L., VEBER, D. F., ANDERSON, P. S., FREIDINGER, R. M. (1986), Design of potent, orally effective, nonpeptidal antagonists of the peptide hormone cholecystokinin, *Proc. Natl. Acad. Sci. USA* **83,** 4918–4922.

FUJII, T., MURAI, M., MORIMOTO, H., NISHIKAWA, M., KIYOTO, S. (1991), Effects of actinomycin D on airway constriction induced by tachykinin and capsaicin in guinea-pigs, *Eur. J. Pharmacol.* **194,** 183–188.

GILBERT, M., KIRIHARA, J., MILLS, J. (1991), Enzyme-linked immunoassay for human immunodeficiency virus type I envelope glycoprotein 120, *J. Clin. Microbiol.* **29,** 142–147.

GOETZ, M. A., LOPEZ, M., MONAGHAN, R. L., CHANG, R. S. L., LOTTI, V. J., CHEN, T. B. (1985), Asperlicin, a novel non-peptidal cholecystokinin antagonist from *Aspergillus alliaceus.* Fermentation, isolation and biological properties, *J. Antibiot.* **38,** 1633–1637.

GOETZ, M. A., MONAGHAN, R. L., CHANG, R. S. L., ONDEYKA, J., CHEN, T. B., LOTTI, V. J. (1988), Novel cholecystokinin antagonists from *Aspergillus alliaceus.* I. Fermentation, isolation, and biological properties. *J. Antibiot.* **41,** 875–877.

HASHIMOTO, M., HAYASHI, K., MURAI, M., FUJII, T., NISHIKAWA, M., KIYOTO, S., OKUHARA, M., KOHSAKA, M., IMANAKA H. (1992), WS9326A, a novel tachykinin antagonist isolated from *Streptomyces violaceusniger* No. 9326. II. Biological characterization and pharmacological characterization of WS9326A and tetrahydro-WS9326A (FK224), *J. Antibiot.* **45,** 1064–1070.

HAYASHI, K., HASHIMOTO, M., SHIGEMATSU, N., NISHIKAWA, M., EZAKI, M., YAMASHITA, M., KIYOTO, S., OKUHARA, M., KOHSAKA, M., IMANAKA, H. (1992), WS9326A, a novel tachykinin antagonist isolated from *Streptomyces violaceusniger* No. 9326. I. Taxonomy, fermentation, isolation, physico-chemical properties and biological activities, *J. Antibiot.* **45,** 1055–1062.

HEDGE, V. R., PATEL, M. G., HORAN, A. C., SCHWARTZ, J. L., HART, R., PUOR, M. S., GULLO, V. P., IYENGAR, S. (1991), Sch42029, a naturally produced dopamine rceptor ligand: taxonomy, fermentation, isolation and structure, *J. Ind. Microbiol.* **8,** 187–192.

HENSENS, O. D., BORRIS, R. P., KOUPAL, L. R., CALDWELL, C. G., CURRIE, S. A., HAIDRI, A. A., HOMIK, C. F., HONEYCUTT, S. S., LINDENMAYER, S. M., SCHWARTZ, C. D., WEISSBERGER, B. A., WOODRUFF, H. B., ZINK, D. L., ZITANO, L., FIELDHOUSE, J. M., ROLLINS, T., SPRINGER, M. S., SPRINGER, J. P. (1991), L-156,602, A C5a antagonist with a novel cyclic hexadepsipeptide structure from *Streptomyces* sp. MA6348. Fermentation, isolation and structure determination, *J. Antibiot.* **44,** 249–254.

HORI, Y., ABE, Y., EZAKI, M., GOTO, T., OKUHARA, M., KOHSAKA, M. (1993a), R1128 substances, novel non-steroidal estrogen-receptor antagonists produced by a *Streptomyces.* I. Taxonomy, fermentation, isolation and biological properties, *J. Antibiot.* **46,** 1055–1062.

HORI, Y., TAKASE, S., SHIGEMATSU, N., GOTO, T., OKUHARA, M., KOHSAKA, M. (1993b), R1128 substances, novel non-steroidal estrogen-receptor antagonists produced by a *Streptomyces.* II. Physico-chemical properties and structure determination. *J. Antibiot.* **46,** 1063–1068.

HORI, Y., ABE, Y., NISHIMURA, M., GOTO, T., OKUHARA, M., KOHSAKA, M. (1993c), R1128 substances, novel non-steroidal estrogen-receptor antagonists produced by a *Streptomyces.* III.

Pharmacological properties and antitumor activities, *J. Antibiot.* **46,** 1069–1075.

HORI, Y., ABE, Y., NAKAJIMA, H., TAKASE, S., FUJITA, T., GOTO, T., OKUHARA, M., KOHSAKA, M. (1993d), WB2838 [3-chloro-4-(2-amino-3-chlorophenyl)-pyrrole]: Non-steriodal androgen-receptor antagonist produced by a *Pseudomonas, J. Antibiot.* **46,** 1327–1333.

HORI, Y., ABE, Y., SHIGEMATSU, N. S., GOTO, T. OKUHARA, M., KOHSAKA, M. (1993e), Napyradiomycins A and B1: Non-steroidal androgen-receptor antagonists produced by a *Streptomyces, J. Antibiot.* **46,** 1890–1894.

HORI, Y., ABE, Y., NAKAJIMA, H., SHIGEMATSU, N., TAKASE, S., GOTO, T., OKUHARA, M., KOHSAKA, M. (1993f), WS9761A and B: New non-steroidal androgen-receptor antagonists produced by a *Streptomyces, J. Antibiot.* **46,** 1901–1903.

HURLEY, T. R., BUNGE, R. H., WILLMER, N. E., HOKANSON, G. C., FRENCH, J. C. (1986), PD124,895 and PD124,966, two new antitumor antibiotics, *J. Antibiot.* **39,** 1651–1656.

IHARA, M., FUKURODA, T., SAEKI, T., NISHIKIBE, M., KOJIRI, K., SUDA, H., YANO, M. (1991), An endothelin receptor (ETA) antagonist isolated from *Streptomyces misakiensis, Biochem. Biophys. Res. Commun.* **178,** 132–137.

IMURA, R., SANO, T., GOTO, J., YAMADA, K., MATSUDA, Y. (1992), Inhibition by HS-142-1, a novel nonpeptide atrial natriuretic peptide antagonist of microbial origin, of atrial natriuretic peptide-induced relaxation of isolated rabbit aorta through the blockade of guanylyl cyclase-linked receptors, *Mol. Pharmacol.* **42,** 982–990.

ITOH, Z., SUZUKI, T., NAKAYA, M., INOUE, M., ARAI, H., WAKABAYASHI, K. (1985), Structure–activity relation among macrolide antibiotics in intitiation of interdigestive migrating contractions in the canine gastrointestinal tract, *Am. J. Physiol.* **248,** G320–525.

JOHNSTON, M. I., HOTH, G. F. (1993), Present status and future prospects for HIV therapies, *Science* **260,** 1286–1293.

KAMIYAMA, T., UMINO, T., FUJISAKI, N., FUJIMORI, K., SATOH, T., YAMASHITA, Y., OHSHIMA, S., WATANABE, J., YOKOSE, K. (1993a), Tetrafibricin, a novel fibrinogen receptor antagonist. I. Taxonomy, fermentation, isolation, characterization and biological activities, *J. Antibiot.* **46,** 1039–1046.

KAMIYAMA, T., ITEZONO, Y., UMINO, T., SATOH, T., NAKAYAMA, N., YOKOSE, K. (1993b), Tetrafibricin, a novel fibrinogen receptor antagonist. II. Structural elucidation, *J. Antibiot.* **46,** 1047–1054.

KANEKO, I., KAMOSHITA, K., TAKAHASHI, S. (1989), Complestatin, a potent anti-complement substance produced by *Streptomyces lavendulae.* I. Fermentation, isolation and biological characterization, *J. Antibiot.* **42,** 236–241.

KING, K., DOHLMAN, H. G., THORNER, J., CARON, M. G., LEFKOWITZ, R. J. (1990), Control of yeast mating signal transduction by a mammalian β_2-adrenergic receptor and Gs α subunit, *Science* **250,** 121–123.

KÖGL, F., COX, H. C., SALEMINK, C. A. (1957), Synthese eines Gemischs von Muscarin und seinen stereoisomeren Formen, *Experienta* **13,** 137–138.

KOJIRI, K., IHARA, M., NAKAJIMA, S., KAWAMURA, K., FUNAISHI, K., YANO, M., SUDA, H. (1991), Endothelial-binding inhibitors, BE-18257A and BE-18257B. 1. Taxonomy, fermentation, isolation and characterization, *J. Antibiot.* **44,** 1342–1347.

KONDO, Y., TORII, K., ŌMURA, S., ITOH, Z. (1988), Erythromycin and its derivatives with motilin-like biological activities inhibit the specific binding of ^{125}I-motilin to duodenal muscle, *Biochem. Biophys. Res. Commun.* **150,** 877–882.

KUBOTA, K., SUNAGANE, N., SUGAYA, K., MATSUOKA, Y., URUNO, T. (1983), Competitive antagonists of cholecystokinin and some benzodiazepines at cholecystokinin receptors of smooth muscle, *Jpn. J. Pharmacol.* **33,** 87.

KUBOTA, K., SUGAYA, K., SUNGANE, N., MATSUDA, I., URANO, T. (1985), Cholecystokinin antagonism by benzodiazepines in the contractile response of the isolated guinea-pig gallbladder, *Eur. J. Pharmacol.* **110,** 225–231.

KUBOTA, K., SUGAYA, K., KOIZUMI, Y., TODA, M. (1989), Cholecystokinin antagonism by anthramycin, a benzodiazepine antibiotic, in the central nervous system in mice, *Brain Res.* **485,** 62–66.

KUEHL, F. A., JR., LEBEL, N., RICHTER, J. W. (1955), Isolation and characterization studies on muscarine, *J. Am. Chem. Soc.* **77,** 6663–6665.

LAM, Y. K. T., BOGEN, D., CHANG, R. S., FAUST, K. A., HENSENS, O. D., ZINK, D. L., SCHWARTZ, C. D., ZITANO, L., GARRITY, G. M., GAGLIARDI, M. M., CURRIE, S. A., WOODRUFF, H. B. (1991), Novel and potent gastrin and brain cholecystokinin antagonists from *Streptomyces olivaceus.* Taxonomy, fermentation, isolation, chemical conversion, and physico-chemical and biological properties, *J. Antibiot.* **44,** 613–625.

LAM, Y. K. T., WILLIAMS, D. L. JR., SIGMUND, J. M., SANCHEZ, M., GENILLOUD, O., KONG, Y. L., STEVENS-MILES, S., HUANG, L., GARRITY, G. M. (1992), Cochinmicins, novel and potent cyclodepsipeptide endothelin antagonists from a

Microbispora sp. I. Production, isolation and characterization. *J. Antibiot.* **45,** 1709–1716.

LAUER, U., ANKE, T., HANSSKE, F. (1991), Antibiotics from basidiomycetes XXXVIII. 2-Methoxy-5-methyl-1,4-benzoquinone, a thromboxane A_2 receptor antagonist from *Lentinus adhaerens, J. Antibiot.* **44,** 59–65.

LIESCH, J. M., HENSENS, O. D., SPRINGER, J. P., CHANG, R. S. L., LOTTI, V. J. (1985), Asperlicin, a novel non-peptidal cholecystokinin antagonist from *Aspergillus alliaceus.* Structure elucidation, *J. Antibiot.* **38,** 1638–1641.

LIESCH, J. M., HENSENS, O. D., ZINK, D. L., GOETZ, M. A. (1988), Novel cholecystokinin antagonists from *Aspergillus alliaceus.* II. Structure determination of asperlicins B, C, D, and E, *J. Antibiot.* **41,** 878–881.

LIFSON, J. D., FEINBERG, M. B., REYES, G. R., RABIN, L., BANAPOUR, B., CHAKRABART, S., MOSS, B., WONG-STALL, F., STEIMER, K. S., ENGELMAN, E. G. (1986), Induction of CD4-dependent cell fusion by the HTLV-III/LAV envelope glycoprotein, *Nature* **323,** 725–728.

MCDOUGAL, J. S., KENNEDY, M. S., SLIGH, J. M., CORT, S. P., MAWLE, C. A., NICHOLSON, J. K. A. (1986), Binding of HTLV-III/LAV to $T4^+$ cells by a complex of the 110K viral protein and the T4 molecule, *Science* **231,** 382–385.

MIYATA, S., HASHIMOTO, M., MASUI, Y., EZAKI, M., TAKASE, S., NISHIKAWA, M., KIYOTO, S., OKUHARA, M., KOHSAKA, M. (1992a), WS-7338, new endothelin receptor antagonists isolated from *Streptomyces* sp. No. 7338. I. Taxonomy, fermentation, isolation, physico-chemical properties and biological activities, *J. Antibiot.* **45,** 74–82.

MIYATA, S., HASHIMOTO, M., FUJIE, K., NISHIKAWA, M., KIYOTO, S., OKUHARA, M., KOHSAKA, M. (1992b), WS-7338, new endothelin receptor antagonists isolated from *Streptomyces* sp. No. 7338. II. Biological characterization and pharmacological characterization of WS-7338B, *J. Antibiot.* **45,** 83–87.

MIYATA, S., FUKAMI, N., NEYA, M., TAKASE, S., KIYOTO, S. (1992c), WS-7338, new endothelin receptor antagonists isolated from *Streptomyces* sp. No. 7338. III. Structures of WS-7338A, B, C, and D and total synthesis of WS-7338B, *J. Antibiot.* **45,** 788–791.

MIYATA, S., OHHATA, H., MURAI, H., MASUI, Y., EZAKI, M., TAKASE, S., NISHIKAWA, M., KIYOTO, S., KOHSAKA, M. (1992d), WS009A and B, new endothelin receptor antagonists isolated from *Streptomyces* sp. No. 89009. I. Taxonomy, fermentation, isolation, physico-chemical properties and biological activities, *J. Antibiot.* **45,** 1029–1040.

MIYATA, S., HASHIMOYO, M., FUJIE, K., SHOUHO, M., SOGABE, K., KIYOTO, S., OKUHARA, M., KOHSAKA, M. (1992e), WS009A and B, new endothelin receptor antagonists isolated from *Streptomyces* sp. No. 89009. II. Biological characterization and pharmacological characterization of WS009A and B, *J. Antibiot.* **45,** 1041–1046.

MORI, Y., ABE, Y., SHIGEMATSU, N., GOTO, T., OKUHARA, M., KOHSAKA, M. (1993c), Neoiradiomycins A and B1: non-steroidal estrogen-receptor antagonists produced by a *Streptomyces, J. Antibiot.* **46,** 1890–1893.

MORISHITA, Y., SANO, T., ANDO, K., SAITOH, Y., KASE, H., YAMADA, K., MATSUDA, Y. (1991a), Microbial polysaccharide, HS-142-1, competitively and selectively inhibits ANP binding to its guanylyl cyclase-containing receptor, *Biochem. Biophys. Res. Commun.* **176,** 949–957.

MORISHITA, Y., TAKAHASHI, M., SANO, T., KAWAMOTO, I., ANDO, K., SANO, H., SAITOH, Y., KASE, H., MATSUDA, Y. (1991b), Isolation and purification of HS-142-1, a novel nonpeptide antagonist for the atrial natriuretic peptide receptor form *Aureobasidium* sp., *Agric. Biol. Chem.* **55,** 3017–3025.

MORISHITA, Y., SANO, T., KASE, H., YAMADA, K., INAGAMI, T., MATSUDA, Y. (1992), HS-142-1, a novel nonpeptide atrial natriuretic peptide (ANP) antagonist, blocks ANP-induced renal responses through a specific interaction with guanylyl cyclase-linked receptors, *Eur. J. Pharmacol.* **225,** 203–207.

NAKAGAWA, A. (1992), Vasoactive substances, in: *The Search for Bioactive Compounds from Microorganisms* (ŌMURA, S., Ed.), pp. 198–212. Berlin–Heidelberg–New York: Springer-Verlag.

NAKAJIMA, S., NIIYAMA, K., IHARA, M., KOJIRI, K., SUDA, H. (1991), Endothelin-binding inhibitors, BE-18257A and BE-18257B. II. Structure determination, *J. Antibiot.* **44,** 1348–1356.

NODA, M., TAKAHASHI, H., TANABE, T., TOYOSATO, M., KIKYOTANI, S., FURUTANI, Y., HIROSE, T., TAKASHIMA, H., INAYAMA, S., MIYATA, T., NUMA, S. (1983), Structural homology of *Torpedo californica* acetylcholine receptor subunits, *Nature* **302,** 528–532.

ODA, S., SANO, T., MORISHITA, Y., MATSUDA, Y. (1992), Pharmacological profile of HS-142-1, a novel nonpeptide atrial natriuretic peptide (ANP) antagonist of microbial origin. II. Restoration by HS-142-1 of ANP-induced inhibition of aldosterone production in adrenal glomerulosa cells, *J. Pharmacol. Exp. Ther.* **263,** 241–245.

OHASHI, H., AKIYAMA, H., NISHIKORI, K., MOCHIZUKI, J. (1992), Asterric acid, a new endothelin binding inhibitor, *J. Antibiot.* **45,** 1684–1685.

OHTSUKA, T., KUDOH, T., SHIMMA, N., KOTAKI, H., NAKAYAMA, N., ITEZONO, Y., FUJISAWA, N., WATANABE, J., YOKOSE, K., SETO, H. (1992), Tetronothiodin, a novel cholecystokinin type-B receptor antagonist produced by *Streptomyces* sp., *J. Antibiot.* **45,** 140–143.

OHTSUKA, T., KOTAKI, H., NAKAYAMA, N., ITEZONO, Y., SHIMMA, N., KUDO, T., KUWAHARA, T., ARISAWA, M., YOKOSE, K., SETO, H. (1993a), Tetronothiodin, a novel cholecystokinin type-B receptor antagonist produced by *Streptomyces* sp. NR0489. II. Isolation, characterization and biological activities. *J. Antibiot.* **46,** 11–17.

OHTSUKA, T., NAKAYAMA, N., ITEZONO, Y., SHIMMA, N., KUWAHARA, T., YOKOSE, K., SETO, H. (1993b), Tetronothiodin a novel cholecystokinin type-B receptor antagonist produced by *Streptomyces* sp. NR0489. III. Structural elucidation, *J. Antibiot.* **46,** 18–24.

OKAMOTO, M., YOSHIDA, K., UCHIDA, I., NISHIKAWA, M., KHOSAKA, M., AOKI, H. (1986a), Studies of platelet activating factor (PAF) antagonists from microbial products. I. Bisdethiobis-(methylthio)gliotoxin and its derivatives, *Chem. Pharm. Bull.* **34,** 340–344.

OKAMOTO, M., YOSHIDA, K., NISHIKAWA, M., ANDO, T., IWAMI, M., KHOSAKA, M., AOKI, H. (1986b), FR-900452, a specific antagonist of platelet activating factor (PAF) produced by *Streptomyces phaeofaciens.* I. Taxonomy, fermentation, isolation and physico-chemical and biological characteristics, *J. Antibiot.* **39,** 198–204.

OLEYNEK, J. J., SEDLOCK, D. M., BARROW, C. J., APPELL, K. C., CASIANO, F., HAYCOCK, D., WARD, S. J., KAPLITA, P., GILLUM, A. M. (1994), WIN64821, a novel neurokinin antagonist produced by an *Aspergillus* sp. II. Biological activity, *J. Antibiot.* **47,** 399–410.

ŌMURA, S., TSUZUKI, K., SUNAZUKA, T., MARUI, S., TOYODA, H., INATOMI, N., ITOH, Z. (1987), Macrolides with gastrointestinal motor stimulating activity, *J. Med. Chem.* **30,** 1941–1943.

ŌMURA, S., TANAKA, H., MATSUZAKI, K., IKEDA, H., MASUMA, R. (1993), Isochromophilones I and II, novel inhibitors against gp120–CD4 binding from *Penicillium* sp., *J. Antibiot.* **46,** 1908–1911.

PEROUTKA, S. J. (1994), *Handbook of Receptors and Channels.* Boca Raton, FL: CRC Press.

PETTIBONE, D. J., CLINESCHMIDT, B. V., ANDERSON, P. S., FREIDINGER, R. M., LUNDELL, G. F., KOUPAL, L. R., SCHWARTS, C. D., WILLIAMSON, J. M., GOETZ, M. A., HENSENS, O. D., LIESCH, J. M., SPRINGER, J. P. (1989), A structurally unique, potent, and selective oxytocin antagonist derived from *Streptomyces silvensis, Endocrinology* **125,** 217–222.

SADROSKI, J., GOH, W. C., ROSEN, C., CAMPBELL, K., HASELTIME, W. A. (1986), Role of HTLV-III/LAV envelope in syncytium formation and cytopathicity, *Nature* **322,** 470–474.

SANO, T., IMURA, R., MORISHITA, Y., MATSUDA, Y., YAMADA, K. (1992a), HS-142-1, a novel polysaccharide of microbial origin, specifically recognizes guanylyl cyclase-linked ANP receptor in rat glomeruli, *Life Sci.* **51,** 1445–1451.

SANO, T., MORISHITA, Y., MATSUDA, Y., YAMADA, K. (1992b), Pharmacological profile of HS-142-1, a novel nonpeptide atrial natriuretic peptide (ANP) antagonist of microbial origin. I. Selective inhibition of the actions of natriuretic peptides in anesthetized rats, *J. Pharmacol. Exp. Ther.* **260,** 825–831.

SANO, T., MORISHITA, Y., YAMADA, K., MATSUDA, Y. (1992c), Effects of HS-142-1, a novel non-peptide ANP antagonist, on diuresis and natriuresis induced by acute volume expansion in anesthetized rats, *Biochem. Biophys. Res. Commun.* **182,** 824–829.

SCHWARTZ, R. E., HIRSCH, C. F., SPRINGER, J. P., PETTIBONE, D. J., ZINK, D. L. (1987), Unusual cyclopropane-containing hapalindolinones from a cultured Cyanobacterium, *J. Org. Chem.* **52,** 3704–3706.

SEDLOCK, D. M., BARROW, C. J., BROWNELL, J. E., HONG, A., GILLUM, A. M., HOUCK, D. R. (1994), WIN64821, a novel neurokinin antagonist produced by an *Aspergillus* sp. I. Fermentation and isolation, *J. Antibiot.* **47,** 391–398.

SHIGEMATSU, N., TUJI, E., KAYAKIRI, N., TAKASE, S., TANAKA, H., TADA, T. (1992), WF11605, an antagonist of leukotriene B_4 produced by a fungus. II. Structure determination, *J. Antibiot.* **45,** 704–708.

STOB, M., BALDWIN, R. S., TUITE, J., ANDREWS, F. N., GILLETTE, K. G. (1962), Isolation of an anabolic, uterotrophic compound from corn infected with *Gibberella zeae, Nature* **196,** 1318.

SUGANO, M., SATO, A., IIJIMA, Y., OSHIMA, T., FURUYA, K., KUWANO, H., HATA, T., HANZAWA, H. (1991), Phomatin A: a novel PAF antagonist from a marine fungus *Phoma* sp., *J. Am. Chem. Soc.* **113,** 5463–5464.

SUNAZUKA, T., TSUZUKI, K., MARUI, S., TOYODA, H., ŌMURA, S., INATOMI, N., ITOH, Z. (1989), Motilides, macrolides with gastrointestinal motor stimulating activity. II. Quaternary *N*-substituted derivatives of 8,9-anhydroerythromycin A 6,9-hemiacetal and 9,9-dihydroerythromycin A 6,9-epoxide, *Chem. Pharm. Bull.* **37,** 2701–2709.

TANAKA, T., ICHIMURA, M., NAKAJO, S., SNAJDAR, R. M., MORISHITA, Y., SANO, T., YAMADA, K., INAGAMI, T., MATSUDA, Y. (1992), HS-142-1, a novel non-peptide antagonist for atrial natriuretic peptide receptor, selectively inhibits particulate guanylyl cyclase and lower cyclic GMP in LLC-PK_1 cells, *Biosci. Biotech. Biochem.* **56,** 1041–1045.

TASKESAKO, K., KURODA, H., UENO, M., SAITO, H., YAMAMOTO, J., NAKAMURA, T. (1988) Screening of muscarine antagonists by radioreceptor assay, *Nippon Nogei Kagaku Kaishi* **62,** 338.

TOKI, S., ANDO, K., YOSHIDA, M., KAWAMOTO, I., SANO, H., MATSUDA, Y. (1992a), ES-242-1, a novel compound from *Verticillium* sp., binds to a site on N-methyl-D-aspartate receptor that is coupled to the channel domain, *J. Antibiot.* **45,** 88–93.

TOKI, S., ANDO, K., KAWAMOTO, I., SANO, H., YOSHIDA, M., MATSUDA, Y. (1992b), ES-242-2, -3, -4, -5, -6, -7, and -8, novel bioxanthracenes produced by *Verticillium* sp., which act on the N-methyl-D-aspartate receptor, *J. Antibiot.* **45,** 1047–1054.

TOKI, S., MORISHITA, Y., SANO, T., MATSUDA, Y. (1992c), HS-142-1, a novel non-peptide ANP antagonist, blocks the cyclic GMP production elicited by natriuretic peptides in PC 12 and NG 108-5 cells, *Neurosci. Lett.* **135,** 117–120.

TOKI, S., TSUKUDA, E., NOZAWA, M., NONAKA, H., YOSHIDA, M., MATSUDA, Y. (1992d), The ES-242s, novel N-methyl-D-aspartate antagonists of microbial origin, interact with both the neurotransmitter recognition site and the ion channel domain, *J. Biol. Chem.* **267,** 14884–14892.

TSUJI, E., TSURUMI, Y., MIYATA, S., FUJIE, K., KAWAKAMI, A., OKAMOTO, M., OKUHARA, M. (1992a), WF11605, an antagonist of leukotriene B4 produced by a fungus. I. Producing strain, fermentation, isolation and biological activity, *J. Antibiot.* **45,** 698–703.

TSUJI, E., SHIGEMATSU, N., HATANAKA, H., YAMASHITA, M., OKAMOTO, M., OKUHARA, M. (1992b), Novobiocin, an antagonist of leukotriene B4, *J. Antibiot.* **45,** 1958–1960.

TSUZUKI, K., SUNAZUKA, T., MARUI, S., TOYODA, H., ŌMURA, S., INATOMI, N., ITOH, Z. (1989), Motilides, macrolides with gastrointestinal motor stimulating activity. I. O-Substituted and tertiary N-substituted derivatives of 8,9-anhydroerythromycin A 6,9-hemiacetal, *Chem. Pharm. Bull.* **37,** 2687–2700.

UENO, M., TAKESAKO, K., IKAI, K., KATAYAMA, K., SHIMANAKA, K., NAKAMURA, T. (1988), Isolation of novel muscarine antagonists, IJ 2702 and PF 6766, *Nippon Nogei Kagaku Kaishi* **62,** 339.

URRY, W. H., WEHRMEISTER, H. L., HODGE, E. B., HIDY, P. H. (1966), The structure of zearalenone, *Tetrahedron Lett.* **27,** 3109–3114.

WATANABE, J., FUJISAKI, N., FUJIMORI, K., ANZAI, Y., OSHIMA, S., SANO, T., OHTSUKA, T., WATANABE, K., OKUDA, T. (1993), Tetronothiodin, a novel cholecystokinin type-B receptor antagonist produced by *Streptomyces* sp. NR0489. I. Taxonomy, yield improvement and fermentation, *J. Antibiot.* **46,** 1–10.

WEBER, W., FISCHLI, W., HOCHULI, E., KUPFER, E., WEIBEL, E. K. (1991), Anantin – a peptide antagonist of the atrial natriuretic factor (ANF). I. Producing organism, fermentation, isolation and biological activity, *J. Antibiot.* **44,** 164–171.

WONG, S.-M., KULLNIG, R., DEDINAS, J., APPELL, K. C., KYDD, G. C., GILLUM, A. M., COOPER, R., MOORE, R. (1993a), Anthrotainin, an inhibitor of substance P binding produced by *Gliocladium catenulatum, J. Antibiot.* **46,** 214–221.

WONG, S.-M., MUSZA, L. L., KYDD, G. C., KULLNIG, R., GILLUM, A. M., COOPER, R. (1993b), Fiscalins: New substance-P inhibitors produced by the fungus *Neosartorya fischeri.* Taxonomy, fermentation, structures, and biological activities, *J. Antibiot.* **46,** 545–553.

WYSS, D. F., LAHM, H. W., MANNEBERG, M., LABHARDT, A. M. (1991), Anantin – a peptide antagonist of the atrial natriuretic factor (ANF). II. Determination of the primary sequence by NMR on the basis of proton assignments, *J. Antibiot.* **44,** 172–180.

ZINK, D., HENSENS, O. D., LAM, Y. K. T., REAMER, R., LIESCH, J. M. (1992), Cochinmicins, novel and potent cyclodepsipeptide endothelin antagonists from a *Microbispora* sp. II. Structure determination, *J. Antibiot.* **45,** 1717–1722.

4 Microbial Lipids

Colin Ratledge

Hull, United Kingdom

1 Introduction

Since the publication of the 1st Edition of "Biotechnology" and the earlier chapter on the biotechnology of lipids in 1986, a considerable number of developments have taken place in this field. Some microbial lipid products have now been produced commercially and prospects for other developments appear to be not too far away. In some cases, as e.g. with the bacterial lipid poly-β-hydroxybutyrate, no counterpart exists from plant or animal sources and consequently the economics of producing this product lie outside the normal oils and fats domain. With most other microbial lipids, these are the equivalent in composition to plant-derived oils and consequently must compete against these in any potential market place. Only the highest valued oils have any chance of being produced by biotechnological means as it is impossible for microorganisms to produce oils and fats as cheaply as the main commodity oils are produced from plant and animal sources. However, there is always the possibility of producing a microbial oil as an adjunct to some waste treatment process in a way similar to that often used to produce microbial proteins (SCP – single cell protein) for animal feed from some unwanted substrate. Microbial oils – which could then be referred to as single cell oils (SCO) – would have the double advantage over SCP in that they could probably sell for a higher price than SCP and, moreover, could be used for a technical purpose should the nature of the substrate prevent the product being returned into the food chain.

The major commercial plant oils continue to be dominated by soybean oil (current 1993 production is about $18 \cdot 10^6$ t); palm oil, though, continues to be the fastest growing market with $14 \cdot 10^6$ t now being produced compared to $6 \cdot 10^6$ t in 1983. If the present rate of expansion in palm oil production continues in Malaysia and Indonesia (BASIRON and IBRAHIM, 1994; LEONARD, 1994), then palm oil will overtake soybean oil production by the end of this decade. Rapeseed oil (now $9 \cdot 10^6$ t in 1993) is also expanding mainly due to increased cultivation in Europe and Canada. The variety now under cultivation is the low- (or zero-) erucic acid (20:1) oil which is then a permitted oil for food manufacture.

Overall production of plant and animal oils is increasing at about 3% per annum; production in 1992/93 was about $85 \cdot 10^6$ t and is expected to reach $105 \cdot 10^6$ t by the year 2000 (MIELKE, 1992). Pricing of these materials remains highly competitive as most products using oils can switch between the various types according to the price of the day. The average price index for the major commodity oils is about US$ 500–550 per t though groundnut oil, e.g., is always significantly higher than the average at $ 800–850 per t. The highest priced commodity oil, excluding the speciality materials, is always olive oil at $ 1,500–2,000 per t. Its price depends on its quality which includes minor, but very important, flavor components. Animal fats (tallow and lard) have steadily declined in consumption over the past decade and are likely to fall even further to about 20% of the total market by 2001 (SHUKLA, 1994). Their prices are therefore usually at or below the average index level.

The trends in world oil and fats supplies are under constant surveillance and are frequently reviewed in various publications: the extensive reviews by SHUKLA (1994) and MIELKE (1992) can be recommended though for current information journals such as *Lipid Technology* (P. T. Barnes & Associates), *Oils and Fats International* (Chase Webb, St Ives PLC), *INFORM* (American Oil Chemists' Society, Illinois) provide invaluable and continuously up-dated information in most areas. There are, in addition, a number of specialized trade reviews that provide weekly prices of the traded oils.

The fatty acid composition of the major commercial oils is given in Tab. 1. The nomenclature of lipids is given in Sect. 1.1. It will be appreciated that, unlike say animal feed protein, the composition of the fats varies considerably from species to species. In all cases, however, the oil or fat is composed almost entirely (>98%) of triacylglycerols (formerly known as triglycerides) – Sect. 1.1. For edible purposes, the oil or fat is retained in this form though the individual fatty acyl groups on the glycerol can be modified – usually by chemical means – without affecting the triacylglycerol structure *per se*. Some en-

Tab. 1. Fatty Acid Composition of Fats and Oils of Animal and Plant Origin

Fats/Oils	Relative Proportion of Fatty Acyl Groups [%(w/w)]											
	4:0–10:0	12:0	14:0	16:0	16:1	18:0	18:1	18:2	18:3	20:0	20:1	Others
Animal Fats												
Butterfat	10	3	11	27	2	12	29	2	–	–	–	15:0+17:0, 3%
Beef tallow	–	–	3	24	4	19	43	3	1	–	–	15:0+17:0, 2%; 14:1+17:1, 2%
Lard	–	–	2	26	3	14	44	10	–	–	–	
Plant Oils												
Coconut oil	15	47	18	9	–	3	6	2	–	–	–	–
Palm kernel oil	8	48	16	8	–	3	15	2	–	–	–	–
Cocoa butter	–	–	–	26	–	35	35	3	–	1	–	–
Olive oil	–	–	–	13	1	3	71	10	1	1	–	–
Rapeseed oil	–	–	–	4	–	2	62	22	10	–	–	–
Groundnut oil[a]	–	–	–	11	–	2	48	32	–	1	2	22:0+24:0, 5%
Sunflower oil	–	–	–	7	–	5	19	68	1	–	–	–
Soybean oil	–	–	–	11	–	4	24	54	7	–	–	–
Corn oil	–	–	–	11	–	2	28	58	1	–	–	–
Cotton seed oil	–	–	1	22	1	3	19	54	1	–	–	–
"Exotic" Plant Oils												
Borage seed oil	–	–	–	11	–	4	16	39	22[b]	–	4.5	22:1, 2.5%; 24:0, 1.5%
Evening primrose seed oil	–	–	–	8	–	2	9	70	9[b]	–	–	
Blackcurrant seed oil	–	–	–	6	–	1	10	48	17[b]	–	–	α-18:3, 13%

[a] Also known as peanut oil.
[b] γ-Linolenic acid, 18:3 (6, 9, 12).

zymatic reformulation of triacylglycerols occurs on an industrial scale using stereospecific lipases to transesterify palm oil fractions into the much more expensive cocoa butter-like triacylglycerols (OWUSU-ANSAH, 1993).

With some technical applications of oils, it is the fatty acid that is required; consequently saponification (hydrolysis) of the triacylglycerol is carried out and the fatty acid used either as such, e.g., with soap manufacture, or is modified to an appropriate derivative which is then used in a multitude of products: from detergents to adhesives.

The aim of all biotechnological processes is to produce products that are either cheaper than can be obtained from other sources, including possible chemical synthesis, or are not available by any other means. Within the field of lipids, the opportunities to produce triacylglycerol lipids are limited to the highest valued materials. The highest priced bulk (commodity) oil is cocoa butter whose price has varied between \$ 8,000 to \$ 3,000 per t over the past decade. At the higher price level, the prospects of producing a cocoa butter equivalent oil by yeast technology have looked favorable. This topic is specifically reviewed later (see Sect. 3.2.1).

Other very high valued oils are those in the health care market and which have had various claims made on their behalf for the amelioration of various diseases and conditions. Of current interest are oils containing the polyunsaturated fatty acids: γ-linolenic acid, 18:3 (ω-6); arachidonic acid, 20:4 (ω-6); eicosapentaenoic acid, 20:5 (ω-3); and docosahexaenoic acid, 22:6 (ω-3). Oils containing such fatty acids are found in a number of microorganisms and are reviewed in Sects. 3.3 and 3.4.

The very highest priced lipids though are probably the prostanoid compounds encompassing the prostaglandins, leukotrienes, and thromboxanes. These are mainly used for treatment of uncommon disorders or for experimental purposes. Consequently, the amounts required per annum are probably at the kilogram stage rather than the ton (or kiloton) stage with other lipid products. Prospects for producing such materials are briefly mentioned in Sect. 6.4.

Thus, if we view microorganisms as a potential source of the widest types of lipids then it is possible to identify a number of potentially attractive products. For the purposes of this article, I have therefore used the broad definition of a lipid as any material that is derived from a (micro)organism, is directly soluble in organic solvents, and is essentially a water-insoluble material. However, as there is still considerable interest in the manner in which microorganisms synthesize large quantities of lipids, much of the review will be taken up with the more conventional types of oils and fats that they produce. The entire subject of microbial lipids, encompassing all aspects and not just biotechnology, has been the above subject of a two-volume monograph by RATLEDGE and WILKINSON (1988a, 1989). The industrial applications of microbial lipids have also been the subject of a monograph edited by KYLE and RATLEDGE (1992). Details concerning the degradation of fats, oils, and fatty acids, including the action of lipases and phospholipases, which are not covered here have been recently reviewed elsewhere by the author (RATLEDGE, 1993).

It will be appreciated, of course, that although microorganisms remain a potential source of oils and fats, there is considerable effort being put into the production of oils and fats from conventional plant sources. Such efforts include the modification of peanut oils (groundnut oil) to produce changes in the fatty acid composition so that the more desirable oils can be produced more cheaply. The application of genetic engineering is now gathering pace as a means of producing "tailor-made" oils and fats in plants and is likely to supersede the traditional plant breeding approach as a means of creating what is wanted more quickly and with greater certainty. This review, however, will not include any detailed review of the current developments in plant genetic engineering as applied to the commodity oils and fats. Readers should though be aware that such advances are now likely to be a major influence in the availability of "improved" oils for everyday use and will undoubtedly ensure that these materials remain highly competitively priced for many years to come. The recent reviews by HARWOOD (1994a, b), MURPHY (1994a) and RATTRAY (1994) and the monographs

Tab. 2. Developments in Genetic Engineering (by *in vitro* Mutagenesis and Gene Transfer) Being Applied for the Modification of Plant Oils (adapted from RATTRAY, 1994)

Plant Target	Fatty Acid	Objective	Application
Soybean and rapeseed	16:0	increase	margarine
	16:0	decrease	edible oil
	18:0	increase	margarine; cocoa butter substitute(?)
	18:1	increase	improved edible oil
	α-18:3	decrease	improved stability and odor
Rapeseed	22:1	increase	erucic acid for oleochemicals
Sunflower	18:1	increase	olive oil substitute
	18:2+18:3	decrease	
Linseed	18:3	increase	oleochemicals
Groundnut	18:1	increase	improved edible oil

edited by RATTRAY (1991) and by MURPHY (1994b) will be found particularly useful in this respect. Tab. 2 summarizes some of the current developments that are now taking place in this area.

The opportunities for microorganisms to produce oils and fats of commercial value for the bulk markets remain doubtful but where the product cannot be obtained from elsewhere then this provides a much better opportunity for a microbial oil than attempting to replicate what is already available from plant sources. Some opportunities nevertheless do exist but they have to be identified with some care. Hopefully, some of the following material may indicate to the astute reader where such opportunities may lie.

1.1 Lipid Nomenclature and Major Lipid Types

Fatty acids are long chain aliphatic acids (alkanoic acids) varying in chain length from, normally, C_{12} to C_{22} though both longer and shorter chain-length acids are known. In most cells (microbial, plant, and animal), the predominant chain lengths are 16 and 18. Fatty acids may be saturated or unsaturated with one or more double bonds which are usually in the *cis* (or *Z*) form. The structure of a fatty acid is represented by a simple notation system $-X{:}Y$, where X is the total number of C atoms and Y is the number of double bonds. Thus, 18:0 is octadecanoic acid, that is stearic acid; 16:1 is hexadecenoic acid, that is palmitoleic acid, with one double bond, and 18:3 would represent octadecatrienoic acid, a C_{18} acid with three double bonds. The position of the double bond(s) is indicated by designating the number of the C atom, starting from the COOH terminus, from which the double bond starts: oleic acid is thus 18:1 (9) signifying the bond is from the 9th (to the 10th) C atom. If it is necessary to specify the isomer, this is added as "*c*" (for "*cis*" = *Z*) or "*t*" for "*trans*" = *E*). In this review, the *cis*/*trans* system is used. Thus, *cis*, *cis*-linoleic acid is 18:2 (*c* 9, *c* 12). Most naturally-occurring unsaturated fatty acids are in the *cis* configuration and, unless it is stated otherwise, this configuration may be assumed.

With polyunsaturated fatty acids (PUFA), the double bonds are normally methylene-interrupted: $-CH{=}CH{-}CH_2{-}CH{=}CH-$. Thus, once the position of one bond is specified all the others are also indicated. In numbering PUFAs, a "reverse" system is used where the position of only the last double bond is given by the number of carbon atoms it is from the CH_3 terminus. To denote the "counting back" system is in operation the notation is given as ω-x or *n*-x; thus the two main isomers of linolenic acid (18:3) are given as 18:3 (ω-3) and 18:3 (ω-6) or as 18:3 (*n*-3) and 18:3 (*n*-6). In some systems, the minus sign may be omitted giving, for example, 18:3 (ω3) or 18:3 (*n*3) and 18:3 (ω6) and 18:3 (*n*6). Respectively, these two isomers are:

$$\overset{18}{CH_3}-\overset{17}{CH_2}-\overset{16}{CH}=\overset{15}{CH}-\overset{14}{CH_2}-\overset{13}{CH}=\overset{12}{CH}-\overset{11}{CH_2}-\overset{10}{CH}=\overset{9}{CH}$$
$$HOOC-H_2\overset{2}{C}-H_2\overset{3}{C}-H_2\overset{4}{C}-H_2\overset{5}{C}-H_2\overset{6}{C}-H_2\overset{7}{C}-H_2\overset{8}{C}$$

$$CH_3-CH_2-CH_2-CH_2-CH_2-CH=CH-CH_2-CH=CH$$
ω ω-1 ω-2 ω-3 ω-4 ω-5 ω-6
n *n*-1 *n*-2 *n*-3 *n*-4 *n*-5 *n*-6
$$HOOC-H_2C-H_2C-H_2C-H_2C-HC=HC-H_2C$$

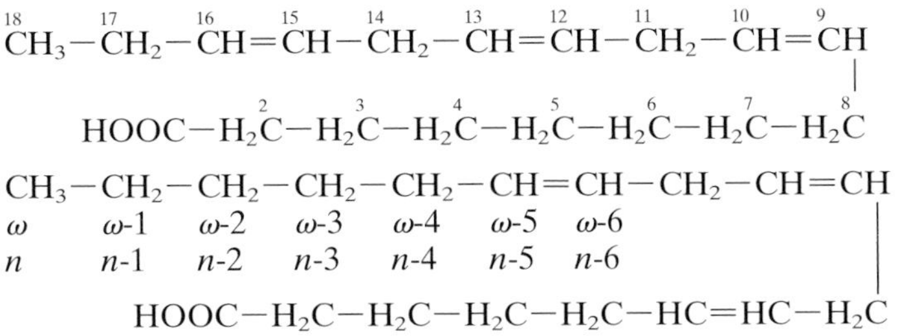

The ω-x system will be used in this review.

When fatty acids are esterified to glycerol, they give a series of esters: mono-, di-, and triacylglycerols (I, II, III). This is the preferred nomenclature to the older mono-, di-, and triglycerides.

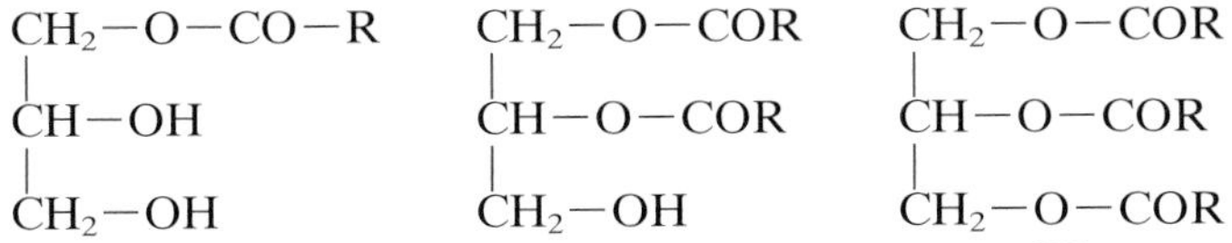

$$\begin{array}{lll} CH_2-O-CO-R & CH_2-O-COR & CH_2-O-COR \\ | & | & | \\ CH-OH & CH-O-COR & CH-O-COR \\ | & | & | \\ CH_2-OH & CH_2-OH & CH_2-O-COR \\ \text{I} & \text{II} & \text{III} \end{array}$$

where R is a long alkyl chain and RCO— is, therefore, the fatty acyl group.

As various isomeric forms are possible, the position of attached acyl group must be specified in most cases. For this, the stereospecific numbering (*sn*-) system is used so that the two prochiral positions of glycerol (IV) can be distinguished as *sn*-1 and *sn*-3.

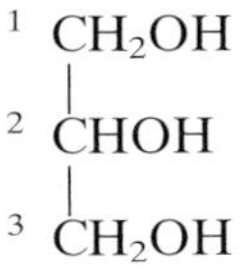

$$\begin{array}{ll} ^1 & CH_2OH \\ & | \\ ^2 & CHOH \\ & | \\ ^3 & CH_2OH \\ & \text{IV} \end{array}$$

With respect to the monoacylglycerols, there are obviously three possible isomers and similarly for the diacylglycerols.

Where different acyl groups are attached to the glycerol moiety, these can then be individually given. For example, 1-stearoyl-2-oleoyl-3-palmitoyl-*sn*-glycerol is the major triacylglycerol of cocoa butter with stearic, oleic, and palmitic acids on the three OH positions.

Phospholipids possess two fatty acyl groups at the *sn*-1 and *sn*-2 positions of glycerol with a phospho group at *sn*-3 which is also linked to a polar head group: choline, serine, ethanolamine, and inositol are the common ones. For the nomenclature and naming of phospholipids and other microbial lipids, the multiauthored treatise *Microbial Lipids*, edited by RATLEDGE and WILKINSON (1988a, 1989), may be helpful though there are numerous text books on lipids that provide similar information.

2 Accumulation of Lipid

2.1 Patterns of Accumulation

Not all microorganisms can be considered as abundant sources of oils and fats, though, like all living cells, microorganisms always contain lipids for the essential functioning of membranes and membranous structures. Those microorganisms that do produce a high content of lipid may be termed "oleaginous" in parallel with the designation given to oil-bearing plant seeds. Of the some 600 different yeast species, only 25 or so are able to accumulate more than 20% lipid; of the 60,000 fungal species fewer than 50 accumulate more than 25% lipid (RATLEDGE, 1989a).

The lipid which accumulates in oleaginous microorganisms is mainly triacylglycerol (see Sect. 1.1). If lipids other than this type are required then considerations other than those expressed here might have to be taken into account to optimize their production. With few exceptions, oleaginous microorganisms are eukaryotes and thus representative species include algae, yeasts, and molds. Bacteria do not usually accumulate significant amounts of triacylglycerol but many do accumulate waxes and polyesters (see Sects. 5.1 and 5.2) which are now of commercial interest.

The process of lipid accumulation in yeasts and molds growing in batch culture was elucidated in the 1930s and 1940s (see WOODBINE, 1959, for a review of the early literature, and RATLEDGE, 1982, for an updated review of these aspects). A typical growth pattern is shown in Fig. 1. This pattern is also found with the accumulation of polyester material in bacteria (see Sect. 5).

The key to lipid accumulation lies in allowing the amount of nitrogen supplied to the

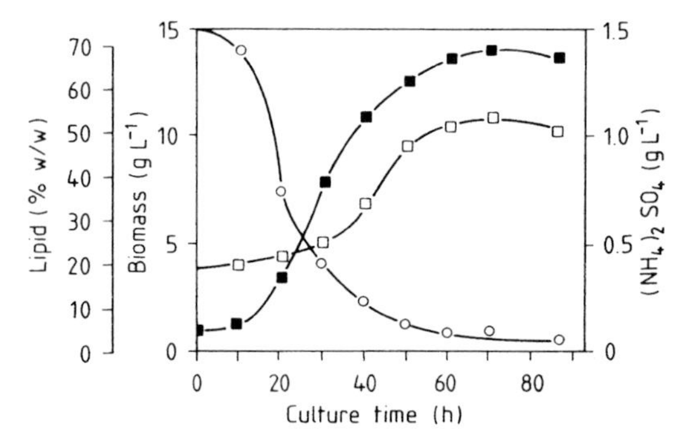

Fig. 1. Typical lipid accumulation pattern for a yeast (*Rhodotorula glutinis* = *R. gracilis*) growing on a high C:N ratio medium in batch culture. Biomass ■, % lipid content □, NH_4^+ in medium ○ (from YOON et al., 1982).

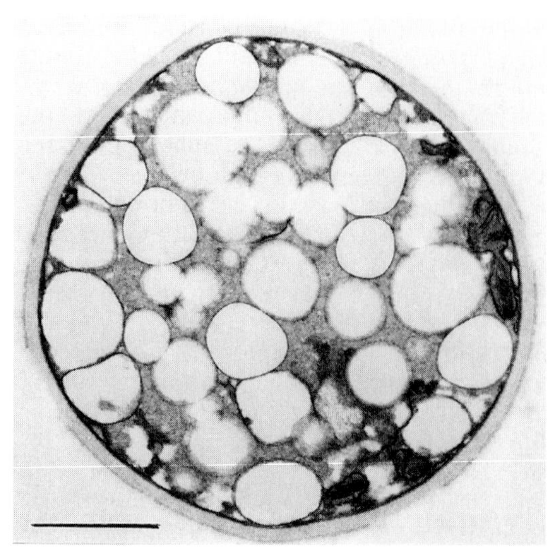

Fig. 2. Electron micrograph of *Cryptococcus curvatus* (= *Candida curvata* = *Apiotrichum curvatum*) strain D grown for 2 days on nitrogen-limiting medium (*viz.* Fig. 1) showing presence of multiple lipid droplets. Total lipid content approx. 40%, marker bar: 1 μm (from HOLDSWORTH et al., 1988).

culture to become exhausted within about 24–48 h. Exhaustion of nutrients other than nitrogen can also lead to the onset of lipid accumulation (see GRANGER et al., 1993, for a recent reference) but, in practice, cell proliferation is most easily effected by using a limiting amount of N (usually NH_4^+ or urea) in the medium. The excess carbon which is available to the culture after N exhaustion continues to be assimilated by the cells and, by virtue of the oleaginous organism possessing the requisite enzymes (see below), is converted directly into lipid.

The essential mechanism which operates is that the organism is unable to synthesize essential cell materials – protein, nucleic acids, etc. – because of nutrient deprivation and thus cannot continue to produce new cells. Because of the continued uptake of carbon and its conversion to lipid, the cells can then be seen to become engorged with lipid droplets (Fig. 2). It is important to appreciate, however, that the specific rate of lipid biosynthesis does not increase; the cells fatten because other processes slow down or cease altogether and, as lipid biosynthesis is not linked to growth, this may continue unabated. The process of lipid accumulation (Fig. 1) can be seen as a two-phase batch system: the first phase consists of balanced growth with all nutrients being available; the subsequent "fattening" or "lipogenic" stage occurs after the exhaustion of a key nutrient other than carbon and, of course O_2. The role of O_2 during lipid formation was discussed briefly in the 1st Edition of "Biotechnology" (RATLEDGE, 1986).

Accumulation of lipid has also been achieved in single stage continuous culture (RATLEDGE et al., 1984) and a typical accumulation profile dependent upon the dilution rate (growth rate) is shown in Fig. 3. As with batch cultivation, the medium has to be formulated with a high carbon-to-nitrogen ratio, usually about 50:1. The culture must be grown at a rate which is about 25–30% of the maximum. Under this condition, the concentration of nitrogen in the medium is virtually nil and the organism then has sufficient residence time within the chemostat to assimilate the excess carbon and convert it into lipid. The rate of lipid production (i.e., g L^{-1} h^{-1}) is usually faster in continuous cultures than in batch ones (EVANS and RATLEDGE, 1983; FLOETENMEYER et al., 1985).

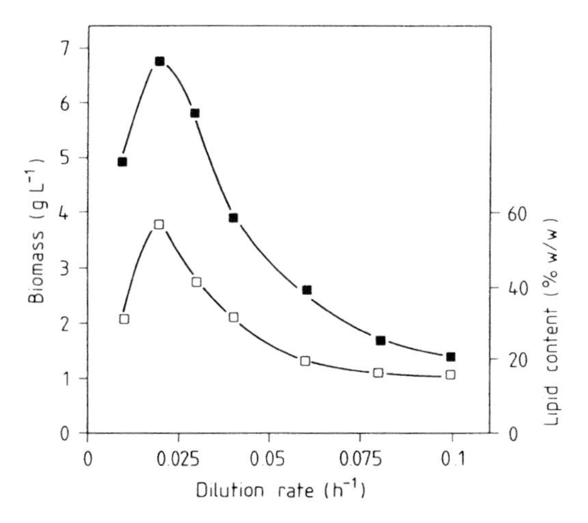

Fig. 3. Typical lipid accumulation pattern for a yeast (*Rhodotorula glutinis*) growing on nitrogen limiting medium in continuous culture. Biomass ■, % lipid content ○ (from YOON and RHEE, 1983).

The exact ratio of C to N chosen for the medium was originally considered to be of little consequence provided N was the limiting nutrient and sufficient carbon remained to ensure good lipid accumulation. However, YKEMA et al. (1986) showed that a range of lipid yields in an oleaginous yeast, *Apiotrichum curvatum* (originally *Candida curvata* but now *Cryptococcus curvatus*; see BARNETT et al., 1990) were traversed in continuous culture by varying the C:N ratio of the growth medium. There was a hyperbolic relationship between the C:N ratio and the maximum growth (dilution) rate that the organism could attain: the lowest growth rate was at the highest C:N ratio of 50:1 and this, in turn, controlled the amount of lipid produced and the efficiency of yield (g lipid per g glucose used) with which it was produced. Although the highest lipid contents of the cell (50% w/w) were obtained with a C:N ratio of 50:1 or over, the optimum ratio for maximum productivity (g L^{-1} h^{-1} lipid) was at a ratio of 25:1 with glucose (YKEMA et al., 1986) and at 30–35:1 when whey permeates were used with same yeast (YKEMA et al., 1988). Similar results for describing the optimum C:N ratio for lipid accumulation have been developed by GRANGER et al. (1993) using *Rhodotorula glutinis*.

Interestingly, YKEMA et al. (1986) commented that *Apiotrichum curvatum* simultaneously accumulated about 20% carbohydrate in the cells along with the 50% lipid. Such a phenomenon of carbohydrate formation had been conjectured by BOULTON and RATLEDGE (1983a) to be a likely event to account for an observed delay in lipid synthesis after glucose assimilation had been initiated. This carbohydrate was also recognized independently by HOLDSWORTH et al. (1988) in the same yeast and was considered to be glycogen. As YKEMA et al. (1986) pointed out, if the biosynthesis of the polysaccharide which, like lipid, is a reserve storage material, could be prevented then this would enhance the total amount of lipid producible with a cell.

Although most studies on microbial lipid accumulation have been conducted using batch cultivation and, for accuracy, in continuous culture, other growth systems have also been explored. In particular, fed-batch culture has proved effective in increasing both the cell density and lipid contents of oleaginous yeasts: YAMAUCHI et al. (1983) used ethanol as substrate with *Lipomyces starkeyi* and achieved a biomass density of 150 g L^{-1} with a lipid content of 54%. Similarly, PAN and RHEE (1986) achieved 185 g (dry wt.) of *Rhodotorula glutinis* per liter with a lipid content of 43% using glucose as the fed-batch substrate. In this latter case, O_2-enriched air (40% O_2 + 60% air) had to be used to sustain the cells. At the density recorded, the packed cell volume was 75% of the total volume of the fermentation medium. Without using additional O_2, it seems likely that cell densities of up to 100 g L^{-1} could be achieved with most oleaginous yeasts (see, e.g., YKEMA et al., 1988) though filamentous molds may pose other problems. Economic considerations, however, would probably be against the use of O_2-enriched air for any commercial process. Interestingly, it is suggested that higher rates of lipid formation may occur with fed-batch techniques than with batch- or continuous-culture approaches (YKEMA et al., 1988).

At the end of the lipid accumulation phase (see Fig. 1), it is essential that the cells are promptly harvested and processed. If glucose, or other substrate, has become exhausted on the end of the fermentation, then the organism will begin to utilize the lipid as the role of the accumulated material is to act as a reserve store of carbon, energy, and possibly even water. HOLDSWORTH and RATLEDGE (1988) showed with a number of oleaginous yeasts that after carbon exhaustion following lipid accumulation, the lipid began to be utilized within 1.5 h thus indicating the dynamic state of storage lipids in these organisms.

2.2 Efficiency of Accumulation

The efficacy of conversion of substrate to lipid has been examined in some detail in both batch and continuous culture. In general, the latter technique offers the better means of attaining maximum conversions as the cells are operating under steady state conditions and carbon is not used with different efficiencies at each stage of the growth cycle. Conversions of glucose and other carbohydrates including lactose and starch, to lipid up to 22% (w/w) have been recorded with a variety of yeasts (RATLEDGE, 1982; YKEMA et al., 1988; DAVIES and HOLDSWORTH, 1992; HASSAN et al., 1993) which compares favorably with the theoretical maximum of about 31–33% (RATLEDGE, 1988). Somewhat lower yields appear to pertain with molds (WOODBINE, 1959; WEETE, 1980). The reason for this difference is not obvious though it may be due to a somewhat slower growth rate of molds than yeasts. It should be said, however, that there has not been the same amount of detailed work carried out with molds as with yeasts. Claims that microorganisms have achieved higher conversions of glucose or other sugars to lipid should be treated with caution: either there will be found to be additional carbon within the medium and not taken into the mass balance or, as may occasionally happen, the "lipid" has been improperly extracted and may contain non-lipid material. However, if experimental data are calculated so that the yield of lipid or fatty acids can be based on the fraction of glucose being used solely for lipid biosynthesis, then values close to the theoretical value have been attained in practice (GRANGER et al., 1993).

When ethanol is used as substrate, the theoretical yield of lipid is 54% (w/w) (RATLEDGE, 1988). Though only a 21% conversion of ethanol to lipid was recorded by YAMAUCHI et al. (1983) in the fed-batch culture of *Lipomyces starkeyi*, higher conversions were recorded by EROSHIN and KRYLOVA (1983) also using a fed-batch system for the cultivation of yeasts on ethanol: conversions of 26%, 27%, and 31% were obtained using, respectively, *Zygolipomyces lactosus* (*Lipomyces tetrasporus*) and two strains of *Cryptococcus albidus* var. *aerius*. The reason for these very high values lies in the efficiency by which ethanol can be converted to acetyl- CoA, the starting substrate for lipid biosynthesis (see below). With glucose, the maximum yield of acetyl-CoA can only be 2 mol per mol utilized whereas with ethanol the yield is 1 mol per mol. On a weight-to-weight basis, therefore, ethanol (MW 46) is almost twice as efficient as glucose (MW 180) in providing C_2 units. GRANGER et al. (1993) have recorded direct conversions of ethanol to lipid (that is excluding ethanol being converted to non-lipid biomass) of 42% (w/w) with *Rhodotorula glutinis*.

While most oleaginous microorganisms accumulate lipid equally well from a number of different carbon sources (see, e.g., EVANS and RATLEDGE, 1983; YOON et al., 1982; DAVIES and HOLDSWORTH, 1992; HAMMOND et al., 1990), and do so without regard to the source of nitrogen used in the medium, a few yeasts are known which only accumulate lipid when an organic source of nitrogen, such as urea, glutamate, or aspartate, is used (WITTER et al., 1974; EVANS and RATLEDGE, 1984a, b). These yeasts appear to be mainly confined to strains of *Trichosporon (Endomycopsis) pullulans* and *Rhodosporidium toruloides*. A biochemical explanation for this has been advanced based on nitrogen catabolite inhibition of phosphofructokinase (see Fig. 4) which is a key enzyme controlling the rate of flux of glucose (as carbon substrate) to acetyl-CoA (EVANS and RATLEDGE, 1984c). The effect, though, is not explicable in terms of a greater pH decrease with inorganic NH_4^+ salts than with glutamate or asparagine, as has been suggested elsewhere (MORETON, 1988a). The same effect was produced in *Rhodotorula gracilis* using ammonium tartrate as was produced with NH_4Cl where the former salt caused little downward pH drift but, without changing the biomass yield, increased the lipid content of the cells from 18% to over 50% (EVANS and RATLEDGE, 1984a, c).

2.3 Biochemistry of Accumulation

The pathway of synthesis of any microbial product, or indeed plant or animal product, needs to be understood so that it then becomes possible to manipulate the cell to enhance the formation of that product. The formation of lipids is not different from any other product in this general concept. The pathway of triacylglycerol formation from glucose is given in Fig. 4. The pathway accounts for lipid formation in oleaginous microorganisms and, with the accompanying regulatory control mechanisms, can explain how lipid accumulation occurs in this group of organisms. In non-oleaginous organisms, one of the key enzymes, ATP-citrate lyase, does not occur and consequently the formation of acetyl-CoA occurs by another route (see SHERIDAN et al., 1989) and does not lead to a lipogenic state being created.

In oleaginous microorganisms, the following sequence of events is considered to happen to cause lipid accumulation:

(1) When the culture has consumed all available N from the medium, a nitrogen-scavenging process is initiated. This takes the form, but there are likely to be other examples, of deaminating AMP via the enzyme AMP-deaminase (see Reaction 1) which becomes activated at the point of N exhaustion (EVANS and RATLEDGE, 1985c).

$$AMP \rightarrow IMP + NH_3 \quad (1)$$

As a result of the activity of AMP deaminase, some NH_3 is provided for the cell to help maintain protein and nucleic acid synthesis but, simultaneously, the concentration of AMP drops rapidly (BOULTON and RATLEDGE, 1983a) and this is then the first major trigger in the lipogenic cascade mechanism.

(2) AMP is required as an activator of the enzyme isocitrate dehydrogenase (Reaction 2) operating in the mitochondrion (EVANS et al., 1983). As a consequence of the absence of AMP the reaction is unable to proceed as part of the tricarboxylic acid cycle.

$$\text{Isocitrate} + NADP^+ \rightarrow \text{2-Oxoglutarate} + NADPH + CO_2 \quad (2)$$

(3) With the cessation or slowing down of isocitrate dehydrogenase, isocitrate cannot be further metabolized and both isocitrate and citrate begin to accumulate. The equilibrium lies in favor of citrate so as the assimilation of glucose continues unabated by the N-limited cells, (BOTHAM and RATLEDGE, 1979), citrate becomes a major product of its metabolism (BOULTON and RATLEDGE, 1983a).

(4) Citrate now exits from the mitochondrion in a malate-mediated citrate translocase reaction (see Fig. 4). Citrate is then cleaved by ATP–citrate lyase (Reaction 3), an enzyme which appears to be uniquely associated with the lipogenic process. The dependency of isocitrate dehydrogenase (Reaction 2) upon

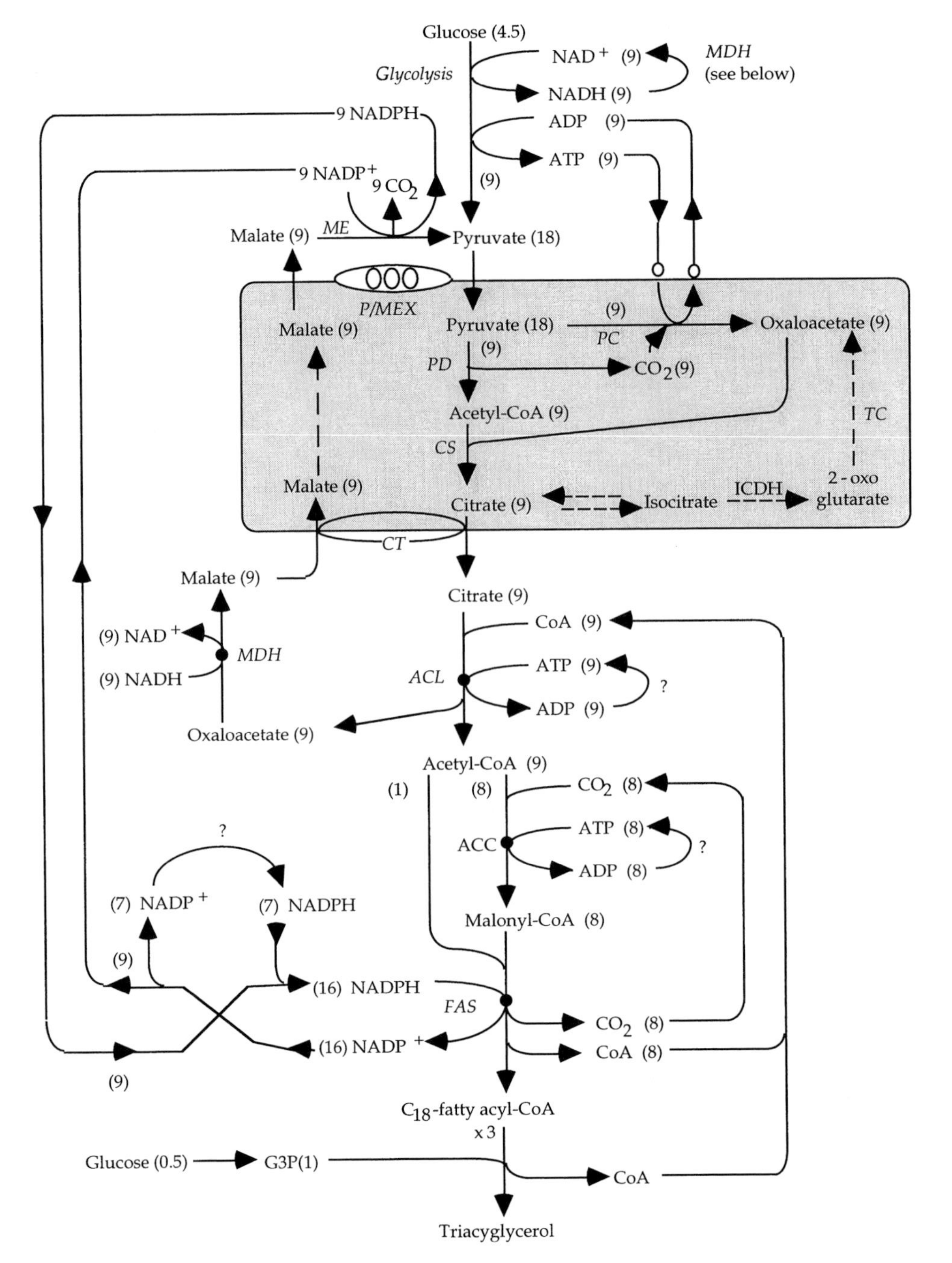

Fig. 4. Pathway of biosynthesis of triacylglycerols from glucose in oleaginous microorganisms. Numbers in parentheses indicate approximate stoichiometry for conversion of glucose to fatty acyl-CoA as elucidated in oleaginous yeasts (see also RATLEDGE, 1988; RATLEDGE and EVANS, 1989).
G3P: Glycerol 3-phosphate (this is derived from 0.5 mol glucose during glycolysis)

AMP is, though, another metabolic feature, possibly peculiar to the oleaginous organisms (EVANS and RATLEDGE, 1985b, 1986).

Citrate + ATP + CoA → Acetyl-CoA + Oxaloacetate + ADP + P_i (3)

(5) The acetyl-CoA from Reaction 3 serves as the primer for fatty acid synthesis. However, in addition to a supply of C_2 units, the cell must also provide NADPH as reductant for fatty acid synthesis. This is provided from the subsequent metabolism of oxaloacetate, first to malate via malate dehydrogenase, and then to pyruvate via malic enzyme (Reaction 4):

Malate + $NADP^+$ → Pyruvate + CO_2 + NADPH (4)

Some NADPH may also be supplied by metabolism of glucose via the pentose phosphate pathway.

The malic acid for the latter reaction is presumed to be by it leaving the mitochondrion in exchange for pyruvate in a series of coupled transport reactions across the mitochondrial membrane (see Fig. 4).

The overall flux of glucose to fatty acyl-CoA and then into triacylglycerol is given in Fig. 4. The overall stoichiometry is approximately:

15 Glucose → Triacylglycerol + 36 CO_2

Therefore, if glucose is not used for the synthesis of any other product the yield of lipid is approximately 32 g per 100 g glucose.

The role of ATP–citrate lyase in lipid accumulation appears to be central. The enzyme itself has been purified and partially characterized from *Rhodotorula gracilis* (SHASHI et al., 1990). It has a M_r of approx. 520 kDa and comprises four identical subunits each about 120-130 kDa in size. Other properties of the enzyme to establish its involvement in lipid accumulation have been presented by BOTHAM and RATLEDGE (1979), BOULTON and RATLEDGE (1981, 1983b), and by EVANS and RATLEDGE (1985a, c). The enzyme appears similar in overall size and structure to mammalian ATP–citrate lyase (HOUSTON and NIMMO, 1984, 1985). Microorganisms lacking ATP–citrate lyase generally do not accumulate lipid above 10–15%. However, it is important to state that the corollary, that microorganisms with ATP–citrate lyase will accumulate lipid, does not necessarily follow because other enzymes that function at key regulatory points such as AMP deaminase (Reaction 1), the AMP-dependent isocitrate dehydrogenase (Reaction 2) and malic enzyme (Reaction 3) may be lacking or be under alternative control mechanisms.

Malic enzyme (ME) itself is absent in some

◀ Enzymes:

PC: Pyruvate carboxylase (Pyruvate + CO_2 + ATP → Oxaloacetate + ADP + P_i
PD: Pyruvate dehydrogenase (Pyruvate + CoA + NAD^+ → Acetyl-CoA + CO_2 + NADH)
CS: Citrate synthase (Acetyl-CoA + Oxaloacetate → Citrate + CoA)
ACL: ATP:citrate lyase
MDH: Malate dehydrogenase
ME: Malic enzyme
ACC: Acetyl-CoA carboxylase
FAS: Fatty acid synthase
P/MEX: Pyruvate/malate exchange (triple-linked system)
ICDH: Isocitrate dehydrogenase (NAD^+ – requiring, AMP-dependent; see text Reaction 5)
CT: Citrate translocase
TC: Tricarboxylic acid cycle reactions
?: unknown (unspecified) reactions

Overall conversion (assuming the 9 mol of NADH required for the *MDH* reaction will be either from glycolysis or the PD reaction):
4.5 Glucose + CoA + 9 NAD^+ + 7 NADPH + 17 ATP → C_{18}-fatty acyl-CoA + 9 CO_2 + 9 NADH + 7 $NADP^+$ + 17 ADP + 17 P_i)
The production of 1 mol triacylglycerol, therefore, requires approx. 5 mol glucose as an additional 0.5 mol glucose is needed to provide glycerol 3-phosphate (G3P) for the triacylglycerol and a further mol of glucose must be oxidized to provide additional energy (ATP) and reducing power (NADPH).

oleaginous yeasts, notably *Lipomyces* spp. (BOULTON, 1982), and here the supply of NADPH is presumably by reactions of the pentose phosphate cycle. ME activity has also been found in oleaginous fungi and has also been implicated as the provider of NADPH for the desaturation of fatty acids (KENDRICK and RATLEDGE, 1992c) in *Mucor circinelloides*. Here, a second form of ME is associated with the membranes of endoplasmic reticulum – the microsomal fraction of the cell – that then serves to provide NADPH directly within the membranes. The transfer of reducing equivalents from the NADPH to the O_2-dependent desaturation reaction (Reaction 5) may not be linked via cytochrome b_5 which is normally associated with such reactions in other tissues (see Reaction 5).

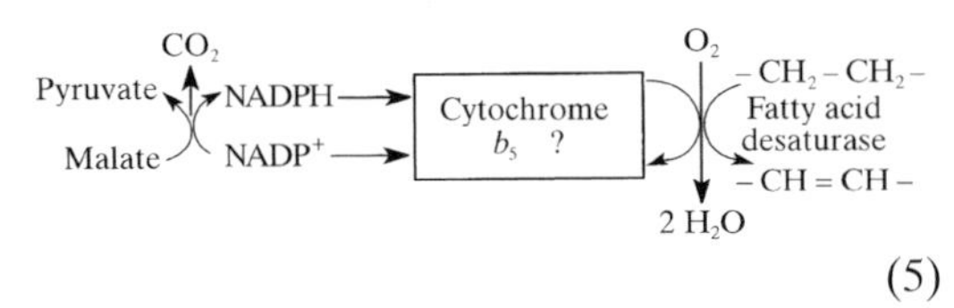

(5)

So far there have been no other reports of the occurrence of malic enzyme in microsomal membranes in other systems.

In non-oleaginous microorganisms, where ATP–citrate lyase does not occur, acetyl-CoA for fatty acid biosynthesis is generated from the intramitochondrial pool by transfer of the acetyl group via carnitine acetyl transferase (CAT). Oleaginous microorganisms also possess this enzyme activity (RATLEDGE and GILBERT, 1985; HOLDSWORTH et al., 1988) though it is not immediately obvious why two routes to acetyl-CoA generation are necessary though it does emphasize that without ATP–citrate lyase cells cannot presumably provide sufficient acetyl-CoA under N-limiting growth conditions to keep lipid biosynthesis fully primed. Non-oleaginous microorganisms, therefore, tend to accumulate carbohydrate reserves when placed under the same growth conditions.

The mechanism of fatty acid biosynthesis itself from acetyl-CoA is well documented in most standard text books. Recent reviews that may be consulted on this topic include those by SCHWEIZER (1989) which covers fatty acid biosynthesis in bacteria, yeasts, and fungi; CARMAN and HENRY (1989) which covers phospholipid biosynthesis in yeast; PIERINGER (1989) covering the biosynthesis of non-terpenoid lipids from fatty acids; and COOLBEAR and THRELFALL (1989) reviewing the biosynthesis of terpenoid lipids.

3 Triacylglycerols and Fatty Acids

Fatty acids do not occur as such in living cells because of their inherent toxicity. Consequently, fatty acids are esterified, usually as triacylglyerols (see Sect. 1.1, Structure III), or occasionally as wax esters (see Sect. 5). Triacylglycerols (TAG) are produced as the major storage product of most oleaginous yeasts and molds. Their proportion of the total lipid is usually over 80% (see RATLEDGE, 1986) and can be over 90%. Clearly, the proportion of TAG will depend upon a number of factors which would include the total amount of lipid in the oleaginous cell (the more accumulated lipid, the greater the amount of TAG) and also the method used for lipid extraction. More severe methods of extraction will remove both free and bound lipids whereas gentler methods will extract only the freely soluble lipid which will be predominantly TAGs. Thus, comparisons of lipid analysis, carried out by different researchers using different methods with different organisms, must be treated with some caution. However, from a biotechnological viewpoint, the TAG content of the cell is usually of paramount importance as this is the form in which an oil will eventually be offered for sale. Should the TAG content of a cell be low, but the fatty acids of interest, then total extraction of all fatty acyl lipids will have to be carried out and the fatty acids then recovered after hydrolysis. In these cases, the fatty acyl profile of the total lipid becomes the major determinant of the potential usefulness of lipid. The fatty acids can be esterified (usually methyl or ethyl derivatives) and then offered for sale.

Lipid extraction is fraught with a number of difficulties and pitfalls for the unwary. The article by RATLEDGE and WILKINSON (1988b) discussed these problems at some length and suggested methods that could be used to minimize post-harvest changes in the lipid composition of microbial cells. The presence of partial acylglycerols (mono- and diacylglycerols) along with free fatty acids in the extracts is indicative of faulty extraction procedures that have failed to subdue the latent activity of lipases and phospholipases of the cells. These enzymes function successfully in the very solvent systems that are used for lipid extraction and require rapid inactivation (usually by heating) to ensure that the lipid remains unchanged. Fractionation of extracted lipids that show the present of significant amounts of these hydrolysis products, and especially free fatty acids, are therefore seldom worth reporting and, as there are so few good examples with oleaginous yeasts and molds (see RATLEDGE, 1986), the erudite reader is therefore referred to the more extensive reviews of RATTRAY (1988) and LÖSEL (1988) who detail the lipid analyses of a large number of yeasts and molds. This present chapter, however, is concerned solely with the biotechnological potential of microorganisms and consequently it is only those species that are prolific in the production of desirable oils or fatty acids that will be considered here.

3.1 Bacteria

Although bacteria do not produce triacylglycerols and are usually considered as sources of novel lipids (see Sect. 5), they nevertheless are of current interest in that some species are known that produce polyunsaturated fatty acids (PUFA) of dietary or even pharmaceutical importance.

3.1.1 Polyunsaturated Fatty Acids in Bacteria

Several bacterial species have been recently isolated from marine sources that synthesize either eicosapentaenoic acid (20:5 ω-3) (EPA) and docosahexaenoic acid (22:6 ω-3) (DHA). EPA has now been identified in the lipids (presumably phospholipids) of a number of species of marine bacteria: *Alteromonas, Shewanella, Flexibacter,* and *Vibrio* (RINGO et al., 1992; AKIMOTO et al., 1990, 1991; YAZAWA et al., 1988, 1992; HENDERSON et al., 1993), though it was first recognized in the lipids of *Flexibacter polymorphus* some years ago by JOHNS and PERRY (1977). YAZAWA et al. (1992) carried out an extensive survey of some 24,000 bacteria isolated from various fish and marine mammals. One isolate, which approximated taxonomically to *Shewanella putrefaciens,* was found that produced EPA at up to 40% of the total fatty acids; the lipid content of the bacterium was between 10 to 15% and the organism grew readily in the laboratory achieving 15 g (dry wt.) L^{-1} in 12–18 h. The content of EPA in dry cells was approx. 2%. Total hydrolysis of the extracted lipids was needed to release the EPA from the phospholipid fraction. EPA was found almost exclusively at the *sn*-2 position of phosphatidylethanolamine and phosphatidylglycerol. It was the only PUFA that was recognized in the total lipids; other unsaturated fatty acids were 16:1 (12%), 18:1 (oleic + *cis*-vaccenic) (2%) and 17:1 (6%). HENDERSON et al. (1993), using a marine *Vibrio* isolate, showed that EPA formation was higher (at 9% of the total fatty acids) when the cells were grown at 5°C than 20°C, whereas YAZAWA et al. (1992) had only used 10–15°C for the growth of their bacterium. The gene(s) responsible for EPA production have been transferred into *E. coli* with the resultant production of EPA in this bacterium (WATANABE and YAZAWA, 1992).

DHA was first recognized in several marine bacteria by DELONG and YAYANOS (1986). Like EPA, DHA was exclusively associated with the phospholipids in all cases. However, DHA was predominant in bacteria grown at 2°C and at very high pressures ($\sim 10^5$ Pa) making them unlikely candidates for biotechnological exploitation. More recently, YANO et al. (1994) have reported similar findings with a further five deep-sea bacterial isolates. Again there was a strong dependency on high pressures and low temperatures for DHA formation though, for the first time, both DHA and EPA, together with traces of 20:4 (ω-3) and 20:3 (ω-3), were noted.

The commercial values of EPA and DHA as individual fatty acids are unclear as they are both readily obtainable as a mixture in fish oils which are relatively cheap. It is clear, however, that the bacterium described above by YAZAWA et al. (1992) – *Shewanella putrefaciens* – could represent a valuable source of EPA but the commercial potential for EPA and DHA would both be considerably enhanced if it should be shown that single fatty acids were required for medical purposes. WATANABE et al. (1994) have recently shown that isolated DHA (and by inference EPA would act similarly) can be readily incorporated into bacterial phospholipids, including those of *E. coli* and also *Rhodopseudomonas capsulata* which is currently used for the production of larvae of marine fish. In this way DHA could be delivered to the developing larvae and potentiate their growth rate. DHA though at present must be obtained from fish oils though mold and algae sources of it are known (see Sects. 3.3.5 and 3.4.4). If recombinant DNA technology could be used to increase the levels of DHA and EPA in more easily cultivatable bacteria (see WATANABE and YAZAWA, 1992), or even yeasts, this could open up new and exciting horizons for these PUFAs. At the moment, however, EPA production by bacteria is someway off and that of DHA would appear almost impossible. Other sources of EPA, however, include both molds and algae (see Sects. 3.3 and 3.4).

3.2 Yeasts

The number of oleaginous yeasts (Tab. 3) is relatively small in comparison to the total number of species – 590 (see BARNETT et al., 1990). Besides producing triacylglycerols, these yeasts also usually produce between 2–10% of their total lipid as phospholipids with smaller amounts of sterol and sterol esters (RATLEDGE, 1986). The fatty acid profile of the yeasts lipids (Tab. 3) is typically that of several commercial plant oils (see Tab. 1) and, as such, would not command a high price: say, maximally $ 600 per t. Consequently, the opportunities for exploiting yeast oils as a commercial possibility are limited and other more expensive targets have had to be identified. One major target that has been identified is to produce a yeast oil as a cocoa butter equivalent (CBE). The fatty acid composition of cocoa butter is given in Tab. 1.

3.2.1 Production of a Cocoa Butter Equivalent Yeast Fat

CBE fats are used extensively in the confectionery business: in the manufacture of cooking chocolate and chocolate-type materials and it can also be included, to an agreed percentage – usually 5% – of cocoa butter itself, in chocolate manufacture in several countries including UK, Ireland, and Denmark (KERNON, 1992). CBE materials have to have similar physical characteristics to cocoa butter itself and currently they are manufactured from palm oil by fractional crystallization though they can also be produced by enzymic transesterification of palm oil and stearic acid or its esters (WILLNER et al., 1993; OWUSU-ANSAH, 1993).

The requirements for a satisfactory CBE is that there should be approximately equal amounts of stearic acid (18:0), oleic acid (18:1), and palmitic acid (16:0) attached to the glycerol moiety with the unsaturated acid being in the *sn*-2 position. This is known as the POSt-TAG (1-palmitoyl-2-oleoyl-3-stearoyl-*sn*-glycerol). The price of cocoa butter itself has varied from $ 8600 per t in the mid 1980s (MORETON, 1988b; SMIT et al., 1992) to about $ 3000 per t in 1994. As CBEs command a price of about 80% of cocoa butter, it can be appreciated that at the higher price level, a yeast CBE would be an attractive economical proposition. Not surprisingly, a number of developments took place in the mid 1980s using appropriate yeast technology with this objective in mind. Simultaneously, the attractive price of cocoa butter and of CBEs stimulated the current interest in lipase-catalyzed transformation of oils and fats. However, with the present low price of commercial cocoa butter and CBEs, the yeast route has been deemed to be non-profitable and even

Tab. 3. Lipid Contents and Fatty Acid Profiles of Oleaginous Yeasts[a]

Yeast Species[b]	Maximum Lipid Content [% (w/w)]	Major Fatty Acyl Residues [Relative % (w/w)]						Others
		16:0	16:1	18:0	18:1	18:2	18:3	
Candida diddensiae	37	19	3	5	45	17	5	18:4 (1%)
Candida sp. 107	42	44	5	8	31	9	1	
Cryptococcus albidus var. *aerius*	65	12	1	3	73	12	–	
C. albidus var. *albidus*	65	16	trace	3	56	–	3	21:0 (7%) 22:0 (12%)
C. curvatus D[c]	58	32	–	15	44	8	–	
C. curvatus R[c]	51	31	–	12	51	6	–	
C. laurentii	32	25	1	8	49	17	1	
Endomyces (*Endomycopsis*) *magnusii*[d]	28	17	19	1	36	25	–	
Galactomyces geotrichum[e]	50	no record[e]						
Williopsis saturnus[f]	28	16	16	–	45	16	5	
Waltomyces lipofer[g]	64	37	4	7	48	3	–	
Lipomyces starkeyi	63	34	6	5	51	3		
L. tetrasporus (*Zygolipomyces lactosus*)	67	31	4	15	43	6	1	
Rhodosporidium toruloides	66	18	3	3	66	–	–	23:0 (3%) 24:0 (6%)
Rhodotorula glutinis	72	37	1	3	47	8	–	
R. graminis	36	30	2	12	36	15	4	
R. mucilaginosa	28				no record			
Trichosporon beigelii[h]	45	12	–	22	50	12	–	
T. fermentans[i]	20	17	1	4	42	34	trace	
T. pullulans[j]	65	15	–	2	57	24	1	
Yarrowia lipolytica[k]	36	11	6	1	28	51	1	

[a] A lower lipd content of cells for inclusion in the list has been taken as 20%. Data taken from RATLEDGE and EVANS (1989) and RATLEDGE (1989b) except for *Cryptococcus albidus* var. *aerius* where original data of ZVYAGINSTEVA et al. (1975) – see RATLEDGE and EVANS (1989) – is used. Not included in the table and of uncertain oleaginicity are:
Candida (*Pichia*) *guilliermondii* (22% lipid), *C. methylica* (20%), *C. stellatoidea* (*C. albicans*) (20%), *C. tropicalis* (23%), *Cryptococcus* (*Filobasidiella*) *neoformis* (22%), *Hansenula* (*Pichia*) *ciferri* (22%), *Lipomyces* spp. (two unspecified species: 59 and 67% lipid), *Schwanniomyces occidentalis* (23%), and *Trigonopsis variabilis* (an unusual yeast which only accumulates lipid – up to 40% – if grown with methionine in the medium).

[b] Nomenclature given accordingly to BARNETT et al. (1990).

[c] Although *Candida curvata* strains D and R are often referred to as *Apiotrichum curvatum,* this name is not recognized as such. BARNETT et al. (1990) consider *C. curvata* to be a synonym of *Cryptococcus curvatus* which is now the recommended name.

[d] *Endomyces magnusii.* This name is of dubious status and may be the imperfect stage of *Trichosporon pullulans* (*q.v.*) according to KREGER-VAN RIJ (1984). It is not listed by BARNETT et al. (1990).

[e] Formerly *Geotrichum candidum.* Although this organism was quoted as containing up to 50% lipid in the 1930s (see HESSE, 1949; WOODBINE, 1959) there have been no recent reports of such levels being repeated. The original oleaginous strains may, therefore, be lost.

[f] Formerly *Hansenula saturnus.*

[g] Formerly *Lipomyces lipofer.*

[h] Formerly *Trichosporon cutaneum.*

[i] Formerly *Trichosporon fermentans.*

[j] *Trichosporon pullulans* possibly includes *Endomycopsis vernalis* (see KREGER-VAN RIJ, 1984) though yeast correspond to this latter description appears to be no longer available.

[k] *Yarrowia lipolytica* is variously referred to as *Candida lipolytica* and *Saccharomycopsis lipolytica* amongst other names. Only a few strains may be oleaginous.

the lipase route must also be under severe financial pressure. Nevertheless, the approach used to achieving a yeast CBE is illustrative of what can be achieved given limited resources and manpower in trying to produce a marketable biotechnological bulk product.

Four different approaches have been tried to produce a satisfactory yeast CBE. The main difficulty to be overcome has been how to increase the inherently low content of stearic acid (maximally from 10 to 12%) in oleaginous yeasts (see Tab. 3) up to the required 30% stearate or more.

3.2.1.1 Direct Feeding of Stearic Acid

The simplest way to increase the stearic acid content of a yeast oil is to feed stearic acid or its ester to the yeast. The fatty acid or its ester is then taken up by the yeast and, although some may be degraded, the bulk seems to be directly esterified into the storage triacylglycerols in the cell. Typical results of such efforts are shown in Tab. 4. Principal commercial concerns that attempted this route were Fuji Oil Co. Ltd. (1979, 1981) and CPC International Inc. (1979, 1982a, b). Best results were obtained by feeding both palmitic acid and stearic acid, as their methyl esters, to *Torulopsis* ATC C 20507 (Fuji Oil Co. Ltd., 1979, 1981). However, even this oil required fractionation to produce a satisfactory CBE and this then significantly added to the costs.

The obvious problem with this approach is the cost of the stearic acid or the stearate – palmitate mixture. Stearic acid itself is usually produced by chemical hydrogenation of oleic acid, which in turn is usually derived most cheaply from animal fats. Unfortunately, this origin of stearic acid then negates any claim that may be made for the yeast CBE being wholly derived from non-animal sources and would make it unacceptable for vegetarians and some religious groups. This situation has now, however, changed with the advent of high-oleic acid sunflower oil (see Tab. 2). For the first time, it is therefore possible to produce stearic acid from oleic acid at a relative-

Tab. 4. Conversions of Fatty Acids or Fatty Acid Esters to Triacylglycerols as Potential Cocoa Butter Equivalents with Various Yeasts

	Candida guilliermondii[a]	*Trichosporon* sp. ATCC 20505[a]	*Torulopsis* sp. ATCC 20507[a]		*Rhodosporidium toruloides*[a]	*Candida* sp. 1-31[b]	*Candida* sp. 1-31[b]	*Rhodosporidium toruloides*[c]	*Lipomyces starkeyi*[c]
Substrate added	Me-16:0 + Me-18:0	Me-16:0 + Me-18:0	Me-18:0	Me-16:0 + Me-18:0	18:0 + glucose	Ethyl-16:0	Ethyl-18:0	16:0/18:0	16:0/18:0
Lipid [% of cells]	34.6	64.5	52.8	NG	NG	51.3	44.8	NG	NG
Fatty acyl groups [%]									
16:0	26.0	38.6	4.2	18.0	17.4	60.8	13.4	22.0	42.0
16:1	4.3	5.2	2.6	2.0	4.2	10.9	0.8	1.3	5.0
18:0	26.2	18.4	40.2	32	32.0	5.2	37.8	30	10.5
18:1	38.8	31.4	47.9	43	31.3	17.4	40.7	37	38.4
18:2	4.6	5.2	4.8	4	5.7	2.2	5.2	5.1	1.0

NG: not given
[a] from Fuji Oil Co. Ltd. (1979, 1981)
[b] from Noguchi et al. (1982)
[c] from CPC International Inc. (1982a)

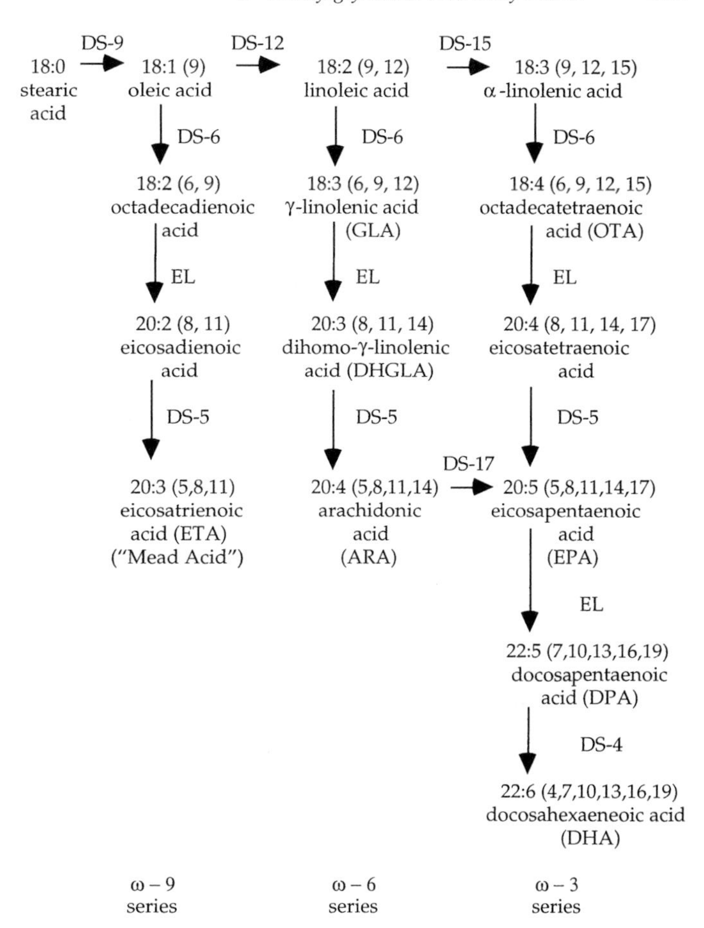

Fig. 5. Formation of polyunsaturated fatty acids in fungi. DS-n: fatty acyl desaturase acting at nth C atom of fatty acid; EL: elongase using acetyl-CoA. The DS-17 is only found in certain filamentous fungi; it is not found in animals. It could though conceivably also convert DHGLA (dihomo-γ-linolenic acid) to eicosatetraenoic acid.

ly low cost and still being classed as a plant oil.

In spite of the development of a cheap, plant source of stearic acid, this approach to a yeast CBE appears to have been abandoned though it did establish that yeast oils could be produced with an unprecedented high content of saturated fatty acids.

The remaining three approaches to produce a yeast CBE have all sought to limit the conversion of stearic acid to oleic acid within the yeast cell. This reaction (see Reaction 5, Sect. 2.3), functions at the level of the coenzyme A derivatives of the fatty acids and requires molecular O_2 as well as a supply of reducing equivalents (NADPH). It is catalyzed by Δ 9-stearoyl desaturase. The subsequent desaturations (see Fig. 5) are carried out with fatty acyl group being detached to a phospholipid (Kendrick and Ratledge 1992c, d)

3.2.1.2 Inhibition of Stearoyl Desaturase

The naturally occurring sterculic acid, *cis*-9,10-methyleneoctadecenoic acid (V), which is found in the seed oil of sterculia and kapok plants is an effective inhibitor of the Δ 9 desaturase.

$$CH_3-(CH_2)_7-\underset{\diagdown\ \diagup \atop CH_2}{C=C}-(CH_2)_7COOH$$

Sterculic acid

V

Tab. 5. Effect of Δ9- and Δ12-Cyclopropene Fatty Acids on the Fatty Acyl Composition of *Rhodosporidium toruloides* IFO 0559 (from MORETON, 1988b)

	Relative Fatty Acyl Composition [% (w/w)]						
	14:0	16:0	16:1	18:0	18:1	18:2	18:3
Control (no additions)	1.7	27.8	0.3	7.3	40.0	17.5	4.6
Sterculia oil[a] 0.3 mL L^{-1}	0.7	14.9	–	48.3	18.5	9.1	3.7
Δ12-Cyclopropene[b] 0.4 mL L^{-1}	1.0	36.0	0.3	8.3	44.7	1.8	–
Sterculia oil[a] + 0.3 mL L^{-1} Δ12-Cyclopropene[b] 0.3 mL L^{-1}	–	19.5	–	46.9	21.8	4.7	1.5

[a] Contains 50% (w/w) of the Δ9-cyclopropene $C_{18:1}$
[b] Δ12-Cyclopropene $C_{18:1}$

MORETON (1985) successfully demonstrated that as little as 100 mg sterculic acid per L of growth medium was effective inhibiting the reaction in a number of yeasts: *Candida* sp. 107, *Rhodosporidium toruloides*, and *Trichosporon cutaneum*. Stearic acid then accumulated up to 48% of the total fatty acids (see Tab. 5). However, the effect of the inhibitor was extremely specific and did not affect the subsequent conversion of oleic acid to linoleic acid (18:2) via the action of the Δ 12 desaturase enzyme. Consequently, the proportion of 18:2 in the yeast oil was unaffected by sterculic acid, and this was detrimental to the properties required of the CBE lipid. MORETON (1988b) subsequently showed that when the *cis*-Δ 12 analog of sterculic acid which had to be chemically synthesized, was added to the yeasts this now had the desired effect of decreasing the linoleic acid content (see Tab. 5). In the presence of both sterculic acid and *cis*-12,13-methyleneoctadecenoic acid, the oil of *R. toruloides* (the best of the yeasts examined) was now almost exactly as required: the three principal fatty acids, palmitic, oleic, and stearic, were present at a ratio of 1:1:2 (MORETON and CLODE, 1985).

Although this approach of MORETON (1985, 1988b) was scientifically very successful in meeting its objectives, the costs of the cyclopropene inhibitors were to prove beyond what the economics of the process could accommodate. Further, the acceptability of using known metabolic inhibitors in a biotechnology process designed to produce an edible oil was very uncertain. Clearly, regulatory authorities would be extremely cautious in allowing a yeast CBE to be used in foods that had some possibility of containing any residual inhibitor. This approach, which had been pioneered by Cadbury-Schweppes plc, the large UK-based chocolate manufacturer, was abandoned in 1986.

3.2.1.3 Mutation

Mutation and genetic manipulation of bacteria are now commonplace; haploid yeasts and molds are also similarly mutatable though, of course, many yeasts are diploid or aneuploid and would thus be not amenable to simple mutational strategies. Nevertheless and without ever apparently assessing whether their chosen yeast was haploid, aneuploid, diploid, or polypoid SMIT et al. at the Free University of Amsterdam embarked upon an ambitious project to delete the stearoyl desaturase from the yeast *Cryptococcus curvatus* (YKEMA et al., 1989, 1990; VERWOERT et al., 1989; SMIT et al., 1992). (This yeast was originally isolated by HAMMOND et al. from dairy wastes as a yeast that would readily grow on lactose, the principal carbohydrate of whey, with whey being judged to be a potential cheap and plentiful substrate (MOON et al., 1978). Initially, the yeast was named as *Candida curvata* and then reclassified as *Apiotri-*

chum curvatum which is now recognized as *Cryptococcus curvatus* (BARNETT et al., 1990).)

In the work of SMIT et al., *C. curvata* was treated with chemical mutagens and a number of auxotrophic mutants isolated that required oleic acid for growth. Such mutants would thus be unable to produce unsaturated fatty acyl groups for incorporation into their phospholipid membrane structures and would be unable to maintain membrane fluidity and growth. It was presumed, but never shown by direct enzyme assay, that these mutants would have been affected in the gene coding for the Δ 9 desaturase. One of these mutants, Ufa 33, now contained 50% stearic acid (see Tab. 6). However, as the mutant required the addition of oleic acid to the medium, it was obvious from the fatty acid analysis that the Δ12- and Δ15-desaturases (forming linoleic and then linolenic acids, respectively) were unaffected in this mutant. Although small amounts of the 18:2 and 18:3 fatty acids can be tolerated in CBE preparations their amounts need to be very low in order to maintain the sharp melting point transition of a CBE. To decrease the content of 18:2 and 18:3 in the yeast oil, a further mutation program would have been required but this was never carried out.

From the initial oleate-auxotroph, Ufa 33, a number of partial revertants and hybrids were subsequently produced (YKEMA et al., 1990; VERWOERT et al., 1989) that no longer required oleic acid to be added to the growth medium: in other words the Δ9-desaturase was now partially functional allowing a small amount of oleic acid to be synthesized inside the cell. The fatty acid profile of these yeasts (Tab. 6) all contained substantial amounts of stearic acid and, in some cases, notably that of the hybrid F33.10, with excellent similarities to cocoa butter itself.

Similar mutational programs have been reported by BEAVAN et al. (1992), working for Diversified Research Laboratories as a subsidiary company of G. Weston Ltd., Canada, and by HASSAN et al. (1993, 1994a) working in the group of GÉRARD GOMA, Toulouse, France. BEAVAN et al. (1992) isolated 6,725

Tab. 6. Cocoa Butter Fatty Acids and the Fatty Acyl Compsition of Triacylglycerols from *Cryptococcus curvatus* wild type (WT) strain, two unsaturated fatty acid auxotrophic mutants (Ufa 33 and Ufa M3), a revertant mutant (R22.72), and a hybrid (F33.10), both derived from Ufa 33 compared to the best results obtained with the wild-type strain grown with limited O_2 supply do diminish desaturase activity

	Major Fatty Acyl Groups [Relative % (w/w)]					
	16:0	18:0	18:1	18:2	18:3	24:0
Cocoa butter	23–30	32–37	30–37	2–4	–	trace
Yeast						
WT	17	12	55	8	2	1
Ufa 33[a]	20	50	6	11	4	4
Ufa M3[b]	26	37	22	8	4	2
R22.72[c]	16	43	27	7	1	2
F33.10[d]	24	31	30	6	–	4
WT-NZ[e]	18	24	48	3	1	2

[a] Grown with 0.2 g L^{-1} oleic acid; from YKEMA et al. (1990).
[b] Grown with 0.2 g L^{-1} oleic acid; from HASSAN et al. (1993, 1994a).
[c] Grown without oleic acid; from YKEMA et al. (1990).
[d] Grown without oleic acid; from VERWOERT et al. (1989).
[e] Wild type grown on whey lactose in a 500 L bubble column fermenter with a restricted O_2 supply; from DAVIES et al. (1990).

yeasts (which would include many identical organisms) of which one organism, DRL-D221, was taken as the best with respect to the rate of lactose utilization, lipid production, yield, and composition. The organism was tentatively identified as *Trichosporon cutaneum* which was originally suggested as an identification of *C. curvata*. A mutant of this yeast, DRL-JF34, contained 19% 16:0, 34% 18:0, and 23% 18:1. The work of HASSAN et al. (1993, 1994a) used the same yeast as SMIT et al.; their results are also shown in Tab. 6.

Although all three groups engaged on this work produced greatly increased proportions of stearic acid in the yeast oils, without diminution of the lipid content of the cells and without significant decrease in the overall growth performance of the yeasts, none of the mutants have yet been used in commercial trials to produce CBE oils. Large-scale trials with mutants are, however, essential as the characteristics of these organisms are often not fully revealed until they are grown in large fermenters. Thus, the use of mutants to produce yeast oil CBEs remains a potential, rather than a proven, approach to this goal.

3.2.1.4 Metabolic Manipulation

The final approach to increasing the stearic acid content of yeast oils has been to control the amount of O_2 entering the fermentation. As mentioned above (see Reaction 5, Sect. 2.3), O_2 is an essential co-reactant in the stearoyl desaturase reaction. Accordingly, DAVIES et al. (1990) carried out a series of fermentation runs with *C. curvata* in which the oxygen uptake rate was progressively decreased by the simple expedient of restricting the air supply (Tab. 6). Best results were obtained with a 500 L fermenter in which the O_2 supply could be effectively regulated. (The effect of O_2 deprivation is extremely difficult to demonstrate with conventional 5 L laboratory fermenters as the amount of air that needs to be supplied to produce the effect is so small that the usual air control systems are unsuitable; C. RATLEDGE and D. GRANTHAM, unpublished work.) The highest levels of stearic acid so obtained by DAVIES et al. (1990) were not far from the required cocoa butter composition (see Tab. 6). Significantly and interestingly, this approach by DAVIES et al. (1990) also served to decrease the contents of the polyunsaturated fatty acids (18:2 and 18:3) which, of course, also require O_2 in their formation. Thus by a single and obvious metabolic manipulation, the goal of a high-stearate yeast oil was almost perfectly achieved.

3.2.1.5 Conclusions

The pursuit of yeast oil as a CBE has been now ceased after a decade of intensive activity. DAVIES (1984) was the first to appreciate the potential of this as a commercial target and to carry out a sustained program to this end. Of key significance was the appreciation that a cheap substrate would be essential for success as, from the previous sections (Sect. 2.2), approximately 5 t of glucose or equivalent carbohydrate are needed to produce 1 t of oil. DAVIES, by working in New Zealand, quickly realized that the waste whey generated from the extensive dairy industry of that country could represent such a source of fermentable carbohydrate. The earlier discovery of the yeast *C. curvata* (MOON et al., 1978) that could be readily grown on lactose, which is the major carbohydrate of whey, quickly indicated that it was probably the most likely one to use in this process. DAVIES et al. (see DAVIES, 1988, 1992a, b, c; DAVIES and HOLDSWORTH, 1992) then developed a process up to a pilot-plant level of 200 m^3 using a bubble column fermenter with *Candida curvata* and casein (milk) whey as its substrate. From those extensive trials, DAVIES was able to calculate the probable economics of a yeast CBE process. This was based on the use of six, similar sized reactors which would operate continuously and with a cell recycling mode of operation to allow yeast densities of 50 g dry cells per L to be achieved. Recovery of the oil from the yeast was also examined and the economics of this process taken into account in the final calculations. A summary of DAVIES' costings is given in Tab. 7. The assumption is made that the yeast CBE product would sell for approximately 80% of the price of cocoa butter and that no significant

Tab. 7. Yeast CBE Process: Estimated Operating Budget (from DAVIES and HOLDWORTH, 1992)

Basis of process	
Available whey	200,000 m^3/a^{-1}
Lactose content	39% (w/v)
Cost of whey (in New Zealand)	\$ 0.5 per m^3
CBE yield	0.16 kg kg^{-1} lactose
CBE production	1,250 t a^{-1}
Value of CBE	\$ 2,400 t^{-1}
Total sales value of CBE	\$ 3,000,000
Costs	
Direct manufacturing costs (untilities, substrate costs, downstream processing, wages, etc.)	\$ 1,000,000
Manufacturing overheads (laboratory costs, site charges, effluent disposal, maintenance, insurance, service overheads, etc.)	\$ 460,000
Finance and sales (distribution, research, and development)	\$ 300,000
Plant depreciation over 10 years on capital of \$ 5.4 M	\$ 540,000
Interest at 12%	\$ 650,000
Total costs	= \$ 2.95 M
Total sales	= \$ 3.0 M
Profit	= \$ 50,000

costs would be involved in toxicological trials before the CBE could be offered for sale.

DAVIES (1992a) and DAVIES and HOLDSWORTH (1992) assumed that a fermentation plant would have to be built specifically for the process. Clearly, if a fermentation plant already existed and could be used without major modification, then this would significantly decrease the indirect costs and the whole process might then show a significant annual profit.

Although DAVIES' cost analysis will obviously change from country to country and will be heavily influenced by the cost of substrate, the overall conclusion is that this yeast CBE process is a process that is waiting for its day to come rather than being another uneconomic biotechnology pipe dream. Clearly, a number of other substrates can be used besides whey (see BEDNARSKI et al., 1986; GLATZ et al., 1985, VEGA et al., 1988; FALL et al., 1984; GUERZONI et al., 1985; DOSTALEK, 1986; HASSAN et al., 1994b) though it is essential that these be available throughout the year to enable continual operation of the plant. Costs of labor and utilities are obviously cheaper in some countries than in others and, consequently, it would not be surprising if this yeast CBE process or a similar one was not adopted somewhere in the world over the next decade. Only the continuing low price of cocoa butter will prevent it from becoming an operating reality.

It has recently been reported by ROUX et al. (1994) that some species of *Mucor*, especially *M. circinelloides*, will simultaneously produce a CBE-SCO as well as producing a valuable polyunsaturated fatty acid (γ-linolenic acid). This is described later in Sect. 3.3.1.

3.3 Molds

A list of oleaginous species of mold is given below.

Lipid contents of oleaginous molds grown in

the vegetative mycelial state are given in parentheses (taken from RATLEDGE, 1986; 1989a, 1993; LÖSEL, 1988 and additional references as indicated).

Zygomycetes

Entomophthorales

Conidiobolus nanodes (26); *Entomophthora conica* (38); *E. coronata* (43); *E. obscura* (34); *E. thaxteriana* (32); *E. virulenta* (26); *Glomus caledonius* (72)

Mucorales

Absidia corymbifera (27); *A. spinosa* (28); *Blakesleea trispora* (37); *Cunninghamella echinulata* (45); *C. japonica* (60); *C. elegans* (44, 56); *C. homothallica* (38); *Mortierella alpina* (33) (TOTANI et al., 1992); *M. elongata* (34) (BAJPAI et al., 1992); *M. isabellina* (63–86); *M. pusilla* (59); *M. vinacea* (66); *Mucor albo-ater* (42); *M. circinelloides* (65); *M. hiemalis* (42) (KENNEDY et al., 1993); *M. miehei* (25); *M. mucedo* (51); *M. plumbeus* (63); *M. pusillus* (26); *M. ramanniana* (56); *M. spinosus* (47); *M. vinacea* (25) (HANSSON und DASTALEK, 1988); *Phycomyces blakesleeanus* (33); *Rhizopus arrhizus* (32–57); *R. delemar* (32–45); *R. oryzae* (57); *Zygorhynchus moelleri* (40)

Peronosporales

Pythium irregulare (42); *P. ultimum* (48)

Ascomycotina

Ascomycetes

Aspergillus fisheri (53); *A. flavus* (35); *A. minutus* (35); *A. nidulans* (51, 25); *A. ochraceus* (48); *A. oryzae* (37–57); *A. terreus* (57); *A. ustus* (28); *Chaetomium globosum* (54); *Fusarium bulbigenum* (50); *F. equiseti* (48); *F. graminearum* (31); *F. lini* (35); *F. lycopersicum* (35); *F. oxysporum* (29, 34); *Fusarium* sp. *N11* (39); *Gibberella fujikuroi* (*F. moniliforme*) (48); *Humicola lanuginosa* (75); *Penicillium gladioli* (32); *P. javanicum* (39); *P. lilacinum* (51, 56); *P. soppii* (40); *P. spinulosum* (64); *Stilbella thermophila* (38)

Clavicipitacae

Claviceps purpurea (31–60)

Tulasuellales

Pellicularia practicola (39)

Hyphomycetes

Cladosporium herbarum (49); *Malbranchea pulchella* (27); *Myrothecium* sp. (30); *Sclerotium bataticola* (46)

Hymenomycetes

Lepista (Tricholoma) nuda (48)

Ustilaginomycetes

Ustilaginales

Sphacelotheca reiliana (41); *Tilletia controversa* (35); *Tolyposporium ehrenbergii* (41); *Ustilago zeae* (59)

Urediniomycetes

Cronartium fusiforme (28); *Puccinia coronata* (37)

For inclusion in the list, the lower cut-off of a lipid content of 25% has been used. There are numerous molds that could have been listed if the limit had been dropped to 20% and, indeed, the commercial value of mold lipids, as will be explained below, could still be high even with molds having lipid contents of less than 20%. Commercialization of mold oils, therefore, unlike yeast oils, does not depend so much on the amount of oil that a mold produces but on the quality of that oil. The quality of the oil is determined by its fatty acid profile (Tab. 8): some fatty acids, particularly the polyunsaturated ones (PUFA) that have nutritional and some medical importance, are therefore select targets for current developments in this field.

The number of oleaginous species of mold is greater than the number of oleaginous yeasts (cf. Tab. 3 and the list given above) but this is hardly surprising considering that there are some 100 times more molds (60,000 approx.) than yeasts. Although most of the 590 species of yeast have probably been assessed for oleaginicity, it is likely that most molds have not. Therefore, one may expect the list of oleaginous molds will be considerably increased as further work continues to be carried out. Extensive reviews of fungal lipids have been prepared by LÖSEL (1988) and WEETE (1980).

As a generalization, molds produce higher levels of the polyunsaturated acids 18:2 and 18:3, than yeasts. Although there are various isomeric possibilities, the fatty acids in yeasts, molds, and algae appear to be the same as those found in plants. With linolenic acid (18:3), two isomers occur in both plants and microorganisms. The more common isomer is

Tab. 8. Fatty Acid Analyses of Lipid from Selected Molds (data taken from RATLEDGE (1986, 1989a) and LÖSEL (1988) and additional references indicated)

Organism	Major Fatty Acyl Groups [Relative % (w/w)] of						Others
	14:0	16:0	18:0	18:1	18:2	18:3	
Entomophthorales							
Conidiobolus nanodes[a]	1	23	15	25	1	4[f]	20:1 (13%) 22:1 (8%) 20:4 (4%)
Entomophthora coronata	31	9	2	14	2	1[f]	12:0 (40%)
E. obscura	8	37	7	4	trace	trace	12:0 (41%)
Mucorales							
Absidia corymbifera	1	24	7	46	8	10[f]	–
Cunninghamella japonica	trace	16	14	48	14	8[f]	–
Mortierella alpina[b]	–	19	8	28	9	8[f]	20:3 (7%) 20:4 (21%)
M. elongata[c]	–	7	2	18	12	25[f]	20:4 (16%) 20:5 (15%)
M. isabellina	1	29	3	55	3	3[f]	–
Rhizopus arrhizus	19	18	6	22	10	12[f]	–
Mucor alpina-peyron[a]	10	15	7	30	9	1[f]	20:0 (8%) 20:3 (6%) 20:4 (5%)
Peronosporales							
Pythium ultimum[d]	7	15	2	20	16	1*	20:1 (4%) 20:4 (15%) 20:5 (12%)
P. irregulare[e]	8	17	2	14	18	–	20:1 (5%) 20:4 (11%) 20:5 (14%)
Ascomycetes							
Aspergillus terreus	2	23	trace	14	40	21	–
Fusarium oxysporum	trace	17	8	20	46	5	–
Pellicularia practicola	trace	8	2	11	72	2	–
Pennicillium spinulosum	–	15	7	42	31	1	–
Hyphomycetes							
Cladosporium herbarum	trace	31	12	35	18	1	–
Ustilaginales							
Tolyposporium ehrenbergii	1	7	5	81	2	–	–
Calvicipitacae							
Claviceps purpurea	trace	23	2	19	8	–	12-HO-18:1 (42%)

[a] KENRICK and RATLEDGE (1992a)
[b] YAMADA et al. (1992)
[c] BAJPAI et al. (1992)
[d] GANDHI and WEETE (1991)
[e] O'BRIEN et al. (1993)
[f] γ-linolenic acid, 18:3 (ω-6)

Tab. 9. Fatty Acid Profiles of a Commercial Fungal Oil Product Compared with Evening Primrose Oil, Borage Oil, and Blackcurrant Oils Containing γ-Linolenic Acid

	Mucor circinelloides[a]	Evening Primrose Seed Oil	Borage Seed Oil	Blackcurrant Seed Oil
Oil content [% (w/w)]	20	16	30	30
Fatty acid				
16:0	22–25	6–10	9–13	6
16:1	0.5–1.5	–	–	–
18:0	5–8	1.5–3.5	3–5	1
18:1	38–41	6–12	15–17	10
18:2	10–12	65–75	37–41	48
γ-18:3	15–18	8–12	19–25	17
α-18:3	0.2	0.2	0.5	13
20:1	–	0	4.5	–
22:1	–	–	2.5	–
24:0	–	–	1.5	–

[a] Production organism used by J & E Sturge Ltd., Selby, Yorkshire, UK, from 1985–1990.

α-linolenic acid, 18:3 (*c* 9, *c* 12, *c* 15), or 18:3 (ω-3), – see Sect. 1.1 – which is found in most plants seed oils, yeasts, and the majority of molds. The more unusual isomer is γ-linolenic acid – 18:3 (*c* 6, *c* 9, *c* 12) or 18:3 (ω-6) – which occurs in the seed oils of *Oenothera* (evening primrose), *Ribes* (blackcurrant, redcurrant, etc.) and the borage family (Boraginaceae) (see Tab. 9). It also occurs throughout the lower fungi, also known as phycomycetes (see Tab. 8). Both isomers occur simultaneously in some algae, though not in molds nor with *Oenothera* and borage plants. In *Ribes*, however, both isomers occur in equal amounts. Longer chain polyunsaturated fatty acids, up to 22:6, have been detected in the lipids of many species of the phycomycetes and these are discussed separately below.

Unusual fatty acids such as hydroxy fatty acids or branched fatty acids are found, respectively, in a few species of *Claviceps* and *Conidiobolus* (TYRRELL and WEATHERSTONE, 1976). The high content of ricinoleic acid, 12-hydroxyoleic acid, in *Claviceps* spp. (see Tab. 8) occurs only in the sclerotial tissue of the fungus and is absent from the vegetative mycelium (see LÖSEL, 1988). It has, therefore, no potential as an alternative source of castor oil which is the major source of this acid. Epoxy and dihydroxy fatty acids are also found in relative abundance in the lipids extracted from the spores of some basidiomycetes. Some acetylenic acids may also occur (LÖSEL, 1988). The occurrence of hydroxy fatty acids in fungi has been recently reviewed by VAN DYK et al. (1994).

Although molds contain an exceptional diversity of fatty acids, current commercial interest centers principally upon the formation of particular PUFAs that have dietary or medical applications. The role of such PUFAs, in both healthy and dysfunctional patients, has been the subject of considerable research and investigation. For a more than adequate statement of the current status of this work, the reader is referred to a recent Congress whose proceedings are given in a volume edited by SINCLAIR and GIBSON (1992). A discussion, or even *précis*, of the various conditions and aliments that seemingly benefit by a dietary intake of PUFA is beyond the scope of this review but consultation of this symposium volume should give adequate information on most topics. The following sections review the current work being carried out to produce PUFAs from fungal sources. The formation of PUFAs is set out in Fig. 5.

3.3.1 γ-Linolenic Acid (GLA, 18:3 ω-6)

Lower fungi, that is the "phycomycetes", invariably produce the γ-isomer of 18:3 instead of the more common α-isomer. SHAW (1966) was the first to point out this distinction amongst fungi, though the occurrence of GLA in fungal lipids was first reported by BERNHARD and ALBRECHT (1948) who had examined the lipid from *Phycomyces blakesleeanus*.

Interest in oils containing this acid has a long history and the oil from the seeds of evening primrose (*Oenothera biennis*) have been used for many centuries, being described as the "King's Cure-All", as a remedy for a number of disorders. The efficacy of evening primrose oil has been attributed to its content of GLA (about 8–10% of the total fatty acids). GLA itself has been reported as suppressing acute and chronic inflammations, decreasing blood cholesterol concentrations, and improving atopic eczema (HORROBIN, 1992), however, SCHÄFER and KRAGBALLE (1991) found no clear evidence in support of GLA being an efficacious treatment for atopic dermatitis. Alternative plant sources to evening primrose have been identified more recently and include borage (*Borago officinalis*) and *Ribes* spp. (see Tab. 7).

In view of the known occurrence of GLA in fungi, steps to develop a biotechnological route to GLA were first initiated in the author's laboratory in 1976. A strain of *Mucor circinelloides* (*M. javanicus*) was identified from a large screening program as being a suitable production organism. The first sales of the GLA-rich oil were in 1985, the process being run by J & E Sturge Ltd. (now Harmann & Reimer) of Selby, Yorkshire, UK, at the 220 m^3 level. The process ran until 1990 when production ceased following transfer of the company to its present owners who are part of the Bayer industrial group.

The specifications of the fungal oil were equal or better than evening primrose oil in almost every respect including its much higher content of GLA (see Tab. 9). A similar process to the Sturge one, was considered to have been developed by Idemitsu Petroleum Co. Ltd. of Japan (see RATLEDGE, 1989b) which used *Mortierella isabellina*. However, a recent report from NAKAJIMA and IZU (1992) of that company makes no mention of any previous large-scale industrial process for GLA production and, indeed, highlights only strains of *Mucor circinelloides* as potential candidate organisms. The largest scale of operation was only 30 L but, optimistically, the authors consider that scale-up to 200 m^3 was now a feasible proposition, but no reports of work at this level were given. One strain of the mold was found that could produce 25% GLA in the total fatty acid with, however, only a 6% oil content of cells. The highest oil producer (30% of the cells) only produced 10% GLA in the fatty acid. This reverse correlation between oil and GLA content has been previously noted by RATLEDGE (1989b) and now been documented in some detail by KENNEDY et al. (1993).

KENNEDY et al. (1993) showed that even in a single organism (they used *Mortierella ramanniana*, *Mucor hiemalis,* and *Mucor circinelloides*) the GLA content of the oil could range from 5%–32% with oil contents from 43%–4% with *Mucor hiemalis* with a narrower range for *Mortierella rammaniana*. Maximum productivity of GLA (that is $g\ L^{-1}\ h^{-1}$) was calculated for *Mucor hiemalis* with a GLA content of the oil at 8–10%. For *Mucor circinelloides*, maximum productivity was with a GLA content of 14–16% which is close to the commercial process outlined above and shown in Tab. 9.

High proportions of GLA of 20–26% in the total fatty acids, together with oil contents of 15–25%, appear to be attainable by a number of species of *Mucor* and *Mortierella* (HANSSON and DOSTALEK, 1988; DAVIES, 1992a; KENNEDY et al., 1993; ROUX et al., 1994). It is, therefore, a matter of simple screening to identify a likely candidate for large-scale production. However, because of the high costs of fermentation processing, other parameters have to be taken into account besides GLA content: these include (1) the density to which the cells can be grown – values of up to 25 $g\ L^{-1}$ are not uncommon (see KENNEDY et al., 1993; NAKAJIMA and IZU, 1992) but are still far from optimal (about 40–80 $g\ L^{-1}$ should be attainable); (2) the rate of growth –

full growth and lipid accumulation should be attained within 72–96 h; and (3) the ability to extract the oil from the cells. The oil should obviously be free from any deleterious material, including free fatty acids and partial acylglycerols, but, as toxicological trials may be called for prior to the release of the oil for human consumption, it is an obvious advantage if the organism being used already has an established record of safe usage in foods. For these many reasons, *Mucor circinelloides* seems to have been an excellent choice for a GLA-production organism.

ROUX et al. (1994) have recently reported that some *Mucor* spp. when grown on acetic acid also produce a high content of stearic acid (18:0) besides GLA (up to 38 mg per g dry biomass was attained) and have suggested that, by appropriate fractionation, it should be possible to produce both a GLA-rich oil and a cocoa butter equivalent fat. Although some strains of *Mucor* produced a high content of stearic acid when grown in glucose (27% of the total fatty acids with *M. flavus*), the highest combined yields of stearic acid and GLA were with *M. circinelloides* grown in a pH-stat, fed-batch culture using acetic acid as sole carbon source. Stearic acid was up to 19% of the neutral lipid with GLA at 8%.

3.3.2 Dihomo-γ-Linolenic Acid (DHGLA, 20:3 ω-6)

DHGLA is produced biosynthetically from GLA by chain elongation (see Fig. 5). It is the precursor of the Group 1 of prostaglandins and thromboxanes and is often a minor component of lipids from fungi and algae but is also found in animals. There is probably no large market for DHGLA. Nevertheless, various attempts have been made to develop a process for its production as, undoubtedly, small amounts of oils containing high amounts of DHGLA would command a very high price if only for exploratory trials and for experimental laboratory work.

Small amounts of DHGLA, up to 5% of the total fatty acids have been found in oils from *Conidiobolus* spp. (NAKAJIMA and IZU, 1992) and in *Mortierella* spp. of the subgenus *Mortierella* (AMANO et al., 1992). The highest amounts occurred with *M. alpina* strain 1S-4 which had been grown in the presence of sesame oil (SHIMIZU et al., 1989a; YAMADA et al., 1992). The concept behind adding the sesame oil to the fungal culture had been to see if exogenous oils could be taken up by the cells and then desaturated to particular fatty acids. With sesame oil there was an apparent inhibition in the formation of arachidonic acid (20:4), which being produced directly from DHGLA (see Fig. 5), then led to the accumulation of DHGLA. The inhibitor was identified as a minor component of sesame oil, sesamin, which acted specifically against the $\Delta 5$ desaturase (SHIMIZU et al., 1991). DHGLA was produced up to 23% of the total fatty acids and at a yield of 2.2 g L^{-1}.

NAKAJIMA and IZU (1992) have similarly shown that a number of anisole derivatives when presented to *Conidiobolus nanodes* also led to the accumulation of DHGLA. Like sesame seedoil with *M. alpina*, these compounds had only a minor effect on cell growth and lipid accumulation. Maximum effect was produced with *tert*-butylhydroxyanisole (BHA) giving 18% DHGLA in the total lipid fraction which was about 35% of the cells.

As both BHA and sesamin appeared to act as specific inhibitors of the $\Delta 5$ desaturase, the next logical step was to delete this enzyme by mutational techniques. JAREONKITMONGKOL et al. (1992a, c) reported the results of such a study using *M. alpina* and succeeded in isolating a mutant that produced 3.2 g L^{-1} DHGLA, that is 123 mg per g cells, and accounting for 23% of the total fatty acids. In comparison, the wild type produced less than a quarter of this amount. Other mutants of the same organism have been reported that accumulated increased amounts of other PUFAs (see Sect. 3.3.6, eicosatrienoic acid).

The approach of deleting various desaturases at the genetic level that are involved in conversion of the PUFAs is obviously a very powerful one. Simple mutational techniques coupled with extensive screening for the correct mutants can clearly pay handsome commercial rewards.

3.3.3 Arachidonic Acid (ARA, 20:4 ω-6)

Arachidonic acid (ARA) and eicosapentaenoic acid (EPA, see below) are intermediates in the formation of several key prostaglandins and leukotrienes which exert profound physiological control over various bodily functions and are the subject of much nutritional and medical research (see SINCLAIR and GIBSON, 1992).

ARA has a much more restricted distribution than GLA and it is clear that many molds do not synthesize fatty acids beyond C_{18} in length (SHAW, 1966). However, phycomycetes molds of the subdivision Mastigomycotina synthesize fatty acids up to C_{22} and formation of ARA has been recorded, for example, in several *Pythium* spp. (GANDHI and WEETE, 1991), *Saprolegnia parasitica* (GELLERMAN and SCHLENK, 1979), in several *Conidiobolus* and *Entomophthora* spp. (TYRRELL, 1967, 1968, 1971; KENDRICK and RATLEDGE, 1992a, b, c; NAKAJIMA and IZU, 1992) and in a number of species of *Mortierella* subgenus *Mortierella* (Totani and OBA, 1987, 1988; YAMADA et al., 1987a; SHIMIZU et al., 1988a; SHINMEN et al., 1989; AMANO et al., 1992). The subject has been recently reviewed by RADWAN (1991) and BAJPAI and BAJPAI (1992), the latter recorded the ARA contents in 27 species of phycomycetes fungi as well as in 42 species of marine algae (see also below). The more general review of LÖSEL (1988) records a number of fungi that contain ARA, though without regard to any biotechnological potential.

All researchers and reviewers to date have been generally agreed that the family of ω-6 PUFA, which includes ARA as well as GLA, are confined to the "lower fungi" or phycomycetes. RADWAN and SOLIMAN (1988), however, reported that they had found ARA in the lipids of a number of ascomycetes (or higher) fungi: *Aspergillus versicolor, A. niger, A. oryzae, A. ustus, A. fumigatus, Paecilomyces lilacinus, Penicillium* sp., *Fusarium oxysporum,* and another *Fusarium* sp. In all cases, the fungi had been cultivated on single, shorter chain fatty acids, either saturated or monounsaturated, i.e., 14:0, 16:0, 18:0, or 18:1. As the identity of the ARA was not confirmed by capillary GC or by CG-MS, only by argentation TLC and by packed column GC, there must be grave doubts about the authenticity of these claims. Biochemically, it would be an unprecedented reaction that could convert a fatty acid of the ω-3 series, which are invariably produced by these fungi (LÖSEL, 1988), into the ω-6 series (see Fig. 5) as this would involve a saturation reaction of a double bond at the ω-3 position which has never yet been recorded in any aerobically-growing microorganism. It is, therefore, more than likely that the 20:4 PUFA reported by RADWAN and SOLIMAN (1988) as arachidonic acid was not the ω-6 isomer (i.e., ARA) but was the ω-3 isomer, that is 20:4 (8, 11, 14, 17) which would have behaved in both the GC and TLC analyses as ARA. Nevertheless, it is still quite exceptional for these higher fungi to be recorded as producing any fatty acid beyond C_{18} in length (LÖSEL, 1988) and one can only conclude that it was the cultivation of these fungi on shorter chain fatty acids that led to this most unusual result, a result, though, which has yet to be confirmed in another laboratory.

The highest ARA contents have been recorded in *Mortierella alpina* with up to 79% ARA in the total fatty acids which represented 26% of the cell dry weight (TOTANI and OBA, 1987). Further work with this organism has been developed up to the 300 L scale (TOTANI et al., 1992) with some slight diminution of yield. Significantly, exceptionally long fermentation times up to 16 d were needed to produce the greatest yields. Such lengthy times would invariably increase the costs of any large-scale process.

Both YAMADA et al. (1992) and BAJPAI et al. (1991c) have reported some possibly interesting developments in which the mold of choice is first grown for up to 10 d then allowed to stand at a lower temperature without further aeration. Under these conditions, the content of ARA (and EPA) increased up to 70% of the total fatty acids. However, mass balances were not carried out with these stored cells so it is not immediately apparent where the extra ARA or lipid might have originated; although the lipid content of *M. alpina* increased from 14–15% to 33–45% during the aging of the cells for 6 d (BAJPAI et al., 1991c) it is not clear from where this extra lipid could have arisen. One simple explanation is that biomass other than lipid was self-utilized. Thus the total amount of lipid may not have changed but only appeared to increase as the remainder of the biomass was consumed.

3.3.4 Eicosapentaenoic Acid (EPA, 20:5 ω-3)

EPA exerts a number of physiological effects when fed to experimental animals including lowering of blood triacylglycerol concentration. Consequently, this would decrease the potential for a coronary heart attack and, therefore, the consumption of EPA is encouraged by many advocates (SIMOPOULOS, 1989). EPA, like ARA is the precursor of prostaglandins and leukotrienes. Currently, the major source of EPA is fish oil where it occurs, usually in low concentrations along with docosahexaenoic acid (DHA) (see below).

In molds, EPA frequently occurs along with ARA. The conversion of ARA to EPA occurs directly in fungi (see Fig. 5) but not in animals (GELLERMAN and SCHLENK, 1979); the necessary $\Delta 17$ desaturase for this conversion is thus apparently unique to fungi. Small amounts (usually <10% of the fatty acids) of EPA have been recorded in a number of lower fungi (LÖSEL, 1988) as well as in the marine fungi *Thraustochytrium* and *Schizochytrium* spp. (ELLENBOGEN et al., 1969) though these latter fungi also contain higher amounts of DHA (see below).

SHIMIZU et al. (1988a) were the first research group to search specifically for the occurrence of EPA at high levels in fungi though its presence had long been known as minor component amongst many of arachidonic acid-containing fungi (SHAW, 1966). SHIMIZU et al. (1988a, b) having earlier screened fungi for their ARA contents (YAMADA et al., 1987a, b) observed that many of these fungi showed enhanced contents of EPA if they were grown at a lower temperature (12°C) than that used for ARA formation (28°C) (SHIMIZU et al., 1988b). Highest yields were attained with *Mortierella alpina* and *M. hygrophila* which produced 29 and 41 mg EPA per g cells. Subsequent work showed that exogenously added α-linolenic acid (18:3 ω-3) was converted by *M. alpina* into EPA eventually pushing up the yield to 67 mg per g dry cells (YAMADA et al., 1992; SHIMIZU et al., 1989b). This was an unusual finding as most fungi will not modify exogenously added fatty acids (RATLEDGE, 1989b).

Other groups have not been as successful as the Japanese in finding productive fungi for EPA. BAJPAI et al., (1992) found the highest yields with *Mortierella elongata* were 15 mg per g dry cells and when α-linolenic acid was added this increased to only 36 mg per g. In *Pythium ultimum* the maximum content was 34 mg EPA per g dry weight which was attained only after careful selection of the strain and its culture conditions (GANDHI and WEETE, 1991). O'BRIEN et al. (1993) reported a maximum yield of 25 mg EPA per g dry cells using *Pythium irregulare* and have recently described a pilot-plant process using a colloid mill for the recovery of EPA at 96% yield from this fungus (O'BRIEN and SENSKE, 1994).

3.3.5 Docosahexaenoic Acid (DHA, 22:6 ω-3)

DHA is abundant in the phospholipids of retina and brain tissues and is usually regarded as an essential fatty acid for humans and other animals (THOMAS and HOLUB, 1994). It occurs in the oils of many fish where, along with EPA, it may account for over 50% of the total fatty acids (ACKMAN, 1994; NICHOLS et al., 1994). However, fish do not synthesize either EPA or DHA but acquire them by ingestion of planktonic algae. DHA is considered important in the development of brain tissue of babies but conclusive evidence for the nutritional role of DHA is lacking as it is always administered during feeding trials along with EPA as fish oil is always used in such studies. A number of different nutritional roles for DHA have been proposed (THO-

MAS and HOLUB, 1993; see also YADWAD et al., 1991).

Although algae are clearly a potential source of DHA (see below), there are several fungi that have been considered as possible candidate organisms for its production though it must be said that prospects for a biotechnological route to DHA via fungi seems remote at this stage.

The presence of DHA has been recorded in small amounts in the lipids of *Conidiobolus* and *Entomophthora* spp. (TYRRELL, 1967, 1968, 1971) but more abundantly in the lipids of the marine fungi *Thraustochytrium* and *Schizochytrium* (ELLENBOGEN et al., 1969), in which the high content of DHA and EPA were considered to have a role in maintaining membrane fluidity in the organism whilst at low temperatures and in saline conditions. BAJPAI et al. (1991a, b) and KENDRICK and RATLEDGE (1992b) both independently re-examined the marine fungi as potential sources of DHA. All these fungi grew slowly and generally to low growth yields giving only low contents of lipid. None contained the key enzyme for oleaginicity, ATP–citrate lyase (see Sect. 2.3) and KENDRICK and RATLEDGE, (1992b) concluded that these fungi would be extremely difficult to exploit directly for the production of DHA. Maximum amounts of DHA were produced by *T. aureum* ATCC 34304 at up to 50% of its total fatty acids but with a lipid content of less than 15% (BAJPAI et al., 1991a, b) and usually not more than 10% (KENDRICK and RATLEDGE, 1992b). About two-thirds of the lipid was neutral lipid (triacylglycerols) and this still had a high content (30%) of DHA. However, the greatest difficulty with the exploitation of this fungus was its low growth yield: KENDRICK and RATLEDGE (1992b) obtained only 4 g biomass per L over 72 h under conditions which yielded up to 12 g L^{-1} of other fungi.

3.3.6 Eicosatrienoic Acid (ETA, 20:3 ω-9, "Mead Acid")

This comparatively rare PUFA occurs in small amounts in lipids of animal tissues and arises by direct elongation and desaturation of oleic acid (18:1 ω-9) (see Fig. 5). It was first described by MEAD and SLATON (1956). Its nutritional status still remains unclear though it may be converted to the 12-hydroxy derivative which can affect blood platelet aggregration (LAGARDE et al., 1985). It is presumably produced because animal tissues are unable to desaturate fatty acids between the existing double bond, in this case at the $\Delta 9$ position, and the terminal CH_3 group.

There has only been one report of the formation of ETA in fungi: JAREONKITMONGKOL et al. (1992b), following their work on the deletion of various fatty acid desaturases in *Mortierella alpina* (see Sects. 3.3.2 and 3.3.3) which led to increased production of DHGLA and ARA, found another mutant that was no longer able to convert oleic acid (18:1 ω-9) to linoleic acid (18:2 ω-6). This mutant accumulated several fatty acids of the ω-9 series: 18:2 (6,9), 20:2 (8,11) and also ETA. Under optimal conditions, in a fed-batch submerged cultivation for 10 d at 20°C, ETA reached 56 mg per g dry biomass = 0.8 g L^{-1}. Growth of the mutant itself reached about 15 g L^{-1} indicating that a small scale process might be possibly developable to produce this acid if sufficient commercial demand for it was forthcoming.

3.3.7 Conclusions

Although molds contain a large number of fatty acids, their biotechnological potential lies in their ability to produce a few selected polyunsaturated fatty acids (PUFA) in some quantity. Already we have seen the commercial production of γ-linolenic acid (GLA) during the 1980s and it is likely that the production of arachidonic acid (ARA) may not be far off. A number of industrial companies are known to be developing processes for this PUFA and at least one company (Martek Corp. Inc, Maryland, USA) now offers a triacylglycerol oil containing 48% ARA for sale at $ 2,000 per kg. Free ARA itself at 80% purity is offered at $ 20 per g (i.e., $ 20,000 per kg). Demand may increase for this particular fatty acid if it can be unequivocally demonstrated that it is beneficial when added to milk destined for neonatal babies. The absence of ARA, DHA, and GLA in cow's milk (but not mother's milk) has suggested that these acids may fulfil important nutritional roles in the development of the early brain of children. Current sources of ARA are from animals and as long as there is continuing and developing concern over the presence of undetected viruses and prions in animal products, the greater will be the driving force to identify alternative and safe sources of these acids. PUFAs derived from molds by fermentation technology are, of course, free from such infectious agents and can be produced to higher levels of quality control than plant oils. Furthermore, they do not contain herbicide or pesticide residues that would occur in oils derived from plant crops that have been treated with these chemical agents as part of the usual agricultural regimen of routine crop spraying. Experience with GLA from *Mucor circinelloides* has indicated it to be a very high quality oil free from all deleterious substances. Once a high market demand for an ARA-rich oil has been developed, the price should fall dramatically from the level quoted above. Production costs for a finished mold oil should lie in the region of $ 25–35 per kg. This price would include refinement (removal of non-TAG lipids) and decolorization or deodorization if needed by passage through charcoal which are standard procedures for producing all high quality oils.

Demand for PUFAs other than GLA and ARA from molds is less certain although DHGLA and ETA ("Mead acid") could be produced if needed but these tend to be regarded as "rare" PUFAs which are probably only required in small amounts (say 10 kg annually) for experimental purposes. With EPA and DHA, mold sources are not as good as algae or bacteria for these acids: EPA rarely exceeds 20% of the total fatty acids in a mold oil and is often much less though the recent description of a pilot-plant extraction process for the recovery of EPA from *Pythium irregulare* (O'Brien and Senske, 1994) may indicate possible future developments in this area. Although DHA can occur at up to 50% of the total fatty acids in a few molds, these species are slow-growing and may be difficult to develop on a large scale though their exploitation appears under active consideration.

As with all biotechnological products, what is produced will be dictated by market forces. Increased demand for a commodity invariably pushes up the price: should demand begin to increase for any of the PUFAs then rapid exploitation of molds could then be anticipated. The high quality of the oils ensures that molds are realistic alternative sources to either plant or animal products.

3.4 Algae

The term "algae" covers 14 distinct biological groups and includes both the macro- and microalgae. The macroalgae are the seaweeds and related families; the microalgae are the equivalent of eukaryotic microorganisms but also include the cyanobacteria, formerly termed the blue-green algae, which are part of the prokaryotic eubacteria. It is the microalgae that are the subject of most research for the production of designated lipids and will therefore be covered here.

Microalgae have long been used as sources of protein for use in animal and human foods. Their potential as sources of biomass and fine chemicals has been the subject of several recent major monographs (Becker, 1993; Cresswell et al., 1989; Borowitzka and Borowitzka, 1988; Stadler et al., 1988). Prospects for the production of oils and fatty acids by algae biotechnology have been reviewed, in general by Kyle (1991), Yongmanitchai and Ward (1989), Volkmann (1989), and Borowitzka (1988), and, with respect to the production of specific fatty acids by Kyle (1992), Seto et al. (1992), Cohen and Heimer (1992), Boswell et al. (1992), and Kyle et al. (1992).

The main problem in the biotechnological exploitation of algae is their cultivation. Most algae will only grow photosynthetically and, therefore, require illumination and, although their carbon source, being CO_2, is regarded as free, growth is usually limited by the CO_2 content of the air. When algae are grown outdoors in ponds or lagoons, a warm ambient temperature is needed besides high illumination which limits their geographical development but many algae are susceptible to contamination by other algae or even bacteria or may be attacked by predatory protozoa. Consequently, to maintain a pure monoculture of an alga, high-cost illuminated fermenters may be needed which become impractical because of cost for large-scale growth. Although numerous devices, including clear plastic tubular fermenters, have been suggested for photosynthetic algae growth (see, e.g., LEE, 1986), no satisfactory system has yet been developed. Those commercial algae units that do exist, mainly for the production of carotenoids (see Sect. 4.2), use robust algae that have a particular nutritional advantage that prevents contaminants or predators affecting the culture. For example, *Dunaliella salina* grows in hypersaline ponds or lakes and little else can survive in such environments. However, an alternative to autotrophic growth (sunlight and CO_2) of algae may be possible in some cases. Thus KYLE (1991, 1992) has described the heterotrophic growth (darkness and glucose) of several algae that continue to produce high amounts of lipid in the absence of light. These examples are described below (see Sects. 3.4.3 and 3.4.4).

Current attention on microalgae as sources of lipids is mainly because of their high contents of PUFAs. However, there is also interest in algae as potential sources of essential fatty acids for marine animals, particularly by developing fish larvae, mollusks, and crustacea (VOLKMANN, 1989). In these cases, the actual microalgae itself is of importance as the whole cells, not the isolated lipid, becomes the feed.

With PUFA production, the choice of algae is less critical as the lipid itself will be extracted and used as the source of fatty acids. This, though, highlights a second major problem with algal lipids. Unlike oleaginous yeasts and molds, algae produce a large number of lipids many of which have functional roles in connection with the photosynthetic process. Only a relatively small proportion (10–40%) of the total lipid may be triacylglycerol (see RATLEDGE, 1986) with the remainder being phospholipids, other polar lipids, and a variety of glycolipids. In the context of algae being used for animal feeding, the type of lipid does not matter as long as it is accessible and digestible. With oils for human consumption, the commercial emphasis is on producing an oil which is acceptable in appearance: a clear, pale yellow oil with no taste, or aftertaste, is usually needed. This means that algae lipids may have to be fractionated or alternatively the constituent fatty acids removed from the total lipid by hydrolysis, either chemically or enzymatically, and then re-esterified to ethanol or glycerol. Ethyl esters of PUFAs are acceptable alternatives to the natural TAGs. Such additional processing though increases the costs of the final product quite considerably.

A list of oleaginous microalgae is given below. Maximum reported lipid contents as % biomass dry weight are given in parentheses (from ROESSLER, 1990 and RATLEDGE, 1989a).

Prokaryota (Cyanobacteria, blue-green algae)
Anabaena cylindrica (9); *Calothrix castelli* (10); *Nostoc* sp. (8); *Oscillatoria* ssp. (18); *Spirulina maxima* (2); *S. platensis* (17).

Eukaryota
Amphiprora pyalina (30); *Ankistrodesmus* sp. (40); *Biddulphia aurita* (40); *Botryococcus braunii* (53–70); *Chlorella minutissima* (23); *C. pyrenoidosa* (36, 72); *C. vulgaris* (40); *Chlamydomonas applanta* (33); *Chrysochromulina* ssp. (33–48); *Crypthecodinium cohnii* (25); *Cyclotella cryptica* (37); *Dunaliella bardawil* (*D. salina*) (47); *Euglena gracilis* (14–20); *Isochyrysis galbana* (22); *Monalanthus salina* (72); *Nannochloris* sp. (48, 55); *N. oculata* (42); *Navicula acceptata* (38); *N. pelliculosa* (22–45); *Neochloris oleoabundans* (35–54); *Nitzschia palea* (40); *Ochromonas danica* (39–71); *Oocystis polymorpha* (35); *Ourococcus* sp. (50); *Peridinium cinctum* (36); *Phaeodactylum tricornutum* (14); *Porphyridium cruentum* (14, 22); *Prymnesium parvum* (22–38); *Radiosphaera negevensis* (43); *Scenedesmus acutus* (26); *S. dimorphus* (16–40); *S. obliquus* (49); *Scotiella* sp. (16–35).

The fatty acid profiles of selected species of microalgae are given in Tab. 10. The cyanobacteria (blue-green algae) tend to have low lipid contents and do not produce fatty acids longer than C_{18}. Although some species contain γ-linolenic acid (see below), the amount is too low to warrant recovery though the

Tab. 10. Fatty Acid Profiles of Selected Microalgae

	Major Fatty Acyl Residues in Lipids [Relative %]											Others	Reference
	14:0	16:0	16:1 (ω-7)	18:1 (ω-9)	18:2 (ω-6)	18:3 (ω-6)	18:3 (ω-3)	20:3 (ω-6)	20:4 (ω-6)	20:5 (ω-3)	22:6 (ω-3)		
Prokaryota	8	63	2	4	9	12	–	–	–	–	–		HUDSON and KARIS (1974)
Spirulina maxima													
S. platensis	1	26	5	23	10	21	–	–	–	–	–		TORNABENE et al. (1985)
S. platensis (SRS-1h)	–	41	5	3	25	24	–	–	–	–	–		COHEN et al. (1992)
Eukaryota													
Chlorella minutissima	12	13	21	1	2	–	–	–	3	45	–		SETO et al. (1984)
Chlorella vulgaris	–	16	2	58	9	–	14	–	–	–	–		SHIFRIN (1984)
Chlorella NKG042401	<1	22	3	8	28	11	14	–	–	–	–		HIRANO et al. (1990)
Chlorella CHLOR-1	<1	35	<1	44	7	<1	9	–	–	–	–		GUCKERT and COOKSEY (1990)
Crypthecodium cohrii	47	19	1	5	–	–	–	–	–	–	9	12:0, 16%	HENDERSON et al. (1988)
Isochrysis galbana	12	10	11	3	2	–	–	–	<1	25	11	18:4, 11%	MOLINA GRIMA et al. (1993)
Nannochloropsis oculata (849/1)	4	15	22	3	1	–	–	1	4	38	–		HODGSON et al. (1991)
Nannochloropsis sp.	5	14	21	4	3	–	–	–	7	38	–		SETO et al. (1992)
Phaeodactylum tricornutum	–	10	21	1	4	1	–	–	1	33	4		YONGMANITCHAI and WARD (1992)
Porphyridium cruentum (SRP-7)	–	30	5	<1	5	1	–	<1	16	–	–		COHEN et al. (1992)
Unspecified													
Martek isolate MK 8908	22	29	1	26	3	–	1	–	–	5	–		BOSWELL et al. (1992)
Martek isolate MK 8805	18	14	2	11	–	–	–	–	–	–	30	12:0, 6%	KYLE et al. (1992)

presence of this PUFA in dry biomass may have a marginal nutritional benefit if algae are used as a dietary supplement. However, the cyanobacteria are not readily digested and may possess some toxicity (TORNABENE et al., 1985). *Spirulina* spp. have though been used as a source of supplementary dietary protein in both Mexico and Chad where blooms of the algae occur on Lake Texcoco and Lake Chad (RATLEDGE, 1989a). These species, therefore, may be presumed to have a safe health record. Most attention on microalgae has, though, focussed on the eukaryotic species as sources of PUFA.

3.4.1 γ-Linolenic Acid (GLA, 18:3 ω-6)

Spirulina platensis and *S. maxima* have been known to contain GLA for some time (NICHOLS and WOOD, 1968) and have occasionally been considered as potential sources of this PUFA. HIRANO et al. (1990), who appear to be the last research group to look at this source of GLA in any seriousness, were only able to achieve a maximum content of GLA in dry biomass of *S. platensis* of 12 mg per g and this was after heterotrophic cultivation for 7 d at 30°C. The total fatty acid content of the cells was less than 5%. These values are not substantially different from those recorded by previous workers (see RATLEDGE, 1989a) using these and other cyanobacteria. HIRANO et al. (1990) also screened a large number (>300) of marine eukaryotic algae for GLA formation and found that the highest production was with a *Chlorella* sp. (NKG 042401) that contained about 10% total fatty acids with a 10% content of GLA.

In an attempt to improve GLA production in *S. platensis*, COHEN et al. (1992) isolated a number of cell lines that were resistant to the herbicide known as SAN 9785 which is a substituted pyridazinone that selectively inhibits fatty acid desaturation. Slight increases in total fatty acid contents (4–6% dry wt.) were obtained in the resistant cells with GLA increasing from 21.5%–23.5%. GUCKERT and COOKSEY (1990) used an alkaline stress cultivation regimen with a *Chlorella* isolate and found that above pH 11 the triacylglycerol (TAG) fraction of the total lipid increased to over 20% whereas below this pH it was less than 3%. GLA was only about 10% of the total fatty acids in the TAG fraction but was over 40% in the glycolipid fraction which was between 50 and 60% of the total lipids.

Thus no readily recognizable, useful source of GLA has been found in any microalgae whether prokaryotic or eukaryotic. The plethora of lipid types (see, e.g., GUCKERT and COOKSEY, 1990) means that direct production of a GLA-TAG oil from algae is an uneconomic proposition.

3.4.2 Arachidonic Acid (ARA, 20:4 ω-6)

Surveys of the lipids of macro- and microalgae (WOOD, 1988; YONGMANITCHAI and WARD, 1989; ROESSLER, 1990) indicate that the red alga (Rhodophyceae) *Porphyridium cruentum* is superior to all other species for the formation of ARA. Small amounts of ARA do, though, occur in many of the marine microalgae but only in *P. cruentum* does its content exceed 30% of the total fatty acids. Very high amounts of ARA in the alga were originally reported by AHERN et al. (1983) at up to 60% of the total fatty acids. COHEN and HEIMER (1992) have confirmed the potential of this alga for large-scale production of ARA which can be grown satisfactorily in large outdoor ponds (VONSHAK et al., 1985). However, under such conditions, the alga produces mainly a reserve polysaccharide at 40% of the biomass; ARA is only 1.5% of the cell dry weight. According to how this alga is cultivated, it may also contain equal amounts of EPA (*q.v.*) to ARA (COHEN et al., 1988) and may, therefore, be considered as a potential source of either or both PUFAs. A method for fractionating the glycolipid fraction of fatty acids, which is the major lipid fraction, has been described and has yielded ARA at 80% purity and EPA at 97% purity (COHEN and COHEN, 1991).

Improvements to this alga are still being sought by the selection of herbicide-resistant cell lines so that they will have higher propor-

Tab. 11. Cultivation of *Porphyridium curentum* in Open (Outdoor) Ponds for the Production of ARA and EPA (from COHEN and HEIMER, 1992)

Cell Concentration	ARA [% dry wt.[a]]	EPA [% dry wt.[a]]	Output Rates $[g\ m^{-2}\ d^{-1}]$		
			Biomass	ARA	EPA
Winter[b]					
Low	0.76	2.2	3.6	0.03	0.08
Medium	0.75	2.3	5.0	0.04	0.12
High	0.73	2.2	2.1	0.02	0.05
Summer[c]					
Low	0.71	1.3	19.7	0.14	0.25
Medium	1.2	1.2	24.0	0.28	0.28
High	1.3	1.0	13.0	0.17	0.17

[a] Ash-free dry wt.
[b] Maximum daily temperature 16–18 °C.
[c] Maximum daily temperature 28–31 °C.

tions of desirable PUFAs (COHEN and HEIMER, 1992). The best results so far achieved in outdoor cultivation (see Tab. 11), which is the only route available for large-scale production, still indicate that there is little chance of this being an economic route to ARA, or EPA, production. Contents of ARA at less than 2% of the cell dry weight compare unfavorably with yields attained with molds which can achieve 5% and possibly 6% ARA contents (*q.v.*). Moreover, molds produce TAG oils requiring the minimum of downstream processing and can be easily grown to produce zero EPA. Thus fractionation and separation of the ω-6 and ω-3 fatty acids would be unnecessary using mold technology.

3.4.3 Eicosapentaenoic Acid (EPA, 20:5 ω-3)

SETO et al. (1984) reported that the high contents of EPA in *Chlorella minutissima* made this alga an attractive source of this acid for both health foods and for the pharmaceutical industry. Maximum contents of EPA of the total fatty acids reached 45% but were only 2.7% of the dry biomass. Cell yields, moreover, were very low at 300 mg L^{-1} (compared to molds which reach cell densities of over 50 g L^{-1}). Nevertheless, interest in this and other alga as a source of EPA has continued (YONGMANITCHAI and WARD, 1991).

Tab. 12 summarizes the principal photosynthetically-grown algal species that have been considered of some potential in the production of EPA. Although there are other algae with higher contents of EPA in their total fatty acids (see YONGMANITCHAI and WARD, 1989, 1991), the lipid content of these algae may not be very high and furthermore, they are mostly macroscopic algae which are not easily cultivatable under controlled conditions.

A separate approach to PUFA production by algae has been developed by the Martek Corp. Inc. (Maryland, USA) in which selected algae are grown heterotrophically using glucose as carbon and energy source. Thus, photobioreactors are unnecessary and the organism behaves, and performs, as a eukaryotic yeast or mold (KYLE, 1992). Of several thousand algae that were screened for PUFA production, one – MK8909 – was selected for EPA production. (Another isolate was subsequently exploited for DHA production, see below.) This isolate has been described as a apochlorotic diatom, that is not possessing chlorophyll and, therefore, incapable of photosynthesis. The alga may be related to *Navicula saprophilla* which KYLE et

Tab. 12. Photosynthetically-Grown Algae as Potential Sources of EPA

Alga	Lipid Content of Biomass[a] [%]	EPA in Fatty Acids [%]	Reference
Phaeodactylum tricornutum	15	35	YONGMANITCHAI and WARD (1992)
Nannochloropsis oculata	14	45	SETO et al. (1992)
Isochrysis galbana	22	26	MOLINA GRIMA et al. (1992, 1993, 1994)
Porphyridium cruentum	5.6[b]	41	COHEN et al. (1992)
Chlorella minutissima	15	45	SETO et al. (1984)

[a] The lipids are usually composed of glycolipids, phospho- and other polar lipids with triacylglycerol as a minority component. The total fatty acid content of the lipids may be less than 50%.
[b] Total fatty acids.

al. (1988) had earlier described as having some potential for PUFA production.

The content of EPA in MK8909 was only 5% of the total fatty acids but the oil content was 50% (w/w) of the cells. The organism grew readily and yielded 50 g dry cells per L in 3 d. As the alga produced an easily extractable oil, the EPA was considered of potential commercial value as there was a complete absence of any other PUFA: 18:2 was present at only 3% and 18:1 was at 26%. Thus enrichment of EPA from the oil is a relatively easy proposition and such oils are now offerred for sale by Martek.

The overall problem with EPA production from any microbial source has already been discussed (Sect. 3.3.4) in that EPA is usually easily obtainable from fish oils though as a mixture with DHA. Unless, and until, someone demonstrates the clear need for EPA (or DHA) as a single PUFA, there seems little prospect that EPA will be required for anything more than experimental work.

3.4.4 Docosahexaenoic Acid (DHA, 22:6 ω-6)

DHA has a very limited distribution in most algae though it occurs throughout the family of dinoflagellated algae (Dinophyceae) along with EPA and also 18:4 (YONGMANITCHAI and WARD, 1989). YONGMANITCHAI and WARD (1989) highlighted three species that could warrant further attention for DHA production: *Crypthecodinium cohnii, Gonyaulax catenella,* and *Gymnodinium nelsonii* as all contained over 30% DHA in their total fatty acids. HENDERSON et al. (1988) have examined the first algae in some details growing the cells non-photosynthetically. The harvested biomass had a total lipid content of 25% of which the triacyglycerol fraction constituted over 50% though this fraction only contained 9% DHA. The major DHA-containing component was phosphatidylcholine (PC) having 57% DHA in its total fatty acids and PC itself constituted 18% of the lipid. The complete fatty acid profile is given in Tab. 9.

Of the other photosynthetic algae, only the species of *Isochrysis* have been reported as containing more than a small amount of DHA. This marine alga has been recently examined by MOLINA GRIMA et al. (1992, 1993, 1994) for its potential to produce DHA and EPA. Following screening of a number of isolates, one was selected for further work (LOPEZ ALONSO et al., 1992). This strain, when grown in chemostat culture with illumination, produced up to 2% of the cell dry weight as DHA (MOLINA GRIMA et al., 1993). EPA though reached up to 6% of the cells (see above and Tab. 12). As a source of DHA, the alga is, therefore, no better than many fish oils; DHA would be difficult to purify as a single PUFA and its presence in all lipid fractions (MOLINA GRIMA et al., 1994) at about 10% of the total fatty would not be regarded as particularly encouraging.

On the other hand, the concept developed by the Martek Corp. Inc. (see Sect. 3.4.3) of using algae growing heterotrophically has succeeded in identifying an unknown (i.e., undeclared) phytoplankton species (MK 8805) that produced DHA as the sole PUFA at 50% of its total fatty acids (KYLE et al., 1992). The only other unsaturated fatty acids in the algae oil were 18:1 (at 11%) and 16:1 (at 2%). The only drawback with this alga was its low lipid content at between 10–15%. Nevertheless, it could be grown to high cell densities (about 25 g L^{-1}) in 84 h in a conventional (non-illuminable) fermenter. The oil and DHA-enriched fractions of it with up to 80% purity are now offered for sale by Martek. This oil is, therefore, the only one which is commercially available that does not contain EPA.

3.4.5 Conclusions

Algae are undoubtedly a rich source of PUFA. They are the sources of PUFA in fish, which do not carry out *de novo* synthesis of these acids, but rely on the ingestion of algae for them. Nevertheless, algae are not exceptional sources of PUFA. ARA is possibly producible from *Porphyridium cruentum* being grown in outdoor ponds but the economics do not look as attractive as obtaining ARA from fungi. EPA and DHA may occur together in algae in which case the lipid would be no better than fish oil as a source of either acid. Some algae though only produce EPA and, therefore, might be useful sources of this acid if it were not for the fact that EPA is distributed throughout all the many lipid classes that occur in algae. This plethora of lipid classes necessitates complete hydrolysis of the total lipid and recovery of the EPA as the free acid.

The more realistic commercial approach to PUFA production by algae has been to grow them heterotrophically: treating them as yeasts or lower fungi and growing them without illumination in conventional bioreactors. Martek Corp. Inc. in the USA have pioneered this approach and since 1992 have on sale oils containing EPA and DHA as single PUFA not containing any other PUFA or even diunsaturated fatty acids. It is understood, however, that demand for these oils has been very modest principally because there is, as yet, no clear dietary or clinical indication that one acid or the other would be useful in the treatment of any disorder or nutritional imbalance. In view of the current high costs of these oils, not unnaturally the public have preferred to buy various fish oils and PUFA concentrates derived from fish oils that contain both EPA and DHA together. The possible inclusion of DHA in infant food formulations (see Sect. 3.3.7), however, is likely to stimulate interest to find an alternative source to fish oil for this PUFA. The algae and fungi already mentioned will be the principal candidates.

4 Sterols, Carotenoids, and Polyprenes

4.1 Sterols

Although steroids, including their precursor squalene, have been reported occasionally from the prokaryotic microorganisms, these are only of academic interest as the quantities involved are extremely small. Sterols, however, have been produced commercially from eukaryotic microorganisms though only as a by-product by the extraction of either spent fungal mycelium recovered from a fermentation process or from spent brewer's yeast. Sterols for commercial use are usually obtained by extraction of plant materials which is not only a cheaper route than from microorganisms but also provides the correct substituents on the molecule for easy transformation into the commercially lucrative steroid hormone market. Most microbial processes for steroid formation now appear to be defunct but some extraction of brewer's yeast for the production of ergosterol (VI) does appear to continue in a few locations though the exact number and scale of the various operations is not generally revealed by industrialists.

Ergosterol (VI) is the commonest microbial sterol and it occurs both in the free form and as its fatty acyl ester in algae, yeasts, and molds. Highest levels of sterol are found in the yeasts, especially in *Saccharomyces, Kluyveromyces, Metschnikowia, Pichia,* and *Torulaspora* (RATTRAY, 1988). While the major steroid is ergosterol, over 14 other sterols have also been identified (WEETE, 1980), the principal ones of which are lanosterol (VII), zymosterol (VIII), and ergosta-5,7,22,24(28)-tetraen-3β-ol (IX).

Although the usual range of extractable sterols (free sterols plus sterol esters) in yeasts is generally in the range of 0.03–6% of

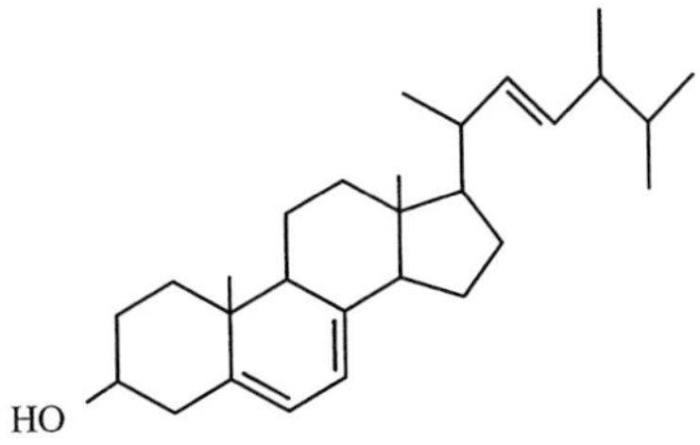

HO
Ergosterol
VI

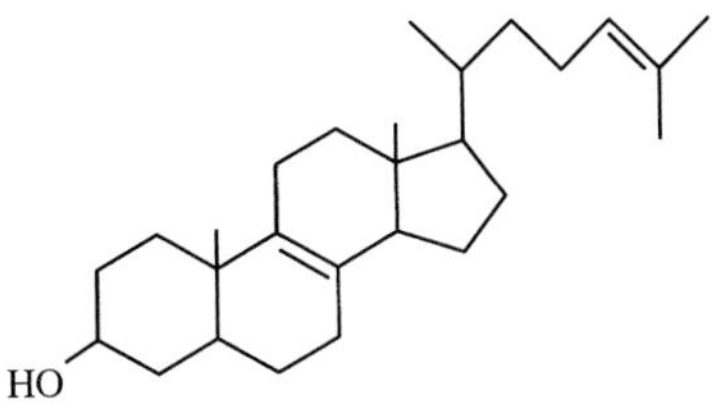

HO
Lanosterol
VII

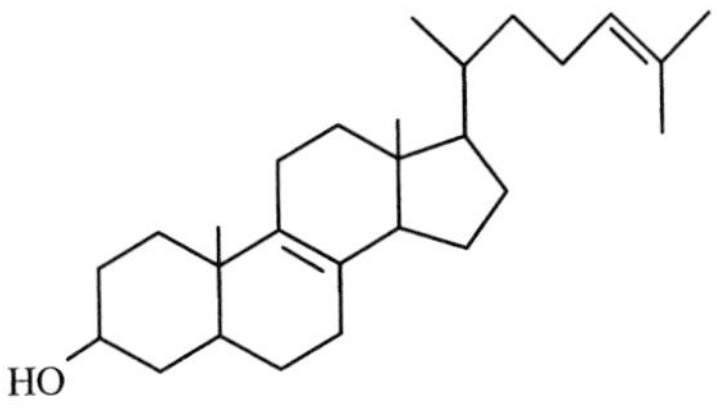

HO
Zymosterol
VIII

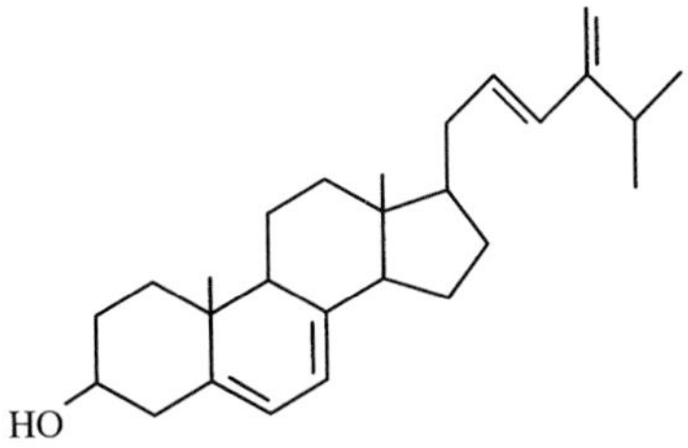

HO
Ergosta-5,7,22,24(28)tetraen-3β-ol
IX

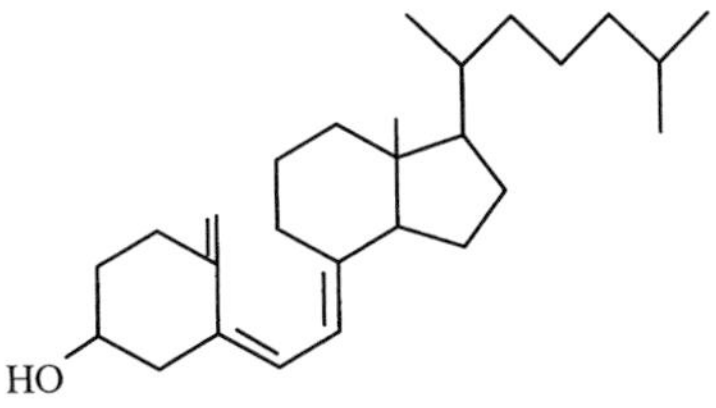

HO
Cholecalciferol
X

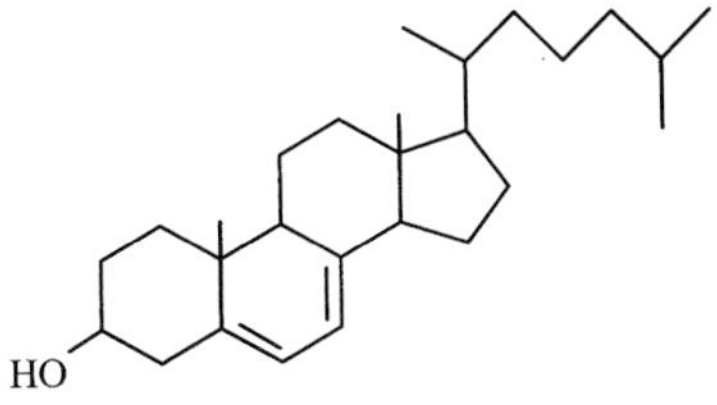

HO
7-Dehydrocholesterol
XI

the cell dry weight (RATTRAY, 1988), DULANEY et al., (1954) found several strains of *S. cerevisiae* which produced ergosterol up to 10% of the biomass. When the yeasts were grown in submerged culture yields of sterol of up to 4 g L^{-1} were obtained. A process for the production of ergosterol was patented by DULANEY (1957). Other workers have been also able to achieve similar results: e.g., EL-REFAI and EL-KADY (1968a, b; 1969) recorded up to 23% sterol contents in *Saccharomyces fermentati* (now *Torulaspora delbrueckii*) if grown in the presence of potassium persulphate, hydroquinone, or indigocarmine.

Increased accumulation of sterols in *S. cerevisiae* occurs at low specific growth rates under nitrogen-limited, aerobic growth conditions (NOVOTNY et al., 1988) and by exploiting such information BEHALOVA and VORISEK (1988) increased the total sterol content of this yeast to 7.2% of the dry biomass. The total lipid content of the cells, including the sterols, reached 31% making the yeast a candidate for inclusion in the list of oleaginous yeasts (see Tab. 3). However, such lipid contents in *S. cerevisiae* are quite exceptional indicating that these authors may be using an atypical strain.

Ergosterol is not of major economic significance though it has some commercial value. MARGALITH (1989) has outlined the various attempts to produce it using yeasts. Its main application is that of an analog for cholecalciferol, vitamin D_3, (X), which arises from 7-dehydrocholesterol (XI) by the action of ultraviolet light; likewise ergosterol (VI) is converted to ergocalciferol, vitamin D_2 (XII). However, the latter is only 10% as nutritionally effective in chickens as is vitamin D_3. The prospects of being able to produce mutants of *S. cerevisiae* which would accumulate 7-dehydrocholesterol rather than ergosterol are, therefore, extremely attractive (PARKS et al., 1984).

In molds, a large range of sterols has been recognized (WEETE, 1980; LÖSEL, 1988), though once more ergosterol is the major constituent in many, though not every, species. Cholesterol (XIII) has been noted as the major sterol in a number of Mucorales and also, unexpectedly, in *Penicillium funiculosum*. The highest level of sterol reported would appear to be that found by OSMAN et al. (1969) for *Aspergillus fumigatus* which contained 5% of its dry biomass as ergosterol. SHIMIZU et al. (1992) have recently reported the occurrence of a novel sterol, 24,25- methylenecholest-5-en-3β-ol (XIV) in *Mortierella alpina* being grown for arachidonic acid production (Sect. 3.3.3). This sterol contains a cyclopropane ring in its side chain. Similar sterols occur in sponges but this is the first example of a sterol with a cyclopropane in this particular position in any organism.

4.2 Carotenoids

Carotenoids occur through the whole of the microbial world and, of course, throughout nature. Their occurrence and structure has been comprehensively reviewed by GOODWIN (1980, 1983). The review of LÖSEL (1988) on fungal lipids also includes coverage of the carotenoids of fungi. A useful monograph on the various pigments of microorganisms has been prepared (MARGALITH, 1992) and contains a short but informative chapter on carotenoids, their chemistry, biochemistry, and functions. The biosynthesis of carotenoids has been described by HARRISON (1986), BRAMLEY (1985), and BRAMLEY and MACKENZIE (1988).

Although all algae contain carotenoids, these are part of the chloroplast photosynthetic apparatus and, therefore, do not usually form a major constituent of the biomass. In the marine brown macroalgae (the Phaeophyta), the total annual biosynthesis of carotenoids through the oceans of the world has been calculated as $1.2 \cdot 10^7$ t (JENSEN, 1966) – but, of course, none of this is harvested for the carotenoids. The major carotenoids, in order of abundance, in these seaweeds are fucoxanthin (XV), violaxanthin (XVI), and β-carotene (XVII).

Amounts of carotene in microalgae do not usually exceed a few mg per g dry weight (GOODWIN, 1980). However, *Spirulina platensis*, which is used as a source of food by inhabitants around Lake Chad in Africa and, which will be grown on brackish water, contains about 0.4% of its dry weight as β-caro-

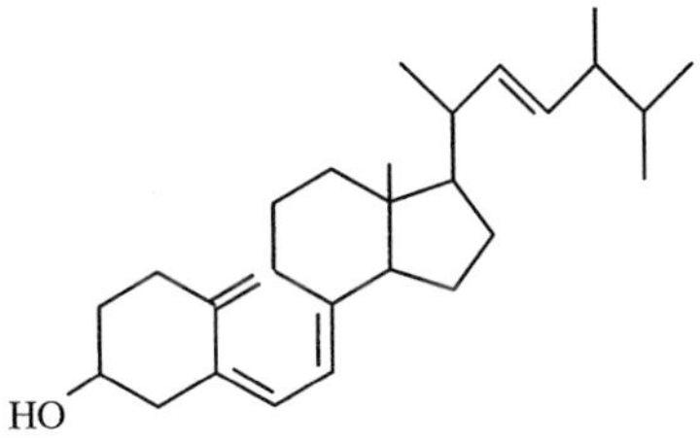

HO

Ergocalciferol

XII

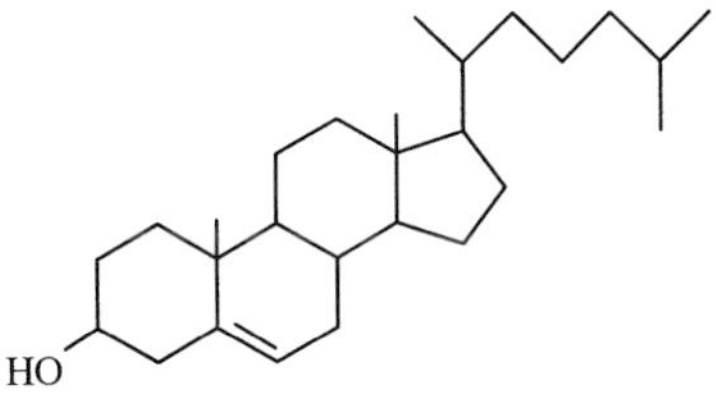

HO

Cholesterol

XIII

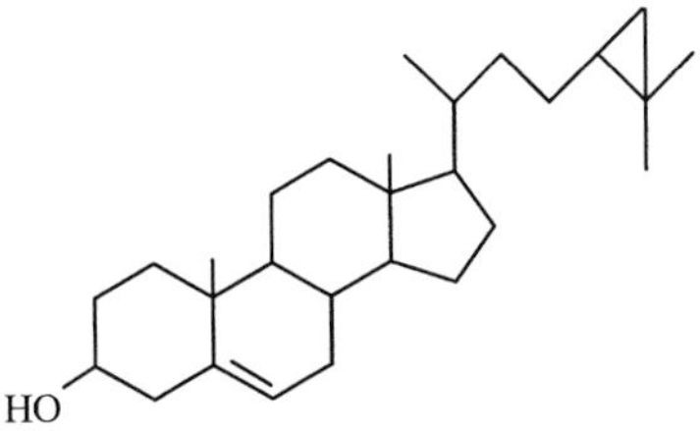

HO

24,25-methylenecholest-5-en-3β-ol

XIV

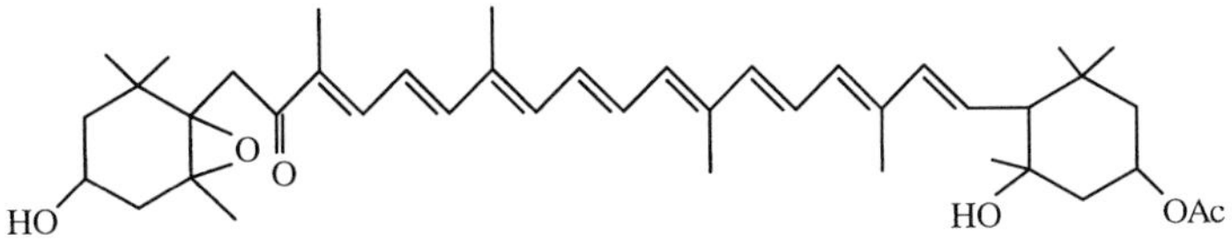

HO O O HO OAc

Fucoxanthin

XV

OH O O HO

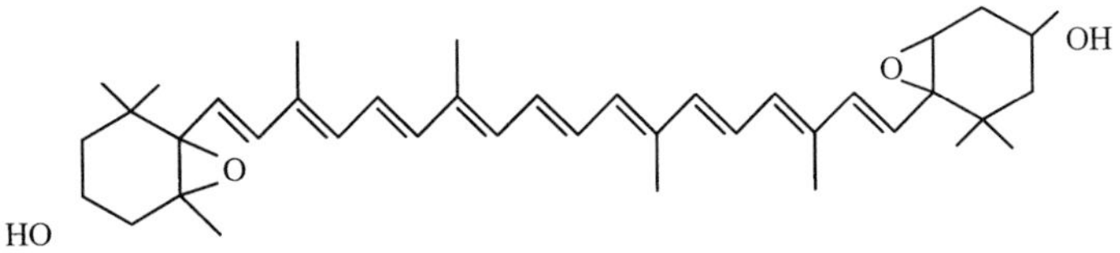

Violaxanthin

XVI

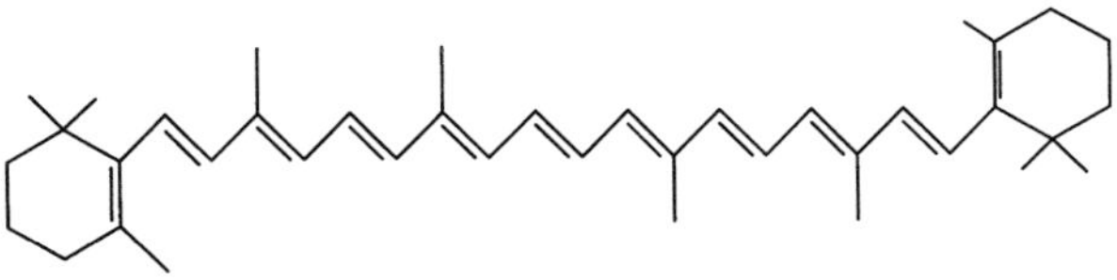

β-Carotene

XVII

tene and other xanthophylls (CLEMENT, 1975). The halophilic *Dunaliella salina* or sometimes referred to as *D. bardawil*, which can grow in waters of high salinity and produce glycerol in some abundance (DUBINSKY et al., 1978), can produce up to 400 mg β-carotene m^{-2} d^{-1} (BEN-AMOTZ and AVRON, 1980). β-Carotene may be up to 10% of the dry weight of *D. bardawil* (MARGALITH, 1992) and with such yields has been grown commercially in Australia by Western Biotechnology Ltd. at Hutt Lagoon in Western Australia, in Israel (Nature Beta Technologies), and in the USA (Microbio Resources).

All systems use outdoor lagoons: that in Australia, e.g., has a coverage of 50 ha (10 × 5 ha ponds) and was scheduled to add a further 25 ha pond in 1994. (The author is grateful to Dr. M. A. BOROWITZKA for this information.) Full descriptions of this process have been provided by BOROWITZKA and BOROWITZKA (1989, 1990) and BOROWITZKA (1992). In Israel, *Dunaliella* is grown in a section of the Dead Sea.

Algal β-carotene is comprised of 60% 9-*cis* isomer (BEN-AMOTZ and AVRON, 1983; BEN-AMOTZ et al., 1988). This appears to be assimilated by experimental animals more rapidly than the all-*trans* isomer (BEN-AMOTZ et al., 1989). From a commercial viewpoint, the most important carotenoid is β-carotene (XVII) which, besides being an important foodstuffs colorant, has provitamin A activity (vitamin A: XVIII). It has also been suggested that β-carotene may also act through its role as an antioxidant as a tumor-suppressing agent and be useful in chemoprevention of cancer (MARGALITH, 1992). However it is the β-*cis* isomer which appears to be effective rather than the all-*trans*, chemically-produced β-carotene. Demand may consequently shift towards the more expensive natural β-carotene if these claims for its therapeutic effectiveness are confirmed.

β-Carotene is also the predominant carotenoid in many fungi and is a minor constituent

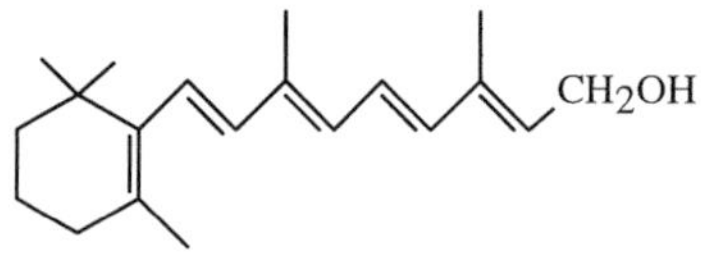

CH_2OH

Vitamin A

XVIII

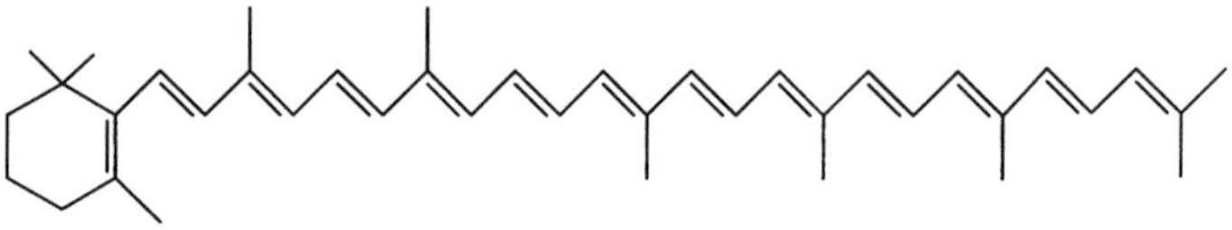

Torulene

XIX

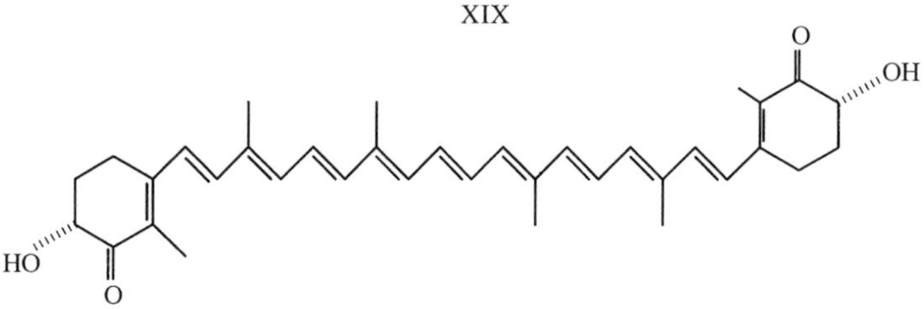

O OH HO O

Astaxanthin (3*R*, 3*R*' isomer)

XX

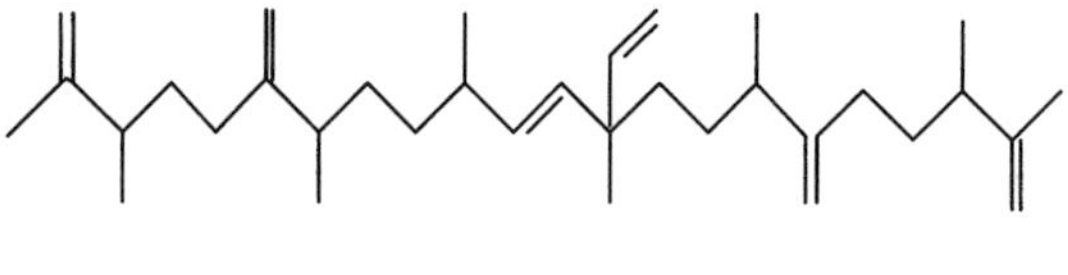

Botryococcene

XXI

in most others, though not all fungi do contain carotenoids. The most abundant production of β-carotene has been achieved with the phycomycete fungi, *Blakesleea trispora* and *Phycomyces blakesleeanus* (MURILLO et al., 1978; NINET and RENAUT, 1979; CERDÁ-OLMEDO, 1989) where yields of carotene in mutant strains of up to 2% of the biomass have been recorded. Commercial processes for the production of β-carotene using *B. trispora* have been developed and at least one, in the Ukraine, is still in operation. SHLOMAI et al. (1991) have recently re-examined the β-carotene produced by *P. blakesleeanus* and, contrary to the original supposition that the all-*trans* isomer would be found (BRAMLEY and MACKENZIE, 1988), 15% of the total β-carotene was the 9-*cis* isomer. The remaining 85% was though the all-*trans* form. Nevertheless, this result was considered of considerable interest as further research should be able to increase the proportion of 9-*cis*-β-carotene. Such formation of the highly desirable isomer for cancer prevention (see above) could now re-stimulate interest in the exploitation of fungi for β-carotene production. Cultivation of heterotrophic fungi appears to be a better commercial proposition than having to grow algae autotrophically where the costs of land, harvesting and the necessity for high light intensities and high ambient temperatures pose many severe limitations (MARGALITH, 1992).

Torulene (XIX), which has only half the provitamin A capability as β-carotene, is the major carotenoid pigment of the red *Rhodotorula* and *Rhodosporidium* yeasts, of which several species are oleaginous (see Tab. 3). Other carotenoids also occur in these species (GOODWIN, 1980). However, the amounts are far less than needed for any commercial interest and these carotenoids remain of academic interest.

The pink yeast, *Phaffia rhodozyma* produces astaxanthin (XX) which is the carotenoid giving the characteristic pink color to salmon, crabs, lobsters, and other crustaceans (and indirectly flamingos and other birds living off these life forms). It is now produced commercially by Gist-Brocades, Netherlands, for use in feed formulations for poultry but mainly for pen-reared salmonids. The demand for astaxanthin for fish feed may exceed 100 t a^{-1} by the end of this century (JOHNSON and AN, 1991). The yeast carotenoid has, however, the opposite chirality at $3R$ and $3R'$ (i.e., the hydroxyl group on the cyclohexene end groups) to the lobster astaxanthin (see GOODWIN, 1980; 1983) but this does not affect its acceptability. Synthetic astaxanthin which is the all-*trans* isomer, is produced by Hoffman-La Roche Ltd. but awaits approval from the FDA for use in the USA (JOHNSON and AN, 1991). Consequently the emphasis is now on *P. rhodozyma* as a potential worldwide source.

Initial yields of astaxanthin by *P. rhodozyma* were less than 500 $\mu g\ g^{-1}$ (JOHNSON et al., 1977; 1980). Whilst some increase is possible by careful selection of the growth medium and conditions (HAARD, 1988; NELIS and DE LEENHEER, 1989), high yields can only be achieved after mutagenesis (AN et al., 1989; MEYER et al., 1993). Even these yields though can be enhanced by selecting individual cells by a cell sorter using the fluorescence of astaxanthin as the indicator (AN et al., 1991). Yields of astaxanthin of 2.5 mg per g dry cells have been achieved by using a mutant and carefully selected growth conditions (MEYER et al., 1993). However, it is considered that commercial production probably needs to reach 4 to 5 mg g^{-1} in order to be economic.

An alternative microbial source of astaxanthin is the freshwater green alga, *Haematococcus pluvialis* (BOROWITZKA, 1992). Outdoor cultivation of this alga so far has proved unreliable and commercialization seems unlikely at the moment even though astaxanthin may reach up to 20 mg per g cell dry weight.

4.3 Polyprenoids

The only polyprenoid of possible biotechnological significance is botryococcene (XXI) which occurs as the principal lipid component of the alga *Botryococcus braunii*. This alga has been reported as containing up to 85% of its biomass as a mixture of polyprene hydrocarbons when it has been isolated from coal deposits, which appears to be its natural ecological niche. The alga can be grown in the laboratory, though yields of botryococcene and related C_{34} hydrocarbons are then only from 24 to 45% of the biomass (HILLEN et al., 1982; YAMAGUCHI et al., 1987; CASADEVALL et al., 1985). The hydrocarbons can be thermally cracked to give gasoline and other fuels of commercial value (HILLEN et al., 1982) and thus have been suggested as possible sources of energy. However, it seems unlikely that this would represent a commercially exploitable source of fuel hydrocarbons particularly in view of the slow rate of hydrocarbon production at about 0.15 g L^{-1} d^{-1} (CASADEVALL et al., 1985) though a slight increase occurs when the cells are immobilized (BAILLIEZ et al., 1985; 1986) and used in an air lift, illuminated bioreactor (BAILLIEZ et al., 1988).

The impracticality of scaling-up such a system would, though, preclude any serious biotechnological application but SAWAYAMA et al. (1992) have recently reported that *B. braunii* can be usefully cultivated on secondarily treated sewage from domestic wastewaters so that it not only removes N and P but still produces about 50% of its biomass weight as the hydrocarbon. Thus, the only hope for commercial take-up would be to use the alga for a dual purpose: sewage cleanup and hydrocarbon production. As the hydrocarbon would be used as a fuel and not as a food supplement, there is obviously no restriction on what the alga may be grown on as none of it would be returned into the food chain.

5 Wax Esters and Polyesters

5.1 Wax Esters

Wax esters of the type RCOOR′ where R and R′ are long alkyl chains occur in bacteria, algae, and yeast. Their route of biosynthesis is given in Fig. 6. The fatty acid and alcohol moieties may be saturated or unsaturated. The diunsaturated wax ester is desirable as a substitute for jojoba oil. With algae, it is the protozoan, *Euglena gracilis*, that has been the most studied (see KAWABATA and KANEYANA, 1989) for wax ester production but the

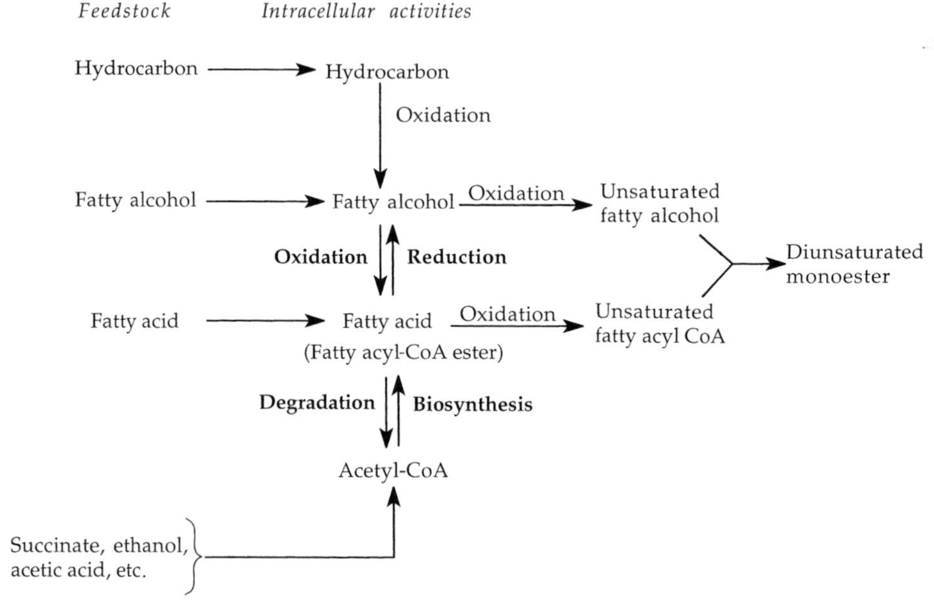

Fig. 6. Pathways to a diunsaturated wax ester (jojoba oil type substitute) from various substrates.

amounts are less than 10% of the cell biomass (about 100 μg per 10^6 cells). With yeasts, wax esters are not usual components of the lipid fraction but SEKULA (1992) has reported formation of them in several yeasts when grown in fatty alcohols. The amounts formed were not disclosed but appeared, from TLC evidence, to be equal in amount to the triacylglycerol fraction. Evidence was presented that the fatty acid moiety of the ester may be oxidized to a ω-1 keto fatty acid which could then be incorporated into the wax ester (see Fig. 6).

The greatest amounts of monoester appear to occur in bacteria of the genera *Acinetobacter, Micrococcus, Nocardia, Mycobacteria*, and *Corynebacterium*. In the latter three genera, the fatty acids involved may be long-chained ($>C_{24}$) and/or methyl branched; the waxes may also be associated with virulence and thus biotechnological exploitation is unlikely. *Acinetobacter* and *Micrococcus*, which are now probably synonymous at least as far as the wax-producing strains are concerned, have been studied by various groups including that of Cetus Corp., California, USA and numerous patents (see RATLEDGE, 1986) taken out. The aim of this work had been to achieve production of a jojoba oil-like material which is essentially a 20:1–20:1 fatty acid/fatty alcohol ester. Yields, however, have remained low (<1 g L^{-1}) and no take-up of the process has occurred. Indeed, prospects for a microbial route to wax ester production would now seem to have receded even further with the recent description of how oleoyl oleate can be chemically synthesized using oleic acid, oleoyl alcohol, and a zeolite catalyst (SANCHEZ et al., 1992).

5.2 Polyesters – Poly-β-Hydroxyalkanoates

The major microbial polyesters of commercial interest are the poly-β-hydroxy alkanoates (PHA) of which poly-β-hydroxybutyrate (PHB, $R=CH_3$ in XXII) is of major importance. PHB and PHA are found principally in bacteria though related molecules have been found in small amounts in the membranes of eukaryotic cells (ANDERSON and DAWES, 1990). The subject has been the topic of a number of monographs and international symposia (DAWES, 1990; DOI, 1990; SCHLEGEL and STEINBÜCHEL, 1992; VERT et al., 1992) and more are to follow, see below. A typical electron micrograph of a PHB-containing cell is given in Fig 7.

Present interest in PHB/PHA arises from its use as a biodegradable plastic. PHB itself is considered too brittle to be conveniently molded into appropriate shapes and so is con-

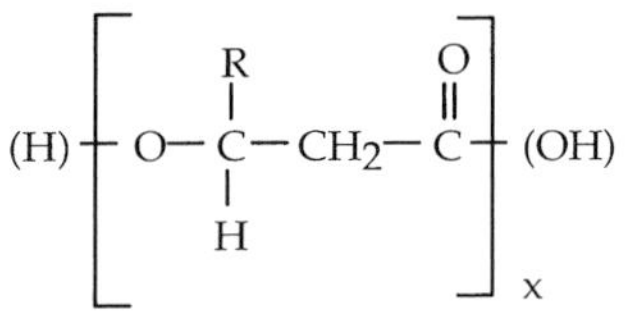

$$(H)\left[O-\overset{\overset{R}{|}}{\underset{\underset{H}{|}}{C}}-CH_2-\overset{\overset{O}{\|}}{C} \right]_x (OH)$$

x = 10 000 to 20 000

R = $-CH_3$ for poly-β-hydroxybutyrate (PHB)

R = $-C_2H_5$ for poly-β-hydroxyvalerate (PHV)

R = $-C_nH_{2n-1}$ for poly-β-hydroxyalkanoates (PHA) up to n = 9

Poly-β-hydroxybutyrate and alkanoates

XXII

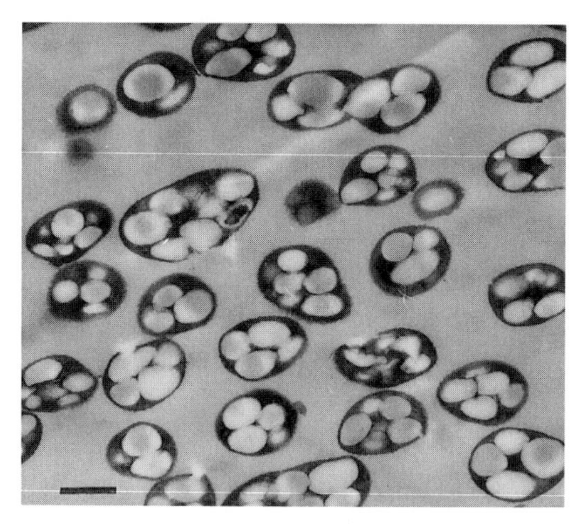

Fig. 7. Electron micrograph of poly-β-hydroxybutyrate granules in *Alcaligenes eutrophus;* marker bar: 1 μm (photograph kindly supplied by Dr. A. J. ANDERSON, University of Hull, UK).

sequently produced as a heteropolymer along with β-hydroxyvalerate (V, $R = -C_2H_5$ in XXII) as the other monomeric unit to β-hydroxybutyrate. Whilst β-hydroxybutyrate is synthesized from acetate–acetate condensation, β-hydroxyvalerate is produced from acetate–propionate condensation. This requires propionic acid to be presented to the bacterial cultures as a cosubstrate along with glucose. PHB and the co-polymer, PHB/V, are produced by Zeneca plc, UK, (formerly ICI plc) and uses *Alcaligenes eutrophus* as production organism. Yields are up to 80% of the total cell biomass. For commercial purposes the molecular weight of the polymer should be as high as possible and certainly in excess of 10^6 Da. The process has been described, with perhaps understandable perfunctoriness, by BYROM (1990, 1992). The commercial product is sold under the trade name of Biopol®. A rival industrial process operated by Chemie Linz GmbH, Austria, has been described by HRABAK (1992) but it is uncertain whether this operates other than as a demonstration unit.

Accumulation of PHB in bacteria is favored by much the same environmental factors that are needed for triacylglycerol accumulation in yeasts and molds: that is a depletion of N (or other nutrient) from the culture with the provision of excess carbon to ensure continued formation of the polymer (see Sect. 2.1). As bacteria do not readily synthesize triacylglycerols, PHB and PHA may be regarded as the bacterial equivalent to triacylglycerols serving the same metabolic functions as a (chemically) reduced storage compound; that is, it is accumulated under conditions of nutrient deficiency and utilized during periods of carbon starvation. The pathway of biosynthesis of PHB (Fig. 8) is less com-

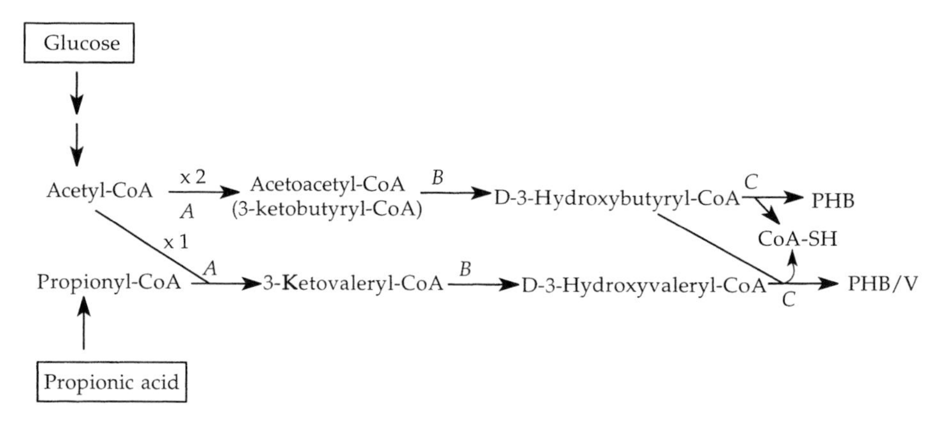

Fig. 8. Biosynthesis of poly-β-hydroxybutyrate (PHB) and the copolymer of poly-(β-hydroxybutyrate/β-hydroxyvalerate) (PHB/V). PHB, $R = CH_3$ in Fig. 10; PHV, $R = C_2H_5$ in Fig. 10. Enzymes: *A:* 3-ketothiolase; *B:* acetoacetyl-(3-ketoacyl-)CoA reductase (NADPH-dependent); *C:* PHB synthase (or polymerase); 1×: one mol; 2×: two mol.

plex than that of triacylglycerols. Indeed, the biosynthesis requires only three additional enzymes to those already present for fatty acid biosynthesis: 3-ketothiolase, acetoacetyl-CoA reductase, and PHB (PHA) synthase or polymerase. All three enzymes have been studied in some detail (see ANDERSON and DAWES, 1990) and the genes for each of them have been identified, sequenced (STEINBÜCHEL et al., 1992), and now cloned and expressed in plants (POIRIER et al., 1992). (The PHA polymerase is probably a different enzyme from that of PHB polymerase and the corresponding gene awaits to be described.)

This latter work has considerable commercial potential. Whilst the economics of PHB/V production by large-scale bacterial fermentation are regarded as only just favorable for commercial exploitation, the economics of production would be considerably enhanced if the same production could be achieved in plants. Just as plants produce triacylglycerol oils and fats at a tenth, or even less, of the cost of producing the same materials biotechnologically, so the costs of producing PHB would be considerably decreased by switching production into plants. For the work to be successful, yields of PHB will have to equal those currently achieved by oilseed crops for the production of triacylglycerol oils, i.e., up to at least 30% of the harvested crop. Present results indicate that PHB formation is only about 100 μg in the hybrid plants (POIRIER et al., 1992) which places the work as being still in the preliminary experimental stages.

For successful production of PHB in plants, the most suitable plant would appear to be one which has the essential machinery for accumulation of a (plant) product already in place. Moreover, not only will the genes for PHB biosynthesis have to be inserted but the genes for producing the existing product will have to be deleted so that the flux of carbon can then be diverted wholly into PHB:

$$CO_2 \rightarrow \text{Sugar} \begin{cases} \nearrow \text{Old product (oil or fat)} \\ \searrow \text{PHB} \end{cases}$$

As PHB is synthesized by direct condensation of acetyl-CoA units (Fig. 8), it is fairly obvious that the ideal plant for PHB production would be an oil-producing one as such a plant would already have the necessary biochemical machinery present to produce acetyl-CoA in some abundance. Thus, if fatty acid biosynthesis could be prevented, say by deletion of acetyl-CoA carboxylase or even impairment of the fatty acid synthase complex, then, with the three PHB genes being successfully introduced and fully expressed in the plant, carbon should now flow into PHB production.

Such scenarios are now being developed for oilseed rape by Zeneca Seeds in the UK (SMITH et al., 1994) and by the Carnegie Institution in conjunction with Procter & Gamble in the USA using *Arabidopsis* as as a model plant system (POIRIER et al., 1992). Alternative plants to oilseed rape could clearly include sunflower, which is clearly amenable to genetic modification (see Tab. 2), and possibly even the palm oil tree. Yields, however, at the moment are very far from any practical value, and considerable technical work, if not scientific innovation, will be necessary to achieve commercially viable yields.

The future demand for PHB and related molecules is currently seen to depend on their value as a biodegradable plastic. Its "environmentally friendly" nature has been repeatedly stressed. However, progress to producing other non-PHB, biodegradable plastics using chemical synthesis is now proceeding apace (VERT et al., 1992). Should large-scale chemical synthesis of an alternative, but still "environmentally friendly" polymer be achieved in the near future, this would have serious consequences for the economic viability of PHB/V. Conceivably though, if production of PHB could be achieved in plants to the same yield that they now produce oils, then this could remain an alternative route to production. Long term prospects for the production of PHB by microbial fermentation would appear uncertain. However, just as microorganisms have been dismissed as alternative producers of oils and fats that are already commercially available, the way forward for microbial PHBs may be to identify higher valued products for niche markets.

The range of polyalkanoates produced by microorganisms extends far beyond the simple PHB and PHV polymers. Other potentially useful polymers could include the poly-β-hydroxyalkanoates where the side chain (R in XXII) could be an alkyl chain up to C_9. These are known as the medium chain length PHAs. Their occurrence has been described mainly in *Pseudomonas oleovorans* (WITHOLT et al., 1990; ANDERSON and DAWES, 1990). Formation is considerably enhanced by growing the bacteria on long chain alkanes, fatty alcohols, and fatty acids. However, growing the bacterium on a long chain fatty acid, such as oleic acid (18:1), only induces formation of the C_{12} monomer ($R=C_9$ in XXII) though, by using it, PHAs are produced with an unsaturated side chain (WITHOLT et al., 1990) but still no longer than C_7. The properties of these PHAs have been described (DE KONING, 1995) and have indicated some potential for producing, after chemical modification, a rubber latex material that still retained its biodegradability.

Considerably further work with the PHAs may be anticipated over the next few years. Although there is currently only one industrial producer of a PHA (see above), a recent (1994) conference held in Montreal, Canada, attracted over 300 delegates. The proceedings of this meeting were published in the *Canadian Journal of Microbiology,* probably during 1995.

6 Other Lipids

The range of microbial lipids is, of course, extensive (see RATLEDGE and WILKINSON, 1988a, 1989). The biotechnological exploitation of any particular lipid will depend upon the perception of the purpose to which that lipid can be put. In some instances the formation of a particular lipid may have been known for many years but how it may be of commercial benefit is not so obvious. An example of this would be the formation of the sophorolipids by several *Candida* spp. which, although excellent surfactants, have not proved sufficiently superior to chemically-produced materials to warrant full scale production (see below). In other cases, the lipid may be a minor component of apparently only academic curiosity but then in the hands of an appropriate industrial company is scaled up into being a significant product. In the following some examples of the range of microbial lipids are given; in some cases commercial exploitation has occurred, in others the lipids remain academic curiosities.

6.1 Biosurfactants

Most microorganisms produce a range of surfactants that seemingly allow them to become attached to water-insoluble substrates or to surfaces of leaves and other parts of plants or soils. Production of surfactants may be enhanced by growing the organism on hydrocarbons or vegetable oils and, for a while, this was considered to be a prerequisite for growth to take place. It is now not certain if this is the case as some surfactants are produced in some abundance even when water-soluble substrates are used.

A wide diversity of chemical types is known ranging from glycolipids which usually involve one or more fatty acyl residues attached glycosidically to a mono- or di-saccharide. Examples include the sophorolipid from *Candida bombicola* (XXIII) and the rhamnolipids from *Pseudomonas aeruginosa* (XXIV). In some cases, macromolecular surfactants are formed with a M_r of up to 10^6 Da in which the lipid component may not be the major one. An extensive monograph on the entire subject has been recently published (KOSARIC, 1993). Readers are therefore referred to this book for full details of the range and potential of these molecules.

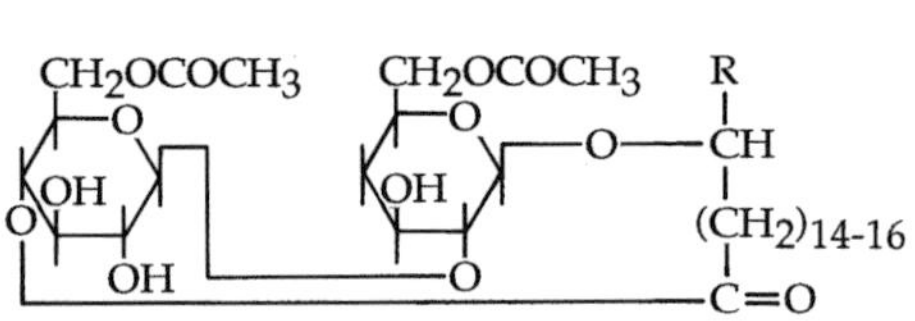

Lactone form
R = H or CH_3

Sophorolipid

XXIII

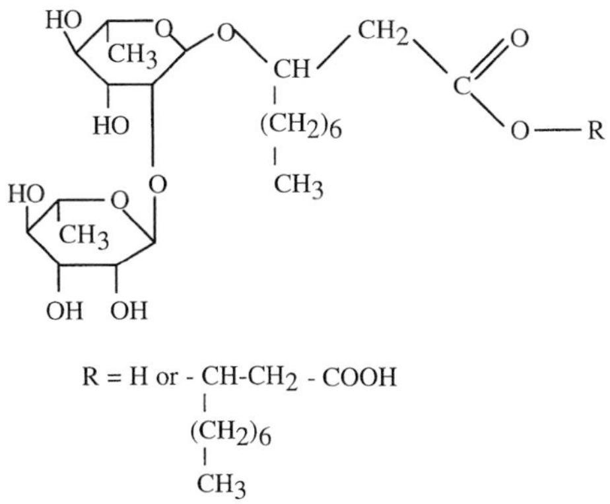

R = H or - CH-CH$_2$ - COOH (with (CH$_2$)$_6$, CH$_3$ side chain)

Rhamnolipid

XXIV

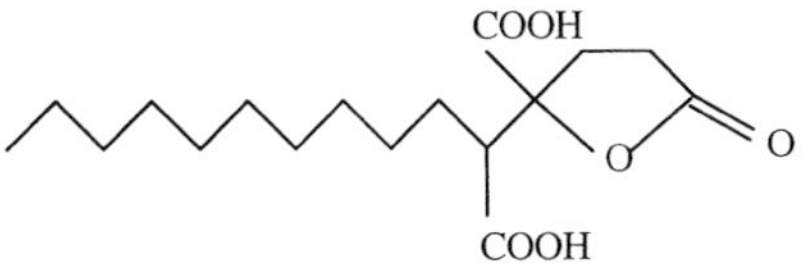

Spiculisporic acid

XXV

Of potential commercial relevance in this field but not included in KOSARIC's monograph, is spiculisporic acid (XXV) which is produced at up to 110 g L^{-1} by *Penicillium spiculisporum* and whose possible uses have included acting as a surfactant or as a synthetic intermediate for such materials (ISHIGAMI, 1993). Other fatty acid derivatives may also have a similar potential: the review by ISHIGAMI (1993) may, therefore, prove helpful in delineation of these different molecules.

6.2 Ether (Archaebacterial) Lipids

The newly-created bacterial domain of Archaea, which has arisen from the appreciation of the chemical and biochemical distinctiveness of these organisms from conventional bacteria, contain a plethora of novel lipid types not seen elsewhere in either prokaryotic or eukaryotic organisms. The organisms which are regarded as the most ancient of all life forms, are amongst the most resistant of all organisms to the extremes of environment. They therefore include the thermoacidophiles, the extreme halophiles, and the methanogens.

The lipids of these bacteria are characterized by being ether, rather than ester, derivatives of glycerol. The alkyl groups, however, are not derived from fatty acids but are formed by the mevalonate pathway which leads to the formation of the isopentenyl lipids. (The isopentenyl-derived lipids include carotenoids, sterols, and other terpenoid lipids which are of ubiquitous distribution but the archaebacterial ether lipids are confined to the Archaea.) Some typical examples of these lipids are given in Fig. 9. Although these lipids also use glycerol as the polyol unit to which the acyl groups are attached, they are attached to the *sn*-2 and *sn*-3 positions (c.f. Sect. 1.1) rather than the *sn*-1 and *sn*-2 positions of the conventional diacylglycerols and phospholipids. The ether lipids may be classed as di- or tetraethers depending on the number of ether linkages that are present in a particular molecule.

Excellent reviews of these highly unusual lipids have been written by DE ROSA and GAMBACORTA (1988), SMITH (1988), KATES (1992, 1993), and by GAMBACORTA et al., (1995). A comprehensive monograph has also recently been published (KATES et al., 1993). The much greater chemical stability of these lipids than the acylated glycerol ester lipids of Eubacteria and the Eukaryota (the other two domains of the microbial world) has suggested the means whereby the archaebacteria are able to withstand temperatures of over 110°C, salinities approaching that of saturated salt solution, and pH values of 1 or even less. The biotechnological applications of the lipids are not immediately apparent though they may present interesting ideas to chemists for synthesis of novel compounds. The biotechnological applications of the Archaea themselves, however, are the subject of much current speculative research; this area has been

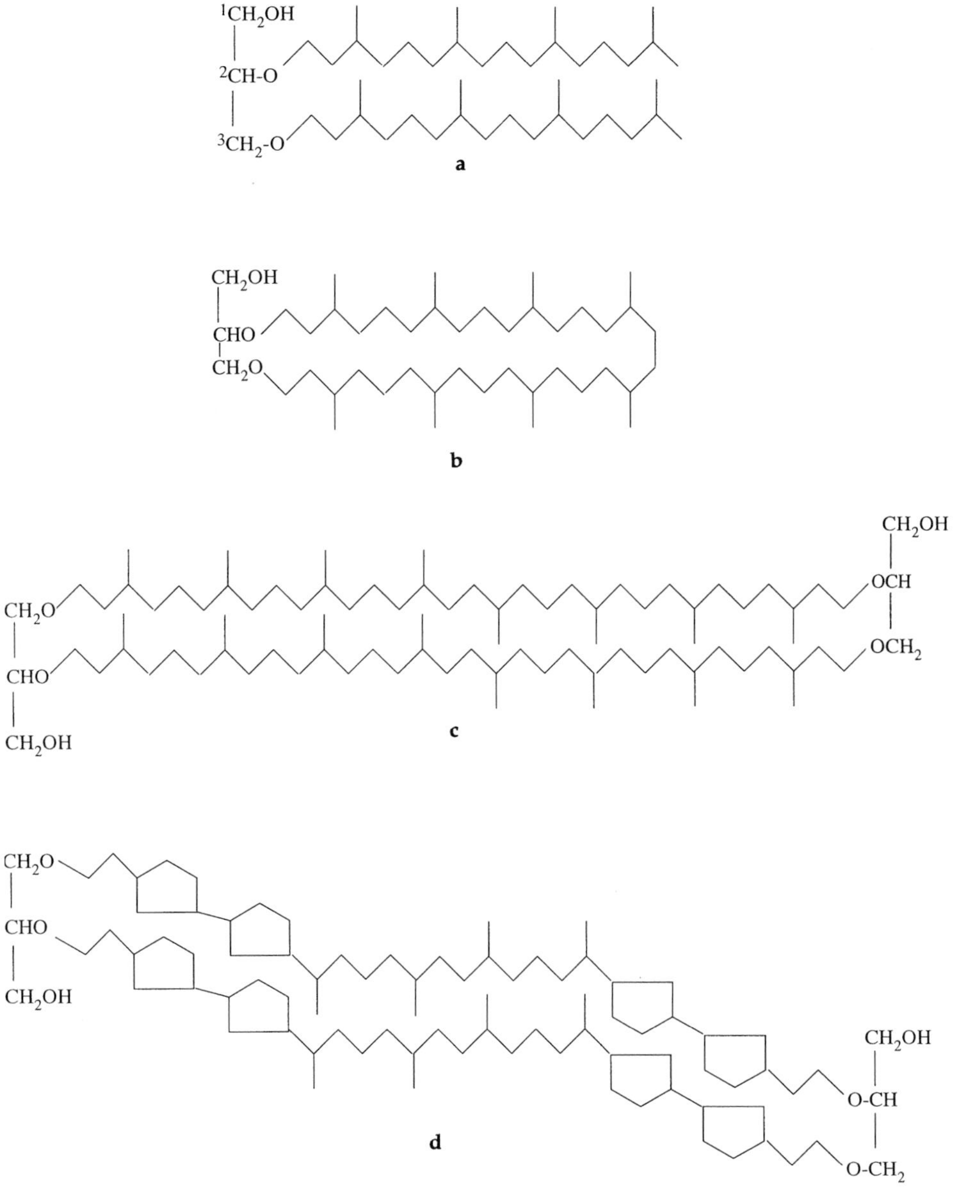

Fig. 9. Isoprenoid lipids of Archaea.
a 2,3-di-*O*-phytanyl-*sn*-glycerol (archaeol) found in several genera; *b* 2,3-di-*O*-biphytanyl-*sn*-glycerol, a macrocyclic diether found in the thermophilic methanogen, *Methanococcus, jannadschii; c* Glycerol-dialkyl-glycerol tetraether (caldarchaeol) found in *Sulfobolus* and other genera; *d* Tetracyclized glycerol dialkylnonitol tetraether (cyclized nonitolcaldarchaeol) found in thermoacidophiles and some methanogens.

recently reviewed by VENTOSA and NIETO (1995) and LEUSCHNER and ANTRANIKAN (1995) amongst others (see AGUILAR, 1995).

6.3 Phospholipids and Sphingolipids

Phospholipids, specially *sn*-1,2-diacyl glycerol-3-phosphorylated compounds, are found in all living cells save for the Archaea where alternative phospholipids occur (see Fig. 14). The range of phospholipids in microorganisms is extensive and has been reviewed by the present author (RATLEDGE, 1987, 1989b). Very few novel biotechnological applications have been identified as the amounts are usually less than 5% of the dry microbial biomass and growth of a microorganism specifically to produce a phospholipid would be clearly uneconomic. An attempt though was made in the early 1980s to extract the phospholipid from bacteria being grown on methanol as a source of single cell protein (SMITH, 1981). Some yeast "lecithin" (which is the unfractionated phospholipid fraction of which phosphatidylcholine is the predominant type) is produced for the health food market by extraction of spent brewer's yeast.

In most cases, large-scale commercial demand for phospholipids is satisfied by using

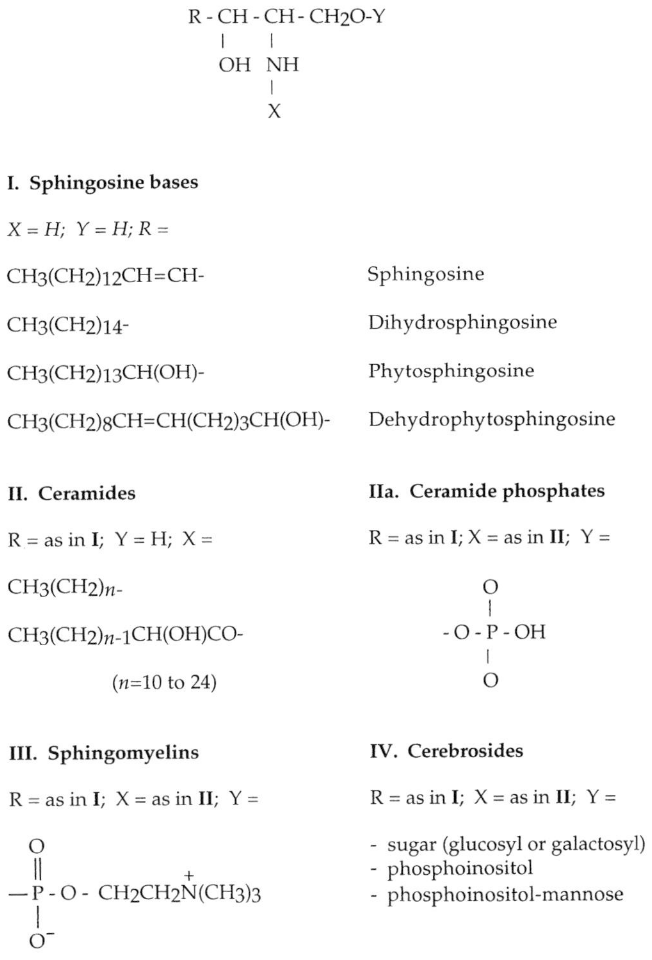

```
R - CH - CH - CH2O-Y
    |    |
    OH   NH
         |
         X
```

I. Sphingosine bases

X = H; Y = H; R =

$CH_3(CH_2)_{12}CH{=}CH$-	Sphingosine
$CH_3(CH_2)_{14}$-	Dihydrosphingosine
$CH_3(CH_2)_{13}CH(OH)$-	Phytosphingosine
$CH_3(CH_2)_8CH{=}CH(CH_2)_3CH(OH)$-	Dehydrophytosphingosine

II. Ceramides

R = as in **I**; Y = H; X =

$CH_3(CH_2)_n$-

$CH_3(CH_2)_{n-1}CH(OH)CO$-

(n=10 to 24)

IIa. Ceramide phosphates

R = as in **I**; X = as in **II**; Y =

```
      O
      |
- O - P - OH
      |
      O
```

III. Sphingomyelins

R = as in **I**; X = as in **II**; Y =

```
 O
 ||           +
—P - O - CH2CH2N(CH3)3
 |
 O⁻
```

IV. Cerebrosides

R = as in **I**; X = as in **II**; Y =

- sugar (glucosyl or galactosyl)
- phosphoinositol
- phosphoinositol-mannose

Fig. 10. Structures and nomenclature of sphingolipids found in yeasts and fungi.

plant-derived materials, which are described loosely as lecithin, and are recovered as by-products following the refining of plant seed oils (see Szuhaj, 1989, for their applications in the food industry). However, phospholipids also have important applications in the pharmaceutical industries and can be used for a number of functions (Hanin and Ansell, 1987). Such functions include the use of highly purified phosphatidylcholine as a lung surfactant for neonatal children (Bangham, 1992) and its uses to improve biocompatibility of various medical devices including contact lenses, implants, and disposable devices used in human health care. It is now sold by a number of commercial companies. The material may be produced by large-scale chromatographic purification but is more likely to be synthesized from glycero-*sn*-3-phosphocholine, which can be produced from plant lecithin, being reacted with ethyl esters of highly purified individual fatty acids in the presence of appropriate phospholipases (see Ratledge, 1994). Other uses for phospholipids could include preparation of artificial liposomes which may be used as a means of delivery of chemotherapeutic agents to selected tissues of a patient (Hanin and Ansell, 1987). If such liposomes require the presence of specific acyl groups or polar head groups then it may be possible to identify these in certain microorganisms: however, for reasons already given it is unlikely that such lipids could be produced cost-effectively as the sole microbial product from a biotechnological process.

Sphingolipids (see Fig. 10) are usually only minor components in microbial lipids. They occur in bacteria (Wilkinson, 1988), in yeasts (Rattray, 1988; Ratledge and Evans, 1989), and in some fungi (Lösel, 1988). All types of sphingolipid shown in Fig. 5, have been recognized in some microorganism. Their contents in cells are usually less than 0.5% of the dry biomass. The four sphingosine bases (see Fig. 5) do not usually occur free but usually as acylated derivatives, or ceramides.

The major microorganism of interest in this area is *Hansenula ciferri* (now known as *Pichia ciferri*). This yeast appears to be unique in producing both intra- and extracellular ceramides in which the base is either phytosphingosine or dehydrosphingosine. Kulmacz and Schroepfer (1978) observed that addition of pentadecanoic acid ($C_{15:0}$) to the growth medium of *Pichia ciferri* increased the production of the extracellular materials. This technique is now used to produce a number of ceramides with this yeast. Patent applications have been filed by Gist-Brocades (Netherlands) with respect to the production of *N*-stearoylphytosphingosine (XXVI) and other related compounds. The ceramides, after purification, are used in cosmetic industry for the controlled release of dermatologically active (or beneficial) compounds into the epidermal, dermal, and subcutaneous tissues of the skin. The yeast ceramides are regarded as identical to the ceramides that occur naturally in the human skin; this includes the correct chirality at the three chiral centers in the molecule (see XXVI). This ensures optimal performance in various skin-care formulations. (I am grateful to Dr. H. Streekstra of Gist-Brocades for supplying the above information.)

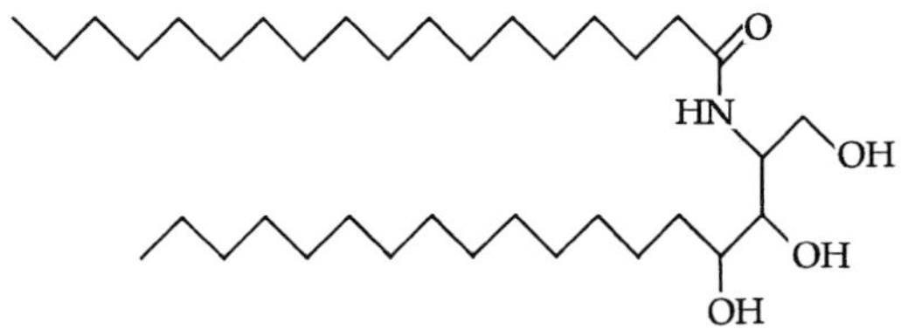

O
HN
OH
OH
OH

Stearoylphytosphingosine

XXVI

6.4 Prostanoid-Type Lipids

Prostaglandins, and the related leucotrienes and thromboxanes, exert unique physiological control over many key metabolic sequences in animals, including humans (Sinclair and Gibson, 1992). Such materials are used clinically for induction of childbirth and are also required by a large number of medical research groups. Prospects for producing such materials microbiologically may be remote but some indication has been given that the yeast *Dipodascopsis uninucleata* may be

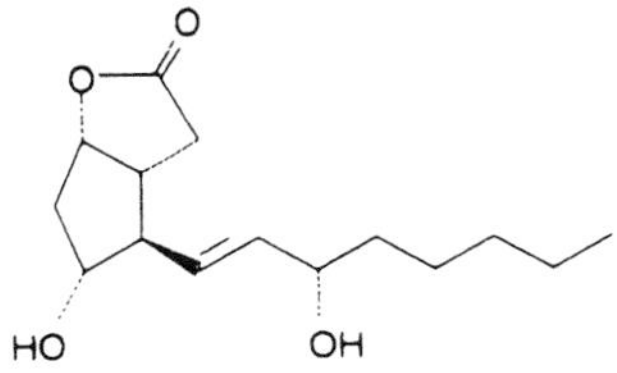

Fig. 11. ARA metabolites produced by *Dipodascopsis uninucleata; a* α-pentanor PGF_{2a} γ-lactone; *b* 3-HETE: 3-hydroxyarachidonic acid (from VAN DYK et al., 1994).

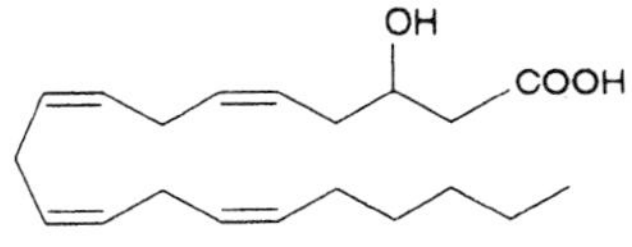

able to convert exogenously supplied arachidonic acid to a compound or compounds (see Fig. 11) showing prostaglandin-like activities when administered to experimental animals (KOCK et al., 1991; VAN DYK et al., 1991; KOCK and RATLEDGE, 1993). Current ongoing research should be able to determine within the next two or three years if such prospects of producing prostanoid lipids in this way are realistic. A recent review (VAN DYK et al., 1994) has indicated that the well-characterized oxygenase systems that occur in plants and animals for the formation of hydroxy PUFAs, which become the immediate precursors of the prostaglandins, may be found in some lower fungi (Saprolegniales and Lagenidiales) as well as species of *Dipodascopsis*.

7 Conclusions

Microbial lipids seemingly offer an almost bewildering array of possibilities for biotechnological exploitation. However, careful examination reveals that in many cases the oil or fat that is produced by a microorganism is not essentially different from that found in a plant oil, or occasionally, an animal fat. In these circumstances, the cost of the microbial route of production is likely to be many times that of the existing route. The only way that such processes could, therefore, be economically viable would be if the microbial oil was being produced as an adjunct to some waste disposal process. Such concepts were developed widely in the 1960s and 70s for the production of single cell proteins (SCP) which, although just selling, literally, as chicken feed, nevertheless produced a positive income to offset the cost of waste disposal via a biotechnological route. With respect to a microbial oil – or single cell oil (SCO) – this too could be similarly produced. There are two obvious additional advantages with this alternative strategy to producing SCP. Firstly, the oil would, if carefully selected, be worth considerably more than just chicken-feed SCP. This is nicely illustrated by the attempt to produce a cocoa butter equivalent (CBE) in New Zealand using deproteinized whey as substrate (see Sect. 3.2.1.5). Even here though, given the comparatively high price of a CBE, there was still insufficient profit margin to proceed against rival plant products selling for about $ 2,000 per t.

The second potential advantage of SCO over SCP is that the SCO need not be used for food or feed. SCP must be put back with the food chain by being fed to animals which, in turn, will be consumed, in whole or in part, by ourselves. This then requires that the substrate, or feedstock, is of an acceptable food-grade quality or poses no long-term toxicological problems. SCO can, of course, when produced from food-acceptable substrates also be used in food materials (see, e.g., the yeast SCO–CBE product already cited), but an SCO could also be used for some technical purpose. In this case, the requirement for the substrate to be of food-grade quality no longer applies. In the extreme case, SCO could be produced from any fermentable substrate if it were subsequently put to some non-food use such as incorporation into paints, lubricants, detergents, plasticizers, etc. However, the cost-effectiveness of this scenario would have to rely almost entirely on the biological process being used in some form of waste removal or environmental cleanup process as the selling price of the non-food SCO would probably be quite low. The advantage would be that the SCO-producing microorganism achieves environmental cleanup and simultaneously produces a saleable end product more valuable than SCP.

As it appears at the present time, the future of microbial oils probably lies outside these areas and the best chance for producing a commercially viable product is probably with high valued materials that are difficult to produce from plant or animal sources. This approach began in the 1980s with the use of fungi to produce polyunsaturated fatty acids (PUFA) and especially γ-linolenic acid (GLA) that was available from only one or two plant sources at exceedingly high prices. However, as with any high-priced material which appears to generate considerable profit for one or two producers, other producers quickly began rival processes of the GLA-producing plant crops so that in a few short years the price of GLA-rich oils was halved and fell even lower. Profit margins which were sufficient to maintain commercial interest in producing SCO-GLA in the mid to late 1980s then disappeared under this downward commercial pressure.

The scene in PUFAs was now moved on beyond GLA with other PUFAs, especially ARA and DHA, becoming the principal targets for microbial production. Whether these other PUFAs will follow the example of GLA and become more abundant, and thus less expensive, from other non-microbial sources will be a risk that the biotechnologist must, therefore, try to assess. In the longer term, though, it is probably these selected, very high-priced fatty acids, or derivatives from them, that will become the next SCOs to enter the marketplace.

Certain lipids, besides those based on the triacylglycerols, are already being produced biotechnologically where there is a market opportunity at the correct (i.e., profitable) price level. Carotenoids, such as β-carotene and astaxanthin, are produced commercially (see Sect. 4.2) but need to be marketed very shrewdly to convince the purchaser that these microbial products are superior to the chemically-produced, and thus cheaper, products. As microorganisms tend to produce only one chiral isomer – and usually this is equivalent to the existing plant or animal material – then this gives the microbial product a significant edge over the chemical product that rarely has the right chirality. Thus where stereospecificity is an important attribute of a product, a biotechnological route will usually be found to be superior to a chemical process. This is also seen in the formation of ceramides by yeast technology (Sect. 6.3).

The range of microbial lipids is enormous. Which are of potential commercial value and which are of academic interest is hard to assess. Insight of a particular field may bring an appreciation to one person that a microbial lipid is of value but this may not be apparent to most others. Therefore, the perusal of the range of types of lipid molecules that are available from microbial sources could bring rich rewards to the shrewd reader. However, let me conclude by saying I have touched on only some of the microbial lipids; there are many others that have not been included directly here but may just have been referred to obliquely or *en passant*. Nevertheless, I hope that I have given sufficient references for the reader to pursue some of these more esoteric opportunities for themselves: fortune will always favor the prepared mind.

8 References

ACKMAN, R. (1994), Animal and marine lipids, in: *Technological Advances in Improved and Alternative Sources of Lipids* (KAMEL, B. S., KAKUDA, Y., Eds.), pp. 292–328. London: Blackie.

AGUILAR, A. (Ed.) (1995), Special Review Topic: Biotechnology of Extremophile Microorganisms, *World J. Microbiol. Biotechnol.* **11**, issue No. 1, pp. 7–131.

AHERN, T. J., KATOH, S., SADA, E. (1983), Arachidonic acid production by the red alga *Porphyridium cruentum, Physiol. Plant.* **16**, 36–643.

AKIMOTO, M., ISHII, T., YAMAGAKI, K., OHTAGUCHI, K., KOIDE, K., YAZAWA, K. (1990), Production of eicosapentaenoic acid by a bacterium isolated from mackerel intestines, *J. Am. Oil Chem. Soc.* **67**, 911–915.

AKIMOTO, M., ISHII, T., YAMAGAKI, K., OHTAGUCHI, K., KOIDE, K., YAZAWA, K. (1991), Metal salts requisite for the production of eicosapentaenoic acid by a marine bacterium isolated from mackerel intestines, *J. Am. Oil Chem. Soc.* **68**, 504–508.

Amano, N., Shinmen, Y., Akimoto, K., Kawashima, H., Amachi, T., Shimizu, S., Yamada, H. (1992), Chemotaxonomic significance of fatty acid composition in the genus *Mortierella* (Zygomycetes, Mortierellaceae), *Mycotaxon* **44**, 257–265.

An, G.-H., Schuman, D. B., Johnson, E. A. (1989), Isolation of *Phaffia rhodozyma* mutants with increased astaxanthin content, *Appl. Environ. Microbiol.* **55**, 116–123.

An, G. H., Bielich, J., Auerbach, R., Johnson, E. A. (1991), Isolation and characterization of carotenoid hyperproducing mutants of yeast by flow cytometry and cell sorting, *Biotechnology* **9**, 70–73.

Anderson, A. J., Dawes, E. A. (1990), Occurrence, metabolism, metabolic role, and industrial uses of bacterial polyhydroxyalkanoates, *Microbiol. Rev.* **54**, 450–472.

Bailliez, C., Largean, C., Casadevall, E. (1985), Growth and hydrocarbon production of *Botryococcus braunii* immobilized in calcium alginate gel, *Appl. Microbiol. Biotechnol.* **23**, 99–105.

Bailliez, C., Largean, C., Berkaloff, C., Casadevall, E. (1986), Immobilization of *Botryococcus braunii* in alginate: influence on chlorophyll content, photosynthetic activity and degeneration during batch cultures, *Appl. Microbiol. Biotechnol.* **23**, 361–366.

Bailliez, C., Largean, C., Casadevall, E., Yang, L. W., Berkaloff, C. (1988), Photosynthesis, growth and hydrocarbon production of *Botryococcus braunii* immobilized by entrapment and adsorption in polyurethane foams, *Appl. Microbiol. Biotechnol.* **29**, 141–147.

Bajpai, P., Bajpai, P. K. (1992), Arachidonic acid production by microorganisms, *Biotechnol. Appl. Biochem.* **15**, 1–10.

Bajpai, P. K., Bajpai, P., Ward, O. P. (1991a), Optimization of production of docosahexaenoic acid (DHA) by *Thraustochytrium aureum* ATCC 34304, *J. Am. Oil Chem. Soc.* **68**, 509–514.

Bajpai, P., Bajpai, P. K., Ward, O. P. (1991b), Production of docasohexaenoic acid by *Thraustochytrium aureum, Appl. Microbiol. Biotechnol.* **35**, 706–710.

Bajpai, P., Bajpai, P. K., Ward, O. P. (1991c), Effects of ageing *Mortierella* mycelium on production of arachidonic and eicosapentaenoic acids, *J. Am. Oil Chem. Soc.* **68**, 775–780.

Bajpai, P. K., Bajpai, P., Ward, O. P. (1992), Optimization of culture conditions for production of eicosapentaenoic acid by *Mortierella elongata* NRRL 5513, *J. Ind. Microbiol.* **9**, 11–18.

Bangham, A. (1992), "Surface tension" in lungs, *Nature* (London) **359**, 110.

Barnett, J. A., Payne, R. W., Yarrow, D. (1990) *Yeasts; Characteristics and Identification,* 2nd Edn. Cambridge: University Press.

Basiron, Y., Ibrahim, A. (1994), Palm oil: old myths new facts, *INFORM* **5**, 977–980.

Beavan, M., Kligerman, A., Droniuk, R., Drouin, C., Goldenberg, B., Effio, A., Yu, P., Giuliany, B., Fein, J. (1992), Production of microbial cocoa butter equivalents, in: *Industrial Applications of Single Cell Oils* (Kyle, D. J., Ratledge, C., Eds.), pp. 156–184. Champaign, IL: American Oil Chemists' Society.

Becker, E. W. (1993), *Microalgae: Biotechnology and Microbiology.* Cambridge: University Press.

Bednarski, W., Leman, J., Tomasik, J. (1986), Utilization of beet molasses and whey for fat biosynthesis by a yeast, *Agric. Wastes* **18**, 19–26.

Behalova, B., Vorisek, J. (1988), Increased sterol formation in *Saccharomyces cerevisiae.* Analysis of cell components and ultrastructure of vacuoles, *Folia Microbiol.* **33**, 292–297.

Ben-Amotz, A., Avron, M. (1980), Glycerol, beta-carotene and dry algal meal production by commercial cultivation of *Dunaliella,* in *Algae Biomass* (Shelef, G., Soeder, C. J., Eds.), pp. 603–610. Amsterdam: Elsevier.

Ben-Amotz, A., Avron, M. (1983), On factors which determine massive β-carotene accumulation in the halotolerant alga *Dunaliella bardawil, Plant Physiol.* **72**, 593–597.

Ben-Amotz, A., Lers, A., Avron, M. (1988), Steroisomers of β-carotene and phytoene in the alga *Dunaliella bardawil, Plant Physiol.* **86**, 1286–1291.

Ben-Amotz, A., Mokady, S., Edelstein, S., Avron, M. (1989), Bioavailability of natural isomer mixture as compared with all-*trans*-β-carotene in rats and chicks, *J. Nutr.* **119**, 1013–1019.

Bernhard, K., Albrecht, H. (1948), Die Lipide aus *Phycomyces blakesleeanus, Helv. Chim. Acta* **31**, 977–988.

Borowitzka, M. A. (1988), Fats, oils and hydrocarbon, in: *Micro-Algal Biotechnology* (Borowitzka, M. A., Borowitzka, L. J., Eds.), pp. 257–287. Cambridge: University Press.

Borowitzka, M. A. (1992), Comparing carotenogenesis in *Dunalliella* and *Haematococcus:* implications for commercial production strategies, in: *Profiles on Biotechnology* (Villa, T. G., Abalde, J., Eds.), pp. 301–310. Servicio de Publications, Universidade de Santiago, Spain.

Borowitzka, M. A., Borowitzka, L. J. (Eds.) (1988), *Micro-Algal Biotechnology,* Cambridge: University Press.

Borowitzka, L. J., Borowitzka, M. A. (1989),

β-Carotene (provitamin A) production with algae, in: *Biotechnology of Vitamins, Pigments and Growth Factors* (VANDAMME, E. J., Ed.), pp. 15–26. London: Elsevier Applied Science.

BOROWITZKA, L. J., BOROWITZKA, M. A. (1990), Commercial production of β-carotene by *Dunaliella salina* in open ponds, *Bull. Marine Sci.* **47**, 244–252.

BOSWELL, K. D. V., GLADUE, R., PRIMA, B., KYLE, D. J. (1992), SCO production by fermentative microalgae, in: *Industrial Applications of Single Cell Oils* (KYLE, D. J., RATLEDGE, C., Eds.), pp. 274–286. Champaign, IL: American Oil Chemists' Society.

BOTHAM, P. A., RATLEDGE, C. (1979), A biochemical explanation for lipid accumulation in *Candida* 107 and other oleaginous micro-organisms, *J. Gen. Microbiol.* **114**, 351–375.

BOULTON, C. A. (1982), The biochemistry of lipid accumulation in oleaginous yeasts, *Ph. D. Thesis,* University of Hull, UK.

BOULTON, C. A., RATLEDGE, C. (1981), Correlation of lipid concentration in yeasts with possession of ATP:citrate lyase, *J. Gen. Microbiol.* **127**, 169–176.

BOULTON, C. A., RATLEDGE, C. (1983a), Use of transition studies in continuous cultures of *Lipomyces starkeyi,* an oleaginous yeast, to investigate the physiology of lipid accumulation, *J. Gen. Microbiol.* **129**, 2871–2876.

BOULTON, C. A., RATLEDGE, C. (1983b), Partial purification and some properties of ATP:citrate lyase from the oleaginous yeast *Lipomyces starkeyi, J. Gen. Microbiol.* **129**, 2863–2869.

BRAMLEY, P. M. (1985), The *in vitro* biosynthesis of carotenoids, *Adv. Lipid Res.* **21**, 243–279.

BRAMLEY, P. M., MACKENZIE, A. (1988), Regulation of carotenoid biosynthesis, *Curr. Top. Cell. Regul.* **29**, 291–343.

BYROM, D. (1990), Industrial production of copolymer from *Alcaligenes eutrophus,* in: *Novel Biodegradable Microbial Polymers* (DAWES, E. A., Ed.), pp. 113–117 (NATO ASI Series No 186). Dordrecht: Kluwer.

BYROM, D. (1992), Production of poly-β-hydroxybutyrate: poly-β-valerate copolymers, *FEMS Microbiol. Rev.* **103**, 247–250.

CARMAN, G. M., HENRY, S. A. (1989), Phospholipid biosynthesis in yeast, *Annu. Rev. Biochem.* **58**, 635–669.

CASADEVALL, E., DIF, D., LARGEAU, C., GUDIN, C., CHANMON, D., DESANTI, O. (1985), Studies on batch and continuous cultures of *Botryococcus braunii,* hydrocarbon production in relation to physiological state, cell ultrastructure and phosphate nutrition, *Biotechnol. Bioeng.* **27**, 286–295.

CERDÁ-OLMEDO, E. (1989), Production of carotenoids by fungi, in: *Biotechnology of Vitamins, Pigments and Growth Factors* (VANDAMME, E. J., Ed.), pp. 27–42. London: Elsevier Applied Science.

CIFERRI, O., TIBONI, O. (1985), The biochemistry and industrial potential of *Spirulina, Ann. Rev. Microbiol.* **39**, 503–526.

CLEMENT, G. (1975), Producing *Spirulina* with CO_2, in: *Single Cell Protein,* Vol. 2 (TANNENBAUM, S. R., WANG, D. I. C., Eds.), pp. 79–89. Cambridge, MA: M.I.T. Press.

COHEN, Z., COHEN, S. (1991), Preparation of eicosapentaenoic acid (EPA) concentrate from *Porphyridium cruentum, J. Am. Oil Chem. Soc.* **68**, 16–19.

COHEN, Z., HEIMER, Y. M. (1992), Production of polyunsaturated fatty acid (EPA, ARA and GLA) by the microalgae *Porphyrium* and *Spirulina,* in: *Industrial Applications of Single Cell Oils* (KYLE, D. J., RATLEDGE, C., Eds.), pp. 243–273, Champaign, IL: American Oil Chemists' Society.

COHEN, Z., VONSHAK, A., RICHMOND, A. (1988), Effect of environmental conditions on fatty acid composition of the red alga *Porphyridium cruentum:* correlation to growth rate, *J. Phycol.* **24**, 328–332.

COHEN, Z., DIDI, S., HEIMER, Y. M. (1992), Overproduction of γ-linolenic acid and eicosapentaenoic acids by algae, *Plant Physiol.* **98**, 569–572.

COOLBEAR, T., THRELFALL, D. R. (1989), Biosynthesis of terpenoid lipids, in: *Microbial Lipids,* Vol. 2 (RATLEDGE, C., WILKINSON, S. G., Eds.), pp. 115–254, London: Academic Press.

CPC International INC. (1979), Process for the microbiological production of oil, *Eur. Patent* 0005277.

CPC International INC. (1982a), Multistage process for the preparation of fats and oils, *Brit. Patent* 2091285A.

CPC International INC. (1982b), Preparation of fats and oils, *Brit. Patent* 2091286A.

CRESSWELL, R. C., REES, T. A., SHAH, N. (Eds.) (1989), *Algal and Cyanobacterial Technology.* London: Longmans.

DAVIES, R. (1984), Oil from whey, *Food Technol. N. Z.* (June), 33–37.

DAVIES, R. J. (1988), Yeast oil from cheese whey – process development, in: *Single Cell Oil* (MORETON, R. S., Ed.), pp. 99–145. Harlow: Longman.

DAVIES, R. J. (1992a), Advances in lipid processing in New Zealand, *Lipid Technol.* **4**, 6–13.

DAVIES, R. J. (1992b), Scale up of yeast oil technology, in: *Industrial Applications of Single Cell*

Oils (KYLE, D. J., RATLEDGE, C., Eds.), pp. 196–218. Champaign, IL: American Oil Chemists' Society.

DAVIES, R. J., HOLDSWORTH, J. E. (1992), Synthesis of lipids in yeasts: biochemistry, physiology, production, *Adv. Appl. Lipid Res.* **1**, 119–159.

DAVIES, R. J., HOLDSWORTH, J. E., READER, S. L. (1990), The effect of low oxygen uptake rate on the fatty acid profile of the oleaginous yeast, *Apiotrichum curvatum, Appl. Microbiol. Biotechnol.* **33**, 569–573 (*erratum: ibid.* 1991, **34**, 832–833).

DAWES, E. A. (Ed.) (1990), *Novel Biodegradable Microbial Polymers,* (NATO ASI Series No. 186). Dordrecht: Kluwer.

DE KONING, G. (1995), Physical properties of bacterial poly((R)-3-hydroxyalkanoates), *Can. J. Microbiol.* **41** (Suppl. 1), 303–309.

DE LONG, E. F., YAYANOS, A. A. (1986), Biochemical function and ecological significance of novel bacterial lipids in deep-sea prokaryotes, *Appl. Environ. Microbiol.* **51**, 730–737.

DE ROSA, M., GAMBACORTA, A. (1988), The lipids of Archaebacteria, *Prog. Lipid Res.* **27**, 153–175.

DELONG, E. F., YAYANOS, A. A. (1986), Biochemical function and ecological significance of novel bacterial lipids in deep-sea prokaryotes, *Appl. Environ. Microbiol.* **51**, 730–737.

DOI, Y. (1990), *Microbial Polyesters.* Weinheim: VCH.

DOSTALEK, M. (1986), Production of lipid from starch by a nitrogen-controlled mixed culture of *Saccharomyces fibuliger* and *Rhodosporidium toruloides, Appl. Microbiol. Biotechnol.* **24**, 19–23.

DUBINSKY, Z., BERNER, T., AARONSON, S. (1978), Potential of large-scale algal culture for biomass and lipid production in arid lands, *Biotechnol. Bioeng. Symp.* **8**, 51–68.

DULANEY, E. L. (1957), Preparation of ergosterol containing yeasts, *U.S. Patent* 2817624.

DULANEY, E. L., STAPLEY, E. O., SIMPF, K. (1954), Studies on ergosterol production by yeasts, *Appl. Microbiol.* **2**, 371–379.

EL-REFAI, A. E. M., EL-KADY, I. A. (1968a), Sterol production of yeast strains, *Z. Allg. Mikrobiol.* **8**, 355–360.

EL-REFAI, A. E. M., EL-KADY, I. A. (1968b), Sterol biosynthesis in *Saccharomyces fermentati, Z. Allg. Mikrobiol.* **8**, 361–366.

EL-REFAI, A. E. M., EL-KADY, I. A. (1969), Utilization of some industrial by-products for the microbiological production of sterols from *Saccharomyces fermentati, J. Bot. U.A.R.* **12**, 55–66.

ELLENBOGEN, B. A., AARONSON, S., GOLDSTEIN, S., BELSKY, M. (1969), Polyunsaturated fatty acids of aquatic fungi: possible phytogenetic significance, *Comp. Biochem. Physiol.* **29**, 805-811.

EROSHIN, V. K., KRYLOVA, N. I. (1983), Efficiency of lipid synthesis by yeasts, *Biotechnol. Bioeng.* **25**, 1693–1700.

EVANS, C. T. (1983), The control of metabolism in lipid-accumulating yeasts, *Ph. D. Thesis,* University of Hull, UK.

EVANS, C. T., RATLEDGE, C. (1983), A comparison of the oleaginous yeast, *Candida curvata,* grown on different carbon sources in continuous and batch culture, *Lipids* **18**, 623–629.

EVANS, C. T., RATLEDGE, C. (1984a), Effect of nitrogen source on lipid accumulation in oleaginous yeasts, *J. Gen. Microbiol.* **130**, 1693–1704.

EVANS, C. T., RATLEDGE, C. (1984b), Influence of nitrogen metabolism on lipid accumulation by *Rhodosporidium toruloides* CBS 14, *J. Gen. Microbiol.* **130**, 1705–1710.

EVANS, C. T., RATLEDGE, C. (1984c), Phosphofructokinase and the regulation of the flux of carbon from glucose to lipid in the oleaginous yeast *Rhodosporidium toruloides, J. Gen. Microbiol.* **130**, 3251–3264.

EVANS, C. T., RATLEDGE, C. (1985a), Partial purification and properties of pyruvate kinase and its regulatory role during lipid accumulation by the oleaginous yeast *Rhodosporidium toruloides* CBS 14, *Can. J. Microbiol.* **31**, 479–484.

EVANS, C. T., RATLEDGE, C. (1985b), The role of the mitochondrial NAD^+:isocitrate dehydrogenase in lipid accumulation by the oleaginous yeast *Rhodosporidium toruloides* CBS 14, *Can. J. Microbiol.* **31**, 845–850.

EVANS, C. T., RATLEDGE, C. (1985c), Possible regulatory roles of ATP:citrate lyase, malic enzyme and AMP deaminase in lipid accumulation by *Rhodosporidium toruloides* CBS 14, *Can. J. Microbiol.* **31**, 1000–1005.

EVANS, C. T., RATLEDGE, C. (1986), The physiological significance of citric acid in the control of metabolism in lipid-accumulating yeasts, *Biotechnol. Gen. Eng. Rev.* **3**, 349–375.

EVANS, C. T., SCRAGG, A. H., RATLEDGE, C. (1983), A comparative study of citrate efflux from mitochondria of oleaginous and non-oleaginous yeasts, *Eur. J. Biochem.* **130**, 195–204.

FALL, R., PHELPS, P., SPINDLER, D. (1984), Bioconversion of xylan to triglycerides by oil-rich yeasts, *Appl. Environ. Microbiol.* **47**, 1130–1134.

FLOETENMEYER, M. D., GLATZ, B. A., HAMMOND, E. G. (1985), Continuous culture fermentation of whey permeate to produce microbial oil, *J. Dairy Sci.* **68**, 633–637.

Fuji Oil CO. LTD. (1979), Method for producing cacao butter substitute, *Brit. Patent* 1555000.

Fuji Oil CO. LTD. (1981), Method for producing cacao butter substitute, *U.S. Patent* 4268527.

GAMBACORTA, A., GLIOZZI, A., DE ROSA, A. (1995), Archaeal lipids and their biotechnological applications, *World J. Microbial Biotechnol.* **12**, 115–131.

GANDHI, S. R., WEETE, J. D. (1991), Production of the polyunsaturated fatty acids arachidonic acid and eicosapentaenoic acid by the fungus *Phytium ultimum, J. Gen. Microbiol.* **137**, 1825–1830.

GELLERMAN, J. L., SCHLENK, H. (1979), Methyl-directed desaturation of arachidonic to eicosapentaenoic acid in the fungus *Saprolegnia parasitica, Biochim. Biophys. Acta* **573**, 23–30.

GLATZ, B. A., FLOETENMEYER, M. D., HAMMOND, E. G. (1985), Fermentation of bananas and other food wastes to produce microbial lipid, *J. Food Prot.* **48**, 574–577.

GOODWIN, T. W. (1980), *The Biochemistry of the Carotenoids,* 2nd Edn., Vol. 1. London: Chapman & Hall.

GOODWIN, T. W. (1983), *The Biochemistry of the Carotenoids,* 2nd Edn., Vol. 2. London: Chapman & Hall.

GRANGER, L. M., PERLOT, P., GOMA, G., PAREILLEUX, A. (1993), Effect of various nutrient limitations on fatty acid production by *Rhodotorula glutinis, Appl. Microbiol. Biotechnol.* **38**, 784–789.

GUCKERT, J. B., COOKSEY, K. E. (1990), Triglyceride accumulation and fatty acid profile changes in *Chlorella* (Chlorophyta) during high pH-induced cell cycle inhibition, *J. Phycol.* **26**, 72–79.

GUERZONI, M. E., LAMBERTINI, P., LERCKER, G., MARCHETTI, R. (1985), Technological potential of some starch degrading yeasts, *Starch/Stärke* **37**, 52–57.

HAARD, N. F. (1988), Astaxanthin formation by the yeast *Phaffia rhodozyma* on molasses, *Biotechnol. Lett.* **10**, 609–614.

HAMMOND, E. G., GLATZ, B. A. (1990), Biotechnology applied to fats and oils, in: *Food Technology,* Vol. 2 (KING, R. D., CHEETHAM, P. S. J., Eds.), pp. 173–217. London: Elsevier Applied Science.

HANIN, I., ANSELL, G. B. (Eds.) (1987), *Lecithin: Technological, Biological and Therapeutic Aspects.* New York: Plenum Press.

HANSSON, L., DOSTALEK, M. (1988), Effect of culture conditions on mycelial growth and production of γ-linolenic acid by the fungus *Mortierella ramanniana, Appl. Microbiol. Biotechnol.* **28**, 240–246.

HANSSON, L., DOSTALEK, M. (1989), Effect of culture conditions on mycelial growth and production of γ-linolenic acid by the fungus *Mucor ramanniana, Appl. Microbiol. Biotechnol.* **28**, 240–246.

HARRISON, D. M. (1986), The biosynthesis of carotenoids, *Nat. Prod. Rep.* **3**, 205–215.

HARWOOD, J. L. (1994a), Environmental effects on plant lipid metabolism, *INFORM,* **5**, 835–839.

HARWOOD, J. L. (1994b), Environmental factors which can alter lipid metabolism, *Prog. Lipid Res.* **33**, 193–202.

HASSAN, M., BLANC, P. J., GRANGER, L. M., PAREILLEUX, A., GOMA, G. (1993), Lipid production by an unsaturated fatty acid auxotroph of the oleaginous yeast *Apiotrichum curvatum* grown in single-stage continuous culture, *Appl. Microbiol. Biotechnol.* **40**, 483–488.

HASSAN, M., BLANC, P. J., PAREILLEUX, A., GOMA, G. (1994a), Selection of fatty acid auxotrophs from the oleaginous yeast *Cryptococcus curvatus* and production of cocoa butter equivalents in batch culture, *Biotechnol. Lett.* **16**, 819–824.

HASSAN, M., BLANC, P. J., PAREILLEUX, A., GOMA, G. (1994b), Production of single cell oil from prickly pear juice fermentation by *Cryptococcus curvatus* grown in batch culture, *World J. Microbiol. Biotechnol.* **10**, 534–537.

HENDERSON, R. J., LEFTLEY, J. W., SARGENT, J. R. (1988), Lipid composition and biosynthesis in the marine dinoflagellate *Crypthecodium cohnei, Phytochemistry* **27**, 1679–1683.

HENDERSON, R. J., MILLAR, R. M., SARGENT, J. R., JOSTENSEN, J. P. (1993), *Trans*-neonoenoic and polyunsaturated fatty acids in phospholipids of a *Vibrio* species in relation to growth conditions, *Lipids,* **28**, 389–396.

HESSE, A. (1949) Industrial biosyntheses, part I. fats, *Adv. Enzymol.* **9**, 653–704.

HILLEN, L. W., POLLARD, G., WAKE, L. V., WHITE, N. (1982), Hydrocarbon of the oils of *Botryococcus braunii* to transport fuels, *Biotechnol. Bioeng.* **24**, 193–205.

HIRANO, M., MORI, H., MIURA, Y., MATSUNAGA, N., NAKAMURA, N., MATSUNAGA, T. (1990), γ-Linolenic acid production by microalgae, *Appl. Biochem. Biotechnol.* **24/25**, 183–191.

HODGSON, P. A., HENDERSON, K. J., SARGENT, J. R., LEFTLEY, J. W. (1991), Patterns of variation in the lipid class and fatty acid composition of *Nannochloropsis oculata* (Eustigmatophyceae) during batch culture, I. The growth cycle, *J. Appl. Phycol.* **3**, 169–181.

HOLDSWORTH, J. E., RATLEDGE, C. (1988), Lipid turnover in oleaginous yeasts, *J. Gen. Microbiol.* **134**, 339–346.

HOLDSWORTH, J. E., VEENHUIS, M., RATLEDGE, C. (1988), Enzyme activities in oleaginous yeasts accumulating and utilizing exogenous or endogenous lipids, *J. Gen. Microbiol.* **134**, 2907–2915.

HORROBIN, D. F. (1992), Clinical applications of n-6 essential fatty acids: atopic eczema and inflammation, diabetic neuropathy and retinopathy, breast pain and viral infections, in: *Essential Fatty Acids and Eicosanoids,* 3rd Int. Congr. (SINCLAIR, A., GIBSON, R., Eds.), pp. 367–372. Champaign: IL: American Oil Chemists' Society.

HOUSTON, B., NIMMO, H. G. (1984), Purification and some kinetic properties of rat liver ATP:citrate lyase, *Biochem. J.* **224**, 437–443.

HOUSTON, B., NIMMO, H. G. (1985), Effects of phosphorylation on the kinetic properties of rat liver ATP:citrate lyase, *Biochim. Biophys. Acta* **884**, 233–239.

HRABAK, O. (1992), Industrial production of poly-β-hydroxybutyrate, *FEMS Microb. Rev.* **103**, 251–256.

HUDSON, B. J. F., KARIS, I. G. (1974), The lipids of the alga *Spirulina, J. Sci. Food Agric.* **25**, 759–763.

ISHIGAMI, Y. (1993), Biosurfactants face increasing interest, *INFORM* **4**, 1156–1165.

JAREONKITMONGKOL, S., KAWASHIMA, H., SHIRASAKA, N., SHIMIZU, S., YAMADA, H. (1992a), Production of dihomo-γ-linolenic acid by a Δ5-desaturase-defective mutant of *Mortierella alpina* 1S-4, *Appl. Environ. Microbiol.* **58**, 2196–2200.

JAREONKITMONGKOL, S., KAWASHIMA, H., SHIMIZU, S., YAMADA, H. (1992b), Production of 5,8,11-*cis*-eicosatrienoic acid by a Δ12-desaturase-defective mutant of *Mortierella alpina* 1S-4, *J. Amer. Oil Chem. Soc.* **69**, 939–944.

JAREONKITMONGKOL, S., SHIMIZU, S., YAMADA, H. (1992c), Fatty acid desaturation-defective mutants of an arachidonic acid-producing fungus, *Mortierella alpina* 1S-4, *J. Gen. Microbiol.* **138**, 997–1002.

JENSEN, A. (1966), *Report No. 31,* Norwegian Institute of Seaweed Research.

JOHNS, R. B., PERRY, G. J. (1977), Lipids of the marine bacterium *Flexibacter polymorphus, Arch. Microbiol.* **114**, 267–271.

JOHNSON, E. A., AN, G. H. (1991), Astaxanthin from microbial sources, *CRC Crit. Rev. Biotechnol.* **11**, 297–321.

JOHNSON, E. A., CONKLIN, D. E., LEWIS, M. J. (1977), The yeast *Phaffia rhodozyma* as a dietary pigment source for salmonids and crustaceans, *J. Fish. Res. Board Can.* **34**, 2417–2421.

JOHNSON, E. A., VILLA, T. G., LEWIS, M. J. (1980), *Phaffia rhodozyma* as an astaxanthin source in salmonid diets, *Aquaculture* **20**, 123–134.

KATES, M. (1992), Archaebacterial lipids: structure, biosynthesis and function, *Biochem. Soc. Symp.* **58**, 51–72.

KATES, M. (1993), Membrane lipids of extreme halophytes: biosynthesis, function and evolutionary significance, *Experientia* **49**, 1027–1036.

KATES, M., KUSHNER, D. J., MATHESON, A. T. (Eds.) (1993), *The Biochemistry of Archaea (Archaebacteria).* Amsterdam: Elsevier Science Publishers B.V.

KAWABATA, A., KANEYAMA, M. (1989), The effect of growth temperature on wax ester content and composition of *Euglena gracilis, J. Gen. Microbiol.* **135**, 1461–1467.

KENDRICK, A., RATLEDGE, C. (1992a), Lipids of selected molds grown for the production of n-3 and n-6 polyunsaturated fatty acids, *Lipids* **27**, 15–20.

KENDRICK, A., RATLEDGE, C. (1992b), Lipid formation in the oleaginous mold *Entomophthora exitalis* grown in continuous culture: effects of growth rate, temperature and dissolved oxygen tension on polyunsaturated fatty acids, *Appl. Microbiol. Biotechnol.* **37**, 18–22.

KENDRICK, A., RATLEDGE, C. (1992c), Desaturation of polyunsaturated fatty acids in *Mucor circinelloides* and the involvement of a novel membrane-bound malic enzyme, *Eur. J. Biochem.* **209**, 667–673.

KENDRICK, A., RATLEDGE, C. (1992d), Phospholipid fatty acyl distribution of three fungi indicates positional specificity for n-6 vs. n-3 fatty acids, *Lipids* **27**, 505–508.

KENNEDY, M. J., READER, S. L., DAVIES, R. J. (1993), Fatty acid production characteristics of fungi with particular emphasis on γ-linolenic acid, *Biotechnol. Bioeng.* **42,** 625–634.

KERNON, J. (1992), Vegetable fats in chocolate – EC and US legislation issues, *Lipid Technol.* **4,** 113–114.

KOCK, J. L. F., RATLEDGE, C. (1993), Changes in lipid composition and arachidonic turnover during the life cycle of the yeast *Dipodascopsis uninucleata, J. Gen. Microbiol.* **139**, 459–464.

KOCK, J. L. F., COETZEE, D. J., VAN DYK, M. S., TRUSCOTT, M., CLOETE, F. C., VAN WYK, V., AUGUSTYN, O. P. H. (1991), Evidence for pharmacologically active prostaglandin in yeasts, *South Afr. J. Sci.* **87**, 73–76.

KORITALA, S. (1989), Microbiologial synthesis of wax esters by *Euglena gracilis, J. Am. Oil Chem. Soc.* **66**, 133–134.

KOSARIC, N. (Ed.) (1993), *Biosurfactants: Production, Properties, Applications.* New York: Marcel Dekker.

KREGER-VAN RIJ, N. J. W. (Ed.) (1984), *The Yeasts. A Taxonomic Study,* 3rd Edn. Amsterdam: Elsevier Science Publishers.

KULMACZ, R. J., SCHROEPFER, G. J. (1978), Sphingolipid base metabolism, *J. Am. Chem. Soc.* **100**, 3963–3964.

KYLE, D. J. (1991), Speciality oils from microalgae: new perspectives, in: *Biotechnology of Plant Fats and Oils* (RATTRAY, J. Ed.), pp. 130–143. Champaign, IL: American Oil Chemists' Society.

KYLE, D. J. (1992), Production and use of lipids from microalgae, *Lipid Technol.* **4**, 59–64.

KYLE, D. J., RATLEDGE, C. (1992), *Industrial Applications of Single Cell Oils,* Champaign, IL: American Oil Chemists' Society.

KYLE, D. J., BEHRENS, P., BINGHAM, S., ARNETT, K., LIEBERMAN, D. (1988), Microalgae as a source of EPA-containing oils, in: *Biotechnology for the Fats and Oils Industry* (APPLEWHITE, T. H., Ed.), pp. 117–121. Champaign, IL: American Oil Chemists' Society.

KYLE, D. J., SICOTTE, V. J., REEB, S. E. (1992), Bioproduction of docosahexaenoic acid (DHA) by microalgae, in: *Industrial Applications of Single Cell Oils* (KYLE, D. J., RATLEDGE, C., Eds.), pp. 287–300. Champaign, IL: American Oil Chemists' Society.

LAGARDE, M., BURTIN, M., RIGAND, M., SPRECHER, H., DECHAVANNE, M., RENAND, S. (1985), Prostaglandin E_2 – like activity of 20: 3n-9 platelet lipoxygenase end-product, *FEBS Lett.* **181**, 53–56.

LEE, Y. K. (1986), Enclosed bioreactors for the mass cultivation of photosynthetic microorganisms: the future trend, *Trends Biotechnol.* **4**, 186–189.

LEONARD, E. C. (1994), Salim group – major factor in oil palm trade, *INFORM* **5**, 987–992.

LEUSCHNER, C., ANTRANIKAN, G. (1995), Heat-stable enzymes from extreme thermophilic and hyperthermophilic microorganisms, *World J. Microbiol. Biotechnol.* **11**, 95–114.

LOPEZ ALONSO, D., MOLINA GRIMA, E., SANCHEZ PEREZ, J. A., GARCIA SANCHEZ, J. L., GARCIA CAMACHO, F. (1992), Fatty acid variation among different isolates of a single strain of *Isochrysis galbana, Phytochemistry* **31**, 3901–3904.

LÖSEL, D. M. (1988), Fungal lipids, in: *Microbial Lipids,* Vol. 2 (RATLEDGE, C., WILKINSON, S. G. Eds.), pp. 699–806. London: Academic Press.

MARGALITH, P. (1989), Vitamin D: the biotechnology of ergosterol, in: *Biotechnology of Vitamins, Pigments and Growth Factors* (VANDAMME, E. J., Ed.), pp. 81–104. London: Elsevier Applied Science.

MARGALITH, P. Z. (1992), *Pigment Microbiology.* London: Chapman & Hall.

MEAD, J. F., SLATON, W. H. (1956), Metabolism of essential fatty acids. III. Isolation of 5,8,11-eicosatrieonic acid from fat-deficient rats, *J. Biol. Chem.* **219**, 705–709.

MEYER, P. S., DU PREEZ, J. C., KILIAN, S. G. (1993), Selection and evaluation of astaxanthin-overproducing mutants of *Phaffia rhodozyma, World J. Microbiol. Biotechnol.* **9**, 514–520.

MIELKE, S. (1992), Trends in supply, consumption and prices, in: *Oils and Fats in the Nineties* (SHUKLA, V. K. S., GUNSTONE, F. D., Eds.), pp. 10–22. Lystrup, Denmark: International Food Science Centre A/S.

MILLER, M., KOCK, J. L. F., BOTES, P. J. (1989a), The significance of long-chain fatty acid compositions and other phenotypic characteristics in the yeast *Pichia* Hansen *emend.* Kurtzman, *Syst. Appl. Microbiol.* **12**, 70–79.

MOLINA GRIMA, E., SANCHEZ PEREZ, J. A., GARCIA SANCHEZ, J. L. (1992), EPA from *Isochrysis galbana.* Growth conditions and productivity, *Process Biochem.* **27**, 299–305.

MOLINA GRIMA, E., SANCHEZ PEREZ, J. A., GARCIA CAMACHO, F., GARCIA SANCHEZ, J. L., LOPEZ ALONSO, D. (1993), n-3 PUFA productivity in chemostat cultures of microalgae, *Appl. Microbiol. Biotechnol.* **38**, 599–605.

MOLINA GRIMA, E., SANCHEZ PEREZ, J. A., GARCIA CAMACHO, F., FERNANDEZ SEVILLA, J. M., ACIEN FERNANDEZ, F. G. (1994), Effect of growth rate on the eicosapentaenoic acid and docasahexaenoic acid content of *Isochrysis galbana* in chemostat culture, *Appl. Microbiol. Biotechnol.* **41**, 23–27.

MOON, N. J., HAMMOND, E. G., GLATZ, B. A. (1978), Conversion of cheese whey and whey permeate to oil and single-cell protein, *J. Dairy Sci.* **61**, 1537–1547.

MORETON, R. S. (1985), Modification of fatty acid composition of lipid accumulating yeasts with cyclopropene fatty acid desaturase inhibitors, *Appl. Microbiol. Biotechnol.* **33**, 41–45.

MORETON, R. S. (1988a), *Single Cell Oil.* Harlow: Longman.

MORETON, R. S. (1988b), The physiology of lipid accumulating yeasts, in: *Single Cell Oil* (MORETON, R. S., Ed.), pp. 1–32. Harlow: Longman.

MORETON, R. S., CLODE, D. M. (1985), Microbial desaturase enzyme inhibitors and their use in a method of producing lipids. *Brit. Patent* 8407195.

MURILLO, F. J., CALDERON, I. L., LOPEZ-DIAZ, I., CERDA-OLMENDO, E. (1978), Carotene-superproducing strains of *Phycomyces, Appl. Environ. Microbiol.* **36**, 477–479.

MURPHY, D. J. (1994a), Transgenic plants – a future source of novel edible and industrial oils, *Lipid Technol.* **6**, 84–91.

MURPHY, D. J. (Ed.) (1994b), *Designer Crop Oils.* Weinheim: VCH.

NAKAHARA, T., YOKOCKI, T., KAMISAKA, Y., SUZUKI, O. (1992), Gamma-linolenic acid from genus *Mortierella,* in: *Industrial Applications of Single Cell Oils* (KYLE, D. J., RATLEDGE, C., Eds.), pp. 61–97. Champaign, IL: American Oil Chemists' Society.

NAKAJIMA, T., IZU, S. (1992), Microbial production and purification of w-6 polyunsaturated fatty acids, in: *Essential Fatty Acids and Eicosanoids* (SINCLAIR, A., GIBSON, R., Eds.), pp. 57–64. Champaign, IL: American Oil Chemists' Society.

NEIDLEMAN, S., HUNTER-CEVERA, J. (1992), Wax ester production by *Acinetobacter* sp. HO1-N, in: *Industrial Applications of Single Cell Oils* (KYLE, D. J., RATLEDGE, C., Eds.), pp. 16–28. Champaign, IL: American Oil Chemists' Society.

NELIS, H. J., DE LEENHEER, A. P. (1989), Microbial production of carotenoids other than β-carotene, in: *Biotechnology of Vitamins, Pigments and Growth Factors* (VANDAMME, E. J., Ed.), pp. 43–80. London, New York: Elsevier Applied Science.

NICHOLS, B. W., WOOD, B. J. B. (1968), The occurrence of biosynthesis of γ-linolenic acid in a blue-green alga, *Spirulina platensis, Lipids* **3**, 46–50.

NICHOLS, P. D., NICHOLS, D. S., BAKES, M. J. (1994), Marine oil products in Australia, *INFORM* **5**, 254–261.

NINET, L., RENAUT, J. (1979), Carotenoids, in: *Microbial Technology,* 2nd Edn., Vol. 1 (PEPPLER, H. J., PERLMAN, D., Eds.), pp. 529–544. San Diego: Academic Press.

NOGUCHI, Y., KAME, M., IWAMOTO, H. (1982), Studies on lipid production by yeasts. Fatty acid composition of lipid from glucose and fatty acid esters by *Rhodotorula* sp. and *Candida* sp., *Yukagawa* **31**, 431–437.

NOVOTNY, C., BEHALOVA, B., STRUZINSKY, R., NOVAK, M., ZAJICEK, J. (1988), Sterol composition of a Δ5, 7-sterol-rich strain of *Saccharomyces cerevisiae* during batch growth, *Folia Microbiol.* **33**, 377–385.

O'BRIEN, D. J., SENSKE, G. E. (1994), Recovery of eicosapentaenoic acid from fungal mycelia by solvent extraction, *J. Am. Oil Chem. Soc.* **71**, 947–950.

O'BRIEN, D. J., KURANTZ, M. J., KWOCZAK, R. (1993), Production of eicosapentaenoic acid by the filamentous fungus *Pythium irregulare, Appl. Microbiol. Biotechnol.* **40**, 211–214.

O'LEARY, W. M., WILKINSON, S. G. (1988), Gram-positive bacteria, in: *Microbial Lipids,* Vol. 1 (RATLEDGE, C., WILKINSON, S. G., Eds.), pp. 117–201. London: Academic Press.

OSMAN, H. G., MOSTAFA, M. A., EL-REFAI, A. H. (1969), Production of lipids and sterols by *Aspergillus fumigatus.* Some culture conditions favouring the formation of lipids and sterols, *J. Chem. U.A.R.* **12**, 185–197.

OWUSU-ANSAH, Y. J. (1993), Enzymes in lipid technology and cocoa butter substitutes, in: *Technological Advances in Improved and Alternative Sources of Lipids* (KAMEL, B. S., KAKUDA, Y., Eds), pp. 360–389. Glasgow: Blackie Academic & Professional.

PAN, J. G., RHEE, J. S. (1986), Biomass yields and energetic yields of oleaginous yeasts in batch culture, *Biotechnol. Bioeng.* **18**, 112–114.

PARKS, L. W., RODRIQUEZ, R. J., MCCAMMON, M. T. (1984), Sterols of yeast: a model for biotechnology in the production of fats and oils, in: *Biotechnology for the Oils and Fats Industry* (RATLEDGE C., DAWSON, P., RATTRAY, J., Eds.), pp. 177–187. Champaign, IL: American Oil Chemists' Society.

PIERINGER, R. A. (1989), Biosynthesis of non-terpenoid lipids, in: *Microbial Lipids,* Vol. 2 (RATLEDGE, C., WILKINSON, S. G., Eds.), pp. 51–114. London: Academic Press.

POIRIER, Y., DENNIS, D., KLOMPARENS, K., NAWRATH, C., SOMERVILLE, C. (1992), Perspectives on the production of polyalkanoates in plants, *FEMS Microbiol. Rev.* **103**, 237–246.

POIRIER, Y., NAWRATH, C., SCHECHTMAN, L. A., SATKOWSKI, M. M., NODA, I., SOMERVILLE, C. (1994), PHB production in plants: some recent advances, *Abst. Intern. Symp. on Bacterial PHA, Montreal, Canada,* p. 44.

RADWAN, S. S. (1991), Sources of C_{20}-polyunsaturated fatty acids for biotechnological use, *Appl. Microbiol. Biotechnol.* **35**, 421–430.

RADWAN, S. S., SOLIMAN, A. H. (1988), Arachidonic acid from fungi utilizing fatty acids with shorter chains as sole sources of carbon and energy, *J. Gen. Microbiol.* **134**, 387–393.

RATLEDGE, C. (1982), Microbial oils and fats: an assessment of their commercial potential, *Prog. Ind. Microbiol.* **16**, 119–206.

RATLEDGE, C. (1986), Lipids, in: *Biotechnology,* 1st Edn., Vol. 4 (REHM, H. J., REED, G., Eds.), pp. 185–213. Weinheim: VCH.

RATLEDGE, C. (1987), Microorganisms as sources of phospholipids, in: *Lecithin: Technological, Biological and Therapeutic Aspects* (HANIN, I.,

Ansell, G. B., Eds.), pp. 17–35. New York: Plenum Press.

Ratledge, C. (1988), Biochemistry, stoichiometry, substrates and economics, in: *Single Cell Oil* (Moreton, R. S., Ed.), pp. 33–70. Harlow: Longman.

Ratledge, C. (1989a), Biotechnology of oils and fats, in: *Microbial Lipids* (Ratledge, C., Wilkinson, S. G., Eds.), pp. 567–668. London: Academic Press.

Ratledge, C. (1989b), Microbiological sources of phospholipids, in: *Lecithins: Sources, Manufacture and Uses* (Szuhaj, B. F., Ed.), pp. 72–96. Champaign, IL: American Oil Chemists' Society.

Ratledge, C. (1994), Biodegradation of oils, fats and fatty acids, in: *Biochemistry of Microbial Degradation* (Ratledge, C., Ed.), pp. 89–141. Dordrecht: Kluwer Academic Publisher.

Ratledge, C., Evans, C. T. (1989), Lipids and their metabolism, in: *The Yeasts,* 2nd Edn., Vol. 3 (Rose, A. H., Harrison, J. S., Eds.), pp. 367–455. London: Academic Press.

Ratledge, C., Gilbert, S. C. (1985), Carnitine acetyltransferase activity in oleaginous yeasts, *FEMS Microbiol. Lett.* **27**, 273–275.

Ratledge, C., Wilkinson, S. G. (1988a), *Microbial Lipids,* Vol. 1. London: Academic Press.

Ratledge, C., Wilkinson, S. G. (1989), *Microbial Lipids,* Vol. 2. London: Academic Press.

Ratledge, C., Wilkinson, S. G. (1988b), An overview of microbial lipids, in: *Microbial Lipids,* Vol. 1 (Ratledge, C., Wilkinson, S. G., Eds.), pp. 3–22. London: Academic Press.

Ratledge, C., Boulton, C. A., Evans, C. T. (1984), Continuous culture studies of lipid production by oleaginous microorganisms, in: *Continuous Culture 8: Biotechnology, Medicine and the Environment* (Dean, A. C. R., Ellwood, D. C., Evans, C. G. T., Eds.), pp. 272–291. Chichester: Ellis Horwood.

Rattray, J. B. M. (1988), Yeasts, in: *Microbial Lipids,* Vol. 1 (Ratledge, C., Wilkinson, S. G., Eds.), pp. 555–697. London: Academic Press.

Rattray, J. B. M. (1991), Plant biotechnology and the oils and fats industry, in: *Biotechnology of Plant Fats and Oils* (Rattray, J. B. M., Ed.), pp. 1–35. Champaign, IL: American Oil Chemists' Society.

Rattray, J. B. M. (1994), Biotechnological advances in improved and alternative sources of lipids, in: *Technological Advances in Improved and Alternative Sources of Lipids* (Kamel, B. S., Kakuda, Y., Eds.), pp. 50–92. London: Blackie.

Ringo, E., Jostensen, J. P., Olsen, R. E. (1992), Production of eicosapentaenoic acid by freshwater *Vibrio, Lipids* **27**, 564–566.

Roessler, P. G. (1990), Environmental control of glycerolipid metabolism in microalgae: commercial implication and future research directions, *J. Phycol.* **26**, 393–399.

Roux, M. P., Kock, J. L. F., Betha, A., du Preez, J. C., Wells, G. V., Botes, P. J. (1994), Mucor – a source of cocoa butter and gamma-linolenic acid, *World J. Microbiol. Biotechnol.* **10**, 417–422.

Sanchez, N., Martinez, M., Aracil, J., Corma, A. (1992), Synthesis of oleyl oleate as a jojoba oil analog, *J. Am. Oil Chem. Soc.* **69**, 1150–1153.

Sawayama, S., Minowa, T., Dote, Y., Yokoyama, S. (1992), Growth of the hydrocarbon-rich microalga *Botryococcus braunii* in secondarily treated sewage, *Appl. Microbiol. Biotechnol.* **38**, 135–138.

Schäfer, L., Kragballe, K. (1991), Supplementation with evening primrose oil in atopic dermatitis: effect of fatty acids in neutrophils and epidermis, *Lipids* **26,** 557–560.

Schlegel, H. G., Steinbüchel, A. (Eds.) (1992), Int. Symp. Bact. Polyalkanoates ISBP '92, *FEMS Microbiol. Rev.,* Vol. **103** (special issue).

Schweizer, E. (1989), Biosynthesis of fatty acids and related compounds, in: *Microbial Lipids,* Vol. 2 (Ratledge, C., Wilkinson, S. G., Eds.), pp. 3–50. London: Academic Press.

Sekula, B. (1992), Wax ester production by yest, in: *Biotechnology of Plant Fats and Oils* (Rattray, J. B. M., Ed.), pp. 162–176. Champaign, IL: American Oil Chemists' Society.

Seto, A., Wang, H. L., Hesseltine, C. W. (1984), Culture conditions affect eicosapentaenoic acid content of *Chlorella minutissima, J. Am. Oil Chem. Soc.* **61,** 892–894.

Seto, A., Kumasaka, K., Hosaka, M., Kojima, E., Kashiwakura, M., Kato, T. (1992), Production of eicosapentaenoic acid by a marine microalgae and its commercial utilization for aquaculture, in: *Industrial Applications of Single Cell Oils* (Kyle, D. J., Ratledge, C., Eds.), pp. 219–234. Champaign, IL: American Oil Chemists' Society.

Shashi, K., Bachhawat, A. K., Joseph, R. (1990), ATP:Citrate lyase of *Rhodotorula gracilis:* purification and properties, *Biochim. Biophys. Acta* **1033,** 23–30.

Shaw, R. (1966), The polyunsaturated fatty acids of microorganisms, *Adv. Lipid Res.* **4,** 107–174.

Sheridan, R., Ratledge, C., Chalk, P. A. (1989), Pathways to acetyl-CoA formation in *Candida albicans, FEMS Microbiol. Lett.* **69,** 165–170.

SHIFRIN, N. S. (1984), Oils from microalgae, in: *Biotechnology for the Oils and Fats Industry* (RATLEDGE, C., DAWSON, P., RATTRAY, J., Eds.), pp. 145–163. Champaign, IL: American Oil Chemists' Society.

SHIMIZU, S., KAWASHIMA, H., SHINMEN, Y., AKIMOTO, K., YAMADA, H. (1988a), Production of eicosapentaenoic acid by *Mortierella fungi, J. Am. Oil Chem. Soc.* **65,** 1455–1459.

SHIMIZU, S., SHINMEN, Y., KAWASHIMA, H., AKIMOTO, K., YAMADA, H. (1988b), Fungal mycelia as a novel source of eicosapentaenoic acid. Activation of enzyme(s) involved in eicosapentaenoic acid production at low temperature, *Biochem. Biophys. Res. Commun.* **150,** 335–341.

SHIMIZU, S., AKIMOTO, K., KAWASHIMA, H., SHINMEN, Y., YAMADA, H. (1989a), Production of dihomo-γ-linolenic acid by *Mortierella alpina* 1S-4, *J. Am. Oil Chem. Soc.* **66,** 237–241.

SHIMIZU, S., KAWASHIMA, H., AKIMOTO, K., SHINMEN, Y., YAMADA, H. (1989b), Microbial conversion of an oil containing α-linolenic acid to an oil containing eicosapentaenoic acid, *J. Am. Oil Chem. Soc.* **66,** 342–437.

SHIMIZU, S., AKIMOTO, K., SHINMEN, Y., KAWASHIMA, H., SUGANO, M., YAMADA, H. (1991), Sesamin is a potent and specific inhibitor of Δ5-desaturase in polyunsaturated fatty acid biosynthesis, *Lipids* **26,** 512–516.

SHIMIZU, S., KAWASHIMA, H., WADA, M., YAMADA, H. (1992), Occurrence of a novel sterol, 24, 25-methylenecholest-5-en-3β-ol, in *Mortierella alpina* 1S-4, *Lipids* **27,** 481–483.

SHINMEN, Y., SHIMIZU, S., AKIMOTO, K., KAWASHIMA, H., YAMADA, H. (1989), Production of arachidonic acid by *Mortierella* fungi, *Appl. Microbiol. Technol.* **31,** 11–16.

SHLOMAI, P., BEN-AMOTZ, A., MARGALITH, P. (1991), Production of carotene stereoisomers by *Phycomyces blakesleeanus, Appl. Microbiol. Biotechnol.* **34,** 458–462.

SHUKLA, V. K. S. (1994), Present and future outlook of the world fats and oil supply, in: *Technological Awareness in Improved and Alternative Sources of Lipids* (KAMEL, B. S., KAKUDA, Y., Eds.), pp. 1–15. London: Blackie.

SIMOPOULOS, A. P. (1989), Summary of the NATO Advanced Research workshop on dietary ω3 and ω6 fatty acids: biological effects and nutritional essentiality, *J. Nutr.* **119,** 521–528.

SINCLAIR, A., GIBSON, R. (Eds.) (1992), *Essential Fatty Acids and Eicosanoids,* 3rd Int. Congr. Champaign, IL: American Oil Chemists' Society.

SMIT, H., YKEMA, A., VERBREE, E. C., VERWOERT, I. I. G. S., KATER, M. M. (1992), Production of cocoa butter equivalents by yeast mutants, in: *Industrial Applications of Single Cell Oils* (KYLE, D. J., RATLEDGE, C., Eds.), pp. 185–195. Champaign, IL: American Oil Chemists' Society.

SMITH, E., WHITE, K. A., HOLT, D., FENTEM, P. A., BRIGHT, S. W. J. (1994), Expression of polyhydroxybutyrate in oilseed rape, *Abstr. Int. Symp. Bacterial PHA,* Montreal, Canada, p. 44.

SMITH, P. F. (1988), Arachaebacteria and other specialized bacteria, in: *Microbial Lipids*, Vol. 1 (RATLEDGE, C., WILKINSON, S. G., Eds.), pp. 489–553. London: Academic Press.

SMITH, S. R. L. (1981), Some aspects of ICI's Single Cell Protein Process, in: *Microbial Growth on* C_1 *Compounds* (DALTON, H., Ed.), pp. 343–348. London: Heyden and Sons.

STADLER, T., MOLLION, J., VERDUS, M. C., KARAMANOS, Y., MORVAN, H., CHRITIAEN, D. (Eds.) (1988), *Algal Biotechnology.* Amsterdam: Elsevier.

STEINBÜCHEL, A., HUSTEDE, E., LIEBERGESELL, M., PIEPER, U., TIMM, A., VALENTIN, H. (1992), Molecular basis for biosynthesis acids in bacteria, *FEMS Microbiol. Rev.* **103,** 217–230.

SZUHAJ, B. F. (Ed.) (1989), *Lecithins: Sources, Manufacture and Uses.* Champaign, IL: American Oil Chemists' Society.

THOMAS, L. M., HOLUB, B. J. (1994), Nutritional aspects of fats and oils, in: *Technological Awareness in Improved and Alternative Sources of Lipids* (KAMEL, S., KAKUDA, Y., Eds.), pp. 16–49. London: Blackie.

TORNABENE, T. G., BOURNE, T. F., RAZIUDDIN, S., BEN-AMOTZ, A. (1985), Lipid and lipopolysaccharide constituents of cyanobacterium *Spirulina platensis, Mar. Ecol. Prog. Ser.* **22,** 121–125.

TOTANI, N., OBA, K. (1987), The filamentous fungus *Mortierella alpina,* high in arachidonic acid, *Lipids* **22,** 1060–1062.

TOTANI, N., OBA, K. (1988), A simple method for production of arachidonic acid by *Mortierella alpina, Appl. Microbiol. Biotechnol.* **28,** 135–137.

TOTANI, N., SOMEYA, K., OBA, K. (1992), Industrial production of arachidonic acid by *Mortierella,* in: *Industrial Applications of Single Cell Oils* (KYLE, D. J., RATLEDGE, C., Eds.), pp. 52–60. Champaign, IL: American Oil Chemists' Society.

TYRRELL, D. (1967), The fatty acid composition of 17 *Entomophthora* isolates, *Can. J. Microbiol.* **13**, 755–760.

TYRRELL, D. (1968), The fatty acid composition of some Entomophthoraceae II, *Lipids* **3**, 368–372.

TYRRELL, D. (1971), The fatty acid composition of some Entomophthoraceae III, *Can. J. Microbiol.* **17,** 1115–1118.

TYRRELL, D., WEATHERSTONE, J. (1976), The fatty acid composition of some Entomophthoraceae IV. The occurrence of branched-chain fatty acids in *Conidiobolus* species, *Can. J. Microbiol.* **22**, 1058–1060.

VAN DYK, M. S., KOCK, J. L. F., COETZEE, D. J., AUGUSTYN, O. P. H., NIGAM, S. (1991), Isolation of a novel arachidonic acid metabolite 3-hydroxy-5,8,11,14-eicosatetraenoic acid (3-HETE) from the yeast *Dipodascopsis uninucleata* UOFS-Y128, *FEBS Lett.* **283**, 195–198.

VAN DYK, M. S., KOCK, J. L. F., BOTHA, A. (1994), Hydroxy long chain fatty acids in fungi, *World J. Microbiol. Biotechnol.* **10**, 495–504.

VEGA, E. Z., GLATZ, B. A., HAMMOND, E. G. (1988), Optimization of banana juice fermentation for the production of microbial oil, *Appl. Environ. Microbiol.* **54**, 748–752.

VENTOSA, A., NIETO, J. J. (1995), Biotechnological applications and potentialities of halophilic microorganisms, *World J. Microbiol. Biotechnol.* **11**, 85–94.

VERT, M., FEIJEN, J., ALBERTOSON, A., SCOTT, G., CHIELLINI, E. (Eds.) (1992), *Biodegradable Polymers and Plastics.* Cambridge: Royal Society of Chemistry.

VERWOERT, I. I. G. S., YKEMA, A., VALKENBURG, J. A. C., VERBREE, E. C., NIJKAMP, H. J. J., SMIT, H. (1989), Modification of the fatty acid composition in lipids of the oleaginous yeast *Apiotrichum curvatum* by intraspecific spheroplast formation, *Appl. Microbiol. Biotechnol.* **32**, 327–333.

VOLKMANN, J. K. (1989), Fatty acids of microalgae used as feedstocks in aquaculture, in: *Fats for the Future* (CAMBIE, R. C., Ed.), pp. 263–283. Chichester: Ellis Horwood.

VONSHAK, A., COHEN, Z., RICHMOND, A. (1985), The feasibility of mass cultivation of *Porphyridium, Biomass* **8**, 13–25.

WATANABE, K., YAZAWA, K. (1992), Cloning of DNA coding for EPA – biosynthesis from a marine bacterium into *E. coli, J. Jpn. Oil Chem. Soc.* **42**, 42–49.

WATANABE, K., ISHIKAWA, C., INOVE, H., CENHUA, D., YAZAWA, K., KONDO, K. (1994), Incorporation of exogenous docosahexaenoic acid into various bacterial phospholipids, *J. Am. Oil Chem. Soc.* **71**, 325–330.

WEETE J. D. (1980), Lipid Biochemistry of Fungi and other Organisms. New York: Plenum Press.

WEETE, J. D., GANDHI, S. R. (1992), Enhancement of C_{20} polyunsaturated fatty acid production in *Pythium ultimum,* in: *Industrial Applications of Single Cell Oils* (KYLE, D. J., RATLEDGE, C., Eds.), pp. 98–117. Champaign, IL: American Oil Chemists' Society.

WILKINSON, S. G. (1988), Gram-negative bacteria, in: *Microbial Lipids,* Vol. 1 (RATLEDGE, C., WILKINSON, S. G., Eds.), pp. 299–488. London: Academic Press.

WILLNER, T., SITZMANN, W., WEBER, K. (1993), Dry fractionation for cocoa butter replacers, in: *Oil and Fats in the Nineties,* pp. 162–175. Lystrup: International Food Science Centre A/S.

WITHOLT, B., HUISMAN, G. W., PREUSTING, H. (1990), Bacterial poly(3-hydroxyalkanoates) in: *Novel Biodegradable Microbial Polymers* (NATO ASI Series No. 186), (DAWES, E. A., Ed.), pp. 161–173. Dordrecht: Kluwer.

WITTER, B., DEBUCH, H., STEINER, M. (1974), Die Lipide von *Endomycopsis vernalis* bei verschiedener Stickstoff-Ernährung, *Arch. Microbiol.* **101**, 321–335.

WOOD, B. J. B. (1988), Lipids of algae and protozoa, in: *Microbial Lipids,* Vol. 1 (RATLEDGE, C., WILKINSON, S. G., Eds.), pp. 807–867. London: Academic Press.

WOODBINE, M. (1959), Microbial fat: microorganisms as potential producers, *Prog. Ind. Microbiol.* **1**, 179–245.

YADWAD, V. B., WARD, O. P., NORONHA, L. C. (1991), Application of lipase to concentrate the docosahexaenoic acid (DHA) fraction of fish oil, *Biotechnol. Bioeng.* **38**, 956–959.

YAMADA, H., SHIMIZU, S., SHINMEN, Y. (1987a), Production of arachidonic acid by *Mortierella elongata* IS-5, *Agric. Biol. Chem.* **51**, 785–790.

YAMADA, H. S., SHIMIZU, S., SHINMEN, Y., KAWASHIMA, H., AKIMOTO, K. (1987b), Production of arachidonic acid and eicosapentaenoic acid by microorganisms in: *Proc. World Conf. Biotechnol. for the Oils and Fats Industry* (APPLEWHITE, T. H., Ed.), pp. 173–177. Champaign, IL: American Oil Chemists' Society.

YAMADA, H., SHIMIZU, S., SHINMEN, Y., AKIMOTO, K., KAWASHIMA, H., JAREONKITMONGKOL, S. (1992), Production of dihomo-γ-linolenic acid, arachidonic acid and eicosapentaenoic acid by filamentous fungi, in: *Industrial Applications of Single Cell Oils* (KYLE, D. J., RATLEDGE, C., Eds.), pp. 118–138. Champaign, IL: American Oil Chemists' Society.

YAMAGUCHI, K., NAKANO, H., MURAKAMI, M., KONOSU, S., NAKAYAMA, O., KANDA, M., NAKANUIRA, A., IWAMOTO, H. (1987), Lipid composition of a green alga, *Botryococcus braunii, Agric. Biol. Chem.* **51**, 493–498.

YAMAUCHI, H., MORI, H., KOBAYASHI, T., SHIMIZU, S. (1983), Mass production of lipids by *Lipomyces starkeyi* in microcomputer aided fed-batch culture, *Ferment. Technol.* **61**, 275–280.

YANO, Y., NAKAYAMA, A., SAITO, H., ISHIHARA, K. (1994), Production of docosahexaenoic acid by marine bacteria isolated from deep sea fish, *Lipids* **29,** 527–528.

YAZAWA, K., ARAKI, K., OKAZAKI, N., WATANABE, K., ISHIKAWA, C., INOVE, A., NUMAO, N., KONDO, K. (1988), Production of eicosapentaenoic acid by marine bacteria, *J. Biochem.* **103**, 5–11.

YAZAWA, K., WATANABE, K., ISHIKAWA, C., KONDO, K., KIMURA, S. (1992), Production of eicosapentaenoic acid from marine bacteria, in: *Industrial Applications of Single Cell Oils* (KYLE, D. J., RATLEDGE, C., Eds.), pp. 29–51. Champaign, IL: American Oil Chemists' Society.

YKEMA, A., VERBREE, E. C., VAN VERSEVELD, H. W., SMIT, H. (1986), Mathematical modelling of lipid production by oleaginous yeasts in continuous culture, *Antonie van Leeuwenhoek* **52,** 491–506.

YKEMA, A., VERBREE, A. C., KATER, M. M., SMIT, H. (1988), Optimization of lipid production in the oleaginous yeast *Apiotrichum curvatum* in whey permeate, *Appl. Microbiol. Biotechnol.* **29**, 211–218.

YKEMA, A., VERBREE, E. C., NIJKAMP, H. J. J., SMITH, H. (1989), Isolation and characterization of fatty acid auxotrophs from the oleaginous yeast *Apiotrichum curvatum, Appl. Microbiol. Biotechnol* **32**, 76–84.

YKEMA, A., VERBREE, E. C., VERWOERT, I. I. G. S., VAN DER LINDEN, K. H., NIJKAMP, H. J. J, SMIT, H. (1990), Lipid production of revertants of Ufa mutants from the oleaginous yeast *Apiotrichum curvatum, Appl. Microbiol. Biotechnol.* **33**, 176–182.

YONGMANITCHAI, W., WARD, O. P. (1989), Omega-3 fatty acids: alternative sources of production, *Process Biochem.* **24**, 117–125.

YONGMANITCHAI, W., WARD, O. P. (1991), Screening of algae for potential alternative sources of eicosapentaenoic acid, *Phytochemistry* **30**, 2963–2967.

YONGMANITCHAI, W., WARD, O. P. (1992), Growth and eicosapentaenoic acid production by *Phaeodactylum tricornutum* in batch and continuous culture systems, *Am. Oil Chem. Soc.* **69**, 584–590.

YONGMANITCHAI, W., WARD, O. P. (1993), Positional distribution of fatty acids, and molecular species of polar lipids, in the diatom *Phaeodactylum tricornutum, J. Gen. Microbiol.* **139**, 465–472.

YOON, S. H., RHEE, J. S. (1983), Quantitative Physiology of *Rhodotorula glutinis* for microbial lipid production, *Process Biochem.* **18**, 2–4.

YOON, S. H., RHIM, J. W., CHOI, S. Y., RYU, D. D. W., RHEE, J. S. (1982), Effect of carbon and nitrogen sources of lipid production of *Rhodotorula gracilis, J. Ferment. Technol.* **60**, 243–246.

ZVGAGINTSEVA, I. S., PITRYUK, I. A., BAB'EVA, I. P., RUBAN, E. L. (1975), Fatty acid composition of lipids of soil and epiphytic yeasts, *Mikrobioloiya USSR* (Int. Edn.) **44**, 625–631.

5 Microbial Siderophores

GÜNTHER WINKELMANN
HARTMUT DRECHSEL

Tübingen, Federal Republic of Germany

1 Introduction

Iron nutrition in microbes and higher organisms has become a most fascinating topic in microbiology and biotechnology. Extensive studies in a variety of microorganisms have shown that low-molecular weight, ferric-specific ligands, named siderophores, are essential for growth and survival in natural environments. The biosynthesis of siderophores and their cognate membrane transport systems in bacteria are regulated at the transcriptional level by internal iron sensors named ferric uptake regulation proteins (Fur) which respond to the external iron concentration and function as iron-loaded repressors. The literature on siderophores has increased considerably during the past five years and several comprehensive books offer detailed descriptions of structures and functions of the various siderophores (WINKELMANN et al., 1987; WINKELMANN, 1991a; BARTON and HEMMING, 1993; WINKELMANN and WINGE, 1994). Attention has been paid primarily to the particular functions of siderophores, their occurrence and distribution among bacteria and fungi. Thus, plenty of information on siderophores already exists. From a biotechnological point of view iron is an extremely important element for fermentation processes in order to enhance growth and production yields. Thus, concentration, presence or absence of iron and other ions or medium constituents may greatly affect synthesis and excretion of fermentation products.

The present review is intended to update the existing structural diversity of siderophores and to specifically address some biotechnological aspects of siderophores which had already been the aim of the previous review in Vol. 4 of the first edition of this series published about ten years ago (WINKELMANN, 1986). Since then, a number of novel siderophore structures have been described which are not only variations of known structures but even represent novel types or classes of siderophores. Overall, the topic of siderophores seems to be a never ending story that continues to inspire the community of microbiologists. The biotechnological value of these results still is not fully appreciated. It may be predicted that a variety of fermentation processes will be optimized and precisely controlled by the addition or withdrawal of iron and siderophores. Other examples will be briefly addressed at the end of this review.

2 General Aspects

The majority of aerobic and facultative anaerobic microorganisms respond to a decreasing iron content in their environment by the expression of siderophores and their cognate uptake systems. Because of the highly specific ferric iron binding, the ligands have attracted the attention of chemists who developed strategies for the synthesis of a number of natural siderophores (LEE and MILLER, 1983; BERGERON and MCMANIS, 1991) and biomimetic analogs (SHANZER and LIBMAN, 1991). The isolation of siderophores from biological fluids is based on various extraction and detection methods (NEILANDS and NAKAMURA, 1991). However, yields are very low sometimes because of the fact that production and excretion of siderophores are negatively regulated by iron. Therefore, mutants defective in regulation of siderophore biosynthesis or in the expression of outer membrane receptors have been used for the production of siderophores (YOUNG and GIBSON, 1979; WINKELMANN et al., 1994). The biosynthesis and excretion of many siderophores can be detected in culture filtrates by the appearance of color due to metal-to-ligand charge transfer bands. Thus, ferric hydroxamate complexes show absorption maxima at 420–440 nm, resulting in a yellow-brown color, while ferric catecholate complexes are red-blue in color due to an absorption maximum at 495–520 nm. Ferric carboxylate siderophores show only a weak absorption at 335 nm (DRECHSEL et al., 1992) which results in a pale yellow color.

Survival of microorganisms in an aerobic atmosphere where the concentration of soluble ferric ions in solution at neutral pH is smaller than 10^{-17} M has been a challenge for about 3.5 billion years and enabled the evolution of siderophores and their cognate membrane transport systems. Moreover, siderophore-mediated iron acquisition has been correlated with the ability of various microorganisms to establish and maintain infection of a host, although examples of virulence enhancement by siderophores are still scarce (BAGG and NEILANDS, 1987; HESEMANN, 1987; ACTIS et al., 1988; ENARD et al., 1988; STOJILJKOVIC and HANTKE, 1992).

The history of siderophores started with the discovery of "growth factors" for mycobacteria and fungi (reviewed in: WINKELMANN, 1991a). NEILANDS (1952) was the first to isolate a siderophore in crystalline form and thus opened an era of intensive search for new iron-binding compounds, which were previously named sideramines or siderochromes and are now collectively designated as siderophores. Siderophores are designed for the solubilization, transport, and storage of iron in microorganisms. It is now generally agreed that siderophores including their biosynthesis and the expression of the corresponding membrane-located transport systems are carefully regulated by iron. Iron-binding compounds that are not regulated seem to occur but do not fall into the category of natural siderophores. Moreover, the term siderophore should be reserved for the metal-free ligand, although in some cases the names had been given to the metal complex earlier (ferrichrome, ferrioxamine, coprogen). The corresponding iron-free compounds then need the prefix desferri- or deferri-; siderophores which represent the iron-free form (enterobactin, aerobactin, staphyloferrin, rhizoferrin) need the prefix ferri- or ferric when iron is bound to the ligand.

The ligands involved in iron(III) binding are either phenolates or catecholates, hydroxamates, oxazolines, α-hydroxy carboxylates (e.g., citrate derivatives), or keto hydroxy bidentates. Ferric siderophores are octahedral complexes in which the coordinated metal ion is d^5, high spin, and rapidly exchangeable. After reduction, the ferrous ion shows little affinity for the ligand, and this is obviously the mechanism by which iron is removed from siderophores within the cell or at the cell surface of microorganisms. Non-reducible aluminum and gallium complexes and kinetically stable chromic complexes have been used to study the metabolism of siderophores in various microorganisms. The physicochemical properties of natural and synthetic siderophores have been compiled in a recent review by CRUMBLISS (1991). The reader is also referred to a comprehensive description devoted solely to the solution and structural chemistry of siderophores (MATZANKE et al., 1989).

3 Bacterial Siderophores

3.1 Catecholate Siderophores

Enterobactin (Fig. 1), also referred to as enterochelin, is the prototype catecholate siderophore of enteric bacteria. It forms highly stable octahedral complexes with ferric iron ($K_f = 10^{52}$). The three catecholamide-binding subunits of enterobactin are attached to a tri-L-serine lactone ligand backbone. While the ligand is uncharged at neutral pH, the ferric form is a trianionic complex. Convenient methods for the isolation of enterobactin (enterochelin) have been described by YOUNG

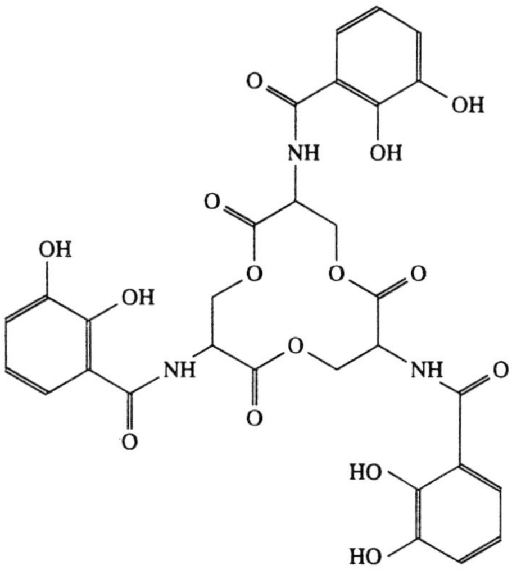

Fig. 1. Enterobactin.

and Gibson (1979), Neilands and Nakamura (1991), and Berner et al. (1991). HPLC separation and isolation from culture fluids have been described which also allow the simultaneous detection of degradation products of linear 2,3-dihydroxybenzoyl serine derivatives (Winkelmann et al., 1994). The production procedure of Young and Gibson (1979) employs a mutant strain (*fepA*) of *Escherichia coli* which is unable to transport the ferric enterobactin complex into the cell. Whereas wild-type strains produce enterobactin only during severe iron deficiency (<0.2 μM) and terminate production after iron transport into the cell, *fepA* mutants continue to produce high amounts of enterobactin in a medium not necessarily iron-deficient.

Enterobactin can be estimated from aqueous solutions (4 mL) after acidification with conc. H_2SO_4 and extracted with an equal volume of ethyl acetate. If required, the extract may be washed with an equal volume of sodium phosphate buffer (0.1 M, pH 7) to remove any of the hydrolysis products of enterobactin which might be present in varying amounts (Winkelmann et al., 1994), and the concentration is determined at 316 nm. The concentration of ferri-enterobactin is determined using the molar extinction coefficient ($\varepsilon_{495\,nm} = 5600\ M^{-1}\ cm^{-1}$).

The transport of ferric enterobactin into *E. coli* requires the expression of the FepA outer membrane protein (81 kDa), which has been cloned, sequenced (Lundrigan and Kadner, 1986), and also crystallized (Jalal and van der Helm, 1989). Enterobactin is the most powerful natural iron sequestering agent (reviewed by Crumbliss, 1991). Physicochemical properties, analogs as well as mechanistic aspects of transport have been discussed in detail (Matzanke, 1991; Shanzer and Libman, 1991). The FepA receptor recognizes the ferric enterobactin complex present in a *delta-cis* configuration as shown by comparison of natural enterobactin and synthetic *lambda-cis-enantio*-enterobactin (Neilands et al., 1981). Purification of FepA by FPLC has been reported from the plasmid-harboring *E. coli* strain UT5600/pBB2 (Zhou et al., 1995). Successful crystallization of the purified FepA protein had been reported earlier by this group (Jalal and van der Helm, 1989) and crystal structures are to be expected in the near future. Five additional gene products are required to complete the transport into the cells, of which only FepB has been sequenced so far (Elkins and Earhardt, 1989). The actual mechanism of enterobactin transport is still unsettled. Due to the very low redox potential (-790 mV vs. NHE at pH 7.4) a direct reduction model could be excluded and an esterolytic degradation prior to reduction has been proposed. Alternative mechanisms involving protonated molecular species of the ferric enterobactin complex during membrane transport have been discussed (Cass et al., 1989; Matzanke, 1991). An additional internal redox reaction has also been suggested as ^{55}Fe-enterobactin is taken up while ^{67}Ga-enterobactin is not. Recent HPLC data from the authors' laboratory have shown that the esterase seems to attack ferric enterobactin after entrance via FepA, producing several linear dihydroxybenzoyl serine products (Winkelmann et al., 1994) which then can be easily reduced (-350 mV vs. NHE). These results are in favor of a combined esterase–reductase mechanism and thus obviate any protonation or internal redox reaction of the ferric enterobactin complex.

Chrysobactin

Chrysobactin (Fig. 2), a simple catechol-type siderophore containing only one 2,3-dihydroxybenzoyl group connected to a dipep-

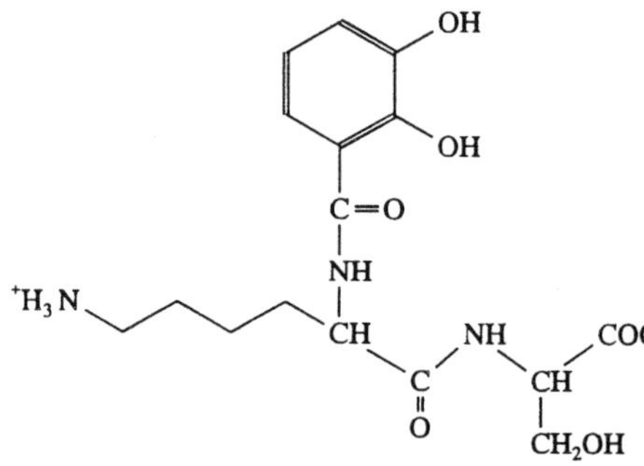

Fig. 2. Chrysobactin.

tide was isolated from *Erwinia chrysanthemi* which is a phytopathogenic species in *Saintpaulia* plants (PERSMARK et al., 1989; PERSMARK and NEILANDS, 1992). Structure elucidation revealed that chrysobactin is *N*-[N^2-(2,3-dihydroxybenzoyl)-D-lysyl]-L-serine.

While for similar natural monocatechol siderophores, like 2,3-dihydroxybenzoyl serine and 2,3-dihydroxybenzoyl lysine, an L-configuration was determined, the lysyl residue in chrysobactin has a D-configuration. However, as shown by transport experiments, the configuration of the amino acids in chrysobactin seems to be of minor importance compared to that of the catechol–iron center (PERSMARK et al., 1992).

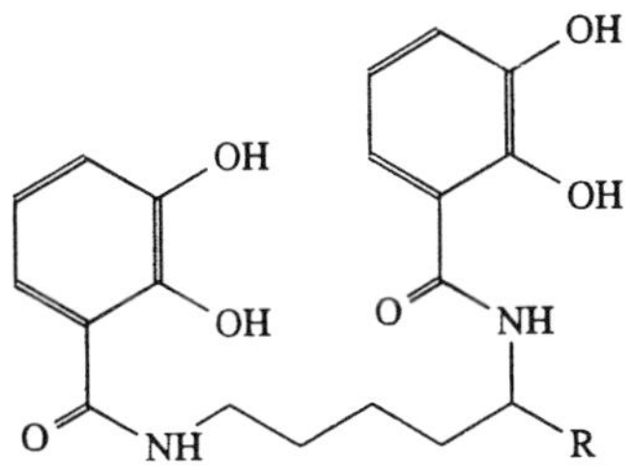

Aminochelin, Azotochelin, Myxochelin A, and Protochelin

Azotobacter vinelandii produces several catecholate siderophores under iron-limiting conditions (FUKASAWA et al., 1972; PAGE and TIGERSTROM, 1988; CORBIN and BULEN, 1969; DEMANGE et al., 1988; BUDZIKIEWICZ et al., 1992): 2,3-dihydroxybenzoic acid (DHBA), aminochelin (Fig. 3), and azotochelin (Fig. 4). These three compounds seem to be biogenetically related. Aminochelin and

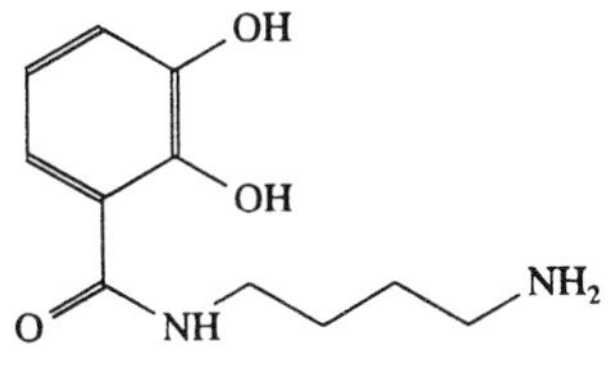

OH OH O NH NH_2

Fig. 3. Aminochelin.

OH OH OH OH O NH O NH R

Fig. 4. Azotochelin: R=COOH; myxochelin A: R=CH_2OH.

azotochelin can be extracted from the culture filtrate using ethyl acetate. Aminochelin is 4-N-2,3-dihydroxybenzoyl-1,4-diaminobutane, and azotochelin is *N,N'*-bis-(2,3-dihydroxybenzoyl)-lysine. It was anticipated that these compounds may possibly represent siderophore precursors or degradation products originating from a more complex siderophore which indeed has been found recently and was named protochelin (Fig. 5) (TARAZ et al., 1990). Although recently detected in *Azotobacter,* (CORNISH and PAGE, 1995), protochelin has been previously shown to occur in a bacterium (DSM 5746) which is able to grow on methanol as the sole carbon source. The structure of protochelin is a direct combination of aminochelin and azotochelin. As shown by PAGE and HUYER (1984) DHBA is produced constitutively by *Azotobacter,* whereas azotochelin and azotobactin are produced only under iron limitation.

Myxochelin A (Fig. 4) is a new catecholate siderophore isolated from *Angiococcus disciformis* (Myxobacteria). The structure is *N,N'*-bis-(2,3-dihydroxybenzoyl)-lysinol (KUNZE et al., 1989) which is related to aminochelin (*N,N'*-bis-(2,3-dihydroxybenzoyl)-lysine) pro-

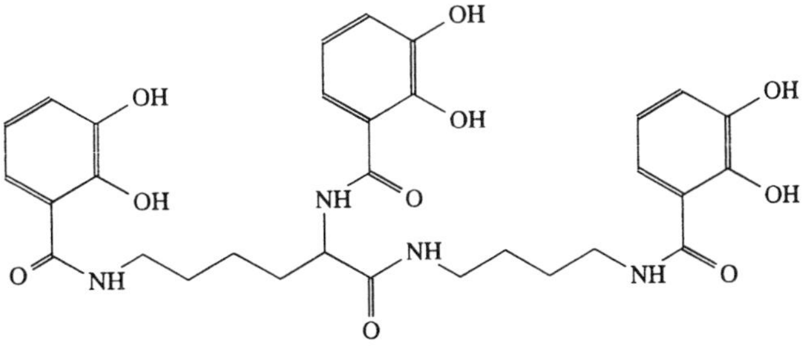

OH OH OH OH OH OH NH O NH O NH NH O O

Fig. 5. Protochelin.

duced by *Azotobacter vinelandii* (PAGE and TIGERSTROM, 1988). Myxochelin A exerts weak antibiotic activity on various gram-positive bacteria, e.g., *Bacillus brevis, B. cereus, B. subtilis, B. megaterium, Micrococcus luteus, Staphylococcus aureus, Arthrobacter simplex, Brevibacterium linens, Corynebacterium fascians,* and *Nocardia corallina* (KUNZE et al., 1989).

Agrobactin

Agrobactin (Fig. 6a) is the characteristic siderophore of *Agrobacterium tumefaciens* (ONG et al., 1979). In low-iron media *A. tumefaciens* B6 was found to produce a neutral, ethyl acetate-extractable substance. The UV spectrum revealed absorption maxima at 316 and 252 nm. Agrobactin was shown by its blue fluorescence in the ultraviolet range and positive Arnow reaction to belong to the catechol-type siderophores. Acid hydrolysis yielded DHBA, spermidine, and L-threonine. The carboxyl group of DHBA as well as the amino group and the β-hydroxyl group of threonine are linked to an oxazoline ring. The crystal structure of the ligand was published by ENG-WILMOT and VAN DER HELM (1980). Exposure to acid leads to the formation of the oxazoline ion which opens to yield the ester and subsequently under neutral conditions undergoes the N-O-acyl shift to give the amide. This open form was named agrobactin A (Fig. 6b). Whereas enterobactin adopts a *delta-cis* configuration about the iron center, the configuration of agrobactin is *lambda-cis.* Agrobactin and several other polyamine catecholamide chelators, such as parabactin and vibriobactin (see below), have been chemically synthesized (reviewed by BERGERON and MCMANIS, 1991). Recently, the structure of serratiochelin from *Serratia marcescens* has been elucidated, which is a derivative of agrobactin (Fig. 6a) lacking the tetramethylenedihydroxybenzoylamide (EHLERT et al., 1994).

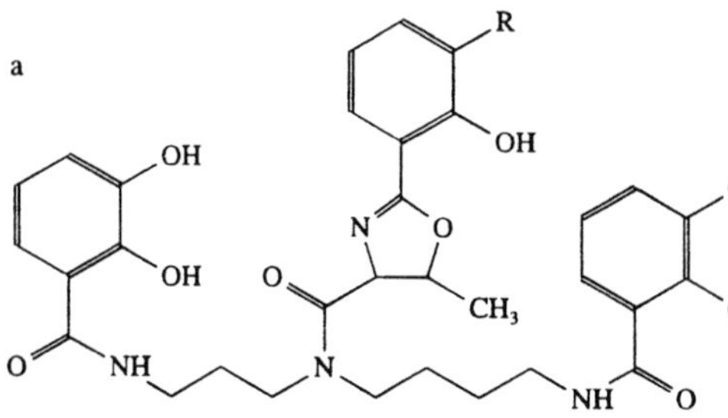

Parabactin

In 1975, TAIT isolated a siderophore from low-iron cultures of *Micrococcus denitrificans* (now *Paracoccus denitrificans*) which he named compound III. A reinvestigation (PETERSON and NEILANDS, 1979) of the structure (Fig. 6a) revealed an analogous structure to agrobactin lacking the OH group of the central catechol in position 3. This compound was named parabactin. Again two forms are conceivable: one possessing a closed oxazoline ring (parabactin) and an open form (parabactin A).

Inspection of the molecular models of agrobactin and parabactin revealed that the tertiary N-atom of the oxazoline ring and the *o*-hydroxyphenyl function are involved in the six-coordinate ferric complex. Because of the optically active substituent L-threonine, a particular coordination isomer of parabactins and agrobactins can be expected. The CD spectra revealed a *lambda-cis* configuration as in ferrichrome. In addition, steric constraints ruled out the possible presence of geometrical isomers of the *trans* variety. The binding strength of parabactins and agrobactins is comparable to that of enterobactin which is approximately 10^{52} at pH 7.4.

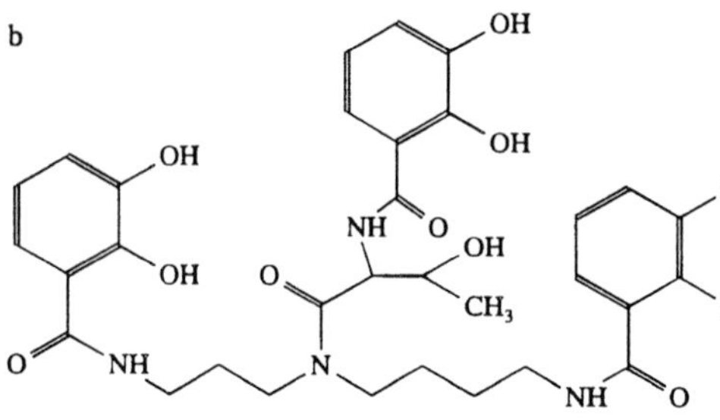

Fig. 6a. Agrobactin: R=OH; parabactin: R=H; **b** agrobactin A.

The study of biological activities of agrobactins and parabactins in *Agrobacterium tumefaciens* and *Paracoccus denitrificans* showed that only the closed but not the open oxazoline forms counteract growth retardation caused by EDTA. Using both labeled metal and labeled ligand complexes, BERGERON et al. (1985) presented evidence that parabactin operates by the "iron taxi" mechanism. Iron was delivered to the cell, but the ligand was not taken up in higher amounts. The same authors also presented evidence that *enantio*-parabactin possessing a *delta* coordination isomer was unable to transfer higher amounts of labeled metal to *Paracoccus denitrificans.* Therefore, the parabactin receptor stereoselectively recognizes the natural *lambda* coordination isomer.

Vibriobactin

From low-iron cultures of *Vibrio cholerae* the siderophore vibriobactin (Fig. 7a) was isolated by GRIFFITH et al. (1984). *V. cholerae* (serovar O1 and O139) is known to have caused several severe pandemic diarrheal diseases, and the latest one (the 7th) is still existing in Asia, Australia, and South America (WACHSMUTH et al., 1994). Vibriobactin, like agrobactin, contains three 2,3-dihydroxybenzoyl residues and two residues of L-threonine per molecule both of which form an oxazoline ring. The polyamine backbone proved to be N-(3-aminopropyl)-1,3-diamino propane (norspermidine), in contrast to spermidine which is found in agrobactin and parabactin.

Cells of *V. cholerae* Lou15 were used for the original isolation of vibriobactin (GRIFFITH et al., 1984). The yield of the preparation varied between 10–25 mg per 6 L batch. Thin layer chromatography (TLC) on silica gel plates using chloroform–methanol 4:1 gave one spot detected by its blue fluorescence under UV light or by spraying with ferric chloride or iodine. The UV absorption spectrum is qualitatively the same as that of agrobactin. Biological activity testing revealed that the growth of the producing *V. cholerae* strain Lou15, which is only weakly pathogenic, was stimulated by vibriobactin and also by agrobactin with nearly equal efficiency. Vibriobactin is not the only siderophore utilized by *V. cholerae* as mutants defective in vibriobactin synthesis can alternatively use ferric citrate (SIGEL et al., 1985) or obtain iron from heme (STOEBNER and PAYNE, 1988). As shown below, additional carboxylate transport systems for vibrioferrin may exist. The outer membrane receptor for vibriobactin has been identified (STOEBNER et al., 1992) and the gene (*viuA*) encoding the outer membrane receptor (74 kDa) has recently been cloned and shown to be negatively regulated by the *fur* gene at the transcriptional level (BUTTERTON et al., 1992). The role of *fur* in the iron-regulated expression of the outer membrane receptor for vibriobactin (ViuA) and the outer membrane virulence protein (IrgA) has been demonstrated recently (LITWIN and CALDERWOOD, 1994). The fact that further proteins were observed which were negatively regulated by iron but independent of Fur suggested that the gene

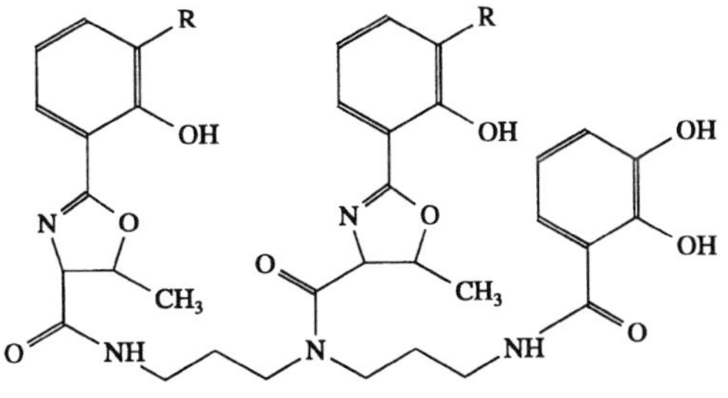

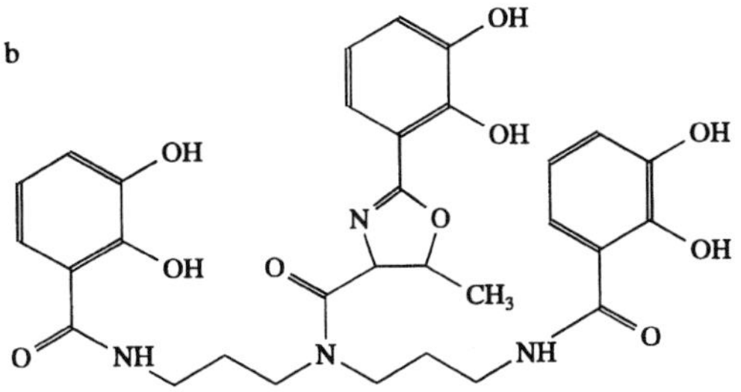

Fig. 7a. Vibriobactin: R = OH; vulnibactin: R = H; **b** fluvibactin.

regulation in *V. cholerae* by Fur and iron is much more complex than previously assumed.

Vulnibactin

The group of YAMAMOTO has also reported on the structures of further polyamine-containing catecholate siderophores from the human pathogen *Vibrio vulnificus* (OKUJO et al., 1994a). The principal compound isolated contained two oxazoline-bound salicyl moieties and one outer DHBA residue and was named vulnibactin (Fig. 7a).

Fluvibactin

Fluvibactin (Fig. 7b) from *Vibrio fluvialis* (YAMAMOTO et al., 1993) resembles agrobactin with respect to the three DHBA residues and the central oxazoline ring. However, fluvibactin, like vibriobactin and vulnibactin, contains norspermidine as a backbone (identical propane spacer between the two nitrogens). This is consistent with the finding that *Vibrio* species contain norspermidine in its free form as a major polyamine (YAMAMOTO et al., 1991).

Amonabactins

Several amonabactins have been isolated from the fish pathogen *Aeromonas hydrophila* (BARGHOUTHI et al., 1989). Recent structure elucidation of the amonabactins have shown that four structural varieties of amonabactins exist (TELFORD et al., 1994). Amonabactins contain 2 mol DHBA, glycine and lysine and 1 mol of either tryptophan or phenylalanine. Two of them are glycine-deleted forms (Fig. 8). A survey on various *Aeromonas* strains revealed that although most strains produced amonabactins, some produced an enterobactin-like or a so far uncharacterized siderophore (ZYWNO et al., 1992). In strains producing amonabactins a biosynthetic gene (*amoA*) was identified in a *Sau*3A gene library of *Aeromonas hydrophila* 495A2 chromosomal DNA (BARGHOUTHI et al., 1991). This gene seems to correspond to the first 2,3-DHBA biosynthetic enzyme (isochorismate synthetase) in *E. coli* (MASSAD et al., 1994).

Anguibactin

While the characteristic moiety of the siderophores described above are oxazoline rings, the following structures contain a thiazoline ring. The first member of this series is anguibactin (Fig. 9a). Isolated from the fish pathogen *Vibrio anguillarum* (ACTIS et al., 1986), it was one of the first siderophores regarded as an important plasmid-encoded virulence factor (CROSA, 1989). Structure elucidation based on single-crystal X-ray diffraction studies revealed that anguibactin consists of 2,3-dihydroxybenzoic acid linked to cysteine by a thiazoline ring which in turn is linked by a hydroxamic acid bound to ω-*N*-hydroxy histamine (JALAL et al., 1989).

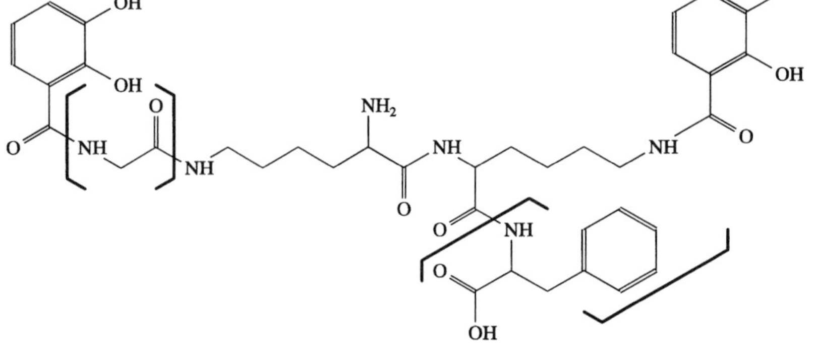

Fig. 8. Amonabactins. Phe can be substituted by Trp, Gly-deleted forms occur.

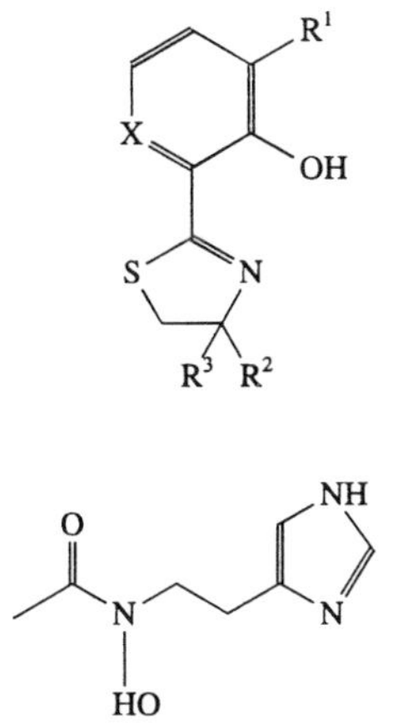

Fig. 9. Anguibactin: $R^1=OH$, $R^2=A$, $R^3=H$, $X=CH$; desferri-ferrithiocin: $R^1=H$, $R^2=CH_3$, $R^3=COOH$, $X=N$; aeruginic acid: $R^1=R^2=H$, $R^3=COOH$, $X=CH$.

Desferrithiocin

Besides the indicator antibiotics β- and γ-rubromycins the iron-binding compound desferri-ferrithiocin is produced by *Streptomyces antibioticus* (Tü 1998; Fig. 9b) (NAEGELI and ZÄHNER, 1980) which resembles the structure of aeruginic acid published previously (Fig. 9c). Desferri-ferrithiocin contains a hydroxy pyridine residue linked to a thiazoline ring. Although a function in iron transport has never been shown in the producing strain, it has been suggested to be a potent, orally available iron chelator.

Pyochelin

Pyochelin (Fig. 10a) is an iron-chelating compound isolated from low-iron cultures of *Pseudomonas aeruginosa* strain PAO-1. Pyochelin has also been found in clinical isolates. The structure, 2-(2-*o*-hydroxyphenyl-2-thiazolin-4-yl)-3-methylthiazolidine-4-carboxylic acid, is presumed to be biosynthesized from salicylcysteinyl cysteine by cyclization and hydration of the thiazolidine ring. Pyochelin was purified by preparative TLC on silica gel G (COX et al., 1981) and detected both by fluo-

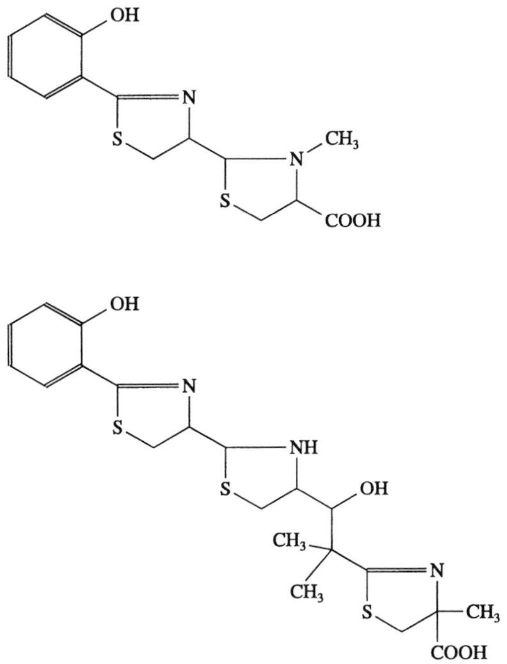

Fig. 10a. Pyochelin; **b** yersiniabactin.

rescence and a red reaction using an iron spray. It forms a stable, red complex with ferric iron. Although the ligand groups involved in iron chelation have not been identified, it may be assumed that iron is bound to the phenyl OH group, the carboxyl group, and one nitrogen in a 2:1 ligand-to-iron ratio. The simpler compound 2-*o*-hydroxyphenyl-2′-thiazoline-4-carboxylic acid, named aeruginic acid, is known to occur as a fermentation product of *P. aeruginosa.* Ferrithiocin isolated from *Streptomyces antibioticus* is related to aeruginic acid, but contains hydroxy pyridine instead of phenyl and a methyl group in the vicinity of the carboxyl group resulting in an additional chiral center.

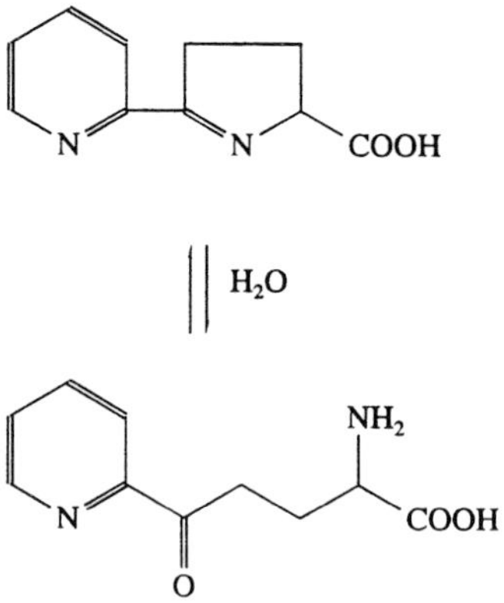

Yersiniabactin

From cultures of *Yersinia enterocolitica,* an iron-complexing and iron-transporting compound named yersiniabactin was isolated (HAAG et al., 1993; CHAMBERS and SOKOL, 1994). The structure of the siderophore was determined by a variety of spectroscopic methods, including 2D NMR experiments on the metal-free ligand as well as its gallium complex (DRECHSEL et al., 1995b). The novel siderophore contains a benzene and a thiazolidine ring as well as two thiazoline rings (Fig. 10b) and thus resembles the thiazoline-containing siderophores pyochelin, anguibactin, and desferri-ferrithiocine. The molecule contains five chiral centers, four of them inherent in the connectivity and substituents of the thiazoline rings. The compound forms stable complexes with trivalent cations such as ferric iron and gallium. Iron binding constants have not yet been determined. From pH dependent UV and CD measurements it is clear, however, that the yersiniabactin–iron(III) complexes are stable at least between pH 4–11. The complexes with Ga^{3+} and Fe^{3+} are monomeric, 1:1 complexes. Yersiniabactin constitutes the principal siderophore of *Yersinia enterocolitica* sought for a long time. The same siderophore has been isolated as aluminium complex and has been termed yersiniophore (CHAMBERS et al., 1996).

N N COOH

H_2O

NH_2 N COOH O

Fig. 11. Proferrorosamine A.

Ferrorosamine A

Erwinia rhapontici has been shown to produce proferrorosamine A, L-(2-pyridyl)-1-pyrroline-5-carboxylic acid, a Fe(II)-complexing agent (Fig. 11) which after complexing gives a reddish pigment named ferrorosamine A (FEISTNER et al., 1983). While proferrorosamine A in neutral solution exists in the bicyclic form, the pyrroline ring opens under acidic conditions to give the α-amino acid residue which no longer complexes ferrous ions. The proferrorosamine producing *Pseudomonas* sp. strain GH (LMG 11358), has later been reclassified as *E. rhaptontici* (VANDE WOESTYNE et al., 1991; DE VOS et al., 1993).

Siderochelin A

Siderochelin A (Fig. 12) isolated from fermentation broths of *Nocardia* sp. SC 11340 was identified as *trans*-3,4-dihydro-4-hydroxy-

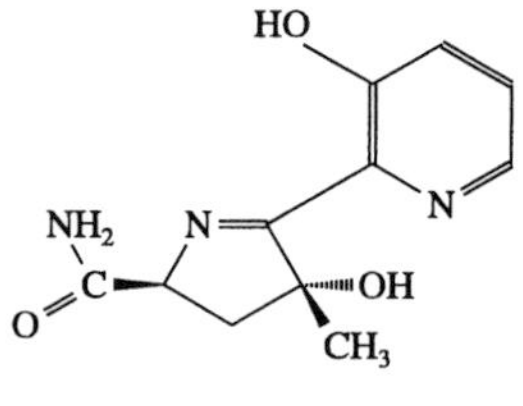

HO N NH_2 N C OH O CH_3

Fig. 12. Siderochelin A.

5-(3-hydroxy-2-pyridinyl)-4-methyl-2H-pyrrole-2-carboxamide (LIU et al., 1981).

3.2 Hydroxamate Siderophores

Generally, hydroxamate siderophores in bacteria can be divided into two main groups: (1) exclusively containing hydroxamate groups as iron-binding bidentates, (2) containing additional iron-binding ligands, e.g., α-hydroxycarboxylate residues, like citrate or catecholate groups as found in the pyoverdins. Members of the first group comprise the ferrioxamines originally isolated from the gram-positive streptomycetes and are now known to occur also in the gram-negative genus *Pseudomonas* and in Enterobacteria, such as *P. stutzeri* (MEYER and ABDALLAH, 1980) and *Erwinia, Enterobacter, Hafnia* (BERNER and WINKELMANN, 1990; REISSBRODT et al., 1990).

Ferrioxamines

Streptomyces pilosus produces a group of closely related desferrioxamines (A_1, A_2, B, C, D_1, D_2, E, F, G, and H) among which desferrioxamine B or E generally predominate (BICKEL et al., 1960; KELLER-SCHIERLEIN et al., 1964). Although commonly isolated in their ferric forms as ferrioxamines, the stuctures shown in Fig. 13 represent the iron-free forms. Desferrioxamine B is produced industrially by large-scale fermentation of *S. pilosus* strain A 21748 as the methane sulfonate salt (Desferal®). Desferal® still is the only drug approved for treatment of iron overload diseases although several other useful iron chelators are currently being developed (DIONIS et al., 1991).

Desferrioxamine B is a linear trihydroxamate. Acid hydrolysis yields acetic acid, succinic acid, and 1-amino-5-hydroxylamino pentane in a ratio of 1:2:3. After addition of fer-

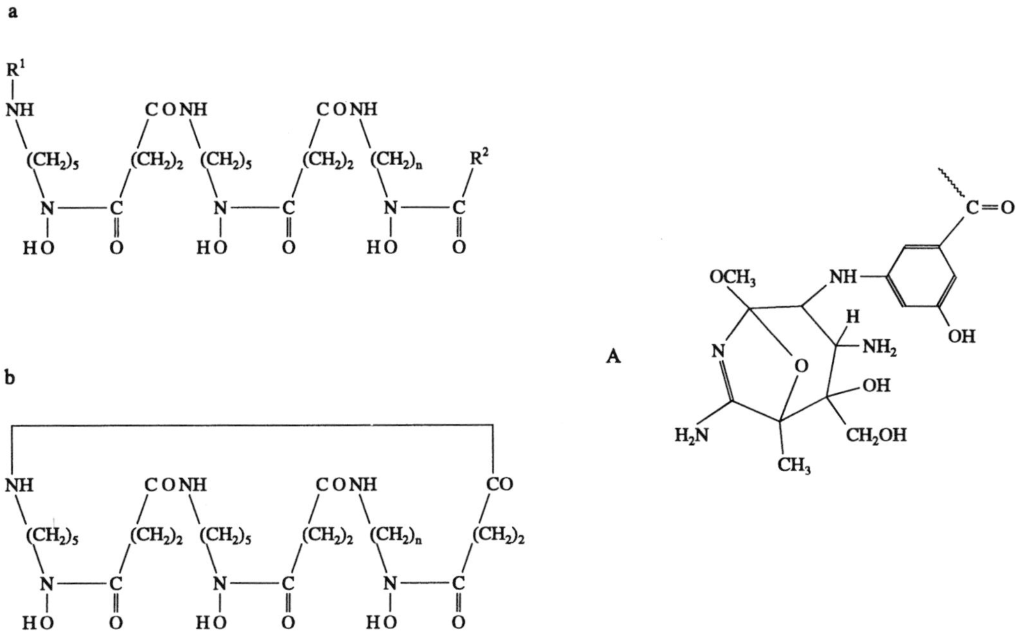

Fig. 13. Ferrioxamines (iron-free).
a Ferrioxamine B: $R^1=H$, $R^2=CH_3$, n=5; ferrioxamine D_1: $R^1=COCH_3$, $R^2=CH_3$, n=5; ferrioxamine G_1: $R^1=H$, $R^2=CH_2CH_2COOH$, n=5; ferrioxamine G_2: $R^1=H$, $R^2=CH_2CH_2COOH$, n=4; **b** ferrioxamine E; ferrimycin A_1: $R^1=A$,
$R^2=CH_3$.

ric iron, a 1:1 complex is formed with a stability constant of $K_f=10^{30.6}$ showing a red-brown color due to the metal-to-ligand charge transfer band at $\lambda_{max}=428$ nm ($\varepsilon=2800$ M^{-1} cm^{-1}). Since ferrioxamine B possesses a free amino group (pK=9.74), it migrates as a cation during electrophoresis in a weak acid buffer. Ferrioxamine D_1 (N-acetyl-ferrioxamine B) was isolated as a minor constituent of the ferrioxamine fraction and can be prepared synthetically by acetylation of ferrioxamine B with acetic acid anhydride. The ligand of ferrioxamine E is identical with the antibiotic "nocardamine" isolated from *Nocardia.* The crystal structure of ferrioxamine E was determined (VAN DER HELM and POLING, 1976), and the coordination about the iron center is *cis.* The ligand is a cyclic structure consisting of 3 mol succinic acid and 3 mol 1-amino-5-hydroxylamino pentane. Formal hydrolysis of one peptide bond in ferrioxamine E leads to ferrioxamine G which in turn can be converted to ferrioxamine E by treatment with dicyclohexyl carbodiimide. Some of the isolated ferrioxamines contained a mixture of 1-amino-5-hydroxylamino pentane and 1-amino-4-hydroxylamino butane (molar ratio 2:1) and were designated as ferrioxamine A (related to ferrioxamine B), and ferrioxamine A_1, (related to ferrioxamine D_1) and ferrioxamine D_2 (related to ferrioxamine E). Recent investigations in the authors' laboratory have shown that ferrioxamines are not confined to the genus *Streptomyces* but are also characteristic siderophores of several enterobacterial genera, e.g., *Erwinia, Pantoea, Enterobacter, Ewingella,* and *Hafnia* (BERNER et al., 1988; BERNER and WINKELMANN, 1990; REISSBRODT et al., 1990). Earlier reports have shown that *Pseudomonas stutzeri* is able to produce desferri-ferrioxamine E, previously named nocardamine (MEYER and ABDALLAH, 1980). Although the name proferrioxamine has been suggested for the ligand instead of desferrioxamine, deferrioxamine, or deferoxamine (FEISTNER et al., 1993), the new terminology of pFOs has not been accepted in the biological literature so far but seems to be of great value in distinguishing the vast number of structurally differing ligand derivatives by a simple subscript (e.g., desferrioxamine B=$pFO_{555\,Ac}$) indicating number and size of the consituent diamines, cyclic structure, and C-terminal acetyl residues (FEISTNER, 1995).

Ferrioxamine B is the most intensely studied ferrioxamine. Today it is used in the mesylate form (Desferal®) for the treatment of iron storage diseases in man. The biological activity of ferrioxamine B has been studied using growth promotion tests with iron-auxotrophic bacteria, such as *Microbacterium lacticum* ATCC 8181, *Arthrobacter flavescens* JG9 ATCC 29091, and *A. terregens* as reviewed by WINKELMANN (1991a). Transport experiments with iron-labeled ferrioxamines in *Streptomyces pilosus* (ATCC 19797) described by MÜLLER and RAYMOND (1984) and MÜLLER et al. (1984) showed that ferrioxamine B has the highest ratio of iron uptake ($K_m=0.2$ μM). Chromic and gallium complexes of ferrioxamine B were transported at similar rates indicating that the intact complexes were taken up. Differences in *cis* and *trans* isomers were not observed. In contrast, ferrioxamine transport studies using *in vivo* Mössbauer spectroscopy and radioactive labeling in Enterobacteria, such as *Erwinia herbicola* (now called *Pantoea herbicola*), revealed that the ligand is not accumulated inside the cells (MATZANKE et al., 1991).

Bisucaberin and Alcaligin

Bisucaberin (Fig. 14a), a dihydroxamate siderophore isolated from low-iron cultures of the marine bacterium *Alteromonas haloplanctis* is a cyclic dimer of succinyl-(N-hydroxy)-cadaverine. It thus resembles ferrioxamine E which is a cyclic molecule composed of succinyl diaminopentane residues.

From the heterotrophic marine bacterium *Alcaligenes denitrificans* a similar cyclic dihydroxamate siderophore named alcaligin (Fig. 14b) was isolated (NISHIO et al., 1988). Alcaligin is a cyclic molecule consisting of two N^1-hydroxy-3(OH)-putrescine residues linked by two succinic acid residues. Alcaligin has recently been found in the obligate pathogens *Bordetella pertussis* and *B. bronchiseptica* (MOORE et al., 1995; BRICKMAN et al., 1996).

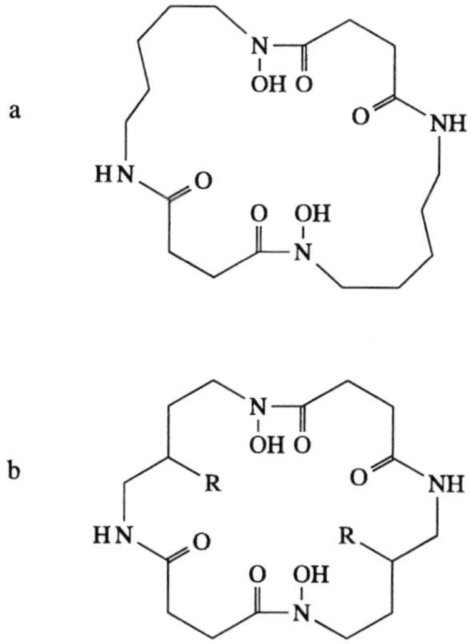

Fig. 14a. Bisucaberin; **b** alcaligin, R=OH; putrebactin, R=H.

A related cyclic dihydroxamate named putrebactin (Fig. 14c), containing putrescine instead of hydroxy-putrescine, was isolated from *Shewanella putrefaciens* (ALISON BUTLER, personal communication).

Ferrimycins and Albomycins

The ferrimycins (Fig. 13f) represent ferrioxamine-type siderophores possessing antibiotically active residues. Ferrimycins (A_1, A_2, B) and the compound A-22765 are ferrioxamine-type antibiotics isolated from *Streptomyces griseoflavus* and other species. The ferrimycins inhibit growth of gram-positive bacteria, such as *Staphylococcus aureus* and *Bacillus subtilis.* A revised structure of ferrimycin A was published by KELLER-SCHIERLEIN et al. (1984). Agar diffusion tests revealed that the antibiotic activity of the ferrimycins is antagonized by the presence of ferrioxamines. This has been explained by the existence of a transport antagonism in which the ferrioxamines and ferrimycins compete for the same target in the transport system (ZÄHNER et al., 1977). It remains an open question whether the antibiotically active group is split off from the transport moiety or whether ferrimycin acts as a whole after entering the cell.

Albomycins (δ_1, δ_2, ε) (Fig. 15) have also been isolated from *Streptomyces* strains, although the triornithine peptide is a characteristic feature of the fungal siderophores, e.g., ferrichromes. The originally published albomycin structure has been revised (BENZ et al.,

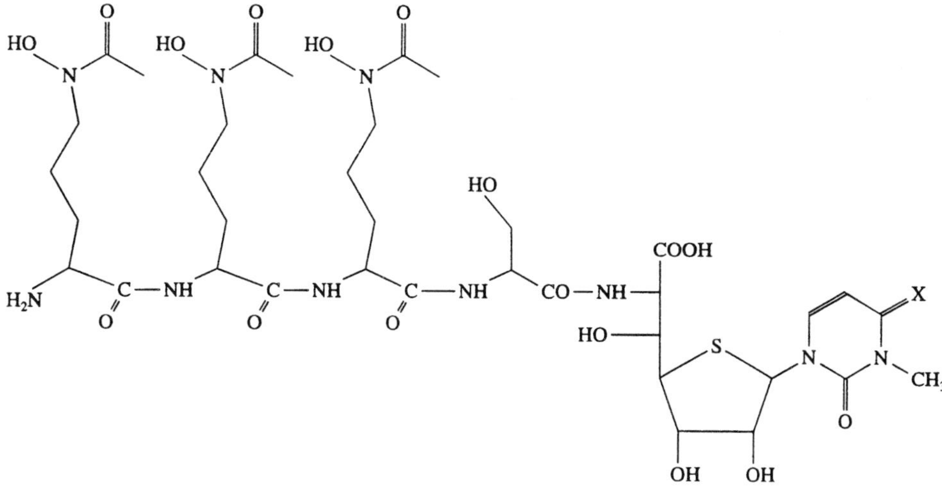

Fig. 15. Albomycins.
Albomycin δ_1: X=O; albomycin δ_2: X=$NCONH_2$; albomycin ε: X=NH.

1982) and was shown to consist of a linear peptidonucleoside (Orn_1-Orn_2-Orn_3-Ser-nucleoside). Albomycins are highly active against gram-negative and gram-positive bacteria. Ferrimycins and albomycins use the iron transport systems of ferrioxamine and ferrichrome, respectively. Albomycin is taken up in *E. coli* via the FhuA receptor protein and the *tonB* gene product, as shown with *fhuA* and *tonB* mutants (HARTMANN et al., 1979). Moreover, using ^3H- and ^{35}S-labeled albomycin it was confirmed that, contrary to the iron atom, the ligand is not accumulated inside the cell. From the small portion of labeled sulfur detected within the cells it was concluded that the nucleoside residue was split off to exert its antibiotic activity. Although inhibition of protein biosynthesis was demonstrated, it was also suggested that the inhibition of oxidative phosphorylation is the initial inhibitory effect of albomycins. Growth of *E. coli* is inhibited at very low concentrations (10^{-8} M). Because of the high frequency of resistant mutants, ferrimycins and albomycins have not found any application in the therapy of bacterial infections so far.

3.3 Peptide Siderophores

Alterobactin

Alterobacin, a peptide siderophore containing the sequence cycl.-(DHBA-Ser-Gly-Arg-β-OH-Asp-β-OH-Asp) (Fig. 16), was isolated from a marine *Pseudomonds*-like

Fig. 16. Alterobactin.

bacterium, *Alteromonas luteoviolacea* (REID et al., 1993). It has been shown to possess an exceptionally high formation constant ($K = 10^{49}$–10^{53}). Alterobactin A is a cyclic siderophore containing a lactone ester bond, while alterobactin B is the corresponding open-chain form. So far, only two siderophores from *Alteromonas* have been isolated – alterobactin and bisucaberin from *Alteromonas haloplanctis* (TAKAHASHI et al., 1987).

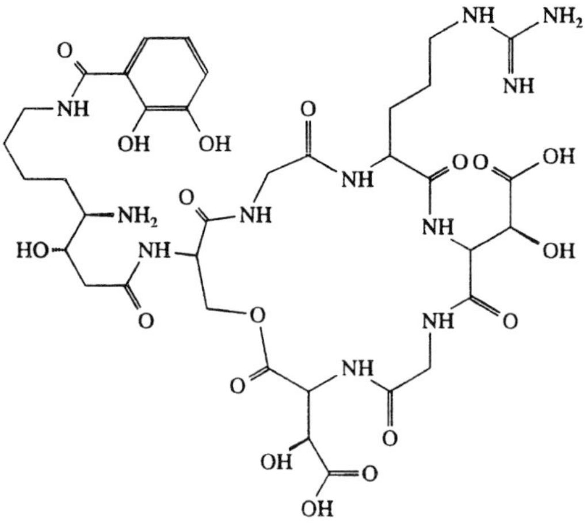

Ferrocins

Ferrocins (Fig. 17) represent a new group of iron-containing peptide antibiotics pro-

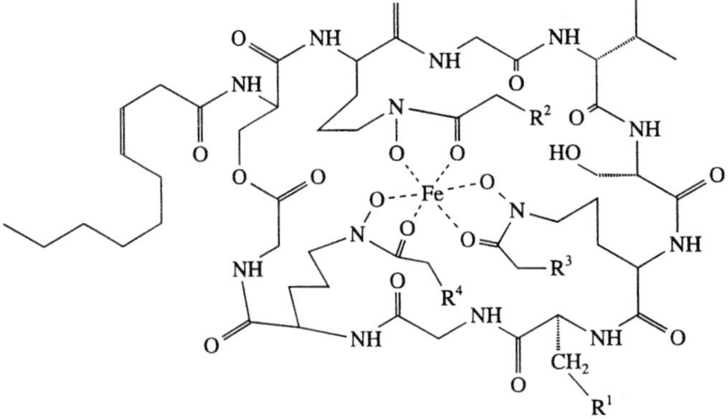

Fig. 17. Ferrocins. Ferrocin A: $R^1 = H$, $R^2 = R^3 = R^4 = H$; ferrocin B: $R^1 = OH$, $R^2 = R^3 = R^4 = H$; ferrocin C and D: $R^1 = H$, R^2, R^3, $R^4 = 2 \cdot H$, $1 \cdot CH_3$.

duced by *Pseudomonas fluorescens* YK-310 (KATAYAMA et al., 1993). These siderophore antibiotics showed antibacterial activity against *E. coli* and *P. aeruginosa* and strong therapeutic effects on *P. aeruginosa.* The structure elucidation revealed that the ferrocins are cyclic decapeptides containing three hydroxamate residues (TSUBOTANI et al., 1993). Four different ferrocins (A, B, C, D) have been isolated. Ferricrocin A is a cyclic peptide containing a fatty acid residue and a lactone ring between the C-terminal Gly and the N-terminal Ser: (Z)-3-Decenoic acid-cycl.-Ser-X-Gly-Val-Ser-X-Ala-Gly-X-Gly. X may represent *N*-Acetyl-*N*-OH-Orn or *N*-propionyl-*N*-OH-Orn. Because of the three ornithine residues and the peptidic nature of the ferrocins they resemble the albomycins, although the latter contain a nucleoside residue.

Pseudobactins and Pyoverdins

Strains of the *Pseudomonas fluorescens* group (*P. fluorescens, P. putida, P. aeruginosa, P. syringae,* and related species) produce yellow-green fluorescent siderophores. Typical structures are the linear pseudobactin (Fig. 18) (TEINTZE et al., 1981) and pyoverdin Pfl2, containing a cyclic substructure (Fig. 19) (BUDZIKIEWICZ, 1993). Pseudobactins and pyoverdins contain a common structural

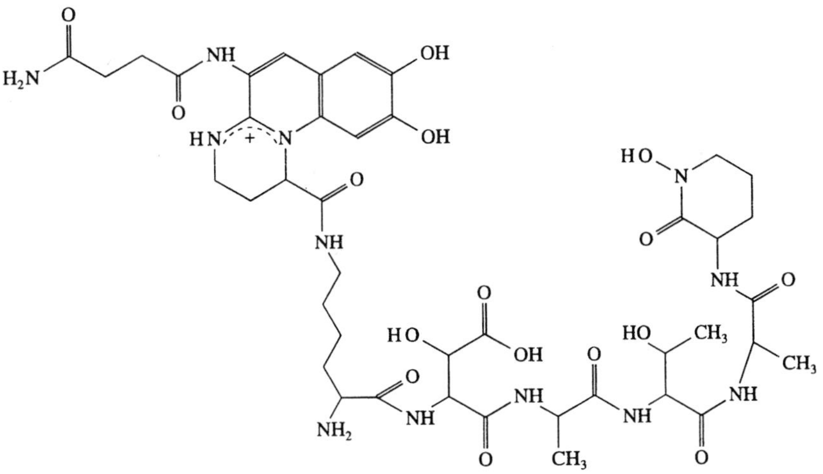

Fig. 18. Pseudobactin.

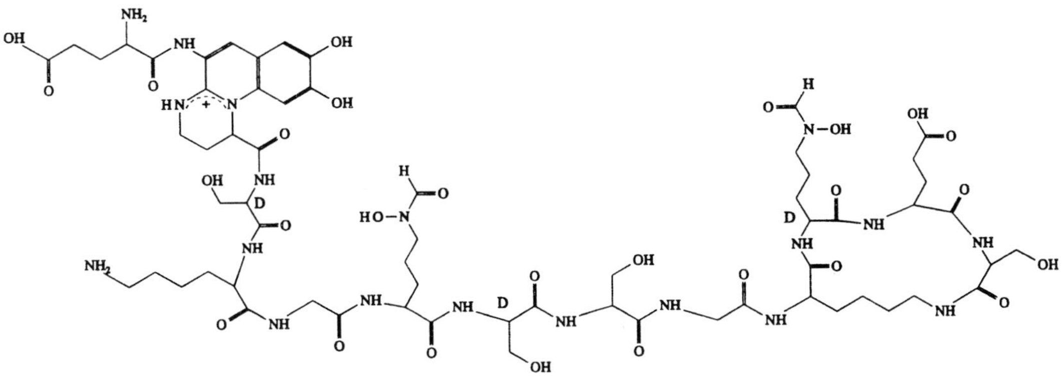

Fig. 19. Pyoverdin Pfl12.

element, the chromophore which is a derivative of 2,3-diamino-6,7-dihydroxyquinoline, possessing an additional 1-carboxyl-pyrimidino ring (*S* configuration) and an amino group acylated with different dicarboxylic acid residues: succinic acid, succinamide, 2-oxo-glutaric acid, glutamic acid, malic acid (amide). The chromophore is connected via the C-1 carboxyl group predominantly (for exceptions, see BUDZIKIEWICZ, 1993) to the N-terminus of a peptide chain of varying chain length (6–12 amino acids), half of which possess D configuration. In addition to the catecholate bidentate of the quinoline residue two further amino acids within the peptide chain function as iron bidentates among which N^5-acyl-N^5-hydroxy ornithine is present in all compounds studied so far. The N^5-acyl group of ornithine may be formyl, acetyl, or β-hydroxybutyryl or may even be connected to its own carboxyl group (cyclic OH-ornithine). The second iron-binding amino acid is generally *threo*-β-hydroxy aspartic acid. However, some of the pyoverdins lack β-hydroxy aspartic acid and possess two N^5-acyl-N^5-hydroxy ornithine residues instead. In addition to the cyclic N^5-OH-Orn the peptide chain may possess larger internal cyclic substructures where ε-amino groups (Lys) may be connected via amide bonds to distant C-terminal carboxyl groups or where Ser and Thr form additional tetrahydropyrimidine rings. Although internal rings have been detected in a variety of pyoverdins, pseudobactin is a linear molecule as confirmed by X-ray analysis (TEINTZE et al., 1981). A list of the currently known pyoverdin structures is shown in Fig. 20. After chelation with ferric iron the complex adopts a red-brown color. At pH 7 the Fe complexes show UV–VIS absorption maxima at 400 (log ε=4.2), 320 and 280 nm and additional charge transfer bands at 470 and 550 nm. Formation constants have been reported to be in the range of 10^{24}–10^{26}.

Ferribactin

During the isolation of iron-binding pigments from *P. fluorescens* a non-fluorescent precursor named ferribactin was isolated (MAURER et al., 1968). The studies of BUDZIKIEWICZ et al. (1992) and LINGET et al. (1992) have generally confirmed the composition of ferribactin. Both compounds, ferribactin and pyoverdin, isolated from the same strain have been shown to differ in the chromophore part but are identical in their peptide sequence (BUDZIKIEWICZ et al., 1992). Thus, the composition of ferribactin isolated from *Pseudomonas aptata* was similar to pyoverdin Pap (*P. aptata*) but lacked the fluorescent chromophore and contained Glu and Tyr/Dbu instead. Ferribactin is a much weaker iron-binding compound than the pyoverdins as the catechol group provided by the chromophore is absent. However, the phenol group of Tyr and the carboxylate group of Glu in ferribactin may possibly substitute the chromophore ligand binding sites.

Because of the diversity of the pyoverdin structures produced by the various *Pseudomonas* species a pronounced specificity during uptake of ferric pyoverdins can be expected and has indeed been observed by HOHNADEL and MEYER (1988) and MEYER (1992).

The structural gene for the ferric pseudobactin receptor (PupA) in *Pseudomonas putida* WCS358 has been cloned and sequenced (MARUGG et al., 1989; BITTER et al., 1991). An inducible ferric pseudobactin receptor (PupB) of *P. putida* WCS358 has also been identified and characterized (KOSTER et al., 1993).

Different types of pyoverdin-defective (*pvd*) mutants have been isolated and characterized (VISCA et al., 1992). The *pvd-1* mutant is an L-N^5-hydroxy ornithine auxotroph unable to oxygenate L-ornithine and requiring L-N^5-hydroxy ornithine for pyoverdin production. Other types of mutants appear to be blocked in further steps of the biosynthetic pathway leading to pyoverdin, the acylation of L-N^5-hydroxy ornithine (*pvd-2*) and chromophore biosynthesis (*pvd-3*). The oxygenase gene *pvd-A* (previously named *pvd-1*) from *P. aeruginosa* has been cloned (VISCA et al., 1994). From a gene bank using the broad-host range cosmid pLAFR3 mobilized in a pvdA-defective mutant a *trans*-complementing cosmid pPV4 was obtained which restored pyoverdine synthesis and oxygenase activity in the *pvdA* mutant.

Bacteria	Structures of pyoverdins and related compounds
Pseudomonas putida ATCC 12633	**Chr**-L-Asp-L-Lys-D-**OHAsp**-(D,L)Ser-L-Thr-D-Ala-D-Glu-(D,L)Ser-L-**cOHOrn**
Pseudomonas putida CFBP 2461	**Chr**-L-Asp-(D,L)-Lys-D-**OHAsp**-D-Ser-L-Thr-D-Ala-(D,L)-Lys-L-Thr-L-**cOHOrn**
Pseudomonas chlororaphis ATCC 9446	**Chr**-D-Ser-L-Lys--Gly-(D,L)-**FoOHOrn**-c[L-Lys-(D,L)-**FoOHOrn**-L-Ser]
Pseudomonas fluorescens ATCC 13525	**Chr**-D-Ser-L-Lys-Gly-(D,L)-**FoOHOrn**-c[L-Lys-(D,L)-**FoOHOrn**-L-Ser]
Pseudomonas fluorescens SB 83	**Chr**-D-Ala-L-Lys-L-Thr-D-Ser-L-**AcOHOrn**-L-**cOHOrn**
Pseudomonas fluorescens ATCC 17400 (Demange et al., 1990a)	**Chr**-D-Ala-D-Lys-Gly-Gly-L-**OHAsp**-D-GlnCTHPMD-L-Ser-D-Ala-L-**cOHOrn**
Pseudomonas fluorescens CCM 2798 (Demange et al., 1990b)	**Chr**-SerCTHPMD-Gly-L-Ser-D-**OHAsp**-L-Ala-Gly-D-Ala-Gly-L-**cOHOrn**
Pseudomonas aeruginosa ATCC 15692 (Briskot et al., 1989)	**Chr**-D-Ser-L-Arg-D-Ser-L-**FoOHOrn**-c[L-Lys-L-**FoOHOrn**-L-Thr-L-Thr]
Pseudomonas tolaasii NCPPB 2192 (Demange et al., 1990b)	**Chr**-D-Ser-L-Lys-L-Ser-D-Ser-L-Thr-D-Ser-L-**AcOHOrn**-L-Thr-D-Ser-D-**cOHOrn**
Azotobacter vinelandii CCM 289 (azotobactin D) (Demange et al., 1988)	**ChrA**-L-Asp-D-Ser-L-Hse-Gly-D-**OHAsp**-L-Ser-D-Cit-L-Hse-D-**AcOHOrn**-L-Hse
Pseudomonas fluorescens ATCC 13525 (desferribactin) (Linget et al., 1992)	L-Glu-D-TyrCTHPMD-D-Ser-L-Lys-Gly-(D,L)-**FoOHOrn**-c[L-Lys-(D,L)-**FoOHOrn**-L-Ser]
Azomonas macrocytogenes ATCC 12334 (azoverdin) (Bernardini et al., 1996)	**Chr**-L-Hse-D-**AcOHOrn**-D-Ser-L-**AcOHOrn**-D-Hse-L-CTHPMD

For additional structures see: Abdallah 1991 and Budzikiewicz 1993.

Chr = Chromophore, ChrA = chromophore of azotobactin, OHOrn and OHAsp = N^{δ}-hydroxyornithine and ß-hydroxyaspartic acid, c = cyclic, cOHOrn = cyclo-N^{δ}-hydroxyornithine, Fo and Ac = formyl and acetyl present at the N^{δ}-nitrogen atom of hydroxyornithines, SerCTHPMD, TyrCTHPMD and GlnCTHPMD represent the amino acids derived from tetrahydropyrimidine resulting from ring formation of L-2,4-diaminobutyric acid with Ser, Gln or Tyr:

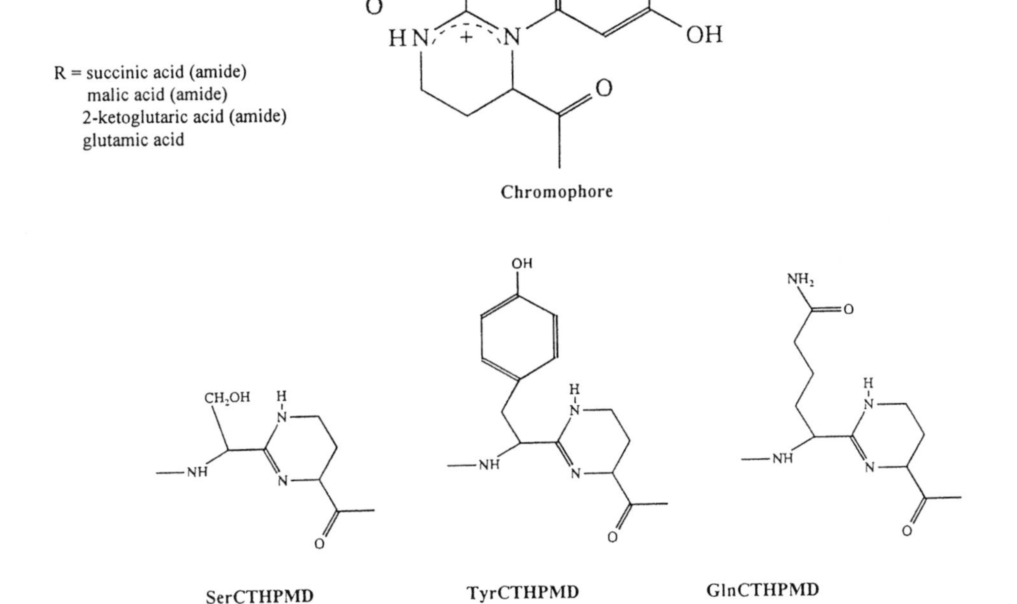

Fig. 20. Structures of pseudobactins and pyoverdins.

Azotobactin

The peptide siderophore azotobactin (Fig. 21) seems to be the most prominent siderophore of the Azotobacteriaceae showing a striking similarity to the pyoverdins. The main structural difference between the azotobactins and pyoverdins is that azotobactins form an extra imidazolone ring connecting the amino group at C-5 to the chromophore N-4 by an urea unit, whereas the pyoverdins possess an amidically bound dicarboxylic acid at that position (CORBIN et al., 1970). As this is a minor structural difference, some authors have included the azotobactins in the pyoverdin group (MENHARDT et al., 1991).

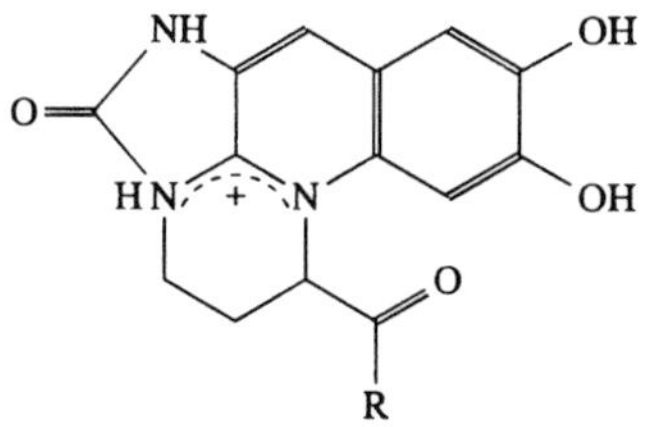

Fig. 21. Chromophore of azotobactin.

FUKASAWA et al. (1972) described the first structure of the yellow-green fluorescent peptide, azotobactin 0, produced by *Azotobacter vinelandii* (strain 0) which contained only two bidentate chelating groups. Two further structures have been elucidated by now: azotobactin D (DEMANGE et al., 1988) and azotobactin from *A. vinelandii* DSM 87 (ATCC 12837), the latter containing an N-hydroxy ornithine acylated with 3-hydroxy butyric acid (BUDZIKIEWICZ et al., 1992).

Using ^{55}Fe-labeled compounds it was shown that azotochelin and azotobactin do function as siderophores in *A. vinelandii* (KNOSP et al., 1984). However, siderophore-mediated uptake of iron was not observed until substantial non-specific binding of iron to the cells was eliminated. Transport of azotobactin and azotochelin was also decreased by high concentrations of cations, such as Na^+, K^+, Li^+, or Mg^{2+}, due to some interference with the siderophore transport system.

VISWANATHA and his group (MENHARDT et al., 1991) have recently discussed the heterogeneity of the pyoverdins and azotobactins observed during growth under iron limitation. From their pulse labeling experiments it was inferred that an amino acid deficiency or a lack of stringency may be responsible for the observed heterogeneity of the produced azotobactins and pyoverdins.

Ornibactins

Ornibactins (Fig. 22) represent modified tetrapeptide siderophores containing *N*-terminal 3-hydroxy-acyl residues of different chain lengths (C4, C6, C8). The peptide contains two L-Orn (N^5-OH, N^5-acyl), one Asp (β-OH), one L-Ser and a C-terminal 1,4-diamino butane residue (STEPHAN et al., 1993a, b). Thus the ornibactins resemble the pyoverdins but they possess an acyl residue instead of a chromophore and are much smaller peptides. They have been isolated from *Pseudomonas cepacia*-like strains. The *"cepacia* group"

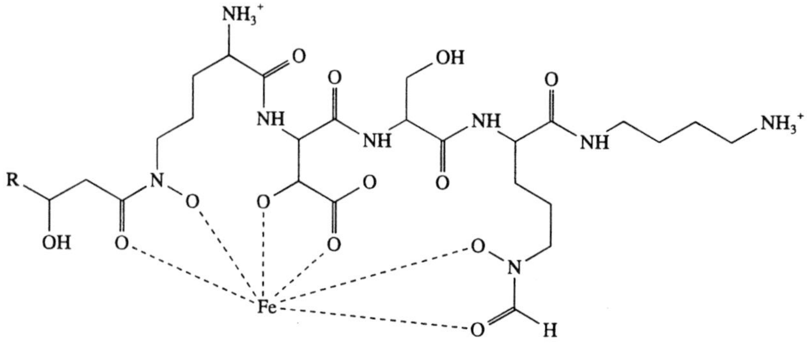

Fig. 22. Ornibactins. Ornibactin C4 (R = methyl); ornibactin C6 (R = propyl); ornibactin C8 (R = pentyl).

(rRNA homology group II) has been recently transferred to the genus *Burkholderia* (YABUUCHI et al., 1992). Ornibactin production has been found in clinical isolates as well as in nitrogen-fixing bacterial isolates from the rhizosphere of rice plants. Moreover, ornibactin-mediated iron uptake was observed in all *Burkholderia* strains but was absent in *Pseudomonas aeruginosa, P. fluorescens,* and *P. stutzeri* (MEYER et al., 1995).

3.4 Mycobactins and Related Siderophores

Mycobacteria produce a series of lipid-soluble, iron-binding compounds termed mycobactins (Fig. 23a) which have been described in previous reviews by SNOW (1970) and RATLEDGE (1987). Among the different mycobactins (A, F, H, M, N, P, R, S, T), mycobactin P, isolated from *Mycobacterium phlei,* and mycobactin T, isolated from *M. tuberculosis,* were analyzed in more detail (SNOW, 1965). After splitting the ester link, two products known as mycobactic acid and cobactin were obtained. Regarding the P type mycobactin, chemical analysis of the mycobactic acid unit yields (1) an aromatic acid, either salicylic acid or 6-methyl salicylic acid, the latter undergoing further degradation to *m*-cresol and carbon dioxide; (2) a β-hydroxy amino acid, either serine or threonine; (3) a mixture of homologous long-chain fatty acids, usually with a double bond adjacent to the carboxyl group; (4) N^6-hydroxy lysine. The cobactin unit yields a hydroxy acid, either 3-hydroxy butyric acid or 3-hydroxy-2-methyl pentanoic acid, and a second molecule of N^6-hydroxyl lysine.

The mycobactins are associated with the mycobacterial cell walls and have to be extracted by organic solvents. In order to avoid extraction of undesired fatty material, the most convenient method is to suspend the cells in cold ethanol. HALL and RATLEDGE (1982) have reported a simple method for the production and isolation of mycobactins: Mycobacteria were grown on solid media and scraped from the agar after growth. The bacteria were then held for 24 h in ethanol. $FeCl_3$ in ethanol was added to convert the mycobactins completely to the ferric form. An equal volume of chloroform was added, followed by water until two layers were formed. The chloroform layer was washed three times with water to remove excess iron; it was then dried with $MgSO_4$, evaporated to dryness, and the residue was dissolved in methanol. As pointed out by SNOW (1970), the characteristic UV absorption spectra of the iron-free mycobactins arise from two parts of the molecule, the 2-(*o*-hydroxyphenyl)-oxazoline structure and the acyl hydroxamic acid. Depending on the substitution of the benzene ring the following λ_{max} values in methanol are given: (no methyl group) 243, 249, and 304 nm with inflection at 258 nm; (with methyl group at position 6) 250 and 311 nm with a shoulder at 254 nm and an inflection at 265 nm.

Growth of *Mycobacterium paratuberculosis* (formerly named *M. johnei*) was stimulated in the presence of 5–15 ng mycobactin P (FRANCIS et al., 1953). However, other mycobactins also gave significant effects so that there seems to be little specificity. *M. paratuberculosis* and some Mycobacteria related to *Mycobacterium avium* have lost the ability to synthesize mycobactins and are, therefore, stimulated by the addition of exogenous mycobactins. Nocobactins have been isolated from *Nocardia* species as compounds equivalent in function to the mycobactins and similar in structure (RATLEDGE and PATEL, 1976).

Mycobactins are extremely lipophilic compounds and are generally not excreted into the medium. Therefore, RATLEDGE (1987) suggested that mycobactins function as an iron shuttle within the lipid boundary of the cell surface. Although salicylic acid is excreted into the medium, its contribution to iron solubilization seems to be insignificant. Instead, water-soluble exochelins are thought to function as iron scavengers and to transport iron to the mycobactins (STEPHENSON and RATLEDGE, 1979). The mechanism proposed for iron uptake into *Mycobacterium smegmatis* involves the mediation of mycobactin in shuttling iron across the boundary layers of the cell (RATLEGE and MARSHALL, 1972). The loading of iron into mycobactin is presumed to occur via exochelins which are

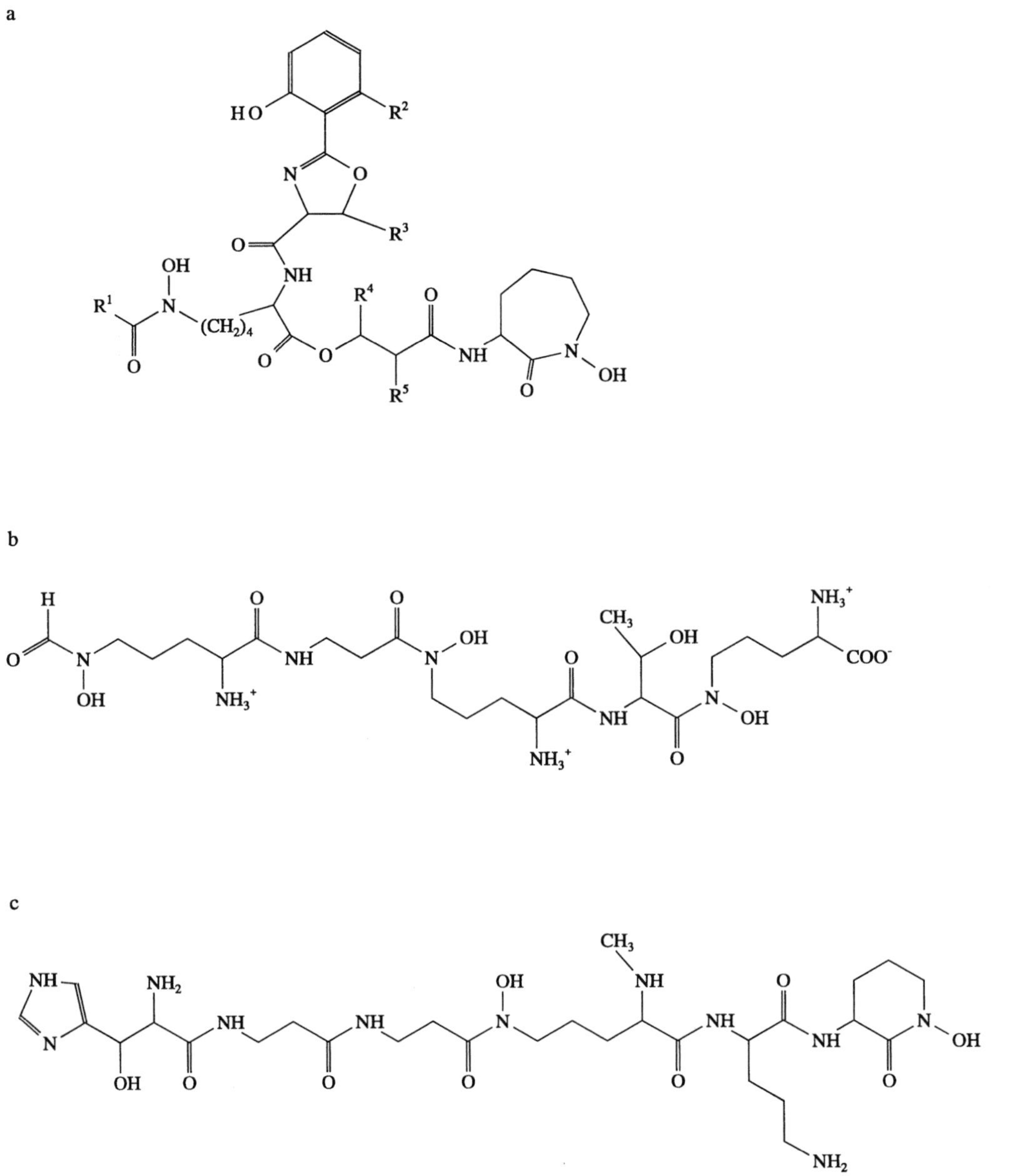

Fig. 23a–c. General structure of **a** mycobactins (for side chains and nomenclature see SNOW, 1970); **b** exochelin MS; **c** exochelin MN.

able to solubilize iron from the environment (MACHAM et al., 1975; STEPHENSON and RATLEDGE, 1979). Structure elucidation of exochelins from *M. smegmatis* has shown the presence of a modified linear pentapeptide containing three N^{δ}-hydroxy-ornithines (Fig. 23b) (SHARMAN et al., 1995a). Another novel water-soluble iron-binding linear hexapeptide named exochelin MN has recently been isolated from *M. neoaurum* (SHARMAN et al.,

1995b) containing β-hydroxyhistidine, 2 β-alanines, N^{α}-methyl-$^{\delta}$N-hydroxyornithine, ornithine, and cyclic $^{\delta}$N-hydroxyornithine (Fig. 23c). In addition, exochelins showing structural similarities with mycobactins have been isolated from *M. tuberculosis* (GOBIN et al., 1995, LANE et al., 1995; RATLEDGE and EWING, 1996). These exochelins contain dicarboxylic acid methyl esters instead of the long chain fatty acids occurring in the mycobactins.

Acinetobactin

From low-iron cultures of *Acinetobacter baumannii* ATCC19606 another compound with both catecholate and hydroxamate functional groups was isolated containing 2,3-dihydroxy benzoic acid and threonine combined to an oxazoline ring and ω-N-hydroxy histamine (YAMAMOTO et al., 1994a). The structure was elucidated by chemical degradation, fast atom bombardment mass spectrometry, and ^{1}H and ^{13}C NMR spectroscopy. Acinetobactin (Fig. 24) was identified in 4 of

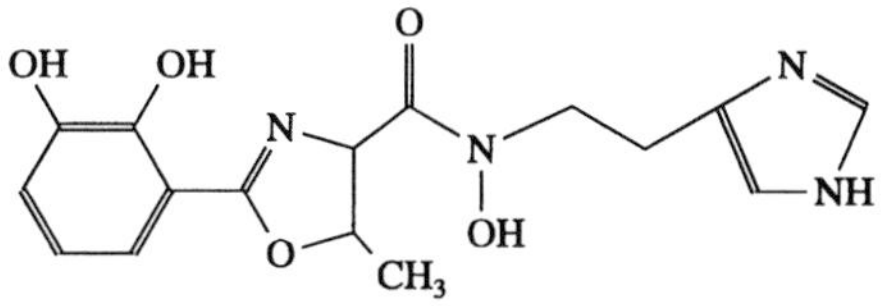

Fig. 24. Acinetobactin.

12 clinical isolates of *A. baumannii.* Acinetobactin is structurally related to anguibactin, a siderophore isolated earlier from *Vibrio anguillarum* which has a thiazoline ring instead of an oxazoline ring (JALAL et al., 1989). Another *Acinetobacter* species, *A. haemolyticus,* has previously been reported to produce a novel citrate-based siderophore, acinetoferrin (OKUJO et al., 1994b).

Frankobactin

The siderophore of *Frankia* (strain 52065), named francobactin, has recently been isolated and shown to be a novel hydroxamate-containing peptide siderophore (Fig. 25). The following components have been tentatively identified (MW 731): a phenyl oxazoline ring, serine, glycine, ornithine, and β-HO-aspartate (BOYER and ARONSON, 1994; ARONSON and BOYER, 1994). The actinomycete *Frankia* is a gram-positive, aerobic, filamentous, sporulating, and nitrogen-fixing bacterium that exists as a free-living saprophyte or as a symbiont in the roots of the actinorhizal plants including alder trees (e.g., *Alnus glutinosa*), *Ceanothus americanus, Myrica gale,* and several species of *Casuarina.* Unlike the *Rhizobium* strains, *Frankia* strains are also able to fix nitrogen in aerobic culture which makes *Frankia* an interesting organism to study siderophore production under symbiotic and non-symbiotic conditions.

Maduraferrin

An *Actinomadura madurae* strain (Actinomycetes) produces the novel siderophore maduraferrin (KELLER-SCHIERLEIN et al., 1988). Maduraferrin (Fig. 26) is the iron complex of an oligopeptide siderophore composed of salicylic acid, β-alanine, glycine, L-serine, N^5-hydroxy-N^2-methyl-L-ornithine, and L-hexahydropyridazine-3-carboxylic acid. Although most *Actinomadura* strains produced various ferrioxamines, some strains, like strain DSM 43067, were found to pro-

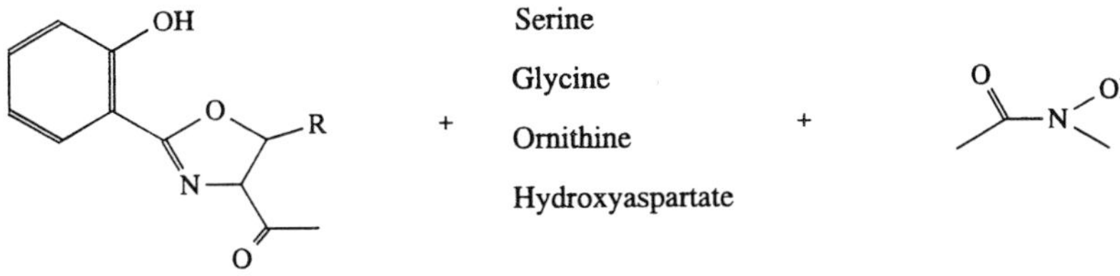

Serine
Glycine
Ornithine
Hydroxyaspartate

Fig. 25. Frankobactin.

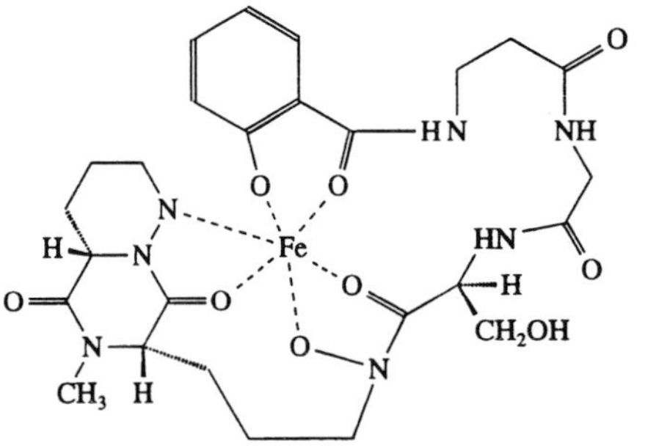

Fig. 26. Maduraferrin.

duce desferri-maduraferrin. Because of the low growth in chemically defined media, maduraferrin production was performed in a complex medium (soy, yeast extract) with the addition of $AlCl_3$ which mimics iron deficiency. Transport of iron mediated by maduraferrin was demonstrated in the producing strains but was absent in *E. coli* or *Staphylococcus aureus*, suggesting a special uptake route in *Actinomadura* strains.

3.5 Citrate Hydroxamate Siderophores

Schizokinen

Schizokinen (Fig. 27a), originally described as a factor that reduced the division lag of *Bacillus megaterium, B. cereus,* and *B. subtilis,* was isolated from a culture of *B. megaterium* ATCC 19213 (BYERS et al., 1967). The growth factor activity was confirmed with a schizokinen-auxotroph of *B. megaterium.* The structure of schizokinen was elucidated by MULLIS et al. (1971). Hydrolysis of schizokinen yielded 1-amino-3-(N-hydroxyamino) propane, and oxidation with periodate yielded acetate as the acyl moiety of the hydroxamic acid bonds. The central backbone is composed of a citric acid residue. Titration and CPK models confirmed that both the central hydroxyl and carboxyl group can bind to the metal ion to give an octahedral six-oxygen coordination sphere. At neutral pH the ferric complex is an anion indicating the presence of ionized citrate hydroxyl. The absorption

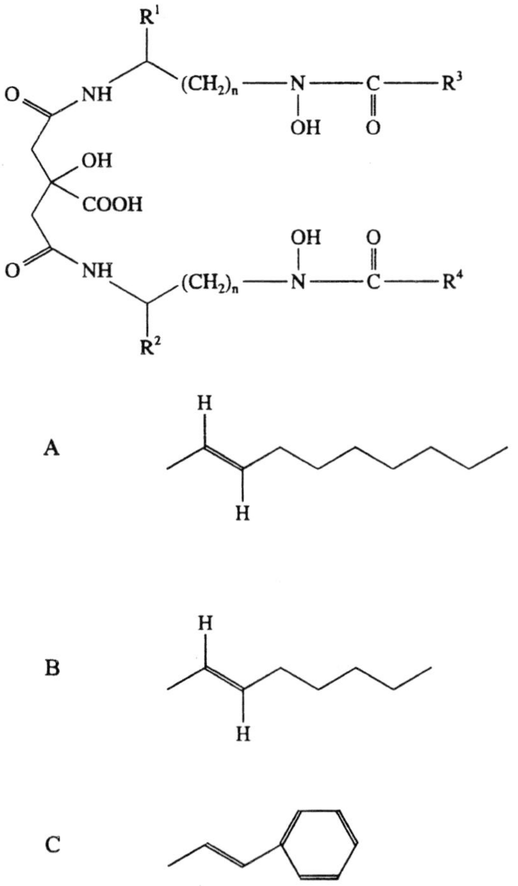

Fig. 27. Citrate hydroxamates.
Schizokinen: $R^1=R^2=H$, $R^3=R^4=CH_3$, n=2; rhizobactin 1021: $R^1=R^2=H$, $R^3=A$, $R^4=CH_3$, n=2; acinetoferrin: $R^1=R^2=H$, $R^3=R^4=B$, n=2; arthrobactin: $R^1=R^2=H$, $R^3=R^4=CH_3$, n=4; aerobactin: $R^1=R^2=COOH$, $R^3=R^4=CH_3$, n=4; nannochelin A: $R^1=R^2=COOCH_3$, $R^3=R^4=C$, n=4; nannochelin B: $R^1=COOCH_3$, $R^2=COOH$, $R^3=R^4=C$, n=4; nannochelin C: $R^1=R^2=COOH$, $R^3=R^4=C$, n=4.

($\lambda_{max}=390$ nm) was pH independent between pH 5 and 9.5, but shifted to red below pH 5.

The cultures were grown in a chemically defined low-iron medium at 35 °C with strong aeration for one week (3% inoculum). After removing the cells, the supernatant was concentrated to 1/10 of the original volume, acidified with conc. HCl to pH 2, extracted with chloroform–phenol (1:1), and transferred with ether to the aqueous phase. This was then concentrated by evaporation and applied to a column of the acetate form of Dowex AG-2-X10. The column was washed with distilled water and schizokinen was eluted with 0.2 M ammonium chloride and further purified on Bio-Gel P-2. TLC on silica gel with methanol ($R_f=0.60$) using tetrazolium chloride or $FeCl_3$ in 0.05 N HCl as sprays was performed to confirm purity. Schizokinen was also found in *Anabaena* strains (LAMMERS and SANDERS-LÖHR, 1982) and may be a siderophore in other cyanobacteria.

A derivative of schizokinen, schizokinen A (Fig. 28), has recently been isolated from alcalophilic *Bacillus* strains containing a cyclic side chain due to the condensation of the carboxyl at the quarternary C atom with one of the amide NH groups (STEPHAN et al., unpublished results). Schizokinen A has been synthesized by LEE and MILLER (1983). A similar compound was isolated earlier from *B. megaterium*. In the original publication, however, this product was incorrectly described to have a six-membered ring (MULLIS et al., 1971). It is still unclear whether or not cyclization to schizokinen A is an artifact arising during the isolation procedure of schizokinen or if it is a real natural product. The occurrence of imido forms in citrate containing siderophores has also been reported for staphyloferrin A (KONETSCHNY-RAPP et al., 1990) and rhizoferrin (DRECHSEL et al., 1992). Rhizoferrin, e.g., tends to give both, imidorhizoferrin and bis-imidorhizoferrin (DRECHSEL et al., 1992).

BYERS and co-workers were the first to investigate the transport of iron in *B. megaterium,* using double-labeled ^{59}Fe–^{3}H-schizokinen. The results indicated an initial temperature-independent binding of ferric schizokinen, followed by a temperature-dependent (active) transport of the chelate into the cell, and an enzyme-catalyzed separation of iron from the chelate (ARCENEAUX et al., 1973). Moreover, a mutant of *B. megaterium* SK11 unable to synthesize schizokinen showed recognition capacity for the chemical structure of schizokinen. Ferrioxamine B (Desferal®) was taken up at lower rates and aerobactin did not support uptake of iron (HAYDON et al., 1973). A recent investigation has shown that *B. subtilis* utilizes three types of hydroxamate siderophores: ferrichromes, ferrioxamines, and schizokinen (SCHNEIDER and HANTKE, 1993). Moreover, the transport system showed significant homology to the binding protein-dependent components of *E. coli.* As a periplasm is missing in gram-positive bacteria, the FhuD-corresponding protein is anchored as a lipoprotein in the cytoplasmic membrane.

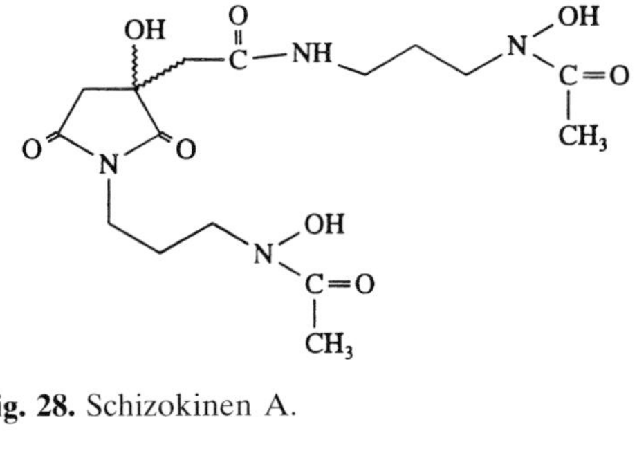

OH O C NH N OH C=O CH3 O N O N OH C=O CH3

Fig. 28. Schizokinen A.

Rhizobactin 1021

Rhizobactin 1021 (Fig. 27b) is a novel asymmetric citrate-based dihydroxamate isolated from *Rhizobium meliloti* strain 1021. While acinetoferrin contains two identical hydroxamic acid acyl groups as (*E*)-2-octenoic acids, rhizobactin 1021 contains two different acyl residues, (*E*)-2-octenoic acid and (*E*)-2-decenoic acid (PERSMARK et al., 1993). Ferric rhizobactin 1021 predominantly exists in solution in the *lambda* configuration in an apparent equilibrium of a monomeric and a dimeric species.

Acinetoferrin

Acinetoferrin (Fig. 27c) has been isolated from *Acinetobacter haemolyticus* and shown to belong to the citrate-based dihydroxamates (OKUJO et al., 1994b). Structure elucidation revealed that acinetoferrin is a structural analog of schizokinen as it contains a citric acid backbone linked amidically to two 1-amino-3-(N-hydroxy propane) residues which in turn are condensed by hydroxamic acid bonds to two (*E*)-2-octenoic acid moieties. Fast atom bombardment mass spectroscopy revealed a quasi molecular ion peak (MH^+) at $m/z=585$ corresponding to a molecular mass of 584 amu. Maximal production of acinetoferrin occurred when *A. haemolyticus* ATCC 17906 was grown in a chemically defined medium containing approximately 0.1 μM iron. Absorption spectra of acinetoferrin showed an absorption maximum at 210 nm ($2.9 \cdot 10^4$ M^{-1} cm^{-1}) with a shoulder at 250 nm, indicative of unsaturated bonds. After addition of iron a red-brown color develops that is characteristic for a ligand-to-metal charge transfer band ($\lambda_{max}=486$ nm; $\varepsilon=9.8 \cdot 10^2$ M^{-1} cm^{-1}).

Arthrobactin

Arthrobactin (Fig. 27d), previously known as the "*terregens* factor", is produced by strains of *Arthrobacter* and supported the growth of the iron-auxotrophs *Arthrobacter terregens* and *A. flavescens* JG9 ATCC 29091. The structure of arthrobactin was elucidated by LINKE et al. (1972). It contained two 1-amino-5-(*N*-acetyl-*N*-hydroxy)amino pentane residues linked symmetrically to citric acid by peptide bonds. Arthrobactin was isolated from low-iron cultures of *Arthrobacter pascens* (ATCC 13346) in an asparagine–salts medium. After inoculation with a spore suspension the culture was incubated for 24 h at 27 °C. Then $FeCl_3$ was added, the culture filtrate was saturated with ammonium sulfate, and extracted with benzyl alcohol. Diethylether was added to transfer the ferric arthrobactin complex into the aqueous phase. The crude material was purified by countercurrent extraction and by chromatography and gel filtration. Arthrobactin and schizokinen were synthesized chemically (LEE and MILLER, 1983).

Aerobactin

Examination of supernatants from cultures of *Aerobacter aerogenes* 62-I (now called *Enterobacter aerogenes*) and related strains showed the presence of a hydroxamic acid for which the trivial name, aerobactin (Fig. 27e), was suggested (GIBSON and MAGRATH, 1969). Aerobactin consists of a conjugate of N^6-acetyl-N^6-hydroxy lysine and citric acid. After complexing with ferric iron the internal citrate carboxyl and the α-carboxyls of the amino acid residues remain free resulting in a ferric complex with three negative charges at neutral pH. The synthesis of aerobactin was described by MAURER and MILLER (1982). Coordination chemistry, stability constants, ligand protonation, and metal ligand equilibria were determined by HARRIS et al. (1979). The ferric aerobactin complex shows an overall formation constant of $\log K_f=22.93$ and an absorption of $\lambda_{max}=398$ nm ($\varepsilon=2170$ M^{-1} cm^{-1}). Based on the p*M* value (stability under specified concentration and pH) aerobactin can effectively compete with transferrin. The comparatively low redox potential (-0.336 V) allows reduction via physiological reductants and reuse of the ligand, which seems to be an advantage for the producing organism. The CD spectrum revealed that ferric aerobactin predominantly exists as the *lambda* optical isomer in aqueous solution.

Aerobactin can be isolated from strains of *Enterobacter aerogenes* using chemically defined media without added iron as described by GIBSON and MAGRATH (1969). After inoculation and incubation for 20 h at 37 °C with stirring and aeration, the cells are separated and the supernatant is passed through a Dowex-1 (Cl^-) column. All the material giving a red color with $FeCl_3$ is held on the column. The column is then eluted with an ammonium chloride gradient (0.4–1 M). 2,3-Dihydroxy benzoate derivatives are eluted first, followed by aerobactin. The fraction containing aerobactin is then evaporated to dryness, dissolved in water, and further purified on a Dowex 50W-X8 (H^+) (200–400 mesh) column with water. $FeCl_3$-positive fractions are collected, freeze-dried, and dried over P_2O_5 for several days.

Aerobactin is also produced by *Enterobacter cloacae* and *Shigella flexneri* (PAYNE, 1980), *Erwinia carotovora* (ISHIMARU and LOPER, 1992), and by *Escherichia coli* strains containing the plasmid ColV (BRAUN, 1981). It was shown that aerobactin uptake requires the expression of an outer membrane receptor (Iut A, 74 kDa) which serves as a common binding site for aerobactin and the bacteriocin cloacin (VAN TIEL-MENKVELD et al., 1982; BINDEREIF et al., 1982). The genes coding for aerobactin synthesis were analyzed by GROSS et al. (1985) and DELORENZO et al. (1986). Aerobactin production also seems to be widespread in other bacterial genera. Besides in enterobacteria, it has been detected in cultures of a halophilic pseudomonad (BUYER et al., 1991) and in a culture of a gram-positive DSM 4640 strain (WINKELMANN, unpublished data).

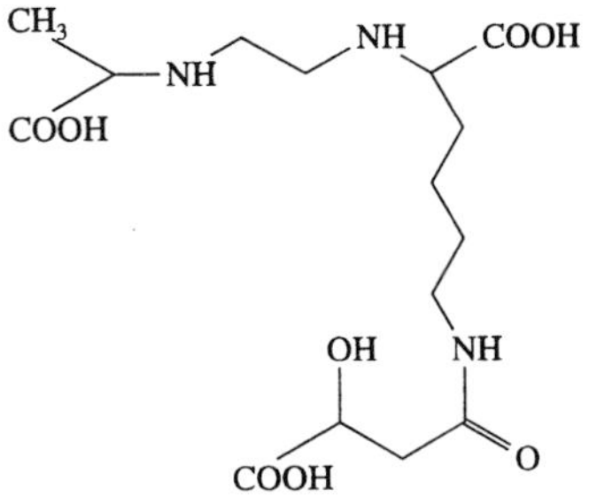

Nannochelins

Citrate hydroxamate siderophores named nannochelins have been isolated from low-iron cultures of *Nannocystis excedens* (Myxobacteria) (KUNZE et al., 1992) (Fig. 27f). Nannochelins resemble aerobactin, but contain two *N*-ε-cinnamoyl residues instead. While nannochelin C contains two non-esterified lysyl carboxyl groups, nannochelin B has one and nannochelin A two methyl-esterified carboxyl groups.

3.6 Carboxylate Siderophores

Because of their high stability constants and their pronounced colored ferric complexes, catecholate and hydroxamate siderophores were the first microbial iron chelates detected. However, with novel chemical and biological assay methods being available, e.g., the chrome azurole S test (CAS test), additional carboxylate-type siderophores have been detected (SCHWYN and NEILANDS, 1987).

Rhizobactin DM4

Rhizobactin DM4 (Fig. 29) isolated from *Rhizobium meliloti* DM4 (SMITH et al., 1985) should not be confused with rhizobactin 1021 isolated from *R. meliloti* 1021 (PERSMARK et al., 1993), although both have been isolated from isolates of the same species. Rhizobactin DM4 contains a characteristic ethylene diamine bridge between a (non-amidically linked) alanine and lysine residue, the latter being N^6-acylated by malic acid. Thus, rhizobactin DM4 is regarded as a complexone type or carboxylate type siderophore, while rhizobactin 1021 is a citrate-based dihydroxamate. It was anticipated that the iron complex of rhizobactin DM4 is hexacoordinated involving the two ethylene diamine nitrogens, the two amino acid carboxylates, and the α-hydroxy acid function of malic acid (NEILANDS, 1993). There is structural similarity of rhizobactin DM4 and the mugineic acids isolated as phytosiderophores from grasses having a diamine backbone with (α-hydroxy)carboxylate side functions. Mugineic acids may be also regarded as complexone-type siderophores (NOMOTO et al., 1987).

Fig. 29. Rhizobactin DM4.

Staphyloferrin A

Staphyloferrin A (Fig. 30) isolated from staphylococci is a carboxylate type siderophore consisting of two citric acid residues linked amidically to D-ornithine (KONETSCHNY-RAPP et al., 1990; MEIWES et al., 1990). *Staphylococcus hyicus* DSM 20459 was used to produce staphyloferrin A in larger amounts. In addition, three dehydration products were obtained, two mono forms and one diimido form, as a result of intramolecular condensation of the central carboxyl groups with the amide NH groups. Interestingly feeding with D- and L-ornithine led to incorporation and synthesis of D-ornithine-containing staphyloferrin A. Transport experiments with ^{55}Fe-labeled compounds

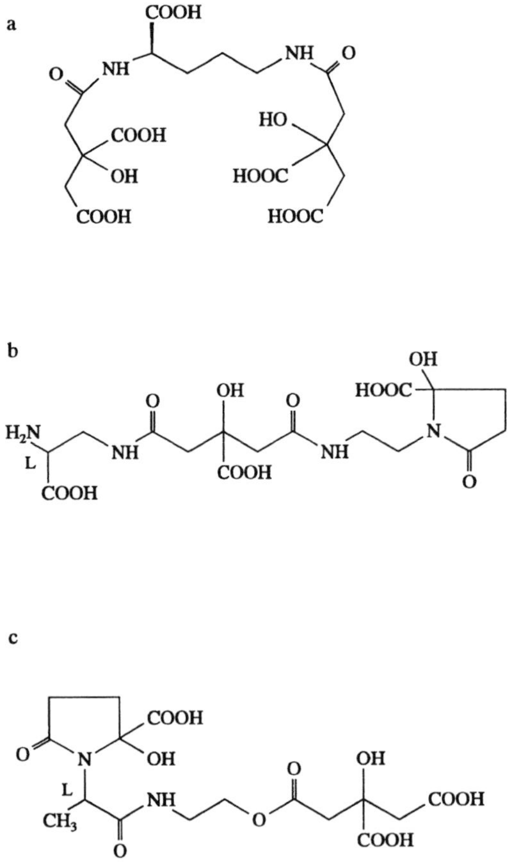

Fig. 30a. Staphyloferrin A; **b** staphyloferrin B; **c** vibrioferrin.

showed that ferric staphyloferrin A acts as a siderophore in staphylococci.

Staphyloferrin B

Staphyloferrin B has also been isolated from Staphylococci, however, the structural components differ completely from staphyloferrin A (DRECHSEL et al., 1993a) in that citric acid is linked by a 2,3-diamino propionic acid residue at one side and by a diamino ethane and a 2-oxo glutaric acid residue at the other side (Fig. 30b). The molecular mass was determined by pneumatically assisted electro spray from aqueous solution and indicates hydration of the 2-oxo function. The 2-oxoglutaric acid residue can easily form a five-membered ring by nucleophilic addition of the amide nitrogen to the carbonyl carbon. NMR experiments showed NOE contacts confirming the existence of a cyclization product as shown in the structural formula of staphyloferrin B. Whether this shift of equilibrium from the open to the cyclized form is caused by sample preparation or measurement conditions is still unclear.

Vibrioferrin

A structurally novel siderophore was isolated by successive column and TLC from *Vibrio parahaemolyticus* (YAMAMOTO et al., 1992). This organism is known to occur in marine and estuarine environments and causes diarrhea associated with seafood consumption. Vibrioferrin was negative in the chemical assays of ARNOW and CSÁKY (for description, see NEILANDS and NAKAMURA, 1991) excluding the presence of catecholates and hydroxamates. The constituents of the siderophore found after hydrolysis were L-alanine, ethanolamine, citric acid, and 2-oxoglutaric acid. Fast atom bombardment mass spectrometry revealed a mass of 446 amu. A complete structure has recently been reported (YAMAMOTO et al., 1994b). 2-oxoglutaric acid is amidically linked to L-alanine and, like in staphyloferrin B, exists in equilibrium with a five-membered hydroxylactam (Fig. 30c).

Cepabactin

Cepabactin (Fig. 31) is a natural bacterial pyridinone siderophore (1-hydroxy-5-methoxy-6-methyl-2(1H)-pyridinone) isolated from *Pseudomonas cepacia* ATCC 25416 (MEYER et al., 1989). Cepabactin can be re-

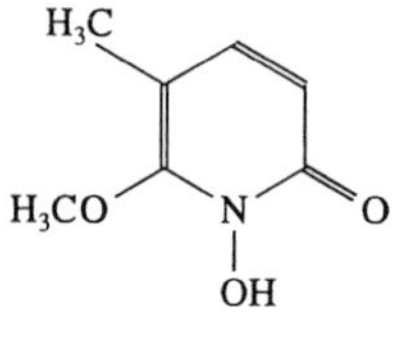

Fig. 31. Cepabactin.

garded as a cyclic hydroxamate or as a keto hydroxy bidentate.

3.7 Keto Hydroxy Bidentates

There seems to be a wide range of compounds in nature possessing keto hydroxy bidentate ligands, most of which have not been shown to be involved in iron transport. However, some selected keto hydroxy bidentate ligands proved to be very effective in iron transport, as demonstrated recently (THIEKEN and WINKELMANN, 1993) using bacteria of the group Proteeae (*Proteus, Providencia, Morganella*) all of which are able to produce α-keto acids as a result of oxidative deamination of amino acids (DRECHSEL et al., 1993b). Deamination to 2-oxo acids is a characteristic trait of the Proteeae and is only of minor importance in other enterobacterial genera. Because of the high production rates in Proteeae and their activity in iron nutrition, α-keto acids have been assigned a siderophore function in Proteeae and in other genera (DRECHSEL et al., 1993b; REISSBRODT et al., 1994). Interestingly, strains of *Proteus mirabilis* were found to be unable to transport iron via the carboxylate type siderophore rhizoferrin but still were able to transport ferric complexes of keto hydroxy bidentate ligands (THIEKEN and WINKELMANN, 1993), suggesting that the transport systems for ferric rhizoferrin and ferric complexes of keto hydroxy bidentate ligands are different. Although rhizoferrin is a fungal siderophore (see Sect. 4.5), uptake kinetics and characterization of the genes involved in *Morganella morganii* have recently been described (CARRANO et al., 1996; KÜHN et al., 1996).

4 Fungal Siderophores

4.1 Ferrichromes

Ferrichrome (Fig. 32), first isolated in 1952 from *Ustilago sphaerogena* (NEILANDS, 1952), was found later in other *Ustilago* species (DEML, 1985) as *U. maydis* (BUDDE and LEONG, 1989), in *Neovossia indica* (Tilletiales) (DEML et al., 1984), Lipomycetaceae (VAN der WALT et al., 1990) and the dermatophyte *Trichophyton mentagrophytes* (MOR et al., 1992). Besides ferrichrome, *Ustilago* species also produce ferrichrome A, the structure of which is closely related to ferrichrome. Crystal structures of ferrichrome A and ferrichrome were determined by ZALKIN et al. (1966) and VAN DER HELM et al. (1980) showing a cyclic, modified Orn-Orn-Orn-Gly-Gly-Gly hexapeptide. Ferrichrome may be regarded as a prototype siderophore from which a variety of other "ferrichromes" are derived, e.g., ferricrocin, ferrichrysin, ferrichrome C, ferrichrome A, ferrirubin, ferrirhodin, malonichrome. An updated detailed

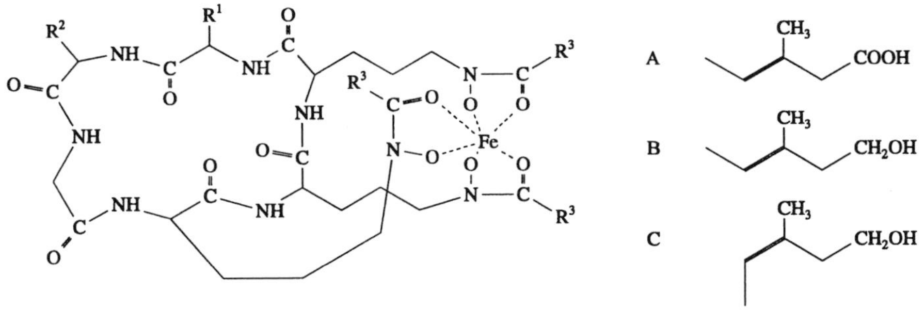

Fig. 32. Ferrichromes.
Ferrichrome: $R^1=R^2=H$, $R^3=CH_3$; ferrichome A: $R^1=R^2=H$, $R^3=A$; ferricrocin: $R^1=H$, $R^2=CH_2OH$, $R^3=CH_3$; ferrichrysin: $R^1=R^2=CH_2OH$, $R^3=CH_3$; ferrirubin: $R^1=R^2=CH_2OH$, $R^3=B$; ferrirhodin: $R^1=R^2=CH_2OH$, $R^3=C$; asperchromes: $R^1=R^2=CH_2OH$, R^3=mixture of C and CH_3.

description is given by VAN DER HELM and WINKELMANN (1994). With the exception of tetraglycyl ferrichrome, a heptapeptide (DEML et al., 1984), all other members of the ferrichrome family possess a cyclic hexapeptide backbone. Ferrichrome type siderophores contain a characteristic tripeptide sequence of N^5-acyl-N^5-hydroxy-L-ornithine and a variable tripeptide sequence of short amino acids (Gly, Ser, Ala). Variations in the latter tripeptide sequence and in the nature of the *N*-acyl substituents of the ornithine residues account for the number of distinct ferrichrome-type siderophores mentioned above. X-ray diffraction studies and CD measurements in solution have revealed that the configuration of all ferrichromes is *lambda-cis* (VAN DER HELM et al., 1980; LEONG and RAYMOND, 1974). The stability of desferri-ferrichrome with ferric ions is in the range of $K_f = 10^{29}$–10^{31}.

VAN DER HELM and co-workers have isolated the so-called "asperchromes" as minor constituents in low-iron culture filtrates of *Aspergillus ochraceus,* containing a common cyclic Orn-Orn-Orn-Ser-Ser-Gly hexapeptide sequence as in ferrichrysin, ferrirhodin, and ferrirubin but different ornithyl-N^5-acyl groups (JALAL et al., 1984). Interestingly, a ferrichrome-type siderophore with an incomplete, open-chain peptide was detected and named des(diserylglycyl)ferrirhodin (JALAL et al., 1985). The latter has recently also been found in the edible mushroom *Agaricus bisporus* (ENG-WILMOT et al., 1992).

Ferrichrome was originally isolated from *Ustilago sphaerogena* grown in iron-containing medium. A key observation was made when the production of ferrichrome could be enhanced by growing *Ustilago* in an iron-deficient medium. The composition of the medium for the production of ferrichrome and ferrichrome A is a modified Grimm–Allen medium. To isolate ferrichromes, iron is added under stirring to the cell-free supernatant of a low-iron culture after a week of growth with strong aeration. The ferrichromes can easily be adsorbed on a column of XAD-2 resin and, after washing with distilled water to remove salts and other polar constituents, the ferrichromes can be desorbed with acetone–water (1:1) or methanol. Lipids can be removed on a Sephadex LH 20 column and a final separation on a silica gel–silica gur (1:1) column with dichloro methane–methanol (2:1) as an eluting solvent (DEML et al., 1984). HPLC separation using acetonitrile–water mixtures (KONETSCHNY-RAPP et al., 1988) or TLC on silica gel plates using dichloro methane–methanol–water (65:25:4) as a solvent system are recommended for identification and isolation. A detailed survey on the methods of isolation and identification of fungal siderophores is given by JALAL and VAN DER HELM (1991).

The biological properties of siderophores can be studied by growth promotion tests or by transport experiments with radio-labeled siderophores as discussed in previous reviews (WINKELMANN, 1991a, 1993). Originally ferrichrome and ferrichrome A were characterized by their growth-stimulating effect on *Pilobolus* and *Arthrobacter* species. *A. flavescens* (now reclassified as *Aureobacterium flavescens*) JG9 (ATCC 25091) is a siderophore-auxotroph and requires trace amounts of ferrichrome or other hydroxamate siderophores but does not respond to catechols (RABSCH and WINKELMANN, 1991; SHANZER et al., 1988). EMERY (1971) was the first to show that ferrichrome indeed functions as a siderophore in *Ustilago sphaerogena.* Ferrichrome is actively taken up. The iron-free ligand is excreted again and can be used for another round of transport. Contrary to that, ferrichrome A does not enter the cells but donates iron to the external face of the cell membrane where it is removed by a reduction process (ECKER et al., 1982a). A mutant of *Neurospora crassa,* unable to synthesize siderophores in the absence of ornithine, has been used to study uptake of ferrichrome-type siderophores and coprogen (WINKELMANN and ZÄHNER, 1973). Uptake of ferrichromes in *N. crassa* and *Penicillium parvum* has been shown to be highly stereospecific (WINKELMANN, 1979; WINKELMANN and BRAUN, 1981). Thus *enantio*-ferrichromes (NAEGELI and KELLER-SCHIERLEIN, 1978) were not recognized by the fungal siderophore transport systems (HUSCHKA et al., 1985, 1986).

4.2 Coprogens

Coprogen (Fig. 33) has been isolated from *Penicillium* species, from *Neurospora crassa, Curvularia,* and a variety of other fungi (JALAL and VAN DER HELM, 1991). A method for coprogen production and isolation from *N. crassa* was described by WONG et al. (1983). The structure of coprogen was elucidated by KELLER-SCHIERLEIN and DIEKMANN (1970) and revealed a linear trihydroxamate siderophore composed of 3 mol of N^5-acyl-N^5-hydroxy-L-ornithine, 3 mol of anhydro mevalonic acid and 1 mol of acetic acid. Desacetyl coprogen, named coprogen B, was isolated from *Fusarium dimerum* (DIEKMANN, 1970). A further coprogen, triornicin, was found in low-iron culture filtrates of *Epicoccum purpurescens* (FREDERICK et al., 1981). *Curvularia lunata* produces neocoprogen I and neocoprogen II, in which the two outer anhydromevalonic acid residues are substituted by one or two acetyl groups, respectively. The CD spectrum of coprogen showed a *delta* configuration in solution (WONG et al., 1983), and the crystal structure of neocoprogen I revealed a *delta*-trans configuration (HOSSAIN et al., 1987). Further coprogen derivatives, such as hydroxy coprogen, N^{α}-dimethyl coprogen as well as their corresponding neoderivatives have been isolated from *Alternaria* sp. (for review, see JALAL and VAN DER HELM, 1991).

Analysis of coprogen transport in wild type and mutant strains of *Neurospora crassa* using single-labeled ^{55}Fe-coprogen, ^{14}C-coprogen, or double-labeled ^{55}Fe–^{14}C-coprogen (WINKELMANN and ZÄHNER, 1973; HUSCHKA et al., 1985) indicated that the intact coprogen molecule can enter the cell. Furthermore, Mössbauer spectroscopic studies (MATZANKE and WINKELMANN, 1981) confirmed that after uptake a major portion of intracellular coprogen remains in the iron complex form and iron is removed by reduction only slowly. Moreover, an additional ferrichrome-type siderophore, ferricrocin, was found to be synthesized only as an intracellular siderophore functioning as an iron storage compound from which iron is gradually removed by reduction. Uptake and competition experiments have shown that *N. crassa* possesses two different siderophore recognition sites, one for ferrichrome-type siderophores one for coprogen, both of which donate their siderophores to a common transport system located in the cytoplasmic membrane (HUSCHKA et al., 1985; CHUNG et al., 1986). Details of the actual transport process are still unresolved. However, there is evidence that the transport of siderophores in fungi is functionally connected to the membrane potential

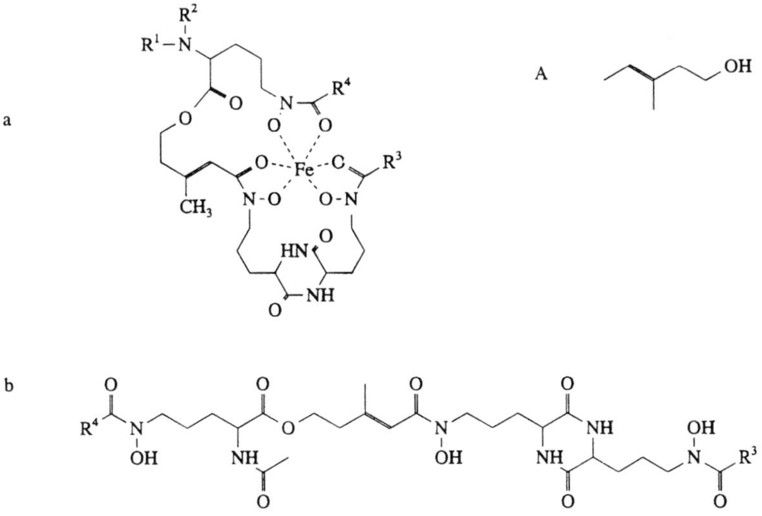

Fig. 33. Coprogens.
a Coprogen: $R^1=H$, $R^2=COCH_3$, $R^3=R^4=A$.
Coprogen B: $R^1=R^2=H$, $R^3=R^4=A$.
Neocoprogen I: $R^1=H$, $R^2=COCH_3$, $R^3=CH_3$, $R^4=A$.
Neocoprogen II: $R^1=H$, $R^2=COCH_3$, $R^3=R^4=CH_3$.
N^{α}-dimethyl coprogen: $R^1=R^2=CH_3$, $R^3=R^4=A$.
b Desferricoprogen: $R^3=R^4=A$.

generated by the plasma membrane ATPase (HUSCHKA et al., 1983).

4.3 Rhodotorulic Acid

A compound with the properties of a secondary hydroxamic acid strongly binding iron(III) was isolated from supernatants of iron-deficient cultures of *Rhodotorula pilimanae* (ATKIN and NEILANDS, 1968). This compound, named rhodotorulic acid (RA) (Fig. 34a), was characterized as the diketo piperazine of N-acetyl-L-N-hydroxy ornithine and has subsequently been found in species of *Leucosporidium, Rhodosporidium, Sporidiobolus,* and *Sporobolomyces* (ATKIN et al., 1970). The diketopiperazine moiety is also present in coprogen as dimerum acid (Fig. 34b). Thus, rhodotorulic acid (RA) is structurally related to the coprogen family. The iron complex of rhodotorulic acid at neutral pH has been found to be dimeric with the formula $Fe_2(RA)_3$, where both iron atoms have the *delta-cis* configuration (CARRANO and RAYMOND, 1978). Below pH 3 this complex dissociates to give the monomer, $FeRA^+$, in which each iron is bound to two hydroxamate ligands. Ferric RA has an absorption maximum at 425 nm (ε=2700 M^{-1} cm^{-1}) at pH 7 and becomes red at pH 2 with an absorption maximum at 480 nm (ε=1750 M^{-1} cm^{-1}). Ferric RA functions as an iron transport agent in *Rhodotorula pilimanae* and probably in other RA-producing heterobasidiomycetous yeasts. RA mediates iron transport to the cell but does not actually transport iron into the cell (CARRANO and RAYMOND, 1978). Citrate was found to be as effective in iron transport as RA in this fungus.

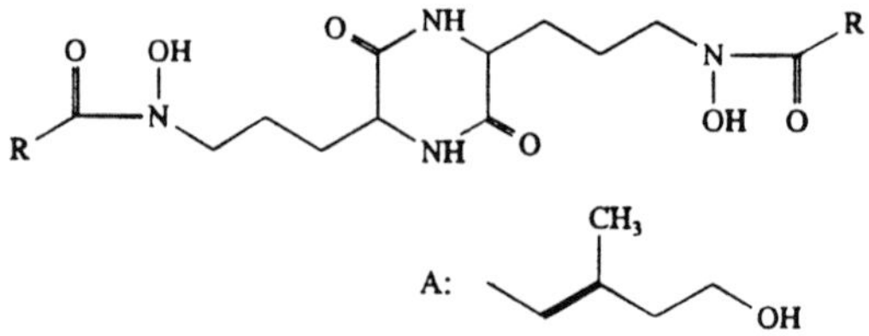

Fig. 34. Rhodotorulic acid: R=CH_3; dimerum acid: R=A.

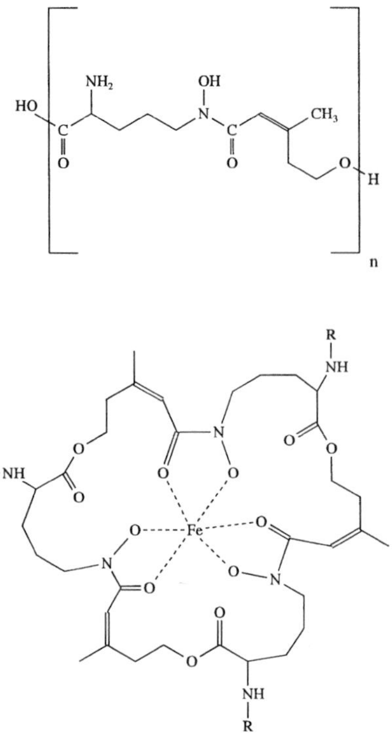

Fig. 35. Fursarinines.
a Fusarinine A: n=2, fusarinine B: n=3; **b** fusarinine C (Fusigen), R=H; triacetyl fusarinine C (triacetyl fusigen), R=$COCH_3$.

4.4 Fusarinines (Fusigens)

Fusarinines (Fig. 35) had originally been isolated from *Fusarium cubense* (DIEKMANN, 1967) and *F. roseum* (SAYER and EMERY, 1968) and were later shown to occur in other genera, e.g., *Giberella, Aspergillus, Penicillium,* and a variety of other fungi (DIEKMANN, 1968; JALAL et al., 1986, 1987). Recently, fusarinine C was described as an endogenous siderophore in *Agaricus bisporus* (ENG-WILMOT et al., 1992). Alkaline degradation yielded 3 mol of the monohydroxamic acid *cis*-fusarinine (Fig. 35a) in which N^5-hydroxy-L-ornithine is *N*-acylated by *cis*-5-hydroxy-3-methyl-2-pentenoic acid. The three fusarinine molecules are esterified head-to-tail to build the cyclic triester fusigen (Fig. 35b). The free amino groups show a pK of 7.1

and the ferric iron complex is stable up to pH 1. In addition to the cyclic triester, the linear trimer (fusarinine B), the linear dimer (fusarinine A), and the monomer (fusarinine) can be found in the culture filtrate. Another member of the fusigen family, *N,N′,N″*-triacetyl fusarinine C (TAFC) or triacetyl fusigen (Fig. 35c), has been isolated from *Aspergillus fumigatus* strains (Diekmann and Krezdorn, 1975) and from *Penicillium* sp. (Moore and Emery, 1976). The ferric triacetyl fusarinine C is a relatively flat molecule with a total thickness of about 45 pm as determined from the crystal structure (Hossain et al., 1980). Two different coordination isomers of the triacetyl fusigen molecule have been observed. Crystals from ethanol/benzene adopted a *lambda-cis* absolute configuration. However, in solution and in the morphologically different crystals from chloroform, the molecule predominantly existed as a *delta-cis* isomer, as determined by circular dichroism.

Fusigen can be isolated from low-iron cultures of *Fusarium cubense* in an asparagine–salts medium (Diekmann and Zähner, 1967). The medium is inoculated with a suspension of spores and incubated at 27 °C on a rotary shaker or in a fermenter with aeration. After 3–5 d $FeCl_3$ solution is added, and the siderophores are adsorbed on Servachrome XAD-2, washed with three volumes of distilled water, and desorbed with one volume of methanol. The crude siderophore solution is evaporated to dryness and dissolved in 0.01 M ammonium acetate buffer (pH 5) and passed through a CM-Sephadex column equilibrated and eluted with the same buffer. Bound fusigen is then eluted with 0.1 M ammonium acetate buffer, and again adsorbed on XAD-2, washed, and desorbed as described before. A further purification can be achieved on Sephadex LH-20 in methanol.

Triacetyl fusigen can be isolated from *Penicillium* and *Aspergillus* strains using the same medium as described for the production of fusigen (Diekmann and Krezdorn, 1975). Transport experiments with ^{55}Fe-labeled fusigen in *Aspergillus fumigatus* revealed high uptake rates, suggesting rapid utilization of the chelated iron by the producing strain (Wiebe and Winkelmann, 1975). Emery (1976) reported *Fusarium roseum* to contain an enzyme which hydrolyzes the ornithine ester bonds of fusigen (fusarinine C) but not the ferric chelate of fusigen. *Penicillium* sp. was found to hydrolyze triacetyl fusigen and was fully active on the ferric trihydroxamate chelate. The iron-free form of triacetyl fusigen was also reported to act as an antibiotic on a variety of bacteria grown on minimal media (Anke, 1977). *Bacillus brevis, Clostridium pasteurianum, Pseudomonas fluorescens,* and *Streptomyces viridochromogenes* were found to be sensitive even on complex media.

4.5 Rhizoferrins

Rhizoferrin (Fig. 36) is a novel carboxylate-type siderophore first isolated from low-iron cultures of *Rhizopus microsporus* var. *rhizopodiformis* (Drechsel et al., 1991, 1992; Winkelmann, 1992). Further studies have shown that rhizoferrin seems to be the characteristic siderophore in the order of Mucorales (Zygomycetes) as a variety of other strains from different genera (*Mucor, Phycomyces, Chaetostylum, Absidia, Cokeromyces, Cunninghamella, Mycotypha,* and *Mortierella*) produce rhizoferrin as their only siderophore (Thieken and Winkelmann, 1992). Hydroxamates could not be found. However, recent results from the authors' laboratory point to the existence of hydroxamate siderophores in strains of *Basidiobolus* and *Conidiobolus* (Thieken and Winkelmann, unpublished results) which are regarded as a separate genus of Zygomycetes (Zycha and Siepmann, 1969).

The stability constant of rhizoferrin with ferric iron has been determined to be $K_f = 10^{24}$ (Albrecht-Gary, personal communication). The existence of rhizoferrins as a novel carboxylate siderophore in Mucorales

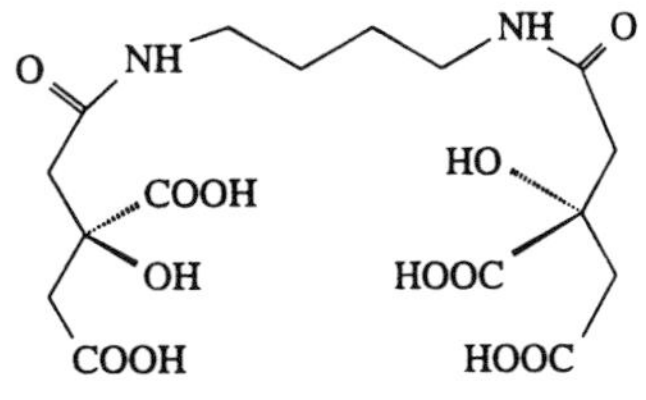

Fig. 36. Rhizoferrin.

is especially interesting, since the Zygomycetes are the only fungal class where ferritins as iron storage proteins have been found yet. Ferritins have not been detected so far in Ascomycetes and Basidiomycetes (MATZANKE, 1994). Therefore, rhizoferrin iron transport and ferritin iron storage might coexist without metabolic disturbance.

The siderophore activity has been investigated using ^{55}Fe-labeled rhizoferrin (DRECHSEL et al., 1991; CARRANO et al., 1996). Furthermore, a comparison with ferrioxamines revealed that iron uptake rates mediated by rhizoferrin and ferrioxamines in *R. microsporus* were similarly effective, suggesting that fungi of the Mucorales are also able to use ferrioxamines as an iron source (DRECHSEL et al., 1991). This has important consequences for their role as pathogens, since some members of this group have been repeatedly isolated during fatal cases of mucormycosis in dialysis patients (BOELAERT et al., 1993, 1994). Thus, a substantial number of patients treated with desferrioxamine (Desferal®) for either aluminium or iron overload have developed mucormycosis caused by *Rhizopus* (BOELAERT et al., 1991).

A simple growth promotion assay using bacteria of the *Proteus* group has been developed to detect bioactive siderophores containing keto hydroxy bidentate ligands including rhizoferrin and simple keto and hydroxy acids (THIEKEN and WINKELMANN, 1993). This bioassay not only allows the detection of further fungal carboxylate siderophores but also discriminates between carboxylate and hydroxamate producing fungi. Several derivatives of rhizoferrin have been prepared by directed fermentation (DRECHSEL et al., 1995a). Thus the addition of analogous diamino acids or diamines to the culture medium yielded rhizoferrin analogs with either shorter (norrhizoferrin) or longer diamino residues (homorhizoferrin). Even substituted amines were incorporated leading to 2-ketorhizoferrin and oxarhizoferrin. Variation of the citryl residues of rhizoferrin by addition of tricarballylic acid to the culture medium yielded the corresponding mono- and didesoxy rhizoferrin derivatives. As the latter alterations affect the iron-liganding properties of the rhizoferrins, the chelate stability with iron should be reduced. Uptake of iron rhizoferrin and its metal analogs (Cr, Rh, Ga) was studied in *Absidia spinosa* (Mucorales) and in *Morganella morganii* (Proteeae) (CARRANO et al., 1996). While uptake kinetics in the fungus could be explained by an active transport system via a shuttle mechanism, growth promotion assays with *M. morganii* could not easily be explained and were dependent on a number of factors including the nature of the chelating agents used to induce iron deficiency. Two genes, *rumA* and *rumB* (**r**hizoferrin **u**ptake into ***M**organella*), encoding an outer membrane protein and a periplasmic protein, respectively, have been identified in *M. morganii* (KÜHN et al., 1996).

5 Miscellaneous Compounds

Several miscellaneous compounds have been isolated from fungi containing iron-binding catecholate, phenolate, or keto hydroxy bidentate ligands. Most of these compounds originate from the shikimate–chorismate pathway and represent condensation products of tyrosine. They have been characterized earlier as pigments from fungi, mainly of the Boletales, e.g., *Boletus, Suillus, Paxillus, Hydnellum, Polyporus,* and *Xerocomus* (GILL and STEGLICH, 1987). The blueing of members of the Boletaceae, and the red and yellow stains of several *Agaricus* and *Cortinarius* species have long been used as taxonomic characteristics by fungal taxonomists. In many cases quinones react by the combined action of alkali, oxygen, or ferric choride.

Thus compounds like the terphenyl quinones (polysporic acid, atromentin, leucomelone, and variegatin) (Fig. 37) possess both keto hydroxy and phenolic iron binding functionalities. Internal cyclization leads to cycloleucomelone, cyclovariegatin, and thelephoric acid (Fig. 38) having only phenolic groups. Another series of compounds are the pulvinic acids (vulpinic acid, atromentic acid, xerocomic acid, and variegatic acid) (Fig. 39) which, depending on the kind and number of

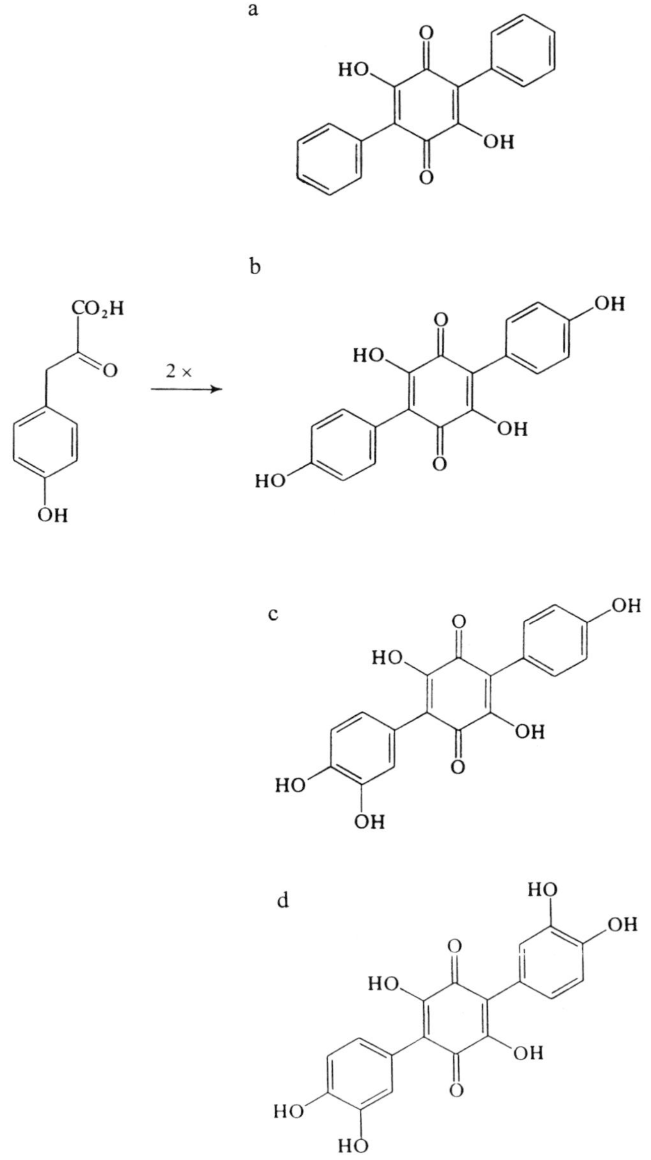

Fig. 37a. Polysporic acid; **b** atromentin; **c** leucomelone; **d** variegatin.

functional groups, possess good iron-binding properties. Because of the quinoid and polyphenolic structure these compounds can be regarded as iron-binding pigments, but their involvement in iron transport has still to be proven.

A novel pyridinone-type iron-binding compound has recently been isolated from strains of the fungus *Tolypocladium* named tolypocin (HL) (Fig. 40) and terricolin (FeL_3), respectively (JEGOROV et al., 1993). Members of the entomopathogenic genus *Tolypocladium* are known for their ability to produce cyclosporin A and a variety of pigments. Although the iron complex (terricolin) has been characterized by X-ray structure determination, iron transport properties have not been reported. The absolute configuration of terri-

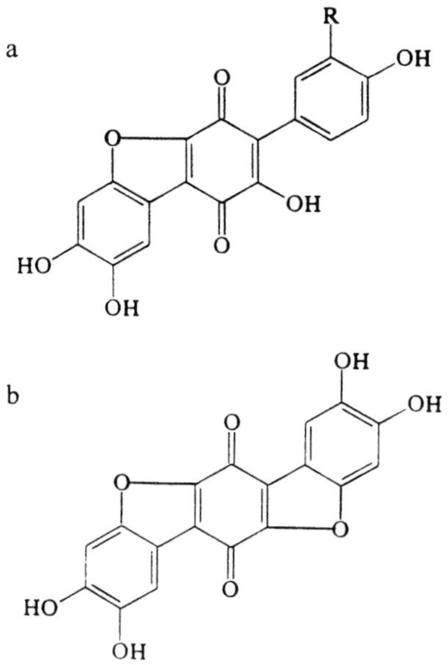

Fig. 38a. Cycloleucomelone: R=H; cyclovariegatin: R=OH; **b** thelephoric acid.

colin has been shown to be *lambda-cis* both in solution and in the solid state.

Pulcherriminic acid (Fig. 41a) and its iron complex pulcherrimin have been isolated from *Candida pulcherrima* (KLUYVER et al., 1953). Pulcherrimin, obtained by extraction of the cells with methanolic KOH as an amorphous red powder represents a polymer of dihydroxy-pyrazine dioxide which is insoluble in water and may be regarded as an iron storage compound rather than a siderophore.

Several related compounds, like the antibiotic mycelianamide (Fig. 41b) from *Penicillium griseofulvum* (OXFORD and RAISTRICK, 1948) or aspergillic acid (Fig. 41c), hydroxy aspergillic acid, muta-aspergillic acid, and neo-aspergillic acid from culture fluids of *Aspergillus flavus* and *A. oryzae* (DUTCHER, 1958) were isolated. The latter compounds are water-soluble and showed growth promotion activity with *A. flavescens* JG. However, siderophore activity in the producing strains has not been demonstrated.

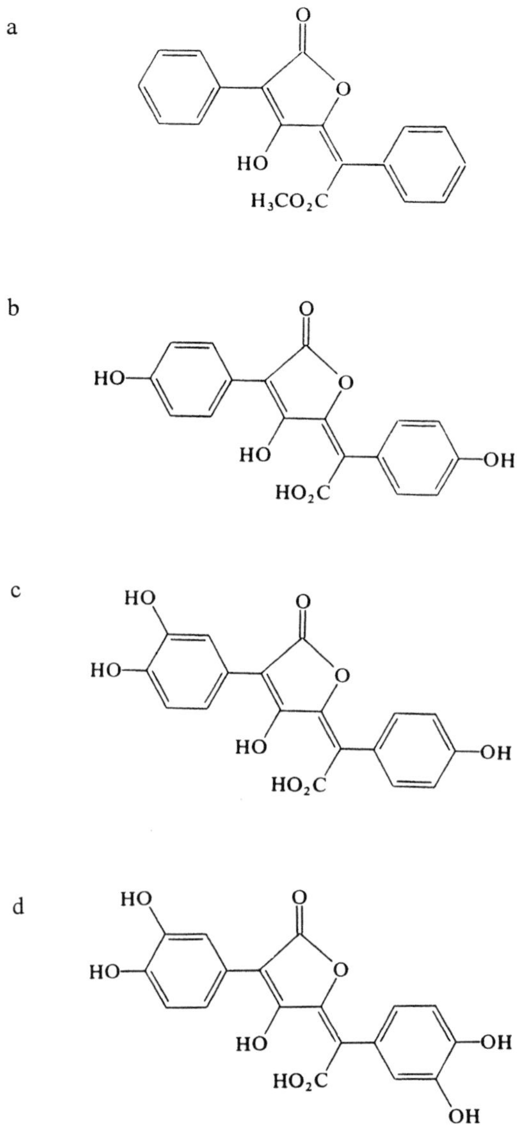

Fig. 39a. Vulpinic acid; **b** atromentic acid; **c** xerocomic acid; **d** variegatic acid.

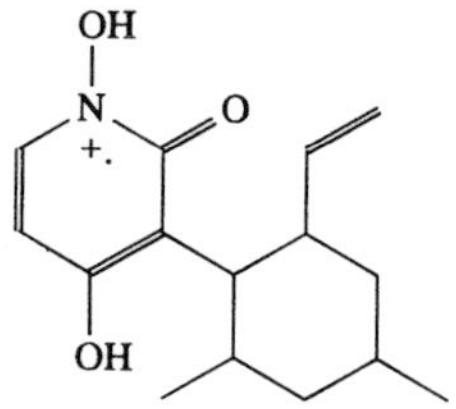

Fig. 40. Tolypocin.

a

b

c

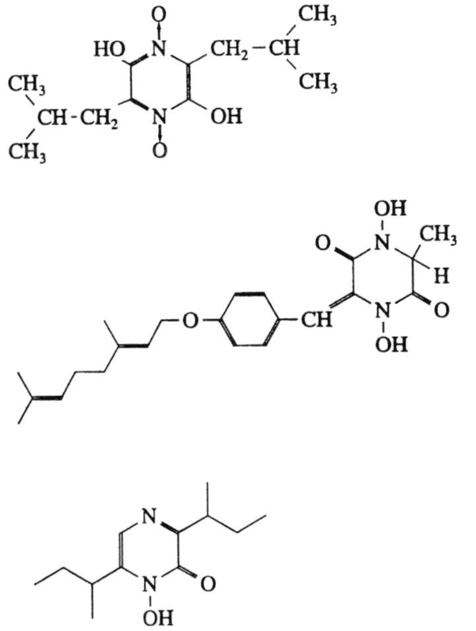

Fig. 41a. Pulcherrimic acid; **b** mycelianamide; **c** aspergillic acid.

6 Transport Mechanisms

To determine whether or not an iron-binding compound can be regarded as a siderophore, its physiological function and genetic regulation must be studied. An increasing number of microbial iron transport systems are currently being identified and characterized at the molecular level. Siderophore transport in enterobacteria and pseudomonads has been reviewed by BRAUN and HANTKE (1991) and MARUGG and WEISBEEK (1991). Fungal iron transport, although mainly characterized in terms of structure–function relationship (VAN DER HELM and WINKELMANN, 1994) is coming to the same molecular level of understanding that has been achieved in bacterial systems (MEI and LEONG, 1994; LESUISSE and LABBE, 1994). Siderophore-mediated iron transport has been studied in bacteria and fungi using radio-labeled siderophores in wild type strains and mutants. In principle, three mechanisms of uptake into microorganisms are conceivable:

(1) Diffusion of siderophores across the membranes along a diffusion gradient.
(2) Active transport of the entire iron chelate into the cell and intracellular reductive removal of iron with degradation or expulsion of the free ligand.
(3) Transport of siderophores only to the membrane surface, with or without reduction of iron, and donation of iron to the cell interior without penetration of the complex or ligand into the cell (taxi model). Indeed, all three modes of siderophore uptake have been shown to occur in certain microorganisms under certain experimental conditions (WINKELMANN, 1991a).

Siderophore iron transport in fungi has recently been reviewed with emphasis on the conditions and precautions of kinetic measurements (WINKELMANN, 1993). Generally, there are several approaches to unravel the mechanisms of fungal iron transport: radioactive labeling, photoaffinity labeling, and spectroscopy. Using ^{55}Fe- and ^{14}C-labeled siderophores four types of transport have been defined (VAN DER HELM and WINKELMANN, 1994): (1) a shuttle mechanism (e.g., ferrichromes in *Neurospora*); (2) a taxi cab mechanism (e.g., Fe-rhodotorulate in *Rhodotorula*); (3) a hydrolytic mechanism (e.g., fusarinines in *Mycelia sterilia*); and (4) a reductive mechanism (siderophores and ferric ions in *Saccharomyces* and other non-siderophore containing yeasts). Using a photoaffinity label (*p*-azidobenzoyl coprogen) 50% irreversible inhibition of coprogen and ferrichrome uptake was measured in *Neurospora crassa* (BAILEY et al., 1986) – an argument in favor of a common transport system for these siderophores. These results correspond to earlier transport kinetics showing a common transport system but different siderophore receptors (HUSCHKA et al., 1985). Although a variety of siderophores may be recognized by fungal siderophore receptors, recognition is highly stereoselective as determined by comparative studies with ferrichrome and *enantio*-ferrichrome (WINKELMANN, 1979; WINKELMANN and BRAUN, 1981). These results confirm that fungal siderophore transport is not a diffusion-controlled process but requires specific interactions with components of the transport system, although proteins of fungal siderophore transport systems have not been identified so far.

There are several reports on the application of EPR and Mössbauer spectroscopic measurements to fungal siderophore iron uptake (MATZANKE and WINKELMANN, 1981; ECKER et al., 1982; MATZANKE et al., 1988) supporting the view that iron undergoes subsequent reductive removal and that under certain conditions siderophores can also serve as iron storage compounds (see also the review of MATZANKE, 1994). Interestingly, siderophores are also found in conidiospores and seem to have an important function during sporulation and germination (HOROWITZ et al., 1976; MATZANKE et al., 1987; WINKELMANN, 1991b). Up to now only few reports on the transport of carboxylate type siderophores, like rhizoferrin, in fungi are available (DRECHSEL, et al., 1991; THIEKEN and WINKELMANN, 1992). The mechanism of uptake in fungi of the Mucorales (Zygomycetes) is still under investigation.

The mechanism of hydroxamate iron transport in enterobacteria was examined in early studies with *E. coli* K-12 and *Salmonella typhimurium* LT-2 mutants defective in enterobactin synthesis using ^{55}Fe- and ^{3}H-labeled ferrichrome and their chromic analogs (LEONG and NEILANDS, 1976). These investigations already showed that ferrichrome is taken up as an intact chelate in both *E. coli* and *Salmonella.* Later studies revealed that the ferrichrome ligand is modified by acylation of the hydroxamate *N*-OH groups after iron delivery (HARTMANN and BRAUN, 1980) thereby reducing the stability of ferrichrome with iron within the cell. Our current knowledge of ferrichrome uptake in *E. coli* has increased considerably by a detailed genetic analysis of the various membrane components involved (BRAUN and HANTKE, 1991). In gram-negative bacteria, the transport of iron bound to hydroxamate siderophores generally depends on the expression of outer membrane receptor proteins. Ferrichrome, e.g., is taken up via the FhuA receptor protein (78 kDa) localized in the outer membrane of gram-negative bacteria. FhuA also binds the phages T1, T5, Φ80, to colicin M, and the antibiotic albomycin (HANTKE and BRAUN, 1975). Although ferrichrome is first bound to the FhuA receptor, it has to be processed further by additional gene products (FhuB, C, D) to permit completion of siderophore transport into the cell (KÖSTER, 1991). This mechanism of nutrient translocation is known as a periplasmic binding-dependent transport which earlier was described for other transport processes (AMES, 1986).

Different ferric hydroxamate uptake (Fhu) outer membrane receptors of *E. coli* have been shown to mediate the transport of various fungal siderophores, e.g., ferrichromes, coprogen, and rhodotorulic acid (HANTKE, 1983). Surprisingly, the fungal siderophores coprogen and Fe rhodotorulate enter the cells of *E. coli* by the same outer membrane receptor protein (FhuE) as the bacterial ferrioxamines, whereas members of enterobacterial *Erwina herbicola* (*Pantoea herbicola*) use a separate outer membrane receptor FoxA to transport ferrioxamines (BERNER and WINKELMANN, 1990).

While FhuD is located in the periplasmic space, FhuB and FhuC are integrated in the cytoplasmic membrane. All *fhu* genes are organized in an operon; they have been sequenced, and the deduced molecular weights of the polypeptides are in accordance with those found by SDS-PAGE. Although the hydroxamate siderophores enter the cells by different outer membrane receptors, e.g., ferrichromes (FhuA), coprogens (FhuE) and Fe rhodotorulate (FhuE), aerobactin (Iut), and ferrioxamines (FoxA), their transport across the cytoplasmic membrane is accomplished by the same transport proteins (FhuD, B, C). In addition, a TonB protein is essential for an "energized state" of the outer membrane receptors as shown by *tonB* mutations which prevent correct interaction of the TonB protein with outer membrane receptor proteins (BRAUN et al., 1991). Thus TonB is regarded as a coupling device between the outer and inner membrane. The interaction of TonB with the receptor proteins requires the presence of a common binding domain which is characterized on the genetic level by a consensus sequence, the so-called "TonB box", present in all TonB-dependent receptor proteins (BRAUN and HANTKE, 1991).

ERNST et al. (1978) were the first to observe a constitutive production of siderophores in a *S. typhimurium* mutant and coined the term *fur* (ferric uptake regulation) for the gene responsible for the regulation of siderophore biosynthesis and expression of the cognate receptor proteins. The *fur* gene later was also found in *E. coli* and was subsequently cloned and sequenced (HANTKE, 1984). The corresponding protein was characterized by WEE et al. (1988). It turned out that all *fhu* genes and other siderophore receptor genes (enterobactin, aerobactin, ferrioxamines) are regulated by iron via the *fur* gene product. This also applies to the biosynthetic genes of siderophores.

Uptake of ferrichrome in fungi is stereospecific, as proven with synthetic *enantio*-ferrichrome which has the opposite chirality (WINKELMANN and BRAUN, 1981). However, uptake of *enantio*-ferrichrome (*delta* configuration) in *E. coli* was still 50% as compared to uptake of the natural ferrichrome (*lambda* configuration), indicating either an incomplete recognition or additional routes for iron uptake. Enantioselectivity seems to be more pronouced with the cloned *Yersinia* ferrichrome receptor protein (FcuA) accepting ferrichrome but excluding *enantio*-ferrichrome completely (BÄUMLER and HANTKE, 1992). A recent analysis of the siderophore specificity of different Fhu receptors in *E. coli* has revealed that certain ferrichromes like ferrichrysin and ferrirubin may not only enter via FhuA but may also use the FhuE receptor (KILLMANN and BRAUN, 1992).

7 Conclusion and Perspectives

The knowledge of siderophores and their cognate iron transport systems in microorganisms is crucial for an understanding of basic events, such as growth, metabolic activity, host invasion, and virulence. In all cases iron nutrition is a prerequisite. The diversity of siderophores is remarkable and can only be discussed in relation to the structural requirements of receptor and binding proteins involved. There is a structure–function relationship between iron-containing siderophores and the surface of transport proteins. Conformation of siderophores seems to be the most important factor. In gram-negative bacteria two membranes and a periplasmic space have to be penetrated before iron can be utilized by the cell metabolism. Although molecular genetics has become the most powerful tool in analyzing the transport of siderophores, many open questions remain, e.g.: Which part of the siderophore molecule is recognized and what kind of interaction is to be expected? Should the outer membrane receptor protein be a channel protein through which the siderophore can pass with specific interaction, as has been inferred from recent results (KILLMANN et al., 1993; RUTZ et al., 1993)? How is the outer membrane receptor gated or energized by the TonB protein? What kind of interactions prevail at the periplasmic binding proteins (FhuD) and, finally, how does the translocation through the inner cytoplasmic membrane proceed?

In fungi similar structure–function relationships have been documented and allocated to putative cytoplasmic membrane proteins. However, since siderophore transport mutants are not available, membrane proteins involved in siderophore binding and transport have never been identified. However, the principal mechanisms of interaction with specific membrane proteins also seem to be valid in fungal siderophore transport. The basic question as to whether the entire chelate molecule or only the iron atom enters the cell has been solved with double-labeled siderophores in various fungi. According to our present knowledge only some siderophores, mainly ferrichromes, can enter the fungal cell and remain intact. Several other siderophores need to interact with membrane receptors to allow exchange of iron. In every case, however, specific interaction is observed. Thus the essential features of structure–function relationship among bacterial and fungal siderophore systems seem to be identical.

Future applications of siderophores are manifold. The lag phase of slowly growing microorganisms, e.g., might be shortened to a certain extent by simply adding very small amounts of genus-specific siderophores. Moreover, addition of siderophores to biological fluids or food may overcome iron restriction due to the iron-withholding properties of iron-binding proteins (SOUKKA et al., 1992) and may thus allow outgrowth and detection of otherwise dormant microorganisms. The use of genetically engineered organisms that possess improved iron transport systems recognizing specific siderophores may be an alternative approach to enhance growth and metabolite production. Also the use of non-utilizable or antibiotic-linked siderophores (NIKAIDO and ROSENBERG, 1990) designed for reducing the bioavailablity of iron by contaminating bacteria might be an interesting future biotechnological application. The best examples of siderophore application can be found in nature where biocontrol and host defense mechanisms based on nutritional iron deficiency are active in a finely tuned manner without negative effects on the surrounding ecosystems. The biotechnological applications of siderophores to medicine, agriculture, the environment, and the food industry are numerous. For example, the affinity of siderophores for toxic metals, such as chromium and plutonium, suggests that siderophores may be useful agents for remediation of polluted environments. Likewise, the development of orally effective siderophores suitable for iron chelation therapy through directed fermentation is a virtually unexplored area of investigation. A possible role of bacterial siderophores in reducing inflammation (COFFMANN et al., 1990) as well as protecting cells from free radical damage has also been suggested.

However, addressing of possible applications of siderophores in more detail is beyond the scope of this chapter. The authors are convinced that the future will bring exciting advances in this area. The present review aims at understanding the wealth of the structural diversity of siderophores, their production, and mode of action and will hopefully provide a stimulus for future research in microbial physiology, ecology, and fields of applied biotechnology.

8 References

ABDALLAH, M. A. (1991), Pyoverdins and pseudobactins, in: *Handbook of Microbial Iron Chelates* (WINKELMANN, G., Ed.), pp. 139–153, Boca Raton, FL: CRC Press.

ACTIS, L. A., FISH, W., CROSA, J. H., KELLERMAN, K., ELLENBERGER, S. R., HAUSER, F. M., SANDERS-LÖHR, J. (1986), Characterization of anguibactin, a novel siderophore from *Vibrio anguillarum* 775 (pJM1), *J. Bacteriol.* **167**, 57–65.

ACTIS, L. A., TOMALSKY, M. E., FARREL, D. H., CROSA, J. H. (1988), Genetic and molecular characterization of essential components of the *Vibrio anguillarum* plasmid-mediated iron transport system, *J. Biol. Chem.* **263**, 2853–2860.

AMES, G. F. L. (1986), Bacterial periplasmic transport systems: structure, mechanism, and evolutio, *Ann. Rev. Biochem.* **55**, 397–425.

ANKE, H. (1977), Metabolic products of microorganisms: desferritriacetylfusigen, an antibiotic from *Aspergillus deflectus, J. Antibiot.* **30**, 125–128.

ARCENEAUX, J. E. L., DAVIS, W. B., DOWNER, D. N., HAYDON, A. H., BYERS, B. R. (1973), Fate of labeled hydroxamates during iron transport from hydroxamate-iron chelates, *J. Bacteriol.* **115**, 919–927.

ARONSON, D. B., BOYER, G. L. (1994), Growth and siderophore formation in six iron-limited strains of *Frankia, Soil Biol. Biochem.* **265**, 561–567.

ATKIN, C. L., NEILANDS, J. B. (1968), Rhodotorulic acid, a diketopiperazine dihydroxamic acid with growth factor activity: isolation and characterization, *Biochemistry* **7**, 3734–3739.

ATKIN, C. L., NEILANDS, J. B., PHAFF, H. J. (1970), Rhodotorulic acid from species of *Leucosporidium, Rhodosporidium, Rhodotorula, Sporidiobolus,* and *Sporobolomyces* and a new alanine containing ferrichrome from *Cryptococcus melibiosum, J. Bacteriol.* **103**, 722–733.

BAGG, A., NEILANDS, J. B. (1987), Molecular mechanism of regulation of siderophore-mediated iron assimilation, *Microbiol. Rev.* **51**, 509–518.

BAIIEY, C. T., KIME-HUNT, E. M., CARRANO, C. J., HUSCHKA, H., WINKELMANN, G. (1986), A photoaffinity label for the siderophore-mediated iron transport system in *Neurospora crassa, Biochim. Biophys. Acta* **883**, 299–305.

BARGHOUTHI, S., YOUNG, R., ARCENEAUX, J. E. L., BYERS, B. R. (1989), Physiological control of amonabactin biosynthesis in *Aeromonas hydrophila, BioMetals* **2**, 155–160.

BARGHOUTHI, S., PAYNE, S. M., ARCENEAUX, J. E. L., BYERS, B. R. (1991), Cloning, mutagenesis, and nucleotide sequence of a siderophore biosynthetic gene (*amoA*) from *Aeromonas hydrophila, J. Bacteriol.* **173**, 5121–5128.

BARTON, L. L., HEMMING, B. C. (Eds.) (1993), *Iron Chelation in Plants and Soil Microorganisms.* San Diego, CA: Academic Press.

BÄUMLER, A. J., HANTKE, K. (1992), Ferrioxamine uptake in *Yersinia enterocolitica:* characterization of the receptor protein FoxA, *Mol. Microbiol.* **6**, 1309–1321.

BENZ, G., SCHRÖDER, T., KURZ, J., WÜNSCHE, C. K. W., STEFFENS, G., PFITZNER, J., SCHMIDT, D. (1982), Konstitution der Desferriform der Albomycine δ_1, δ_2, ε, *Angew. Chemie* **94**, 552–553.

BERGERON, R. J., MCMANIS, J. S. (1991), Synthesis of catecholamide and hydroxamate siderophores, in: *Handbook of Microbial Iron Chelates* (WINKELMANN, G., Ed.), Boca Raton, Fl: CRC Press.

BERGERON, R. J., DIONIS, J. B., ELLIOT, G. T., KLINE, S. J. (1985), Mechanism and stereospecificity of the parabactin-mediated iron transport system in *Paracoccus denitrificans, J. Biol. Chem.* **260**, 7936–7944.

BERNER, I., WINKELMANN, G. (1990), Ferrioxamine transport mutants and the identification of the ferrioxamine receptor protein (FoxA) in *Erwinia herbicola (Enterobacter agglomerans), BioMetals* **2**, 197–202.

BERNER, I., KONETSCHNY-RAPP, S., JUNG, G., WINKELMANN, G. (1988), Characterization of ferrioxamine E as the principal siderophore of *Erwinia herbicola (Enterobacter agglomerans), BioMetals* **1**, 51–56.

BERNER, I., GREINER, M., METZGER, J., JUNG, G., WINKELMANN, G. (1991), Identification of enterobactin and linear dihydroxybenzoylserine compounds by HPLC and ion spray mass spectrometry (LC/MS and MS/MS), *BioMetals* **4**, 113–118.

BERNHARDINI, J., LINGET-MORICE, C., HOH, F., COLLINSON, S. K., KYSLIK, P., PAGE, W. J., DELL, A., ABDALLAH, M. A. (in press), Bacterial siderophores: Structure elucidation and ^{1}H, ^{13}C, ^{15}N two-dimensional NMR assignments of azotoverdin and related siderophores synthesized by *Azomonas macrocytogenes* ATCC 12334, *BioMetals* **9**, 107–120.

BICKEL, H., BOSSHARDT, R., GÄUMANN, E., REUSSER, P., VISCHER, E., VOSER, W., WETTSTEIN, A., ZÄHNER, H. (1960), Stoffwechselprodukte von Actinomyceten. Über die Isolierung und Charakterisierung der Ferrioxamine A–F, neuer Wuchsstoffe der Sideramin-Gruppe, *Helv. Chim. Acta* **43**, 2118–2128.

BINDEREIF, A., BRAUN, V., HANTKE, K. (1982), The cloacin receptor of ColV-bearing *Escherichia coli* is part of the Fe^{3+}-aerobactin transport system, *J. Bacteriol.* **150**, 1472–1475.

BITTER, W., MARUGG, J. D., DE WEGER, L. A., TOMMASSEN, J., WEISBEEK, P. J. (1991), The ferric-pseudobactin receptor PupA of *Pseudomonas putida* WCS 358: homology to TonB-dependent *Escherichia coli* receptors and specificity of the protein, *Mol. Microbiol.* **5**, 647–655.

BOELAERT, J. R., FENVES, A. Z., COBURN, J. W. (1991), Deferoxamine therapy and mucormycosis in dialysis patients: report of an international registry, *Am. J. Kidney Dis.* **18**, 660–667.

BOELAERT, J. R., DELOCHT, M., VAN CUTSEM, J., KERBELS, V., CANTINIEAUX, B., VERDONCK, A., VAN LANDUYT, SCHNEIDER, Y.-J. (1993), Mucormycosis during desferoxamine therapy is a siderophore-mediated infection. *In vitro* and *in vivo* animal studies, *J. Clin. Invest.* **91**, 1979–1986.

BOELAERT, J. R., VAN CUTSEM, J., DELOCHT, M., SCHNEIDER, Y.-J., CRICHTON, R. R. (1994), Deferoxamine augments growth and pathogenicity of *Rhizopus,* while hydroxypyridinone chelators have no effect, *Kidney Int.* **45**, 667–671.

BOYER, G. L., ARONSON, D. B. (1994), Iron uptake and siderophore formation in the actinorhizal symbiont *Frankia,* in: *Biochemistry of the Metal Micronutrients in the Rhizosphere* (MANTHEY, J. A., CROWLEY, D. E., LUSTER, D. G., Eds.), pp. 41-54. Boca Raton, FL: CRC Press.

BRAUN, V. (1981), *Escherichia coli* cells, containing the plasmid ColV produce the iron ionophore aerobactin, *FEMS Microbiol. Lett.* **11**, 225–228.

BRAUN, V., HANTKE, K. (1991), Genetics of bacterial iron transport, in: *Handbook of Microbial Iron Chelates* (WINKELMANN, G., Ed.), pp. 107–138. Boca Raton, FL: CRC Press.

BRAUN, V., GÜNTER, C., HANTKE, K. (1991), Transport of iron across the outer membrane, *BioMetals* **4**, 14–22.

BRICKMAN, T. J., HANSEL, J.-G., MILLER, M. J., ARMSTRONG, S. K. (1996), Purification spectroscopic analysis, and biological activity of the macrocyclic dihydroxamate siderophore alcaligin produced by *Bordetella pertussis* and *Bordetella bronchiseptica, BioMetals* **9**, 191–203.

BRISKOT, G., TARAZ, K., BUDZIKIEWICZ, H. (1989), Pyoverdin-type siderophores from *Pseudomonas aeruginosa, Liebigs Ann. Chem.* **1989**, 375–384.

BUDDE, A. D., LEONG, S. A. (1989), Characterization of siderophores from *Ustilago maydis, Mycopathologia* **108**, 125–133.

BUDZIKIEWICZ, H. (1993), Secondary metabolites from fluorescent pseudomonads, *FEMS Microbiol. Rev.* **104**, 209–228.

BUDZIKIEWICZ, H., SCHAFFNER, E. M., TARAZ, K. (1992), A novel azotobactin from *Azotobacter vinelandii, Nat. Prod. Lett.* **1**, 9–14.

BUTTERTON, J. R., STOEBNER, J. A., PAYNE, S. M., CALDERWOOD, S. B. (1992), Cloning, sequencing, and transcriptional regulation of *vizA*, the gene encoding the ferric vibriobactin receptor of *Vibrio cholerae, J. Bacteriol.* **174**, 3729–3738.

BUYER, J. S., DELORENZO, V., NEILANDS, J. B. (1991), Production of the siderophore aerobactin by a halophilic Pseudomonad, *Appl. Environ. Microbiol.* **57**, 2246–2250.

BYERS, B. R., POWELL, M. V., LANKFORD, C. E. (1967), Iron-chelating hydroxamic acid (schizokinen) active in initiation of cell division in *Bacillus megaterium, Bacteriol.* **93**, 286–294.

CARRANO, C. J., RAYMOND, K. N. (1978), Coordination chemistry of microbial iron transport compounds: Rhodotorulic acid and iron uptake in *Rhodotorula pilimanae, J. Bacteriol.* **136**, 69–74.

CARRANO, C., THIEKEN, A., WINKELMANN, G. (1996), Specificity and mechanism of rhizoferrin mediated metal ion uptake, *BioMetals* **9**, 185–189.

CASS, M. E., GARRETT, T. M., RAYMOND, K. N. (1989), The salicylate mode of bonding in protonated ferric enterobactin analogs, *J. Am. Chem. Soc.* **111**, 1677–1682.

CHAMBERS, C. E., MCINTYRE, D. D., MOUCK, M., SOKOL, P. (1996), Physical and structural characterization of yersiniophore, a siderophore produced by clinical isolates of *Yersinia enterocolitica, BioMetals* **9**, 157–168.

CHAMBERS, C. E., SOKOL, P. A. (1994), Comparison of siderophore production and utilization in pathogenic and environmental isolates of *Yersinia enterocolitica, J. Clin. Microbiol.* **32**, 32–39.

CHUNG, T. D. Y., MATZANKE, B. F. M., WINKELMANN, G., RAYMOND, K. N. (1986), The inhibitory effect of the partially resolved coordination isomers of chromic desferricoprogen on coprogen uptake in *Neurospora crassa, J. Bacteriol.* **165**, 283–287.

COFFMAN, T., COX, C. D., EDECKER, B. L., BRITIGAN, B. E. (1990), Possible role of bacterial siderophores in inflammation, *J. Clin. Invest.* **86**, 1030–1037.

CORBIN, J. L., BULEN, W. A. (1969), The isolation and identification of 2,3-dihydroxybenzoic acid and 2-N, 6-N-di (2,3-dihydroxybenzoyl)-L-lysine formed by iron-deficient *Azotobacter vinelandii, Biochemistry* **8**, 757–762.

CORBIN, J. L., KARLE, I. L., KARLE, J. (1970), Crystal structure of the chromophore from the fluorescent peptide produced by iron-deficient *Azotobacter vinelandii, J. Chem. Soc., Chem. Commun.* D 186–187.

CORNISH, A. S., PAGE, W. J. (1995), Production of the tricatecholate siderophore protochelin by *Azotobacter vinelandii, BioMetals* **8**, 332–338.

COX, C. D., RINEHART, K. L., MOORE, M. L., COOK, J. C. (1981), Pyochelin: novel structure of an iron-chelating growth promotor for *Pseudomonas aeruginosa, Proc. Nat. Acad. Sci. USA* **78**, 4256–4260.

CROSA, J. H. (1989), Genetics and molecular biology of siderophore-mediated iron transport in bacteria, *Microbiol. Rev.* **53**, 517–530.

CRUMBLISS, A. L. (1991), Aqueous solution equilibrium and kinetic studies of iron siderophore and model siderophore complexes, in: *Handbook of Microbial Iron Chelates* (WINKELMANN, G., Ed.), pp. 177–233. Boca Raton, FL: CRC Press.

DELORENZO, V., BINDEREIF, A., PAW, B. H., NEILANDS, J. B. (1986), Aerobactin biosynthesis and transport genes of plasmid ColV-K30 in *Escherichia coli* K-12, *J. Bacteriol.* **165**, 570–578.

DEMANGE, P., BATEMAN, A., DELL, A., ABDALLAH, M. A. (1988), Structure of azotobactin D, a siderophore of *Azotobacter vinelandii* strain D (CCM 289), *Biochemistry* **27**, 2745–2752.

DEMANGE, P., BATEMAN, A., MACLEOD, J. K., DELL, A., ABDALLAH, M. A. (1990a), Bacterial siderophores: Unusual 3,4,5,6-Tetrahydropyrimidine-based amino acids in pyoverdins from *Pseudomonas fluorescens, Tetrahedron Lett.* **31**, 7611–7614.

DEMANGE, P., BATEMAN, A., MERZ, C., DELL, A., PIÉMONT, Y., ABDALLAH, M. (1990b), Bacterial siderophores: Structures of pyoverdines Pt, siderophores of *Pseudomonas tolaassii* NCPPB 2192, and pyoverdins Pf, siderophores of *Pseudomonas fluorescens* CCM 2789. Identification of an unusual natural amino acid, *Biochemistry* **29**, 11041–11051.

DEML, G. (1985), Studies in heterobasidiomycetes, Part 34. A survey on siderophore formation in low-iron cultured smuts from the floral parts of Polygonaceae, *Syst. Appl. Microbiol.* **6**, 23–24.

DEML, G., VOGES, K., JUNG, G., WINKELMANN, G. (1984), Tetraglycylferrichrome – the first heptapeptide ferrichrome, *FEBS Lett.* **173**, 53–57.

DEVOS, P., VANDE WOESTYNE, M., VANCANNEYT, M., VERSTRAETE, W., KERSTERS, K. (1993), Identification of proferrorosamine producing *Pseudomonas* sp. strain GH (LMG 11358) as *Erwinia rhapontici, System. Appl. Microbiol.* **16**, 252–255.

DIEKMANN, H. (1967), Fusigen – ein neues Sideramin aus Pilzen, *Arch. Microbiol.* **58**, 1–5.

DIEKMANN, H. (1968), Stoffwechselprodukte von Mikroorganismen. 68. Die Isolierung von *trans*-5-Hydroxy-3-methylpenten-(2)-säure, *Arch. Microbiol.* **62**, 322–327.

DIEKMANN, H. (1970), Stoffwechselprodukte von Mikroorganismen, 81. Mitteilung. Vorkommen und Strukturen von Coprogen B und Dimerumsäure, *Arch. Microbiol.* **73**, 65–76.

DIEKMANN, H., KREZDORN, E. (1975), Stoffwechselprodukte von Mikroorganismen. 150. Ferricrocin, Triacetylfusigen und andere Sideramine aus Pilzen der Gattung *Aspergillus,* Gruppe *fumigatus, Arch. Microbiol.* **106**, 191–194.

DIEKMANN, H., ZÄHNER, H. (1967), Konstitution von Fusigen und dessen Abbau zu *delta*-2-Anhydromevalonsäurelacton, *Eur. J. Biochem.* **3**, 213–218.

DIONIS, J. B., JENNY, H., PETER, H. H (1991), Therapeutically useful iron chelators, in: *Handbook of Microbial Iron Chelates* (WINKELMANN, G., Ed.), pp. 339–356. Boca Raton, FL: CRC Press.

DRECHSEL, H., METZGER, J., FREUND, S., JUNG, G., BOELAERT, J., WINKELMANN, G. (1991), Rhizoferrin – a novel siderophore from the fungus *Rhizopus microsporus* var. *rhizopodiformis, BioMetals* **4**, 238–243.

DRECHSEL, H., JUNG, G., WINKELMANN, G. (1992), Stereochemical characterization of rhizoferrin and identification of its dehydration products, *BioMetals* **5**, 141–148.

DRECHSEL, H., FREUND, S., NICHOLSON, G., HAAG, H., JUNG, O., ZÄHNER, H., JUNG, G. (1993a), Purification and chemical characterization of staphyloferrin B, a hydrophilic siderophore from staphylococci, *BioMetals* **6**, 185–192.

DRECHSEL, H., THIEKEN, A., REISSBRODT, R., JUNG, G., WINKELMANN, G. (1993b), α-Keto acids are novel siderophores in the genera *Proteus, Providencia,* and *Morganella* and are produced by amino acid deaminases, *J. Bacteriol.* **175**, 2727–2733.

DRECHSEL, H., TSCHIERSKE, M., THIEKEN, A., ZÄHNER, H., JUNG, G., WINKELMANN, G. (1995), The carboxylate-type siderophore rhizoferrin and its analogs produced by directed fermentation, *J. Ind. Microbiol.* **14**, 105–112.

DRECHSEL, H., STEPHAN, H., LOTZ, R., HAAG, H., ZÄHNER, H., HANTKE, K., JUNG, G. (1995), Structure elucidation of Yersiniabactin, a siderophore from highly virulent *Yersinia* strains, *Liebigs Ann. Chem.*, 1727–1733.

DUTCHER, J. D. (1958), Aspergillic acid: An antibiotic substance produced by *Aspergillus flavus.* III. The structure of hydroxyaspergillic acid, *J. Biol. Chem.* **232**, 785.

ECKER, D. J., LANCASTER, J. R., EMERY, T. (1982), Siderophore iron transport followed by electron paramagnetic resonance spectroscopy, *J. Biol. Chem.* **257**, 8623–8626.

EHLERT, G., TARAZ, K., BUDZIKIEWICZ, H. (1994), Serratiochelin, a new catecholate siderophore from *Serratia marcescens, Z. Naturforsch.* **49**c, 11–17.

ELKINS, M. F., EARHARDT, C. F. (1989), Nucleotide sequence and regulation of the *Escherichia coli* gene for ferrienterobactin transport protein FepB, *J. Bacteriol.* **171**, 5443–5451.

EMERY, T. (1971), Role of ferrichrome as a ferric ionophore in *Ustilago sphaerogena, Biochemistry* **10**, 1483–1488.

EMERY, T. (1976), Fungal ornithine esterases: relationship to transport, *Biochemistry* **15**, 2723–2728.

ENARD, C., DIOLEZ, A., EXPERT, D. (1988), Systemic virulence of *Erwinia chrysanthemi* on *Saintpaulia chrysanthemi* 3937 requires a functional iron assimilation system, *J. Bacteriol.* **170**, 2419–2426.

ENG-WILMOT, L. D., VAN DER HELM, D. (1980), Molecular and crystal structure of the linear tricatechol siderophore, agrobactin, *J. Am. Chem. Soc.* **102**, 7719–7725.

ENG-WILMOT, L. D., ADJIMANI, J.P., VAN DER HELM, D. (1992), Siderophore mediated iron-(III) transport in the mycelia of the cultivated fungus, *Agaricus bisporus, J. Inorg. Biochem.* **48**, 183–195.

ERNST, J. F., BENNET, R. L., ROTHFIELD, L. I. (1978), Constitutive expression of the iron-enterochelin and ferrichrome uptake systems in a mutant strain of *Salmonella typhimurium, J. Bacteriol.* **135**, 928–934.

FEISTNER, G. F. (1995), Suggestion for a new, semirational nomenclature for the free chelators of ferrioxamines, *BioMetals* **8**, 193–196.

FEISTNER, G., KORTH, H., KO, H., PULVERER, G., BUDZIKIEWICZ, H. (1983), Ferrorosamine A from *Erwinia rhapontici, Curr. Microbiol.* **8**, 239–243.

FEISTNER, G. J., STAHL, D. C., GABRIK, A. H. (1993), Proferrioxamine siderophores of *Erwinia amylovora.* A capillary liquid chromatographic/electrospray tandem mass spectrometric study, *Org. Mass Spectrom.* **28**, 163–175.

FRANCIS, J., MACTURK, H. M., MADINAVEITIA, J., SNOW, G. A. (1953), Mycobactin, a growth factor for *Mycobacterium johnei.* I. Isolation from *Mycobacterium phlei, Biochem. J.* **55**, 596–607.

FREDERICK, C. B., BENTLEY, M. D., SHIVE, W. (1981), The structure of triomicin, a new siderophore, *Biochemistry* **20**, 2436–2438.

FUKASAWA, K., GOTO, M., SASAKI, K., HIRATA,

Y., Sato, S. (1972), Structure of the yellow-green fluorescent peptide produced by iron-deficient *Azotobacter vinelandii* strain 0, *Tetrahedron* **28**, 5359–5365.

Gibson, F., Magrath, D. I. (1969), The isolation and characterization of a hydroxamic acid (aerobactin) formed by *Aerobacter aerogenes* 62-1, *Biochim. Biophys. Acta* **192**, 175–184.

Gill, M., Steglich, W. (1987), Pigments of fungi (Macromycetes), in: *Progr. Chem. of Organic Natural Products*, Vol. 51 (Herz, W., Grisebach, H., Kirby, G. W., Tamm, C., Eds.). Heidelberg: Springer-Verlag.

Gobin, J., Moore, C. H., Reeve, J. R. Wong, D. K., Gibson, B. W., Horwitz, M. A. (1995), Iron acquisition by *Mycobacterium tuberculosis:* isolation and characterization of a family of iron-binding exochelins, *Proc. Natl. Acad. Sci. USA* **92**, 5189–5193.

Griffith, G. L., Sigel, S. P., Payne, S. M., Neilands, J. B. (1984), Vibriobactin, a siderophore from *Vibrio cholerae, J. Biol. Chem.* **259**, 383–385.

Gross, R., Engelbrecht, Braun, V. (1985), Identification of the genes and their polypeptide products responsible for aerobactin synthesis by ColV plasmids, *Mol. Gen. Genet.* **201**, 204–212.

Haag, H., Hantke, K., Drechsel, H., Strojilkovic, I., Jung, G., Zähner, H. (1993), Purification of yersiniabactin: a siderophore and possible virulence factor of *Yersinia enterocolitica, J. Gen. Microbiol.* **139**, 2159–2165.

Hall, R. M., Ratledge, C. (1982), Mycobactins as chemotaxonomic characters for some rapidly growing mycobacteria, *J. Gen. Microbiol.* **130**, 1883–1892.

Hantke, K. (1983), Identification of an iron uptake system specific for coprogen and rhodotorulic acid in *Escherichia coli* K12, *Mol. Gen. Genet.* **191**, 301–306.

Hantke, K. (1984), Cloning of the repressor protein gene of iron regulated systems in *E. coli* K12, *Mol. Gen. Genet.* **197**, 337–341.

Hantke, K., Braun, V. (1975), A function common to iron-enterochelin transport and action of colicins B, I, V in *Escherichia coli, FEBS Lett.* **59**, 277–281.

Harris, W. R., Carrano, C. J., Raymond, K. N. (1979), Coordination chemistry of microbial iron transport compounds. 16. Isolation, characterization, and formation constants of ferric aerobactin, *J. Am. Chem. Soc.* **101**, 2722–2727.

Hartmann, A., Braun, V. (1980), Iron transport in *E. coli:* uptake and modification of ferrichrome, *J. Bacteriol.* **143**, 246–255.

Hartmann, A., Fiedler, H.-P., Braun, V. (1979), Uptake and conversion of the antibiotic albomycin by *Escherichia coli* K-12, *Eur. J. Biochem.* **99**, 517–524.

Haydon, A. H., Davis, W. B., Arceneaux, J. E. L., Byers, B. R. (1973), Hydroxamate recognition during iron transport from hydroxamate-iron chelates, *J. Bacteriol.* **115**, 912–918.

Hesemann, J. (1987), Chromosomal-encoded siderophores are required for mouse virulence of enteropathogenic *Yersinia* species, *FEMS Microbiol. Lett.* **18**, 229–233.

Hohnadel, D., Meyer, J.-M. (1988), Specificity of pyoverdin-mediated iron uptake among fluorescent *Pseudomonas* strains, *J. Bacteriol.* **170**, 4865–4873.

Horowitz, N. H., Charlang, G., Horn, G., Williams, N. P. (1976), Isolation and identification of the conidial germination factor of *Neurospora crassa, J. Bacteriol.* **127**, 135–140.

Hossain, M. B., Eng-Wilmot, D. L., Loghry, R. A., van der Helm, D. (1980), Circular dichroism, crystal structure, and absolute configuration of the siderophore ferric *N, N', N"*-triacetylfusarinine, *J. Am. Chem. Soc.* **102**, 5766–5773.

Hossain, M. B., Jalal, M. A. F., Benson, B. A., Barnes, C. L., van der Helm, D. (1987), Structure and conformation of two coprogen-type siderophores: Neocoprogen I and Neocoprogen II, *J. Am. Chem. Soc.* **109**, 4948–4954.

Huschka, H., Müller, G., Winkelmann, G. (1983), The membrane potential is the driving force for siderophore iron transport in fungi, *FEMS Microbiol. Lett.* **20**, 125–129.

Huschka, H., Naegeli, H., Leuenberger-Ryf, H., Keller-Schierlein, W., Winkelmann, G. (1985), Evidence for a common siderophore transport system but different siderophore receptors in *Neurospora crassa, J. Bacteriol.* **162**, 715–721.

Huscka, H., Jalal, M. A. F., van der Helm, D., Winkelmann, G. (1986), Molecular recognition of siderophores in fungi: Role of iron-surrounding N-acyl residues and peptide backbone during membrane transport in *Neurospora crassa, J. Bacteriol.* **167**, 1020–1024.

Ishimaru, C. A., Loper, J. E. (1992), High-affinity iron uptake systems present in *Erwinia carotovora* subsp. *carotovora* include the hydroxamate siderophore aerobactin, *J. Bacteriol.* **174**, 2993–3003.

Jalal, M. A. F., van der Helm, D. (1989), Purification and crystallization of ferric enterobactin receptor protein, FepA, from outer membranes of *Escherichia coli* UT5600/pBB2, *FEBS Lett.* **243**, 366–370.

Jalal, M. A. F., van der Helm, D. (1991), Isolation and spectroscopic identification of fungal siderophores, in: *Handbook of Microbial Iron Chelates* (Winkelmann, G., Ed.), pp. 235–269. Boca Raton, FL: CRC Press.

Jalal, M. A. F., Mocharla, R., Barnes, C. L., Hossain, M. B., Powell, D. R., Eng-Wilmot, D. L., Grayson, S. L., Benson, B. A., van der Helm, D. (1984), Extracellular siderophores from *Aspergillus ochraceus, J. Bacteriol.* **158**, 683–688.

Jalal, M. A. F., Galles, J. L., van der Helm, D. (1985), Structure of des(diserylglycyl)ferrirhodin, DDF, a novel siderophore from *Aspergillus ochraceus, J. Org. Chem.* **50**, 5642–5645.

Jalal, M. A. F., Love, S. K., van der Helm, D. (1986,) Siderophore mediated iron(III) uptake in *Gliocladium virens.* 1. Properties of *cis*-fusarinine, *trans*-fusarinine, dimerum acid, and their ferric complexes, *J. Inorg. Biochem.* **28**, 417–430.

Jalal, M. A. F., Love, S. K., van der Helm, D. (1987), Siderophore mediated iron(III) uptake in *Gliocladium virens.* 2. Role of ferric mono- and dihydroxamates as iron transport agents, *J. Inorg. Biochem.* **29**, 259–267.

Jalal, M. A. F., Hossain, M. B., van der Helm, D., Sanders-Löhr, J., Actis, L. A., Crosa, J. H. (1989), Structure of anguibactin, a unique plasmid-related bacterial siderophore from the fish pathogen *Vibrio anguillarum, J. Am. Chem. Soc.* **111**, 292–296.

Jegorov, A., Matha, V., Husak, M., Kratochvil, B., Stuchlik, J., Sedmera, P., Havlicek, V. (1993), Iron uptake system of some members of the genus *Tolypocladium:* Crystal structure of the ligand and its iron(III) complex, *J. Chem. Soc. Dalton Trans.* 1993, 1287–1293.

Katayama, N., Nozaki, Y., Okonogi, K., Harada, S, Ono, H. (1993), Ferrocins, new iron-containing peptide antibiotics produced by bacteria. Taxonomy, fermentation and biological activity, *J. Antibiot.* **46**, 65–70.

Keller-Schierlein, W., Diekmann, H. (1970), Zur Konstitution des Coprogens, *Helv. Chim. Acta* **53**, 2035–2044.

Keller-Schierlein, W., Prelog, V., Zähner, H. (1964), Siderochrome, in: *Fortschritte der Chemie organischer Naturstoffe* (Zechmeister, L., Ed.), pp. 279–322. Heidelberg: Springer-Verlag.

Keller-Schierlein, W., Huber, P., Kawaguchi, H. (1984), Chemistry of danomycin, an iron-containing antibiotic, in: *Natural Products and Drug Development* (Krogsgaard-Larsen, P., Brogger Christensen, S., Kofod, H., Eds.), pp. 213–227. Copenhagen: Munksgaard.

Keller-Schierlein, W., Hagmann, L., Zähner, H., Huhn, W. (1988), Stoffwechselprodukte von Mikroorganismen. 167. Maduraferrin, ein neuartiger Siderophor aus *Actinomadura madurae, Helv. Chim. Acta* **71**, 1528–1540.

Killmann, H., Braun, V. (1992), An aspartate deletion mutation defines a binding site of the multifunctional FhuA outer membrane receptor of *Escherichia coli, J. Bacteriol.* **174**, 3479–3486.

Killmann, H., Benz, R., Braun, V. (1993), Conversion of the FhuA transport protein into a diffusion channel through the outer membrane of *Escherichia coli, EMBO J.* **12**, 3007–3016.

Kluyver, A. J. J., van der Walt, J. P., van Triet, A. J. (1953), Pulcherrimin, the pigment of *Candida pulcherrima, Proc. Natl. Acad. Sci. U.S.A.* **39**, 583.

Knosp, O., Tigerstrom, M., Page, W. J. (1984), Siderophore-mediated uptake of iron in *Azotobacter vinelandii, J. Bacteriol.* **159**, 341–347.

Konetschny-Rapp, S., Huschka, H., Winkelmann, G., Jung, G. (1988), High-performance liquid chromatography of siderophores from fungi, *BioMetals* **1**, 9–17.

Konetschny-Rapp, S., Jung, G., Meiwes, J., Zähner, H. (1990), Staphyloferrin A: a structurally new siderophore from staphylococci, *Eur. J. Biochem.* **191**, 65–74.

Koster, M., van de Vossenberg, J., Leong, J., Weisbeek, P. J. (1993), Identification and characterization of the *pupB* gene encoding an inducible ferric-pseudobactin receptor of *Pseudomonas putida* WCS358, *Mol. Microbiol.* **8**, 591–601.

Köster, W. (1991), Iron(III) hydroxamate transport across the cytoplasmic membrane of *Escherichia coli, BioMetals* **4**, 23–32.

Kunze, B., Bedorf, N., Kohl, W., Höfle, G., Reichenbach, H. (1989), Myxochelin A, a new iron-chelating compound from *Angiococcus disciformis* (Myxobacteriales), *J. Antibiot.* **42**, 14–17.

Kunze, B., Trowitzsch, W., Höfle, G., Reichenbach, H. (1992), Nannochelins A, B and C, new iron-chelating compounds from *Nannocystis excedens* (Myxobacteria), *J. Antibiot.* **45**, 147–150.

Kühn, S., Braun, V., Köster, W. (1996), Ferric rhizoferrin uptake into *Morganella morganii:* characterization of genes involved in the uptake of a polycarboxylate siderophore, *J. Bacteriol.* **178**, 496–504.

Lammers, P. J., Sanders-Löhr, J. (1982), Active transport of ferric schizokinen in *Anabaena* sp., *J. Bacteriol.* **151**, 288–294.

Lane, S. J., Marshall, P. S., Upton, R. J., Ratledge, C., Ewing, M. (1995), Novel extracellular mycobactins, the carboxymycobactins from *Mycobacterium avium, Tetrahedron Lett.* **36**, 4129–4132.

LEE, B. H., MILLER, M. J. (1983), Natural ferric ionophores: Total synthesis of schizokinen, schizokinen A, and arthrobactin, *J. Org. Chem.* **48**, 24–31.

LEONG, J., NEILANDS, J. B. (1976), Mechanisms of siderophore iron transport in enteric bacteria, *J. Bacteriol.* **126**, 823–830.

LEONG, J., RAYMOND, K. N. (1974), Coordination isomers of biological iron transport compounds. II. The optical isomers of chromic desferriferrichrome and desferriferrichrysin, *J. Am. Chem. Soc.* **96**, 6628–6630.

LESUISSE, E., LABBE, P. (1994), Reductive iron assimilation in *Saccharomyces cerevisiae,* in: *Metal Ions in Fungi* (WINKELMANN, G., WINGE, D. R., Eds.), pp. 149–178. New York: Marcel Dekker.

LINGET, C., STYLIANOU, D. G., DELL, A., WOLFF, R. E., PIEMONT, Y., ABDALLAH, M. (1992), Bacterial siderophores: the structure of a desferribactin produced by a *Pseudomonas fluorescens* ATCC 13525, *Tetrahedron Lett.* **33**, 3851–3854.

LINKE, W. D., CRUEGER, A., DIEKMANN, H. (1972), Stoffwechselprodukte von Mikroorganismen. 106. Zur Konstitution des Terregens-Faktors, *Arch. Microbiol.* **85**, 44–50.

LITWIN, C. M., CALDERWOOD, S. B. (1994), Analysis of the complexity of gene regulation by Fur in *Vibrio cholerae, J. Bacteriol.* **176**, 240–248.

LIU, W., FISHER, S., WELLS, J. S., RICCA, C. S., PRINCIPE, P. A., TREJO, W. H., BONNER, D. P., GOUGOUTOS, J. Z., POEPLITZ, B., SYKES, R. B. (1981), Siderochelin, a new ferrous-ion chelating agent produced by *Nocardia, J. Antibiot.* **34**, 791–799.

LUNDRIGAN, M. D., KADNER, R. J. (1986), Nucleotide sequence of the gene for the ferrienterochelin receptor FepA in *Escherichia coli, J. Biol. Chem.* **261**, 10797–10801.

MACHAM, L. P., RATLEDGE, C., NOCTON, J. C. (1975), Extracellular iron acquisition by mycobacteria: Role of the exochelins and evidence against the participation of mycobactin, *Infect. Immun.* **12**, 1242–1251.

MARUGG, J. D., WEISBEEK, P. J. (1991), Molecular genetics of siderophore biosynthesis in fluorescent pseudomonads, in: *Handbook of Microbial Iron Chelates* (WINKELMANN, G., Ed.), pp. 155–175. Boca Raton, FL: CRC Press.

MARUGG, J. D., DE WEGER, L. A., NIELANDER, H. B., OORTHUIZEN, M., RECOURT, K., LUGTENBERG, B., VAN DER HOFSTAD, G. A. J. M., WEISBEEK, P. J. (1989), Cloning and characterization of a gene encoding an outer membrane protein required for siderophore-mediated uptake of Fe^{3+} in *Pseudomonas putida* WCS358, *J. Bacteriol.* **171**, 2819–2826.

MASSAD, G., ARCENEAUX, J. E. L., BYERS, B. R. (1994), Diversity of siderophore genes encoding biosynthesis of 2,3-dihydroxybenzoic acid in *Aeromonas* spp., *BioMetals* **7**, 227–236.

MATZANKE, B. F. (1991), Structures, coordination chemistry and functions of microbial iron chelates, in: *Handbook of Microbial Iron Chelates* (WINKELMANN, G., Ed.) pp. 15–64. Boca Raton, FL: CRC Press.

MATZANKE, B. F. (1994), Iron storage in fungi, in: *Metal Ions in Fungi* (WINKELMANN, G., WINGE, D. R., Eds.), pp. 179–214. New York: Marcel Dekker.

MATZANKE, B. F., WINKELMANN, G. (1981), Siderophore transport followed by Mössbauer spectroscopy, *FEBS Lett.* **130**, 50–53.

MATZANKE, B. F., BILL, E., TRAUTWEIN, A. X., WINKELMANN, G. (1987), Role of siderophores in iron storage in spores of *Neurospora crassa* and *Aspergillus ochraceus, J. Bacteriol.* **169**, 5873–5876.

MATZANKE, B. F., BILL, E., TRAUTWEIN, A. X., WINKELMANN, G. (1988), Ferricrocin functions as the main intracellular iron-storage compound in mycelia of *Neurospora crassa, BioMetals* **1**, 18–25.

MATZANKE, B. F., MÜLLER, G., RAYMOND, K. N. (1989), Siderophore mediated iron transport, in: *Iron Carriers and Iron Proteins* (LOEHR, T., Ed.), pp. 1–121. Weinheim: VCH.

MATZANKE, B. F., BERNER, I., BILL, E., TRAUTWEIN, A. X., WINKELMANN, G. (1991), Transport and utilization of ferrioxamine-E-bound iron in *E. herbicola (Pantoea herbicola), BioMetals* **4**, 181–185.

MAURER, P. J., MILLER, M. J. (1982), Microbial iron chelators: Total synthesis of aerobactin and its constituent amino acid N^6-acetyl-N^6-hydroxylysine, *J. Am. Chem. Soc.* **104**, 3096–3101.

MAURER, B., MÜLLER, A., KELLER-SCHIERLEIN, W., ZÄHNER, H. (1968), Ferribactin, ein Siderophor aus *Pseudomonas fluorescens* Migula, *Arch. Microbiol.* **60**, 326–339.

MEI, B., LEONG, S. A. (1994), Molecular biology of iron transport, in: *Metal Ions in Fungi* (WINKELMANN, G., WINGE, D. R., Eds.), pp. 117–148. New York: Marcel Dekker.

MEIWES, J., FIEDLER, H., HAAG, H., ZÄHNER, H., KONETSCHNY-RAPP, S., JUNG, G. (1990), Isolation and characterization of staphyloferrin A, a compound with siderophore activity from *Staphylococcus hyicus* DSM 20459, *FEMS Microbiol. Lett.* **67**, 201–206.

MENHARDT, N., THARIATH, A., VISWANATHA, T. (1991), Characterization of the pyoverdins of

Azotobacter vinelandii ATCC 12837 with regard to heterogeneity, *BioMetals* **4**, 223–232.

MEYER, J. M. (1992), Exogenous siderophore-mediated iron uptake in *Pseudomonas aeruginosa:* possible involvement of porin OprF in iron translocation, *J. Gen. Microbiol.* **138**, 951–958.

MEYER, J. M., ABDALLAH, M. A. (1980), The siderochromes of non-fluorescent pseudomonads: production of nocardamine by *Pseudomonas stutzeri, J. Gen. Microbiol.* **118**, 125–129.

MEYER, J. M., HOHNADEL, D., HALLÉ, F. (1989), Cepabactin from *Pseudomonas cepacia,* a new type of siderophore, *J. Gen. Microbiol.* **135**, 1479–1487.

MEYER, J.-M., TRAN VAN, V., STINTZI, A., BERGE, O., WINKELMANN, G. (1995), Ornibactin production and transport properties in strains of *Burkholderia vietnamiensis* and *Burkholderia cepacia* (formerly *Pseudomonas cepacia*), *BioMetals* **8**, 309–317.

MOORE, R. E., EMERY, T. (1976), N^{δ}-Acetylfusarinines: isolation, characterization, and properties, *Biochemistry* **15**, 2719–2722.

MOORE, C. H., FOSTER, L.-A., GERBIG JR., D. G., DYER, D. W., GIBSON, B. (1995), Identification of alcaligin as the siderophore produced by *Bordetella pertussis* and *B. bronchiseptica, J. Bacteriol.* **7**, 1116–1118.

MOR, H., KASHMAN, Y., WINKELMANN, G., BARASH, I. (1992), Characterization of siderophores produced by different species of the dermatophytic fungi *Microsporum* and *Trichophyton, BioMetals* **5**, 213–216.

MÜLLER, G., RAYMOND, K. N. (1984), Specificity and mechanism of ferrioxamine-mediated iron transport in *Streptomyces pilosus, J. Bacteriol.* **160**, 304–312.

MÜLLER, G., MATZANKE, B. F., RAYMOND, K. N. (1984), Iron transport in *Streptomyces pilosus* mediated by ferrichrome siderophores, rhodotorulic acid, and *enantio*-rhodotorulic acid, *J. Bacteriol.* **160**, 313–318.

MULLIS, K. B., POLLACK, J. R., NEILANDS, J. B. (1971), Structure of schizokinen, an iron transport compound from *Bacillus megaterium, Biochemistry* **10**, 4894–4898.

NAEGELI, H., KELLER-SCHIERLEIN, W. (1978), Stoffwechselprodukte von Mikroorganismen. Eine neue Synthese des Ferrichroms; *enantio*-Ferrichrom, *Helv. Chim. Acta* **61**, 2088–2095.

NAEGELI, H., ZÄHNER, H. (1980), Stoffwechselprodukte von Mikroorganismen. Ferrithiocin, *Helv. Chim. Acta* **63**, 1400–1406.

NEILANDS, J. B. (1952), A crystalline organo-iron pigment from a rust fungus (*Ustilago sphaerogena*), *J. Am. Chem. Soc.* **74**, 4846–4847.

NEILANDS, J. B. (1993), Overview of bacterial iron transport and siderophore systems in *Rhizobia,* in: *Iron Chelation in Plants and Soil Microorganisms* (BARTON, L. L., HEMMING, B. C., Eds.), pp. 179–195. San Diego, CA: Academic Press.

NEILANDS, J. B., NAKAMURA, K. (1991), Detection, determination, isolation, characterization and regulation of microbial iron chelates, in: *Handbook of Microbial Iron Chelates* (WINKELMANN, G., Ed.), pp. 1–14. Boca Raton, FL: CRC Press.

NEILANDS, J. B., ERICSON, T. J., RASTETTER, W. H. (1981), Stereospecificity of the ferric enterobactin receptor of *Escherichia coli* K-12, *J. Biol. Chem.* **256**, 3831–3832.

NIKAIDO, H., ROSENBERG, E. Y. (1990), Cir and Fiu proteins in the outer membrane of *Escherichia coli* catalyze transport of monomeric catechols: Study with β-lactam antibiotics containing catechol and analogous groups, *J. Bacteriol.* **172**, 1361–1367.

NISHIO, T., TANAKA, N., HIRATAKE, JUN., KATSUBE, Y., ISHIDA, Y., ODA, J. (1988), Isolation and structure of the novel dihydroxamate siderophore alcalignin, *J. Am. Chem. Soc.* **110**, 8733–8734.

NOMOTO, K., SUGIURA, Y., TAKAGI, S. (1987), Mugineic acids, studies on phytosiderophores, in: *Iron Transport in Microbes, Plants and Animals* (WINKELMANN, G., VAN DER HELM, D., NEILANDS, J. B., Eds.), pp. 401–425, Weinheim: VCH.

OKUJO, N., SAITO, M., YAMAMOTO, S., YOSHIDA, T., MIYOSHI, S., SHINODA, S. (1994a), Structure of vulnibactin, a new polyamine-containing siderophore from *Vibrio vulnificus, BioMetals* **7**, 109–116.

OKUJO, N., SAKAKIBARA, Y., YOSHIDA, T., YAMAMOTO, S. (1994b), Structure of acinetofferin, a new citrate-based dihydroxamate siderophore from *Acinetobacter haemolyticus*, *BioMetals* **7**, 170–176.

ONG, S. A., PETERSON, T., NEILANDS, J. B. (1979), Agrobactin, a siderophore from *Agrobacterium tumefaciens, J. Biol. Chem.* **254**, 1860–1865.

OXFORD, A. E., RAISTRICK, H. (1948), Studies on the biochemistry of microorganisms. 76. Mycelianamide, a metabolic product of *Penicillium griseofulvum.* Part I. Preparation, properties and breakdown products, *Biochem. J.* **42**, 323.

PAGE, W. J., HUYER, M. (1984), Derepression of the *Azotobacter vinelandii* siderophore system using iron-containing minerals to limit iron repletion, *J. Bacteriol.* **158**, 496–502.

PAGE , W. J., TIGERSTROM, M. (1988), Aminochelin, a catecholamine siderophore produced by *Azotobacter vinelandii, J. Gen. Microbiol.* **134**, 453–460.

PAYNE, S. M. (1980), Synthesis and utilization of siderophores by *Shigella flexneri*, *J. Bacteriol.* **143**, 1420–1424.

PERSMARK, M., NEILANDS, J. B. (1992), Iron(III) complexes of chrysobactin, a siderophore of *Erwinia chrysanthemi, BioMetals* **5**, 29–36.

PERSMARK, M., EXPERT, D., NEILANDS, J. B. (1989), Isolation, characterization, and synthesis of chrysobactin, a compound with siderophore activity from *Erwinia chrysanthemi, J. Biol. Chem.* **264**, 3187–3193.

PERSMARK, M., EXPERT, D., NEILANDS, J. B. (1992), Ferric iron uptake in *Erwinia chrysanthemi* mediated by chrysobactin and related catechol-type compounds, *J. Bacteriol.* **174**, 4783–4789.

PERSMARK, M., PITTMAN, P., BUYER, J. S., SCHWYN, B., GILL, P. R., NEILANDS, J. B. (1993), Isolation and structure of rhizobactin 1021, a siderophore from the alfalfa symbiont *Rhizobium meliloti* 1021, *J. Am. Chem. Soc.* **115**, 3950-3956.

PETERSON, T., NEILANDS, J. B. (1979), Revised structure of a catecholamide spermidine siderophore from *Paracoccus denitrificans, Tetrahedron Lett.* **50**, 4805–4808.

RABSCH, W., WINKELMANN, G. (1991), The specificity of bacterial siderophore receptors probed by bioassays, *BioMetals* **4**, 244–250.

RATLEDGE, C. (1987), Iron metabolism in mycobacteria, in: *Iron Transport in Microbes, Plants and Animals* (WINKELMANN, G., VAN DER HELM, D., NEILANDS, J. B., Eds.), pp. 207–233, Weinheim: VCH.

RATLEDGE, C., EWING, M. (1996), The occurrence of carboxymycobactin, the siderophore of pathogenic mycobacteria, as a second extracellular siderophore in *Mycobacterium smegmatis, Microbiology* **142**, 2207–2212.

RATLEDGE, C., MARSHALL, B. J. (1972), Iron transport in *Mycobacterium smegmatis*: the role of mycobactin, *Biochim. Biophys. Acta* **279**, 58–74.

RATLEDGE, C., PATEL, P. V. (1976), The isolation, properties and taxonomic relevance of lipid-soluble iron-binding compounds (the nocobactins) from *Nocardia, J. Gen. Microbiol.* **93**, 141–152.

REID, R. T., LIVE, D. H., FAULKNER, D. J., BUTLER, A. (1993), A siderophore from a marine bacterium with an exceptional ferric ion affinity constant, *Nature* **366**, 455–458.

REISSBRODT, R., RABSCH, W., CHAPEAUROUGE, A., JUNG, G., WINKELMANN, G. (1990), Isolation and identification of ferrioxamine G and E in *Hafnia alvei, BioMetals* **3**, 54–60.

REISSBRODT, R., ERLER, W., WINKELMANN, G. (1994), Iron supply of *Pasteurella multocida* and *Pasteurella haemolytica, J. Basic Microbiol.* **34**, 61–63.

RUTZ, J. M., LIU, J., LYONS, J., GORANSON, J., ARMSTRONG, S. K., MCINTOSH, M. A., FEIX, J. B., KLEBBA, P. E. (1993), Formation of a gated channel by a ligand-specific transport protein in the bacterial outer membrane, *Science* **258**, 471–475.

SAYER, J. M., EMERY, T. (1968), Structures of the naturally occurring hydroxamic acids, fusarinines A and B, *Biochemistry* **7**, 184–190.

SCHNEIDER, R., HANTKE, K. (1993), Iron-hydroxamate uptake systems in *Bacillus subtilis:* identification of a lipoprotein as part of a binding protein-dependent transport system, *Mol. Microbiol.* **8**, 111–121.

SCHWYN, B., NEILANDS, J. B. (1987), Universal chemical assay for the detection and determination of siderophores, *Anal. Biochem.* **160**, 47–56.

SHANZER, A., LIBMAN, J. (1991), Biomimetic siderophores, in: *Handbook of Microbial Iron Chelates* (WINKELMANN, G., Ed.), pp. 309–338. Boca Raton, FL: CRC Press.

SHANZER, A., LIBMAN, J., LAZAR, R., TOR, Y., EMERY, T. (1988), Synthetic ferrichrome analogues with growth promotion activity for *Arthrobacter flavescens, Biochem. Biophys. Res. Commun.* **157**, 389–394.

SHARMAN, G., WILLIAMS, D. H., EWING, D. F., RATLEDGE, C. (1995a), Isolation, purification and structure of exochelin MS, the extracellular siderophore from *Mycobacterium smegmatis, Biochem. J.* **305**, 187–196.

SHARMAN, G., WILLIAMS, D. H., EWING, D. F., RATLEDGE, C. (1995b), Determination of the structure of exochelin MN the extracellular siderophore from *Mycobacterium neoaurum, Chem. Biol.* **2**, 553–561.

SIGEL, S. P., STOEBNER, J. A., PAYNE, S. M. (1985), Iron-vibriobactin transport system is not required for virulence of *Vibrio cholerae, Infect. Immun.* **47**, 360–362.

SMITH, M. J., SHOOLERY, J. N., SCHWYN, B., HOLDEN, I., NEILANDS, J. B. (1985), Rhizobactin, a structurally novel siderophore from *Rhizobium meliloti, J. Am. Chem. Soc.* **107**, 1739–1743.

SNOW, G. A. (1965), Isolation and structure of mycobactin T, a growth factor from *Mycobacterium tuberculosis, Biochem. J.* **97**, 166–175.

SNOW, G. A. (1970), Mycobactins: iron chelating growth factors from Mycobacteria, *Microbiol. Rev.* **34**, 99–125.

SOUKKA, T., TENOVUO, J., LENANDER-LUMIKARI, M. (1992), Fungicidal effect of human lactoferrin against *Candida albicans, FEMS Microbiol. Lett.* **90**, 223–228.

STEPHAN, H., FREUND, S., BECK, W., JUNG, G., WINKELMANN, G. (1993a), Ornibactins – a new family of siderophores from *Pseudomonas, BioMetals* **6**, 93–100.

STEPHAN, H., FREUND, S., MEYER, J.-M., WINKELMANN, G., JUNG, G. (1993b), Structure elucidation of the gallium-ornibactin complex by 2D-NMR spectroscopy, *Liebigs Ann. Chem.* 1993, 43–48.

STEPHENSON, M. C., RATLEDGE, C. (1979), Iron transport in *Mycobacterium smegmatis:* Uptake of iron from ferriexochelin, *J. Gen. Microbiol.* **110**, 193–202.

STOEBNER, J. A., PAYNE, S. M. (1988), Iron-regulated hemolysin production and utilization of heme and hemoglobin by *Vibrio cholerae, Infect. Immun.* **56**, 2891–2895.

STOEBNER, J. A., BUTTERTON, J. R., CALDERWOOD, S. B., PAYNE, S. M. (1992), Identification of the vibriobactin receptor of *Vibrio cholerae, J. Bacteriol.* **174**, 3270–3274.

STOJILJKOVIC, I., HANTKE, K. (1992), Hemin uptake system of *Yersinia enterocolitica:* similarities with other TonB-dependent systems in gram-negative bacteria, *EMBO J.* **11**, 4359–4367.

TAKAHASHI, A., NAKAMURA, H., KAMEYAMA, T., KURASAWA, S., NAGANAWA, H., OKAMI, Y., TAKEUCHI, S, UMEZAWA, H. (1987), Bisucaberin, a new siderophore, sensitizing tumor cells to macrophage-mediated cytolysis. II. Physicochemical properties and structure determination, *J. Antibiot.* **40**, 1671–1676.

TARAZ, K., EHLERT, G., GEISEN, K., BUDZIKIEWICZ, H., KORTH, H., PULVERER, G. (1990), Protochelin, ein Catecholat-Siderophor aus einem Bakterium (DSM 5746), *Z. Naturforsch.* **45b**, 1327–1332.

TEINTZE, M., HOSSAIN, M. B., BARNES, C. L., LEONG, J., VAN DER HELM, D. (1981), Structure of ferric pseudobactin, a siderophore from a plant-deleterious *Pseudomonas, Biochemistry* **20**, 6446–6457.

TELFORD, J. R., LEARY, J. A., TUNSTAD, L. M. G., BYERS, B. R., RAYMOND, K. N. (1994), Amonabactin: Characterization of a new series of siderophores from *Aeromonas hydrophila, J. Am. Chem. Soc.* **116**, 4499–4500.

THIEKEN, A., WINKELMANN, G. (1992), Rhizoferrin: A complexone type siderophore of the Mucorales and Entomophthorales, *FEMS Microbiol. Lett.* **94**, 37–42.

THIEKEN, A., WINKELMANN, G. (1993), A novel bioassay for the detection of siderophores containing keto-hydroxy bidentate ligands, *FEMS Microbiol. Lett.* **111**, 281–286.

TSUBOTANI, S., KATAYAMA, N., FUNABASHI, Y., ONO, H., HARADA, S. (1993), Ferrocins, new iron-containing peptide antibiotics produced by bacteria. Isolation, characterization and structure elucidation, *J. Antibiot.* **46**, 287–293.

VAN DER HELM, D., POLING, M. (1976), The crystal structure of ferrioxamine E, *J. Am. Chem. Soc.* **98**, 82–86.

VAN DER HELM, D., BAKER, J. R., ENG-WILMOT, D. L., HOSSAIN, M. B., LOGHRY, R. A. (1980), Crystal structure of ferrichrome and comparison with the structure of ferrichrome A, *J. Am. Chem. Soc.* **102**, 4224–4231.

VAN DER HELM, D., WINKELMANN, G. (1994), Hydroxamates and polycarboxylates as iron transport agents (siderophores) in fungi, in: *Metal Ions in Fungi* (WINKELMANN, G., WINGE, D. R., Eds.), pp. 39–98. New York: Marcel Dekker.

VAN DER WALT, J. P., BOTHA, A., EICKER, A. (1990), Ferrichrome production by Lipomycetaceae, *Syst. Appl. Microbiol.* **13**, 131–135.

VAN TIEL-MENKVELD, G.J., MENTJOX-VERVUURT, M., OUDEGA, B., DEGRAAF, F. K. (1982), Siderophore production by *Enterobacter cloacae* and a common receptor protein for the uptake of aerobactin and cloacin DF13, *J. Bacteriol.* **150**, 490–497.

VANDE WOESTYNE, M., BRUYNEEL, B., MERGEAY, M., VERSTRAETE, W. (1991), The Fe^{2+} chelator proferrorosamine A is essential for the siderophore-mediated uptake of iron by *Pseudomonas roseus fluorescens, Appl. Environ. Microbiol.* **57**, 949–954.

VISCA, P., SERINO, L., ORSI, N. (1992), Isolation and characterization of *Pseudomonas aeruginosa* mutants blocked in the biosynthesis of pyoverdin, *J. Bacteriol.* **174**, 5727–5731.

VISCA, P., CIERVO, A., ORSI, N. (1994), Cloning and nucleotide sequence of the *pvdA* gene encoding the pyoverdin biosynthetic enzyme L-ornithine N^5-oxygenase in *Pseudomonas aeruginosa, J. Bacteriol.* **176**, 1128–1140.

WACHSMUTH, I. K., BLAKE, P. A., OLSVIK, O. (Eds.) (1994), *Vibrio cholerae and Cholera – Molecular to global perspectives.* Washington, DC: Am. Soc. Microbiol.

WEE, S., NEILANDS, J. B., BITTNER, M. L., HEMMING, B. C., HAYMORE, B. L., SEETHARAM, R. (1988), Expression, isolation and properties of Fur (ferric uptake regulation) protein of *Escherichia coli* K12, *BioMetals* **1**, 62–68.

WIEBE, C., WINKELMANN, G. (1975), Kinetic studies on the specificity of chelate iron uptake in *Aspergillus, J. Bacteriol.* **123**, 837–842.

WINKELMANN, G. (1979), Evidence for stereospecific uptake of iron 14 chelates in fungi, *FEBS Lett.* **97**, 43–46.

WINKELMANN, G. (1986), Iron complex products (siderophores), in: *Biotechnology*, 1st Edn., Vol. 4, (REHM, H.-J., REED, G., Eds), pp. 215–243. Weinheim: VCH.

WINKELMANN, G. (Ed.) (1991a), *Handbook of Microbial Iron Chelates.* Boca Raton, FL: CRC Press.

WINKELMANN, G. (1991b), Importance of siderophores in fungal growth, sporulation and spore germination, in: *Frontiers in Mycology* (HAWKSWORTH, D. L., Ed.), pp. 49–65. Wallingford: CAB International.

WINKELMANN, G. (1991c), Specificity of iron transport in bacteria and fungi, in: *Handbook of Microbial Iron Chelates* (WINKELMANN, G., Ed.), pp. 65–105. Boca Raton, FL: CRC Press.

WINKELMANN, G. (1992), Structures and functions of fungal siderophores containing hydroxamate and complexone type iron binding ligands, *Mycol. Res.* **96**, 529–534.

WINKELMANN, G. (1993), Kinetics, energetics, and mechanisms of siderophore iron transport in fungi, in: *Iron Chelation in Plants and Soil Microorganisms* (BARTON, L. L., HEMMING, B. C., Eds.), pp. 219–239. Boca Raton, FL: CRC Press.

WINKELMANN, G., BRAUN, V. (1981), Stereoselective recognition of ferrichrome by fungi and bacteria, *FEMS Microbiol. Lett.* **11**, 237–241.

WINKELMANN, G., WINGE, D. R. (Eds.) (1994), *Metal Ions in Fungi.* New York: Marcel Dekker.

WINKELMANN, G., ZÄHNER, H. (1973), Stoffwechselprodukte von Mikroorganismen. 115. Eisenaufnahme bei *Neurospora crassa.* Zur Spezifität des Eisentransportes, *Arch. Microbiol.* **88**, 49–60.

WINKELMANN, G., VAN DER HELM, D., NEILANDS, J. B. (Eds.) (1987), *Iron Transport in Microbes, Plants and Animals.* Weinheim: VCH.

WINKELMANN, G., CANSIER, A., BECK, W., JUNG, G. (1994), HPLC separation of enterobactin and linear 2,3-dihydroxybenzoylserine derivatives: a study on mutants of *Escherichia coli* defective in regulation (*fur*), esterase (*fes*) and transport (*FepA*), *BioMetals* **7**, 149–154.

WONG, G. B., KAPPEL, M. J., RAYMOND, K. N., MATZANKE, B., WINKELMANN, G. (1983), Coordination chemistry of microbial iron transport compounds. 24. Characterization of coprogen and ferricrocin, two ferric hydroxamate siderophore, *J. Am. Chem. Soc.* **105**, 810–815.

YABUUCHI, E., KOSAKO, Y., OYAIZU, H., YANO, I., HOTTA, H., HASHIMOTO, Y., EAZAKI, T., ARAKAWA, M. (1992), Proposal of *Burkholderia* gen. nov. and transfer of seven species of the genus *Pseudomonas* homology group II to the new genus, with the type species *Burkholderia cepacia* (PALLERONI and HOLMS 1981) comb. nov., *Microbiol. Immunol.* **36**, 1251–1275.

YAMAMOTO, S., CHOWDHURY, M. A. R., KURODA, M., et al. (1991), Further study on polyamine compositions in Vibrionaceae, *Can. J. Microbiol.* **37**, 148–153.

YAMAMOTO, S., FUJITA, Y., OKUJO, N., MINAMI, C., MATSUURA, S., SHINODA, S. (1992), Isolation and partial characterization of a compound with siderophore activity from *V. parahaemolyticus, FEMS Microbiol. Lett.* **94**, 181–186.

YAMAMOTO, S., OKUIO, N., FUJITA, Y., SAITO, M., YOSHIDA, T., SHINODA, S. (1993), Structures of two polyamine-containing catecholate siderophores from *Vibrio fluvialis, J. Biochem.* **113**, 538–544.

YAMAMOTO, S., OKUJO, N., SAKAKIBARA, Y. (1994a), Isolation and structure elucidation of acinetobactin, a novel siderophore from *Acinetobacter baumannii, Arch. Microbiol.* 249–254.

YAMAMOTO, S., OKUJO, N., YOSHIDA, T., MATSUURA, S., SHINODA, S. (1994b), Structure and iron transport activity of Vibrioferrin, a new siderophore from *Vibrio parahaemolyticus, J. Biochem.* **115**, 868–874.

YOUNG, I. G., GIBSON, F. (1979), Isolation of enterobactin from *Escherichia coli, Methods Enzymol.* **56**, 394–398.

ZÄHNER, H., DIDDENS, H., KELLER-SCHIERLEIN, W., NAEGELI, H. (1977), Some experiments with semisynthetic sideromycin, *J. Antibiot.* **30** (Suppl.), S-201–206.

ZALKIN, A., FORRESTER, J. D., TEMPLETON, D. H. (1966), Ferrichrome A tetrahydrate. Determination of crystal and molecular structure, *J. Am. Chem. Soc.* **88**, 1810–1814.

ZHOU, X. H., VAN DER HELM, D., VENKATRAMANI, L. (1995), Binding characterization of the iron transport receptor from the outer membrane of *Escherichia coli* (FepA): differentiation between FepA und FecA, *BioMetals* **8**, 129–136.

ZYCHA, H., SIEPMANN, R. (Eds.) (1969), *Mucorales – Eine Beschreibung aller Gattungen und Arten dieser Pilzgruppe. Mit einem Beitrag zur Gattung Mortierella von G. Linnemann,* Lehre: Verlag von Cramer.

ZYWNO, S. R., ARCENEAUX, J. E. L., ALTWEG, M., BYERS, B. R. (1992), Siderophore production and DNA hybridization groups of *Aeromonas* spp., *J. Clin. Microbiol.* **30**, 619–622.

6 Advances in the Molecular Genetics of β-Lactam Antibiotic Biosynthesis

PAUL L. SKATRUD

Indianapolis, IN, USA

TORSTEN SCHWECKE

Cambridge, UK

HENK VAN LIEMPT

Bonn, Germany

MATTHEW B. TOBIN

Indianapolis, IN, USA

1 Introduction

120 years ago scientists observed growth inhibition when certain microbes were cultured together and they recognized the potential therapeutic value of this antagonism (PASTEUR and JOUBERT, 1877). Half a century later a similar observation in the British laboratory of SIR ALEXANDER FLEMING piqued interest in this phenomenon (FLEMING, 1929). Over a decade later, these discoveries and other observations led SIR EDWARD ABRAHAM and his colleagues to the first amazingly successful clinical use of penicillin in 1941 (ABRAHAM et al., 1941). In the ensuing years, the clinical use of β-lactam antibiotics (i.e., penicillins, cephalosporins, and cephamycins) became the most successful foray of mankind in his on-going battle against infectious diseases. Indeed, it has been suggested that in the United States the use of antibiotics has added approximately ten years to life expectancy (MCDERMOTT and ROGERS, 1982). Due to their intrinsically high therapeutic index and clinical success, β-lactam antibiotics have been and continue to be the most intensely used class of antibacterial compound.

The mode of action of β-lactam antibiotics on bacteria has been extensively studied since the identification of penicillin (reviewed by WAXMAN and STROMINGER, 1982). WISE and PARK (1965) and TIPPER and STROMINGER (1965) demonstrated that penicillin inhibits the transpeptidation reaction required for peptidoglycan crosslinking, and hypothesized that penicillin acts as a structural analog of the acyl-D-ala-D-ala terminus of the stem peptide (Park nucleotide) found in peptidoglycan. Other studies have demonstrated that penicillin also binds at the same active-site serine in two bacterial D-alanine carboxypeptidases (YOCUM et al., 1979). Regeneration of these enzymes requires cleavage of the acyl-enzyme complex by hydrolysis (of carboxypeptidases) or transaminolysis (of transpeptidases). The scissile bond of the β-lactam ring replaces the natural D-ala-D-ala substrate at the enzyme active site. Acylation of the catalytically active amino acid would occur with the opening of the β-lactam ring, forming an inactive penicilloyl enzyme resistant to hydrolysis or aminolysis.

Why the inhibition of cell wall biosynthesis should ultimately lead to cell death is a complex question (reviewed by SHOCKMAN et al., 1982). Peptidoglycan assembly is performed by several specific enzymes forming glycosidic, amide, and peptide bonds, and as such requires some kind of coordinate regulation of these enzymes. Early studies conducted by LEDERBERG (1957) assumed that inhibition of cell wall biosynthesis led directly to a shortage of building blocks, so that as growing cells increased in size, the limited or weakened cell wall was unable to contain the contents and lysed. However, more recent observations indicate that not all bacterial species undergo lysis after treatment with cell wall inhibitors such as penicillins. Indeed, the historically accepted dogma concerning that bacteriolysis is the direct result of penicillin treatment of susceptible bacteria has recently been challenged. GIESBRECHT et al. (1993) were able to distinguish between the onset of killing and the onset of bacteriolysis after treatment of *S. aureus* with β-lactam antibiotics, suggesting that the penicillin-induced bacteriolysis of sensitive bacteria is not the cause of death, but rather its consequence.

The unparalleled clinical success of β-lactam antibiotics stimulated intense interest and research into these compounds as well as the organisms that produce them. For decades, industrial research was focused on improving overall productivity and on understanding the biosynthesis of penicillin G in *Penicillium chrysogenum* and of cephalosporin C in *Cephalosporium acremonium* (syn. *Acremonium chrysogenum*). Strain improvement studies utilized natural selection and extensive mutagenesis programs coupled with biochemical selections. These strain improvement studies resulted in strains vastly improved in their capacity to produce β-lactam antibiotics. By 1985, the central reaction pathways for the biosynthesis of β-lactams had been described (O'SULLIVAN and SYKES, 1983). Most of the intermediates had been isolated and the order of reactions could be assigned (QUEENER and NEUSS, 1982) (see Fig. 1). The stage was set for the complete elucidation of biosynthetic pathway mechanisms, including the catalytic entities and the corresponding genes. Specifically the production of modified β-lactam antibiotics seemed feasible and this spurred research in both industrial groups and academic centers. Initial recombinant DNA studies on β-lactam producing fungi were done in the industrial setting (QUEENER et al., 1985; PENALVA et al., 1985) and as collaborative studies between industry (Eli Lilly and Company, U.S.) and academia (Oxford University, U.K.) (SAMSON et al., 1985). Since that time both communities have contributed heavily to the expansion of our knowledge in this arena. In the years since 1983, two lines of research have come together to prepare the way for a rationally designed exploitation of β-lactam antibiotic producing organisms. The first major step was the identification and cloning of the β-lactam biosynthetic genes which engendered a means to influence the catalytic properties of the corresponding enzymes. And second, the development of transformation protocols for achieving stable expression of foreign DNA in β-lactam producing strains led to an encompassing array of methods for drug design and production.

Several reviews have been written in the past few years which address progress made in understanding the molecular biology of β-lactam biosynthesis as well as the genetic manipulation and characterization of organisms producing these compounds. The reader is referred to several of these reviews for more extensive coverage of topics discussed in this chapter (MILLER and INGOLIA, 1989; INGOLIA and QUEENER, 1989; SKATRUD, 1991, 1992; MARTÍN et al., 1991; AHARONOWITZ et al., 1992; VICHITSOONTHONKUL et al., 1994; MARTÍN and GUTIÉRREZ, 1995). On the following pages, the developments in these lines of research are highlighted and examples of the use of these advances in basic knowledge for the improvement of antibiotic production are given.

2 Organisms which Produce β-Lactam Antibiotics

A variety of prokaryotic and eukaryotic microorganisms are able to synthesize bicyclic and monocyclic β-lactams (Tab. 1). Of these

Tab. 1. Representative β-Lactam-Producing Organisms (modified from TURNER, 1992)

β-Lactam Structure	Fungi	Bacteria	
		Actinomycetes	Eubacteria
Penam	*Aspergillus nidulans* *Penicillium chrysogenum* *Epidermophyton* *Trichophyton* *Polypaecilum* *Malbranchea* *Microsporum*		
Cephem	*Cephalosporium acremonium* *Paecilomyces* *Scopulariopsis* *Diheterospora* *Spiroidium*	*Streptomyces* *Nocardia*	*Flavobacterium* *Lysobacter* *Xanthomonas*
Clavam		*Streptomyces*	*Erwinia* *Serratia*
Carbapenem		*Streptomyces*	

compounds, the bicyclic β-lactams possess the most potent antimicrobial activity. Thus far, only *P. chrysogenum* and *C. acremonium* have been used extensively at the industrial scale for production of bicyclic β-lactam compounds. Despite their relatively potent antimicrobial activity, only a minority of these compounds are directly used in the clinical setting. However, chemical modification of bicyclic β-lactam fermentation products has produced a wide array of clinically useful semi-synthetic β-lactams. For the purposes of this chapter we will focus on the biosynthesis, molecular genetics, and enzymology of bicyclic β-lactams.

3 Biosynthesis of β-Lactam Antibiotics

A clear understanding of β-lactam biosynthesis (QUEENER and NEUSS, 1982) enabled investigators to efficiently dissect and modify this pathway through the use of molecular biology. All β-lactams contain the four-membered azetidinone ring system in which antimicrobial activity resides (Fig. 1). Bicyclic β-lactam antibiotics fall into two natural classes: penicillins with the second ring system being a five-membered thiazoladine ring and cephalosporins/cephamycins in which the second ring system is a six-membered dihydrothiazine ring. The biosynthetic pathways leading to the production of penicillin G in *P. chrysogenum*, cephalosporin C in *C. acremonium*, and cephamycin C in *Streptomyces clavuligerus* are illustrated in Fig. 1 as a branched pathway. Designations for the genes encoding enzymes of this pathway were established by INGOLIA and QUEENER (1989) and SKATRUD (1991); *pcb* refers to genes common to all three end products, *pen* defines genes found only in penicillin biosynthesis, *cef* refers to genes found only in cephalosporin C biosynthesis, and *cmc* denotes genes specific for cephamycin C biosynthesis. The so-called early genes (*pcbAB* and *pcbC*) encode the δ-(L-α-aminoadipoyl)-L-cysteinyl-D-valine synthetase (ACVS) and isopenicillin N synthase (IPNS) enzymes, respectively. In *Nocardia lactamdurans* and *S. clavuligerus*, α-aminoadipic acid is formed from L-lysine by lysine-ε-aminotransferase, encoded by the *lat* gene. In the fungi, α-aminoadipate is recruited from L-lysine biosynthesis. Subsequent to the formation of isopenicillin N (IPN), the penicillin and cephalosporin/cephamycin biosynthetic pathways diverge. In *P. chrysogenum* and *A. nidulans*, hydrophobic penicillins (e.g., penicillin G) are formed by a transacylation reaction catalyzed by acyl-coenzyme A:isopenicillin N acyltransferase (AT), encoded by the *penDE* gene. In cephalosporin/cephamycin producing organisms (e.g., *C. acremonium*, *N. lactamdurans*, and several strains of Streptomycetes), IPN is converted to penicillin N by an epimerase, encoded by the *cefD* gene. The thiazolidine ring of the penicillin is then ring-expanded to form deacetoxycephalosporin C (DAOC), containing the dihydrothiazine ring characteristic of the cephalosporins. DAOC is subsequently hydroxylated at the C-3 position, forming deacetylcephalosporin C (DAC). In the filamentous fungus *C. acremonium*, the *cefEF* gene product (DAOC/DAC synthase, expandase/hydroxylase, REX/H) catalyzes both of these reactions. In the prokaryotes *S. clavuligerus* and *N. lactamdurans*, these two reactions are catalyzed by separate enzymes, DAOC synthase (REX, expandase) and DAC synthase (hydroxylase), encoded by two separate genes which bear striking homology to one another (*cefE* and *cefF*, respectively). At this stage, the biosynthetic pathways of cephalosporins and cephamycins diverge. In *C. acremonium*, DAC is converted to cephalosporin C, the end product of cephalosporin biosynthesis, by an acetyltransferase reaction at the C-3 position. The enzyme catalyzing this reaction is encoded by the *cefG* gene. The formation of cephamycin C in *S. clavuligerus* involves further modification of DAC. A 3′-hydroxymethylceph-3-em-*O*-carbamoyltransferase, encoded by the *cmcH* gene, catalyzes the carbamoylation at the C-3 position of DAC to form *O*-carbamoyldeacetylcephalosporin C. The last two steps in cephamycin C biosynthesis involve 7′-α-hydroxylase and methyltransferase reactions, encoded by the *cmcI* and *cmcJ* genes, respectively.

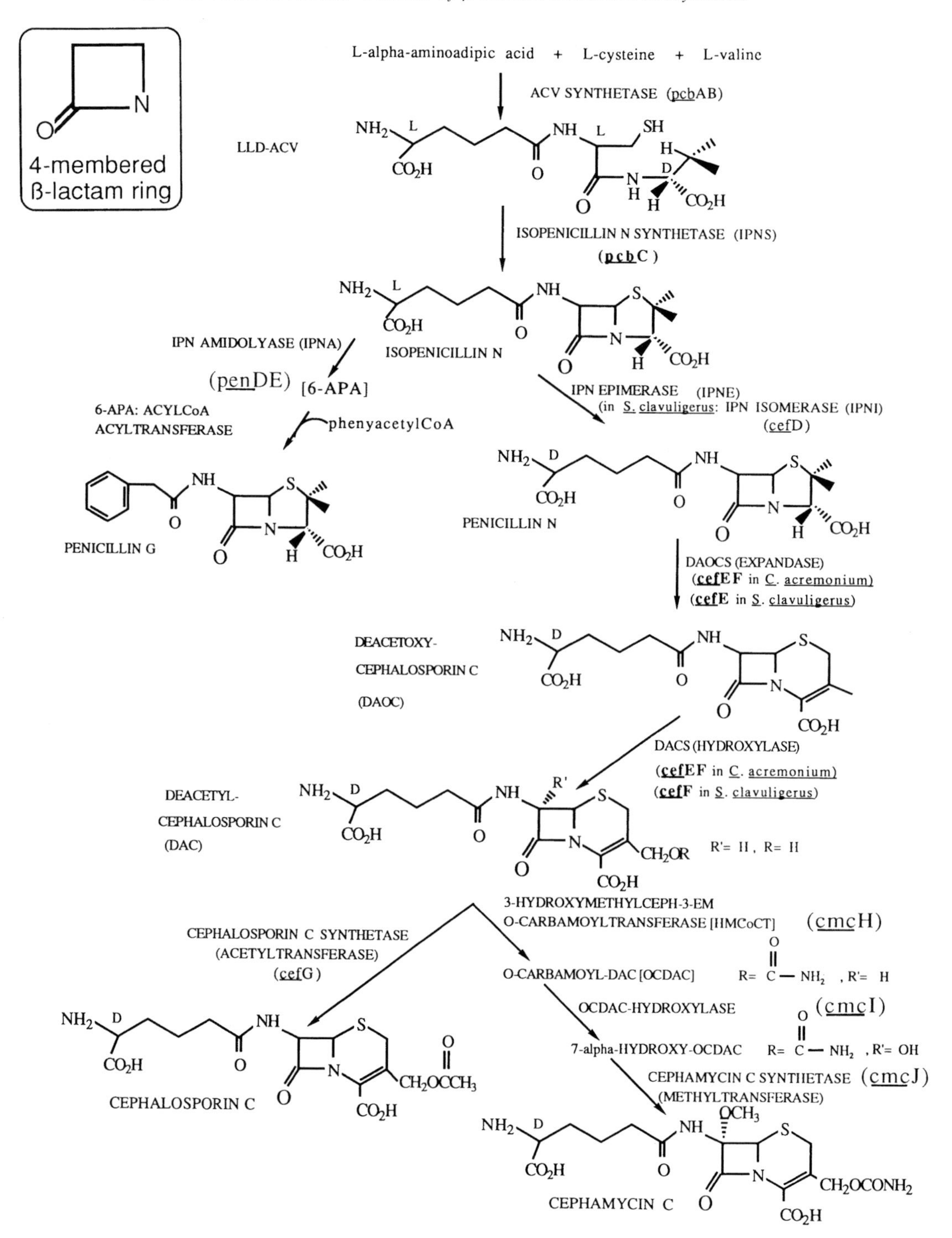

Fig. 1. Biosynthesis of β-lactam antibiotics.

4 Development of Genetic Transformation Systems

The application of recombinant DNA technology was based on the development of genetic transformation systems for β-lactam producing organisms and cloning of biosynthetic genes. The ability to transform industrially relevant organisms (specifically the filamentous fungi *P. chrysogenum* and *C. acremonium*) would allow investigators to manipulate the biosynthetic pathway in a precise manner and provide an avenue for practical applications (e.g., gene dosage studies for rate-limiting steps, gene disruptions to alter end products, or the addition of novel genes to modify the end product). Experimental conditions utilized in the development of the first fungal transformation system (i.e., *Saccharomyces cerevisiae,* HINNEN et al., 1978) were relied upon heavily in establishing conditions for transformation of β-lactam producing fungi. The first fungal β-lactam producing organism to be transformed was *Aspergillus nidulans* (YELTON et al., 1984), however, this organism is not used on the industrial scale for production of β-lactam antibiotics. *C. acremonium* was transformed the following year (QUEENER et al., 1985; PENALVA et al., 1985). While the transformation of *A. nidulans* relied on auxotrophic complementation, transformation of *C. acremonium* made use of a dominant selectable marker, hygromycin B phosphotransferase. Mutagenesis of a highly developed production strain frequently resulted in a decline in overall productivity, thus the use of a dominant selectable marker which permitted transformation of a broad range of *C. acremonium* strains without mutagenesis was a significant advance. Several laboratories also developed transformation systems for *P. chrysogenum* shortly thereafter (BERI and TURNER, 1987; CANTORAL et al., 1987; SANCHEZ et al., 1987; SKATRUD et al., 1987a). Typical for filamentous fungi, the transformation frequencies for these organisms are low (i.e., 1–10 transformants per μg of transforming DNA). Subsequently, utilization of a homologous promoter (*pcbC* promoter) attached to the hygromycin B phosphotransferase (HPT) gene significantly increased the rate of transformation for *C. acremonium* (SKATRUD et al., 1987b). Transformation occurs by integration rather than replication – in contrast to *S. cerevisiae* where both modes of transformation are readily available. In *C. acremonium* and *P. chrysogenum,* most transformation events occurred by ectopic integration into genomic DNA; only a small percentage of the transformants harbored homologous integration events. Despite these limitations, transformation systems such as these opened the door for the application of recombinant DNA technology to these organisms.

Transformation systems for prokaryotes which produce β-lactam antibiotics have also been developed. Recently, an efficient genetic transformation system was developed for the prokaryote *Nocardia lactamdurans,* a cephamycin C-producing organism, using an endogenous plasmid found in a species of *Amycolatopsis* (KUMAR et al., 1994). This development made possible the genetic manipulation of *Nocardia* species and the related genus *Amycolatopsis,* some of which are used on an industrial scale for the production of antibiotics (e.g., efrotomycin, vancomycin, and rifamycin).

5 Enzymes Involved in Penicillin Biosynthesis

While the basic biosynthetic routes reading to the different β-lactam antibiotics (see Fig. 1) have been known for several decades, the enzymes catalyzing the reactions have only recently been isolated and studied in detail. Knowledge about primary and tertiary structures and catalytic properties of these rather unique enzymes provided a rationale for planning new biosynthetic routes and for the design of novel biocatalysts. Here, we will focus on the three enzymes leading to the formation of penicillin G. As stated above, isopenicillin N is a precursor for both penicillin G and cephalosporins, thus the first two enzymes are common to all organisms producing penicillin and cephalosporin-type compounds.

5.1 δ-(L-α-Aminoadipyl)-L-Cysteinyl-D-Valine Synthetase (ACVS)

δ-(L-α-Aminoadipyl)-L-cysteinyl-D-valine (ACV), the first common intermediate in the formation of all penicillins and cephalosporins by eukaryotic and prokaryotic microorganisms, was initially thought to be synthesized by two different enzymes in analogy to the structurally similar tripeptide glutathione. This type of process would predict that a dipeptide existed as a free intermediate. However, BANKO et al. (1986, 1987) showed that an extract of *C. acremonium* catalyzed the synthesis of ACV at a substantially higher rate from the individual amino acids than from δ-(L-α-Aminoadipyl)-L-cysteine (AC) and L-valine, and also faster than AC was formed from first component amino acids. The authors concluded that ACV synthesis in *C. acremonium* involved the action of a single multifunctional enzyme. JENSEN et al. (1988) reported on a cell-free extract of *S. clavuligerus* capable of synthesizing ACV from the appropriate component amino acids. The actual demonstration of catalysis by a multienzyme was achieved by VAN LIEMPT et al. (1989) with the isolation of a large synthetase from *A. nidulans* which catalyzed the formation of ACV from the L-amino acid precursors. The corresponding gene was identified and found to direct the biosynthesis of a protein of 422 kDa (MACCABE et al., 1991). To date several other genes encoding ACV synthetase from prokaryotes as well as from eukaryotes have been cloned and characterized. The deduced sizes of the encoded polypeptides were between 404 and 424 kDa (AHARONOWITZ et al., 1993). A common feature found in all these enzymes was the presence of three repeated modules, each comprised of more than 500 amino acids. These modules exhibited extensive sequence homology with nonribosomal peptide synthetases such as firefly luciferase and acyl-CoA ligases (AHARONOWITZ et al., 1992). So far, the multienzymes from *A. nidulans* (VAN LIEMPT et al., 1989) *S. clavuligerus* (BALDWIN et al., 1990; JENSEN et al., 1990; SCHWECKE et al., 1992a; ZHANG et al., 1992), and *C. acremonium* (BALDWIN et al., 1990; ZHANG and DEMAIN, 1990a) have been purified to homogeneity and enzymatically characterized (ZHANG, 1991; ZHANG and DEMAIN, 1992). The *S. clavuligerus* ACVS was a monomeric protein (JENSEN et al., 1990; SCHWECKE et al., 1992b; ZHANG and DEMAIN, 1990b) whereas the *C. acremonium* ACV synthetase appeared to be a homodimer with weak physical interactions between the two subunits (ZHANG and DEMAIN, 1990a, b, 1992; ZHANG et al., 1992; ZHANG, 1991). In analogy to many known peptide synthetases, ACVS activates the three substrate amino acids at the expense of the α-β-phosphate bond of ATP (SCHWECKE et al., 1992a; VAN LIEMPT et al., 1989). Formation of aminoacyl adenylates with all three amino acids was demonstrated by measuring formation of labeled ATP from $[^{32}P]$-PP_i in the presence of the single amino acids. One mole of ATP was consumed per peptide bond formed (KALLOW et al., 1994). Substrate specificity studies have demonstrated that ACVS accepts a broad range of amino acid analogs (SHIAU et al., 1995; ZHANG and DEMAIN, 1992). The α-aminoadipate dependent ATP/PP_i exchange reaction catalyzed by the *S. clavuligerus* enzyme was inhibited by Tris buffer (SCHWECKE et al., 1992a) as well as tripeptide formation (ZHANG et al., 1992). ZHANG and DEMAIN (1992) compared several β-lactam biosynthetic enzymes in crude cell-free extracts of *C. acremonium* and *S. clavuligerus* and found that ACVS possessed only 1–10% of the specific activity of isopenicillin synthase, isopenicillin epimerase, and deacetoxycephalosporin C-synthetase. After being activated as aminoacyl adenylates, the substrate amino acids were covalently bound to the cysteamine moieties of 4g-phosphopantetheine (ROLAND et al., 1975) which were attached to each activation domain of the enzyme by a posttranslational modification (BALDWIN et al., 1991; SCHLUMBOHM et al., 1991). In the final product formed by ACVS, valine occurred in the D-configuration, however, only the L-configuration of covalently bound valine could be released from the enzyme indicating that epimerization took place upon or after peptide bond formation (SHIAU et al., 1995; SCHWECKE and VON DÖHREN, unpublished data). The putative thioesterase function located in the C-terminal region of ACVS has been proposed to specifically release the LLD-tripeptide (KLEINKAUF and VON DÖHREN, 1996).

The order of the partial reactions involved in the biosynthesis of the complete tripeptide is still not completely resolved. A particulate preparation of lysed protoplasts from a *Cephalosporium* sp. incorporated DL-[^{14}C]-valine into penicillin when incubated with the dipeptide AC and ATP (LODER and ABRAHAM, 1971b). BANKO and colleagues obtained ACV from AC and valine (BANKO et al., 1986, 1987). However, SHIAU et al. (1995) recently reported on the formation of cysteinyl-valine (CV) as an intermediate in ACV biosynthesis implicating a different mechanism, which implies the initial formation of the second peptide bond. Nevertheless, the yield was low and no biosynthesis of ACV occurred from α-aminoadipate and the dipeptide CV (SHIAU et al., 1995).

Such results do not resolve the issue of whether the order of peptide bond formation is ACV or VCA. A critical requirement for analyzing and manipulating ACVS is, therefore, the ability to characterize each individual module or domain thereof. TAVANLAR and co-workers used limited proteolysis to isolate and analyze ACVS fragments from *C. acremonium* (TAVANLAR, SCHWECKE, VAN LIEMPT, and VON DÖHREN, unpublished data, cited in KLEINKAUF and VON DÖHREN, 1996). Three major bands with a Mr of approximately 116 kDa were observed and purified after digestion of a homogenous ACVS sample using either subtilisin or proteinase K. The digestion products were investigated by sequence analysis and biochemical assays. The activation of cysteine as aminoacyl adenylate could undoubtedly be attributed to the middle fragment by means of ATP/PP$_i$ exchange assay. Surprisingly, the N-terminal fragment activated valine to a higher degree than α-aminoadipate. These results were in contrast to the assumption that the order of modules are colinear with the order of building blocks in the final product, as has been convincingly shown for other peptide synthetases (HORI et al., 1991; KRAUSE et al., 1985; KRAUSE and MARAHIEL, 1988; KUROTSO et al., 1991; PIEPER et al., 1995; SAITO et al., 1995; STACHELHAUS et al., 1995; TURGAY et al., 1992; VAN SINDREN et al., 1993; WEBER et al., 1994).

Heterologous expression of individual modules in *E. coli* has so far been hampered by the fact that the fragments expressed were insoluble within the bacterium (SCHWECKE et al., 1992b; V. UHLMANN, Masters Thesis, Technical University Berlin, cited in KLEINKAUF and VON DÖHREN, 1996). After renaturation of protein from unfolded inclusion bodies containing the N-terminal module of the *S. clavuligerus* ACVS, SCHWECKE et al. (1992b) demonstrated adenylation of all three constituent amino acids with valine showing highest activity in the ATP/[32]PP$_i$ exchange reaction followed by cysteine and α-aminoadipate. This somewhat surprising result showed that this approach to determine kinetic constants is of limited use in intact ACVS. However, upon incubation with all constituent ^{14}C-labeled amino acids only α-aminoadipate gave rise to radioactivity covalently bound to the refolded fragment, supporting the idea of a modular arrangement of activation units on the multienzyme that reflects the order of amino acids in the product. No investigation into the nature of the covalent bond was conducted. Therefore, it cannot be ruled out that the observed enzyme-bound radioactivity was at least partially due to an unidentified contaminant known to be present in the ^{14}C-α-aminoadipate (SCHWECKE et al., 1992a).

While cysteine and valine are available in all cells, α-aminoadipate has to be synthesized exclusively for β-lactam formation from lysine in actinomycetes (KERN et al., 1980; KIRKPATRICK et al., 1973; MADDURI et al., 1989). Cephalosporin and cephamycin producing strains possess an enzyme, L-lysine ε-aminotransferase (LAT), which mediates the formation of α-aminoadipate by removal of the ε-amino group from lysine (MADDURI et al., 1989, 1991; VINING et al., 1990). The resulting 1-piperidine-6-carboxylic acid was postulated to be oxidized by a yet unknown dehydrogenase in analogy to a pipecolic acid pathway found in *Pseudomonas* sp. (CALVERT and RODWELL, 1966). The gene encoding LAT has been mapped to the *pcb/cef* gene cluster in β-lactam antibiotic producing strains of *Streptomyces* (COQUE et al., 1991b; MADDURI et al., 1991) and is absent in non-producers (MADDURI et al., 1989). MALMBERG and colleagues found 4-fold elevated levels of LAT and 2–5 times increased cephamycin production after introduction of an additional copy of the gene into the chromosome of *S. clavuligerus* (MALMBERG et al., 1993, 1995). Fungal β-lactam producing strains make α-aminoadipate as an intermediate in the synthesis of lysine and do not seem to require such an enzyme.

5.2 Isopenicillin N Synthase (IPNS)

A key step in the biosynthesis of β-lactam antibiotics is the conversion of the tripeptide ACV into the bicyclic fused β-lactam–thiazolidine ring structure. This particular step has been the subject of many studies over several decades. The structure of the β-lactam ring was correctly established through the work of ABRAHAM and colleagues over 50 years ago. The structure of the direct precursor (ACV) was determined in 1971 (LODER and ABRAHAM, 1971a). The enzymatic conversion of the tripeptide ACV into a β-lactam compound was finally established in 1979 (O'SULLIVAN et al., 1979). ABRAHAM et al. (1981) noted that IPNS required Fe^{2+} and molecular oxygen for activity. The catalytic activity of IPNS was greatly stimulated by the presence of ascorbate and thiol compounds such as DTT and by catalase (ABRAHAM et al., 1981). The first purification of IPNS was reported from *C. acremonium* in 1984 (PANG et al., 1984). Analysis of purified IPNS enzyme predicted a protein of 38 kDa, which required ferrous iron and ascorbate and used dioxygen as a co-substrate during catalysis. The nature of the predicted iron-binding site within the active site cleft was addressed by CHEN and colleagues (KRIAUCIUNAS et al., 1991). They suggested that the metal ion was bound to two histidine residues, rendering the otherwise highly reactive iron less damaging to the cell. The IPNS enzyme catalyzed both cyclizations found in the structure of isopenicillin N. Since the initial purification of IPNS, the enzyme has been isolated from many different sources, including fungi, gram-positive and gram-negative bacteria. In each case, IPNS was characterized as a protein with a molecular mass very close to that of IPNS from *C. acremonium.*

The IPNS reaction mechanism has been addressed in elaborate studies by BALDWIN and coworkers (BALDWIN and ABRAHAM, 1988; BALDWIN and BRADLEY, 1990). By analyzing kinetic isotope effects, they were able to show that the reaction proceeds in a two-step mechanism, the intermediate being a monocyclic β-lactam in which the sulphur is bound to iron dioxygen of the enzyme. Based on cell-free catalytic systems using purified recombinant IPNS, HUFFMAN and colleagues demonstrated that IPNS could recognize and utilize substrates similar to ACV. Such studies led to the formation of many novel β-lactam compounds, some of which possessed antibiotic activity (HUFFMAN et al., 1992).

The three-dimensional structure of IPNS has recently been elucidated by BALDWIN and coworkers (ROACH et al., 1995). In a complex with manganese, the recombinant IPNS protein from *A. nidulans* was crystallized, and the structure was determined at a resolution of 2.5 Å. Structural analysis of crystalized IPNS revealed the presence of 10 helices and 16 β-strands, eight of which fold into a jelly-roll motif. The active site was buried within the β-barrel. A glutamine residue which is conserved in all IPNS primary structures available to date interacted with the metal ion (manganese in the case of the crystalline form of IPNS) as well as two histidines and one aspartate residue. Furthermore, two water molecules were coordinated to the manganese. Based on these findings, a reaction mechanism was proposed in which ACV and dioxygen bind to the coordination sites occupied by the water molecules and glutamate.

IPNS is now the most well understood enzyme in the β-lactam biosynthetic pathway. The elucidation of the reaction mechanism aided by knowledge of the three dimensional crystal structure combined with the known amino acid sequences of IPNS molecules from many different sources may provide an avenue for rational design of IPNS proteins with new desired activities.

5.3 Acyl-Coenzyme A: Isopenicillin N Acyltransferase (AT)

The final step in the biosynthesis of hydrophobic penicillins in *P. chrysogenum* and *A. nidulans* involves the removal of the L-α-aminoadipoyl side chain from isopenicillin N and its exchange with one of many coenzyme A-derived monosubstituted acetic acids (e.g., phenylacetyl, forming benzylpenicillin) (BRUNNER et al., 1968; FAWCETT et al., 1975; ALVAREZ et al., 1987, 1993; WHITEMAN et al., 1990). Side chain exchange either occurs directly or as a two-step process, forming 6-aminopenicillanic acid (6-APA) as an intermediate (QUEENER and NEUSS, 1982) (see Fig. 3).

A single multifunctional enzyme, acyl-coenzyme A:isopenicillin N acyltransferase (AT), catalyzes all of the possible steps involved in this transacylation, including 6-APA formation *via* IPN side chain removal (IPN amidohydrolase, IAH), 6-APA acylation (acyl-CoA:6-APA acyltransferase, AAT), IPN transacylation (acyl-CoA:IPN acyltransferase, IAT), and acyl-CoA hydrolysis (ALVAREZ et al., 1993). The kinetic parameters reported for each of these reactions suggests that removal of the α-aminoadipoyl side chain from IPN is the rate-limiting step in the acyltransferase reaction (ALVAREZ et al., 1993).

The ability of the AT enzyme to accept a variety of acyl-CoA derivatives (LUENGO et al., 1986; ALONSO et al., 1988), together with the capacity of *P. chrysogenum* to form a number of acyl-CoA thioesters, has allowed the *in vivo* production of more than 100 different penicillins in *P. chrysogenum* fermentations containing the corresponding precursor acid (COLE, 1966). The production of penicillin G or V, e.g., is accomplished by precursing fermentations with phenylacetic acid or phenoxyacetic acid, respectively.

The composition of the active form of AT has been the subject of some controversy. Initial purifications of the AT enzyme suggested that it was a monomeric protein of approximately 30 kDa (ALVAREZ et al., 1987; ALONSO et al., 1988). However, subsequent protein purifications (e.g., WHITEMAN et al., 1990) have indicated that AT is a heterodimer composed of two dissimilar subunits of 11 kDa (α subunit) and 29 kDa (β subunit).

The eventual cloning of the gene encoding AT (*penDE*) from both *P. chrysogenum* (BARREDO et al., 1989c; TOBIN et al., 1990) and *A. nidulans* (MONTENEGRO et al., 1990; TOBIN et al., 1990), demonstrated that the *penDE* gene is composed of four exons encoding 357 amino acids, translated as a 40 kDa proenzyme, and posttranslationally processed to generate the two subunits (Fig. 2). Based on NH_2-terminal amino acid sequences, proenzyme cleavage occurs between Gly^{102} and Cys^{103} (BARREDO et al., 1989c; WHITEMAN et al., 1990). However, the presence of the ~11 kDa subunit in active AT has been questioned by some investigators (e.g., BARREDO et al., 1989c).

Recent electrospray mass spectrometric analysis of both recombinant and native *P. chrysogenum* AT has verified the location of the proenzyme cleavage site between Gly^{102} and CyS^{103} (APLIN et al., 1993a, b). These reports also indicated that both the 11 kDa (α subunit) and the 29 kDa (β subunit) proteins are present in purified AT forming an α,β-heterodimer, in agreement with the report of WHITEMAN et al. (1990). To further investigate the requirement for both of these subunits, TOBIN et al. (1993) separated the regions of *penDE* encoding each subunit. Independent production of either subunit in *E. coli* did not yield active AT, and post-expression complementation by the mixing of cell sonicates containing separately produced subunits did not regenerate activity. However, reconstitution of AT activity was accomplished by coproduction of the two subunits produced from separate plasmids in the same cell. Further, *in vitro* refolding of separately produced 11 kDa and 29 kDa subunits only resulted in active AT when mixed together

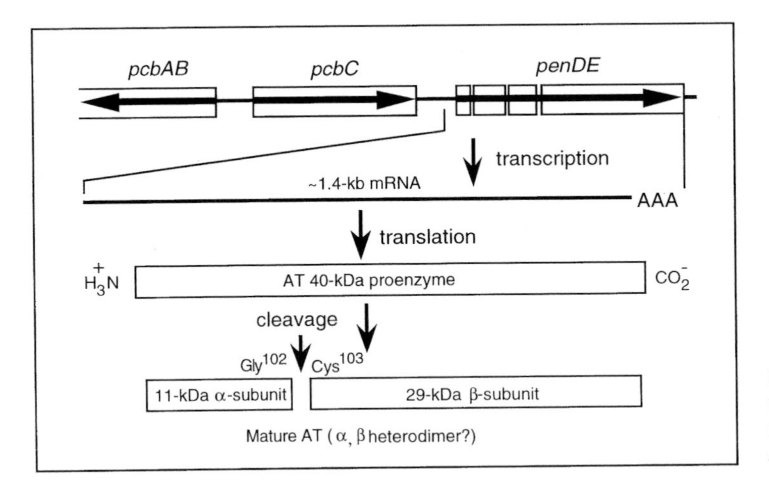

Fig. 2. Transcription, translation, and posttranslational modification of AT in *P. chrysogenum.*

prior to refolding (TOBIN, 1994). These results indicate that the 11 kDa and the 29 kDa proteins interact at an intermediate step in the folding pathway to form AT enzyme.

Studies of the posttranslational cleavage of the proenzyme have been hampered by the inability to repeatably isolate the ~40 kDa AT proenzyme from *P. chrysogenum.* Using a heterologous *penDE* expression system for *E. coli* APLIN et al. (1993a) observed the soluble production of heterodimeric AT, suggesting that other specific *P. chrysogenum* factors were not required for proenzyme cleavage. Using insolubly produced, resolubilized, and purified 40 kDa recombinant protein, TOBIN (1994) further demonstrated cleavage of the AT proenzyme by *in vitro* refolding. Cleavage occurred between Gly^{102} and Cys^{103}, as observed for native and solubly produced recombinant AT, suggesting that this event is autolytic.

Mutational analysis and recombinant expression of the *penDE* gene has identified three amino acid residues essential for AT activity. Substitution of several residues in the proenzyme cleavage site region demonstrated that Cys^{103} was absolutely required for AT proenzyme cleavage and AT activity. The presence of Ser or Ala at this position yielded uncleaved and inactive recombinant AT (TOBIN et al., 1995). Ser^{227}, residing in an S-Q-N motif in AT and other β-lactam biosynthetic enzymes was also examined for a potential role in AT activity (TOBIN et al., 1994). Substitution of this residue with Cys did not destroy AT activity, however, Ala at this position did not result in either proenzyme cleavage or activity. Further, a thioesterase-like domain (G-X-S^{309}-X-G) has been identified in AT (ALVAREZ et al., 1993). TOBIN et al. (1994) observed that $Ser^{309}\rightarrow$Cys retained activity, however, $Ser^{309}\rightarrow$Ala was not active. AT enzyme containing either of these mutations was posttranslationally cleaved, indicating that this residue is involved in the AT enzyme mechanism itself. Together with the sensitivity of AT activity to chloromercuribenzoate and iodoacetamide and the requirement for reducing agents (e.g., DTT) for AT activity (ALVAREZ et al., 1993), these results imply that a thioesterase activity may be involved in the AT enzyme mechanism. Although there are five conserved Cys residues in *P. chrysogenum* and *A. nidulans* AT, the requirement for Cys^{103} tempts speculation that this residue may bind the CoA-derived acyl group, subsequently transferred to 6-APA or IPN by a thioesterase activity. The apparent absolute requirement for proenzyme cleavage to produce active AT additionally suggests that proenzyme cleavage may be an enzyme activation mechanism.

Substrate specificity studies have demonstrated that, in addition to IPN and 6-APA, AT will hydrolyze and acylate penicillin N forming 6-APA and penicillin G, respectively (ALVAREZ et al., 1987; TOBIN, 1994). Further, it has recently been shown that the cephem nucleus 7-aminodeacetoxycephalosporanic acid (7-ADCA) is poorly acylated by recombinant AT, forming deacetoxycephalosporin G (TOBIN, 1994). Transacylation of cephalosporins containing aliphatic side chains (e.g., deacetoxycephalosporin C, deacetylcephalosporin C, and cephalosporin C) has not been observed. However, the acceptance of 7-ADCA and penicillin N (the latter containing the D-configured α-aminoadipoyl side chain) as a substrate would suggest that AT could be used to produce hydrophobic cephalosporins for industrial application to semi-synthetic cephalosporin production, pending mutagenesis, and screening for the optimization of these activities.

6 Cloning of the Genes Involved in the Biosynthesis of Penicillin G, Cephalosporin C, and Cephamycin C

A combination of approaches has been used to identify, clone, and characterize genes involved in β-lactam biosynthesis including genetic complementation, gene disruption, genetic linkage, and reverse genetics (MILLER and INGOLIA, 1989; SKATRUD, 1991). Over the past decade, a substantial array of β-lactam biosynthetic genes have been cloned and sequenced from several different organisms (Tab. 2). A striking similarity exists between corresponding genes and enzymes isolated from various organisms, despite their phylogenetic diversity. This characteristic has been exploited to permit the isolation of genes from several different organisms by cross-hybridization techniques.

7 Clustering of β-Lactam Biosynthetic Genes

The isolation of antibiotic biosynthetic genes from *Streptomyces* species revealed that these genes tend to be organized in clusters, and that they are often associated with antibiotic resistance genes (MALARTIDA and HOPWOOD, 1984; SENO and BALTZ, 1989). This clustering was also to be true for β-lactam biosynthetic genes when BAILEY et al. (1984) observed the production of clavulanic acid by introducing a cosmid clone containing a segment of *S. clavuligerus* DNA into a non-producing mutant. BURNHAM et al. (1987) provided evidence that the *cefE, cefF,* and *cmc* genes were clustered in *S. clavuligerus.* KOVACEVIC et al. (1989) later isolated a cosmid clone from *S. clavuligerus* containing both the *pcbC* and *cefE* genes involved in cephalosporin C/cephamycin C biosynthesis. This clone was subsequently used to isolate the *cefD* (KOVACEVIC et al., 1990), *cefF* (KOVACEVIC and MILLER, 1991), *lat*, and *pcbAB* genes (TOBIN et al., 1991).

Both DIÉZ et al. (1989) and BARREDO et al. (1989b) demonstrated that the *pcbC* and *penDE* genes were adjacent to one another in the filamentous fungus *P. chrysogenum.* A study employing genetic complementation and heterologous cloning techniques later indicated that all of the genes involved in the biosynthesis of penicillins resided in close proximity to one another in this organism (SMITH et al., 1990b). In addition to demonstrations of linkage in *S. clavuligerus* and *P. chrysogenum,* clustering of β-lactam biosynthetic genes has been demonstrated in *Aspergillus nidulans* (SMITH et al., 1990a) as well as in the bacteria *Flavobacterium* and *N. lactamdurans* (COQUE et al., 1991a, b). In addition, a β-lactamase gene has been identified within the *Flavobacterium* (*Lysobacter*) and *N. lactamdurans* clusters (KIMURA et al., 1990; COQUE et al., 1993). Fig. 3 illustrates the arrangement of β-lactam biosynthetic genes in various representative organisms. Recent results have demonstrated that the clavulanic acid biosynthetic gene cluster in *S. clavuligerus* is adjacent to the gene cluster for β-lac-

Tab. 2. DNA Sequences Reported for β-Lactam Biosynthetic Genes

Gene (Enzyme Endocded)	Organism	Reference
lat (lysine-ε-aminotransferase)	*S. clavuligerus*	TOBIN et al. (1991)
	N. lactamdurans	COQUE et al. (1991a)
pcbAB (ACV synthetase)	*P. chrysogenum*	SMITH et al. (1990c), DÍEZ et al. (1990)
	A. nidulans	MACCABE et al. (1991)
	C. acremonium	GUTIÉRREZ et al. (1991)
		HOSKINS et al. (1990)
	S. clavuligerus (partial)	TOBIN et al. (1991)
pcbC (IPN synthase)	*C. acremonium*	SAMSON et al. (1985)
	P. chrysogenum	CARR et al. (1986), BARREDO et al. (1989a)
	A. nidulans	RAMÓN et al. (1987), WEIGEL et al. (1988)
	S. lipmanii	WEIGEL et al. (1988)
	S. clavuligerus	LESKIW et al. (1988)
	S. jumonjinensi	SHIFFMAN et al. (1988)
	S. griseus	GARCIA-DOMINGUEZ et al. (1991)
	Flavobacterium sp. 12154	SHIFFMAN et al. (1990)
	N. lactamdurans	COQUE et al. (1991b)
penDE (IPN acyltransferase)	*P. chrysogenum*	BARREDO et al. (1989c)
	A. nidulans	MONTENEGRO et al. (1990), TOBIN et al. (1990)
cefD (IPN epimerase)	*S. clavuligerus*	KOVACEVIC et al. (1990)
	N. lactamdurans	COQUE et al. (1993)
cefEF (REX/H)	*C. acremonium*	SAMSON et al. (1987)
cefE (expandase)	*S. clavuligerus*	KOVACEVIC et al. (1989)
	N. lactamdurans	COQUE et al. (1993)
cefF (hydroxylase)	*S. clavuligerus*	KOVACEVIC and MILLER (1991)
cefG (acyltransferase)	*C. acremonium*	GUTIÉRREZ et al. (1992)
cmcH (carbamoyltransferase)	*N. lactamdurans*	COQUE et al. (1995a)
cmcI (7′-α-hydroxylase)	*N. lactamdurans*	COQUE et al. (1995b)
cmcJ (methyltransferase)	*N. lactamdurans*	COQUE et al. (1995b)

tam antibiotic biosynthesis (WARD and HODGSON, 1993).

An anomaly regarding linkage of biosynthetic genes occurs in the filamentous fungus *C. acremonium*. The early genes (*pcbAB*, *pcbC*) are linked (HOSKINS et al., 1990), however, they reside on a different chromosome from the late (*cefEF, cefG*) genes. SKATRUD and QUEENER (1989) reported that the early genes were located on chromosome VI and the late genes on chromosome II in an industrial strain of *C. acremonium* (strain 394-4). However, a later report (FIERRO et al., 1994) demonstrated that in *C. acremonium* ATCC 28901 the early and late genes resided in chromosomes VII and I, respectively. In both

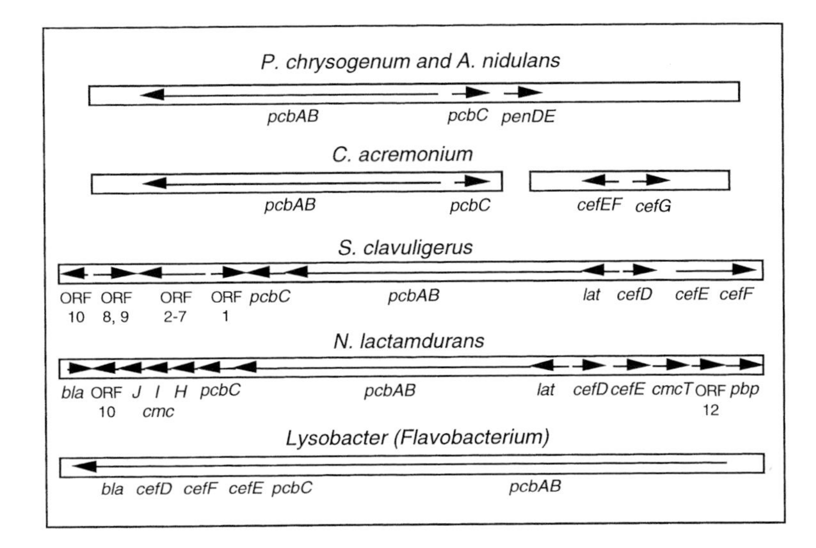

Fig. 3. Clustering of β-lactam biosynthetic genes in various organisms.

studies the early and late genes were in separate clusters. It is conceivable that these apparently conflicting reports are due to differences between the strains. The location of the *cefD* gene in *C. acremonium* has not been determined.

In *A. nidulans* ATCC 28901 the penicillin biosynthetic pathway is located on chromosome VI (3.0-Mb) (MONTENEGRO et al., 1992). In *P. chrysogenum* AS-P-78 and P2 the cluster is located on chromosome I (FIERRO et al., 1994). In *P. notatum,* the organism originally used for penicillin production, the biosynthetic genes reside on chromosome II (FIERRO et al., 1994).

The orientation of transcription of the β-lactam biosynthetic genes relative to one another is not entirely conserved. In the prokaryotes (*Flavobacterium, N. lactamdurans,* and *S. clavuligerus*) the early genes tend to be transcribed in the same direction. PETRICH and JENSEN (1994) reported that the sequential arrangement of the *lat, pcbAB,* and *pcbC* genes affords coordinate regulation at the level of transcription, involving transcription units of various lengths. In addition to a short transcript for *pcbC* alone, a polycistronic message encoding LAT, ACVS, and IPNS has been suggested by these investigators. Whether or not the *lat* gene may be transcribed by itself is not clear (S. JENSEN, University of Alberta, personal communication). In addition, KOVACEVIC et al. (1989) reported that the *cefD* and *cefE* genes are cotranslated. Thus, portions of the biosynthetic pathway constitute operons. In the fungi, however, the early genes are transcribed in opposite directions. While clustering of biosynthetic genes in fungi is common separate genes have their own promoter, so relative orientation is not important in a regulatory sense (TURNER, 1992).

8 Compartmentalization of β-Lactam Biosynthetic Enzymes

The role of subcellular compartmentalization in the biosynthesis of penicillin in *P. chrysogenum* has been elucidated in some detail. Early studies indicated that ACVS activity was associated with a particular fraction of the crude cell homogenate (FAWCETT and ABRAHAM, 1976). IPNS is apparently a soluble enzyme, although its activity in cell-free extracts appears to be stimulated by Triton X-100 or sonication (SAWADA et al., 1980). More recently, KURYLOWICZ et al. (1987) suggested that penicillin G is synthesized in the Golgi system. Pursuant to this finding, MÜLLER et al. (1991) determined that the last step in penicillin biosynthesis (acyl transfer) is accomplished in organelles of 200–800 nm in diameter. A subsequent report by this group determined that a short signal peptide (Ala-Arg-Leu) located at the COOH-terminus of the AT enzyme is required for its localization in organelles ("microbodies") and for benzylpenicillin biosynthesis (MÜLLER et al., 1992). This signal sequence is similar to that found at the COOH-terminus of several proteins known to be transported to peroxisomes (GOULD et al., 1989). In addition, evidence for the compartmentalization of the amino acid precursors required for ACV formation has been obtained (AFFENZELLER and KUBICEK, 1991; LEDENFELD et al., 1993) and confirms the suspicions of earlier studies regarding the limitation of ACVS activity *in vivo* by one of its substrates (HÖNLINGER et al., 1988; HÖNLINGER and KUBICEK, 1989). LEDENFELD et al. (1993) ultimately concluded that the α-aminoadipate, L-cysteine, and L-valine precursors for ACV formation are derived from the vacuole, and ACVS is located either within or bound to the vacuolar membrane. A current model (as suggested by LEDENFELD et al., 1993) regarding the compartmentalization of events involved in penicillin biosynthesis in *P. chrysogenum* is illustrated in Fig. 4.

In light of the compartmentalization of the penicillin biosynthetic pathway it seems likely that the alleviation of substrate limitations, particularly with regard to the formation of ACV, would be an effective route to strain improvement.

9 Evolution of the β-Lactam Biosynthetic Pathway: The Horizontal Transfer Hypothesis

The improbable identity demonstrated between particular genes and enzymes involved in β-lactam biosynthesis from a wide diversity of microorganisms has generated a great deal of speculation concerning the horizontal transfer of the β-lactam biosynthetic pathway from a prokaryote to a eukaryote. The first theory – as put forward by T. D. INGOLIA in CARR et al. (1986) and later expanded (WEIGEL et al., 1988; SKATRUD, 1991) – was designed to explain the identity of the *pcbC* genes of *P. chrysogenum* and *C. acremonium* and the presence of IPNS activity in gram-positive bacteria of the *Streptomyces* genus. As

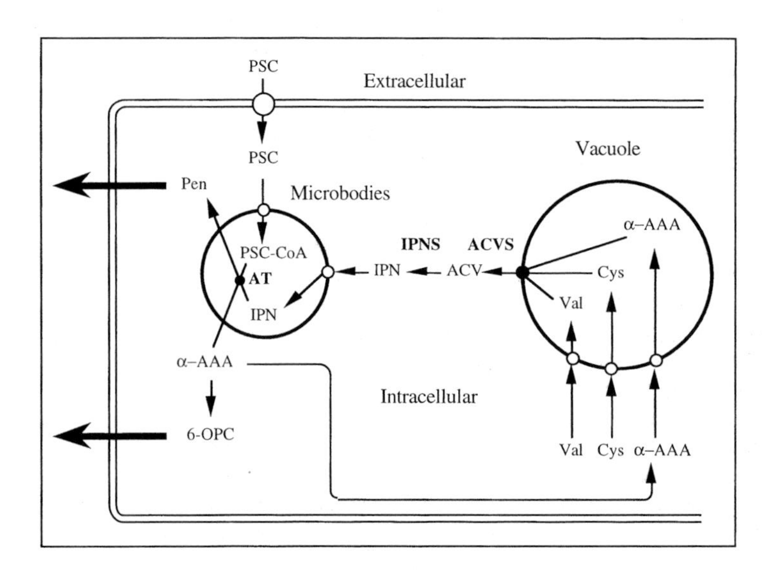

Fig. 4. Compartmentalization of the penicillin biosynthetic pathway in *P. chrysogenum* (adapted from LENDENFELD et al., 1993). Open circles represent possible transport steps, filled circles represent enzymatic steps; α-AAA: α-aminoadipic acid; PSC: precursor side chain; PSC-CoA: CoA-activated PSC; Pen: penicillin; 6-OPC: 6-oxopiperidine-2-carboxylic acid; enzymes are indicated in bold letters.

originally stated, the so-called "horizontal transfer hypothesis" argued for the transfer of the *pcbC* genes from *Streptomyces* to a eukaryotic organism. Since then, the DNA sequences of several more *pcbC* and other β-lactam biosynthetic genes have been elucidated, and the theory remains consistent with the new information. Supportive evidence from other genes, particularly the *pcbAB, penDE, cefD, cefE, cefF,* and *cefEF* genes, has lent credence to the theory of horizontal transfer (SKATRUD, 1991). The close linkage of the genes in a variety of organisms makes a single transfer event plausible. However, the DNA sequence of the *pcbC* gene from a gram-negative bacterium, *Flavobacterium,* has produced an extension/modification of the suggested transfer theory. Based on this DNA sequence AHARONOWITZ et al. (1992) suggested that more than one horizontal transfer event may have occurred during the evolution of this biosynthetic pathway.

It is virtually impossible to accurately predict or suggest how this horizontal transfer might have occurred. However, there are several known mechanisms of DNA transfer, including conjugative plasmids, DNA transformation, protoplast fusion, and bacteriophage transduction. In one study plasmids in β-lactam producing streptomyces were detected recently (KINASHI et al., 1995) The plasmids ranged in size from 12 kb to 450 kb – an adequate size to contain the entire β-lactam biosynthetic pathway in at least some cases. However, hybridization analysis did not provide evidence for the genes on these plasmids (KINASHI et al., 1995). The ability to conjugate *E. coli* to both *Saccharomyces cerevisiae* (HEINEMANN and SPRAGUE, 1989) and *Schizosaccharomyces pombe* (SIKORSKI et al., 1990) in the laboratory is a direct demonstration of DNA transfer between a prokaryote and a eukaryote. BORK and DOOLITTLE (1992) have proposed that the fibronectin type III domain of animal proteins has been acquired by bacteria through horizontal transfer from mammalian cells. In addition, bacterial acquisition of antibiotic resistance by plasmids, transposable elements, and gene transfers is well known (DAVIES, 1994) and recent examples of horizontal DNA transfers of PBP genes to *Streptococcus pneumoniae* and *Staphylococcus aureus* from unknown sources have been proposed (DOWSON et al., 1989; MATSUHASHI et al., 1986). Thus, in a mechanistic sense, a horizontal transfer of the β-lactam biosynthetic genes from a prokaryote to a eukaryote appears plausible.

While a great deal of attention has been paid to the acquisition of the biosynthetic pathway by various organisms, little notice has been given to the explanation of how a number of genes recognizing similar substrates appeared and how genetic linkage arose. There are two general scenarios: first, it is conceivable that several of the genes in the pathway arose *via* convergent evolution of substrate specificity to perform the various reactions. The second possibility is that some of the genes may have arisen by divergent evolution, requiring (1) a gene duplication, and (2) divergent evolution of function. There does not appear to be a great deal of DNA or amino acid homology between most genes/enzymes for different steps in the pathway, with the exception of a conserved motif (S...Q...N) identified by KOVACEVIC and MILLER (1991) and TOBIN et al. (1994). This motif is surrounded by a block of similar amino acids in all the enzymes examined. However, the case of the expandase and hydroxylase functions provides an elegant testimonial for the second possibility. In *C. acremonium,* one gene (*cefEF*) encodes both activities, however, in *S. clavuligerus,* these two activities are encoded by separate genes (*cefE* and *cefF*). In the latter organism, each enzyme retains a residual level of the other activity, however, the DNA and amino acid sequences are strikingly similar (KOVACEVIC and MILLER, 1991). The level of identity indicates that the separate *cefE* and *cefF* genes arose by gene duplication in *S. clavuligerus* (KOVACEVIC and MILLER, 1991). An extended account of the horizontal transfer hypothesis has been given in SKATRUD (1991).

10 β-Lactam Antibiotics and Recombinant DNA Technology: Practical Applications

The use of gene dosage of genes involved in penicillin biosynthesis for industrial strain improvement of *P. chrysogenum* was unsuccessful (SKATRUD et al., 1987a). Shortly thereafter, evidence emerged which explained such results. Analysis of industrial strains of *P. chrysogenum* revealed that the cluster of genes responsible for penicillin production was reiterated several times (BARREDO et al., 1989b; SMITH et al., 1989; TOBIN, unpublished data). Thus gene dosage was already utilized by these strains as a means of increasing overall productivity. To further increase yields in such strains, it may be more fruitful to study and alter regulation of gene expression within the penicillin biosynthetic cluster.

In contrast, gene dosage studies in *C. acremonium* did meet with success. In the first example, characterization of an industrial strain of *C. acremonium* suggested that the expandase/hydroxylase step was rate-limiting in the biosynthesis of cephalosporin C due to the accumulation of isopenicillin N (see Fig. 1). To overcome this rate limitation, this strain was transformed with a plasmid containing the *cefEF* gene encoding expandase/hydroxylase. Isolate LU4-79-6 contained one extra copy of *cefEF* ectopically integrated into its genome. In fermentations utilizing strain LU4-79-6, isopenicillin N no longer accumulated, suggesting that this rate-limitation was overcome. Further analysis revealed that this strain possessed twice the amount of expandase/hydroxylase activity as compared to the recipient and produced approximately 15% more cephalosporin C when scaled up to industrial level (SKATRUD et al., 1989). In another example, the *cefG* gene which is directly adjacent to *cefEF* was cloned from *C. acremonium* (MATHISON et al., 1993). This gene encodes the enzyme which carries out the last step in cephalosporin C production (Fig. 1). These investigators returned multiple copies of this gene *via* genetic transformation to a different strain of *C. acremonium*. They were able to show a direct correlation between *cefG* copy number, message level, and cephalosporin C titer suggesting a different rate-limitation in their strain. It is interesting to note that in the study by SKATRUD et al. (1989) the fragment containing the *cefEF* gene was large enough to include the *cefG* gene as well. Thus *cefG* encoded activity may have been augmented as well, depending on the site of integration within the transforming DNA, and may have exerted some influence on the results in that study.

Several steps in the biosynthesis of cephalosporin C are oxygen dependent. DEMODENA et al. (1993) reasoned that increasing the intracellular availability of oxygen might positively affect overall antibiotic productivity. To accomplish this goal they expressed a bacterial hemoglobin gene in *C. acremonium* (strain C10). Fermentation results demonstrated increased cephalosporin C production in some transformants carrying the bacterial hemoglobin gene. It should be noted, however, that the strain transformed was not a highly developed strain in terms of cephalosporin C production. Also, fermentation analyses were performed in shake flasks (baffled and unbaffled) in which oxygen limitation is an inherent problem. It is not clear that the same type of effect would be observed in large-scale fermentations.

Another exciting area for application of recombinant DNA technology to β-lactam producing organisms was the production of intermediate compounds useful in the manufacture of semi-synthetic cephalosporins (SKATRUD, 1992). Medically important cephalosporins such as cephalexin, cephradine, and cefadroxil are made by the addition of different side-chains to the 7-amino group of 7-aminodeacetoxycephalosporin C (7-ADCA) (BUNNELL et al., 1986). This procedure normally includes several chemical modification steps with production of organic solvent waste products. A biosynthetic/enzymatic process to produce 7-ADCA was proposed by CANTWELL et al. (1990) which would eliminate three chemical modifications and reduce pollution. The proposed process involved *in vivo* ring expansion of penicillin V in *P. chrysogenum* via a genetically engineered *cefE* gene to form cephalosporin V and subsequent enzymatic deacylation resulting in the desired end product, 7-ADCA. The initial step in establishing feasibility of such a process was accomplished by these investigators. The *cefE* gene from *S. clavuligerus* was modified by fusing it in frame with the *pcbC* promoter of *P. chrysogenum*. The modified gene was transformed into an industrial strain of *P. chrysogenum*. Transformants were recovered which maintained their capacity to produce penicillin and possessed expandase activity (CANTWELL et al., 1990). In order for this work to achieve maximal impact, the amount of *cefE* gene product produced in *P. chrysogenum* transformants would have to be increased and the substrate specificity of *cefE* would have to be modified to efficiently act on penicillin V.

CRAWFORD et al. (1995) took a somewhat different approach to achieve the production of 7-ADCA in *P. chrysogenum*. Their process involved the expression of the *S. clavuligerus cefE* or the *C. acremonium cefEF* gene in genetic transformants of *P. chrysogenum*, and subsequent fermentation in the presence of adipic acid. Transformants produced adipoyl-6-APA (presumably by acyltransferase-catalyzed transacylation of IPN and adipoyl-CoA), which is subsequently ring-expanded, producing high titers of adipoyl-7-aminodeacetoxycephalosporanic acid (ad-7-ADCA) (*cefE*-catalyzed), or adipoyl-7-aminodeacetylcephalosporanic acid (ad-7-ACA) (*cefEF*-catalyzed). The adipic acid side chain from ad-7-ADCA is removed by the *Proteus* SY77-1 cephalosporin acylase, producing 7-ADCA in high yields. Current work is directed at attaching this acylase to a resin to enable the removal of the adipic acid from the acylase reaction. One apparent drawback to this approach may involve isolation of ad-7-ADCA or ad-7-ACA from the fermentation broth, however, these products may prove to be extractable by an organic solvent. It is interesting to note that the observed *in vivo* ring expansion of adipoyl-6-APA could not be duplicated *in vitro* with cell-free extracts of *S. clavuligerus* (MAEDA et al., 1995). YEH et al. (1994) similarly reported no *in vitro* ring expansion of adipoyl-6-APA with expandase or expandase/hydroxylase produced by recombinant *E. coli* cells. In contrast, BALDWIN et al. (1987), working with cell-free extracts from *C. acremonium*, were able to detect ring expansion of this substrate.

11 Regulation of β-Lactam Biosynthetic Gene Expression

Investigations of fungal strains used in the production of β-lactam antibiotics have revealed two general strategies to enhance productivity. In general terms, increasing the amount of biosynthetic enzymes present in an organism should increase productivity as long as primary metabolites or other factors utilized in biosynthesis are not depleted. Two routes for increasing the concentration of biosynthetic enzymes are readily apparent. An increase in the rate of transcription (i.e., up-regulation) or an increase in the number of genes (i.e., gene dosage) or a combination of both, should achieve enhanced production of β-lactams. As detailed earlier, genes involved in the biosynthesis of β-lactam antibiotics are closely linked in several organisms. Such close physical proximity of the biosynthetic genes entails obvious benefits regarding the inheritance of these genes and may simplify the possibility of coordinate transcriptional regulation.

In *P. chrysogenum*, physical linkage of the penicillin biosynthetic genes proved to be an advantage in terms of improving overall penicillin productivity by gene dosage. Two groups independently demonstrated that the penicillin biosynthetic cluster in *P. chrysogenum* was amplified several-fold in strains highly developed for penicillin production (SMITH et al., 1990a; DÍEZ et al., 1990). The three genes formally recognized as a part of the biosynthetic pathway are contained within an approximately 15-kb region. The total amplified region was about 35 kb in size. In a more recent study, FIERRO et al. (1995) found a 106.5-kb region repeated several-fold in tandem on chromosome I in a different strain of *P. chrysogenum* which was highly developed for the production of penicillin. Another strain they investigated contained a ~58-kb repeat containing the penicillin biosynthetic pathway. These investigators also noted a hexanucleotide repeat (TTTACA) which bounded the amplified units in all strains examined, regardless of copy number or size of amplified unit. FIERRO et al. (1995) suggested that amplification occurred by mutation-induced site-specific recombination at the conserved hexanucleotide sequence.

A salient point regarding the amplified unit of genomic DNA in *P. chrysogenum* is that the length of the *pcbAB, pcbC,* and *penDE* genes is only a fraction of the amplified region. It is possible that other genes involved in penicillin biosynthesis and the regulation thereof are encoded by this region. Indeed, a recent report suggests that the *pcbAB* and *pcbC* genes are subject to *trans*-acting regulation (CHU et al., 1995). A transcriptional regulatory protein which binds to the upstream regions of both *pcbAB* (−32 to −200 and −188 to −344) and *pcbC* (−10 to −131) was identified by electromobility shift assay. Binding of this regulatory protein to these regions results in a repression of transcription of both genes. A similar situation has been observed in *A. nidulans* (PÈREZ-ESTEBAN et al., 1993). Whether these elements effect *penDE* transcription, either directly or indirectly by repression of *pcbAB* and *pcbC* transcription, has not been reported. FENG et al. (1995) reported several nuclear proteins which bind to the intergenic region between *pcbAB* and *pcbC.* One particularly abundant nuclear protein, nuclear factor A (NF-A), recognizes and binds to a specific 8 base pair sequence (GCCAAGCC) found in the intergenic region. BRAKHAGE and VANDENBRULLE (1995) utilized a reporter gene system in *A. nidulans* to identify recessive *trans*-acting mutations residing in two complementation groups (prgA1 and prgB1) which regulate β-lactam biosynthesis. Mutations in these complementation groups resulted in a depression of penicillin production.

Although not clearly understood at the molecular level, it has been known for years that penicillin biosynthesis is sensitive to environmental factors such as nitrogen source, carbon source, and growth stage in *P. chrysogenum* suggesting the presence of several regulatory factors. ESPESO and PENALVA (1992) described a temporal pattern of expression of penicillin biosynthetic genes in *A. nidulans* which was dependent upon carbon source. An enhancement of that knowledge came when FENG et al. (1994) verified that expression of the penicillin biosynthetic genes in *P. chrysogenum* was regulated by these factors at the level of transcription. Using a reporter gene analysis, their study revealed that promoters found within the intergenic region between *pcbAB* and *pcbC* were sensitive to nitrogen repression, glucose repression, and growth stage. The gene (NRE) encoding the major nitrogen regulatory protein has recently been cloned from *P. chrysogenum* (HAAS et al., 1995). This gene resembles other major nitrogen regulatory genes (e.g., 60% identical to *areA* from *A. nidulans* and 30% identical to *nit2* from *Neurospora crassa*). HAAS and MARZLUF (1995) carried these studies further by demonstrating that the NRE encoded protein binds specifically to the intergenic promoter regions of nitrate assimilation and penicillin biosynthetic gene clusters of *P. chrysogenum.* All three of the above mentioned major nitrogen regulatory proteins are members of the GATA protein family – as such they are DNA binding proteins characterized by the presence of a zinc finger motif (Cys-X_2-Cys-X_{17}-Cys-X_2-Cys) and they recognize the sequence GATA which is present in the intergenic region of *pcbAB*/*pcbC*. ESPESO et al. (1993) and TILBURN et al. (1995) demonstrated that the *pcbC* gene was regulated by PACC, a regulatory protein which mediates pH response of several other *A. nidulans* genes as well. THEN BERGH et al. (1996) identified a major *cis*-acting DNA element, located between *pcbAB* and *pcbC* in *A. nidulans* which regulates expression of both genes. In this organism 872 base pairs separate these two divergently transcribed genes. Deletion of nucleotides −353 to −432 upstream of *pcbAB* lead to an altered expression of both genes. Further, analysis revealed the sequence CCAAT (designated PENR1) found within this region, was bound specifically by a protein or protein complex. Evidence was also provided which suggested that the corresponding genes in *C. aremonium* and *P. chrysogenum* contained a similar regulatory circuit.

The regulation of genes involved in cephalosporin C production in *C. acremonium* or cephamycin C production in the streptomycetes has not been studied as extensively as those involved in penicillin production. It is interesting to note that the first two genes of the biosynthetic pathway (*pcbAB* and *pcbC*) in *C. acremonium* are physically arranged in the same manner as in *A. nidulans* and *P. chrysogenum* (see Fig. 2). Despite a similar physical arrangement, gene amplification for the cephalosporin C biosynthetic pathway has not been observed. This observation suggests that alteration in gene regulation produced enhanced cephalosporin C production in industrial strains of *C. acremonium.* The relative strength of the *pcbAB* and *pcbC* promoters was recently studied with a reporter gene system (MENNE et al., 1994). Results of that study suggested that the *pcbC* promoter was approximately five times stronger than the *pcbAB* promoter. Such a contrast in the relative strength of promoters in the same biosynthetic pathway might suggest differences in protein stability, protein localization (i.e., compartmentalization vs. cytoplasmic), or specific activity of the enzymes involved in these steps. Regulation of the rate of transcription for the cephalosporin C biosynthetic pathway can be positively influenced by the addition of exogenous methionine to the medium (VELASCO et al., 1994).

12 Summary

The last decade has produced numerous significant advances in the β-lactam antibiotic field. Most noteworthy perhaps has been the application of recombinant DNA technology to the organisms which produce these lifesaving compounds. This application of technology has opened new doors to many avenues of research which were in some instances becoming less fruitful (e.g., strain improvement studies). Currently, strain improvement may be approached not only from the standpoint of enhanced productivity but also biosynthetic pathway engineering resulting in production of valuable new end products useful in the manufacture of semisynthetic β-lactam compounds (SKATRUD, 1992).

In the same time frame, the clinical effectiveness of β-lactam antibiotics was once again threatened by wide-spread emergence of serious resistance problems (e.g., methicillin-resistant *Staphylococcus aureus* [MRSA], penicillin-resistant *Streptococcus pneumoniae*). The primary mode of resistance in these organisms was not due to the elaboration of β-lactamases which has been a major clinical problem in the past. Rather, the targeted penicillin-binding proteins (PBPS) were either modified by mutation (PBP2x of *S. pneumoniae*) or replaced with a novel PBP (PBP2a in MRSA) resulting in a target with low affinity for β-lactam antibiotics (SPRATT, 1989; HARTMAN and TOMASZ, 1984). Once again recombinant DNA technology may have a dramatic impact on the use of β-lactam antibiotics. However, in this case the application of this technology will also be on the organisms which are killed by these antibiotics, not on the producing organisms. One novel approach to development of new antibiotics targeted at the modified PBPs may involve structure-based drug design in which the three-dimensional structure of a low-affinity PBP will be determined (WU et al., 1992). Based on that information, modifications to existing compounds may lead to new more effective β-lactam antibiotics and novel approaches for rapid diagnostic detection of problematic organisms (ÜNAL et al., 1992).

13 References

ABRAHAM, E. P., CHAIN, E, FLETCHER, C. M., GARDNER, A. D., HEATLEY, N. G., JENNINGS, M. A., FLOREY, H. W. (1941), Further observations on penicillin, *Lancet* **241** (2), 177–188.

ABRAHAM, E. P., HUDDLESTON, J. A., JAYATILAKE, G. S., O'SULLIVAN, J., WHITE, R. L. (1981), in: Recent Advances in the Chemistry of β-Lactam Antibiotics, *R. Soc. Chem.,* Special Publication No. 38, pp. 125–134.

AFFENZELLER, K., KUBICEK, C. P. (1991), Evidence for a compartmentation of penicillin biosynthesis in a high- and a low-producing strain of *Penicillium chrysogenum, J. Gen. Microbiol.* **137**, 1653–1660.

AHARONOWITZ, Y., COHEN, G., MARTÍN, J. F. (1992), Penicillin and cephalosporin biosynthetic genes: structure, organization, regulation and evolution, *Annu. Rev. Microbiol.* **46**, 461–495.

AHARONOWITZ, Y., BERGMEYER, H., CANTORAL, J. M., COHEN, G., DEMAIN, A. L., FINK, U., KINGHORN, J., LANGE, M., MCCABE, A., PALISA, H., PFEIFER, E., SCHWECKE, T., VAN LIEMPT, H., VON DÖHREN, H., ZHANG, J. (1993), δ-(L-α-Aminoadipyl)-L-cysteinyl-D-valine synthetase, the multienzyme integrating the four primary reactions in β-lactam biosynthesis, as a model peptide synthetase, *Bio/Technology* **11**, 807–810.

ALONSO, M. J., BERMEJO, F., REGLERO, A., FERNÁNDEZ-CAÑÓN, J. M., GONZÁLEZ DE BUITRAGO, G., LUENGO, J. M. (1988), Enzymatic synthesis of penicillins, *J. Antibiot.* **41**, 1074–1084.

ALVAREZ, E., CANTORAL, J. M., BARREDO, J. L., DÍEZ, B., MARTÍN, J. F. (1987), Purification to homogeneity and characterization of acyl-coenzyme A: 6-aminopenicillanic acid acyltransferase of *Penicillium chrysogenum, Antimicrob. Agents Chemother.* **31**, 1675–1682.

ALVAREZ, E., MEESSCHAERT, B., MONTENEGRO, E., GUTIÉRREZ, S., DÍEZ, B., BARREDO, J. L., MARTÍN, J. F. (1993), The isopenicillin-N acyltransferase of *Penicillium chrysogenum* has isopenicillin-N amidohydrolase, 6-aminopenicillanic acid acyltransferase and penicillin amidase activities, all of which are encoded by the single *penDE* gene, *Eur. J. Biochem.* **215**, 323–332.

APLIN, R. T., BALDWIN, J. E., COLE, S. C. J., SUTHERLAND, J. D., TOBIN, M. B. (1993a), On the production of α,β-heterodimeric acyl-coenzyme A:isopenicillin N acyltransferase of *Penicillium chrysogenum:* studies using a recombinant source, *FEBS Lett.* **319**, 166–170.

APLIN, R. T., BALDWIN, J. E., ROACH, P. L., ROBINSON, C. V., SCHOFIELD, C. J. (1993b), Investigations into the post-translational modification and mechanism of acyl-coenzyme A:isopenicillin N acyltransferase using electrospray mass spectrometry, *Biochem. J.* **294**, 357–363.

BAILEY, C. R., BUTLER, M. J., NORMANSELL, I. D., ROWLANDS, R. T., WINSTANLEY, D. J. (1984), Cloning a *Streptomyces clavuligerus* genetic locus involved in clavulanic acid biosynthesis, *Bio/Technology* **2**, 808–811.

BALDWIN J. E., ABRAHAM, E. P. (1988), The biosynthesis of penicillin and cephalosporins, *Nat. Prod. Rep.* **5**, 129–145.

BALDWIN, J. E., BRADLEY, M. (1990), Isopenicillin N synthase mechanistic studies, *Chem. Rev.* **90**, 1079–1088.

BALDWIN, J. E., ADDLINGTON, R. M., COATES, J. B., CRABBE, M. J., CROUCH, N. P., KEEPING, J. W., KNIGHT, B. C., SCHOFIELD, C. J., TING, H.-H., VALLEJO, C. A., THORNILEY, M., ABRAHAM, E. P. (1987), Purification and initial characterization of an enzyme with deacetoxycephalosporin C synthase and hydroxylase activities, *Biochem. J.* **245**, 831–841.

BALDWIN, J. E., BIRD, J. W., FIELD, R. A., O'CALLAGHAN, N. M., SCHOFIELD, C. J. (1990), Isolation and partial characterization of ACV synthetase from *Cephalosporium acremonium* and *Streptomyces clavuligerus, J. Antibiot.* **43**, 1055–1057.

BALDWIN, J. E., BIRD, J. W., FIELD, R. A., O'CALLAGHAN, N. M., SCHOFIELD, C. J., WILLIS, A. C. (1991), Isolation and partial characterization of ACV synthetase from *Cephalosporium acremonium* and *Streptomyces clavuligerus:* evidence for the presence of phosphopantothenate in ACV synthetase, *J. Antibiot.* **44**, 241–248.

BANKO, G., WOLFE, S., DEMAIN, A. L. (1986), Cell-free synthesis of δ-(L-α-aminoadipyl)-L-cysteine, the first intermediate of penicillin and cephalosporin biosynthesis, *Biochem. Biophys. Res. Commun.* **137**, 528–535.

BANKO, G., DEMAIN, A. L., WOLFE, S. (1987), δ-(L-α-Aminoadipyl)-L-cysteinyl-D-valine synthetase (ACV synthetase): a multifunctional enzyme with broad substrate specificity for the synthesis of penicillin and cephalosporin precursors, *J. Am. Chem. Soc.* **109**, 2858–2860.

BARREDO, J. L., CANTORAL, J. M., ALVAREZ, E., DÍEZ, B., MARTÍN, J. F. (1989a), Cloning, sequence analysis and transcriptional study of the isopenicillin N synthase of *Penicillium chrysogenum* AS-P-78, *Mol. Gen. Genet.* **216**, 91–98.

BARREDO, J. L., DIÉZ, B., ALVAREZ, E., MARTÍN, J. F. (1989b), Large amplification of a 35-kb DNA fragment carrying two penicillin biosynthetic genes in high penicillin producing strains of *Penicillium chrysogenum, Curr. Genet.* **16**, 453–459.

BARREDO, J. L., VAN SOLINGEN, P., DÍEZ, B., ALVAREZ, E., CANTORAL, J. M., KATTEVILDER, A., SMAAL, E. B., GROENEN, M. A. M., VEENSTRA, A. E., MARTÍN, J. F. (1989c), Cloning and characterization of the acyl-coenzyme A:6-aminopenicillanic acid acyltransferase gene of *Penicillium chrysogenum, Gene* **83**, 291–300.

BERI, R. K., TURNER, G. (1987), Transformation

of *P. chrysogenum* using the *A. nidulans amdS* gene as a dominant selective marker, *Curr. Genet.* **11**, 639–641.

BORK, P., DOOLITTLE, R. F. (1992), Proposed acquisition of an animal protein domain by bacteria, *Proc. Natl. Acad. Sci. USA* **89**, 8990–8994.

BRAKHAGE, A. A., VANDENBRULLE, J. (1995), Use of reporter genes to identify recessive *trans*-acting mutations specifically involved in the regulation of *Aspergillus nidulans* penicillin biosynthesis genes, *J. Bacteriol.* **177**(10), 2781–2788.

BRUNNER, R., RÖHR, M., ZINNER, M. (1968), Zur Biosynthese des Penicillins, *Hoppe-Seyler's Z. Physiol. Chem.* **349**, 95–103.

BUNNEL, C. A., LUKE, W. D., PERRY, F. M., JR. (1986), in: *β-Lactam Antibiotics for Clinical Use* (QUEENER, S. F., WEBER, J. A., QUEENER, S. W., Eds.), pp. 255–284. New York: Marcel Dekker.

BURNHAM, M. K. R., HODGSON, J. E., NORMANSELL, I. D. (1987), Isolation and expression of genes involved in the biosynthesis of β-lactams, *Eur. Patent Application* No. 0233715.

CALVERT, A. F., RODWELL, V. W. (1966), Metabolism of pipecolic acid in a *Pseudomonas sp.*, *J. Biol. Chem.* **241**, 409–414.

CANTORAL, J. M., DIEZ, B., BARREDO, J. L., ALVAREZ, E., MARTÍN, J. E. (1987), High-frequency transformation of *P. chrysogenum, Bio/Technology* **5**, 494–497.

CANTWELL, C. A., BECKMANN, R. J., DOTZLAF, J. E., FISHER, D. L., SKATRUD, P. L., YEH, W.-K., QUEENER, S. W. (1990), Cloning and expression of a hybrid *S. clavuligerus cefE* gene in *P. chrysogenum, Curr. Genet.* **17**, 213–221.

CARR, L. G., SKATRUD, P. L., SCHEETZ, M. E., QUEENER, S. W., INGOLIA, T. D. (1986), Cloning and expression of the isopenicillin N synthetase gene from *Penicillium chrysogenum, Gene* **48**, 257–266.

CHU, Y. W., RENNO, D., SAUNDERS, G. (1995), Detection of a protein which binds specifically to the upstream region of the *pcbAB* gene in *Penicillium chrysogenum, Curr. Genet.* **28**(2), 184–189.

COLE, M. (1966), Microbial synthesis of penicillins: Part 1, *Process Biochem.* **1**, 334–338.

COQUE, J. J. R., LIRAS, P., LAIZ, L., MARTÍN, J. F. (1991a), A gene encoding lysine 6-aminotransferase, which forms the β-lactam precursor α-aminoadipic acid, is located in the cluster of cephamycin biosynthetic genes in *Nocardia lactamdurans, J. Bacteriol.* **173**, 6258–6264.

COQUE, J. J. R., MARTÍN, J. F., CALZADA, J. G., LIRAS, P. (1991b), The cephamycin biosynthetic genes *pcbAB,* encoding a large multidomain peptide synthetase, and *pcbC* of *Nocardia lactamdurans* are clustered together in an organization different from the same genes in *Acremonium chrysogenum* and *Penicillium chrysogenum, Mol. Microbiol.* **5**, 1125–1133.

COQUE, J. J. R., LIRAS, P., MARTÍN, J. F. (1993), Genes for a β-lactamase, a penicillin-binding protein and a transmembrane protein are clustered with the cephamycin biosynthetic genes in *Nocardia lactamdurans, EMBO J.* **12**, 631– 639.

COQUE, J. J. R., LLARENA, F. J. P., ENGUITA, F. J., DE LA FUENTE, J. L., MARTÍN, J. F., LIRAS, P. (1995a), Characterization of the *cmcH* genes of *Nocardia lactamdurans* and *Streptomyces clavuligerus* encoding a functional 3′-hydroxymethylcephem O-carbamoyltransferase for cephamycin biosynthesis, *Gene* **162**, 21–27.

COQUE, J. J. R., ENGUITA, F. J., MARTÍN, J. F., LIRAS, P. (1995b), A two-protein component 7α-cephem-methoxylase encoded by two genes of the cephamycin C cluster converts cephalosporin C to 7-methoxycephalosporin C, *J. Bacteriol.* **177** (8), 2230–2235.

CRAWFORD, L., STEPAN, A. M., MCADA, P. C., RAMBOSEK, J. A., CONDER, M. J., VINCI, V. A., REEVES, C. D. (1995), Production of cephalosporin intermediates by feeding adipic acid to recombinant *Penicillium chrysogenum* strains expressing ring expansion activity, *Bio/Technology* **13**, 58–62.

DAVIES, J. (1994), Inactivation of antibiotics and the dissemination of resistance genes, *Science* **264**, 375–382.

DEMODENA, J. A., GUTIERREZ, S., VELASCO, J., FERNANDEZ, F. J., FACHINI, R. A., GALAZZO, J. L., HUGHES, D. E., MARTÍN, J. F. (1993), The production of cephalosporin C by *Acremonium chrysogenum* is improved by the intracellular expression of a bacterial hemoglobin, *Bio/Technology* **11**, 926–929.

DÍEZ, B., BARREDO, J. L., ALVAREZ, E., CANTORAL, J. M., VAN SOLINGEN, P., GROENEN, M. A. M., VEENSTRA, A. E., MARTÍN, J. F. (1989), Two genes involved in penicillin biosynthesis are linked in a 5.1 kb *Sal* I fragment in the genome of *Penicillium chrysogenum, Mol. Gen. Genet.* **218**, 572–576.

DÍEZ, B., GUTIÉRREZ, S., BARREDO, J. L., VAN SOLINGEN, P., VAN DER VOORT, L. H. M., MARTÍN, J. F. (1990), The cluster of penicillin biosynthetic genes: Identification and characterization of the *pcbAB* gene encoding the α-aminoadipyl-cysteinyl-valine synthetase and linkage to the *pcbC* and *penDE* genes, *J. Biol. Chem.* **265**, 16358–16365.

DOWSON, C. G., HUTCHISON, A., BRANNIGAN, J. A., GEORGE, R. C., HANSMAN, D., LIÑARES, J., TOMASZ, A., MAYNARD SMITH, J., SPRATT, B.

G. (1989), Horizontal transfer of penicillin binding protein genes in penicillin-resistant clinical isolates of *Streptococcus pneumoniae, Proc. Natl. Acad. Sci. USA* **86**, 8842–8846.

ESPESO, E. A., PENALVA, M. A. (1992), Carbon catabolite repression can account for the temporal pattern of expression of a penicillin biosynthetic gene in *Aspergillus nidulans, Mol. Microbiol.* **6** (11), 1457–1465.

ESPESO, E. A., TILBURN, J., ARST, H. N., JR., PENALVA, M. A. (1993), pH regulation is a major determinant in expression of a fungal biosynthetic gene, *EMBO J.* **12**, 3947–3956.

FAWCETT, P. A., ABRAHAM, E. P. (1976), δ-(L-α-Aminoadipyl)cysteinylvaline synthetase, *Methods Enzymol.* **43**, 471–473.

FAWCETT, P. A., USHER, J. J., ABRAHAM, E. P. (1975), Behavior of tritium-labelled isopenicillin N and 6-aminopenicillanic acid acyltransferase from *Penicillium chrysogenum, Biochem. J.* **151**, 741–746.

FENG, B., FRIEDLIN, E., MARZLUF, G. A. (1994), A reporter gene analysis of penicillin biosynthesis gene expression in *Penicillium chrysogenum* and its regulation by nitrogen and glucose catabolite repression, *App. Environ. Mirobiol.* **60** (12), 4432–4439.

FENG, B., FRIEDLIN, E., MARZLUF, G. A. (1995), Nuclear DNA-binding proteins which recognize the intergenic control region of penicillin biosynthetic genes, *Curr. Genet.* **27**, 351–358.

FIERRO, F., MONTENEGRO, E., MARCOS, A. T., MARTÍN, J. F. (1994), Chromosomal location of the β-lactam biosynthetic genes in fungal overproducing strain, presented at *7th Int. Symp. Genet. Ind. Microorg.*, Montreal, Quebec, Canada, June 26–July 1, 1994.

FIERRO, F., BARREDO, J. L., DÍEZ, B., GUTIÉRREZ, S., FERNÁNDEZ, F. J., MARTÍN, J. F. (1995), The penicillin cluster is amplified in tandem repeats linked by conserved hexanucleotide sequences, *Proc. Natl. Acad. Sci. USA* **92**, 6200–6204.

FLEMING, A. (1929), On the antibacterial action of cultures of a *Penicillium,* with special reference to their use in the isolation of *B. influenzae, Br. J. Exp. Pathol.* **10**, 226–236.

GARCIA-DOMINGUEZ, M., LIRAS, P., MARTÍN, J. F. (1991), Cloning and characterization of the isopenicillin N synthase gene of *Streptomyces griseus* NRRL 3851 and studies of expression and complementation of the cephamycin pathway in *Streptomyces clavuligerus, Antimicrob. Agents Chemother.* **35**, 44–52.

GIESBRECHT, P., FRANZ, M., KRÜGER, D., LABISCHINSKI, H., WECKE, J. (1993), Bacteriolysis of Staphylococci is only a side-effect of penicillin-induced death, in: *50 Years of Penicillin Application: History and Trends* (KLEINKAUF, H., VON DÖHREN, H., Eds.), Technische Universität Berlin, Germany (printed by Public Ltd., Czech Republic), pp. 353–363.

GOULD, S. J., KELLER, G.-A., HOSKEN, N., WILKINSON, J., SUBRAMANI, S. (1989), A conserved tripeptide sorts proteins to peroxisomes, *J. Cell Biol.* **108**, 1655–1664.

GUTIÉRREZ, S., DÍEZ, B., MONTENEGRO, E., MARTÍN, J. F. (1991), Characterization of the *Cephalosporium acremonium pcbAB* gene encoding α-aminoadipyl-L-cysteinyl-D-valine synthetase, a large multidomain synthetase: linkage to the *pcbC* gene as a cluster of early cephalosporin biosynthetic genes and evidence of multiple functional domain, *J. Bacteriol.* **173**, 2354–2365.

GUTIÉRREZ, S., VELASCO, J., FERNANDEZ, F. J., MARTÍN, J. F. (1992), The *cefF* gene of *Cephalosporium acremonium* is linked to the *cefEF* gene and encodes a deacetylcephalosporin C acetyltransferase closely related to homoserine *O*-acetyltransferase, *J. Bacteriol.* **174**, 3056–3064.

HAAS, H., MARZLUF, G. A. (1995), NRE, the major nitrogen regulatory protein of *Penicillium chrysogenum,* binds specifically to elements in the intergenic promoter regions of nitrate assimilation and penicillin biosynthetic gene clusters, *Curr. Genet.* **28**, 177–183.

HAAS, H., BAUER, B., REDL, B., STOFFLER, G., MARZLUF, G. A. (1995), Molecular cloning and anlysis of *nre,* the major nitrogen regulatory gene of *P. chrysogenum, Curr. Genet.* **27**, 150–158.

HARTMAN, B. J., TOMASZ, A. (1984), Low-affinity penicillin-binding protein associated with β-lactam resistance in *Staphylococcus aureus, J. Bacteriol.* **158**, 513–516.

HEINEMANN, J. A., SPRAGUE, G. F. (1989), Bacterial conjugative plasmids mobilize DNA transfer between bacteria and yeast, *Nature* **340**, 205–209.

HINNEN, A., HICKS, J. B., FINK, J. R. (1978), Transformation of yeast, *Proc. Natl. Acad. Sci. USA* **75**, 1929–1933.

HÖNLINGER, C., KUBICEK, C. P. (1989), Regulation of δ-(L-α-aminoadipyl)-L-cysteinyl-D-valine and isopenicillin N biosynthesis in *Penicillium chrysogenum* by the α-aminoadipate pool size, *FEMS Microbiol. Lett.* **65**, 71–76.

HÖNLINGER, C., HAMPEL, W. A., RÖHR, M., KUBICEK, C. P. (1988), Differential effects of general amino acid control of lysine biosynthesis on penicillin formation in strains of *Penicillium chrysogenum, J. Antibiot.* **41**, 255–257.

HORI, K., YAMAMOTO, Y., TOKITA, K., SAITO, F., KUROTSU, T., KANDA, M., OKAMURA, K., FURUYAMA, J., SAITO, Y. (1991), The nucleotide-sequence for a proline-activating domain of gramicidin-S-synthetase-2 gene from *Bacillus brevis, J. Biochem.* **110**, 111–119.

HOSKINS, J., O'CALLAGHAN, N., QUEENER, S. W., CANTWELL, C. A., WOOD, J. S., CHEN, V. J., SKATRUD, P. L. (1990), Gene disruption of the *pcbAB* gene encoding ACV synthetase in *Cephalosporium acremonium, Curr. Genet.* **18**, 523–530.

HUFFMAN, G. W., GESELLSCHEN, P. D., TURNER, J. R., ROTHENBERGER, R. B., OSBORNE, H. E., MILLER, F. D., CHAPMAN, J. L., QUEENER, S. W. (1992), Substrate specificity of isopenicillin N synthetase, *J. Med. Chem.* **35**(10), 1897–1914.

INGOLIA, T. D., QUEENER, S. W. (1989), β-Lactam biosynthetic genes, *Med. Res. Rev.* **9** (2), 245–264.

JENSEN, S. E. (1985), Biosynthesis of cephalosporins, *Crit. Rev. Biotechnol.* **3**, 277–301.

JENSEN, S. E., WESTLAKE, D. W. S., WOLFE, S. (1988), Production of the penicillin precursor δ-(L-α-aminoadipyl)-L-cysteinyl-D-valine (ACV) by cell-free extracts from *Streptomyces clavuligerus, FEMS Microbiol. Lett.* **49**, 213–218.

JENSEN, S. E., WONG, A., ROLLINS, M. J., WESTLAKE, D. W. S. (1990), Purification and partial characterization of δ-(L-α-aminoadipyl)-L-cysteinyl-D-valine synthetase from *Streptomyces clavuligerus, J. Bacteriol.* **172**, 7269–7271.

KALLOW, W., VON DÖHREN, H., KENNEDY, J., TURNER, G. (1994), Structure and functional study of ACV synthetase, *7th Int. Congr. Bacteriol. Appl. Microbiol. Mycol.* Prague, Czech Republic, p. 426.

KERN, B. A., HENDLIN, D., INAMINE, E. (1980), L-Lysine-ε-aminotransferase involved in cephamycin biosynthesis, in *Streptomyces lactamdurans, Antimicrob. Agents Chemother.* **17**, 679–685.

KIMURA, H., MIYANOSHITA, H., SUMINO, Y. (1990), DNA and applications thereof, *Jpn. Patent Application* No. 2-3762.

KINASHI H., DOI, M., NIMI, O. (1995), Isolation of large linear plasmids from β-lactam producing actinomycete strains, *Biotechnol. Lett.* **17**(3), 243–246.

KIRKPATRICK, J. R., DOOLIN, L. E., GODFREY, L. W. (1973), Lysine biosynthesis in *Streptomyces lipmannii:* implications in antibiotic biosynthesis, *Antimicrob. Agents Chemother.* **4**, 542–550.

KLEINKAUF, H., VON DÖHREN, H. (1996), A nonribosomal system of peptide biosynthesis, *Eur. J. Biochem.* **236**, 335–351.

KOVACEVIC, S., MILLER, J. R. (1991), Cloning and sequencing of the β-lactam hydroxylase gene (*cefEF*) from *Streptomyces clavuligerus:* gene duplication may have led to separate hydroxylase and expandase activities in the actinomycetes, *J. Bacteriol.* **173**, 398–400.

KOVACEVIC, S., WEIGEL, B. J., TOBIN, M. B., INGOLIA, T. D., MILLER, J. R. (1989), Cloning, characterization, and expression in *Escherichia coli* of the *Streptomyces clavuligerus* gene encoding deacetoxycephalosporin C synthetase, *J. Bacteriol.* **171**, 754–760.

KOVACEVIC, S., TOBIN, M. B., MILLER, J. R. (1990), The β-lactam biosynthesis genes for isopenicillin N epimerase and deacetoxycephalosporin C synthetase are expressed from a single transcript in *Streptomyces clavuligerus, J. Bacteriol.* **172**, 3952–3958.

KRAUSE, M., MARAHIEL, M. A. (1988), Organization of the biosynthesis genes for the peptide antibiotic gramicidin S, *J. Bacteriol.* **170**, 4669–4674.

KRAUSE, M., MARAHIEL, M. A., VON DÖHREN, H., KLEINKAUF, H. (1985), Molecular cloning of an ornithine-activating fragment of the gramicidin-S-synthetase-2-gene from *Bacillus brevis* and its expression in *Escherichia coli, J. Bacteriol.* **162**, 1120–1125.

KRIAUCIUNAS, A., FROLIK, C. A., HASSELL, T. C., SKATRUD, P. L., JOHNSON, M. G., HOLBROOK, N. L., CHEN, V. J. (1991), The functional role of cysteines in isopenicillin N synthase, *J. Biol. Chem.* **266**, 11779–11788.

KUMAR, C. V., COQUE, J.-J. R., MARTÍN, J. F. (1994), Efficient transformation of the cephamycin C producer *Nocardia lactamdurans* and development of shuttle and promoter-probe cloning vectors, *Appl. Environm. Microbiol.* **60**(11), 4068–4093.

KUROTSO, T., HORI, K., SAITO, Y. (1991), Characterization and location of the L-proline activating fragment from the multifunctional gramicidin-S synthetase-2, *J. Biochem.* **109**, 763–769.

KURYLOWICZ, W., KURZATKOWSKI, W., KURZATKOWSKI, J. (1987), Biosynthesis of benzylpenicillin by *Penicillium chrysogenum* and its Golgi apparatus, *Arch. Immunol. Ther. Exp.* **35**, 699–724.

LEDERBERG, J. (1957), Mechanism of action of penicillin, *J. Bacteriol.* **73**, 144.

LENDENFELD, T., GHALI, D., WOLSCHEK, M., KUBICEK-PRANZ, E. M., KUBICEK, C. P. (1993), Subcellular compartmentation of penicillin biosynthesis in *Penicillium chrysogenum:* the amino acid precursors are derived from the vacuole, *J. Biol. Chem.* **268**, 665–671.

LESKIW, B. K., AHARONOWITZ, Y., MEVARECH, M., WOLFE, S., VINING, L. C., WESTLAKE, D. W. S., JENSEN, S. E. (1988), Cloning and nucleotide sequence of the IPNS gene from *Streptomyces clavuligerus, Gene* **62**, 87–196.

LODER, P. B., ABRAHAM, E. P. (1971a), Isolation and nature of intracellular peptides from a cephalosporin C-producing *Cephalosporium* sp., *Biochem. J.* **123**, 471–476.

LODER, P. B., ABRAHAM, E. P. (1971b), Biosynthesis of peptides containing α-aminoadipic acid and cysteine in extracts of a *Cephalosporium* sp., *Biochem. J.* **123**, 477–482.

LUENGO, J. M., IRISO, J. L., LÓPEZ-NIETO, M. J. (1986), Direct enzymatic synthesis of natural penicillins using phenylacetyl-CoA:6-APA phenylacetyl transferase of *Penicillium chrysogenum:* minimal and maximal side chain requirements, *J. Antibiot.* **39**, 1754–1759.

MACCABE, A. P., VAN LIEMPT, H., PALISSA, H., UNKLES, S. E., RIACH, M. B. R., PFEIFER, E., VON DÖHREN, H., KINGHORN, J. R. (1991), δ-(L-α-Aminoadipyl)-L-cysteinyl-D-valine synthetase from *Aspergillus nidulans, J. Biol. Chem.* **266**, 12646–12654.

MADDURI, K., STUTTARD, C., VINING, L. C. (1989), Lysine catabolism in *Streptomyces* spp. is primarily through cadverine: β-lactam producers also make α-aminoadipate, *J. Bacteriol.* **171**, 299–302.

MADDURI, K., STUTTARD, C., VINING, L. C. (1991), Cloning and location of a gene governing lysine ε-aminotransferase, an enzyme initiating β-lactam biosynthesis in *Streptomyces* sp., *J. Bacteriol.* **173**, 985–988.

MAEDA, K., LUENGO, J. M., FERRERO, O., WOLFE, S., LEBEDEV, M. Y., FANG, A., DEMAIN, A. L. (1995), The substrate specificity of deacetoxycephalosporin C synthase ("expandase") of *Streptomyces clavuligerus* is extremely narrow, *Enzyme Microb. Technol.* **17**, 231–234.

MALMBERG, L. H., HU, W.-S., SHERMAN, D. H. (1993), Precursor flux control through targeted chromosomal insertion of the lysine ε-aminotransferase (LAT) gene in cephamycin C biosynthesis, *J. Bacteriol.* **175**, 6916–6924.

MALMBERG, L. H., HU, W. S., SHERMAN, D. H. (1995), Effects of enhanced lysine ε-aminotransferase activity on cephamycin biosynthesis in *Streptomyces clavuligerus, Appl. Microbiol. Biotechnol.* **44**, 198–205.

MALPARDITA, F., HOPWOOD, D. A. (1984), Molecular cloning of the whole biosynthetic pathway of a *Streptomyces* antibiotic and its expression in a heterologous host, *Nature* **309**, 462–464.

MARTÍN, J. F., GUTIÉRREZ, S. (1995), Genes for β-lactam antibiotic biosynthesis, *Antonie van Leeuwenhoek* **67**, 181–200.

MARTÍN, J. F., INGOLIA, T. D., QUEENER, S. W. (1991), Molecular genetics of penicillin and cephalosporin antibiotic biosynthesis, in: *Molecular Industrial Mycology: Systems and Applications for Filamentous Fungi* (LEONG, S. A., BERKA, R. M., Eds.), pp. 149–196. New York: Marcel Dekker.

MATHISON, L., SOLIDAY, C., STEPAN, T., ALDRICH, T., RAMBOSEK, J. (1993), Cloning, characterization, and use in strain improvement of the *C. acremonium* gene *cefG* encoding acetyl transferase, *Curr. Genet.* **23**, 33–41.

MATSUHASHI, M., SONG, M. D., ISHINO, F., WACHI, M., DOI, M., INOUE, M., UBUKATA, K., YAMASHITA, N., KONNO, M. (1986), Molecular cloning of the gene of a penicillin-binding protein supposed to cause high resistance to β-lactam antibiotics in *S. aureus, J. Bacteriol.* **167**, 975–980.

MCDERMOTT, W., ROGERS, D. E. (1982), Social ramifications of control of microbial disease, *Johns Hopkins Med. J.* **151**, 302–312.

MENNE, S., WALZ, M., KÜCK, U. (1994), Expression studies with the bidirectional *pcbAB-pcbC* promoter region from *Acremonium chrysogenum* using reporter gene fusions, *Appl. Microbiol. Biotechnol.* **42**, 57–66.

MILLER, J. R., INGOLIA, T. D. (1989), Cloning β-lactam genes from *Streptomyces* spp. and fungi, in: *Genetics and Molecular Biology of Industrial Microorganisms* (HERSHBERGER, C. L., QUEENER, S. W., HEGEMAN, G., Eds.), pp. 246–255. Washington, DC: American Society for Microbiology.

MONTENEGRO, E., BARREDO, J. L., GUTIÉRREZ, S., DÍEZ, B., ALVAREZ, E., MARTÍN, J. F. (1990), Cloning, characterization of the acyl-CoA:6-aminopenicillianic acid acyltransferase gene of *Aspergillus nidulans* and linkage to the isopenicillin N synthase gene, *Mol. Gen. Genet.* **221**, 322–330.

MONTENEGRO, E., FIERRO, F., FERNANDEZ, F. J., GUTIÉRREZ, S., MARTÍN, J. F. (1992), Resolution of chromosomes III and VI of *Aspergillus nidulans* by pulse-field gel electrophoresis shows that the penicillin biosynthetic pathway genes *pcbAB, pcbC,* and *penDE* are clustered on chromosome VI (3.0 Megabases), *J. Bacteriol.* **174**, 7063–7067.

MÜLLER, W. H., VAN DER KRIFT, T. P., KROUWER, A. J. J., WÖSTEN, H. A. B., VAN DER VOORT, L. H. M., SMAAL, E. B., VERKLEIJ, A. J. (1991), Localization of the pathway of the penicillin biosynthesis in *Penicillium chrysogenum, EMBO J.* **10**, 489–495.

MÜLLER, W. H., BOVENBERG, R. A. L., GROOTHUIS, M. H., KATTEVILDER, F., SMAAL, E. B., VAN DER VOORT, L. H. M., VERKLEIJ, A. J. (1992), Involvement of microbodies in penicillin biosynthesis, *Biochim. Biophys. Acta* **1116**, 210–213.

O'SULLIVAN, J., SYKES, R. B. (1986), β-Lactam antibiotics, in: *Biotechnology* 1st. Edn., Vol. 4 (REHM, H.-J., REED, G., Eds.), pp. 247–273. Weinheim: VCH.

O'SULLIVAN, J., BLEANEY, R. C., HUDDLESTON, J. A., ABRAHAM, E. P. (1979), Incorporation of 3H from δ-(L-α-amino(4,5-3H)adipyl)-L-cysteinyl-D-(4,4-3H)valine into isopenicillin N, *Biochem. J.* **184**, 421–426.

PANG, C. P., CHAKRAVARTI, B., ADLINGTON, R. M., TING, H. H., WHITE, R. L., JAYATILAKE, G. S., BALDWIN, J. E., ABRAHAM, E. P. (1984), Purification of isopenicillin N synthetase, *Biochem. J.* **222**, 789–795.

PASTEUR, L., JOUBERT, J. (1877), Charbon et septicémie, *C. R. Acad. Sci. Paris* **85**, 101–115.

PENALVA, M. A., TOURINO, A., PATINO, C., SANCHEZ, F., FERNANDEZ SOUSA, J. M., RUBIO, V. (1985), Studies of transformation of *Cephalosporium acremonium,* in: *Molecular Genetics of Filamentous Fungi* (TIMBERKLAKE, W. E., Ed.), pp. 59–68. New York: Alan Liss.

PÈREZ-ESTEBAN, B., OREJAS, M., GÓMEZ-PARDO, E., PENALVA, A. (1993), Molecular characterization of a fungal secondary metabolism promoter: transcription of the *Aspergillus nidulans* isopenicillin N synthetase gene is modulated by upstream negative elements, *Mol. Microbiol.* **9** (4), 881–895.

PETRICH, A. K., JENSEN, S. E. (1994), Transcriptional regulation of the genes involved in the early steps of the cephamycin biosynthetic pathway in *Streptomyces clavuligerus,* presented at *7th Int. Symp. Genet. Ind. Microorg.* Montreal, Canada, June 26–July 1, 1994.

PIEPER, R., HAESE, A., SCHRIDER, W., ZOCHER, R. (1995), Arrangement of catalytic sites in the multifunctional enzyme enniatin synthetase, *Eur. J. Biochem.* **230**, 119–126.

QUEENER, S. W., NEUSS, N. (1982), The biosynthesis of β-lactam antibiotics, in: *The Chemistry and Biology of β-Lactam Antibiotics* (MORIN, R. B., MORGAN, M., Eds.), pp. 1–81. London: Academic Press.

QUEENER, S. W., INGOLIA, T. D., SKATRUD, P. L., CHAPMAN, J. L., KASTER, K. R. (1985), A system for genetic transformation of *Cephalosporium acremonium,* in: *Microbiology 1985* (LIEVE, L., Ed.), pp. 468–472. Washington, DC: American Society for Microbiology.

RAMÓN, D., CARRAMOLINO, L., PATIÑO, C., SANCHEZ, F., PEÑALVA, M. A. (1987), Cloning and characterization of the isopenicillin N synthetase gene mediating the formation of the β-lactam ring in *Aspergillus nidulans, Gene* **57**, 71–181.

ROACH, P. L., CLIFTON, I. J., FÜLÖP, V., HARLOS, K., BARTON, G. J., HAJDU, J., ANDERSSON, I., SCHOFIELD, C. J., BALDWIN, J. E. (1995), Crystal structure of isopenicillin N synthase is the first from a new structural family of enzymes, *Nature* **375**, 700–704.

ROLAND, I., FROYSHOV, O., LALAND, G. (1975), On the presence of pantothentic acid in the three complementary enzymes of bacitracin synthetase, *FEBS Lett.* **60**, 305–308.

SAITO, F., HORI, K., KANDA, M., KUROTSU, T., SAITO, Y. (1995), Entire nucleotide sequence for *Bacillus brevis* Nagano Grs2 gene encoding gramicidin S synthetase 2: a multifunctional peptide synthetase, *J. Biochem.* **116**, 357–367.

SANCHEZ, F., LOZANO, M., RUBIO, V., PENALVA, M. A. (1987), Transformation of *P. chrysogenum, Gene* **51**, 97–102.

SAMSON, S. E., BELAGAJE, R., BLANKENSHIP, D. T., CHAPMAN, J. L., PERRY, D., SKATRUD, P. L., VAN FRANK, R. M., ABRAHAM, E. P., BALDWIN, J. E., QUEENER, S. W., INGOLIA, T. D. (1985), Isolation, sequence determination and expression in *E. coli* of the isopenicillin N synthetase gene from *Cephalosporium acremonium, Nature* (London) **318**, 191–194.

SAMSON, S. M., DOTZLAF, J. E., SLISZ, M. L., BECKER, G. W., VAN FRANK, R. M., VEAL, L. E., YEH, W. K., MILLER, J. R., QUEENER, S. W., INGOLIA, T. D. (1987), Cloning and expression of the fungal expandase/hydroxylase gene involved in cephalosporin biosynthesis, *Bio/Technology* **5**, 1207–1214.

SAWADA, Y., BALDWIN, J. E., SINGH, P. D., SOLOMON, N. A., DEMAIN, A. L. (1980), Cell-free cyclization of δ-(L-α-aminoadipyl)-L-cysteinyl-D-valine to isopenicillin N, *Antimicrob. Agents Chemother.* **18**, 456–470.

SCHLUMBOHM, W., STEIN, T., ULLRICH, C., VATER, J., KRAUSE, M., MARAHIEL, M. A., KRUFT, V., WITTMANN-LIEBOLD, B. (1991), An active serine is involved in covalent substrate amino acid binding at each reaction center of gramicidin S synthetase, *J. Biol. Chem.* **266**, 23135–23141.

SCHWECKE, T., AHARONOWITZ, Y., PALISSA, H., VON DÖHREN, H., KLEINKAUF, H., VAN LIEMPT, H. (1992a), Enzymatic characterization of the multifunctional enzyme δ-(L-α-aminoadipyl)-L-cysteinyl-D-valine synthetase from *Streptomyces clavuligerus, Eur. J. Biochem.* **205**, 687–694.

SCHWECKE, T., TOBIN, M., KOVACEVIC, S., MILLER, J. R., SKATRUD, P. L., JENSEN, S. E. (1992b), The A module of ACV synthetase: expression, renaturation and functional characterization, in: *Industrial Microorganisms: Basic and*

Applied Molecular Genetics (BALTZ, R. H., HEGEMAN, G. E., SKATRUD, P. L., Eds.), p. 291. Washington DC: American Society for Microbiology.

SENO, E. T., BALTZ, R. H. (1989), Structural organization and regulation of antibiotic synthesis and resistance genes in Actinomycetes, in: *Regulation of Secondary Metabolism in Actinomycetes* (SHAPIRO, S., Ed.), pp. 1–48. Boca Raton, FL: CRC Press.

SHIAU, C. Y., BALDWIN, J. E., BYFORD, M. F., SOBEY, W. J., SCHOFIELD, C. J. (1995), δ-(L-α-Aminoadipyl)-L-cysteinyl-D-valine synthetase: the order of peptide bond formation and timing of the epimerisation reaction, *FEBS Lett.* **358**, 97–100.

SHIFFMAN, D., MEVARECH, M., JENSEN, S. E., COHEN, G., AHARONOWITZ, Y. (1988), Cloning and comparative sequence analysis of the gene coding for isopenicillin N synthase in *Streptomyces, Mol. Gen. Genet.* **214**, 562–569.

SHIFFMAN, D., COHEN, G., AHARONOWITZ, Y., PALISSA, H., VON DÖHREN, H., KLEINKAUF, H., MEVARECH, M. (1990), Nucleotide sequence of the isopenicillin N synthase gene (*pcbC*) of the gram-negative *Flavobacterium* sp. SC 12,154, *Nucleic Acids Res.* **18**, 660.

SHOCKMAN, G. D., DANEO-MOORE, L., MCDOWELL, T. D., WONG, W. (1982), The relationship between inhibition of cell wall synthesis and bacterial lethality, in: *The Chemistry and Biology of β-Lactam Antibiotics,* Vol. 3 (MORIN, R. B., GORMAN, M., Eds.), pp. 303–338. London: Academic Press.

SIKORSKI, R. S., MICHAUD, W., LEVIN, H. L., BOEKE, J. D., HIETER, P. (1990), Trans-kingdom promiscuity, *Nature* **345**, 581–582.

SKATRUD, P. L. (1991), Molecular biology of the β-lactam producing fungi, in: *More Gene Manipulations in Fungi* (BENNET, J. W., LASURE, L. L., Eds.), pp. 364–395. New York: Academic Press.

SKATRUD, P. L. (1992), Genetic engineering of β-lactam antibiotic biosynthetic pathways in filamentous fungi, *Trends BioTechnol.* **10**(9), 324–329.

SKATRUD, P. L., QUEENER, S. W. (1989), An electrophoretic molecular karyotype for an industrial strain of *Cephalosporium acremonium, Gene* **78**, 331–338.

SKATRUD, P. L., FISHER, D. L., CHAPMAN, J. L., CANTWELL, C. A., QUEENER, S. W. (1987a), Strain improvement studies in *P. chrysogenum* using the cloned *P. chrysogenum* isopenicillin N synthetase gene and the *amdS* gene of *Aspergillus nidulans, SIM News* **37**(4), 77.

SKATRUD, P. L., QUEENER, S. W., CARR, L. G., FISHER, D. L. (1987b), Efficient integrative transformation of *C. acremonium, Curr. Genet.* **12**, 337–348.

SKATRUD, P. L., TIETZ, A. J., INGOLIS, T. D., CANTWELL, C. A., FISHER, D. L., CHAPMAN, J. L., QUEENER, S. W. (1989), Use of recombinant DNA to improve production of cephalosporin C in *C. acremonium, Bio/Technology* **7**, 411–485.

SMITH, D. J., BULL, J. H., EDWARDS, J., TURNER, G. (1989), Amplification of the isopenicillin N synthetase gene in a strain of *Penicillium chrysogenum* producing high levels of penicillin, *Mol. Gen. Genet.* **216**, 492–497.

SMITH, D. J., BURNHAM, M. K. R., BULL, J. H., HODGSON, J. E., WARD, J. M., BROWNE, P., BROWN, J., BARTON, B., EARL, A. J., TURNER, G. (1990a), β-Lactam antibiotic biosynthetic genes have been conserved in prokaryotes and eukaryotes, *EMBO J.* **9**, 741–747.

SMITH, D. J., BURNHAM, M. K. R., EDWARDS, J., EARL, A. J., TURNER, G. (1990b), Cloning and expression of the penicillin biosynthetic gene cluster from *Penicillium chrysogenum, Bio/Technology* **8**, 39–41.

SMITH, D. J., EARL, A. J., TURNER, G. (1990c), The multifunctional peptide synthetase performing the first step in penicillin biosynthesis in *Penicillium chrysogenum* is a 421037 Dalton protein similar to *Bacillus brevis* peptide antibiotic synthetases, *EMBO J.* **9**, 2743–2750.

SPRATT, B. G. (1989), Resistance to β-lactam antibiotics mediated by alterations of penicillin-binding proteins, in: *Handbook of Experimental Pharmacology,* Vol. 91 (BRYAN, L. E., Ed.), pp. 77–100. Berlin: Springer-Verlag.

STACHELHAUS, T., SCHNEIDER, A., MARAHIEL, M. A. (1995), Rational design of peptide antibiotics by targeted replacement of bacterial and fungal domains, *Science* **269**, 69–72.

THEN BERGH, K., LITZKA, O., BRAKHAGE, A. A. (1996), Identification of a major *cis*-acting DNA element controlling the bidirectionally transcribed penicillin biosynthesis genes *acvA* (*pcbAB*) and *ipnA* (*pcbC*) of *Aspergillus nidulans, J. Bacteriol.* **178**(13), 3908–3916.

TILBURN, J., SARKAR, S., WIDDICK, D. A., ESPESA, E. A., OREJAS, M., MUNGROO, J., PENALVA, M. A., ARST, H. N., JR. (1995), The *Aspergillus* PacC zinc finger transcription factor mediates regulation of both acidic- and alkaline-expressed genes by ambient pH, *EMBO J.* **14**, 779–790.

TIPPER, D. J., STROMINGER, J. L. (1965), Mechanism of action of penicillins: a proposal based on their structural similarity to acyl-D-alanyl-D-alanine, *Proc. Natl. Acad. Sci. USA* **54**, 1133–1141.

TOBIN, M. B. (1994), Genetic engineering of the acyl-coenzyme A:isopenicillin N acyltransferase from *Penicillium chrysogenum, Thesis,* University of Oxford.

TOBIN, M. B., FLEMING, M. D., SKATRUD, P. L., MILLER, J. R. (1990), Molecular characterization of the acyl-coenzyme A:isopenicillin N acyltransferase gene (*penDE*) from *Penicillium chrysogenum* and *Aspergillus nidulans* and activity of recombinant enzyme in *Escherichia coli, J. Bacteriol.* **172**, 5908–5914.

TOBIN, M. B., KOVACEVIC, S., MADDURI, K., HOSKINS, J., SKATRUD, P. L., VINING, L. C., STUDDARD, C., MILLER, J. R. (1991), Localization of the lysine ε-aminotransferase (*lat*) and δ-(L-α-aminoadipyl)-L-cysteinyl-D-valine synthetase (*pcbAB*) genes from *Streptomyces clavuligerus* and production of lysine ε-aminotransferase activity in *E. coli, J. Bacteriol.* **173**, 6223–6229.

TOBIN, M. B., BALDWIN, J. E., COLE, S. C. J., MILLER, J. R., SKATRUD, P. L., SUTHERLAND, J. D. (1993), The requirement for subunit interaction in the production of *Penicillium chrysogenum* acyl-coenzyme A:isopenicillin N acyltransferase in *Escherichia coli, Gene* **132**, 199–206.

TOBIN, M. B., COLE, S. C. J., KOVACEVIC, S., MILLER, J. R., BALDWIN, J. E., SUTHERLAND, J. D. (1994), Acyl-coenzyme A:isopenicillin N acyltransferase from *Penicillium chrysogenum:* effect of amino acid substitutions at Ser^{227}, Ser^{230} and Ser^{309} on proenzyme cleavage and activity, *FEMS Microbiol. Lett.* **121**, 9–46.

TOBIN, M. B., MILLER, J. R., BALDWIN, J. E., SUTHERLAND, J. D. (1995), Amino acid substitutions in the cleavage site of acyl-coenzyme A:isopenicillin N acyltransferase from *Penicillium chrysogenum:* effect on proenzyme cleavage and activity, *Gene* **162**, 29–35.

TURGAY, K., KRAUSE, M., MARAHIEL, M. A. (1992), Four homologous domains in the primary structure of GrsB are related to domains in a superfamily of adenylate forming enzymes, *Mol. Microbiol.* **6**, 529–546.

TURNER, G. (1992), Genes for the biosynthesis of β-lactam compounds in microorganisms, in: *Secondary Metabolites: Their Function and Evolution* (CHADWICK, D. J., WHELAN, J., Eds.), pp. 113–128. Chichester: Wiley and Sons (Ciba Foundation Symposium 171).

ÜNAL, S., HOSKINS, J., FLOKOWITSCH, J. E., WU, C. Y. E., PRESTON, D. A., SKATRUD, P. L. (1992), Detection of methicillin-resistant staphylococci by using the polymerase chain reaction, *J. Clin. Microbiol.* **30**, 1685–1691.

VAN LIEMPT, H., VON DÖHREN, H., KLEINKAUF, H. (1989), δ-(L-α-Aminoadipyl)-L-cysteinyl-D-valine synthetase from *Aspergillus nidulans, J. Biol. Chem.* **264**, 3680–3684.

VAN SINDREN, D., GALLI, G., COSMINA, F., DEFERRA, S., WITTHOFF, S. VENEMA, G., GRANDI, G. (1993), Characterization of the *srfA* locus of *Bacillus subtilis:* only the valine-activating domain of *srfA* is involved in the establishment of genetic competence, *Mol. Microbiol.* **8**, 833–841.

VELASCO, J., GUTIÉRREZ, S., FERNANDEZ, F. J., MARCOS, A. T., ARENOS, C., MARTÍN, J. F. (1994), Exogenous methionine increases levels of mRNAs transcribed from *pcbAB, pcbC,* and *cefEF* genes, encoding enzymes of the cephalosporin biosynthetic pathway, in *Acremonium chrysogenum, J. Bacteriol.* **176**(4), 985–991.

VICHITSOONTHONKUL, T., CHU, Y. W., SODHI, H. S., SAUNDERS, G. (1994), β-Lactam antibiotics produced by genetically engineered filamentous fungi, in: *Recombinant Microbes for Industrial and Agricultural Applications* (MUROOKA, Y., IMANAKA, E., Eds.), pp. 119–135. New York: Marcel Dekker.

VINING, L. C., SHAPIRO, S., MADDURI, K., STUTTARD, C. (1990), Biosynthesis and control of β-lactam antibiotics: the early steps in the "classical" tripeptide pathway, *Biotechnol. Adv.* **8**, 159–183.

WARD, J. M., HODGSON, J. E. (1993), The biosynthetic genes for clavulanic acid and cephamycin production occur as a "super-cluster" in three *Streptomyces, FEMS Microbiol. Lett.* **110**, 239–242.

WAXMAN, D. J., STROMINGER, J. L. (1982), β-Lactam antibiotics: biochemical modes of action, in: *The Chemistry and Biology of β-Lactam Antibiotics*, Vol. 3 (MORIN, R. B., GORMAN, M., Eds.), pp. 209–285. London: Academic Press.

WEBER, G., SCHÖRGENDORFER, K., SCHNEIDER-SCHERZER, E., LEITNER, E. (1994), The peptide synthetase catalyzing cyclosporine production in *Tolypocladium niveum* is encoded by a giant 45.8-kilobase open reading frame, *Curr. Genet.* **26**, 120–125.

WEIGEL, B. J., BURGETT, S. G., CHEN, V. J., SKATRUD, P. L., FROLIK, C. A., QUEENER, S. W., INGOLIA, T. D. (1988), Cloning and expression in *Escherichia coli* of isopenicillin N synthetase genes from *Streptomyces lipmanii* and *Aspergillus nidulans, J. Bacteriol.* **170**, 3817–3826.

WHITEMAN, P. A., ABRAHAM, E. P., BALDWIN, J. E., FLEMING, M. D., SCHOFIELD, C. J., SUTHERLAND, J. D., WILLIS, A. C. (1990), Acyl coenzyme A:6-aminopenicillanic acid acyltransferase from *Penicillium chrysogenum* and *Aspergillus nidulans, FEBS Lett.* **262**, 342–344.

WISE, E. M., JR., PARK, J. T. (1965), Penicillin: its basic site of action as an inhibitor of a peptide cross-linking reaction in cell wall mucopeptide synthesis, *Proc. Natl. Acad. Sci. USA* **54**, 75–81.

WU, C. Y. E., HOSKINS, J., BLASZCZAK, L. C., PRESTON, D. A., SKATRUD, P. L. (1992), Construction of a water-soluble form of penicillin-binding protein 2a from a methicillin-resistant *Staphylococcus aureus* isolate, *Antimicrob. Agents Chemother.* **36**, 533–539.

YEH, W. K., DOTZLAF, J. E., HUFFMAN, G. W. (1994), Biochemical characterization and evolutionary implication of β-lactam expandase/hydroxylase, expandase and hydroxylase, in: *50 Years of Penicillin Application; History and Trends* (KLEINKAUF, H., VON DÖHREN, H., Eds.), pp. 208–223. Prague: Public.

YELTON, M. M., HAMER, J. E., TIMBERLAKE, W. E. (1984), Transformation of *Aspergillus nidulans* using a *trpC* plasmid, *Proc. Natl. Acad. Sci. USA* **81**, 1470–1474.

YOCUM, R. R., WAXMAN, D. J., RASMUSSEN, J. R., STROMINGER, J. L. (1979), Mechanism of penicillin action: penicillin and substrate bind covalently to the same active site serine in two bacterial D-alanine carboxypeptidases, *Proc. Natl. Acad. Sci. USA* **76**, 2730–2734.

ZHANG, J. (1991), ACV synthetase in cephalsporin biosynthesis, *Thesis,* Massachussets Institute of Technology, Cambridge, MA.

ZHANG, J., DEMAIN, A. L. (1990a), Purification from *Cephalosporium acremonium* of the initial enzyme unique to the biosynthesis of penicillins and cephalosporins, *Biochem. Biophys. Res. Commun.* **137**, 528–535.

ZHANG, J., DEMAIN, A. L. (1990b), Purification of ACV synthetase from *Streptomyces clavuligerus, Biotechnol. Lett.* **12**, 649–654.

ZHANG, J., DEMAIN, A. L. (1992), ACV synthetase, *Crit. Rev. Biotechnol.* **12**, 245–260.

ZHANG, J., WOLFE, S., DEMAIN, A. L. (1992), Biochemical studies of ACV synthetase activity from *Streptomyces clavuligerus, Biochem. J.* **383**, 691–698.

7 Peptide Antibiotics

HORST KLEINKAUF
HANS VON DÖHREN

Berlin, Germany

List of Abbreviations

Note: If no D- or D-prefix is given, all amino acids are in the L-configuration.

1-Cl-D-vinylGly	1-chloro-D-vinylglycine
2a3h4buOA	2-amino-3-hydroxy-4-butyl-octanoic acid
2a4m4HEA	2-amino-4-methyl-hex-4-enoic acid
2a3h4mOA	2-amino-3-hydroxy-4-methyl-octanoic acid
2a3h4,8m$_2$NA	2-amino-3-hydroxy-4,8-dimethyl-nonanoic acid
2-Cl-DAla	2-chloro-D-alanine
2-F-DAla	2-fluoro-D-alanine
2hAsn	2-hydroxy-asparagine
2hBu	2-hydroxy-butyric acid
2hiCap	2-hydroxy-isocaproic acid
2hPhe	2-hydroxy-phenylalanine
2hVal	2-hydroxy-valine
2h3mVa	2-hydroxy-3-methyl-valeric acid
2,3h3mP	2,3-hydroxy-3-methyl-pentanoic acid
2,4h3mP	2,4-hydroxy-3-methyl-pentanoic acid
2,5h3mP	2,5-hydroxy-3-methyl-pentanoic acid
3hAsn	3-hydroxy-asparagine
3hCHA	3-hydroxy-cyclohexyl-alanine
3hC$_{14}$	3-hydroxy-tetradecanoic acid
3hLeu	3-hydroxy-leucine
3S-Pro	3-thioproline
3,4ΔPro	3,4-dehydro-proline
4hPro	4-hydroxy-proline
4S-Pro	4-thioproline
4,5hIle	4,5-hydroxy-isoleucine
6′hTrp	6′-hydroxy-tryptophan
α-Aad	α-aminoadipic acid
βAla	β-alanine
ΔAla	dehydro-alanine
ΔAbu	dehydro-aminobutyric acid
Abu	aminobutyric acid
Aeo	2-amino-8-oxo-9,10-epoxi-decanoic acid
aIle	*allo*-isoleucine
allylGly	allyl-glycine
AOC	amino-octanoic acid
aThr	*allo*-threonine
Aze	azetidine-2-carboxylic acid
Bmt	(4*R*)-4-[(4*E*)-2-butenyl]-4-methyl-L-threonine
<u>C</u>	cysteine linked as thioether
Cit	citrulline
CPG	cyclopropyl-glycine
cyclodihydroBmt	cyclodehydro-Bmt (see Bmt)
cyclopropylGly	cyclopropyl-glycine
deoxyBmt	deoxy-Bmt (see Bmt)
Dbu	diaminobutyric acid
fIle	formyl-isoleucine
fVal	formyl-valine
fOrn	f-ornithine
Himv	2-hydroxy-4-methyl-valeric acid
Hiv	hydroxy-isovaleric acid
Hmp	hydroxy-methyl-pentanoic acid
hPro	4-hydroxy-proline
Hyp	4-hydroxy-proline
Lac	lactic acid
LYSA	lysergic acid
mFPhe	*meta*-fluoro-phenylalanine
Mha	4-methyl-hydroxy-anthranilic acid
Mtz	methyloxazole (formed by cyclization of threonyl side chain)
N2MeAsp	N,2-methyl-aspartic acid
Nle	norleucine
Nva	norvaline
oFPhe	*ortho*-fluoro-phenylalanine
OMeSer	O-methyl-serine
Oxz	oxazole (formed by cyclization of seryl side chain)
pGlu	pyro-glutamic acid
PhSer	phenyl-serine
PPT	phosphinothricine
Q	quinaldic acid
Qaa	quinaldinic acid
Qoa	quinazol-4-one-3-acetic acid
Qxa	quinoxaline-2-carboxylic acid
Sar	sarcosine
Ser*	modified serine contained in tetrahydropyridine moiety of thiopeptides
Spro	thioproline

T̲	dehydrated threonyl side chain linked as thioether
tbuAla	t-butyl-alanine
tbuGly	t-butyl-glycine
Thz	thizole (formed by side chain cyclization of cysteine)
vinylGly	vinyl-glycine

D-prefix for D-configuration, used without-
Me-prefix used for N-methyl-
Nme-prefix used for N-methyl-

1 Introduction

Peptides as a chemically defined group of metabolites have attracted considerable attention for a variety of reasons. The first antibiotics discovered – penicillins, tyrocidines, and gramicidins – are peptides, and new peptides emerge from screenings at a steady rate. However, in addition to the enzymatic pathways of biosynthesis which, e.g., produce polyketides or terpenoids, there is the additional route of ribosomal peptide synthesis. The number of publications on peptide structures reflects that increasing attention is paid to this route. About a half of the 100 references on *peptide antibiotics* traced from December 1995 until December 1996 deal with peptides of ribosomal origin. Those detected via non-antibiotic properties like hormone functions are not even part of this cluster. In recent treatises on commercially important microbial products peptides have an increasing share (VINING and STUTTARD 1995; STROHL, 1996). In this volume, further chapters deal with highly significant fields: the established β-lactams (SKATRUD et al., Chapter 6), the promising antibacterials called dalbaheptides (LANCINI and CAVALLERI, Chapter 9), and the ribosomally formed lantibiotics (JACK et al., Chapter 8). The currently predominant immunosuppressor cyclosporin is treated in Chapter 12 by FLIRI et al. Other compounds of peptidic origin are contained in the chapters on siderophores (WINKELMANN and DRECHSEL, Chapter 5) and on antitumor agents (GRÄFE et al., Chapter 14).

Main applications of peptides still are pharmacological formulations applied in both human and animal health care. Besides antibacterial uses, increasing attention is paid to the introduction of antifungal compounds while immunomodulators have been firmly established. A considerable number of proteinase inhibitors and other enzyme inhibitors are continuously emerging and find various applications. In Sect. 5 a compilation of recently described peptides can be consulted for current screening targets and their results.

As will be shown below, antibiotics evolved in the ribosomal or enzymatic nonribosomal system may have similar structures and properties. Similar genetic backgrounds for the manufacturing of complex metabolites can be followed from prokaryotes to lower and higher eukaryotes, and boundaries between fields of microbial, plant or animal origin may disappear. This becomes obvious in the screening of marine organisms where in some instances metabolites isolated from sponges could be traced to associated microorganisms. Such possibly symbiotic relationships may indeed trigger the production of useful metabolites. Likewise, experimental simulations of microbial invasions lead to the production of antimicrobials, and this principle has been applied successfully, e.g., in the detection of ribosomally produced antibiotics of animal origin.

The aim of these new approaches is to understand the ecological significance of such metabolites and to try to trace the concepts of the evolution of structures. These concepts will be beneficial in the design of selection screens in biosynthetic approaches utilizing the natural combinatorial potential. In addition, boundaries between compound structures and their biosynthetic origins show the tendency of opening up due to the unifying concept of genetic structures. This is particularly evident from the similar genetic background of various polyketide structures and their combination with peptide forming systems, e.g., in the prominent immunomodulators FK506 and rapamycin (SCHWECKE et al., 1995).

2 Structures

2.1 Ribosomal Origin

Ribosomally encoded peptide antibiotics in microorganisms have been a topic throughout the history of antibiotic research. The promising field of lanthionine-containing peptides for which the term "lantibiotics" has been coined is treated in Chapter 8 of this volume by JACK et al. These peptides are used in the food industry. As animal antibiotics of ribosomal origin they have led to very promising fields of research.

Defenses against microbial infections – besides the highly specific system of adaptive immunity in vertebrates – were detected in a variety of targets. Thus compounds with antimicrobial activity, mainly peptides, were found in animals, plants, and bacteria. Animal antibiotics are located at surfaces like skin or mucosal surfaces, e.g., epithelial cells, and are components of body fluids like neutrophils, macrophages, and killer cells in mammals and haemocytes in insects. These antibacterials function as a primary defense system which acts faster and simpler and saves energy as well as information compared to the slow and highly specific memory-based system of adaptive immunity (BOMAN, 1995). Fundamental problems are the growth rates of bacteria compared to lymphocytes which differ 50- to 100fold.

Focusing on peptides, molecular genetics have opened up research to the complicated network of innate immunity. Current work describes the structure of gene clusters, regulation of the expression of peptide genes, their posttranslational processing and modification, and their fate in the respective environment. Obviously, these approaches are not limited to antimicrobial peptides and apply as well to, e.g., hormones or toxins. It has been pointed out that peptides signaling defense reactions in plants exhibit analogies to defense signaling in animals (BERGEY et al., 1996). Thus the release of the 18-residue peptide systemin from a 200-amino acid precursor is induced by mechanical wounding of tomato plants. Systemin induces a lipid-based signaling cascade leading to the expression of wound response proteins. Signaling functions have now been reported in a variety of cases including the involvement of peptide factors like the 45-residue ComS in competence in *Bacillus subtilis* (D'SOUZA et al., 1994; HAMOEN et al., 1995; MAGNUSON et al., 1994; SOLOMON et al., 1995).

In this chapter several aspects of mainly antimicrobial peptides are presented. As the field has been reviewed recently (LEHRER et al., 1993; BOMAN et al., 1994; BOMAN, 1995), we focus on biotechnological aspects.

2.1.1 Structural Classifications

There have been several general approaches to group or classify peptides of ribosomal origin. Two groups of linear peptides with or without disulfide linkages is a common scheme, which is further refined into respective subgroups of *dominating residues, antibacterial actions,* or *processed from precursors,* or two, three, or four *disulfides* (BOMAN, 1995). Recently, HANCOCK et al. introduced the group of cationic bactericidal peptides (HANCOCK et al., 1995) which are defined as polypeptides with less than 100 amino acid residues carrying a net charge of at least +2. It is obvious that there will be various overlaps within such restricted systems as many compounds show several properties, including linear portions, disulfides and antibacterial properties. Their linear parts may be dominated by certain residues or they may include cationic regions. Schemes of this kind will be useful in some respects, but some compounds will be part of several subgroups. As will be shown below, there is a similar situation in the classification of nonribosomally generated peptides.

2.1.2 Similar Structures of Ribosomal and Nonribosomal Peptides – Monocyclic Structures

Many cyclic and branched cyclic peptides of microbial origin containing nonprotein constituents are well known antimicrobials. Cyclic peptides of ribosomal origin were characterized from bovine neutrophils (bactenecins (I); ROMEO et al., 1988), and from the skin of bullfrogs (brevinin 1 (II), brevinin 2 (III), *Rana brevipode*; esculentin (IV), *Rana esculenta*; MORIKAWA et al., 1992; SIMMACO et al., 1993; ranalexin (V), *Rana catesbeiana*; CLARK et al., 1994).

three identical 198-base pair open reading frames which encode identical 66-amino acid peptides, but differ in their presumably regulatory untranslated 5′-regions. The amino terminal prepropeptide region is composed of a signal sequence separated by a prohormone processing site from the acidic region preced-

```
      R→L→C →R→I→V→V
I            \        ↓
             R←C←C←R←I

      FLPVLAGIAAKVVPALF→C→K→I→T
II                       \     ↓
                          C←K←K

      GIMDTLKNLAKTAGKGALQSLLNKAS→C→A→V→T
III                                \     ↓
                                    C←K←K

      GIFSKLGRKKIKNLLISGLKNVGKEVGMDVVRTGIDIAG→C→K→I→K
IV                                              \     ↓
                                                 C←E←G

      FLGGLIKIVPAMI→C→A→V→T
V                    \     ↓
                      C←K←K
```

The cycloheptapeptide moiety appears to be a common structural element of the *Rana* peptides, carrying positively charged side chains as found in polymyxin (VI), a nonribosomal antibacterial from *Bacillus polymyxa.* Cyclononapeptide structures as in bactenecins are found, e.g., in nonapeptidolactones like lysobactin/katanosin (VII, *Lysobacter* sp. and *Cytophaga* sp.).

ing the antibiotic peptide. Similar sequences are known from other propeptides (BEVINS and ZASLOFF, 1990). These signal sequences resemble sequences found in amphibian opioid peptides including dermorphins and deltorphins (VIII and IX). These opioid receptor-binding heptapeptides were first isolated from the skin of South American frogs (*Phyllomedusa* sp.) and have been shown to

```
      acyl→Dbu→Thr→DDbu→Dbu→Dbu→DLeu→Thr
VI                       |              ↓
                         └──── Thr←Dbu←Dbu

      DLeu→Leu→PhSer→3hLeu→Leu→Arg→Val
VII                  |               ↓
                     Ser←3hAsn←Gly←aThr
```

Variable side chains point to additional functional features, possibly even linked to the processing of precursor structures. CLARK et al. (1994) isolated the gene for ranalexin using the cDNA approach and detected an acidic amino acid-rich propeptide sequence at the amino terminal end. The ranalexin structutral gene has been found to be contained in

contain D-amino acids (ERSPAMER et al., 1989).

VIII Tyr-**DAla**-Phe-Gly-Tyr-Pro-SerNH_2

IX Tyr-**DAla**-Phe-Asp/Glu-Val-Val-GlyNH_2

X pGlu-Asp-Tyr(HSO_3)-Thr/Leu/Met-Gly-Trp-Met-Asp-PheNH_2

XI pGlu-Leu-Tyr-Glu-Asn-Lys-Pro-Arg-Arg-Pro-Tyr-Ile-Leu

Both types of bioactive peptides are produced in granular glands of the skin, compounds like dermorphin, caeruleins (X), and neurotensin (XI) being noxious to predators. The common processing/transport signals indicate their common secretory pathway.

2.1.3 D-Amino Acids in Ribosomally Produced Peptides

Epimerizations of amino acid residues in gene-encoded peptides have been verified for dermorphin (VIII) and dermenkephalin (XII). Dermorphin binds exclusively to the μ-opioid receptor and is 1000 times more potent than morphine upon intracerebroventricular injection. The D-residue is of crucial importance for this high affinity. Isolation of the respective gene revealed the Ala codon as precursor of D-Ala (RICHTER et al., 1987). The predicted biosynthetic precursor contains 4 dermorphin sequences together with one for dermenkephalin flanked by putative cleavage sites. In addition, D-amino acids have been found in the neuroexcitatory snail peptides from *Achatia fulica* achatin I (XIII) and fulicin (XIV; FUJITA et al., 1995), and the 48 amino acid-containing Ca^{2+} antagonist ω-AgaIVC from the venom of the *Agelopsis aperta* spider (XV). This venom has been the source of a cofactor-independent isomerase acting on peptides with the common sequence Leu-*Xaa*-Phe-Ala. This enzyme has been shown to isomerize Ser, Cys, OMeSer, and Ala residues in a reversible reaction (HECK et al., 1995). At least in the case of dermorphins the conversion to D-Ala was shown to proceed at the prepropeptide stage (MOR et al., 1991).

XII Tyr-**DMet**-Phe-His-Leu-Met-AspNH_2

XIII Gly-**DPhe**-Ala-Asp

VIV Phe-**DAsn**-Glu-Phe-ValNH_2

A different route of conversion is found in lantibiotics where Ser or Thr residues are initially reduced to the dehydro derivatives (SKAUGEN et al., 1994). These may further react with Cys side chains to thioethers, but isolated dehydro residues are known, e.g., in all lantibiotics (JACK et al., Chapter 8, this volume). These reactive side chains may be involved in covalent binding of peptides to target proteins, as has been shown for the cyanobacterial cyclopeptide microcystin and protein phosphatase 1 (MACKINTOSH et al., 1995). A desdehydro derivative of nisin A shows reduced antibacterial activity (ROLLEMA et al., 1996).

2.1.4 Posttranslational Modifications of Side Chains

Non-disulfide cyclic peptides include lantibiotics (JACK et al., Chapter 8, this volume; JACK et al., 1995) with stable thioether links forming multiple 3- to 6-membered cyclic peptide structures. These unique antibacterials, which are approved as food components, have been found exclusively in bacteria so far. Examples given are epidermin (XVI) and subtilin (XVII). Several respective gene clusters have been characterized and, besides the structural genes of the propeptides, they contain modifying enzymes and proteins involved in export. Structural modifications to direct specificity and improved stability are under investigation (for details, see Chapter 8). In contrast to disulfide bonds, enzymes opening thioether bonds have not yet been identified.

Side chain modifications of cysteine and serine residues to oxazoles and thiazoles, established for nonribosomal products for a long time, have recently been detected in ribosomal peptides as well, including microcins (MORENO et al., 1995) and the rhizobial peptide trifolitoxin (BREIL et al., 1996; TRIPLETT, personal communication). A respective transforming enzyme complex has been identified by the analysis of the microcin B17 (XVIII)

XV EDNCIAEDIGKCTWGGTKCCRGRPCRCSMIGTNCECTPRLIMEGL-D-Ser-SFA

(XVI) IA**S**KFI**CT**PG**C**AK**T**G**S**FN**SYCC**

(XVII) WK**S**E**T**L**CT**PG**C**VTGALQ**TC**FLQ**T**L**TC**N**C**KISK

gene cluster and it contains three proteins (LI et al., 1996). Microcin B17 is the first known gyrase inhibitor of peptidic nature. Its 69-residue precursor peptide is modified at Gly-Cys and Gly-Ser segments, respectively, or at Gly-Ser-Cys or Gly-Cys-Ser segments, with the formation of bicyclic thiazol-oxazole derivatives. Replacement of oxazoles by thiazoles leads to an inactive product (BAYER et al., 1995). All respective examples reported so far seem again to be restricted to bacterial sources, including thiopeptides like thiostrepton (STROHL and FLOSS, 1995), the antitumor peptide bleomycin, bacitracin, and various marine peptides like ulithiacylamide (XIX). These cyclic peptides isolated from sponge extracts are likely to originate from associated microorganisms, although this has not been generally established (BEWLEY et al., 1996).

XVIII VGIGGGGGGGGG-Oxz-Thz-GGQGGG-Thz-GG-Thz-SNG-Thz-Oxz-GGNGG-Thz-GG-Thz-GSHI

XIX Leu-Thz-Cys-Mtz
S-S
Mtz-Cys-Thz-Leu

bosomal origin has now been verified by the detection of the structural gene during the *Bacillus subtilis* genome sequencing program.

(XXI) 4,5hIle-6′hTrp-Gly
4hPro SO Ile
Asn—Cys—Gly

(XXII) Qxa
Qaa
Qoa-DSer-Ala-Cys-NMeVal
S
NMeVal-Cys-Ala-DSer-Qoa
Qaa
Qxa

(XXIII) Qxa
Qaa
Qoa-DSer-Ala-Cys-NMeVal
S-S
NMeVal-Cys-Ala-DSer-Qoa
Qaa
Qxa

(XX)
[Q]-Ile-Ala-ΔAla-Ala-[Ser*]-Thz-Thr-ΔAbu-DCys-3,4hIle-Thz-Thr-Thz-[Ser*]-Thz-ΔAla-ΔAla(NH_2)

2.1.5 Multicyclic Peptides

Prominent examples of multicyclic peptides from microorganisms include thiostrepton (XX), amanitins (XXI), and quinomycin (XXII). For the latter *in vivo* transformation of triostins (XXIII) to quinomycins was demonstrated (WILLIAMSON et al., 1982). More recently, the structural elucidation of the *Bacillus subtilis* antibacterial subtilosin (XXIV) introduced a new level of complexity. Its ri-

The principle of multiple rings in gene-encoded peptides is well established for numerous structures containing up to 4 disulfide links (see compilation in Sect. 5). Common cyclization patterns emerge and point to a similar conserved enzymology. Properties of the peptides identified so far are mainly antimicrobial, cytotoxic, or like trifoil peptides involved in tissue repair and cell proliferation (CHINERY and COFFEY, 1996). The rat intestinal trifoil factor rITF has even been shown to form covalent dimers (CHINERY and COF-

(XXIV) GLGLWGNKGCATCSIGAACLVDGPIPDG*IAGAX
S-S X

FEY, 1996). Recent excitement has been provoked by the discovery of microbial multicyclics with promising antiviral properties like aborycin (RP71955, XXV) (POTTERAT et al., 1994; FRÉCHET et al., 1994).

(XXV)

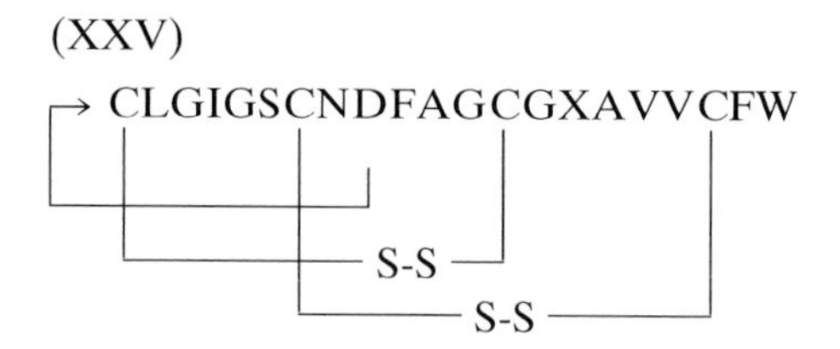

2.1.6 Linear Peptides

A number of peptides without any side chain modifications or disulfide bonds with diverse biological activities are known. For antimicrobial peptides at least several groups have been described of (1) amphipathic α-helical peptides (e.g., magainin) and (2) two α-helices linked by a hinge region (e.g., cecropins). Three other groups show high contents of (1) glycine (e.g., diptericin), (2) proline and arginine (e.g., apedaecin), or (3) tryptophan (e.g., indolicidin) (LEHRER et al., 1993; BOMAN et al., 1994; BOMAN, 1995). The tryptophan-rich indolicidin (XXVI) shows antibacterial and hemolytic activity. Replacement of tryptophan by phenylalanine abolishes only hemolytic activity (SUBBALAKSHMI et al., 1996) while antibacterial activities remain, even when proline residues are substituted by alanine.

(XXVI) ILPWKWPWWPWRRNH_2

A compilation of structures can be found in Sect. 5.

2.1.7 Production of Ribosomally Formed Peptides

Besides solid-phase synthesis production by fermentation is in an early phase. However, regioselective formation of several disulfide bonds still remains a challenge for chemists and overall yields are in the range of a few percent (KELLENBERGER et al., 1995). Microbial peptides produced in their hosts do not present a special problem. The production of animal or plant antibacterial peptides, in bacterial species, however, poses the problem of a suicide situation. This could be overcome by simultaneously introducing resistance or export genes. So far, either the expression of fusion constructs has been practiced (PIERS et al., 1993) or insensitive yeast cells have been employed (REICHHART et al., 1992).

2.1.8 Similar Open Questions in Gene-Encoded and Enzymatically Formed Peptides

The findings on multicyclic peptides illustrate well that the formation of constrained disulfide linkages to stabilize and functionalize peptide chains is not restricted to higher eukaryotes. The production of various multicyclic peptides may thus be well achieved by microbial cultures. However, a number of questions might be relevant, brought up recently in connection with defensins (SELSTED and OULETTE, 1995):

- How can the spontaneous lysis of granules producing cytolytic peptides be prevented? The same question is unresolved for the many microbial compounds with lytic properties; there might be natural antagonists or storage molecules.
- Are there extracellular roles for neutrophil defensins, or are there intracellular roles for enteric defensins? This question applies to most metabolites with, e.g., antimicrobial properties and it has not yet been answered for any secondary metabolite convincingly.
- Do defensin isoforms cooperate to form heteromer assemblies in target cell mem-

branes? The question of assembly is of considerable importance since dimerization or multimerization properties cannot be predicted yet. The human defensin HNP-3 forms dimers, and dimer formation is essential for the function of many microbial peptides like the linear gramicidin pore formation or vancomycin binding to peptidoglycan precursors.

2.2 Nonribosomal Origin

In contrast to those of ribosomal origin, more unusual structures and compounds are found in nonribosomally formed peptides. Systematic approaches to peptide structures are difficult, since in reports which are basically oriented towards structural chemistry the compounds may be found as peptide alkaloids (LEWIS, 1996), as indole alkaloids (IHARA and FUKUMOTO, 1996), or they may be contained in source-oriented reviews (e.g., marine metabolites; FAULKNER, 1996, and references therein). Peptide structures linked with polyketides and terpenoids may raise additional problems of compound classification. In a biosynthetic approach simply the types of enzyme structures involved are used. Since all peptides originate from linear precursors followed by various processing reactions, several approaches to systematics are possible (VON DÖHREN, 1990). However, in many cases there are several structural features involved and convenient approaches should then permit the search for a structural profile.

3 Biotechnology of Peptides

In the biotechnology of peptides fermentation of pure microbial cultures is used traditionally. An increasing number of processes deals with cell culture techniques that are still dominated by larger peptides and proteins as products. Obviously, the manufacturing of most bioactive peptides, especially antibiotics, needs self-protective systems. Such systems of self-immunity are inherent in producer cells but may limit the levels of production. Problems arise when non-self products are to be generated, e.g., the expression of animal-derived antibacterials in bacterial cultures. Solutions of such problems might be the use of *in vitro* production systems or the introduction of resistance genes into the expression organism.

3.1 Biosynthesis

The essential questions to be addressed refer to the biosynthetic origin of the compound and to the availability of precursors. It has to be decided whether a structural gene or a protein template is involved. This decision can be made upon inspection of the structural features. A collection of modified amino acids found in peptides of ribosomal origin is given in Tab. 1. A respective collection of amino acids encountered in nonribos-

Tab. 1. Unusual Features of Ribosomally Derived Peptides

Modified Amino Acid	Modification	Process
Glu	pyroGlu	terminal modification
Ala	D-Ala	in-chain epimerization
Met	D-Met	
Asp	D-Asp	
Ser	D-Ala	enzymatic dehydration
Thr	D-Abu	
Cys/Ser	lanthionine	enzymatic dehydration and
Cys/Thr	methyl-lanthionine	thioether linkage
Ser/Thr	oxazole-derivatives	enzymatic cyclization
Cys	thiazole-derivatives	

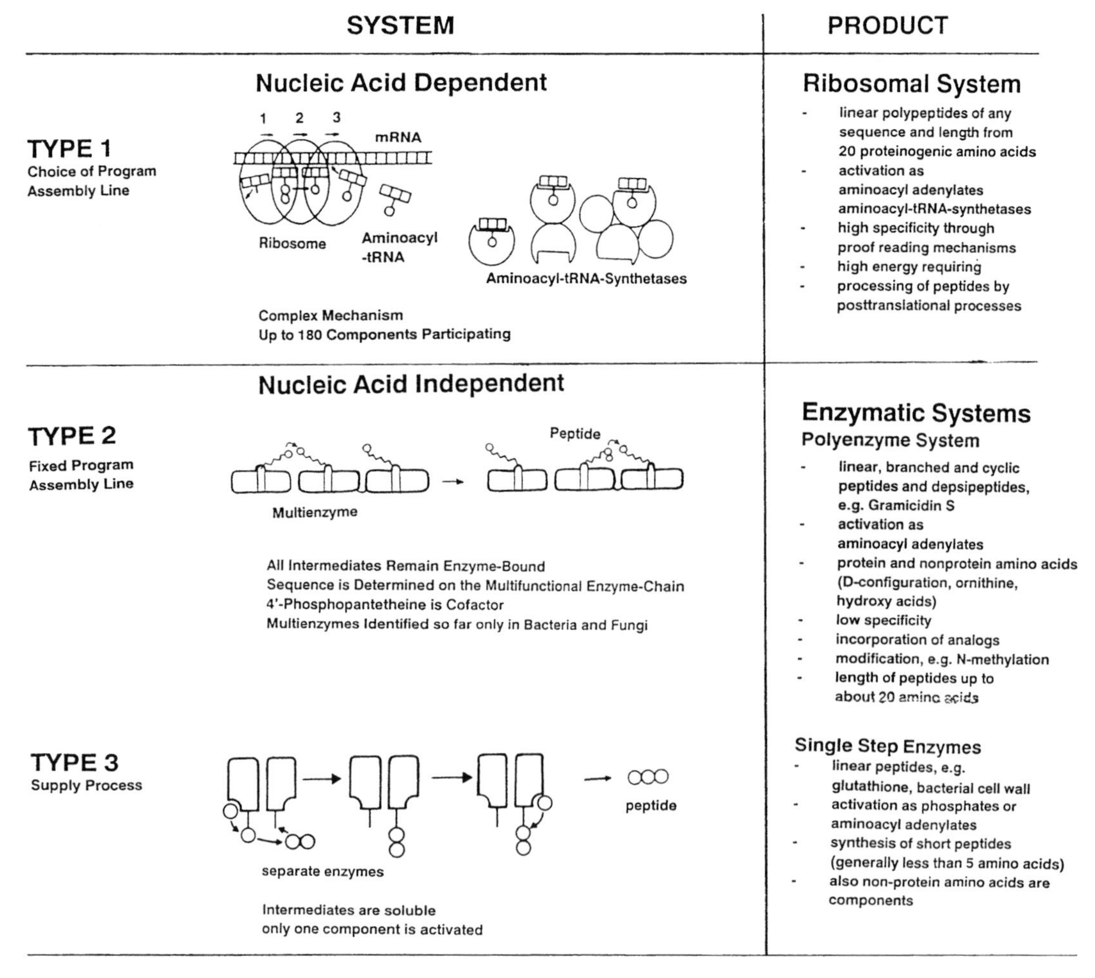

Fig. 1. Mechanisms of peptide biosynthesis. Scheme comparing ribosomal (type 1) and nonribosomal systems (type 2 – multistep systems, type 3 – single-step systems).

omally formed peptides is presented in Sect. 5. The principal differences of peptide forming systems are shown in Fig. 1. Nonribosomal peptides are formed either by single-step enzymes with free intermediates or by multistep systems. Well-known examples of peptides made by single-step enzymes are glutathione, pantetheine, or muropeptides. Thiol peptides termed "phytochelatins" which upon exposure to Cd^{2+}, Cu^{2+}, or Zn^{2+} are induced in the fission yeast *Schizosaccharomyces pombe* (KONDO et al., 1984) or in cultured plant cells (GRILL et al., 1989) are composed of Glu, Cys, and Gly of the composition (γGlu-Cys)$_n$Gly, with $n = 2$–11 (RAUSER, 1990; STEFFENS, 1990). Besides Gly also β-Ala or Ser may function as C-terminal residues (GEKELER et al., 1989; KLAPHECK et al., 1994). In maize seedlings two additional peptide families of the type (γGlu-Cys)$_n$ and (γGlu-Cys)$_n$Glu have been observed upon exposure to Cd^{2+} (MEUWLY et al., 1995). Biosynthetic systems for these peptides are supposed to resemble glutathione forming enzymes.

The modification of peptide structures by alteration of their genetic background is an emerging approach (see Sect. 3.2.2). In general, the structural genes for peptides or peptide synthetases are clustered together with those for modifying enzymes, regulatory signals, and – if secreted – for export proteins.

3.1.1 Biosynthetic Gene Clusters

Clusters identified for peptide biosynthesis are listed in Tab. 2. So far, only microbial clusters have been characterized in detail. Genome sequencing has revealed that there are no nonribosomal peptide biosynthetic activities in yeast, *Haemophilus influenzae, Mycoplasma genitalium, Synechocystis,* and *Methanococcus jannaschii.* However, evidence for multiple peptide forming clusters has been obtained in *Bacillus subtilis, Microcystis aeruginosa, Amycolatopsis mediterranei,* and *Bacillus brevis.*

Gene structures generally permit the identification of the involved biosynthetic reactions. Structural genes for ribosomally encoded peptides are a direct proof of their origin (e.g., subtilosin). A respective nonribosomal system is readily identified by the presence of very large genes (up to 45 kb) with modular structures. The most prominent structural element is the amino acid activating module or adenylate domain. Domains like this can be identified by a number of highly conserved motifs (KLEINKAUF and VON DÖHREN, 1996). The signature sequence SGTTGxPKG is most prominent and is also contained in acyl-CoA synthetases. Peptide forming systems are recognized by the repeated occurrence of these modules directly related to the number of amino acids contained in the respective peptide.

3.1.2 Biosynthetic Modules

The term "module" has been introduced for a gene-encoded functional unit performing an elongation step in either polyketide or peptide biosynthesis (DONADIO et al., 1991; KLEINKAUF and VON DÖHREN, 1996). During the process the particular residue entering the elongation cycle may be structurally altered, which implies epimerization, methylation, or hydroxylation in peptide forming systems, or reductions and dehydration in the polyketide systems. A module is thus comprised of submodules. Boundaries of these submodules have been determined at the protein level by limited proteolysis studies and sequence alignments. Therefore, the respective DNA modules have corresponding functional protein modules, which contain domains and subdomains of defined structures and functions. All modules can be readily identified by the presence of highly conserved core sequences (PFEIFER et al., 1995; STEIN and VATER, 1996) (Fig. 2); a selection is given in Tab. 3.

Arrangement of modules

Modules and their corresponding functional domains show a linear arrangement corresponding to the sequence of catalytic events. As these sequential reactions are catalyzed by protein surfaces this strict linearity is not obvious. Domains which are not directly linked need to interact frequently, e.g., in condensation reactions (see below). Domains may be reused for repeated sequences as in the formation of the cyclodepsipeptide enniatin (XXVII). The respective synthetase contains only two, not six activation modules.

```
          NMeVal-Hiv-NMEVal
             |           |
(XXVII)   Hiv-NMeVal-Hiv
```

Adenylate forming domains

In contrast to polyketide forming systems peptide synthetases activate their building blocks, i.e., amino acids, imino acids, and hydroxy acids, as adenylates on an integrated activation domain. These activation domains show structural similarities to the carboxyl-activating acyl-CoA synthetases of the polyketide systems. The positioning of activating domains corresponds in sequence to the amino acid sequence of the product in either wholly or partially integrated interacting enzyme systems (Fig. 3). The reaction sequence includes binding of carboxyl substrate and $MATP^{2-}$ (with $M = Mg^{2+}$, Mn^{2+}) and formation of the

Tab. 2. Peptide Biosynthetic Clusters Identified

Compound	Type[a]	Organism	Reference
Ribosomally Formed Peptides:			
Cytolysins	P-?-M	*Enterococcus faecalis*	SAHL et al., 1995
Epidermin	P-21-M	*Staphylococcus epidermidis*	SCHNELL et al., 1992
Lactocin	P-37-M	*Lactobacillus sake*	SKAUGEN et al., 1994
Lactococcin	P-27-M	*Lactococcus lactis*	RINCE et al., 1994
Microcin B17	P-43-M	*Escherichia coli*	LI et al., 1996
Nisin	P-34-M	*Lactococcus lactis*	JACK et al., 1995
Pep-5	P-34-M	*Staphylococcus epidermidis*	KALETTA et al., 1989
Subtilin	P-32-M	*Bacillus subtilis*	HANSEN, 1993
Trifolitoxin	P-1l-M	*Rhizobium leguminosarum*	BREIL et al., 1996
Nonribosomally Formed Peptides:			
A54145	R-P-13-L-10	*Streptomyces fradiae*	BALTZ, 1996
Actinomycin	R-(L-5)$_2$	*Streptomyces chrysomallus*	SCHAUWECKER et al., 1996
Anguibactin	R-P-2-M	*Vibrio anguillarum*	CHEN et al., 1996
Ardacin	P-7-M	*Kibdelosporangium aridum*	PIECQ et al., 1994
Bacilysin	P-2-M	*Bacillius subtilis*	SAKAJOH et al., 1987; HILTON et al., 1988
Bacitracin	R-P-12-C-7	*Bacillus licheniformis*	HERZOG-VELIKONJA et al., 1994
Bialaphos	P-3	*Streptomyces viridochromogenes*	SCHWARTZ et al., 1996
CDA	R-P-11-L-10	*Streptomyces coelicolor*	CHONG et al., 1996
Clavulanic acid	modified amino acid	*Streptomyces clavuligerus*	HODGSON et al., 1995
Cephalosporin	P-3-M	*Acremonium chrysogenum*	MARTÍN and GUTIERREZ, 1995
		Lysobacter lactamgenus	KIMURA et al., 1996
Cephamycin	P-3-M	*Nocardia lactamdurans*	COQUE et al., 1995a, b
Coronatin	acyl amino acid	*Pseudomonas syringae*	BENDER et al., 1996
Cyclosporin	C-11	*Tolypocladium niveum*	WEBER et al., 1994
Daptomycin	R-P-13-L-10	*Streptomyces roseosporus*	BALTZ, 1996
Destruxin	L-6	*Metarhizium anisopliae*	BAILEY et al., 1996
Enniatin	D-6	*Fusarium scirpi*	HAESE et al., 1993
Enterochelin	P-C-E-3	*Escherichia coli* *Bacillus subtilis*	REICHERT et al., 1992
Fengymycin	R-P-10-?	*Bacillus subtilis*	LIU et al., 1996
Ferrichrome	C-6	*Ustilago maydis*	MEI et al., 1993
Gramicidin S	C-(P-5)$_2$	*Bacillus brevis*	TURGAY and MARAHIEL, 1995
HC-toxin	C-4	*Helminthosporium carbonum*	PITKIN et al., 1996
Immunomycin	modified polyketide	*Streptomyces* sp.	MOTAMEDI, 1996
Iturin	C-8	*Bacillus subtilis*	HUANG et al., 1993
Lysobactin	P-11-L-9	*Lysobacter* sp.	BERNHARD et al., 1996
Microcystin	C-7	*Microcystis aeruginosa*	MEISSNER et al., 1996
Nikkomycins	modified peptides	*Streptomyces tendae*	BORMANN et al., 1996
Nosiheptide	R-P-13-C-10-M	*Streptomyces actuosus*	STROHL and FLOSS, 1995
Penicillin	P-3-M	*Aspergillus nidulans* *Penicillium chrysogenum*	SMITH et al., 1990; MACCABE et al., 1990; DÍEZ et al., 1990

Tab. 2. Continued

Compound	Type[a]	Organism	Reference
Phaseolotoxin	P-4-M	*Pseudomonas syringae*	TURGAY and MARAHIEL, 1995
Pristinamycin A	R-C-6	*Streptomyces pristinaespiralis*	BLANC et al., 1997
Pristinamycin M	polyketide/peptide		
Pyoverdin	R-P-8-M	*Pseudomonas fluorescens*	STINTZI et al., 1996
Rapamycin	modified polyketide	*Streptomyces hygroscopicus*	SCHMECKE et al., 1995; MOLNAR et al., 1996; APARICIO et al., 1996
Saframycin	P-4-M	*Myxococcus xanthus*	POSPIECH et al., 1996
SDZ214-103	L-11	*Cylindrotrichum oligospermum*	BERNHARD et al., 1996
Surfactin	L-8	*Bacillus subtilis*	COSMINA et al., 1993
Syringomycin	R-L-9	*Pseudomonas syringae*	MO et al., 1995
Syringostatin			
Thiostrepton	R-P-17-C-10-M	*Streptomyces laurentii*	STROHL and FLOSS, 1995
Tolaasin	R-P-18-L-5	*Pseudomonas tolaasi*	RAINEY et al., 1993
Tyrocidine	C-10	*Bacillus brevis*	MITTENHUBER et al., 1989
Yersiniabactin	R-P-3-M	*Yersinia enterocolitica*	GUILVOUT et al., 1993

[a] The abbreviations used are: P peptide, C cyclopeptide, L lactone, D depsipeptide, E ester, R acyl, M modified. The structural types are defined by the number of amino-, imino- or hydroxy acids in the precursor chain. The ring sizes of cyclic structures are indicated by the number following C, L, D, or E, defining the type of ring closure.

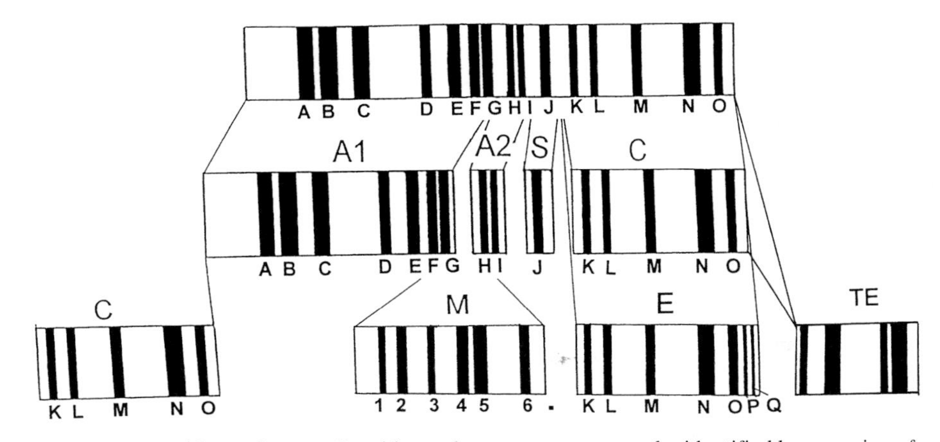

Fig. 2. Domain construction of peptide synthetases. Peptide synthetase sequences can be identified by a spacing of motifs, which have been assigned to adenylate formation (A1 and A2), aminoacylation of pantetheine (S), N-methylation of aminoacyl residues (M), condensation (peptide bond formation) (C), epimerization (E), and thioesterase (TE). The respective consensus sequences are given in Tab. 3.

acyladenylate which, as a highly reactive mixed anhydride, is stabilized against hydrolysis by the two subdomains. The released MPP_i^{2-} may reverse the activation reaction leading to the two substrates. This reaction employing [P-32]-PP_i usually is used as a sensitive assay for peptide synthetases and their activation domains. In the presence of high concentrations of ATP the adenylate may react to form AP_4A, but it is not known if this reaction has any physiological significance.

Tab. 3. Consensus Sequences of Peptide Synthetase Motifs

Adenylate Domains

A: LTxxELxxxAxxLxR
B: AVxxAxAxYVxIDxxYPxER
C: YSTGTTGxPKG
D: IIxxYGxT
E: GELxIxGxxVAR
F: RLYRTGDL
G: IEYLGRxDxQVKIRxxRIELGEIE
H: LxxYMVP
I: LTxxGKLxRKAL

Acyl Carrier Domains

J: LGGxSIxAI

Elongation (1) Epimerization[a] (2) Domains

K	(B1): YPSVxxQxRMYIL,	(B2): LxPIQxWF
L	(B1): LIxRHExL,	(B2): LxxxHD
M	(B1): DMHHIIxDGxSxxI	(B2) HHxxVDxVSWxIL
N	(B1): LSKxGQxDIIxGTPxAGR	(B2): VxxEGHGRE
O	(B1): IxGMFVNTxLALR,	(B2): TVGWFTxxxPxxL
		P: PxxGxGy
		Q:

N-Methyl-Transferases[b]

M1: GxDFxxWTSMYDG;
M2: LEIGTGTGMVLFNLxxxxGL;
M3: VxNSVAQYFP;
M4: ExxEDExLxxPAFF;
M5: HVExxPKxMxxxNELSxYRYxAV
N6: GxxVExSxARQxGxLD.

[a] Derived from *Bacillus* synthetases only.
[b] Derived from fungal domains.

Aminoacylation or carrier domains

Adenylates formed on acyl-CoA synthetases are transferred to CoA, and these thioesters are the stable transport form of aliphatic carboxylic acids. Amino acids are transferred to acyl carrier modules integrated within the multienzyme structures. These carrier modules have a structure similar to acyl carrier proteins involved in various polyketide forming systems including fatty acid biosynthesis (STEIN and VATER, 1996; STACHELHAUS et al., 1996a). Their essential functional group, 4′-phosphopantetheine, has to be introduced by posttranslational modification from CoA. The respective transferases have recently been identified with the aid of the holo-acyl carrier protein synthase structure from *Escherichia coli* (LAMBALOT et al., 1996). Respective genes have been detected in several biosynthetic gene clusters, and for each cluster a specific transferase to activate the pathway has been proposed. The finding that apoenzymes form unusually tight complexes with the transferases in the absence of CoA suggests an additional control mechanism (LAMBALOT and WALSH, 1995; PFEIFER et al., unpublished data).

As in the case of the acyl transfer to CoA the adenylate is cleaved by the phosphopantetheine thiol to release AMP and an enzyme-bound intermediate thioester is produced (PFEIFER et al., 1995).

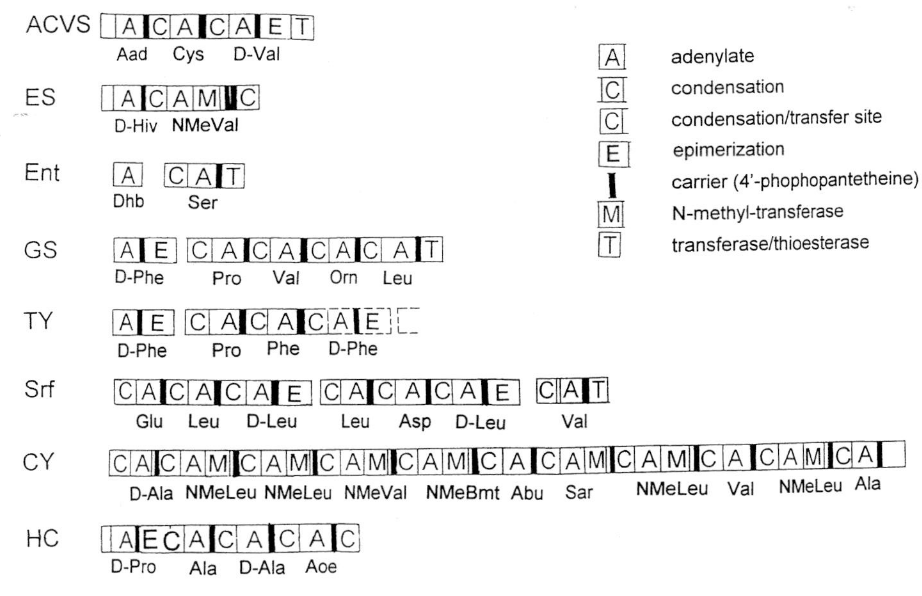

Fig. 3. Domains in peptide forming multienzyme systems. ACVS: ACV synthetase, ES: enniatin synthetase, Ent: enterobactin biosynthetic enzymes EntE and EntF, GS: gramicidin S synthetases 1 and 2, TY: tyrocidine synthetases 1 and 2, Srf: surfactin synthetases 1, 2 and 3, CY: cyclosporin synthetase, HC: HC-toxin synthetase. Amino or hydroxy acid specificity of the domains is indicated. Note that two different types of condensation domains are found, slightly deviating in their consensus sequences, and one very closely related epimerization domain. The E-C domain in the HC-system catalyzes both epimerization and condensation.

Condensation and epimerization domains

Adjacent to the carrier domains condensation or epimerization domains with similar structures are found. These are structurally stable regions of about 50 kDa (DIECKMANN et al., unpublished results). Epimerization domains are readily identified by two additional motifs (see Tab. 3). Epimerizations are catalyzed at the aminoacyl or peptidyl stage resulting in a thioester-bound mixture of both isomers (KLEINKAUF and VON DÖHREN, 1997). In the following step the selection of the stereospecific intermediate is controlled. The current model of peptide bond formation assumes that aminoacyl and peptidyl intermediates of two adjacent modules are directed towards peptide bond formation at the enclosed condensation domain. In order to explain the direction of bond formation aminoacyl and peptidyl binding sites have been postulated in analogy to the ribosomal system (KLEINKAUF and VON DÖHREN, 1997) (Fig. 4).

In several systems D-amino acid residues are formed during a transfer reaction between two multienzymes. In these cases an epimerization domain terminates the first enzyme while the second starts with a condensation domain (DE CRÉCY-LAGARD et al., 1995). If the transfer occurs within an enzyme system and is thus an intraenzymatic transfer, a third type of domain with a size of about 100 kDa is found resembling a fusion of both types of domains. Such a domain was detected first in HC toxin synthetase, a multienzyme forming the plant pathogenic cyclotetrapeptide HC toxin (SCOTT-CRAIG et al., 1992).

N-methylation domains

Methylation of amino groups is catalyzed by a transferase domain at the aminoacyl-

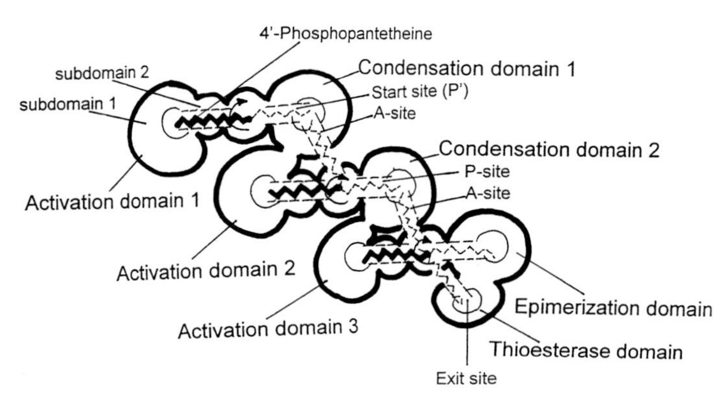

Fig. 4. Schematic view of peptide biosynthesis using a tripeptide model system, ACV synthetase. The three amino acids are activated at the activation domains 1–3. Each activation domain is composed of two subdomains. Adjacent to each activation domain is a carrier domain, which contains the cofactor 4′-phosphopantetheine. This factor may swing into three positions: binding the activated amino acid (activation domain), transport of the activated amino acid to the condensation site (A-site), and transport of the starter unit or peptidyl intermediate to the condensation P-site. In the terminating reaction epimerization occurs and the peptide is stereospecifically released at the thioesterase domain.

thioester stage (BILLICH and ZOCHER, 1987, 1990). This domain is inserted within the adenylate domain and has a size of about 50 kDa (HAESE et al., 1993; WEBER et al., 1994). The highly conserved core motifs (see Tab. 3) are not found in other known methyl transferases which all share the substrate S-adenosyl-methionine (BURMESTER et al., 1995). The reaction has been investigated in some detail and methods to detect methylation domains at the protein or gene level have been developed (ZOCHER et al., unpublished data).

Thioesterase domains

Domains with similarities to thioesterases have been detected in peptide synthetases and polyketide synthases as well. Their positioning in the C-terminal region of terminating multienzymes indicates their functions in terminating a catalytic cycle. Such functions have been demonstrated in several polyketide forming enzyme systems. Experimental work on peptide synthetases is only available so far on ACV synthetase producing the tripeptide precursor of β-lactam antibiotics (for details, see Chapter 6, this volume). Since a linear peptide is released the thioesterase function seems to be obvious in this case. Upon change of a highly conserved serine residue contained in the motif GXSXG in the thioesterase domain, δ-(L-α-aminoadipyl)-L-cysteinyl-L-valine is released as the main product instead of exclusively δ-(L-α-aminoadipyl)-L-cysteinyl-D-valine (KALLOW et al., 1996). This result indicates that the thioesterase catalyzes the stereospecific release of the product. Similarily condensation domains adjacent to epimerization domains are thought to catalyze the stereospecific condensation of only one of the peptidyl or aminoacyl intermediates.

Enzymes still to be identified

A number of reactions known to be involved in enzymatic peptide biosynthesis have not been studied yet. It could not be established so far, if these proceed within the context of integrated multienzymes or if separate single-step enzymes interact with the respective systems. Such reactions include the modification of amino acids (like hydroxylation or O-methylation), side chain modifications (like oxazole and thiazole ring formation of serine, threonine and cysteine side chains; cf. Sect. 2.1.4), and ether linkages or direct linkages of aromatic side chains.

3.1.3 Biosynthetic Enzyme Systems

Multienzyme systems for most types of peptides have been characterized (KLEINKAUF and VON DÖHREN, 1996) (Tab. 4). It seems evident that systems of eukaryotic origin are fully integrated while prokaryotic systems contain up to three interacting multienzymes. The only system deviating from this rule is ACV synthetase forming the β-lactam precursor tripeptide. This enzyme is recovered in a fully integrated state from both prokaryotic and eukaryotic sources. Isopenicillin N synthase forming the bicyclic penam precursor from the tripeptide is located adjacent to the synthetase in the respective gene clusters and has been implicated in horizontal gene transfer (AHARONOWITZ et al., 1992; BUADES and MOYA, 1996). Transfer, however, is thought to proceed from prokaryotic to eukaryotic hosts.

The integration of all sequential condensation functions does not rule out various interactions with other enzymes involved. In cyclosporin formation D-alanine and the complex amino acid (4*R*)-4-[(E)-2-butenyl]-4-methyl-L-threonine (Bmt) have to be supplied as direct precursors. Thus a respective alanine racemase and an enzyme system forming a polyketide precursor which has to be transaminated are involved in the process as well (OFFENZELLER et al., 1996). Likewise, acylated peptides can be assumed to have specific polyketide synthases required for the respective acyl-CoA starter unit contained in or coregulated with their respective peptide biosynthetic genes.

Biosynthetic systems can be identified by the respective alignments identifying the various modules. For isolation of genes convenient use can be made of the highly conserved core sequences to PCR-specific gene segments (BORCHERT et al., 1992). Adjacent regions are explored then, and the correlation is attempted by gene disruption and agreement of the peptide structure with the predicted modular arrangement (BERNHARD et al., 1996; MEISSNER et al., 1996). These procedures are in complete analogy to polyketide systems (SCHWECKE et al., 1995). In the case of *Bacillus subtilis* genome sequencing has revealed a peptide biosynthetic cluster with a yet unidentified product (TOGNONI et al., 1995). The conditions of expression of this cluster are unknown. Tentatively, from correlating activation modules with known structures, as has been discussed by COSMINA et al. (1993), the already identified peptide fengycin could be the product.

3.2 Production

The best studied system for peptide production is the penicillin fermentation where relevant parameters have been investigated in considerable detail. These include precursor concentrations and transport, intracellular compartmentation, levels and activity of biosynthetic enzymes, their regulation, expression, and stability in relation to the gene copy number, their posttranslational modification and intracellular localization, their possible inhibition by metabolites or products, and finally their export. This still uncomplete list of factors illustrates the complexity of tasks to be addressed. Many of the details are discussed in Chapter 6. For an overview of the state of metabolic flow studies in *Penicillium chrysogenum* the reader is referred to the recent publications of NIELSEN et al. (JORGENSEN et al., 1995a, b; NIELSEN and JORGENSEN, 1995). Comparable work on the fermentation of other peptides is scarce, but precursor-directed feeding work has been reviewed recently (THIERICKE and ROHR, 1993). Some of the results are discussed in Sect. 3.2.2.1.

3.2.1 Fermentation Procedures

It is beyond the scope of this chapter to discuss data on fermentation procedures which actually are similar in secondary metabolite production in general. Therefore, just a few remarks on recent developments are added: The control of parameters in metabolite fermentations has been improved by measuring not only dissolved oxygen but also redox potential, dissolved carbon dioxide (DAHOD, 1993), and culture fluorescence (NIELSEN et

Tab. 4. Peptide Synthetases Currently Studied

Peptide	Organism	Structural Type[a]	Multi-Enzymes (size kDa)[b]	Activation Modules
Linear Peptides:				
ACV	*Aspergillus nidulans*	P-3[c]	425[b]	3
	Acremonium chrysogenum		415[b]	3
	Streptomyces clavuligerus		420	3
Ergopeptines	*Claviceps purpurea*	R-P-3[d]	140	1
			370	3
Gramicidin	*Bacillus brevis*	P-15-M[e]	300	2
			700	5
				?
Cyclopeptides:				
HC-toxin	*Helminthosporium carbonum*	P-4[f]	570[b]	4
Gramicidin S	*Bacillus brevis*	C-$(P\text{-}5)_2$[g]	127[b]	1
			512[b]	4
Tyrocidine	*Bacillus brevis*	C-10[h]	123[b]	1
			400	3
			800	6
Cyclosporin	*Beauveria niveum*	C-11[i]	1700[b]	11
Lactones:				
Destruxin	*Metarhizium anisopliae*	L-6-M[j]	800	6
Actinomycin	*Streptomyces chrysomallus*	R-$(L\text{-}5)_2$[k]	45[b]	1
			280	2
			480	3
SDZ215-104	*Cylindrotrichum oligospermum*	L-11[l]	1650	11
Surfactin	*Bacillus subtilis*	R-L-7[m]	402[b]	3
			401[b]	3
			144[b]	1
Branched Cyclopeptides:				
Lysobactin	*Lysobacter* sp.	P-11-L-9[h]	450	3
			900	8
Bacitracin	*Bacillus licheniformis*	P-12-C-7[o]	700	5
			350	2
			750	5
Cyclodepsipeptides:				
Enniatin	*Fusarium scirpi*	C-$(D\text{-}2)_3$[p]	347[b]	2

[a] The abbreviations used are: P peptide, C cyclopeptide, L lactone, D depsipeptide, R acyl, M modified. The number of residues activated by each (multi)enzyme is indicated. 3×2 stands for the trimerization of 2 residues.
[b] Size verified by gene sequencing.
[c] α-Aad-Cys-DVal
[d] DLYSA-Ala-Phe-Pro
[e] f-Val-Gly-Ala-DLeu-Ala-DVal-Val-DVal-Trp-(DLeu-Trp)3-ethanolamine
[f] c(DPro-Ala-DAla-Aeo)
[g] c(DPhe-Pro-Val-Orn-Leu)2
[h] c(DPhe-Pro-Phe-DPhe-Asn-Gln-Val-Orn-Leu)
[i] c(DAla-NMeLeu-NMeLeu-NMeVal-NMeBmt-Abu-Sar-NMeLeu-Val-NMeLeu-Ala)
[j] c(DHimv-Pro-Ile-NMeVal-NMeAla-βAla)
[k] Mha-c(Thr-DVal-Pro-Sar-Val)
[l] c(DHiv-NMeLeu-Leu-NMeVal-NMeBmt-Thr-Sar-NMeLeu-Leu-NMeLeu-Ala)
[m] c(3hC_{14}-Glu-Leu-DLeu-Val-Asp-DLeu-Leu)
[n] DLeu-Leu-c(2hPhe-2hLeu-Leu-DArg-Ile-aThr-Gly-2hAsn-Ser)
[o] (Ile-Cys)-Leu-DGlu-Ile-c(Lys-DOrn-Ile-Phe-Asn-DAsp-His)
[p] c(NMeVal-DHiV)$_3$

al., 1994). Monitoring of the state of mycelia in fermentations permits more detailed investigations of morphology and vacuolation (PAUL and THOMAS, 1996; VANHOUTTE et al., 1995; NIELSEN et al., 1995; NIELSEN and KRABBEN, 1995). Oils have been used with advantage as carbon sources in cephamycin fermentations (PARK et al., 1994a, b). Solid-state fermentation for the production of surfactin and iturin by *Bacillus subtilis* has been exploited (OHNO et al., 1995, 1996). The separation of products is a promising approach to avoid feedback problems. Attempts with an aqueous two-phase system have been made in subtilin fermentation with *Bacillus subtilis* (KUBOI et al., 1994). Alternatively, a microfiltration module was used in nisin production with *Lactococcus lactis* (TANIGUCHI et al., 1993).

The discovery of the role of autoregulators in metabolite production (CHATER and BIBB, Chapter 2, this volume) has led to production processes now. Product yields are enhanced by the controlled addition of an autoregulators at high cell concentrations (YANG et al., 1996).

3.2.2 Structural Alterations

In order to achieve structural alterations two approaches are available, either to screen for analog producers or to manipulate the peptide composition by precursor feeding. Genetic manipulations are less obvious because the respective selection processes are still unclear. It seems obvious that peptides encountered in various sources show a unique design with respect to structural variations. These variations are evident from the observation of peptide families. It is fascinating to observe variations in certain residues while others are invariant. Remarkable invariance can be achieved solely by selection within the enzymatic steps of peptide synthesis and may reflect proof-reading mechanisms known from the ribosomal machinery. It can be assumed that structural selections are connected with biological targets which still need to be identified. Once biological selection processes are understood selection against defined targets under conditions of enhanced recombination could become a valuable tool.

3.2.2.1 Family Exploitations

The concept of peptide families is illustrated in Fig. 5 using the examples of gramicidin S, gramicidin, cyclosporin, and aureobasidin. Analogs of gramicidin S have not been detected for a long time. Modern mass spectrometric techniques have considerably enhanced analytical resolution (NOZAKI and MURAMATSU, 1987; THIBAULT et al., 1992). Both bacterial peptides show quite limited structural variations. The only remarkable feature is that the intiating valine in linear gramicidin has been found to be replacable by isoleucine. Likewise, only tryptophan in position 11 of gramicidin has been exchanged by other aromatic amino acids. From large-scale fermentations of the fungal metabolites cyclosporin and aureobasidin 32 and 29 analogs have been purified and characterized. In general, the number of tolerated substitutions is not more than two. This indicates a sieving mechanism presumably exerted by the biosynthetic process. The combinatorial game as played in chemical synthesis of analogs thus meets only limited success in the enzymatic system. In fact, only a small amount of predictable analogs is detected. 18 variable positions should permit the formation of more than 10^4 cyclosporin analogs, but only 32 have been detected (FLIRI et al., Chapter 12, this volume). The respective abortion products, however, have not been investigated yet. Likewise, 14 replacements of aureobasidins permit the synthesis of more than 10^3 analogs, 29 of which have been found (IKAI et al., 1991a, b; AWAZU et al., 1995).

Within a notably small range of variations analogs have been detected by screening methods. The scope of this approach is illustrated by LANCINI and CAVALLERI in Chapter 9 on dalbaheptides. Screening for cyclosporin analogs gave rise to the peptidolactone analog SDZ204-125 (Tab. 4), where D-alanine is replaced by D-hydroxyisovalerate, the variable position 2 is dominated by threonine, and one N-methylation is missing. In the case of the cyclodepsipeptide enniatin different

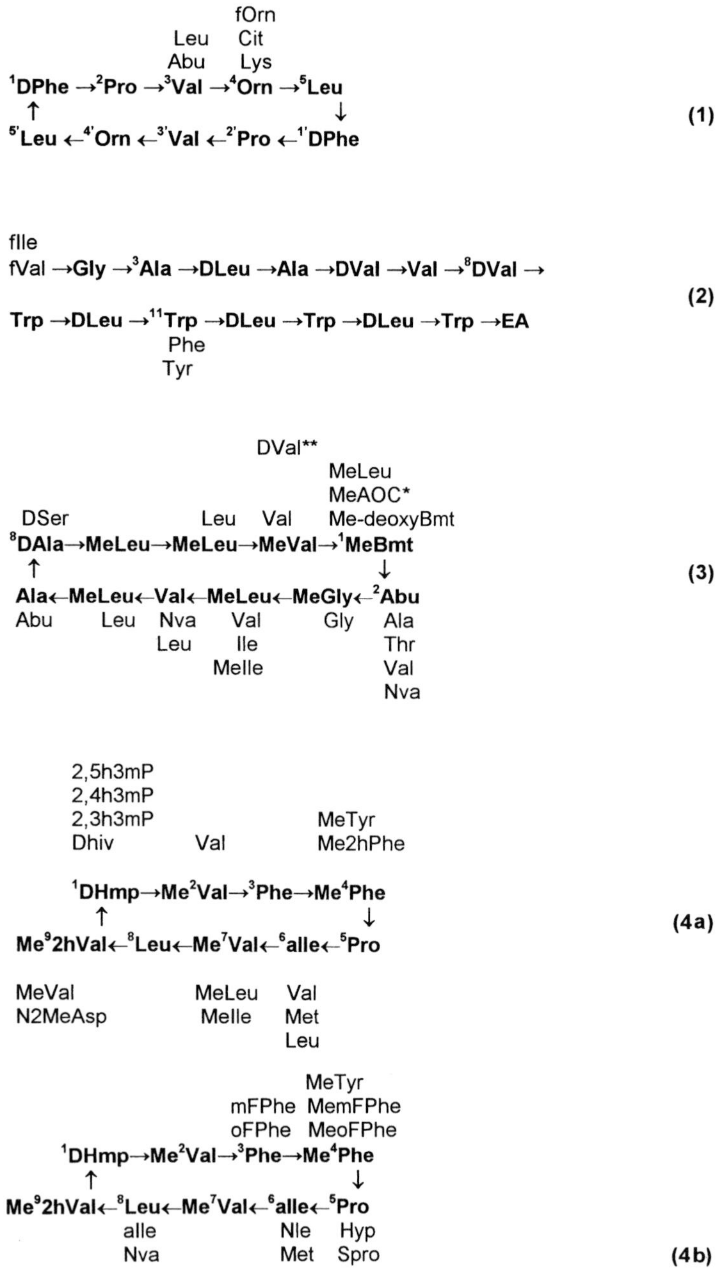

fOrn
Leu Cit
Abu Lys
^{1}DPhe →^{2}Pro →^{3}Val →^{4}Orn →^{5}Leu
↑ ↓ **(1)**
$^{5'}$Leu ←$^{4'}$Orn ←$^{3'}$Val ←$^{2'}$Pro ←$^{1'}$DPhe

fIle
fVal →**Gly →^{3}Ala →DLeu →Ala →DVal →Val →^{8}DVal →**
(2)
Trp →DLeu →^{11}Trp →DLeu →Trp →DLeu →Trp →EA
Phe
Tyr

DVal**
MeLeu
MeAOC*
DSer Leu Val Me-deoxyBmt
^{8}DAla→MeLeu→MeLeu→MeVal→^{1}MeBmt
↑ ↓ **(3)**
Ala←MeLeu←Val←MeLeu←MeGly←^{2}Abu
Abu Leu Nva Val Gly Ala
Leu Ile Thr
MeIle Val
Nva

2,5h3mP
2,4h3mP
2,3h3mP MeTyr
Dhiv Val Me2hPhe

^{1}DHmp→Me^{2}Val→^{3}Phe→Me^{4}Phe
↑ ↓ **(4a)**
Me^{9}2hVal←^{8}Leu←Me^{7}Val←^{6}aIle←^{5}Pro

MeVal MeLeu Val
N2MeAsp MeIle Met
Leu

MeTyr
mFPhe MemFPhe
oFPhe MeoFPhe
^{1}DHmp→Me^{2}Val→^{3}Phe→Me^{4}Phe
↑ ↓
Me^{9}2hVal←^{8}Leu←Me^{7}Val←^{6}aIle←^{5}Pro
aIle Nle Hyp
Nva Met Spro **(4b)**

Fig. 5. Peptide analogs isolated from fermentation broths: (1) gramicidin S analogs, (2) linear gramicidins, (3) cyclosporins, and (4) aureobasidins. Only two positions are variable in the *Bacillus* peptides (1) and (2). Note that only one of the four tryptophans has been found to be exchanged. The fungal peptides show conserved and less restricted positions. In all analogs found usually one of the variations is present, and in rare cases two positions are exchanged compared to the main products. More exchanges apparently reduce the rate of synthesis so that such products become extremely rare. Aureobasidin analogs (4b) were obtained by directed biosynthesis.

Fusaria have been shown to have altered profiles of branched-chain analogs. A study of the respective peptide synthetases revealed altered substrate binding sites (PIEPER et al., 1992).

3.2.2.2 Biosynthetic Manipulations

The potential of biosynthetic manipulations has to address either the precursor situation or the enzymatic template. A convenient procedure includes feeding of substrates or substrate analogs. Template alterations are still in their infancy and they will become of special importance when selection against new targets not encountered in natural or ecological systems becomes available.

Directed biosynthesis

Work in directed biosynthesis has been successful in various cases including tyrocidines, actinomycins, viridogriseins (where the natural D-hydroxyproline is replaced by feeding of proline to the nonhydroxylated analog which is a more effective antibiotic; OKUMURA, 1990), ergot peptides, enniatins, cyclosporins, and aureobasidins (for a review, see THIERICKE and ROHR, 1993). The remarkable extension of products is illustrated with aureobasidin analogs which are not available without feeding (Fig. 5). It is possible to replace proline by hydroxyproline and thioproline (TAKESAKO et al., 1996). However, systems may be surprisingly restrictive. Attempts to replace 4-hydroxyproline in the echinocandin analog L-671329 by feeding a culture of the producer *Zalerion arboricola* were not successful (ADEFARATI et al., 1991). It has been established later that the hydroxy group is essential for the antifungal activity. Such results imply a target adaptation of biosynthetic systems.

An extension of *in vivo* feeding approaches is the *in vitro* enzymatic synthesis of peptides employing the isolated multienzyme systems (KLEINKAUF and VON DÖHREN, 1997). The potential of the method is illustrated in Fig. 6 using gramicidin S, cyclosporin, and the peptolide SDZ214-125 (KLEINKAUF and VON DÖHREN, 1982; LAWEN and TRABER, 1993). Rates of synthesis of gramicidin S analogs have been studied with respect to multiple

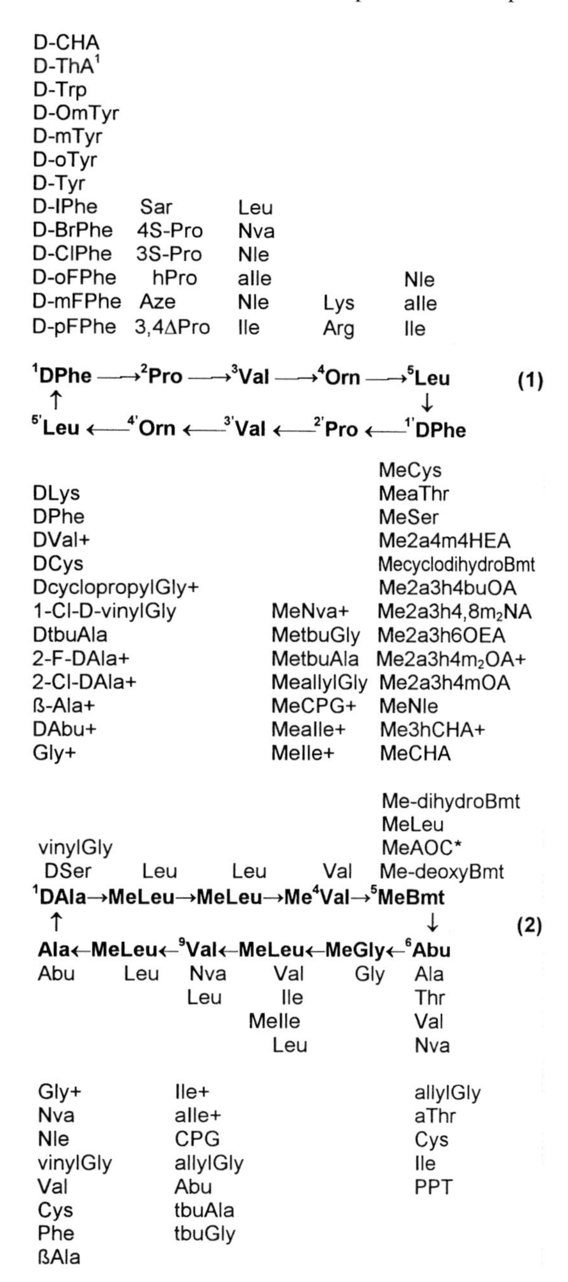

Fig. 6. Peptides formed by *in vitro* synthesis with supply of amino acids, ATP, and SAM if required; (1) gramicidin S, (2) cyclosporin, and (3) SDZ214-125. For abbreviations see p. 278.

2hiCap+
D2h3mVa+ MetbuAla
D2hVa+ MeallylGly Me-dihydroBmt
D2hBu+ MeCPG Me2a3h4mOA
vinylGly Mealle+ MeLeu*
DLac MeAbu+ MeCHA

^{1}DHiv→MeLeu→Leu→Me^{4}Val→Me^{5}Bmt
↑ ↓ **(3)**
Ala←MeLeu←^{9}Leu←MeLeu←MeGly←^{6}Thr

Abu+ Leu 3hNva
vinylGly Abu
Nva Nva
Cys Nle

+ molecular mass by FAB-MS

Fig. 6. Continued.

substitutions. It has been shown that rates are roughly additive and overall rates decrease drastically when poor substrates are combined. In the evaluation of properties of cyclosporin and SDZ213-125 analogs a large number of cyclopeptides have been produced with these giant multienzymes in the mg range. A comparison revealed that the peptolide synthetase had a much more restricted substrate profile (LAWEN and TRABER, 1993).

In vitro investigation of enzyme properties is essential for the efficient utilization of substrate feeding and the evaluation of structural alterations of enzymes.

Genetic approaches

Metabolic engineering could approach both substrate pools and composition as well as the substrate selection properties of enzymes involved. Should the latter approach attempt to change the amino acid sequence of the considered peptides the term "reprogramming" has been introduced. The idea of reprogramming is to exchange biosynthetic modules in order to arrive at new structures. This exchange has been demonstrated in several cases for the polyketide forming systems. As polyketide building blocks are restricted to acetate, propionate, or butyrate residues which may be modified by reductions and dehydration, the reprogramming approach is quite limited. In the peptide field a much larger reservoir of structures seems to be available. Using a double recombination approach STACHELHAUS et al. (1995, 1996b) succeeded in replacing the last amino acid module of surfactin synthetase by substituting the terminal leucine domain with bacterial ornithine and phenylalanine domains and with fungal cysteine and valine domains. This approach offers exciting opportunities of varying residues within a given structural type. Finally, even new peptide structures should be available from designed enzyme systems.

4 Future Prospects

4.1 Peptides of Ribosomal Origin

Impressive examples of antimicrobial properties of peptides of ribosomal origin have been demonstrated. So far uses have been limited to topical applications (a magainin analog is in clinical trial for cutaneous application in skin and eye infections), and no data are available on the uptake in tissues. An obvious limitation seems to be the degradation, especially by trypsin-like proteases. Promising fields of research include the use of precursor peptides and sequence alterations to improve proteolytic stability. Efforts for the exploitation of peptide structures are justified by the ease of engineering these peptide structures compared to enzymatic systems and research has been directed towards multidrug resistant bacteria, sepsis, and even anticancer drug development (BOMAN, 1995).

A fascinating area of research aims at the *in vivo* production of peptide factors to counteract insect-borne diseases like malaria, trypanosomiasis, and filariasis. It has been proposed to modify the insect immune system of the respective vectors or symbionts in order to eliminate the parasites (HAM et al., 1994).

4.2 Peptides of Nonribosomal Origin

Peptides of enzymatic origin continue to emerge from various screening programs. New promising sources are, e.g., marine organisms like sponges as a rich source of peptidic structures, harboring various kinds of microorganisms. The exploitation of such microbes that are often surface-associated has just begun and is mainly hampered by the lack of experience in cultivation.

The extensive understanding of biosynthetic reactions permits various structural modifications to be introduced into the genetic background, thus facilitating the biotechnological production of compounds. New attempts will be made by the implementation of strategies from ribosomal antibiotic induction approaches to various microbial, plant, and animal sources. A lot can be learned from the interaction of organisms in ecosystems, and genome sequencing will certainly enhance this understanding considerably. The advancing analytical techniques will permit a more rapid access to even more complicated natural products. The production of such compounds will remain a challenge for biotechnology.

5 Compilation of Compounds

It is the purpose of this compilation covering most peptides and related structures published during the last five years (1992–1996) to illustrate trends in sources, screening targets, variability of structures within groups of organisms, and to address some problems of source identification and stability. This list updates and extends the table of peptides presented in Vol. 4 of the First Edition of *"Biotechnology"* (KLEINKAUF and VON DÖHREN, 1986). A more extensive compilation of peptide structures including publications until 1989 can be found in VON DÖHREN (1990). Data on specific compounds are retrieved best from the available data banks (Chemical Abstracts; Kitasato Microbial Chemistry Database, Usako Corp., Tokyo; Antibase, Database, Chemical Concepts, Heidelberg). Data on ribosomal peptides are easily accessed from the standard gene banks. For retrieval of additional information the given names of compounds, organisms, authors, and years of publication are sufficient.

Peptides of ribosomal origin have been grouped in Tab. 5. This compilation illustrates the historically grown pathways of the studies which originated mainly in the insect field, then achieved spectacular results with frog peptides, and now identify many antimicrobial peptides from mammalian sources. For details, the reader is referred to recent reviews (HANCOCK et al., 1995; LEHRER et al., 1993; BOMAN et al., 1994; BOMAN, 1995). Microbial peptides of ribosomal origin are the subject of Chapter 8 on lantibiotics (JACK et al., this volume) while the fields of neuropeptides and hormones are not within the scope of the chapter.

Peptides of nonribosomal origin are listed in Tab. 6. A number of observations can be made concerning the sources especially suited for peptides. Thus the actinomycetes are still the leading group for new products, and here extensive variations in the structural types of compounds are found. The groups of dalbaheptides and thiopeptides in particular have benefitted from directed screening approaches. Bacilli have received less attention, but the potential of various cyclopeptides from well-known strains and also from marine Bacilli with slightly deviating structures has not yet been explored in depth. A surprising group are the blue-green algae, mainly cyanobacteria, which show an impressive variability of peptide structures with various properties. This variability implies frequent horizontal transfer of biosynthetic genes, a promising field to be exploited in order to make use of natural combinatorial approaches. Fungi continue to provide interesting compounds, most peptides being rather small. Large fungal peptides are mainly assembled in the family of peptaibols.

Plant peptides of presumably nonribosomal origin often are not modified and could as well originate from ribosomal precursors by

Tab. 5. Compilation of Peptides of Ribosomal Origin

Peptide	Source	Sequence[a]	Properties	Reference
Microbial Sources				
Leukocin A Ual 187	*Leuconostoc gelidum*	KYYGNGVHCTKSGCSVNWGEAFSAGVHR-LANGGNGFW	antibacterial (G –)	HASTING et al. (1991)
Nisin	*Lactococcus lactis*	ITSISLCTPGCKTGALMGCNMKTATCHC-SIHVSK	antibacterial (G*)	HURST (1981)
Pep-5	*Staphylococcus epidermidis*	TAGPAIRASVKQCQKTLKATRLFTVSCK-GKNGCK	antibacterial (G*)	KALETTA et al. (1987)
Plantaricin A	*Lactobacillus plantarum*	AYSLQMGATAIKQVKKLFKKW	antibacterial	NISSEN-MEYER et al. (1993)
Sillucin	*Rhizomucor pusillus*	ACLPNSCVSKGCCCGBSGYWCRQCGIKYTC	antibacterial (G*)	BRADLEY and SOM-kuti (1979)
Subtilin	*Bacillus subtilis*	MSKFDDFDLDVVKVSKQDSKITPQWKSES-LCTPGCVTGALQTCFLQTLTCNCKISK	antibacterial	BANERJEE and HANSEN (1988)
Plants				
AFP1	Rape (*Brassica napus*)	QKLCERPSGTWSGVCGNNNACKNQCINLE-KARHGSCNYVFPAHK	antifungal	TERRAS et al. (1992)
AFP2	Turnip (*Brassica rapa*)	QKLCERPSGTXSGVCGNNNACKNQCIR	antifungal	TERRAS et al. (1992)
Ac-AMP	Amaranth (*Amaranthus caudatus*)	VGECVRGRCPSGMCCSQFGYCGKGP-KYCGR	antibacterial, antifungal	BROEKAERT et al. (1992)
Crambin	Crambe plants (*Crambe abyssinica*)	TTCCPSIVARSNFNVCRIPGTPEAICATYTG-CIIIPGATCPGDYAN		TEETER et al. (1981)
MBP-1	Maize (*Zea mays*)	RSGRGECRRQCLRRHEGQPWET-QECMRRCR	antifungal	DUVICK et al. (1992)
Mj-Ampl (2 types)	*Mirabilis jalapa*	QCIGNGGRCNENVGPPYCCSGFCLRQPGQY-GYCKNR	antibacterial (G*); anti-fungal	CAMMUE et al. (1992)
Rs-AFP1	Radish (*Raphanus sativus*)	QKLCERPSGTWSGVCGNNNACKNQCINLE-KARHGSCNYVFPAHK	antibacterial (G*); anti-fungal	TERRAS et al. (1990)
Thionin BTH6	Barley (*Hordeum vulgare*, leaf)	KSCCKDTLARNCYNTCRFAGGSRPVCA-GACRCKIISGPKCPSDYPK		BOHLMANN et al. (1988)

Insects				
Abaecin	Honey bee (*Apis mellifica*)	YVPLPNVPQPGRRPFPTFPGQGPFNP-KIKQPQGY	antibacterial	Castreels et al. (1990)
Andropin	Fruit fly (*Drosophila melanogaster*)	VFIDILDKVENAIHNAAQVGIGFAKPFEK-LINPK		
Apedaecin IA	Honey bee (*Apis mellifica*, lymph fluid)	GNNRPVYIPQPRPPHPRI	antibacterial (G−)	Castreels et al. (1989)
Bactericidin B2	Tobacco hornworm (*Manduca sexta*, larvae hemolymph)	WNPFKELERAGQRVRDAVISAAPAVATVG-QAAAIARG	antibacterial, cytotoxic	Dickinson et al. (1988)
Bombolitin	Bumblebee (*Megabombus pennsylvanicus*, venom)	IKITTMLAKLGKVLAHVa	antibacterial, cytotoxic	Argiolas and Pisano (1985)
Cecropin	Silk moth (*Bombyx mori*)	RWKIFKKIEKVGQNIRDGIVKAGPAVAVVG-QAATI	antibacterial	Qu et al. (1987)
Crabrolin	Hornet (*Vespa crabro*, venom)	FLPLILRKIVTALa	cytotoxic	Argiolas and Pisano (1984)
Drosocin	Fruit fly (*Drosophila melanogaster*)	GKPRPYSPRPTSIHPRPIRV	antibacterial	Bulet et al. (1993)
Insect defensin	Dragonfly (*Aeschna cyanea*, larvae)	GFGCPLDQMQCHRHCQTITGRSG-GYCSGPLKLTCTCYR	antibacterial (G+)	Bulet et al. (1992)
Lepidopteran C	Silkworm (*Bombyx mori*)	RWKLFKKIEKVGRNVRDGLIKAGPAIAVIG-QAKSL		Teshima et al. (1987)
Mastoparan	Wasp (*Vespula lewisii*, venom)	INLKALAALAKKIL	antibacterial (G+)	Bernheimer and Rudi (1986)
Melittin	Bee (*Apis mellifica*, venom)	GIGAVLKVLTTGLPALISWIKRKRQQ	antibacterial, cytostatic, antifungal	Tosteson and Tosteson (1984)
Phormicin A (B)	Blowfly (*Phormia terranova*)	ATCDLLSGTGINHSACAAHCKLRGNRG-GYCNGKGVCVCRN	antibacterial (G+)	Lambert et al. (1989)
Royalisin	Bee, royal jelly (*Apis mellifica*)	VTCDLLSFKGQVNDSACAANCLGKAGGH-CEKGVCICRKTSFKDLWDKYF	antibacterial (G+)	Fujiwara et al. (1990)
Sapecin	Fleshfly (*Sarcophaga peregrina*)	ATCDLLSGTGINHSACAAHCLLRGNRG-GYCNGKAVCVCRN	antibacterial	Hanzawa et al. (1990)
Sarcotoxin (3 types)	Fleshfly (*Sarcophaga peregrina*)	GWLKKIGKKIERVGQHTRDATIQGLGIAQ-QAANVAATARa	antibacterial	Okada and Natori (1985)
Spiders and Scorpions				
Charybdotoxin	Scorpion (*Leiurus quinquestriatus hebraeus*, venom)	ZFTNVSCTTSKECWSVCQRLHNTSRGKCMN-KKCRCYS		Schweitz et al. (1989)
Defensin 4K	Scorpion (*Leiurus quinquestrianus*)	GFGHCPLNQGACHRHCRSIRRRGGY-CAGFFKQTCTCYRN	antibacterial (G+)	Cociancich et al. (1993)

Tab. 5. Continued

Peptide	Source	Sequence[a]	Properties	Reference
Polyphemusin	Horseshoe crab (*Limulus polyphemus*)	RRWCFRVCYRGFCYRKCRa	antibacterial	Lambert et al. (1989)
Tachyplesin (3 types)	Horseshoe crab (*Limulus polyphemus*)	KWCFRVCYRGICYRRCR	antibacterial; antifungal	Nakamura et al. (1988)
Toxin 2	Sahara scorpion (*Androctonus australis Hector*)	VKDGYIVDDVNCTYFCGRNAYC-NEECTKLKGESGYCQWASPYGNA-CYCKLPDHVRTKGPGRCH		Bontems et al. (1991)
Molluscs				
CARP	*Mytilus edulis*	AMPMLRLa	catch-relaxing peptide (muscle)	Hirata et al. (1987)
MIP (10 types)	Land snail (*Achatina fulica*)	AAPKFVGRRGAPYFV	contraction inhibitor	Ikeda et al. (1992)
MIP (12 types)	Land snail (*Helix pomatia*)	GAPAFV	contraction inhibitor	Ikeda et al. (1992)
Frogs/Toads				
Adenoegulin	Two-colored leaf frog (*Phyllomedusa bicolor*)	GLWSKIKEVGKEAAKAAAKAAGKAAL-GAVSEAV	antibacterial, antifungal	Daly et al. (1992)
Bombinin	Yellow-bellied toad (*Bombina variegata*)	GIGALSAKGALKGLAKGLAEHFANa	anibacterial	Csordas and Michl (1970)
BLP-1 (4 types)	Asian toad (*Bombina orientalis*)	GIGASELSAGKSALKGLAKGLAEHFANa	antibacterial	Gibson et al. (1991)
Brevinin	European frog (*Rana esculenta*)	FLPLLAGLAANFLPKIFCKITRKC	antibacterial, cytotoxic	Simmaco et al. (1993)

Dermaseptin 1 (6 types)	Sauvage's leaf frog (*Phyllomedusa sauvagii*)	ALWKTMLKKLGTMALHAGKAALKAAAD-TISQGTQ	antibacterial, antifungal	MOR et al. (1991)
Esculentin	European frog (*Rana esculenta*)	GIFSKLGKKIKNLLISGLKNVG-KEVGMDVVRTGIDIAGCKIKGEC	antibacterial	SIMMACO et al. (1993)
Magainin I	*Xenopus laevis*, skin	GIGKFLHSAGKFGKAFVGEIMKS	anibacterial, antifungal	ZASLOFF (1987)
PGLa	*Xenopus laevis*, skin	GMASKAGAIAGKIAKVALKAL	antibacterial	KUCHLER et al. (1989)
PGQ	*Xenopus laevis*, stomach	GVLSNVIGYLKKLGTGALNAVLK	antibacterial; antifungal	MOORE et al. (1991)
Ranalexin	Bullfrog (*Rana catesbeiana*, skin)	FLGGLIKIVPAMICAVTKKC	antibacterial	CLARK et al. (1994)
XPF	*Xenopus laevis*, skin	GWASKIGQTLGKIAKVGLKELIQPK	antibacterial	SURES and CRIPPA (1984)
Snakes				
Toxin 1	Waglers pit viper (*Trimeresurus wagleri*, venom)	GGKPDLRPCHPPCHYIPRPKPR		SCHMIDT et al. (1992)
Mammals				
Bactenecin	bovine neutrophils	RLCRIVVIRVCR	antibacterial	ROMEO et al. (1988)
Bac5	bovine neutrophils	RFRPPIRRPPIRPPFYPPFRPPIRPPIFP-PIRPPFRPPLRFP	antibacterial (G –)	FRANK et al. (1990)
BNBD-2 (13 types)	bovine neutrophils	VRNHVTCRINRGFCVPIRCPGRTRQIGTCFG-PRIKCCRSW	antibacterial	SELSTED et al. (1993)
Cecropin P1	Pig (*Sus scrofa*, small intestine)	SWLSKTAKKLENSAKKRISEGIAIAIQGGPR	antibacterial (G –)	LEE et al. (1989)
Cryptdin (5 types)	Mouse (*Mus musculus*, intestine)	LRDLVCYCRSRGCKGRERMNGTCRKGHL-LYTLCCR		SELSTED et al. (192)
Endozepine	Pig (*Sus scrofa domestica*)	KQATVGDINTERPDILDKGKAKW-DAWNGLKGTSKEDAMKAYINKVEELKK-KYGI	antibacterial	AGERBERTH et al. (1993)
Gastric inhibitory peptide (GIP)	Pig (*Sus scrofa domestica*)	ISDYSIAMDKIRQQDFVNWL-LAQKGKKSDWKHNITQ	antibacterial	AGERBERTH et al. (1993)
GNCP-1	Guinea pig (*Cavia cutteri*)	RRCICTTRTCRFPYRRLGTCIFQNRVYTFCC	antibacterial	YAMASHITA and SAITO (1988)

Tab. 5. Continued

Peptide	Source	Sequence[a]	Properties	Reference
Hiastadin	*Macaca fascicularis*	DSHEERHHGRHGHHKYGRKFHEKHHSHR-GYRSNYLYDN	antifungal	Xu et al. (1990)
HNP-1 (6 types)	Human neutrophils	ACYCRIPACIAGERRYGTCIYQGRLWAFCC	antibacterial, antifungal, cytotoxic	Lehrer et al. (1991)
Indolicidin	Bovine neutrophils	ILPWKWPWWPWRR	antibacterial	Selsted et al. (1992)
Lacto-ferricin	Bovine (N-terminal of lacto-ferrin)	FKCRRWQWRMKKLGAPSITCVRRAF	antibacterial	Bellamy et al. (1992)
MCPI (2 types; NP-1 (5 types)	Rabbit (*Oryctolagus cuniculus,* macrophage; neutrophils)	VVCACRRALCLPRERRAGFCRIR-GRIHPLCCRR	antibacterial, antifungal, cytotoxic	Selssted et al. (1983)
Peptide 3910	Pig (*Sus scrofa domestica*)	RADTQTYQPYNKDWIKEKIYVLLRRQAQ-QAGK	antibacterial	Agerberth et al. (1993)
RatNP-1 (4 types)	Rat (*Rattus norvegicus*)	VICYCRRTRCGFRERLSGACGYR-GRIYRLCCR	antifungal, antibacterial	Eisenhauer et al. (1989)
Sarcolipin	Rabbit (*Oryctolagus cuniculus,* skeletal muscle)	MERSTRELCLNFTVVLITVILIWLLVRSYQY	proteolipid-like, ionophore?	Wawrzynow et al. (1992)
Seminal-plasmin	Bovine (*Bos taurus,* seminal plasma)	SDEKASPDKHHRFSLSRYAKLANRLANPKL-LETFLSKWIGDRGNRSV	antibacterial, antifungal, cytotoxic	Reddy and Bhar-gava (1979)
TAP	Bovine (*Bos taurus,* tracheal mucosa)	NPVSCVRNKGICVPIRCPGSMKQIGTCV-GRAVKCCRKK	antibacterial, antifungal	Diamond et al. (1991)

[a] Only plain sequences are given, no disulfide links are indicated: a: C-terminal amidated; most accession numbers of the gene sequences can be retrieved from Hancock et al. (1995).

Tab. 6. Compilation of Peptides of Nonribosomal Origin (Bacteria/Fungi/Plants/Animals)

Compound	Organism	Structural Type[a]	Properties	Reference
Bacteria – Bacilli				
FR901537	*Bacillus* sp.	AA-M	antitumor	OOHATA et al. (1995)
Fusaricidin	*Bacillus polymyxa*	R-L-6	antibacterial	KAJIMURA and KANEDA (1996)
N-4909	*Bacillus* sp.	L-8	stimulates apolipoprotein E secretion	HIRAMOTO et al. (1996)
Halobacillin	*Bacillus* sp. (marine)	L-8(2)	cytotoxic	TRISCHMANN et al. (1994)
Isohalobacillin B	*Bacillus* sp.	L-8	ACAT inhibitor	HASUMI et al. (1995)
Surfactins	*Bacillus subtilis natto*	L-8(2)		OKA et al. (1993)
Surfactin-like	*Bacillus pumilus* (marine)	L-8(2		KALINOVSKAYA et al. (1995)
Bacillopeptin	*Bacillus subtilis*	C-8(2)	antifungal (phytopathol.)	KAJIMURA et al. (1995)
Cereulide	*Bacillus cereus*	C-D-12	K^+-complex	SUWAN et al. (1995)
Homocereulide	*Bacillus cereus* (marine)	C-D-12	extremely cytotoxic	WANG et al. (1995)
Bacitracins	*Bacillus* sp.	P-12-C-7-M(2)	antibacterials	IKAI et al. (1995)
Actinomycetes				
Monamidocin	*Streptomyces* sp.	P-2	fibrinogen receptor antagonist	KAMIYAMA et al. (1995)
Matylstatin	*Actinomadura atramentaria*	R-P-2-M	type IV collagenase inhibitor	OGITA et al. (1992)
Phevalin	*Streptomyces* sp.	C-2-M	calpain inhibitor	ALVAREZ et al. (1995)
Thaxtomin	*Streptomyces scabies*	C-2-M	phytotoxin	GELIN et al. (1993)
Cutinostatin	Actinomycete	R-P-2-M	cutinase inhibitor	HIGASHI et al. (1996)
Amonobactins	*Aeromonas hydrophila*	R-P-3	siderophore	TELFORD et al. (1994)
Napsamycin	*Streptomyces* sp.	R-P-3-M(U)	antibacterial (*Pseudomonas*)	CHATTERJEE et al. (1994)
Nerfilin I	*Streptomyces halstedii*	R-P-3	neurite outgrowth inducer	HIRAO et al. (1995)
Pepticinnamins	*Streptomyces* sp.	R-P-3-M (C2)	farnesyl-protein transferase inhibitor	ŌMURA et al. (1993)
Epoxomicin	Actinomycete	R-P-4-M	antitumor	HANADA et al. (1992)
Respiranin	*Streptomyces* sp.	R-C-D-4 (3,E)	insecticidal	URUSHIBATA et al. (1993)
Enamidonin	*Streptomyces* sp.	R-P-5-C-4(2)	antitumor, detransforming activity	KOSHINA et al. (1995)

Tab. 6. Continued

Compound	Organism	Structural Type[a]	Properties	Reference
α-MAPI	*Streptomyces* sp. *Micromonospora* sp. *Nocardia* sp.	P-4(U)	HIV-1 protease inhibitor	Stella et al. (1991)
Mer-N5075A	*Streptomyces chromofuscus*	P-4-M(U)	HIV-1 protease inhibitor	Kaneto et al. (1993)
MTMTLA	*Streptomyces griseus*	P-4-M	calpain inhibitor	Alvarez et al. (1994)
GE20372	*Streptomyces* sp.	P-4-M	HIV-1 protease inhibitor	Stefanelli et al. (1995)
MR-387	*Streptomyces neyagawaensis*	P-4-M	aPase inhibitor	Chung et al. (1996)
Formobactin	*Nocardia* sp.	R-D-4	radical scavenger, neuronal protecting	Murakami et al. (1996)
Muredomycins	*Streptomyces flavidovirens*	R-P-4-M(U)	antibiotic	Isono et al. (1993)
Echinoserine	*Streptomyces tendae*	$(R-P-4)_2$-M	antibacterial (weak)	Blum et al. (1995)
BE-22179	*Streptomyces gangtokensis*	$(R-P-4)_2$-M	topoisomerase II inhibitor	Okada et al. (1994)
Poststatin	*Streptomyces viridochromogenes*	P-5-(CO)	pro-endopeptidase inhibitor	Aoyagi et al. (1991)
Cyclothialidine	*Streptomyces filipinensis*	P-5-M	DNA gyrase inhibitor	Kamiyama et al. (1994)
WS 1279	*Streptomyces willmorei*	R-P-5	bone marrow growth stimulating	
Sandramycin	*Nocardioides* sp.	$(R-P-5)_2$	antitumor	Matson et al. (1993)
Rotihibin	*Streptomyces graminofaciens*	R-P-5-AA-M	plant growth regulator	Fukuchi et al. (1995)
WS-7338	*Streptomyces* sp.	C-5	endothelin receptor antagonist	Miyata et al. (1992)
BE-18257	*Streptomyces misakiensis*	C-5	enothelin binding inhibitor	Kojiri et al. (1991)
Aurantimycin	*Streptomyces aurantiacus*	C-6	antibacterial, cytotoxic	Gräfe et al. (1995)
WF11899A/B/C	*Coleophoma emperri*	R-C-6(4)	antifungal	Iwamoto et al. (1994)
Protactin	*Streptomyces cucumerosporus*	R-L-6	antibacterial, antitumor	Hanada et al. (1992)
WS1279	*Streptomyces willmorei*	R-P-6-M	mitogenic	Kurimura et al. (1993)
Verucopeptin	*Actinomadura* sp.	R-L-6-M	antitumor	Nishiyama et al. (1993)
Xanthostatin	*Streptomyces spiroverticillatus*	R-L-6-M	antibiotic	Kim et al. (1992)

Salinamide	*Streptomyces* sp. (marine)	R-L-6-M	anti-inflammatory	TRISCHMANN et al. (1994)
IC101	*Streptomyces* sp.	R-L-6-M	extracellular matrix antagonist	UENO et al. (1993)
RPI-856	*Streptomyces* sp.	P-6(7)-M	HIV protease inhibitor	ASANO et al. (1994)
Bottromycin A2	*Streptomyces bottropensis*	P-7-C-4-M		KANEDA (1992)
Cochinmicins	*Microbispora* sp.	P-7-L-5	endothelin antagonist	LAM et al. (1992)
Piperastatin A	*Streptomyces lavendofoliae*	R-P-7	sprine carboxypeptidase inhibitor	MURAKAMI et al. (1996)
WS9326A	*Streptomyces violaceusniger*	R-L-7	tachykinin antagonist	SHIGEMATSU et al. (1993)
Kistamicins	*Microtetraspora parvosata*	P-7-M	antiviral	NARUSE et al. (1993)
MM 55266/8	*Amycolatopsis* sp.	R-P-7-M (GLYC)	antibacterial	BOX et al. (1991)
Galacardins	*Saccharothrix* sp.	R-P-7-M (GLYC)	antibacterial	TAKEUCHI et al. (1992)
Balhimycin	*Amycolatopsis* sp.	R-P-7-M (GLYC)	antibacterial	NADKARNI et al. (1994)
Chloropeptins	*Streptomyces* sp.	P-7-M	gp120-CD4 binding inhibitor	MATSUZAKI et at. (1994)
WS9326A	*Streptomyces violaceusniger*	R-C-7	tachykinin inhibitor	HAYASHI et al. (1992)
A21459	*Actinoplanes* sp.	C-8	antibacterial	SELVA et al. (1996)
Himastatin	*Streptomyces hygroscopicus*	$(L\text{-}6)_2$	antitumor	LEET et al. (1996)
Quinoxapeptin	MA7095 (nocardioform)	R-$(P\text{-}5)_2$-M	HIV-1,2 reverse transcriptase inhibitor	LINGHAIN et al. (1996)
Cypemycin	*Streptomyces* sp.	P-11-M	antibiotic	MINAMI et al. (1994)
Promotiocin	*Streptomyces* sp.	P-13-C-10 [PYR]-M	tip A promoter inducing	YUN et al. (1994)
Berninamycin	*Streptomyces bernensis*	P-14-C-12 [PYR]-M		LAU and RINEHART (1994)
Geninthiocin	*Streptomyces* sp.	P-15-C-13 [PYR]-M	tip A promoter inducing	YUN et al. (1994)
GE37, 468	*Streptomyces* sp.	P-15-C-10-M [PYR]	antibacterial	FERRARI et al. (1995)
RES 701	*Streptomyces* sp.	P-16-C-9	endothelin B receptor antagonist	OGAWA et al. (1995)
Thiotipin	*Streptomyces* sp.	P-16-C-12 [PYR]-M	tipA promoter inducing	YUN et al. (1994)

Tab. 6. Continued

Compound	Organism	Structural Type[a]	Properties	Reference
A10255B/G/J	*Streptomyces gardneri*	P-17-C-14 [PYR]-M		DEBONO et al. (1992)
Promoinducin	*Streptomyces* sp.	P-18-C-14 [PYR]-M		YUN and SETO (1995)
Cyanobacteria				
Aeruginosin	*Microcystis aeruginosa*	R-P-3	trypsin inhibitor	MURAKAMI et al. (1995)
Aeruginosin 298A	*Microcystis aeruginosa*	P-4	thrombin and trypsin inhibitor	MURAKAMI et al. (1994)
Muscoride A	*Nostoc muscorum*	P-4-M (TERP)		NAGATSU et al. (1995)
Cryptophycin	*Nostoc* sp.	L-4(4)		BARROW et al. (1995)
Antillatoxin	*Lyngbya majuscula*	L-4-M(4)	ichthyotoxic	TAKIZAWA et al. (1995)
Linear peptides	*Microcystis* sp.	R-P-4(2,3)		CHOI et al. (1993)
Microginin	*Microcystis aeruginosa*	P-5	angiotensin converting enzyme inhibitor	OKINO et al. (1993)
Microcolins	*Lyngbya majuscula*	R-P-5-M	immunosuppressive	KOEHN et al. (1992)
Nodularins	*Nodularia spumignea*	C-5(2,3)		NAMIKOSHI et al. (1994)
Anabaenopeptins	*Anabaena flos-aquae*	P-6-C-5(4)		HARADA et al. (1995)
Nostocyclamide	*Nostoc* sp.	C-6	antialgal, anticyano-bacterial	TODOROVA et al. (1995)
Westiellamide	*Westiellopsis prolifica*	C-6-M	cytotoxic (moderate)	PRINSEP et al. (1992)
Oscillamide	*Oscillatoria agardhii*	C-6-M(4)	chymotrypsin inhibitor	SANO and KAYA (1995)
Micropeptin 90	*Microcystis aeruginosa*	R-L-6-M	plasmin/trypsin inhibitor	ISHIDA et al. (1995)
DHB-microcystin-RR	*Oscillatoria agardhii*	C-7-M(2)		SANO and KAYA (1995)
Nostocyclin	*Nostoc* sp.	R-P-7-L-5-M	protein phosphatase-1 inhibitor	KAYA et al. (1996)
Microcystilide A	*Microcystis aeruginosa*	R-P-7-L-6-M	cell differentiation promoting	TSUKAMOTO et al. (1993)
Cyanopeptolins A–D	*Microcystis* sp.	R-P-7-L-6-M		MARTIN et al. (1993)
Micropeptins A, B	*Microcystis aeruginosa*	R-P-7-L-6-M	plasmin/trypsin inhibitor	OKINO et al. (1993)
Oscillapeptin	*Oscillatoria agardhii*	R-P-7-L-6-M	elastase/chymotrypsin inhibitor	SHIN et al. (1995)
A90720A	*Microchaete loktakensis*	R-P-7-L-6	serine protease inhibitor	BONJOUKLIAN et al. (1996)

Aeruginopeptins	*Microcystis aeruginosa*	R-P-8-L-6-M		HARADA et al. (1993)
Calophycin	*Calothrix fusca*	C-10-M	fungicidal	MOON et al. (1992)
Puwainaphycins	*Anabaena* sp.	R-C-10(2)	cardioactive	GREGSON et al. (1992)
Laxaphycins	*Anabaena laxa*	C-11	antifungal	FRANKMOLLE et al. (1992)
Hormothamnin	*Hormothamnion entero-morphoides*	C-11		GERWICK et al. (1992)
Schizotrin	*Schizothrix* sp.	R-P-13-C-12(2)	antimicrobial	PERGAMENT and CARMELI (1994)
Microviridins	*Microcystis aeruginosa*	R-P-14-(C-4/L-4)	elastase inhibitor	OKINO et al. (1995)
Bacteria – Various				
Epothilon	*Mucor hiemalis*	AA-M(PK)	antifungal, cytotoxic	GERTH et al. (1995)
Fluvibactin	*Vibrio fluvialis*	R3-A	siderophore	YAMAMOTO et al. (1993)
WS75624	*Saccharothrix* sp.	R-P-2-M	endothelin converting enzyme inhibitor	TSURUMI et al. (1995)
CI-4	*Pseudomonas* sp. (marine)	C-2	chitinase inhibitor	IZUMIDA et al. (1996)
TAN-057	*Flexibacter* sp.	P-3-M	antibacterial (*S. aureus* methicillin resistant)	KATAYAMA et al. (1993)
Methanofurans	*Methanobacterium thermoautotrophicum*	R-P-3-M	natural cofactor	SULLIONS et al. (1993)
Cyclotetrapeptides	*Lactobacillus* sp.	C-4	metal binding, melanin inhibitor; tyrosinase inhibitor	KURANARI et al. (1993), KAWAGISHI et al. (1993)
Rakicidins	*Micromonospora* sp.	L-4-M	cytotoxic	MCBRIEN et al. (1995)
Glycopeptidolipid antigen	*Mycobacterium sene-galense*	R-P-4-M (GLYC)		LOPEZ MARIN (1993)
Thiangole	*Polyangium* sp.	R-P-4-M	HIV-1 inhibitor	BOYCE et al. (1994)
FR-901, 228	*Chromobacterium violaceum*	C-R-P-4-M	antitumor	LI et al. (1996)
Thiangazole	*Polyangium* sp.	R-P-5-M		JANSEN et al. (1993)
Aletrobactin A	*Alteromonas luteoviolacea*	R-P-6-L-5	siderophore	DENG et al. (1995)
Azoverdin	*Azomonas macrocytogenes*	R-P-7-M	siderophore	LINGET et al. (1992)
YM47141/2	*Flexibacter* sp.	P-7-L-5	elastase inhibitor	ORITA et al. (1995)
Myxochromide A	*Myxococcus virescens*	R-L-6		TROWITZSCH-KIENSAT et al. (1993)
TAN-1511	*Streptosporangium amethystogenes*	R-S-R′-P-7	induction of cytokines	TAKIZAWA et al. (1995)
WLIP	*Pseudomonas reactans*	R-P-9-L-6	white line inducing principle	HAN et al. (1992)

Tab. 6. Continued

Compound	Organism	Structural Type[a]	Properties	Reference
Vioprolides	*Cystobacter violaceus*	L-9-M(2)	antifungal, cytotoxic	SCHUMMER et al. (1996)
Syringomycin	*Pseudomonas syringae*	R-L-9	phytotoxin	FUKUCHI et al. (1992)
Ferrocins	*Pseudomonas fluorescens*	R-L-10-M	siderophore	TSUBOTANI et al. (1993)
Pholipeptin	*Pseudomonas* sp.	R-P-11-L-9(2)		UI et al. (1995)
Bu-2841	unidentified gram-negative	R-P-12-L-11	antibacterial	FUKAI et al. (1995)
Fungi				
Bassiatin	*Beauveria bassiana*	c-R-AA	platelet aggregation inhibitor	KAGAMIZONO et al. (1995)
Cathestatin	*Microascus longiirostris*	R-AA-A	cysteine protease inhibitor	YU et al. (1996)
AM4299A	*Chromelosporium fulvum*	R-AA-M	thiol protease inhibitor	MORISHITA et al. (1994)
Lipoxamycin	*Aspergillus fumigatus*	R-AA	palmitoyl-transferase inhibitor	MANDALA et al. (1994)
NK-374200	*Talaromyces*	P-2	insecticidal	
WIN 64821	*Aspergillus* sp.	P-2-M		POPP et al. (1994)
OPC-15161	*Thielavia minor*	C-2-M	superoxide anion generation inhibitor	KITA et al. (1994)
Gypsetin	*Nannizia gypsea*	C-2-M	ACAT inhibitor	SHINOHARA et al. (1994)
Tryprostatins	*Aspergillus fumigatus*	C-2-M (TERP)	cell cycle inhibitor	CUI et al. (1996)
PEDT	*Aspergillus flavus*	$(C\text{-}2)_2$-M	substance P inhibitor	BARROW and SEDLOCK (1994)
Sch52900/1	*Gliocladium* sp.	$(C\text{-}2)_2$-M	oncogene inhibitor	CHU et al. (1995)
Benzomalvins	*Penicillium* sp.	C-3-M	substance P inhibitor	SUN et al. (1994)
Ustiloxins	*Ustilaginoidea virens*	L-3-M(2,5)	antimitotic, cancerostatic	KOISO et al. (1994)
Citreoindole	*Penicillium citreo-viride*	P-3-C-2-M		MATSUNAGA et al. (1991)
Fellutamide	*Penicillium fellutanum* (marine)	R-P-3-M	nerve growth factor promoter	SHIGEMORI et al. (1991), TSUJI et al. (1994)
Ro 09-1679	*Mortierella alpina*	R-P-3-M	thrombin inhibitor	KAMIYAMA et al. (1992)
Ergobine	*Claviceps purpurea*	R-P-3-M		PERELLINO et al. (1993)
Ergogaline	*Claviceps purpurea*	R-P-3-M		CVAK et al. (1994)
CAPPA	*Verticillium coccosporum*	C-4	phytotoxic	GUPTA et al. (1994)
Beauverolides I, II	*Beauveria* sp.	L-4	insecticidal (I)	MOCHIZUKI et al .(1993)
Beauverolides L/La	*Beauveria tenella* *Paecilomyces fumo-soroseus*	L-4(2)		JEGOROV et al. (1994)

Stevastelins	*Penicillium* sp.	L-4(2)	immunosuppressant	Morino et al. (1996)
Beauvericins	*Beauveria bassina*	C-D-4-M	insecticidal	Gupta et al. (1995)
YF-044P-D	*Candida albicans*	R-P-5	aspartic proteinase inhibitor	Sato et al. (1994)
Plactin	F-165 (Fungi Imperfecti)	C-5	stimulates fibrinolytic activity	Inoue et al. (1996)
Leualacin	*Hapsidospora irregularis*	L-5-M-M(2)	Ca^{2+}-blocker	Hamano et al. (1992)
Malformin	*Aspergillus niger*	C-5-M	phytotoxic	Kim et al. (1993)
Pithomycolide	*Pithomyces chartarum*	C-D-5		Moussa and Le Quesne (1996)
Aselacins	*Acremonium* sp.	R-P-5-L-4(3)	endothelin receptor antagonist	Hochlowski et al. (1994)
Pneumocandins	*Zalerion arboricola*	R-C-6(3)	antifungal	Hensens et al. (1992), Morris et al. (1994)
Deoxymulundocandin	*Aspergillus sydowii*	R-C-6(3)	antifungal	Mukhopadhay et al. (1992)
Destruxin A4/5	*Aschersonia* sp.	L-6-M	insecticidal	Krasnoff et al. (1996)
Desmethyldestruxin	*Metarhizium anisopliae*	L-6	suppresses hepatitis B virus surface antigen production	Chen et al. (1995)
Bursaphelocides	perfect fungus	L-6-M	nematicidal	Kawazu et al. (1993)
Enniatins D-F	*Fusarium* sp.	C-D-6		Tomoda et al. (1992)
Enniatin A2	*Fusarium avenaceum*	C-D-6		Kastic et al. (1992)
WIN 66306	*Aspergillus* sp.	C-7-M (TERP)	neurokinin antagonist	Barrow et al. (1994)
PF1022	*Mycelia sterilia*	C-D-8	antihelmintic	Scherkenbeck et al. (1995)
BZR-cotoxin IV	*Bipolyris zeicola*	L-8-M		Ueda et al. (1995)
BZR-cotoxin I/II	*Bipolyris zeicola*	C-D-9-M	phytotoxic?	Ueda et al. (1992, 1994)
Leucinostatins, P-168	*Paecilomyces* sp.	R-P-9-M	antibiotic	Kuwata et al. (1992), Isogai et al. (1992)
Helioferins	*Mycogone rosea*	R-P-10-M	antifungal	Gräfe et al. (1995)
LP237-F8	*Tolypocladium geodes*	P-10-M	antibiotic	Tsantrizos et al. (1996)
Trichogin A	*Trichoderma longibrachiatum*	P-10-M		Auvin-Guette et al. (1992)
FR901459	*Stachybotrys chartarum*	C-11-M	immunosuppressive	Sakamoto et al. (1993)
Petriellin A	*Petriella sordida*	L-13-M	antifungal	Lee et al. (1995)
Harzianin	*Trichoderma harzianum*	R-P-13-M	antibiotic	Rebuffat et al. (1995)
Trichkindins	*Trichoderma harzianum*	P-17-M	induce Ca^{2+}-dependent catecholamine secretion	Iida et al. (1994)

Tab. 6. Continued

Compound	Organism	Structural Type[a]	Properties	Reference
Hypelcins	*Hypocrea peltata*	R-P-19-M	antibiotic	MATSURA et al. (1993, 1994)
Trichosporin B	*Trichoderma polysporum*	R-P-19-M	ion channel forming	IIDA et al. (1993), NAGAOKA et al. (1994)
Plants				
p-Coumaroyl-tryptophan	*Coffea canephora*	R-AA		MURATA et al. (1995)
Aurantiamide	*Piper aurantiacum* *Clematis hepaulensis*	R-P-2-M		BANERJEE et al. (1993)
Cyclopeptide alkaloids	*Zizyphus mucronata* (roots)	R-P-3-M(E)		BARBONI et al. (1994)
Mucronine J	*Zizyphus mucronata* (root, bark)	R-P-4-M(C)		AUVIN et al. (1996)
Astin A	*Aster tataricus*	C-5(2)	antitumor	MORITA et al. (1993)
Astins D, E	*Aster tataricus*	C-5		MORITA et al. (1993)
Astin J	*Aster tataricus* (roots)	P-5		MORITA et al. (1995)
Asterinins A–C	*Aster tataricus* (roots)	P-5		CHENG et al. (1994)
Asterinins D, E	*Aster tataricus*	P-5-M		CHENG et al. (1996)
Astericins D–F	*Aster tataricus* (roots)	P-5		CHENG et al. (1996)
Astins	*Aster tataricus*	C-5-(U)	antitumor	MORITA et al. (1996)
Astins	*Aster tataricus*	C-5(2)		MORITA et al. (1994)
Pseudostellarins A–C	*Pseudostellaria heterophylla*	C-5	tyrosinase inhibitor	MORITA et al. (1994)
Mucronine J	*Zizyphus mucronata* (bark)	R-P-4-M		AUVIN et al. (1996)
Stellarins B, C	*Stellaria yunnanensis* (roots)	C-6		ZHAO et al. (1995)
Dichotomins	*Stellaria dichotoma*	C-6		MORITA et al. (1996)
Segetalin A/B–D	*Vaccaria segetalis*	C-6/C-5		MORITA et al. (1995)
Citrusins	*Citrus unshiu, Citrus sinensis, Citrus natsudaidai* peeling	C-7		MATSUBARA et al. (1992)
Heterophyllins	*Pseudostellaria heterophylla* (root)	C-7		TAN et al. (1993)
Yunnanins C–F	*Stellaria yunnanensis*	C-7		MORITA et al. (1996)

Cyclogossine A	*Jatropha gossypifolia* (latex)	C-7		HORSTEN et al. (1996)
Stellarins F, G	*Stellaria yunnanensis*	C-8		ZHAO et al. (1995)
Pseudostellarin D/H	*Pseudostellaria heterophylla*	C-7/8		MORITA et al. (1995)
Podacycline A	*Jatropha podagrica* (latex)	C-9		VAN DEN BERG et al. (1996)
Segetalin A	*Vaccaria segetalis* (seeds)	C-6		MORITA et al. (1994)
RA VII	*Rubia cordifolia*	C-6	eukaryotic translation inhibitor	ITOKAWA et al. (1993)
Stellarins D, E	*Stellaria yunnanensis*	C-7		ZHAO et al. (1995)
Segetalin E	*Vaccaria segetalis* (seeds)	C-7		MORITA et al. (1996)
Lyciumin A	*Lycium chinense*	P-8-C-5(CN)		MORITA et al. (1996)
Cycloneoluripeptides	*Leonurus heterophylus*	C-9		MORITA et al. (1996)
Sponges				
Geodiamolide G	*Cymbastela* sp.	L-5-M(7)	cytotoxic	COLEMAN et al. (1995)
Cyclotheonamide	*Theonella* sp.	C-5(CO,3)	serine protease inhibitor	LEE et al. (1993)
Microsclerodermins	*Microscleroderma* sp.	C-6(2,3)	antifungal	BEWLEY et al. (1994)
Konbamide	*Theonella* sp.	P-6(U)-C-5	calmodulin antagonist	SCHMIDT and WEINBRENNER (1996)
Orbiculamide	*Theonella* sp.	R-P-7-C-6-M(3)	cytotoxic	FUSETANI et al. (1991)
Discobahamins	*Discodertnia* sp.	R-P-7-C-6-M(3,4)	antifungal (*Candida albicans*)	GUNASEKERA et al. (1994)
Pseudoaxinellin	*Pseudoaxinella massa*	C-7		KONG et al. (1992)
Aciculitins	*Acucilites orientalis*	C-7-M	cytotoxic, antifungal	BEWLEY et al. (1996)
Stylostatin	*Stylotella aurantium*	C-7	cytotoxic	PETTIT et al. (1992)
Stylopeptide I	*Stylotella* sp./*Phakellia costata*	C-7		PETTIT et al. (1995)
Phakellistatin	*Phakellia costata* and *Stylotella aurantium*	C-7	cancer cell growth inhibitor	PETTIT et al. (1993, 1995)
Axinastatins	*Axinella* spp.	C-7	cytotoxic	PETTIT et al. (1993, 1994), KONAT et al. (1995)
Malaysiatin	*Pseudoaxinyssa* sp.			
Cyclodidemnamide	*Didemnum molle*	C-7-M		TOSKE and FENICAL (1995)
Hymenamides	*Hymeniacidon* sp.	C-7/8		TSUDA et al.(1993), KOBAYASHI et al. (1996)
Perthamide B	*Theonella* sp.	C-8		GULAVITA et al. (1994)

Tab. 6. Continued

Compound	Organism	Structural Type[a]	Properties	Reference
Keramamides	*Theonella* sp.	R-P-8-C-6-M(3,CO)	superoxide generation response of human neutrophils inhibitor	KOBAYASHI et al. (1991, 1995)
Callipeltins	*Callipelta* sp.	L-8-C-7-M		D'AURIA et al. (1996)
Majusculamide C	*Ptilocaulis trachys*	L-9(3)		WILLIAMS et al. (1993)
Callipeltin A	*Callipelta* sp.	R-P-9-L-6-M	anti-HIV	ZAMPELLA et al. (1996)
Didemnin B*	*Trididemnum cyanophorum*	R-P-9-L-7(3)		ABOU-MANSOUR et al. (1995)
Keramamide F	*Theonella* sp.	R-P-9-C-7-M(CO)		ITAGAKI et al. (1992)
Theonellamides	*Theonella* sp.	C-12(2,3)	cytotoxic	MATSUNAGA and FUSETANI (1995)
Theonegramide	*Theonella swinhoei*	C-12(2,3)-M (GLYC)	antifungal	BEWLEY and FAULKNER (1994)
Polydiscamide A	*Discoderma* sp.	R-P-13-L-5-M		GULAVITA et al. (1992)
Theonellapeptolide IId	*Theonella swinhoei*	R-P-13-L-11(M)		KOBAYASHI et al (1994)
Discodermins F–H	*Discodermia kiiensis*	R-P-14-L-6	antimicrobial	RYU et al. (1994)
Halicylindramides	*Halichondria cylindrata*	R-P-14-L-6	antifungal, cytotoxic	LI et al. (1995)
Polytheonamides	*Theonella swinhoei*	P-48		HAMADA et al. (1994)
Ascidians				
Bistratamides C/D	*Lissoclinum bistratum*	C-6-M	CNS depressant (D)	FOSTER et al. (1992)
Patellins	*Lissoclinum* sp.	C-6-M(terp), C-7/C8		CAROLL et al. (1996)
Trunkamide A				
Patellamides	*Lissoclinum patella*	C-6-M	cytotoxic	ISHIDA et al. (1995)
Discokilides	*Discoderma kiiensis*	L-7-M	cytotoxic	TADA et al. (1992)
Mollamide	*Didemnum molle*	C-7-M	cytotoxic	CAROLL et al. (1994)
Nairaiamides	*Lissoclinum bistratum*	C-7-M (TERP)		FOSTER and IRELAND (1993)
Patellamide E	*Lissoclinum patella*	C-8-M		MACDONALD and IRELAND (1992)
Tawicyclamide B	*Lissoclinum patella*	C-8-M	cytotoxic	MACDONALD et al. (1992)
Didemnins	*Tridemnum solidum*	D-12-L-7-M(4)		SAKAI et al. (1995)

Sea Hares				
Aurilide	*Dolabella auricularia*	C-D-7(5)		SUENAGA et al. (1996)
Dolastatin H	*Dolabella auricularia*	R-P-4-M	cytotoxic	SONE et al. (1996)
Dolastatins C, D	*Dolabella auricularia*	R-D-4	cytotoxic (D)	SONE et al. (1993)
Molluscs				
Keenamide	*Pleurobranchus forskalii*	C-6-M(terp)	cytotoxic	WESSON and HAMANN (1996)
Kahalide F	*Elysia rufescens* and *Bryopsis* sp.	R-P-13-L-5		HAMANN and SCHEUER (1993)
Spiders/Scorpions				
Nephilatoxins	*Nephila clavata*	P2-A-AA	neurotoxins	MIYASHITA et al. (1992)
Nephilatoxin 7	*Nephila clavata*	R-P-2-A	neurotoxins	MATSUSHITA et al. (1995)
Clavamine	*Nephila clavata*	P-2-A-P-3	insecticidal	YOSHIOKA et al. (1992)
Spidamine	*Nephila clavata*	R-P-2-A	toxin	CHIBA et al. (1996)

[a] Abbreviations used in the description of the structural types are: A amine, AA amino acid, P peptide, C cyclopeptide, L peptidolactone (cyclic structure closed with an ester bond), D depsipeptide (compound containing both peptide and ester bonds); the number associated with each structural type (P-n, C-n, L-n, D-n) gives the number of amino acids or hydroxy acids contained in the structure; branched cyclic compounds are given in the total length with the ring size included (P-10-C-5 is thus a branched decapeptide with a pentapeptide cyclic region); acyl compounds, amidations or modifications by aminoalcohols are not included. Additional information is provided by R acylation by aliphatic or aromatic carboxylic acids, M modifications (largely N-methylations), (GLYC) glycosylation, (TERP) modification by terpenoids, [PYR] ring formation by a pyridine moiety formed by two modified serine residues (thioipeptides), U urea-type of linkage in the peptide chain, CO an additional CO-residue in the peptide chain, E ether bond cyclization, S thioether link, CN direct C-N linkage for cyclization, (n) gives information on cyclization types differing from the common α-carboxyl-α-amino-link; 2, 3, and 4 stand for bonds involving β, γ or δ positioned amino, carboxylic, or hydroxy groups.

yet unidentified cleavage and cyclization reactions. The restricted number of sources indicates that this field investigations have almost begun.

Animals as sources have been restricted mainly to marine organisms. The large group of sponges gives rise to the problem of the origin of compounds, since sponges are known to contain large numbers of associated microbes. Recently, the production of a complex sponge metabolite has been traced to an associated actinomycete (BEWLEY et al., 1996). In the future of screenings more attention will be paid to the interactions of organisms. In addition, search for biosynthetic genes involved in peptide formation is a promising pathway to detect, generate, and produce new metabolites.

6 References

ADEFARATI, A. A., GIACOBBE, R. A., HENSENS, O. D., TKACZ, J. S. (1991), Biosynthesis of L-671329, an echinocandin-type antibiotic produced by *Zalerion arboricola*: origins of some of the unusual amino acids and the dimethylmyristic acid side chain, *J. Am. Chem. Soc.* **113**, 3542–3545.

AHARONOWITZ, Y., COHEN, G., MARTÍN, J. F. (1992), Penicillin and cephalosporin biosynthetic genes: structure, organization, regulation, and evolution, *Annu. Rev. Microbiol.* **46**, 461–495.

APARICIO, J. F., MOLNAR, I., SCHWECKE, T., KÖNIG, A., HAYDOCK, S. F., KHAW, L. E., STAUNTON, J., LEADLAY, P. F. (1996), Organisation of the biosynthetic gene cluster for rapamycin in *Streptomyces hygroscopicus* – analysis of the enzymatic domains in the modular polyketide synthase, *Gene* **169**, 9–16.

AWAZU, N., IKAI, K., YUMAMOTO, J., NISHIMURA, K., MIZUTANI, S., TAKESAKO, K., KATO, I. (1995), Structures and antifungal activities of new aureobasidins, *J. Antibiot.* **48**, 525–527.

BAILEY, A. M., KERSHAW, M. J., HUNT, B. A., PATERSON, I. C., CHARNLEY, A. K., REYNOLDS, S. E., CLARKSON, J. M. (1996), Cloning and sequence analysis of an intron-containing domain from a peptide synthetase of the entomopathogenic fungus *Metarhizium anisopliae, Gene* **173**, 195–197.

BALTZ, R. H. (1996), Daptomycin and A54145, in: *Biotechnology of Industrial Antibiotics,* 2nd Edn. (STROHL, W., Ed.). New York: Marcel Dekker.

BAYER, A., FREUND, S., JUNG, G. (1995), Post-translational heterocyclic backbone modifications in the 43-peptide antibiotic microcin B17, *Eur. J. Biochem.* **234**, 414–426.

BENDER, C., PALMER, D., PENALOZAVAZQUEZ, A., RANGASWAMY, V., ULLRICH, M. (1996), Biosynthesis of coronatine, a thermoregulated phytotoxin produced by the phytopathogen *Pseudomonas syringae, Arch. Microbiol.* **166**, 71–75.

BERGEY, D. R., HOWE, G. A., RYAN, C. A. (1996), Polypeptide signaling for plant defensive genes exhibits analogies to defense signaling in animals, *Proc. Natl. Acad. Sci.* **93**, 12053–12058.

BERNHARD, F., DEMEL, G., SOLTANI, K., VON DÖHREN, H., BLINOV, V. (1996), Identification of genes encoding for peptide synthetases in the gram-negative bacterium *Lysobacter* sp. ATCC 53042 and the fungus *Cyclidrotrichum oligospermum, DNA Sequence* **6**, 319–330.

BEVINS, C. L., ZASLOFF, M. (1990), Peptide from frog skin, *Annu. Rev. Biochem.* **59**, 395–414.

BEWLEY, C. A., HOLLAND, N. D., FAULKNER, D. J. (1996), Two classes of metabolites from *Theonelle swinhoei* are localized in distinct populations of bacterial symbionts, *Experientia* **52**, 716–722.

BILLICH, A., ZOCHER, R. (1987), N-methyltransferase function of the multifunctional enzyme enniatin synthetase, *Biochemistry* **26**, 8417–8423.

BILLICH, A., ZOCHER, R. (1990), Biosynthesis of N-methylated peptides, in: *Biochemistry of Peptide Antibiotics* (KLEINKAUF, H., VON DÖHREN, H., Eds.), pp. 57–80. Berlin: de Gruyter.

BLANC, V., GIL, P., BAMAS-JACQUES, N., LORENZON, S., ZAGOREC, M., SCHLEUNIGER, J., BISCH, D., BLANCHE, F., DEBUSSCHE, L., CROUZET, J., THIBAUT, D. (1997), Identification and analysis of genes from *Streptomyces pristinaspiralis* encoding enzymes involved in the biosynthesis of the 4-dimethylamino-L-phenylalanine precursor of pristinamycin I, *Mol. Microbiol.* **23**, 191–202.

BOMAN, H. G. (1995), Peptide antibiotics and their role in innate immunity, *Annu. Rev. Immunol.* **13**, 62–92.

BOMAN, H. G., MARSH, J., GOODE, J. A. (1994), *Antimicrobial Peptides.* Chichester, New York: John Wiley & Sons.

BORCHERT, S., PATIL, S. S., MARAHIEL, M. A. (1992), Identification of putative multifunctional peptide synthetase genes using highly conserved oligonucleotide sequences derived from known synthetases, *FEMS Microbiol. Lett.* **92**, 175–180.

BORMANN, C., MÖHRLE, V., BRUNTNER, C. (1996), Cloning and heterologous expression of the entire set of structural genes for nikkomycin synthesis from *Streptomyces tendae* Tü901 in *Streptomyces lividans, J. Bacteriol.* **178**, 1216–1218.

BREIL, B. T., BORNEMAN, J., TRIPLETT, E. W. (1996), A newly discovered gene, TfuA, involved in the production of the ribosomally synthesized peptide antibiotic trifolitoxin, *J. Bacteriol.* **178**, 4150–4156.

BUADES, C., MOYA, A. (1996), Phylogenetic analysis of isopenicillin N synthetase horizontal gene transfer, *J. Mol. Evol.* **42**, 537–542.

BURMESTER, J., HAESE, A., ZOCHER, R. (1995), Highly conserved N-methyltransferases as integral part of peptide synthetases, *Biochem. Mol. Biol. Int.* **37**, 201–207.

CHEN, Q., WERTHEIMER, A. M., TOLMASKY, M. E., CROSA, J. H. (1996), The AngR protein and the siderophore anguibactin positively regulate the expression of iron-transport genes in *Vibrio anguillarum, Mol. Microbiol.* **22**, 127–134.

CHINERY, R., COFFEY, R. J. (1996), Trefoil peptides: less clandestine in the intestine, *Science* **274**, 204.

CHONG, P. P., PODMORE, S., KIESER, H., REDENBACH, M., TURGAY, K., MARAHIEL, M. A., HOPWOOD, D., SMITH, C. (1996), Physical identification of a chromosomal locus encoding biosynthetic genes for the calcium-dependent antibiotic of *Streptomyces coelicolor* A3(2). *Abstracts* Symp. Enzymol. Biosynth. Nat. Prod., Technical University, Berlin.

CLARK, D. P., DURELL, S., MALOY, W. L., ZASLOFF, M. (1994), Ranalexin, *J. Biol. Chem.* **269**, 10849–10855.

COQUE, J. J. R., ENGUITA, F. J., MARTÍN, J. F., LIRAS, P. (1995a), A two protein component 7-cephem-methoxylase encoded by two genes of the cephamycin C cluster converts cephalosporin to 7-methoxycephalosporin C, *J. Bacteriol.* **177**, 2230–2235.

COQUE, J. J. R., PÉREZ-LLARENA, F. J., ENGUITA, F. J., FUENTE, J. L., MARTÍN, J. F., LIRAS, P. (1995b), Characterization of the *cmcH* genes of *Nocardia lactamdurans* encoding a functional 3′-hydroxymethylcephem O-carbamoyltransferase for cephamycin biosynthesis, *Gene* **162**, 21–27.

COSMINA, P., RODRIGUEZ, F., DE FERRA, F., GRANDI, G., PEREGO, M., VENEMA, G., VAN SINDEREN, D. (1993), Sequence and analysis of the genetic locus responsible for surfactin synthesis in *Bacillus subtilis, Mol. Microbiol.* **8**, 821–831.

DAHOD, S. K. (1993), Dissolved carbon dioxide measurement and its correlation with operating parameters in fermentation process, *Biotechnol. Progr.* **9**, 655–660.

DE CRÉCY-LAGARD, V., MARLIÉRE, P., SAURIN W. (1995), Multienzymatic nonribosomal peptide biosynthesis: identification of the functional domains catalysing peptide elongation and epimerization, *C. R. Acad. Sci. Paris* **318**, 927–936.

DÍEZ, B., GUTIÉRREZ, S., BARREDO, J. L., VAN SOLINGEN, P., VAN DER VOORT, L. H. M., MARTÍN, J. F. (1990), The cluster of penicillin biosynthetic genes. Identification and characterization of the *pcbAB* gene encoding α-aminoadipyl-cysteinyl-valine synthetase and linkage to the *pcbC* and *penDE* genes, *J. Biol. Chem.* **265**, 16358–16365.

DONADIO, S., STAVER, M. J., MCALPINE, J. B., SWANSON, J. B., KATZ, L. (1991), Modular organization of genes required for complex polyketide biosynthesis, *Science* **252**, 675–679.

D'SOUZA, C., NAKANO, M. M., ZUBER, P. (1994), Identification of *comS,* a gene of the *srfA* operon that regulates the establishment of genetic competence in *Bacillus subtilis, Proc. Natl. Acad. Sci. USA* **91**, 9397–9401.

ERSPAMER, V., MELCHIORRI, P., FALCONIERI-ERSPAMER, G., NEGRI, L., CORSI, R., SEVERINI, C., BARRA, D., SIMMACO, M., KREIL, G. (1989), Deltorphins: a family of naturally occurring peptides with high affinity and selectivity for δ-opioid binding sites, *Proc. Natl. Acad. Sci. USA* **86**, 5188–5192.

FAULKNER, D. J. (1996), Marine natural products, *Nat. Prod. Rep.* **13**, 75–125.

FRÉCHET, D., GUITTON, J. D., HERMAN, F., FAUCHER, D., HELYNCK, G., MONEGIER DE SORBIER, B., RIDOUX, J. P., JAMES-SURCOUF, E., VUILHORGNE, M. (1994), Solution structure of RP-71955, a new 21 amino acid tricyclic peptide active against HIV-1 virus, *Biochemistry* **33**, 42–50.

FUJITA, K., MINAKATA, H., NOMOTO, K., FURUKAWA, Y., KOBAYASHIE, M. (1995), Structure–activity relations of fulicin, a peptide containing a D-amino acid residue, *Peptides* **16**, 565–568.

GEKELER, W., GRILL, E., WINNACKER, E.-L., ZENK, M. H. (1989), Survey of the plant kingdom for the ability to bind heavy metals through phytochelatin, *Z. Naturforsch.* **44c**, 361–369.

GRILL, E., LOEFFLER, E., WINNACKER, E.-L., ZENK, M. H. (1989), Phytochelatins, a class of heavy-metal binding peptides from plants, are synthesized from glutathione by a specific γ-glutamylcysteine dipeptididyl transpeptidase (phytochelatin synthase), *Proc. Natl. Acad. Sci. USA* **86**, 6838–6842.

GUILVOUT, I., MERCEREAU-PUIJALON, O., BONNEFOY, S., PUGSLEY, A. P., CARNIEL, E. (1993),

High-molecular-weight protein 2 of *Yersinia enterocolitica* is homologous to AngR of *Vibrio anguillarum* and belongs to a family of proteins involved in nonribosomal peptide synthesis, *J. Bacteriol.* **175**, 5488–5504.

HAESE, A., SCHUBERT, M., HERRMANN, M., ZOCHER, R. (1993), Molecular characterization of the enniatin synthetase gene encoding a multifunctional enzyme catalysing N-methyl-depsipeptide formation in *Fusarium scirpi, Mol. Microbiol.* **7**, 905–914.

HAM, P. J., CHALK, R., SMITHIES, B., ALBUQUERQUE, C., HAGEN, H. (1994), Antibacterial peptides in insect vectors of tropical parasitic diseases, in: *Antimicrobial Peptides* (J. GOODE, Ed.), pp. 140–150, Ciba Foundation Series. New York: John Wiley & Sons.

HAMOEN, L. W., ESHUIS, H., JONGBLOED, J., VENEMA, G., VAN SINDEREN, D. (1995), A small gene, designated *comS,* located within the coding region of the fourth amino acid-activation domain of *srfA,* is required for competence development in *Bacillus subtilis, Mol. Microbiol.* **15**, 55–63.

HANCOCK, R. E. W., FALLA, T., BROWN, M. (1995), Cationic bactericidal peptides, *Adv. Microb. Physiol.* **37**, 135–175.

HANSEN, J. N. (1993), Antibiotics synthesized by posttranslational modification, *Annu. Rev. Microbiol.* **47**, 535–564.

HECK, S. D., FARACI, W. S., KELBAUGH, P. R., SACCOMANO, N. A., THADEIO, P. F., VOLKMANN, R. A. (1995), Posttranslational amino acid epimerization – enzyme catalyzed isomerization of amino acid residues in peptide chains, *Proc. Natl. Acad. Sci. USA* **93**, 4036–4039.

HERZOG-VELIKONJA, B., PODLESEK, Z., GRABNAR, M. (1994), Isolation and characterization of Tn917-generated bacitracin deficient mutants of *Bacillus licheniformis, FEMS Microbiol. Lett.* **121**, 147–152.

HILTON, M. D., ALAEDDINOGLU, N. G., DEMAIN, A. L. (1988), *Bacillus subtilis* mutant deficient in the ability to produce the dipeptide antibiotic bacilysin: isolation and mapping of the mutation, *J. Bacteriol.* **170**, 1018–1021.

HODGSON, J. E., FOSBERRY, A. P., RAWLINSON, N. S., ROSS, H. N. M., NEAL, R. J., ARNELL, J. C., EARL, A. J., LAWLOR, E. J. (1995), Clavulanic acid biosynthesis in *Streptomyces clavuligerus*: gene cloning and characterization, *Gene* **166**, 49–55.

HUANG, C. C., ANO, T., SHODA, M. (1993), Nucleotide sequence and characteristics of the gene, *Lpa-14*, responsible for biosynthesis of the lipopeptide antibiotics iturin a and surfactin from *Bacillus subtilis* Rb14, *J. Ferment. Bioeng.* **76**, 445–450.

IHARA, M., FUKUMOTO, K. (1996), Recent progress in the chemistry of non-terpenoid indole alkaloids, *Nat. Prod. Rep.* **13**, 241–261.

IKAI, K., TAKESAKO, K., SHIOMI, K., MORIGUCHI, M., UMEDA, Y., YAMAMOTO, J., KATO, I., NAGANAWA, H. (1991a), Structure of aureobasidin A, *J. Antibiot.* **44**, 925–933.

IKAI, K., SHIOMI, K., TAKESAKO, K., MIZUTANI, S., YAMAMOTO, J., OGAWA, Y., UENO, M., KATO, I. (1991b), Structures of aureobasidins B to R, *J. Antibiot.* **44**, 1187–1198.

JACK, R., TAGG, J. R., RAY, B. (1995), Bacteriocins of gram-positive bacteria, *Microbiol. Rev.* **59**, 171–200.

JORGENSEN, H., NIELSEN, J., VILLADSEN, J., MOLLGAARD, H. (1995a), Metabolic flux distributions in *Penicillium chrysogenum* during fedbatch cultivations, *Biotechnol. Bioeng.* **46**, 117–131.

JORGENSEN, H., NIELSEN, J., VILLADSEN, J., MOLLGAARD, H. (1995b), Analysis of penicillin V biosynthesis during fed-batch cultivations with a high-yielding strain of *Penicillium chrysogenum, Appl. Microbiol. Biotechnol.* **43**, 123–130.

KALETTA, C., ENTIAN, K.-D., KELLNER, R., JUNG, G., REIS, M., SAHL, H.-G. (1989), Pep-5, a new lantibiotic: structural gene isolation and prepeptide sequence, *Arch. Microbiol.* **152**, 16–19.

KALLOW, W., VON DÖHREN, H., KENNEDY, J., TURNER, G. (1996), *Abstract,* 3rd Eur. Conf. Fungal Genet., Münster, p. 159.

KELLENBERGER, C., HIETTER, H., LUU, B. (1995), Regioselective formation of the three disulfide bonds of a 35-residue insect peptide, *Peptide Res.* **8**, 321–327.

KIMURA, H., MIYASHITA, H., SUMINO, Y. (1996), Organization and expression in *Pseudomonas putida* of the gene cluster involved in cephalosporin biosynthesis from *Lysobacter lactamgenus* YK90, *Appl. Microbiol. Biotechnol.* **45**, 490–501.

KLAPHECK, S., FLIEGNER, W., ZIMMER, I. (1994), Hydroxymethyl-phytochelatins (γ-glutamylcysteine)$_n$-serine) are metal-induced peptides of the Pocaceae, *Plant Physiol.* **104**, 1325–1332.

KLEINKAUF, H., VON DÖHREN, H. (1982), A survey of enzymatic biosynthesis of peptide antibiotics, in: *Trends in Antibiotic Research,* pp. 220–232. Tokyo: Japan Antibiotics Res. Ass.

KLEINKAUF, H., VON DÖHREN, H. (1986), Peptide antibiotics, in: *Biotechnology* 1st Edn., Vol. 4 (REHM, H.-J., REED, G., Eds.), pp. 283–307. Weinheim: VCH.

KLEINKAUF, H., VON DÖHREN, H. (1996), A nonribosomal system of peptide biosynthesis, *Eur. J. Biochem.* **236**, 335–351.

KLEINKAUF, H., VON DÖHREN, H. (1997), Enzymatic generation of complex peptides, *Progr. Drug Res.* **48**, 27–51.

KONDO, N., IMAI, K., ISOBE, M., GOTO, T., MURASUGI, A., WADA-NAKAGAWA, C., HAYASHI, Y. (1984), Cadystin A and B, major unit peptides comprising cadmium binding peptides induced in a fission yeast-separation, revision of structure and synthesis, *Tetrahedron Lett.* **25**, 3869–3872.

KUBOI, R., MARUKI, T., TANAKA, H., KOMASAWA, I. (1994), Fermentation of *Bacillus subtilis* ATCC 6633 and production of subtilin in polyethylene glycol/phosphate aqueous two-phase systems, *J. Ferment. Bioeng.* **78**, 431–436.

LAMBALOT, R. H., WALSH, C. T. (1995), Cloning, overproduction, and characterization of the *Echerichia coli* holo-acyl carrier protein synthase, *J. Biol. Chem.* **270**, 24658–24661.

LAMBALOT, R. H., GEHRING, A. M., FLUGEL, R. S., ZUBER, P., LACELLE, M., MARAHIEL, M. A., REID, R., KHOSLA, C., WALSH, C. T. (1996), A new enzyme superfamily – the phosphopantetheinyl-transferases, *Chemistry & Biology* **3**, 923–936.

LAWEN, A., TRABER, R. (1993), Substrate specificities of cyclosporin synthetase and peptolide SDZ 214-103 synthetase, *J. Biol. Chem.* **268**, 20452–20465.

LEHRER, R. I., LICHTENSTEIN, A. K., GANZ, T. (1993), Defensins: antimicrobial and cytotoxic peptides of mammalian cells, *Annu. Rev. Immunol.* **11**, 105–128.

LEWIS, J. R. (1996), Muscarine, imidazole, oxazole, thiazole and peptide alkaloids, and other miscellaneous alkaloids, *Nat. Prod. Rep.* **13**, 435–467.

LI, Y. M., MILENE, J. C., MADISON, L. L., KOLTER, R., WALSH, C. T. (1996), From peptide precursors to oxazole and thiazole-containing peptide antibiotics – microcin B17 synthase, *Science* **274**, 1188–1193.

LIU, G.-H., CHEN, I.-L., HSIEH, J.-C., LIU, S.-T. (1996), Cloning and sequencing of the peptide synthetase genes encoding the biosynthesis of fengycin, *Abstracts,* Symp. Enzymol. Biosynth. Nat. Prod., Technical University, Berlin.

MACCABE, A. P., RIACH, M. B. R., UNKLES, S. E., KINGHORN, J. R. (1990), The *Aspergillus nidulans npeA* locus consists of three contiguous genes required for penicillin biosynthesis, *EMBO J.* **9**, 279–287.

MACKINTOSH, R. W., DALBY, K. N., CAMPBELL, D. G., COHEN, P. T., COHEN, P., MACKINTOSH, C. (1995), The cyanobacterial toxin microcystin binds covalently to cysteine-273 on protein phosphatase 1, *FEBS Lett.* **371**, 236–240.

MAGNUSON, R., SOLOMON, J., GROSSMAN, A. D. (1994), Biochemical and genetic characterization of a competence pheromone from *B. subtilis, Cell* **77**, 207–216.

MARTÍN, J. F., GUTIERREZ, S. (1995), Genes for beta-lactam antibiotic biosynthesis, *Antonie van Leeuwenhoek Int. J. Gen. Mol. Microbiol.* **67**, 181–200.

MEI, B., BUDDE, A. D., LEONG, S. A. (1993), *sid1,* a gene intiating siderophore biosynthesis in *Ustilago maydis*: molecular characterization, regulation by iron, and role in phytopathogenicity, *Proc. Natl. Acad. Sci. USA* **90**, 903–907.

MEISSNER, K., DITTMANN, E., BÖRNER, T. (1996), Toxic and nontoxic strains of the cyanobacterium *Microcystis aeruginosa* contain sequences homologous to peptide synthetase genes, *FEMS Microbiol. Lett.* **135**, 295–303.

MEUWLY, P., THIBAULT, P., SCHWAN, A. L., RAUSER, W. E. (1995), Three families of thiol peptides are induced by cadmium in maize, *Plant J.* **7**, 391–400.

MITTENHUBER, G., WECKERMANN, R., MARAHIEL, M. A. (1989), Gene cluster encoding the genes for tyrocidine synthetase 1 and 2 from *Bacillus brevis.* Evidence for an operon, *J. Bacteriol.* **171**, 4881–4887.

MO, Y. Y., GEIBEL, M., BONSALL, R. F., GROSS, D. C. (1995), Analysis of sweet cherry (*Prunus avium* L.) leaves for plant signal molecules that activate the *syrB* gene required for synthesis of the phytotoxin, syringomycin, by *Pseudomonas syringae* pv. *syringae, Plant Physiol.* **107**, 603–612.

MOLNAR, I., APARICIO, J. F., HAYDOCK, S. F., KHAW, L. E., SCHWECKE, T., KÖNIG, A., STAUNTON, J., LEADLAY, P. F. (1996), Organisation of the biosynthetic gene cluster for rapamycin in *Streptomyces hygroscopicus* – analysis of genes flanking the polyketide synthase, *Gene* **169**, 1–7.

MOR, A., DELFOUR, A., NICOLAS, P. (1991), Identification of a D-alanine-containing polypeptide precursor for the peptide opioid, dermorphin, *J. Biol. Chem.* **266**, 6264–6270.

MORIKAWA, N., HAGIWARA, K., HAKAJIMA, T. (1992), Brevinin-1 and -2, unique antimicrobial peptides from the skin of the frog, *Rana brevipoda porsa, Biochem. Biophys. Res. Commun.* **189**, 184–190.

MOTAMEDI, H. (1996), FK506 biosynthetic gene cluster. *Abstracts*, Symp. Enzymol. Biosynth. Nat. Prod., Technical University, Berlin.

NIELSEN, J., JORGENSEN, H. S. (1995), Metabolic control analysis of the penicillin biosynthetic pathway in a high-yielding strain of *Penicillium chrysogenum, Biotechnol. Progr.* **11**, 299–305.

NIELSEN, J., KRABBEN, P. (1995), Hyphal growth and fragmentation of *Penicillium chrysogenum* in submerged cultures, *Biotechnol. Bioeng.* **46**, 588–598.

NIELSEN, J., JOHANSEN, C. L., VILLADSEN, J. (1994), Culture fluorescence measurements during batch and fed-batch cultivations with *Penicillium chrysogenum, J. Biotechnol.* **38**, 51–62.

NIELSEN, J., JOHANSEN, C. L., JACOBSEN, C. L., KRABBEN, P., VILLADSEN, J. (1995), Pellet formation and fragmentation in submerged cultures of *Penicillium chrysogenum* and its relation to penicillin production, *Biotechnol. Progr.* **11**, 93–98.

NOZAKI, S., MURAMATSU, I. (1987), Natural homologs of gramicidin S, *J. Antibiot.* **37**, 689–90.

OFFENZELLER, M., SANTER, G., TOTSCHNIG, K., SU, Z., MOSER, H., TRABER, R., SCHNEIDER-SCHERZER, E. (1996), Biosynthesis of the unusual amino acid (4*R*)-4-[(E)-2-butenyl]-4-methyl-L-threonine of cyclosporin A: enzymatic analysis of the reaction sequence including identification of the methylation precursor in a polyketide pathway, *Biochemistry* **35**, 8401–8412.

OHNO, A., ANO, T., SHODA, M. (1995), Production of a lipopeptide antibiotic, surfactin, by recombinant *Bacillus subtilis* in solid state fermentation, *Biotechnol. Bioeng.* **47**, 209–214.

OHNO, A., ANO, T., SHODA, M. (1996), Use of soybean curd residue, okara, for the solid state substrate in the production of a lipopeptide antibiotic, iturin a, by *Bacillus subtilis* NB22, *Process Biochem.* **31**, 801–806.

OKUMURA, Y. (1990), Bioysynthesis of viridogrisein, in: *Biochemistry of Peptide Antibiotics* (KLEINKAUF, H., VON DÖHREN, H., Eds.), pp. 365–378. Berlin: de Gruyter.

PARK, Y. S., INOUE, K., YAHIRO, K., OKABE, M. (1994a), Improvement of cephamycin C production by a mutant resistant to linoleic acid, *J. Ferment. Bioeng.* **78**, 88–92.

PARK, Y. S., MOMOSE, I., TSUNODA, K., OKABE, M. (1994b), Enhancement of cephamycin C production using soybean oil as the sole carbon source, *Appl. Microbiol. Biotechnol.* **40**, 773–779.

PAUL, G. C., THOMAS, C. R. (1996), A structured model for hyphal differentiation and penicillin production using *Penicillium chrysogenum, Biotechnol. Bioeng.* **51**, 558–572.

PFEIFER, E., PAVELA-VRANCIC, M., VON DÖHREN, H., KLEINKAUF, H. (1995), Characterization of tyrocidine synthetase 1 (TY1): Requirement of post-translational modification for peptide biosynthesis, *Biochemistry* **34**, 7450–7459.

PIECQ, M., DEHOTTAY, P., BIOT, A., DUSART, J. (1994), Cloning and nucleotide sequence of a region of the *Kibdelosporangium aridum* genome homologous to polyketide biosynthetic genes, *J. DNA Sequ. Mapping* **4**, 219–229.

PIEPER, R., KLEINKAUF, H., ZOCHER, R. (1992), Enniatin synthetases from different Fusaria exhibiting distinct amino acid specificities, *J. Antibiot.* **45**, 1273–1277.

PIERS, K. L., BROWN, M. H., HANCOCK, R. E. W. (1993), Recombinant DNA procedures for producing small antimicrobial cationic peptides in bacteria, *Gene* **134**, 7–13.

PITKIN, J. W., PANACCIONE, D. G., WALTON, J. D. (1996), A putative cyclic peptide efflux pump encoded by the *toxA* gene of the plant pathogenic fungus *Cochliobolus carbonum, Microbiology* **142**, 1557–1565.

POSPIECH, A., BIETENHADER, J., SCHUPP, T. (1996), Two multifunctional peptide synthetases and an O-methyltransferase are involved in the biosynthesis of the DNA-binding antibiotic and antitumour agent saframycin Mx1 from *Myxococcus xanthus, Microbiology* **142**, 741–746.

POTTERAT, O., STEPHAN, H., METZGER, J. W., GNAU, V., ZÄHNER, H., JUNG, G. (1994), Aborycin – a tricyclic-21-peptide antibiotic isolated from *Streptomyces griseoflavus, Liebigs Ann. Chem.* **1994**, 741–743.

RAINEY, P. B., BRODEY, C. L., JOHNSTONE, K. (1993), Identification of a gene cluster encoding three high-molecular weight proteins, which is required for synthesis of tolaasin by the mushroom pathogen *Pseudomonas tolaasii, Mol. Microbiol.* **8**, 643–652.

RAUSER, W. E. (1990), Phytochelatins, *Annu. Rev. Biochem.* **59**, 61–86.

REICHERT, J., SAKAITANI, M., WALSH, C. T. (1992), Characterization of EntF as a serine-activating enzyme, *Protein Sci.* **1**, 549–555.

REICHHART, J.-M., PETIT, I., LEGRAIN, M., DIMARCQ, J. L., KEPPI, E., LECOCQ, J. P., HOFFMANN, J. A., ACHSTETTER, T. (1992), Expression and secretion in yeast of active insect defensin, an inducible antibacterial peptide from the fleshfly *Formia terranova, Invert. Reprod. Dev.* **21**, 15–24.

RICHTER, K., EGGER, R., KREIL, G. (1987), D-Alanine in the frog skin dermorphin is derived from L-alanine in the precursor, *Science* **238**, 200–202.

RINCE, A., DUFOUR, A., LEPOGAM, S., THUAULT, D., BOURGEOIS, C. M., LE PENNEC, J. P. (1994), Cloning, expression and nucleotide sequence of genes involved in production of lactococcin DR, a bacteriocin from *Lactococcus lactis, Appl. Environ. Microbiol.* **60**, 1652–1657.

ROLLEMA, H. S., METZGER, J. W., BOTH, P., KUIPERS, O. P., SIEZEN, R. J. (1996), Structure and biological activity of chemically modified nisin species, *Eur. J. Biochem.* **241**, 716–722.

ROMEO, D., SKERLAVAJ, B., BOLOGNESI, M., GENNARO, R. (1988), Structure and bactericidal activity of an antibiotic dodecapeptide purified from bovine neutrophils, *J. Biol. Chem.* **263**, 9573–9575.

SAHL, H.-G., JACK, R. W., BIERBAUM, G. (1995), Lantibiotics – biosynthesis and biological activities of peptides with unique posttranslational modifications, *Eur. J. Biochem.* **230**, 827–853.

SAKAJOH, M., SOLOMON, N. A., DEMAIN, A. L. (1987), Cell-free synthesis of the dipeptide antibiotic bacilysin, *J. Ind. Microbiol.* **2**, 201–208.

SCHAUWECKER, F., PFENNIG, F., KELLER, U. (1996), Molecular cloning of the actinomycin synthetase genes *acmA* and *acmB* from *Streptomyces chrysomallus, Abstracts,* Symp. Enzymol. Biosynth. Nat. Prod., Technical University, Berlin.

SCHNELL, N., ENGELKE, G., AUGUSTIN, J., ROSENSTEIN, R., UNGERMANN, V., GÖTZ, F., ENTIAN, K.-D. (1992), Analysis of genes involved in the biosynthesis of the lantibiotic epidermin, *Eur. J. Biochem.* **204**, 57–68.

SCHWARTZ, D., ALIJAH, R., NUSSBAUMER, B., PELZER, S., WOHLLEBEN, W. (1996), The peptide synthetase gene *phsA* from *Streptomyces virdochromogenes* is not juxtaposed with other genes involved in nonribosomal biosynthesis of peptides, *Appl. Environ. Microbiol.* **62**, 570–577.

SCHWECKE, T., APARICIO, J. F., MOLNAR, I., KÖNIG, A., KHAW, L. E., HAYDOCK, S. F., OLIYNYK, M., CAFFREY, P., CORTES, J., LESTER, J. B., BÖHM, G. A., STAUNTON, J., LEADLAY, P. F. (1995), The biosynthetic gene cluster for the polyketide immunosuppressant rapamycin, *Proc. Natl. Acad. Sci. USA* **92**, 7839–7843.

SCOTT-CRAIG, J. S., PANACCIONE, D. G., PACARD, J. A., WALTON, J. D. (1992), The cyclic peptide synthetase catalysing HC toxin production in the filamentous fungus *Cochliobolus carbonum* is encoded by a 15.7 kilobase open reading frame, *J. Biol. Chem.* **267**, 26044–26049.

SELSTED, M. E., OUTLETTE, A. J. (1995), Defensins in granules of phagocytic and non-phagocytic cells, *Trends Cell. Biol.* **5**, 114–119.

SIMMACO, M., MIGNOGNA, G., BARRA, D., BOSSA, F. (1993), Novel antimicrobial peptides from skin secretion of the European frog *Rana esculenta, FEBS Lett.* **324**, 159–161.

SKAUGEN, M., NISSEN-MEYER, J., JUNG, G., STEVANOVIC, S., SLETTEN, K., INGER, C., ABILDGAARD, M., NES, I. F. (1994), *In vivo* conversion of L-serine to D-alanine in a ribosomally synthesized poylpeptide, *J. Biol. Chem.* **269**, 27183–27185.

SMITH, D. J., BURNHAM, M. K. R., BULL, J. H., HODGSON, J. E., WARD, J. M., BROWNE, P., BROWN, J., BARTON, B., EARL, A. J., TURNER, G. (1990), β-Lactam antibiotic biosynthetic genes have been conserved in clusters in prokaryotes and eukaryotes, *EMBO J.* **9**, 741–747.

SOLOMON, J. M., MAGNUSON, R., SRIVASTAVA, A., GROSSMAN, A. D. (1995), Convergent sensing pathways mediate response to two extracellular competence factors in *Bacillus subtilis, Genes Dev.* **9**, 547–558.

STACHELHAUS, T., SCHNEIDER, A., MARAHIEL, M. A. (1995), Rational design of peptide antibiotics by targeted replacement of bacterial and fungal domains, *Science* **269**, 69–72.

STACHELHAUS, T., HUSER, A., MARAHIEL, M. A. (1996a), Biochemical characterization of peptides carrier protein (PCP), the thiolation domain of multifunctional peptide synthetases, *Chemistry & Biology* **3**, 913–921.

STACHELHAUS, T., SCHNEIDER, A., MARAHIEL, M. A. (1996b), Engineered biosynthesis of peptide antibiotics, *Biochem. Pharmacol.* **52**, 177–186.

STEFFENS, J. C. (1990), The heavy metal-binding peptides of plants, *Annu. Rev. Plant Physiol. Plant Mol. Biol.* **41**, 533–575.

STEIN, T., VATER, J. (1996), Amino acid activation and polymerization at modular multienzymes in nonribosomal peptide biosynthesis, *Amino Acids* **10**, 201–227.

STINTZI, A., CORNELIS, P., HOHNADEL, D., MEYER, J.-M., DEAN, C., POOLE, K., KOURAMBAS, S., KRISHNAPILLAI, V. (1996), Novel pyoverdine biosynthesis gene(s) of *Pseudomonas aeruginosa* PAO, *Microbiology* **142**, 1181–1190.

STROHL, W. R. (Ed.) (1996), *Biotechnology of Industrial Antibiotics,* 2nd Edn. New York: Marcel Dekker.

STROHL, W. R., FLOSS, H. G. (1995), Thiopeptides, in: *Genetics and Biochemistry of Antibiotic Production* (VINING, L. C., STUTTARD, C., Eds.), pp. 223–238. Boston, MA: Butterworth-Heinemann.

SUBBALAKSHMI, C., KRISHNAKUMARI, V., NAGARAJ, R., SITARAM, N. (1996), Requirements for antibacterial and hemolytic activities in the bovine neutrophil derived peptide indolicidin, *FEBS Lett.* **395**, 48–52.

TAKESAKO, K., MIZUTANI, S., SAKAKIBARA, H., ENDO, M., YOSHIKAWA, Y., MSUDA, T., SONO-KOYAMA, E., KATO, I. (1996), Precursor directed biosynthesis of aureobasidins, *J. Antibiot.* **49**, 676–681.

TANIGUCHI, M., HOSHINO, K., URASAKI, H., FUJII, M. (1993), Continuous production of an antibiotic polypeptide (nisin) by *Lactococcus lactis* using a bioreactor coupled to a microfiltration module, *J. Ferment. Bioeng.* **78**, U122.

THIBAULT, P., FAUBERT, D., KARUNANITHY, S., BOYD, R. K., HOLMES, C. F. B. (1992), Isolation, mass spectrometric characterization, and protein phosphatase inhibition properties of cyclic peptide analogs of gramicidin S from *Bacillus brevis* (Nagano strain), *Biol. Mass Spectrom.* **21**, 367–379.

THIERICKE, R., ROHR, R. (1993), Biological variation of microbial metabolites by precursor-directed biosynthesis, *Nat. Prod. Rep.* **10**, 263–289.

TOGNONI, A., FRANCHI, E., MAGISTRELLI, C., COLOMBO, E., COSMINA, P., GRANDI, G. (1995), A putative new peptide synthase operon in *Bacillus subtilis*: Partial characterization, *Microbiology* **141**, 645–648.

TURGAY, K., MARAHIEL, M. A. (1995), The *gtcRS* operon coding for two-component system regulatory proteins is located adjacent to the *grs* operon of *Bacillus brevis, DNA Sequence* **5**, 283–290.

VANHOUTTE, B., PONS, M. N., THOMAS, C. R., LOUVEL, L., VIVIER, H. (1995), Characterization of *Penicillium chrysogenum* physiology in submerged cultures by color and monochrome image analysis, *Biotechnol. Bioeng.* **48**, 1–11.

VINING, L. C., STUTTARD, C. (Eds.) (1995), *Genetics and Biochemistry of Antibiotic Production.* Boston, MA: Butterworth-Heinemann.

VON DÖHREN, H. (1990), Compilation of peptide structures – a biogenetic approach, in: *Biochemistry of Peptide Antibiotics* (KLEINKAUF, H., VON DÖHREN, H., Eds.), pp. 411–507. Berlin, New York: de Gruyter.

WEBER, G., SCHÖRGENDORFER, K., SCHNEIDER-SCHERZER, E., LEITNER, E. (1994), The peptide synthetase catalyzing cyclosporine production in *Tolypocladium niveum* is encoded by a giant 45.8 kilobase open reading frame, *Curr. Genet.* **26**, 120–125.

WILLIAMSON, M. P., GAUVREAU, D., WILLIAMS, D. H., WARING, M. J. (1982), Structure and conformation of fourteen antibiotics of the quinoxaline group determined by ^{1}H-NMR, *J. Antibiot.* **35**, 62–66.

YANG, Y. K., MORIKAWA, M., SHIMIZU, H., SHIOYA, S., SUGA, K., NIHIRA, T., YAMADA, Y. (1996), Maximum virginiamycin production by optimization of cultivation conditions in batch culture with autoregulator addition, *Biotechnol. Bioeng.* **49**, 437–444.

8 Lantibiotics

Ralph Jack
Friedrich Götz
Günther Jung

Tübingen, Germany

1 Introduction

The search for novel pharmaceutical compounds with potent new biological activities has only relatively recently begun to look inward and away from complex in-lab synthetic processes toward the huge array of "natural" compounds produced by bacteria, offering a plethora of new possibilities. Such studies of bacterial-derived proteins in particular have unearthed a vast array of naturally produced compounds, not the least of which is the group of highly modified antimicrobial and enzyme-inhibitory peptides collectively referred to as "lantibiotics". Originally coined to describe the rapidly expanding group of an**tibiotic**-like peptides which were found to contain the non-protein amino acids **lan**thionine and 3-methyl lanthionine (SCHNELL et al., 1988), the name lantibiotics belies the full extent and complexity of this class of bacterially synthesized peptides. For example, as the number of characterized lantibiotics increases, novel modified amino acids of interest to chemists and biochemists alike are constantly being identified, including those with unusual crosslinks or unsaturated R-groups, while analysis of the solution structure of the peptides concerned is offering new perspectives on the relationship between peptide structure and biological function.

Perhaps more strikingly though, a number of recent studies of the biosynthesis of lantibiotics has revealed that they are produced by ribosomal biosynthesis and subsequently modified to generate the mature, biologically active compound (SCHNELL et al., 1988; BANERJEE and HANSEN, 1988; BUCHMANN et al., 1988; KALETTA and ENTIAN, 1989). What this means is that there are proteins within the lantibiotic-producing cells which are able to carry out specific transformations of amino acids, converting them to novel structures which may play a role in the biological activity, stability, etc. of the peptide and, more importantly, may prove useful to those biotechnologists interested in creating similar modifications of other peptides.

It is this last feature of ribosomal biosynthesis which sets the lantibiotics apart from the classical peptide antibiotics which are typically synthesized by large multi-enzyme complexes in the cell and for which there exists no structural gene (KATZ and DEMAIN, 1977; KLEINKAUF and VON DÖHREN, 1986, 1987, 1990; NAKANO and ZUBER, 1990). Studies of the lantibiotics have revealed that they are genetically encoded and are uniformly produced on the ribosome as a precursor peptide which is subsequently modified at specific points to give rise to the large number of modified amino acids found in these peptides. In addition, the peptides are produced with a leader peptide which is removed during maturation and are transported by specific transport-related proteins from the cell. Still other lantibiotic-specific proteins are involved in the genetic regulation of biosynthesis and generation of the specific producer cell self-protection mechanism(s) frequently observed.

The mature lantibiotics have also found a number of potential applications including medical and veterinary antibiosis, food, beverage, and cosmetic preservation and as regulators of both human immune function and blood pressure (HURST, 1981; JUNG, 1991a, b; DELVES-BROUGHTON, 1990; MOLITOR and SAHL, 1991; BIERBAUM and SAHL, 1993; DE VUYST and VANDAMME, 1993; JACK et al., 1995; SAHL et al., 1995). In the following chapter we wish to provide an overview of the novel structures, mechanism(s) of biosynthesis, genetic regulation, biological activities and current as well as potential applications for this fascinating, novel class of bacterial-derived, biologically-active peptides.

2 The Unique Chemistry and Structure of Lantibiotics

In order to appreciate the potential of the lantibiotics as biologically active peptides, as well as the possibilities for the application of their respective biosynthetic machinery, it is first necessary to discuss the structure of both the peptides themselves and the modified

Tab. 1. A Compilation of the Currently Described Lantibiotics[a] Including Several of their Respective Chemical and Physical Characteristics

Lantibiotic	Producing Organism	Mass [Da]	Net Charge[b] (at pH 7.0)	Number of Rings[c]	% Modified Residues[d]	Reference
Type A:						
Nisin A	*Lactococcus lactis*	3353	+3	5	38	GROSS and MORRELL (1971)
Nisin Z	*Lactococcus lactis*	3330	+3	5	38	MULDERS et al. (1991)
Subtilin	*Bacillus subtilis*	3317	+2	5	40	GROSS et al. (1973)
Epidermin	*Staphylococcus epidermidis*	2164	+3	4	41	ALLGAIER et al. (1986)
Gallidermin	*Staphylococcus gallinarum*	2164	+3	4	41	KELLNER et al. (1988)
[1V,6L]-Epidermin	*Staphylococcus epidermidis*	2151	+3	4	41	SAHL et al. (1995)
Pep5	*Staphylococcus epidermidis*	3488	+7	3	26	KELLNER et al. (1989)
Epilancin K7	*Staphylococcus epidermidis*	3032	+5	3	32	VAN DE KAMP et al. (1995a)
Lactocin S	*Lactobacillus sake*	3764	−1	2	24	SKAUGEN et al. (1994)
SA-FF22	*Streptococcus pyogenes*	2795	+1	3	27	JACK et al. (1994a)
Lactococcin DR	*Lactococcus lactis*	2901	0	3	26	RINCE et al. (1994)
Salivaricin A	*Streptococcus salivarius*	2315	0	3	27	ROSS et al. (1993)
Cytolysin L1	*Enterococcus faecalis*	4164	0	N.R.[e]	N.R.	GILMORE et al. (1994)
Cytolysin L2	*Enterococcus faecalis*	2631	0	N.R.	N.R.	GILMORE et al. (1994)
Carnocin UI49	*Carnobacterium piscicola*	4635	N.R.	N.R.	N.R.	STOFFELS et al. (1994)
Mutacin	*Streptococcus mutans*	3245	N.R.	3	N.R.	NOVAK et al. (1994)
Type B:						
Cinnamycin	*Streptomyces cinnamoneus*	2042	0	4	47	FREDENHAGEN et al. (1991[f])
Duramycin	*Streptomyces cinnamoneus*	2014	0	4	47	FREDENHAGEN et al. (1991[f])
Duramycin B	*Streptoverticillium* spp.	1951	0	4	47	FREDENHAGEN et al. (1991)
Duramycin C	*Streptomyces griseoluteus*	2008	−1	4	47	FREDENHAGEN et al. (1991)
Ancovenin	*Streptomyces* spp.	1959	0	3	37	WAKAMIYA et al. (1985)
Mersacidin	*Bacillus subtilis*	1825	−1	4	42	KOGLER et al. (1991)
Actagardine	*Actinoplanes* spp.	1890	0	4	45	ZIMMERMANN (1995[f])

[a] Adapted from JACK and SAHL (1995) and SAHL et al. (1995).
[b] Includes those amino- and carboxy termini which remain unmodified following posttranslational modification.
[c] Total number of covalent ring structures, including both thioether (lanthionine, β-methyllanthionine, aminovinyl cysteine, aminovinyl-methyl cysteine, lanthionine sulphoxide) and aminoether (lysinolalanine) bridges.
[d] Derived from the total number of modified amino acid residues per total number of residues in the respective peptide.
[e] N.R.: not reported
[f] Although these structures have been previously described, the references given are for the corrected structures after revisions.

amino acids which they contain. Tab. 1 lists all of the currently characterized lantibiotics, the bacterium which produces them as well as a number of their physicochemical properties relevant to the proceeding sections (SAHL et al., 1995; JACK and SAHL, 1995).

In addition, an earlier review of the lantibiotics has suggested that the peptides can be divided into two distinct groups (JUNG, 1991a, b), a subdivision scheme which has been maintained in the following chapter. The type A lantibiotics can be defined as elongated, helical peptides whose primary mode of action appears to be directed at the depolarization of the bacterial cytoplasmic membrane through the formation of voltage-dependent pores. Alternatively, type B lantibiotics are generally somewhat smaller, compact, globular structures which either inhibit bacterial cell wall replication, interact with specific enzymes, or activate T-cell proliferation (see also Sect. 5).

2.1 Modified Amino Acids Found in the Lantibiotics

While the lantibiotics take their name from the observation that they contain lanthionine (SCHNELL et al., 1988), this is not the only modified amino acid that they possess. Currently, a number of modified amino acids and other residues have been found in lantibiotic structures including: *meso*-lanthionine (Lan), lanthionine sulphoxide, *threo*-β-methyllanthionine (MeLan), S-[(*Z*)-2-aminovinyl]-D-cysteine (AviCys), S-[(*Z*)-2-aminovinyl]-3-methyl-D-cysteine, 2,3-didehydroalanine (Dha), 2,3-didehydrobutyrine (Dhb), D-alanine, 2-oxopyruvate, 2-oxobutyrate, hydroxypyruvate, *erythro*-3-hydroxyaspartate and (2*S*,8*S*)-lysinoalanine, the structures of which are presented for comparison in Fig. 1. Some of these residues such as Lan, MeLan, Dha, and Dhb occur in most, if not all, lantibiotics so-

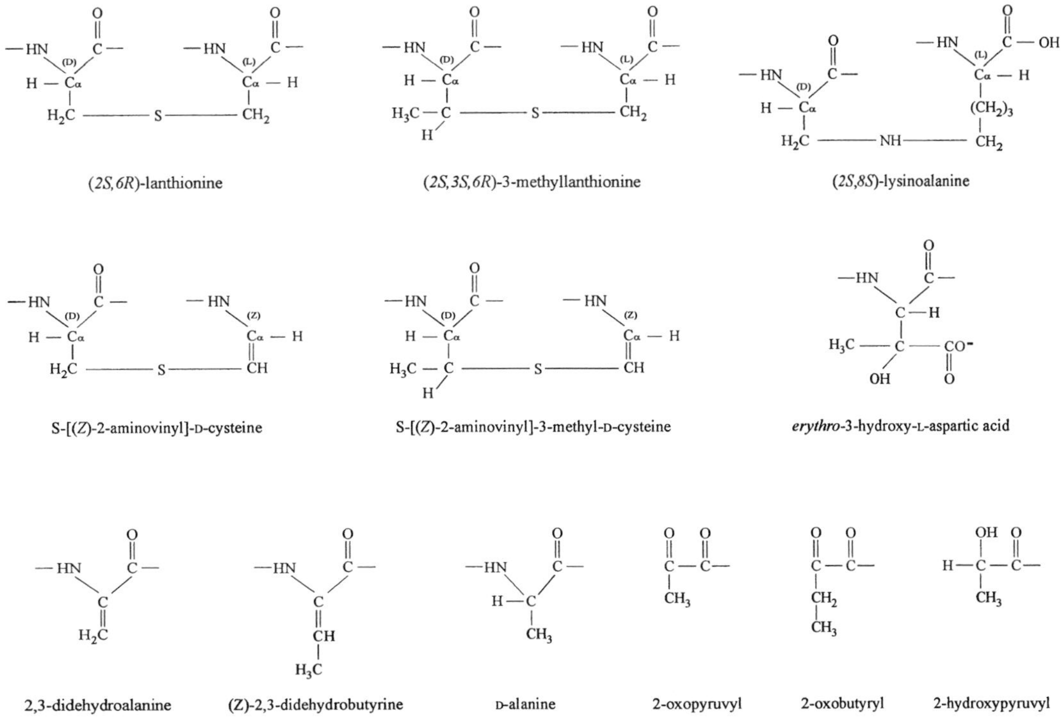

(2*S*,6*R*)-lanthionine

(2*S*,3*S*,6*R*)-3-methyllanthionine

(2*S*,8*S*)-lysinoalanine

S-[(*Z*)-2-aminovinyl]-D-cysteine

S-[(*Z*)-2-aminovinyl]-3-methyl-D-cysteine

erythro-3-hydroxy-L-aspartic acid

2,3-didehydroalanine

(Z)-2,3-didehydrobutyrine

D-alanine

2-oxopyruvyl

2-oxobutyryl

2-hydroxypyruvyl

Fig. 1. The structures of the modified amino acids and other residues found in lantibiotics. The chirality of various structures is indicated (where appropriate) above the respective α-carbon atom.

far characterized, while others occur only in individual cases. However, due in part to their novel nature and properties, many of these residues may be of great interest to the researchers involved in biotechnology.

2.1.1 Lanthionine (Lan) and 3-Methyllanthionine (MeLan)

It was the pioneering research of E. GROSS and co-workers that finally demonstrated that the peptide antibiotics nisin and subtilin actually contained Lan and MeLan (as well as Dha and Dhb), confirming previously held beliefs (BERRIDGE et al., 1952; GROSS and MORRELL, 1971; GROSS et al., 1973). However, it was not clear at that time how these amino acid structures should arise. Later studies showed that inhibitors of protein biosynthesis prevented nisin and subtilin production (HURST, 1981), whilst the isolation of the structural genes for a number of lantibiotics has finally proved conclusively that these residues arise from posttranslational modification of a ribosomally synthesized precursor peptide (SCHNELL et al., 1988; BUCHMANN et al., 1988; BANERJEE and HANSEN, 1988; KALETTA and ENTIAN, 1989). In addition, these studies also showed that Ser, Thr and Cys residues found in the prepeptides but not in the mature lantibiotics, must be the precursors for the formation of Lan, MeLan, Dha and Dhb.

While it is not clear how Lan and MeLan are synthesized in the cell, the structure of these thioether-linked, *di*-carboxy, *di*-amino acids has been studied in detail and a model (Fig. 2) for their biosynthesis has been proposed (SCHNELL et al., 1988; JUNG, 1991a, b). The first step involves the site-specific dehydration of the α-amino-β-hydroxy acids Ser and/or Thr in the propeptide part of the prelantibiotic; such dehydration results in the formation of Dha or Dhb, respectively. Presumably, this reaction is carried out by a specific enzyme within the lantibiotic-producing cells, however, neither the enzyme nor the molecular mechanism responsible have yet been identified. Also, the enzyme(s) capable of such a reaction should be novel since, currently characterized serine and threonine dehydratases appear to act on free amino groups via pyridoxal phosphate-dependent Schiff base reactions, suggesting that Ser and Thr residues incorporated into a peptide chain would not be appropriate as substrates for these enzymes.

The subsequent step in lanthionine biosynthesis (Fig. 2) involves the addition of the sulphydryl group of a neighboring Cys residue to the α,β-unsaturated amino acid to form a covalent thioether bridge (SCHNELL et al., 1988; JUNG, 1991a, b). Again, while it is not clear how this step occurs *in vivo* (i.e., spontaneous addition or specific protein-directed addition), it has some interesting ramifications for the stereospecificity of the final product. Since lantibiotics prepeptides are synthesized on the ribosome, they contain only L-amino acids; despite this, Lan and MeLan in the mature lantibiotic are consistently found in the *meso*(i.e., D/L-isomer)-configuration with the Ser/Thr-derived "half" in the D-form while the Cys-derived "half" remains in the L-configuration. In addition, in the case of type A lantibiotics the Ser/Thr-derived "half" of the Lan and MeLan is always closer to the N-terminus than is the Cys-derived "half" of the amino acid, however, this is not always true amongst the type B lantibiotics (JUNG, 1991a, b).

2.1.2 2,3-Didehydroalanine (Dha) and 2,3-Didehydrobutyrine (Dhb)

The α,β-unsaturated amino acids Dha and Dhb (formed as described above for the first step in Lan/MeLan biosynthesis) are stable within the mature lantibiotic, however, their appearance at the amino terminus of the peptide, such as occurs during amino acid sequencing by Edman-degradation, can prove problematic. The structure elucidation of a number of lantibiotics has previously been hampered by the presence of these amino acids (GROSS and MORRELL, 1971; ALLGAIER et al., 1986; KELLNER et al., 1988, 1989; JUNG, 1991a, b; JACK et al., 1994a; VAN DE KAMP et al., 1995b), the problems resulting from the spontaneous oxidative deamina-

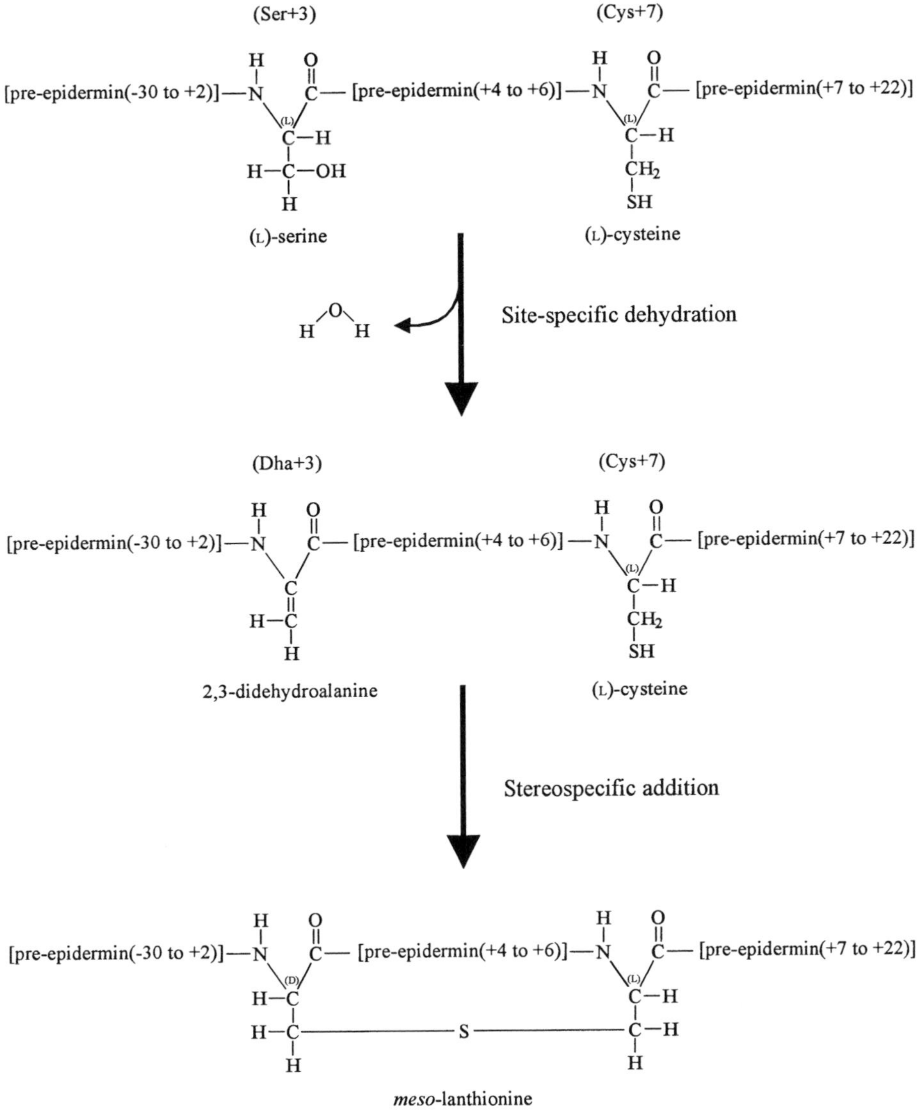

(Ser+3) (Cys+7)

[pre-epidermin(-30 to +2)]—NH—(L)CH(CH_2OH)—C(=O)— [pre-epidermin(+4 to +6)]—NH—(L)CH(CH_2SH)—C(=O)— [pre-epidermin(+7 to +22)]

(L)-serine (L)-cysteine

H_2O ← Site-specific dehydration

(Dha+3) (Cys+7)

[pre-epidermin(-30 to +2)]—NH—C(=CH_2)—C(=O)— [pre-epidermin(+4 to +6)]—NH—(L)CH(CH_2SH)—C(=O)— [pre-epidermin(+7 to +22)]

2,3-didehydroalanine (L)-cysteine

Stereospecific addition

[pre-epidermin(-30 to +2)]—NH—(D)CH(CH_2—S—CH_2)(L)CH—NH— ... C(=O)— [pre-epidermin(+4 to +6)]—N ... C(=O)— [pre-epidermin(+7 to +22)]

meso-lanthionine

Fig. 2. Proposed pathway for the chemical reactions occurring during the biosynthesis of lanthionine from the precursor amino acids serine and cysteine using, as an example, the formation of ring A of epidermin. The chirality of the respective compounds is indicated (where appropriate) above the α-carbon atom.

tion of N-terminally situated Dha or Dhb to produce a sequence-blocking pyruvyl or butyryl group, respectively (GROSS and MORELL, 1971; JUNG, 1991a, b). During the elucidation of the structure of Pep5 (KELLNER et al., 1989) this problem was circumvented by first reducing the two Dhb residues with benzylthioalcohol, however this has proved unsuccessful with other lantibiotics such as SA-FF22 (JACK et al., 1994a). Recently, MEYER (1994) has developed a convenient, new amino acid sequencing regimen for the detection of suitably derivatized dehydroamino acids and thioethers which involves thiol addition, oxidation with *per*-trifluoroacetic acid and a second round of thiol addition. Similar oxidative deaminations of N-terminal Dhb and Dha leads to the formation of the novel sequence blocking 2-oxobutyryl and 2-oxopyruvyl residues found at the N-terminus of Pep5 and lactocin S, respectively (KELLNER et al., 1989; SKAUGEN et al., 1994) and will be discussed in subsequent sections.

2.1.3 D-Alanine

Amino acids in bacterially-synthesized proteins and peptides are normally present as L-isomers, however, the type A lantibiotic lactocin S, which is produced by *Lactobacillus sake* strain Lb45, contains D-alanine (SKAUGEN et al., 1994). Interestingly, analysis of the prepeptide sequence showed that the precursor for this amino acid was L-serine, rather than alanine as might be expected. Thus, it seems that specific serine residues are dehydrated to form Dha and then hydrogenated in a stereospecific fashion to yield D-alanine during lactocin S biosynthesis (Fig. 3). Since the Dha residues found in other lantibiotics are generally stable, it seems likely that such a reaction should be catalyzed by a hitherto unknown enzyme. In addition, the identification of an enzyme or enzyme system able to create D-isomers of alanine in peptides from serine precursors offers considerable hope for new approaches in biotechnology and peptide engineering.

(Ser+7)

[lactocin S(1 to 6)]—NH—(L)CH(CH_2OH)—C(=O)— [lactocin S(8-37)]

(L)-serine

H_2O ← Site-specific dehydration

[lactocin S(1 to 6)]—NH—C(=CH_2)—C(=O)— [lactocin S(8-37)]

2,3-didehydroalanine

Hydrogenation

[lactocin S(1 to 6)]—NH—(D)CH(CH_3)—C(=O)— [lactocin S(8-37)]

(D)-alanine

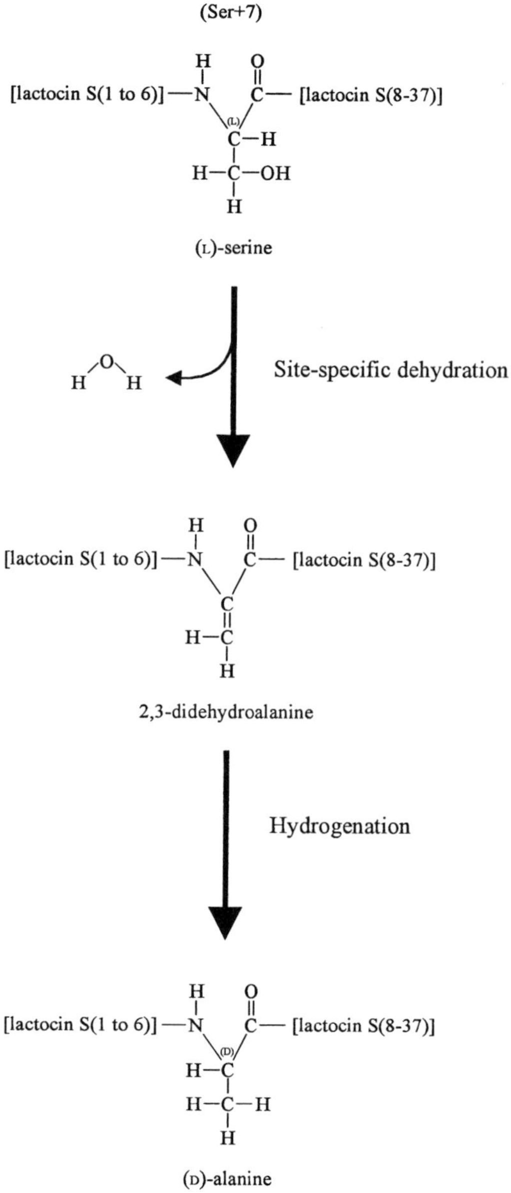

Fig. 3. Proposed pathway for the chemical reactions occurring during the biosynthesis of D-alanine from a precursor serine residue at position 7 of the type A lantibiotic lactocin S. The chirality of the respective compounds is indicated above the α-carbon atom.

2.2 Type A Lantibiotic Primary Structures

2.2.1 Nisin

Since the structure of nisin was the first lantibiotic described (GROSS and MORRELL, 1971) and nisin was probably the first lantibiotic ever identified (ROGERS and WHITTIER, 1928), it is fitting to begin any treatise of lantibiotic structure with this particular example. Nisin is produced by the cheese starter culture organism *Lactococcus lactis* ssp. *lactis* and is antimicrobial for a broad variety of gram-positive bacteria (HURST, 1981; MOLI-

TOR and SAHL, 1991). The peptide itself (Fig. 4) is a 34 amino acid, 3353 Da, pentacyclic structure formed from its content of one Lan and 4 MeLan residues (GROSS and MORRELL, 1971). While rings A, B, and C are separate the C-terminal pair of rings (D and E) form a bicycle (in all cases rings have been assigned ascending alphabetic characters, starting with the A ring closest to the N-terminus). In addition, nisin contains one Dha and 2 Dhb residues.

In addition to nisin, a number of nisin fragments have also been isolated and characterized, many of which have altered antimicrobial, activity. BERRIDGE et al. (1952) first reported the isolation of nisin fragments generated by long-term storage under acidic conditions which have subsequently been isolated and characterized by CHAN et al. (1989a, b). Nisin1–32(amide), which results from the hydrolysis of the Val33-Dha34 peptidyl bond, has similar antimicrobial activity to native nisin, suggesting that the C-terminal two amino acids are dispensable and that Dha33 plays no particular role in the biological activity of the peptide (ROLLEMA et al., 1991). However nisin1–32(amide), in which the Dha5-Leu6 peptidyl bond had also been hydrolyzed and ring A was therefore opened, was virtually inactive.

Subsequently these studies have been confirmed by site-directed mutagenesis (DODD and GASSON, 1994); the exchange Dha33Ala resulted in only slightly decreased activity while the double exchange Dha5Ala/Dha33Ala essentially destroyed the antimicrobial activity of the peptide. Similarly, the presumably conservative exchange Dha5Dhb led to a substantial decrease in activity (between 2- and 10-fold) against some indicators (KUIPERS et al., 1992), perhaps suggesting that these residues are not only important for antimicrobial activity but may also alter the spectrum of bacterial strains affected. In addition, these authors also altered ring C of nisin with the dual exchange Met17Gln/Gly18Thr. Two products were resolved from this exchange; intact nisin containing either Thr18 or Dhb18. The latter product not only contains a ring with the same sequence as the structurally similar lantibiotic subtilin (Fig. 4), but also demonstrates that the biosynthetic machinery of the cell was able to cope with forming a novel dehydrated residue at a position not normally occupied by such an amino acid. Interestingly, [Met17Gln/Gly18Dhb]-nisin also had improved antimicrobial activity against *Bacillus cereus* and decreased activity against *Micrococcus flavus*, while [Met17Gln/Gly18/Thr]-nisin had heightened activity against *M. flavus* and reduced activity against *B. cereus* (KUIPERS et al., 1992).

Recently, MULDERS et al. (1991) identified a naturally occurring analog of nisin which they named nisin Z. The peptide was characterized by a single amino acid exchange (His27Asn), resulting from a single base change in the 3rd base of the 27th codon of the structural gene *nisZ*. Interestingly, while this exchange appears to have little (if any) effect on the biological activity of the mutant nisin, it seems to make the peptide more soluble at or around pH 7, presumably as a result of the increased hydrophilicity of Asn at neutral pH. Thus, nisin Z might prove to be useful for applications where the low solubility of nisin at neutral pH (HURST, 1981) has traditionally proved problematic (see also Sect. 6).

2.2.2 Subtilin

Like nisin, subtilin is a pentacyclic type A lantibiotic (Fig. 4); in fact, the arrangement and nature of all of the thioethers are conserved between nisin and subtilin, even though the amino acid sequences of the two peptides varies significantly (GROSS and MORRELL, 1971; GROSS et al., 1973). First described by JANSEN and HIRSCHMANN (1944), subtilin is produced by *Bacillus subtilis* and is active against both the vegetative cells of a number of gram-positive bacteria as well as the outgrowth of endospores of *Bacillus* spp. and *Clostridium* spp. (HURST, 1981). Interestingly, the two activities are distinguishable, apparently dependent on saturation of Dha5 which leads to loss of subtilin antispore activity, the half-life of which is only 0.8 d (HANSEN et al., 1991; LIU and HANSEN, 1992; HANSEN, 1993). Using site-directed mutagenesis, the exchange Dha5Ala was shown to result in the loss of antispore activity, confirm-

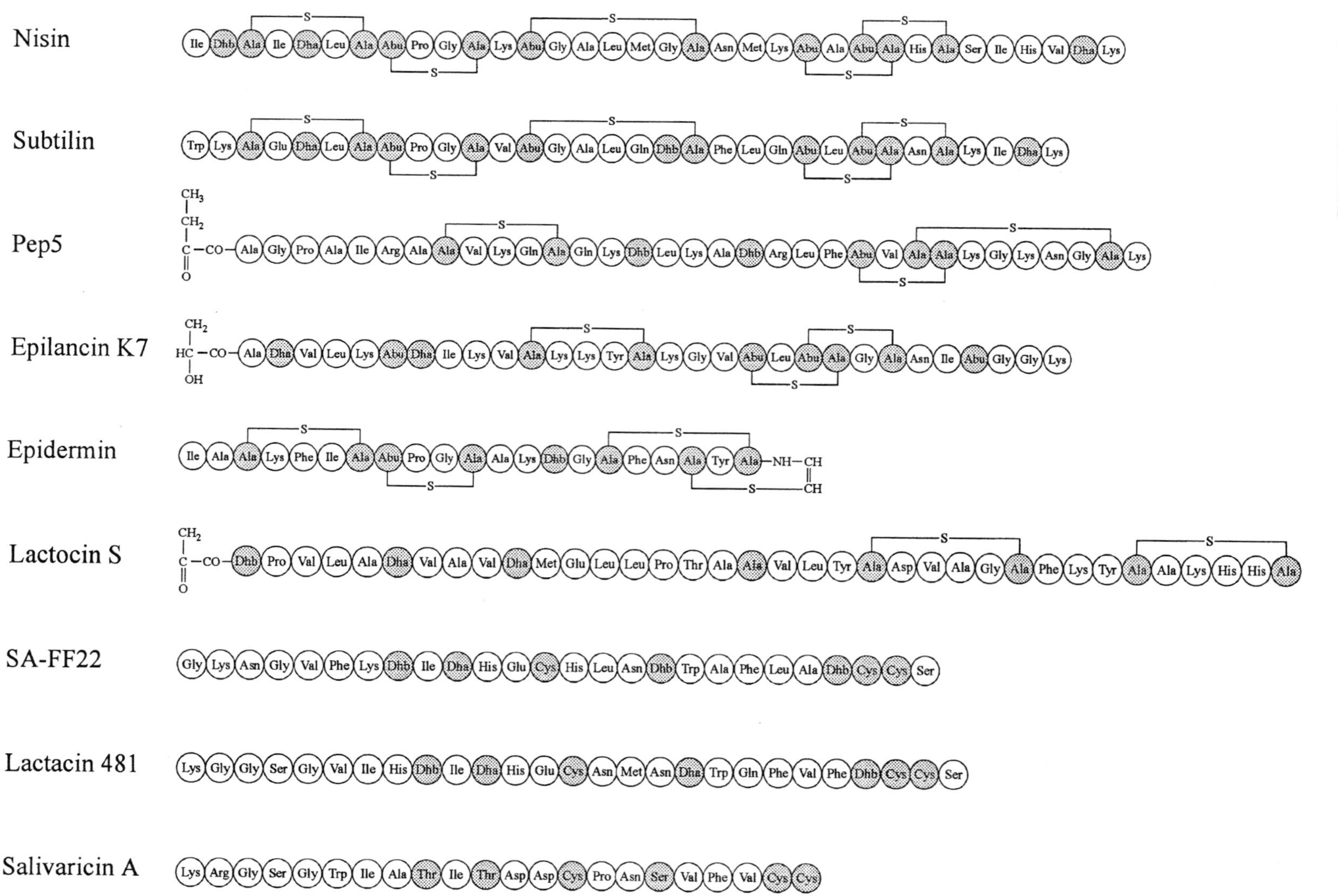
Nisin
Subtilin
Pep5
Epilancin K7
Epidermin
Lactocin S
SA-FF22
Lactacin 481
Salivaricin A

◀ **Fig. 4.** The structure of several type A lantibiotics. Amino acid residues involved in posttranslational modifications have been identified by highlighting while additional modified residues are indicated by their respective chemical structures. Dha, 2,3-didehydroalanine; Dhb, 2,3-didehydrobutyrine; Abu, α-aminobutyric acid; Ala-S-Ala, (2*S*,6*R*)-lanthionine; Abu-S-Ala (2*S*,3*S*,6*R*)-3-methyllanthionine. The arrangement of the thioether bonds in the lantibiotics SA-FF22, lacticin 481/lactococcin DR, and salivaricin A remains unreported.

ing that saturation of Dha5 was the primary cause of subtilin instability (LIU and HANSEN, 1992). In addition, these authors also showed that the exchange Glu4Ile (i.e., the same as is found in this position of nisin; Fig. 4) not only increased the half-life of the antispore activity by 57-fold but also increased the sporicidal activity by a factor of 3. From these results, they suggest that Glu4 may be involved in the spontaneous saturation of subtilin, probably acting as a Michael-type acceptor. In similar structure–function studies, subtilin with a succinylated N-terminus has been isolated; this peptide also had reduced biological activity (CHAN et al., 1993).

2.2.3 Epidermin and Gallidermin

Elucidation of the structure of the *Staphylococcus epidermidis* lantibiotic epidermin (ALLGAIER et al., 1986) has proved important in terms of our understanding of lantibiotic biosynthesis, as it assisted in the isolation of the structural gene (*epiA*), proving conclusively that lantibiotics were ribosomally synthesized (SCHNELL et al., 1988). Epidermin (Fig. 4) is a 22 amino acid (2164 Da), tetracyclic peptide which contains one residue each of Dhb and MeLan along with two residues of Lan. The 4th cyclic structure results from the novel C-terminal *mono*-carboxy, *di*-amino acid, Cys(Avi) (Fig. 1) between residues 19 and 22, the formation of which results from oxidative decarboxylation of the C-terminal Cys residue by the enzyme EpiD (see also Sect. 4.3.2). In addition, ring B is identical to ring B of both nisin and subtilin, as is the arrangement of both rings A and B among these 3 lantibiotics.

Subsequently, a structural analog of epidermin, called gallidermin and produced by *Staphylococcus gallinarum*, has also been isolated and characterized (KELLNER et al., 1988). While gallidermin appears to have a slightly different spectrum of biological activity to epidermin (especially against *Propionibacterium acnes*), it differs only by the exchange of Ile6 in epidermin for Leu, an exchange which is conservative and does not alter the molecular mass. In all other respects, the two structures are identical.

2.2.4 Pep5

The type A lantibiotic Pep5 was first isolated from *Staphylococcus epidermidis* strain 5 (SAHL and BRANDIS, 1981) and has subsequently been thoroughly characterized (KELLNER et al., 1989, 1991). The peptide (Fig. 4) is the largest characterized type A lantibiotic (3488 Da) and consists of 34 amino acids, 8 of which are basic. In addition, Pep5 is tricyclic as a result of its containing one MeLan and two Lan residues; ring A is separate while, rings B and C overlap. The N-terminus of Pep5 is occupied by a 2-oxobutyryl residue (KELLNER et al., 1989) and analysis of the sequence of the structural gene, *pepA* (KALETTA et al., 1989), shows that the +1 codon of the prepeptide encodes a threonine residue which is subsequently converted to Dhb during modification and maturation (WEIL et al., 1990). It appears that once this Dhb residue is located at the N-terminus (i.e., following removal of the leader peptide) it undergoes oxidative deamination to form 2-oxobutyryl (Fig. 5), probably by a spontaneous mechanism (KELLNER et al., 1989).

2.2.5 Epilancin K7

Also produced by *Staphylococcus epidermidis*, epilancin K7 is somewhat similar to Pep5 (Fig. 4). Recently, VAN DE KAMP et al. (1995a, b) solved the structure of the peptide by a variety of chemical and homonuclear NMR-based strategies and showed that it is a 3052 Da, tricyclic, type A lantibiotic containing one Lan and two MeLan residues. In addition, epilancin contains two residues each of Dha and Dhb as well as 6 lysine residues (and

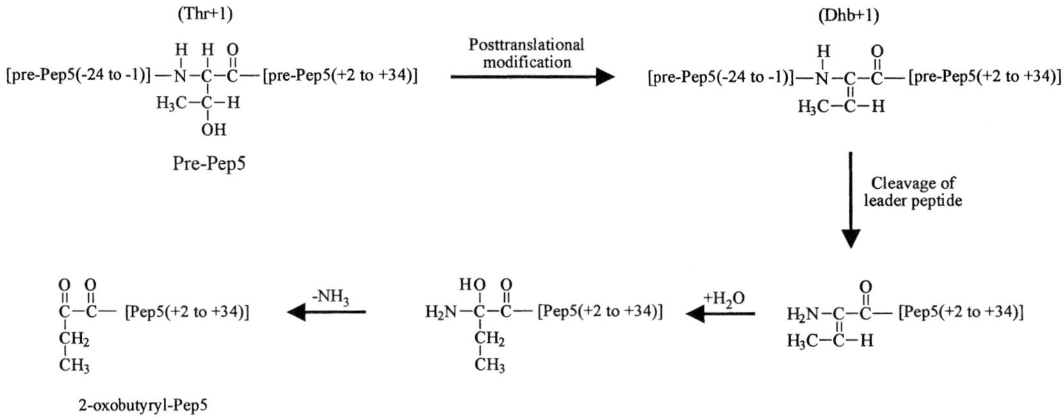

Fig. 5. Proposed pathway for the spontaneous oxidative deamination of dehydrobutyrine located at the N-terminus of Pep5 after cleavage of the leader peptide.

no acidic amino acids), accounting for its extremely basic nature. Again, like Pep5 (KELLNER et al., 1989), the N-terminus of epilancin is blocked; however, while spontaneous deamination of the N-terminal Dha should result in the formation of 2-oxopyruvyl (in analogy to that shown for Pep5; Fig. 5), VAN DE KAMP et al. (1995a) suggest that the N-terminal residue is occupied by 2-hydroxypyruvate (Fig. 1). The formation of this group at the N-terminus probably suggests that an additional, novel enzyme function is required for epilancin biosynthesis. In addition, this study also isolated a deletion peptide resulting from hydrolysis of the Ala2-Dha3 peptidyl bond; epilancin K7(3–33) is also blocked, in this case by the expected 2-oxopyruvyl group, presumably formed by oxidative deamination of the N-terminal Dha residue. Interestingly, this peptide appears to be approximately as active as native epilancin K7, suggesting these two N-terminal amino acids are not essential for its biological activity.

2.2.6 Streptococcin A-FF22, Lacticin 481 (Lactococcin DR), and Salivaricin A

At around the same time as the structure of nisin was being reported (GROSS and MORRELL, 1971), TAGG et al. (1973a, b) reported that a clinical isolate of *Streptococcus pyogenes* designated strain FF22 produced "a diffusible inhibitor of bacterial growth" which they called streptococcin A-FF22. Subsequent partial purification and characterization showed that the inhibitory agent was essentially proteinaceous, of low molecular weight (TAGG et al., 1973a, b), and was rather similar to nisin (TAGG and WANNAMAKER, 1978; JACK and TAGG, 1992). Recently, streptococcin A-FF22 was purified to homogeneity and chemically characterized (JACK and TAGG, 1991); the lantibiotic was shown to be a 27 amino acid, 2795 Da type A lantibiotic (Fig. 4) which contains one Lan and two MeLan residues as well as one residue of Dhb (JACK et al., 1994a). Using the sequence information gained from these studies, the structural gene *scnA* has been cloned and sequenced and confirms the correct sequence for the peptide (HYNES et al., 1993). Unfortunately, because of the overlapping nature of the thioether rings in streptococcin A-FF22, assignment of the ring order by enzymic and chemical digestions and modifications has proved impossible (JACK et al., 1994a).

In addition, JACK and TAGG (1991) also isolated a naturally-occurring derivative of streptococcin A-FF22 devoid of detectable antibacterial activity, apparently as a result of the loss of the N-terminal 4 amino acids which are not involved in formation of ring structures. While C-terminal amino acid-deficient derivatives have not yet been found, it would seem that the antibacterial activity of streptococcin A-FF22 is dependent on (at least) an intact N-terminus, in direct contrast to observations made with the type A lantibiotic epilancin K7 (VAN DE KAMP et al., 1995a).

Another type A lantibiotic which shows striking similarities to the structure of streptococcin A-FF22 has been isolated from *Lactococcus lactis* ssp. *lactis* and called lacticin 481 (PIARD et al., 1992). In addition, RINCE et al. (1994) have recently isolated a lantibiotic from *Lactococcus lactis* ssp. *lactis* DR which proved to be identical to lacticin 481. Preliminary amino acid sequence analysis of lacticin 481 has allowed the isolation of the structural gene, *lctA* (PIARD et al., 1993) and complete analysis of the primary structure of the peptide has shown it to be a 28 amino acid, 2901 Da type A lantibiotic with one MeLan, one Dhb and two Lan residues (Fig. 4). Interestingly the peptide shares both sequence and structural similarities with streptococcin A-FF22 (JACK et al., 1994a), especially in the presence of the 3 overlapping thioether rings and the single residue of unmodified serine at the C-terminus (PIARD et al., 1993).

Salivaricin A (Fig. 4) is a lantibiotic produced by *Streptococcus salivarius* strain 20P3 which acts against a number of pathogenic streptococci including the principal human pathogen, *Streptococcus pyogenes* (DEMPSTER and TAGG, 1982). Purification and microcharacterization of the peptide responsible showed that it is a small (2315 Da) type A lantibiotic with one Lan and two MeLan residues (ROSS et al., 1993). In addition, this particular lantibiotic is apparently the only one so-far characterized which does not contain either Dha or Dhb. Also, although no attempts appear to have been made to determine the order of the thioether bridges, it seems likely that all three should overlap.

2.2.7 Lactocin S

Lactocin S (Fig. 4) is a large, 37 amino acid lantibiotic produced by *Lactobacillus sake* which has been purified and partially characterized (MØRTVEDT and NES, 1990; MØRTVEDT et al., 1991). In addition, the structural gene encoding the prepeptide has been isolated and characterized in order to gain an insight into the likely structure of this lantibiotic (SKAUGEN et al., 1994). Taken together, these studies have demonstrated that the peptide is the largest lantibiotic so-far characterized (3764 Da) and probably contains two Lan residues. In addition, in a situation analogous to that observed in Pep5 (Fig. 5; KELLNER et al., 1989) the N-terminus is blocked, probably by a 2-oxopyruvyl group which would arise from the spontaneous oxidative deamination of a N-terminally located Dha residue. However, the nature of the blocking group as well as the arrangement of the thioether bridges has yet to be resolved. Finally, lactocin S has also been shown to contain D-alanine at positions which contain Ser in the prepeptide sequence, prompting speculation that these residues arise following stereospecific hydrogenation of Dha residues at these sites in the propeptide domain (see also Sect. 2.1.3).

2.2.8 Enterococcal Cytolysin/Bacteriocin

The cytolysin/bacteriocin of *Enterococcus faecalis* has long-since been shown to be a virulence determinant which also has the ability to kill certain gram-positive bacterial strains (BROCK and DAVIE, 1963), however, the recent isolation of the genetic elements responsible for its biosynthesis have demonstrated the relationship between this toxin and the lantibiotics (GILMORE et al., 1994). The cytolysin/bacteriocin system appears to comprise two peptides both of which are lantibiotics and both of which are required for antibacterial activity; Cyl1 is a 4163 Da peptide while Cyl2 has a mass of only 2631 Da, however, while lanthionine has been identified in these peptides the exact number of residues etc. is not yet clear (SAHL et al., 1995). From the derived amino acid sequences obtained through analysis of the genes, it appears that the two peptides have regions of identity, suggesting that they may have arisen through gene duplication and partial deletion (GILMORE et al., 1994).

2.2.9 Carnocin UI49 and Mutacin

STOFFELS et al. (1992) have reported the isolation of a large lantibiotic from *Carnobacterium piscicola* which they named carnocin UI49. The peptide is particularly hydrophobic as judged by its chromatographic behavior, has a mass of 4635 Da and a novel N-terminal amino acid sequence of G-S-E-I-Q-P-R which is blocked to further sequencing at the 8th residue. While the exact number and nature of amino acids present in this lantibiotic remains to be determined, preliminary results suggest it is somewhat similar to lactocin S (SAHL et al., 1995). In addition, the peptide appears to have a propensity for membrane perturbation and seems to interact with nisin-producing strains in a specific way, suggesting there may be an interaction of carnocin UI49 with both the cytoplasmic membrane and parts of the nisin-synthesizing machinery of the cell (STOFFELS et al., 1994).

Mutacin, a lantibiotic from *Streptococcus mutans* T8, has also been isolated and characterized (NOVAK et al., 1994). This lantibiotic has a mass of 3245 Da, contains one MeLan and two Lan residues and has the unique N-terminal amino acid sequence N-R-W-W-Q-G-V-V-X, where X indicates a sequence blocking group.

2.3 Type B Lantibiotic Primary Structures

2.3.1 Cinnamycin, Duramycin, Duramycin B, Duramycin C, and Ancovenin

Many of the type B lantibiotics described so-far have gained considerable interest, not so much because of their unusual structures nor their antimicrobial activity (which in most cases is only moderate), but because they act as inhibitors of essential enzymes involved in both the human immune and circulatory blood pressure regulatory systems, including: phospholipase A_2 and angiotensin-converting enzyme (see also Sect. 5). Over a period of a number of years, searches for such inhibitors of immune function and blood pressure regulation have identified a number of these compounds (some of which have subsequently been shown to be identical) including: cinnamycin (formerly also known as lanthiopeptin or Ro09–0198), duramycin (formerly also known as leucopeptin), duramycin B, duramycin C, and ancovenin (BENEDICT et al., 1952; SHOTWELL et al., 1958; GROSS, 1977; WAKAMIYA et al., 1985, 1988; KESSLER et al., 1987, 1988, 1991; NARUSE et al., 1989; FREDENHAGEN et al., 1990, 1991; HAYASHI et al., 1990), the structures of which are shown in Fig. 6. Each of these compounds has been isolated from different strains of *Streptomyces* spp. or *Streptoverticillium* spp. and all show limited antimicrobial activity, principally directed against *Bacillus* spp. (see also Sect. 5). Since the structures of cinnamycin, duramycin, duramycin B, duramycin C, and ancovenin are so similar, they will be regarded here as the cinnamycin-group of type B lantibiotics and described together. Indeed, SAHL et al. (1995) have suggested that since these type B lantibiotics share such an extent of structural and sequence similarity (only 7 conservatively exchanged amino acids overall), that they ought to be viewed as structural variants of the same compounds.

As would be expected, the type B lantibiotics of the cinnamycin group all contain lanthionine; indeed, the ring structures and arrangements are conserved among all these peptides (Fig. 6). However, while each contains one Lan and two MeLan residues (FREDENHAGEN et al., 1991; KESSLER et al., 1991; JUNG, 1991a, b), there are some notable differences between type A and type B lantibiotics. Firstly, these type B lantibiotics all contain a "head-to-tail" bridge, formed by a MeLan residue spanning residues 1–18 and this Melan, along with that formed between residues 5 and 11 is in the opposite orientation to that observed in all type A lantibiotics (i.e., the Cys-derived "half" of the *di*-amino acid is located toward the N-terminus). In addition, cinnamycin-group type B lantibiotics (with the exception of ancovenin) also con-

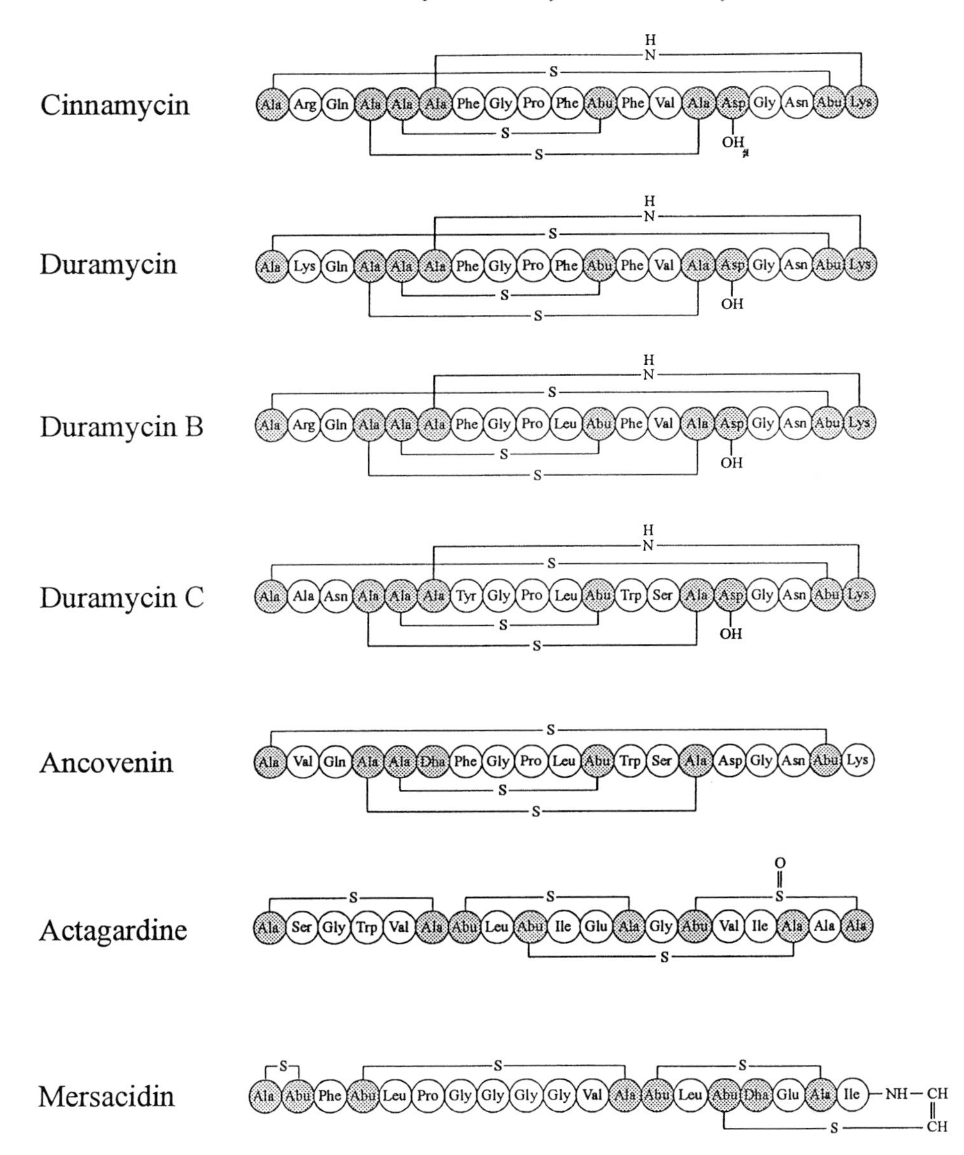

Fig. 6. The structure of several type B lantibiotics. Amino acid residues involved in posttranslational modifications have been identified by highlighting while additional modified residues are indicated by their respective chemical structures. Abu, α-aminobutyric acid; Asp-OH, *erythro*-3-hydroxy-aspartic acid; Ala-NH-Lys, (2*S*,8*S*)-lysinoalanine; Ala-S-Ala, (2*S*,6*R*)-lanthionine; Ala-SO-Ala, lanthionine sulphoxide; Abu-S-Ala (2*S*,3*S*,6*R*)-3-methyllanthionine.

tain a single residue of hydroxy-aspartic acid (Figs. 1 and 6), however, it is not clear from whence or by what mechanism(s) this residue arises (FREDENHAGEN et al., 1991; JUNG, 1991a, b; SAHL et al., 1995). Similarly these type B lantibiotics (again with the exception of ancovenin) contain the aminoether linked, *di*-carboxy, *di*-amino acid, (2*S*,8*S*)-lysinoalanine (Fig. 1), formed between residues 6 and 19 (Fig. 6). It has been proposed (JUNG, 1991a, b) that such a residue could form by the addition of the ε-amino group of Lys19 across the unsaturated Dha6, in a manner rather analogous to that observed for the formation of lanthionine shown in Fig. 2. What is not clear about these additional modifications is what role (if any) they play in the biological activity of their respective lantibiotic; indeed, ancovenin contains neither hydroxyaspartic acid nor lysinoalanine, yet is functionally similar to other type B lantibiotics of this group (WAKAMIYA et al., 1985; SHIBA et al., 1986), suggesting that they may not be essential for biological activity.

2.3.2 Mersacidin and Actagardine

By contrast to the cinnamycin-like type B lantibiotics described in the preceding section, mersacidin and actagardine have been studied extensively, in part, because of their strong antimicrobial activities (see also Sect. 5). Furthermore, as can be seen in Fig. 6 they also differ significantly in their structural characteristics, neither containing the head-to-tail bridging pattern described above.

Mersacidin (Fig. 6) is produced by *Bacillus* spp. (strain HIL Y–85,54728), is the smallest lantibiotic (1825 Da) and consists of 20 amino acids arranged to give a tetracyclic structure with rings C and D forming a bicycle (CHATERJEE et al., 1992a, b). It is hydrophobic with no net charge, contains a single residue of Dha and 3 MeLan residues and is the only lantibiotic to contain the C-terminal unsaturated amino acid (S)-[(*Z*)-2-aminovinyl]-3-methyl-D-cysteine. This last residue is probably formed by a similar process as that responsible for Cys(Avi) formation in epidermin (Fig. 1; ALLGAIER et al., 1986; KUPKE et al., 1992), except that the dehydrogenated decarboxylated C-terminal Cys residue should react with a Dhb residue, rather than a Dha. In addition, KOGLER et al. (1991) have shown that mersacidin also contains a residue of MeLan (between residue 1 and 2) formed by linkage in the opposite direction to that observed with all type A lantibiotics (i.e., the Cys-derived "half" of the first ring of mersacidin is located to the N-terminal side with respect to the Dhb-derived "half").

Actagardine (formerly gardimycin) was isolated from *Actinoplanes* spp. ATCC 31048 and 31049 by PARENTI et al. (1976) and the correct structure of the peptide has been determined by ZIMMERMANN and JUNG (1995) after previous attempts (MALABARBA et al., 1985, 1990; KETTENRING et al., 1990). Actagardine contains 4 sulphide rings (1 Lan and 3 MeLan residues) with the rings B, C and D forming a tricycle. Interestingly, with the exception of a single, conservative amino acid exchange, ring B of actagardine and ring C of mersacidin are identical (Fig. 6), a feature which may be related to their similar biological activity (see also Sect. 5), which is the inhibition of peptidoglycan biosynthesis (SOMMA et al., 1977; BRÖTZ et al., 1995; SAHL et al., 1995). In addition, actagardine is the only lantibiotic shown to contain 3-methyllanthionine sulphoxide (Fig. 6), however, it is not yet clear whether this residue has arisen as an artefact of the purification process (KETTENRING et al., 1990; ZIMMERMANN and JUNG, 1995).

3 Lantibiotic Structures in Solution

The lantibiotics, be they either type A or type B, contain a vast number of modified residues, many of which form bridging structures along various portions of the respective peptides length (c.f. Figs. 5, 6 and Tab. 1). Obviously then, these modified residues and unusual bridging patterns should have some ramifications on the overall structure adopted by the peptides in solution. Initially, unsuccessful attempts were made to obtain crystals of several lantibiotics, in order to determine their spatial structures by X-ray crystallography (JUNG et al., unpublished data). Subsequently, the advent of sensitive, high resolution 2-D, 3-D and 4-D NMR techniques has allowed a number of laboratories to look at the solution conformations of various lantibiotic structures (SLIJPERS et al., 1989; CHAN et al., 1989b; PALMER et al., 1989; VAN DE VEN et al., 1991a, b; LIAN et al., 1991; GOODMAN et al., 1991; FREUND et al., 1991a, b, c; KESSLER et al., 1991; ZIMMERMANN et al., 1993; ZIMMERMANN and JUNG, 1995; VAN DE KAMP et al., 1995a). In addition, a number of these studies have also been able to take advantage of the nature of NMR and assess the effects of solutions of differing lipophilicity, micelles, perturbants (e.g., urea) and phospholipid vesicles on the solution structures obtained.

3.1 Type A Lantibiotics

3.1.1 Gallidermin

The structure of gallidermin (Fig. 7) has been obtained both in trifluoroethanol (TFE)/water (95:5) and dimethylsulphoxide (DMSO) where it adopts an extended, cork-screw-like conformation (FREUND et al., 1991a, b). Ring B (residues 8–11), which is identical to ring B of nisin, forms a β-turn type II, while the central domain (residues 11–15) shows the greatest degree of flexibility over the whole molecule, due to its content of turn-like motifs. Interestingly, this region incorporates a potential trypsin cleavage site (Lys13-Dhb14) which appears to be exposed by this flexibility and has been shown by molecular modelling to fit the active site of the enzyme (FREUND et al., 1991b). Residues

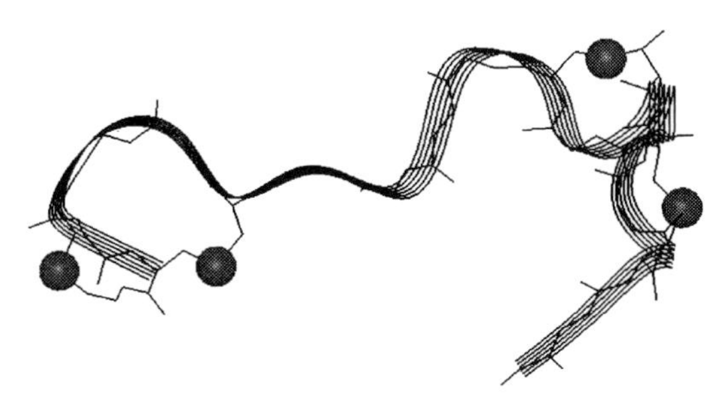

Fig. 7. Stereorepresentation (backbone ribbon) of the solution structure of the type A lantibiotic gallidermin; c.f. Fig. 4 for the primary structure. The dots represent sulphur atoms, and the N-terminus is at the right side.

Asn18, Ala19, Ala21, and the C-terminal Cys(Avi) are oriented inward forming a hydrophilic core to an otherwise hydrophobic cage-like structure consisting of the overlapping rings at the C-terminus which incorporate the outwardly oriented residues Phe17 and Tyr18 (FREUND et al., 1991a, b).

Overall, gallidermin adopts a distorted helix-like structure which has some degree of flexibility in the central region. In addition, it is amphiphilic with the hydrophobic C-terminal residues aligned on one "face", while the N-terminal part contains the hydrophilic residues also oriented towards the opposite "face" of the cork-screw (FREUND et al., 1991a, b). In TFE/water the peptide has an overall length of about 3 nm, a diameter of approximately 1 nm and a net dipole moment of around 75 Debye (JUNG, 1991a, b). The amphiphilicity, high dipole moment and membrane spanning length observed for gallidermin in solution may help to explain its biological activity, which is the formation of voltage-dependent pores in biological membranes (SAHL, 1991; BENZ et al., 1991).

3.1.2 Nisin and Subtilin

A number of research groups have independently investigated the solution structure of nisin (and, to some extent, subtilin) and found that in aqueous solution the peptide adopts a relatively random, unordered structure (VAN DE VEN et al., 1991a, b; GOODMAN et al., 1991; LIAN et al., 1991, 1992; CHAN et al., 1992). Some degree of restraint is shown in the structures of both ring B and the overlapping rings D and E, however, ring A and C, several of the residues located at the N- and C-termini of the peptide as well as those in the central part show considerable conformational freedom.

By contrast, in mixed, lipophilic solvents such as TFE/water or DMSO the structure of nisin becomes somewhat more stabilized and shows that the peptide adopts overall α-helical structure. Taken together, nisin can be viewed as a pair of helical structures (residues 3–19 and 23–28), separated by a flexible hinge region between amino acids 20–22 (VAN DE VEN et al., 1991a, b; LIAN et al., 1992); the remaining residues at the N-terminus (amino acids 1–2) and C-terminus (amino acids 29–34) are extremely flexible and could not be defined, even in the stabilizing, lipophilic solutions. In addition, while it is not so obvious in aqueous solution, nisin, like gallidermin, is amphiphilic. Furthermore, a study by GOODMAN et al. (1991) of nisin in DMSO showed that it formed a kinked rod-like structure with an overall length of ca. 5 nm, diameter of ca. 2 nm and calculated dipole moment of at least 80 Debye, all features which are consistent with its mode of action (see also Sect. 5).

3.1.3 Pep5

Although they are not yet complete, studies of the lantibiotic Pep5 in solution suggest that it has a similar structure to the type A lantibiotics gallidermin, nisin, and subtilin previously characterized (FREUND et al., 1991c; FREUND and JUNG, 1992). Analysis of the circular dichroism of Pep5 in water suggests it forms a random, disordered structure while, on the addition of lipophilic solvents there is an increasing propensity toward formation of helical elements. Similarly NMR studies have shown that in aqueous solution Pep5 has no defined structure except in the region of the four-membered ring B and that the unbridged regions (amino acids 1–8) and the central region (amino acids 14–23) are particularly flexible and unstructured (FREUND et al., 1991c; FREUND and JUNG, 1992). However, as with the CD experiments, NMR of Pep5 in 90% TFE shows that the peptide is able to adopt an overall amphiphilic helical structure with most of the charged, hydrophilic amino acids oriented to one face of the rod. In addition, the central region is not bridged however, the presence of the two dehydroamino acids appear to stabilize local structure in this region (FREUND and JUNG, 1992).

3.2 Type B Lantibiotics

3.2.1 Cinnamycin and the Duramycins

In many respects the type B lantibiotics cinnamycin, duramycin, duramycin B, duramycin C, and ancovenin can be considered structural variants of one another (DE VOS et al., 1991) as they have the same principal bridging pattern (apart from ancovenin which lacks the lysinoalanine bridge) and amino acid exchanges are relatively conservative (FREDENHAGEN et al., 1991; SAHL et al., 1995). Therefore, it is not unreasonable to expect that their solution structures should be somewhat similar. In addition, all of these lantibiotics contain a head-to-tail MeLan bridge which reduces the possibilities for conformational freedom when compared to the type A lantibiotics (c.f. Figs. 4 and 6).

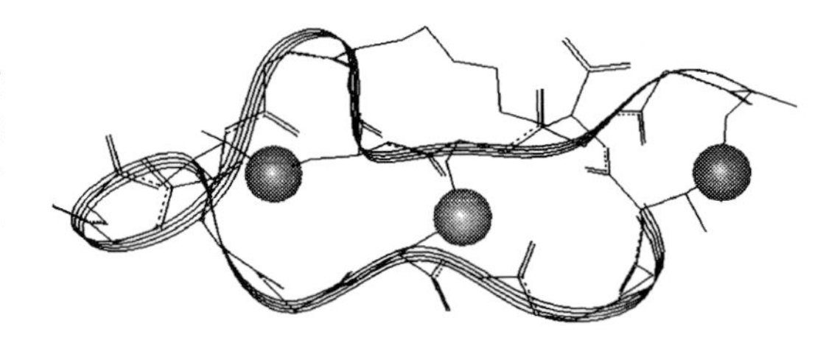

Fig. 8. Stereorepresentation (backbone ribbon, sulphur atoms represented as dots) of the type B lantibiotic duramycin B. For the primary structure see Fig. 6.

The currently published structures of cinnamycin, duramycin B, and duramycin C (Fig. 8) as well as the preliminary structure of ancovenin are all in general agreement as to the overall structure of these lantibiotics (KESSLER et al., 1991; ZIMMERMANN et al., 1993; NISHIKAWA et al., 1988). The peptides are bent into a U-shape by a turn induced by Pro9 and stabilized by the three thioether rings. In addition, the structure is further stabilized by antiparallel β-sheets in the N- and C-terminal regions and the planar nature of the backbone is slightly distorted by the lysinoalanine ring between residues 6 and 19 (ZIMMERMANN et al., 1993). Furthermore, the amino acid exchanges have little or no influence on the overall structures, but do appear to influence the degree of mobility of local structural elements and the overall hydrophobicity of the peptides. Like the type A lantibiotics, the type B lantibiotics are highly amphiphilic with all of the hydrophobic amino acids clustered into the bend of the "U" and the hydrophilic residues localized at the termini. Interestingly, the C-terminal region of cinnamycin has considerable structural similarity to cyclo-TP5, a thymopoietin analog able to stimulate T-cell maturation (KESSLER et al., 1991). The possibility that cinnamycin could stimulate T-lymphocytes may help to explain the previous observation that cinnamycin has *in vivo* anti-*Herpes simplex* activity (WAKAMIYA et al., 1988).

3.2.2 Mersacidin and Actagardine

So-far at least, the spatial structure of mersacidin has not been reported. However, preliminary results for actagardine (ZIMMERMANN and JUNG, unpublished data) show that it is a very rigid peptide, with well-defined structure, even though it lacks the head-to-tail bridge of the other type B lantibiotics. In addition, the structure appears to have an amphiphilic nature.

4 The Genes/Proteins Involved in Lantibiotic Biosynthesis and Genetic Regulation

4.1 The Requirement for Multiple Gene Products in Lantibiotic Biosynthesis

Lantibiotics are encoded by specific gene sequences and are synthesized on the ribosome hence, the prepeptides are limited to containing only the 20 amino acids allowed for in the genetic code (SCHNELL et al., 1988; SAHL et al., 1995; JACK et al., in press; JACK and SAHL, 1995). Therefore, in order to mature into the biologically active forms which have been isolated outside the producing cells, a number of posttranslational modification events must occur including: specific amino acid modifications, formation of intrapeptide thioether and/or aminoether rings, removal of the leader peptide from the N-terminus and transport of the peptide out of the cell to the extracellular matrix. In addition to these functions other controlled events are necessary for lantibiotic production, such as the regulation of their biosynthesis and the generation of immunity mechanism(s) to protect the producing cell from the action of its own lantibiotic. Although it is not possible to state a definitive order of events, the generalized chain of reactions leading to lantibiotic biosynthesis are shown for epidermin in Fig. 9 (see also Sect. 4.8).

In light of these observations, it is not surprising that following the discovery of the first specific genes encoding lantibiotic structural genes (SCHNELL et al., 1988; BUCHMANN et al., 1988; BANERJEE and HANSEN, 1988; KALETTA and ENTIAN, 1989), subsequent analysis of the DNA flanking these genes has revealed the presence of a number of other open reading frames, often clustered together, which seem to be involved in the biosynthesis, genetic regulation and immunity required during lantibiotic production. A number of these lantibiotic-synthesizing gene clusters are compared in Fig. 10; wherever possible, genes/gene products with similar function have been given the same letter designation (DE VOS et al., 1991) and the prefix *lan*/Lan (lantibiotic-related) has been used to indicate groups of genes (SAHL et al., 1995). Thus, e.g., the general term *lanA* indicates all lantibiotic structural genes while the general term LanA indicates all pre-lantibiotics.

4.2 Structural Genes and Pre-Lantibiotics

4.2.1 The Location and Features of the Structural Genes and their Transcription

Much debate has surrounded the location of the structural gene encoding pre-nisin, with various research groups suggesting that it was located on either the chromosome or a plasmid (KOZAK et al., 1974; TSAI and SANDINE, 1987; BUCHMANN et al., 1988; KALETTA and ENTIAN, 1989; GIREESH et al., 1992). However, it is now clear that the entire operon responsible for nisin production, consisting of *nisABTCIPRKFEG* (KUIPERS et al., 1993a; VAN DER MEER et al., 1993; ENGELKE et al., 1994; SIEGERS and ENTIAN, 1995) is encoded on a 70 kbp conjugative transposon, either called *Tn*5301 (DODD et al., 1990, 1991; HORN et al., 1991) or *Tn*5276 (RAUCH et al., 1990, 1991; RAUCH and DE VOS, 1992), depending on the strain of *Lactococcus lactis*

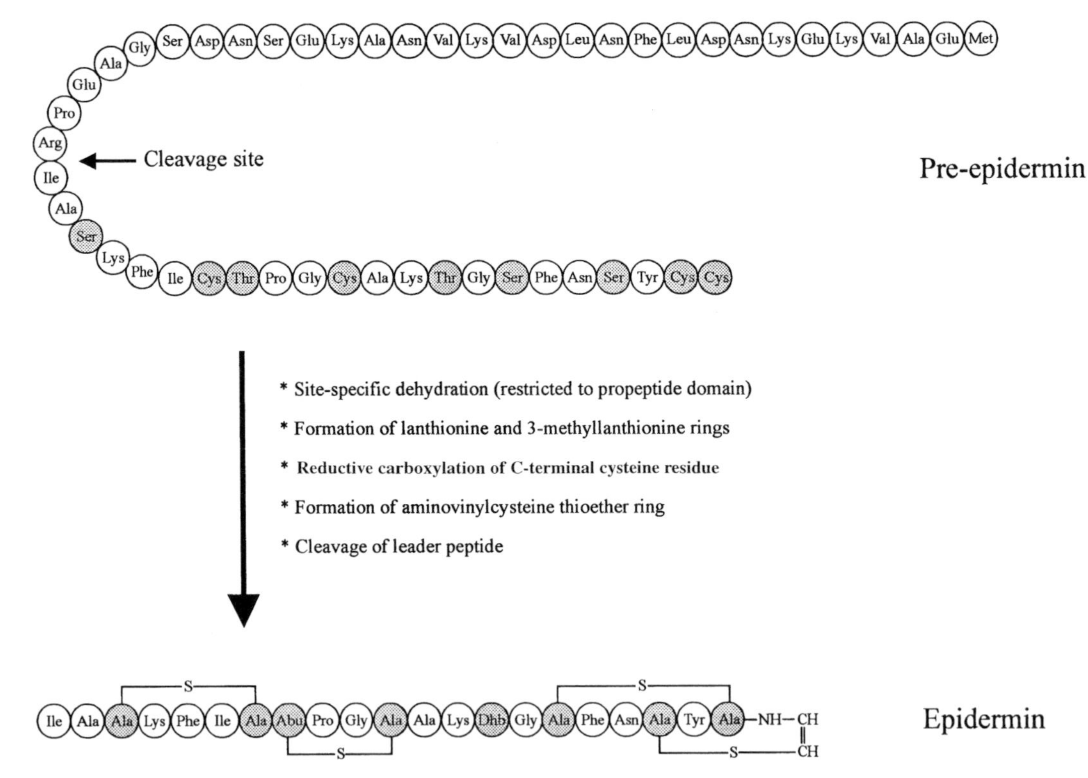

Fig. 9. General schema for the events occurring during the biosynthesis of the type A lantibiotic epidermin.

from which it was isolated. Furthermore, these authors have shown that these transposons also encode several of the genes necessary for sucrose metabolism in lactococci. Therefore, the recent observations that sucrose metabolism-related genes can be found immediately proceeding the *nisABTCIPRKFEG* gene cluster suggests that this represents all of the genes necessary for the production of nisin (SIEGERS and ENTIAN, 1995).

The nisin structural gene, *nisA* (Fig. 10), is the first gene in the cluster and is located very close to the 5' terminus of the transposon (DODD et al., 1990). BUCHMAN et al. (1988) identified a putative ρ-independent terminator immediately following the structural gene and subsequent studies have mapped the transcription start-site and identified an appropriate promoter for transcription of *nisA* (KUIPERS et al., 1993a; ENGELKE et al., 1992). In addition, it has been suggested that *nisA* and *nisB* may be co-transcribed and appropriate sized transcripts have been identified along with a terminator after *nisB* (STEEN et al., 1991; ENGELKE et al., 1992; KUIPERS et al., 1993a). Little is known of the events surrounding transcription of the genes downstream of *nisAB*, however, a tandem promoter immediately preceding *nisT* could suggest that the remaining genes are transcribed as part of a polycistronic message (STEEN et al., 1991).

The genes involved in the biosynthesis of epidermin have also been studied extensively. The gene cluster responsible for epidermin production consists of *epiT'T''ABCDQP* and is carried on the 54 kbp plasmid pTü32 (SCHNELL et al., 1991, 1992; AUGUSTIN et al., 1991, 1992). The structural gene, *epiA* (Fig. 10), is very likely transcribed along with *epiBCD*, since *epiB* does not appear to possess its own promoter and transcripts of ap-

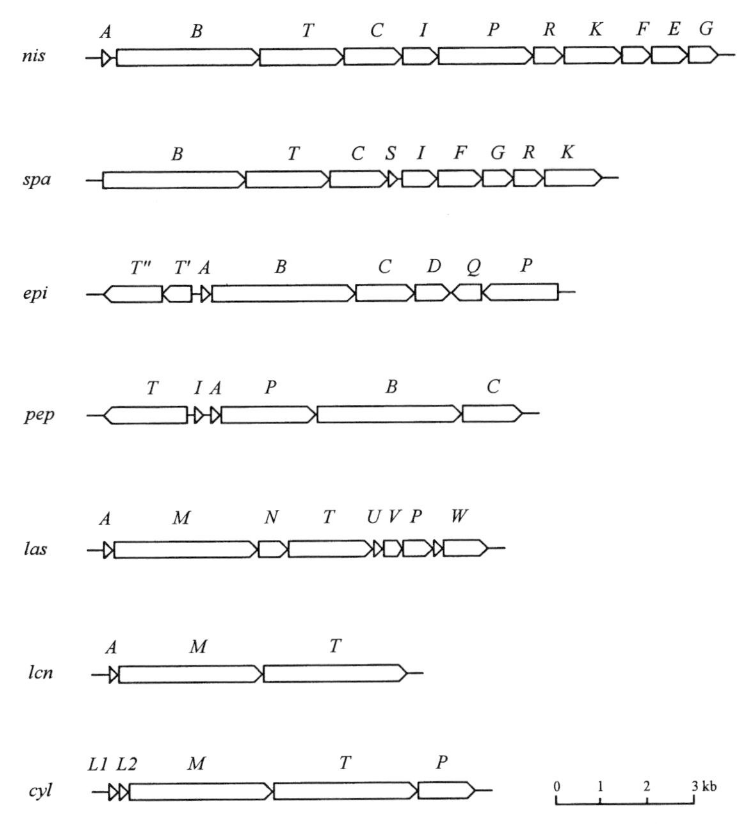

Fig. 10. The arrangement of the gene clusters involved in the biosynthesis of the the lantibiotics nisin (*nis*), subtilin (*spa*), epidermin (*epi*), Pep5 (*pep*), lactocin S (*las*), lactococcin DR (*lcn*), and the enterococcal cytolysin/bacteriocin (*cyl*). In general, homologous genes have been given the same alphabetic suffix. Thus, *A*-genes are the structural genes encoding the prepeptides (with the exception of *spaS*), *B*, *C*, and *M*-genes code for putative modification enzymes, *T*, *E*, and *F*-genes encode ABC superfamily translocator proteins, *P*-genes are responsible for the production of leader peptidases, *R* and *K*-genes produce histidine kinase/response regulator proteins (with the exception of *epiQ* which is equivalent to the *R*-genes of other gene clusters), and *I*-genes encode specific proteins involved in producer self-protection. In addition, the function of the alternatively named genes in the lactocin S-producing gene cluster are unknown. The arrows represent the relative direction of transcription of the respective genes.

propriate size for co-transcription (ca. 5 kb) can be found (SCHNELL et al., 1992; AUGUSTIN et al., 1992). Furthermore, these studies have speculated that a putative terminator between *epiA* and *epiB* may be "leaky", allowing some readthrough and thereby regulating the levels of transcription of *epiB*, *epiC*, and *epiD*. In addition, the start-site for *epiA* transcription along with an appropriate promoter have been identified; the promoter is activated by the regulatory protein EpiQ which probably binds to an inverted repeat located immediately preceding the –35 region (PESCHEL et al., 1993).

Alternatively, the subtilin structural gene, *spaS* (Fig. 10), is located in the middle of the subtilin-producing gene cluster *spaBTCSIRKFG*, preceded by an appropriate transcriptional promoter (BANERJEE and HANSEN, 1988). These authors were also able to map the transcriptional start-site and identify a 0.5 kb mRNA transcript corresponding to *spaS* which had an unusually long half-life of approximately 45 min. In addition, a pair of typical ρ-independent terminator structures have been identified immediately proceeding the structural gene (BANERJEE and HANSEN, 1988; HANSEN et al., 1991). While transcriptional data for the genes downstream of *spaS* has not been reported, the identification of a transcriptional start-site for *spaB* (the first gene in the subtilin-synthesizing gene cluster) has prompted speculation that upstream genes are produced as a large polycistronic mRNA transcript (CHUNG and HANSEN, 1992; CHUNG et al., 1992; HANSEN, 1993).

PepA, the structural gene for Pep5 (Fig. 10) can be localized to the 18 kbp plasmid pED503, harbored within *Staphylococcus epidermidis* strain 5 (ERSFELD-DREßEN et al., 1984; KALETTA et al., 1989). The gene cluster responsible for Pep5 production consists of *pepTIAPBC*, all of which, with the exception of the putative transporter gene (*pepT*), appear to be essential for Pep5 production and immunity (BIERBAUM et al., 1994; MEYER et al., 1995). Both *pepI* (responsible for immunity) and *pepA* are preceded by promoters, however, while both can be transcribed independently, PepI is not produced in the absence of *pepA* (REIS et al., 1994). In addition, REIS et al. (1994) were able to identify a weak ρ-independent terminator immediately after *pepA* and suggested that, analogous to the situation observed in the epidermin-synthesizing gene cluster, read-through may occur and regulate downstream transcription. This view would be further supported by the observation that the proceeding ORF, *pepP*, appears to lack a promoter. Recent analysis of the genes responsible for production of epilancin K7 have suggested that it also has a similar arrangement of genes to that observed for the Pep5-producing gene cluster except that no immunity gene (*elkI*) could be identified between *elkT* and *elkA*, the structural gene (VAN DE KAMP et al., 1995b).

The structural genes for a number of other lantibiotics including the bacteriocin/cytolysin of enterococci, lactococcin DR (also called lacticin 481), SA-FF22 and lactocin S have also been identified and sequenced (GILMORE et al, 1994; RINCE et al., 1994; PIARD et al., 1993; HYNES et al., 1993; SKAUGEN, 1994). The structural genes for pre-lactococcin DR, *lcnDR1* (Fig. 10), was located on a 70 kbp plasmid in *Lactococcus lactis* ssp. *lactis* DR along with two other genes (RINCE et al., 1994). Interestingly, this study showed that only *lcnDR1* and *lcnDR2* were required for the production of lactococcin DR, while the 3rd gene *lcnDR3* which has strong homology with transport proteins was dispensable. Similar genes to *lcnDR2*, designated *lasM* and *cylM* have been located in the gene clusters responsible for production of lactocin S and cytolysin/bacteriocin, respectively (SKAUGEN, 1994; GILMORE et al., 1994). In addition, *lasM* is surrounded by a number of additional ORFs which do not appear to have homology with known proteins, one (or more) of which may encode the novel enzyme responsible for catalyzing the stereospecific conversion of Dha to D-Ala (SKAUGEN, 1994). Analysis of the DNA surrounding *scnA*, the structural gene encoding pre-streptococcin A-FF22, has revealed that *scnA* is preceded by a putative terminator and proceeded by an inverted repeat which might act either as a transcriptional terminator or as an mRNA processing site (HYNES et al., 1993).

So-far at least, there is a distinct paucity of information concerning the genetic elements responsible for the production of type B lantibiotics and only the structural genes for cinnamycin and mersacidin have yet been reported (ENTIAN and KALETTA, 1991; KALETTA et al., 1991a; BIERBAUM et al., 1995). In the case of cinnamycin, the structural gene (*cinA*) was identified, cloned and sequenced from the chromosome of *Streptoverticillium griseoverticillatum* (ENTIAN and KALETTA, 1991; KALETTA et al., 1991a) while, the mersacidin structural gene (*mrsA*) was identified, cloned and sequenced from the producing strain *Bacillus subtilis* HIL Y–85,54728 (BIERBAUM et al., 1995). In both cases it is apparent that the products (pre-cinnamycin and pre-mersacidin) are quite different from the structural gene products of the type A lantibiotics; both have extremely long leader peptides with cleavage sites resembling those found in signal sequences for secreted proteins. So-far at least, no information concerning additional genes involved in the biosynthesis of the type B lantibiotics appears to have been reported.

4.2.2 The General Structure of Prepeptides

In general, the information currently available concerning the structure of the lantibiotic prepeptides has been deduced from their gene sequences. Overall, lantibiotic prepeptides (Fig. 11) consist of two domains: a leader peptide and a propeptide domain, surrounding an appropriate processing site. In addition, the leader peptide domains are generally acidic while the propeptide domains are generally basic or neutral, both domains have predicted helical propensity and are separated by a predicted turn structure encompassing the cleavage site which may allow the processing protease access to this region (JUNG, 1991a, b; JACK et al., 1995; SAHL et al., 1995). Exceptions to these observations can be found in the two type B lantibiotic prepeptide sequences so-far identified (ENTIAN and KALETTA, 1991; KALETTA et al., 1991a; BIERBAUM et al., 1995); the leader peptides are considerably longer and do not contain the structural features identified for the type A lantibiotics (Fig. 11).

Physical evidence for the structure of the pre-type A lantibiotics has come from the isolation of pre-Pep5 from the producing strain *Staphylococcus epidermidis* strain 5 (WEIL et al., 1990). In this study, pre-Pep5 was isolated, albeit in very small quantities due to the apparent raid turnover within the cell, and shown to contain modified amino acids only within the propeptide domain, even though the leader peptide contains hydroxyl amino acids which could potentially be dehydrated. Similar results have subsequently been obtained with pre-nisin (VAN DER MEER et al., 1993). In addition, using a mutant *S. epidermidis* strain 5 it has subsequently been possible to isolate prepeptides in different stages of dehydration, however, the leader peptide alone was never detected, suggesting it is rapidly destroyed after cleavage (SAHL et al., 1991).

Chemical synthesis of appropriate pre-lantibiotics has also provided useful insight into the structure of their fully unmodified form. Both the propeptide and leader peptide domains (as well as segments of these) of gallidermin and Pep5 have been synthesized, as well as the complete pre-gallidermin (BECK-

Fig. 11. The primary structure of several representative lantibiotic prepeptides, i.e., the primary translation product of the respective structural genes. The sequences are shown centered around the cleavage site for leader peptide processing.

SICKINGER and JUNG, 1991). Circular dichroism analysis of these segments confirmed that the leader peptide formed predominantly helical structures. In addition, both synthetic pre-gallidermin and naturally isolated unmodified pre-Pep5 showed stronger helical propensity than either of their respective domains alone, suggesting that the prepeptide helix may be stabilized by interaction between the pro- and leader peptide domains (BECK-SICKINGER and JUNG, 1991; SAHL et al., 1991).

Similarly, both pre-nisin and a large number of their fragments have been synthesized and studied (BYCROFT et al., 1991; SUROVOY et al., 1992). Initial results from the NMR analysis of pre-nisin in water suggests that it is highly flexible with little or no preference for particular conformations (FREUND and JUNG, 1992). In addition, synthetic pre-nisin has been shown to stoichiometrically bind Zn^{2+} ions, an event which may act to stabilize the conformation of the prepeptide and may have ramifications in other biosynthetic reactions (SUROVOY et al., 1992).

4.2.3 The Possible Role(s) of the Leader Peptide

A number of roles for the leader peptide have been suggested and should be considered. Firstly, the leader peptide may serve to keep the prepeptide in an inactive form within the cell. This view may be supported by the observations that leader peptide cleavage is generally the last step in lantibiotic biosynthesis and that fully modified but unprocessed pre-nisin and pre-pep5 are not biologically active (WEIL et al., 1990; SAHL et al., 1991; VAN DER MEER et al., 1994). Alternatively, it may be that the leader peptide signals transport of the lantibiotic out of the cell. However, the leader peptides of lantibiotics share little similarity with the signal peptides of the *sec*-dependent export systems (PUGSLEY, 1993) and the identification of specific transport systems of the ABC-superfamily (FATH and KOLTER, 1993) within lantibiotic-producing gene clusters would suggest that *sec*-dependent transport is not used, although the leader peptide might still direct the immature lantibiotic to the identified transporters. Finally, the conserved properties of the lantibiotic leader peptides might suggest that they contain specific sequences or motifs which direct the biosynthetic enzymes to the appropriate site for amino acid modification. Alternatively, they may stabilize the conformation of the propeptide domain such that the biosynthetic machinery can access it and carry out specific posttranslational modifications (JUNG, 1991a, b).

Two recent studies may support this last proposal. Firstly, VAN DER MEER et al. (1994) have shown that a number of lantibiotics have conserved motifs at particular positions among their respective leader peptides. In this study, they identified the conserved residues Phe–18, Asn or Asp–17, Leu–16, Asp or Glu–15, Ser–10, Asp-7, Ser-6, Pro–2, and Arg or Gln–1 and systematically set about altering these residues in the structural gene for nisin by site-directed mutagenesis. Alterations at Arg–1 resulted in failure of the protease to cleave the leader peptide while, somewhat surprisingly, alterations to the highly conserved Pro–2 had no effect on either biosynthesis or processing. Similarly, changing the Ala–4 resulted in production of unprocessed, mature nisin, suggesting that the peptide no longer fit the active site of the leader peptidase. Alternatively, exchanges with Asp–7 had little effect while those at Ser–10 resulted in improved production of nisin. However, exchanges in positions –6, –15, –16 and –18 could all be shown to completely inhibit nisin biosynthesis, suggesting that (at least) these residues may be essential for the biosynthesis of this lantibiotic.

Secondly, gene fusion of the nisin propeptide domain to the leader domain for subtilin production resulted in the production of fully modified but unprocessed nisin, suggesting that the biosynthetic machinery of the cell (with the exception of the processing protease) recognized the hybrid protein as a suitable substrate (KUIPERS et al., 1993b). Alternatively, a similar hybrid produced by fusion of the nisin leader domain to the subtilin propeptide region was neither matured nor processed in *Bacillus subtilis* and only a leader peptide domain which contained the first 7 amino acids of the subtilin leader fused to the remaining 17 amino acids from the nisin leader sequence was sufficient to get production of subtilin (RINTALA et al., 1993). Taken together, these results seem to suggest that there may be structural motifs in the leader peptide which aid in creating a suitable substrate for subsequent modifications.

4.3 Amino Acid Modifying Proteins

4.3.1 LanB/LanC and LanM Proteins

The most obvious and striking feature of the structure of the lantibiotics is their content of the non-protein amino acids lanthionine and 3-methyllanthionine. Since neither codons nor tRNAs for these amino acids are found in the cell, they must arise by the action of specific enzymes acting on the precursor amino acids found in the prepeptide. Isolation and genetic analysis of the gene clusters responsible for the production of a number of lantibiotics has revealed the presence of a number of open reading frames; although the role of most can be predicted by homology with other proteins, several genes have been identified which have no homology with proteins of known function and may be the gene products responsible for the formation of the novel amino acids found in the lantibiotics (SCHNELL et al., 1991; AUGUSTIN et al., 1991; KALETTA et al., 1991). Indeed it has been clearly shown by gene disruption and complementation of both *spaB* and *spaC* as well as with *epiB* and *epiC* that lantibiotic biosynthesis is dependent on the production of these proteins (KLEIN et al., 1992). In addition, the appearance of subtilin in the culture supernatant correlates directly with the expression of SpaB (GUTOWSKI-ECKEL et al., 1994).

The *lanB* genes so-far characterized encode proteins of approximately 1000 amino acids (Fig. 10) which are principally hydrophilic, but which also have several hydrophobic regions which might indicate membrane spanning domains. Indeed, using antibodies raised to NisB, ENGELKE et al. (1992) were able to demonstrate that NisB associates with the membranes of fractionated nisin-producing cells, however they were not able to determine whether NisB was an integral membrane protein or merely associated loosely with this fraction. Similar results have subsequently been obtained with SpaB analysis, suggesting that the site of lantibiotic biosynthesis might be organized and localized to the cytoplasmic membrane (GUTOWSKI-ECKEL et al., 1994).

The deduced size of the proteins produced by the *lanC* genes (Fig. 10) is somewhat smaller at around 450 amino acids (SAHL et al., 1995) and a number of conserved residues can be identified within their sequences. Although localization studies have not yet been reported, the LanC proteins appear to consist of regularly alternating regions of hydrophobic and hydrophilic nature (ENGELKE et al., 1992; GUTOWSKI-ECKEL et al., 1994).

Recently a number of lantibiotic-producing gene clusters have been identified which do not contain either the *lanB* or *lanC* genes, including lactococcin DR (also known as lacticin 481), lactocin S and the enterococcal cytolysin/bacteriocin (RINCE et al., 1994; PIARD et al., 1993; SKAUGEN, 1994; GILMORE et al., 1994). Interestingly, these gene clusters contain instead a different gene, designated *lanM* (Fig. 10), which share homology in their C-terminal region with the *lanC* genes. Although they do not appear to have any homology with the *lanB* gene products, it is possible that the LanM proteins represent hybrid modifying proteins, able to carry out the functions of both LanB and LanC. This hypothesis may be further supported by the observation that the leader peptides of the respective lantibiotics synthesized by *lanM*-containing gene clusters are quite different to those synthesized by *lanB/LanC*-containing operons (Fig. 10). What must still be investigated is the molecular mechanism(s) involved in the catalysis of dehydration and thioether ring formation, regardless of whether or not it is carried out by a LanB/LanC or LanM system or even by some hitherto unidentified protein.

4.3.2 EpiD

The gene cluster responsible for the biosynthesis of epidermin contains the gene *epiD* (Fig. 10), which is not found in any of the other lantibiotic-producing operons (AUGUSTIN et al., 1991, 1992; SCHNELL et al., 1991, 1992); similarly, only epidermin and its structural analog gallidermin possess the C-terminal modified amino acid (S)-[(*Z*)-2-aminovinyl]-D-cysteine (Figs. 1 and 4; ALLGAIER et al., 1986; KELLNER et al., 1988). Furthermore, expression of *epiD* has been shown to be essential for epidermin production (SCHNELL et al., 1991, 1992; AUGUSTIN et al., 1991, 1992). Subsequently, the protein EpiD has been extensively studied and represents the first example of a novel enzyme responsible for amino acid modification isolated from a lantibiotic-producing strain. KUPKE et al. (1992) were able to overexpress EpiD as a fusion protein with maltose-binding protein (MBP) and recover the EpiD after removal of the MBP; analysis of EpiD revealed that is a 21 kDa flavoprotein enzyme which requires the cofactor flavinmononucleotide (FMN) and is able to catalyze the oxidative decarboxylation of the C-terminal cysteine residue of pre-epidermin (Fig. 12). Subsequently, both purified pre-epidermin (KUPKE et al., 1993) and synthetic pro-epidermin have been used as substrates to characterize the catalytic activity of this enzyme by mass spectrometry (KUPKE et al., 1994). From this study it could be concluded that EpiD is responsible for, at least, the oxidation of the C-terminal Cys residue of epidermin (the decarboxylation shown in Fig. 12 might occur simultaneously and that, since EpiD acted both on pre-epidermin and the synthetic propeptide domain, its activity is not "directed" via the leader peptide region of the prepeptide form.

4.4 Transport Proteins

Genes which are likely to encode proteins responsible for the transport of lantibiotics out of the cell (designated *lanT* genes) have so-far been identified in each of the characterized, lantibiotic-producing operons (Fig. 10). Interestingly, all belong to the superfamily of transport complexes known as the ATP-binding cassette (ABC) transporters (HIGGINS, 1992; FATH and KOLTER, 1993). In general terms, ABC transporters consist of a dimeric complex consisting of two duplicated domains, one domain of each monomer containing the cytoplasmic ATP-binding region characterized by a consensus Gly-Xaa-Gly-Lys-Ser-Thr sequence (where Xaa indicates any amino acid) and the other a membrane spanning region. In some cases, each domain may be encoded by separate genes, although, with the exception of *epiT'T"*, all of the characterized lantibiotic transporters are encoded by a single gene encoding both domains (FATH and KOLTER, 1993; SAHL et al., 1995).

In the case of epidermin production, it is not clear that *epiT'T"* is a functional translocator since, a frame shift would be required to transcribe both proteins. Furthermore, heterologous expression of epidermin in *Staphylococcus carnosus* suggests that transport is not dependent on *epiT'T"* and that host-encoded transport systems can efficiently substitute for this putative transport system (SCHNELL et al., 1992). Similarly, BIERBAUM et al. (1994) were able to show that disruption of *pepT* reduced yields of Pep5 by about 90% compared to the wild type. The observations of both of these studies indicate that the gene cluster-encoded transporters in these staphylococci may well be complemented, at least in part, by other cellular transporters. Similar results have been obtained with other transporters of the ABC-superfamily, which are generally thought to have some degree of flexibility in their substrate specificity (FATH and KOLTER, 1993). In direct contrast, interruption of the *spaT* gene results in abolition of the production of subtilin by *Bacillus subtilis* (CHUNG et al., 1992; KLEIN and ENTIAN, 1994).

4.5 Leader Peptidases (LanP)

Three genes encoding putative peptidases (Fig. 10) which are probably responsible for the cleavage of the leader peptide of the respective lantibiotics have so far been se-

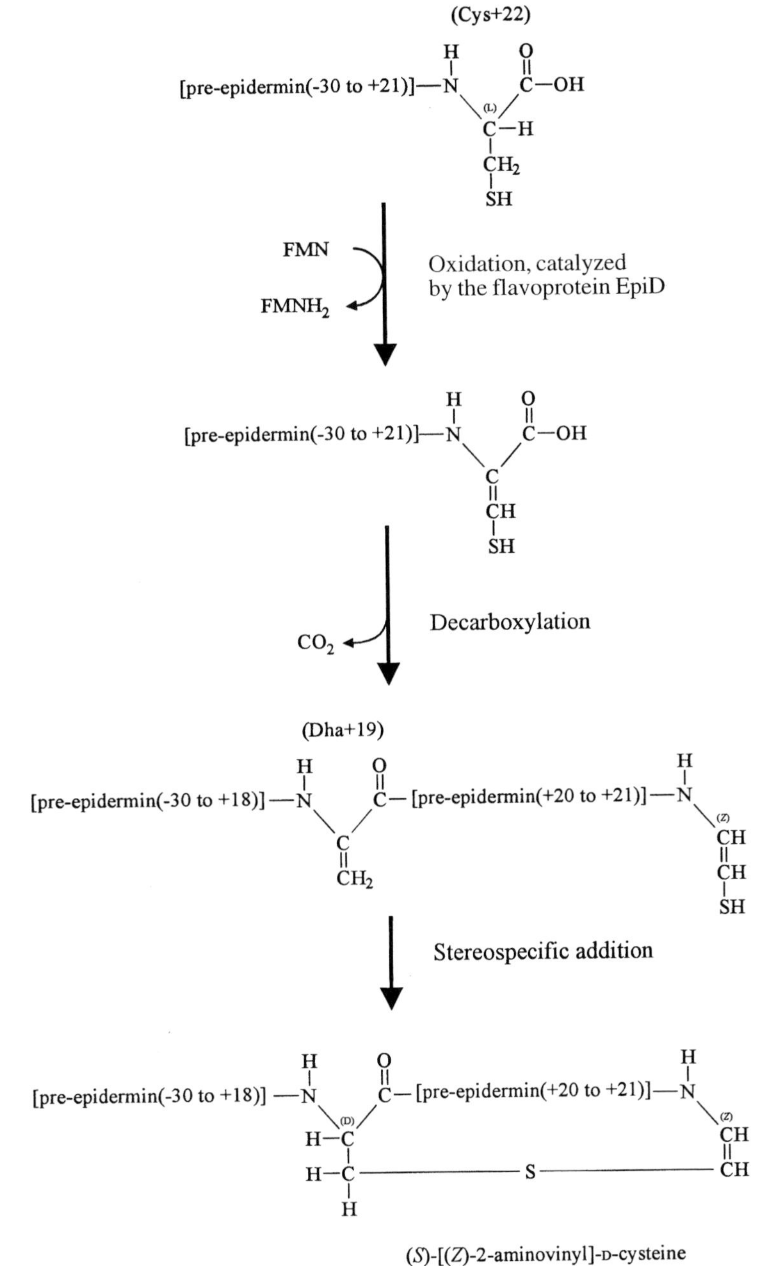

Fig. 12. Proposed pathway for the formation of the aminovinyl cysteine residue at the C-terminus of epidermin by the flavoprotein enzyme EpiD. The chirality of the respective residues is indicated (where appropriate) above the α-carbon atom. FMN, flavinmononucleotide; $FMNH_2$, reduced FMN.

quenced: *nisP* (VAN DER MEER et al., 1993; ENGELKE et al., 1994), *epiP* (SCHNELL et al., 1992), and *pepP* (MEYER et al., 1995), and VAN DE KAMP et al. (1995b) have identified the partial sequence of *elkP*, probably responsible for cleavage of the leader peptide of epilancin K7. Each of the proteins shares homology with subtilisin-like serine proteases, showing conserved active-site residues and a putative oxyanion hole.

NisP (Fig. 10) encodes a putative 74.7 kDa protein which, when expressed in *E. coli*, can be detected as a 54 kDa protein, consistent with processing of a prepropeptide to generate a mature protease (VAN DER MEER et al., 1993). Furthermore, NisP contains a C-terminal extension not found on most other proteases of this class and which could serve as a membrane anchor, suggesting that NisP is both exported from the cell and anchored to the outside of the cytoplasmic membrane. This observation would suggest that cleavage of the nisin leader peptide should be the ultimate step in maturation, a conclusion confirmed since modified but unprocessed nisin can be found in the culture media after growth of a NisP-deficient mutant of a nisin-producing *Lactococcus lactis* (VAN DER MEER et al., 1993). In addition, this study showed that NisP produced in *E. coli* was able to cleave this accumulated unprocessed nisin to release the mature, biologically active form.

The putative protease required for processing of epidermin (EpiP) is also encoded within the epidermin-synthesizing gene cluster (Fig. 10) and is a 461 amino acid protein with calculated mass of 51 kDa (SCHNELL et al., 1992). Interestingly, heterologous expression of epidermin in *Staphylococcus carnosus* is not dependent on the presence of a functional *epiP* gene, suggesting that host-encoded protease(s) are capable of substituting for EpiP (AUGUSTIN et al., 1992). Furthermore, no equivalent protease has been found in the subtilin-synthesizing gene cluster, suggesting that an intrinsic *Bacillus subtilis*-encoded protease is able to process immature subtilin (SAHL et al., 1995).

By contrast, *pepP* (Fig. 10), encoding the leader peptidase involved in Pep5 maturation, encodes a smaller protease which is devoid of a preprosequence and is therefore not exported from the cell (MEYER et al., 1995). Thus the site of leader peptide cleavage for Pep5 must be intracellular and is therefore not the ultimate step in lantibiotic maturation in *Staphylococcus epidermidis* 5. In addition PepP lacks some of the conserved residues typical of subtilisin-like serine proteases however, this study showed, using both gene disruption and site-directed mutagenesis of specific residues, that PepP is essential for processing of the immature Pep5. Similarly, the partial sequence of *elkP* suggests that epilancin K7 processing is carried out inside the cell, since the putative peptidase lacks a preprosequence required for export (VAN DEN KAMP et al., 1995b).

4.6 Proteins Regulating Lantibiotic Biosynthesis

Except in the cases of subtilin and mersacidin production, lantibiotics are generally produced most abundantly during logarithmic growth when energy sources are at their maximum, suggesting constitutive production as opposed to regulated biosynthesis (SAHL and BRANDIS, 1981; HÖRNER et al., 1990; DE VUYST and VANDAMME, 1991; JACK and TAGG, 1992); however, this may not be the case since the lantibiotic-producing gene clusters appear to encode specific regulatory elements (Fig. 10). Like the transport proteins which are members of a larger superfamily of transport proteins, so too the gene products responsible for the regulation of lantibiotic biosynthesis appear to part of a large group of proteins with similar function, that of the two-component response/regulatory elements. In general (MSADEK et al., 1993), these regulatory elements are made up of two proteins the first of which is a membrane-bound histidine kinase able to respond to an extracellular signal by autophosphorylation of a specific histidine residue in its cytoplasmic domain. Subsequently, the phosphate group is transferred to the second component, an intracellular regulatory protein which is usually a transcriptional activator, able to bind DNA. Genes with homology to these two-component regulators have been found in the gene clusters encoding biosynthesis of both nisin and subtilin (KLEIN et al., 1993; KUIPERS et al., 1993a; VAN DER MEER et al., 1993; ENGELKE et al., 1994). Both the insertional inactivation of *spaRK* followed by complementation of the *sapR* (KLEIN et al., 1993) as well as deletion of *nisR* (VAN DER MEER et al., 1993) have been used to demonstrate that both of these genes are essential for regulation of subtilin and nisin biosynthesis, respectively.

However, so-far at least, none of the respective proteins have been studied in detail, nor have potential activator binding regions been identified in the gene clusters.

In contrast, considerable study of the regulation of epidermin production, both at the DNA and protein level, has been carried out. The gene cluster responsible for epidermin biosynthesis (Fig. 10) encodes EpiQ, which appears to be equivalent to the regulator element and shares significant homology with both SpaR and NisR (AUGUSTIN et al., 1992; SCHNELL et al., 1992; PESCHEL et al., 1993). While a respective histidine kinase has not been identified, overexpression of EpiQ leads to increased epidermin production and the gene appears to be essential for heterologous expression of epidermin in *Staphylococcus carnosus*, suggesting that this heterologous host may provide a suitable response element. In addition, the purified EpiQ has also been studied; EpiQ is a 25 kDa DNA-binding transcriptional activator able to bind to one or more of several putative operator sites upstream of the *epiA* promoter (PESCHEL et al., 1993).

Interestingly, no response/regulator gene pair has yet been identified associated with the gene cluster responsible for Pep5 biosynthesis which is carried on the plasmid pED503 in *Staphylococcus epidermidis* 5 (ERSFELD-DREßEN et al., 1984; MEYER et al., 1995). However, a vector containing the genes *epiABCDQ*, which are insufficient for epidermin production in *Staphylococcus carnosus*, converted a strain of *S. epidermidis* 5 devoid of pED503 into an epidermin producer (AUGUSTIN, 1991). This gene cluster is insufficient for epidermin production in *S. carnosus* since *epiQ* is transcribed from the same promoter as one of the deleted genes (*epiP*; Fig. 10). In addition, a chromosomal fragment isolated from *S. epidermidis* 5 and containing a response/regulator could complement this vector and give epidermin production in *S. carnosus* (AUGUSTIN, 1991). Thus, it may be that since these chromosomal genes from *S. epidermidis* 5 can regulate heterologous epidermin production, they may also be capable of directing the biosynthesis of Pep5.

What remains unclear is: What is the external "signal" to which the response regulatory elements are responding? In the case of nisin, DE VUYST and VANDAMME (1991, 1992) have suggested that nisin production may be associated with carbon source control, since nisin production and sucrose metabolism phenotypes are both encoded on the same transposon and that high phosphate levels could stimulate nisin biosynthesis (DE VUYST and VANDAMME, 1993). However, HÖRNER et al. (1990) demonstrated that high phosphate was detrimental for production and that the carbon source had not effect on biosynthesis of both epidermin and Pep5. Thus, the signal remains enigmatic however, lantibiotic biosynthesis seems, in general, programmed to occur when there is an abundance of energy-providing substrate in the growth medium.

4.7 Producer Self-Protection ("Immunity") Mechanisms

Clearly, lantibiotic synthesizing cells are producing substances which are potentially detrimental to their own continued well-being, however, such strains generally show a high degree of resistance to the action of their own lantibiotic. Recent studies have shown that this resistance is provided by specific producer self-protection (immunity) mechanisms. A number of early studies using co-elimination or co-transfer demonstrated that lantibiotic production and immunity were linked (ERSFELD-DREßEN et al., 1984; GASSON, 1984; TAGG and WANNAMAKER, 1978). More recently, the genes encoding specific immunity-related polypeptides have been identified (Fig. 10) at least in the producers of nisin, subtilin, and Pep5 (KUIPERS et al., 1993a; KLEIN and ENTIAN, 1994; REIS and SAHL, 1991; REIS et al., 1994). Interestingly, these studies have shown that, while there is a great deal of similarity between SpaI and NisI, the producing strains do not share cross immunity with each other, nor are they immune to the effects of Pep5. Indeed, cross immunity between lantibiotic producers has only been demonstrated where the lantibiotics produced can be considered natural variants of one another, such as in the case of nisin A and nisin Z or epidermin and [Val1Ile]-epidermin (DE VOS et al., 1993; SAHL et al., 1995).

PepI, the immunity protein responsible for protection against the action of Pep5, is a 69 amino acid peptide with a strongly hydrophilic C-terminal region and a strongly hydrophobic N-terminal domain; expression of PepI of which is absolutely essential for the immune phenotype (REIS and SAHL, 1991; REIS et al., 1994). These studies also showed that production of the immune phenotype is dependent on co-expression of both *pepA* and *pepI* and that even partial deletions in *pepA*, the structural gene, resulted in complete abolition of immunity, although the reasons for this are not apparent. In addition, since PepI appears to be accessible to exogenously added proteases, it is likely to be located outside the membrane, perhaps loosely associated with it (REIS et al., 1994). Furthermore, in this study it was observed that the membranes of cells expressing PepI were not depolarized by Pep5, suggesting that PepI and Pep5 might interact with PepI preventing formation of transmembrane pores. However, analysis of the possible *in vitro* interactions between synthetic PepI and Pep5 was unable to demonstrate any direct interaction between these two peptides (SUROVOY and JUNG, unpublished data; SAHL et al., 1995).

In contrast to PepI, the immunity proteins NisI and SpaI (Fig. 6) are considerably larger at 245 and 163 amino acids, respectively (KUIPERS et al., 1993a; ENGELKE et al., 1994). In addition, both are typical of bacterial lipoproteins, possessing both an N-terminal signal sequence and a membrane-anchoring Cys residue immediately proceeding the cleavage site. Moreover, both have relatively hydrophilic C-terminal domains, suggesting that this region might be exposed outside the cytoplasmic membrane. At least in the case of NisI, partially immune phenotypes are possible, however, as was observed for PepI, full immunity to nisin was only achieved when both *nisA* and *nisI* were co-expressed (KUIPERS et al., 1992; REIS and SAHL, 1991; REIS et al., 1994). In addition, expression of NisI in *E. coli* with disrupted outer membranes protected the cells from the action of exogenous nisin (KUIPERS et al., 1993a).

Potential additional mechanism(s) involved in immunity to lantibiotics have been identified in both nisin-producing and subtilin-producing bacterial strains. Recently, SIEGERS and ENTIAN (1995) have identified the genes *nisE, nisF* and *nisG* in *Lactococcus lactis* 6F3 and shown that disruption of these genes leads to increased sensitivity of the producing strain to exogenous nisin. Analysis of the gene sequences obtained suggest that NisE and NisF together form another ABC-transport system but, this transporter is apparently not involved in the transport of nisin during maturation. In addition, NisE/NisF shares homology with McbE/McbF, the transporter reported to be responsible for immunity to microcin B17 in certain strains of *E. coli* (GARRIDO et al., 1988). NisG on the other hand is an hydrophobic protein with significant similarities to many of the colicin immunity proteins, which are thought to interact directly with the channel-forming colicins (PUGSLEY, 1988; SONG and CRAMER, 1991).

Similar results have also been obtained with analysis of subtilin immunity determinants (except that the NisE/NisF homolog in *Bacillus subtilis* is called spaF/SpaG), prompting speculation that the E/F and G-proteins of these two strains are involved in providing an additional degree of protection to that generated by the immunity proteins NisI and SpaI (KLEIN and ENTIAN, 1994; SIEGERS and ENTIAN, 1995). However, how these proteins might achieve this is anything but clear. Observations of similarities between NisE/NisF and the microcin B17 immunity proteins McbE/McbF further confuse the issue because this bacteriocin/colicin apparently has an intracellular target, specifically inhibiting DNA gyrase (GARRIDO et al., 1988). However, the primary mode of action of nisin and subtilin involves depolarization of the cytoplasmic membrane through the formation of voltage-dependent pores (see also Sect. 5). Thus, some authors have speculated that the additional transport systems might act either to transport external nisin into the cell for degradation or transport internalized subtilin out of the cell, thus generating the observed additional degree of immunity (KLEIN and ENTIAN, 1994; SIEGERS and ENTIAN, 1995). In any case, it is now apparent that multiple systems are involved in self-protection of *Lactococcus lactis* and *Bacillus subtilis* against nisin and subtilin, respectively. So-far no reports of such additional immunity mechanism(s) have been given for other type A lantibiotic-producing strains.

4.8 The Chain of Events Leading to Lantibiotic Biosynthesis and Maturation

From an historical point of view it is interesting to note that even at the same time as the structures of nisin and subtilin were being determined (GROSS and MORRELL, 1971; GROSS et al., 1973) other researchers were proposing that these lantibiotics were synthesized by posttranslational modification. Initially, HURST (1966) showed that nisin biosynthesis was prevented by inhibitors of protein synthesis. On the basis of the incorporation of radiolabeled amino acids INGRAM (1969, 1970) suggested that didehydroamino acids should result from dehydration of Ser and Thr and that the thioether rings might arise from addition of the sulphydryl groups of Cys residues. Later, NISHIO et al. (1983) used anti-subtilin antibodies to isolate precursor subtilin and then used cell extracts from subtilin-producing *B. subtilis* to convert this precursor into a substance with the same activity and electrophoretic mobility as natural subtilin. However, the isolation of the structural genes encoding the precursor peptides for epidermin, nisin, and subtilin provided the final evidence that both didehydroamino acid and thioether rings arose from modifications to genetically encoded Ser, Thr, and Cys residues (SCHNELL et al., 1988; BANERJEE and HANSEN, 1988; BUCHMANN et al., 1988; KALETTA and ENTIAN, 1989).

Subsequently the isolation of pre-peptides of Pep5 and nisin has shed a great deal of light on the sequence and subcellular localization of modification events required for lantibiotic biosynthesis. Cytoplasmic pre-Pep5 has been shown to consist of a mixture of dehydrated forms (6-fold dehydrated, 5-fold dehydrated, etc.) which do not contain Lan/MeLan rings (WEIL et al., 1990; SAHL et al., 1991). Thus, it could be concluded that the primary translation product has a very short half-life, site-specific dehydration is the first step in biosynthesis and that thioether rings are formed in a separate step. In the case of nisin biosynthesis VAN DER MEER et al. (1993) showed that nisin secretion is the penultimate step in processing and that the last step involves cleavage of the leader peptide. However, this sequence of events may not hold for all lantibiotics since Pep5 (at least) appears to be processed inside the cell with transport the ultimate step in production of this lantibiotic (MEYER et al., 1995).

Several modification reactions are likely to occur spontaneously, including the formation of the N-terminal 2-oxopyruvyl and 2-oxobutyryl groups of lactocin S and Pep5, respectively (MØRTVEDT et al., 1991; SKAUGEN et al., 1994; KELLNER et al., 1989). These reactions would require the removal of the leader peptide and should therefore proceed processing. Similarly, since the N-terminal hydroxypyruvyl group at the N-terminus of epilancin K7 probably requires enzymatic reduction, this reaction must proceed processing and may constitute one of the last steps in biosynthesis of this lantibiotic (VAN DE KAMP et al., 1995a, b; SAHL et al., 1995). Alternatively, since EpiD is able to act on the C-terminal Cys residue of both pre-epidermin, synthetic pro-epidermin, and other synthetic peptides not related to epidermin, oxidative decarboxylation may well precede thioether ring formation during Cys(Avi) biosynthesis (SCHNELL et al., 1988; KUPKE et al., 1992, 1993, 1994, 1995).

5 Biological Activities of Lantibiotics

5.1 Type A Lantibiotics

5.1.1 Primary Mode of Action

As earlier mentioned, the type A lantibiotics have been defined as those lantibiotic peptides whose mode of action is principally directed towards the killing of bacterial cells (JUNG, 1991a, b). In general their activity is confined to gram-positive bacteria. However, provided that the outer membrane is first disrupted, several type A lantibiotics have been shown to also act against a number of gram-negative bacteria, including *E. coli* and *Salmonella* spp. (STEVENS et al., 1991). Thus, it would appear that the lipid-rich outer membrane protects gram-negative bacteria from type A lantibiotic action, probably by preventing access to the inner membrane, the site of their action. In addition, the range of bacteria affected by the different type A lantibiotics varies considerably. Some, such as salivaricin A exhibit a relatively limited antimicrobial spectrum (ROSS et al., 1993), while others such as nisin inhibit a broad range of gram-positive bacteria (e.g., many strains of micrococci, streptococci, lactococci, pediococci, staphylococci, lactobacilli, *Listeria* spp., and mycobacteria) as well as both the vegetative cells and spores of *Bacillus* spp. and *Clostridium* spp. (HURST, 1981; DELVES-BROUGHTON, 1990; MOLITOR and SAHL, 1991; BIERBAUM and SAHL, 1993; DE VUYST and VANDAMME, 1993).

Historically, it is interesting that while nisin has been used in biopreservation for more than 35 years, the mechanism by which it and other type A lantibiotics kill bacterial cells has only recently been clarified (HURST, 1982; MOLITOR and SAHL, 1991; JACK et al., in press; SAHL et al., 1995). In early studies, RAMSEIER (1960) suggested that nisin might be acting as a detergent both because of its highly basic p*I* and the observation that nisin treatment of cells induced leakage of UV-absorbing intracellular constituents. Later, GROSS and MORRELL (1971) determined the structure of nisin and observed that the unsaturated amino acids Dha and Dhb found in nisin might be able to interact with the sulphydryl groups of specific enzymes within a cell, while other studies suggested that nisin could interfere with the biosynthesis of the bacterial cell wall (LINNETT and STROMINGER, 1973; REISINGER et al., 1980). Subsequently, it has been suggested that type A lantibiotics such as nisin kill cells by interfering with energy transduction (SAHL, 1985); this conclusion was based on the observation that nisin treatment of sensitive cells resulted in the cessation of most macromolecular biosyntheses (e.g., protein, DNA, RNA, and polysaccharides) and that nisin activity was dependent on external factors such as pH, temperature and the phase of growth of the target cells (HURST, 1981; SAHL and BRANDIS, 1981; SAHL, 1991). Based on these studies, subsequent analysis of the mechanism of action of the type A lantibiotics has revealed much about the way in which they exert their antimicrobial activity.

Treatment of susceptible bacterial cells with micromolar concentrations of type A lantibiotics such as nisin, Pep5, subtilin, epidermin, gallidermin, or streptococcin A-FF22 arrests amino acid uptake and induces rapid efflux of preaccumulated amino acids (SAHL and BRANDIS, 1983; RUHR and SAHL, 1985; SCHÜLLER et al., 1989; SAHL, 1991; BENZ et al., 1991; JACK et al., 1994b) and, at least in some cases, has also been shown to cause the efflux of the potassium analog Rb^+ (SAHL and BRANDIS, 1983; RUHR and SAHL, 1985). In addition, following Pep5-treatment of cells, ATP could be found in the external medium (SAHL and BRANDIS, 1983); since there are no known transport systems for ATP, these results, along with the observed efflux of other low molecular-weight intracellular macromolecules, suggest that type A lantibiotics form discrete pores in the cytoplasmic membrane of susceptible bacteria. This mechanism is in contrast to the generalized disruption that might be expected from the action of a surfactant. In addition, both ATP and Rb^+ efflux as well as the arrest of macromolecule biosynthesis is effected by the growth phase of the target cells (SAHL and BRANDIS, 1983; SAHL, 1991; JACK et al., 1994b), suggesting that the pore formation occurs in an energy-dependent manner.

Further evidence for the mode of action of type A lantibiotics come from experiments with cytoplasmic membrane vesicles. Artificially-energized vesicles, which have been treated with type A lantibiotics, rapidly efflux preaccumulated radiolabeled amino acids; by contrast, pre-treatment of the vesicles (prior to energization) induced little or no efflux until after the cells were sufficiently energized (SAHL, 1985; RUHR and SAHL, 1985; SAHL et al., 1987; Kordel et al., 1988; Schüller et al., 1989; Jack et al., 1994b). These results further demonstrate that type A lantibiotics form transmembrane pores in an energy-dependent fashion and allow efflux of preaccumulated intracellular components. However, these same studies also showed that lantibiotic treatment of either whole cells, artificial vesicles which had been energized with valinomycin-induced potassium gradients or artificially energized liposomes (Gao et al., 1991; Abee et al., 1991) resulted in dissipation of the membrane potential, suggesting that the pores formed are non-specific and allow influx of extracellularly accumulated protons and (probably) other ions and small molecules. The dissipation of the membrane potential accounts for the observed arrest of energy-dependent macromolecular biosynthesis and is different to that of the protonophores since pore formation is energy-dependent, potentiated by the membrane potential (Sahl, 1991; Garcia-Garcera et al., 1993).

Further support for the prediction that type A lantibiotics kill cells by disruption of energy transduction through the formation of energy-dependent pores in the cytoplasmic membrane have been provided by analysis of the pores formed in artificial bilayers such as the black-lipid membranes (BLM); in addition these studies have allowed the measurement of several of the physical properties of the formed pores (Benz et al., 1978, 1991). Artificial BLMs can be formed across a small whole separating two chambers or wells in a teflon block, both of which are filled with a conducting salt buffer solution. Using electrodes, it is then possible to apply a defined potential difference across an artificial bilayer and measure lantibiotic-induced current flows through the normally insulative model membrane. In such a system, the type A lantibiotics can be shown to induce discrete pores since, application of a sufficiently high potential induces current flow; reduction of the potential allows the pores to close and restores the insulative properties of the bilayer, indicating that membrane disruption is not generalized (Sahl et al., 1987; Kordel et al., 1988; Schüller et al., 1989; Benz et al., 1991; Jack et al., 1994b). In addition, these studies have also shown that nisin and Pep5 form pores only in the presence of a *trans*-negative membrane potential, while subtilin, epidermin, gallidermin, and streptococcin A-FF22 form pores irrespective of the orientation of the applied potential. Analysis of the current voltage curves obtained from these studies also allowed determination of the threshold potential for lantibiotic-induced pore formation; epidermin and gallidermin required ca. 50 mV, Pep5, nisin and subtilin require ca. 80 mV, while streptococcin A-FF22 required ca. 100 mV.

Further physical properties, including the mean diameter and the lifetime of type A lantibiotic channels can be determined from the analysis of single channels formed in BLMs. If the pore is assumed to be a cylinder with a length equivalent to the thickness of the membrane which is filled with the same solution as bathes the membrane (of known conductance), then from the observed conductance it is possible to calculate the diameter of the pore. Streptococcin A-FF22 pores appear to be relatively unstable, appearing as millisecond time scale "bursts" and a with a mean diameter of about 0.5–0.6 nm (Jack et al., 1994b). Alternatively, nisin and Pep5 pores are somewhat more stable (tens to hundreds of milliseconds) and have a diameter of ca. 1 nm (Sahl et al., 1987; Kordel et al., 1988), while subtilin pores are larger again at ca. 2 nm diameter (Schüller et al., 1989). Epidermin and gallidermin pores are somewhat different; their mean lifetimes may extend up to 30 s and their diameter appears to increase with the applied potential (Benz et al., 1991).

The determination of many of the physical properties of the type A lantibiotic peptides has allowed some understanding of how they might form pores in the bacterial cytoplasmic membrane. In general, type A lantibiotics have both sufficient length and a sufficiently high dipole moment to be consistent with voltage-dependent channel formation in phospholipid bilayers (Freund et al., 1991a b; Benz et al., 1991) as well as the central flexible region thought to be essential for stabilization of the transmembrane structure (Vogel et al., 1993). In addition, NMR studies have shown that the peptides themselves appear to form amphiphilic helical conformations in solution and present their hydrophobic residues on one "face" of the helix and the charged, hydrophilic residues on the opposite "face" (FREUND et al., 1991a, b, c; GOODMAN et al., 1991; LIAN et al., 1991; PALMER et al., 1989; SLIJPER et al., 1989; VAN DE VEN et al., 1991a, b). Clearly, a single lantibiotic molecule is insufficient to induce pore formation and multiple peptides must somehow coalesce in the bilayer to form an aggregate with the properties observed above. In addition, since type A lantibiotics can form pores in artificial vesicles and bilayers, it is clear they have no requirements for specific membrane-associated "receptors" as has been suggested for several other channel-forming antibacterial peptides including lactococcin A and pediocin PA–1 (VAN BELKUM et al., 1991; CHIKINDAS et al., 1993).

Although the exact mechanism remains unclear, current models (Fig. 13) for type A lantibiotic-induced pore formation suggest that the peptides accumulate at the cytoplasmic membrane, perhaps attracted to the bilayer by ionic interactions (SAHL, 1991; BENZ et al., 1991; JACK et al., 1994b). In the absence of a membrane potential they should remain oriented lateral to the membrane but apparently with their hydrophobic "face" intimately associated with the bilayer (SCHÜLLER et al., 1989; SAHL, 1991; BENZ et al., 1991; JACK et al., 1994b; DRIESSEN et al., 1995); on application of a sufficiently high membrane potential the peptides apparently adopt a transmembrane orientation and create a pore. However, a number of questions remain unanswered; e.g., it remains to be seen whether the peptides aggregate before or after adoption of the transmembrane orientation, how many pores are the minimum required for transmembrane conductance and which terminus of the peptide inserts through the membrane. In addition, the observations that pore diameters fluctuate and the short pore lifetimes observed with most type A lantibiotics suggest that the number of peptide involved in pore formation is not static, but rather more dynamic.

5.1.2 Secondary Mode of Action

In addition to forming pores in phospholipid bilayers, both Pep5 and nisin have been shown to induce autolysis in *Staphylococcus simulans* cells (BIERBAUM and SAHL, 1991). Characterization of the mechanism by which they achieve this suggests that cationic lantibiotics such as nisin and Pep5 are able to competitively release cell wall autolytic enzymes normally bound to, and regulated by, polyanionic cell wall constituents such as lipoteichoic-, teichoic-, and teichuronic acids, probably by an ion exchange-like mechanism (SAHL, 1985; BIERBAUM and SAHL, 1987,

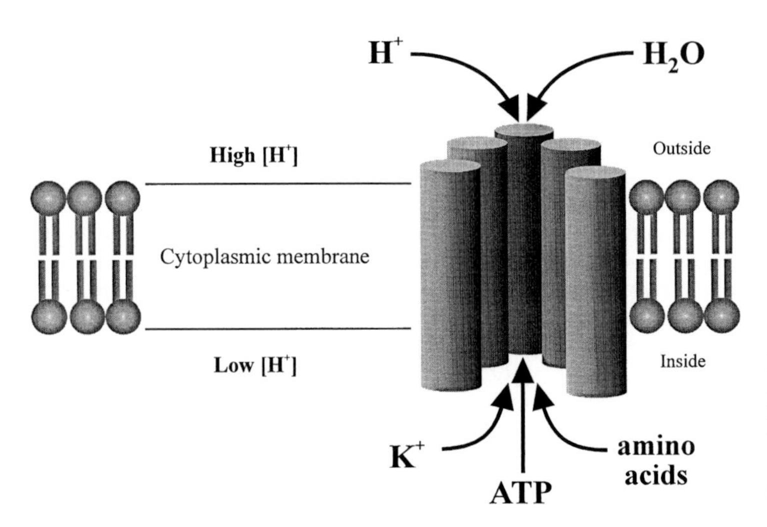

Fig. 13. Model for the mode of action of the type A lantibiotics, involving the formation of voltage-dependent pores in the cytoplasmic membrane of sensitive bacteria.

1988, 1991). In addition, it has been shown that activation of the autolytic enzymes occurred most markedly in the area of the septa between dividing daughter cells (BIERBAUM and SAHL, 1985, 1991) and that the activation was not specific since similar results could be obtained using synthetic cationic peptides, such as poly-lysine or poly-arginine (BIERBAUM and SAHL, 1991). Since only small solutes may pass outward (Fig. 13) pores formed by these lantibiotics in the cytoplasmic membrane should allow an influx of water into the cell, increasing osmotic pressure. This increase in intracellular pressure combined with the interference in energy transduction (resulting from depolarization of the membrane) which should prevent repair of the weakened cell wall, probably results in the observed lysis (BIERBAUM and SAHL, 1991).

5.2 Type B Lantibiotics

5.2.1 Mersacidin and Actagardine

The type B lantibiotics mersacidin and actagardine also kill bacterial cells rather efficiently, however, their primary mechanism of activity appears to be vastly different to that described above for the type A lantibiotics, directed primarily at the cell wall rather than the cytoplasmic membrane. Mersacidin acts principally against streptococci and staphylococci; this particular type B lantibiotic has demonstrated *in vivo* activity against methicillin-resistant *Staphylococcus aureus* strains and could therefore offer an alternative to vancomycin treatment of such infections (CHATERJEE et al., 1992b). Alternatively, actagardine is primarily effective against obligate anaerobes and streptococci and has been used experimentally to treat *Streptococcus pyogenes* infections (ARIOLI et al., 1976; MALABARBA et al., 1990).

In the late 1970s, SOMMA et al. (1977) showed that low concentrations of actagardine inhibited the incorporation of N-acetylglucosamine and L-alanine into *Bacillus subtilis* cell wall peptidoglycan. In these experiments it was observed that other intermediates also failed to be incorporated into cross-linked peptidoglycan in the actagardine-treated bacilli and isolation of a labeled intermediate (UDP-acetylmuramyl-N-acetylglucosamine pentapeptide linked to a C55-isoprenylphosphate carrier) suggested that actagardine treatment was able to inhibit the transfer of cell wall precursor to the peptidoglycan receptor during cell wall biosynthesis.

When staphylococci are incubated in the presence of mersacidin their growth is first inhibited, following which the cells begin to lyse; cessation of growth is accompanied by an inability to incorporate D-alanine into the cell wall and by a marked decrease in the thickness of the cell wall (BRÖTZ et al., 1995). In addition, these authors observed that mersacidin treatment had no effect on other macromolecular biosynthetic processes such as DNA, RNA, and protein biosynthesis, in direct contrast to the effects observed with similar cells treated with type A lantibiotics. Furthermore, whereas both mersacidin and the glycopeptide antibiotic vancomycin have approximately the same MIC and both act on cell wall biosynthesis, mersacidin was not inhibited by the tripeptide L-Lys-D-Ala-D-Ala which is a potent inhibitor of vancomycin action. This last result suggests that, while the molecular target of mersacidin is unclear, it differs from that of vancomycin.

5.2.2 Cinnamycin and the Duramycins

Among other type B lantibiotics, at least cinamycin and duramycin have been shown to inhibit the growth of *Bacillus* spp.; treated cells show increased membrane permeability, undergo a reduction in ATP-dependent protein translocation and ATP-dependent calcium uptake, show marked slowing in the rate of chloride uptake and show significantly reduced potassium and sodium ATPase activities (RACKER et al., 1983; STONE et al., 1984; NAVARRO et al., 1985; CHEN and TAI, 1987). Duramycin-resistant bacilli remain unaffected and have been shown to possess cytoplasmic membranes with altered phospholipid compositions; whereas sensitive cells contain phosphatidyl ethanolamine, resistant cells do not (NAVARRO et al., 1985; DUNKLEY et al., 1988). These results suggest that duramycin recognizes and interacts with specific phospholipids in the cytoplasmic membrane of sensitive cells and that alterations to this composition, while conferring duramycin resistance, do not seem to effect other membrane-associated cellular activities (DUNKLEY et al., 1988; CLEJAN et al., 1989).

In addition to inhibiting the growth of bacilli, duramycin also inhibits a number of metabolic properties of isolated mitochondria and specifically increases the permeability of the inner membrane, the major component of which is phosphatidyl ethanolamine (SOKOLOVE et al., 1989). Similarly, cinnamycin has been shown to induce hemolysis in isolated erythrocytes and phosphatidyl ethanolamine-containing liposomes; cinnamycin had no effect on erythrocytes if it was first incubated with phosphatidyl ethanolamine or on liposomes which were prepared from phosphatidyl inositol, phosphatidyl serine, or cardiolipin (CHOUNG et al., 1988a, b). The latter results suggest that cinnamycin interacts with phospholipids containing both a glyceryl backbone and a primary amino group. Using electron-spin resonance, these studies also showed that the interaction of cinnamycin and phosphatidyl ethanolamine results in reorganization of the membrane bilayer. Taken together, these results have further confirmed the direct interaction of type B lantibiotics with specific phospholipids present in the membrane bilayer.

Recently, SHETH et al. (1992) have suggested that the increases in membrane permeability associated with duramycin treatment could result from the formation of ion channels. They observed that both the membranes of cultured epithelial cells and artificial BLMs treated with duramycin showed complex conductance states, some of which were discrete, and that pores were weakly anion selective. In addition, this study showed that pore diameter increased with time, perhaps explaining duramycin-induced increases in membrane permeability.

Duramycin, duramycin B, duramycin C, and cinnamycin may also act as immunoregulators as they are able to influence the activity of phospholipase A2, an important enzyme of the immune system involved in the production of prostaglandins and leucotrienes (MÄRKI and FRANSON, 1986; FREDENHAGEN et al., 1990, 1991; MÄRKI et al., 1991). The principal substrate of phospholipase A2 is phosphatidyl ethanolamine; interaction of these type B lantibiotics with the substrate renders it unavailable for lipolysis, thus interfering with immune function.

6 Applications of Lantibiotics

6.1 Applications as a Food/Beverage Preservative

Since the type A lantibiotic nisin has been used as a biopreservative for at least 35 years, there is currently a great deal of information regarding its application to foods and beverages for which readers are referred to the excellent reviews of HURST (1981), DELVES-BROUGHTON (1990), MOLITOR and SAHL (1991), and DE VUYST and VANDAMME (1993). However, it is worth briefly mentioning that already this peptide has been utilized in a number of areas including: fish and low-temperature processed canning, brewing, winemaking, cheese and other fermented dairy product manufacture as well as in the preservation of cosmetics and deodorants. Nisin has proved particularly useful in a number of these areas because of its characteristics. Nisin, e.g., demonstrates good activity against a number of food-borne pathogens and spoilage bacteria including *Clostridium botulinum, Listeria monocytogenes, Lactobacillus* spp. and *Leuconostoc* spp. In addition, the peptide is particularly heat and acid stable making it particularly suitable for addition to products with low pH (e.g., fermented products) and products which subsequently undergo some sort of pasteurization.

6.2 Medical/Paramedical and Veterinary Applications

Interestingly, some of the lantibiotics may also prove to be useful in medicinal preparations for both human and animal application. The type A lantibiotics epidermin and gallidermin have previously been shown to be active against *Propionibacterium acnes*, the causative agent of the acnes disease. These observations have led to speculation that these lantibiotics might be used in the external treatment of acnes as a replacement therapy for the currently used erythromycin/vitamin A creams. Such application of a lantibiotic should have several advantages: (1) so-far at least, acquired resistance to these lantibiotics has not been observed, (2) because of their size and nature, the peptides should not be absorbed into the body and they should be of low toxicity, and (3) it should be possible to develop large-scale, low cost production schemes for these bacterially-derived peptides (ALLGAIER et al., 1991; JUNG, 1991a, b; UNGERMANN et al., 1991). Similarly, the notable acid stability of nisin (HURST, 1981) has led several recent studies to suggest that it may have efficacy in the treatment of gastric ulcers, due to its antimicrobial activity against *Helicobacter pylori* (DE VUYST and VANDAMME, 1993). In addition, alternative suggested applications of nisin include: as a mouthrinse for the prevention of gingivitis and plaque or as a germicidal preparation for the prevention of mastitis in dairy cattle (SEARS et al., 1992; HOWELL et al., 1993).

More significantly perhaps, the cinnamycin-group type B lantibiotics have been shown to effect human immune function, leucocyte proliferation and may play a role in protection against *Herpes simplex* virus (MÄRKI and FRANSON, 1986; FREDENHAGEN et al., 1990, 1991; MÄRKI et al., 1991); thus the potentials for application of these compounds in human health care are obvious. In addition, the observation that mersacidin displays antibacterial activity against methicillin-resistant *Staphylococcus aureus* strains as well as multidrug resistant enterococci by novel mechanisms offers the greatest hope for the development of new therapeutic agents to combat the threat of these potentially fatal microorganisms (BRÖTZ et al., 1995).

7 Conclusions and Future Perspectives

From the biotechnological standpoint then, the lantibiotics represent an exciting class of peptides for two reasons. Firstly, as antimicrobial compounds they may offer not only new opportunities in areas such as food and beverage preservation (HURST, 1981; DELVES-BROUGHTON, 1990; MOLITOR and SAHL, 1991; DE VUYST and VANDAMME, 1993), but may also represent (at least in the case of mersacidin and gallidermin) novel therapeutic agents with potential for medicinal use (BRÖTZ et al., 1995; SAHL et al., 1995). Secondly, it is clear that the lantibiotics are produced by novel biosynthetic mechanisms; the enzymes which are capable of transforming the 20 "protein" amino acids into such highly modified forms may well prove useful for the generation of new peptide compounds and mimetics, previously only produced through time-consuming and costly in-lab chemical syntheses (SAHL et al., 1995; JACK et al., in press; JACK and SAHL, 1995).

Thus, greater understanding of the biosynthetic mechanisms involved in lantibiotic production may one day allow us to introduce non-peptide amino acids and enantiomeric amino acids into useful proteins, perhaps improving their stability, resistance to proteolytic activity or even their biological and catalytic activity. The continued study of novel peptides and biosynthetic enzymes, such as those involved in lantibiotic production, give the highest prospects for the development of new, useful, peptide-based compounds through biotechnology.

8 References

ABEE, T., GAO, F. H., KONINGS, W. N. (1991), The mechanism of the lantibiotic nisin in artificial membranes, in: *Nisin and Novel Lantibiotics* (JUNG, G., SAHL, H.-G., Eds.), pp. 373–385. Leiden: ESCOM Scientific Publishers BV.

ALLGAIER, H., JUNG, G., WERNER, R. G., SCHNEIDER, U., ZÄHNER, H. (1986), Epidermin: sequencing of a heterodet tetracyclic 21-peptide amide antibiotic, *Eur. J. Biochem.* **160**, 9–22.

ALLGAIER, H., WALTER, J., SCHLÜTER, M., WERNER, R. G. (1991), Strategy for the purification of lantibiotics, in: *Nisin and Novel Lantibiotics* (JUNG, G., SAHL, H.-G., Eds.), pp. 422–433. Leiden: ESCOM Scientific Publishers BV.

ARIOLI, V., BERTI, M., SILVESTRI, L. G. (1976), Gardimycin, a new antibiotic from *Actinoplanes*. III. Biological properties, *J. Antibiot.* **29**, 511–515.

AUGUSTIN, J. (1991), Identifizierung und Regulation der Biosynthesegene des Lantibiotikums Epidermin aus *Staphylococcus aureus* Tü 3298. *PhD Thesis*, University of Tübingen, Germany.

AUGUSTIN, J., ROSENSTEIN, R., KUPKE, T., SCHNEIDER, U., SCHNELL, N., ENGELKE, G., ENTIAN, K.-D., GÖTZ, F. (1991), Identification of epidermin biosynthetic genes by complementation studies and heterologous expression, in: *Nisin and Novel Lantibiotics* (JUNG, G., SAHL, H.-G., Eds.), pp. 277–286. Leiden: ESCOM Scientific Publishers BV.

AUGUSTIN, J., ROSENSTEIN, R., WIELAND, B., SCHNEIDER, U., SCHNELL, N., ENGELKE, G., ENTAIN, K.-D., GÖTZ, F. (1992), Genetic analysis of epidermin biosynthetic genes and epidermin-negative mutants of *Staphylococcus epidermidis, Eur. J. Biochem.* **204**, 1149–1154.

BANERJEE, S., HANSEN, J. N. (1988), Structure and expression of a gene encoding the precursor of subtilin, a small peptide antibiotic, *J. Biol. Chem.* **263**, 9508–9514.

BECK-SICKINGER, A. G., JUNG, G. (1991), Synthesis and conformational analysis of lantibiotic leader-, pro- and pre-peptides, in: *Nisin and Novel Lantibiotics* (JUNG, G., SAHL, H.-G., Eds.), pp. 218–230. Leiden: ESCOM Scientific Publishers BV.

BENEDICT, R. G., DVONCH, W., SHOTWELL, O. L., PRIDHAM, T. G., LINDENFELSER, L. A. (1952), Cinnamycin, an antibiotic from *Streptomyces cinnamonensis* nov. sp, *Antibiot. Chemother.* **2**, 591–594.

BENZ, R., JANKO, K., BOOS, W., LÄUGER, P. (1978), Formation of large ion-permeable membrane Channels by the matrix protein (porin) of *Escherichia coli, Biochim. Biophys. Acta* **511**, 305–319.

BENZ, R., JUNG, G., SAHL, H.-G. (1991), Mechanism of channel formation by lantibiotics in black lipid membranes, in: *Nisin and Novel Lantibiotics* (JUNG, G., SAHL, H.-G., Eds.), pp. 359–372. Leiden: ESCOM Scientific Publishers BV.

BERRIDGE, N. J., NEWTON, G. G., ABRAHAM, E. P. (1952), Purification and nature of the antibiotic nisin, *Biochem. J.* **52**, 529–535.

BIERBAUM, G., SAHL, H.-G. (1985), Induction of autolysis of staphylococci by the basic peptide antibiotics Pep5 and nisin and their influence on the activity of autolytic enzymes, *Arch. Microbiol.* **141**, 249–254.

BIERBAUM, G., SAHL, H.-G. (1987), Autolytic system of *Staphylococcus simulans* 22: influence of cationic peptides on activity of *N*-acetylmuramoyl-L-alanine amidase, *J. Bacteriol.* **169**, 5452–5458.

BIERBAUM, G., SAHL, H.-G. (1988), Influence of cationic peptides on the activity of the autolytic endo-β-*N*-acetylglucosaminidase of *Staphylococcus simulans* 22, *FEMS Microbiol. Rev.* **58**, 223–228.

BIERBAUM, G., SAHL, H.-G. (1991), Induction of autolysis of *Staphylococcus simulans* 22 by Pep5 and nisin and influence of the cationic peptides on the activity of the autolytic enzymes, in: *Nisin and Novel Lantibiotics* (JUNG, G., SAHL, H.-G., Eds.), pp. 386–396. Leiden: ESCOM Scientific Publishers BV.

BIERBAUM, G., SAHL, H.-G. (1993), Lantibiotics – unusually modified bacteriocin-like peptides from Gram-positive bacteria, *Zentralbl. Bakteriol.* **278**, 1–22.

BIERBAUM, G., REIS, M., SZEKAT, C., SAHL, H.-G. (1994), Construction of an expression system for engineering of the lantibiotic Pep5, *Appl. Environ. Microbiol.* **60**, 4332–4338.

BIERBAUM, G., BRÖTZ, H., KOLLER, K.-P., SAHL, H.-G. (1995), Cloning, sequencing and production of the lantibiotic mersacidin, *FEMS Microbiol. Lett.* **127**, 121–126.

BROCK, T. D., DAVIE, J. M. (1963), Probable identity of a group D hemolysin with a bacteriocine, *J. Bacteriol.* **86**, 708–712.

BRÖTZ, H., BIERBAUM, G., MARKUS, A., MOLITOR, E., SAHL, H.-G. (1995), Mode of action of mersacidin – inhibition of peptidoglycan synthesis via a novel mechanism? *Antimicrob. Agents Chemother.* **39**, 714–719.

BUCHMAN, W. B., BANERJEE, S., HANSEN, J. N. (1988), Structure, expression and evolution of a gene encoding the precursor of nisin, a small protein antibiotic, *J. Biol. Chem.* **263**, 16260–16266.

BYCROFT, B. W., CHAN, W. C., ROBERTS, G. C. K. (1991), Synthesis of pro- and prepeptides related to nisin and subtilin, in: *Nisin and Novel Lantibiotics* (JUNG, G., SAHL, H.-G., Eds.), pp. 204–217. Leiden: ESCOM Scientific Publishers BV.

CHAN, W. C., LIAN, L.-Y., BYCROFT, B. W., ROBERTS, G. C. K. (1989a), Isolation and characterization of two degradation products derived from the peptide antibiotic nisin, *FEBS Lett.* **252**, 29–36.

CHAN, W. C., LIAN, L.-Y., BYCROFT, B. W., ROBERTS, G. C. K. (1989b), Confirmation of the complete structure of nisin by complete ^{1}H-NMR resonance assignments in aqueous and dimethylsulphoxide solution, *J. Chem. Soc. Perkin Trans.* **1**, 2359–2367.

CHAN, W. C., BYCROFT, B. W., LEYLAND, M. L., LIAN, L.-Y., YANG, J. C., ROBERTS, G. C. K. (1992), Sequence-specific resonance assignment and conformational analysis of subtilin by 2D-NMR, *FEBS Lett.* **300**, 56–62.

CHAN, W. C., BYCROFT, B. W., LEYLAND, M. L., LIAN, L.-Y., YANG, J. C., ROBERTS, G. C. K. (1993), A novel posttranslational modification of the peptide antibiotic subtilin: isolation and characterization of a natural variant from *Bacillus subtilis* ATCC 6633, *Biochem. J.* **291**, 23–27.

CHATERJEE, S., CHATERJEE, S., LAD, S. J., PHANSALKAR, M. S., RUPP, R. H., GANGULI, B. N. FEHLHABER, H.-W., KOGLER, H. (1992a), Mersacidin, a new antibiotic from *Bacillus*: fermentation, isolation, purification and chemical characterization, *J. Antibiot.* **45**, 832–838.

CHATERJEE, S., CHATERJEE, D. K., JANI, R. H., BLUMBACH, J., GANGULI, B. N. KLESEL, N., LIMBERT, M., SEIBERT, G. (1992b), Mersacidin, a new antibiotic from *Bacillus*: *in vitro* and *in vivo* antibacterial activity, *J. Antibiot.* **45**, 839–845.

CHEN, L. L., TAI, P. C. (1987), Effects of antibiotics and other inhibitors on ATP-dependent protein translocation into membrane vesicles, *J. Bacteriol.* **169**, 2372–2379.

CHIKINDAS, M. L., GARCIA-GARCERA, M. J., DRIESSEN, A. J. M., LEDERBOER, A. M., NISEEN-MEYER, J., NES, I. F., ABEE, T., KONING, W. N., VENEMA, G. (1993), Pediocin PA-1, a bacteriocin from *Pediococcus acidilactici* PAC1.0 forms hydrophilic pores in the cytoplasmic membrane of target cells, *Appl. Environ. Microbiol.* **59**, 3577–3584.

CHOUNG, S.-Y., KOBAYASHI, T., INOUE, J., TAKEMOTO, K., ISHITSUKA, H., INOUE, K. (1988a), Haemolytic activity of a cyclic peptide Ro 09–0198 isolated from *Streptoverticillium, Biochim. Biophys. Acta* **940**, 171–179.

CHOUNG, S.-Y., KOBAYASHI, T., TAKEMOTO, K., ISHITSUKA, H., INOUE, K. (1988b), Interaction of a cyclic peptide, Ro 09–0198, with phosphatidylethanolamine in liposomal membranes, *Biochim. Biophys. Acta* **940**, 180–187.

CHUNG, Y.-J., HANSEN, J. N. (1992), Determination of the sequence of *spaE* and identification of a promoter in the subtilin (*spa*) operon in *Bacillus subtilis, J. Bacteriol.* **174**, 6699–6702.

CHUNG, Y.-J., STEEN, M. T., HANSEN, J. N. (1992), The subtilin gene of *Bacillus subtilis* ATCC 6633 is encoded in an operon that contains a homologue of the hemolysin B transport protein, *J. Bacteriol.* **174**, 1417–1422.

CLEJAN, S., GUFFANTI, A. A., COHEN, M. A., KRULWICH, T. A. (1989), Mutations of *Bacillus firmus* OF4 to duramycin resistance results in substantial replacement of membrane lipid phosphatidylethanolamine by its plasmalogen form, *J. Bacteriol.* **171**, 1744–1746.

DELVES-BROUGHTON, J. (1990), Nisin and its uses as a food preservative, *Food Technol.* **44**, 100–112.

DEMPSTER, R. P., TAGG, J. R. (1982), The production of bacteriocin-like substances by the oral bacterium *Streptococcus salivarius*, *Arch. Oral Biol.* **27**, 151–157.

DE VOS, W. M., JUNG, G., SAHL, H.-G. (1991), Appendix: definitions and nomenclature of lantibiotics, in: *Nisin and Novel Lantibiotics* (JUNG, G., SAHL, H.-G., Eds.), pp. 457–463. Leiden: ESCOM Scientific Publishers BV.

DE VOS, W. M., MULDERS, J. W. M., SIEZEN, R. J., HUGENHOLTZ, J., KUIPERS, O. P. (1993), Properties of nisin Z and distribution of its gene *nisZ* in *Lactococcus lactis, Appl. Environ. Microbiol.* **59**, 213–218.

DE VUYST, L., VANDAMME, E. J. (1991), Microbial manipulation of nisin biosynthesis and fermentation, in: *Nisin and Novel Lantibiotics* (JUNG, G., SAHL, H.-G., Eds.), pp. 397–409. Leiden: ESCOM Scientific Publishers BV.

DE VUYST, L., VANDAMME, E. J. (1992), Influence of the carbon source on nisin production in *Lactococcus lactis* subspp. *lactis* batch fermentations, *J. Gen. Microbiol.* **138**, 571–578.

DE VUYST, L., VANDAMME, E. J. (1993), Nisin, a lantibiotic produced by *Lactococcus lactis* subspp. *lactis*: properties, biosynthesis, fermentation and applications, in: *Bacteriocins of Lactic Acid Bacteria: Microbiology, Genetics and Applications* (DE VUYST, L., VANDAMME, E., Eds.), pp. 151–221. London: Chapman and Hall.

DODD, H. M, GASSON, M. J. (1994), Bacteriocins of lactic acid bacteria, in: *Genetics and Biotechnology of Lactic Acid Bacteria* (GASSON, M. J., DE VOS, W. M., Eds.), pp. 211–251. London: Chapman and Hall.

DODD, H. M., HORN, N., GASSON, M. J. (1990), Analysis of the genetic determinant for production of the peptide antibiotic nisin, *J. Gen. Microbiol.* **136**, 555–566.

DODD, H. M., HORN, N., SWINDELL, S., GASSON, M. J. (1991), Physical and genetic analysis of the chromosomally located transposon Tn*5301*, responsible for nisin biosynthesis, in: *Nisin and Novel Lantibiotics* (JUNG, G., SAHL, H.-G., Eds.), pp. 231–242. Leiden: ESCOM Scientific Publishers BV.

DRIESSEN, A. J. M., VAN DEN HOOVEN, H. W., KUIPER, W., VAN DE KAMP, M. M., SAHL, H.-G., KONINGS, R. N. H, KONINGS, W. N. (1995), Mechanistic studies of lantibiotic-induced permeabilization of phospholipid vesicles, *Biochemistry* **34**, 1606–1614.

DUNKLEY, E. A., JR., CLEJAN, S., GUFFANTI, A. A., KRULWICH, T. A. (1988), Large decreases in membrane phosphatidylethanolamine and diphosphatidylglycerol upon mutation to duramycin resistance do not change the protonophore resistance of *Bacillus subtilis, Biochim. Biophys. Acta* **943**, 13–18.

ENGELKE, G., GUTOWSKI-ECKEL, Z., HAMMELMANN, M., ENTIAN, K.-D. (1992), Biosynthesis of the lantibiotic nisin: genomic organization and membrane localization of the NisB protein, *Appl. Environ. Microbiol.* **58**, 3730–3743.

ENGELKE, G., GUTOWSKI-ECKEL, Z., KIESAU, P., SIEGERS, K., HAMMELMANN, M., ENTIAN, K.-D. (1994), Regulation of nisin biosynthesis and immuntiy in *Lactococcus lactis* 6F3, *Appl. Environ. Microbiol.* **60**, 814–825.

ENTIAN, K.-D., KALETTA, C. (1991), Isolation and characterization of the cinnamycin structural gene, in: *Nisin and Novel Lantibiotics* (JUNG, G., SAHL, H.-G., Eds.), pp. 303–308. Leiden: ESCOM Scientific Publishers BV.

ERSFELD-DREßEN, H., SAHL, H.-G., BRANDIS, H. (1984), Plasmid involvement in production of and immunity to the staphylococcin-like peptide Pep5, *J. Gen. Microbiol.* **130**, 3029–3035.

FATH, M. J., KOLTER, R. (1993), ABC transporters: bacterial exporters, *Microbiol. Rev.* **57**, 995–1017.

FREDENHAGEN, A., FENDRICH, G., MÄRKI, F., MÄRKI, W., GRUNER, J., RASCHDORF, F., PETER, H. H. (1990), Duramycins B and C, two new lanthionine-containing antibiotics as inhibitors of phospholipase A2, *J. Antibiot.* **43**, 1403–1412.

FREDENHAGEN, A., MÄRKI, F., FENDRICH, G., MÄRKI, W., GRUNER, J., VAN OOSTRUM, J., RASCHDORF, F., PETER, H. H. (1991), Duramycin B and C, two new lanthionine-containing antibiotics as inhibitors of phospholipase A2 and structural revision of duramycin and cinnamycin, in: *Nisin and Novel Lantibiotics* (JUNG, G., SAHL, H.-G., Eds.), pp. 131–140. Leiden: ESCOM Scientific Publishers BV.

FREUND, S., JUNG, G. (1992), Lantibiotics: an overview and conformational studies on gallidermin and Pep5, in: *Bacteriocins, Microcins and Lantibiotics* (JAMES, R., LAZDUNSKI, C., PATTUS, F., Eds.), pp. 75–92. Berlin: Springer-Verlag.

FREUND, S., JUNG, G., GUTBROD, O., FOLKERS, G., GIBBONS, W. A., ALLGAIER, H., WERNER, R. (1991a), The solution structure of the lantibiotic gallidermin, *Biopolymers* **31**, 803–811.

FREUND, S., JUNG, G., GUTBROD, O., FOLKERS, G., GIBBONS, W. A. (1991b), The three-dimensional solution structure of gallidermin determined by NMR-based molecular graphics, in: *Nisin and Novel Lantibiotics* (JUNG, G., SAHL, H.-G., Eds.), pp. 91–102. Leiden: ESCOM Scientific Publishers BV.

FREUND, S., JUNG, G., GIBBONS, W. A., SAHL, H.-G. (1991c), NMR and circular dichroism studies on Pep5, in: *Nisin and Novel Lantibiotics* (JUNG, G., SAHL, H.-G., Eds.), pp. 103–112. Leiden: ESCOM Scientific Publishers BV.

GAO, T., ABEE, T., KONINGS, W. N. (1991), Mechanism of the peptide antibiotic nisin in liposomes and cytochrome c oxidase-containing proteoliposomes, *Appl. Environ. Microbiol.* **57**, 2164–2170.

GARCIA-GARCERA, M. G. J., ELFERINK, M. G. L., DRIESSEN, A. J. M., KONINGS, W. N. (1993), *In vitro* pore-forming activity of the lantibiotic nisin: role of proton motive force and lipid composition, *Eur. J. Biochem.* **212,** 417–422.

GARRIDO, M. C., HERRERO, M., KOLTER, R., MORRENO, P. (1988), The export of the DNA replication inhibitor microcin B17 provides immunity for the host cell, *EMBO J.* **7,** 1853–1862.

GASSON, M. J. (1984), Transfer of sucrose fermenting ability, nisin resistance and nisin production into *Lactococcus lactis* 712, *FEMS Microbiol. Lett.* **21**, 7–10.

GILMORE, M. S., SEGARRA, R. A., BOOTH, M. C., BOGIE, C. P., HALL, L. R., CLEWELL, D. B. (1994), Genetic structure of the *Enterococcus faecalis* plasmid pAD1-encoded cytolytic toxin system and its relationship to lantibiotic determinants, *J. Bacteriol.* **176**, 7335–7344.

GIREESH, T., DAVIDSON, B. E., HILLIER, A. J. (1992), Conjugal transfer in *Lactococcus lactis* of a 68-kilobase-pair chromosomal fragment containing the structural gene for the peptide bacteriocin nisin, *Appl. Environ. Microbiol.* **58**, 1670–1676.

GOODMAN, M., PALMER, D. E., MIERKE, D., RO, S., NUNAMI, K., WAKAMIYA, T., FUKASE, K.,

HORIMOTO, S., KITAZAWA, M., FUJITA, H., KUBO, A., SHIBA, T. (1991), Conformations of nisin and its fragments using synthesis, NMR and computer simulations, in: *Nisin and Novel Lantibiotics* (JUNG, G., SAHL, H.-G., Eds.), pp. 59–75. Leiden: ESCOM Scientific Publishers BV.

GROSS, E. (1977), α,β-Unsaturated and related amino acids in peptides and proteins, *Adv. Exp. Med. Biol.* **86**, 131–153.

GROSS, E., MORELL, J. L. (1971), The structure of nisin, *J. Am. Chem. Soc.* **93**, 4634–4635.

GROSS, E., KILTZ, H. H., NEBELIN, E. (1973), Subtilin. VI. Die Struktur des Subtilins. *Hoppe-Seyler's Z. Physiol. Chem.* **354**, 810–812.

GUTOWSKI-ECKEL, G., KLEIN, C., SIEGERS, K., BOHM, K., HAMMELMANN, M., ENTIAN, K.-D. (1994), Growth phase-dependent regulation and membrane localization of SpaB, a protein involved in biosynthesis of the lantibiotic subtilin, *Appl. Environ. Microbiol.* **60**, 1–11.

HANSEN, J. N. (1993), Antibiotics synthesized by posttranslational modification, *Ann. Rev. Microbiol.* **47**, 535–564.

HANSEN, J. N., CHUNG, Y. J., LIU, W., STEEN, M. T. (1991), Biosynthesis and mechanism of action of nisin and subtilin, in: *Nisin and Novel Lantibiotics* (JUNG, G., SAHL, H.-G., Eds.), pp. 287–302. Leiden: ESCOM Scientific Publishers BV.

HAYASHI, F., NAGASHIMA, K., TERUI, Y., KAWAMURA, Y., MATSUMOTO, K., ITAZAKI, H. (1990), The structure of PA48009: the revised structure of duramycin, *J. Antibiot.* **43**, 1421–1430.

HIGGINS, C. F. (1992), ABC-transporters: from microorganisms to man, *Ann. Rev. Cell Biol.* **8**, 67–113.

HORN, N., SWINDELL, S., DODD, H. M., GASSON, M. J. (1991), Nisin biosynthesis genes are encoded by a novel conjugative transposon, *Mol. Gen. Genet.* **228**, 129–135.

HÖRNER, T., UNGERMANN, V., ZÄHNER, H., FIEDLER, H.-P., UTZ, R., KELLNER, R., JUNG, G. (1990), Comparative studies on the fermentative production of lantibiotics by staphylococci, *Appl. Microbiol. Biotechnol.* **32**, 511–517.

HOWELL, T. H., FIORELLINI, J. P., BLACKBURN, P., PROJAN, S. J., DE LA HARPE, J., WILLIAMS, R. C. (1993), The effect of a mouthrinse based on nisin, a bacteriocin, on developing plaque and gingivitis in beagle dogs, *J. Clin. Periodontol.* **20**, 335–339.

HURST, A. (1966), Biosynthesis of the antibiotic nisin by whole *Streptococcus lactis* organisms, *J. Gen. Microbiol.* **44**, 209–220.

HURST, A. (1981), Nisin, *Adv. Appl. Microbiol.* **27**, 85–123.

HYNES, W. L., FERRETTI, J. J., TAGG, J. R. (1993), Cloning of the gene encoding streptococcin A-FF22, a novel lantibiotic produced by *Streptococcus pyogenes* and determination of its nucleotide sequence, *Appl. Environ. Microbiol.* **59**, 1969–1971.

INGRAM, L. (1969), Synthesis of the antibiotic nisin: formation of lanthionine and β-methyllanthionine, *Biochim. Biophys. Acta* **184**, 216–219.

INGRAM, L. (1970), A ribosomal mechanism for synthesis of peptides related to nisin, *Biochim. Biophys. Acta* **224**, 263–265.

JACK, R. W., SAHL, H.-G. (1995), Unique postranslational modifications involved in lantibiotic biosynthesis, *Trends Biotechnol.* **13**, 269–278.

JACK, R. W., TAGG, J. R. (1991), Isolation and partial structure of streptococcin A-FF22, in: *Nisin and Novel Lantibiotics* (JUNG, G., SAHL, H.-G., Eds.), pp. 171–179. Leiden: ESCOM Scientific Publishers BV.

JACK, R. W., TAGG, J. R. (1992), Factors affecting the production of the group A *Streptococcus* bacteriocin SA-FF22, *J. Med. Microbiol.* **36**, 132–138.

JACK, R. W., CARNE, A., METZGER, J., STEVANOVIĆ, S., SAHL, H.-G., JUNG, G., TAGG, J. R. (1994a), Elucidation of the structure of SA-FF22, a lanthionine-containing antibacterial peptide produced by *Streptococcus pyogenes* strain FF22, *Eur. J. Biochem.* **220**, 455–462.

JACK, R. W., BENZ, R., TAGG, J. R., SAHL, H.-G. (1994b), The mode of action of SA-FF22 a lantibiotic isolated from *Streptococcus pyogenes* strain FF22, *Eur. J. Biochem.* **219**, 699–705.

JACK, R. W., TAGG, J. R., RAY, B. (1995), Bacteriocins of Gram-positive bacteria, *Microbiol. Rev.* **59**, 171–200.

JACK, R. W., BIERBAUM, G., HEIDRICH, C., SAHL, H.-G. (in press), The genetics of lantibiotic biosynthesis, *BioEssays*.

JANSEN, E. F., HIRSCHMANN, D. J. (1944), Subtilin – an antibacterial product of *Bacillus subtilis*: culturing conditions and properties, *Arch. Biochem.* **4**, 297–304.

JUNG, G. (1991a), Lantibiotics – ribosomally synthesized biologically active polypeptides containing sulphide rings and α,β-didehydroamino acids, *Angew. Chem.* (Int. Edn. Engl.) **30**, 1051–1068.

JUNG, G. (1991b), Lantibiotics: a survey, in: *Nisin and Novel Lantibiotics* (JUNG, G., SAHL, H.-G., Eds.), pp. 1–34. Leiden: ESCOM Scientific Publishers BV.

KALETTA, C., ENTIAN, K.-D. (1989), Nisin, a peptide antibiotic: cloning and sequencing of the *nisA* gene and posttranslational processing of its peptide product, *J. Bacteriol.* **171**, 1597–1601.

KALETTA, C., ENTIAN, K.-D., KELLNER, R., JUNG, G., REIS, M., SAHL, H.-G. (1989), Pep5, a new lantibiotic: structural gene isolation and prepeptide sequence, *Arch. Microbiol.* **152**, 16–19.

KALETTA, C., ENTIAN, K.-D., JUNG, G. (1991a), Prepeptide sequence of cinnamycin (Ro 09–0198): the first structural gene of a duramycin-type lantibiotic, *Eur. J. Biochem.* **199**, 411–415.

KALETTA, C., KLEIN, C., SCHNELL, N., ENTIAN, K.-D. (1991b), An operon-like structure of the genes involved in subtilin biosynthesis, in: *Nisin and Novel Lantibiotics* (JUNG, G., SAHL, H.-G., Eds.), pp. 309–319. Leiden: ESCOM Scientific Publishers BV.

KATZ, E., DEMAIN, A. L. (1977), The peptide antibiotics of *Bacilllus*: chemistry, biogenesis and possible functions, *Bacteriol. Rev.* **41**, 499–474.

KELLNER, R., JUNG, G., HÖRNER, T., ZÄHNER, H., SCHNELL, N., ENTIAN, K.-D., GÖTZ, F. (1988), Gallidermin, a new lanthionine-containing polypeptide antibiotic, *Eur. J. Biochem.* **177**, 53–59.

KELLNER, R., JUNG, G., JOSTEN, M., KALETTA, C., ENTIAN, K.-D., SAHL, H.-G. (1989), Pep5: structure elucidation of a large lantibiotic, *Angew. Chem.* **28**, 616–619.

KELLNER, R., JUNG, G., SAHL, H.-G. (1991), Structure elucidation of the tricyclic lantibiotic Pep5 containing eight positively charged amino acids, in: *Nisin and Novel Lantibiotics* (JUNG, G., SAHL, H.-G., Eds.), pp. 141–158. Leiden: ESCOM Scientific Publishers BV.

KESSLER, H., STEUERNAGEL, S., GILLESSEN, D., KAMIYAMA, T. (1987), Complete sequence determination and localization of one imino and three sulphide bridges of the nonadecapeptide Ro 09–0198 by homonuclear 2D-NMR spectroscopy: the DQF-RELAYED-NOESY-experiment, *Helv. Chim. Acta* **70**, 726–741.

KESSLER, H., STEUERNAGEL, S., WILL, M., JUNG, G., KELLNER, R., GILLESSEN, D., KAMIYAMA, T. (1988), The structure of the polycylic nonadecapeptide Ro 09–0198, *Helv. Chim. Acta* **71**, 1924–1929.

KESSLER, H., SEIP, S., WEIN, T., STEUERNAGEL, S., WILL, M. (1991), Structure of cinnamycin (Ro 09–0198) in solution, in: *Nisin and Novel Lantibiotics* (JUNG, G., SAHL, H.-G., Eds.), pp. 141–158. Leiden: ESCOM Scientific Publishers BV.

KETTENRING, J., MALABARABA, A., VEKEY, K., CAVALLERI, B. (1990), Sequence determination of actagardine, a novel lantibiotic, by homonuclear 2D NMR spectroscopy, *J. Antibiot.* **43**, 1082–1088.

KLEIN, C., ENTIAN, K.-D. (1994), Genes involved in self-protection against the lantibiotic subtilin produced by *Bacillus subtilis* ATCC 6633, *Appl. Environ. Microbiol.* **60**, 2793–2801.

KLEIN, C., KALETTA, C., SCHNELL, N., ENTIAN, K.-D. (1992), Analysis of the genes involved in the biosynthesis of the lantibiotic subtilin, *Appl. Environ. Microbiol.* **58**, 132–142.

KLEIN, C., KALETTA, C., ENTIAN, K.-D. (1993), Biosynthesis of the lantibiotic subtilin is regulated by a histidine kinase/response regulator system, *Appl. Environ. Microbiol.* **59**, 296–303.

KLEINKAUF, H., VON DÖHREN, H. (1986) Peptide antibiotics, in: *Biotechnology*, 1st Edn., Vol. 4 (REHM, H. J., REED, G., Eds.), pp. 283–307, Weinheim: VCH.

KLEINKAUF, H., VON DÖHREN, H. (1987), Biosynthesis of peptide antibiotics, *Ann. Rev. Microbiol.* **41**, 259–289

KLEINKAUF, H., VON DÖHREN, H. (1990), Non-ribosomal biosynthesis of peptide antibiotics, *Eur. J. Biochem.* **192**, 1–15.

KOGLER, H., BAUCH, M., FEHLHABER, H.-W., GRIESINGER, C., SCHUBERT, W., TEETZ, V. (1991), NMR-spectroscopic investigations on mersacidin, in: *Nisin and Novel Lantibiotics* (JUNG, G., SAHL, H.-G., Eds.), pp. 159–170. Leiden: ESCOM Scientific Publishers BV.

KORDEL, M., BENZ, R., SAHL, H.-G. (1988), Mode of action of the staphylococcin-like peptide Pep5: voltage-dependent depolarization of bacterial and artificial membranes, *J. Bacteriol.* **170**, 84–88.

KOZAK, W., RAJCHERT-TRZPIL, M., ZAJDEL, J., DOBZANSKI, W. T. (1974), The effect of proflavin, ethidium bromide and elevated temperature on the appearance of nisin-negative clones in nisin-producing strains of *Streptococcus lactis, J. Gen. Microbiol.* **83**, 295–302.

KUIPERS, O. P., ROLLEMA, H. S., YAP, W. M. G. J., BOOT, H. J., SIEZEN, R. J., DE VOS, W. M. (1992), Engineering dehydrated amino acid residues in the antimicrobial peptide nisin, *J. Biol. Chem.* **267**, 24340–24346.

KUIPERS, O. P., BEERTHUYZEN, M. M., SIEZEN, R. J., DE VOS, W. M. (1993a), Characterization of the nisin gene cluster *nisABTCIPR* of *Lactococcus lactis*: requirement of expression of *nisA* and *nisI* genes for development of immunity, *Eur. J. Biochem.* **216**, 281–292.

KUIPERS, O. P., ROLLEMA, H. S., DE VOS, W. M., SIEZEN, R. J. (1993b), Biosynthesis and secretion of a precursor of nisin Z by *Lactococcus lactis* directed by the leader peptide of the homologous lantibiotic subtilin from *Bacillus subtilis, FEBS Lett.* **330**, 23–27.

KUPKE, T., STEVANOVIĆ, S., SAHL, H.-G., GÖTZ, F. (1992), Purification and characterziation of EpiD, a flavoprotein involved in the biosynthesis of the lantibiotic epidermin, *J. Bacteriol.* **174**, 5354–5361.

Kupke, T., Stevanović, S., Ottenwälder, B., Metzger, J. W., Jung, G., Götz, F. (1993), Purification and characterization of EpiA, the peptide substrate for posttranslational modifications involved in epidermin biosynthesis, *FEMS Microbiol. Lett.* **112**, 43–48.

Kupke, T., Kempter, C., Gnau, V., Jung, G., Götz, F. (1994), Mass spectroscopic analysis of a novel enzymatic reaction: oxidative decarboxylation of the lantibiotic precursor peptide EpiA catalyzed by the flavoprotein EpiD, *J. Biol. Chem.* **269**, 5653–5659.

Kupke, T., Kempter, C., Jung, G., Götz, F. (1995), Oxidative decarboxylation of peptides catalyzed by flavoprotein EpiD: determination of substrate specificity using peptide libraries and neutral loss mass spectrometry, *J. Biol. Chem.* **270**, 11282–11289.

Lian, L.-Y., Chan, W. C., Morley, S. D., Roberts, G. C. K., Bycroft, B. W., Jackson, D. (1991), NMR studies of the solution structure of nisin A, in: *Nisin and Novel Lantibiotics* (Jung, G., Sahl, H.-G., Eds.), pp. 43–58. Leiden: ESCOM Scientific Publishers BV.

Lian, L.-Y., Chan, W. C., Morley, S. D., Roberts, G. C. K., Bycroft, B. W., Jackson, D. (1992), Solution structures of nisin and its two major degradation products determined by NMR, *Biochem. J.* **283**, 413–420.

Linnett, P. E., Strominger, J. L. (1973), Additional inhibitors of peptidoglycan synthesis, *Antimicrob. Agents Chemother.* **4**, 231–236.

Liu, W., Hansen, J. N. (1992), Enhancement of the chemical and antimicrobial properties of subtilin by site-directed mutagenesis, *J. Biol. Chem.* **267**, 25078–25085.

Malabarba, A., Landi, M., Pallanza, R., Cavalleri, B. (1985), Physicochemical and biological properties of actagardine and some acid hydrolysis products, *J. Antibiot.* **38**, 1506–1511.

Malabarba, A., Pallanza, R., Berti, M., Cavalleri, B. (1990), Synthesis and biological activity of some amide derivatives of the lantibiotic actagardine, *J. Antibiot.* **43**, 1089–1097.

Märki, F., Franson, R. (1986), Endogenous suppression of neutral-active and calcium-dependent phospholipase A2 in human polymorphonuclear leucocytes, *Biochim. Biophys. Acta* **879**, 149–156.

Märki, F., Hänni, E., Fredenhagen, A., van Oostrum, J. (1991), Mode of action of the lanthionine-containing peptide antibiotics duramycin, duramycin B, duramycin C and cinnamycin as direct inhibitors of phospholipase A2, *Biochem. Pharmacol.* **42**, 2027–2035.

Meyer, H. E. (1994), Analyzing posttranslational protein modifications, in: *Microcharacterization of Proteins* (Kellner, R., Lottspeich, F., Meyer, H. E., Eds.). Weinheim: VCH.

Meyer, C., Bierbaum, G., Heidrich, C., Reis, M., Süling, J., Iglesias-Wind, M. I., Kempter, C., Molitor, E., Sahl, H.-G. (1995), Nucleotide sequence of the lantibiotic Pep5 biosynthetic gene cluster and functional analysis of PepP and PepC: evidence for a role of PepC in thioether formation, *Eur. J. Biochem.* **232**, 478–489.

Molitor, E., Sahl, H.-G. (1991), Applications of nisin: a literature survey, in: *Nisin and Novel Lantibiotics* (Jung, G., Sahl, H.-G., Eds.), pp. 434–439. Leiden: ESCOM Scientific Publishers BV.

Mørtvedt, C. I., Nes, I. F. (1990), Plasmid-associated bacteriocin production by *Lactobacillus sake, J. Gen. Microbiol.* **136**, 1601–1607.

Mørtvedt, C. I., Nissen-Meyer, J., Sletten, K., Nes, I. F. (1991), Purification and amino acid sequence of lactocin S, a bacteriocin produced by *Lactobacillus sake* L45, *Appl. Environ. Microbiol.* **57**, 1829–1834.

Msadek, T., Kunst, F., Rapoport, G. (1993), Two component regulatory systems, in: *Bacillus subtilis and Other Gram-Positive Bacteria* (Sonenshein, A., Hoch. L., Losick, R., Eds.), pp. 729–746. Washington, DC: ASM.

Mulders, J. W. M., Boerrigter, I. J., Rollema, H. S., Siezen, R. J., de Vos, W. M. (1991), Identification and characterization of the lantibiotic nisin Z, a natural nisin variant, *Eur. J. Biochem.* **201**, 581–584.

Nakano, M. M., Zuber, P. (1990), Molecular biology of antibiotic production in *Bacillus, Crit. Rev. Biotechnol.* **10**, 223–240.

Naruse, N., Tenmyo, O., Tomita, K., Konishi, M., Miyaki, T., Kawaguchi, H., Fukase, K., Wakamiya, T., Shiba, T. (1989), Lanthiopeptin, a new peptide antibiotic. Production, isolation and properties of lanthiopeptin, *J. Antibiot.* **42**, 837–845.

Navarro, J., Chabot, J., Sherrill, K., Aneja, R., Zahler, S. A., Racker, E. (1985), Interaction of duramycin with artificial and natural membranes, *Biochem.* **24**, 4645–4650.

Nishikawa, M., Teshima, T., Wakamiya, T., Shiba, T., Kobayashi, Y., Okubo, T., Kyogoku, Y., Kido, Y. (1988), Chemistry of lantibiotics, in: *Peptide Chemistry 1987* (Shiba, T., Sakakibara, S., Eds.), pp. 71–74. Osaka: Protein Research Foundation.

Nishio, C., Komura, S., Kurahashi, K. (1983), Peptide antibiotic subtilin is synthesized via precursor proteins, *Biochem. Biophys. Res. Comm.* **116**, 751–758.

NOVAK, J., CAULFIELD, P. W., MILLER, E. J. (1994), Isolation and biochemical characterization of a novel lantibiotic mutacin from *Streptococcus mutans, J. Bacteriol.* **176**, 4316–4320.

PALMER, D. E., MIERKE, D. F., PATTARONI, C., GOODMAN, M., WAKAMIYA, T., FUKASE, K., FUJITA, H., SHIBA, T. (1989), Interactive NMR and computer simulation studies of lanthionine-ring structures, *Biopolymers* **28**, 397–408.

PARENTI, F., PAGANI, H., BERETTA, G. (1976), Gardimycin, a new antibiotic from *Actinoplanes.* I. Description of the producer strain and fermentation studies, *J. Antibiot.* **24**, 501–506.

PESCHEL, A., AUGUSTIN, J., KUPKE, T., STEVANOVIĆ, S., GÖTZ, F. (1993), Regulation of epidermin biosynthetic genes by EpiQ, *Mol. Microbiol.* **9**, 31–39.

PIARD, J.-C., MURIANA, P. M., DESMAZEAUD M. J., KLAENHAMMER, T. R. (1992), Purification and partial characterization of lacticin 481, a lanthionine-containing bacteriocin produced by *Lactococcus lactis* subspp. *lactis* CNRZ 481, *Appl. Environ. Microbiol.* **58**, 279–284.

PIARD, J.-C., KUIPERS, O. P., ROLLEMA, H. S., DESMAZEAUD, M. J., DE VOS, W. M. (1993), Structure, organization and expression of the *lct* gene for lacticin 481, a novel lantibiotic produced by *Lactococcus lactis, J. Biol. Chem.* **268**, 16361–16368.

PUGSLEY, A. P. (1988), The immunity and lysis genes of ColN plasmid pCHAP4, *Mol. Gen. Genet.* **211**, 335–341.

PUGSLEY, A. P. (1993), The complete general secretory pathway in Gram-negative bacteria, *Microbiol. Rev.* **57**, 50–108.

RACKER, E., RIEGLER, C., ABDEL-GHANY, M. (1983), Stimulation of glycolysis by placental polypeptides and inhibition by duramycin, *Cancer, Res.* **44**, 1364–1367.

RAMSEIER, H. R. (1960), Die Wirkung von Nisin auf *Clostridium butyricum, Arch. Microbiol.* **37**, 57–94.

RAUCH, P. J. G., DE VOS, W. M., (1992), Characterization of the novel nisin-sucrose conjugative transposon Tn*5276* and its insertion in *Lactococcus lactis, J. Bacteriol.* **174**, 1280–1287.

RAUCH, P. J. G., BEERTHUYZEN, M. M., DE VOS, W. M. (1990), Nucleotide sequence of IS*904* from *Lactococcus lactis* subspp. *lactis* NIZO R5, *Nucleic Acids Res.* **18**, 4253.

RAUCH, P. J. G., BEERTHUYZEN, M. M., DE VOS, W. M. (1991), Molecular analysis and evolution of conjugative transposons encoding nisin production and sucrose metabolism in *Lactococcus lactis*, in: *Nisin and Novel Lantibiotics* (JUNG, G., SAHL, H.-G., Eds.), pp. 243–249. Leiden: ESCOM Scientific Publishers BV.

REIS, M., SAHL, H.-G. (1991), Genetic analysis of the producer self-protection mechanism (immunity) against Pep5, in: *Nisin and Novel Lantibiotics* (JUNG, G., SAHL, H.-G., Eds.), pp. 320–331. Leiden: ESCOM Scientific Publishers BV.

REIS, M., ESCHBACH-BLUDAU, M., IGLESIAS-WIND, M. I., KUPKE, T., SAHL, H.-G. (1994), Producer immunity towards the lantibiotic Pep5: identification of the immunity gene *pepI* and localization and functional analysis of its gene product, *Appl. Environ. Microbiol.* **60**, 4332–4338.

REISINGER, P. SEIDEL, H., TSCHESCHE, H., HAMMES, W. P. (1980), The effect of nisin on murein synthesis, *Arch. Microbiol.* **127**, 187–193.

RINCE, A., DUFOUR, A., LE POGAM, S., THUAULT, D., BOURGEOIS, C. M., LE PENNEC, J. P. (1994), Cloning, expression and nucleotide sequence of genes involved in production of lactococcin DR, a bacteriocin from *Lactococcus lactis, Appl. Environ. Microbiol.* **60**, 1652–1657.

RINTALA, H., GRAEFFE, T., PAULIN, L., KALKKINEN, N., SARIS, P. E. J. (1993), Biosynthesis of nisin in the subtilin producer *Bacillus subtilis* ATCC 6633, *Biotechnol. Lett.* **15**, 991–996.

ROGERS, L. A., WHITTIER, E. O. (1928), Limiting factors in lactic fermentation, *J. Bacteriol.* **16**, 211–214.

ROLLEMA, H. S., BOTH, P., SIEZEN, R. J. (1991), NMR and activity studies of nisin degradation products, in: *Nisin and Novel Lantibiotics* (JUNG, G., SAHL, H.-G., Eds.), pp. 123–130. Leiden: ESCOM Scientific Publishers BV.

ROSS, K. F., RONSON, C. W., TAGG, J. R. (1993), Isolation and characterization of the lantibiotic salivaricin A and its structural gene *salA* from *Streptococcus salivarius* 20P3, *Appl. Environ. Microbiol.* **59,** 2014–2021.

RUHR, E., SAHL, H.-G. (1985), Mode of action of the peptide antibiotic nisin and influence on the membrane potential of whole cells and on artificial membrane vesicles, *Antimicrob. Agents Chemother.* **27,** 841–845.

SAHL, H.-G. (1985), Influence of the staphylococcin-like peptide Pep5 on membrane potential of bacterial cells and cytoplasmic membrane vesicles, *J. Bacteriol.* **162,** 833–836.

SAHL, H.-G. (1991), Pore formation in bacterial membranes by cationic lantibiotics, in: *Nisin and Novel Lantibiotics* (JUNG, G., SAHL, H.-G., Eds.), pp. 347–358. Leiden: ESCOM Scientific Publishers BV.

SAHL, H.-G., BRANDIS, H. (1981), Production, purification and chemical properties of an antistaphylococcal agent produced by *Staphylococcus epidermidis, J. Gen. Microbiol.* **127**, 377–384.

SAHL, H.-G., BRANDIS, H. (1983), Efflux of low Mr substances from the cytoplasm of sensitive cells caused by the staphylococcin-like agent Pep5, *Zentralbl. Bakteriol. Hyg. I. Abt. Orig. A.* **252**, 166–175.

SAHL, H.-G., KORDEL, M., BENZ, R. (1987), Voltage-dependent depolarization of bacterial membranes and artificial lipid bilayers by the peptide antibiotic nisin, *Arch. Microbiol.* **149**, 120–124.

SAHL, H.-G., REIS, M., ESCHBACH, M., SZEKAT, C., BECK-SICKINGER, A. G., METZGER, J., STEVANOVIĆ, S., JUNG, G. (1991), Isolation of Pep5 prepeptides in different stages of modification, in: *Nisin and Novel Lantibiotics* (JUNG, G., SAHL, H.-G., Eds.), pp. 332–346. Leiden: ESCOM Scientific Publishers BV.

SAHL, H.-G., JACK, R. W., BIERBAUM, G. (1995), Lantibiotics: Biosynthesis and biological activities of peptides with unique posttranslational modifications, *Eur. J. Biochem.* **230**, 827–853.

SCHNELL, N., ENTIAN, K.-D., SCHNEIDER, U., GÖTZ, F., ZÄHNER, H., KELLNER, R., JUNG, G. (1988), Prepeptide sequence of epidermin, a ribosomally-synthesized antibiotic with four sulphide rings, *Nature* (London) **333**, 276–278.

SCHNELL, N., ENGELKE, G., AUGUSTIN, J., ROSENSTEIN, R., GÖTZ, F., ENTIAN, K.-D. (1991), The operon-like organization of lantibiotic epidermin biosynthesis genes, in: *Nisin and Novel Lantibiotics* (JUNG, G., SAHL, H.-G., Eds.), pp. 269–276. Leiden: ESCOM Scientific Publishers BV.

SCHNELL, N., ENGELKE, G., AUGUSTIN, J., ROSENSTEIN, R., UNGERMANN, V., GÖTZ, F., ENTIAN, K.-D. (1992), Analysis of genes involved in the biosynthesis of the lantibiotic epidermin, *Eur. J. Biochem.* **204**, 57–68.

SCHÜLLER, F., BENZ, R., SAHL, H.-G. (1989), The peptide antibiotic subtilin acts by formation of voltage-dependent multi-state pores in bacterial and artificial membranes, *Eur. J. Biochem.* **182**, 181–186.

SEARS, P. M., SMITH, B. S., STEWART, W. K., GONZALEZ, R. N., RUBINO, S. D., GUSIK, S. A., KULISEK, E. S., PROJAN, S. J., BLACKBURN, P. (1992), Evaluation of a nisin-based germicidal formulation on teat skin of live cows, *J. Dairy Sci.* **75,** 3185–3190.

SHETH, T. R., HENDERSON, R. M., HLADKY, S. B., CUTHBERT, A. W. (1992), Ion-channel formation by duramycin, *Biochim. Biophys. Acta* **1107,** 179–185.

SHIBA, T., WAKAMIYA, T., FUKASE, K., SANO, A., SHIMBO, K., UEKI, Y. (1986), The chemistry of lanthionine-containing peptides, *Biopolymers* **25,** S11-S19.

SHOTWELL, O. L., STODOLA, F. H., MICHAEL, W. R., LINDENFELSER, L. A., DWORSCHAK, G., PRIDHAM, T. G. (1958), Antibiotics against plant disease. III. Duramycin, a new antibiotic from *Streptomyces cinammonensis* forma *azacoluta, J. Am. Chem. Soc.* **80,** 3912–3915.

SIEGERS, K., ENTIAN, K.-D. (1995), Genes involved in immunity to the lantibiotic nisin produced by *Lactococcus lactis* 6F3, *Appl. Environ. Microbiol.* **61,** 1082–1089.

SKAUGEN, M., NISSEN-MEYER, J., JUNG, G., STEVANOVIĆ, S., SLETTEN, K., MORTVEDT-ABILDGAARD, C. I., NES, I. (1994), *In vivo* conversion of L-serine to D-alanine in a ribosomally-synthesized polypeptide, *J. Biol. Chem.* **269**, 27183–27185.

SLIJPERS, M., HILBERS, C. W., KONINGS, R. N. H., VAN DE VEN, F. J. M. (1989), NMR studies of lantibiotics: assignment of the ^{1}H-NMR spectrum of nisin and identification of interresidual contacts, *FEBS Lett.* **252**, 22–28.

SOKOLOVE, P. M., WESTPHAL, P. A., KESTER, M. B., WIERWILE, R., VAN METER, K. S. (1989), Duramycin effects on the structure and function of heart mitochondria. I. Structural alterations and changes in membrane permeability, *Biochim. Biophys. Acta* **983**, 15–22.

SOMMA, S., MERATI, W., PARENTI, F. (1977), Gardamycin, a new antibiotic inhibiting peptidoglycan synthesis, *Antimicrob. Agents Chemother.* **11**, 396–401.

SONG, H. Y., CRAMER, W. A. (1991), Membrane topography of ColE1 gene products: the immunity protein, *J. Bacteriol.* **173**, 2935–2943.

STEEN, M. T., CHUNG, Y. J., HANSEN, J. N. (1991), Characterization of the nisin gene as part of a polycistronic operon in the chromosome of *Lactococcus lactis* ATCC 11454, *Appl. Environ. Microbiol.* **57**, 1181–1188.

STEVENS, K. A., SHELDON, B. W., KLAPES, N. A., KLAENHAMMER, T. R. (1991), Nisin treatment for inactivation of *Salmonella* species and other Gram-negative bacteria, *Appl. Environ. Microbiol.* **537**, 3613–3615.

STOFFELS, G., NISSEN-MEYER, J., GUDMUNDSDOTTIR, A., SLETTEN, K., HOLO, H., NES, I. F. (1992), Purification and characterization of a new bacteriocin isolated from a *Carnobacterium* spp., *Appl. Environ. Microbiol.* **58**, 1417–1422.

STOFFELS, G., GUDMUNDSDOTTIR, A., ABEE, T. (1994), Membrane-associated proteins encoded by the nisin gene cluster may function as a receptor for the lantibiotic carnocin UI49, *Microbiology* **140**, 1443–1450.

STONE, D. K., XIE, X. S., RACKER, E. (1984), Inhibition of clathrin-coated vesicle acidification by duramycin, *J. Biol. Chem.* **259**, 2701–2703.

SUROVOY, A., WEIDELICH, D., JUNG, G. (1992), Electrospray mass spectroscopic analysis of metal-peptide complexes, in: *Peptides 1992, Proceedings of the 22nd European Peptide Symposium* (SCHNEIDER, C. H., EBERLE, A. N., Eds.), pp. 563–564. Leiden: ESCOM Scientific Publishing BV.

TAGG, J. R., WANNAMAKER, L. W. (1978), Streptococcin A-FF22: nisin-like antibiotic substance produced by a group A *Streptococcus, Antimicrob. Agents Chemother.* **14**, 31–39.

TAGG, J. R., DAJANI, A. S., WANNAMAKER, L. W., GRAY, E. D. (1973a), Group A streptococcal bacteriocin: production, purification and mode of action, *J. Exp. Med.* **138**, 1168–1183.

TAGG, J. R., READ, R. S. D., MCGIVEN, A. R. (1973b), Bacteriocin of group A *Streptococcus*: Partial purification and properties, *Antimicrob. Agents Chemother.* **4**, 214–221.

TSAI, H.-J., SANDINE, W. E. (1987), Conjugal transfer of nisin plasmid genes from *Streptococcus lactis* 7962 to *Leuconostoc dextranicum* 181, *Appl. Environ. Microbiol.* **53**, 352–357.

UNGERMANN, V., GOEKE, K., FIEDLER, H.-P., ZÄHNER, H. (1991), Optimization of fermentation and purification of gallidermin and epidermin, in: *Nisin and Novel Lantibiotics* (JUNG, G., SAHL, H.-G., Eds.), pp. 410–421. Leiden: ESCOM Scientific Publishers BV.

VAN BELKUM, M. J., KOK, J., VENEMA, G., HOLO, H., NES, I. F., KONINGS, W. N., ABEE, T. (1991), The bacteriocin lactococcin A specifically increases the permeability of lactococcal cytoplasmic membranes in a voltage-independent, protein-mediated manner, *J. Bacteriol.* **173**, 7934–7941.

VAN DE KAMP, M., HORSTINK, L. M., VAN DEN HOOVEN, H., KONING, R. N. H., HILBERS, C. W., FREY, A., SAHL, G.-G., METZGER, J., VAN DE VEN, F. J. M. (1995a), Sequence analysis by NMR spectroscopy of the peptide lantibiotic epilancin K7 from *Staphylococcus epidermidis* K7, *Eur. J. Biochem.* **227**, 757–771.

VAN DE KAMP, M., VAN DE VEN, F. J. M., KONINGS, R. H. H., HILBERS, C. W., METZGER, J. W., JUNG, G., KUIPERS, O. P., BIERBAUM, G., SAHL, H.-G. (1995b), Elucidation of the primary structure of the peptide lantibiotic epilancin K7 from *Staphylococcus epidermidis*: cloning of the epilancin K7-encoding gene and Edman degradation of the mature peptide, *Eur. J. Biochem.* **230**, 587–600.

VAN DE VEN, F. J. M., VAN DEN HOOVEN, H. W., KONINGS, R. N. H., HILBERS, C. W. (1991a), NMR-studies of lantibiotics: the structure of nisin in aqueous solution, *Eur. J. Biochem.* **202**, 1181–1188.

VAN DE VEN, F. J. M., VAN DEN HOOVEN, H. W., KONINGS, R. N. H., HILBERS, C. W. (1991b), The spatial structure of nisin in aqueous solution, in: *Nisin and Novel Lantibiotics* (JUNG, G., SAHL, H.-G., Eds.), pp. 35–42. Leiden: ESCOM Scientific Publishers BV.

VAN DER MEER, J. R., POLMAN, J., BEERTHUYZEN, M. M., SIEZEN, R. J., KUIPERS, O. P., DE VOS, W. M. (1993), Characterization of the *Lactococcus lactis* nisin A operon genes *nisP* encoding a subtilisin-like serine protease involved in precursor processing and *nisR* encoding a regulatory protein involved in nisin biosynthesis, *J. Bacteriol.* **175**, 2578–2588.

VAN DER MEER, J. R., ROLLEMA, H. S., SIEZEN, R. J., BEERTHUYZEN, M. M., KUIPERS, O. P., DE VOS, W. M. (1994), Influence of amino acid substitutions in the nisin leader peptide on biosynthesis and secretion of nisin by *Lactococcus lactis, J. Biol. Chem.* **269**, 3555–3562.

VOGEL, H., NILSSON, L., RIGLER, R., MEDER, S., BOHEIM, G., BECK, W., KURTH, H.-H., JUNG, G. (1993), Structural fluctuations between two conformational states of a transmembrane helical peptide are related to its channel-forming properties in planar lipid membranes, *Eur. J. Biochem.* **212**, 305–313.

WAKAMIYA, T., UEKI, Y., SHIBA, T., KIDO, Y., MOTOKI, Y. (1985), The structure of ancovenin, a new peptide inhibitor of angiotensin I converting enzyme, *Tetrahedron Lett.* **26**, 665–668.

WAKAMIYA, T., FUKASE, K., NARUSE, N., KONISHI, M., SHIBA, T. (1988), Lanthiopeptin, a new peptide effective against *Herpes simplex* virus: structural determination and comparison with Ro 09–0198, an immunopotentiating peptide, *Tetrahedron Lett.* **29**, 4771–4772.

WEIL, H.-P., BECK-SICKINGER, A. G., METZGER, J., STEVNANOVIĆ, S., JUNG, G., JOSTEN, M., SAHL, H.-G. (1990), Biosynthesis of the lantibiotic Pep5: Isolation and characterization of a prepeptide containing dehydroamino acids, *Eur. J. Biochem.* **194**, 217–223.

ZIMMERMANN, N. (1995), Raumstrukturaufklärung der Lantibiotika Duramycin B, Duramycin C und Actagardin durch mehrdimensionale Kernresonanzspektroskopie, *PhD Thesis*, University of Tübingen, Germany.

ZIMMERMANN, N., JUNG, G. (1995), The tetracyclic lantibiotic actagardine. ^{1}H-NMR and ^{13}C-NMR assignments and revised primary structure, *Eur. J. Biochem.* **228**, 786–797.

ZIMMERMANN, N., FREUND, S., FREDENHAGEN, A., JUNG, G. (1993), Solution structures of the lantibiotics duramycin B and C, *Eur. J. Biochem.* **216**, 419–428.

9 Glycopeptide Antibiotics (Dalbaheptides)

GIANCARLO LANCINI

BRUNO CAVALLERI

Gerenzano, Italy

1 Introduction

The term "glycopeptide antibiotics" is commonly used to indicate a family of microbial metabolites, active on gram-positive bacteria, and closely related in their chemical structure and biological activity to the antibiotics vancomycin and ristocetin – the first members of the family isolated in the early 1950s. The expression "glycopeptide antibiotics of the vancomycin–ristocetin family" is more precise and is also often used.

The characteristics common to all members of the family are:

- a structure composed of a linear heptapeptide in which at least five of the amino acid residues are aromatic, the rings linked to form a triphenyl ether moiety and a diphenyl group;
- a unique mechanism of action, i.e., inhibition of bacterial growth by binding to the D-Ala-D-Ala terminus of peptidoglycan precursors, thus inhibiting cell wall formation.

Taking into account their chemical characteristics and unusual mechanism of action, the name Dalbaheptides, from **Dal** (anyl-D-alanine) **B**(inding) **A**(ntibiotics with) **Hept**(apept)**ide** (structure), was proposed for these antibiotics by PARENTI and CAVALLERI (1989).

Several reviews have been published reporting different aspects of glycopeptide antibiotics: discovery, isolation, and purification (CASSANI, 1989; SITRIN and FOLENA-WASSERMAN, 1989; CAVALLERI and PARENTI, 1992), mechanism of action (BARNA and WILLIAMS, 1984; REYNOLDS, 1989), fermentation and biosynthesis (LANCINI, 1989; LANCINI and CAVALLERI, 1990), chemistry and chemical derivatives (MALABARBA et al., 1993a), structure–activity relationship (NAGARAJAN, 1993), pharmacology (CASSETTA et al., 1991; PHILLIPS and GOLLEDGE, 1992; BROGDEN and PETERS, 1994). Biological and chemical aspects are comprehensively discussed in a multiauthor book published recently (NAGARAJAN, 1994a).

Although many natural and semisynthetic glycopeptides are provided with relevant antibacterial activity, only vancomycin and teicoplanin are used in human medicine, due to their effect on gram-positive pathogens refractory to established antibiotics, such as multiresistant *Staphylococcus aureus*, coagulase-negative staphylococci, clostridia, and enterococci. Eremomycin, a relatively recent natural product is under clinical evaluation. Avoparcin is commercially available as growth promoter in animal feed. Finally, ristocetin is a diagnostic agent for a particular disorder of genetic origin (von Willebrand's disease) due to the ability to aggregate blood platelets.

2 Descriptive Chemistry

The first glycopeptide antibiotic isolated from microbial fermentation was ristocetin in 1953, followed by vancomycin in 1955. Almost immediately their composition of sugar carrying small peptides was defined, but the determination of their structures was completed only 20 years later, mainly by studies with high-field NMR. In the 1960s and 1970s the isolation of a small number of new glycopeptides including avoparcin and teicoplanin was reported, but only in the 1980s the introduction into screening programs of specific methods for the detection of these antibiotics in fermentation broth resulted in the discovery of the majority of glycopeptides now available. Modern techniques of isolation and purification such as reverse-phase chromatography and affinity chromatography and availability of powerful spectroscopic techniques such as 2D-NMR, Fast Atom Bombardment Mass Spectrometry (FAB-MS), and Ion Spray MS have now made it possible to determine the structure of new products in a relatively short time.

A list of natural glycopeptides is shown in Tabs. 1–4. More detailed references and the chemical structures are provided in the previously cited reviews by LANCINI and CAVALLERI (1990) and CAVALLERI and PARENTI (1992). It should be noted that in some cases different code numbers or names were assigned to identical chemical entities produced by different strains, before their structure had been elucidated. A striking example is that of eremomycin that was successively reported as A82846A, MM45289, and LY264826. In addition to the products listed in Tabs. 1–4 the following compounds with unknown structure were reported: AM374 from *Streptomyces ebureosporeus* (KUNSTMANN and PORTER, 1974), A477 from *Actinoplanes* sp. (HAMILL et al., 1973), AB65 from *Saccharomonospora viride* (TAMURA and TAKEDA, 1975).

Many of the microorganisms listed produce families (complexes) of strictly related compounds. The components (factors) of a complex usually differ in the level of methylation or chlorination of the peptidic skeleton or in the presence of additional sugars. Analogs lacking some or all sugar units present in the parent antibiotic (therefore designated as "pseudoaglycones" or aglycones, respectively) are either the result of incomplete glycosylation or generated by chemical or enzymatic deglycosylation during fermentation or recovery and purification. In several cases, further studies on complex producing strains have revealed the presence of additional minor components in the fermentation broth that had initially been overlooked (BORGHI et al., 1989; NAGARAJAN, 1993).

The glycopeptide skeleton is shown in Fig. 1. The numbering of the amino acid residues is that proposed by BARNA et al. (1984); it has been widely used because it allows an easy comparison of NMR data of glycopeptides having different core structures. It should be

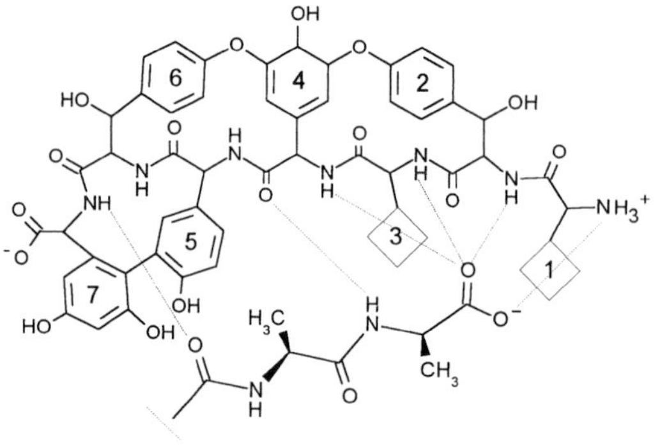

Fig. 1. General heptapeptide structure of dalbaheptides and interaction with the D-alanyl-D-alanine peptide terminus. Hydrogen bonds are indicated by dottet lines.

Tab. 1. Naturally Occurring Dalbaheptides – Ristocetin Type

Name	Producing Strain[a]	Company, Year[b]	References
Ristocetin A, B	*Nocardia lurida,* NRRL 2430	Abbott, 1953	SZTARICSKAI and BOGNÁR, 1984 KATRUKHA and SILAEV, 1986
Ristomycin A, B	*Proactinomyces fructiferi*	Inst. New Antibiotics, Moscow, 1962	SZTARICSKAI and BOGNÁR, 1984 KATRUKHA and SILAEV, 1986
Actaplanin (A4696 complex)	*Actinoplanes missouriensis,* ATCC 23342	Lilly, 1971	DEBONO et al., 1984 HUNT et al., 1984a
A35512 complex	*Streptomyces candidus,* NRRL 8156	Lilly, 1976	MICHEL et al., 1980 DEBONO et al., 1980
A41030 complex	*Streptomyces virginiae,* NRRL 15156	Lilly, 1982	BOECK et al., 1985 HUNT et al., 1985
A47934	*Streptomyces toyocaensis,* NRRL 15009	Lilly, 1982	BOECK and MERTZ, 1986
UK-68597	*Actinoplanes* sp., ATCC 53533	Pfizer, 1987	SKELTON and WILLIAMS, 1990 HOLDOM et al., 1988
UK-69542	*Saccharothrix aerocolonigenes,* ATCC 53829	Pfizer, 1991	HOLDEN et al., 1991 SKELTON et al., 1991

[a] The name of the producing strain is according to the original publication; note that several strains have been subsequently reclassified into different genera.
[b] Year of first paper or patent publication.

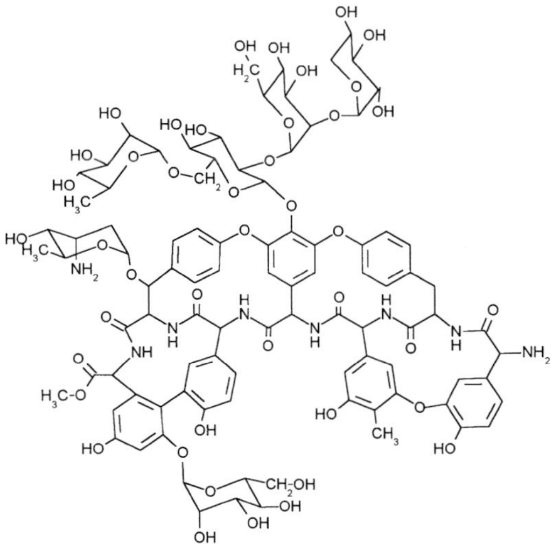

Fig. 2. Ristocetin A.

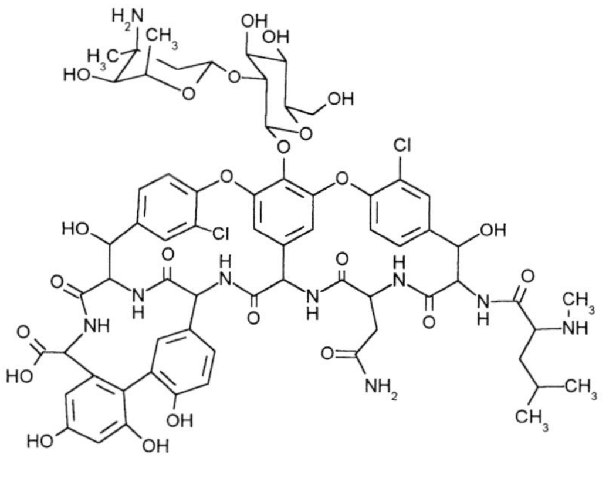

Fig. 3. Vancomycin.

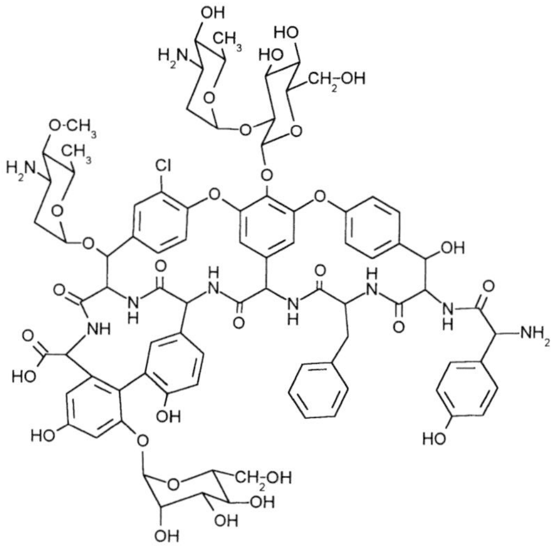

Fig. 4. Actinoidin A.

noted that the residues are numbered from the amino terminus, in contrast to the usual system of peptide numbering starting from the carboxyl terminus.

Phenylamino acids 2, 4, 5, 6, and 7 are present in all glycopeptides and a classification based on the remaining amino acids 1 and 3 is commonly used. The compounds wherein the phenylic moieties of amino acids 1 and 3 are linked by an oxygen belong to the ristocetin type (Tab. 1); their structure is exemplified by that of ristocetin A in Fig. 2. Vancomycin type includes the glycopeptides listed in Tab. 2, where amino acids 1 and 3 are aliphatic. Synmonicin, also included in Tab. 2, is an exception, since amino acid 3 is a thioamino acid and amino acid 1 is aromatic. The structure of vancomycin is shown in Fig. 3. In the actinoidin type (Tab. 3) amino acid 1 is always a *p*-hydroxyphenylglycine whereas amino acid 3 is a phenylalanine or *p*-hydroxyphenylglycine (Fig. 4).

Within a group other characteristics account for the large variety of glycopeptide structures with chlorine atoms, methyl and hydroxy groups at different positions of the phenyl residues. Amino acid 6 and occasionally amino acid 2 carry a hydroxyl group in β-position. Common or unusual sugars are linked at different positions through glycosidic bonds. The sugars vancosamine (and the related epivancosamine and 4-ketovancosamine), ristosamine, actinosamine, and acosamine were first isolated from glycopeptides. Only A41030A, B, E and A47934 do not contain any sugar. In a few compounds a phenolic hydroxyl is esterified as a sulfate monoester (A47934, UK-68597, UK-69542). The terminal carboxyl is often a methyl ester and the terminal amino group can be methylated.

In some ristocetin-type glycopeptides linear or branched (exceptionally unsaturated) fatty

Tab. 2. Naturally Occurring Dalbaheptides – Vancomycin Type

Name	Producing Strain[a]	Company, Year[b]	References
Vancomycin	*Streptomyces orientalis,* NRRL 2450	Lilly, 1955	HARRIS et al., 1983 SZTARICSKAI and BOGNÁR, 1984
K-288	*Streptomyces haranomachiensis*	Tohoku University, 1961	MATSUMOTO, 1961
OA-7653 A, B	*Streptomyces hygroscopicus* ssp. *hiwasaensis,* ATCC 31613	Otsuka Pharm., 1978	KAMOGASHIRA et al., 1983 ANG et al., 1988
A51568A (N-demethyl vancomycin)	*Nocardia orientalis,* NRRL 15232	Lilly, 1982	BOECK et al., 1984 HUNT et al., 1984b
M43 complex	*N. orientalis,* NRRL 2450	Lilly, 1984	HIGGINS et al., 1985
Izupeptin A, B	*Nocardia* sp., FERM P-8656	Kitasato Institute, 1986	SPIRI-NAKAGAWA et al., 1986
Synmonicin A, B, C (CWI-785)	*Synnemomyces mamnoorii,* ATCC 53296	Smith, Kline & French, 1986	ARJUNA RAO et al., 1986
Orienticin (PA-42867-A, B, C, D)	*N. orientalis,* PA-42867	Shionogi, 1987	TSUJI et al., 1988b
Eremomycin	*Actinomyces* sp., INA-238	Inst. New Antibiotics, Moscow, 1987	BRAZHNIKOVA et al., 1989 GAUSE et al., 1989
A42867	*Nocardia* sp., ATCC 53492	Lepetit, 1987	RIVA et al., 1989
A82846 A, B, C	*Amycolatopsis orientalis,* NRRL 18098, NRRL 18099	Lilly, 1987	HAMILL et al., 1988 NAGARAJAN et al., 1989a
Chloroorienticin A, B, C, D, E (PA-45052)	*A. orientalis,* PA-45052	Shionogi, 1988	TSUJI et al., 1988a NAGARAJAN et al., 1989b
A80407 A, B	*Kibdelosporangium philippinensis,* NRRL 18198	Lilly, 1989	DOOLIN et al., 1989
MM 45289, MM 47756	*A. orientalis,* NCIB 12531	Beecham, 1989	GOOD et al., 1990
MM 47761, MM 49721	*A. orientalis,* NCIB 12608	Beecham, 1989	BOX et al., 1990
Decaplanin (M 86-1410)	*Kibdelosporangium deccaensis,* DSM 4763	Hoechst, 1990	FRANCO et al., 1990
UK-72051	*A. orientalis*	Pfizer, 1990	SKELTON et al., 1990
MM 55270, MM 55271, MM 55272	*Amycolatopsis* sp., NCIB 40086	Beecham, 1990	COATES et al., 1990a
Balhimycin	*Amycolatopsis* sp., DSM 5908	Hoechst, 1992	NADKARNI et al., 1994
A83850 A, B	*Amycolatopsis albus,* NRRL 18532	Lilly, 1993	HAMILL and YAO, 1993

See footnotes in Tab. 1

acids are linked as amides to the amino group of a glucosamine (or 2-aminoglucuronic acid) moiety. These antibiotics listed in Tab. 4 are thus complexes the components of which differ in the nature of the aliphatic side chains as illustrated by the structure of teicoplanin (Fig. 5). The aliphatic moieties cause a certain lipophilic character of these "lipoglycopep-

Tab. 3. Naturally Occurring Dalbaheptides – Actinoidin Type

Name	Producing Strain[a]	Company, Year[b]	References
Actinoidin A, B	*Proactinomyces actinoides*	Inst. New Antibiotics, Moscow, 1956	BERDNIKOVA et al., 1982
Avoparcin (LL-AV290 complex)	*Streptomyces candidus,* NRRL 3218	American Cyanamid, 1966	MCGAHREN et al., 1980 MCGAHREN et al., 1983
Chloropolysporin A, B, C	*Faenia interjecta,* FERM BP-583	Sankyo, 1983	OKAZAKI et al., 1987 TAKATSU et al., 1987a
Actinoidin A_2	*Nocardia* sp., SKF-AAJ-193	Smith, Kline & French, 1987	DINGERDISSEN et al., 1987 HEALD et al., 1987
Helvecardin A, B	*Pseudonocardia compacta* ssp. *helvetica,* SANK 65185	Sankyo, 1988	TAKEUCHI et al., 1991a TAKEUCHI et al., 1991b
MM 47766, MM 47767, MM 55256, MM 55260	*Amycolatopsis orientalis,* NCIB 40011	Beecham, 1989	ATHALYE et al., 1989
Galacardin A, B	*Saccharothrix* sp., SANK 64289	Sankyo, 1990	TAKEUCHI et al., 1992

See footnotes in Tab. 1

Tab. 4. Naturally Occurring Dalbaheptides – Lipoglycopeptides

Name	Producing Strain[a]	Company, Year[b]	References
Teicoplanin	*Actinoplanes teichomyceticus,* ATCC 31121	Lepetit, 1975	PARENTI et al., 1978 CORONELLI et al., 1987
Ardacin (aridicin, AAD-216 complex)	*Kibdelosporangium aridum,* ATCC 39323	Smith, Kline & French, 1983	SHEARER et al., 1985 SITRIN et al., 1985
A40926 complex	*Actinomadura* sp., ATCC 39727	Lepetit, 1984	SELVA et al., 1986 GOLDSTEIN et al., 1987
Kibdelin (AAD-609 complex)	*K. aridum,* ATCC 39922	Smith, Kline & French, 1985	SHEARER et al., 1986 FOLENA-WASSERMAN et al., 1986
Parvodicin (AAJ-271)	*Actinomadura parvosata,* ATCC 53463	Smith, Kline & French, 1986	CHRISTENSEN et al., 1987
MM 49728, MM 55266, MM 55267, MM 55268	*Amycolatopsis* sp., NCIB 40089	Beecham, 1990	COATES et al., 1990b BOX et al., 1991
A84575 complex	*Streptosporangium carneum,* NRRL 18437, NRRL 18505	Lilly, 1991	MICHEL and YAO, 1991
MM 56597, MM 56598	*Amycolatopsis* sp., NCIB 40089	Beecham, 1991	COATES et al., 1991

See footnotes in Tab. 1

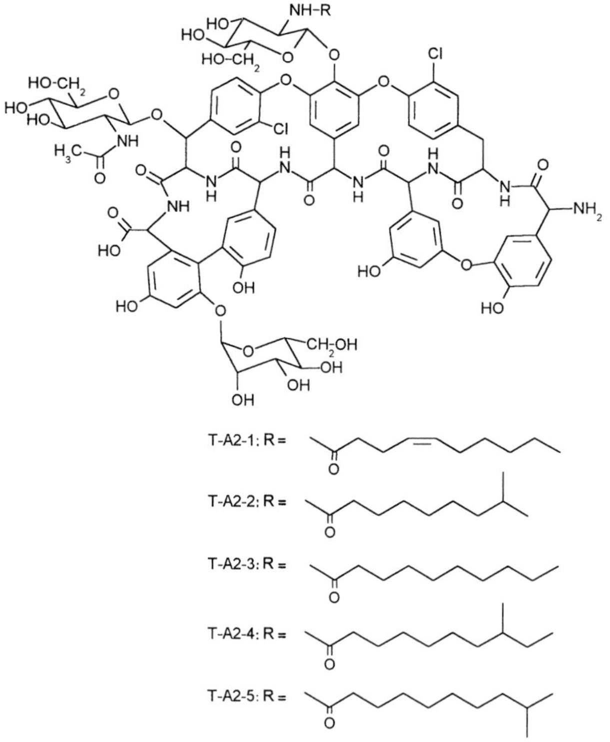

Fig. 5. Teicoplanin (teichomycin complex).

tides" that directly influences their biological properties and in particular their pharmacokinetic behavior.

2.1 Physicochemical Properties

Glycopeptides are colorless or whitish powders, generally water-soluble, that usually strongly retain water or crystallization solvents. Their molecular weights range from 1150–2300 Da. Some of them as, e.g., teicoplanin can be isolated as an internal salt or as a partial monoalkaline (sodium) salt, depending on the pH value of the aqueous medium in the final purification step. Others can be isolated as acidic salts: vancomycin and actaplanin as hydrochlorides, and ristocetin A, avoparcin, and eremomycin as sulfates. Many pharmaceutically acceptable basic and acidic addition salts are claimed in patent applications. The net charge of several glycopeptides has been determined by electrofocusing. The isoelectric points reported in Tab. 5 range

Tab. 5. Isoelectric Points (I. P.) of Some Dalbaheptides[a]

Antibiotic	I. P.
A47934	3.2
A40926	3.7–3.9
Ardacin	3.9
A41030	4.9
OA-7653	4.9
Teicoplanin	5.0
Chloropolysporin	7.5
A35512B	7.5–7.8
Actaplanin	7.6–8.5
AM374	7.7–8.0
Avoparcin	7.7–8.1
Vancomycin	7.7
Ristocetin	8.1
A42867	8.1
A477	7.9–8.1

[a] Determined by electrofocusing (RIVA, E., SOFFIENTINI, A., personal communication)

from 3.2 (A47934) to 8.1 (ristocetin, A477, and A42867).

Retention times from reverse-phase HPLC give an approximate indication of the relative lipophilicity. In Tab. 6 the values obtained with representative compounds are listed.

All glycopeptides exhibit similar UV absorption spectra with maxima at about 280 nm in acidic or neutral media that shift to 292–300 nm under alkaline conditions. NMR experiments led to complete assignment of each hydrogen, carbon, and nitrogen atom for the majority of the glycopeptides and greatly contributed to structure elucidation of both natural and semisynthetic compounds. Although almost 1000 natural and semisynthetic glycopeptides are presently known no crystalline forms suitable for X-ray analysis have been reported, except for a degradation product of vancomycin (designated as CDP-I) that allowed the determination of the absolute configuration of vancomycin (SHELDRICK et al., 1978). The configuration of the seven amino acids is 1*R*, 2*R*, 3*S*, 4*R*, 5*R*, 6*S*, and 7*S*, respectively. The NMR spectra of vancomycin and of glycopeptides discovered later show that the stereochemistry and the overall three-dimensional conformation is the same in all compounds except for a few epimers at amino acid residue 1. In teicoplanin five of the amide bonds are *trans*, the 5-6 bond is *cis*: modification of the bond conformation often occurs during chemical reactions. As an example, the three-dimensional structure of teicoplanin aglycone is presented in Fig. 6.

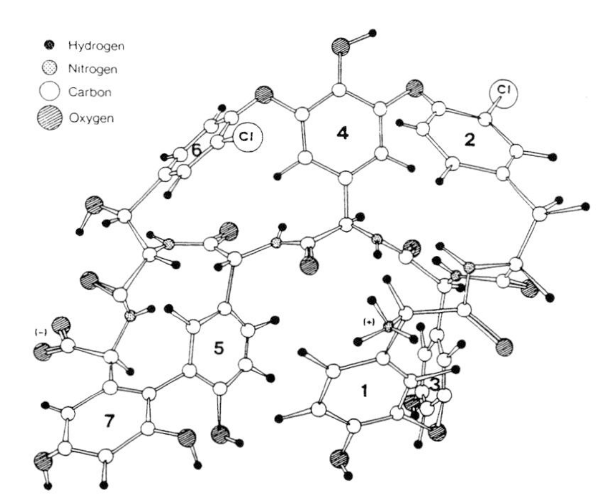

Fig. 6. Stereo model of teicoplanin aglycone.

Tab. 6. Reverse-Phase HPLC Retention Times (t_R) of Some Dalbaheptides[a]

Antibiotic	t_R [min]
AM374	7
Vancomycin	10.5
Ristocetin	11
Chloropolysporin	12
A35512B	12.5
Avoparcin	13
Actaplanin	14
A47934	17
OA-7653	20
A41030	21
A477	23
Ardacin	24–26
Teicoplanin	24–28
A40926	30

[a] On Ultrasphere ODS 5 μm (250–4.6 mm) column (Beckman);
mobile phases:
(A) 0.02 M aq NaH_2PO_4/CH_3CN 9:1, pH 6;
(B) 0.02 M aq NaH_2PO_4/CH_3CN 3:7, pH 6;
gradient from 5%–60% of eluent (B) in 40 min;
flow rate 1.6 mL·min^{-1}.
(RIVA, E. and SOFFIENTINI, A., personal communication)

3 Biological Activity

3.1 *In vitro* Antibacterial Activity

Glycopeptide antibiotics are active on most gram-positive aerobic and anaerobic bacteria (CAMPOLI-RICHARDS et al., 1990). Gram-negative bacteria are generally insensitive except for *Neisseria, Gardnerella,* and *Branhamella* strains on which several natural or semisynthetic glycopeptides exert a weak effect. However, most strains of the genera *Lactobacillus, Leuconostoc,* and *Pediococcus* – although gram-positive – are insensitive (JOHNSON et al., 1990). The minimal inhibitory concentrations (MIC) of some glycopep-

Tab. 7. Comparison of Antibacterial Activity (MIC, $\mu g \cdot mL^{-1}$) of Some Dalbaheptides (GOLDSTEIN et al., 1987)

Antibiotic	*Staphylococcus aureus* TOUR	*Staphylococcus epidermidis* ATCC 12228	*Streptococcus pyogenes* C 203	*Enterococcus faecalis* ATCC 7080	*Propionibacterium acnes* ATCC 6919	*Bacteroides fragilis* ATCC 23745	*Neisseria gonorrhoeae* ISM 68/126
A40926	0.06	0.06	0.06	0.06	0.016	64	2
A47934	0.06	0.03	0.13	0.13	nd	nd	8
Teicoplanin	0.13	0.13	0.06	0.13	0.13	128	32
Vancomycin	0.25	0.5	0.13	0.5	nd	nd	32
Ristocetin	4	2	0.25	1	nd	nd	64
A35512B	1	0.5	0.13	0.5	nd	nd	64
A41030A	0.03	0.008	0.13	0.13	nd	nd	64
Aridicin	1	4	0.13	2	nd	nd	64
Avoparcin	2	2	0.25	0.25	nd	nd	128
Actaplanin	1	2	0.13	0.5	nd	nd	>128

nd: not determined

tides on representative pathogens are reported in Tab. 7. Since – as discussed later – the activity of these antibiotics is due to their ability to bind peptidoglycan precursors that are similar in gram-positive and gram-negative bacteria, the insensitivity of the latter is attributed to a lack of penetration through the outer membrane. It is, however, noteworthy that some semisynthetic derivatives exhibit a certain activity against *Escherichia coli, Proteus vulgaris,* and *Pseudomonas aeruginosa* (MALABARBA et al., 1993a).

The effect of glycopeptides on staphylococci and streptococci is clearly bactericidal at concentrations slightly higher than the MICs. Their bactericidal effect is weaker in *Enterococcus faecium* or *E. faecalis* (CAMPOLI-RICHARDS et al., 1990). Lethality can be accompanied by cell lysis depending on the strain and on environmental conditions.

3.2 Mechanism of Action

A most interesting property of glycopeptide antibiotics is their ability to block cell wall formation in sensitive bacteria by binding to precursors of peptidoglycan synthesis (REYNOLDS, 1989; NICAS and ALLEN, 1994).

Inhibition of cell wall synthesis in growing bacterial populations is supported by several experimental data (WALLAS and STROMINGER, 1963; SOMMER et al., 1984) the most relevant of which are:

- Addition of a glycopeptide to a bacterial culture blocks the uptake of peptidoglycan precursors such as acetyl glucosamine or diamino pimelic acid several minutes before any effect can be noticed on the uptake of precursors of DNA, RNA, or proteins.
- Inhibition of uptake is accompanied by accumulation of intermediates of peptidoglycan biosynthesis, notably UDP-muramylpentapeptide, indicating that these antibiotics do not interfere with early steps of peptidoglycan synthesis but affect a later stage of the process.

The observation that vancomycin activity is reversed by cell wall fragments or by late intermediates of the peptidoglycan biosynthetic pathway gave the first indication of the action of glycopeptides on a molecular level. The classical work of NIETO and PERKINS (1971a, b) established clearly that glycopeptides bind with relatively high affinity to peptides having a D-Ala-D-Ala carboxyl terminus. A lower binding affinity is observed when other amino acids substitute for one of these residues. The tripeptide N,N-diacetyl-L-Lys-D-Ala-D-Ala was used by several authors as a model of the natural ligand to determine the binding affinity of different antibiotics. An association constant of $1.5 \cdot 10^6$ L·mol^{-1} was calculated for vancomycin. The values obtained with ristocetin and teicoplanin were $5.9 \cdot 10^5$ and $2.6 \cdot 10^6$ L·mol^{-1}, respectively (SOMMA et al., 1984).

Experiments performed to establish which biochemical reaction is specifically inhibited in the peptidoglycan biosynthetic pathway have not achieved decisive results so far. Glycopeptides are most probably unable to cross the bacterial cytoplasmic membrane as was experimentally shown with iodinated vancomycin (PERKINS and NIETO, 1970); therefore, their action appears to be limited to events occurring at the membrane surface, such as transglycosylation or transpeptidation reactions (PERKINS, 1982). Vancomycin and teicoplanin were reported to inhibit transglycosylation in a cell free membrane system. However, the inhibitory effect was observed only at antibiotic concentrations substantially higher than the MICs of these drugs (SOMMA et al., 1984).

Another aspect difficult to be interpreted is the often observed poor correlation between the degree of affinity to model tripeptides and the antimicrobial activity of the antibiotic. An extreme case is that of eremomycin, consistently more active than vancomycin but demonstrating a significantly lower affinity to Ac_2-Lys-D-Ala-D-Ala (GOOD et al., 1990).

3.3 Resistance

To date, despite many years of clinical use, vancomycin resistance has not been reported for *Streptococcus* sp. or *Staphylococcus aureus* clinical isolates. *In vitro* selection for vancomycin or teicoplanin resistance in *S. aureus* is difficult and strains with only modest increases in MIC are obtained (JOHNSON et al., 1990). A decreased susceptibility to vancomycin and more frequently to teicoplanin has been observed with coagulase-negative staphylococci, such as *S. epidermidis* or *S. haemolyticus*. The biochemical mechanism determining this low-level resistance has not been elucidated yet (ARDUINO and MURRAY, 1993).

Enterococci resistant to vancomycin or teicoplanin have been isolated with increasing frequency in the last years, especially in intensive care units. Three major phenotypes termed VanA, VanB, VanC can be distinguished in resistant *Enterococcus* species (ARTHUR and COURVALIN, 1993; NICAS and ALLEN, 1994). Enterococci carrying the *VanA* gene are characterized by high-level resistance to vancomycin and teicoplanin. A cluster of genes the expression of which is associated with the VanA phenotype has been identified in these strains on a transposon designated Tn*1546*. Biochemical characterization of the gene products has revealed that the resistance is due to an unusual pathway of peptidoglycan biosynthesis in which UDP-muramyl-tetrapeptide-D-lactate is synthesized instead of the normal precursor UDP-muramyl-pentapeptide. Apparently, all enzymes catalyzing the subsequent reactions of the pathway accept as substrates the intermediates carrying this modification. The result is a peptidoglycan structure in which the residue D-Ala-D-Lac substitutes for D-Ala-D-Ala. As glycopeptides do not form complexes with peptides ending with D-Ala-D-Lac the strain results resistant to their action.

Phenotype VanB is characterized by moderate or high-level resistance to vancomycin and sensitivity to teicoplanin. The mechanism of resistance is the same as of the VanA phenotype. However, in VanB strains expression of resistance genes is inducible by vancomycin but not by teicoplanin which explains the different susceptibility to the two antibiotics.

The VanC phenotype comprises species such as *Enterococcus gallinarum* in which the pentapeptides of peptidoglycan and its biosynthetic precursors constitutively possess a D-Ala-D-Ser ending. These species show low-level resistance to vancomycin and sensitivity to teicoplanin.

3.4 *In vivo* Efficacy and Pharmacology

The efficacy of the principal glycopeptides in curing experimental infections has been assessed in septicemia models in mice (PAL-

Tab. 8. Experimental Septicemia in Mice[a]

Antibiotic	Infecting Organism (Clinical Isolates)					
	S. aureus L 165		*S. pyogenes* L 49		*S. pneumoniae* L 44	
	MIC [μg·mL^{-1}]	ED_{50} [mg·kg^{-1}]	MIC [μg·mL^{-1}]	ED_{50} [mg·kg^{-1}]	MIC [μg·mL^{-1}]	ED_{50} [mg·kg^{-1}]
Teicoplanin	0.4	0.72	0.05	0.11	0.1	0.41
Vancomycin	0.8	7.2	0.8	0.58	0.8	1.9
Ampicillin	0.1	8.1	0.02	0.1	0.02	4.1
Cephaloridine	0.05	2.8	0.01	0.03	0.02	0.93
Erythromycin	0.5	28	0.05	0.44	0.01	26

[a] Animals treated subcutaneously once daily for three days (PALLANZA et al., 1983).

LANZA et al., 1983). The results obtained with teicoplanin and vancomycin are reported in Tab. 8 in comparison with other antibiotics. In general, a higher *in vivo* efficacy corresponds to a higher *in vitro* activity although no simple correlation is apparent. Eremomycin appears to be more active than vancomycin both *in vitro* and in experimental infections. Ardacins are less active *in vivo* than vancomycin although having a comparable antibacterial activity. This also holds for A40926 when compared to teicoplanin.

A few products have been tested in more complex models of experimental infection. Teicoplanin and vancomycin were found effective in reducing the bacterial load in experimentally induced endocarditis in rats and rabbits (GOLDSTEIN et al., 1994).

A clinical overview of vancomycin has been published by ZECKEL and WOODWORTH (1994). Toxicology, pharmacokinetics, pharmacology, and therapeutic use of teicoplanin have been extensively reviewed by GOLDSTEIN et al. (1994) and BROGDEN and PETERS (1994).

4 Producing Organisms

All known glycopeptides are produced by microorganisms of the order of *Actinomycetales* isolated from soil samples collected from almost everywhere. Surprisingly, only a few glycopeptides are produced by streptomycetes and practically all of them are of the ristocetin type, although ristocetin itself was isolated from strains originally considered as *Nocardia* or *Proactinomyces* and later classified as *Amycolatopsis* (LECHEVALIER et al., 1986). A large proportion of known glycopeptide is generated by this genus. In fact, the vancomycin producing strain has been in turn classified as *Streptomyces,* then as *Nocardia*, and at present as *Amycolatopsis;* several glycopeptides of the vancomycin type are produced by *A. orientalis* and most probably several other strains should be reclassified into the same genus. Actinoidin producers are probably also *Amycolatopsis* although originally considered as *Proactinomyces* or *Nocardia.* Lipoglycopeptides are produced by a variety of rare actinomycetes such as *Actinomadura, Actinoplanes, Streptosporangium,* and the new genus *Kibdelosporangium.*

Interestingly, glycopeptides that are minor components of complexes produced by a microbial strain are the main products of other soil isolates. Partially purified vancomycin produced by *A. orientalis* NRRL 2450 contains minor quantities of at least nine structurally related compounds (NAGARAJAN, 1993). One of them, A51586A, is the main product of *N. orientalis* NRRL 15232 (HUNT et al., 1984b).

5 Methods of Screening

The first glycopeptide antibiotics were discovered by conventional screening procedures such as growth inhibition of a test organism on agar plates. Later, specific tests were devised to detect inhibitors of cell wall synthesis. In the Lepetit Laboratories, fermentation broths were tested for differential inhibition of a *S. aureus* strain and an L-form derived from it. In this way teicoplanin was discovered. Similarly, izupeptin was found on the basis of its differential activity on cell wall possessing bacteria and on mycoplasma (SPIRI-NAKAGAWA et al., 1986).

Subsequently, very efficient screening tests were implemented exploiting the unique mechanism of action of the antibiotics. One of them was based on affinity chromatography; matrix bound D-Ala-D-Ala was prepared by forming a peptide bond between aminocaproyl-Sepharose and the amino group of the dipeptide (CORTI and CASSANI, 1985). A different antimicrobial activity of culture filtrates before and after passage on the affinity resin was taken as evidence for the presence of a glycopeptide that could easily be isolated by elution. By this method 72 strains (about 0.3% of the fermentation broth tested) were identified as producers of glycopeptides in a screening campaign (CASSANI, 1989); about 60% were ristocetin producers. Among the others teicoplanin, avoparcin or actaplanin producing strains were identified, and two new glycopeptides (A40926 and A42867) were discovered as well.

RAKE et al. (1986) applied an assay based on the affinity of glycopeptides to the tripeptide N,N-diacetyl-L-Lys-D-Ala-D-Ala to 1936 cultures. Reversion of antibacterial activity by addition of this receptor analog indicated the presence of a glycopeptide. The test proved to be very specific: 42 glycopeptides were isolated, 6 of which were novel ones including kibdelin, parvodicin, and actinoidin A_2.

A colorimetric assay (SPERA) was proposed (CORTI et al., 1985) based on the competition of horseradish peroxidase bound teicoplanin and the putative glycopeptide for a D-Ala-D-Ala peptide attached to the test tube. Glycopeptide A82846 was discovered by testing fermentation broths against vancomycin in a polyclonal antibody assay (ELISA) (YAO et al., 1988).

6 Fermentation

Culture media and fermentation conditions reported for production of glycopeptide antibiotics either at the laboratory level or in industrial manufacturing do not substantially differ from those described for other antibiotics. As usual, precultures (vegetative cultures) are prepared in one or more stages using rich media composed of soluble proteins and rapidly utilizable carbon sources. Fermentation conditions must be adjusted individually for the different producing strains. In general, optimal temperatures are between 28 and 30°C, and the optimal pH is around 7.0. All producing organisms are strictly aerobic and a consistent oxygen supply must be provided by adequate aeration and agitation systems.

6.1 Fermentation Media – Carbon Sources

Carbon sources frequently used in fermentation media include glucose, glycerol, starch, or dextrin. The use of oleate or other lipids is less common.

Antibiotic production is often repressed by the presence of rapidly utilizable carbon sources in the production medium. A few studies of this effect have been reported for glycopeptide fermentation. There is evidence that antibiotic production by *Amycolatopsis* species is subjected to this repression, since higher yields of ristomycin (ristocetin) are obtained when galactose or glycerol are used as carbon sources in *Amycolatopsis lurida* fermentations (TOROPOVA et al., 1982). Similarly the production of demethyl vancomycin is increased when galactose or dextrin substitutes for glucose in the fermentation medium (BOECK et al., 1984); dextrin or starch are the preferred carbon sources in the fermentations of vancomycin and of the related antibiotic izupeptin (MERTZ and DOOLIN, 1973; SPIRI-NAKAGAWA et al., 1986). All these antibiotics are produced by *Amycolatopsis* strains. A marked negative effect of glucose is observed in the production of the antibiotic complex A41030 by cultures of *Streptomyces virginiae* (BOECK et al., 1985).

In contrast, *Actinoplanes* species appear to be less affected by the nature of carbon sources. Only minor differences in actaplanin yields are observed when dextrin or other slowly metabolized carbon sources substitute for glucose in *Actinoplanes missouriensis* fermentations. A medium composed of glucose and yeast extract appears suitable for teicoplanin production by cultures of *A. teichomyceticus*.

A substantial increase in yields of ardacins and kibdelins, complexes produced by *Kibdelosporangium* strains, is observed up on addition of oleic acid (as methyl ester) to the fermentation medium (SHEARER et al., 1985). Oleic acid is a good source of acetyl CoA – one of the starting materials of these antibiotic biosyntheses. Higher availability of acetyl CoA was therefore proposed to explain the oleate effect on yields. However, since ardacins and kibdelins are lipoglycopeptides oleic acid possibly is a direct precursor of their fatty acid chain as it was shown for teicoplanin, a lipoglycopeptide (see Sect. 8.2).

6.2 Fermentation Media – Nitrogen Sources

Soybean meal or similar soy products are almost universally used as nitrogen providing ingredients in production media; yeast extract and corn steep liquor are also used frequently. Ammonium salts are never found among the suitable nitrogen sources. Although specific studies are lacking it can be assumed that – similarly to other antibiotics – glycopeptide production is repressed by high concentrations of ammonium ions. Nitrates can be utilized; at least in the case of *S. virginiae* fermentations addition of sodium nitrate to a soybean and corn steep containing medium increased production of A41030 by 80% (BOECK et al., 1985).

6.3 Fermentation Media – Effect of Phosphates

Phosphate concentration is an important parameter in antibiotic fermentation processes as phosphates are necessary for optimal growth, but high concentrations inhibit production.

In vancomycin fermentations antibiotic production is depressed by phosphate concentrations above $0.1\ g \cdot L^{-1}$ (MERTZ and DOOLIN, 1973). Even lower concentrations are sufficient to inhibit production of demethyl vancomycin or ristocetin (BOECK et al., 1984; TOROPOVA et al., 1974). The fact that these antibiotics belong to different structural types but are all produced by *Amycolatopsis* species suggests that phosphate control is characteristic for this genus.

Production of ardacins by *Kibdelosporangium aridum* and of A47934 by *S. toyocaensis* is also inhibited by relatively low concentrations of phosphates (BOECK and MERTZ, 1986). In contrast, production of the complex A41030 by *S. virginiae* was enhanced by increasing phosphate concentrations up to $1\ g \cdot L^{-1}$ (BOECK et al., 1985).

6.4 Inhibition by the Final Product

Several examples of inhibition of antibiotic biosynthesis by the final fermentation product are reported in the literature. This phenomenon has also been observed in some glycopeptide fermentations. Ristocetin production, e.g., is inhibited by low concentrations of this antibiotic (TOROPOVA et al., 1974). It was observed in the authors' laboratory that A40926 inhibits its own production when present in the growth phase of cultures.

Well documented is the effect of actaplanin on its producing strain. *A. missouriensis* cultures start producing actaplanin about 40 h after inoculation at the end of mycelial growth. Depending on the fermentation medium 60–90% of the antibiotic produced is tightly bound to the mycelium. When mycelium was centrifuged and resuspended in saline no further growth was observed but production continued normally. When mycelium was centrifuged, washed to completely eliminate soluble actaplanin, and resuspended, growth was resumed and no production was observed. Thus, even low concentrations of actaplanin in the medium apparently inhibit growth and also have a killing effect as demonstrated by viable colony counting in a parallel experiment. Surprisingly the cell bound antibiotic that can be as much as 50 mg per gram of cells has no effect on either growth or production (HUBER et al., 1987).

6.5 Control of Complex Composition

An antibiotic composed of several factors has to be produced during fermentation in constant relative amounts to ensure uniformity of the finished product. Alternatively, the relative amount of the major component can be increased to obtain as a final product a single substance rather than a complex. The factors composing the lipoglycopeptide antibiotics teicoplanin and A40926 differ in the structure of their acyl chains, in which the branched type, either *iso* or *anteiso,* predominates. In biosynthesis of branched fatty acids the initiator molecule determines the type of the final product: isobutyric acid gives rise to *iso* chains with an even number of carbon atoms, 2-methylbutyric acid and isovaleric acid give rise to chains with an odd number of carbon atoms of the *anteiso* and the *iso* type, respectively. Addition of one of these precursors to the fermentation broth can selectively increase the amount of the fatty acid that it is initiating and thus of the corresponding component of the complex. It has been shown that in this way the complex composition of both teicoplanin and A40926 can be substantially altered (BORGHI et al., 1991a; SELVA et al., 1992). In practice, better results could be achieved by adding their natural precursors, i.e., valine for isobutyric acid, isoleucine for 2-methylbutyric acid, and leucine for isovaleric acid, to the cultures instead of the acids mentioned above.

7 Recovery and Purification

Although glycopeptides are relatively water-soluble, often a large proportion of the product is found in the harvest broth bound to mycelium. Solubilization and release can normally be obtained by adjusting the pH to an appropriate value (according to the isoelectric point of the product acidic or alkaline conditions are required). Alternatively extraction with acetone or methanol has been used. Recovery from the filtered broth may include adjustment of pH and use of water immiscible organic solvents (butanol for teicoplanin), or ion-pair extraction (avoparcin). Other isolation schemes described use ion exchange matrices (Dowex, Amberlite IR9), acidic alumina, cross-linked polymeric adsorbents (Diaion HP, Amberlite XAD), cation exchange dextran gel (Sephadex) and polyamides in various sequences (SITRIN and FOLENA-WASSERMAN, 1994).

Reverse-phase chromatography with semipreparative and preparative columns, packed with silanized silica gel, allowed purification and separation of glycopeptides and of the single factors of complexes on the basis of differential hydrophobicity. Isocratic and gradient elutions were carried out. Examples are given by SITRIN and FOLENA-WASSERMAN (1989) and in almost every publication listed in Tab. 1–4.

The elucidation of the mechanism of action at a molecular level allowed the development of affinity chromatography both as a specific discovery tool and as a powerful method of isolation and purification of glycopeptides from fermentation broths (CORTI and CASSANI, 1985; FOLENA-WASSERMAN et al., 1987). The affinity adsorbent is prepared by immobilizing D-Ala-D-Ala on commercially available activated supports. The affinity constant for the sepharose-D-Ala-D-Ala resin, as determined by equilibrium binding experiments, is $8.08 \cdot 10^5$ $L \cdot mol^{-1}$ for teicoplanin, $1.57 \cdot 10^5$ $L \cdot mol^{-1}$ for vancomycin and $3.89 \cdot 10^5$ $L \cdot mol^{-1}$ for ristocetin A (CORTI and CASSANI, 1985).

An aqueous solution of the glycopeptide is contacted with the adsorbent in batch mode or loaded on columns. After washing to remove impurities the glycopeptides are eluted with a small volume of an aqueous buffer solution and an organic solvent (acetonitrile, methanol, ethylene glycol) to disrupt the interaction. By varying the pH of the buffer, mixtures of different glycopeptides or components of a complex can be resolved, with the additional advantage of obtaining a concentrated solution. The final pure material can be obtained after desalting by lyophilization.

8 Biosynthesis

Glycopeptides are complex molecules the biosynthesis of which must include several steps:

(1) synthesis of the uncommon amino acids constituting the heptapeptide,
(2) polymerization of the amino acids,
(3) formation of the ether and carbon–carbon bonds between the phenylic groups,
(4) synthesis of the unusual sugars and, if present, of fatty acid chains,
(5) glycosylation and other final modifications.

The sequence of these steps may, to some extent, differ from that indicated; chlorination of the phenyl rings, e.g., may occur before or after the amino acid assembly and glycosylation may precede the oxidative ring linking.

Experimental studies to elucidate the biosynthetic pathway are scanty. In fact, although it appears to be most probable that assembly is performed through the well known thiotemplate system, no experimental evidence is available on this most important aspect. The origin of the unusual sugars and the reactions leading to the formation of the triphenyl ether and diphenyl groups have never been studied. This lack of information is, in part, due to the fact that several laboratories failed to isolate blocked mutants accumulating biosynthetic intermediates. In addition, the most intensely studied antibiotics, i.e., vancomycin, teicoplanin and ristocetin, are produced by *Actinomyces* strains for which a system of transformation with exogenous DNA is not available. This has severely hampered studies based on molecular genetic methods.

8.1 Origin of the Uncommon Amino Acids

A few uncommon amino acids are found as building blocks of almost all glycopeptide antibiotics: β-hydroxytyrosine, 3-chloro-β-hydroxytyrosine, *p*-hydroxyphenylglycine and *m*-dihydroxyphenylglycine (Fig. 7).

Analysis of ^{13}C NMR spectra of avoparcin produced by fermentations of *Streptomyces candidus* which had been supplied with 2-^{13}C-

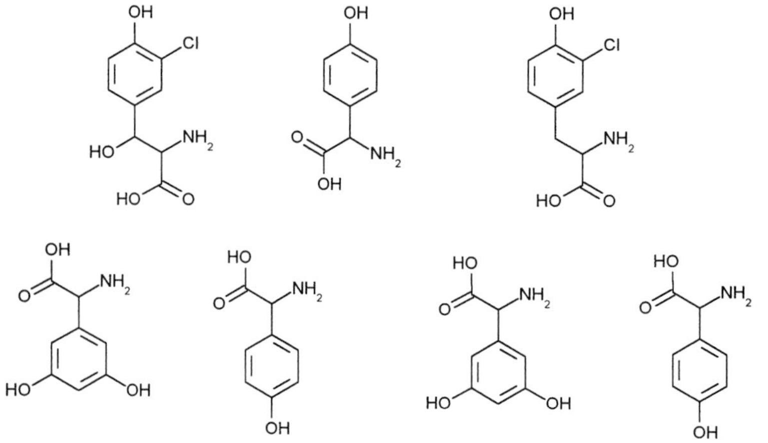

Fig. 7. Amino acids composing teicoplanin heptapeptide.

tyrosine demonstrated that tyrosine is the precursor of the *β*-hydroxytyrosine and *p*-hydroxyphenylglycine residues of the peptide (McGahren et al., 1980). These results were confirmed and extended by studies on vancomycin. Experiments with ^{13}C and ^{3}H labeled tyrosine showed that both L- and D-tyrosine are precursors of the two 3-chloro-*β*-hydroxytyrosine units, although these differ in their configuration at C-2. Moreover, it was found that *β*-hydroxylation occurs with retention of configuration at C-3 (Hammond et al., 1982). Similar results were obtained by adding D,L-2-^{13}C-tyrosine to ristocetin fermentations. As expected, the label was incorporated into C-2 of the *β*-hydroxytyrosine and C-1 of the *p*-hydroxyphenylglycine residues (Hammond et al., 1983).

Although the origin of *p*-hydroxyphenylglycine from tyrosine is clearly demonstrated, the sequence of reactions and the intermediates of this conversion can be deduced only from indirect evidence. *p*-Hydroxyphenylglyoxylic acid appears to be the direct precursor of *p*-hydroxyphenylglycine, since demethyl vancomycin and A47934 yields are increased by addition of either of these compounds to *A. orientalis* or *S. toyocaensis* cultures, respectively (Boeck et al., 1984; Boeck and Mertz, 1986). In *K. aridum* fermentations, addition of cold *p*-hydroxyphenylglyoxylic acid or *p*-hydroxy mandelic acid depresses the incorporation of radioactivity from L-U^{14}C-tyrosine into ardacin indicating dilution of labeled intermediates (Chung et al., 1986a). It is noteworthy that *p*-hydroxyphenyl acetic acid has no effect, providing evidence that this compound is not an intermediate in the conversion. These results and the proof that the producing organisms are able to convert tyrosine into *β*-hydroxytyrosine are consistent with the following reaction sequence (Fig. 8):

Tyrosine → *β*-hydroxytyrosine →
β-*p*-dihydroxyphenylpyruvic acid →
p-hydroxymandelicacid →
p-hydroxyphenylglyoxylic acid →
p-hydroxyphenylglycine.

The origin of *m*-dihydroxyphenylglycine units has been demonstrated by experiments in which 1,2-^{13}C-acetate was added either to vancomycin or ristocetin fermentations (Hammond et al., 1982). The carbons of the *m*-dihydroxyphenylglycine residues resulted specifically labeled, indicating that this amino acid backbone originates from cyclization of a polyketomethylene chain. A tetraketide chain was originally suggested as the immediate precursor. However, this requires a ring closure involving the initial methyl group of the chain, a feature never observed in biosynthesis of polyketide antibiotics. It was, therefore,

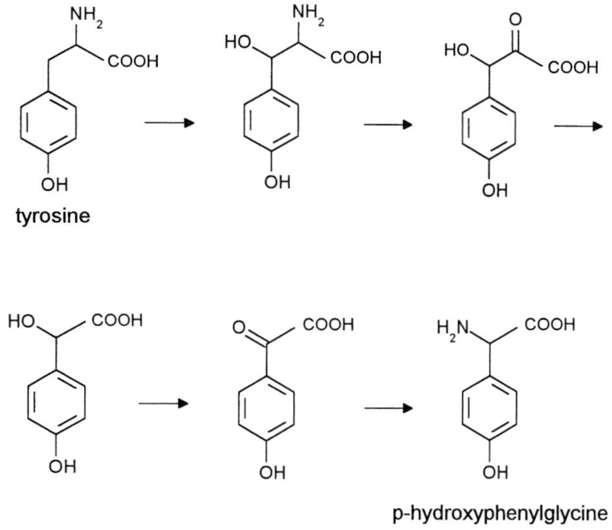

tyrosine

p-hydroxyphenylglycine

Fig. 8. Presumptive pathway of *p*-hydroxyphenyl glycine biosynthesis from tyrosine.

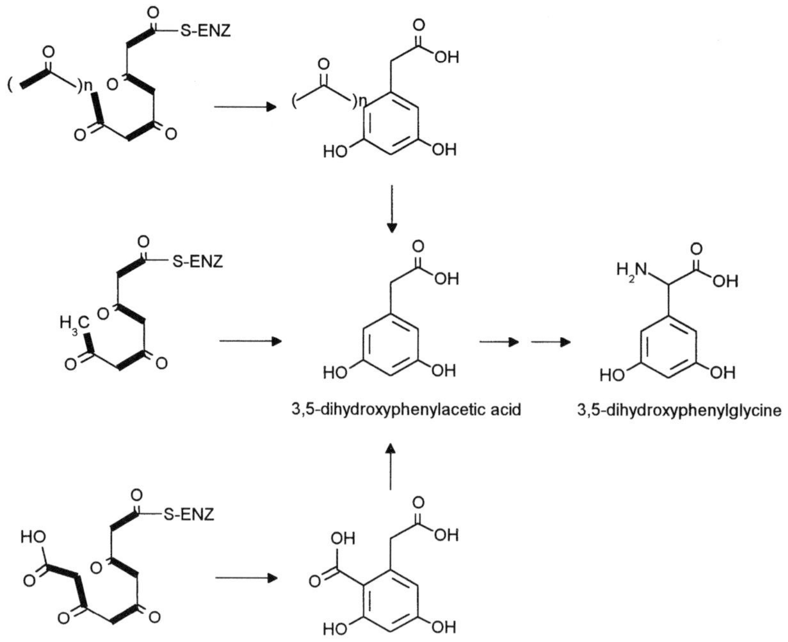

3,5-dihydroxyphenylacetic acid

3,5-dihydroxyphenylglycine

Fig. 9. Hypothetical polyketide chains that can give rise to 3,5-dihydroxyphenyl acetic acid and to 3,5-dihydroxyphenyl glycine.

proposed that a longer chain is formed first, part of which could be degraded after cyclization (HAMMOND et al., 1983). An alternative hypothesis can be considered assuming that malonate rather than acetate is the initiator molecule of the polymerization process. The resulting product, 6-carboxy-3,5-dihydroxyphenylacetic acid, could be easily converted into *m*-dihydroxyphenylacetic acid by decarboxylation (Fig. 9).

8.2 Origin of Fatty Acids of Lipoglycopeptides

All glycopeptides listed in Tab. 4 are complexes of factors characterized by the presence of different acyl chains linked, as amides, to amino sugars. The origin of the fatty acids constituting the acyl moieties of the teicoplanin components (see Fig. 5) has been extensively investigated (BORGHI et al., 1991a). Altogether the results obtained indicate that these chains are not synthesized *de novo* but derived from degradation of long-chain fatty acid components of cell lipids or from those present in the fermentation medium. This conclusion is based on the following experimental evidence:

- Production of teicoplanin factor T-A2-1 by *A. teichomyceticus* characterized by a 4-decenoyl moiety is entirely dependent on the presence of linoleic acid in the fermentation medium.
- The relative amount of factor T-A2-3 produced that is characterized by a linear decanoyl chain is substantially increased by addition of oleic acid to the medium.
- Factors T-A2-2, T-A2-4, and T-A2-5 bear branched acyl chains, namely 8-methylnonanoic acid (*iso*-C10:0), 8-methyldecanoic acid (*anteiso*-C11:0) and 9-methyldecanoic acid (*iso*-C11:0). Analysis of fatty acid constituents of cell lipids revealed the presence of three major components, 14-methylpentadecanoic acid (*iso*-C16:0), 14-methylhexadecanoic acid (*anteiso*-C17:0) and 13-methyltetradecanoic acid (*iso*-C15:0). These appear to be the logical precursors of the *iso*-C10:0, *anteiso*-C11:0, and *iso*-C11:0 moieties of T-A2-2, T-A2-4, and T-A2-5, respectively, assuming the loss of acetate units by the common β-oxidation mechanism of fatty acid degradation.

- A mutant strain of *A. teichomyceticus* produces a novel teicoplanin factor characterized by a *n*-nonanoic moiety. Correspondingly, cell lipids contain heptadecenoic acid not present in parent strain cells.
- Addition of ^{14}C-acetate to the culture medium at the time of inoculation resulted in substantial labeling of fatty acid moieties. When ^{14}C-acetate was added to grown mycelium resuspended in saline, radioactivity of acyl chains was negligible in comparison to that of teicoplanin aglycone, demonstrating that the fatty acid moieties derive from molecules formed during the growth phase.

A similar correspondence between acyl chains of antibiotic factors and cell lipid composition was found for A40926 and its producer *Actinomadura* strain (Zerilli et al., 1992).

8.3 Glycosylation and Final Modifications

The last steps of the biosynthetic pathway were studied in some detail in ardacin production by *K. aridum.* Sugar substituents in these antibiotic molecules are mannose and a N-acyl-2-aminoglucuronic unit. A related complex of antibiotics, kibdelins, differs from ardacins in the presence of N-acylglucosamine instead of acylaminoglucuronic acid. Kibdelins are readily converted into ardacins by cultures of *K. aridum* indicating that oxidation of C-6 of glucosamine could be the last biosynthetic step (Chung et al., 1986a).

Less clear are the results of experiments aimed at defining the glycosylation sequence. When ^{14}C labeled ardacin aglycone or mannosyl aglycone were added to *K. aridum* cultures, labeled ardacins were produced (Chung et al., 1986a). However, since transformation required a long incubation period and only a small fraction of the added radioactivity was incorporated, the possibility of a degradation of the molecule and recycling of the fragments cannot be ruled out.

Similar attempts were carried out by adding teicoplanin aglycone to *A. teichomyceticus* cultures. Conversion into the final molecule was never achieved; the aglycone was rapidly transformed into mannosyl aglycone but no further glycosylation step was observed (Borghi et al., 1991b).

Antibiotic A47934 produced by *S. toyocaensis* has a chemical structure similar to that of ardacin aglycone from which it differs in the presence of a sulfate ester on an aromatic ring. Extensive experiments with labeled substrates, blocked mutants, and biochemical inhibitors indicate that the sulfate is added prior to the formation of intermediates that possess antimicrobial activity. These results exclude that sulfate esterification of the aglycone that is antimicrobially active is the last reaction of the biosynthetic pathway (Zmijewski et al., 1987).

9 Chemical Modifications

Early investigations on glycopeptide chemistry were carried out mainly for structural studies. Hydrolysis reactions yielded the aglycones and pseudoaglycones, some of which exhibited a higher antimicrobial activity than the parent antibiotic (Philip et al., 1960). Later, knowledge of the mechanism of action on a molecular level, elucidation of the structure of several compounds, and establishment of some correlations between pharmacokinetic and physicochemical properties (Pitkin et al., 1986) allowed a more rational design of chemical derivatives. In particular, the efficacy of teicoplanin in experimental infections pointed out the importance of lipophilic side chains and opened the way to the oriented synthesis of a large number of derivatives both of teicoplanin and other glycopeptides. Products with improved antimicrobial activity or better pharmacokinetics have been reported, but none of them is in clinical use so far. Structure–activity relationships of natural and semisynthetic glycopeptides have been reviewed by Nagarajan (1994b). The ideal glycopeptide has been outlined in a recent article by Felmingham (1993).

An extensive review of chemical modifications is out of the scope of this chapter, therefore only a condensed summary is given.

Reaction conditions are those conventional in peptide chemistry including suitable protection of active functions not involved in the reaction. However, their application sometimes requires specific conditions since different glycopeptides treated with the same reagent under relatively similar conditions yield different results depending on the nature of amino acids 1 and 3, on steric factors, and on the presence of sugars or chlorine atoms.

A common modification of glycopeptides is the selective hydrolysis of sugars leading to the preparation of aglycones and pseudoaglycones. Reports are available, e.g., on avoparcin (McGAHREN et al., 1983), teicoplanin (MALABARBA et al., 1984), parvodicin (CHRISTENSEN et al., 1987), ardacin (SITRIN et al., 1986), and eremomycin (KOBRIN et al., 1988). The hydrolyses were performed mainly for investigational purposes but also provided compounds of interest for their intrinsic activity or as substrates for further chemical transformations.

Since some components of certain glycopeptide complexes differ only in the number or position of Cl atoms, partial or complete removal of chlorine converts one glycopeptide into another, e.g., A82846B into orienticin A (NAGARAJAN et al., 1989a). Selective dechlorination of vancomycin (HARRIS et al., 1985) and teicoplanin (MALABARBA et al., 1989b) and its effect on the affinity of the antibiotics for model peptides and on their antibacterial activity have been reported. Iodovancomycin (HARRIS et al., 1986) and ^{125}I-labeled ristocetin (KIM et al., 1989) were also described.

The terminal carboxyl group which may be free as in teicoplanin or methylated as in ristocetin has been submitted to esterification, reduction to alcohol, or amidation. Teicoplanin, its pseudoaglycones, and its aglycone have been amidated with various amines and alkyldiamines (MALABARBA et al., 1989c), α-amino acids (MALABARBA et al., 1989a), or polyamines (MALABARBA et al., 1992).

A40926, an antibiotic characterized like ardacins by an N-acylglucuronic moiety shows a good antibacterial activity on several pathogens, including *N. gonorrhoeae* (GOLDSTEIN et al., 1987). The glucuronic carboxyl group was reduced to hydroxymethyl to obtain a molecule with physicochemical properties similar to those of teicoplanin, and the peptidic carboxyl group was condensed with a number of amines. Among the derivatives obtained the $-NH(CH_2)_3N(CH_3)_2$ amide (MDL 63,246) exhibits a higher antimicrobial activity than the parent compound on several pathogens, including vancomycin and teicoplanin resistant strains (MALABARBA et al., 1993b).

It was suggested that a protonated amino group plays an important role in the binding process through an electrostatic interaction with the carboxyl group of the target peptide (WILLIAMSON et al., 1984). The terminal amino group of several glycopeptides such as teicoplanin, vancomycin, ristocetin, and eremomycin has been variously alkylated, and acylated. In vancomycin the higher basicity of the vancosamine amino group compared to that of the terminal N-methylleucine allowed a selective acylation of the former (KANNAN et al., 1988; NAGARAJAN et al., 1988).

To understand the role of the terminal "carboxylate binding pocket" in glycopeptide–substrate interactions binding constants of glycopeptide fragments (produced by controlled degradations) were determined with peptide models. NAGARAJAN and SCHABEL (1988) prepared the "vancomycin hexapeptide" by removal of the terminal N-methylleucine, and MALABARBA and CIABATTI (1992) reported a tetrapeptide fragment from teicoplanin.

On the other hand, relevant peptide fragments were obtained by stereospecific syntheses with a view of total synthesis of vancomycin (RAMA RAO et al., 1992) and teicoplanin (CHAKRABORTY and VENKAT REDDY, 1992). Both these fragments and those prepared by controlled degradation may be used as building blocks for the synthesis of non-natural glycopeptides.

The various approaches to the synthesis of vancomycin and ristocetin aglycones have been reviewed by EVANS and DEVRIES (1994).

10 Biotransformations

10.1 Deglycosylation

Selective hydrolysis of certain sugar substituents that could not be performed chemically was carried out enzymatically.

Using naranginase chloropolysporin B and α- or β-avoparcin were converted to the corresponding derhamnosyl derivatives with a 40% and 80% yield, respectively. Although the commercial enzyme used contained β-glucosidase the glucose moiety was not hydrolyzed, probably because of steric hindrance. α-Mannosidase converted chloropolysporin B and C to the corresponding demannosyl pseudoaglycones with a 71–50% yield. With this enzyme it was also possible to obtain the demannosyl derivative of β-avoparcin which was found identical with ε-avoparcin (TAKATSU et al., 1987b).

Enzymatic deglycosylation clarified the relationship between galacardin A and β-avoparcin. Treatment of galacardin A with α-galactosidase selectively removed the galactose units linked by α-glycosidic bonds. After incubation either the galactose at the hydroxyl of amino acid 3 or the galactose linked to the L-rhamnose unit at amino acid 1 were removed leaving the sugar units linked by β-glycosidic bonds unaffected (TAKEUCHI et al., 1992).

The acyl glucosamine moiety of teicoplanin is selectively hydrolyzed by mild acid treatment. More acidic conditions are required to remove the other sugar substituents of the molecule. Therefore, a selective, chemical hydrolyzation of the mannose moiety was not possible without removing the acyl glucosamine unit. The single components of the teicoplanin complex were converted into the corresponding demannosyl derivatives by treatment with cultures of *N. orientalis* NRRL 2450 (the producer of vancomycin-type M43 complex) or *S. candidus* NRRL 3218 (the producer of actinoidin-type avoparcin). The time course of the transformation with washed mycelium of *N. orientalis* showed a 40% conversion in 72 h at a concentration of 0.25 $g \cdot L^{-1}$ (BORGHI et al., 1991b).

10.2 Glycosylation

Cultures of *A. teichomyceticus*, the producer of teicoplanin, and of *K. aridum*, the producer of ardacins, easily transform the aglycones of their antibiotics into the mannosyl derivatives (CHUNG et al., 1986a; BORGHI et al., 1991b). Transformation is almost quantitative after 24 h of incubation.

Mannosylation is independent of antibiotic biosynthesis, since with *K. aridum* it occurs also in the presence of glyphosate, a biosynthesis inhibitor, and with *A. teichomyceticus* it is performed by washed mycelium resuspended in saline. The reaction is not very specific with respect to the substrate: pseudoaglycones are also efficiently transformed by both microorganisms.

Protoplast preparations of *K. aridum* were tested for their ability to convert ardacin aglycone into the complete antibiotic. In this case the prevalent transformation compound was also mannosyl aglycone, accompanied by smaller amounts of poorly defined derivatives carrying other neutral sugars (CHUNG et al., 1986b).

10.3 Deacylation

At present, microbial deacylation seems to be the only method to remove the fatty acid residues from the N-acylamino sugars of lipoglycopeptides. The main component of A40926 complex is characterized by the presence of a 10-methylundecanoyl chain that acylates the amino group of the glucuronic moiety. A few hundred microorganisms were tested for their ability to selectively hydrolyze the amide bond. Positive results were obtained with cultures of *A. teichomyceticus* ATCC 31121 (the producer of teicoplanin), *A. missouriensis* NRRL 15646 and NRRL 15647 (SELVA et al., 1988).

Ardacins and the components of AAD 609 complex are also deacylated by *A. teichomyceticus* cultures (CHUNG et al., 1986b). Teicoplanin is not deacylated by its producer – which is somewhat surprising in view of the structure similarity with kibdelins.

10.4 Other Biotransformations

Actaplanin single components or its pseudoaglycone are bioconverted by pure cultures of *A. missouriensis* NRRL 15646 or *A. missouriensis* NRRL 15647 into the corresponding CUC/CSV glycopeptides in which the $-CO-CH(NH_2)-$ group on the terminal amino acid 1 is oxidatively deaminated into a $-CO-CO-$ group (CLEM et al., 1985). Antibiotic CUC/CSV is also produced by cofermentation of *A. missouriensis* NRRL 15646 and NRRL 15647. They are blocked mutants of *A. missouriensis* ATCC 31683 (producer of actaplanin components B_3, E_1, and pseudoaglycone) which in turn derives from mutation of *Actinoplanes* ATCC 23342 (actaplanin producer). The two strains NRRL 15646 and NRRL 15647 do not produce actaplanin (HERSHBERGER, 1985).

Bromoactaplanins carrying a Br atom instead of the Cl on the phenyl residue of amino acid 6 were obtained by supplying a spontaneous mutant of the producing strain with NaBr (HUBER et al., 1988). As discussed in Sect. 8.2 the acyl chains of the teicoplanin factors originate from degradation of long-chain fatty acids. This notion was exploited to biosynthesize new antibiotic derivatives. Addition of ricinoleic acid (12-hydroxyoleic acid) to cultures of *A. teichomyceticus* resulted in the formation of a novel teicoplanin component characterized by a 4-hydroxydecanoyl chain. Similarly, addition of linolenic acid (which has an 18 carbon linear chain with three double bonds) gives raise to a teicoplanin derivative with a 4,7-decadienoyl acyl chain (unpublished data from the authors' laboratory).

Selective removal of amino acid 1 of vancomycin (to obtain "vancomycin hexapeptide") was carried out by biotransformation, as an alternative to the chemical method. Screening of soil microorganisms for their ability to grow on a model compound (D-leucyl-D-tyrosine) as the sole carbon source led to the isolation of a strain *(Actinomadura citrea)* that converted vancomycin B into the hexapeptide in 90% yields (ZMIJEWSKI et al., 1989).

11 References

ANG, S-G., WILLIAMSON, M. P., WILLIAMS, D. H. (1988), Structure elucidation of a glycopeptide antibiotic, OA-7653, *J. Chem. Soc. Perkin Trans.* **I,** 1949–1956.

ARDUINO, R. C., MURRAY, B. E. (1993), Vancomycin resistance in Gram-positive organisms, *Curr. Op. Infect. Dis.* **6,** 715–724.

ARJUNA RAO, V., RAVISHANKAR, D., SADHUKHAN, A. K., AHMED, S. M., GOEL, A. K., PRABHU, N. S., VERMA, A. K., VENKATESWARLU, A., ALLAUDEEN, H. S., HEDDE, R. D., NISBET, L. J. (1986), Synmonicins: a novel antibiotic complex produced by *Synnemomyces mamnoorii* gen. et sp. nov. I. Taxonomy of the producing organism, fermentation and biological properties, Abstract No. 939, *26rd Interscience Conference on Antimicrobial Agents and Chemotheraphy,* New Orleans: Am. Soc. Microbiol.

ARTHUR, M., COURVALIN, P. (1993), Genetics and mechanisms of glycopeptide resistance in enterococci, *Antimicrob. Agents Chemother.* **37,** 1563–1571.

ATHALYE, M., COATES, N. J., MILNER, P. H. (1989), *Eur. Patent Appl.* 339982.

BARNA, J. C. J., WILLIAMS, D. H. (1984), The structure and mode of action of glycopeptide antibiotics of the vancomycin group, *Ann. Rev. Microbiol.* **38,** 339–357.

BARNA, J. C. J., WILLIAMS, D. H., STONE, D. J. M., LEUNG, T.-W. C., DODDRELL, D. M. (1984), Structure elucidation of the teicoplanin antibiotics, *J. Am. Chem. Soc.* **106,** 4895–4902.

BERDNIKOVA, T. F., LOMAKINA, N. N., POTAPOVA, N. P. (1982), Structure of actinoidins A and B, *Antibiotiki* **27,** 252–258.

BOECK, L. D., MERTZ, F. P. (1986), A47934, a novel glycopeptide-aglycone antibiotic produced by a strain of *Streptomyces toyocaensis.* Taxonomy and fermentation studies, *J. Antibiot.* **39,** 1533–1540.

BOECK, L. D., MERTZ, F. P., WOLTER, R. K., HIGGENS, C. E. (1984), N-demethylvancomycin, a novel antibiotic produced by a strain of *Nocardia orientalis.* Taxonomy and fermentation, *J. Antibiot.* **37,** 446–453.

BOECK, L. D., MERTZ, F. P., CLEM, G. M. (1985), A41030, a complex of novel glycopeptide antibiotics produced by a strain of *Streptomyces virginiae.* Taxonomy and fermentation studies, *J. Antibiot.* **38,** 1–8.

BORGHI, A., ANTONINI, P., ZANOL, M., FERRARI, P., ZERILLI, L. F., LANCINI, G. C. (1989), Isolation and structure determination of two new analogs of teicoplanin, a glycopeptide antibiotic, *J. Antibiot.* **42,** 361–366.

BORGHI, A., EDWARDS, D., ZERILLI, L. F., LANCINI, G. C. (1991a), Factors affecting the normal and branched-chain acyl moieties of teicoplanin components produced by *Actinoplanes teichomyceticus, J. Gen. Microbiol.* **137,** 587–592.

BORGHI, A., FERRARI, P., GALLO, G. G., ZANOL, M., ZERILLI, L. F., LANCINI, G. G. (1991b), Microbial demannosylation and mannosylation of teicoplanin derivatives, *J. Antibiot.* **44**, 1444–1451.

BOX, S. J., ELSON, A. L., GILPIN, M. L., WINSTANLEY (1990), MM 47761 and MM 49721, glycopeptide antibiotics produced by a new strain of *Amycolatopsis orientalis.* Isolation, purification and structure determination, *J. Antibiot.* **43,** 931–937.

BOX, S. J., COATES, N. J., DAVIS, C. J., GILPIN, M. L., HOUGE-FRYDRYCH, C. V. S., MILNER, P. H. (1991), MM-55266 and MM-55268, glycopeptide antibiotics produced by a new strain of *Amycolatopsis:* isolation, purification and structure determination, *J. Antibiot.* **44,** 807–813.

BRAZHNIKOVA, M. G., LOMAKINA, N. N., BERDNIKOVA, T. F. (1989), Eremomycin: a new glycopeptide antibiotic, in: *Bioactive Metabolites from Microorganisms* (BUSHELL, M. E., GRAEFE, U., Eds.), Vol. 27, pp. 163–165. Amsterdam: Elsevier.

BROGDEN, R. N., PETERS, D. H. (1994), Teicoplanin. A reappraisal of its antimicrobial activity, pharmacokinetic properties and therapeutic efficacy, *Drugs* **47,** 823–854.

CAMPOLI-RICHARDS, D. M., BROGDEN, R. N., FAULDS, D. (1990), Teicoplanin. A review of its antibacterial activity, pharmacokinetic properties and therapeutic potential, *Drugs* **40,** 449–486.

CASSANI, G. (1989), Glycopeptides: antibiotic discovery and mechanism of action, in: *Bioactive Metabolites from Microorganisms* (BUSHELL, M. E., GRAEFE, U., Eds.), pp. 221–235. Amsterdam: Elsevier.

CASSETTA, A., BINGEN, E., LAMBERT-ZECHOVSKY, N. (1991), La vancomycine en 1991: actualité et perspectives, *Path. Biol.* **39,** 700–708.

CAVALLERI, B., PARENTI, F. (1992), Glycopeptides (dalbaheptides), in: *Kirk-Othmer Encyclopedia of Chemical Technology,* 4/e, Vol. 2, pp. 995–1018. New York: J. Wiley and Sons.

CHAKRABORTY, T. K., VENKAT REDDY, G. (1992), Studies directed toward the synthesis of glycopeptide antibiotic teicoplanin: first synthesis of the N-terminal 14-membered ring, *J. Org. Chem.* **57,** 5462–5469.

CHRISTENSEN, S. B., ALLAUDEEN, H. S., BURKE, M. R., CARR, S. A., CHUNG, S. K., DEPHILLIPS, P., DINGERDISSEN, J. J., DIPAOLO, M., GIOVENELLA, A. J., HEALD, S. L., KILLMER, L. B., MICO, B. A., MUELLER, L., PAN, C. H., POEHLAND, B. L., RAKE, J. B., ROBERTS, G. D., SHEARER, M. C., SITRIN, R. D., NISBET, L. J., JEFFS, P. W. (1987), Parvodicin, a novel glycopeptide from a new species, *Actinomadura parvosata:* discovery, taxonomy, activity and structure elucidation, *J. Antibiot.* **40,** 970–990.

CHUNG, S. K., TAYLOR, P., OH, Y. K., DEBROSSE, C., JEFFS, P. W. (1986a), Biosynthetic studies on aridicin antibiotics. I. Labeling patterns and overall pathways, *J. Antibiot.* **39,** 642–651.

CHUNG, S. K., OH, Y. K., TAYLOR, P., GERBER, R., NISBET, L. J. (1986b), Biosynthetic studies on aridicin antibiotics. II. Microbial transformations and glycosylations by protoplasts, *J. Antibiot.* **39,** 652–659.

CLEM, G. M., BOECK, L. D., ANDERSON, M. T., MICHEL, F. H. (1985), *Eur. Patent Appl.* 142285.

COATES, N. J., ELSON, A. L., ATHALYE, M., CURTIS, L. M., MOORES, L. V. (1990a), *Eur. Patent Appl.* 375213.

COATES, N. J., SYKES, R., CHRISTOPHER, D. J., CURTIS, L. M. (1990b), *Eur. Patent Appl.* 375448.

COATES, N. J., DAVIS, C. J., CURTIS, L. M., SIKES, R. (1991), *WO Patent* 91/16346.

CORONELLI, C., GALLO, G. G., CAVALLERI, B. (1987), Teicoplanin: chemical, physico-chemical and biological aspects, *Il Farmaco, Ed. Scient.* **42,** 767–786.

CORTI, A., CASSANI, G. (1985), Synthesis and characterization of D-alanyl-D-alanine-agarose: a new bioselective adsorbent for affinity chromatography of glycopeptide antibiotics, *Appl. Biochem. Biotechnol.* **11,** 101–109.

CORTI, A., RURALI, L., BORGHI, A., CASSANI, G. (1985), Solid-phase enzyme–receptor assay (SPERA): A competitive binding assay for glycopeptide antibiotics of the vancomycin class, *Clin. Chem.* **31,** 1606–1610.

DEBONO, M., MOLLOY, R. M., BARNHART, M., DORMAN, D. E. (1980), A35512, a complex of new antibacterial antibiotics produced by *Streptomyces candidus.* II. Chemical studies on A35512B, *J. Antibiot.* **33,** 1407–1416.

DEBONO, M., MERKEL, K. E., MOLLOY, R. M., BARNHART, M., PRESTI, E., HUNT, A. H., HAMILL, R. L. (1984), Actaplanin, new glycopeptide antibiotics produced by *Actinoplanes missouriensis.* The isolation and preliminary chemical characterization of actaplanin, *J. Antibiot.* **37,** 85–95.

DINGERDISSEN, J. J., SITRIN, R. D., DEPHILLIPS, P. A., GIOVENELLA, A. J., GRAPPEL, S. F., MEHTA, R. J., OH, Y. K., PAN, C. H., ROBERTS,

G. D., SHEARER, M. C., NISBET, L. J. (1987), Actinoidin A_2, a novel glycopeptide: production, preparative HPLC separation and characterization, *J. Antibiot.* **40,** 165–172.

DOOLIN, L. E., GALE, R. M., GODFREY JR., O. W., HAMILL, R. L., MAHONEY, D. F., YAO, R. C-F. (1989), *Eur. Patent Appl.* 299707.

EVANS, D. E., DEVRIES, K. M. (1994), Approaches to the synthesis of the vancomycin aglycones, in: *Glycopeptide Antibiotics* (NAGARAJAN, R., Ed.), pp. 63–104. New York: Marcel Dekker.

FELMINGHAM, D. (1993), Towards the ideal glycopeptide, *J. Antimicrob. Chemother.* **32,** 663–666.

FOLENA-WASSERMAN, G., POEHLAND, B. L., YEUNG, E. W-K., STAIGER, D., KILLMER, L. B., SNADER, K., DINGERDISSEN, J. J., JEFFS, P. W. (1986), Kibdelins (AAD-609), novel glycopeptide antibiotics. II. Isolation, purification and structure, *J. Antibiot.* **39,** 1395–1406.

FOLENA-WASSERMAN, G., SITRIN, R., CHAPIN, F., SNADER, K. (1987), Affinity chromatography of glycopeptide antibiotics, *J. Chromatogr.* **392,** 225–238.

FRANCO, C. M. M., CHATTERJEE, S., VIJAKUMAR, E. K. S., CHATTERJEE, D. K., GANGULI, B. N., RUPP, R. H., FEHLHABER, H-W., KOGLER, H., SELBERT, G., TEETZ, V. (1990), *Eur. Patent Appl.* 356894.

GAUSE, G. F., BRAZHNIKOVA, M. G., LOMAKINA, N. N., BERDNIKOVA, T. F., FEDOROVA, G. B., TOKAREVA, N. L., BORISOVA, V. N., BATTA, G. Y (1989), Eremomycin – New glycopeptide antibiotic: chemical properties and structure, *J. Antibiot.* **42,** 1790–1799.

GOLDSTEIN, B. P., SELVA, E., GASTALDO, L., BERTI, M., PALLANZA, R., RIPAMONTI, F., FERRARI, P., DENARO, M., ARIOLI, V., CASSANI, G. (1987), A40926, a new glycopeptide antibiotic with anti-*Neisseria* activity, *Antimicrob. Agents Chemother.* **31,** 1961–1966.

GOLDSTEIN, B. P., ROSINA, R., PARENTI, F. (1994), Teicoplanin, in: *Glycopeptide Antibiotics* (NAGARAJAN, R., Ed.), pp. 273–307. New York: Marcel Dekker.

GOOD, V. M., GWYNN, M. N., KNOWLES, D. J. C. (1990), MM 45289, a potent glycopeptide antibiotic which interacts weakly with diacetyl-L-lysyl-D-alanyl-D-alanine, *J. Antibiot.* **43,** 550–555.

HAMILL, R. L., YAO, R. C. (1993), *U.S. Patent* 5187082.

HAMILL, R. L., HANEY, M. E., JR., STARK, W. M. (1973), *Ger. Patent Appl.* 2252937.

HAMILL, R. L., MAHONEY, D. F., NAKASUKASA, W. M., YAO, R. C-F. (1988), *Eur. Patent Appl.* 265701.

HAMMOND, S. J., WILLIAMSON, M. P., WILLIAMS, D. H., BOECK, L. D., MARCONI, G. G. (1982), On the biosynthesis of the antibiotic vancomycin, *J. Chem. Soc., Chem. Commun.* **1982**, 344–346.

HAMMOND, S. J., WILLIAMS, D. H., NIELSEN, R. V. (1983), The biosynthesis of ristocetin, *J. Chem. Soc., Chem. Commun.* **1983**, 116–117.

HARRIS, C. M., KOPECKA, H., HARRIS, T. M. (1983), Vancomycin: Structure and transformation to CDP-I, *J. Am. Chem. Soc.* **105,** 6915–6922.

HARRIS, C. M., KANNAN, R., KOPECKA, H., HARRIS, T. M. (1985), The role of chlorine substituents in the antibiotic vancomycin: preparation and characterization of mono- and didechlorovancomycin, *J. Am. Chem. Soc.* **107,** 6652–6658.

HARRIS, C. M., FESIK, S. W., THOMAS, A. M., KANNAN, R., HARRIS, T. M. (1986), Iodination of vancomycin, ristocetin A, and ristocetin pseudoaglycon, *J. Org. Chem.* **51,** 1509–1513.

HEALD, S. L., MUELLER, L., JEFFS, P. W. (1987), Actinoidins A and A_2: Structure determination using 2D NMR methods, *J. Antibiot.* **40,** 630–645.

HERSHBERGER, C. L. (1985), *Eur. Patent Appl.* 159436.

HIGGINS, H. M., JR., MCCORMICK, M. H., MERKEL, K. E., MICHEL, K. H. (1985), *Eur. Patent Appl.* 159180.

HOLDEN, K. S., RUDDOCK, J. C., TONE, J., MAEDA, H. (1991), *Brit. Patent* 2243610.

HOLDOM, K. S., MAEDA, H., RUDDOCK, J. C., TONE, J. (1988), *Eur. Patent Appl.* 265143.

HUBER, F. M., PIEPER, R. L., TIETZ, A. J. (1987), Characterization of the process of the biosynthesis of the actaplanin complex by *Actinoplanes missouriensis, J. Ferment. Technol.* **65,** 85–89.

HUBER, F. M., MICHEL, K. L., HUNT, A. H., MARTIN, J. W., MOLLOY, R. M. (1988), Preparation and characterization of some bromine analogs of the glycopeptide antibiotic actaplanin, *J. Antibiot.* **41,** 798–801.

HUNT, A. H., ELZEY, T. K., MERKEL, K. E., DEBONO, M. (1984a), Structures of actaplanins, *J. Org. Chem.* **49,** 641–645.

HUNT, A. H., MARCONI, G. G., ELZEY, T. K., HOEHN, M. M. (1984b), A51568A: N-demethylvancomycin, *J. Antibiot.* **37,** 917–919.

HUNT, A. H., DORMAN, D. E., DEBONO, M., MOLLOY, R. M. (1985), Structure of antibiotic A41030A, *J. Org. Chem.* **50,** 2031–2035.

JOHNSON, A. P., UTTLEY, A. H. C., WOODFORD, N., GEORGE, R. C. (1990), Resistance to vancomycin and teicoplanin: an emerging clinical problem, *Clin. Microbiol. Rev.* **3,** 280–291.

KAMOGASHIRA, T., NISHIDA, T., SUGAWARA, M. (1983), A new glycopeptide antibiotic, OA-7653, produced by *Streptomyces hygroscopicus* ssp. *hiwasaensis, Agric. Biol. Chem.* **47,** 499–506.

KANNAN, R., HARRIS, C. M., HARRIS, T. M., WALTHO, J. P., SKELTON, N. J., WILLIAMS, D. H. (1988), Function of the amino sugar and N-terminal amino acid of the antibiotic vancomycin in its complexation with cell wall peptides, *J. Am. Chem. Soc.* **110,** 2946–2953.

KATRUKHA, G. S., SILAEV, A. B. (1986), The chemistry of glycopeptide antibiotics of the vancomycin group, in: *Chemistry of Peptides and Proteins* (VOELTER, W., BAYER, E., OVCHINNIKOV, Y. A., IVANOV, V. T., Eds.), Vol. 3., pp. 289–306. Berlin: de Gruyter.

KIM, K-H., MARTIN, Y., OTIS, E., MAO, J. (1989), Inhibition of ^{125}I-labeled ristocetin binding to *Micrococcus luteus* cells by the peptides related to bacterial cell wall mucopeptide precursors: quantitative structure–activity relationships, *J. Med. Chem.* **32,** 84–93.

KOBRIN, M. B., FEDOROVA, J. B., KATRUKHA, G. S. (1988), Study on deglycosylation of certain antibiotics belonging to vancomycin group, *Antibiotiki i Chemother.* **33,** 331–335.

KUNSTMANN, M. P., PORTER, J. N. (1974), *U.S. Patent* 3803306.

LANCINI, G. C. (1989), Fermentation and biosynthesis of glycopeptide antibiotics, in: *Bioactive Metabolites from Microorganisms* (BUSHELL, M. E., GRAEFE, U., Eds.), pp. 283–296. Amsterdam: Elsevier.

LANCINI, G. C., CAVALLERI, B. (1990), Glycopeptide antibiotics of the vancomycin group, in: *Biochemistry of Peptide Antibiotics, Recent Advances in the Biotechnology of β-Lactams and Microbial Bioactive Peptides* (KLEINKAUF, H., VON DÖHREN, H., Eds.), pp. 159–178. Berlin: de Gruyter.

LECHEVALIER, M. P., PRAUSER, H., LABEDA, D. P., RUAN, J.-S. (1986), Two new genera of nocardioform actinomycetes: *Amycolata* gen. nov. and *Amycolatopsis* gen. nov., *Int. J. Syst. Bacteriol.* **36,** 29–37.

MALABARBA, A., CIABATTI, R. (1992), *Int. Patent* WO 92/10517.

MALABARBA, A., STRAZZOLINI, P., DEPAOLI, A., LANDI, M., BERTI, M., CAVALLERI, B. (1984), Teicoplanin, antibiotics from *Actinoplanes teichomyceticus* nov. sp. VI. Chemical degradation: physico-chemical and biological properties of acid hydrolysis products, *J. Antibiot.* **37,** 988–999.

MALABARBA, A., FERRARI, P., CIETTO, G., PALLANZA, R., BERTI, M. (1989a), Synthesis and biological activity of N^{63}-carboxypeptides of teicoplanin and teicoplanin aglycone, *J. Antibiot.* **42,** 1800–1816.

MALABARBA, A., SPREAFICO, F., FERRARI, P., KETTENRING, J., STRAZZOLINI, P., TARZIA, G., PALLANZA, R., BERTI, M., CAVALLERI, B. (1989b), Dechloro teicoplanin antibiotics, *J. Antibiot.* **42,** 1684–1697.

MALABARBA, A., TRANI, A., STRAZZOLINI, P., CIETTO, G., FERRARI, P., TARZIA, G., PALLANZA, R., BERTI, M. (1989c), Synthesis and biological properties of N^{63}-carboxamides of teicoplanin antibiotics. Structure–activity relationships, *J. Med. Chem.* **32,** 2450–2460.

MALABARBA, A., CIABATTI, R., KETTENRING, J., SCOTTI, R., CANDIANI, G., PALLANZA, R., BERTI, M., GOLDSTEIN, B. P. (1992), Synthesis and antibacterial activity of a series of basic amides of teicoplanin and deglucoteicoplanin with polyamines, *J. Med. Chem.* **35,** 4054–4060.

MALABARBA, A., CIABATTI, R., CAVALLERI, B. (1993a), Semisynthetic dalbaheptides: chemistry and biological activity, *Curr. Topics Med. Chem.* **1,** 359–376.

MALABARBA, A., CIABATTI, R., PANZONE, G., MARAZZI, A. (1993b), *WO Patent* 93/03060.

MATSUMOTO, K. (1961), A vancomycin-related antibiotic from *Streptomyces* sp. K-288. Studies on *Streptomyces* antibiotics, XXXVIII, *J. Antibiot., Ser. A* **14,** 141–146.

MCGAHREN, W. J., MARTIN, J. H., MORTON, G. O., HARGREAVES, R. T., LEESE, R. A., LOVELL, F. M., ELLESTAD, G. A., O'BRIEN, E., HOLKER, J. S. E. (1980), Structure of avoparcin components, *J. Am. Chem. Soc.* **102,** 1671–1674.

MCGAHREN, W. J., LEESE, R. A., BARBATSCHI, F., MORTON, G. O., KUCK, N. A., ELLESTAD, G. A. (1983), Components and degradation compounds of the avoparcin complex, *J. Antibiot.* **36,** 1671–1682.

MERTZ, F. P., DOOLIN, L. E. (1973), The effect of inorganic phosphate on the biosynthesis of vancomycin, *Can. J. Microbiol.* **19,** 263–270.

MICHEL, K. H., SHAH, R. M., HAMILL, R. L. (1980), A35512, a complex of new antibacterial antibiotics produced by *Streptomyces candidus.* I. Isolation and characterization, *J. Antibiot.* **33,** 1397–1406.

MICHEL, K. J., YAO, R. C-F. (1991), *Eur. Patent Appl.* 424051.

NADKARNI, S. R., PATEL, M. V., CHATTERJEE, S., VIJAYAKUMAR, E. K. S., DESIKAN, K. R., BLUMBACH, J., GANGULI, B. N., LIMBERT, M. (1994), Balhimycin, a new glycopeptide antibiotic produced by *Amycolatopsis* sp. Y-86-21022. Taxonomy, production, isolation and biological activity, *J. Antibiot.* **47,** 334–341.

NAGARAJAN, R. (1993), Structure–activity relationships of vancomycin-type glycopeptide antibiotics, *J. Antibiot.* **46,** 1181–1195.

NAGARAJAN, R. (Ed.) (1994a), *Glycopeptide Antibiotics.* New York: Marcel Dekker.

NAGARAJAN, R. (1994b), Structure–activity relationships of vancomycin antibiotics, in: *Glycopeptide Antibiotics* (NAGARAJAN, R., Ed.), pp. 195–218. New York: Marcel Dekker.

NAGARAJAN, R., SCHABEL, A. A. (1988), Selective cleavage of vancosamine, glucose, and N-methyl leucine from vancomycin and related antibiotics, *J. Chem. Soc., Chem. Commun.* **1988**, 1306–1307.

NAGARAJAN, R., SCHABEL, A. A., OCCOLOWITZ, J. L., COUNTER, F. T., OTT, J. L. (1988), Synthesis and antibacterial activity of N-acyl vancomycins, *J. Antibiot.* **41,** 1430–1438.

NAGARAJAN, R., BERRY, D. M., HUNT, A. H., OCCOLOWITZ, J. L., SCHABEL, A. A. (1989a), Conversion of antibiotic A82846B to orienticin A and structural relationships of related antibiotics, *J. Org. Chem.* **54,** 983–986.

NAGARAJAN, R., BERRY, D. M., SCHABEL, A. A. (1989b), The structural relationships of A82846B and its hydrolysis products with chloroorienticins A, B and C, *J. Antibiot.* **42,** 1438–1440.

NICAS, T. I., ALLEN, N. E. (1994), Resistance and mode of action, in: *Glycopeptide Antibiotics* (NAGARAJAN, R., Ed.), pp. 219–241. New York: Marcel Dekker.

NIETO, M., PERKINS, H. R. (1971a), Modification of the acyl-D-alanyl-D-alanine terminus affecting complex-formation with vancomycin, *Biochem. J.* **123,** 789–803.

NIETO, M., PERKINS, H. R. (1971b), The specificity of combination between ristocetins and peptides related to bacterial cell wall mucopeptide precursors, *Biochem. J.* **124,** 845–852.

OKAZAKI, T., ENOKITA, R., MIYAOKA, H., TAKATSU, T., TORIKATA, A. (1987), Chloropolysporins A, B and C, novel glycopeptide antibiotics from *Faenia interjecta* sp. nov. I. Taxonomy of producing organism, *J. Antibiot.* **40,** 917–923.

PALLANZA, R., BERTI, M., GOLDSTEIN, B. P., MAPELLI, E., RANDISI, E., SCOTTI, R., ARIOLI, V. (1983), Teichomycin: *in vitro* and *in vivo* evaluation in comparison with other antibiotics, *J. Antimicrob. Chemother.* **11,** 419–425.

PARENTI, F., CAVALLERI, B. (1989), Proposal to name the vancomycin–ristocetin like glycopeptides as dalbaheptides, *J. Antibiot.* **42,** 1882–1883.

PARENTI, F., BERETTA, G., BERTI, M., ARIOLI, V. (1978), Teichomycins, new antibiotics from *Actinoplanes teichomyceticus* nov. sp. I. Description of the producer strain, fermentation and biological properties, *J. Antibiot.* **31,** 276–283.

PERKINS, H. R. (1982), Vancomycin and related antibiotics, *Pharmacol. Ther.* **16,** 181–197.

PERKINS, H. R., NIETO, M. (1970), The preparation of iodinated vancomycin and its distribution in bacteria treated with the antibiotic, *Biochem. J.* **116,** 83–92.

PHILIP, J. E., SCHENCK, J. R., HARGIE, M. P., HOLPER, J. C., GRUNDY, W. E. (1960), The increased activity of ristocetins A and B following acid hydrolysis, *Antimicrob. Agents Ann.* **1960,** 10–16.

PHILLIPS, G., GOLLEDGE, C. L. (1992), Vancomycin and teicoplanin. Something old, something new, *Med. J. Aust.* **156,** 53–57.

PITKIN, D. H., MICO, B. A., SITRIN, R. D., NISBET, L. J. (1986), Charge and lipophilicity govern the pharmacokinetics of glycopeptide antibiotics, *Antimicrob. Agents Chemother.* **29,** 440–444.

RAKE, J. B., GERBER, R., METHA, R. J., NEWMAN, D. K., OH, Y. K., PHELEN, C., SHEARER, M. C., SITRIN, R. D., NISBET, L. J. (1986), Glycopeptide antibiotics: a mechanism based screen employing a bacterial cell wall receptor mimetic, *J. Antibiot.* **39,** 58–67.

RAMA RAO, A. V., CHAKRABORTY, T. K., JOSHI, S. P. (1992), The first synthesis of C-terminal biphenyl moiety of vancomycin, *Tetrahedron Lett.* **33,** 4045–4048.

REYNOLDS, P. E. (1989), Structure, biochemistry and mechanism of action of glycopeptide antibiotics, *Eur. J. Clin. Microbiol. Infect. Dis.* **8,** 943–950.

RIVA, E., GASTALDO, L., BERETTA, M. G., FERRARI, P., ZERILLI, L. F., CASSANI, G., GOLDSTEIN, B. P., BERTI, M., PARENTI, F., DENARO, M. (1989), A42867, a novel glycopeptide antibiotic, *J. Antibiot.* **42,** 497–505.

SELVA, E., GASTALDO, L., BERETTA, G., BORGHI, A., GOLDSTEIN, B. P., ARIOLI, V., CASSANI, G., PARENTI, F. (1986), *Eur. Patent Appl.* 177882.

SELVA, E., BERETTA, G., BORGHI, A., DENARO, M. (1988), *WO Patent Appl.* 88/02755.

SELVA, E., GASTALDO, L., CASSANI, G., PARENTI, F. (1992), *Eur. Patent Appl.* 259781.

SHEARER, M. C., ACTOR, P., BOWIE, B. A., GRAPPEL, S. F., NASH, C. H., NEWMAN, D. J., OH, Y. K., PAN, C. H., NISBET, L. J. (1985), Aridicins, novel glycopeptide antibiotics. I. Taxonomy, production and biological activity, *J. Antibiot.* **38,** 555–560.

SHEARER, M. C., GIOVENELLA, A. J., GRAPPEL, S. F., HEDDE, R. D., MEHTA, R. J., OH, Y. K., PAN, C. H., PITKIN, D. H., NISBET, L. J. (1986), Kibdelins, novel glycopeptide antibiotics. I. Discovery, production, and biological evaluation, *J. Antibiot.* **39,** 1386–1394.

SHELDRICK, G. M., JONES, P. G., KENNARD, O., WILLIAMS, D. H., SMITH, G. H. (1978), Structure of vancomycin and its complex with acetyl-D-alanyl-D-alanine, *Nature* (London) **271,** 223–225.

SITRIN, R. D., FOLENA-WASSERMAN, G. (1989), Affinity and HPLC purification of glycopeptide antibiotics, in: *Natural Products Isolation* (WAGMAN, G. H., COOPER, R., Eds.), pp. 111–152. Amsterdam: Elsevier.

SITRIN, R. D., FOLENA-WASSERMAN, G. (1994), Separation methodology, in: *Glycopeptide Antibiotics* (NAGARAJAN, R., Ed.), pp. 29–61. New York: Marcel Dekker.

SITRIN, R. D., CHAN, G. W., DINGERDISSEN, J. J., HOLL, W., HOOVER, J. R. E., VALENTA, J. R., WEBB, L., SNADER, K. M. (1985), Aridicins, novel glycopeptide antibiotics. II. Isolation and characterization, *J. Antibiot.* **38,** 561–571.

SITRIN, R. D., CHAN, G. W., CHAPIN, F., GIOVENELLA, A. J., GRAPPEL, S. F., JEFFS, P. W., PHILLIPS, L., SNADER, K. M., NISBET, L. J. (1986), Aridicins, novel glycopeptide antibiotics. III. Preparation, characterization, and biological activity of aglycone derivatives, *J. Antibiot.* **39,** 68–75.

SKELTON, N. J., WILLIAMS, D. H. (1990), Structure elucidation of the novel glycopeptide antibiotic UK-68597, *J. Org. Chem.* **55,** 3718–3723.

SKELTON, N. J., WILLIAMS, D. H., RANCE, M. J., RUDDOCK, J. C. (1990), Structure elucidation of UK-72051, a novel member of the vancomycin group of antibiotics, *J. Chem. Soc. Perkin Trans.* **1,** 77–81.

SKELTON, N. J., WILLIAMS, D. H., RANCE, M. J., RUDDOCK, J. C. (1991), Structure elucidation of a novel antibiotic of the vancomycin group. The influence of ion–dipole interactions on peptide backbone conformation, *J. Am. Chem. Soc.* **113,** 3757–3765.

SOMMA, S., GASTALDO, L., CORTI, A. (1984), Teicoplanin, a new antibiotic from *Actinoplanes teichomyceticus* nov. sp., *Antimicrob. Agents Chemother.* **26,** 917–923.

SPIRI-NAKAGAWA, P., FUKUSHI, Y., MAEBASHI, K., IMAMURA, N., TAKAHASHI, Y., TANAKA, Y., TANAKA, H., ŌMURA, S. (1986), Izupeptins A and B, new glycopeptide antibiotics produced by an actinomycete, *J. Antibiot.* **39,** 1719–1723.

SZTARICSKAI, F., BOGNÁR, R. (1984), The chemistry of the vancomycin group of antibiotics, in: *Recent Developments in the Chemistry of Natural Carbon Compounds* (BOGNÁR, R., SZANTAY, Cs., Eds.), Vol. X, pp. 91–201. Budapest: Akadémiai Kiadó.

TAKATSU, T., TAKAHASHI, S., NAKAJIMA, M., HANEISHI, T., NAKAMURA, T., KUWANO, H., KINOSHITA, T. (1987a), Chloropolysporins A, B and C, novel glycopeptide antibiotics from *Faenia interjecta* sp. nov. III. Structure elucidation of chloropolysporins, *J. Antibiot.* **40,** 933–940.

TAKATSU, T., TAKAHASHI, S., TAKAMATSU, Y., SHIOIRI, T., IWADO, S., HANEISHI, T. (1987b), Chloropolysporins A, B and C, novel glycopeptide antibiotics from *Faenia interjecta* sp. nov. IV. Partially deglycosylated derivatives, *J. Antibiot.* **40,** 941–945.

TAKEUCHI, M., ENOKITA, R., OKAZAKI, T., KAGASAKI, T., INUKAI, M. (1991a), Helvecardins A and B, novel glycopeptide antibiotics. I. Taxonomy, fermentation, isolation and physico-chemical properties, *J. Antibiot.* **44,** 263–270.

TAKEUCHI, M., TAKAHASHI, S., INUKAI, M. (1991b), Helvecardins A and B, novel glycopeptide antibiotics. II. Structural elucidation, *J. Antibiot.* **44,** 271–277.

TAKEUCHI, M., TAKAHASHI, S., ENOKITA, R., SAKAIDA, Y., HARUYAMA, H., NAKAMURA, T., KATAYAMA, T., INUKAI, M. (1992), Galacardins A and B, new glycopeptide antibiotics, *J. Antibiot.* **45,** 297–305.

TAMURA, A., TAKEDA, I. (1975), Antibiotic AB-65, a new antibiotic from *Saccharomonospora viride, J. Antibiot.* **28,** 395–397.

TOROPOVA, E. G., EGOROV, N. S., OBRAZTSOVA, A. Y. (1974), Effect of ristomycin and excess of phosphorus in the medium on activity of enzymes of *Proactinomyces fructiferi* var. *ristomycini, Antibiotiki* **19,** 418–422.

TOROPOVA, E. G., EGOROV, N. S., TKHAKER, V., BURAKAEVA, A. D. (1982), Effect of various sources of carbon and nitrogen on biosynthesis of ristomycin, protease and pigment by *Nocardia fructiferi* var. *ristomycini, Antibiotiki* **27,** 749–753.

TSUJI, N., KAMIGAUCHI, T., KOBAYASHI, M., TERUI, Y. (1988a), New glycopeptide antibiotics. II. The isolation and structures of chloroorienticins, *J. Antibiot.* **41,** 1506–1510.

TSUJI, N., KOBAYASHI, M., KAMIGAUCHI, T., YOSHIMURA, Y., TERUI, Y. (1988b), New glycopeptide antibiotics. I. The structures of orienticins, *J. Antibiot.* **41,** 819–822.

WALLAS, C. H., STROMINGER, J. L. (1963), Ristocetins, inhibitors of cell wall synthesis in *Staphylococcus aureus, J. Biol. Chem.* **238,** 2264–2266.

WILLIAMSON, M. P., WILLIAMS, D. H., HAMMOND, S. J. (1984), Interactions of vancomycin and ristocetin with peptides as a model for protein binding, *Tetrahedron* **40,** 569–577.

YAO, R. C., MAHONEY, D. F., BAISDEN, D. K., MERTZ, F. P., MABE, J. A., NAKATSUKASA, W.

M. (1988), A82846, a new glycopeptide antibiotic complex, produced by *Amycolatopsis orientalis.* I. Discovery, taxonomy and fermentation studies, Abstract No. 974, *28th Interscience Conference on Antimicrobial Agents and Chemotherapy,* Los Angeles: Am. Soc. Microbiol.

ZECKEL, M. L., WOODWORTH, R. (1994), Vancomycin, a clinical overview, in: *Glycopeptide Antibiotics* (NAGARAJAN, R., Ed.), pp. 309–409. New York: Marcel Dekker.

ZERILLI, L. F., EDWARDS, D. M., BORGHI, A., GALLO, G. G., SELVA, E., DENARO, M., LANCINI, G. C. (1992), Determination of the acyl moieties of the antibiotic complex A40926 and their relation with the membrane lipids of the producer strain, *Rapid Commun. Mass. Spectrom.* **6,** 109–114.

ZMIJEWSKI, M. J., BRIGGS, B., LOGAN, R., BOECK, L. D. (1987), Biosynthetic studies on antibiotic A47934, *Antimicrob. Agents Chemother.* **31,** 1497–1501.

ZMIJEWSKI, M. J., LOGAN, R. M., MARCONI, G., DEBONO, M., MOLLOY, R. M., CHADWELL, F., BRIGGS, B. (1989), Biotransformation of vancomycin B to vancomycin hexapeptide by a soil microorganism, *J. Nat. Prod.* **52,** 203–206.

10 Aminoglycosides and Sugar Components in Other Secondary Metabolites

WOLFGANG PIEPERSBERG, JÜRGEN DISTLER

Wuppertal, Germany

1 Introduction

There is no clear chemical or biochemical definition of the term "aminoglycoside" which is traditionally reserved for mono- to oligosaccharidic sugar and/or cyclitol derivatives containing amino nitrogen. Therefore, an attempt is made in this chapter to demonstrate that in biological and especially genetic terms there are many parallels between origins and uses of strongly derived sugar components in secondary metabolites in general and with respect to the aminocyclitol aminoglycosides in particular. A list of the major chemical groups of aminoglycosides (see Tab. 1) shows that some new structures have been described in the past decade.

Since the 1st Edition of *"Biotechnology"* (cf. UMEZAWA et al., 1986) the field of aminoglycoside research has undergone a significant change characterized by the following divergent aspects:

(1) ceasing interest in the search for and development of new antibiotics from this particular group of compounds accompanied by stagnant market figures and clinical use;
(2) increasing interest in basic research on the (molecular) biological fundamentals of the production of aminoglycosides and other secondary metabolic carbohydrates, especially initiated by the new methods of genetic engineering in actinomycetes, bacilli, and other groups of biotechnological interest;
(3) evaluation of new classes of related natural products for other types of applications, e.g., inhibitors of glycosidases and other enzymes.

However, this situation might change again due to the currently increasing problems with antibiotic resistance and the appearance of new and reappearence of "old" infectious deseases (see discussion in DAVIES, 1992; HOTTA et al., 1995). Examples are methicillin-resistant staphylococci, bacterial infections in HIV patients, and the worldwide renaissance of tuberculosis as a major health concern. In line with this development the authors of this chapter focus mainly on new aspects of the molecular genetics, biochemistry, and physiology of production of aminoglycosides and the similarly modified sugar components of other chemical classes of secondary metabolites. In addition, future biotechnological applications of this knowledge are discussed.

Many details of the chemistry, mode of action, classical strain improvement, fermentation technology, and pharmacology, e.g., of 2-deoxystreptamine-containing, streptomycin-like and other related aminocyclitol-aminoglycosides are well known from many reviews published in the past two decades (DAVIES and YAGISAWA, 1983; GRÄFE, 1992; LORIAN, 1980; MALLAMS, 1988; UMEZAWA and HOOVER 1982; UMEZAWA et al., 1986; WALLACE et al., 1979) and will not be extensively dealt with here. For the progress made in related fields, such as the chemical and enzymic synthesis of carbohydrate components and their analogs, e.g., cyclitols, hexose and pentose derivatives, sugar mimics, and their biological activity and application, the reader is referred to the following publications: KENNEDY, 1988; HORTON et al., 1989; THIEM, 1990; LUKACS and OHNO, 1990; YAMAGUCHI and KAKINUMA, 1992a; OGURA et al., 1992; BILLINGTON, 1993; HUDLICKY and CEBULAK, 1993; COLEMAN and FRASER, 1993; LOOK et al., 1993; KROHN et al., 1993; TESTA et al., 1993; COOK et al., 1994.

2 Isolation, Distribution, Ecology, and Fermentation

In the last decade new metabolites which clearly belong to the aminoglycoside group were detected and described. Tab. 1 gives an overview of the chronology of publications and the biochemical grouping of the most important aminoglycosides. Most sugar-containing secondary metabolites relevant to this chapter are formed by bacteria, mainly actinomycetes, bacilli, and pseudomonads (Fig. 1). However, the bioactive glycosides

Tab. 1. Chronology of the Detection and Biochemical Classification of the Major Aminoglycosides[a]

Aminoglycoside[b]	Year of First Description	Pathway Formula[c]
Streptomycin (A1)	1944	Ca(4)-6DOH(2)-HA
Neomycins (A11)	1949	HA-(4)Cb(5)-P(3)-HA
5′-Hydroxystreptomycin (A1)	1950	Ca(4)-6DOH(2)-HA
Paromomycins (catenulin) (A11)	1952	HA-(4)Cb(5)-P(3)-HA
Kanamycins (A12)	1957	HA-(4)Cb(6)-HA
Trehalosamines (A19)	1957	HA(1)-H
Hygromycin B (A16)	1958	Cb(5)-H(2,3)-HepA
Streptozotocin (A18)	1960	HA
Spectinomycin (A2)	1961	Ca(4,5)-6DOH
Bluensomycin (glebomycin) (A1)	1962	Ca(4)-6DOH(2)-HA
Gentamicins (A14)	1964	HA-(4)Cb(6)-PA
Kasugamycin (A4)	1965	Ca(4)-6DOH
Destomycins (A16)	1965	Cb(5)-H(2,3)-HepA
Nojirimycin (A18)	1966	HA
Apramycin (A15)	1968	Cb(4)-OctA(8)-HA
Tobramycin (A12)	1968	HA-(4)Cb(6)-HA
Sisomicin (A14)	1970	HA-(4)Cb(6)-PA
Ribostamycins (A13)	1970	HA-(4)Cb(5)-P
Validamycins (A21)	1970	Cb(1)-Cb(4)-H
Lividomycins (A11)	1971	HA-(4)Cb(5)-P(3)-HA
Butirosin (A13)	1971	HA-(4)Cb(5)-P
Acarbose (A22)	1972	Cb(1)-6DOH(l)-$(H)_2$
Myomycin (A8)	1973	Ca(4)-HA
SS-56-C (A16)	1973	Ca(5)-H(2,3)-HepA
Minosaminomycin (A5)	1974	Ca(4)-6DOH
Amylostatins (A22)	1974	$(H)_n$-(4)Cb(1)-6DOH(1)-$(H)_n$
Siastatin (A18)	1974	HA
Verdamicin (A14)	1975	HA-(4)Cb(6)-PA
Sorbistins (A24)	1976	Ca(2)-HA
Seldomycins (A17)	1977	HA-(4)Cb(6)-PA
LL-BM123α (A6)	1977	Ca(4)-H(4)-HA
Fortimicins (A10)	1977	Ca(6)-HA
1-Desoxynojirimycins (A18)	1978	HA
Adiposins (A22)	1978	$(H)_n$-(4)Cb(1)-H(1)-$(H)_n$
Trestatins (A22)	1978	$(H)_n$-(4)Cb(1)-6DOH(1)-$(H)_n$
Istamycins (A10)	1979	Cb(6)-HA
Sporaricins (A10)	1979	Cb(6)-HA
Sannamycin (A10)	1979	Cb(6)-HA
Dactimicin (A10)	1979	Ca(6)-HA
Oligostatins (A22)	1979	$(H)_n$-(4)Cb(1)-6DOH(1)-$(H)_n$
Lysinomicin (A10)	1981	Ca(6)-HA
Spenilomycin (A3)	1982	Ca(4,5)-6DOH
Valiolamine (A21)	1984	Cb
5′-Hydroxy-*N*-demethyldihydrostreptomycin (A1)	1985	Ca(4)-6DOH(2)-HA
1-*N*-Amidino-1-*N*-demethyl-2-hydroxy-destomycin (A16)	1985	Ca(5)-H(2,3)-HepA
Neotrehalosadiamine (A19)	1986	HA(1)-HA
AC4437 (A1)	1986	Ca(4)-6DOH
Allosamidin (A25)	1987	Ca(4)-HA(4)-HA
CV-1 (A18)	1987	HA
Boholmycin (A7)	1988	HA-(4)Ca(6)-PA(4)-Hep
Ashimycins (A1)	1989	Ca(4)-6DOH(2)-HA(4)-$(H)_n$
Trehazolin (A23)	1991	Ca(1,2)-HA

[a] For references before 1986 see UMEZAWA and HOOPER (1982) and UMEZAWA et al. (1986), others are referenced in the text.
[b] Numbers in brackets refer to the figures showing the formulae in Sect. 10.
[c] See text in Sect. 3; only the basic formula is given if a group of compounds is characterized.

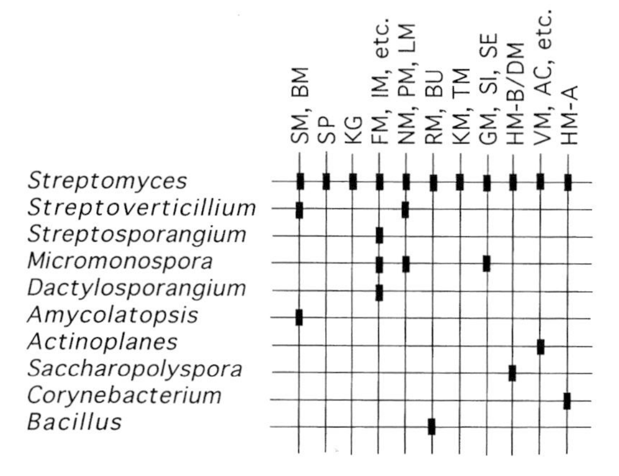

Fig. 1. Aminocyclitol aminoglycoside-producing bacterial genera (extended according to HASEGAWA, 1991).
AC: acarbose and related glycosidase inhibitors; BM: bluensomycin; BU: butirosins; DM: destomycins; FM: fortimicins (astromicins); GM: gentamicins; HM-A,B: hygromycins A or B; IM: istamycins; KG: kasugamycins; KM: kanamycins; LM: lividomycins; NM: neomycins; PM: paromomycins; RM: ribostamycins; SE: seldomycins; SI: sisomicins; SM: streptomycins; SP: spectinomycins; TM: tobramycin; VM: validamycins.

from plants should also be considered. In general, it may be assumed that these glycosides share a common general biochemical basis with those of bacteria and that an equivalent gene pool is used for their cellular production. However, since little is known about their biosynthesis, this chapter concentrates on prokaryotic production systems.

Research covering all aspects of the biology of secondary carbohydrate metabolism in organisms producing secondary extracellular products is still in its initial stages. In contrast, the resistance mechanisms of antibiotic producers for self-protection can be regarded as having been extensively investigated (see Sect. 4.5). Even the basis of physiology in natural biotopes and the ecological role of aminoglycosides and other secondary metabolites are unknown (PIEPERSBERG, 1993; MARSH and WELLINGTON, 1994).

Aminoglycosides have been shown to be produced in soils (WELLINGTON et al., 1993). Therefore, one of the future aims of aminoglycoside aminocyclitol research is to study their ecological role in the biotopes of the producing organisms. There, these substances might be involved in the communication between different organisms and in the hormone-like crosstalk between the differentiated cells of the producer itself (CHADWICK and WHELAN, 1992; PIEPERSBERG, 1993). Also, the routes of dissemination of aminoglycoside resistance mechanisms found in both producers and clinically relevant non-producing bacteria (cf. Sect. 4.5) and the driving selective pressures under natural conditions might become accessible to direct experimental research.

Glycosidic or cyclitol-containing components similar to the so-called secondary metabolites of microbes and plants occur in the structures of many pro- and eukaryotic extracellular polymers, such as polysaccharides, lipopolysaccharides, glycolipids, and other glycoconjugates. So far, in gram-negative unicellular bacteria, only very little evidence has been found for the biosynthesis and secretion of aminoglycosides or other carbohydrate-containing diffusible substances (e.g., pseudomonads produce sorbistins). However, these organisms produce a variable and abundant range of carbohydrate-based extracellular substances, lipopolysaccharides (LPS), and other heteropolymeric polysaccharides (e.g., enterobacterial common antigen: ECA). There is also emerging evidence that the same gene pool is used for the production of these polymers, which in many instances resemble polymerized secondary metabolites, and for the production of the low molecular weight end products of, e.g., actinomycete secondary metabolism itself (see below).

Two recent examples of non-eubacterial aminoglycoside-like compounds are (1) the glucosaminyl archaetidyl-*myo*-inositols (Fig. 2) produced by some methanogenic Archaea (KOGA et al., 1993) and (2) the very similar core structure of the glycosyl phosphatidylinositol protein anchors for outer-surface-attached glycoproteins in eukaryotes (Fig. 3). These are found in a variety of organisms, from trypanosomes to mammals (ENGLUND, 1993). Interestingly, they have the same content of nonacetylated glucosamine and share the 1-phosphoryl-6-glucosami-

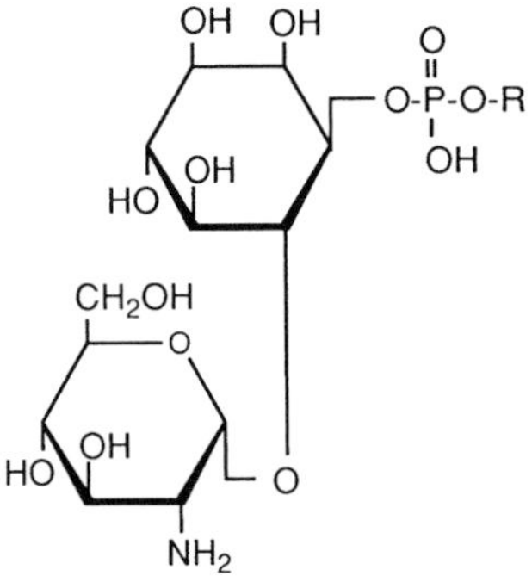

Fig. 2. An aminoglycoside-like glycolipid structure in Archaea.
R: archaetidyl residues (=glyceryl ethers).

nyl *myo*-inositol core unit which is synthesized in the mammalian system from phosphoinositides and UDP-*N*-acetylglucosamine at the inner side of the cytoplasmic membrane. Following deacetylation and acylation at the inositol moiety it is transported to the outer surface where it is condensed to other glycosidic residues (mannosyl, and in some cases galactosyl, *N*-acetyl-galactosaminyl, phosphorylethanolamine). The phosphorylethanolamine unit connects the short mannosyl chain to the C-terminal carboxyl group of the protein via formation of a new amide bond (reviewed by ENGLUND, 1993). These examples might suggest that part of the biochemistry of aminoglycoside biosynthesis and, hence, part of the relevant gene pool are used for different purposes in the metabolism of mainly extracellularly targeted compounds and that they occur in practically all groups of organisms.

3 Biogenesis and Basic Pathways of Cyclitols and Sugar Components

It is unknown what is the general purpose of sugar modification and incorporation into homogenous (sugar-based) or mixed-type ("glycoconjugate-like") *oligomers* – mostly diffusible or membrane-associated substances outside the cells – and *polymers*, which are mostly coating materials for cell surfaces or proteins outside the cells. It might be necessary for cell–cell interactions which could be called "communication", "special purpose", or "individualization metabolism" and which might be a common feature of all types of organisms (CHADWICK and WHELAN, 1992; PIEPERSBERG, 1992, 1993). At least there is good evidence that most of the sugar metabolism beyond glucose, fructose, and some basic pentoses and their incorporation into storage and cell wall materials is specific for the cell type or strain, rather than for a species or higher taxa, and it seems to be based on the same common and highly variable and

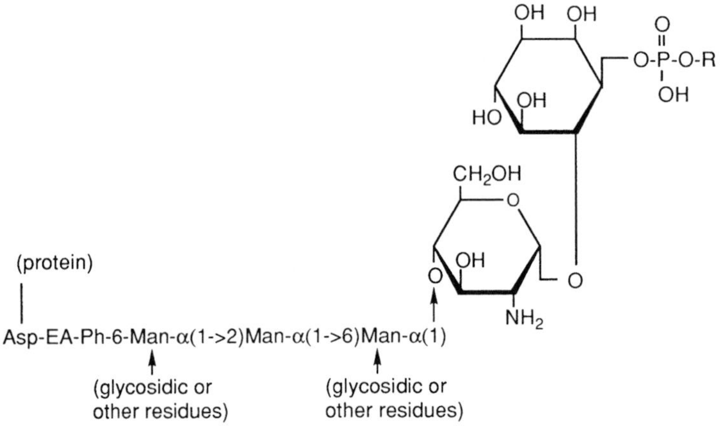

Fig. 3. Structure in glycosyl phophatidylinositols resembling to aminoglycosides. Such structures are protein anchors of proteins bound to the outer surface of eukaryotic cytoplasmic membranes. Asp: aspartyl residue in the protein chain; EA: ethanolamine; Man: mannosyl residues; P: phosphate; R: diacylglycerol.

fluid gene pool. This is reflected by high variations in glycosylation reactions in all organisms and especially in differentiating organisms. This gives rise to the high antigenic variation in microbial and higher eukaryotic cells. Examples of this are the different blood group markers in man and the capsular materials, lipopolysaccharides, and various forms of excreted glycosylated products in bacteria, which are described in this chapter.

3.1 Cyclitols

According to our present knowledge there are two different (amino-)cyclitol pathways (Fig. 4).

(1) The myo-inositol pathway. The first route (designated Ca in this chapter) is via D-*myo-inositol-3-phosphate* which is synthesized by the NAD^+-dependent *myo*-inositol synthase; end products are, e.g., D-*chiro*-inositol, streptidine, bluensidine, actinamine, and D-*myo*-1-deoxy-1-amino-inositol (D-*myo*-inosamine). A compilation of Ca cyclitols is given in Fig. 5 (FLOSS and BEALE, 1989; WALKER, 1975a; SIPOS and SZABO, 1989). The reactions catalyzed by the bacterial and the eukaryotic D-*myo*-inositol-3-phosphate synthases follow that of a typical intramolecular aldol condensation. This involves

- opening of the pyranose ring and rotation of the C-4/C-5 bond to place the C-6 group in an active position,
- the reversible oxidation and reduction of the C-5 position by NAD^+,
- the removal of the *pro-R*-H atom from C-6 and its transfer to the C-1 carbonyl, and
- the use of the β-anomer of D-glucose-6-phosphate as the preferred substrate (WONG and SHERMAN, 1985; FLOSS and BEALE, 1989).

(2) The dehydroquinate pathway. The second route to cyclitols (Cb) follows a NAD^+-dependent dehydroquinate-synthase-like mechanism on open chain hexose-6- or heptulose-7-phosphates yielding nonphosphorylated *1-keto-2-deoxy-cyclitols.* End products are, e.g., 6-deoxystreptamine and valienamine/valiolamine (Fig. 6; WIDLANSKI et al., 1989; GODA and AKHTAR, 1992; RINEHART et al., 1992; YAMAUCHI and KAKINUMA 1992c, 1993). This mechanism involves

- removal of electrons at C-4 of hexosephosphates (C-5 of heptulose derivatives);
- elimination of the phosphoester hydroxyl as a leaving group would create a potential for the formation of a carbanion at C-6 (C-7);
- finally, after the condensation reaction the keto group at C-4 (C-5) is reduced to restore the hydroxy group (cf. Fig. 4).

This step is similar to that which occurs on a C_7 sugar acid in the ring closure reaction during the initial biosynthetic pathway leading to the aromatic amino acids. The main difference in the intitial intermediates of the Ca and Cb pathways is that in Cb a first transamination step can follow immediately whereas in Ca a phosphatase and an oxidase (dehydrogenase) reaction has to first create a suitable keto intermediate for a primary aminotransfer (see below, Sect. 4.1.3). The further processing of cyclitols to aminocyclitols might even be at least in part very similar, i.e., the first aminotransfer and other later intermediate steps, and might be catalyzed by related or even almost identical enzymes. This can be predicted although the enzymes involved are not yet known for most of representative groups of (amino-)cyclitol pathways. However, there is one detail in which also the later stages of Ca and Cb cyclitol conversion differ: in 2-deoxystreptamine (2DOS) formation the incorporation of amino groups stereochemically proceeds in a direction opposite to that shown to occur in streptidine biosynthesis (WALKER, 1975a; see Sects. 4.1 and 4.3).

The chemical concept of aliphatic and aromatic C_7 units which basically constitute cyclohexane derivatives with a nitrogenous group in the *meta* position (m-C_7N) dominates the literature on the biogenesis of building blocks in many secondary metabolites (RINEHART et al., 1992; FLOSS and BEALE, 1989; CHIAO et al., 1995; KIM et al., 1996). These are formally all cyclitol derivatives, most probably formed via the Cb pathways. An aliphatic m-C_7N unit is found in valienamine/validamine-containing substances, such as validoxylamines and validamycins (cf. Sect. 10, Fig. A21), in acarbose, and in related glycosidase inhibitors (see Fig. A22; TRUSCHEIT et al., 1984; MÜLLER, 1989).

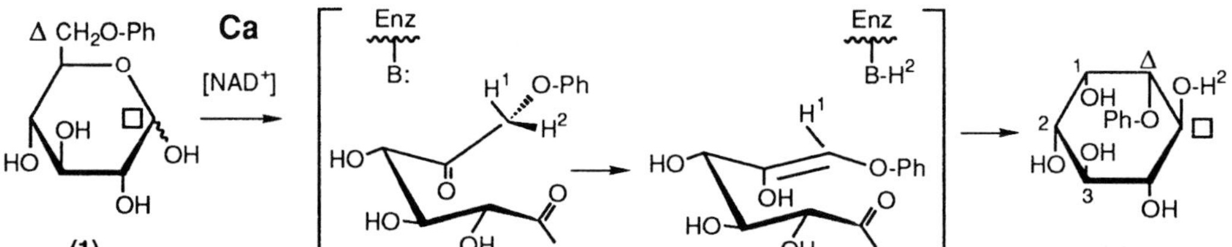

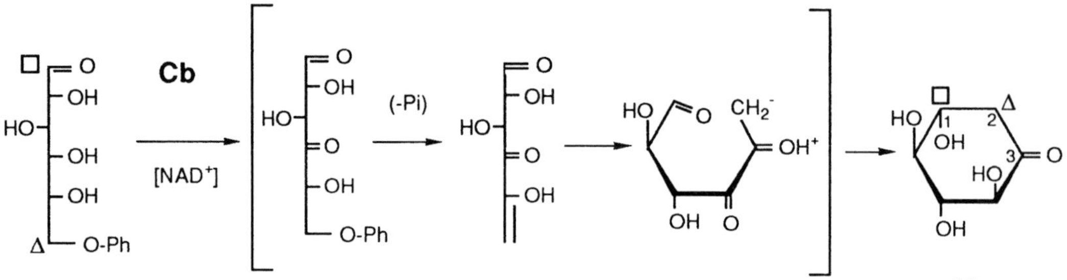

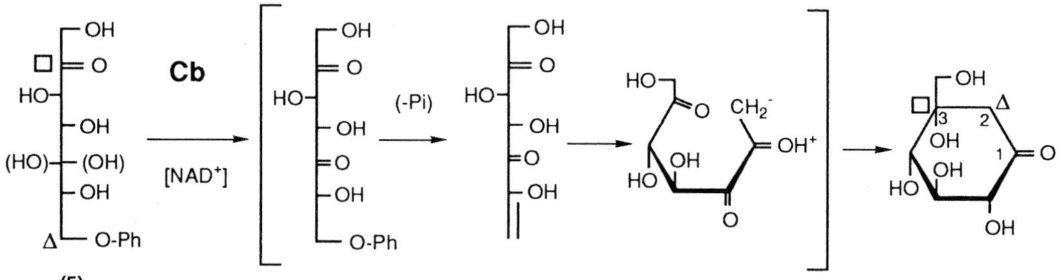

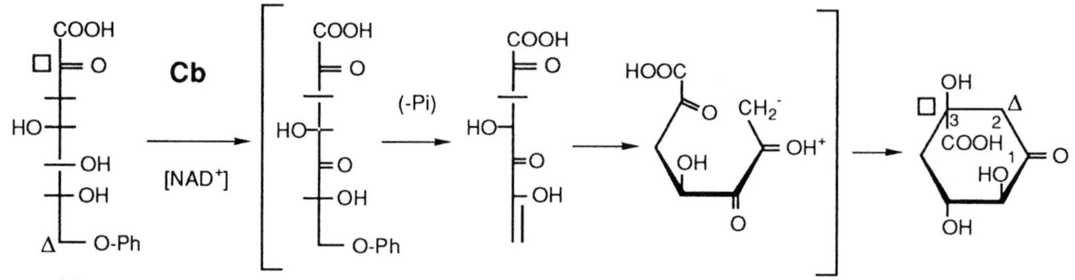

The common dehydroquinate/shikimate pathway of many aromatic compounds should be regarded as a Cb pathway and is involved in the formation of many aromatic m-C_7N units (cf. Fig. 4). These units occur in many

◀ **Fig. 4.** Routes to the formation of (amino-)cyclitols via *myo*-inositol-D-3-phosphate synthase (Ca) and a dehydroquinate synthase-like enzyme mechanism (Cb; cf. reactions 8-7). Substrates can be D-glucose-6-phosphate (1 and 3), sedoheptulose-7-phosphate (5, or its 5-epimer), or 3-deoxyarabinoheptulosonic acid (7). The products of Ca or Cb pathways are either cyclitolphosphates (2) which have to undergo dephosphorylation and further oxidation before transamination or 1-keto-2-deoxycyclitols (4, 6, and 8) which can be transaminated directly, respectively. The numbering of ring atoms in (2) and (4) is according to the counting system used in streptamine derivatives (streptidine, actinamine, 2-deoxystreptamine); numbers in (6) and (8) are given according to the nomenclature in valienamine and dehydroquinate, respectively. □ (C-1 or C-2) and △ (C-6 or C-7) mark the original carbohydrate atoms forming the new C—C bond in cyclitol products. The dehydroquinate pathway (7–8) can also be started with a 5-amino-3,5-deoxyarabinoheptulosonic acid or an intermediate becomes transaminated after cyclization to yield an aromatic m-C_7N unit as is found in many secondary metabolites (for detail, see FLOSS and BEALE, 1989; RINEHART et al., 1992; YAMAUCHI and KAKINUMA, 1992c, 1993).

well-known secondary metabolites, such as many ansamycins, e.g., rifamycins, streptovaricins, geldanamycin (not shown), in candicidins (not shown), and in pactamycin (cf. Figs. 5 and A30), which is unusual in that it contains components seemingly derived from the two alternate cyclitol pathways, one following the Ca and the other an aromatic Cb route different from that of the ansamycin family (RINEHART et al., 1981, 1992; FLOSS and BEALE, 1989). A detailed account of the aromatic Cb pathway is beyond the scope of this chapter; the reader is referred to the literature (CHIAO et al., 1995; FLOSS, in press; KIM et al., 1996).

The cyclitol moieties generally may be regarded as aglyca in glycosylation reactions, e.g., in streptomycins (see Sect. 4.1), fortimicins, and istamycins (see Sect. 4.2), most 2-deoxystreptamine-(2DOS-) containing aminoglycosides (see Sect. 4.3), and some other products such as the nucleoside antibiotic adenomycin (cf. Fig. A27 and Sect. 5.2). Alternatively, they are (co-)substrates in other types of condensation reactions, e.g., with another molecule derived from the same pathway (e.g., validamycin, acarbose, and other related compounds; see Sect. 4.4). In extreme cases, such as in pactamycin (see Fig. A30), they can constitute the central building block which is condensed with four or more side groups of different origin and varying complexity. In pactamycin they consist of two aromates (one derived from the dehydroquinate pathway, a m-C_7N unit, the other from the polyacetate/polyketide pool), a dimethylated urea group, a hydroxyethyl group derived from the methyl groups of methionine, and a C-bound methyl group. In addition, four of the five C atoms of the pentitol moiety which seems to be a *myo*-inositol derivative are substituted with nitrogenous group.

However, it is doubtful whether some of the representatives of a common chemical group of aminoglycosides share the same cyclitol pathway. For instance, the destomycin group contains two compounds with aminocyclitols derived from streptamine, one of which is identical to 1-*N*-amidinostreptamine (see Fig. A16; IKEDA et al., 1985b), an intermediate in the streptomycin pathway (Ca pathway; see Sect. 4.1.1). The other members of this family have 2-deoxy-*scyllo*-streptamine (2DOS) as cyclitol moieties (Cb pathway). Similarly, the fortimicins (Ca) and the istamycins (Cb) could be put into separate groups (see Fig. A10 and Sect. 4.2). This might suggest even two different initial pathways for cyclitol formation resulting in very similar end products or, alternatively, a subsequent modification by oxidoreductases (dehydroxylation of Ca derivatives or hydroxylation of Cb products) as optional routes to produce the alternate category of building blocks. However, the latter possibility is less likely since in no case were both pathways found in the same producer.

3.2 Sugar Components

The sugar derivatives which can be attached to the cyclitols of both the Ca and the Cb type, may stem either from nucleotidyl-activated monomeric precursors (glycosyltransfer reactions) or from the pool of polymerized D-glucose directly (e.g., maltodextrins;

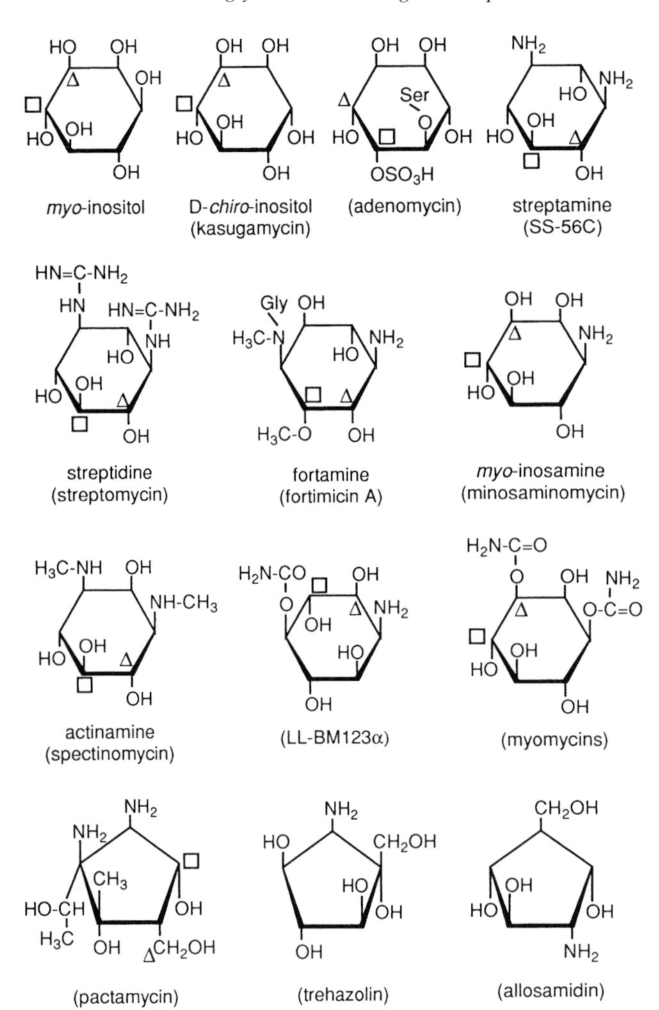

Fig. 5. Structures of (amino-)hexitol and pentitol components in secondary metabolites known or assumed to be formed via a Ca pathway (see Fig. 4). Some examples of their occurrence are given in brackets. Known or assumed positions derived from the original C-1 (□) and C-6 (△) of the glucose precursor are labeled; in the pentitols of trehazolin and allosamidin no suggestion can be given without any direct experimental evidence.

transglycosylations). For a systematic analysis of the pathways involved we define here the following general routes by which carbohydrate building blocks are preformed:

(1) the 6-deoxyhexose pathway (6DOH);
(2) the pyranosidic or furanosidic pentose (P), hexose (H), heptose (Hep), octose (Oct), or the hexosamine (HA) pathway;
(3) some glucosyl residues ($(H)_n$) are possibly incorporated via transglycosylation from di- or oligosaccharides such as maltodextrins, especially where α-1,4-bound glucose or maltose moieties are encountered;
(4) spacing non-carbohydrate-derived residues are unspecified (X).

Thus, for the purposes of the discussion here and for simplification a "pathway formula" is defined for each carbohydrate-containing compound considered. The postulated pathway for the derivation of the mono- and oligosaccharidic structures is given by connecting the above abbreviations for the individual precursor pathways where the figures in brackets indicate the points of glycosidic (or other) substitution. For example, streptomycin (Ca(4)-6DOH(2)-HA) or mannosylmannosidostreptomycin (Ca(4)-6DOH(2)-

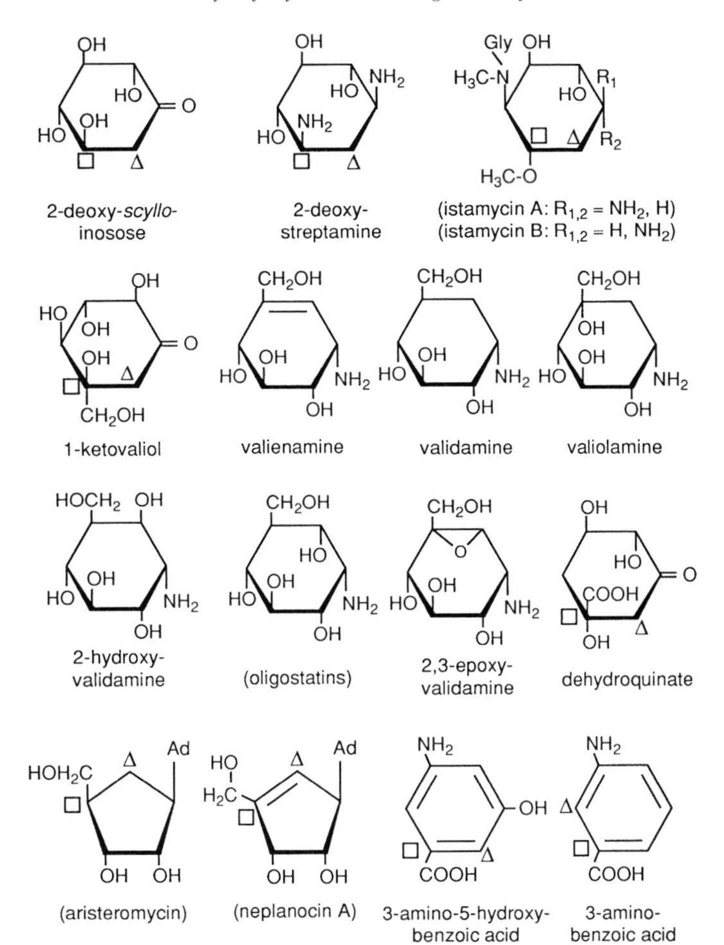

Fig. 6. Structures of (amino-)hexitols and pentitols or benzoic acid derivatives in secondary metabolites known or assumed to be formed via a Cb pathway (see Fig. 4). Some examples of their occurrence are given in brackets. Known or assumed positions derived from the original C-1 or C-2 (□) and C-6 or C-7 (△) of the glucose or heptulose (heptulosonic acid) precursors, respectively, are labeled.

HA(4)-$(H)_2$) would be classified together with spectinomycin (Ca(4,5)-6DOH) and set apart from neomycins (HA-(4)Cb(6)-P(3)-HA), validamycins (Cb(1)-Cb), and amylostatins ($(H)_n$-(4)Cb(1)-6DOH(l)-$(H)_n$) (see Sect. 10 and Tab. 1 for a compilation).

3.2.1 6-Deoxy- and Other Deoxyhexoses

D-glucose, a precursor of most or all 6-deoxyhexoses (6DOH) in prokaryote-formed (antibiotic-like) secondary metabolites, seems to be generally activated by deoxythymidine diphosphate (dTDP). Catalysis of the first two steps in their pathway is accomplished by the enzymes dTDP-D-glucose synthetase and dTDP-D-glucose 4,6-dehydratatase yielding 4-keto-6-deoxyhexose intermediates (Fig. 7). The 4-keto compounds can be used as a common precursor for branching the further pathways into the D- and L-series of hexose derivatives (Figs. 8 and 9). The L-6DOHs are isomerized from the D-configurated precursors by a 3,5-epimerase (see Fig. 7). In many cases both biosynthetic routes are followed in the same producing cell. Frequently, by such branching routes both D- and L-6DOH derivatives are formed and are specifically incorporated into a particular complex end product, e.g., macrolides and some angucyclins. In aminoglycosides, 6DOH components are re-

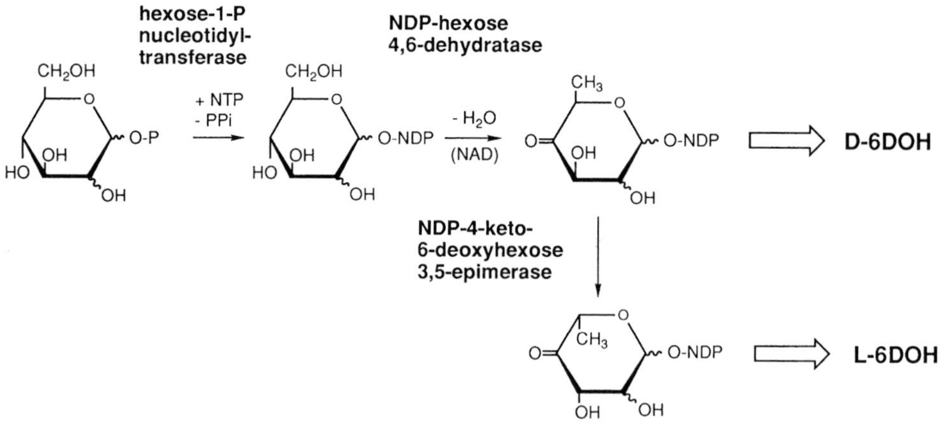

Fig. 7. General pathway for the biosynthesis of 6-deoxyhexoses. Hexose-1-phosphate precursors are either D-glucose-1-phosphate or D-mannose-1-phosphate.
P: phosphate; NDP: dTDP, CDP, or GDP.

latively rare compared with the dominance and variability among modifying sugar side chains in other chemical classes of microbial natural products.

Important and interesting types of further modification are (1) the deoxygenations (formally: dehydroxylations) at C-2, C-3, and C-4, (2) the transaminations (formally: exchange of hydroxyl for amino groups) at C-2, C-3, and C-4, and (3) the isomerization and epimerization steps. Other types of modifications, such as C, N, O, and S methylations or transfer reactions for more complex side groups are also common in the 6DOH pathways. Here we briefly describe only the first mechanism; the transaminations and other types of isomerization (e.g., epimerization) reactions are discussed below. A mechanism for the deoxygenation steps was elucidated by the recent studies of the CDP-3,6-deoxyhexose pathway (REEVES, 1993; SHNAITMAN and KLENA, 1993; THORSON et al. 1993; THORSON and LIU, 1993a, b; LIU and THORSON, 1994). This is an alternative pathway yielding 6DOHs which is abundantly used in gram-negative bacteria besides the dTDP-pathway, e.g., in the biosynthesis of lipopolysaccharide O-chains. The enzyme system involved consists of two iron–sulfur proteins one of which (E_1) catalyzes the pyridoxamine phosphate (PMP)-dependent dehydration of the 4-keto-6-deoxy hexose via a radical mechanism and by forming a covalently bound PMP–hexose intermediate (Fig. 10). The electrons for this process are delivered via a second enzyme (E_2) which contains FAD in addition to the [2Fe–2S] cluster and uses NADH as an electron donor. This type of mechanism could easily be envisaged also to be involved in the 2-deoxygenation of many other 6DOHs in secondary metabolites, such as the sugar constituents in the daunorubicin–cytorhodin–rhodomycin group of anthracyclines or some of the 6DOHs occurring in chromomycins (cf. Figs. 8 and 9).

3.2.2 Other Sugar Components

As outlined above biogenesis studies in early phases of antibiotic research have often misleadingly suggested that D-glucose, D-glucosamine, glycerol, or related carbon sources are directly incorporated into sugar constituents in secondary metabolites. This was in most cases interpreted as an indication that the regular routes of sugar activation and further processing are adopted from primary metabolic routes, e.g., the UDP-hexosamine pathway in bacterial cell wall biosynthesis. However, our rapidly increasing knowledge of the extreme specificity of secondary metabolic traits does not support this view. Unexpected modes of sugar activation (e.g., nucleotidylation by CTP instead of UTP), 2-amino-hexose derivation from precursors other than D-glucosamine, and unknown mechanisms of modification and condensation have been found (some examples are given below; see Sects. 4.1, 4.2, and 5.1). Some of the known or hypothetical general routes for the biosynthesis of hexosamines are outlined in Fig. 11; it should be emphasized that 6-amino-6-deoxyhexoses and 2-amino-2-deoxyhexoses also could be formed in the same way as the 3- and 4-isomers and that the same principal route also should yield amino-group-containing pentoses, heptoses, and octoses. Further variation may come from the transamination step, before or after sugar activation by nucleotidylation or even after condensation into complex molecules.

Several of the unusual C-5 to C-9 sugar derivatives encountered in many groups of microbial products are found in the aminoglycoside-related and other microbial secondary metabolites. Some examples comprise pyranosidic 2- or 3-pentosamines (e.g., in seldomycins and gentamicins, respectively; cf. Figs. A14 and A17), 2-deoxyhexoses (e.g., in cytosaminomycins; see Fig. A27), 3-deoxyhexosamines (e.g., in lividomycins; Fig. A11), 4-deoxyhexosamines (e.g., in seldomycins; see Fig. A17), L-hexoses (e.g., L-mannose in desertomycin; not shown), and L-hexosamines (e.g., in streptomycin; see Sect. 4.1 and Fig. A1). These moieties are most probably all activated and modified via nucleotidylated intermediates. However, it is thought that simple glucosylations and mannosylations involve UDP-D-glucose and GDP-D-mannose, respectively. Furanosidic pentoses, such as ribose (e.g., in neomycins; see Fig. A11) or arabinose moieties (e.g., in nucleoside antibiotics), might be introduced from 5-phosphoribosyl-1-diphosphate (PRPP) as activated precursors or, again, via an unknown activation and transfer mechanism. The dehydroxylation reactions of pyranosidic hexoses (other than 6DOH) at positions C-2, C-3, and C-4 are especially interesting from a mechanistic point of view. Several examples are known in aminoglycosides (see, e.g., the discussion in Sect. 4.2 on the fortimicin group) and other hexose-derived moieties in many chemical groups of heterogenously composed secondary metabolites.

4 Genetics and Biochemistry of the Biosynthesis and Functions of Aminoglycosides

Aminoglycosides are largely actinomycete products though rare occurrences in other bacterial groups are known (e.g., butirosin-producing *Bacillus circulans,* sorbistin in pseudomonads, and *N*-methyl-*scyllo*-inosamine in rhizobia; see Fig. 1). Aminoglycoside production is mainly a property of members of the filamentous actinomycete genera *Streptomyces* spp., *Streptoverticillium* spp., and *Saccharopolyspora* spp., and of *Micromonospora* spp., *Amycolatopsis* spp., and *Actinoplanes* sp. (WILLIAMS et al., 1989). The biology, biochemistry, mode of action, biotechnology, and clinical applications of the aminoglycoside antibiotics in addition to aminoglycoside resistance have been often reviewed (e.g., KORZYBSKI et al., 1978; DAVIES and SMITH, 1978; WALLACE et al., 1979; PIEPERSBERG et al., 1980; WALKER, 1980; PEARCE and RINEHART, 1981; UMEZAWA and HOOPER, 1982; DAVIES and YAGISAWA, 1983; FOSTER, 1983; UMEZAWA et al., 1986; CUNDLIFFE, 1989, 1990; GRÄFE, 1992; PIEPERSBERG, 1995; HOTTA et al. 1995). Recent investigations have concentrated mainly on the molecular biology of the clinically relevant aminoglycoside resistance and molecular aspects of aminocyclitol aminoglycoside biosynthesis, resistance, and regulation in the producing organisms (mainly with respect to streptomycins and fortimicins; see Sects. 4.1 and 4.2). During the past decade, semisynthetic modification, new screening methods, and new fields of application have resulted in the discovery of several new aminoglycoside

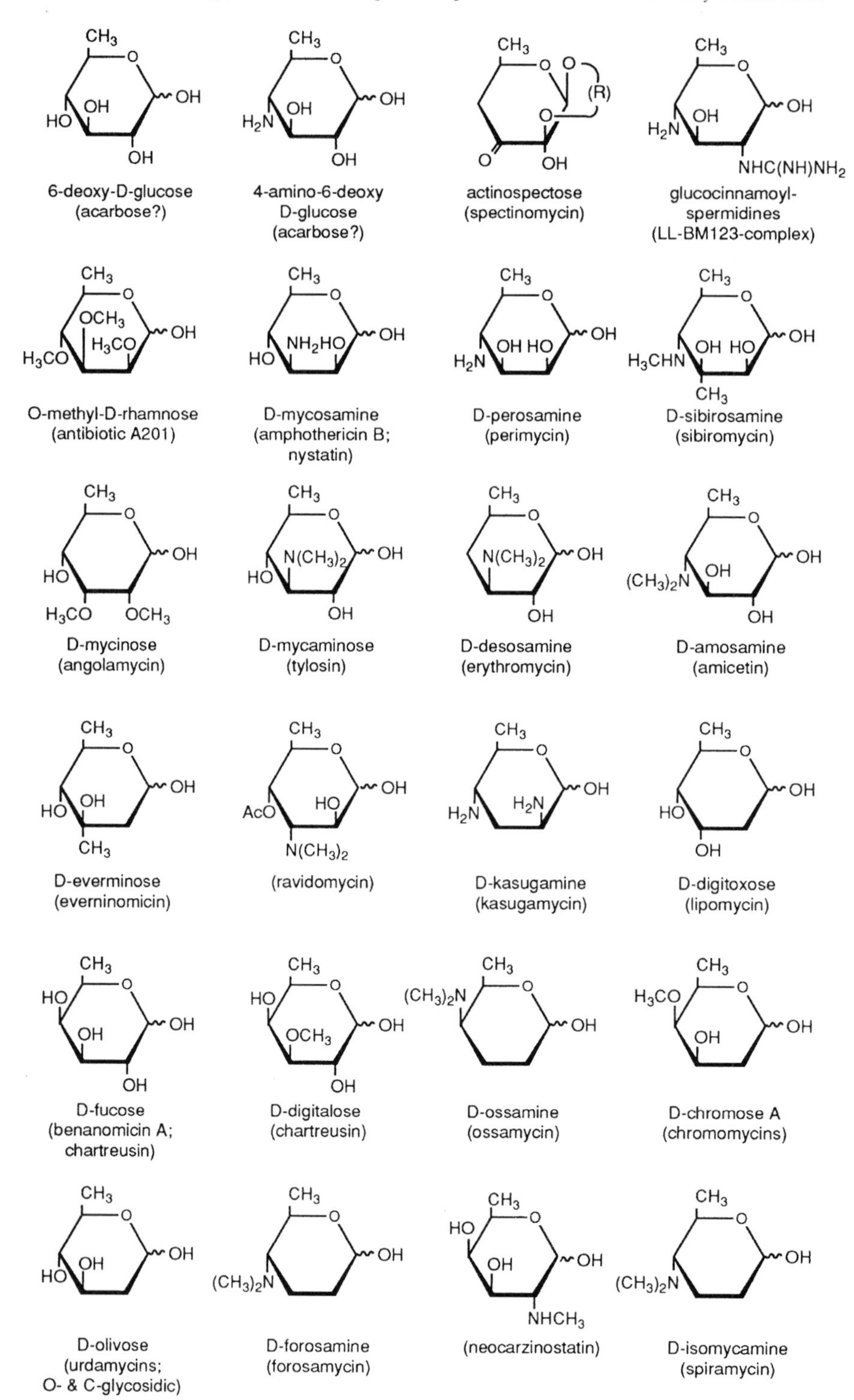
6-deoxy-D-glucose
(acarbose?)
4-amino-6-deoxy
D-glucose
(acarbose?)
actinospectose
(spectinomycin)
glucocinnamoyl-
spermidines
(LL-BM123-complex)
O-methyl-D-rhamnose
(antibiotic A201)
D-mycosamine
(amphothericin B;
nystatin)
D-perosamine
(perimycin)
D-sibirosamine
(sibiromycin)
D-mycinose
(angolamycin)
D-mycaminose
(tylosin)
D-desosamine
(erythromycin)
D-amosamine
(amicetin)
D-everminose
(everninomicin)
(ravidomycin)
D-kasugamine
(kasugamycin)
D-digitoxose
(lipomycin)
D-fucose
(benanomicin A;
chartreusin)
D-digitalose
(chartreusin)
D-ossamine
(ossamycin)
D-chromose A
(chromomycins)
D-olivose
(urdamycins;
O- & C-glycosidic)
D-forosamine
(forosamycin)
(neocarzinostatin)
D-isomycamine
(spiramycin)

Fig. 8

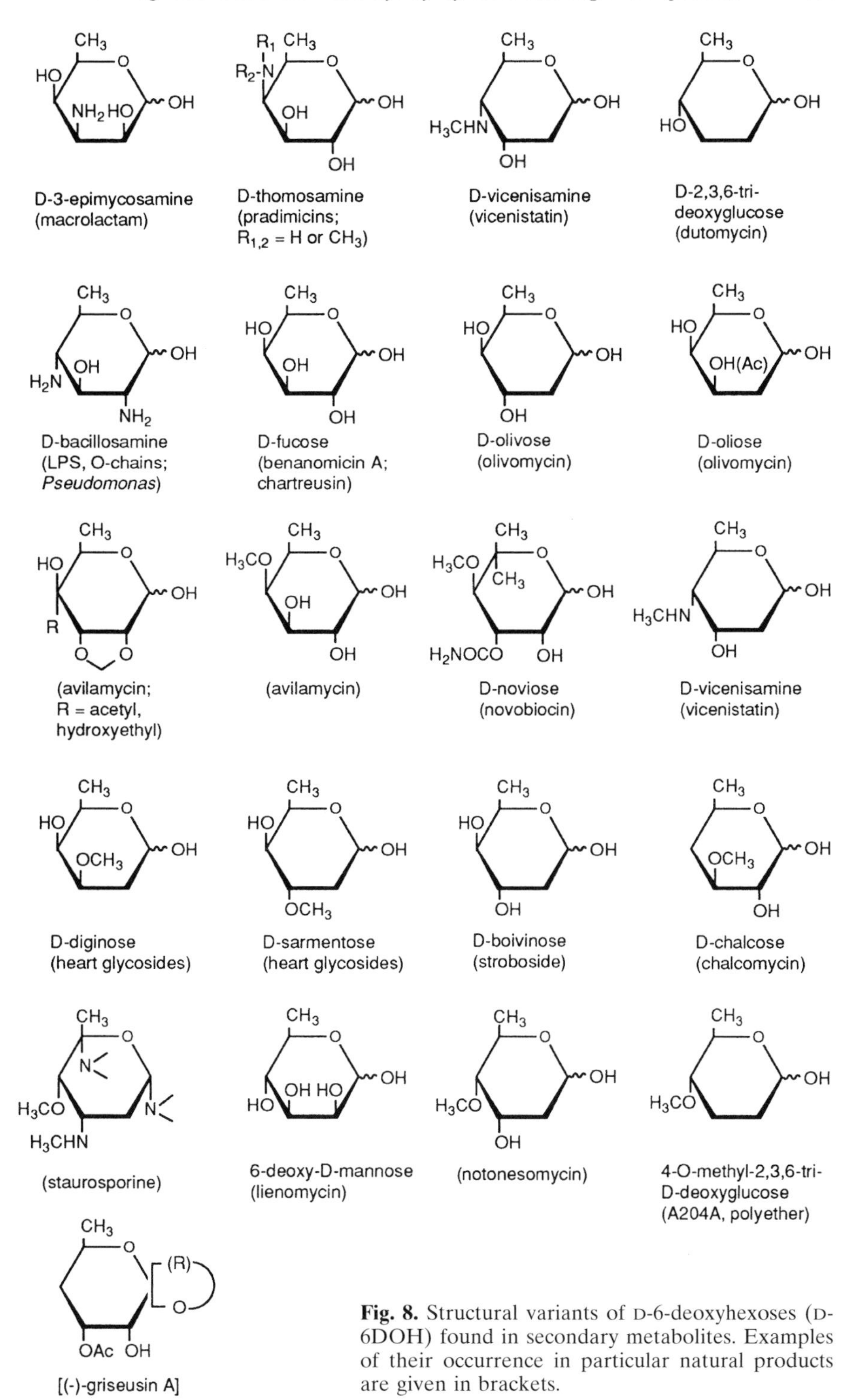

Fig. 8. Structural variants of D-6-deoxyhexoses (D-6DOH) found in secondary metabolites. Examples of their occurrence in particular natural products are given in brackets.

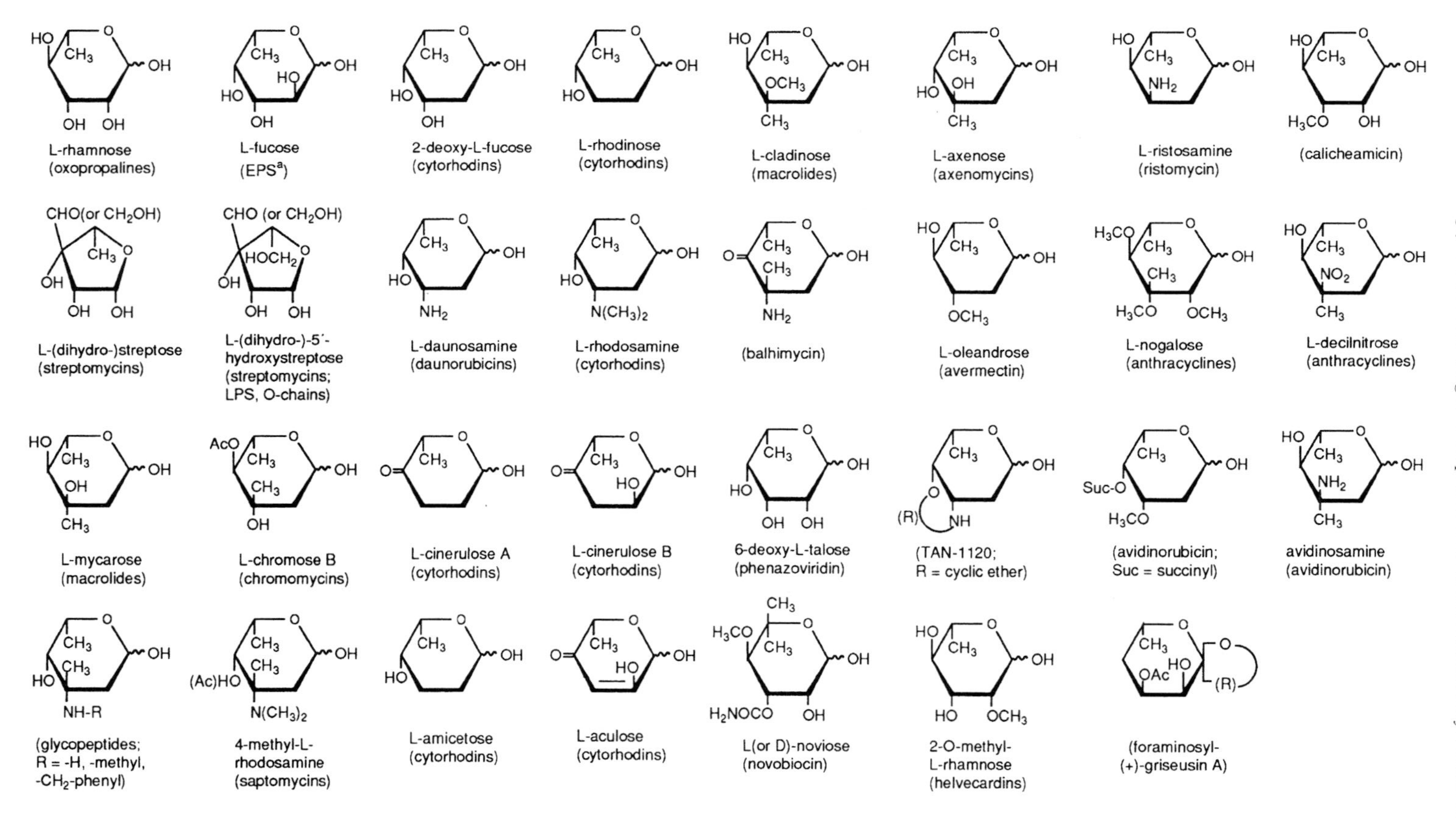
L-rhamnose (oxopropalines)
L-fucose (EPS[a])
2-deoxy-L-fucose (cytorhodins)
L-rhodinose (cytorhodins)
L-cladinose (macrolides)
L-axenose (axenomycins)
L-ristosamine (ristomycin)
(calicheamicin)
L-(dihydro-)streptose (streptomycins)
L-(dihydro-)5'-hydroxystreptose (streptomycins; LPS, O-chains)
L-daunosamine (daunorubicins)
L-rhodosamine (cytorhodins)
(balhimycin)
L-oleandrose (avermectin)
L-nogalose (anthracyclines)
L-decilnitrose (anthracyclines)
L-mycarose (macrolides)
L-chromose B (chromomycins)
L-cinerulose A (cytorhodins)
L-cinerulose B (cytorhodins)
6-deoxy-L-talose (phenazoviridin)
(TAN-1120; R = cyclic ether)
(avidinorubicin; Suc = succinyl)
avidinosamine (avidinorubicin)
(glycopeptides; R = -H, -methyl, -CH2-phenyl)
4-methyl-L-rhodosamine (saptomycins)
L-amicetose (cytorhodins)
L-aculose (cytorhodins)
L(or D)-noviose (novobiocin)
2-O-methyl-L-rhamnose (helvecardins)
(foraminosyl-(+)-griseusin A)

Fig. 9

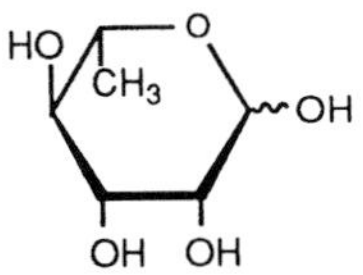

L-rhamnose (oxopropalines)

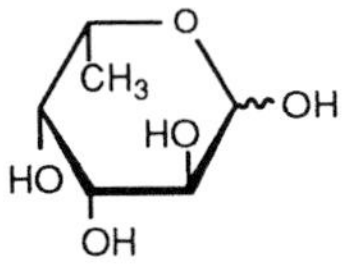

L-fucose (EPS[a])

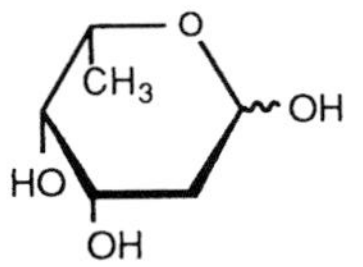

2-deoxy-L-fucose (cytorhodins)

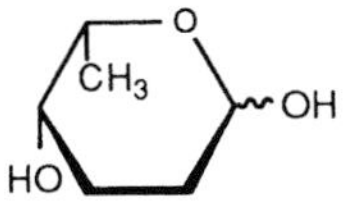

L-rhodinose (cytorhodins)

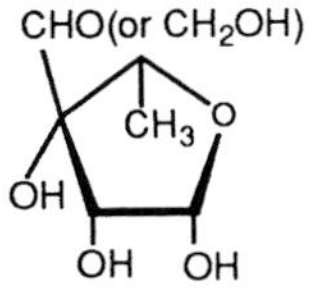

L-(dihydro-)streptose (streptomycins)

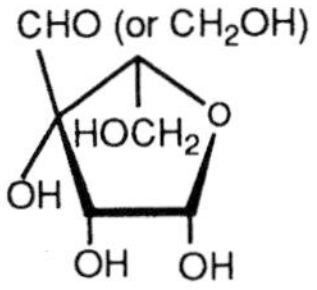

L-(dihydro-)-5′-hydroxystreptose (streptomycins; LPS, O-chains)

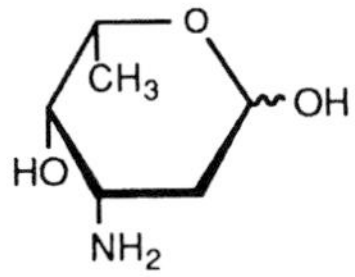

L-daunosamine (daunorubicins)

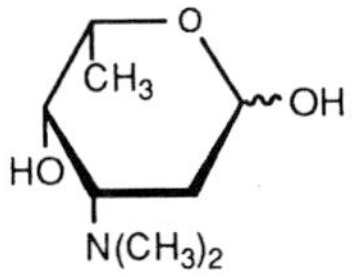

L-rhodosamine (cytorhodins)

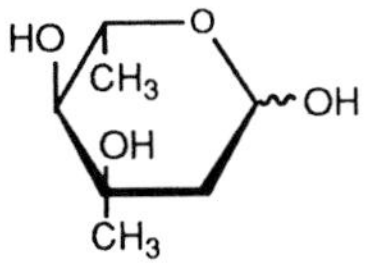

L-mycarose (macrolides)

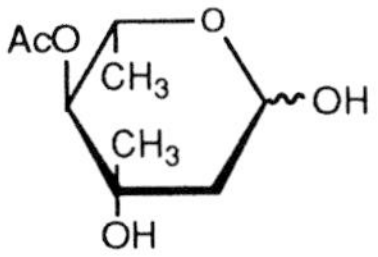

L-chromose B (chromomycins)

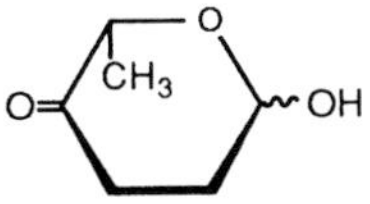

L-cinerulose A (cytorhodins)

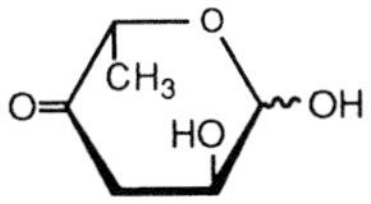

L-cinerulose B (cytorhodins)

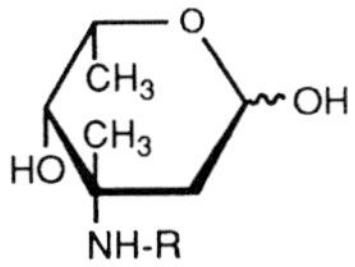

(glycopeptides; R = -H, -methyl, $-CH_2$-phenyl)

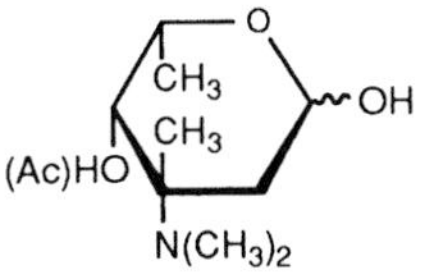

4-methyl-L-rhodosamine (saptomycins)

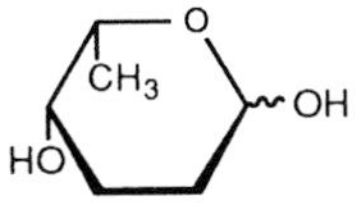

L-amicetose (cytorhodins)

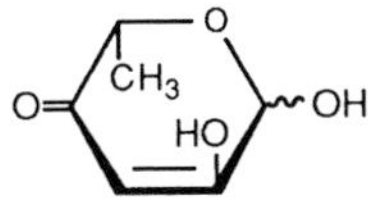

L-aculose (cytorhodins)

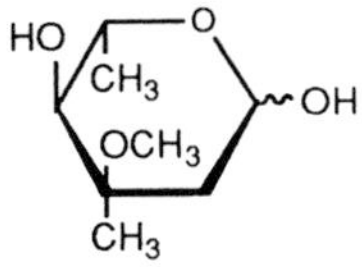

L-cladinose (macrolides)

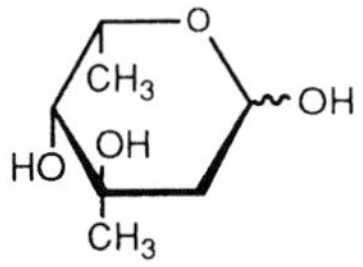

L-axenose (axenomycins)

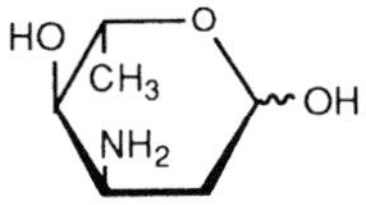

L-ristosamine (ristomycin)

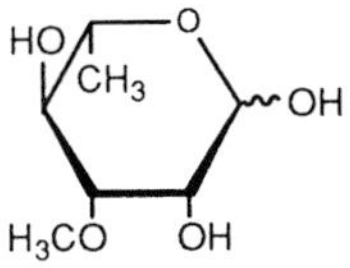

(calicheamicin)

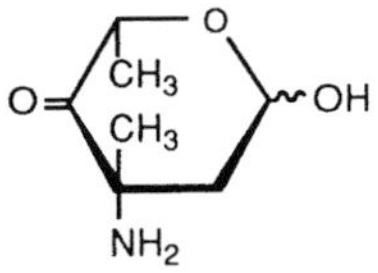

(balhimycin)

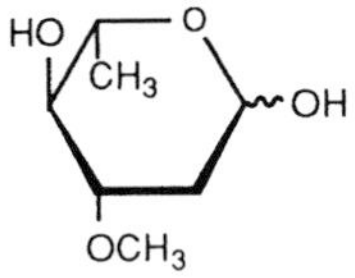

L-oleandrose (avermectin)

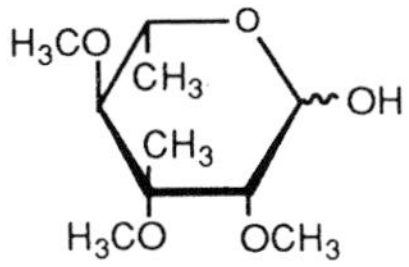

L-nogalose (anthracyclines)

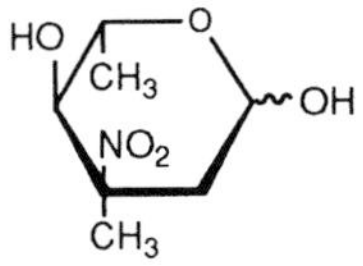

L-decilnitrose (anthracyclines)

Fig. 9

6-deoxy-L-talose (phenazoviridin)

(TAN-1120; R = cyclic ether)

(avidinorubicin; Suc = succinyl)

avidinosamine (avidinorubicin)

L(or D)-noviose (novobiocin)

2-O-methyl-L-rhamnose (helvecardins)

(foraminosyl-(+)-griseusin A)

Fig. 9. Structural variants of L-6-deoxyhexoses (L-6DOH) found in secondary metabolites. Examples of their occurrence in particular natural products are given in brackets.

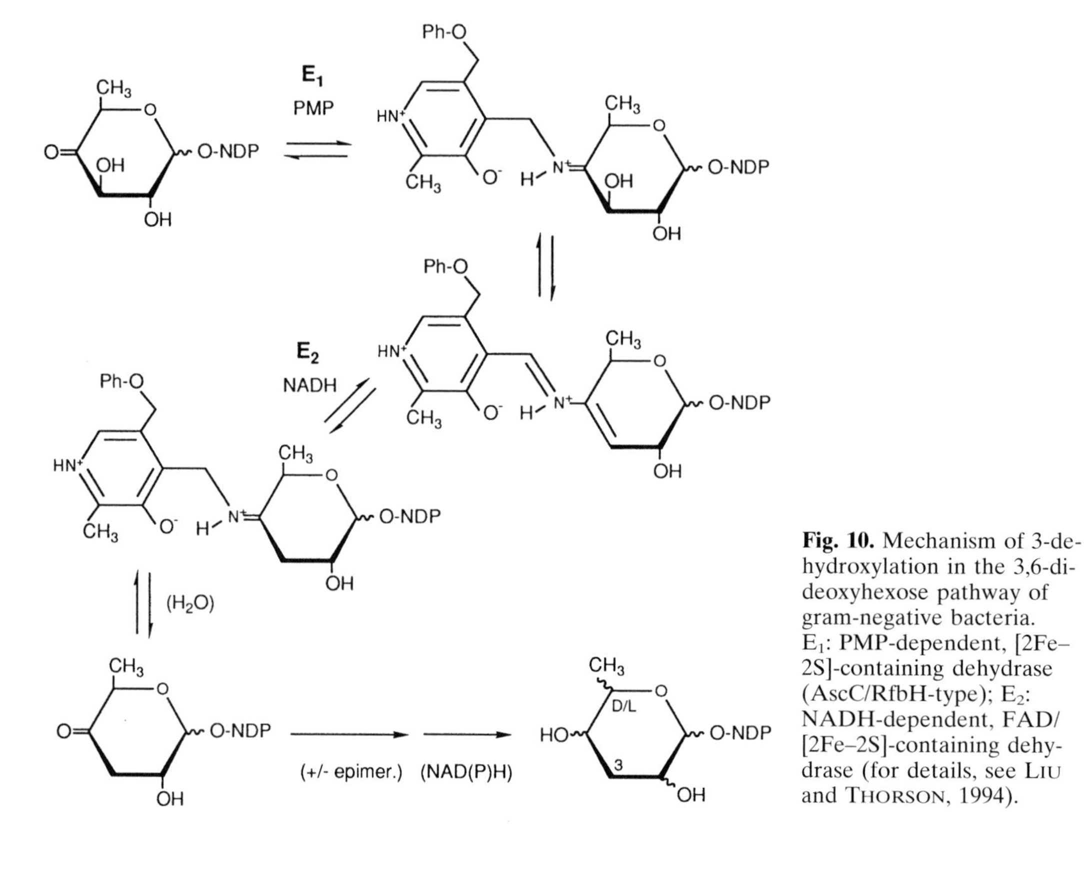

Fig. 10. Mechanism of 3-dehydroxylation in the 3,6-dideoxyhexose pathway of gram-negative bacteria. E_1: PMP-dependent, [2Fe–2S]-containing dehydrase (AscC/RfbH-type); E_2: NADH-dependent, FAD/[2Fe–2S]-containing dehydrase (for details, see LIU and THORSON, 1994).

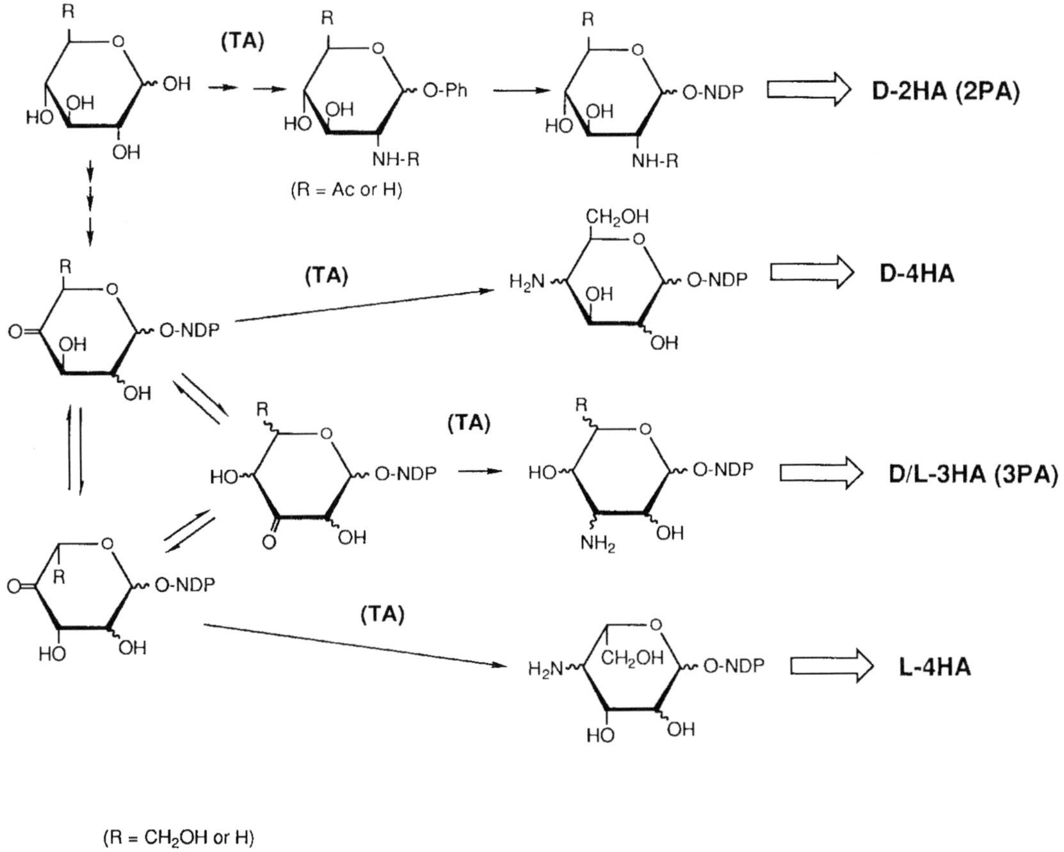

(R = CH_2OH or H)

Fig. 11. General scheme for the derivation of 2-, 3-, or 4-aminated hexoses and pentoses. Also, 2- and 6-aminohexoses (HA), 2-aminopentoses (PA), or aminoheptoses and aminooctoses could be formed in their furanosidic or pyranosidic forms via the oxidation and transamination pathway.

structures and, in at least one case, in the introduction into pharmaceutical use. This chapter will focus mainly on the latter aspects.

4.1 Streptomycins and Related Ca Aminoglycosides

The streptomycins and bluensomycin are a relatively homogenous group of basically pseudotrisaccharidic aminoglycosides (Fig. A1) which can be elongated by one or two further glycosidic residues (mannosyl or ashimosyl; IKEDA et al., 1985a; TOHMA et al., 1989) or be cleaved into two hexose derivatives as in AC4437 (dihydrostreptosyl-streptidine; AWATA et al., 1986). Compounds of this group are produced by a broad range of species (KORZYBSKI et al., 1978). They have been found in the genera *Streptomyces, Streptoverticillium, Amycolatopsis,* and may be present in others. A survey of streptomycin-producing strains from collections and a taxonomic study of new isolates that was recently started (PHILLIPS et al., 1992; MARSH and WELLINGTON, 1994) indicated that they can mainly be grouped into three clusters within the family of Streptomycetaceae. These clusters are represented by *S. griseus, S. hygroscopicus,* and *Streptoverticillium mashuense.*

Therefore, the presently accepted taxonomy of the family Streptomycetaceae (WILLIAMS et al., 1989; EMBLEY and STACKEBRANDT, 1994) assigns those streptomycin-producing species that have been investigated in detail to clearly separated species clusters. Recently, this view was confirmed and extended by sequencing the 16S rRNA genes from several streptomycin-producers (MEHLING et al., 1995). As with other secondary metabolites, the reason for this taxonomically scattered, but relatively stable distribution in indiviual strains worldwide is not known.

4.1.1 Streptomycins

The biosynthesis of streptomycins, including bluensomycin, was intensively studied by feeding labeled primary metabolic precursors (DEMAIN and INAMINE, 1970; MUNRO et al., 1975) and analyzing streptomycin biosynthetic reactions in cell-free systems and with purified biosynthetic enzymes (WALKER, 1975a, b; GRISEBACH, 1978; RINEHART, 1980). Although not all of the biosynthetic reactions have been clarified, it is known that the compounds of the streptomycin family are synthesized via 25–30 enzyme-catalyzed steps (WALKER, 1975a; RINEHART and STROSHANE, 1976; GRISEBACH, 1978; RINEHART, 1980; OKUDA and ITO, 1982; PIEPERSBERG, 1995). The currently hypothesized streptomycin and bluensomycin biosynthesis pathways are summarized in Fig. 12. They involve the formation of the activated precursors, streptidine (bluensidine)-6-phosphate (cf. Sect. 4.1.1.1), dTDP-dihydrostreptose (cf. Sect. 4.1.1.2), and NDP-*N*-(methyl)-L-glucosamine (cf. Sect. 4.1.1.3) which are made most likely from D-glucose-6-phosphate. However, it has been demonstrated that D-glucosamine can be incorporated without breakage of its C–C bonds into the *N*-methyl-L-glucosamine subunit of streptomycin (reviewed by OKUDA and ITO, 1982). In a second phase, the precursors are condensed forming dihydro-streptomycin-6-phosphate (cf. Sect. 4.1.1.4) which is secreted and probably oxidized while passing through the cytoplasmic membrane to give streptomycin-6-phosphate outside the cell (cf. Sect. 4.1.1.4). The biologically active strepto-

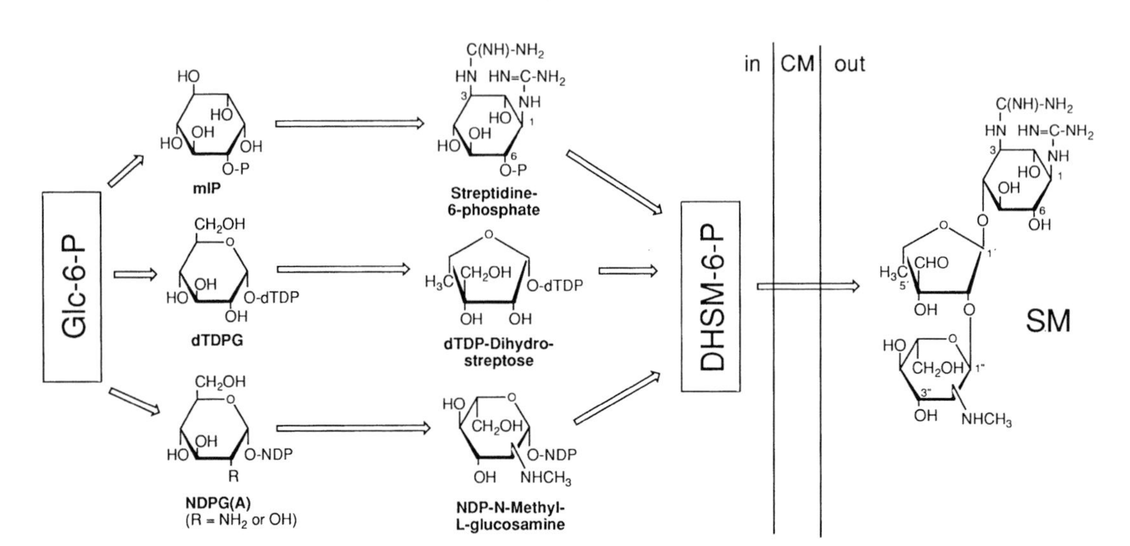

Fig. 12. Gerneral outline of the streptomycin (SM) pathway. The three activated intermediates principally formed from glucose-6-phosphate (G-6-P) are condensed in the cytoplasm to dihydro-SM-6-phosphate (DHSM-6-P) which is oxidized and dephosphorylated to SM during or after transport through the cytoplasmic membrane (CM).
mIP: *myo*-inositolphosphate; dTDPG: deoxythymidinediphosphate-glucose; NDPG(A): nucleosidediphosphate-glucose (or -glucosamine) (NDP = CDP or UDP).

mycins are liberated finally by a specific phosphatase (cf. Sect. 4.1.1.4), and after re-uptake they can be phosphorylated by a streptomycin 6-phosphotransferase representing the resistance mechanism which protects the producers from their own products (cf. Sect. 4.5).

Genetic studies of streptomycin biosynthesis in the various producers started with the cloning of resistance genes (reviewed in CUNDLIFFE, 1989; PIEPERSBERG, 1995). Two genes were initially cloned, *strA* (*aphD*) and *aphE*, which encode two different streptomycin phosphotransferases, APH(6) and APH(3″), respectively (WALKER, 1975b; WALKER and WALKER, 1975; DISTLER and PIEPERSBERG, 1985; DISTLER et al., 1987a; HEINZEL et al., 1988). Only the *strA* gene could be localized up to now in the production gene clusters of *S. griseus* and *S. glaucescens* GLA 0. The *aphE* gene, however, occurs only in *S. griseus* strains and does not appear to be linked to the *str/sts* cluster. Subsequently, further streptomycin biosynthetic genes were identified by chromosome walking and genetic complementation of mutants from different *S. griseus* strains deficient in streptomycin biosynthesis (DISTLER et al., 1985; OHNUKI et al., 1985a, b). About 30 genes for (5′-hydroxy-)streptomycin (*str/sts*) and bluensomycin (*blu*) production have been cloned and analyzed from various strains of *S. griseus*, from *S. glaucescens* GLA.0 (ETH 22794), and from other *Streptomyces* spp., and all were found to be clustered in one region of about 30–40 kb of genomic DNA (Tab. 2; Fig. 13; MANSOURI et al., 1989; DISTLER et al., 1990, 1992; RETZLAFF et al., 1993; PIEPERSBERG, 1995). Comparison of the primary structures of the homologous *str/sts* genes and their gene products in *S. griseus* and *S. glaucescens* revealed identity values varying between 58% and 86%. In contrast, the corresponding nonencoding intercistronic DNA sections have much less or no significant homology. The order of the *str/sts* genes identified so far within the respective operons is similar although the arrangement of the operons differs considerably among producing species. Usually, the genes encoding enzymes for the synthesis of a subunit of streptomycin (e.g., streptidine) are not arranged in subpathway-specific operons. Instead, mixed operons are generally found which may reflect the need for a strictly coordinated regulation of *str/sts* gene expression in order to guarantee a coordinated supply of the activated precursors in streptomycin synthesis. The phenomena involved in genetic regulation of streptomycin biosynthesis are discussed below (cf. Sect. 4.6).

4.1.1.1 Biosynthesis of Streptidine and Bluensidine

The biochemistry of the eleven steps involved in streptidine synthesis was evaluated by WALKER et al. (WALKER, 1975a, 1990) by means of enzymology and led to the proposed pathway outlined in Fig. 14. The initial step is the formation of D-*myo*-inositol-3-phosphate from D-glucose-l-phosphate via the Ca route (cf. Sect. 3.1) by an ATP-dependent *myo*-inositol phosphate synthase (WALKER, 1975a; SIPOS and SZABO, 1989). Up to now, no *str/sts* gene could be identified which encodes the *myo*-inositol synthetase. Also attempts to purify the enzyme failed because of its instability (SIPOS and SZABO, 1989). The end product is streptidine (SD)-6-phosphate, which also seems to be the first intermediate made from externally supplied SD in SD$^-$ mutants (OHNUKI et al., 1985a, b; DISTLER et al., 1985). SD-6-phosphate is synthesized by two sets of five parallel enzymatic reactions each presumably catalyzed by individual enzymes: two cyclitol phosphate phosphatases, two cyclitol dehydrogenases, two aminotransferases, two phosphotransferases, and two amidinotransferases in streptomycin producers (cf. Tab. 2, Fig. 14).

In contrast to this the producer of bluensomycin *S. hygroscopicus* forma *glebosus* lacks steps 8 to 11 (see Fig. 14; WALKER, 1990) which are replaced by carbamoylation and phosphorylation reactions at positions 5 and 4, respectively, yielding bluensidine-6-phosphate (see Fig. 15). The product of the *strO* gene has significant similarity to eukaryotic inositolmonophosphate phosphatases (RETZLAFF et al., 1993). Therefore, this protein is assumed to be the enzyme which catalyzes step 2 in the streptidine pathway (RETZLAFF

Tab. 2. Gene Products and Enzymes Presumed to be Involved in Streptomycin Production

Gene Product[a]	Molecular Data[b]		Enzymatic Function[c] (Preliminary Assignment)	Remarks[d]	References[e]
	MW [kDa]	aa			
Unkown	216 (native)	–	D-*myo*-Inositol-3-phosphate synthetase	ATP-dependent	SIPOS and SZABO (1989)
StrO	28	260 (256)	(D-*myo*-Inositol-3-phosphate phosphatase)		A, RETZLAFF et al. (1993)
StrI	37	348	(*scyllo*-Inosose dehydrogenase)	NAD(P)-dependent	MANSOURI and PIEPERSBERG (1991)
StsC	45	424	*scyllo*-Inosose aminotransferase	Rel. to StrS and StsA; PLP-dependent	A, RETZLAFF et al. (1993)
StrN	36 (35)	320 (316)	(*scyllo*-Inosamine 4-phosphotransferase)		PISSOWOTZKI et al. (1991)
StrB1	39	347	*scyllo*-Inosamine-4-phosphate amidinotransferase		DISTLER et al. (1987b)
Unknown			*N*-Amidino-*scyllo*-inosamine-4-phosphate phosphatase		(cf. WALKER, 1975a)
StsB	52	490	(*N*-Amidino-*scyllo*-inosamine dehydrogenase)	NAD(P)-dependent	A
StsA	43	410	(3-Keto-*N*-amidino-*scyllo*-inosamine aminotransferase)	Rel. to StsC and StrS; PLP-dependent	A, RETZLAFF et al. (1993)
StsE	32	312	(*N*-Amidino-streptamine 6-phosphotransferase)		A
StrB2	38 (35)	349 (319)	*N*-Amidino-streptamine-6-phosphate amidinotransferase		PISSOWOTZKI et al. (1991)
StrD	38	355	dTDP-Glucose synthetase		PISSOWOTZKI et al. (1991)
StrE	36	328	dTDP-Glucose 4,6-dehydratase	NAD-dependent	PISSOWOTZKI et al. (1991)
StrM	22	200	dTDP-4-Keto-6-deoxyglucose 3,5-epimerase		PISSOWOTZKI et al. (1991)
StrL	32	304	dTDP-4-Keto-L-rhamnose dehydrogenase [=dTDP-L-dihydrostreptose synthase]	NADP-dependent	PISSOWOTZKI et al. (1991)
StrH	43	384	(Streptidine-6-phosphate dihydrostreptosyltransferase)		MANSOURI and PIEPERSBERG (1991)
StrQ	34	297	CDP-Hexose synthetase	Rel. to StrD	A
StrP	38	358	(NDP-Hexose 4-dehydrogenase)	Rel. to GalE; NAD(P)-dependent	A, MAYER (1994)

StrX	20	182	(NDP-Hexose 3,5-epimerase)	Rel. to StrM	A, BEYER et al. (1996)
StrU	46	428	(NDP-Hexose oxidoreductase)	NAD(P)-dependent	A, BEYER et al. (1996)
StrF	32	281	(NDP-Hexose epimerase)		MANSOURI and PIEPERSBERG (1991)
StrG	23	199	(NDP-Hexose epimerase)		MANSOURI and PIEPERSBERG (1991)
StsG	27	246	(N-Methyltransferase)		A
StrS	40	378 (377)	(Aminotransferase; unknown function)	Rel. to StsC and StsA; PLP-dependent	A, RETZLAFF et al. (1993)
StrT	33	300	Unknown function		A
StsD	24	213	Unknown function		A
StsF	26	236	Unknown function		A
StrV	(ca. 45–50)	(inc.)	(Exporter for streptomycin 6-(or 3″-)phosphates)	ATP-dependent (ABC-transporter)	A, BEYER et al. (1996)
StrW	63	592	(Exporter for streptomycin 6-(or 3″-)phosphates)	ATP-dependent (ABC-transporter)	A, BEYER et al. (1996)
StrK	46	449 (462)	Streptomycin 6-(or 3″-)phosphate phosphatase	Extracellular protein	MANSOURI and PIEPERSBERG (1991)
StrA (AphD)	33	307	Streptomycin 6-phosphotransferase	ATP-dependent	DISTLER et al. (1987a)
StrR	38 (46)	350 (424)	DNA-binding protein, activator of gene expression		A, DISTLER et al. (1987b); RETZLAFF and DISTLER (1995)
AphE	29	272	streptomycin 3″-phosphotransferase	ATP-dependent	HEINZEL et al. (1988)

[a] Cf. Fig. 13
[b] Genetic data from *Streptomyces griseus* N2-3-11 and/or *S. glaucescens* GLA.0 (from the latter strain given in brackets if different); inc.: incomplete sequence
[c] Cf. Figs. 14–17
[d] PLP: pyridoxalphosphate
[e] A: J. AHLERT, S. BEYER, J. DISTLER, K. MANSOURI, G. MAYER, and W. PIEPERSBERG, unpublished data

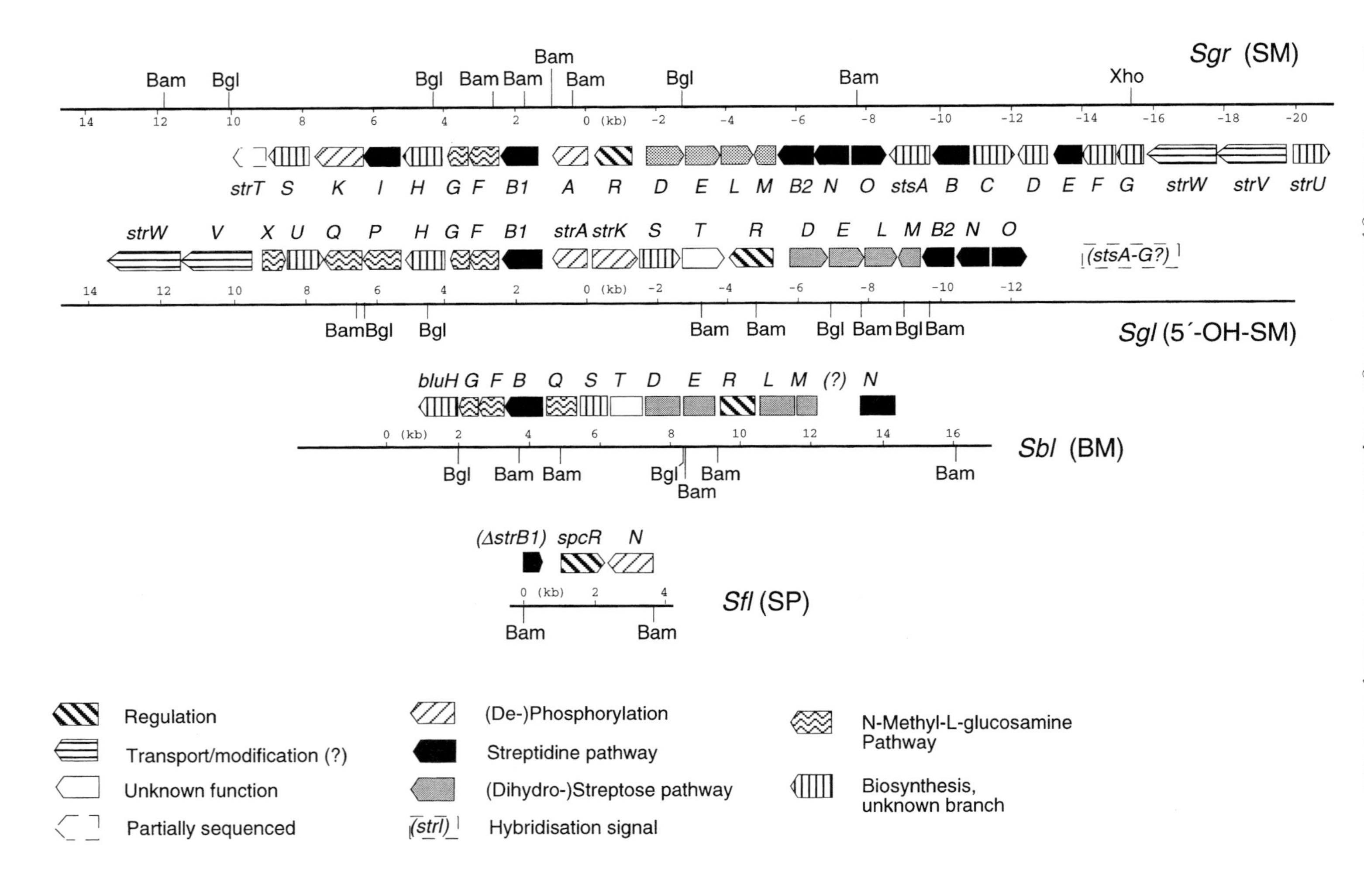
Sgr (SM)
strT S K I H G F B1 A R D E L M B2 N O stsA B C D E F G strW strV strU
strW V X U Q P H G F B1 strA strK S T R D E L M B2 N O
(stsA-G?)
Sgl (5´-OH-SM)
bluH G F B Q S T D E R L M (?) N
Sbl (BM)
(ΔstrB1) spcR N
Sfl (SP)
Bam
Bgl
Xho
Regulation
Transport/modification (?)
Unknown function
Partially sequenced
(De-)Phosphorylation
Streptidine pathway
(Dihydro-)Streptose pathway
(strI) Hybridisation signal
N-Methyl-L-glucosamine Pathway
Biosynthesis, unknown branch

◄ **Fig. 13.** Gene clusters for the production of streptomycins (SM) and the related Ca aminoglycosides bluensomycin (BM) and spectinomycin (SP). The streptomycete producers investigated most intensively are *S. griseus* (*Sgr*) strains N2-3-11 and DSM40236, *S. glaucescens* (*Sgl*) GLA.O (ETH 22794), *S. bluensis* (*Sbl*) DSM40564, and *S. flavopersicus* (*Sfl*) NRRL 2820. Restriction maps for a few enzymes are given for orientation; the clusters are aligned according to their homologous amidinotransferase [*strB*(1)] genes.

et al., 1993; see Fig. 14). The StrI and StsB proteins are clearly members of the oxidoreductase class with an N-terminal dinucleotide coenzyme binding site (MANSOURI and PIEPERSBERG, 1991). Therefore, they are good candidates for the step-3 and -8 cyclitol oxidoreductases which form the keto groups preceeding the two transamination steps (see Fig. 14). The StrI protein significantly resembles the *myo*-inositol-2-dehydrogenase enzyme from *Bacillus subtilis* (FUJITA et al., 1992) and is the enzyme which catalyzes the first cyclitol dehydrogenation step (J. AHLERT, W. PIEPERSBERG; unpublished data).

The StsC protein is the *scyllo*-inosose aminotransferase and catalyzes the transfer of the α-amino group of glutamine to *scyllo*-inosose yielding *scyllo*-inosamine and α-ketoglutaramate, an unusual transamination reaction in bacterial cells (cf. step 4 in Fig. 14; WALKER, 1975a; LUCHER et al., 1989; J. AHLERT, J. DISTLER, and W. PIEPERSBERG, unpublished data). This enzyme, like the *str*/*sts*-gene products StrS and StsA, is a member of a new class

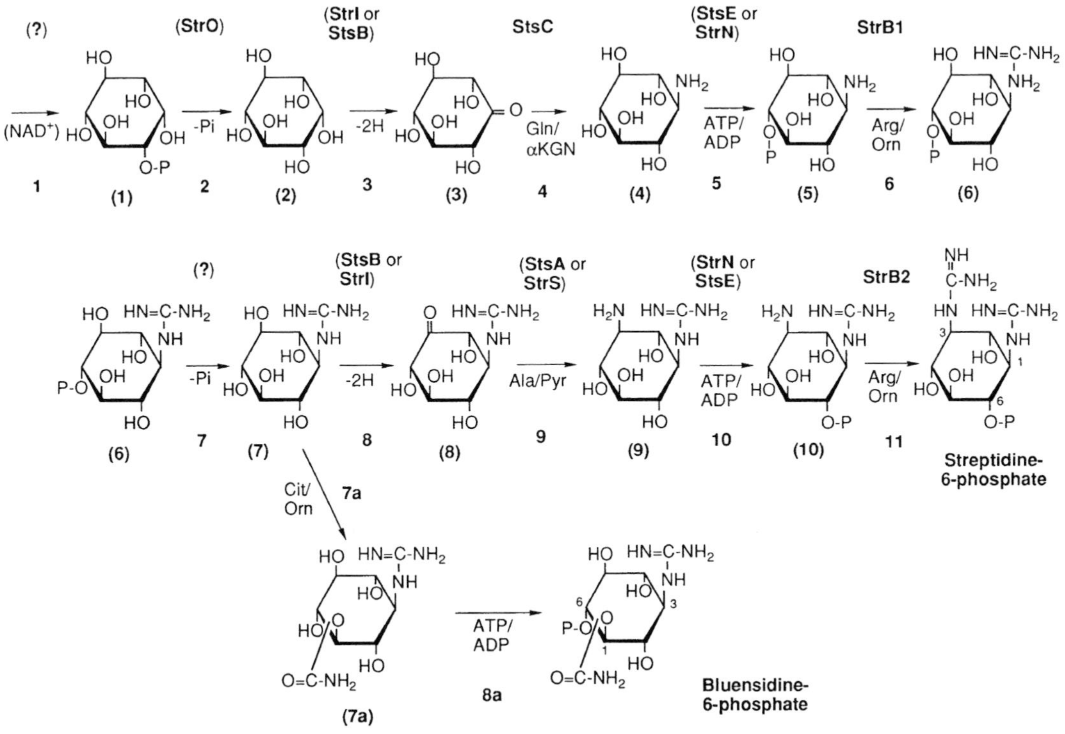

Fig. 14. The streptidine and bluensidine pathways. Enzymatic steps are numbered; where a particular gene product is known or postulated to be involved this is given (cf. Tab. 2 and Fig. 13). Known or postulated intermediates (numbered in brackets) are (1) D-*myo*-inositol-3-phosphate, (2) *myo*-inositol, (3) *scyllo*-inosose, (4) *scyllo*-inososamine, (5) *scyllo*-inososamine-4-phosphate, (6) N^1-amidino-*scyllo*-inososamine-4-phosphate, (7) N^1-amidino-*scyllo*-inososamine, (8) 3-keto-N^1-amidino-*scyllo*-inososamine, (9) N^1-amidino-streptamine, (10) N^1-amidino-streptamine-6-phosphate, (7a) bluensidine; P: phosphate residues.

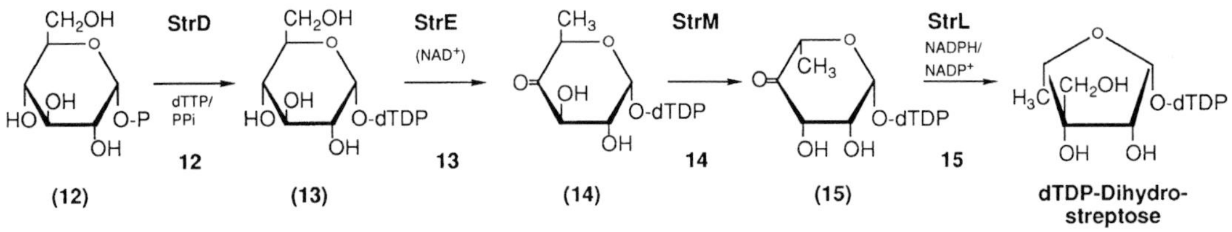

Fig. 15. The dTDP-dihydrostreptose pathway. For the numbering and labeling system, see legend of Fig. 14; intermediates are (12) D-glucose-1-phosphate, (13) dTDP-D-glucose, (14) dTDP-4-keto-6-deoxy-D-glucose, (15) dTDP-4-keto-L-rhamnose.

of pyridoxalphosphate(PLP)-dependent aminotransferases, the so-called secondary metabolic aminotransferases (SMAT). Several other antibiotic biosynthetic aminotransferases (see the compilation and discussion in PIEPERSBERG, 1994), the protein MosB of *Rhizobium meliloti* involved in the metabolism of L-3-*O*-methyl-*scyllo*-inosamine (MURPHY et al., 1993), and the PMP-dependent enzyme E_1 catalyzing the 3-dehydroxylation during the formation of 3,6-dideoxyhexoses in gram-negative bacterial LPS biosynthesis (THORSON et al., 1993; LIU and THORSON, 1994) belong to the SMAT protein family. Therefore, it seems obvious that one of the genes, either *stsA* or *strS*, encodes the *N*-amindino-*scyllo*-inosamine L-alanine aminotransferase, the second transaminase necessary for streptidine-6-phosphate formation (cf. Fig. 14, step 9). Alternatively, one of the remaining SMAT enzymes, StrS or StsA, may be involved as a biosynthetic enzyme in the synthesis of the NDP-*N*-methyl-L-glucosamine (NMLGA) subunit (cf. Sect. 4.1.1.3). The enzymes catalyzing the phosphotransfer (cf. Steps 5 and 10, Fig. 14) seem to be StrN or StsE proteins. Although this function has not yet been proven by *in vitro* assays of the individually expressed enzymes there is much evidence for this assumption. Both proteins contain in their C-terminal portion the characteristic signature motifs which are typical for aminoglycoside phosphotransferases and eukaryotic protein kinases (cf. Sect. 4.5; PIEPERSBERG et al., 1988; HEINZEL et al., 1988). Recently, a gene homologous to *strN* was found on a DNA fragment from *S. flavopersicus,* which confers spectinomycin resistance to *S. lividans* (J. ALTENBUCHNER and D. LYUTZKANOVA, unpublished data; see below Sect. 4.1.2). In addition, a spectinomycin phosphorylating activity was detected in this recombinant *S. lividans* strain (J. DISTLER, J. ALTENBUCHNER, and D. LYUTZKANOVA, unpublished data). These findings support the assumption that *strN* encodes a phosphotransferase. Two closely related genes, *strB1* and *strB2*, were identified which encode the amidinotransferase engaged in the biosynthesis of the streptidine moiety (OHNUKI et al., 1985a, b; DISTLER et al., 1987b; TOHYAMA et al., 1987; MAYER et al., 1988).

Both amidinotransferases of *S. griseus* cloned in *S. lividans* were active in a nonspecific assay which did not differentiate between the first and second transamidination steps (OHNUKI et al., 1985a, b; DISTLER et al., 1987b; TOHYAMA et al., 1987; S. EHRIG-FRANTZKE, and W. PIEPERSBERG, unpublished data). Since the enzymes encoded by the cloned genes have not as yet been tested with their postulated substrates (WALKER, 1975a, see steps 6 and 11, Fig. 14) it remains uncertain whether the assumptions regarding their functions (cf. StrB1, step 6, StrB2, step 11, Fig. 14) are correct (PISSOWOTZKI et al., 1991) or whether StrB1 carries out both steps (OHNUKI et al., 1985b). The latter is supported by the puzzling finding that extracts of *S. bluensis* seemed to contain both enzymatic activities although only StrB1 is used in the bluesidine pathway (WALKER, 1990) and only the *strB1* gene could be detected by hybridization in two bluensomycin producers, *S. bluensis* DSM 40564 and *S. hygroscopicus* ssp. *glebosus* DSM 40823 (cf. Fig. 13; G. MAYER, A. MEHLING, and W. PIEPERSBERG, unpublished data).

4.1.1.2 The L-Dihydrostreptose Pathway

GRISEBACH (1978) postulated a route for dTDP-L-dihydrostreptose similar to that leading to activated L-rhamnose in gram-negative bacteria: activation of D-glucose in form of dTDP-D-glucose (step 12, Fig. 15), dehydratation to dTDP-4-keto-6-deoxyglucose (step 13), epimerization to dTDP-4-keto-L-rhamnose (step 14), and reduction coupled to a rearrangement of the hexose carbon chain yielding dTDP-L-dihydrostreptose (step 15). The 4 enzymes catalyzing the synthesis of dTDP-L-dihydrostreptose are encoded by the genes *strD*, *strE*, *strM*, and *strL* (see Tab. 2 and Fig. 14; DISTLER et al., 1987b; PISSOWOTZKI et al., 1991; PIEPERSBERG, 1994). Evidence in support of this is the high level of homology which they share with the respective genes encoding the L-rhamnose biosynthesis enzymes in salmonellae, *rfbA*, *B*, *C*, *D*, and their activity in *Escherichia coli* (REEVES, 1993; PIEPERSBERG, 1994; S. VERSECK and W. PIEPERSBERG, unpublished data). StrD is related to NDP-hexose synthases (pyrophosphorylases) (DISTLER et al., 1987b). Genes similar to all or part of the *strDELM* cluster are present in many other gene clusters which encode enzymes for the production of streptomycete secondary metabolites containing a 6DOH sugar moiety (PIEPERSBERG, 1994; LIU and THORSON, 1994; VINING and STUTTARD, 1994). In a screening of streptomycete strains more than 50 of which produced 6-deoxyhexose-containing secondary metabolites a majority seemed to possess genes which hybridize to *strD*, *E*(*L*,*M*) (STOCKMANN and PIEPERSBERG, 1992). Thus, this part of the *str* gene cluster appears to be widespread among antibiotic-producing streptomycetes and other bacterial groups (cf. Sects. 5.3 and 6).

4.1.1.3 Hexosamine Pathway

The synthesis of the third moiety of streptomycin, (NDP-activated) *N*-methyl-L-glucosamine (NMLGA), has been intensively studied. Several speculative pathways have been postulated, including nucleotidylation, deacetylation of one of the intermediates (if UDP-*N*-acetyl-hexosamines are formed), and at least three epimerization steps, one of which could be divided into separate oxidation and reduction steps at C-4 of the hexoseamine, followed by N-methylation (RINEHART and STROSHANE, 1976; GRISEBACH, 1978; OKUDA and ITO, 1982; HIROSE-KUMAGAI et al., 1982; KUMADA et al., 1986). The direct precursor still remains unknown but could perhaps be D-glucose-1-phosphate, D-glucosamine-1-phosphate, *N*-acetyl-D-glucosamine-1-phosphate, or *N*-methyl-D-glucosamine (S. BEYER and W. PIEPERSBERG, unpublished data). Also, the published structures, UDP-activated and phosphorylated hexosamines (HIROSE-KUMAGAI et al., 1982; KUMADA et al., 1986), of putative intermediates of the *V*-methyl-L-glucosamine (NMLGA) pathway which were accumulated in the wild type and in a mutant blocked in the NMLGA pathway cannot be explained by the currently postulated pathway. Another unsolved problem is the formation of the *N*-methyl group in *N*-methyl-L-glucosamine which might be introduced at the D-glucosamine level or later in the pathway, perhaps even after condensation of the three streptomycin moieties (OKUDA and ITO, 1982, op. lit.).

There is indirect evidence that the genes *strPQX*, isolated and sequenced from *S. glaucescens*, *strFG*, and *stsG*, analyzed from *S. griseus*, are involved in the *N*-methyl-L-glucosamine pathway (MANSOURI and PIEPERSBERG, 1991; J. AHLERT, G. MAYER, S. BAYER, and W. PIEPERSBERG, unpublished data). StrQ is a CDP-D-glucose pyrophosphorylase which could catalyze the activating step of the *N*-methyl-L-glucosamine pathway (Fig. 16, step 16; S. BAYER and W. PIEPERSBERG, unpublished data). The StrP protein shares a higher degree of similarity with UDP-glucose 4-epimerases than with NDP-hexose 4,6-dehydratases, which are distantly related, and it could, therefore, be a C-4 dehydrogenase (or epimerase; Fig. 16, steps 17, 19, 18a, or 20) for NDP-hexose derivatives (PISSOWOTZKI et al., 1991). Alternatively, StrP could be an oxidoreductase, introducing a keto group at C-2 of a CDP-activated hexose (Fig. 16, step 20). This could then be the substrate for an aminotransferase, encoded by either *strS* or *stsA*,

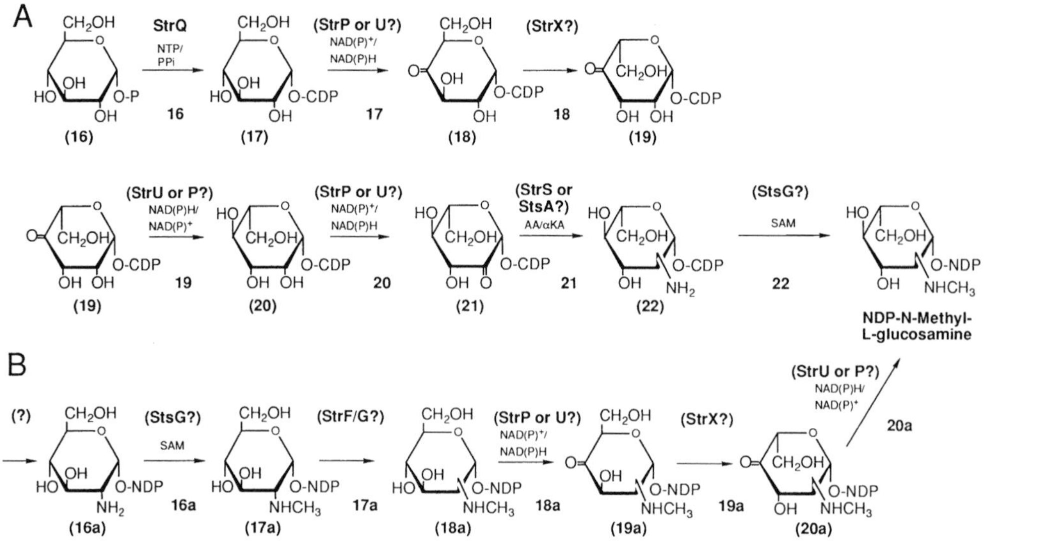

Fig. 16. The NDP-*N*-methyl-L-glucosamine (NMLGA) pathway. For the numbering and labeling system, see legend of Fig. 14. The exact route of formation of NMLGA is unknown and could either procede via CDP-glucose (A) or an unknown derivative of NDP-D-glucosamine (B) as precursors (for details, see text). Possible intermediates in (A) are (16) D-glucose-1-phosphate, (17) CDP-D-glucose, (18) CDP-4-keto-D-glucose, (19) CDP-4-keto-L-mannose, (20) CDP-L-mannose, (21) CDP-2-keto-L-glucose, (22) CDP-L-glucosamine; possible intermediates in (B) are (16a) NDP-D-glucosamine, (17a) NDP-*N*-methyl-D-glucosamine, (18a) NDP-*N*-methyl-D-mannosamine, (19a) NDP-4-keto-*N*-methyl-D-mannosamine, (20a) NDP-4-keto-*N*-methyl-L-glucosamine.

which introduces the 2-amino group (Fig. 16, step 21). The *strFG* genes were mapped in a region which complemented a mutant blocked in the *N*-methyl-L-glucosamine pathway (KUMADA et al., 1986). Protein comparisons suggest that both StrF and StrG could be members of the group of dinucleotide-independent (non-oxidoreductase type) sugar isomerases (or epimerases) and might even form a heterodimeric enzyme (MANSOURI and PIEPERSBERG, 1991). A candidate for a 3,5-epimerase (Fig. 16, steps 18 and 19a) in the *N*-methyl-L-glucosamine pathway is StrX because of its significant similarity to other 3,5-epimerases such as StrM (see Sect. 4.1.1.2; S. BEYER and W. PIEPERSBERG, unpublished data). The StsG protein has three conserved motifs which are generally found in methyltransferases (KAGAN and CLARKE, 1994). Therefore, *stsG* could encode the *N*-methyltransferase necessary for streptomycin formation (Fig. 16, steps 22 and 16a). Based on these genetic data, the following routes for the synthesis of *N*-methyl-L-glucosamine can be proposed among several other possibilities:

(1) CDP activation of D-glucose-1-phosphate (StrQ), epimerization and oxidoreduction followed by a transamination and *N*-methylation yielding CDP-L-glucosamine (Fig. 16A);
(2) activation and modification of D-glucosamine (or a derivative thereof) which would account for the earlier observations that D-glucosamine is preferentially incorporated into *N*-methyl-L-glucosamine (Fig. 16B) (OKUDA and ITO, 1982; KUMADA et al., 1986).

However, the sequence of enzymic steps or the still unknown reactions remain even more speculative. Corresponding to the cyclitol transaminase reactions of the cyclitol an additional phosphotransferase could also be involved in the *N*-methyl-L-glucosamine pathway forming the identified, additionally phosphorylated NDP-hexosamine (HIROSE-KUMAGAI et al., 1982; KUMADA et al., 1986).

4.1.1.4 Condensation of Subunits, Processing, and Export

The condensation of the activated precursors requires two glycosyltransferase steps which are catalyzed by two enzymes localized in the soluble cytoplasmic fraction resulting in dihydro-streptomycin-6-phosphate, which is the last soluble intermediate detected inside producing cells (KNIEP and GRISEBACH, 1976, 1980). The dihydrostreptosyl transferase was partially purified and found to be a dimeric enzyme (KNIEP and GRISEBACH, 1980) with a likely subunit molecular weight of 35kDa. Evidence that the *strH* gene could be one of the two genes needed for glycosyltransfers is weak (OHNUKI et al., 1985a, b; MANSOURI and PIEPERSBERG, 1991). The condensation product dihydro-streptomycin-6-phosphate is converted, probably coupled with the active transport, to streptomycin-6-phosphate by a membrane-associated dehydrogenase (step 26, Fig. 17; MAIER and GRISEBACH. 1979). The proteins encoded by *strVW* and *strU* recently detected in both *S. glaucescens* and *S. griseus* (PIEPERSBERG, 1994; BEYER et al., 1996) are able to form this membrane-bound transport/dehydrogenase complex. The StrU protein has significant similarity to the alcoholic hydroxyl-group-oxidizing dehydrogenases, and the *strV(W)* gene product(s) represent a new member of the family of the so-called ABC transporters (BEYER et al., 1996) suggesting that they could be engaged in the oxidation and export of dihydro-streptomycin-6-phosphate. However, StrU clearly is a member of the dinucleotide coenzyme-dependent dehydrogenase family and does not contain any transmembrane domains nor membrane-association sites. Therefore, the suggested involvement of StrU in the oxidation of dihydrostreptomycin-6-phosphate does not easily correlate to the earlier findings of H. GRISEBACH's group (MAIER and GRISEBACH, 1979) that this oxidase is in the particulate (membrane) fraction and does not require the addition of an electron acceptor such as $NAD(P)^+$. Nevertheless, these phenomena could be explained by a hypothetical oxidase/exporter complex formation between the StrU protein and the cytoplasmic domain of the StrU/W transmembrane complex and a strongly bound dinucleotide coenzyme in StrU. This would also explain the coupling of oxidation and transport steps which have been interpreted as being a functional unit (MAIER and GRISEBACH, 1979). An attractive speculation could be that the oxidation is coupled to both the ATP-driven export of a phosphorylated aminoglycoside and a membrane-bound electron transport.

The final dephosphorylation to release the biologically active antibiotic is catalyzed by StrK, a streptomycin-6-phosphate specific extracellular phosphatase (step 27, Fig. 17; WALKER, 1975a; MANSOURI and PIEPERSBERG, 1991). The gene products of the *strK* genes of *S. griseus* and *S. glaucescens* (*orfI* of VÖGTLI and HÜTTER, 1987) are highly homologous to the alkaline phosphatase (PhoA) of *E. coli* (MANSOURI and PIEPERSBERG, 1991). The properties of the StrK phosphatase of *S. griseus* when expressed in *S. lividans* are similar to those reported earlier for the streptomycin phosphate phosphatase (WALKER and SKORVAGA, 1973; WALKER, 1975a; MANSOURI and PIEPERSBERG, 1991). Each of the five steps (two glycosyltransfers, dehydrogenation, phosphatase reaction, and transport) should have an equivalent in 5′-hydroxy-streptomycin producers and, except for the dehydrogenation, also in the dihydrostreptomycin and bluensomycin producers (cf. Fig. A1). The 5′-hydroxylation reaction forming 5′-hydroxy derivatives of streptomycins is still obscure. It could take place by a hydroxylase reaction during the release at the cytoplasmic membrane or at the stage of the dTDP-hexose. However, it seems unlikely that the original 6-hydroxy group of D-glucose remains in the intermediates since the following steps would then require enzymes with altered substrate specificity or even altered reaction mechanisms. D-Mannosylation at position 4″ of the *N*-methyl-L-glucosamine

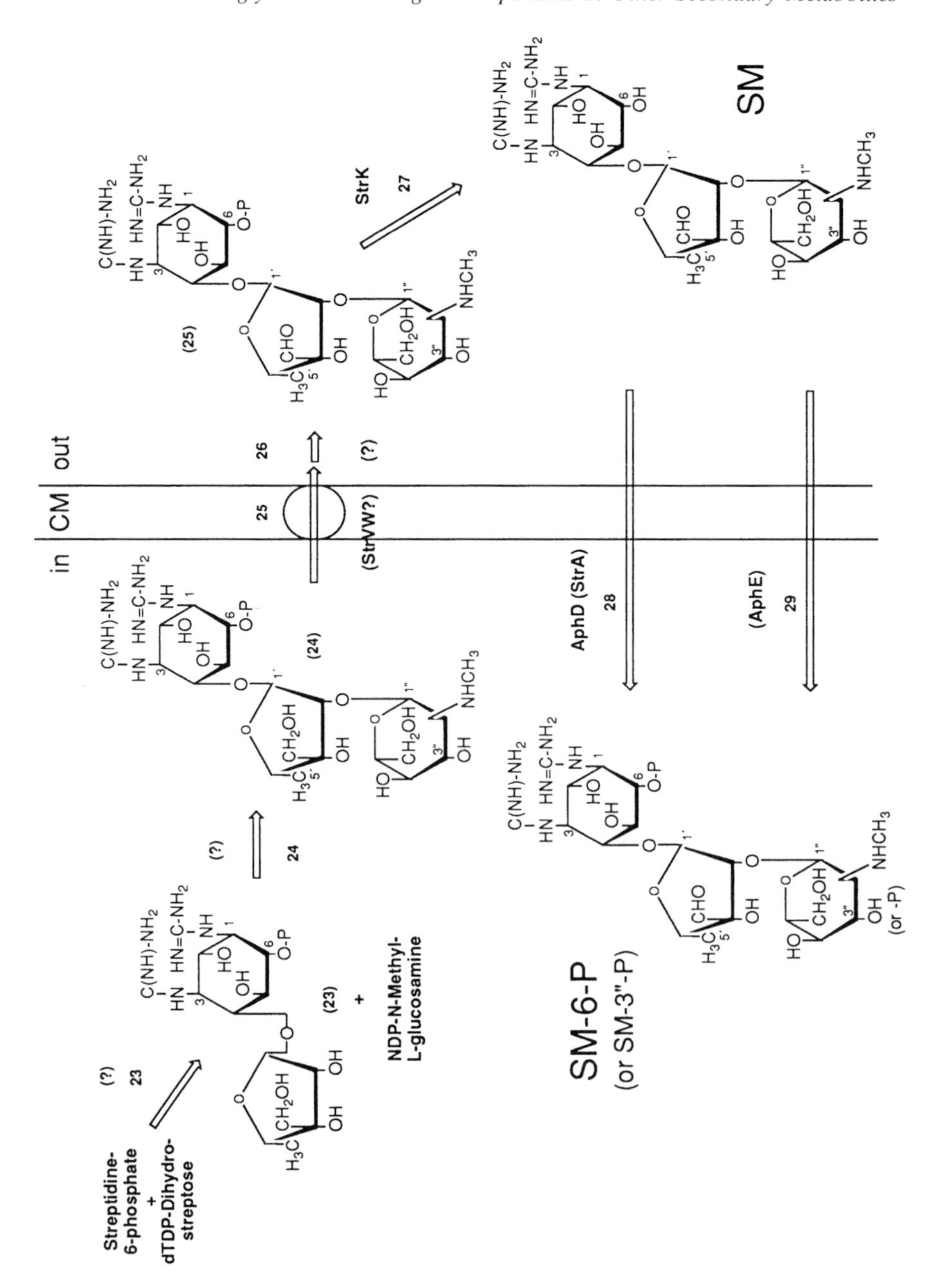

Fig. 17. Pathway for the condensation, export, oxidation, and release of streptomycin (SM) and SM resistance mechanisms. For the numbering and labeling system, see legend of Fig. 14; intermediates are (23) dihydrostreptosylstreptidine-6-phosphate, (24) dihydro-SM-6-phosphate, (25) SM-6-phosphate (SM-6-P). After re-uptake into *S. griseus* the antibiotic is re-phosphorylated, either at the 6- or 3″-hydroxyls by the resistance enzymes AphD or AphE, respectively.

moiety occurs in some streptomycin-producing biovars of *S. griseus*. This appears to be a nonspecific peripheral reaction rather than an integral step in the biosynthetic pathway, since in the same strains a mannosidohydrolase is formed under carbon catabolite depletion (INAMINE and DEMAIN, 1975). Later, also a dimannosylated product was detected in *S. griseus* strains (IKEDA et al., 1985a), in addition to *N*-methyl-L-glucosamine amino group modifications as in the ashimycins produced by *S. griseus* FT3-4 (TOHMA et al., 1989). Also, pseudodisaccharidic end products of the streptomycin family missing the *N*-methyl-L-glucosamine moiety have antibiotic activity and occur as natural products in fermentations of the *Streptomyces* sp. strain AC4437 (AWATA et al., 1986).

4.1.2 Streptomycin-Related Ca Aminoglycosides

Spectinomycins (Actinospectacin). The spectinomycin molecule (see Fig. A2) is composed of two double condensed hexose derivatives, a *myo*-inositol and a 6DOH-derivative (actinospectose, a 4,6-dideoxyhexose; cf. Fig. 8), and therefore is a Ca(4,5)-6DOH compound as is the dihydro-streptosylstreptidine (AC4437) produced by some streptomycetes (AWATA et al., 1986). The incorporation of radioactively and stable isotope-labeled glucose into both moieties clearly supports this interpretation (OTSUKA et al., 1980). However, no further biochemical studies on the pathway have been reported, except for the recent demonstration in a spectinomycin producer of an L-glutamine:*scyllo*-inosose aminotransferase similar to that found in streptomycin producers (cf. Sect. 4.1.1.1; WALKER, 1995). Recently, a spectinomycin-resistance-conferring DNA segment (3.65 kb) was cloned from the spectinomycin producer *S. flavopersicus* NRRL 2820 (J. ALTENBUCHNER and D. LYUTZKANOVA, personal communication). The DNA sequence suggested that it was derived from a rearranged and degenerated streptomycin gene cluster since it showed striking similarity with three genes, *strB1*, *strR*, and *strN* in the respective clusters of *S. griseus* and *S. glaucescens* (cf. Fig. 14). The *strB1*-related gene is deleted and its truncated protein product (80 aa) is certainly nonfunctional. The deletion was probably accompanied or preceded by an insertion of an IS112-like IS element part of which has been retained. Also, the similarity pattern among the reading frames is puzzling: the product of the *strB1*-related gene shows 83.8% and 92.5%, the product of the *strR*-related gene 45.9% and 47.5%, and the product of the *strN*-related gene 29.5% and 29.4% identity to the StrB1, StrR, and StrN proteins, respectively, of *S. griseus* and *S. glaucescens*. It is not yet known whether there are other production genes for spectinomycin adjacent to this cluster. However, this finding strongly supports the hypothesis that an ancestral production gene cluster for streptomycin-like aminoglycosides could have developed into another variant by divergent evolution after degeneration, modification, and, later on, frequent recombination with an intact *str/sts* gene cluster thereby creating a new pathway which yields simpler but still effective end products. More recently, a new spectinomycin derivative, spenolimycin (Fig. A3), has been described (KARWOWSKI et al., 1984). Its 6DOH moiety is altered in that the 3′ position contains an oxymethyl group and the C-3′/4′-bond is unsaturated, reminiscent of modifications which occur in other 6DOH residues, e.g., in anthracyclines (see Figs. 8 and 9 and Sect. 5.3).

Kasugamycins. Not much work has been done to elucidate the biochemical pathway for the production of kasugamycin (Fig. A4) and related Ca(4)-6DOH compounds (minosaminomycin; Fig. A5). However, it is reasonable to postulate basically streptomycin-like pathways also for these aminoglycosides. The cyclitols are again clearly derived from *myo*-inositol. However, the postulate by UMEZAWA et al. (1986) that the 6DOH moiety kasugamine could be derived from UDP-*N*-acetyl-D-glucosamine seems unlikely in view of what has been learned from the streptomycin pathway and from other 6DOH or hexosamine pathways in the recent past. Rather, it seems likely that a dTDP- or a CDP-hexose biosynthetic route is used, since the final product is a 2,3,4,6-tetradeoxy-2,4-diaminohexose derivative (cf. Sects. 3.2.1 and 5.3).

Myomycin. Myomycin, besides being a possible Ca(4)-HA (Fig. A8) compound modified by a varying number of β-lysyl residues, is another streptomycin-related product in that it contains some distant structural similarities to bluensomycin, such as carbamoyl groups in the cyclitol moiety in addition to a guanidino group (in the hexosamine moiety) and apparently a mode of action and binding site which are identical to those of the streptomycins (DAVIES et al., 1988). Since these aminoglycosides are produced by nocardioforms (or coryneforms; FRENCH et al., 1973) it will be interesting to see whether genes related to known *str/sts* genes in the streptomycin or bluensomycin producers are also present in the lower actinomycetes. Examples of such genes which could be found in the myomycin producer are those encoding enzymes of the following families (see Tab. 2): amidinotransferases (e.g., StrB1, StrB2), carbamoyltransferases (bluensomycin pathway), and those related to some of the enzymes involved in the synthesis of the cyclitol (e.g., StrO, StrN) and in other parts of the pathway, e.g., activation, synthesis, and transfer of the 3-aminohexose precursor (e.g., StrQ, StrS, or StsA), or involved in export and resistance phenomena (e.g., StrV, StrW, StrA). Genetic relationships to producers of antibiotics containing β-lysine tails (e.g., streptothricin; cf. Fig. A27; or viomycin) may also become apparent. The 3-guanidinomannose moiety in myomycin could be derived from an unusual NDP-glucose or NDP-mannose pathway similar to the NMLGA unit in streptomycin (cf. Sect. 4.1.1.3).

Boholmycin. The last, structurally new class of aminoglycosides to be described is represented by the pseudotetrasaccharidic compound boholmycin (Fig. A7; SAITOH et al., 1988) and is produced by a strain of *Streptomyces hygroscopicus.* It clearly belongs to the Ca aminoglycosides according to our definitions. It is composed of a dicarbamoyl *scyllo*-inositol, two amino sugars condensed via glycosidic bonds to the 4- and 6-positions of the cyclitol, and a heptose (Ha-(4)Ca(6)-PA-Hep). Its clear structural relationships with both myomycin (in the cyclitol(4)-3-aminohexose pseudodisaccharidic unit; see above) and seldomycins (in the cyclitol(6)-2-aminopentose unit; cf. Sect. 4.3) gives it a clear bridging role between the Ca and the (2DOS-)Cb aminoglycosides, and it will be interesting to see whether this is reflected by similarities at the genetic/enzymic level. The incorporation of a D-mannoheptose makes it unique among the cyclitol-containing aminoglycosides and could be another indication for the hypothesis that the production of cell wall components of gram-negative bacteria (e.g., lipopolysaccharide) and many antibiotics in streptomycetes may be based on the same gene pool (PIEPERSBERG, 1992, 1993).

Ca Aminoglycosides with Monoaminocyclitols. Three monoaminocyclitol aminoglycosidic antibiotics with no importance in pharmaceutical application but with interesting structures have been reported which should be mentioned here.

(1) Minosaminomycin, a kasugamycin-related Ca(4)-6DOH product of *Streptomyces* sp., has already been mentioned above. Its unique structural feature is the presence of a 1-D-1-amino-1-deoxy-*myo*-inositol to which a histidinyl-valine dipeptide is bound via an amide bond (Fig. A5). The introduction of the amino group into the cyclitol could principally follow the same route as in the streptidine pathway, although with altered stereoselectivity in the steps catalyzed by the first-step dehydrogenase and transaminase.

(2) Hygromycin A (Fig. A9) was detected prior to hygromycin B in the same strain of *Streptomyces hygroscopicus* in the early 1950s and was later also found as a product of other *Streptomyces* sp. and of *Corynebacterium equi.* It is also a Ca(1)-[X]-6DOH compound and is interesting in several respects:

- it has an aminocyclitol unit, a 2-amino-*neo*-inosamine, the derivation of which is totally obscure, but suggesting, however, that both a Ca and a Cb (for the biosynthesis of the 2DOS moiety in hygromycin B) pathway is functional in this strain, and that these two pathways can only coexist because of their completely different stereoselectivity of oxidation/transamination steps;

- two of the *cis*-hydroxyl groups are bridged by a methylene residue in the cyclitol moiety;
- the furanosidic 5-keto-6-deoxysugar could be another rare 6DOH unit derived from a dTDP-glucose (cf. Sects. 3.2.1 and 5.3).

(3) The aminoglycosides of the LL-BM123 series (Fig. A6), produced by *Nocardia* sp., are again compounds of biosynthetically mixed origin since they also contain amino acid residues such as minosaminomycin. Their unique pathway formula is Ca(4)-H(4)-HA, where the disaccharide D-glucosaminyl-(β-1,4)-D-mannose is glycosidically linked to the 4 position of the 2-amino-2-deoxy-*myo*-inositol moiety. The latter aminocyclitol should be derived from *myo*-inositol, which again can only be formed via an aminotransferase with different substrate selectivity relative to the StsC protein involved in the biosynthesis of streptidine (cf. Sect. 4.1.1.1).

4.2 Fortimicins, Istamycins

This group of very similar Ca(6)-HA or Cb(6)-HA compounds (Fig. A10) is widespread among filamentous actinomycete genera (cf. Fig. 1), namely *Micromonospora* spp. (fortimicins: FTM; SF-2052), *Dactylosporangium* spp. (dactimicins), *Streptomyces* spp. (istamycins: ISM; sannamycins), and *Saccharopolyspora* spp. (sporaricins). In fact, the only major biosynthetic difference between the two groups containing either fortamine (FTM, dactimicins) or 2-deoxyfortamine (ISM, sporaricins, and sannamycins) as aminocyclitols seems to be the formation of the cyclitol via *myo*-inositol phosphate or *scyllo*-inosose, respectively (cf. Sect. 3.1). The FTM pathway has been most extensively studied in *Micromonospora olivasterospora* ATCC 21819 (FTM-A producer). It has been found to be almost congruent in substrate specificity with that of the producers of SF-2051 *Micromonospora* sp. SF-2089 ATCC 31580, dactimicin *Dactylosporangium matsuzakiense* ATCC 31570, sannamycin *Streptomyces sannanensis* IFO 14239, istamycin *S. tenjimariensis* ATCC 31603, and sporaricin *Saccharopolyspora hirsuta* ATCC 20501 (Fig. 18 and Tab. 3; ITOH et al., 1984; ODAKURA et al., 1984; DAIRI and HASEGAWA, 1989; HASEGAWA, 1991, 1992; DAIRI et al., 1992a, b, c; OHTA et al., 1992a, b, 1993a, b; OHTA and HASEGAWA, 1993a, b; HOTTA et al., 1995). Of the roughly 20 steps of FTM-A biosynthesis, 14 have been identified by various methods, including induction and analysis of blocked mutants, gene cloning, and feeding of intermediates. Most of the production genes (*fms*) seem to be clustered on a DNA fragment of ca. 30 kb or more in *M. olivasterospora.* The order of identified genes in *M. olivasterospora* ATCC 21819 is *fms10, 13, 3, 4, 5, 12, 8, 7, 14, 1, 11,* (*orf2*), *fmrO*, (*orf4*). This DNA segment as a whole only hybridizes with the genomic DNA from *Micromonospora* sp. SF-2089 and *D. matsuzakiense* (DAIRI et al., 1992b). However, the restriction pattern of the hybridizing bands were almost identical only in the strain *Micromonospora* sp. SF-2089. In the DNA of the other three producers investigated no hybridization with the 30 kb fragment was observed, but when individual genes for conserved functions were taken as probes, e.g., the *fms13* (*sms13*; encoding the N-glycyltransferase) genes, significant hybridization was seen with the DNA from all six producers (OHTA et al., 1992b). Thus, it seems likely that all producers of FTM-like aminoglycosides contain highly related gene clusters originating from a common evolutionary source with some minor modifications, such as the use of a different pathway for the formation of the (2-deoxy)-*scyllo*-inosose precursor.

The difference between the two more distant groups (from the hybridization data; see above) is also reflected by the aquisition of two types of resistance genes: (1), *fmrO* (*fmrM*, *fmrD*, from *M. olivasterospora* ATCC 21819, *Micromonospora* sp. SF-2098, and *Dactylosporangium matsuzakiense* ATCC 31570, respectively); (2) *fmrT* (*fmrS*, *fmrH*, from *Streptomyces tenjimariensis* ATCC 31603, *S. sannanensis* IFO 14239, and *Saccharopolyspora hirsuta* ATCC 20501, respectively). Both encode members of the 16S rRNA methyltransferases but these enzymes methylate different residues (G-1405 and A-1408; CUNDLIFFE, 1989; HASEGAWA, 1991; OHTA

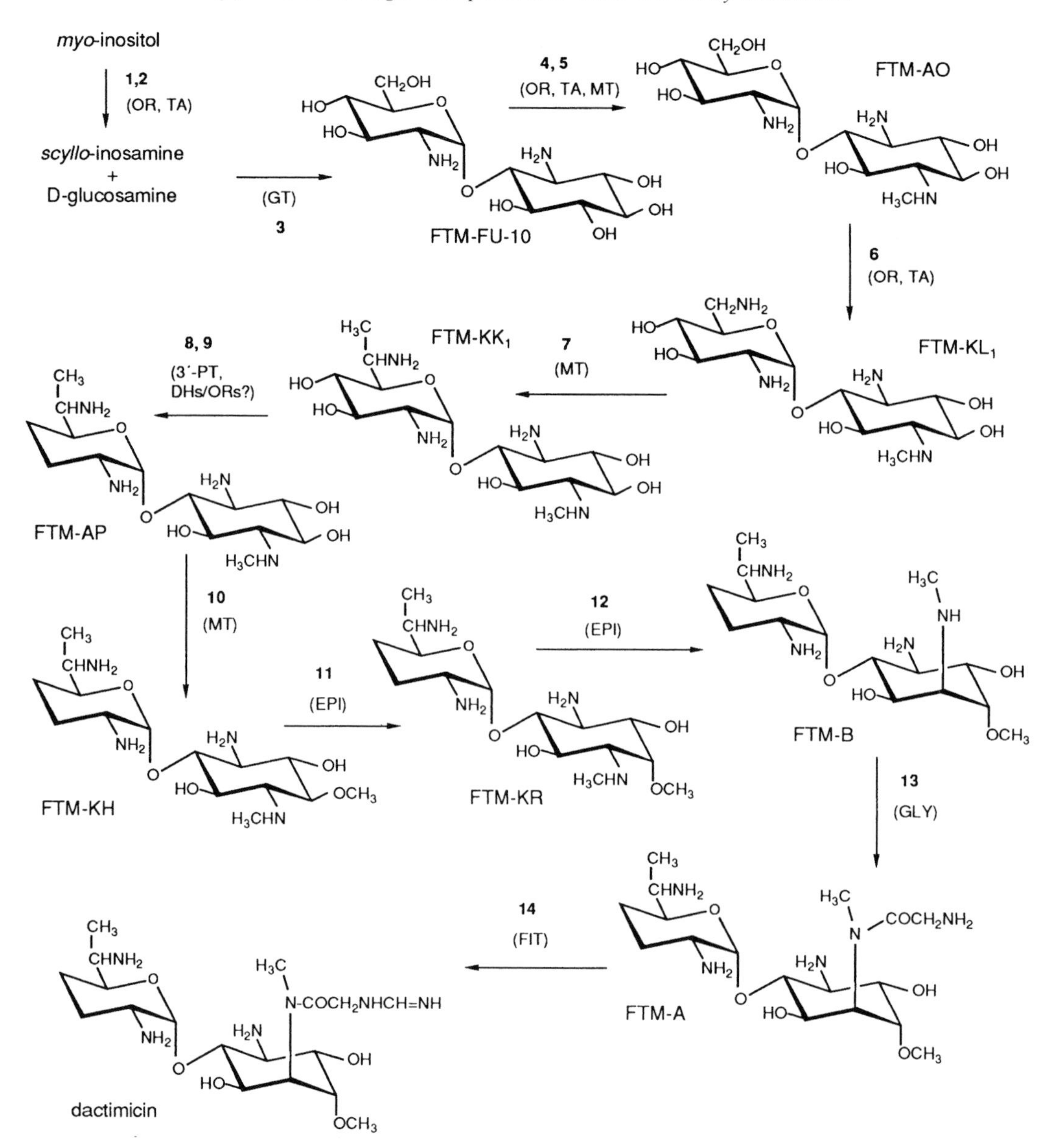

Fig. 18. The fortimicin (FTM, astromicin) pathway. The same pathway starting from 2-deoxy-*scyllo*-inosamine seems to be established in istamycin/sannamycin/sporaricin producers. The known intermediates and postulated enzymatic steps are given; for further details, see HASEGAWA (1992) and HOTTA et al. (1995).

and HASEGAWA, 1993a, b; OHTA et al. 1993a, b; see Sect. 4.5). In each case the single resistance gene seems to reside in the production gene cluster. However, they seem to be differently organized in the gene clusters in both groups (OHTA et al., 1993a). Thus, the set of genes used in the dactimicin producer could be a mixture of the two extreme evolutionary lines found in the other producers of the FTM/ISM group aminoglycosides: (1) it con-

Tab. 3. Gene Products and Enzymes Known or Presumed to be Involved in the Production and Resistance of Fortimicin (FIM)-Group Antibiotics

Gene Product[a]	Coding Capacity of Gene aa[b]	Enzymatic Function or Step[c] (Preliminary Assignment)	Remarks[d]	Organism[e]	References[f]
Fms1		(D-*myo*-Inositol 2-dehydrogenase)	NAD(P)-dependent?	*Mol, Msp, (Dma)*	A
Fms2		(*scyllo*-Inosose aminotransferase)	PLP-dependent?	*Mol, Msp, (Dma)*	A
Fms3		FTM-FU-10 synthesis (D-Glucosaminyltransferase)		*Mol, Msp, (Dma)*	A
Fms4		FTM-AO synthesis		*Mol, Msp, (Dma)*	A
Fms5		FTM-AO synthesis		*Mol, Msp, (Dma)*	A
Fms7		FTM-KK1 synthesis		*Mol, Msp, (Dma)*	A
Fms8		FTM-AP synthesis (FTM-KK1 phosphotransferase)	ATP-dependent, homologous to APH(3′)-II	*Mol, Msp, (Dma)*	A
Fms10		FTM-KH synthesis		*Mol, Msp, (Dma)*	A
Fms11	(126)	FTM-KR synthesis (FTM-KH epimerase)		*Mol, Msp, (Dma)*	A
Fms13 (Sms13)		FTM-B N-glycyltransferase		*Mol, Msp, Dma, Ste, San, Shi*	A, OHTA et al. (1992b)
Fms14	480	N-Formimidoyl FTM-A synthase (oxidase)	FAD, 4-mer	*Mol, Msp, (Dma)*	A, DAIRI et al. (1992c)
FmrO (FmrM, FmrD)	293	16S rRNA (G-1405) methyltransferase	SAM-dependent	*Mol; Msp, Dma*	A, OHTA and HASEGAWA (1993a)
FmrT	211	16S RRNA (A-1408) methyltransferase	SAM-dependent	*Ste, (San)*	A, OHTA and HASEGAWA (1993b)
KamC	156	16S rRNA (A-1408) methyltransferase	SAM-dependent	*Shi*	HOLMES et al. (1991)
Unknown ORF (ORF-2)	99	Unknown (upstream *fmrO*)	(FTM-A synthesis?)	*Mol*	OHTA and HASEGAWA (1993a)
Unknown ORF (ORF-4)	(313)	Unknown (downstream *fmrO*)	(FTM-A synthesis?)	*Mol*	OHTA and HASEGAWA (1993a)
Unknown ORF (ORF-1)	333	(Epoxide hydrolase)	(Sannamycin synthesis?)	*Ste*	OHTA and HASEGAWA (1993b)
Unknown ORF (ORF-3)	(188)	Unknown (downstream *fmrT*)	(Sarmamycin synthesis?)	*Ste*	OHTA and HASEGAWA (1993b)

[a] Cf. Fig. 18
[b] In brackets: partial sequence data
[c] Cf. Fig. 18
[d] PLP: pyridoxalphosphate; SAM: *S*-adenosyl methionine
[e] *Mol*: *Micromonospora olivasterospora*; *Msp*: *M.* sp. SF-2098; *Dma*: *Dactylosporangium matsuzakiense*; *San*: *Streptomyces sannanensis*; *Ste*: *S. tenjimariensis*; *Shi*: *Saccharopolyspora hirsuta*
[f] A: HASEGAWA (1992) and DAIRI et al. (1992a, b)

tains probably a Cb pathway as the ISM (sannamycin) type producers; but (2) it has the resistance gene and profile as well as the stronger DNA sequence similarity to the FTM(SF-2051) type producers.

The last two steps in the formation of FTM-A (glycyltransfer) and FTM-C (N-formimidoylation of the glycyl amino group; cf. Fig. 18), catalyzed by the gene products Fms13 and Fms14, respectively, in *M. olivasterospora* (Tab. 3) represent the biochemically best investigated phase of the biosynthetic pathway for FTM-like aminoglycosides. The genes for these two steps have been cloned, analyzed in part, and found to be present in all producers of FIM/ISM type aminoglycosides either by hybridization or by activity (DAIRI et al., 1992c; OHTA et al., 1992b). In a blocked mutant of the ISM producer *S. tenjimariensis* FTM-B was converted into 1-*epi*-FIM-B, dactimicin, and 1-*epi*-dactimicin; *M. olivasterospora* in turn converted ISM-A_0 and ISM-B_0 into ISM-A3 and ISM-B3, respectively (cf. Fig. A10; HOTTA et al., 1989; DAIRI and HASEGAWA, 1989). The mechanism of the glycyl transfer and the putative activation of the glycyl residue (e.g., aminoacyl-AMP) has not yet been studied. The *N*-formimidoyl group was shown to be derived from glycine, the C-2 group of which is converted via an unusual oxidase mechanism to the formimidoyl group and probably CO_2 in the presence of molecular oxygen only; this is catalyzed by the FAD-containing Fms14 enzyme (DAIRI et al., 1992c). Another interesting gene product is the Fms8 phosphotransferase, which probably catalyzes the 3′-OH phosphorylation of the purpurosamine moiety in the FTM-KK1 intermediate. This enzyme is homologous to the APH(3′) enzymes encoded by the *nmrA* and *aph* genes of neomycin-producing *Micromonospora* sp. and *Streptomyces fradiae,* respectively, and can be replaced by the latter gene products (DAIRI et al., 1992a). However, its involvement in the interesting 3′,4′-dehydroxylation is unclear (cf. Fig. 18, steps 8, 9). For this phase of the pathway probably several steps are required: (1) dehydratation at C-3′could occur via a mechanism similar to that operating in the 3,6-dideoxyhexose pathway in gram-negative bacteria (cf. Sect. 3.2.1 and Fig. 10); (2) a 4′,5′-dehydratase reaction as is suggested by the occurrence of 4′,5′-dehydro-FTM-A; (3) a reductase/dehydrogenase) step reducing the 4′,5′ double bond.

4.3 2-Deoxystreptamine-Containing Aminoglycosides

The large and clinically important group of 2-deoxystreptamine(2DOS)-containing aminoglycosides was extensively studied with regard to its biogenesis in wild type and mutant strains using ^{14}C-, ^{13}C-, ^{3}H-, and ^{15}N-labeled precursors, such as D-glucose, D-glucosamine, and 2DOS. The resulting data obtained mainly with the producers of neomycin, paromomycin, ribostamycin, butirosin, and the gentamicin/sagamicin group antibiotics have been reviewed extensively (RINEHART and STROSHANE, 1976; PEARCE and RINEHART, 1981; KAKINUMA, 1982; KASE et al., 1982; OKUDA and ITOH, 1982; UMEZAWA et al., 1986; GRÄFE, 1992). However, since about 1985 only very few new findings have been published. Therefore, only a brief summary of what is currently known is given here.

The 2DOS moiety which is the basic building block in this family of compounds is made directly from glucose-6-phosphate via the Cb route (see Sect. 3.1; Fig. 19, cf. Fig. 4) as has been clearly demonstrated recently in the neomycin producer *S. fradiae* (YAMAUCHI and KAKINUMA, 1992b, c; 1993; 1995) and earlier postulated (KAKINUMA, 1982). Previously, the biosynthesis of the cyclitol moiety was obscure and was believed to occur either via *myo*-inositol or directly from glucose by an unknown mechanism (RINEHART and STROSHANE, 1976). However, it was already known from the labeling pattern of the C_1 and C_2 positions of glucose and from the incorporation of the first amino group, that 2-deoxy-*scyllo*-inosamine is not derived from D-glucosamine and that the direction of the second-step transamination is opposite to that of the streptidine and actinamine pathways (PEARCE and RINEHART, 1981; UMEZAWA et al., 1986). The first transamination step could be catalyzed by an aminotransferase very similar to the *S. griseus* StsC enzyme (see Sect. 4.1.1.1) since it was shown by WALKER

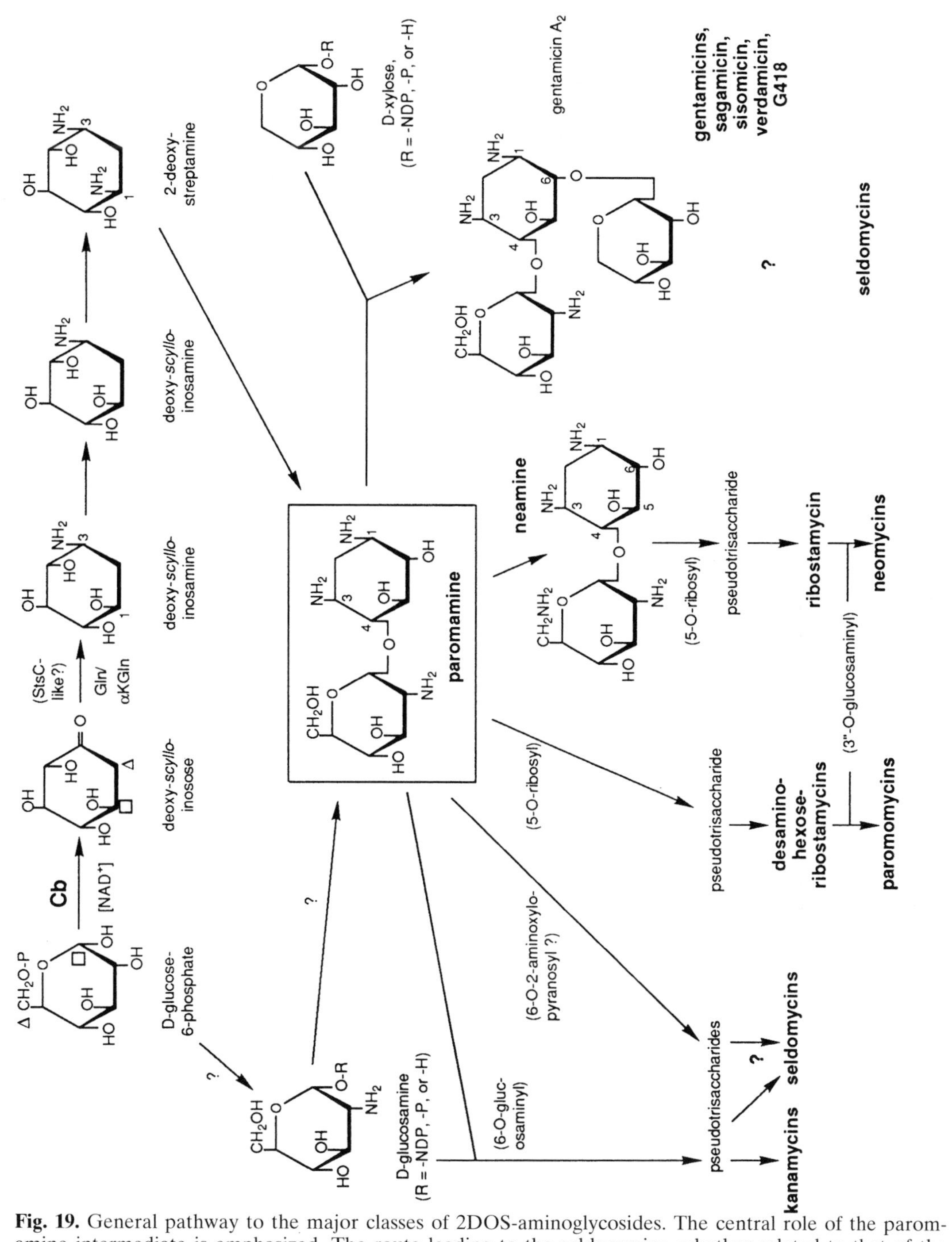

Fig. 19. General pathway to the major classes of 2DOS-aminoglycosides. The central role of the paromamine intermediate is emphasized. The route leading to the seldomycins, whether related to that of the gentamicins or the kanamycins or not, is unknown; for chemical structures, see Sect. 10, Figs. A12 and A14; P: phosphoryl residues.

et al. (CHEN and WALKER, 1977; LUCHER et al., 1989; WALKER, 1995) that ketocyclitols were transaminated by the α-amino group of glutamine in extracts of the neomycin producer *S. fradiae* and the gentamicin producer *Micromonospora purpurea* as in streptomycin and spectinomycin producers. Therefore, it will be interesting to see whether *stsC* probes, e.g., from *S. griseus,* will detect a respective gene in the production gene clusters of 2DOS-producing actinomycetes. The later steps in the formation of 2DOS could also be closely related to the respective steps in the streptidine pathway (cf. Fig. 14), such as dehydrogenation (enzyme StsB- or StrI-like?) and second-step transamination (enzyme StsA-like?). It is unknown whether 2DOS precursors are phosphorylated at some stage or enter the further condensation and secretion steps in a form similar to streptomycin. The presence of aminoglycoside-3′-phosphotransferases as resistance mechanisms in some of the 2DOS producers (see below, Sect. 4.5) would suggest, however, that the phosphorylation does not occur in the cyclitol moiety.

All major 2DOS-containing aminoglycosides, except for the destomycin-hygromycin B group and perhaps also the apramycins, seem to be formed via a common pseudodisaccharidic intermediate, paromamine, which is formed as an early intermediate from 2DOS and a molecule of D-glucosamine, which is probably nucleotide-activated, via the hypothetical pathway outlined in Fig. 19. Whether the precursor formation and the attachment reactions of the D-glucosamine moiety somehow relate to that of the NMLGA unit of streptomycin (see Sect. 4.1) remains to be shown. Further on, the 2DOS pathways branch into various alternate routes resulting in the large variety of end products (see Figs. A11–A17).

4,5-Disubstituted 2DOS Aminoglycosides. The compounds which are 4,5-substituted at the aminocyclitol, such as the neomycins, paromomycins, lividomycins (Fig. A11; basically Ha-(4)Cb(5)-P(3)-HA) or ribostamycins, and butirosins (Fig. A13; basically Ha-(4)Cb(5)-P), are probably synthesized directly from paromamine or via a second pseudodisaccharide, neamine (Fig. 19), followed by the attachment of a furanosidic pentose, xylose (forming xylostasins), or ribose (forming ribostamycins) which finally can be further glycosylated by a 2,6-diamino-2,6-dideoxyhexose (neosamine B or C) in the neomycins and paromomycins. The pseudodisaccharides are clearly intermediates and can be converted directly to the respective end products in mutants blocked in their formation, e.g., in the biosynthetic pathway of the aminocyclitol. Neamine (neomycin A) has measurable antibiotic activity. The pathway of the neosamines B and C is as obscure as that of the first hexosamine unit, though it is clear that D-glucosamine is preferentially incorporated into them. The type of activation, the glycosyltransferase(s), and the route of incorporation of a second amino-N into the C_6 position of the 2-aminohexose remain to be established enzymatically. A similarity to the situation in streptomycin-producing *S. griseus* could exist also here: UDP-(N-acetyl-)D-glucosamine might not be an immediate precursor as in the cell wall biosynthesis. Therefore, it will be necessary to screen also for a StrQ-related nucleotidylating enzyme among the gene products involved in biosynthesis of neomycin-like aminoglycosides. Important in this context are also the results of more recent biogenesis studies with 1-^{13}C- and 6-^{13}C-labeled glucose, which suggest that all building blocks of neomycin seem to be formed from intermediates of the pentose phosphate cycle (Fig. 20; RINEHART et al., 1992). If this "equilibration" of all or most glucose molecules through the pentose phosphate cycle generally occurs for all hexose and other carbohydrate components in secondary metabolites within the production phase, the earlier results of isotope-labeling studies would have to be reinterpreted.

4,6-Disubstituted 2DOS Aminoglycosides. The kanamycins–tobramycin (Fig. A12; HA-(4)Cb(6)-HA), the gentamicins–sagamicin (Fig. A14; HA-(4)Cb(6)-PA), and probably also the seldomycins (Fig. A17; HA-(4)Cb(6)-PA) groups of 2DOS-containing aminoglycosides are also synthesized from paromamine. However, their further substitution at the C_6 position of the cyclitol and the nature and modification of the second glycosidic residue

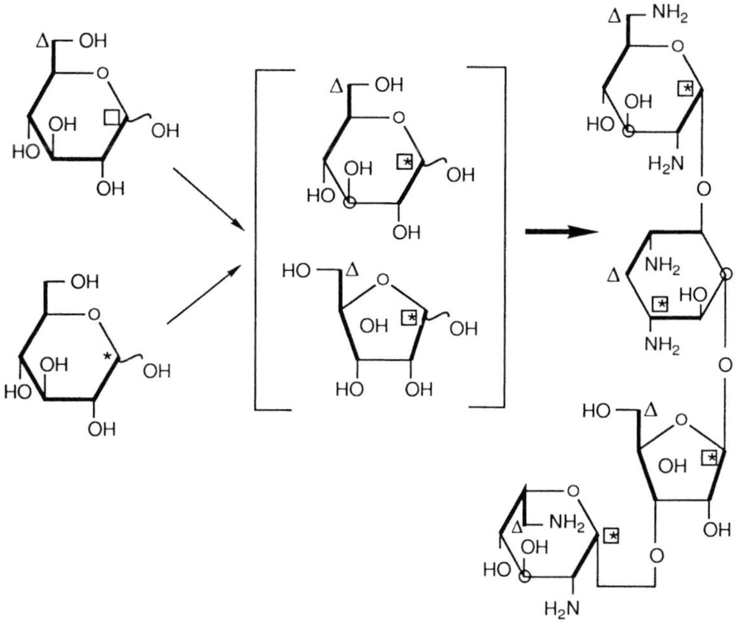

Fig. 20. Biogenesis of the hexose-, pentose- and cyclitol-derived components in neomycin from 6-(C^{13})-D-glucose (according to RINEHART et al., 1992). The labeling patterns measured by (C^{13})-NMR prove that all units are built up preferentially from C_1 (○), C_2, or C_3 units (thick lines) rearranged by transketolase- and transaldolase-catalyzed reactions in passages through the pentosephosphate cycle. The earlier labeling patterns obtained with 1- (*) or 1,6-labeled (□/△) D-glucoses are also given from which a differentiation between direct or indirect incorporation of glucose was not possible.

distinguish the gentamicins from the seldomycins in which a pyranoid C-5 sugar moiety is replaced by a hexosamine. This sugar in the gentamicin-related compounds is D-xylose (the pseudodisaccharide unit formed with the 2DOS cyclitol is called garamine) and could be either D-xylose or 2-amino-2-deoxy-D-xylose in the seldomycins. Since the gentamicins are only produced in *Micromonospora* spp., and the seldomycins are only found in *Streptomyces* spp. it will be of interest to study the mutual relationships between the respective sets of biosynthetic genes/enzymes relative to those involved in the production of the other groups of 2DOS-containing aminoglycosides. The most intensive study on the design of an individual 2DOS pathway was carried out on that of the gentamicin(GM)–sisomicin–sagamicin group (KASE et al., 1982; cf. UMEZAWA et al., 1986) which is one of the most versatile with more than 20 end products identified. Some details should be mentioned for comparison (for formulae see Fig. A 14): (1) a minimum of 20 enzymes is probably involved in the biosynthesis of, e.g., GM-C_1; (2) many of the steps, e.g., dehydrogenation and transamination in positions 3 and 6 of pyranoses, or N-methylation and dehydration, resemble those of other aminoglycoside pathways; (3) the multiply branching pathway proceeds from the first trisaccharidic intermediate GM-A_2 to GM-X_2 and then branches to yield the two unsaturated intermediates sisomicin (via JI-20A) and verdamicin (via G-418, a compound now frequently used for cloning vector selection in plant and animal cells). In some strains these two intermediates can already be the major end products which are released. This biosynthetic phase strongly resembles the 3,4-dehydroxylation steps in the fortimicin pathway (cf. Fig. 18; steps 8, 9); however, the occurrence of a 3,4-unsaturated 2,3,4,6-deoxy-2,6-aminohexose indicates that dehydratation (probably of a 4-hydroxylated precursor after 3-dehydroxylation; cf. LIU and THORSON, 1994) is a step in this process (see also Sect. 4.2). The last phase of GM biosynthesis yielding the final products GM-C_1, GM-C_{1a}, GM-C_2, and sagamicin, which are mainly used in therapeutics variably involves reduction, epimerization, and N-methylation steps, probably by action of the same enzymes on different but structurally related intermediates.

Apramycins and Destomycins. A more distant position relative to the other 2DOS aminoglycosides is taken by apramycin (Fig. A15; **CB(4)-OctA(8)-HA**) and destomycin-hygromycin B (Fig. A16; **Cb/Ca(5)-H(2,3)-HepA**) group. In the apramycins monosubstituted in the 4-position of the 2DOS moiety paromamine could be an intermediate. The further pathway would then require an elongation at the 6′-hydroxymethyl group of the hexosamine moiety by a C-2 unit which, however, seems rather unlikely. An alternative route would be the condensation of 2DOS with an octose derivative, which is reminiscent of the unusual octose pathway in the lincosamides (cf. Sect. 5.1; RINEHART, 1980; see also the discussion in OKUDA and ITO, 1982). Also, the 4-amino-4-deoxy-D-glucose moiety glycosidically linked to the 8′ position is quite unusual among the aminoglycosides; 4 transamination also occurs in the kasugamycin-related compounds, but on a 6DOH unit (see Sect. 4.1). This unit could be synthesized by transamination of a NDP-4-ketoglucose intermediate formed by an enzyme related to the UDP-glucose 4-epimerase (cf. Sect. 3.2). In the destomycins and hygromycin B the 2DOS moiety is 5-substituted with an unusual hexose, D-talose. This in turn is fused via a unique type of chemical bonding, an *ortho*-ester linkage between the 2′- and 3′-hydroxyls and the 1″-position of a sugar acid derived from a 6-amino-6-deoxyheptose, destomic acid. These structural details suggest that there is very little resemblance between the pathways of hygromycin B and destomycin production and those of other aminoglycosides, except for the biosynthesis of the aminocyclitol. Interestingly, strains of *Streptomyces eurocidicus* and of *Saccharopolyspora hirsuta* produce the destomycin derivatives, SS-56-C and 1-*N*-amidino-1-*N*-demethyl-2-hydroxydestomycin A (INOUYE et al., 1973; IKEDA et al., 1985b; cf. Fig. A16), respectively, which are clearly Ca compounds and are related to streptidine in their diaminocyclitol moieties. Here, a DNA recombination event resulting in the fusion of two gene clusters, those for streptomycin and destomycin production, could have created a new mixed pathway. From the streptomycin gene cluster those genes needed for the production of the 1-*N*-amidinostreptamine-6-phosphate intermediate (cf. Fig. 14) could be used. The streptamine unit in SS-56-C could be formed directly from *scyllo*-inosamine or via hydrolysis of the 1-*N*-amidino group from 1-*N*-amidinostreptamine. This also could mean that 6-phosphate intermediates of destomycins are formed inside the cells as in the case of the streptomycin producers.

In summary, compared with the streptomycins and the fortimicins the 2DOS aminoglycosides mentioned so far seem to take an intermediate position with respect to the distribution of modifying steps relative to the condensation reactions. The modifications are practically all finished before condensation in streptomycins, but in the 2DOS antibiotics much more take place after condensation by glycosyltransfer reactions by which only the diaminocyclitol is completely preformed.

4.4 Other Aminoglycosides

Monomeric sugar derivatives. The number of bioactive and stable molecules directly derived from monosaccharidic units or sugar analogs described in the literature is increasing steadily (Fig. A18; cyclitol-related compounds are treated below; cf. Sect. 5.2). Some examples are mentioned here:

(1) A group of glycosidase inhibitors, such as the nojirimycins (inhibit glucosidases and mannosidases), galactostatin (inhibits galactosidases), and siastatin (inhibits sialidases), are sugar analogs with a substituted pyridine ring and can be regarded as a group of bacterial alkaloids derived from aminohexoses or aminopentoses (GRÄFE, 1992). The sugar-like 1-deoxynojirimycin (DNJ) and a similar structured plant alkaloid (castanospermine; not shown) can act as anti-HIV drugs by preventing the maturation of the gp120 envelope glycoprotein. This finding motivated a recent study of the biogenesis of these unusual amino sugars by stable isotope labeling in *Streptomyces subrutilus* (HARDICK et al., 1992). It was found that DNJ is derived from a glucose molecule converted first via fructose, 6-oxidation and reductive 2- or 6-transamination steps to mannonojirimycin (Fig. 21). This intermediate can then be dehydrated and re-

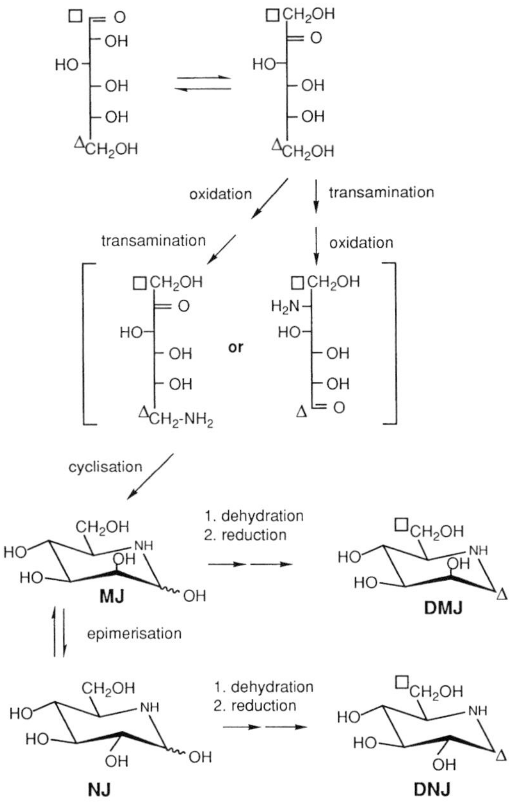

Fig. 21. Proposed biogenesis of nojirimycin and related monosaccharide analogs.
(D)NJ: (1-deoxy-)nojirimycin; (D)MJ: (1-deoxy-)-mannonojirimycin. The labeling patterns obtained from 1,6-labeled D-glucose was adopted from RINECHART et al. (1992) (cf. Fig. 20).

duced to 1-deoxymannonojirimycin or, alternatively, epimerized at C-2 to nojirimycin and subsequently dehydroxylated to DNJ. Thus, several enzymes/genes related to those used in the formation of other amino and/or deoxy sugar components in known antibiotics could be used. The same might hold for other monomeric amino sugars isolated from cultures of microorganisms such as 3-amino-3-deoxy-D-glucose, *N*-carbamoyl-D-glucosamine, prumycin, and streptozotocin (Fig. A18; see also the compilations in UMEZAWA et al., 1986; GRÄFE, 1992). Semisynthetic derivatives of DNJ with improved inhibitory activity on α-glucosidases and pharmacokinetics have also been prepared, two of which, miglitol and emiglitate (Fig. A18), have reached the phase of clinical development (MÜLLER, 1989).

(2) The recently described D-glucosamine derivative CV-1 (Fig. A18), which is produced by a *Streptomyces* spp. and is a weak antibiotic by itself, has a very interesting cooperative effect together with spiramycin on gram-negative bacteria (ICHIMURA et al., 1987). The inhibitory effect of CV-1 was demonstrated to be on LPS synthesis in *E. coli* thereby relieving the barrier effect of the outer membrane for spiramycin which by itself is ineffective on gram-negatives. The biosynthesis of CV-1 involves N-carbamoylation of D-glucosamine, probably from L-citrulline, which conceivably occurs either on a 1-phosphate- or nucleotide-activated precursor or the free sugar. Subsequent reorganization of *N*-carbamoyl-D-glucosamine to the unique open ring hemiaminal was shown to proceed spontaneously (YASUZAWA et al., 1987).

(3) Valiolamine (cf. Fig. A21) is a new monomeric aminocyclitol member of the validamycin group of aminoglycosides (KAMEDA et al., 1984; see below) and has inhibitory activity against α-glucosidases.

Trehalosamines and Other Aminodisaccharides. A larger group of nitrogen-containing and carbohydrate-related actinomycete products are compounds with structural analogy to the disaccharides trehalose or saccharose. Most of these have some biological activity either as antibiotics or as glycosidase inhibitors. The α,α-glycosidic trehalosamines (Fig. A19; HA(1)-H) are known for a long time and were isolated on account of their antibacterial activity which, however, is only weak (cf. UMEZAWA et al., 1986; ASANO et al., 1989). The biosynthesis of 2-trehalosamine and mannosyl glucosaminide from a molecule each of D-glucosamine and either D-glucose or D-mannose, respectively, by enzymes related to trehalose synthases can easily be envisaged. In contrast, the derivation of 3- and 4-trehalosamines which also occur as streptomycete products could follow at least two different routes: (1) formation of an activated 3- or 4-aminohexose which is condensed with D-glucose or (2) modification of preformed trehalose. It would be interesting to know whether these compounds have (auto-)regulatory functions in adaptive processes, such as osmoregulation, cell differentiation, or secondary metabolism; trehalose metabolism could play an important role in those physiological phenomena in streptomycetes (CHAMPNESS and CHATER, 1994). A new member of the trehalosamine family is the α,β-glycosidic antibiotic 3,3′-neotrehalosadiamine (Fig. A19; HA(1)-HA) isolated from a *Bacillus pumilus* for its antibacterial activity on an aminoglycoside-hypersensitive strain of *Klebsiella pneumoniae* (NUMATA et al., 1986; TSUNO et al., 1986).

The 3-amino-3-deoxy analogs of trehalose and sucrose and their epimers, T-I, T-II (3-trehalosamine), and T-III or S-I and S-II (3-sucrosamine) should be regarded as members of this family of compounds (Fig. A20; ASANO et al., 1989). These have been obtained by chemo-enzymatic synthesis from the 3-keto forms of trehalose and sucrose, which were prepared by treatment with D-glucoside 3-dehydrogenase from *Flavobacterium saccharophilum*, and by reductive chemical amination. Except for S-I, these compounds are weak antibiotics; S-I is an inhibitor of invertases.

Validamycins, Acarbose, and Related Oligosaccharidic Aminoglycosides. The cyclitol moieties formally derived from valiolamine, such as valienamine or validamine (cf. Figs. 6 and 22), are most probably products formed via a Cb pathway and may be regarded as aliphatic m-C_7N units, though they are not synthesized using erythrose-4-phosphate and phosphoenolpyruvate as the aromatic m-C_7N-units (cf. Sect. 3; FLOSS and BEALE, 1989; RINEHART et al., 1992). Thus, the suggestion that valienamine is not derived from the shikimate pathway (DEGWERT et al., 1987)

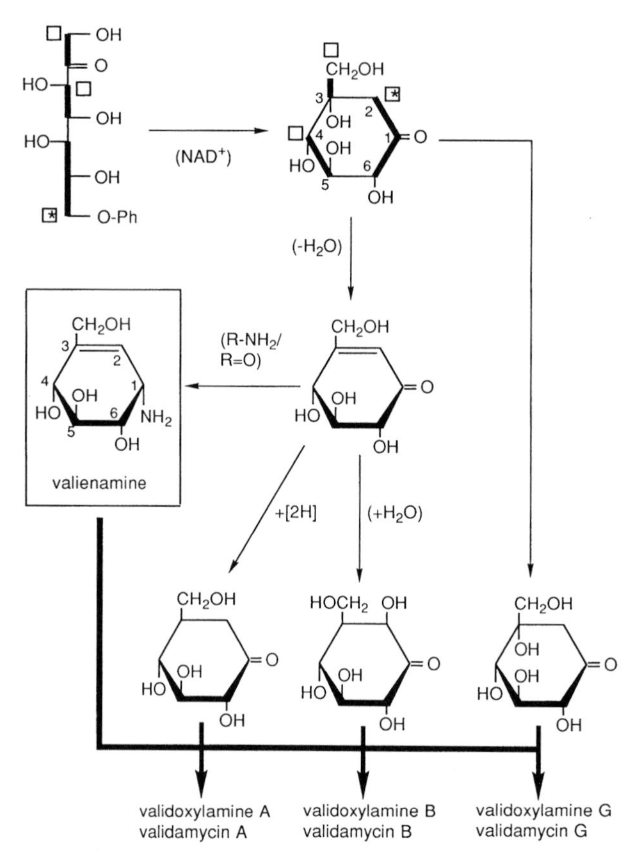

Fig. 22. Proposed scheme of biogenesis of C_7 sugar-derived cyclitols of the validamycin family. The labeling pattern from 1,6- or 6-(C^{13})-D-glucoses was adopted from RINEHART et al. (1992) (cf. Fig. 20).

probably hold true only for the origin of the C_7 precursor (which is either sedoheptulose-7-phosphate or an epimer thereof). However, at least the cyclization and dehydration steps are probably catalyzed by enzymes homologous to dehydroquinate synthase and dehydroquinate dehydrase, respectively. This type of cyclitol is found in validoxylamines and validamycins (Fig. A21; basically Cb(1)-Cb(4)-H compounds), and in acarbose, amylostatin, and other structurally related glycosidase inhibitors (Fig. A22; general formula $(H)_n$-Cb(1)-6DOH/H(1)-$(H)_n$; TRUSCHEIT et al., 1981; MÜLLER, 1989; YOKOSE et al., 1989). Valienamine and the related C_7 cyclitols are all condensed via an imino group with either another molecule of the same origin and similar structure (validamycin family) or at the 4 position with a 4,6-dideoxy-D-glucose or 4-deoxy-D-glucose unit (acarviosine family of α-glucosidase and trehalase inhibitors). This suggests that the aminotransfer reaction takes place on a precursor of the C_7 cyclitol moiety and that valienamine might be an intermediate for all other derivatives (cf. Figs. 6, 21 and A21, A22; cf. RINEHART et al., 1992).

The validamycins are α-D- or β-D-glucosylated at various positions (Fig. A21) which could be achieved by extracellular or cell-wall-associated glucosyltransferases. Also, the acarviosine family of glycosidase inhibitors, comprising acarbose, amylostatins, oligostatins, "amino-oligosaccharides" (epoxy derivatives of amylostatins), adiposins, trestatins, and AI-5662, are all variably glycosylated, mainly by condensation with di- or oligosaccharidic units such as maltose, oligomaltodextrins, trehalose, or an acarviosine units (TRUSCHEIT et al., 1981; MÜLLER, 1989). Again, the mechanism of addition of such blocks of oligosaccharidic sugars could be a membrane-associated extracellular process similar to the bactoprenol-dependent formation of extracellular heteropolysaccharides, e.g., lipopolysaccharide O-chains (GOEKE, 1986; RAETZ, 1996). Alternatively, these compounds could be excreted as inactive precursors by special exporters similar to streptomycin (cf. Sect. 4.1.1.4). This view is supported by the recent identification of an acarbose 7-phosphotransferase in the producing *Actinoplanes* sp. (DREPPER and PAPE, 1996).

Trehazolin. A new type of aminoglycosidic and very specific trehalase inhibitor, trehazolin (or trehalostatin), a product of *Micromonospora* sp. and *Amycolatopsis* sp., and the chemical synthesis of this compound and its β-anomer has recently been reported (ANDO et al., 1991; KOBAYASHI and SHIOZAKI, 1994; Fig. A23; Ca(1,2)-HA). The aminocyclitol moiety in this molecule is unusual in that it could be formed in a Ca pathway (cf. Fig. 4) via *myo*-inositol and later ring contraction as is suggested for pactamycin (see Sect. 5.2) or by direct reductive ring closure from a ketohexose precursor, e.g., fructose-6-phosphate. Also, the presence of a carbamoyl group is reminiscent of other groups of aminoglycosides; the incorporation of this group may involve enzymes similar to those used in pathways of other aminoglycosides. However, the trehazolin molecule exhibits the only known true aminoglycosidic linkage in its bonding to the α-D-glucose moiety.

Sorbistins. The pseudodisaccharidic aminoglycosides of the sorbistin family (Fig. A24) are interesting in two respects: they are produced both by pseudomonads and higher actinomycetes, and they contain an unusual open-chained diaminohexitol, 1,4-diaminosorbitol, of unknown biosynthetic origin (reviewed in UMEZAWA et al., 1986). It is tempting to speculate that these compounds could be derived from a pathway similar to that of the fortimicins followed by a reductive ring cleavage between the C-1 and C-2 positions of an aminocyclitol similar to fortamine later in the pathway (cf. Sect. 4.2; cf. Fig. 18). In fact, a nonacylated and nonmethylated fortamine precursor could be the direct precursor of 1,4-diaminosorbitol with an additional epimerization in position C-5 (C-3 of fortamine) creating the stereochemical configuration of sorbitol. The presence of a 4-amino-4-deoxyglucose can be explained in two ways, as for the biogenesis of 4-amino-4-deoxytrehalose (see the discussion on trehalosamines above). In accordance with the similarity to the fortimicin pathway a preformed (NDP-)4-aminoglucose would be the preferred precursor.

This could be synthesized via reactions similar to the initial steps postulated for the biosynthesis of the NMLGA moiety of streptomycin (cf. Sect. 4.1; cf. Fig. 16), whereby a convenient 4-ketohexose intermediate is formed which could be used as an aminotransferase substrate.

Allosamidin. A new aminoglycosidic product of *Streptomyces* spp., allosamidin and its demethyl and didemethyl derivatives (Fig. A25; Ca(4)-HA(4)-HA), is the first chitinase inhibitor isolated from actinomycetes (SAKUDA et al., 1987; ZHOU et al., 1992, 1993). Allosamidin is composed of unusual components, a branched five-membered C_6 aminocyclitol (allosamizoline) and two *N*-acetyl-D-allosamine units linked via β-1,4-glycosidic bonds which could be synthesized via a novel Ca pathway and via 3-epimerizations from (*N*-acetyl)-D-glucosamine, respectively. The exact route of the synthesis of the allosamizoline cyclitol is unknown, though its C and N atoms were shown to be directly derived from D-glucosamine. Therefore, it could either be synthesized from *myo*-inosamine formed by an enzyme analogous to *myo*-inositolphosphate synthase via successive ring contraction, oxidoreduction/epimerization, and transamination steps. Alternatively, it could be directly cyclized from D-glucosamine-6-phosphate via a Ca type (or other) enzyme forming pentacyclitols by activating C-5 and linking it to C-1. The aminooxazoline ring in the allosamizoline moiety (Fig. A25) is probably introduced via N-amidinotransfer from L-arginine, N-monomethylation, cyclization, and a second N-methylation step (ZHOU et al., 1993). Therefore, also in this pathway several steps could be catalyzed by enzymes related to *str*/*sts* gene products.

4.5 Resistance in Aminoglycoside Producers

Interestingly, aminoglycoside and in particular streptomycin resistance mechanisms by mutation (OZAKI et al., 1969; WALLACE et al., 1979; PIEPERSBERG et al., 1980), and those encoded by specific resistance genes in producers (WALKER, 1975b; COURVALIN et al., 1977) and clinical isolates of gram-negative and gram-positive bacteria, such as obligate pathogens or from opportunistic infections, where they are mostly plasmid-determined (BENVENISTE and DAVIES, 1973; UMEZAWA, 1974; DAVIES and SMITH, 1978; FOSTER, 1983), were the first among all antibiotic resistance phenomena to be identified at the molecular level. In fact, the first resistance gene cloned from an antibiotic producer was the gene encoding butirosin-3′-phosphotransferase, APH(3′)-IV, from *Bacillus circulans* (COURVALIN et al., 1977). Also, the correlation between the phosphorylation of streptomycin and its occurrence in a streptomycin producers, *S. griseus* (MILLER and WALKER 1969; NIMI et al., 1971) and in clinical strains of *Escherichia coli* and *Pseudomonas aeruginosa* (UMEZAWA et al., 1967a, b), for the first time led to the speculation that transferable antibiotic resistance could in general have evolved in the producers of these natural products (BENVENISTE and DAVIES, 1973). Later, ribosomal target site modification via specific 16S rRNA methylation was also identified as a major aminoglycoside resistance mechanism in producers (PIENDL et al., 1984; CUNDLIFFE, 1989, 1992a). For the first time, the transfer of a gene between gram-positive and gram-negative bacteria in nature was demonstrated to occur with an aminoglycoside resistance gene: the gene *aphA-3* (kanamycin/neomycin 3′-phosphotransferase) was transferred between enterococci (also streptococci and staphylococci) and *Campylobacter coli* (reviewed in COURVALIN, 1994). Some of the aminoglycoside-resistance genes might have spread from aminoglycoside producers to other bacteria in earlier evolutionary periods, the mechanisms of which are now becoming apparent (MAZODIER and DAVIES, 1991; COURVALIN, 1994). In this context, it is also interesting that related multifactorial systems exist for the intercellular transport of DNA and protein molecules which function specifically between bacteria and other, unrelated types, such as cells of other bacterial genera, and plant and animal cells (POHLMAN et al., 1994).

Much progress has been made in the analysis of resistance mechanisms in bacterial producers of aminoglycosides and other carbohydrate-containing self-toxic compounds (Tab. 4, Fig. 23). Our present knowledge can be summarized as follows:

(1) There are two basic resistance-conferring biochemical phenomena: first, specific elimination of the inhibitory function inside the producing cell (e.g., inactivation by modification, modification of the target site, or the production of a new, insensitive version of the normal target complex) and second, active transport of the inhibitor out of the cells via energy-driven exporters. The first type could be regarded as a mere self-protection mechanism which employs protective chemical groups or bypass mechanisms, whereas the second has its main function in transporting the compounds out of the cells in order to enable the bioactive end products to reach their natural destinations, namely other cells. We only can speculate on the natural functions of these compounds and the nature of the target cells. In general these target cells could be other cells of the same organism (hormone-like functions: the target cells would be either nonproducing or otherwise differently differentiated stages of the life cycle) or cells of other, e.g., competitive, organisms (DAVIES et al., 1992; PIEPERSBERG, 1993).

(2) Active exporters have been identified so far only for antibiotics such as macrolides, anthracyclines, tetracyclines (see Tab. 4), but not for aminoglycosides. Interestingly, two genes, *strV* and *strW,* have recently been detected in the streptomycin production gene clusters of both *S. glaucescens* and *S. griseus;* these genes encode a new type of ABC transporters (PIEPERSBERG, 1995; BEYER et al., 1996). Since streptomycin is secreted in an inactive, phosphorylated form (WALKER, 1975b; MANSOURI and PIEPERSBERG, 1991; PIEPERSBERG, 1995) these transmembrane exporters, if responsible for secretion, would not give rise to a resistance phenotype (RETZLAFF et al., 1993). There is evidence in support of this hypothesis: *S. lividans* 66 strains carrying a combination of the *strA* [APH(6)] and *strVW* transcription units on plasmids convert added streptomycin to an extracellularly accumulated streptomycin-6-phosphate (unfortunately those clones turned out to be very instable; S. BEYER and W. PIEPERSBERG, unpublished data). Therefore, the above view would be additionally supported by the possible existence of an active export system, also for antibiotically inactive precursors of aminoglycosides from the producing cells. It will be of further interest to investigate the presence of similar transporter genes in other aminoglycoside production gene clusters. A different type of membrane-anchored protein has recently been identified as the product of the *butB* gene in *Bacillus circulans* NRRL B3312 (AUBERT-PIVERT and DAVIES, 1994; HOTTA et al., 1995). The ButB protein is related to the cell-wall-associated S-layer proteins in low-G+C gram-positive bacteria and interruption of its gene blocks butirosin production, thus indicating that it might also be involved in aminoglycoside export.

(3) Some of the aminoglycoside-resistance mechanisms listed in Tab. 4, e.g., phosphorylation and acetylation, are also among those which could give us the biochemical basis for an understanding of the evolution of this type of resistance determinants. These are suspected to be derived from biosynthetic enzymes or serve both purposes at the same time, such as the *pac* and *bar* genes encoding puromycin and phosphinothricin acetyltransferases, respectively (WALKER, 1975a, b; PIEPERSBERG et al., 1988; CUNDLIFE, 1992a; THOMPSON and SETO, 1995; TERCERO et al., 1996). Again, it was the biosynthetic pathway of streptidine which first supported this speculation since intermediates in this pathway are successively dephosphorylated and rephosphorylated twice, the last time at the same position (C-6 of the aminocyclitol) as is phosphorlyated by the resistance enzyme, StrA (AphD or APH(6); see below), in streptomycin producers (cf. Fig. 17; WALKER, 1975b). Since this enzyme also has a phosphorylating activity for streptidine and its immediate precursor, *N*-amidinostreptamine (WALKER, 1975b; however, at much higher K_M values; DISTLER and PIEPERSBERG, 1985) it was suggested to be also a biosynthetic enzyme. However, among the gene products for streptomycin production there are two, StrN and StsE, with peptide motifs similar to those of the catalytic centers of antibiotic and protein kinases, especially the $HxDx_5Nx_{7-14}UD$

Tab. 4. Resistance Mechanisms against Glycosidic Antibiotics in Producers (Examples)

Antibiotic	Resistance Mechanism[a]	Producing Organism[b]	References[c]
Streptomycin	SM-6-PhT; (SM-3″-PhT); (export: ABC?)	*S. griseus, S. glaucescens*	A
Neomycin	neomycin-3-AcT, neom.-6′-AcT, neom.-3′-PhT	*S. fradiae*	A, G
Neomycin	neomycin-3-AcT	*M. chalcea*	B
Paromomycin	paromomycin-3-AcT, par.-3′-PhT	*S. rimosus* forma *paromomycinus*	A
Ribostamycin	ribostamycin-3-AcT, rib.-3′-PhT	*S. ribosidificus*	A
Kanamycin	16S rRNA MT (G-1405)	*S. kanamyceticus*	A
Hygromycin B	hygromycin B-7-PhT	*S. hygroscopicus*	A
Nebramycin	nebramycin-3-AcT, par.-3′-PhT; 16S rRNA MT (A-1408, G-1405); SM-6-PhT	*S. tenebrarius*	A
Gentamicin	16S RRNA MT (?)	*M. purpurea*	C
Fortimicin (astromicin)	(16S rRNA MT?)	*M. olivasterospora*	D
Istamycin	16S RRNA MT (A-1408)	*S. tenjimariensis*	A
Kasugamycin	kasugamycin-2′-AcT; fortimicin/istamycin-2′(+2″)-AcT	*S. kasugaensis*	G
Spectinomycin	spectinomycin-2″-AcT; spectinom.-?-PhT	*S. spectabilis*	G
Pactamycin	16S rRNA MT (A-964)	*S. pactum*	A
Lincomycin	23S rRNA MT (?); export:ABC; export:ΔpH	*S. lincolnensis*	E
Puromycin	puromycin-AcT, export:ΔpH(?)	*S. alboniger*	A
Daunorubicin	export:ABC	*S. peucetius*	F
Erythromycin	23S rRNA MT (A-2058); export:ABC; export:ΔpH	*Saccharopolyspora erythraea*	A
Tylosin	23S rRNA MT (A-2058); export:ABC; export:ΔpH	*S. fradiae*	A
Novobiocin	res. target enzyme: DNA gyrase	*S. sphaeroides*	A

[a] AcT: acetyltransferase; MT: methyltransferase; PhT: phosphotransferase; ABC: ATP-dependent transporter family; ΔpH: pH gradient-dependent transporter family

[b] *S.*: *Streptomyces*; *M.*: *Micromonospora*

[c] A: Cundliffe (1989); B: Salauze et al. (1992); C: Piendl et al. (1984); D: Hasegawa (1991); E: Zhang et al. (1992; and unpublished observations); F: Guilfoile and Hutchinson (1991); G: Hotta et al. (1995)

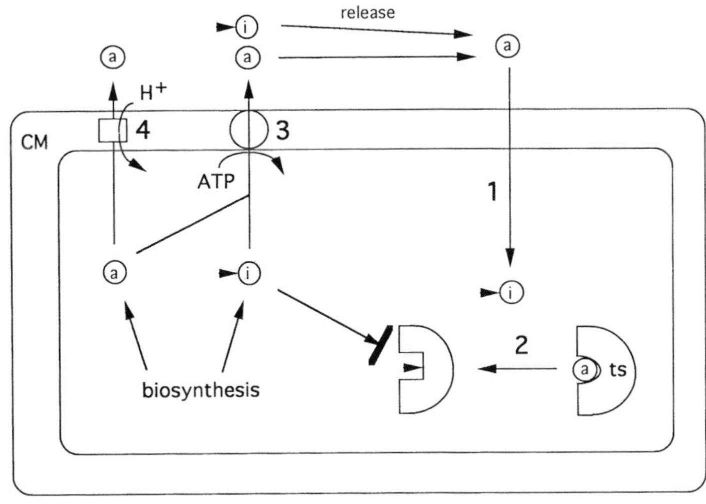

Fig. 23. Schematic representation of self-protecting resistance mechanisms in producers of antibiotically active secondary carbohydrates.

(U = hydrophobic residue, e.g., I, V, L) motif probably involved in the phosphate group transfer (PISSOWOTZKI et al., 1991; RETZLAFF et al., 1993; KNIGHTON et al., 1991; cf. also Sect. 6). These are more likely to be the two phosphotransferases involved in the streptidine pathway (see Sect. 4.1.1.1). Recently, a spectinomycin resistance gene encoding a spectinomycin phosphotransferase, was cloned from *S. flavopersicus* NRRL 2820 which showed striking similarity to StrN of *S. griseus* and *S. glaucescens* (J. ALTENBUCHNER, D. LYUTZKANOVA, and J. DISTLER, personal communication). This finding strongly supports the hypothesis that resistance genes originate from biosynthetic genes. Here, this hypothesis also has to be extended to include the possibility that an ancestral production gene could become a resistance gene by divergent evolution or after degeneration and modification of a pre-existing pathway to yield simpler but still effective end products (this could be the case for spectinomycin; see Sect. 4.1). The kanamycin-6′-acetyltransferase in *S. kanamyceticus* could also be involved primarily in the acetylation of kanamycin precursors for several reasons (CRAMERI and DAVIES, 1986; see also below).

(4) Resistance to aminoglycosides also occurs as a cryptic phenotypic property in streptomycetes, such as a second streptomycin resistance enzyme (streptomycin-3″-phosphotransferase) and kanamycin resistance enzyme (kanamycin-3-N-acetyltransferase) in *S. griseus* (HEINZEL, et al. 1988; HOTTA et al., 1988). It can be used either as a basis for the screening for new producers of aminoglycoside-like antibiotics (ETIENNE et al., 1991) or to stimulate aminoglycoside production, e.g., the 6′-acetyltransferase gene from *S. kanamyceticus* stimulates aminoglycoside production when several copies are introduced into kanamycin and neomycin producers (CRAMERI and DAVIES, 1986).

4.6 Regulation in Streptomycetes

The synthesis of aminoglycosides like the production of other secondary metabolites in *Streptomyces* is regulated by a complicated network of regulator proteins encoded by the individual biosynthetic gene clusters and other pleiotropic regulatory genes. Besides these regulatory genes, hormone-like autoregulators (e.g., A-factor), intracellular signal molecules (e.g., ppGpp), and nutrients (e.g., phosphate, glucose, N sources, etc.) control the production of aminoglycosides at the level of gene expression and/or enzyme activity. Mainly for the streptomycin producing *S. griseus* as a model system we will here describe some of the most important details known or postulated (summarized in Fig. 24).

The depression of the production of aminoglycosides by glucose has been reported for streptomycin, kanamycin, istamycin, and neomycin (VINING and DOULL, 1988; DEMAIN, 1989) and is caused by the repression of antibiotic synthases (e.g., *N*-acetylkanamycin aminohydrolase; DEMAIN, 1989) rather than by an influence on the formation of precursors of secondary metabolites. Carbon catabolite control mechanisms extensively studied in other bacterial systems such as *E. coli* and *Bacillus subtilis* (reviewed by FISCHER, 1992) typically use cAMP/CAP-mediated gene regulation. Although cAMP relieves glucose repression of *N*-acetylkanamycin aminohydrolase in *S. kanamyceticus* (SATOH et al., 1976) there is only weak evidence for the involvement of cAMP in regulating aminoglycoside production in *Streptomyces.* The intracellular cAMP concentration falls sharply in the mid to late vegetative growth phase of *S. griseus* S104 before the onset of secondary metabolism (RAGAN and VINING, 1978). Glucose increases cAMP levels in *S. antibioticus* while simultaneously repressing oleandomycin formation (LISHNEVSKAYA et al., 1986). There is no evidence that cAMP directly affects streptomycin production (NEUMANN et al., 1996) or controls secondary metabolism in general (MARTIN and DEMAIN, 1980). It is more likely that the carbon source repression is mediated by glucose kinase. Glucose kinase-deficient mutants of *S. coelicolor* having normal intracellular levels of cAMP do not exhibit a glucose repression phenotype (ANGELL et al., 1992; KWAKMAN and POSTMA, 1994). In contrast to these results, glucose-insensitive mutants of *S. kanamyceticus* blocked in kanamycin production were isolated which have wild-type levels of glucose kinase (FLORES et al., 1993). Thus, a phosphorylated sugar could mediate carbon source repression of antibiotic production (DEMAIN, 1989).

There is conflicting evidence in the literature concerning the influence of the nitrogen source on the production of streptomycin and other aminoglycosides. Nevertheless, ammonium seems to impede the synthesis of streptomycin, neomycin, and kanamycin (SHAPIRO, 1989) whereas nitrate and some amino acids support the production of aminoglycosides, e.g., kanamycin (BASAK and MAJUMDAR, 1973) and streptomycin. The stimulation of streptomycin production by alanine, arginine, and/or glutamine can be explained in terms of their being direct donors of nitrogenous groups in enzyme-catalyzed steps during streptomycin biosynthesis (cf. Sect. 4.1). The mechanisms underlying the positive effect of asparagine and proline (SHAPIRO, 1989) or the repression of streptomycin synthesis by valine (ENSIGN, 1988; NEUMANN et al., 1996) in *S. griseus* are presently not understood. These compounds could influence the expression of biosynthetic key enzymes at the genetic level. This influence could be mediated by a NtrB/NtrC-like system or by a modulation of the activities of primary metabolic enzymes and streptomycin biosynthetic enzymes at the physiological level via metabolite accumulation and excretion by, e.g., feedback repression or induction of gluconeogenetic pathways or metabolic conditions favoring special routes of intermediary metabolism, such as the pentosephosphate cycle used in the "reverse direction".

The biosynthesis of aminoglycosides (e.g., streptomycin, neomycin, kanamycin) is sensitive to a high concentration (>5 mM) of inorganic phosphate (MARTIN, 1989). The extracellular aminoglycoside phosphate phosphatase which forms the biologically active antibiotic, in the case of streptomycin (see Sect. 4.1), from an inactive phosphorylated precursor is inhibited by phosphate. Streptomycin-6-P is accumulated in cultures of *S. griseus* grown at a high concentration of phosphate or at low pH due to the inhibition of the streptomycin-6-P phosphatase (MANSOURI und PIEPERSBERG, 1991). In addition, there is evidence that some secondary metabolic genes are controlled by a PhoB/PhoR-like system; so-called pho boxes have been detected in phosphate-regulated promoter regions (cf. Fig. 24B; MARTIN, 1989; MARTIN

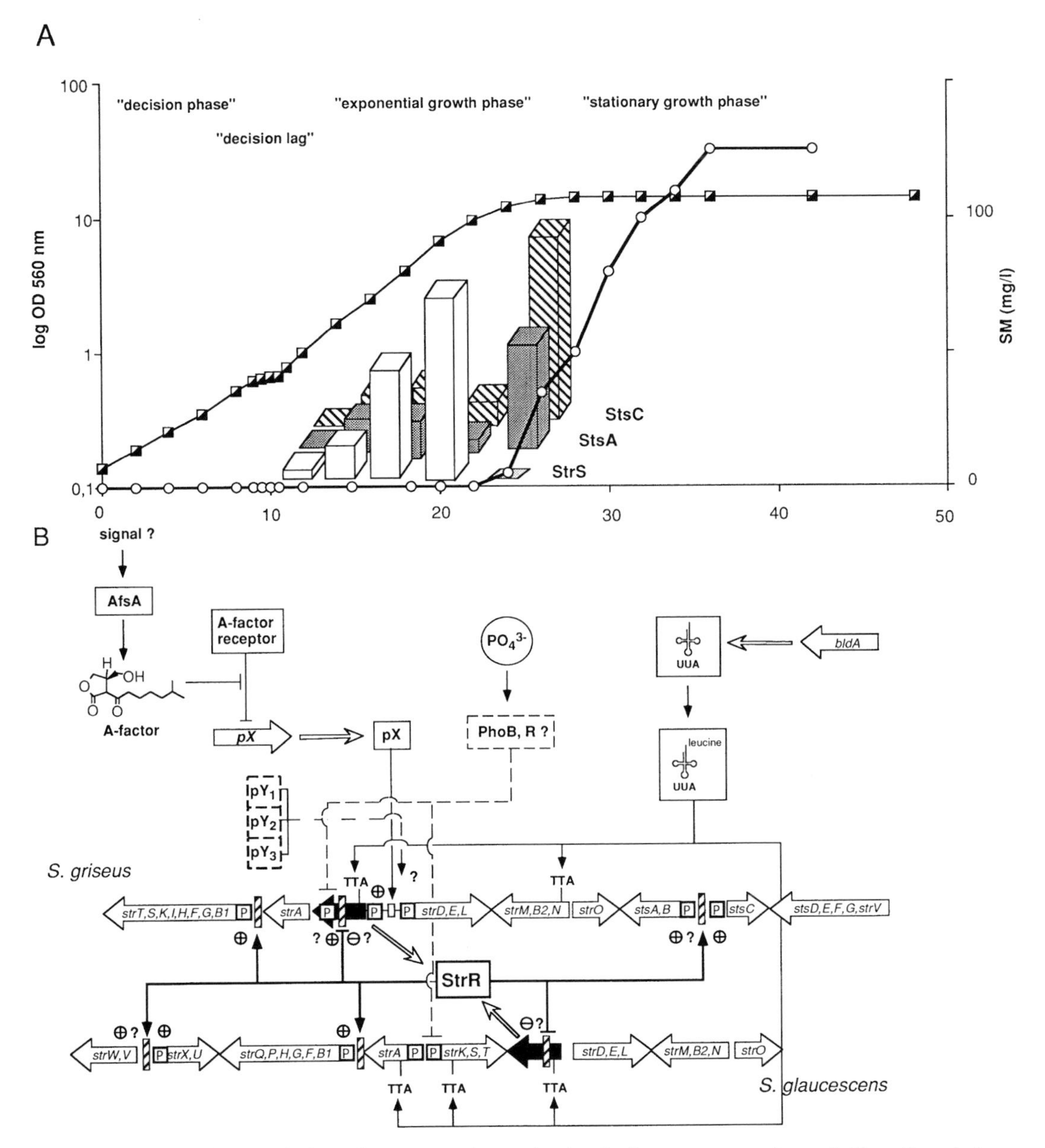

Fig. 24. Model for the regulation of streptomycin production in *Streptomyces griseus.* **A** Growth and streptomycin production of *S. griseus.* The kinetics of biomass accumulation (□) and of streptomycin production (○) are shown. The columns indicate the relative expression rates of StrS, StsA, and StsC as determined by densitometry of the autoradiographed protein patterns from *S. griseus* pulse-labeled with (^{35}S)-methionine after separation on 2-dimensional protein gels. **B** Regulation of the expression of *str/sts* genes by pleiotropic (e.g., A-factor) and pathway-specific (e.g., StrR) factors.
P: promoters; arrows indicate positive regulatory effects; tRNA-molecules are symbolized by cloverleaves. The dominance of the StrR-dependent regulation is indicated by thick lines, for further explanantions see the text.

and LIRAS, 1989; LIRAS et al., 1990). Similar structures were also found in the promoter regions of the *aphDp2* gene of *S. griseus* and of the *strK* (streptomycin-6-P phosphatase) gene of *S. glaucescens* (LIRAS et al., 1990; DISTLER et al., 1990).

The possible role of guanosine tetraphosphate or guanosine pentaphosphate (ppGpp or pppGpp) and GTP pools in controlling secondary metabolism in streptomycin-producing *S. griseus* and other antibiotic producers was extensively studied by OCHI (1987, 1988, 1990). Although *relC* mutants showed a marked reduction in antibiotic production on complete medium it seemed difficult to assess whether the effect on streptomycin production was a direct consequence of the defect in ppGpp formation or an indirect effect of the *relC* mutation (OCHI, 1990). In *S. clavuligerus* there is no relationship between ppGpp and antibiotic production (BASCARAN et al., 1991). The same observation was made for streptomycin synthesis in *S. griseus* on minimal medium; no ppGpp formation could be detected at any growth phase (NEUMANN et al., 1996).

A-factor (2-(6′-methylheptanoyl)-3*R*-hydroxymethyl-4-butanolide), an extracellular diffusible autoregulatory molecule, triggers both streptomycin biosynthesis differentiation and streptomycin biosynthesis in *S. griseus.* This factor, which was discovered by KHOKHLOV et al. (1967, 1988), belongs to a group of chemically similar γ-butyrolactone autoregulators which are synthesized by a variety of *Streptomyces* (HORINOUCHI and BEPPU, 1992, 1994). A-factor induces streptomycin production and sporulation in A-factor-negative mutants of *S. griseus* at concentrations as low as 1 nM. The induction by A-factor exhibits a strict growth phase dependence, in that A-factor has to be present during the first hours of growth ("decision phase" model, see below; PIEPERSBERG, 1995; NEUMANN et al. 1996). A-factor-induced regulation in *S. griseus* depends on an A-factor-binding protein, which is present in very low amounts (ca. 37 molecules per genome) and has a binding constant (K_d) of 0.7 nM (MIYAKE et al., 1989, 1990). Recently, the gene of the A-factor-binding protein, *arpA,* was cloned from *S. griseus* and found to encode a repressor type DNA-binding protein (ONAKA et al., 1995). In a proposed model the A-factor receptor protein acts in the absence of A-factor as a repressor of streptomycin biosynthesis and differentiation (MIYAKE et al., 1990). In addition, A-factor stimulates membrane-bound GTPases in *S. griseus in vivo* and *in vitro* (PENYIGE et al., 1992). A-factor induces the expression of StrR, the activator protein of some of the transcription units in the gene cluster for streptomycin biosynthesis (cf. Sect. 4.1; DISTLER et al., 1987b; RETZLAFF et al., 1993; RETZLAFF and DISTLER, 1995). The A-factor-dependent transcription of *strR* seems to be controlled via regulation by the A-factor receptor of at least one further DNA-binding protein (pX; cf. Fig. 24) which binds to an enhancer-like element located upstream of the *strR* promoter (HORINOUCHI and BEPPU, 1992; VUJAKLIJA et al., 1991). Three additional DNA-binding proteins (pY_{1-3}; cf. Fig. 24B), not dependent on A-factor induction in *S. griseus,* interact with neighboring sites in the same *strR* promoter region, suggesting an even more complex regulatory network governing StrR expression (VUJAKLIJA et al., 1993).

The StrR proteins encoded by the streptomycin biosynthesis gene cluster of *S. griseus* and *S. glaucescens* share an identity of 62.8%. The *strR* gene encodes an activator of streptomycin and OH-streptomycin production. The existence of the StrR activator protein was first postulated in mutants of *S. griseus* and its role was suggested from the mutant phenotype and its suppression by complementation with wild type DNA (OHNUKI et al., 1985a, b). Analysis of the mode of action of StrR in *S. griseus* showed that StrR is a DNA-binding protein which activates the expression of the *str/sts* genes *strB1* and *stsC* at the level of transcription by binding to upstream promoter sequences (hatched boxes in Fig. 24B; RETZLAFF et al., 1993; RETZLAFF and DISTLER, 1995; BEYER et al., 1996). These and a third identified StrR-binding site within the *strR* gene are palindromic sequences with the consensus sequence: GTTCGAnnGn(11)CnnCTCAACG (RETZLAFF and DISTLER, 1995). The function of the StR-binding site within the *str*R gene, e.g., negative feedback regulation of StrR expression or activation of the promoter *aphDp2* located approximately 400 bp downstream of this element, is unclear. In the OH-streptomycin gene cluster of *S. glaucescens* an StrR-binding site was also identified upstream of the *StrB1* and *strX* genes and within the *strR* gene. Recently, a gene homologous to *strR* was identified on a DNA fragment of *S. spectabilis* conferring spectinomycin resistance in *S. lividans* (D. LYUTZKANOVA and J. ALTENBUCHNER, unpublished data). Therefore, it can be speculated that StrR homologous proteins are widely distributed activators of the synthesis of aminoglycoside antibiotics, especially for those using the Ca cyclitol pathway, which could have evolved from a common ancestral gene cluster.

The expression of aminoglycoside biosynthetic genes in *Streptomyces* is regulated by and/or is dependent on additional gene products having pleiotropic effects and influencing both secondary metabolism and differentiation on a more general level. The *bldA* gene encodes a leucine-specific tRNA which recognizes the rare codon UUA (cf. Fig. 24; CHATER, 1989, 1992; LESKIW et al., 1991a, b). *BldA* mutants of *S. griseus* are deficient in aerial mycelium formation and streptomycin production (MCCUE et al., 1992). Recently, it was demonstrated that $tRNA_{UUA}$-dependent control could be a key switching process in the onset of differentiation and secondary metabolism (LESKIW et al., 1991a, b). In *S. griseus, S. glaucescens,* and *S. spectabilis* the *strR* genes contain a respective TTA triplet in conserved positions in the 5′-section of the respective reading frames (DISTLER et al., 1987b; PISSOWOTZKI et al., 1991; MAYER, 1994; D. LUTZKANOVA and J. ALTENBUCHNER, unpublished). In the *strN* genes of *S. griseus* and *S. glaucescens* additional TTA codons were detected. The *strA* (*sph*) gene of *S. glaucescens* possesses also a TTA codon, which, however, is absent from the counterpart of this gene, *strA (aphD)*, in *S. griseus* (VÖGTLI and HÜTTER, 1987; DISTLER et al., 1987a). In *S. coelicolor* seven different RNA polymerase sigma factors were identified and the role of special sigma factors (e.g., WhiG) in controlling differentiation and antibiotic production was elucidated (BUTTNER, 1989). The transcription of aminoglycoside biosynthetic genes could also be dependent on such sigma specific for secondary metabolic genes. In *S. griseus* the similarity of the *str/sts* promoters (*strRp, aphDP,* and *strB1p*) supports this assumption (DISTLER et al., 1987b). There are several reports of the possible involvement of other factors in the regulation of streptomycin production in *S. griseus* strains: the DNA-binding protein ORF1590 necessary for sporulation and probably streptomycin production in *S. griseus* (MCCUE et al., 1992), ADP-ribosylation of proteins (PENYIGE et al., 1992) possibly via influencing membrane-bound G proteins (GTPases; PENYIGE et al., 1992), and C-factor, a cytodifferentiation protein, excreted into the medium by *S. griseus* 45H (SZESZÁK et al., 1991). Recently, it has been found that serine/threonine and tyrosine protein kinases are present in most or all streptomycetes and, therefore, could be part of the complicated regulation network necessary for the induction of antibiotic synthesis and differentiation, also in *S. griseus.* Proteins phosphorylated at a tyrosine residue were detected in *S. griseus* and other *Streptomyces* spp. (WATERS et al., 1994). Specific inhibitors of eukaryotic protein kinases, such as staurosporin (cf. Fig. A29), inhibit sporulation (HONG et al., 1993) and streptomycin production (NEUMANN et al., 1996) of *S. griseus* without affecting vegetative growth. These results suggest that in *Streptomyces* a signal transduction pathway similar to that in eukaryotic organisms may control cell differentiation and secondary metabolism.

S. griseus, like other streptomycetes (HOLT et al., 1992), undergoes a decision-making process during a rather short period of 1–2 h the mid logarithmic growth phase (Fig. 24A; NEUMANN et al., 1996). This "decision phase" depends on pre-existing factors (such as A-factor) and can be suppressed by certain metabolites (e.g., valine) or inhibitors of protein-modifying processes (e.g., 3-aminobenzamide). Various changes in physiology, e.g., a temporal increase in intracellular cAMP levels and gene expression, can be observed during and after this period (DISTLER et al., 1990; NEUMANN et al., 1996), which altogether seem to be an absolute prerequisite for later cell differentiation and streptomycin production. It is notable that the A-factor-dependent expression of some *str/sts* genes (*stsC*, *stsA*, *stsS*) occurs 2–4 h after this decision phase, but 10 h before streptomycin is detectable in the culture medium for the first time (cf. Fig. 24A). This concept of the induction of secondary metabolism during a very early growth phase temporally separated from the production phase demands a global signal transduction network which might include autoregulator molecules, receptors, transducers, protein phosphorylation/dephosphorylation systems, and "second messengers". Although the direct regulation of the *str/sts* genes and the superimposed regulatory cascades in *S. griseus* are still incompletely understood, a model for the regulation of streptomycin production which includes current hypotheses can be proposed (Fig. 24).

4.7 Overview of the Aminoglycoside Pathways

When we summarize what we have learned from the genetics and biochemistry of aminoglycoside synthesis in bacteria and when we compare the respective pathways with each other, several similarities, but also quite divergent traits become apparent. The initial pathways of aminocyclitol formation starting at the (deoxy-)*scyllo*-inosose level, whether via the Ca or Cb routes, and some of the intermediate steps, e.g., the involvement of a D-glucosamine moiety in the formation of a pseudodisaccharide in the 2DOS and fortimicin/istamycin groups, could use very similar enzymes (genes). Also, the resistance mechanisms may be very similar and based on the evolution of common ancestral genes (see also Sect. 4.5 and 6). Whether the Ca or Cb routes for aminohexitol formation are used might depend on the availability of the respective enzymes/genes. However, other more specific factors might force cells to use either one or the other. The **Cb** pathway is more economical because it uses fewer steps (cf. Sect. 3.1). However, the initial step enzyme (D-*myo*-inosotol-3-phosphate synthase) for the **Ca** route might be more wide-spread in streptomycetes because it is probably used also for other, more generally distributed biosynthetic pathways (e.g., for the formation of the recently detected compound mycothiol, which is a major thiol in most actinomycetes; NEWTON et al., 1996). Also, part of the enzymes for the next steps (e.g., an StrI-like oxidoreductase, cf. Sect. 4.1.1.1) could be used in *myo*-inosotol degradation, a frequent property of *Streptomyces* spp. other than *S. griseus*. For these reasons it is an interesting speculation that both the **Ca** and **Cb** pathways are used alternatively when a metabolic sorting mechanism is needed for the alternative use of different anabolic and/or catabolic pathways in the same cell.

The various NDP activation mechanisms of monosaccharides before modification and glycosyltransfers might have a similar metabolic sorting effect. However, another explanation might be that they are designed in order to facilitate the horizontal genetic transfer of whole biosynthetic branches between bacteria and maybe even higher taxa. The 6DOH pathways are used less in aminoglycoside production than are the pathways leading to 6-hydroxylated hexosamines; however, they are preferred over other carbohydrate pathways in the modification of nonaminoglycosidic secondary metabolites (cf. Sect. 3.2 and 5.2). In aminocyclitol aminoglycosides the 6DOH routes are practically exclusively coupled to the Ca type, streptomycin-like compounds (cf. Sect. 4.1). This could reflect a common evolutionary link between both pathways or a tight genetic coupling of both sets of genes required (or both). This coupling also implies that the sets of functionally identical enzymes/genes involved should show a higher degree of relatedness in producers of compounds such as streptomycins, spectinomycins, or kasugamycins than in Cb (e.g., 2DOS) producers. In contrast to this, amino sugars can either be derived from preformed aminated precursors (e.g., D-glucosamine), NDP-activated hexoses, or formed only after condensation of a carbohydrate precursor into (pseudo-)oligosaccharidic compounds and occur in both Ca and Cb aminoglycosides. Therefore, it will be interesting to determine by which alternative route most of the aminoglycoside hexosamine pathways proceed. In particular, the biosynthesis route of the N-methyl-L-glucosamine precursor of streptomycin (mode (2), Sect. 4.1.1.3), the details of which are still unknown, could be an example for a wider distribution of the HA pathway.

The streptomycins and probably some of the more related substances (e.g., spectinomycin, kasugamycin) are made via the condensation of highly modified precursors in the final stage of the pathway. In contrast, the larger group of 2DOS-containing aminoglycosides and the compounds of the fortimicin/istamycin family are condensed first from relatively simple precursors, and the intermediates are strongly modified in the later pathway. Some substances, such as the acarbose-like α-glucosidase inhibitors might be condensed with additional subunits only after export to the cell surface (outer side of the cytoplasmic membrane) with the involvement of membrane anchors and carriers, e.g., undecaprenyl phosphate residues, for the activated intermediates. This could also indicate a more general principle in the formation of secondary metabolites where frequently mixtures of very similar end products formed by a given strain of the producing microorganism are found. Rather than being side products of "inefficient" or "nonspecific" biosynthetic enzymes, complex product mixtures frequently observed in fermentations might be the result of the individual specificities of the export systems which are responsible for the intermediate/product patterns released from the cells.

Thus, the strategies and routes used by the various producers for the production of the chemically relatively homogenous class of aminocyclitol-containing aminoglycosides described so far do not seem to be of similar homogeneity. Instead, a strong selective advantage could exist in the long-term diversification by means of the modulation of the pathway design (e.g., by natural genetic engineering). Hence, in the long-term possibly from time to time a new pathway for an altered end product develops which uniformly serves very similar functions of which the individual producers take advantage and at the same time escape the defense mechanisms of related producers or nonproducers in a competing situation in natural environments. The resistance mechanisms themselves in the aminoglycoside producers do not seem to be similar, although evidence is emerging that some of them are based on biosynthetic functions (e.g., phospho- and acetyltransferases). However, a connection between individual aminoglycoside pathways might exist at a higher level of intracellular regulation or extracellular cell–cell communication between the producing cells and/or other cell types or organisms (see discussion in PIEPERSBERG, 1993).

4.8 Aminoglycoside–Target Site Interactions and General Effects on Bacterial and Eukaryotic Cells

The interaction of the classic aminoglycosides with the small (30S) subunits of eubacteria-type ribosomes (bacterial, mitochondrial, and plastidal ribosomes) has been well investigated and reviewed several times in the past (GORINI, 1974; SCHLESSINGER et al. 1975; WALLACE et al., 1979; VAZQUEZ, 1979; PIEPERSBERG et al., 1980; GALE et al., 1981; HILL et al., 1990; NIERHAUS, 1993). Two functional groups can be distinguished among the aminoglycosidic translational inhibitors: (1) translational misreading-enhancing and bactericidal compounds, e.g., the streptomycins and all 2DOS aminoglycosides; (2) bacteriostatic compounds which do not affect translational accuracy, e.g., spectinomycin and kasugamycin. This difference is also reflected by the alterations in overall translation patterns after aminoglycoside addition to cultures of *E. coli* and *B. subtilis,* where group 1 compounds induce mistranslation and a rapid loss of overall translation ability and group 2 aminoglycosides a "quasi-relaxed" (Rel^-) phenotype similar to that induced by chloramphenicol with an uncoupled residual translation of a special group of proteins for a longer period of time (PIEPERSBERG, 1985). An exceptional position is taken by hygromycin B which exhibits functional traits of both groups but does not seem to induce misreading of the genetic code (BAKKER, 1992).

Originally, it was believed that certain small subunit ribosomal proteins mediate aminoglycoside interaction with the ribosome (OZAKI et al., 1969; PIEPERSBERG et al., 1980). In contrast, an early suggestion of GO-

RINI (1974) that 16S rRNA could be the primary target site of aminoglycosides was proven more recently; evidence is also accumulating that 16S and 23S rRNAs are the catalytically active components in both decoding and peptidyltransfer in the eubacterial ribosome (MOAZED and NOLLER, 1987; DE STASIO et al., 1989; CUNDLIFFE, 1990; NOLLER, 1993). Mistranslation is also observed with some aminoglycosides in some but not all Archaea tested (LONDEI et al., 1988). The decoding process involves an interaction of a short RNA duplex (paired codon–anticodon) with the P and A sites of the 30S subunits, a process similar to that of the self-splicing group I introns (Fig. 25; VON AHSEN and NOLLER, 1993). Therefore, it is not surprising that the RNA molecules of group I introns specifically bind aminoglycosides and that the splicing reaction is inhibited by these antibiotics (VON AHSEN et al., 1991; SCHROEDER et al., 1993). It has also been demonstrated that short RNA analogs of the 16S rRNA decoding site specifically interact with both aminoglycosides and its RNA ligands, tRNA and mRNA, in the absence of ribosomal proteins (PUROHIT and STERN, 1994). These findings initiated speculations that aminoglycosides are "molecular fossils" echoing their possible functions as ligands of catalytically active RNAs in a precellular RNA world (DAVIES et al., 1992). An argument in favor of this hypothesis is the finding that all known translational inhibitors among the aminoglycosides, although they bind at different positions, seem to bind within a common domain in the catalytic center of the 16S rRNA of bacterial ribosomes (CUNDLIFFE, 1990; NOLLER, 1993; BRINK et al., 1994).

Effects other than the direct interaction of the group 1 bactericidal aminoglycosides with the bacterial ribosome have been suggested to be responsible for causing cell death in eubacteria (DAVIS, 1987). Besides mistranslation at actively elongating ribosomes (GORINI, 1974; WALLACE et al., 1979; PIEPERSBERG et al., 1980), two phenomena were observed to be coupled with lethality, mainly in studies with streptomycin and gentamicin: (1) a two-step uptake kinetics of aminoglycosides, the first phase of which is dependent on the $\Delta\Psi$ component of the proton motive force or driven by ATP (HANCOCK, 1981a, b; BRYAN and KWAN, 1983; FRAIMOW et al., 1991), and (2) membrane damage, probably via the induction of membrane channels (DAVIS et al., 1986; BUSSE et al., 1992). Two models could explain the pleiotropy of action of these compounds: (1) incorporation of misread proteins into the cytoplasmic membrane, and (2) mistranslation of proteins with a short half-life involved in DNA replication and/or cell division and made at very few copies in a distinct phase of the cell cycle. The first model would explain also the observed drastic increase of aminoglycoside uptake in the second, killing phase, where the antibiotics are irreversibly accumulated inside the cells in a

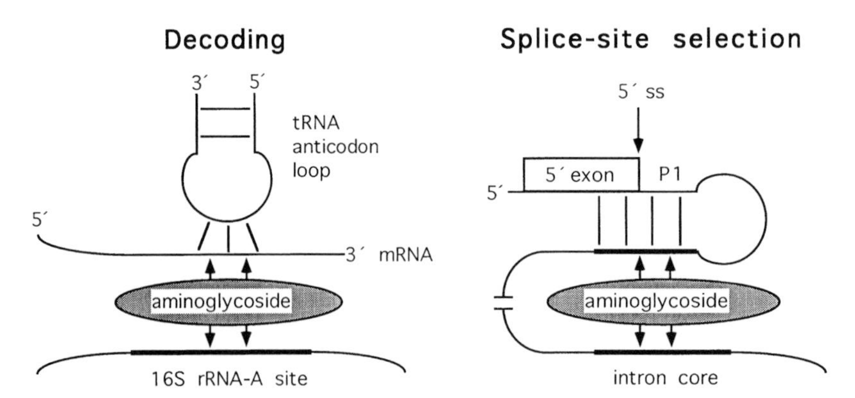

Fig. 25. Similarities between interaction of aminoglycosides in the decoding site of bacterial ribosomes (left) and self-splicing type I introns (modified according to SCHROEDER et al., 1993).

large excess over the number of ribosomes. Aminoglycosides accumulate by binding electrostatically to anionic groups of cytoplasmic macromolecules or by being caged into degradation products of mistranslated proteins (DAVIS, 1987; BUSSE et al., 1992). Also, passage through the outer membrane of gram-negative bacteria could be affected by direct interaction of aminoglycosides with porins and/or lipopolysaccharide components (HANCOCK et al., 1991).

In patients treated with aminoglycosides the most prominent adverse side effects encountered are oto- and nephrotoxicity (PRATT and FEKETY, 1986). Aminoglycoside hypersensitivity in humans is a maternally inherited trait, suggesting a mitochondrion-associated genetic defect, and a molecular mechanism has been proposed recently (HUTCHIN and CORTOPASSI, 1994). Hypersensitive persons have been shown to carry a 1555^{G} mutation, i.e., a replacement of the A residue which is normally in this position of the small (12S) mitochondrial rRNA by a G. This area of small, 16S type rRNAs is known to interact with aminoglycosides and to be involved in the control of codon–anticodon-pairing (see above; cf. Fig. 25). This defect leads to frequent death of hair cells in the ear, which is proposed to be due to enhanced production of superoxide by a mistranslated mitochondrial complex I, the proteins of which account for more than half of the coding capacity of the mitochondrial genome. Nephrotoxicity is primarily accompanied by membrane damage in the renal tubular cells. The biochemical basis of this effect in man is still not fully understood and could be caused either by an effect on lipid metabolism and structure or by mistranslation of proteins inducing malfunctions of membrane traffic (KOHLHEPP et al., 1994; and op. cit.). The nephrotoxic effect is also seen in neonates whose mother was treated with aminoglycosides during gestation (SMAOUI et al., 1993) and can be suppressed by polyaspartic acid (SWAN et al., 1991).

5 Related Sugar Components in Other Secondary Metabolites

5.1 Lincosamides

Characteristic of the lincosamides is an interesting C-8 aminosugar component which is represented by the methylthiolincosaminide (MTL) moiety of lincomycin A (Fig. A26; WRIGHT, 1983). Lincomycins (LM) A and B and celesticetin are members of the lincosamide group of antibiotics produced by *S. lincolnensis* and several other streptomycete species. Intensive biogenesis studies on LM-A involving measurement of stable isotope labeling patterns led to the proposal of a biosynthetic pathway for this antibiotic from an octulose and L-tyrosine. The former which is presumably derived from intermediates of the pentose phosphate cycle is a precursor of the MTL moiety and the latter is a precursor of the propylproline (PPL) subunit of LM-A (Fig. 26; BRAHME et al., 1984a, b). However, for the MTL subunit the genetic record seems to suggest a participation of a nucleotide (probably dTDP) activation step and a series of modification steps, including dehydratation, on NDP-activated sugar intermediates (PESCHKE et al., 1995). Therefore, either a totally different initial pathway starting from D-glucose with the familiar first steps of the 6-deoxyhexose pathways (see Sect. 3.2.1 and below) or an NDP-activation and modification of a C_8 sugar intermediate are the routes for the formation of the MTL subunit which currently can be postulated (cf. Fig. 26; PESCHKE et al., 1995). Also, it was found that L-3,4-dihydroxyphenylalanine (L-DOPA) and 3-propylidene-Δ^1-pyrroline-5-carboxylic acid are probable intermediates in the PPL biosynthetic branch of the pathway (BRAHME et al., 1984a; KUO et al., 1992). In addition to a specific set of biosynthetic enzymes a special coenzyme, the so-called co-synthetic factor, which is structurally identical to the ribodeazaflavin moiety of the F420 coenzyme of methanogenic bacteria is needed for the PPL branch of the LM production pathway in *S.*

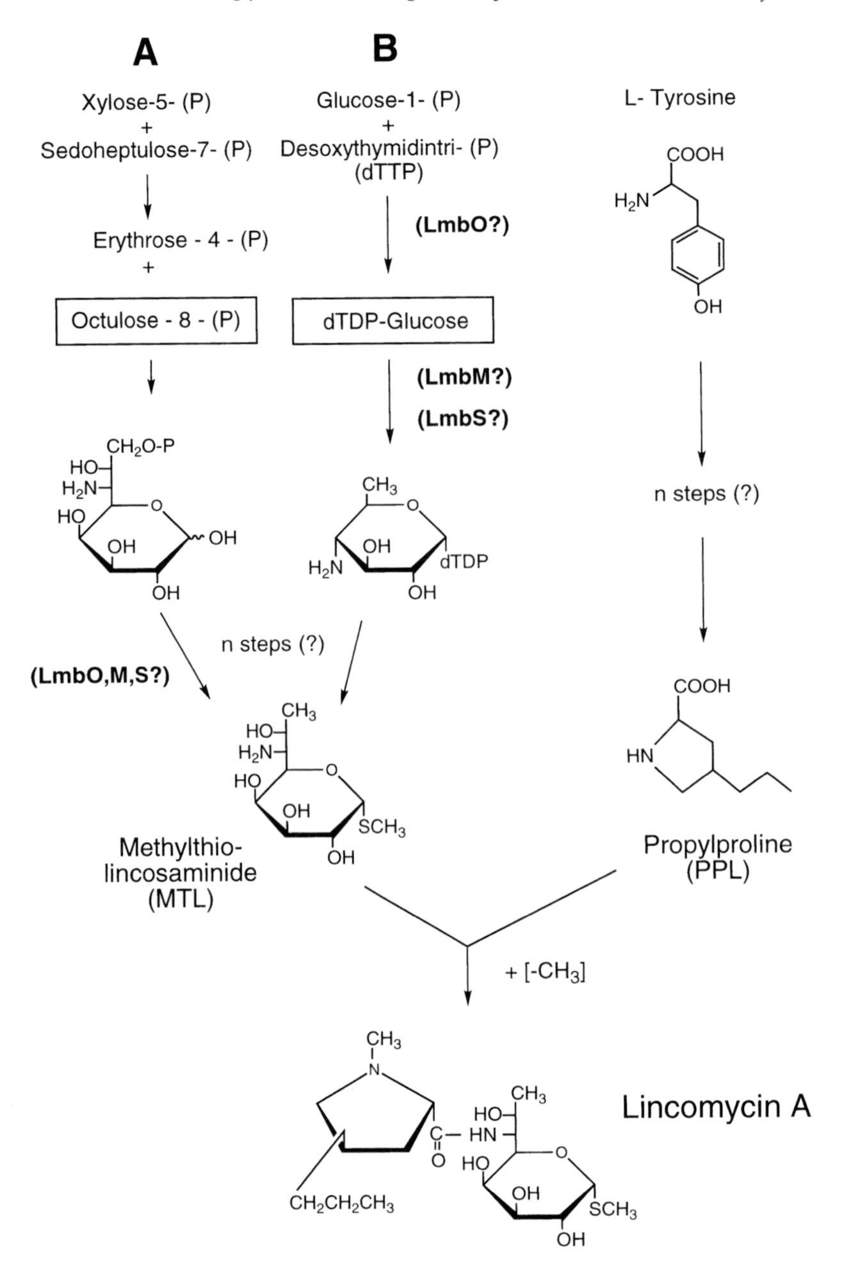

Fig. 26. Hypothetical pathway of lincomycin A. For the biosynthesis of the C_8 sugar moiety, methylthio-lincosaminide (MTL), two alternative pathways are proposed: (A) from intermediates of the pentose phosphate cycle via formation of a C_8 sugar precursor; (B) from dTDP-glucose via a 6DOH pathway an a later extension of the carbon chain. The possible involvement of three of the gene products of the putative MTL genes (cf. Fig. 27) is indicated.

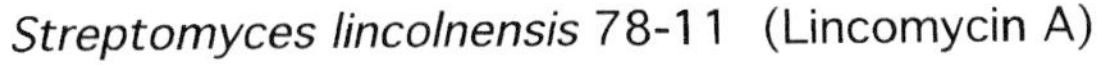

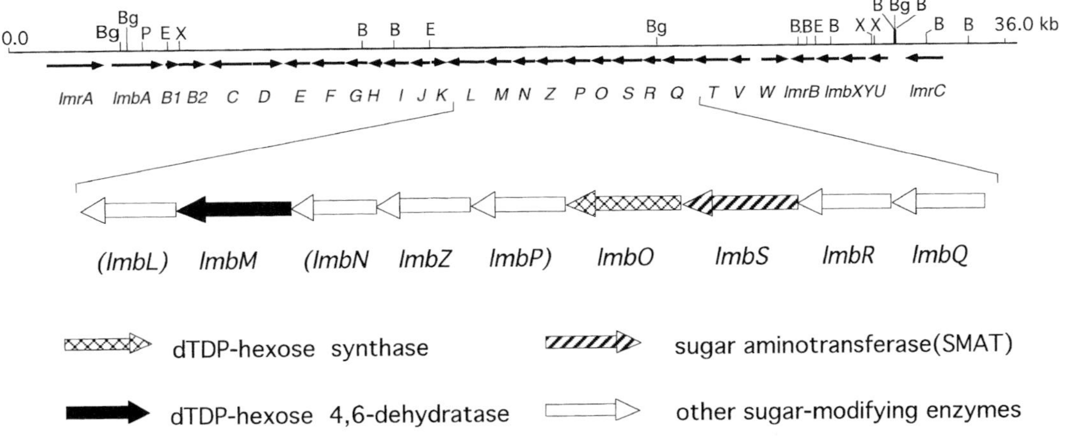

Fig. 27. The, "sugar" subcluster of the biosynthetic gene (*lmb*) cluster for lincomycin A. Genes marked by *lmr* encode resistance (or export) proteins.

lincolnensis (KUO et al., 1989, 1992). In contrast, no intermediates of the MTL subpathway could be identified so far. The biosynthesis of the related antibiotic celesticetin (Fig. A26) probably proceeds via a very similar route.

The LM-A production gene clusters of two overproducing industrial strains derived from *Streptomyces lincolnensis* NRRL2936 were cloned and analyzed by mutagenesis and hybridization; one of them (strain 78-11) has been sequenced (CHUNG and CROSE, 1990; ZHANG et al., 1992; PESCHKE et al., 1995). The *lmb*/*lmr* gene cluster is composed of 27 open reading frames with putative production functions (biosynthetic or regulatory; *lmb* genes) and three resistance (or export; *lmr*) genes, and is flanked by two of them, the *lmrA* and *lmrC* genes (Fig. 27). Compared with the respective genome segment of other lincomycin producers, the *lmb*/*lmr* clusters seem to have a very similar overall organization in the other LM producers, *S. pseudogriseolus* NRRL3985, *Streptomyces* sp. NRRL3890, and *S. vellosus* NRRL8037. However, they are embedded in nonhomologous genomic environments and exhibit polymorphic restriction patterns.

In the wild-type strain (*S. lincolnensis* NRRL2936) the *lmb*/*lmr* cluster is apparently present in a single copy. However, in the industrial strain *S. lincolnensis* 78-11 the gene clusters for the production of LM and melanin (*melC*) are duplicated on a large (450–500 kb) DNA segment by transposition to another genomic region accompanied by deletion events (PESCHKE et al., 1995). This fact indicates that enhanced gene dosage is one of the factors underlying overproduction in the developed industrial strains. Only a minority of the putative Lmb proteins belong to known protein families which among the supposed PPL biosynthetic enzymes include members of the γ-glutamyl transferases (LmbA), an L-tyrosine oxidase (LmbB2), an L-DOPA oxidase (LmbB1), amino acid acyl-adenylate synthetases (LmbC), aromatic amino acid aminotransferases (LmbF), and imidazoleglycerolphosphate dehydratases/histidinolphosphate phosphatases (LmbK) (PESCHKE et al., 1995; U. PESCHKE, D. NEUSSER, S. KASCHABECK, and W. PIEPERSBERG, unpublished data). However, except for LmbB1,2, even these proteins do not easily fit into any hypothetical route from L-Tyr to PPL, suggesting that the whole pathway is largely unknown.

Similarly, in the right hand part of the *lmb*/*lmr* cluster a set of eight genes, *lmbLMNZPOSQ*, encode proteins which to various extents are related to enzymes involved in the sugar converting or modifying metabolism. The existence of these putative enzymes, dTDP-glucose synthase (LmbO), dTDP-glucose 4,6-dehydratase (LmbM), and (NDP-)-ketohexose aminotransferase (LmbS), suggests that in contrast to the earlier proposals of the hypothetical biosynthetic pathway resulting in the MTL moiety (BRAHME et al., 1984b) this branch of the LM-pathway seems to be based on nucleotide-activated sugar intermediates. The stable isotope-labeling pattern found by these authors, suggesting an octulose – phosphate intermediate to be formed first, could also be explained by the same type of "equilibration" of externally applied glucose through the pentose phosphate cycle as was found for the biosyntheses of neomycin and validamycin (RINEHART et al., 1992; cf. Fig. 20). However, this would probably lead to another labeling pattern as that found in the case of the MTL subunit (BRAHME et al., 1984b). At first, for the reaction sequence NDP-pyranose synthesis/4,6-dehydratation (cf. Fig. 7) only a hexose seemed likely as a precursor. Therefore, the alternative pathway given in Fig. 26 (route B) was proposed (PESCHKE et al., 1995; PIEPERSBERG, 1994). The later detection of a gene *lmbR* encoding a transaldolase-like enzyme in the "sugar subcluster" of the *lmb* cluster (cf. Fig. 27) could support an alternative route via a pyranosidic octose intermediate which is NDP-activated and further modified in this form. Finally, a thiomethyl unit to the C1 position of the postulated (NDP-)6-amino-6,8-deoxyoctose intermediate would be added which could be transferred from 5′-thiomethyladenosine, a side product of polyamine biosynthesis from *S*-adenosyl-methionine.

5.2 Cyclitols

Other natural products containing cyclitol moieties or their (e.g., aromatic, pentitol, and cyclohexane carboxylic acid) derivatives are more widespread than is at first obvious. For instance, cyclitol derivatives formed via the Ca (cf. Sect. 3.1) pathway are components in nucleoside type antibiotics, e.g., adenomycin which in addition to an inositol unit contains an L-hexosamine derivative (Fig. A27). Therefore, their formation could have several steps in common with the streptomycin pathway. In this case equivalent enzymes could be used beyond the cyclitol formation, via D-*myo*-inositol-3-phosphate synthase and phosphatase (StrO-like enzyme?). The initial nucleotidylation, e.g., part of the oxidoreduction and epimerization, or transamination steps, which might be catalyzed by the gene products StrQ, StrP, StrX, and StsA (or StrS), respectively, during the formation of the *N*-methyl-L-glucosamine subunit of streptomycin (see Sect. 4.1.1.3), could also be involved in the biosynthesis of the hexosamine moiety.

An example is the cyclopentane ring in pactamycin (Fig. A30) which according to earlier biogenesis studies should also be derived from *myo*-inositol, i.e., via a 1,2-*cis*-diaminocyclitol (RINEHART et al., 1981, 1992; GRÄFE, 1992). Similarly the pentitols in allosamidin and trehazolin (Figs. A23 and A25) are possible products of the Ca route. The ring contraction of a hexitol to a pentitol could be accomplished by a mechanism similar to that operating in the formation of a furanoid ring from a pyranose derivative during dihydrostreptose biosynthesis (ses Sect. 4.1.1.2; cf. Fig. 15). In such a mechanism a dehydrogenase is believed to reduce a keto group and concomitantly introduces a rearrangement in the C chain resulting in a C_1 branch. Other possible routes for the synthesis of pentitols as in pactamycin or in allosamidin could be a new and so far unknown Cb mechanism acting on a ketohexose phosphate (e.g., fructose-6-phosphate) as a precursor and subsequent hydroxylation of the 2-deoxypentitol via 2,3-dehydratation and isomerizing rehydratation such as in the formation of 2-hydroxyvalidamines (cf. Fig. 22).

In contrast to this, as a second example the pentitol moieties of some nucleoside homologs, aristeromycin, neplanocin A (inhibitors of *S*-adenosylcysteine hydrolase; Fig. A27), and adecypenol, which are inhibitors of various enzymes of nucleoside metabolism, are though to be synthesized from a fructose derivative and via pathways either similar to the Ca or analogous to the Cb routes (PARRY et al., 1989; PARRY, 1992). Stable and radioactive isotope labeling experiments have shown that the stereochemistry of the C-6 atom in the cyclopentane ring formation between C-6 and C-2 of the hexose precursor is opposite to that in enzymatic *myo*-inositolphosphate synthesis. Also, the inversion of stereochemistry is unexplained by a mechanism of the Ca type. Therefore, it is more likely that the aristeromycin pentitol is formed by a Cb type enzyme cyclizing fructose-6-phosphate first to a 2-deoxyketopentitol. Successive reduction of the keto group to hydroxyl, and phosphorylation and pyrophosphorylation in the 5′- and 1′-positions, respectively, could result in an analog of 5′-phosphoribosyl-1′-pyrophosphate (PRPP). The addition of the adenine moiety could then be catalyzed by either a salvage enzyme, e.g., purine-RRPP transferase, or a set of enzymes equivalent to the purine biosynthetic enzymes. Indirect evidence has been presented that both could occur in the organism producing aristeromycin and neplanocin, i.e., *Streptomyces citricolor* (PARRY, 1992).

An interesting monoaminocyclitol, *N*-methyl-*scyllo*-inosamine, is found as a so-called rhizopine in rhizobia (MURPHY et al., 1987, 1993). Its anabolic metabolism is largely unknown so far. However, it will be interesting to see whether enzyme-catalyzed steps equivalent to those in the streptidine pathway are used (cf. Sect. 4.1.1.1).

Also, nonglycosylated and neutral cyclitol derivatives, with unknown biological activities and biosynthetic origins, such as 2,3,4-trihydroxy-6-methylcyclohexanone (Fig. A28; MÜLLER et al., 1986), seem to be common and stable actinomycete products. Similar compounds of this group have been found in the chemical screening of actinomycete secondary metabolites (S. GRABLEY, personal communication). Cyclophellitol, produced by the mushroom *Phellinus* sp. and active as a specific inhibitor of almond β-glucosidase (ATSUMI et al., 1990a, b), is another example for this type of compound. It is plausible that this is synthesized by a basic pathway and cyclization mechanism (Cb) similar to that of the valienamine-related C_7 cyclitols (cf. Sect. 4.4). The exact biosynthetic routes and their relationships to the known cyclitol pathways of all these compounds await further clarification, and can now be studied in a more direct way.

5.3 6-Deoxyhexoses

The 6DOH-derivatives, both of D- and L-configuration (cf. Sect. 3.2.1; Figs. 8 and 9), are the most variable group of sugar components in low molecular weight natural products. They are widely distributed especially in the actinomycete antibiotic groups of the aromatic (type I; anthracyclines, angucyclines, granaticin, etc.) and macrolide type polyketides (type II; macrolides, avermectins, polyenemacrolides) and in the glycopeptides (vancomycin family, thiopeptides) (PIEPERSBERG, 1994; LIU and THORSON, 1994). They also are building blocks in highly modified and variable extracellular polysaccharides, such as the lipopolysaccharides of gram-negative bacteria (LIU and THORSON, 1994). Their biosynthetic pathways have not been studied extensively in most cases. However, it was shown that DNA probes taken from the genes encoding enzymes involved in the basic steps of the 6DOH pathway (Sects. 3.2.1 and 4.1.1) detect the gene clusters relevant for 6DOH biosynthesis in many streptomycetes and other actinomycetes (PIEPERSBERG et al., 1991b; STOCKMANN and PIEPERSBERG, 1992). Hybridization signals obtained with these cloned fragments were localized in the predicted production gene clusters. From such experiments it was found that the dTDP-glucose 4,6-dehydratases, StrE, and related enzymes are the most highly conserved ones. In contrast, the genes/enzymes for the first step related to *strD*/StrD from *S. griseus* (dTDP-glucose synthetase) were more divergent and, interestingly, some of them did not hybridize at all, such as *strD* or *tyl*A1 from tylosin-producing *S. fradiae.* The *tylA1* gene is more closely related to gram-negative bacterial genes which encode the same enzyme (such as *rfbA* from salmonellae). Again, this suggested that these genes can be recruited from a common gene pool independent of any taxonomic distances by all eubacteria whenever a particular secondary carbohydrate metabolism is required.

The basic enzyme complement and the gene and protein families involved in 6DOH biosynthesis is becoming increasingly apparent (see reviews by SHNAITMAN and KLENA, 1993; PIEPERSBERG, 1994; LIU and THORSON, 1994; cf. Sect. 3.2). Enzymatic synthesis of the activated sugar components dTDP-L-oleandrose (MACNEIL, 1995; see Fig. 9) and dTDP-L-rhamnose (REEVES, 1993; see Fig. 9) which are naturally derived from 6DOH pathways has been achieved by complete *in vitro* enzymic synthesis from dTDP-D-glucose. Also, total chemical synthesis and introduction into the natural aglyca have been reported for some 6DOH derivatives, e.g., the 6DOHs D-mycosamine (BEAU, 1990) which occurs in polyene macrolides (see Fig. 8) and L-daunosamine (THOMAS, 1990) which occurs in anthracyclines (see Fig. 9).

The types of chemical bonds involved in the linkage between 6DOH moieties and their aglyca also vary widely and are worth some attention. Besides the usual O-glycosidic bonding C- and N-glycosidic or more complex linkages are also observed. C-glycosidic bonds occur with both aromatic and aliphatic C atoms, e.g., in altromycins (BRILL et al., 1990). Examples are the 2,6-dideoxy-D-glucose and 3-amino-3-*N*-methyl-4-*O*-methyl-2,3,6-trideoxy-D-hexose moieties in granaticin and staurosporine, respectively (Fig. A29). In granaticin the incorporation of the double-(1′,4′-)C-bound 6DOH into the benzoisochromanequinone polyketide aglycone could be achieved via two different routes from a dTDP-4-keto-2,6-D-hexose precursor according to earlier proposals (FLOSS and BEALE, 1989): via the formation of a C-glycosidic linkage after dTDP elimination or via a reaction of the aromatic C-10 and the 4′-keto group of the sugar. The latter route is considered more likely since other configurations in the pyrane ring do not alter the regioselectivity of the condensation reaction. In staurosporin and in the staurosporin-related compound K-252a (Fig. A29; KASE et al., 1986) an N-glycosidic linkage is probably formed first between the 1′-position of the amino-6DOH and one of the indole N atoms of the tryptophan-derived indolocarbazole heterocyclic system. This may also be the case for other very similar metabolites since in rebeccamycin as a single bond only the N-glycosidic linkage exists between the 4′-*O*-methylglucose and the indolocarbazole moieties (LAM et al., 1989; Fig. A29). The transferring glycosyltransferases, so far postulated to be involved in 6DOH transfer (LIU and THORSON, 1994; OTTEN et al., 1995; PIEPERSBERG et al., 1995; DICKENS et al., 1996) are related to the macrolide glucosyltransferases (MgtA, MgtB) causing resistance to 14- and 16-membered macrolides (CUNDLIFFE, 1992b; VILCHES et al., 1992).

5.4 Other Pentose, Hexose, and Heptose Derivatives

Nitrogen-containing derivatives of pentoses and hexoses (other than 6DOHs) are found in many microbial secondary metabolites, especially in nucleoside type antibiotics, either as (deoxy-)ribose analogs or as additionally modifying components in products originating from pathways with chemically heterogenous precursors.

Pentosamines. Examples for aminofuranoses are found in some well-studied aminonucleoside antibiotics, e.g., puromycin and the related antibiotic A201A (cf. Fig. A27). In the puromycin-related aminonucleosides the biosynthetic origin of the 3-amino group of the ribose derivative, for which an adenine nucleotide is the precursor, is not yet known. However, intensive investigations of the genetics and the enzymology of these compounds have been initiated recently (LACALLE et al., 1992). Interestingly, the putative products of the two genes *pur3* and *prg1* in the puromycin biosynthetic gene cluster of *Streptomyces alboniger* (TERCERO et al., 1996) are related to the StrO (possible cyclitol-phosphate phosphatase) and StrS/StsA/StsC (SMATs) proteins, respectively, from *S. griseus* (cf. Sect. 4.1). Besides the 3-amino-3-deoxyribose moiety A201A contains two additional rare sugar units: an unusual furanosidic hexose with an unsaturated C–C bond between C-1 and C-2 branching off directly from the furane ring, and a 6DOH, an *O*-methyl-D-rhamnose (cf. Fig. A27). Therefore, a *strE* probe was tested for hybridization and found to give a signal in the genomic DNA of the A201A producer *Streptomyces capreolus* (A. JIMENEZ, personal communication; cf. Sect. 5.3).

Hexosamines. Hexosamines of various structures and positions of amino-N substitution occur in many secondary metabolites (see examples in Figs. A7, A8, A11–15, A17–19, A24, and A27) and many of them may share common biosynthetic traits with aminoglycosides. Most of them have retained a pyranosidic ring structure, but incorporation into other heterocyclic ring systems via linearized intermediates can also occur, e.g., the D-glucosamine-derived building block in mitomycins (Fig. A31; HORNEMANN et al., 1974; OKADA et al., 1988). Examples of aminopyranoses are encountered in some nucleoside type antibiotics, e.g., streptothricins and blasticidin S (cf. Fig. A27). The origin of the 2-amino-D-glucose derivative in streptothricins is still unknown.

The biogenesis of the 4-amino-2,3,4-deoxyglucuronic acid moiety in blasticidin S (cf. Fig. A27), which is produced by *Streptomyces griseochromogenes* and used against rice blast disease, was studied by isotope labeling (GOULD, 1992). The fully retained position-specific labels of fed D-glucose (or even with higher yield with D-galactose) in this compound indicated that it is directly derived from a UDP-D-glucose or UDP-D-galactose precursor with UDP-D-glucuronic acid and cytosylglucuronic acid as intermediates. The latter intermediates were confirmed by direct measurements in cell-free extracts of the enzymes UDP-D-glucose epimerase, UDP-D-glucose oxidase, and cytosylglucuronic acid synthase.

6 Evolutionary Aspects

Most of the aminoglycosides and related compounds considered here are actinomycete products. Therefore, the question of whether the evolution of the respective pathways also occurred in this molecularly defined lineage of bacterial taxonomy arises (DAVIES et al., 1992; PIEPERSBERG, 1993; EMBLEY and STRACKEBRANDT, 1994). Although there are no data at present which could unequivocally prove this hypothesis it nevertheless has a high likelihood. Evidence is accumulating that the genes for these pathways have been transmitted horizontally between the actinomycetes and have also spread to other microbial groups. Hence, the butirosin (Fig. A13) biosynthetic genes are not likely to have evolved independently in the Bacillaceae. Rather, it seems that they have been derived from an actinomycete gene cluster involved in the biosynthesis of a ribostamycin-like aminoglycoside (cf. Fig. A13) which was laterally transferred to *Bacillus circulans* or an ancestor thereof. The elucidation of the structures of the gene clusters which encode enzymes involved in the formation of ribostamycin and butirosin and of other 2DOS aminoglycosides will clarify these evolutionary aspects. The horizontal dissemination of aminoglycoside resistance genes among all major eubacterial groups is particularly well documented (FOSTER, 1983; PIEPERSBERG et al., 1988; CUNDLIFFE, 1989; SHAW et al., 1993). The primary structure similarities, in particular, between the aminoglycoside phosphotransferase enzymes (APH) suggest a common origin of antibiotic phosphotransferases in bacteria and eukaryotic protein kinases (DISTLER et al., 1987a; HEINZEL et al., 1988; PIEPERSBERG et al., 1988, 1991; KIRBY, 1990; RETZLAFF et al., 1993). This common evolutionary origin is also supported by site-directed mutation and gene fusion experiments which yield hybrid APH enzymes. These experiments have shown that the essential amino acid residues and the two-domain structure of the catalytic subunit in the cAMP-dependent protein kinase of eukaryotes are conserved in the APHs (TAYLOR et al., 1990; BLAQUEZ et al., 1991; KNIGHTON et al., 1991; PIEPERSBERG et al., 1991a; RETZLAFF et al., 1993).

The above-mentioned versatile enzyme families, which are involved in secondary sugar metabolism, such as aminocyclitol biosynthesis and hexose activation and modification, are obviously products of a modularly used gene pool, the products of which are mainly involved in the production of highly variable and mostly secreted biomolecules not used in primary cell functions. Instead, they seem to be mainly involved in the extracellular communication of the producing cell with other biological systems (PIEPERSBERG, 1992, 1993). The secreted low molecular weight molecules, e.g., antibiotics, enzyme inhibitors, and autoregulators, could be used for the defense against agonistic or competing organisms or as hormone-like signal transmitters in differentiation processes. The extracellular polysaccharides and other cell surface-bound compounds (LPS, capsular polymers, etc.) are synthesized on the basis of very similar pathways and gene clusters; they may be cell surface "individualizing" material for cell-specific recognition and attachment or protection, i.e., these compounds could serve similar functions and be regarded as cell surface-bound secondary metabolites. In accordance with this, both the genes for secondary metabolites and the genes for LPS biosynthesis have frequently been suggested to be transferred horizontally (PIEPERSBERG, 1993, 1994; REEVES, 1993; SHNAITMAN and KLENA, 1993; LIU and THORSON, 1994). Therefore, the *rfb* genes and the secondary carbohydrate biosynthetic genes in actinomycetes represent an interesting basis for the study of the evolutionary links and dynamics in secondary, highly mobile gene pools.

The biosynthesis pathway of streptidine represents another interesting topic of evolutionary studies. The bluensidine pathway, e.g., was suggested to be ancestral to that of streptidine (WALKER, 1990). Amidinotransferase activities catalyzing both reactions 6 and 11 of the streptidine pathway (cf. Sect. 4.1.1.1 and Fig. 14) were found in the bluensomycin producer *S. hygroscopicus* ssp. *glebosus* ATCC 14607 (WALKER, 1990). However, no *strB2* gene is present in the two bluensomycin producers, *S. bluensis* DSM 40564 and *S. hygroscopicus* ssp. *glebosus* DSM 40823 (MAYER, 1994; G. MAYER, A. MEHLING, and W. PIEPERSBERG, unpublished data.). In these strains only one *strB* gene is conserved which clearly belongs to the *strB1* group, since the adjacent gene downstream is *strF* in both cases (see Fig. 13) and these genes encode proteins with a much higher amino acid sequences identity (ca. 85%) relative to the StrB1 than to StrB2 proteins of streptomycin producers. Therefore, in streptomycin producers gene duplication could have resulted during the evolution of two amidinotransferase with district substrate specificities from one ancestral enzyme exhibiting both activities. Alternatively, the bluensidine pathway could be a degenerated streptidine pathway after loss of the *strB2* gene and changes in the substrate specificity of the StrB1 protein. Also, an open question is whether or not *myo*-inositol is a specific precursor and, therefore, its synthase is an intrinsic enzyme of the streptomycin pathway, since measurements of this enzyme under various culture conditions and in various strains suggested a rather non-specific distribution (SIPOS and SZABO, 1989). The recent finding that probably all actinomycetes contain a major thiol compound called mycothiol (NEWTON et al., 1996) which also could be regarded as an aminoglycoside indicates that the anabolic *myo*-inositol pathway is generally present in this taxonomic group, in contrast to other bacteria. Therefore, the evolution of **Ca** type pathways could be based on a preferred metabolic route in actinomycetes.

The protein similarities between StrS, StsA, and StsC are in the range of 25% identity and, therefore, too low to suggest a gene duplication event during the evolution of the streptomycin pathways which might explain the origin of these proteins. The occurrence of a third possible aminotransferase is puzzling since it was thought that the third amino group introduced during the biosynthesis of streptomycin into NMLGA was derived via the primary metabolic pathway from the D-glucosamine pool (GRISEBACH, 1978). If this precursor is not formed *de novo* under conditions of streptomycin production another explanation for the need of an additional transamination step would be the regeneration of glutamine from α-ketoglutamine, the unusual by-product of step 4 in SD biogenesis. This aminotransferase reaction is not normally observed in prokaryotes (WALKER, 1975a). The N-terminal sequence of the StrS protein is also identical to that of one of the proteins expressed at high concentrations in the streptomycin production phase in *S. griseus* but not in the mutant M881 (see Tab. 2) (DISTLER et al., 1992).

7 Involvement of Primary Metabolism in the Delivery of Carbohydrate Precursors

The involvement of primary metabolic traits in the activation of precursors, the dynamics and alterations of precursor pool sizes, and the use of nutritional sources for precursor formation during the production phase in the producers of aminoglycosides and other secondary carbohydrates has not been studied intensively (SHAPIRO, 1989). The regulation of streptomycin production and its growth phase dependence has already been discussed above (see Sect. 4.6). These aspects will become more accessible for investigation in the future when the genetic programs for the biogenesis of representative members of this group of compounds are analyzed and available for manipulation in different organisms. Biotechnological development and use of production strains in fermentation could then become more efficient; mathematical modeling of the producer's physiology would also greatly facilitate the production processes. It is especially important to gain specific knowledge about the sources and the routes of delivery of the carbohydrate precursors and the nitrogenous groups in the case of the aminoglycosides. The finding that compounds such as neomycins and validamycins are synthesized from a precursor pool related to the pentosephosphate (PP) cycle intermediates (RINEHART et al., 1992), and that the NDP-hexose forming enzymes are encoded by genes in the clusters for aminoglycosides and other sugar-based secondary metabolites (see Sect. 4) in streptomycetes could help to clarify these aspects. If hexose intermediates are derived from the PP cycle mainly, this raises the question of whether compartmentalization or a difference in the efficiencies of the enzymes involved in the conversion of the primary hexosephosphate are responsible for this phenomenon. Alternatively, the inversion of the major route(s) of intermediary C metabolism to the predominant use of the gluconeogenetic pathway could occur during the production phase. In this context it is interesting to note that amino acids seem to be preferred or even essential nutrients in streptomycin or neomycin producers (SHAPIRO, 1989; PIEPERSBERG, 1995).

8 Pathway Engineering and Other Types of Application in Biotechnology

The availability of the genes which encode enzymes necessary for the biosynthesis of secondary metabolites recently has allowed the initiation of experiments designed for the production of new hybrid molecules (HOPWOOD et al., 1990; KATZ and DONADIO, 1993; several articles in VINING and STUTTARD, 1995; PIEPERSBERG, 1994). For the array of aminoglycosides and other carbohydrate-derived natural products this phase of biotechnology has not yet been as successful as it was for the polyketides. The reasons for this are as follows: (1) the pathways involved are composed of multiple steps catalyzed by highly specific and in general monofunctional enzymes; (2) the biochemistry and chemistry of carbohydrates is complicated, and the intermediates of biosynthetic pathways are mostly quite instable and inaccessible to synthetic chemistry; and (3) only relatively few research groups are engaged in the investigation of secondary carbohydrate metabolism. Nevertheless, several fields of biotechnological application of our increasing knowledge of the biochemical and genetic components involved in the formation of activated and modified sugar or cyclitol molecules can be envisaged (cf. also PIEPERSBERG, 1994): (1) genetic engineering of pathways for improved fermentations, with respect to nutrient control and incorporation, optimization of yield, and restriction of product patterns, or for the production of new hybrid end products in the producers themselves (e.g., the hybrid, glycosylated tetracenomycins; DECKER et al.,

1995); (2) development of biotransformation systems for the *in vivo* glycosylation of fed aglycones; (3) *in vitro* enzymatic production of activated sugar derivatives and/or glycosyltransfer to synthetic aglycones; or (4) transfer of the genetic complement for complete pathways or branching subpathways to new host systems and/or their redesign for improved production characteristics or completely new metabolites. However, most of the prerequisites for the planned redesign of the pathways for aminocyclitol aminoglycosides and related secondary carbohydrates are not yet available. These prequisites include the complete genetic and biochemical analyses of several key pathways and knowledge of substrate specificities of key enzymes (especially the glycosyltransferases and other condensing enzymes) and of the bottleneck steps or the flux rates of individual precursors and intermediates through the pathway.

Some interesting goals for pathway engineering in aminocyclitol aminoglycoside producers are as follows:

(1) Transfer of the production and transfer genes for the N_1-α-hydroxy-γ-aminobutyryl moiety of butirosin from *Bacillus circulans* to a kanamycin-producing *Streptomyces kanamyceticus* in order to produce *in vivo* the semisynthetic compound amikacin (=N_1-α-hydroxy-γ-aminobutyryl-kanamycin A), as already suggested much earlier (Fig. 28A; DAVIES and YAGISAWA, 1983). This would require first that the N_1-α-hydroxy-γ-aminobutyryltransferase also recognizes the kanamycin A molecule as a substrate and in its correct amino acceptor group and, second, that the possible kanamycin exporter in *S. kanamyceticus* transports the new end product with equivalent efficiency. Similarly, it might be possible to find enzymic reactions for the designed modification of other aminoglycosides, e.g., group transfers or dehydroxylations such as those used in the synthesis of well-established semisynthetic chemotherapeutics with activity against bacteria with clinically important resistance patterns (cf. SHAW et al., 1993) and with reduced ototoxicity. Examples are (1) netilmicin (=1-*N*-ethylsisomicin; cf. Fig. A14), (2) isepamicin (cf. Fig. 28B), dibekacin (=3′,4′-dideoxykanamycin B; cf. Fig. A12), and arbekacin (=N_1-α-hydroxy-γ-aminobutyryl-dibekacin; cf. Fig. 28B); for most of those modifications, natural counterparts exist as models.

(2) Designed mixing of subpathways could be envisaged. For example, first the Cb pathways could be exchanged for Ca routes and vice versa. In another line of experimental models the exchange of the sugar modifying and transferring subpathways between producers of related groups of aminoglycosides could be attempted, e.g., the PA pathways between the producers of gentamicins (cf. Fig. A14) and seldomycins (cf. Fig. A17) or the 6DOH or HA pathways between the producers of streptomycins (cf. Fig. A1), spectinomycins (cf. Figs. A2 and A3), kasugamycins (cf. Figs. A4 and A5), and boholmycin (cf. Fig. A7).

(3) Some aminoglycosides are produced in genera or strains which are not easily accessible for physiological or genetic manipulation or which produce unwanted side products. The ability to produce could, therefore, be transferred to hosts which can be more easily manipulated. For instance, the aminoglycoside production genes from *Micromospora* spp. could be transferred to *Streptomyces* spp. or even to more distantly related "GRAS"-organisms (GRAS="generally regarded as safe") such as corynebacteria (cf. PIEPERSBERG, 1993).

(4) The regulation of production genes and the nutrient flow in aminoglycoside producers could be further targets of pathway engineering. Because of their complicated physiology and regulation in complex cell differentiation cycles (cf. Sect. 4.6) this would be of particular importance when continuous culture techniques are used in production.

Another application of the new genetic and biochemical data could be the development of genetic screening systems for the search for a particular production ability. For this purpose, we have to find more genes for key functions which are diagnostically relevant in the detection of a specific pathway. An example is the family of *strE*-related genes encoding the dTDP-glucose 4,6-dehydratases characteristic for the **6DOH** pathways (STOCKMANN and PIEPERSBERG, 1992). This type of diagnostic material, together with that typical of other chemical classes of secondary meta-

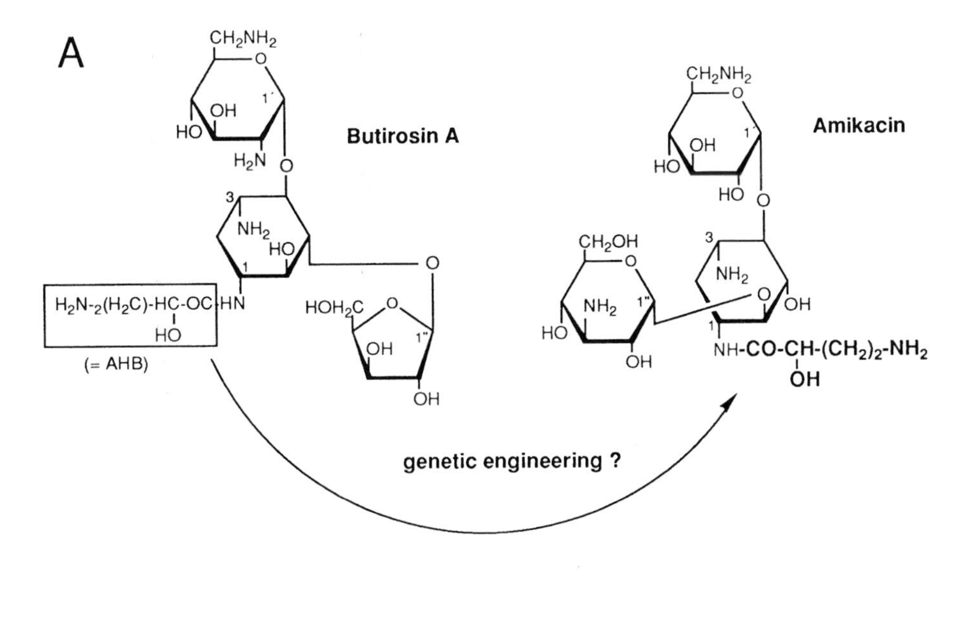

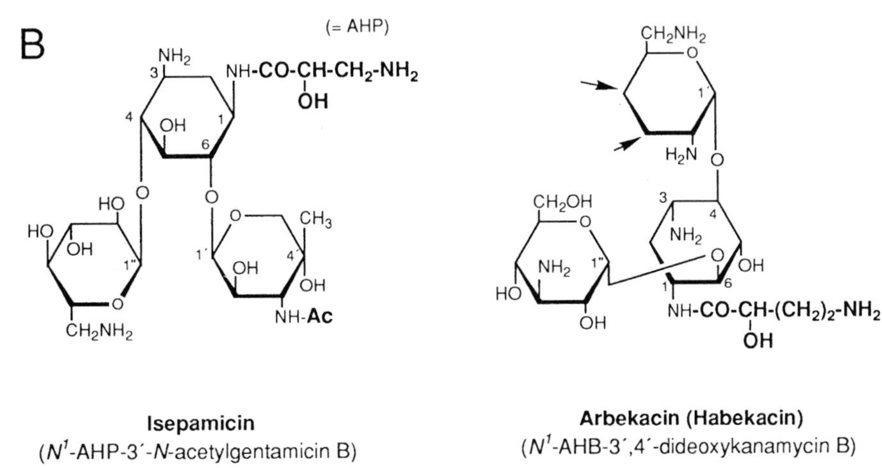

Isepamicin
(N^1-AHP-3´-N-acetylgentamicin B)

Arbekacin (Habekacin)
(N^1-AHB-3´,4´-dideoxykanamycin B)

Fig. 28. Semisynthetic aminoglycosides as models for the production of modified aminoglycosides by pathway engineering. **A** Hypothetical production of the semisynthetic amikacin in a genetically engineered kanamycin A producer; **B** structures of two other semisynthetic aminoglycosides, derivatives of gentamicin B and kanamycin B, with good pharmacological properties and activity against multiple-resistant pathogens. The chemical groups that are introduced by semisynthetic additions (given in bold face) or deleted (arrows) mostly mimic known modifications of related natural components in other molecules.

bolites, could be applied together with classical screening methods for the detection of new compounds. This could be of interest if, e.g., a leading structure is known and new derivatives related to that special target group are searched for, or if in particular microbial groups (e.g., actinomycetes) an unwanted subgroup of producers has to be excluded. In the future, when sufficient and highly predictive pathway-specific gene probes become available this method can also be used for a more rapid prescreening method.

9 Conclusions and Perspectives

The knowledge of all the variants of secondary metabolites composed of (amino-)sugars or their derivatives (secondary carbohydrates) and their biosyntheses is still sparse. Evidence is accumulating, however, that the various pathways of aminocyclitol aminoglycoside production share several common features and that a common gene pool is used for the modular design of these pathways. It can be predicted that the application of molecular genetics and biochemistry on several production systems for secondary carbohydrates will bring some unification to several aspects of this field of research. Concomitantly, also an intensification of applied research on secondary carbohydrates will arise, since these are components in many applied natural products where they are essential for bioactivity. Thus, the data reported in this chapter will influence other fields of research and development, especially of other chemical groups of low-molecular weight bioactive molecules or of polysaccharides and glycoconjugates. Also, the research and development in the field of chemo-enzymatic synthesis will be fertilized by providing the biochemical tools such as enzymes and their substrates (e.g., NDP-activated sugars). The basis for pathway engineering for the designed production of new variants of secondary metabolites will also be available very soon. With respect to the more restricted group of aminocyclitol aminoglycosides in particular, there still seems to be a potential for finding new structural classes as is indicated by the recent detection of new groups via target-directed screenings for glycosidase inhibitors (e.g., acarbose, trestatins, trehazolin, and allosamidins). Other fields of application and new structures will also become available by continuing research efforts and the input of intelligence and scientific skills.

Acknowledgements

We thank all our collaborators for their enthusiastic participation in aminoglycoside research. The work on secondary carbohydrate genetics and metabolism in the laboratory of the authors was generously supported by the Deutsche Forschungsgemeinschaft, the Bundesministerium für Forschung und Technologie, the European Commission, and the pharmaceutical companies Bayer AG and Hoechst AG.

10 Appendix (Chemical Structures)

The figures compiled in the Appendix (Figs. A1–A31) summarize the chemical families of secondary carbohydrates mentioned in the text; their "pathway formulae" are indicated in the legend (cf. Sect. 3.2).

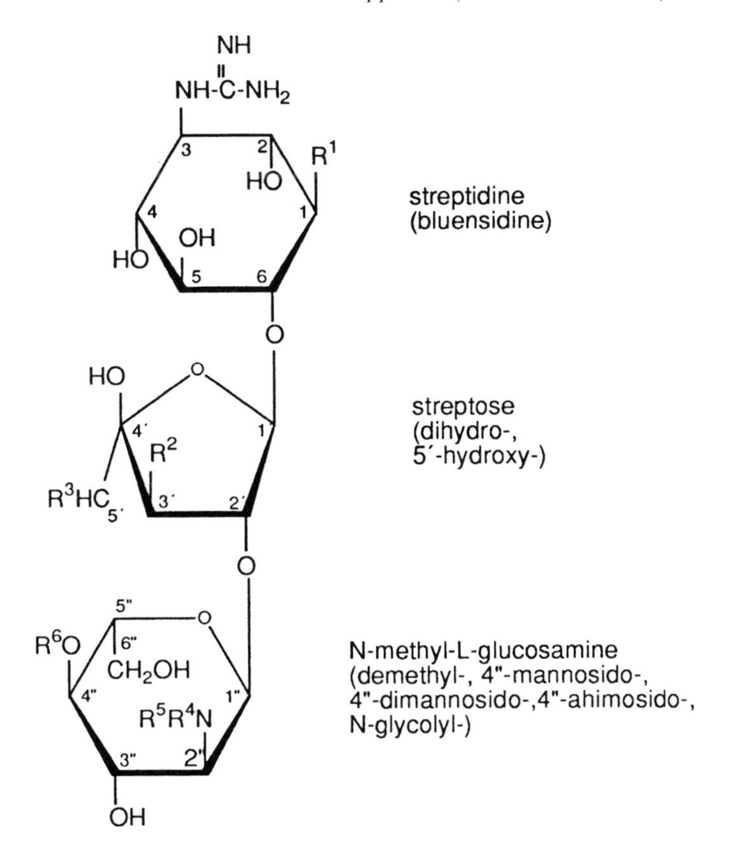

	R^1	R^2	R^3	R^4	R^5	R^6
streptomycin	NH-CNH-NH_2	CHO	CH_3	CH_3	H	H
dihydrostreptomycin	NH-CNH-NH_2	CH_2-OH	CH_3	CH_3	H	H
5′-hydroxystreptomycin	NH-CNH-NH_2	CHO	CH_2-OH	CH_3	H	H
N-demethylstreptomycin	NH-CNH-NH_2	CHO	CH_3	H	H	H
5′-hydroxy-N-demethyl-dihydrostreptomycin	NH-CNH-NH_2	CH_2-OH	CH_2-OH	H	H	H
mannosido-5′-hydroxystreptomycin	NH-CNH-NH_2	CHO	CH_2-OH	CH_3	H	α-D-mannose (= DM)
dimannosidostreptomycin	NH-CNH-NH_2	CHO	CH_3	CH_3	H	α-DM-1,6-α-DM
bluensomycin	O-CO-NH_2	CH_2-OH	CH_3	CH_3	H	H
ashimycin A	NH-CNH-NH_2	CHO	CH_3	CH_3	H	2‴-carboxy-xylo-furanose (ashimose)
ashimycin B	NH-CNH-NH_2	CHO	CH_3	CH_3	CO-CH_2-OH	α-DM-1,6-α-DM

AC4437 = 5′-hydroxystreptomycin lacking NMLGA

Fig. A1. Streptomycins; basic pathway formula Ca(4)-6DOH(2)-HA.

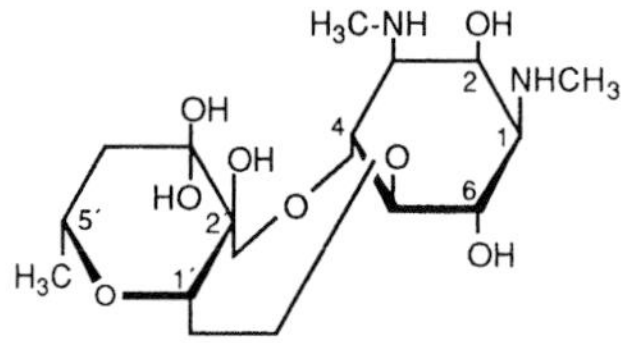

Fig. A2. Spectinomycin; Ca(4,5)-6DOH.

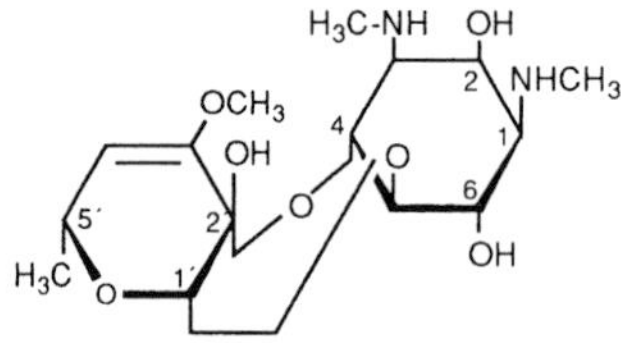

Fig. A3. Spenolimycin; Ca(4,5)-6DOH.

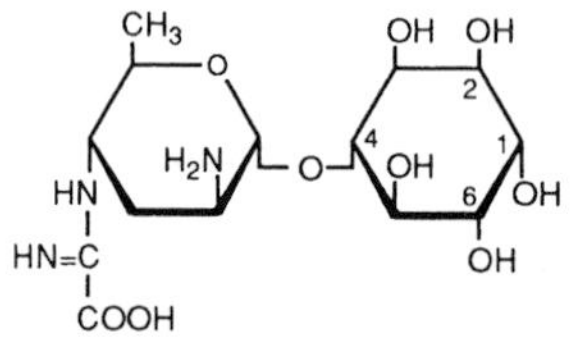

Fig. A4. Kasugamycin; Ca(4)-6DOH.

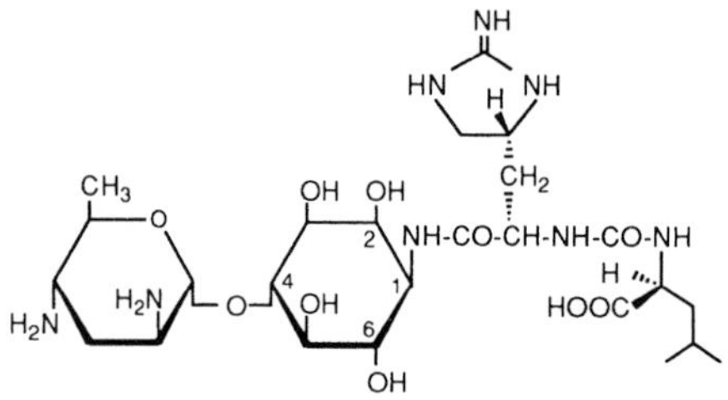

Fig. A5. Minosanimomycin; Ca(4)-6DOH.

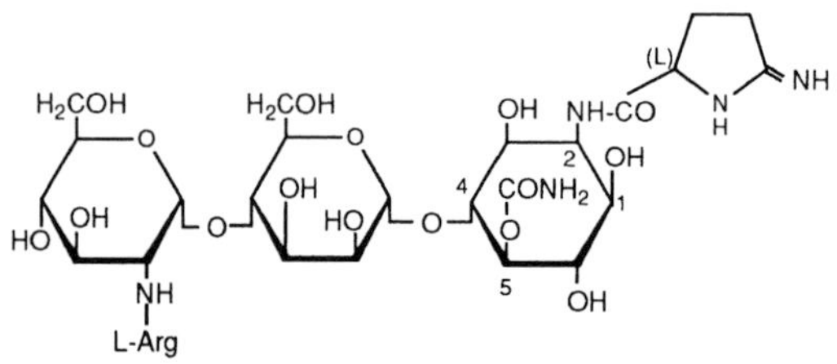

Fig. A6. LL-BM132α; Ca(4)-H(4)-HA.

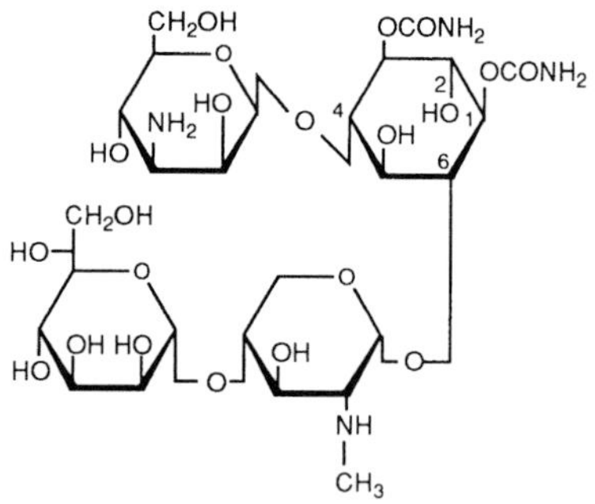

Fig. A7. Boholmycin; HA-(4)Ca(6)-PA(4)-Hep.

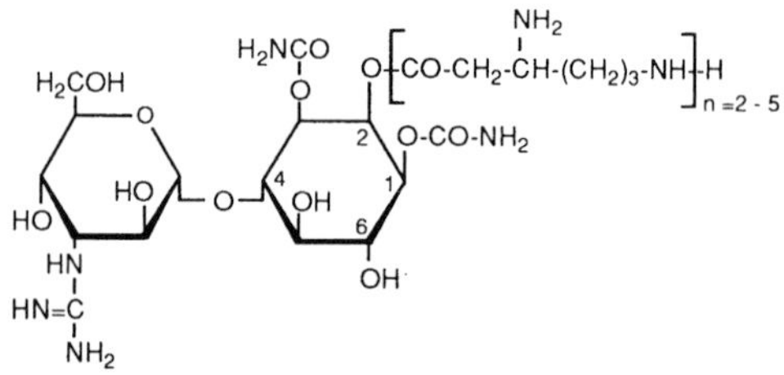

Fig. A8. Myomycins; Ca(4)-HA.

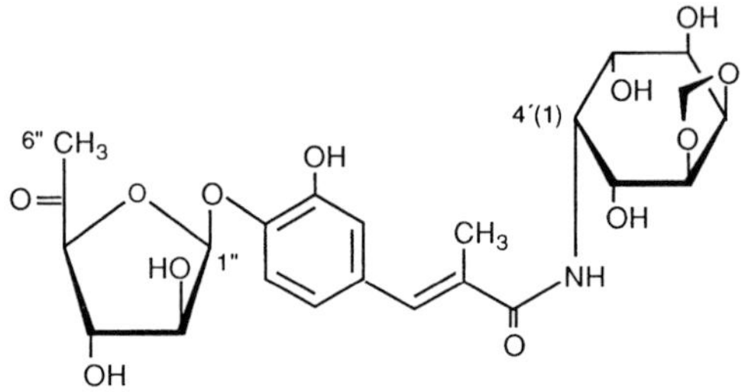

Fig. A9. Hygromycin A; Ca(1)-[X]-6DOH.

A

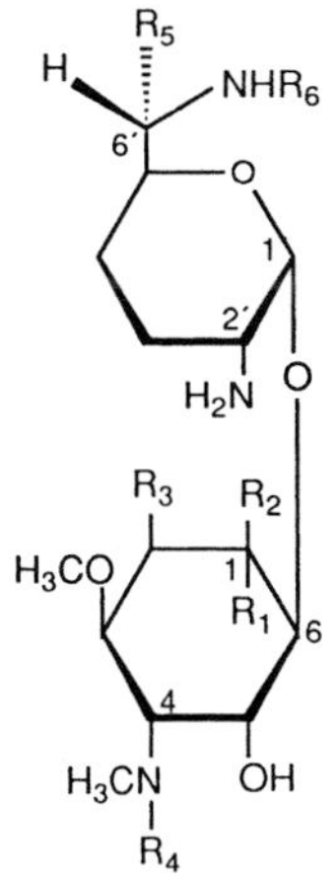

B

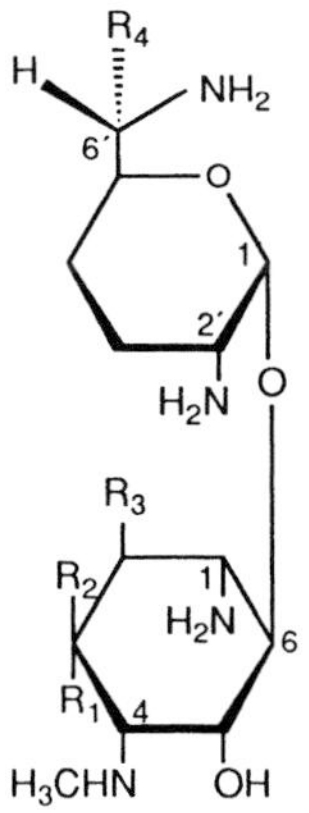

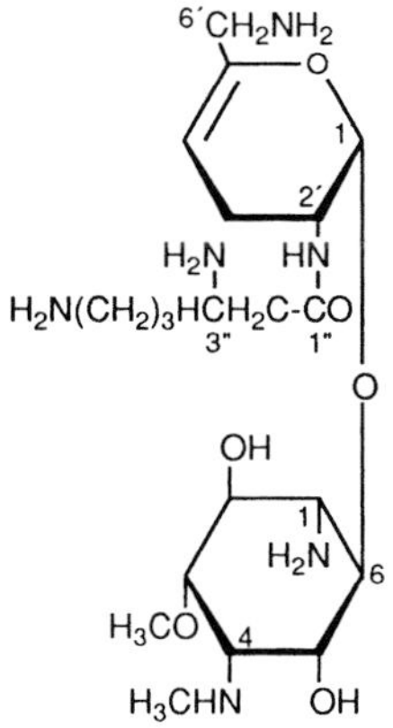

Lysinomicin (**Ca(6)-HA**)

		R_1	R_2	R_3	R_4	R_5	R_6
A	fortimicin A	NH_2	H	OH	$COCH_2NH_2$	CH_3	H
	fortimicin B	NH_2	H	OH	H	CH_3	H
	1-*epi*-fortimicin B	H	NH_2	OH	H	CH_3	H
	fortimicin C	NH_2	H	OH	$COCH_2NHCONH_2$	CH_3	H
	fortimicin D	NH_2	H	OH	$COCH_2NH_2$	H	H
	dactimicin	NH_2	H	OH	$COCH_2NHCH{=}NH$	CH_3	H
	1-*epi*-dactimicin	H	NH_2	OH	$COCH_2NHCH{=}NH$	CH_3	H
	sporaricin A	H	NH_2	H	$COCH_2NH_2$	CH_3	H
	sporaricin B	H	NH_2	H	H	CH_3	H
	istamycin A_0	NH_2	H	H	H	H	CH_3
	istamycin A (= sannamycin)	NH_2	H	H	$COCH_2NH_2$	H	CH_3
	istamycin A_3	NH_2	H	H	$COCH_2NHCH{=}NH$	H	CH_3
	istamycin B_0	H	NH_2	H	H	H	CH_3
	istamycin B	H	NH_2	H	$COCH_2NH_2$	H	CH_3
	istamycin B_3	H	NH_2	H	$COCH_2NHCH{=}NH$	H	CH_3
	istamycin C	NH_2	H	H	$COCH_2NH_2$	H	CH_2CH_3
	istamycin A_2	NH_2	H	H	$COCH_2NHCONH_2$	H	CH_3
	fortimicin KG_3 = 4´,5´-dehydrofortimicin A						
B	fortimicin KH	OCH_3	H	OH	CH_3		
	fortimicin KR	H	OCH_3	OH	CH_3		
	istamycin Y_0	OCH_3	H	H	H		
	istamycin X_0	H	OCH_3	H	H		

Fig. A10. Fortimicin/istamycin family; Ca(6)-HA or Cb(6)-HA.

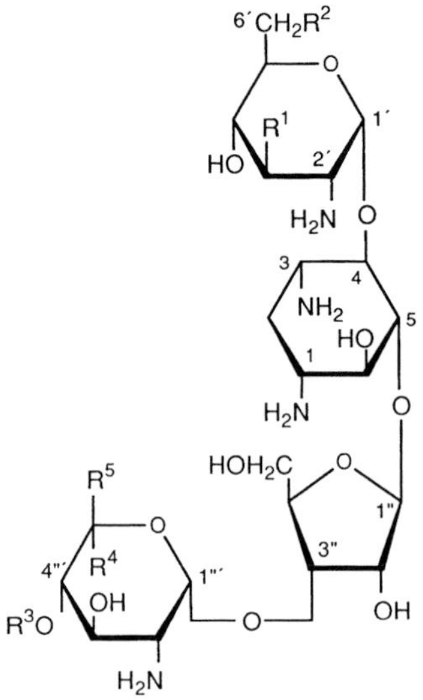

	R^1	R^2	R^3	R^4	R^5
neomycin B	OH	NH_2	H	CH_2NH_2	H
neomycin C	OH	NH_2	H	H	CH_2NH_2
paromomycin I	OH	OH	H	CH_2NH_2	H
paromomycin II	OH	OH	H	H	CH_2NH_2
mannosylparomomycin	OH	OH	α-D-Man	CH_2NH_2	H
lividomycin A	H	OH	α-D-Man	CH_2NH_2	H
lividomycin B	H	OH	H	CH_2NH_2	H

Fig. A11. Neomycin family; HA-(4)Cb(5)-P(3)-HA.

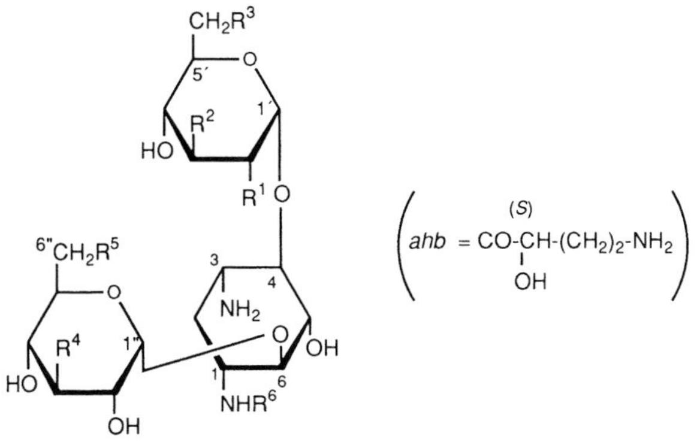

(ahb = CO-CH(OH)-$(CH_2)_2$-NH_2, (S))

	R^1	R^2	R^3	R^4	R^5	R^6
kanamycin	OH	OH	NH_2	NH_2	OH	H
kanamycin B	NH_2	OH	NH_2	NH_2	OH	H
kanamycin C	NH_2	OH	OH	NH_2	OH	H
amikacin	OH	OH	NH_2	NH_2	OH	*ahb*
NK-1001	OH	OH	NH_2	OH	OH	H
NK-1012-1	NH_2	OH	NH_2	OH	OH	H
tobramycin	NH_2	H	NH_2	NH_2	OH	H
nebramycin 4	NH_2	OH	NH_2	NH_2	$OCONH_2$	H
nebramycin 5′	NH_2	H	NH_2	NH_2	$OCONH_2$	H

Fig. A12. Kanamycin family; HA-(4)Cb(6)-HA.

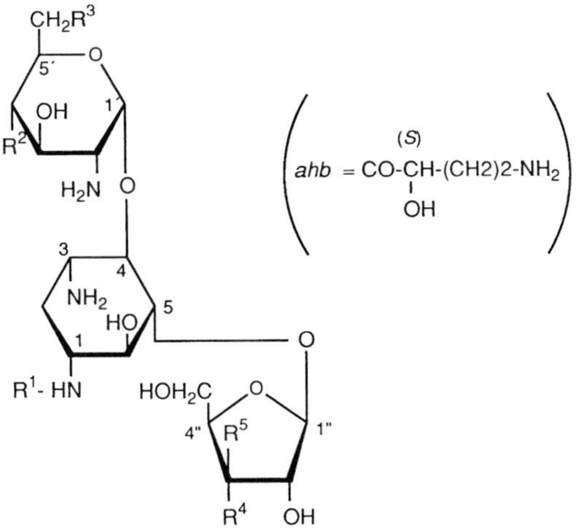

(ahb = CO-CH(OH)-(CH2)2-NH_2, (S))

	R^1	R^2	R^3	R^4	R^5
butirosin A	*ahb*	OH	NH_2	H	OH
butirosin B	*ahb*	OH	NH_2	OH	H
butirosin E_1	*ahb*	OH	OH	H	OH
butirosin E_2	*ahb*	OH	OH	OH	H
butirosin C_1	*ahb*	H	NH_2	H	OH
butirosin C_2	*ahb*	H	NH_2	OH	H
ribostamycin	H	OH	NH_2	OH	H
xylostatin	H	OH	NH_2	H	OH
LL-BM 408α	H	OH	OH	OH	H

Fig. A13. Butirosin/ribostamycin family; HA-(4)Cb(5)-P.

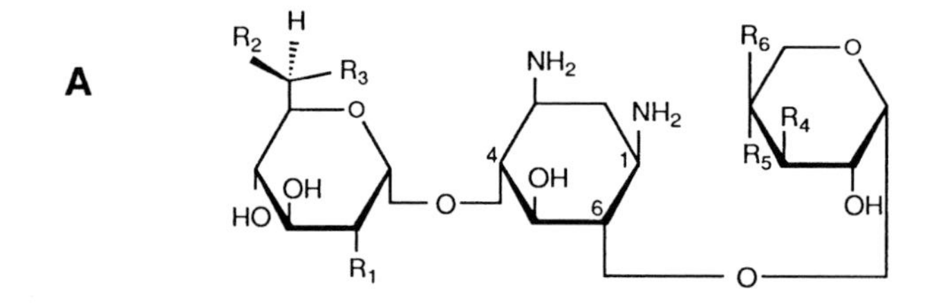

A	R_1	R_2	R_3	R_4	R_5	R_6
gentamicin A	NH_2	H	OH	$NHCH_3$	OH	H
gentamicin A_1	NH_2	H	OH	$NHCH_3$	H	OH
gentamicin A_2	NH_2	H	OH	OH	OH	H
gentamicin A_3	OH	H	NH_2	$NHCH_3$	H	OH
gentamicin A_4	NH_2	H	OH	NCH_3CHO	OH	H
gentamicin B	OH	H	NH_2	$NHCH_3$	CH_3	OH
gentamicin B_1	OH	CH_3	NH_2	$NHCH_3$	CH_3	OH
gentamicin X_2	NH_2	H	OH	$NHCH_3$	CH_3	OH
G-418	NH_2	CH_3	OH	$NHCH_3$	CH_3	OH
JI-20A	NH_2	H	NH_2	$NHCH_3$	CH_3	OH
JI-20B	NH_2	CH_3	NH_2	$NHCH_3$	CH_3	OH

B	R'_1	R'_2	R'_3	R'_4	R'_5	R'_6
gentamicin C_1	CH_3	H	CH_3	CH_3	OH	-
gentamicin C_{1a}	H	H	H	CH_3	OH	-
gentamicin C_2	CH_3	H	H	CH_3	OH	-
gentamicin C_{2a}	H	CH_3	H	CH_3	OH	-
sagamicin	H	H	CH_3	CH_3	OH	-
sisomicin	H	H	H	CH_3	OH	+
verdamicin	CH_3	H	H	CH_3	OH	+
G-52	H	H	CH_3	CH_3	OH	+
66-40B	H	H	H	OH	H	+
66-40D	H	H	H	H	OH	+

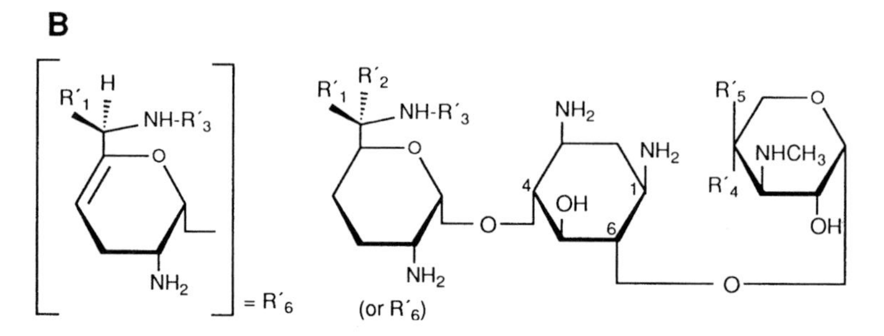

Fig. A14. Gentamicin family; HA-(4)Cb(6)-PA.

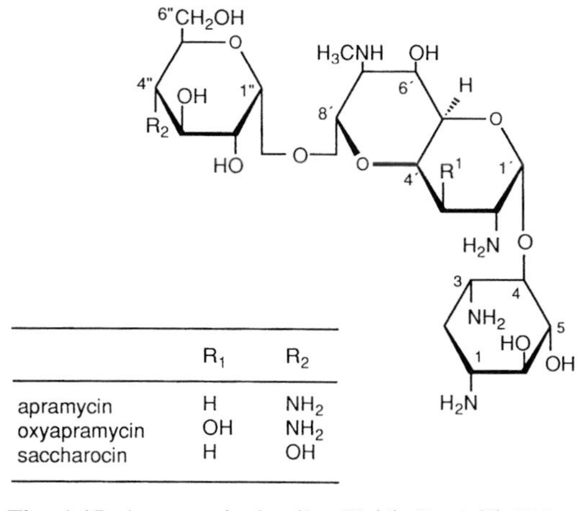

	R_1	R_2
apramycin	H	NH_2
oxyapramycin	OH	NH_2
saccharocin	H	OH

Fig. A15. Apramycin family; Cb(4)-OctA(8)-HA.

	R_1	R_2	R_3	R_4
seldomycin 1	OH	OH	OH	OH
seldomycin 3	OH	NH_2	OH	OH
seldomycin 5	H	NH_2	NH_2	OCH_3

Fig. A17. Seldomycin family; HA-(4)Cb(6)-PA.

	R_1	R_2	R_3	R_4	R_5	R_6	R_7
hygromycin B	H	CH_3	H	OH	H	OH	H
destomycin A	CH_3	H	H	OH	H	OH	H
destomycin B	CH_3	CH_3	H	H	OH	H	OH
destomycin C	CH_3	CH_3	H	OH	H	OH	H
A-396-I	H	H	H	OH	H	OH	H
A-16316-C	CH_3	CH_3	H	H	OH	OH	H
SS-56C	H	H	OH	OH	H	OH	H
1-N-amidino-1-N-demethyl-2-hydroxy-destomycin A	$C(NH)NH_2$	H	OH	OH	H	OH	H

Fig. A16. Hygromycin B/destromycin family; Cb(5)-H(2,3)-HepA or Ca(5)-H(2,3)-HepA.

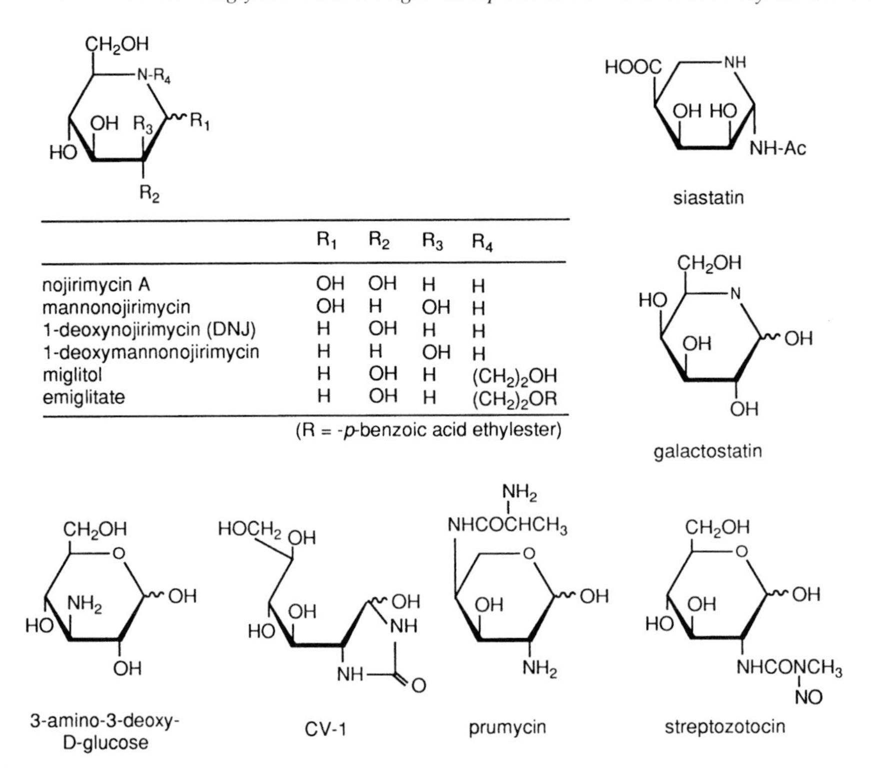

	R_1	R_2	R_3	R_4
nojirimycin A	OH	OH	H	H
mannonojirimycin	OH	H	OH	H
1-deoxynojirimycin (DNJ)	H	OH	H	H
1-deoxymannonojirimycin	H	H	OH	H
miglitol	H	OH	H	$(CH_2)_2OH$
emiglitate	H	OH	H	$(CH_2)_2OR$

(R = -*p*-benzoic acid ethylester)

Fig. A18. Monomeric aminoglycosides which are products of HA or PA pathways.

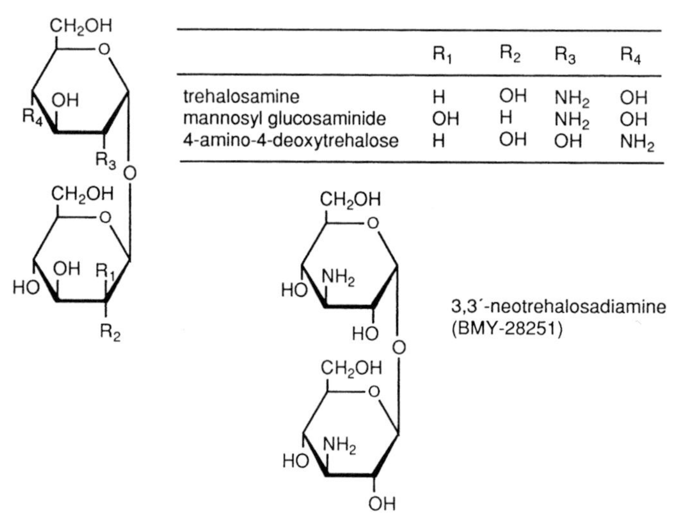

	R_1	R_2	R_3	R_4
trehalosamine	H	OH	NH_2	OH
mannosyl glucosaminide	OH	H	NH_2	OH
4-amino-4-deoxytrehalose	H	OH	OH	NH_2

Fig. A19. Trehalosamine family; HA(1)-H or HA(1)-HA.

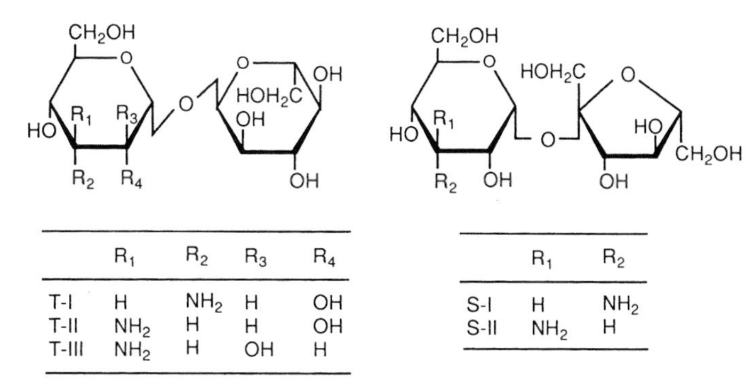

	R_1	R_2	R_3	R_4
T-I	H	NH_2	H	OH
T-II	NH_2	H	H	OH
T-III	NH_2	H	OH	H

	R_1	R_2
S-I	H	NH_2
S-II	NH_2	H

Fig. A20. Trehalosamine-related synthetic disaccharides.

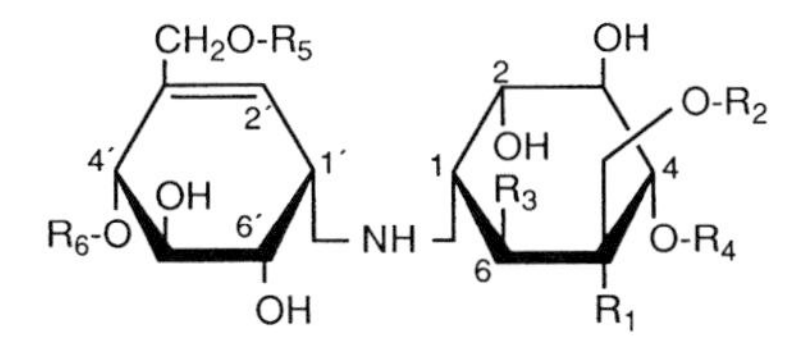

	R_1	R_2	R_3	R_4	R_5	R_6
validoxylamine A	H	H	H	H	H	H
validoxylamine B	H	H	OH	H	H	H
validoxylamine G	OH	H	H	H	H	H
validamycin A	H	H	H	β-D-Glc	H	H
validamycin B	H	H	OH	β-D-Glc	H	H
validamycin C	H	H	H	β-D-Glc	α-D-Glc	H
validamycin D	H	α-D-Glc	H	H	H	H
validamycin E	H	H	H	α-D-Glc(1,4) -β-D-Glc	H	H
validamycin F	H	H	H	β-D-Glc	H	α-D-Glc
validamycin G	OH	H	H	β-D-Glc	H	H
validamycin H	H	H	H	α-D-Glc(1,6) -β-D-Glc	H	H

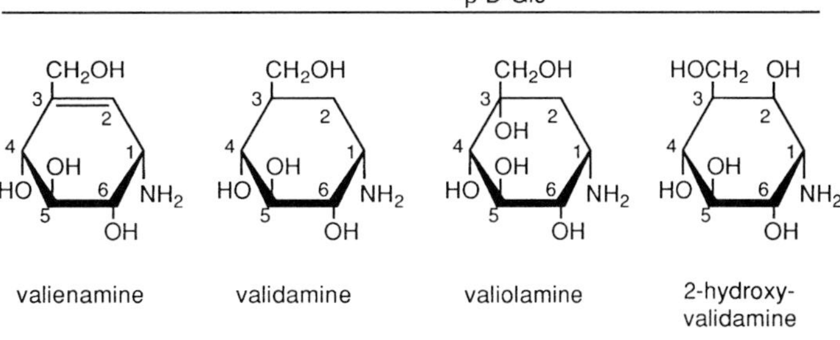

Fig. A21. Validamycin family; Cb(1)-Cb(4)-H. Valiolamine (Cb) was also found as a separate end product.

	R_1	R_2	R_3
acarbose (amylostatins)	H	α-1,4-maltosyl	H, or (α-D-glucopyranosyl)$_n$
adiposins	OH	α-1,4-maltosyl	H, or (α-D-glucopyranosyl)$_n$
trestatin	H	α-1,4-maltosyl-α-1,4-trehalosyl	H, or (core pseudo-trisaccharide)$_n$
2,3-epoxyderivatives of acarbose			
oligostatins = 2,3-dihydro-2-hydroxyderivatives of acarbose			

Fig. A22. A carbose/amylostatin family of glycosidase inhibitors; (H)$_n$-Cb(1)-6DOH(1)-(H)$_n$ or (H)$_n$-Cb(1)-H(1)-(H)$_n$.

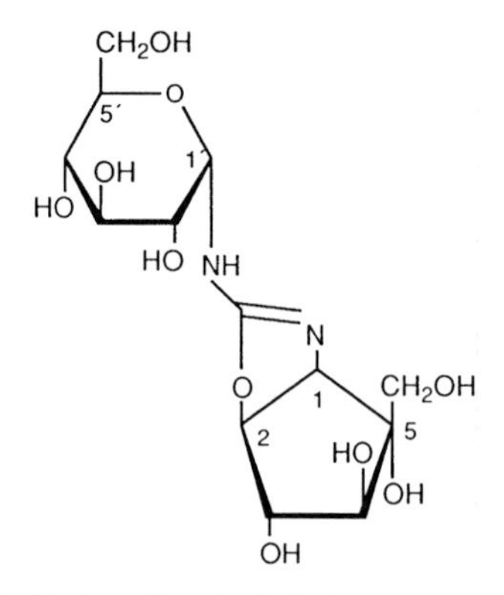

Fig. A23. Trehazolin; Ca(1,2)-HA.

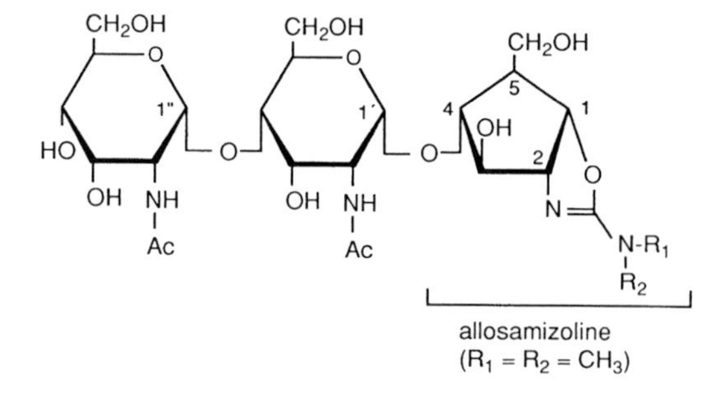

	R_1	R_2
allosamidin	CH_3	CH_3
demethylallosamidin	CH_3	H
didemethylallosamidin	H	H

Fig. A25. Allosamidins; Ca(4)-HA(4)-HA.

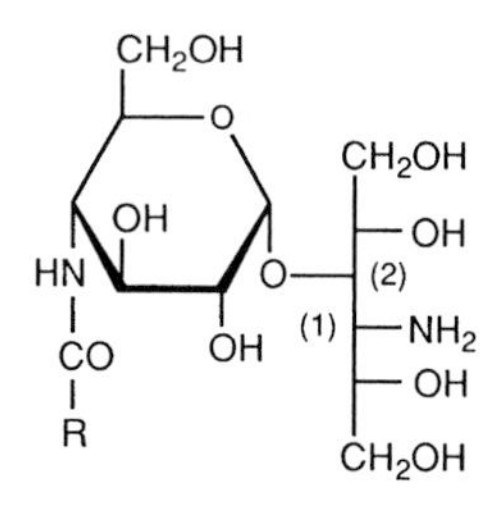

	R
sorbistin A_1	Ethyl
sorbistin A_2	Propyl
sorbistin B	Methyl
sorbistin D	H

Fig. A24. Sorbistins; Ca(2)-HA.

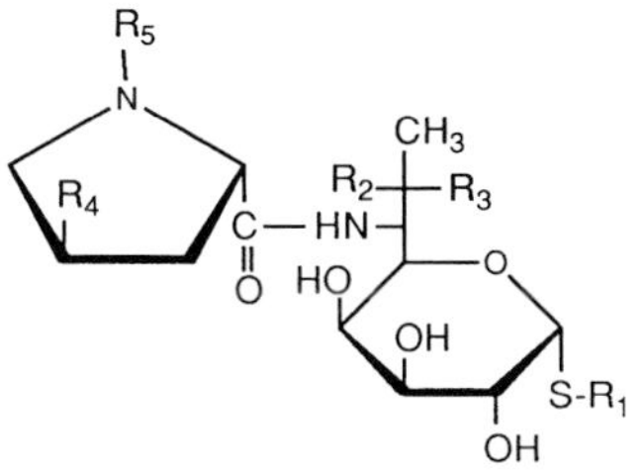

	R_1	R_2	R_3	R_4	R_5
lincomycin A	CH_3	OH	H	CH_2-CH_2-CH_3	CH_3
lincomycin B	CH_3	OH	H	CH_2-CH_3	CH_3
clindamycin	CH_3	H	Cl	CH_2-CH_2-CH_3	CH_3
N-demethyl-lincomycin A	CH_3	OH	H	CH_2-CH_2-CH_3	H
celesticetin	$(CH_2)_2$-O-salicylyl	OCH_3	H	H	CH_3
desalicetin	$(CH_2)_2$-OH	OCH_3	H	H	CH_3
celesticetin B	$(CH_2)_2$-O-isoburyryl	OCH_3	H	H	CH_3
celesticetin C	$(CH_2)_2$-O-anthranilyl	OCH_3	H	H	CH_3
celesticetin D	$(CH_2)_2$-O-acetyl	OCH_3	H	H	CH_3

Fig. A26. Lincosamides; HA or OctA.

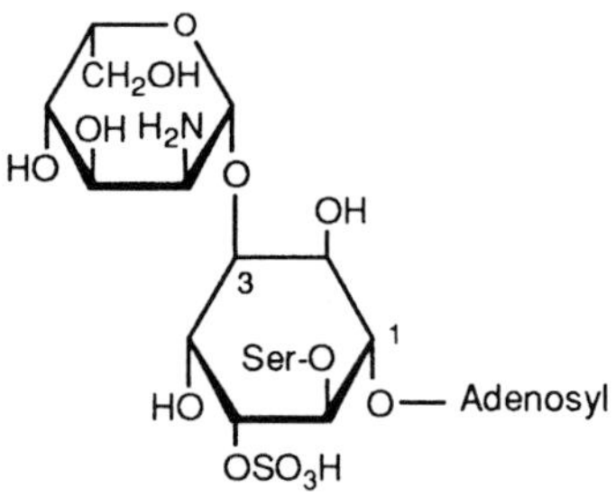

Adenomycin
(**P-(1)Ca(3)-HA**)

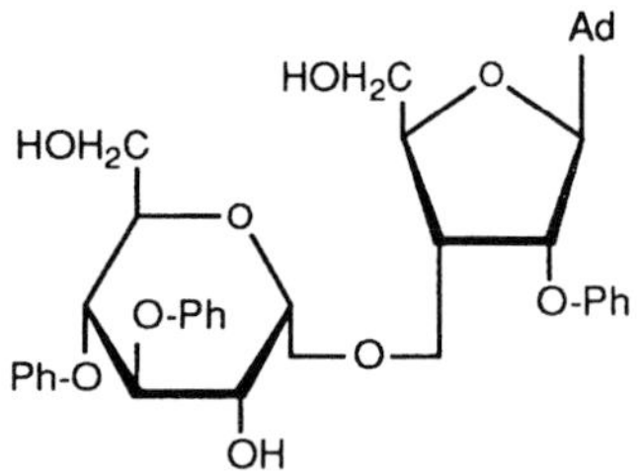

Adenophostin A
(Ad = Adenin;
Ph = PO_3H_2)
(**P(3)-H**)

Cytosaminomycins
(R = acyl residues)
(**6DOH(4)-6DOH**)

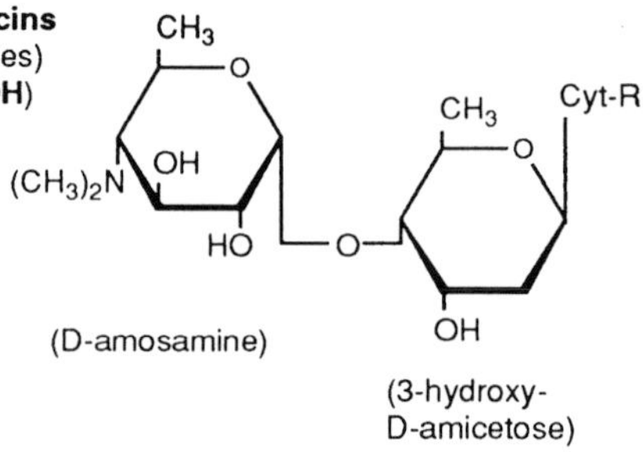

(D-amosamine)

(3-hydroxy-
D-amicetose)

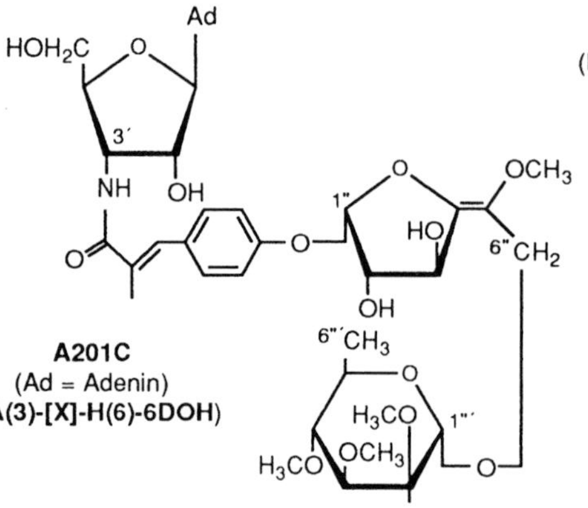

A201C
(Ad = Adenin)
(**PA(3)-[X]-H(6)-6DOH**)

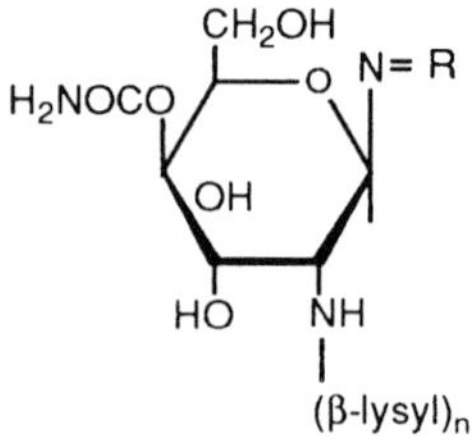

streptothricins
(R = deazapurin)
(**HA**)

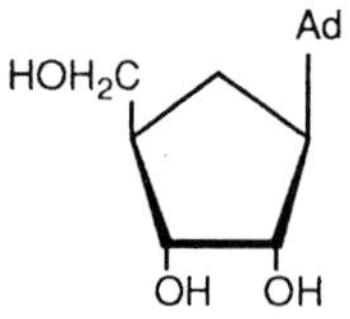

Aristeromycin
(Ad = Adenin)
(**Ca** or **Cb**)

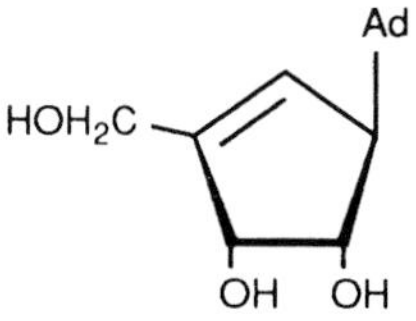

Neplanocin A
(Ad = Adenin)
(**Ca** or **Cb**)

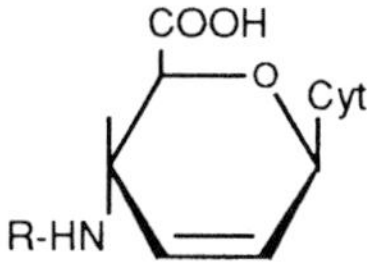

Blasticidin S
(Cyt = cytosin; R =
γ-N-methyl-β-arginine)
(**HA**)

Fig. A27. Nucleoside-type secondary metabolites with unusual sugar components or cyclitols.

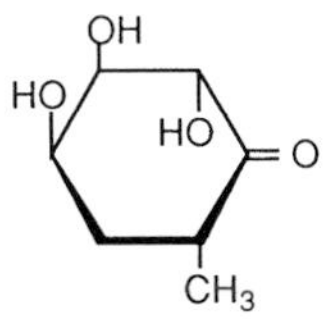

2,3,4-trihydroxy-6-methylcyclohexanone

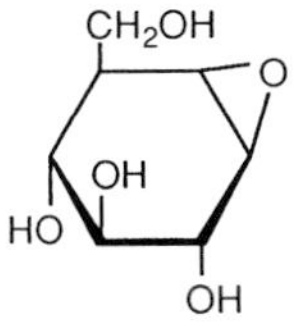

cyclophellitol

Fig. A28. Natural neutral cyclitols; Ca or Cb?

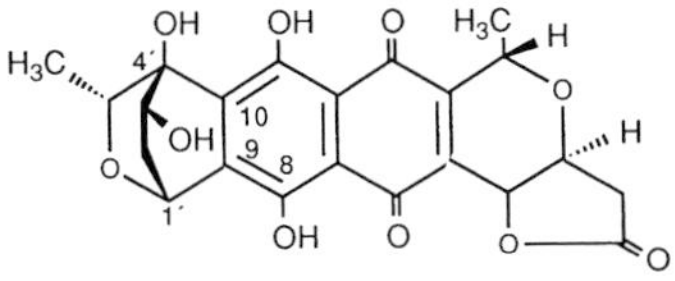

Granaticin

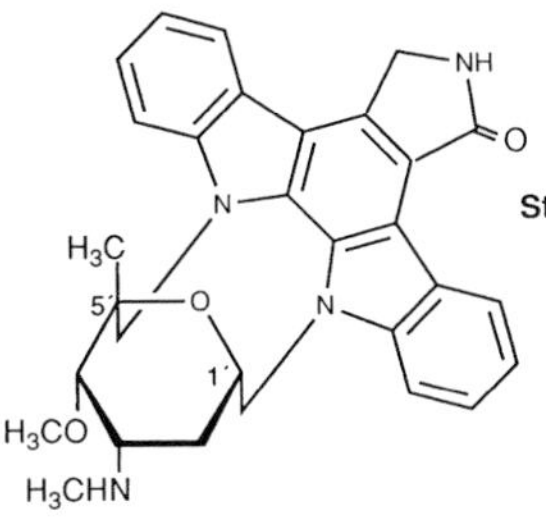

Staurosporine

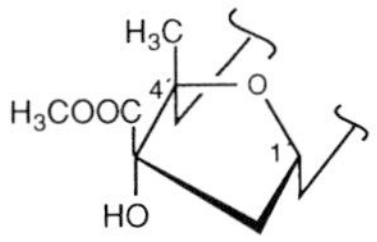

K-252a

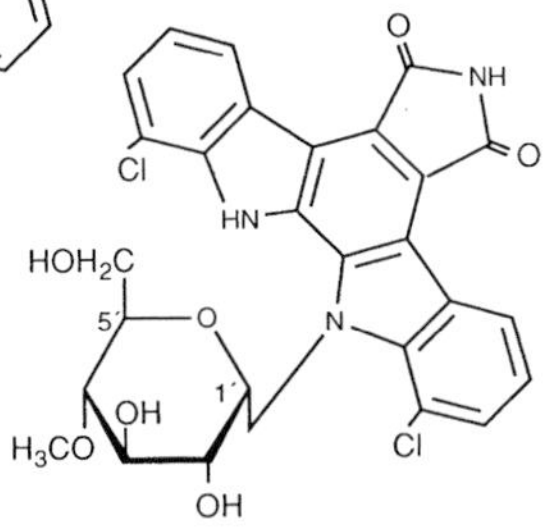

Rebeccamycin

Fig. A29. Granaticin, staurosporine, and staurosporine-related compounds. Unusual bonds in the linkage of 6DOH components in microbial secondary metabolites. C-glycosidic binding in the polyketide granaticin and N-glycosidic binding in the indolocarbazole alkaloids staurosporine, K-252a (same aglycone as staurosporine), and rebeccamycin which are all produced by actinomycetes.

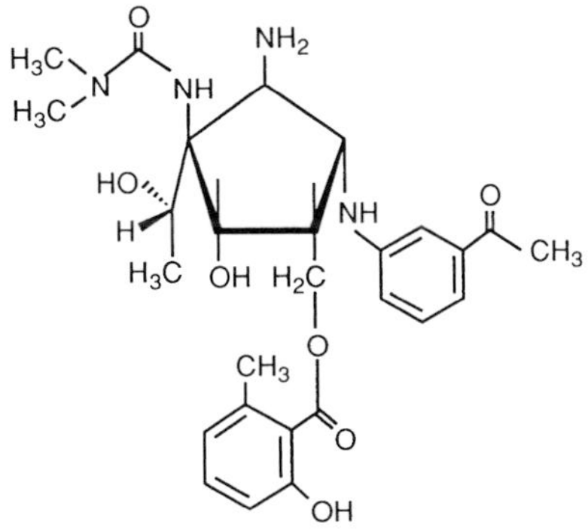

Fig. A30. Pactamycin; Ca?

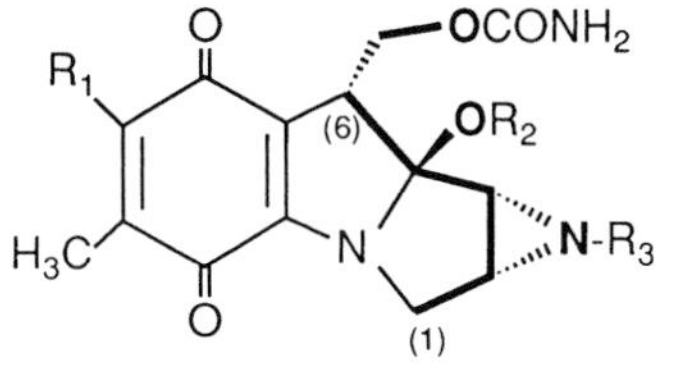

	R_1	R_2	R_3
mitomycin A	OCH_3	CH_3	H
mitomycin B	OCH_3	H	CH_3
mitomycin C	NH_2	CH_3	H
porfiromycin	NH_2	CH_3	CH_3

(thick line, bold face = D-glucosamine unit)

Fig. A31. Mitomycins.

11 References

ANDO, O., SATAKE, H., ITOI, K., SATO A., NAKAJIMA, M., TAKAHASHI, S., HARUYAMA, H., OHKUMA, Y., KINOSHITA, T., ENOKITA, R. (1991), Trehazolin, a new trehalase inhibitor, *J. Antibiot.* **44**, 1165–1168.

ANGELL, S., SCHWARZ, E., BIBB, M. (1992), The glucose kinase gene of *Streptomyces coelicolor* A3(2): its nucleotide sequence, transcriptional analysis and role in glucose repression, *Mol. Microbiol.* **6**, 2833–2844.

ASANO, N., KATAYAMA, K., TAKEUCHI, M., FURUMOTO, T., KAMEDA, Y., MATSUI, K. (1989), Preparation of 3-amino-3-deoxy derivatives of trehalose and sucrose and their activities, *J. Antibiot.* **42**, 585–590.

ATSUMI, S., IINUMA, H., NOSAKA, C., UMEZAWA, K. (1990a), Biological activities of cyclophellitol, *J. Antibiot.* **43**, 1579–1585.

ATSUMI, S., UMEZAWA, K., IINUMA, H., NAGANAWA, H., NAKAMURA, H., IITAKA, Y., TAKEUCHI, T. (1990b), Production, isolation and structure determination of a novel β-glucosidase inhibitor, cyclophellitol, from *Phellinus* sp., *J. Antibiot.* **43**, 49–53.

AUBERT-PIVERT, E., DAVIES, J. (1994), Biosynthesis of butirosin in *Bacillus circulans* NRRL B3312: identification by sequence analysis and insertional mutagenesis of the *butB* gene involved in antibiotic production, *Gene* **147**, 1–11.

AWATA, M., MUTO, N., HAYASHI, M., YAGINUMA, S. (1986), A new aminoglycoside antibiotic, substance AC4437, *J. Antibiot.* **39**, 724–726.

BAKKER, E. P. (1992), Aminoglycoside and aminocyclitol antibiotics: hygromycin B is an atypical bactericidal compound that exerts effects on cells of *Escherichia coli* characteristic for bacteriostatic aminocyclitols, *J. Gen Microbiol.* **138**, 563–569.

BASAK, K., MAJUMDAR, S. K. (1973), Utilization of carbon and nitrogen sources by *Streptomyces kanamyceticus* for kanamycin production, *Antimicrob. Agents Chemother.* **4**, 6–11.

BASCARAN, V., SANCHEZ, L., HARDISSON, C., BRANA, A. F. (1991), Stringent response and initiation of secondary metabolism in *Streptomyces clavuligerus, J. Gen. Microbiol.* **137**, 1625–1634.

BEAU, J.-M. (1990), Polyene macrolides: stereostructural elucidation and synthetic studies of a few members, in: *Recent Progress in the Chemical Synthesis of Antibiotics* (LUKACS, G., OHNO, M., Eds.), pp. 135–182. Berlin: Springer-Verlag.

BENVENISTE, R., DAVIES, J. (1973), Aminoglycoside antibiotic inactivating enzymes in actinomycetes similar to those present in clinical isolates of antibiotic resistant bacteria, *Proc. Natl. Acad. Sci. USA* **70**, 2276–2280.

BEYER, S., DISTLER, J., PIEPERSBERG, W. (1996), The *str* gene cluster for the biosynthesis of 5′-hydroxystreptomycin in *Streptomyces glaucescens* GLA.O (ETH 22794): new operons and evidence for pathway-specific regulation by StrR, *Mol. Gen. Genet.* **250**, 775–784.

BILLINGTON, D. C. (1993), *The Inosotol Phosphates.* Weinheim: VCH.

BLAZQUEZ, J., DAVIES, J., MORENO, F. (1991), Mutations in the *aphA-2* gene of transposon Tn5 mapping with the regions highly conserved in aminoglycoside-phosphotransferases strongly reduce aminoglycoside resistance, *Mol. Microbiol.* **5**, 1511–1518.

BRAHME, N. M., GONZALEZ, J. E., ROLLS, J. P., HESSLER, E. J., MIZSAK, S., HURLEY, L. H. (1984a), Biosynthesis of the Lincomycins. 1. Studies using stable isotopes on the biosynthesis of the propyl- and ethyl-L-hygric acid moieties of lincomycins A and B, *J. Am. Chem. Soc.* **106**, 7873–7878.

BRAHME, N. M., GONZALEZ, J. E., MIZSAK, S., ROLLS, J. R., HESSLER, E. J., HURLEY, L. H. (1984b), Biosynthesis of the Lincomycins. 2. Studies using stable isotopes on the biosynthesis of the methylthiolincosaminide moiety of Lincomycin A, *J. Am. Chem. Soc.* **106**, 7878–7883.

BRILL, G. M., MCALPINE, J. B., WHITTERN, D. N., BUKO, A. M. (1990), Altromycins, novel pluramycin-like antibiotics. II. Isolation and elucidation of structure, *J. Antibiot.* **43**, 229–237.

BRINK, M. F., BRINK, G., VERBEET, M. P., DEBOER, H. A. (1994), Spectinomycin interacts specifically with the residues G1064 and C1192 in 16S rRNA, thereby potentially freezing this molecule into an inactive conformation, *Nucleic Acids Res.* **22**, 325–331.

BRYAN, L. E., KWAN, S. (1983), Roles of ribosomal binding, membrane potential, and electron transport in bacterial uptake of streptomycin and gentamicin, *Antimicrob. Agents Chemother.* **23**, 835–845.

BUSSE, H.-J., WÖSTMANN, K., BAKKER, E. P. (1992), The bactericidal action of streptomycin: membrane permeabilization caused by the insertion of mistranslated proteins into the cytoplasmic membrane of *Escherichia coli* and subsequent caging of the antibiotic inside the cells due to degradation of the protein, *J. Gen Microbiol.* **138**, 551–556.

BUTTNER, M. J. (1989), RNA polymerase heterogeneity in *Streptomyces coelicolor* A3(2), *Mol. Microbiol.* **3**, 1653–1659.

CHADWICK, D. J., WHELAN, J. (Eds.) (1992), *Secondary Metabolites: Their Function and Evolu-*

tion, Ciba Foundation Symposium 171. Chichester: John Wiley & Sons.

CHAMPNESS, W. C., CHATER, K. F. (1994), Regulation and interaction of antibiotic production and morphological differentiation in *Streptomyces* spp., in: *Regulation of Bacterial Differentiation* (PIGGOT, P., MORAN JR., C. P., YOUNGMAN, P., Eds.), pp. 61–94. Washington, DC: American Society for Microbiology.

CHATER, K. F. (1989), Multilevel regulation of *Streptomyces* differentiation, *Trends Genet.* **5**, 372–377.

CHATER, K. F. (1992), Genetic regulation of secondary metabolic pathways in *Streptomyces,* in: *Secondary Metabolites: Their Function and Evolution* (DAVIES, J., CHADWICK, D., WHELAN, J., WIDDOWS, K., Eds.), pp. 144–156. Chichester: John Wiley & Sons.

CHEN, Y.-M., WALKER, J. B. (1977), Transaminations involving keto- and amino-inositols and glutamine in actinomycetes which produce gentamicin and neomycin, *Biochem. Biophys. Res. Commun.* **77**, 688–692.

CHIAO, J. S., XIA, T. H., MEI, B. G., JIN, Z. K., GU, W. L. (1995), Rifamycin SV and related ansamycins, in: *Biochemistry and Genetics of Antibiotic Biosynthesis* (VINING, L., STUTTARD, C., Eds.), pp. 477–498. Boston, MA: Butterworth-Heinemann.

CHUNG, S. T., CROSE, L. L. (1990), Transposon *Tn*4556-mediated DNA insertion and site-directed mutagenesis; in: *Proc. GIM 1990,* Vol. I (HESLOT, H., DAVIES, J., FLORENT, J., BOBICHON, L., DURAND, G., PENASSE, L., Eds.), pp. 207–218. Paris: Société Française de Microbiologie.

COLEMAN, R. S., FRASER, J. R. (1993), Acylketene [4+2]cycloadditions: divergent *de novo* synthesis of 2,6-dideoxy sugars, *J. Org. Chem.* **58**, 385–392.

COOK, G. R., BEHOLZ, L. G., STILLE, J. R. (1994), Construction of hydroxylated alkaloids (+/−)-mannonolactam, (+)-deoxymannojirimycin, and (+/-)-prosopinine through aza-annulation, *J. Org. Chem.* **59**, 3575–3584.

COURVALIN, P. (1994), Transfer of antibiotic resistance genes between gram-positive and gram-negative bacteria, *Antimicrob. Agents Chemother.* **38**, 1447–1451.

COURVALIN, P., WEISBLUM, B., DAVIES, J. (1977), Aminoglycoside-modifying enzyme of an antibiotic-producing bacterium acts as a determinant of antibiotic resistance in *Escherichia coli, Proc. Natl. Acad. Sci. USA* **74**, 999–1003.

CRAMERI, R., DAVIES, J. E. (1986), Increased production of aminoglycosides associated with amplified antibiotic resistance genes, *J. Antibiot.* **39**, 128–135.

CUNDLIFFE, E. (1989), How antibiotic-producing organisms avoid suicide, *Ann. Rev. Microbiol.* **43**, 207–233.

CUNDLIFFE, E. (1990), Recognition sites for antibiotics within rRNA, in: *The Ribosome. Structure, Function, and Evolution* (HILL, W. E., DAHLBERG, A., GARRETT, R. A., MOORE, P. B., SCHLESSINGER, D., WARNER, J. R., Eds.), pp. 479–490. Washington, DC: American Society for Microbiology.

CUNDLIFFE, E. (1992a), Self-protection mechanisms in antibiotic producers, in: *Secondary Metabolites: Their Function and Evolution, Ciba Foundation Symposium 171* (CHADWICK, D, WHELAN, J., Eds.), pp. 199–214. Chichester: John Wiley & Sons.

CUNDLIFFE, E. (1992b), Glycosylation of macrolide antibiotics in extracts of *Streptomyces lividans, Antimicrob. Agents Chemother.* **36**, 348–352.

DAIRI, T., HASEGAWA, M. (1989), Common biosynthetic feature of fortimicin-group antibiotics, *J. Antibiot.* **42**, 934–943.

DAIRI, T., OHTA, T., HASHIMOTO, E., HASEGAWA, M. (1992a), Self cloning in *Micromonospora olivasterospora* of *fms* genes for fortimicin A (astromicin) biosynthesis, *Mol. Gen. Genet.* **232**, 262–270.

DAIRI, T., OHTA, T., HASHIMOTO, E., HASEGAWA, M. (1992b), Organization and nature of fortimicin A (astromicin) biosynthetic genes studied using a cosmid library of *Micromonospora olivasterospora* DNA, *Mol. Gen. Genet.* **236**, 39–48.

DAIRI, T., YAMAGUCHI, K., HASEGAWA, M. (1992c), N-formimidoyl fortimicin A synthase, a unique oxidase involved in fortimicin A biosynthesis: purification, characterization and gene cloning, *Mol. Gen. Genet.* **236**, 49–59.

DAVIES, J. (1992), Another look at antibiotic resistance, *J. Gen. Microbiol.* **138**, 1553–1559.

DAVIES, J., SMITH, D. I. (1978), Plasmid-determined resistance to antimicrobial agents, *Ann. Rev. Microbiol.* **32**, 469–518.

DAVIES, J., YAGISAWA, M. (1983), The aminocyclitol glycosides (aminoglycosides), in: *Biochemistry and Genetic Regulation of Commercially Important Antibiotics* (VINING, L. C., Ed.), pp. 311–354. London: Addison-Wesley Publishing Company.

DAVIES, J., CANNON, M., MAUER, M. B. (1988), Myomycin: mode of action and mechanism of resistance, *J. Antibiot.* **41**, 366–372.

DAVIS, B. D. (1987), Mechanism of bactericidal action of aminoglycosides, *Microbiol. Rev.* **51**, 341–350.

DAVIS, B. D., CHEN, L., TAI, P. C. (1986), Misread protein creates membrane channels: an essential step in the bactericidal action of aminoglycosides, *Proc. Natl. Acad. Sci. USA* **83**, 6164–6168.

DE STASIO, E. A., MOAZED, D., NOLLER, H. F., DAHLBERG, A. E. (1989), Mutations in 16S ribosomal RNA disrupt antibiotic-RNA interactions, *EMBO J.* **8,** 1213–1216.

DECKER, H., HAAG, G., UDVARNORKI, G., ROHR, J. (1995), Novel genetically engineered tetracenomycins, *Angew. Chem.* (Int. Ed Engl.) **34**, 1107–1110.

DEGWERT, U., VAN HÜLST, R., PAPE, H., HERROLD, R. E., BEALE, J. M., KELLER, P. J., LEE, J. P., FLOSS, H. G. (1987), Studies on the synthesis of the α-glucosidase inhibitor acarbose: valienamine, a m-C_7N unit not derived from the shikimate pathway, *J. Antibiot.* **40**, 855–861.

DEMAIN A. L. (1989), Carbon source regulation of idiolite biosynthesis in actinomycetes, in: *Regulation of Secondary Metabolism in Actinomycetes* (SHAPIRO, S., Ed.), pp. 127–134. Boca Raton, FL: CRC Press.

DEMAIN, A. L., INAMINE, E. (1970), Biochemistry and regulation of streptomycin and mannosidostreptomycinase (α-D-mannosidase) formation, *Bacteriol. Rev.* **34**, 1–19.

DICKENS, M. L., YE, J., STROHL, W. R. (1996), Cloning, sequencing and analysis of aklaviketone reductase from *Streptomyces* sp. strain C5, *J. Bacteriol.* **178**, 3384–3388.

DISTLER, J., PIEPERSBERG, W. (1985), Cloning and characterization of a gene from *Streptomyces griseus* coding for a streptomycin-phosphorylating activity, *FEMS Microbiol. Lett.* **28**, 113–117.

DISTLER, J., KLIER, K., PIENDL, W., WERBITZKI, O., BÖCK, A., KRESZE, G., PIEPERSBERG, W. (1985), Streptomycin biosynthesis in *Streptomyces griseus.* I. Characterization of streptomycin-idiotrophic mutants, *FEMS Microbiol. Lett.* **30**, 145–150.

DISTLER, J., BRAUN, C., EBERT, A., PIEPERSBERG, W. (1987a), Gene cluster for streptomycin biosynthesis in *Streptomyces griseus*: Analysis of a central region including the major resistance gene, *Mol. Gen. Genet.* **208**, 204–210.

DISTLER, J., EBERT, A., MANSOURI, K., PISSOWOTZKI, K., STOCKMANN, M., PIEPERSBERG, W. (1987b), Gene cluster for streptomycin biosynthesis in *Streptomyces griseus*: nucleotide sequence of three genes and analysis of transcriptional activity, *Nucleic Acids Res.* **15**, 8041–8056.

DISTLER, J., MAYER, G., PIEPERSBERG, W. (1990), Regulation of biosynthesis of streptomycin, in: *Proc. GIM 1990,* Vol. 1 (HESLOT, H., DAVIES, J., FLORENT, J., BOBICHON, L., DURAND, G., PENASSE, L., Eds.), pp. 379–392. Paris: Société Française de Microbiologie.

DISTLER, J., MANSOURI, K., MAYER, G., STOCKMANN, M., PIEPERSBERG, W. (1992), Streptomycin production and its regulation, *Gene* **115**, 105–111.

DREPPER, A., PAPE, H. (1996), Acarbose 7-phosphotransferase from *Actinoplanes* sp.: purification, properties, and possible physiological function, *J. Antibiot.* **49**, 664–668.

EMBLEY, T. M., STACKEBRANDT, E. (1994), The molecular phylogeny and systematics of the actinomycetes, *Ann. Rev. Microbiol.* **48**, 257–289.

ENGLUND, P. T. (1993), The structure and biosynthesis of glycosyl phosphatidylinositiol protein anchors, *Annu. Rev. Biochem.* **62**, 112–138.

ENSIGN, J. C. (1988), Physiological regulation of sporulation of *Streptomyces griseus,* in: *Biology of Actinomycetes '88* (OKAMI, Y., BEPPU, T., OGAWARA, H., Eds.), pp. 309–315. Tokyo: Japan Scientific Societies Press.

ETIENNE, G., ARMAU, E., DASSIN, M., TIRABY, G. (1991), A screening method to identify antibiotics of the aminoglycoside family, *Antibiot.* **44**, 1357–1366.

FISHER, S. H. (1992), Glutamine synthesis in *Streptomyces* – a review, *Gene* **115**, 13–17.

FLORES, M. E., PONCE, E., RUBIO, M., HUITRÓN, C. (1993), Glucose and glycerol repression of α-amylase in *Streptomyces kanamyceticus* and isolation of deregulated mutants, *Biotechnol. Lett.* **15**, 595–600.

FLOSS, H. G. (in press), Biosynthesis of Antibiotics, in: *Proc. 6th Conf. Genetics and Molecular Biology of Industrial Microorganisms,* Bloomington, IN, Oct. 1996 (HEGEMANN, G. D; Baltz, R. H., BALTZ, R. H., Skatrund, P. L., Eds.).

FLOSS, H. G., BEALE, J. M. (1989), Investigation of the biosynthesis of antibiotics, *Angew. Chem. (Int. Edn.)* **28**, 146–177.

FOSTER, T. J. (1983), Plasmid-determined resistance to antimicrobial drugs and toxic metal ions in bacteria, *Microbiol. Rev.* **47**, 361–409.

FRAIMOW, H. S., GREENMAN, J. B., LEVITON, I. M., DOUGHERTY, T. J., MILLER, M. H. (1991), Tobramycin uptake in *Escherichia coli* is driven by either electrical potential or ATP, *J. Bacteriol.* **173**, 2800–2808.

FRENCH, J. C., BARTZ, Q. R., DION, H. W. (1973), Myomycin, a new antibiotic, *J. Antibiot.* **26**, 272–283.

FUJITA, Y., SHINDO, K., MIWA, Y., YOSHIDA, K. I. (1992), *Bacillus subtilis* inositol dehydrogenase-encoding gene (*idh*): sequencing and expression in *Escherichia coli, Gene* **108**, 121–125.

GALE, E. F., CUNDLIFFE, E., REYNOLDS, P. E., RICHMOND, M. H., WARING, M. J. (1981), *The Molecular Basis of Antibiotic Action,* 2nd Edn. London: Wiley Interscience.

GODA, S. K., AKHTAR, M. (1992), Neomycin biosynthesis: the incorporation of D-6-deoxyglucose derivatives and variously labelled glucose into the 2-deoxystreptamine ring. Postulated involvement of 2-deoxyinosose synthase in the biosynthesis, *J. Antibiot.* **45**, 984–994.

GOEKE, K. (1986), Enzymatische Untersuchungen zum Zuckerstoffwechsel und zur Biosynthese des α-Glucosidase-Inhibitors Acarbose bei *Actinoplanes* spec., *Dissertation,* University of Münster.

GORINI, L. (1974), Streptomycin and misreading of the genetic code, in: *Ribosomes* (NOMURA, M., TISSIERES, A., LENGYEL, P., Eds.), pp. 791–803. New York: Cold Spring Harbor Laboratory.

GOULD, S. J. (1992), Exploring the intricate details of antibiotic biosynthesis, in: *Secondary-Metabolite Biosynthesis and Metabolism* (PETROSKI, R. J., MCCORMICK, S. P., Eds.), pp. 11–25. New York: Plenum Press.

GRÄFE, U. (1992), *Biochemie der Antibiotika.* Heidelberg: Spektrum Akademischer Verlag.

GRISEBACH, H. (1978), Biosynthesis of sugar components of antibiotic substances, *Adv. Carbohydr. Chem. Biochem.* **35**, 81–126.

GUILFOILE, P. G., HUTCHINSON, C. R. (1991), A bacterial analogue of the *mdr* gene of mammalian tumor cells is present in *Streptomyces peucetius,* producer of daunorubicin and doxorubicin, *Proc. Natl. Acad. Sci. USA* **88**, 8553–8557.

HANCOCK, R. E. W. (1981a), Aminoglycoside uptake and mode of action with special reference to streptomycin and gentamicin. I. Antagonists and mutants, *J. Antimicrob. Chemother.* **8**, 249–276.

HANCOCK, R. E. W. (1981b), Aminoglycoside uptake and mode of action with special reference to streptomycin and gentamicin. II. Effects of aminoglycosides on cells, *J. Antimicrob. Chemother.* **8**, 429–445.

HANCOCK, R. E. W., FARMER, S. W., LI, Z., POOLE, K. (1991), Interaction of aminoglycosides with outer membranes and purified lipopolysaccharide and OmpF porin of *Escherichia coli, Antimicrob. Agents Chemother.* **35**, 1309–1314.

HARDICK, D. J., HUTCHINSON, D. W., TREW, S. J., WELLINGTON, E. M. H. (1992), Glucose is a precursor of 1-deoxynojirimycin and 1-deoxymannonojirimycin in *Streptomyces subrutilus, Tetrahedron* **48**, 6285–6296.

HASEGAWA, M. (1991), A gene cloning system in *Micromonospora* which revealed the organization of biosynthetic genes of fortimicin A (astromicin), *Actinomycetol.* **5**, 126–131.

HASEGAWA, M. (1992), A novel, highly efficient gene-cloning system in *Micromonospora* applied to the genetic analysis of fortimicin biosynthesis, *Gene* **115**, 85–91.

HEINZEL, P., WERBITZKY, O., DISTLER, J., PIEPERSBERG, W. (1988), A second streptomycin resistance gene from *Streptomyces griseus* codes for streptomycin-3″-phosphotransferase. Relationship between antibiotic and protein kinases, *Arch. Microbiol.* **150**, 184–192.

HILL, W. E., DAHLBERG, A., GARRETT, R. A., MOORE, P. B., SCHLESSINGER, D., WARNER, J. R. (1990), *The Ribosome. Structure, Function, and Evolution.* Washington, DC: American Society for Microbiology.

HIROSE-KUMAGAI, A., YAGITA, A., AKAMATSU, N. (1982), UDP-*N*-methyl-D-glucoseamine-phosphate. A possible intermediate of *N*-methyl-L-glucosamine moiety of streptomycin, *A. Antibiot* **35**, 1571–1577.

HOLMES, D. J., DROCOURT, D., TIRABY, G., GUNDIFFE, E. (1991), Cloning of an aminoglycoside resistance-endoding gene, *kamC*, from *Saccharopolispora hirsuta:* Comparison with *kamB* from *Streptomyces tenebrarius, Gene* **102**, 19–26.

HOLT, T. G., CHANG, C., LAURENTWINTER, C., MURAKAMI, T., GARRELS, J. I., DAVIES, J. E., THOMPSON, C. J. (1992), Global change in gene expression related to antibiotic synthesis in *Streptomyces hygroscopicus, Mol. Microbiol.* **6**, 969–980.

HONG, S. K., MATSUMOTO, A., HORINOUCHI, S., BEPPU, T. (1993), Effects of protein kinase inhibitors on *in vitro* protein phosphorylation and cellular differentiation of *Streptomyces griseus, Mol. Gen. Genet.* **236**, 347–354.

HOPWOOD, D. A., SHERMAN, D. H., KHOSLA, C., BIBB, M. J., SIMPSON, T. J., FERNANDEZ-MORENO, M. A., MARTINEZ, E., MALPARTIDA, F. (1990), "Hybrid" pathways for the production of secondary metabolites, in: *Proc. 6th Int. Symp. Genet. Ind. Microorg.* (HESLOT, H., DAVIES, J., FLORENT, J., BOBICHON, L., DURAND, G., PENEASSE, L., Eds.), pp. 259–270. Paris: Société Française de Microbiologie.

HORINOUCHI, S., BEPPU, T. (1992), Regulation of secondary metabolism and cell differentiation in *Streptomyces*: A-factor as a microbial hormone and the AfsR protein as a component of a two-component regulator system, *Gene* **115**, 167–172.

HORINOUCHI, S., BEPPU, T. (1994), A-factor as a microbial hormone that controls cellular differentiation and secondary metabolism in *Streptomyces griseus, Mol. Microbiol.* **12**, 859–864.

HORNEMANN, U., KEHRER, J. P., NUNEZ, C. S., RANIERI, A. L. (1974), D-Glucosamine and L-citrulline, precursors in mitomycin biosynthesis in *Streptomyces verticillatus, J. Am. Chem. Soc.* **96**, 320–322.

HORTON, D., HAWKINS, L. D., MCGARVEY, G. J. (1989), Trends in synthetic carbohydrate chemistry, *ACS Symp. Ser.* **386**. Washington, DC: American Chemical Society.

HOTTA, K., ISHIKAWA, J., ICHIHARA, M., NAGANAWA, H., MIZUNO, S. (1988), Mechanism of increased kanamycin-resistance generated by protoplast regeneration of *Streptomyces griseus.* I. cloning of a gene segment directing a high level of an aminoglycoside 3-*N*-acetyltransferase activity, *J. Antibiot.* **41**, 94–103.

HOTTA, K., MORIOKA, M., OKAMI, Y. (1989), Biosynthetic similarity between *Streptomyces tenjimariensis* and *Micromonospora olivasterospora* which produce fortimicin-group antibiotics, *J. Antibiot.* **42**, 745–751.

HOTTA, K., DAVIES, J., YAGISAWA, M. (1995), Aminoglycosides and aminocyclitols (other than streptomycin), in: *Biochemistry and Genetics of Antibiotic Biosynthesis* (VINING, L., STUTTARD, Eds.), pp. 571–595. Stoneham: Butterworth-Heinemann.

HUDLICKY, T., CEBULAK, M. (1993), *Cyclitols and Their Derivatives.* Weinheim: VCH.

HUTCHIN, T., CORTOPASSI, G. (1994), Proposed molecular and cellular mechanism for aminoglycoside ototoxicity, *Antimicrob. Agents Chemother.* **38**, 2517–2520.

ICHIMURA, M., KOGUCHI, T., YASUZAWA, T., TOMITA, F. (1987), CV-1, a new antibiotic produced by a strain of *Streptomyces* sp. I. Fermentation, isolation and biological properties of the antibiotic, *J. Antibiot.* **40**, 723–726.

IKEDA, Y., GOMI, S., YOKOSE, K., NAGANAWA, H., IKEDA, T., MANABE, M., HAMADA, M., KONDO, S., UMEZAWA, H. (1985a), A new streptomycin group antibiotic produced by *Streptomyces sioyaensis, J. Antibiot.* **38**, 1803–1805.

IKEDA, Y., KONDO, S., KANAI, F., SAWA, T., HAMADA, M., TAKEUCHI, T., UMEZAWA, H. (1985b), A new destomycin-family antibiotic produced by *Saccharopolyspora hirsuta, J. Antibiot.* **38**, 436–438.

INAMINE, E., DEMAIN, A. L. (1975), Mannosidostreptomycin hydrolase, *Methods Enzymol.* **43**, 637–640.

INOUYE, S., SHOMURA, T., WATANABE, H., TOTSUGAWA, K., NIIDA, T. (1973), Isolation and gross structure of a new antibiotic SS-56C and related compounds, *J. Antibiot.* **26**, 374–385.

ITOH, S., ODAKURA, Y., KASE, H., SATHO, S., TAKAHASHI, K., IIDA, T., SHIRAHATA, K., NAKAYAMA, K. (1984), Biosynthesis of astromicin and related antibiotics. I. Biosynthetic studies by bioconversion experiments, *J. Antibiot.* **37**, 1664–1669.

KAGAN, R. M., CLARKE, S. (1994), Widespread occurrence of three sequence motifs in diverse S-adenosylmethionine-dependent methyltransferases suggests a common structure for these enzymes, *Arch. Biochem. Biophys.* **310**, 417–427.

KAKINUMA, K. (1982), Biosynthesis of ribostamycin. Application of the deuterium label, in: *Trends in Antibiotic Research. Genetics, Biosyntheses, Actions, and New Substances* (UMEZAWA, H., DEMAIN, A. L., HATA, T., HUTCHINSON, C. R., Eds.), pp. 185–194. Tokyo: Japan Antibiotics Research Association.

KAMEDA, Y., ASANO, N., YOSHIKAWA, M., TAKEUCHI, M., YAMAGUCHI, T., MATSUI, K., SATOSHI, H., FUKASE, H. (1984), Valiolamine, a new α-glucosidase inhibiting aminocyclitol produced by *Streptomyces hygroscopicus, J. Antibiot.* **37**, 1301–1307.

KARWOWSKI, J. P., JACKSON, M., BOBIK, T. A., PROKOP, J. F., THERIAULT, R. J. (1984), Spenolimycin, a new spectinomycin-type antibiotic. I. Discovery, taxonomy and fermentation, *J. Antibiot.* **37**, 1513–1518.

KASE, H., ODAKURA, Y., TAKAZAWA, Y., KITAMURA, S., NAKAYAMA, K. (1982), Biosynthesis of sagamicin and related aminoglycosides, in: *Trends in Antibiotic Research. Genetics, Biosyntheses, Actions, and New Substances* (UMEZAWA, H., DEMAIN, A. L., HATA, T., HUTCHINSON, C. R., Eds.), pp. 195–212. Tokyo: Japan Antibiotics Research Association.

KASE, H., IWAHASHI, K., MATSUDA, Y. (1986), K-252a, a potent inhibitor of protein kinase C of microbial origin, *J. Antibiot.* **39**, 1059–1065.

KATZ, L., DONADIO, S. (1993), Polyketide synthesis: prospects for hybrid antibiotics, *Ann. Rev. Microbiol.* **47**, 875–912.

KENNEDY, J. F. (1988), *Carbohydrate Chemistry.* Oxford: Clarendon Press.

KHOKHLOV, A. S. (1988), Results and perspectives of actinomycete autoregulators studies, in: *Biology of Actinomycetes '88* (OKAMI, Y., BEPPU, T., OGAWARA, H., Eds.), pp. 338–345. Tokyo: Japan Scientific Societies Press.

KHOKHLOV, A. S., TOVAROVA, I. I., BORISOVA, L. N., PLINER, S. A., SHEVCHENKO, L. A., KORNITSKAYA, E. Y., IVKINA, N. S., RAPOPORT, I. A. (1967), A-factor responsible for the biosynthesis of streptomycin by a mutant strain of *Actinomyces streptomycini, Doklady Akad. Nauk SSSR* **177**, 232–235.

KIM, C.-G., KIRSCHNING, A., BERGON, P., ZHOU, P., SU, E., SAUERBREI, B., NING, S., AHN, Y., BREUER, M., LEISTNER, E., FLOSS, H. G. (1996), Biosynthesis of 3-amino-5-hydroxybenzoic acid, the precursor of mC7N units in ansamycin antibiotics, *J. Am. Chem. Soc.* **118**, 7486–7491.

KIRBY, R. (1990), Evolutionary origin of aminoglycoside phosphotransferase resistance genes, *J. Mol. Evol.* **30**, 489–492.

KNIEP, B., GRISEBACH, H. (1976), Enzymatic synthesis of streptomycin. Transfer of L-dihydrostreptose from dTDP-L-dihydrostreptose to streptidine-6-phosphate, *FEBS Lett.* **65**, 44–46.

KNIEP, B., GRISEBACH, H. (1980), Biosynthesis of streptomycin. Purification and properties of a dTDP-L-dihydrostreptose:streptidine-6-phosphate dihydrostreptosyltransferase from *Streptomyces griseus, Eur. J. Biochem.* **105**, 139–144.

KNIGHTON, D. R., ZENG, J., TENEYCK, L. F., ASHFORD, V. A., XUONG, N. H., TAYLOR, S. S. (1991), Crystal structure of the catalytic subunit of cyclic adenosine monophosphate-dependent protein kinase, *Science* **253**, 407–414.

KOBAYASHI, Y., SHIOZAKI, M. (1994), Synthesis of trehazolin β-anomer, *J. Antibiot.* **47**, 243–246.

KOGA, Y., NISHIHARA, M., MORII, H., AKAGAWA-MATSUSHITA, M. (1993), Ether lipids of methanogenic bacteria: structures, comperative aspects, and biosynthesis, *Microbiol. Rev.* **57**, 164–182.

KOHLHEPP, S. J., HOU, L., GILBERT, D. N. (1994), Pig kidney (LLC-PK1) cell membrane fluidity during exposure to gentamicin or tobramycin, *Antimicrob. Agents Chemother.* **38**, 2169–2171.

KORZYBSKI, T., KOWSZYK-GINDIFER, Z., KURYLOWICZ, W. (1978), *Antibiotics. Origin, Nature, and Properties.* Washington, DC: American Society for Microbiology.

KROHN, K., KIRST, H. A., MAAG, H. (1993), *Antibiotics and Antiviral Compounds. Chemical Synthesis and Modification.* Weinheim: VCH.

KUMADA, Y., HORINOUCHI, S., UOZUMI, T., BEPPU, T. (1986), Cloning of a streptomycin-production gene directing synthesis of *N*-methyl-L-glucosamine, *Gene* **43**, 221–224.

KUO, M. S., YUREK, D. A., COATS, J. H., LI, G. P. (1989), Isolation and identification of 7,8-didemethyl-8-hydroxy-5-deazariboflavin, an unusual cosynthetic factor in streptomycetes from *Streptomyces lincolnensis, J. Antibiot.* **42**, 475–478.

KUO, M. S., YUREK, D. A., COATS, J. H., CHUNG, S. T., LI, G. P. (1992), Isolation and identification of 3-propylidene-Δ'-pyrroline-5-carboxylic acid, a biosynthetic precursor of lincomycin, *J. Antibiot.* **45**, 1773–1777.

KWAKMAN, M., POSTMA, W. (1994), Glucose kinase has a regulatory role in carbon catabolite repression in *Streptomyces coelicolor, J. Bacteriol.* **176**, 2694–2698.

LACALLE, R. A., TERCERO, J. A., JIMENEZ, A. (1992), Cloning of the complete biosynthetic gene cluster for an aminonucleoside antibiotic, puromycin, and its regulated expression in heterologous host, *EMBO J.* **11**, 785–792.

LAM, K. S., FLORENZA, S., SCHROEDER, D. R., DOYLE, T. W., PEARCE, C. J. (1989), Biosynthesis of rebeccamycin, a novel antitumor agent, in: *Novel Microbial Products for Medicine and Agriculture* (DEMAIN, A. L., SOMKUTI, G. A., HUNTER-CREVA, J. C., ROSSMOORE, H. W., Eds.), pp. 63–66. Amsterdam: Elsevier Science Publishers.

LESKIW, B. K., BIBB, M. J., CHATER, K. F. (1991a), The use of a rare codon specifically during development, *Mol. Microbiol.* **5**, 2861–2867.

LESKIW, B. K., LAWLOR, E. J., FERNANDEZ-ABALOS, J. M., CHATER, K. F. (1991b), TTA codons in some genes prevent their expression in a class of developmental, antibiotic-negative *Streptomyces* mutants, *Proc. Natl. Acad. Sci. USA* **88**, 2461–2465.

LIRAS, P., ASTURIAS, J. A., MARTIN, J. F. (1990), Phosphate control sequences involved in transcriptional regulation of antibiotic biosythesis, *TIBTECH* **8**, 184–189.

LISHNEVSKAYA, E. B., KUZINA, Z. A., ASINOOVSKAYA, N. K., BELOUSOVA, I. I., MALKOV, M. A., RAVINSKAYA, A. YA. (1986), Cyclic adenosine-3',5'-monophosphoric acid in *Streptomyces antibioticus* and its possible role in the regulation of oleandomycin biosynthesis and culture growth, *Mikrobiologiya* **55**, 350.

LIU, H.-W., THORSON, J. S. (1994), Pathways and mechanisms in the biogenesis of novel deoxysugars by bacteria, *Ann. Rev. Microbiol.* **48**, 223–256.

LONDEI, P., ALTAMURA, S., SANZ, J. L., AMILS, R. (1988), Aminoglycoside-induced mistranslation in thermophilic archaebacteria, *Mol. Gen. Genet.* **214**, 48–54.

LOOK, G. C., FOTSCH, C. H., WONG, C.-H. (1993), Enzyme-catalysed organic synthesis: practical routes to aza sugars and their analogs for use as glycoprocessing inhibitors, *Acc. Chem. Res.* **26**, 182–190.

LORIAN, V. (Ed.) (1980), *Antibiotics in Laboratory Medicine.* Baltimore, MA: Williams & Wilkins.

LUCHER, L. A., CHEN, Y.-M., WALKER, J. B. (1989), Reactions catalysed by purified L-glutamine:keto-scyllo-inositol aminotransferase, an enzyme required for biosynthesis of aminocyclitol antibiotics, *Antimicrob. Agents Chemother.* **33**, 452–459.

LUKACS, G., OHNO, M. (1990), *Recent Progress in the Chemical Synthesis of Antibiotics.* Berlin: Springer-Verlag.

MACNEIL, D. J. (1995), Avermectins, in: *Biochemistry and Genetics of Antibiotic Biosynthesis* (VINING, L., STUTTARD, C., Eds.), pp. 421–442. Boston, MA: Butterworth-Heinemann.

MAIER, S., GRISEBACH, H. (1979), Biosynthesis of streptomycin. Enzymic oxidation of dihydrostreptomycin (6-phosphate) to streptomycin (6-phosphate) with a particulate fraction of *Streptomyces griseus, Biochim. Biophys. Acta* **586**, 231–241.

MALLAMS, A. K. (1988), The carbohydrate-containing antibiotics, in: *Carbohydrate Chemistry* (KENNEDY, J. F., Ed.), pp. 73–133. Oxford: Claendron Press.

MANSOURI, K., PIEPERSBERG, W. (1991), Genetics of streptomycin production in *Streptomyces griseus:* nucleotide sequence of five genes, *strFGHIK*, including a phosphatase gene, *Mol. Gen. Genet.* **228**, 459–469.

MANSOURI, K., PISSOWOTZKI, K., DISTLER, J., MAYER, G., HEINZEL, P., BRAUN, C., EBERT, A., PIEPERSBERG, W. (1989), Genetics of streptomycin production, in: *Genetics and Molecular Biology of Industrial Microorganisms* (HERSHBERGER, C. L., QUEENER, S. W., HEGEMAN, G., Eds.), pp. 61–67. Washington, DC: American Society for Microbiology.

MARSH, P., WELLINGTON, E. M. H. (1994), Molecular ecology of filamentous actinomycetes in soil, in: *Molecular Ecology of Rhizosphere Microorganisms. Biotechnology and the Release of GMOs* (O'GARA, F., DOWLING, D. N., BOESTEN, B., Eds.), pp. 133–149. Weinheim: VCH.

MARTIN, J. F. (1989), Molecular mechanisms for the control by phosphate of the biosynthesis of antibiotics and other metabolites, in: *Regulation of Secondary Metabolism in Actinomycetes* (SHAPIRO, S., Ed.), pp. 213–237. Boca Raton, FL: CRC Press.

MARTIN, J. F., DEMAIN, A. L. (1980), Control of antibiotic biosynthesis, *Microbiol. Rev.* **44**, 230–251.

MARTIN, J. F., LIRAS, P. (1989), Organization and expression of genes involved in the biosynthesis of antibiotics and other secondary metabolites, *Ann. Rev. Microbiol.* **43**, 173–206.

MAYER, G. (1994), Molekulare Analyse und Evolution von 5′-Hydroxystreptomycin- und Bluensomycin-Biosynthesegenen aus *Streptomyces glaucescens* GLA.0 und *Streptomyces bluensis* ISP 5564, *Dissertation,* Bergische Universität GH, Wuppertal.

MAYER, G., VÖGTLI, M., PISSOWOTZKI, K., HÜTTER, R., PIEPERSBERG, W. (1988), Colinearity of streptomycin production genes in two species of *Streptomyces.* Evidence for occurrence of a second amidino-transferase gene, *Mol. Gen.* (Life Sci. Adv.) **7**, 83–87.

MAZODIER, P., DAVIES, J. (1991), Gene transfer between distantly related bacteria, *Annu. Rev. Genet.* **25**, 147–171.

MCCUE, L. A., KWAK, J., BABCOCK, M. J., KENDRICK, K. F. (1992), Molecular analysis of sporulation in *Streptomyces griseus, Gene* **115**, 173–179.

MEHLING, A., WEHMEIER, U. F., PIEPERSBERG, W. (1995), Nucleotide sequence of *Streptomyces* 16S ribosomal DNA: towards a specific identification system for streptomycetes using PCR, *Microbiology* **141**, 2139–2147.

MILLER, A. L., WALKER, J. B. (1969), Enzymatic phosphorylation of streptomycin by extracts of streptomycin-producing strains of *Streptomyces, J. Bacteriol.* **99**, 401–405.

MIYAKE, K., HORINOUCHI, S., YOSHIDA, M., CHIBA, N., MORI, K., NOGAWA, N., MORIKAWA, N., BEPPU, T. (1989), Detection and properties of A-factor-binding protein from *Streptomyces griseus, J. Bacteriol.* **171**, 4298–4302.

MIYAKE, K., KUZUYAMA, T., HORINOUCHI, S., BEPPU, T. (1990), The A-factor-binding protein of *Streptomyces griseus* negatively controls streptomycin production and sporulation, *J. Bacteriol.* **172**, 3003–3008.

MOAZED, D., NOLLER, H. F. (1987), Interactions of antibiotics with 16S rRNA, *Nature* **327**, 389–394.

MÜLLER, L. (1989), Chemistry, biochemistry and therapeutic potential of microbial α-glucosidase inhibitors, in: *Novel Microbial Products for Medicine and Agriculture* (DEMAIN, A. L., SOMKUTI, G. A., HUNTER-CREVA, J. C., ROSSMOORE, H. W., Eds.), pp. 109–116. Amsterdam: Elsevier Science Publishers.

MÜLLER, A., KELLER-SCHIERLEIN, W., BIELECKI, J., RAK, G., STÜMPFEL, J., ZÄHNER, H. (1986), (2*S*,3*R*,4*R*,6*R*) - 2,3,4, - Trihydroxy - 6 - methylhexanon aus zwei Actinomyceten-Stämmen, *Helv. Chim. Acta* **69**, 1829–1832.

MUNRO, M. H. G., TANIGUCHI, M., RINEHART, K. L., GOTTLIEB, D., STOUDT, T. H., ROGERS, T. O. (1975), Carbon-13 evidence for the stereochemistry of streptomycin biosynthesis from glucose, *J. Am. Chem. Soc.* **97**, 4782–4783.

MURPHY, P. J., HEYCKE, N., BANFALVI, Z., TATE, M. E., DE BRUJIN, F., KONDOROSI, A., TEMPE, J., SCHELL, J. (1987), Genes for the catabolism and synthesis of an opine-like compound in *Rhizobium meliloti* are closely linked and on the Sym plasmid, *Proc. Natl. Acad. Sci. USA* **84**, 493–497.

MURPHY P. J., TRENZ, S. P., GRZEMSKI, W., DE BRUIJN, F. J., SCHELL, J. (1993), The *Rhizobium meliloti* rhizopine *mos* locus is a mosaic structure facilitating its symbiotic regulation, *J. Bacteriol.* **175**, 5193–5204.

NEUMANN, T., PIEPERSBERG, W., DISTLER, J. (1996), The decision phase model of the growth phase-dependent regulation of biosynthesis of streptomycin in *Streptomyces griseus, Microbiology* **142**, 1953–1963.

NEWTON, G. L., ARNOLD, K., PRICE, M. S., SHERRILL, C., DELCARDAYRE, S. B., AHARONOWITZ, Y., COHEN, G., DAVIES, J., FAHEY, R. C., DAVIS, C. (1996), Distribution of thiols in microorganisms: mycothiol is a major thiol in most actinomycetes, *J. Bacteriol.* **178**, 1990–1995.

NIERHAUS, K. H. (Ed.) (1993), *The Translational Apparatus.* New York: Plenum Press.

NIMI, O., ITO, G., SUEDA, S., NOMI, R. (1971), Phosphorylation of streptomycin at C6-OH of streptidine moiety by an intracellular enzyme of *Streptomyces griseus, Agr. Biol. Chem.* **35**, 848–855.

NOLLER, H. F. (1993), On the origin of the ribosome: Coevolution of subdomains of tRNA and rRNA, in: *The RNA World* (GESTLAND, R. F., ATKINS, J. F., Eds.), pp. 137–184. New York: Cold Spring Harbor Laboratory Press.

NUMATA, K., YAMAMOTO, H., HATORI, M., MIYAKI, T., KAWAGUCHI, H. (1986), Isolation of an aminoglycoside hypersensitive mutant and its application in screening, *J. Antibiot.* **39**, 994–1000.

OCHI, K. (1987), Metabolic initiation of differentiation and secondary metabolism by *Streptomyces griseus:* significance of the stringent response (ppGpp) and GTP content in relation to A-factor, *J. Bacteriol.* **169**, 3608–3616.

OCHI, K. (1988), Nucleotide pools and stringent response in regulation of *Streptomyces differentiation,* in: *Biology of Actinomycetes '88* (OKAMI, Y., BEPPU, T., OGAWARA, H., Eds.), pp. 330–337. Tokyo: Japan Scientific Societies Press.

OCHI, K. (1990), *Streptomyces griseus,* as an excellent object for studing microbial differentiation, *Actinomycetol.* **4**, 23–30.

ODAKURA, Y., KASE, H., ITOH, S., SATHO, S., TAKASAWA, S., TAKAHASHI, K., SHIRAHATA, K., NAKAYAMA, K. (1984), Biosynthesis of astromicin and related antibiotics. II. Biosynthetic studies with blocked mutants of *Micromonospora olivasteropora, J. Antibiot.* **37**, 1670–1680.

OGURA, H., HASEGAWA, A., SUAMI, T. (1992), *Carbohydrates – Synthetic Methods and Applications in Medicinal Chemistry.* Tokyo: Kodansha.

OHNUKI, T., IMANAKA, T., AIBA S. (1985a), Self-cloning in *Streptomyces griseus* of an *str* gene cluster for streptomycin biosynthesis and streptomycin resistance, *J. Bacteriol.* **164**, 85–94.

OHNUKI, T., IMANAKA, T., AIBA, S. (1985b), Isolation of streptomycin non-producing mutants deficient in biosynthesis of the streptidine moiety or linkage between streptidine-6-phosphate and dihydrostreptose, *Antimicrob. Agents Chemother.* **27**, 367–374.

OHTA, T., HASEGAWA, M. (1993a), Analysis of the self-defense gene (*fmrO*) of a fortimicin A (astromicin) producer, *Micromonospora olivasterospora:* comparison with other aminoglycoside-resistance-encoding genes, *Gene* **127**, 63–69.

OHTA, T., HASEGAWA, M. (1993b), Analysis of the nucleotide sequence of *fmrT* encoding the self-defense gene of the istamycin producer, *Streptomyces tenjimariensis* ATCC 31602; comparison with the sequences of *kamB* of *Streptomyces tenebrarius* NCIB 11028 and *kamC* of *Saccharopolyspora hirsuta* CL102, *J. Antibiot.* **46**, 511–517.

OHTA, T., HASHIMOTO, E., HASEGAWA, M. (1992a), Characterization of sannamycin A-non-producing mutants of *Streptomyces sannanensis, J. Antibiot.* **45**, 289–291.

OHTA, T., HASHIMOTO, E., HASEGAWA, M. (1992b), Cloning and analysis of a gene (*sms*13) encoding sannamycin B-glycyltransferase from *Streptomyces sannanensis* and its distribution among actinomycetes, *J. Antibiot.* **45**, 1167–1175.

OHTA, T., DAIRI, T., HASHIMOTO, E., HASEGAWA, M. (1993a), Use of a heterologous gene for molecular breeding of actinomycetes producing structurally related antibiotics: self-defense genes of producers of fortimicin-A (astromicin)-group antibiotics, *Actinomycetol.* **7**, 145–155.

OHTA, T., DAIRI, T., HASEGAWA, M. (1993b), Characterization of two different types of resistance genes among producers of fortimicin-group antibiotics, *J. Gen. Microbiol.* **139**, 591–599.

OKADA, H., YAMAMOTO, K., TSUTANO, S., NAKAMURA, S. (1988), A new group of antibiotics, hydroxamic acid antimycotic antibiotics. I. Precursor-initiated changes in productivity and biosynthesis of neoenactins NL, and NL_2, *J. Antibiot.* **41**, 869–874.

OKUDA, T., ITO, Y. (1982), Biosynthesis and mutasynthesis of aminoglycoside antibiotics, in: *Aminoglycoside Antibiotics* (UMEZAWA, H., HOOPER, I. R., Eds.), pp. 111–203. Berlin: Springer-Verlag.

ONAKA, H., ANDO, N., NIHIRA, T., YAMADA, Y., BEPPU, T., HORINOUCHI, S. (1995), Cloning and characterization of the A-factor receptor gene from *Streptomyces griseus, J. Bacteriol.* **177**, 6083–6092.

OTSUKA, H., MASKARETTI, O. A., HURLEY, L. H., FLOSS, H. G. (1980), Stereochemical aspects of the biosynthesis of spectinomycin, *J. Am. Chem. Soc.* **102**, 6817–6820.

OTTEN, S. L., LIU, X., FERGUSON, J., HUTCHINSON, C. R. (1995), Cloning and characterization of the *Streptomyces peuceticus dnrQS* genes encoding a daunosamine biosynthesis enzyme and a glycosyltransferase involved in daunorubicin biosynthesis, *J. Bacteriol.* **177**, 6688–6692.

OZAKI, M., MIZUSHIMA, S., NOMURA, M. (1969), Identification and functional characterization of the protein controlled by the streptomycin-resistant locus in *E. coli, Nature* **222**, 333–339.

PARRY, R. J. (1992), Investigation of the biosynthesis of aristeromycin, in: *Secondary-Metabolite Biosynthesis and Metabolism* (PETROSKI, R. J., MCCORMICK, S. P., Eds.), pp. 89–104. New York: Plenum Press.

PARRY, R. J., BORNEMANN, V., SUBRAMANIAN, R. (1989), Biosynthesis of the nucleoside antibiotic aristeromycin, *J. Am. Chem. Soc.* **111**, 5819–5824.

PEARCE, C. J., RINEHART, K. L. (1981), Biosynthesis of aminocyclitol antibiotics, in: *Antibiotics,* Vol. 4, *Biosynthesis* (CORCORAN, J. W., Ed.), pp. 74–100. Berlin: Springer-Verlag.

PENYIGE, A., VARGHA, G., ENSIGN, J. C., BARABÁS, G. (1992), The possible role of ADP ribosylation in physiological regulation of sporulation in *Streptomyces griseus, Gene* **115**, 181–185.

PESCHKE, U., SCHMIDT, H., ZHANG, H.-Z., PIEPERSBERG, W. (1995), Molecular characterization of the lincomycin production gene cluster of *Streptomyces lincolnensis* 78-11, *Mol. Microbiol.* **16**, 1137–1156.

PHILLIPS, L., WELLINGTON, E. M. H., REES, S. B., JUN, L. S., KING, G. P. (1992), The distribution of DNA sequences hybridizing with antibiotic production and resistance gene probes within type strains and wild isolates of *Streptomyces species, J. Antibiot.* **45**, 1481–1491.

PIENDL, W., BÖCK, A., CUNDLIFFE, E. (1984), Involvement of 16S ribosomal RNA in resistance of the aminoglycoside-producers *Streptomyces tenjimariensis, Streptomyces tenebrarius,* and *Micromonospora purpurea, Mol. Gen. Genet.* **197**, 24–29.

PIEPERSBERG, W. (1985), Aminoglycoside Antibiotika: Wichtige Therapeutika und Objekte der Grundlagenforschung, *Forum Mikrobiol.* **8/85**, 153–161.

PIEPERSBERG, W. (1992), Metabolism and cell individualization, in: *Secondary Metabolites: Their Function and Evolution, Ciba Foundation Symposium 171* (CHADWICK, D., WHELAN, J., Eds.), pp. 294–299. Chichester: John Wiley & Sons.

PIEPERSBERG, W. (1993), Streptomycetes and corynebacteria, in: *Biotechology*, 2nd Edn., Vol. 1 (REHM, H.-J., REED, G., Eds.), pp. 433–468. Weinheim: VCH.

PIEPERSBERG, W. (1994), Pathway engineering in secondary metabolite-producing actinomycetes, *Crit. Rev. Biotechnol.* **14**, 251–285.

PIEPERSBERG, W. (1995), Streptomycin and related aminoglycosides, in: *Biochemistry and Genetics of Antibiotic Biosynthesis* (VINING, L., STUTTARD, C., Eds.), pp. 531–570. Boston, MA: Butterworth-Heinemann.

PIEPERSBERG, W., GEYL, D., HUMMEL, H., BÖCK, A. (1980), Physiology and biochemistry of bacterial ribosomal mutants, in: *Genetics and Evolution of RNA Polymerase, tRNA and Ribosomes* (OSAWA, S., OZEKI, H., UCHIDA, H., YURA, T, Eds.), pp. 359–377. Tokyo: University of Tokyo Press.

PIEPERSBERG, W., DISTLER, J., HEINZEL, P., PEREZ-GONZALEZ, J.-A. (1988), Antibiotic resistance by modification: many resistance genes could be derived from cellular control genes in actinomycetes – A hypothesis, *Actinomycetol.* **2**, 83–98.

PIEPERSBERG, W., HEINZEL, P., MANSOURI, K., MÖNINGHOFF, U., PISSOWOTZKI, K. (1991a), Evolution of antibiotic resistance and production genes in streptomycetes, in: *Genetics and Product Formation in Streptomyces* (BAUMBERG, S., KRÜGEL, H., NOVACK, D., Eds.), pp. 161–170. New York: Plenum Press.

PIEPERSBERG, W., STOCKMANN, M., MANSOURI, K., DISTLER, J., GRABLEY, S., SICHEL, P., BRÄU, B. (1991b), *German Patent Application* P41 30 967 – HOE 91/F 300.

PIEPERSBERG, W., JIMENEZ, A., CUNDLIFFE, E., GRABLEY, S., BRÄU, B., MARQUARDT, R. (1994), Glycosyltransferases from streptomycetes as tools in biotransformations, *BRIDGE Progress Report 1994* (CEC) (VASSAROTTI, A., Ed.), pp. 128–133. Brussels: Printeclair.

PISSOWOTZKI, K., MANSOURI, K., PIEPERSBERG, W. (1991), Genetics of streptomycin production in *Streptomyces griseus.* Molecular structure and putative function of genes *strELMB2N, Mol. Gen. Genet.* **231**, 113–123.

POHLMAN, R. F., GENETTI, H. D., WINANS, S. C. (1994), Common ancestry between IncN conjugal transfer genes and macromolecular export systems of plant and animal pathogens, *Mol. Microbiol.* **14**, 655–668.

PRATT, W. B., FEKETY, R. (1986), *The Antimicrobial Drugs.* New York: Oxford University Press.

PUROHIT, P., STERN, S. (1994), Interactions of small RNA with antibiotic and RNA ligands of the 30S subunit, *Nature* **370**, 659–662.

RAETZ, C. R. H. (1996), Bacterial lipopolysaccharides: a remarkable family of bioactive macroamphiphiles, in: *Escherichia coli* and *Salmonella. Cellular and Molecular Biology* (NEIDHARDT, F. C., Ed.), pp. 1035–1063. Washington DC: American Society for Microbiology.

RAGAN, C. M., VINING, L. C. (1978), Intracellular cyclic adenosine 3′,5′-monophosphate levels and streptomycin production in cultures of *Streptomyces griseus, Can. J. Microbiol.* **24**, 1012–1015.

REEVES, P. (1993), Evolution of *Salmonella* O antigen variation by interspecific gene transfer on a large scale, *TIG* **9**, 17–22.

RETZLAFF, L., DISTLER, J. (1995), The regulator of streptomycin gene expression, StrR, of *Streptomyces griseus* is a DNA binding activator protein with multiple recognition sites, *Mol. Microbiol.* **18**, 151–162.

RETZLAFF, L., MAYER, G., BEYER, S., AHLERT, J., VERSECK, S., DISTLER, J., PIEPERSBERG, W. (1993), Streptomycin production in streptomycetes: a progress report, in: *Industrial Microorganisms: Basic and Applied Molecular Genetics* (HEGEMAN, G. D., BALTZ, R. H., SKATRUD, P. L., Eds.), pp. 183–194. Washington: American Society for Microbiology.

RINEHART, K. L., JR. (1980), Biosynthesis and mutasynthesis of aminocyclitol antibiotics, in: *Aminocyclitol Antibiotics, ACS Symp. Ser. 125* (RINEHART, K. L., SUAMI, T., Eds.). Washington, DC: American Chemical Society.

RINEHART, K. L., JR., STROSHANE, R. M. (1976), Biosynthesis of aminocyclitol antibiotics, *J. Antibiot.* **29**, 319–353.

RINEHARD, K. L., POTGIETER, M., DELAWARE, D. L. (1981), Direct evidence from multiple ^{13}C labeling and homonuclear decoupling for the labeling pattern by glucose of the *m*-aminobenzoyl (C_7N) unit in pactamycin, *J. Am. Chem. Soc.* **103**, 2099–2101.

RINEHART, K. L., JR., SNYDER, W. C., STALEY, A. L., LAU, R. C. M. (1992), Biosynthestic studies on antibiotics, in: *Secondary-Metabolite Biosynthesis and Metabolism* (PETROSKI, R. J., MCCORMICK, S. P., Eds.), pp. 41–60. New York: Plenum Press.

SAITOH, K., TSUNAKAWA, M., TOMITA, K., MIAKI, T., KONISHI, M., KAWAGUCHI, H. (1988), Boholmycin, a new aminoglycoside antibiotic. I. Production, isolation and properties, *J. Antibiot.* **41**, 855–861.

SAKUDA, S., ISOGAI, A., MATSUMOTO, S., SUZUKI, A. (1987), Search for microbial insect growth regulators. II. Allosamidin, a novel insect chitinase inhibitor, *J. Antibiot.* **40**, 296–300.

SALAUZE, D., PEREZ-GONZALEZ, J. A., PIEPERSBERG, W., DAVIES, J. (1991), Characterisation of aminoglycoside acetyltransferase-encoding genes of neomycin-producing *Micromonospora chalcea* and *Streptomyces fradiae, Gene* **101**, 143–148.

SATOH, A., OGAWA, H., SATOMURA, Y. (1976), Regulation of *N*-acetylkanamycin amidohydrolase in the idiophase in kanamycin fermentation, *Agric. Biol. Chem.* **40**, 191.

SCHLESSINGER, D., MEDOFF, G. (1975), Streptomycin, dihydrostreptomycin, and the gentamicins, in: *Antibiotics,* Vol. III (CORCORAN, J. W., HAHN, F. E., Eds.), pp. 535–550. Berlin: Springer-Verlag.

SCHROEDER, R., STREICHER, B., WANK, H. (1993), Splice-site selection and decoding: are they related? *Science* **260**, 1443–1444.

SHAPIRO, S. (1989), Nitrogen assimilation in actinomycetes and the influence of nitrogen nutrition on actinomycete secondary metabolism, in: *Regulation of Secondary Metabolism in Actinomycetes* (SHAPIRO, S., Ed.), pp. 135–211. Boca Raton, FL: CRC Press, Inc.

SHAW, K. J., RATHER, P. N., HARE, R. S., MILLER, G. H. (1993), Molecular genetics of aminoglycoside resistance genes and familial relationships of the aminoglycoside-modifying enzymes, *Microbiol. Rev.* **57**, 138–163.

SHNAITMAN, C. A., KLENA, J. D. (1993), Genetics of lipopolysaccharide biosynthesis in enteric bacteria, *Microbiol. Rev.* **57**, 655–682.

SIPOS, L., SZABO, G. (1989), *Myo*-inositol-1-phosphate synthase in different *Streptomyces griseus* variants, *FEMS Microbiol. Lett.* **65**, 339–364.

SMAOUI, H., MALLIE, J.-P., SCHAEVERBEKE, M., ROBERT, A., SCHAEVERBEKE, J. (1993), Gentamicin administered during gestation alters glomerular basement membrane development, *Antimicrob. Agents Chemother.* **37**, 1510–1517.

STOCKMANN, M., PIEPERSBERG, W. (1992), Gene probes for the detection of 6-deoxyhexose metabolism in secondary metabolite-producing streptomycetes, *FEMS Microbiol. Lett.* **90**, 185–190.

SWAN, S. K., KOHLHEPP, S. J., KOHNEN, P. W., GILBERT, D. N., BENNETT, W. M. (1991), Long-term protection of polyaspartic acid in experimental gentamicin nephrotoxicity, *Antimicrob. Agents Chemother.* **35**, 2591–2595.

SZESZÁK, F., VITÁLIS, S., BÉKÉSI, I., SZABÓ, G. (1991), Presence of factor C in streptomycetes and other bacteria, in: *Genetics and Product Formation in Streptomyces* (BAUMBERG, S., KRÜ-

GEL, H., NOVACK, D., Eds.), pp. 11–18. New York: Plenum Press.

TAYLOR, S. S., BUECHLER, J. A., YONEMOTO, W. (1990), cAMP-dependent protein kinase: Framework for a diverse family of regulatory enzymes, *Ann. Rev. Biochem.* **59**, 971–1005.

TERCERO, J. A., ESPINOSA, J. C., LACALLE, R. A., JIMÉNEZ, A. (1996), The biosynthetic pathway of the aminonucleoside antibiotic puromycin, as deduced from the molecular analysis of the *pur* cluster of *Streptomyces alboniger, J. Biol. Chem.* **271**, 1579–1590.

TESTA, B., KYBURZ, E., FUHRER, W., GIGER, R. (1993), *Perspectives in Medicinal Chemistry.* Weinheim: VCH.

THIEM, J. (Ed.) (1990), *Carbohydrate Chemistry. Topics in Current Chemistry 154.* Berlin: Springer-Verlag.

THOMAS, G. J. (1990), Synthetis of anthracyclines related to daunomycin, in: *Recent Progress in the Chemical Synthesis of Antibiotics* (LUKAS, G., OHNO, M., Eds.), pp. 467–496. Berlin: Springer-Verlag.

THOMPSON, C. J., SETO, H. (1995), Bialaphos, in: *Biochemistry and Genetics of Antibiotic Biosynthesis* (VINING, L., STUTTARD, C., Eds.), pp. 197–222. Boston, MA: Butterworth-Heinemann.

THORSON, J. S., LIU, H.-W. (1993a), Characterization of the first PMP-dependent iron-sulfur-containing enzyme which is essential for the biosynthesis of 3,6-dideoxyhexoses, *J. Am. Chem. Soc.* **115**, 7539–7540.

THORSON, J. S., LIU, H.-W. (1993b), Coenzyme B_6 as a redox cofactor: a new role for an old coenzyme? *J. Am. Chem. Soc.* **115**, 12177–12178.

THORSON, J. S., LO, S. F., LIU, H.-W., HUTCHINSON, C. R. (1993), Biosynthesis of 3,6-dideoxyhexoses: New mechanistic reflections upon 2,6-dideoxy, 4,6-dideoxy, and amino sugar construction, *J. Am. Chem. Soc.* **115**, 6993–6994.

TOHMA, S., KONDO, H., YOKOTSUGA, J., IWAMOTO, J., MATSUHASHI, G., ITO, T. (1989), Ashimycins A and B, new streptomycin analogues, *J. Antibiot.* **42**, 1205–1212.

TOHYAMA, H., OKAMI, Y., UMEZAWA, H. (1987), Nucleotide sequence of the streptomycin phosphotransferase and amidinotransferase of *Streptomyces griseus, Nucleic Acids Res.* **15**, 1819–1834.

TRUSCHEIT, E., FROMMER, W., JUNGE, B., MÜLLER, L., SCHMIDT, D. D., WINGEDER, W. (1981), Chemistry and biochemistry of bacterial *alpha*-glucosidase inhibitors, *Angew. Chem.* (Int. Edn.) **20**, 744–761.

TSUNO, T., IKEDA, C., NUMATA, K.-I., TOMITA, K., KONISHI, M., KAWAGUCHI, H. (1986), 3,3′-Neotrehalosadiamine (BMY-28251), a new aminosugar antibiotic, *J. Antibiot.* **39**, 1001–1003.

UMEZAWA, H. (1974), Biochemical mechanisms of resistance to aminoglycoside antibiotics, *Adv. Carbohydr. Chem. Biochem.* **30**, 183–225.

UMEZAWA, H., HOOPER, I. R. (Eds.) (1982), *Aminoglycoside Antibiotics.* Berlin: Springer-Verlag.

UMEZAWA, S., KONDO, S., ITO, Y. (1986), Aminoglycoside antibiotics, in: *Biotechnology,* 2nd Edn., Vol. 4 (REHM, H.-J., REED, G., Eds.), pp. 309–357. Weinheim: VCH.

UMEZAWA, H., OKANISHI, M., UTAHARA, R., MAEDA, K., KONDO, S. (1967a), Isolation and structure of kanamycin inactivated by a cell-free system of kanamycin-resistant *Escherichia coli, J. Antibiot.* **A20**, 136–141.

UMEZAWA, H., OKANISHI, M., KONDO, S., HAMANA, K., UTAHARA, R., MAEDA, K., MITSUHASHI, S. (1967b), Adenylylstreptomycin, a product of streptomycin inactivated by *E. coli* carrying R-factor, *Science* **157**, 1559–1561.

VAZQUEZ, D. (1979), Inhibitors of protein synthesis, in: *Molecular Biology, Biochemistry, and Biophysics,* Vol. 30 (KLEINZELLER, A., SPRINGER, G. F., WITTMANN, H. G., Eds.). Berlin: Springer-Verlag.

VILCHES, C., HERNANDEZ, C., MENDEZ, C., SALAS, J. A. (1992), Role of glycosylation in biosynthesis of and resistance to oleandomycin in the producer organism, *Streptomyces antibioticus, J. Bacteriol.* **174**, 161–165.

VINING, L. C., DOULL, J. L. (1988), Catabolite repression of secondary metabolism in actinomycetes, in: *Biology of Actinomycetes '88* (OKAMI, Y., BEPPU, T., OGAWARA, H., Eds.), pp. 406–411. Tokyo: Japan Scientific Societies Press.

VINING, L., STUTTARD, C. (1995), *Biochemistry and Genetics of Antibiotic Biosynthesis.* Boston, MA: Butterworth-Heinemann.

VÖGTLI, M., HÜTTER, R. (1987), Characterization of the hydroxystreptomycin phosphotransferase gene (*sph*) of *Streptomyces glaucescens:* nucleotide sequencing and promoter analysis, *Mol. Gen. Genet.* **208**, 195–203.

VON AHSEN, U., NOLLER, H. F. (1993), Footprinting the sites of interaction of antibiotics with catalytic group I intron RNA, *Science* **260**, 1500–1503.

VON AHSEN, U., DAVIES, J., SCHROEDER, R. (1991), Antibiotic inhibition of group I ribozyme function, *Nature* **353**, 368–370.

VUJAKLIJA, D., UEDA, K., HONG, S., BEPPU, T., HORINOUCHI, S. (1991), Identification of an A-factor-dependent promotor in the streptomycin biosynthetic gene cluster of *Streptomyces griseus, Mol. Gen. Genet.* **229**, 119–128.

VUJAKLIJA, D., HORINOUCHI, S., BEPPU, T. (1993), Detection of an A-factor-responsive protein that binds to the upstream activation sequence of *strR*, a regulatory gene for streptomycin biosynthesis in *Streptomyces griseus, J. Bacteriol.* **175**, 2652–2661.

WALKER, J. B. (1975a), Pathways of the guanidinated inositol moieties of streptomycin and bluensomycin, *Methods Enzymol.* **43**, 429–470.

WALKER, J. B. (1975b), ATP:streptomycin 6-phosphotransferase, *Methods Enzymol.* **43**, 628–632.

WALKER, J. B. (1980), Biosynthesis of aminoglycoside antibiotics, *Dev. Ind. Microbiol.* **21**, 105–113.

WALKER, J. B. (1990), Possible evolutionary relationships between streptomycin and bluensomycin biosynthetic pathways: Detection of novel inositol kinase and *O*-carbamoyltransferase activities, *J. Bacteriol.* **172**, 5844–5851.

WALKER, J. B. (1995), Enzymatic synthesis of aminocyclitol moieties of aminoglycoside antibiotics from inositol by *Streptomyces* spp.: detection of glutamine–aminocyclitol aminotransferase and diaminocyclitol aminotransferase activities in a spectinomycin producer, *J. Bacteriol.* **172**, 5844–5851.

WALKER, J. B., SKORVAGA, M. (1973), Phosphorylation of streptomycin and dihydrostreptomycin, *J. Biol. Chem.* **248**, 2435–2440.

WALKER, J. B., WALKER, M. S. (1975), ATP:streptomycin 3″-phosphotransferase, *Methods Enzymol.* **43**, 632–634.

WALLACE, B. J., TAI, P.-C., DAVIS, B. D. (1979), Streptomycin and related antibiotics, in: *Antibiotics,* Vol. V-1, *Mechanism of Action of Antibacterial and Antitumor Agents* (CORCORAN, J. W., HAHN, F. E., Eds.), pp. 272–303. Berlin: Springer-Verlag.

WATERS, B., VUJAKLIJA, D., GOLD, M. R., DAVIES, J. (1994), Protein tyrosine phosphorylation in streptomycetes, *FEMS Microbiol. Lett.* **120**, 187–190.

WELLINGTON, E. M. H., MARSH, P., TOTH, I., CRESSWELL, N., HUDDELSTON, L., SCHILHABEL, M. B. (1993), The selective effects of antibiotics in soil, in: *Trends in Microbial Ecology* (GUERRERO, R., PEDRÓS-ALÓ, C., Eds.), pp. 331–336. Madrid: Spanish Society for Microbiology.

WIDLANSKI, T., BENDER, S. L., KNOWLES, J. R. (1989), Dehydroquinate synthase: A sheep in wolf's clothing? *J. Am. Chem. Soc.* **111**, 2299–2300.

WILLIAMS, S. T., SHARPE, M. E., HOLT, J. G. (1989), *Bergey's Manual of Systematic Bacteriology,* Vol. 4. Baltimore, ML: Williams and Wilkins.

WONG, Y.-H. H., SHERMAN, W. R. (1985), Anomeric and other substrate specificity studies with myo-inositol-1-P synthase, *J. Biol. Chem.* **260**, 11083–11090.

WRIGHT, J. L. C. (1983), The lincomycin–celesticetin–anthramycin group, in: *Biochemistry and Genetic Regulation of Commercially Important Antibiotics* (VINING, L. C., Ed.), pp. 311–328. London: Addison-Wesley.

YAMAUCHI, N., KAKINUMA, K. (1992a), Biochemical studies on 2-deoxy-scyllo-inosose an early intermediate in the biosynthesis of 2-deoxystreptamine. I. Chemical synthesis of 2-deoxy-scyllo-inosose and [2,2-$^2H_2$]-2-deoxy-scyllo-inosose, *J. Antibiot.* **45**, 756–766.

YAMAUCHI, N., KAKINUMA, K. (1992b), Biochemical studies on 2-deoxy-scyllo-inosose an early intermediate in the biosynthesis of 2-deoxystreptamine. II. Quantitative analysis of 2-deoxy-scyllo-inosose, *J. Antibiot.* **45**, 767–773.

YAMAUCHI, N., KAKINUMA, K. (1992c), Confirmation of *in vitro* synthesis of 2-deoxy-scyllo-inosose, the earliest intermediate in the biosynthesis of 2-deoxystreptamine, using cell free preparations of *Streptomyces fradiae, J. Antibiot.* **45**, 774–780.

YAMAUCHI, N., KAKINUMA, K. (1993), Biochemical studies on 2-deoxy-scyllo-inosose an early intermediate in the biosynthesis of 2-deoxystreptamine. VI. A clue to the similarity of 2-deoxy-scyllo-inosose synthase to dehydroquinate synthase, *J. Antibiot.* **46**, 1916–1918.

YAMAUCHI, N., KAKINUMA, K. (1995), Enzymatic carbocycle formation in microbial secondary metabolism. The mechanism of the 2-deoxy-scyllo-inosose synthase reaction as a crucial step in the 2-deoxystreptamine biosynthesis in *Streptomyces fradiae, J. Antibiot.* **60**, 5614–5619.

YASUZAWA, T., YOSHIDA, M., ICHIMURA, M., SHIRAHATA, K., SANO, H. (1987), CV-1, a new antibiotic produced by a strain of *Streptomyces* sp. II. Structure determination, *J. Antibiot.* **40**, 727–731.

YOKOSE, K., FURUMAI, T., SUHARA, Y., PIRSON, W. (1989), Trestatin: *alpha*-amylase inhibitor, in: *Novel Microbial Products for Medicine and Agriculture* (DEMAIN, A. L., SOMKUTI, G. A., HUNTER-CREVA, J. C., ROSSMOORE, H. W., Eds.), pp. 117–126. Amsterdam: Elsevier Science Publishers.

ZHANG, H., SCHMIDT, H., PIEPERSBERG, W. (1992), Molecular cloning and characterization of two lincomycin-resistance genes, *lmrA* and *lmrB*, from *Streptomyces lincolnensis* 78-11, *Mol. Microbiol.* **6**, 2147–2157.

ZHOU, Z.-Y., SAKUDA, S., YAMADA, Y. (1992), Biosynthetic studies on the chitinase inhibitor, allosamidin. Origin of the carbon and nitrogen atoms, *J. Chem. Soc. Perkin Trans.* **I** 1992, 1649–1652.

ZHOU, Z.-Y., SAKUDA, S., KINOSHITA, M., YAMADA, Y. (1993), Biosynthetic studies of allosamidin. 2. Isolation of didemethylallosamidin, and conversion experiments of ^{14}C-labeled demethylallosamidin, didemethylallosamidin and their related compounds, *J. Antibiot.* **46**, 1582–1588.

11 Products from Basidiomycetes

GERHARD ERKEL
TIMM ANKE

Kaiserslautern, Germany

1 Introduction

The basidiomycetes (mushrooms) constitute a large class of fungi and are estimated to consist of 30 000 species (MÜLLER and LOEFFLER, 1982) which is approximately one third of all fungi known. Since ancient times many of them were used as food (e.g., boletuses, chantarelles, *Agaricus* spp.) or for cultural purposes (hallucinogenic mushrooms). One of the first to describe pharmacological and toxic activities of fruiting bodies was PLINIUS SECUNDUS (A.D. 23–79). Although a direct use of the fruiting bodies was common practice all over the world, a detailed study of their contents and the metabolites produced by cultured mycelia started only in this century with FLEMING's discovery of the imperfect fungus *Penicillium notatum* as the producer of penicillin, the first antibiotic metabolite for the treatment of bacterial infections in humans. This led to an intensive search for new antibiotics produced by other microorganisms, especially easily available soil-living forms. The pioneers in the search for antibiotics from basidiomycetes are ANCHEL, HERVEY, WILKINS and their coworkers who investigated extracts from fruiting bodies and mycelial cultures of approximately 2000 species (for a review, see FLOREY et al., 1949). Their outstanding work offered a first glance at the basidiomycete chemistry and succeeded in the isolation of pleuromutilin (KAVANAGH et al., 1951), the first basidiomycete metabolite to serve later as a leading structure for the development of a commercial antibiotic. The work on antibiotic producing basidiomycetes was almost completely discontinued following WAKSMAN's discovery of the streptomycetes as most promising antibiotic producers. These bacteria can be obtained from soil samples and grow easily in a variety of technical media. Up to now, the worldwide investigation of *Streptomyces* and related genera resulted in more than 6000 metabolites, many of them being used as antibiotics or for other pharmacological purposes. In spite of the incredible wealth of structures and activities which can to be found in streptomycetes and imperfect fungi it now has become increasingly difficult to find novel chemical entities from these organisms. Because of that and of recent progress in fermentation technology, product recovery, and spectroscopy for structural analysis new investigations of other organisms seem to be attractive again. Among these are rare Actinomycetales, gliding bacteria, marine organisms, and some taxa of higher fungi including basidiomycetes.

In the following chapter an overview of bioactive metabolites from mycelial cultures of basidiomycetes with special emphasis on antibiotics, cytotoxic, and antitumor compounds is given. In some cases secondary metabolites were obtained from fruiting bodies. Normally, basidiomycetes do not form fruiting bodies under laboratory conditions. Therefore, only the larger mushrooms can be used in chemical or biological investigations, and they have been studied intensively for the occurrence of toxins (reviewed by BRESINSKY and BESL, 1985), hallucinogens (reviewed by SCHULTES and HOFMANN, 1980), and pigments (reviewed by GILL, 1994; GILL and STEGLICH, 1987).

2 Cultivation of Basidiomycetes

The life cycle of a typical basidiomycete (e.g., *Agaricus campestris,* which does not form conidia and is devoid of a yeast phase) starts with haploid basidiospores germinating to form haploid mycelia, which – if compatible – fuse to give rise to dikaryotic mycelia from which fruiting bodies are derived. Karyogamy and meiosis take place in the basidia located in the hymenium (e.g., lamellae or pores) of the fruiting body. Usually, 4 haploid basidiospores are formed.

Basidiomycetes are usually collected as fruiting bodies from their natural substrate: dead or living plants, soil or dung. Cultures can be derived either from spores (haploid or dikaryotic mycelia) or tissue plugs (dikaryotic mycelia) which can germinate and grow on complex media, typically containing yeast extract or peptone as a nitrogen source and glucose, maltose, or malt extract as a carbon source. A medium commonly used in the authors' laboratory consists of 4 g glucose, 4 g yeast extract, 10 g malt extract, water to 1 L, pH adjusted to 5.5. The same media usually are suitable for submerged cultivation either in Erlenmeyer flasks or in fermenters. Metabolite diversity and production are mainly dependent on the biosynthetic capabilities of the strain and on the fermentation conditions. From the authors' experience, these conditions can only be varied to a limited extent since many strains are highly sensitive to shear stress imposed by the impellers, and media have to be chosen so as to permit sufficiently fast growth. These problems were addressed by CHENINA et al. (1993), GERMERDONK et al. (1993), and BRAUER and KORN (1993). The modeling of a basidiomycete fermentation was achieved by HASS and MUNACK (1993). An example for a detailed description of a technical-scale process is the production of the antibiotic pleuromutilin (KNAUSEDER and BRANDL, 1975; SCHNEIDER and MOSER, 1987). Several peculiarities in the cultivation of basidiomycetes are commonly encountered: spores of many species, e.g., from the genera *Inocybe* and *Russula,* do not germinate, and no growth from tissue plugs can be observed. Many mycelial cultures grow very slowly on solid media or in submerged cultures. Fermentation times range from one to several weeks. Suitable methods for the preservation of cultures are keeping agar slants at 4°C with periodical transfers (e.g., once a year) or storage in liquid nitrogen. Most cultures lose viability after freeze-drying.

3 Primary and Secondary Metabolites from Basidiomycetes – Biosyntheses and Possible Functions

Secondary metabolites show diverse chemical structures that are often quite different from the primary metabolites (such as amino acids, acetyl coenzyme A, sugars, mevalonic acid, and intermediates from the shikimic acid pathway) from which they are synthesized. The starting point from primary metabolism is the basis of classification according to their biosynthetic precursors (TURNER, 1971; TURNER and ALDRIDGE, 1983). As shown in Fig. 1 the main branching points leading to secondary products are (HERBERT, 1989):

- acetyl coenzyme A, leading to polyketides, polyins, terpenoids, steroids, or carotenoids;
- shikimate, from which aromatic compounds can be derived;
- amino acids, which serve as precursors for peptides and alkaloids;
- glucose, for the biosynthesis of glycosides and aminoglycosides.

Due to the widespread distribution of the biosynthetic pathways mentioned above, related secondary metabolites, like polyketides, steroids, and terpenoids have been isolated from bacteria, plants, fungi, and animals (BEALE, 1990; JANSEN and DE GROOT, 1991).

The secondary metabolism of basidiomycetes is rich in terpenoids, especially sesquiterpenoids (AYER and BROWNE, 1981) and polyacetylenes (JONES and THALLER, 1973). Many of these possess structures which up to now have only been detected in this class of fungi, whereas others closely resemble plant metabolites (FRAGA, 1990).

Illudin M and illudin S (1 and 2, Fig. 2) were two of the first highly antimicrobial and cytotoxic metabolites of basidiomycetes detected by the screenings of ANCHEL et al.

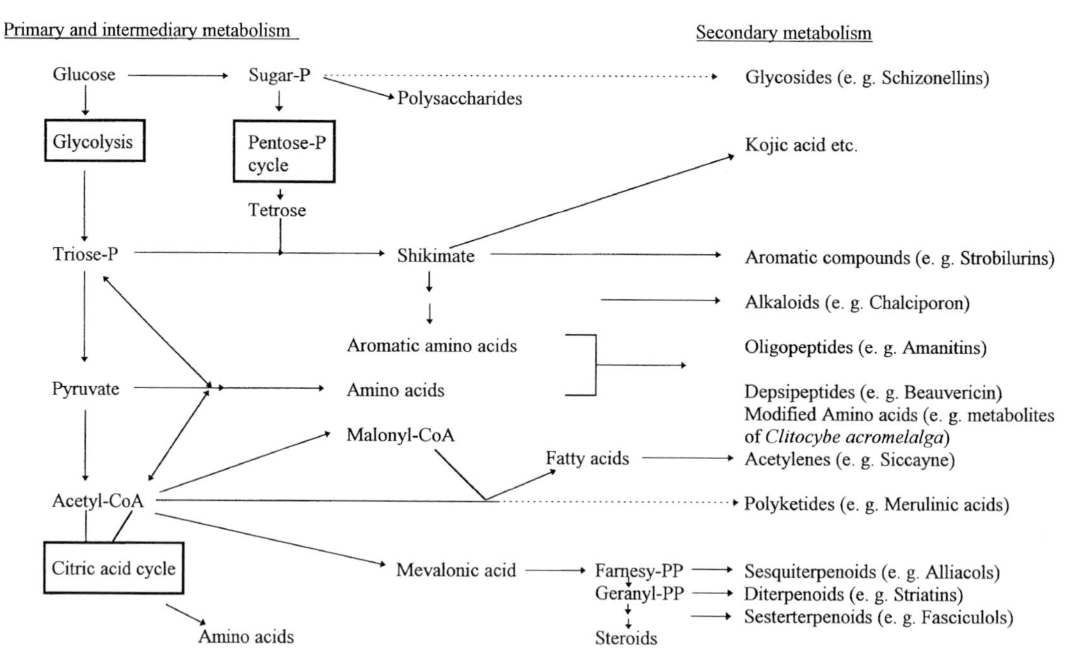

Fig. 1. Interrelationship between primary and secondary metabolism.

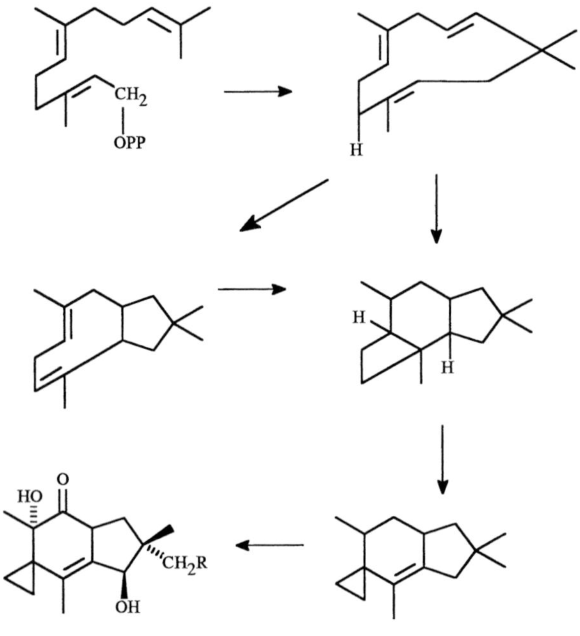

R = H Illudin M (1)
R = OH Illudin S (2)

Fig. 2. Biosynthesis of illudins.

(1950). The producing mushrooms, *Clitocybe illudens* and *Lampteromyces japonicus*, are highly toxic to humans. Illudin S is considered to be the toxic principle. The first information on stages in sesquiterpenoid biosynthesis were obtained by incorporation of radiolabeled mevalonic acid into illudins S and M from a head-to-tail condensation of isopentenyl pyrophosphate and dimethylallyl pyrophosphate by stationary cultures of *Clitocybe illudens* with a humulene-type precursor as the first cyclic intermediate (McMorris and Anchel, 1965; Hanson and Marten, 1973). Price and Heinstein (1978) confirmed this biosynthetic pathway (Fig. 2) by using a cell-free homogenate of *Clitocybe illudens* which incorporated labeled pyrophosphorylated isoprenyl alcohols into illudins.

Another example of metabolites apparently occurring in basidiomycetes only are the diterpenoids – striatins A, B, C, D – and the corresponding striatals (Fig. 3) which are antibiotic and cytotoxic products of *Cyathus striatus, C. poeppigii, C. limbatus, C. montagnei,* and *Gerronema fibula* (Anke et al., 1977a; Hecht et al., 1978). Striatins and striatals consist of a cyathan moiety with an attached pentose unit. In a resting cell system ^{14}C- and ^{13}C-labeled precursors were incorporated into striatins and striatals (Rabe, 1989). Feeding experiments with 1-^{13}C-glucose and 2-^{13}C-glucose and subsequent analysis by NMR spectroscopy revealed that the pentose unit was formed by decarboxylation of C-6 of glucose (70%) and to a smaller extent (30%) via the pentose phosphate cycle (Fig. 3). The labels observed in the cyathan skeleton were consistent with the formation of acetate via glycolysis and subsequent synthesis of mevalonate. Further cyclization of geranyl–geranyl pyrophosphate to the cyathan skeleton occurred according to the mechanism of the cyathine formation as described by Ayer et al. (1979). Herical, an antibiotic metabolite of *Hericium ramosum* consists of a cyathan moiety and D-xylose attached to it by a glycosidic bond. When ^{14}C-herical was prepared and fed to resting cells of *C. striatus* it was readily incorporated into the striatals A and B. Herical is thus considered as a direct precursor of striatals. The formation of the C—C bond between aglycone and xylose involves a loss of protons, intramolecular cyclization, and acylation. Similar mechanisms may be assumed for the formation of the related metabolites dihydrostriatal C and hericin by *H. ramosum.*

It has been proposed that molecules related to secondary metabolites played important roles in biochemical evolution as modulators or effectors, enhancing or controlling biological activities of primitive macromolecules. A number of antibiotics which inhibit translation, such as aminoglycosides, interact with ribosomes by directly binding to specific RNA conformations. It has been suggested that these secondary metabolites served as effectors of translation and other ribozyme-catalyzed reactions in early stages of evolution (Davis et al., 1992). From another viewpoint secondary metabolism might be an evolutionary playground from which new and useful biogenetic pathways could evolve (Zähner, 1982).

The biological role of secondary metabolites is still a matter of debate. Secondary metabolites might be beneficial to producing organisms several ways: improving their ability to grow, reproduce, or disperse under appropriate conditions, or affording protection against competitors or predators. The majority of these compounds fit into these categories (Vining, 1990). Among the growth-supporting substances are the sideramines (ferric ion-chelating compounds) and related metabolites which, in association with specific receptors, play an essential role in solubilization and uptake of iron. Fungal sideramines have been isolated from cultures of *Penicillium* sp., *Neurospora crassa,* some *Fusarium strains, Ustilago* sp., and the basidiomycetous yeast *Rhodotorula pilimanae* (Winkelmann, 1986). Their biosynthesis is regulated by the concentration of soluble iron in the substrate. As has been shown for rhodotorulic acid synthetase (Anke and Diekmann, 1972) and fusigen synthetase (Anke et al., 1973) these key enzymes could not be detected as long as sufficient iron was present in the culture media. The characteristics of fungal sideramine biosynthesis is similar to the non-ribosomal biosynthesis of other peptide antibiotics (Kleinkauf and von Döhren, 1987). Recently, it has been shown that wood decaying basidio-

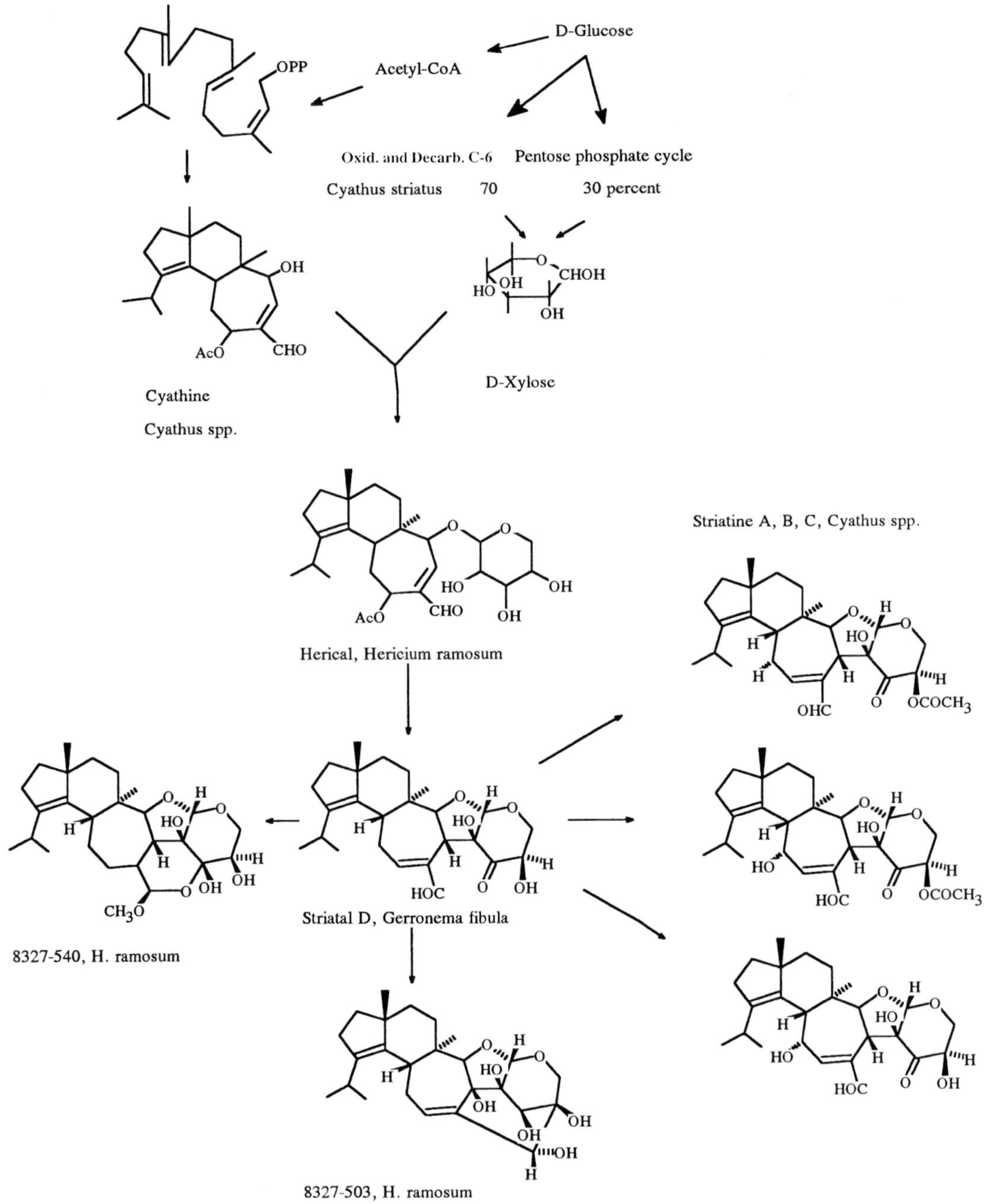

D-Glucose

Acetyl-CoA

Oxid. and Decarb. C-6 Pentose phosphate cycle

Cyathus striatus 70 30 percent

D-Xylose

Cyathine

Cyathus spp.

Herical, Hericium ramosum

Striatine A, B, C, Cyathus spp.

Striatal D, Gerronema fibula

8327-540, H. ramosum

8327-503, H. ramosum

Fig. 3. Biosynthesis of striatals.

mycetes (brown and white rots) produce siderophores of the phenolate type (JELLISON et al., 1990).

The enzymatic transformation of sesquiterpenes in various species of *Russulaceae* is an example for a proposed chemical defense system that preserves the fruiting bodies from attack by parasites and microorganisms (STERNER et al., 1985). The fruiting bodies of *Laccarius vellereus* contain large amounts of stearoylvelutinal (3, Fig. 4) which is transformed to the unsaturated dialdehyde isovelleral a few seconds after injury (4, Fig. 4). While stearoylvelutinal appears to have weak biological activity, isovelleral, like other unsaturated dialdehydes, has strong antimicrobial as well as mutagenic properties (STERNER et al., 1987a). In addition, isovelleral is a potent antifeedant for mammals that normally feed on mushrooms. Further reduction of an aldehyde group converts isovelleral to isovellerol (5, Fig. 4) the mutagenicity, pungency, and antimicrobial activities of which are diminished or lost upon reduction. It seems probable that injured fruiting bodies reduce the unsaturated dialdehydes in order to avoid prolonged contact with their own defense chemicals.

Another example of a chemical defense mechanism in basidiomycetes are azepin derivatives occurring in fruiting bodies of *Chalciporus piperatus*. The main component chalciporon (6, Fig. 4) exhibits antibacterial and antifungal activity. Due to its strong pungency chalciporon is considered to be responsible

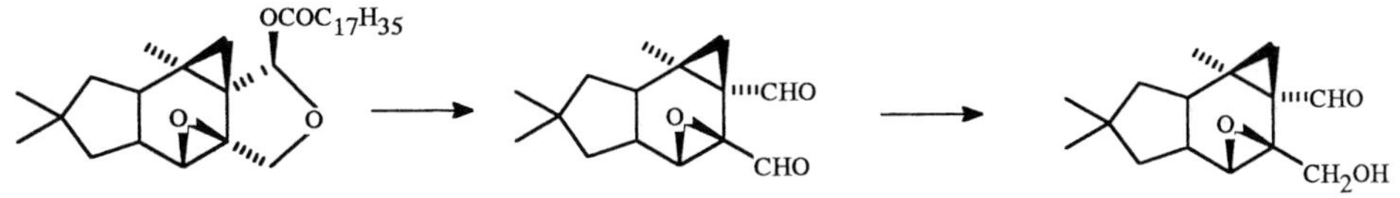

Stearoylvelutinal (3) Isovelleral (4) Isovellerol (5)

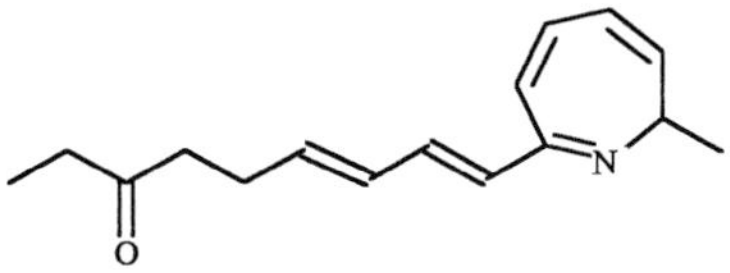

Chalciporon (6) 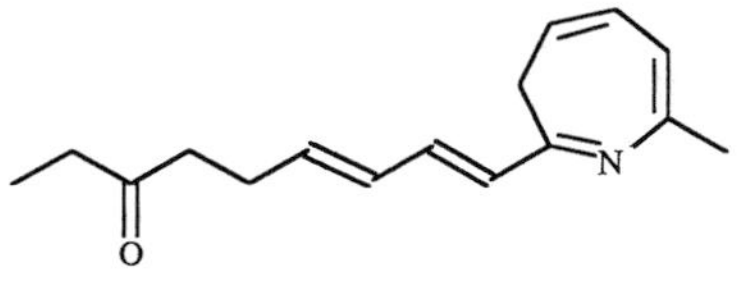 Isochalciporon (7)

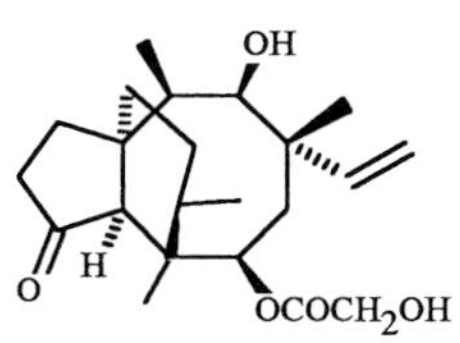

Pleuromutilin (8) 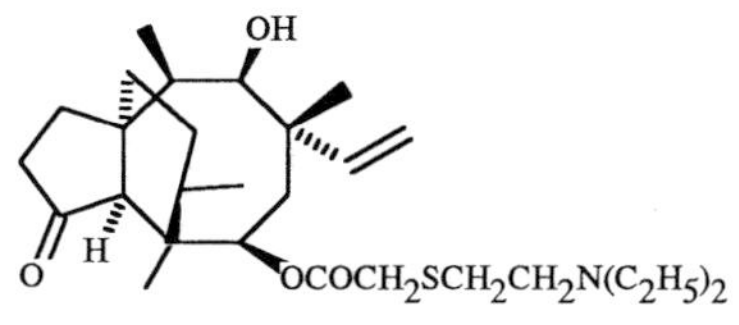 Tiamulin (9)

Fig. 4. Proposed chemical defense mechanisms by *Laccarius vellerus* and *Chalciporus piperatus*.

for the antifeedant acitivity. In solution chalciporon is converted to isochalciporon (7, Fig. 4) which still shows antibiotic activity but has lost pungency (STERNER et al., 1987b).

4 Screening Methods Used for the Detection of Potentially Useful Metabolites

Culture fluids or mycelial extracts of basidiomycetes are amenable to all screening methods applied to other microorganisms. Routine test systems comprise bacteria, filamentous fungi or yeasts, human and rodent cell lines, viruses, plants, and several enzyme assays.

5 Bioactive Metabolites from Basidiomycetes

5.1 Pleuromutilin (Tiamulin)

So far, the only commercial antibiotic produced by a basidiomycete is the diterpene pleuromutilin (8, Fig. 4). Pleuromutilin was first isolated from *Pleurotus mutilus* and *Pleurotus passeckerianus* in a screening for antibacterial compounds (KAVANAGH et al., 1951). The structural formula was elucidated by ARIGONI (1962) and BIRCH et al. (1963, 1966). In 1963, BRANDL et al. (KNAUSEDER and BRANDL, 1976) isolated an antibiotic from *Clitopilus passeckerianus* which was identical with pleuromutilin. Pleuromutilin is active against gram-positive bacteria, but the most interesting biological activity is its high effectiveness against various forms of mycoplasms. The preparation of more than 66 derivatives of pleuromutilin by RIEDL (1976) and EGGER and REINSHAGEN (1976a, b) resulted in the development of tiamulin (9, Fig. 4) which exceeds the activity of the parent compound against gram-positive bacteria and mycoplasms by a factor of 10–50. The minimal inhibitory concentrations (MIC) for different strains of *Mycoplasma* were in the range of 0.0039–6.25 $\mu g/mL^{-1}$ (DREWS et al., 1975). Studies on the mode of action revealed that pleuromutilin and its derivatives act as inhibitors of prokaryotic protein synthesis by interfering with the activities of the 70 S ribosomal subunit. The ribosome-bound antibiotics lead to the formation of inactive initiation complexes which are unable to enter the peptide chain elongation cycle (HÖGENAUER, 1979). In various bacteria drug resistance is developed stepwise. In some *E. coli* mutants the ribosome has lost its binding ability for tiamulin. Because of its outstanding properties tiamulin is currently used for the treatment of mycoplasma infections in animals.

Pleuromutilin can be produced by fermentation in a medium composed of: 50 g glucose, 50 g autolyzed brewer's yeast, 50 g KH_2PO_4, 0.5 g $MgSO_4 \times 7\ H_2O$, 0.5 g $Ca(NO_3)$, 0.1 g NaCl, 0.5 g $FeSO_4 \times 7\ H_2O$, water to 1 L, pH 6.0. The yield after 6 d of growth in a 1000 L fermenter was reported to be 2.2 $g\ L^{-1}$. It could be demonstrated that during fermentation of pleuromutilin derivatives differing in the acetyl portions attached to the 14-OH group of mutilin were formed. The biosynthesis of these derivatives was strongly stimulated by addition of corn oil as a carbon source during fermentation (KNAUSEDER and BRANDL, 1976). Pleuromutilin overproducers were obtained by conventional mutagenesis and selection programs as well as by protoplast fusion and genetic studies (STEWART, 1986).

5.2 The Strobilurins and Oudemansins

Initially, the strobilurins A (10, Fig. 5) and B (11, Fig. 5) were isolated from cultures of *Strobilurus tenacellus,* a small and very common edible mushroom growing on buried pine cones in early spring (ANKE et al., 1977b). Both compounds showed a remarkable activity against a variety of filamentous fungi and yeasts but no antibacterial effects. The structure elucidation by W. STEGLICH'S group (SCHRAMM et al., 1978; ANKE et al., 1984) revealed that both compounds belonged to a new class of antifungal antibiotics. Close similarities between strobilurin A and mucidin, an antifungal antibiotic previously isolated from cultures of the wood-inhabiting basidiomycete *Oudemansiella mucida,* were recognized. Mucidin, however, had been described as a dextrarotatory crystalline compound (MUSILEK, 1969). Fermentations of the *Oudemansiella mucida* strains used in the author's laboratory yielded in addition to strobilurin A a new antifungal antibiotic, oudemansin A (23, Fig. 5) (ANKE et al., 1979). Strobilurin A and mucidin were claimed to be identical by SEDMERA et al. in 1982 and this was finally proven by VON JAGOW et al. (1986) in a direct comparison of both compounds. In the meantime, numerous strobilurins and oudemansins were isolated from many genera of basidiomycetes (Fig. 5), from tropical as well as from temperate regions; among them were many *Mycena* species (BÄUERLE and ANKE, 1980; BÄUERLE, 1981) (Tab. 1).

Surprisingly, several strobilurins were also isolated from an ascomycete, *Bolinea lutea* (FREDENHAGEN et al., 1990a, b). Strobilurins seem to be of worldwide occurrence. Most of their producers grow on wood or decaying plant material.

Strobilurins and oudemansins inhibit fungal growth at very low concentrations (10^{-7}–10^{-8} M) (ANKE et al., 1977b, 1979, 1983; BACKENS et al., 1988; WEBER et al., 1990a, b; ANKE et al., 1990; ZAPF et al., 1994) without any significant antibacterial activity. Weak phytotoxic activity of several strobilurin and oudemansin derivatives have been demonstrated (SAUTER et al., 1995). Insecticidal activity of strobilurin A against adults and larval stages of *Epilachna varivestis* (Mexican bean beetle), *Aphis fabae* (aphid), and *Tetranychus urticae* (mite) were also found at concentrations of 10^{-4}–10^{-5} M (HOLST, University of Gießen, personal communication, 1978).

Reversible cytostatic activity has been described for all strobilurins, with strobilurin E (19, Fig. 5) being the most active. In HeLa S3 cells (human) these are accompanied by a 30% drop of the cellular ATP content and a change in morphology. The observed antiviral effects of strobilurin E (vesicular stomatitis virus in baby hamster kidney cells) are probably due to an inhibition of host cell growth (WEBER et al., 1990a).

5.2.1 Mode of Action – Selective Toxicity

Respiration in fungi and other eukaryotes is completely blocked by strobilurin A and oudemansin A. In Ehrlich ascitic carcinoma cells (ECA, mouse) syntheses of macromolecules (proteins, RNA, DNA) are inhibited due to a depletion of their ATP pool caused by the inhibition of oxidative phosphorylation. Upon addition of glucose this effect is completely reversed since ATP supply by glycolysis seems to be sufficient in these cells. In rat liver mitochondrial oxygen uptake and ATP synthesis were blocked by both α-ketoglutaric acid and succinate as substrates which gave the first evidence of a target within the respiratory chain (ANKE et al., 1979). This molecular target was precisely identified by VON JAGOW and coworkers (BRANDT et al., 1988, 1993). Strobilurins and oudemansins specifically inhibit the ubiquinol oxidation (Qp center) of the mitochondrial bc_1 complex (Fig. 6).

Like other specific inhibitors strobilurins and oudemansins have become valuable tools for the development of a more detailed model of the structure and function of their target. In heterocysts of cyanobacteria the cytochrome b/f complex, which utilizes light energy to generate the proton gradient used for ATP synthesis and transfers electrons to nitrogenase, the key enzyme of nitrogen fixation, plays a central role. Using heterocysts of an *Anabaena* sp. HOUCHINS and HIND (1983) found that strobilurin A inhibited the electron flow from reduced plastoquinone to the cytochrome b/f complex in a similar way as in the mitochondrial bc_1 complex. The similarity of the b/f complexes in cyanobacteria and in higher plants, where the b/f complex plays a

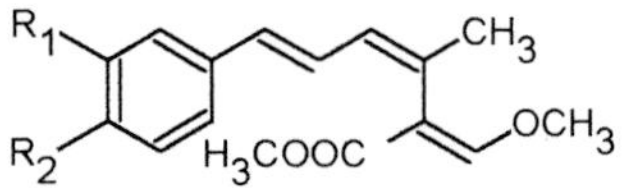

R_1	R_2	
H	H,	Strobilurin A (10)
MeO	Cl,	Strobilurin B (11)
HO	H,	Strobilurin F1 (12)
MeO	H,	Strobilurin H (13)
H	MeO,	Strobilurin X (14)

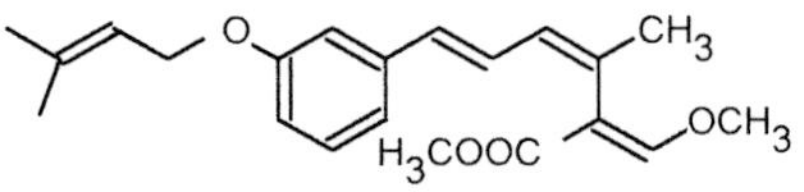

Strobilurin C (15)

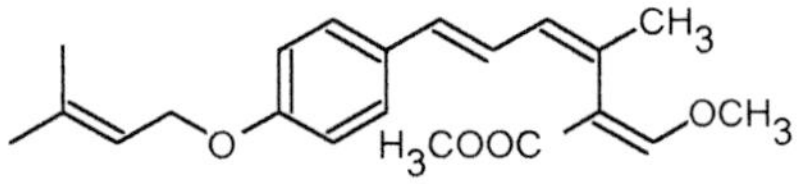

Strobilurin F2 (16)

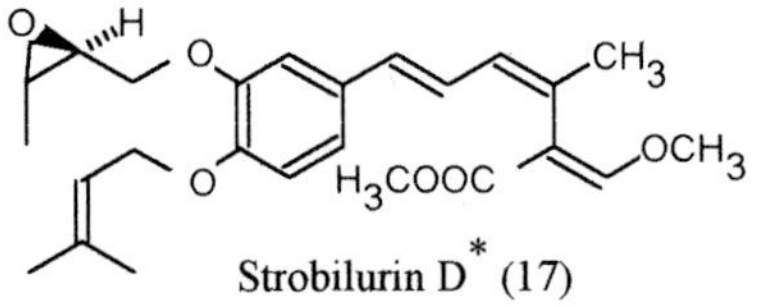

Strobilurin D* (17)

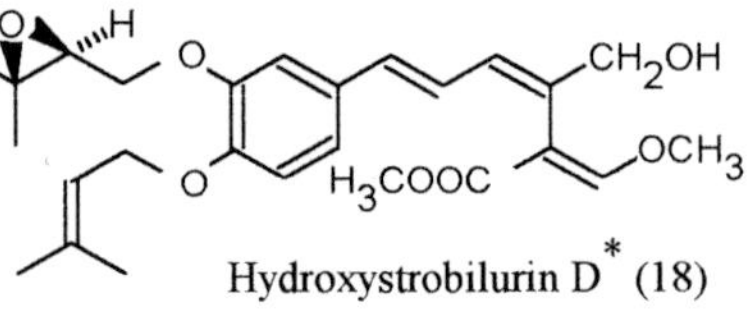

Hydroxystrobilurin D* (18)

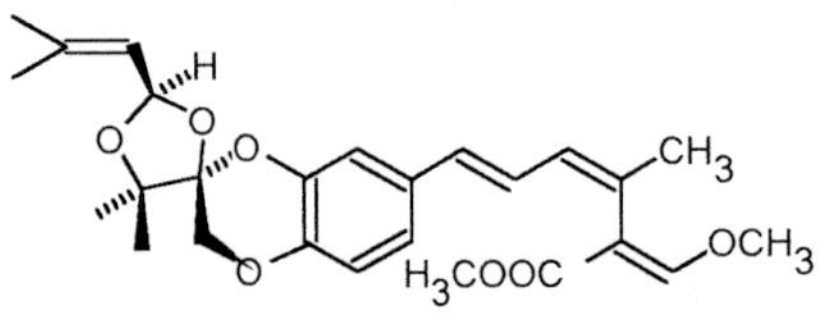

Strobilurin E (19)

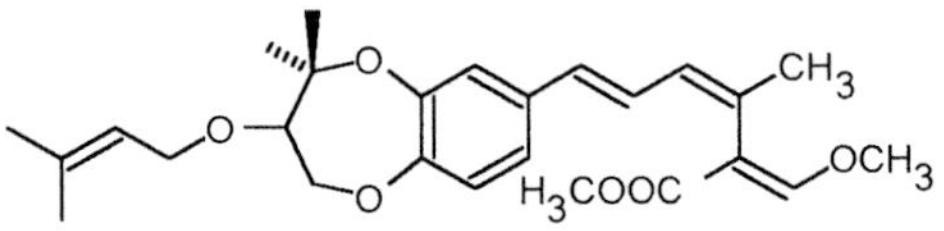

Strobilurin G (20)

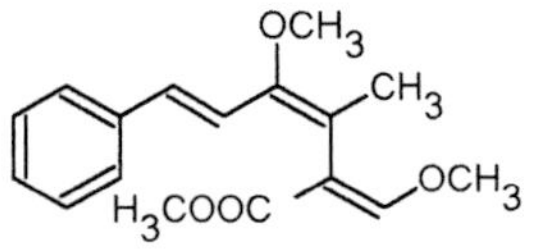

9-Methoxystrobilurin A (21)

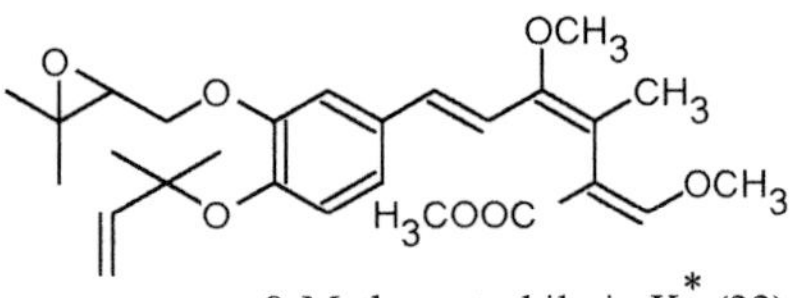

9-Methoxystrobilurin K* (22)

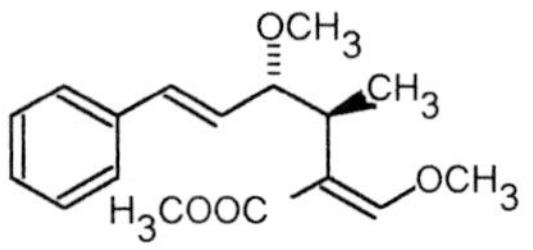

Oudemansin A (23)

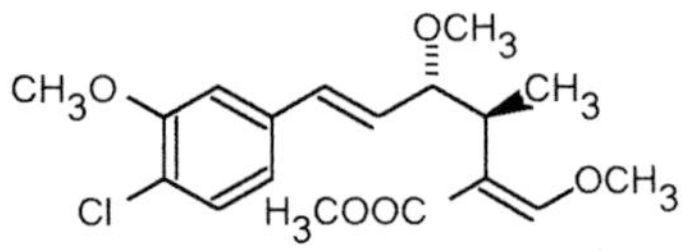

Oudemansin B (24)

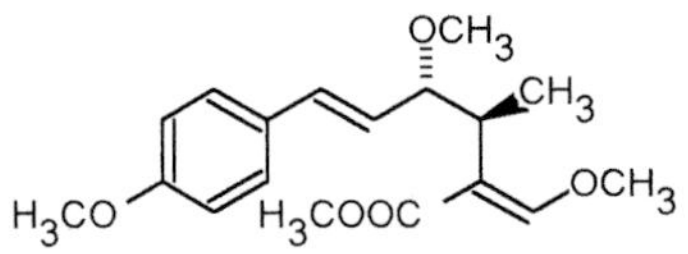

Oudemansin X (25)

Fig. 5. The strobilurins and oudemansins. * The structures of the side chains are currently under investigation.

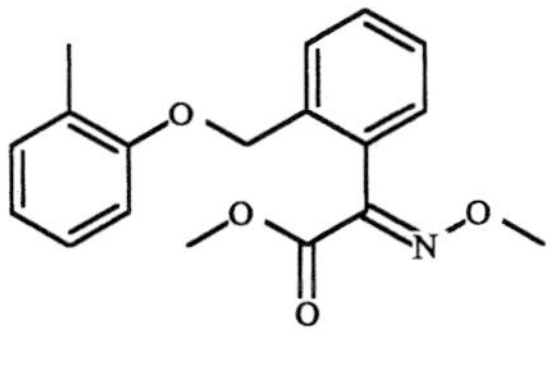

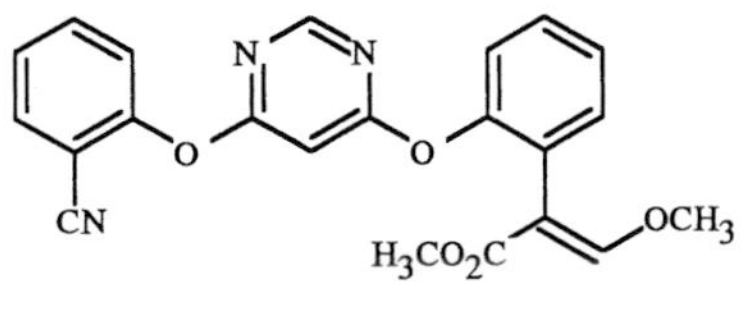

Fig. 5

BAS 490 F (26)

ICIA5504 (27)

Tab. 1. Fungi Producing Strobilurins and Oudemansins

Producer	Compound	References
Basidiomycetes		
Agaricus sp. 89139	10, 12, 17	ZAPF (1994)
Crepidotus fulvomentosus	19	WEBER et al. (1990a)
Cyphellopsis anomala	10, 12, 17	WEBER et al. (1990b)
Favolaschia sp. 87129	10, 12, 17, 19, 21, 22, 23	ZAPF et al. (1995)
Filoboletus sp. 9054	19	SIMON (1994)
Hydropus scabripes	10	BÄUERLE (1981)
Mycena aetites	10	BÄUERLE (1981)
M. alkalina	11	BÄUERLE (1981)
M. atromarginata	10	BÄUERLE (1981)
M. avenacea	11	BÄUERLE (1981)
M. cf. *capillaripes*	10	BÄUERLE (1981)
M. crocata	11	BÄUERLE (1981)
M. fagetorum	10	SCHRAMM et al. (1978)
M. galapoda	10	BÄUERLE (1981)
M. galopoda var. *alba*	10	BÄUERLE (1981)
M. oregonensis	10	BÄUERLE (1981)
M. polygramma	23	BÄUERLE (1981)
M. purpureofusca	10	BÄUERLE (1981)
M. rosella	10	BÄUERLE (1981)
M. sanguinolenta	18	BACKENS et al. (1988)
M. vitilis	11	BÄUERLE (1981)
M. zephirus	10	SCHRAMM et al. (1978)
Oudemansiella mucida	10, 23	ANKE et al. (1979)
O. radicata	10, 25	ANKE et al. (1990)
Strobilurus conigenoides	10	ANKE, unpublished data
S. esculentus	10	ANKE and STEGLICH (1981)
S. tephanocystis	10	ANKE, unpublished data
S. tentacellus	10, 11	ANKE et al. (1977b)
Xerula longipes	11, 15	ANKE et al. (1983)
X. melanotricha	10, 11, 24	ANKE et al. (1983)
Ascomycete		
Bolina lutea	10, 11, 13, 16, 20	FREDENHAGEN et al. (1990a, b)

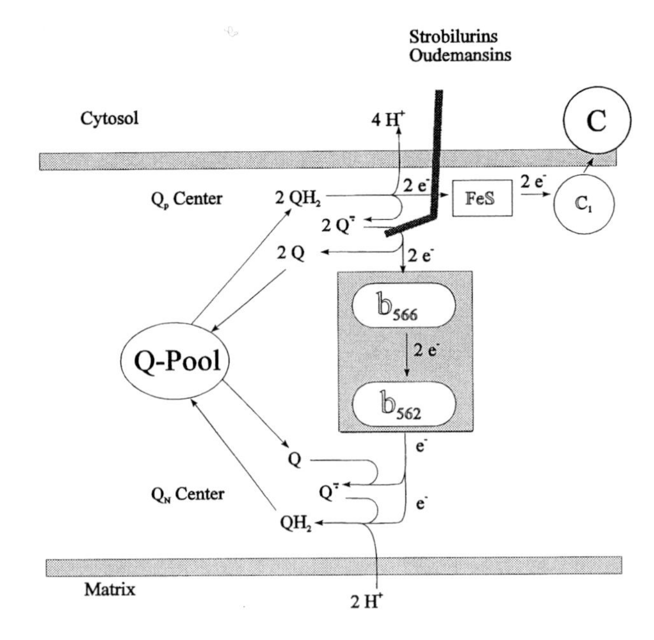

Fig. 6. The Q-cycle mechanism of the bc_1-complex and the mode of action of strobilurins and oudemansis (BRANDT et al., 1993).
Q-Pool: ubichinone pool; Q, $Q^{\dot{-}}$, QH_2: ubiquinone, ubisemiquinone, ubihydroquinone; c, c_1: cytochrome c_1; b_{566}: "low potential" heme b; b_{562}: "high potential" heme b; FeS: iron–sulfur protein.

central role in cyclic photophosphorylation and in coupling photosystems 1 and 2, might contribute, together with an inhibition of mitochondrial respiration, to the phytotoxic activity observed for some strobilurin derivatives (SAUTER et al., 1995).

The mitochondrial bc_1 complex, the target of strobilurins and oudemansins, is common to many eukaryotic taxa. Mitochondrial preparations of rat liver, beef heart, house fly, and corn are all sensitive to strobilurins (ANKE et al., 1979; BRANDT et al., 1993; SAUTER et al., 1995). Surprisingly, strobilurin A and other synthetic mimics exhibited no toxicity to rodents (J. DOUROS, NCI, USA, unpublished; SAUTER et al., 1995). In fact, "mucidermin", a preparation which apparently contains strobilurin A (mucidin) was marketed by Spofa, CSFR, for the treatment of dermatomycoses in humans. This lack of toxicity is probably due to enzymic degradation of the shobilurins by, e.g., mammalian esterases before reaching their target.

5.2.2 Structure–Activity Relationships – Development of Agricultural Fungicides

Extensive synthetic efforts lead to simple mimics and revealed that the *E*-β-methoxyacrylate unit is a prerequisite for the antifungal and respiration inhibiting properties of strobilurins and oudemansins (SCHRAMM, 1980; SCHRAMM et al., 1982; T. ANKE et al., 1988). Continuous efforts by STEGLICH and coworkers (ANKE and STEGLICH, 1989) and by BASF and ICI resulted in compounds with improved activity and light stability and lead to the development of BAS 490 F (26, Fig. 5) (SAUTER et al., 1995) and ICIA5504 (27, Fig. 5) (CLOUGH, 1993), which will be commercialized in the near future by ICI and BASF.

5.2.3 Biosynthesis

The biosynthesis of mucidin (28, Fig. 7) (the *E, E, E* geometry was revised to *E, Z, E* of strobulin A by VON JAGOW et al., 1986) was investigated by NERUD et al. (1982) by feeding isotopically labeled phenylalanine,

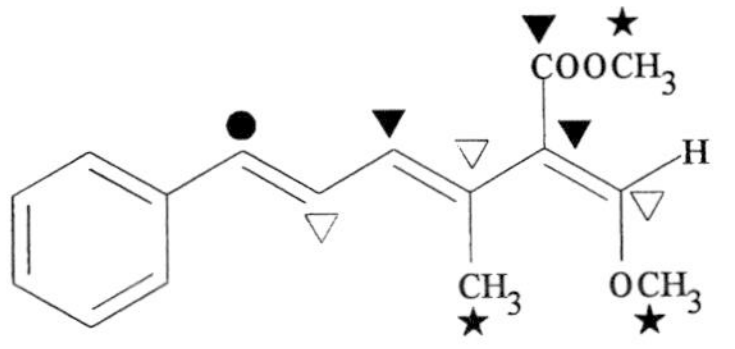

Mucidin (28)

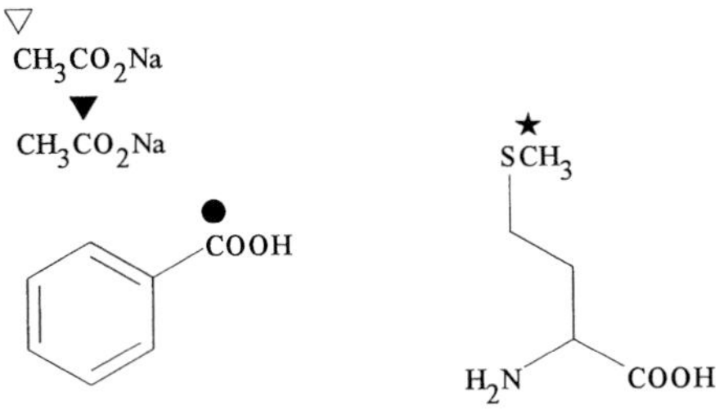

Fig. 7. Biosynthesis of mucidin (NERUD et al., 1982). *E, E, E,* geometry has been revised to *E, Z, E,* of strobilurin A; isotopically labeled carbon atoms in the precursors acetate, benzoic acid, and methionine are marked ▽, ▼, ●, and ★.

benzoic acid, acetate, and methionine. The aromatic part of the molecule and the benzylic carbon atom are derived from the shikimate pathway. The side chain consists of acetate units, and all three methyl groups are derived from methionine (Fig. 7).

5.2.4 Possible Functions of Strobilurins and Oudemansins in the Producing Fungi

The biological activities in the producers of strobilurins and oudemansins suggest a possible role in the defence of habitats and substrates against competing fungi or predatory insects. The use of antifungal antibiotics against competing fungi would require the producing fungus to be resistant to its own product. This was clearly demonstrated for *Strobilurus tenacellus* and *Mycena galapoda* by VON JAGOW and coworkers (BRANDT et al., 1993). In the case of *S. tenacellus*, binding of strobilurin A and oudemansin A to the bc_1 complex was reduced by several orders of magnitude, whereas in the case of *M. galopoda* a markedly increased rate of respiration is likely to confer a greater resistance.

Genetic characterization of the exon–intron organization, the deduced amino acid sequence of the cytochromes b from *S. tenacellus, M. galopoda,* and *M. viriginata* (which does not produce strobilurin A), and a comparative sequence analysis of two regions of cytochrome b contributing to the formation of the Qp center as demonstrated by VON JAGOW and coworkers (KRAICZY et al., 1996) revealed, that the generally lower sensitivity of all three basidiomycetes was due to the replacement of a small amino acid residue in position 127 by isoleucine. For *M. galopoda* replacement of glycine-143 by alanine and glycine-153 by serine, and for *S. tenacellus* replacement of a small residue in position 254 by glutamine and asparagine-261 by aspartate were assumed to cause resistance to *E*-β-methoxyacrylates. The latter exchange is also found in *Schizosaccharomyces pombe* which shows a natural resistance to *E*-β-methoxyacrylates.

On the other hand, it was demonstrated that *Oudemansiella mucida* produces oudemansin A together with strobilurin A on sterilized beech wood which is its natural substrate (SCHWITZGEBEL, 1992). These findings and the observation that many basidiomycetes belonging to different taxa (Tab. 1) use the same and obviously quite effective principle to secure their habitat in very different climates and locations suggest that strobilurins and oudemansis play an important role in the producing fungi.

5.3 Other Antibacterial and Antifungal Metabolites

Lentinellic acid (29, Fig. 8) from *Lentinellus omphalodes* and *Lentinellus ursinus* is a new protoilludine derivative. Interestingly, strains from Europe, USA, and Canada all produce the same antibiotic. Lentinellic acid shows strong antibacterial activity with minimal inhibitory concentrations of 1–5 $\mu g\ mL^{-1}$ for *Bacillus brevis, Aerobacter aerogenes,* and *Corynebacterium insidiosum.* Compared to lentinellic acid its methyl ester exhibits much higher antifungal activity. In

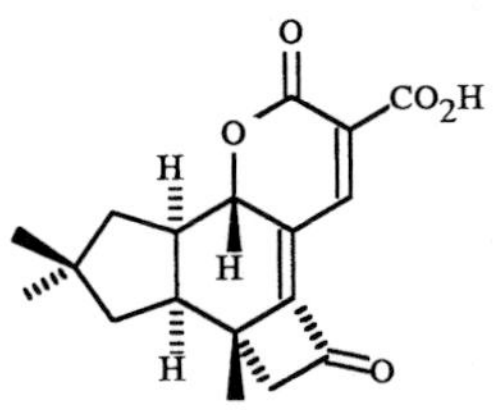

O O CO_2H H H H

Lentinellic acid (29)

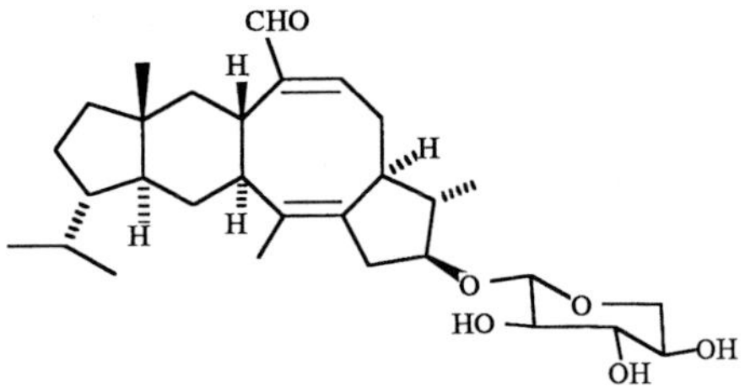

CHO H H H H O HO O OH OH

Aleurodiscal (30)

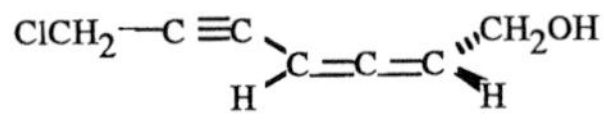

$ClCH_2-C\equiv C$ C=C=C CH_2OH H H

Scorodonin (31)

O OH

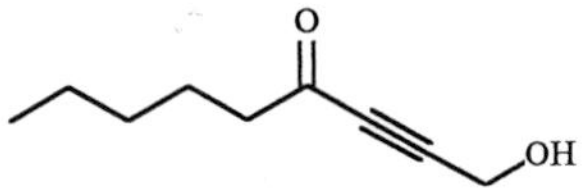

1-Hydroxy-2-nonyn-3-one (32)

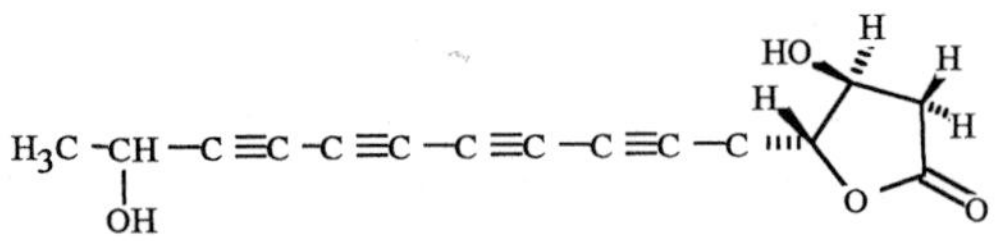

$H_3C-CH-C\equiv C-C\equiv C-C\equiv C-C\equiv C-C$ HO H H H H OH O O

3,4,13-Trihydroxy-tetradeca-5,7,9,11-tetraynic acid-Y-lactone (33)

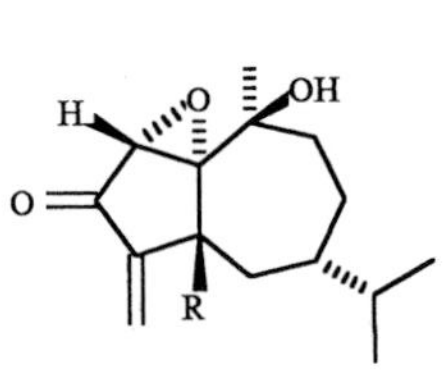

H O OH O R

R = H Fimicolon (34)
R = OH Hydroxyfimicolon (35)

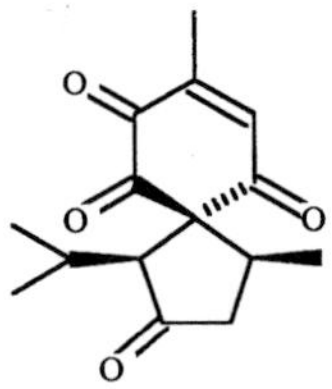

O O O O

Hemimycin (36)

Fig. 8. Other antibacterial and antifungal metabolites.

ECA cells DNA, RNA, and protein syntheses are inhibited by 50% using 20 μg mL^{-1} lentinellic acid (STÄRK et al., 1988). Several related protoilludane orsellinate esters were isolated from cultures of *Armillaria mellea*. These compounds exhibit weak antibacterial and antifungal activity (OBOUCHI et al., 1990; YANG et al., 1991).

Screening for antifungal compounds resulted in the isolation of aleurodiscal (30, Fig. 8) from mycelial cultures of *Aleurodiscus mirabilis* (LAUER et al., 1989). Aleurodiscal, a hydroxysesterterpene aldehyde β-D-xyloside with a novel carbon skeleton, is related to retigeranic acid A which was isolated from lichens (KANEDA et al., 1972). Aleurodiscal possesses weak antibacterial activity and strongly inhibits the growth of several fungi in the agar diffusion assay at concentrations of 2–10 μg per disc. In addition, it causes abnormal branching of apical hyphae of *Mucor miehei* at a concentration of 1 μg mL^{-1}. Acetylenes are strong antifungal metabolites commonly found in basidiomycetes. They also exhibit antibacterial and cytotoxic activity (TURNER, 1983). Examples are scorodonin (31, Fig. 8) from cultures of *Marasmius scorodonius* (ANKE et al., 1980), 1-hydroxy-2-nonyn-4-one (32, Fig. 8) from fermentations of *Ischnoderma benzoinum* (ANKE et al., 1982), and 3,4,13,-trihydroxy-tetradeca-5,7,9,11,-tetraynic acid-γ-lactone (33, Fig. 8) from cultures of *Mycena viridimarginata* (BÄUERLE et al., 1982).

Fimicolon (34, Fig. 8) and hydroxyfimicolon (35, Fig. 8) of *Panaeolus fimicola* and *Psathyrella orbitarium* are antibiotic and cytotoxic guaianes (ANKE et al., 1985a). Guaianes are typical metabolites of higher plants. The structures responsible for the biological activity of the *Pleurotellus* metabolites and fimicolons are similar and consist of a five-membered ring with an exomethylene group adjacent to an oxirane ring.

The antibiotic and cytotoxic compound hemimycin (36, Fig. 8) obtained from *Hemimycena cucullata* and *H. candida* is another example for the occurrence of the same carbon skeleton in basidiomycetes and higher plants (e.g., *Acorus calamus*). Hemimycin is highly oxygenated and contains a double bond which easily reacts with nucleophiles (BÄUERLE et al., 1986). Acoranes have not yet been reported from microbial sources.

5.4 Cytotoxic and Antitumor Metabolites

Illudins isolated from *Clitocybe illudens* and *Lampteromyces japonicus* were two of the first known antitumor metabolites. The lifetime of Ehrlich ascites tumor mice was prolonged by illudin S (2, Fig. 2) when given at a dose of 166 μg kg^{-1} i.p. Enclosure of illudin S into liposomes markedly enhanced this effect, apparently by decreasing the side effects observed under standard experimental conditions (SHINOZAWA et al., 1979). 6-Deoxyilludin M (37, Fig. 9) was isolated from cultures of *Pleurotus japonicus*. This compound is closely related to illudin M and was effective against murine leukemia P388, showing a 24% increase of life span at a daily dose of 5 mg kg^{-1} i.p. (HARA et al., 1987).

Several metabolites of basidiomycetes with strong cytotoxic and antifungal activitiy belong to the sesquiterpenoids with a marasmane or isolactarane skeleton. Marasmic acid (38, Fig. 9) was isolated from cultures of *Marasmius conigenus* in the course of the first extensive screenings for antibacterial compounds conducted by KAVANAGH et al. (1949). The sesquiterpenoid structure of marasmic acid was elucidated by DUGAN et al. in 1966. GREENLEE and WOODWARD achieved the first total synthesis in 1976 and several new synthetic approaches have been published since then (MORISAKI et al., 1980). Marasmic acid was also isolated from cultures of *Lachnella* sp. and *Peniophora laeta* and exhibits pronounced inhibitory action on nucleic acid syntheses in whole mammalian cells and on some enzymes of nucleoside metabolism. In isolated rat liver nuclei the guanylation of mRNA was strongly inhibited by 10 μg mL^{-1} marasmic acid (KUPKA et al., 1983). The life span of P388 lymphocytic leukemia mice was prolonged by marasmic acid when given at a total dose of 3.5 mg kg^{-1} i.p. The LD_{50} for tumor bearing mice was determined to be 28 mg kg^{-1} i.p. (J. DOUROS, National Cancer Institute, USA, personal communication). It was proposed that due to the reactive α,β-un-

6-Deoxyilludin M (37)

Marasmic acid (38)

Pilatin (39)

Merulidial (40)

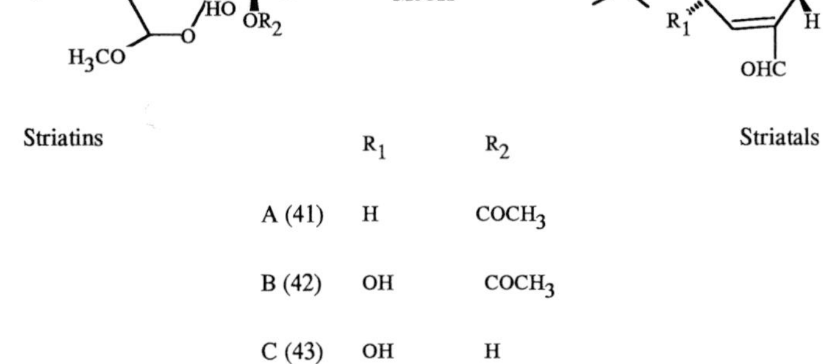

Striatins

Striatals

	R_1	R_2
A (41)	H	$COCH_3$
B (42)	OH	$COCH_3$
C (43)	OH	H

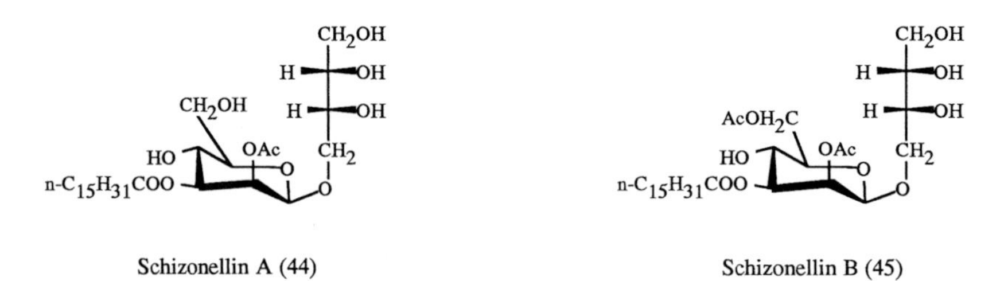

Schizonellin A (44)

Schizonellin B (45)

Fig. 9. Cytotoxic and antitumor metabolites.

Hypnophilin (46)

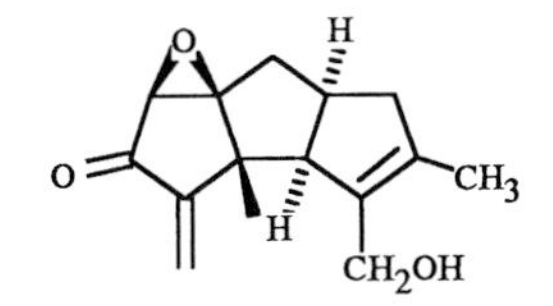

Pleurotellol (47)

Pleurotellic acid (48)

Phellodonic acid (49)

Alliacol A (50)

Alliacol B (51)

Fig. 9 Fulvoferruginin (52)

saturated aldehyde function marasmic acid covalently binds to nucleophilic (e.g., amino) groups of enzymes or to nucleic acids. The hydroxylated derivative of marasmic acid, 9-β-hydroxymarasmic acid, was isolated by H. ANKE et al. (1988). Introduction of a hydroxyl function reduces the biological activity of marasmic acid, but increases mutagenic activity in the Ames test.

Pilatin (39, Fig. 9), a new marasmane derivative, was isolated from fermentations of *Flagelloscypha pilatii,* a cyphelloid fungus (HEIM et al., 1988). Pilatin inhibits the growth of bacteria and fungi at concentrations of 5–50 $\mu g\ mL^{-1}$. It strongly interferes with the DNA and RNA syntheses of ECA cells and both normal and Rous Sarcoma Virus (RSV)-transformed chicken embryo fibroblasts (CEF). In addition, pilatin causes frameshift mutations in *Salmonella typhimurium* TA 98. *In vivo* no significant antitumor activity on P388 lymphocytic leukemia mice was observed for pilatin. The LD_{50} for tumor bearing mice was determined to be 125 $mg\ kg^{-1}$

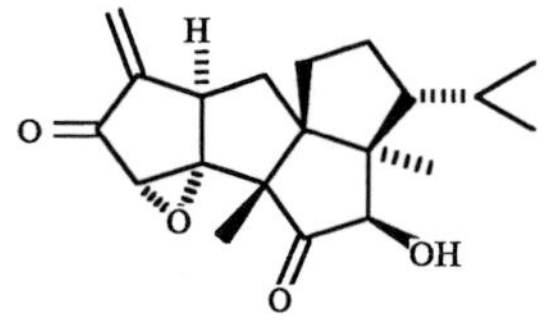

Crinipellin A (53)

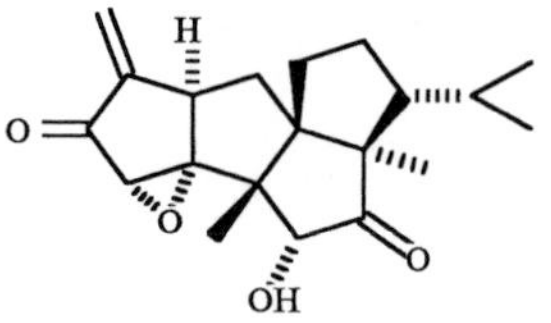

Crinipellin B (54)

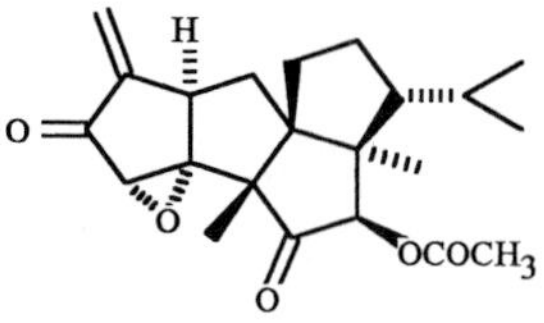

O-acetylcrinipellin A (55)

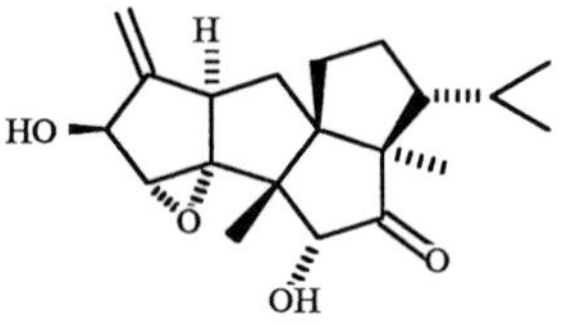

Dihydrocrinipellin B (56)

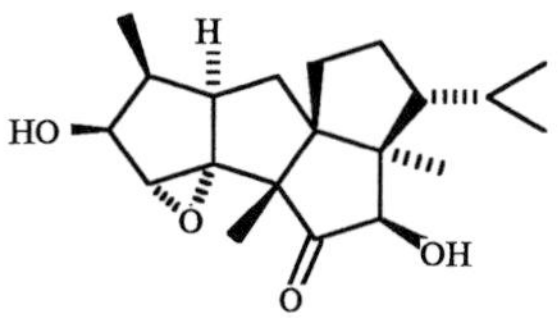

Tetrahydrocrinipellin A (57)

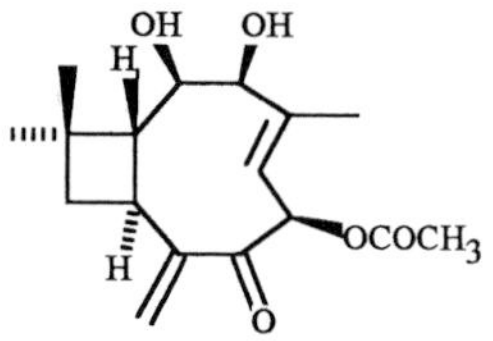

Nematolin (58)

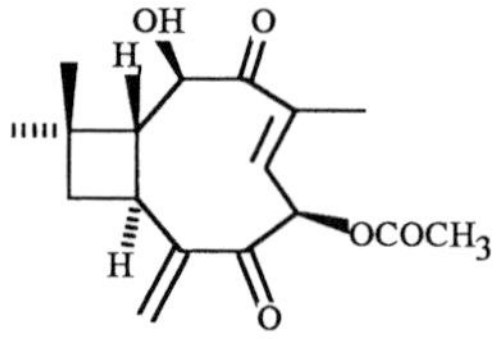

Nematolon (59)

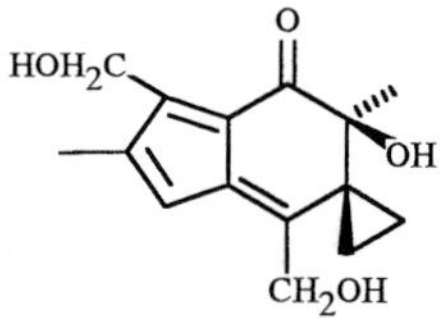

Leianafulvene (60)

Fig. 9

i.p. (J. DOUROS, National Cancer Institute, USA, personal communication).

The isolactarane merulidial (40, Fig. 9) was isolated from submerged cultures of *Merulius tremellosus* (QUACK et al., 1978). The crystalline sesquiterpene dialdehyde very strongly inhibits DNA synthesis in Ehrlich ascitic carcinoma (ECA) cells at 1 μg ml^{-1}. In the assay of AMES et al. (1975) merulidial exhibits mutagenic activity. Comparative studies of merulidial and several hydroxylated and acetylated derivatives revealed that the molecular mechanism responsible for the mutagenicity of merulidial is different from the mechanism resulting in antimicrobial and cytotoxic activity. Acetylation of merulidial to 8-acetylmerulidial, e.g., increases antifungal activity but diminishes mutagenic activity. Hydroxylation of merulidial to 9-α-hydroxymerulidial and 9-β-hydroxymerulidial as well as of acetylmerulidial to 9-α-hydroxyacetylmerulidial does not strongly affect mutagenic activity but dramatically reduces antimicrobial, cytotoxic, and phytotoxic activity (ANKE et al., 1989).

The striatins A, B, and C (41–43, Fig. 9) and the corresponding striatals were isolated from submerged cultures of *Cyathus striatus, C. poeppigii, C. limbatus,* and *C. montagnei.* They were also detected in the fruiting bodies (ANKE et al., 1977a). In ECA cells DNA, RNA, and protein syntheses are completely inhibited by 2 μg mL^{-1} striatins. RSV-transformed CEF were found to be inhibited at lower concentrations as compared to their normal counterparts. Studies on the mode of action revealed that interference with the transport of essential precursors was mainly responsible for their cytotoxic activity (LEE and ANKE, 1979). Striatins and striatals were found to prolong the life span of P388 lymphocytic leukemia mice and to be inhibitory in the system colon xenograft-athymic mouse. The LD_{50} for tumor bearing mice was determined to be 150 mg kg^{-1} i. p. In greenhouse experiments striatins exhibited good fungicidal activity against *Plasmopara viticola* on grape vine, *Phytophtora infestans* on potatoes, *Botrytis cinerea* on green pepper, and *Septoria nodorum* on wheat (ANKE et al., 1986).

The schizonellins A and B (44 and 45, Fig. 9) are glycolipids produced by submerged cultures of the smut fungus *Schizonella melanogramma* (DEML et al., 1980). Like ustilagic acids which are glycolipids of a different type obtained from *Ustilago maidis* (LEMIEUX et al., 1951), schizonellins exhibit weak antibacterial and antifungal activity. In ECA cells the incorporation of leucine, uridine, and thymidine into protein, RNA, and DNA is completely inhibited by 25 μg mL^{-1} of schizonellin A or B. Concomitant lysis of the cells suggests a detergent-like mode of action.

Hypnophilin, pleurotellol, and pleurotellic acid (46–48, Fig. 9) were isolated from fermentations of *Pleurotellus hypnophilus* (KUPKA et al., 1981a). While pleurotellol and pleurotellic acid belong to a new group of sesquiterpenoids, hypnophilin is a new member of the hirsutane family to which a number of typical basidiomycete metabolites belong. All three antibiotics exhibit antimicrobial and very high cytotoxic activity. However, in comparison to normal cells no selective toxicity for RSV-transformed chicken embryo fibroblasts (CEF) could be detected. Hypnophilin and pleurotellol also act as plant growth inhibitors. In the *Avena* coleoptile bioassay they strongly inhibit indole-3-acetic acid-induced growth of coleoptile sections. Like other exomethylene ketones and lactones *Pleurotellus* antibiotics very readily form adducts with cysteine or other thiols and they are mutagenic.

Phellodonic acid (49, Fig. 9), a new hirsutane derivative closely related to hypnophilin, has recently been isolated from cultures of *Phellodon melaleucus* (STADLER et al., 1993b). Like hypnophilin, phellodonic acid exhibits antimicrobial and strong cytotoxic activity. The incorporation of radiolabeled precursors into DNA, RNA, and protein of L 1210 cells is almost completely inhibited at a concentration of 5 μg mL^{-1}.

The alliacols A and B (50 and 51, Fig. 9) from *Marasmius alliaceus* are α,β-unsaturated sesquiterpene lactones which exhibit rather low antimicrobial but highly cytotoxic properties (ANKE et al., 1981). In ECA cells nucleic acid biosyntheses are almost completely inhibited at concentrations of 2–10 μg mL^{-1}. Like other α,β-unsaturated lactones, alliacols readily form adducts with nucleophilic thiols. It is assumed that a rapid reaction with SH groups in enzymes or other proteins is responsible for most of the biological activity of these compounds. Deduced alliacolide shows no antibiotic or cytotoxic properties.

Fulvoferruginin (52, Fig. 9), a sesquiterpenoid carotane derivative has been isolated from *Marasmius fulvoferrugineus* (KLEIN et al., 1990). It is closely related to hercynolactone which was isolated from liverworths (HUNECK et al., 1982). Several carotane sesquiterpenes were isolated from *Ferula* species. Fulvoferruginin exhibits modest antibacterial activity and inhibits the growth of several fungi at 5–50 μg mL^{-1}. In ECA cells the incorporation of leucine, uridine, and thymidine into protein, RNA, and DNA was inhibited by 50% at a concentration of 10–50 μg mL^{-1}.

Crinipellins obtained from fermentations of *Crinipellis stipitaria* are the first known natural tetraquinanes (KUPKA et al., 1979; ANKE et al., 1985b). The crinipellins A and B and *O*-acetylcrinipellin A (53–55, Fig. 9) containing an exomethylene ketone moiety are strong antibacterials and highly cytotoxic metabolites. The reduced compounds dihydrocrinipellin B and tetrahydrocrinipellin A (56 and 57, Fig. 9) are inactive. Like striatins and striatals crinipellins exert their cytotoxic activity mainly by interfering with the transport of essential nutrients and precursors. The cytotoxic activity on RSV-transformed CEF seems to be higher than on normal CEF (KUPKA et al., 1980).

The caryophyllanes nematolin and nematolon (58 and 59, Fig. 9) were isolated from cultures of *Naematoloma capnoides, N. sublateritium, N. fasciculare,* and *N. elongatipes* (BACKENS et al., 1984). Comparison to the caryophyllanes of higher plants these basidiomycete metabolites contain more oxygen functions and one or two α,β-unsaturated carbonyl groups. Nematolon and nematolin are weakly antimicrobial, the cytotoxic activity of nematolon is 5-fold higher than that of nematolin. In ECA cells the incorporation of thymidine into DNA is inhibited by 50% at a nematolon concentration of 2 μg mL^{-1}. *In vivo* no significant antitumor activity (B-16 melanocarcinoma, Lewis lung carcinoma, P-388 lymphocytic leukemia) was found for nematolon. The LD_{50} for tumor bearing mice was determined to be >225 mg kg^{-1} (J. DOUROS, National Cancer Institute, USA, personal communication).

Leaianafulvene (60, Fig. 9), an orange-yellow pigment, was isolated from mycelial cultures of *Mycena leaiana* (HARTTIG et al., 1990). The compound is closely related to the illudins and represents the first example of a natural "isoilludane" derivative which may be formed from an illidane precursor by 1,2-migration of a methyl group. Leaianafulvene exhibits weak antibacterial activity, whereas its cytotoxic activity is quite pronounced. A 50% lysis of ECA cells is observed at 2.5 μg mL^{-1}. DNA and RNA syntheses are inhibited by 50% at a leaianafulvene concentration of 10 μg ml^{-1}. In addition, mutagenic acitvity was also observed.

5.4.1 Antitumor Polysaccharides

Antitumor polysaccharides have been obtained from various kinds of preparations. They include lentinan (CHIHARA et al., 1970), a high-molecular weight β-1,3 glucan isolated from fruiting bodies of *Lentinus edodes* (SASAKI and TAKASUKA, 1976), and schizophyllan (KOMATSU et al., 1969), a high-molecular weight β-1,3 1,6 glucan obtained from cultured mycelia of *Schizophyllum commune.* These compounds inhibit the growth of various transplantable tumors in experimental animals, they increase the survival rate and are considered to exert their antitumor activity by the potentiation of the host animals' defense mechanisms rather than by direct inhibition of tumor cell growth (SUGA et al., 1984). Lentinan in its sulfated form was also used in conjunction with AZT to suppress HIV (DE CLERCQ, 1990). Like several other sulfated polysaccharides, lentinan interferes with syncytium formation resulting from fusion of HIV-infected and uninfected cells. KS-2, a peptide containing α-linked mannose was extracted from the mycelia of *Lentinus edodes* (FUJI et al., 1978). KS-2 suppressed the growth of both Ehrlich tumors and Sarcoma 180 tumors at dose levels of 1 mg kg^{-1} and 100 mg kg^{-1} when administered intraperitoneally or orally. It was also capable of inducing interferone in mice.

PSK (krestin), a polysaccharide preparation isolated from *Coriolus versicolor* predominantly consists of glucan and of ca. 25% tightly bound protein (TSUGAGOSHI et al., 1984). Oral administration of PSK increased the survival rate in several animal cancer models, and PSK is now clinically used in Japan for the treatment of postoperative cancer patients. PSK was also reported to exhibit immunomodulating acitvity by regulating cytokine production and effector cell functions (reviewed by KOBAYASHI, 1993).

5.4.2 Immunosuppressive Metabolites

Only few immunosuppressive compounds from basidiomycetes have been reported. In the course of a screening for metabolites suppressing the proliferation of mouse lymphocytes stimulated with mitogens, three geranylphenols, flavidulol A, B, and C (61–63, Fig. 10) have been isolated from fruiting bodies of *Lactarius flavidulus* (FUJIMOTO et al., 1993). The IC_{50} values for flavidulols A, B, and C were found to be 8.9 μg mL^{-1}, 4.9 μg mL^{-1}, and 36.3 μg mL^{-1}, respectively, in an assay measuring concanavalin A-induced proliferation of mouse lymphocytes, and 6.7 μg mL^{-1}, 3.9 μg mL^{-1}, and 28.3 μg mL^{-1}, respectively, when the cells were stimulated with lipopolysaccharide.

5.5 Antiviral Compounds and Inhibitors of Reverse Transcriptases

The nucleosides 6-methylpurine, 6-methyl-9-β-D-ribofuranosylpurine, and 6-hydroxymethyl-9-β-D-ribofuranosylpurine (64–66, Fig. 11) were isolated from mycelial cultures of *Collybia maculata* in a screening for inhibitors of vesicular stomatitis virus (VSV) multiplication in baby hamster kidney (BHK) cells (LEONHARDT et al., 1987). 6-methylpurine and 6-methyl-9-β-D-ribofuranosylpurine had been obtained before by chemical synthesis. All three nucleosides exhibit modest antifungal and cytotoxic activity; the effect on VSV multiplication in BHK cells is high and compares very favorably with that of *araA*. Besides their antiviral activity 6-methyl-9-β-D-ribofuranosylpurine and 6-hydroxymethyl-9-β-D-ribofuranosylpurine are inhibitors of adenosine desaminase and, therefore, interfere with the nucleoside metabolism.

Other nucleosides that have been reported as secondary metabolites from basidiomycetes are nebularine, described as an antibacterial antibiotic from *Clitocybe nebularis* (LÖFGREN, 1954), lentinacin, a hypercholesterolemic compound from *Lentinus edodes* (CHIBATA et al., 1969), and the insecticidal compound clitocine (67, Fig. 11) from *Clitocybe inversa* (KUBO et al., 1986). A compilation of nucleoside antibiotics from microbial sources and their biological activities was published by ISONO (1988) and by ISAAC et al. (1991).

A screening for inhibitors of avian myeloblastosis virus (AMV) reverse transcriptase resulted in the isolation of clavicoronic acid (68, Fig. 11) from fermentations of *Clavicorona pyxidata* (ERKEL et al., 1992). Clavicoronic acid is a non-competitive inhibitor of AMV (K_i: 130 μM) and Moloney murine leukemia (MMuLV) virus (K_i: 68 μM) reverse transcriptases. In permeabilized cells and isolated nuclei DNA and RNA synthesis are not

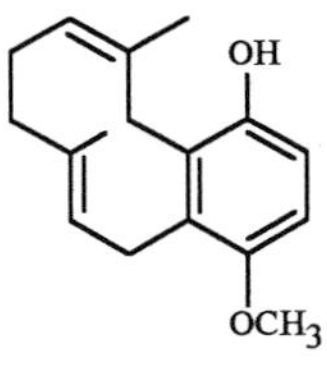

OH OCH3 Flavidulol A (61)

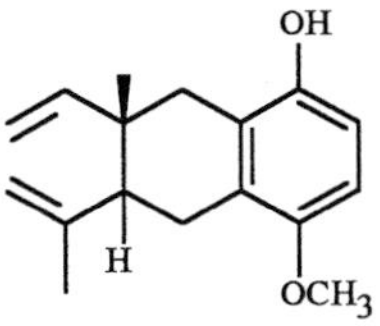

OH H OCH3 Flavidulol B (62)

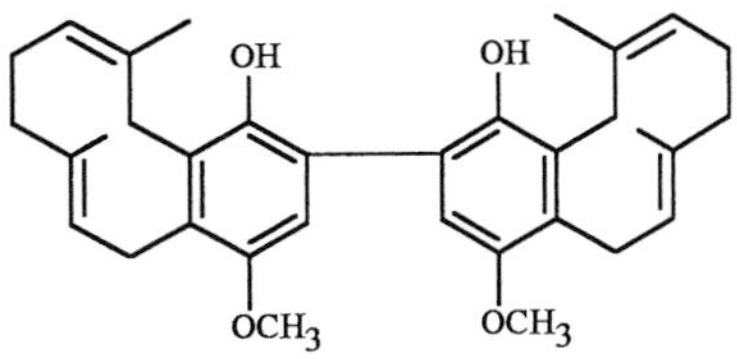

OH OH OCH3 OCH3 Flavidulol C (63)

Fig. 10. Immunosuppressive metabolites.

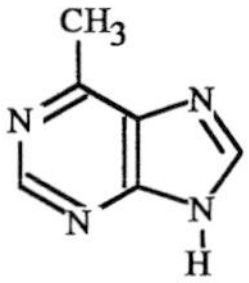

6-Methylpurine (64)

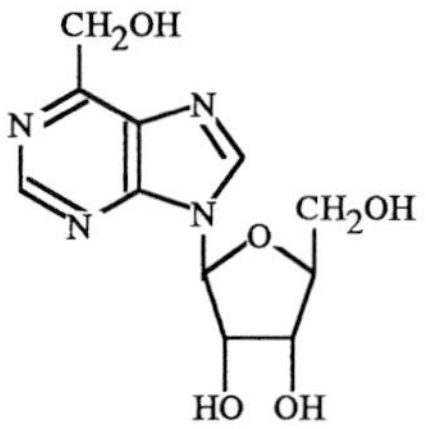

6-Hydroxymethyl-9-ß-D-ribofuranosylpurine (66)

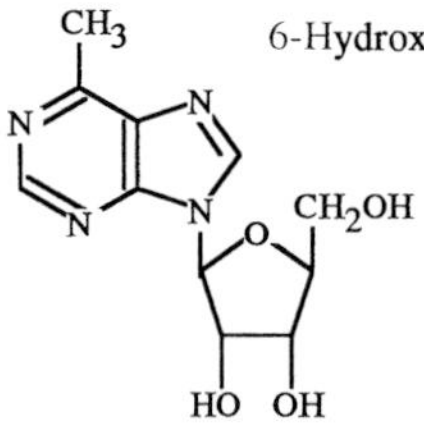

6-Methyl-9-ß-D-ribofuranosylpurine (65)

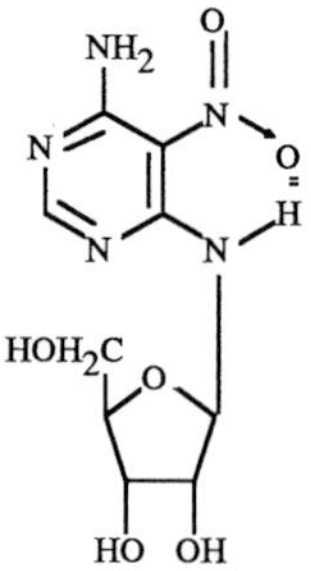

Clitocine (67)

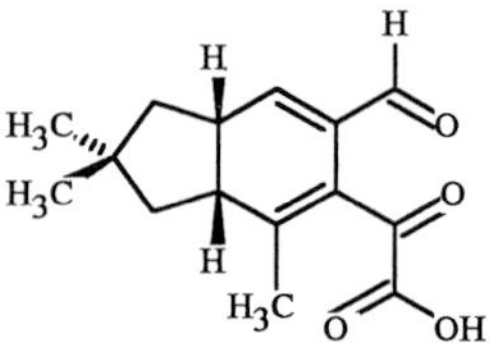

Clavicoronic acid (68)

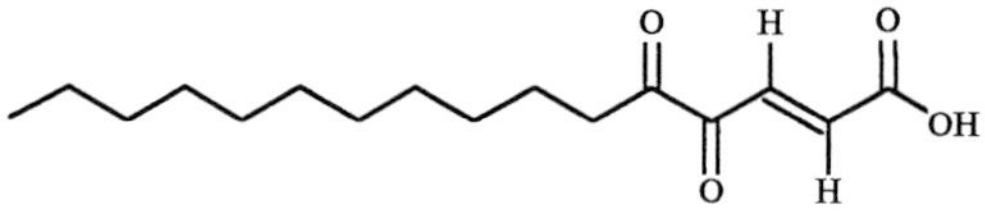

Podoscyphic acid (69)

Fig. 11. Antiviral compounds and inhibitors of reverse transcriptases.

affected. Clavicoronic acid markedly inhibits the multiplication of VSV in BHK cells by interfering with the RNA-directed RNA polymerase of the virus. Clavicoronic acid exhibits no cytotoxic and very weak antimicrobial activity.

Podoscyphic acid, (*E*)-4,5-dioxo-2-hexadecenoic acid (69, Fig. 11), isolated from fermentations of *Podoscypha petalodes,* is a non-competitive inhibitor of AMV and MMuLV reverse transcriptase. The IC_{50} values for the inhibition of AMV reverse transcriptase were 100 $\mu g\ mL^{-1}$ and for the MMuLV reverse transcriptase 10–20 $\mu g\ mL^{-1}$. DNA and RNA syntheses in whole cells and isolated nuclei are not affected by 100 $\mu g\ mL^{-1}$ podoscyphic acid. Comparison of the ethyl ester and the mono-oxo derivative of podoscyphic acid revealed the importance of the free γ-oxoacrylate moiety for its biological activity (ERKEL et al., 1991).

Several drimane sesquiterpenoids from basidiomycetes have been reported as inhibitors of reverse transcriptases. Drimanes had been isolated from a number of other natural sources including higher plants, ascomycetes, mollusks, and sponges (reviewed by JANSEN and DE GROOT, 1991). The mniopetals A, B,

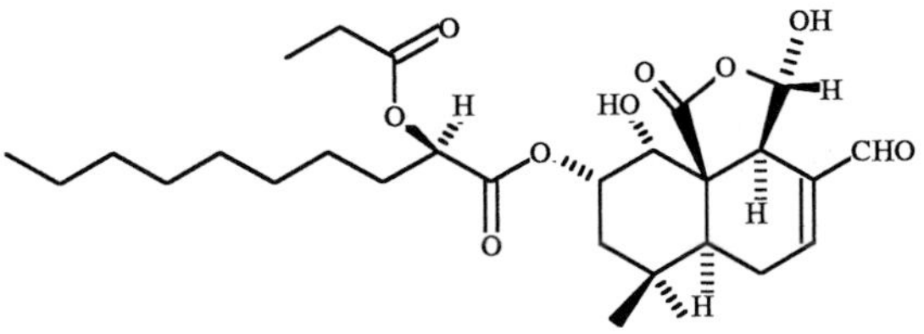

Mniopetal A (70)

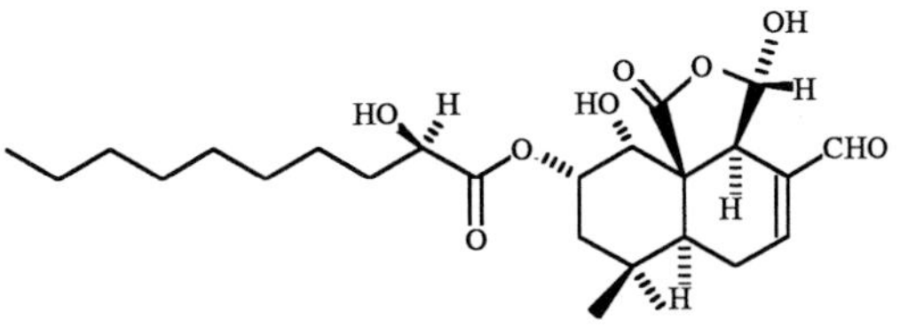

Mniopetal B (71)

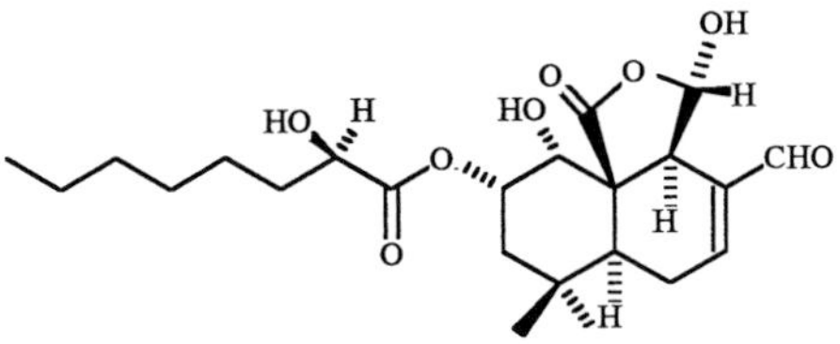

Mniopetal C (72)

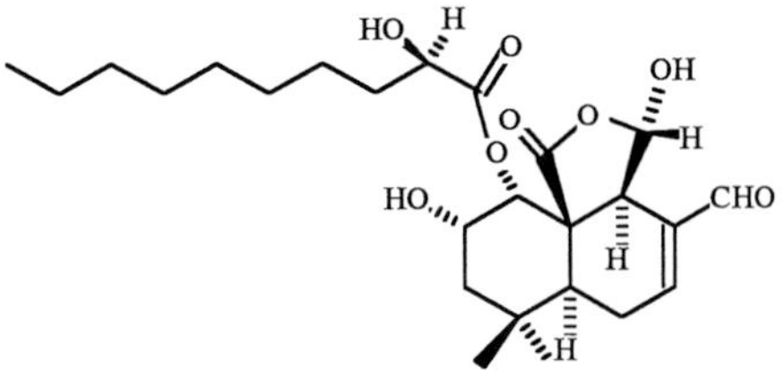

Mniopetal D (73)

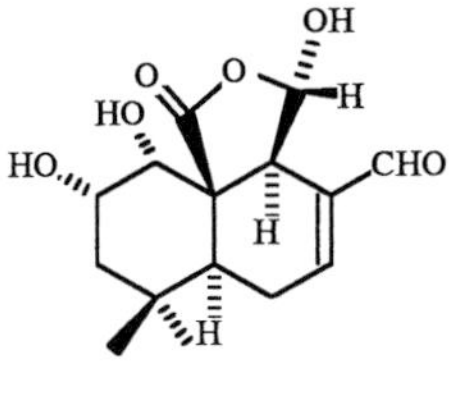

Mniopetal E (74)

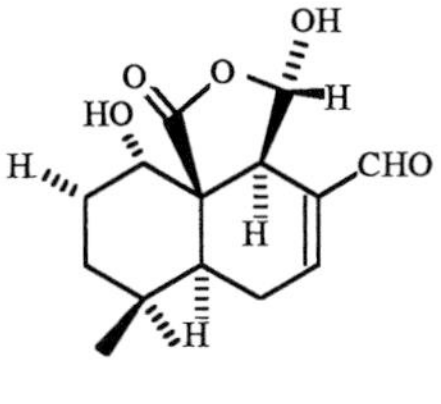

Mniopetal F (75)

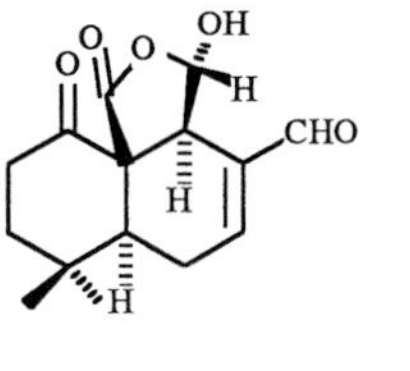

Kuehneromycin A (76)

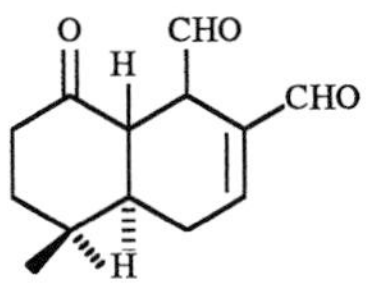

Fig. 11 Kuehneromycin B (77)

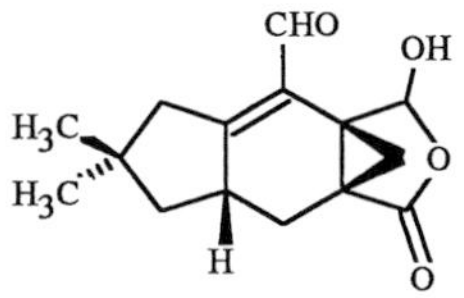

Hyphodontal (78)

C, D, E, and F (70–75, Fig. 11) have been isolated from fermentations of a Canadian *Mniopetalum* species (KUSCHEL et al., 1994). They most strongly inhibit the MMuLV reverse transcriptase at concentrations of 1.7–50 μM, with mniopetal B being the most active compound (IC_{50}: 1.7 μM). The IC_{50} values for the AMV reverse transcriptase are much higher. Inhibition of HIV-1 reverse transcriptase by mniopetals depends on the template primer used. With a natural heteropolymeric template inhibition of HIV-1 reverse transcriptase is most pronounced at concentrations of 30–190 μM. In addition, mniopetals exhibit cytotoxic properties which may be at least partly due to a lyctic action on the cytoplasma membrane. The kuehneromycins A and B (76 and 77, Fig. 11) have been isolated from cultures of a Tasmanian *Kuehneromyces* species (ERKEL et al., 1995; ANKE et al., 1993). Like mniopetals the kuehneromycins A and B preferentially inhibit the MMuLV reverse transcriptase with an IC_{50} of 36 μM, while inhibition of AMV reverse transcriptase is much less pronounced. The activity of HIV-1 reverse transcriptase with the natural heteropolymeric template is reduced to 50% at a concentration of 64 μM. In addition, both compounds exhibit cytotoxic and antimicrobial activity.

Hyphodontal (78, Fig. 11), a new isolactarane sesquiterpenoid, has been isolated from fermentations of a *Hyphodontia* species (ERKEL et al., 1994). Hyphodontal strongly inhibits the growth of several yeasts and is a noncompetitive inhibitor of AMV (K_i: 349 μM) and MMuLV (K_i: 112 μM) reverse transcriptase. The IC_{50} for the HIV-1 reverse transcriptase with the natural heteropolymeric template was determined to be 77 μM (20 μg mL^{-1}). The cytotoxic activity of hyphodontal is mainly due to the interference with DNA and RNA syntheses in whole cells and isolated nuclei.

Up to now, none of the inhibitors described above has been shown to inhibit the multiplication of HIV-1 or HIV-2 viruses in cellular systems.

5.6 Inhibitors of Platelet Aggregation

A screening for antithrombotic compounds using platelet rich plasma from bovine slaughter blood resulted in the isolation of 2-methoxy-5-methyl-1,4-benzochinone (MMBC; 79, Fig. 12) from mycelial cultures of *Lentinus adhaerens* (LAUER et al., 1991).

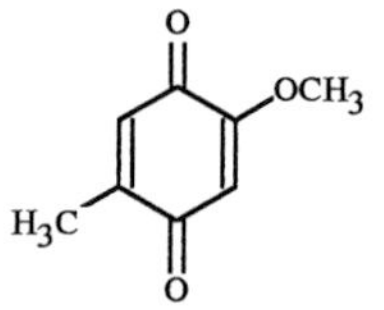

O, OCH3, H3C, O

MMBC (79)

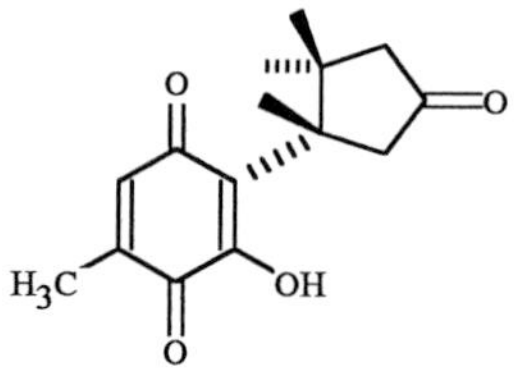

O, O, H3C, OH, O

Lagopodin B (80)

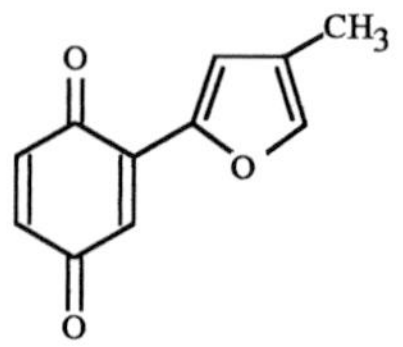

O, CH3, O, O

Omphalone (81)

O, CHO, H, CHO, H

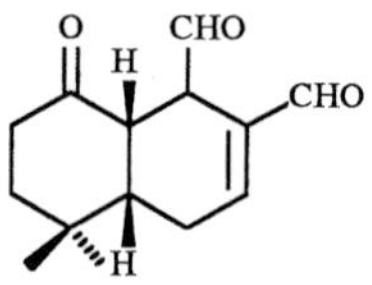

Panudial (82)

Fig. 12. Inhibitors of platelet aggregation.

MMBC inhibits aggregation of human blood platelets induced by U46619 a prostaglandine analog and thromboxane mimic with an IC_{50} of 2.5 μg mL^{-1} (16.45 μM). This effect is completely reversible by the addition of 1.8–2.4 μM U46619. Hence, it MMBC is proposed to act as a competitive thromboxane A_2 receptor antagonist. Similar effects on platelet aggregation were observed with the benzoquinones lagopodin B (80, Fig. 12) obtained from *Coprinus cinereus* and omphalon (81, Fig. 12), an antimicrobial and cytotoxic metabolite isolated from cultures of *Lentinellus omphalodes* (STÄRK et al., 1991).

The drimane panudial (82, Fig. 12) from cultures of a *Panus* species was detected as an inhibitor of platelet aggregation (LORENZEN et al., 1993). Panudial strongly interferes with the ADP-, collagen-, U46619-, ristocetin-, arachidonic acid-, and thrombine-induced aggregation of human and bovine platelets. The IC_{30} values of panudial for all inducers except for thrombine varied between 5 μM and 35 μM. In addition, panudial showed inhibitory activity against type I phospholipase A_2 (IC_{50}=23 μg mL^{-1}) from *Naja mosambique.* In HL-60 cells the incorporation of leucine and uridine into protein and RNA is markedly inhibited by 4–8 μg mL^{-1} panudial.

5.7 Herbicidal Compounds

A screening for inhibitors of the key enzymes of the glyoxylate cycle, which are potential targets for herbicides, resulted in the isolation of mycenon (83, Fig. 13), a novel chlorinated benzoquinone derivative from cultures of a *Mycena* species (HAUTZEL et al., 1990). Mycenon inhibits isocitrate lyase preparations from plants, bacteria, and fungi. The K_i values were determined to be 5.2 μM, 11 μM, and 7.4 μM for the enzymes from *Rhicinus communis, Acinetobacter calcoaceticus,* and *Neurospora crassa,* respectively. Malate synthase, the second key enzyme of the glyoxylate cycle, was not affected. In addition to its phytotoxic activities mycenon exhibits antimicrobial and cytotoxic properties.

Antibiotically active products which are structurally related to mycenon are siccayne (84, Fig. 13) from cultures of the marine basidiomycetes *Halocyphina villosa* (KUPKA et al., 1981b) and *Helminthosporium siccans* (ISHIBASHI et al., 1968), frustulosinol, and frustulosin (85 and 86, Fig. 13) isolated from cultures of *Stereum frustolosum* (NAIR and ANCHEL, 1977). These compounds are hydrochinone derivatives with an isopentenyne side chain but without chlorine substitutions. Other acetylenic substances containing an aromatic ring are the antibacterial and antifungal antibiotics peniophorin B and A (87 and 88, Fig. 13) (GERBER et al., 1980).

The pereniporins A and B (89 and 90, Fig. 13) were isolated from cultures of *Perenniporia medullaepanis.* Pereniporin A inhibits the root elongation of lettuce at 100 μg mL^{-1} and is active against gram-positive bacteria. Both compounds show cytotoxic activity against Friend leukemia cells at 130 μg mL^{-1} and 3.91 μg mL^{-1}, respectively (KIDA et al., 1986).

Fomannosin (91, Fig. 13) was isolated from cultures of the wood-rotting basidiomycete *Fomes annosus* (*Heterobasidion annosum*). *F. annosus* is one of the relatively few basidiomycetes that cause the death of host cells in living trees and an extensive decay of heartwood of contaminated trees. Fomannosin exhibits phytotoxic activity in assays with *Chlorella pyrenoidosa* and against *Pinus taeda* seedlings when applied to the stem base or a lateral root at a concentration of 88 μg per seedling (BASSETT et al., 1967). The absolute configuration and the biosynthesis of fomannosin was elucidated by CAINE and NACHBAR (1978). Fomannoxin (92, Fig. 13), a dihydrobenzofuran which is 100 times more toxic to *Chlorella pyrenoidosa*, was isolated from the same fungus by HIROTANI et al. (1977).

Several plant growth inhibitors were isolated from fruiting bodies of *Naematoloma fasciculare.* The fasciculols A, B, and C (93–95, Fig. 13) are tetracyclic triterpenes with a hydroxylated lanostane skeleton. The fasciculols D, E, and F (96–98, Fig. 13) are the corresponding esters with a novel depsipeptide group consisting of 3-hydroxy-3-methylglutaric acid and glycine (IKEDA et al., 1977). The fasciculols and their depsipeptides inhibited the growth of Chinese cabbage seedlings at concentrations of 100–300 μg mL^{-1}. In addition, fasciculol D exhibited weak antibac-

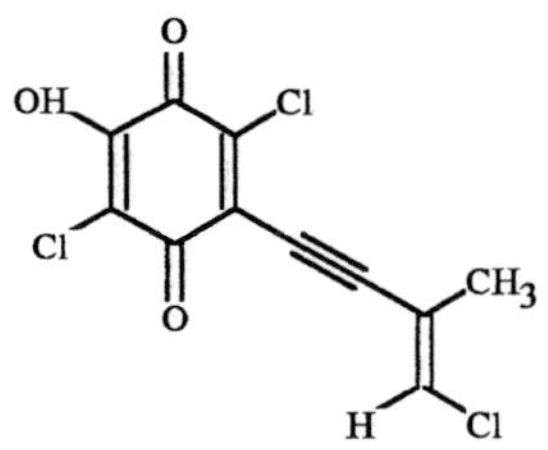

Mycenon (83)

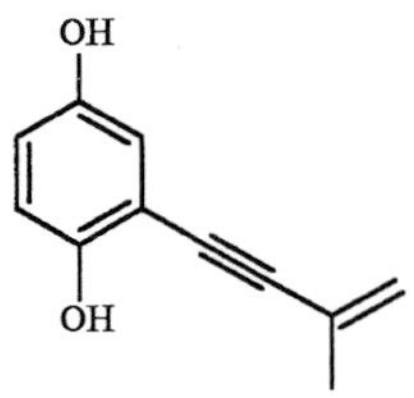

Siccayne (84)

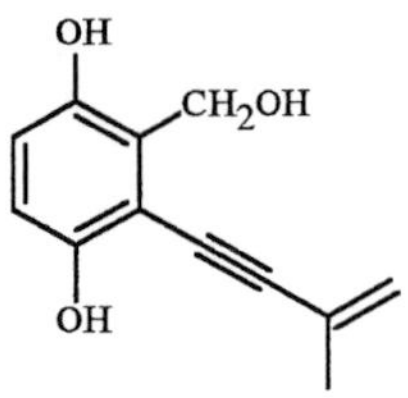

Frustulosinol (85)

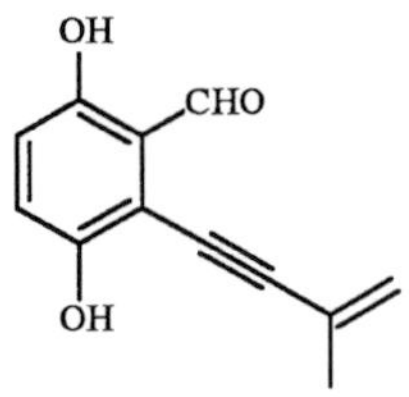

Frustulosin (86)

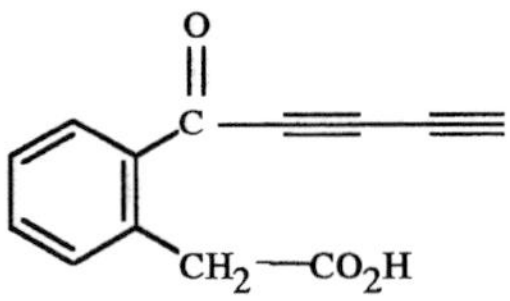

Peniophorin B (87)

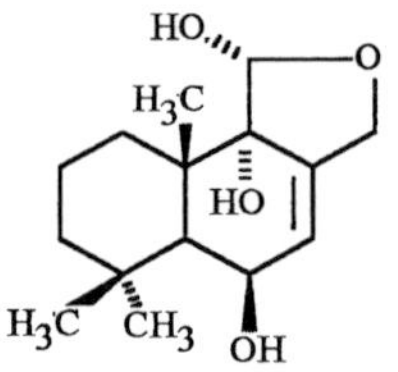

Pereniporin A (89)

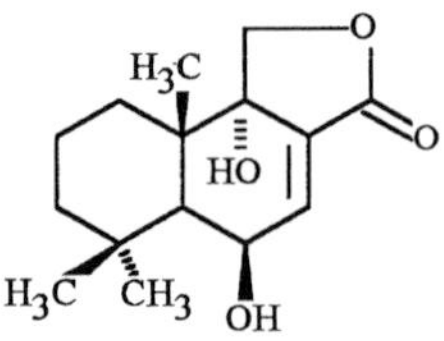

Pereniporin B (90)

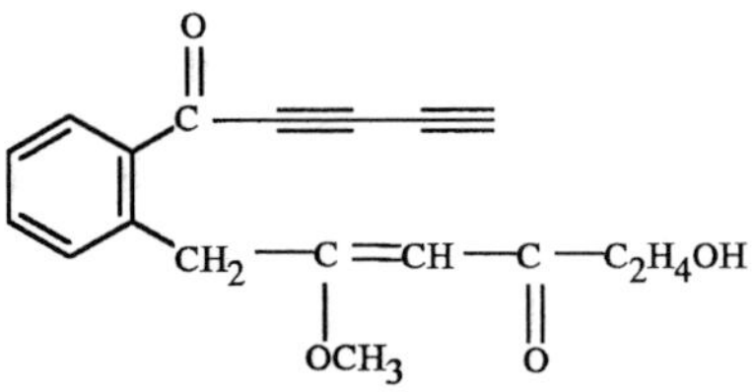

Peniophorin A (88)

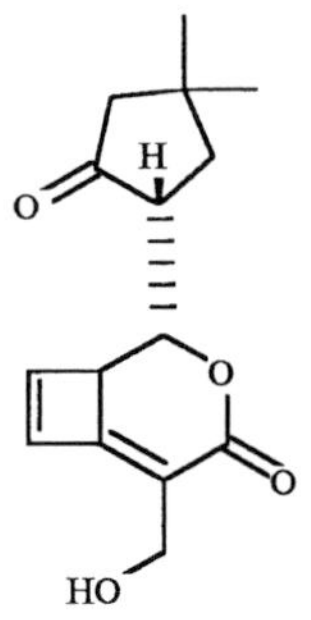

Fomannosin (91)

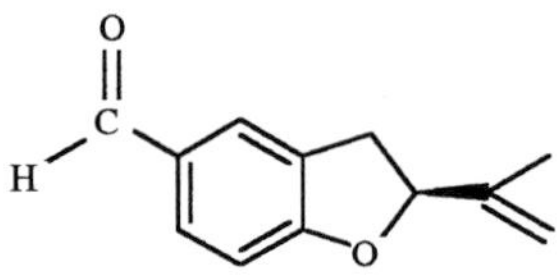

Fomannoxin (92)

Fig. 13. Herbicidal compounds.

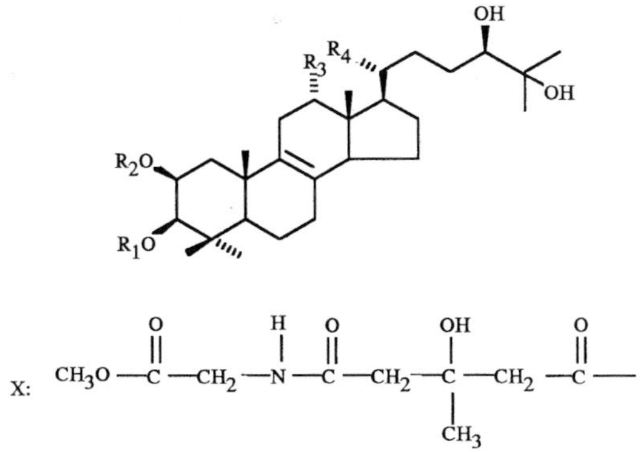

X: $CH_3O-C(=O)-CH_2-NH-C(=O)-CH_2-C(OH)(CH_3)-CH_2-C(=O)-$

	R_1	R_2	R_3	R_4
Fasciculol A (93)	H	H	H	CH_3
Fasciculol B (94)	H	H	OH	CH_3
Fasciculol C (95)	H	H	OH	CH_2OH
Fasciculol D (96)	X	H	OH	CH_3
Fasciculol E (97)	H	X	OH	CH_2OH
Fasciculol F (98)	X	H	OH	CH_2OH

Fig. 13

terial activity against *Staphylococcus aureus* and *Klebsiella pneumoniae.*

Several new related fasciculol esters, the fasciculic acids A, B, and C, were isolated from fruiting bodies of *Naematoloma fasciculare* as inhibitors of a calmodulin-dependent phosphodiesterase (PDE) (TAKAHASHI et al., 1989). The IC_{50} values for the PDE from bovine heart were 6 μM for fasciculic acid B and 10 μM for fasciculic acid A. No data on phytotoxic activity have been published.

Related tetracyclic triterpenoids with a lanostane skeleton have also been isolated from *Gandoderma lucidum, G. applanatum, Piolithus tinctorius, P. arrhizus,* and *Fomes fastuosus* (reviewed by CONNOLLY et al., 1994).

5.8 Insecticidal and Nematicidal Metabolites

Insecticidal activity against houseflies was described for ibotenic acid (99, Fig. 14) and its decarboxylation product muscimol (100, Fig. 14). Ibotenic acid was isolated from fruiting bodies of *Amanita muscaria, A. strobiliformis,* and *A. pantherina* (for a review, see BRESINSKY and BESL, 1985). The cyclodepsipeptide beauvericin (101, Fig. 14) which was isolated from the basidiomycete *Polyporus sulphureus* (DEOL et al., 1978) and from the entomopathogenic fungi *Beauveria bassiana* and *Paecilomyces fumoso-roseus,* exhibits insecticidal activity against mosquito larvae, brine shrimp, houseflies, and cockroach cardiac cells *in vitro* (ROBERTS, 1981). Beauvericin is a ionophore forming complexes with alkali metals. Recently, inhibitory activity on acyl CoA–cholesterol acyltransferase (ACAT) in isolated rat liver microsomes has been described for beauvericin with an IC_{50} value of 3.0 μM (TOMODA et al., 1992). In a cell assay using J774 macrophages the formation of cholesteryl esters was inhibited by 0.17 μM beauvericin.

More than 150 species of nematophagous fungi belonging to Zygomycetes, Ascomycetes, Deuteromycetes, and Basidiomycetes are known to be capable of capturing nematodes, and hence the production of nematicidal toxins has been proposed (BARRON, 1977; SAYRE and WALTER, 1991).

5-Pentyl-2-furaldehyde, 5-(4-pentenyl)-2-furaldehyde, and methyl-3-*p*-anisoloxypropionate (102–104, Fig. 14), were isolated from cultures of *Irpex lacteus* (HAYASHI et al., 1981). All three compounds caused 50% mortality of the nematode *Aphelencoides besseyi* at 25–50 μg mL^{-1}.

2-Pecenedioic acid (105, Fig. 14), a fatty acid toxic to *Panagrellus redivius,* was isolated from a *Pleurotus ostreatus* strain. The compound immobilized 95% of the test nematode at a concentration of 300 μg mL^{-1} within 1 h (KWOK et al., 1992).

A screening using the saprophytic nematode *Caenorhabditis elegans* as a test organism resulted in the isolation of *S*-coriolic acid, linoleic acid, *p*-anisaldehyde, *p*-anisyl alcohol, 1-(4-methoxyphenyl)-1,2-propanediol, and 2-hydroxy-(4-methoxy)-propiophenone (106–111, Fig. 14) from cultures of *Pleurotus pulmonaris* (STADLER et al., 1993b). The most active metabolites were *S*-coriolic acid and linoleic acid with LD_{50} values between 5 and 10 μg mL^{-1}. Interestingly, the nematicidal

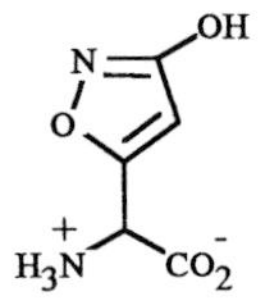

N OH O H_3N^+ CO_2^-

Ibotenic acid (99)

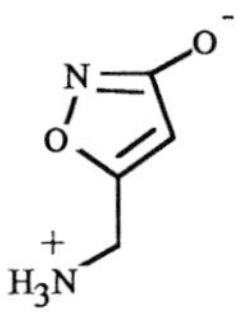

N O^- O H_3N^+

Muscimol (100)

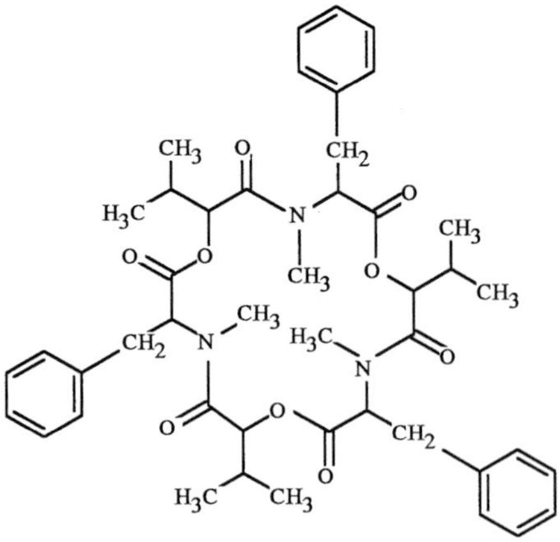

CH_3 O CH_2 H_3C N O CH_3 O O CH_3 O CH_3 CH_3 CH_2 N H_3C N O O O CH_2 O O H_3C CH_3

Beauvericin (101)

O O

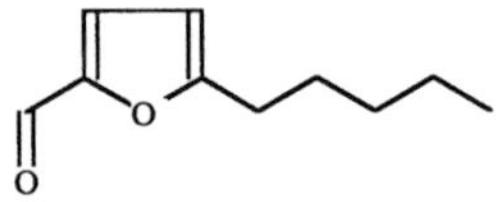

5-Pentyl-2-furaldehyde (102)

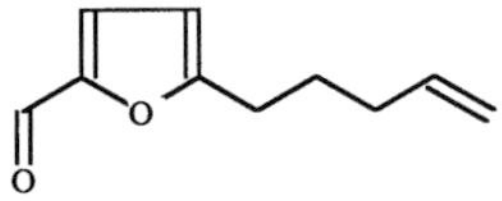

O O

5-(4-Pentenyl)-2-furaldehyde (103)

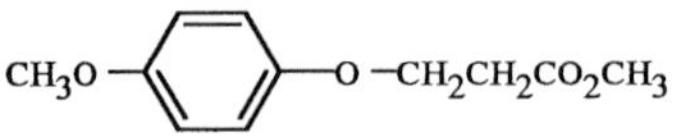

CH_3O—⟨benzene⟩—$O-CH_2CH_2CO_2CH_3$

Methyl-3-*p*-anisoloxypropionate (104)

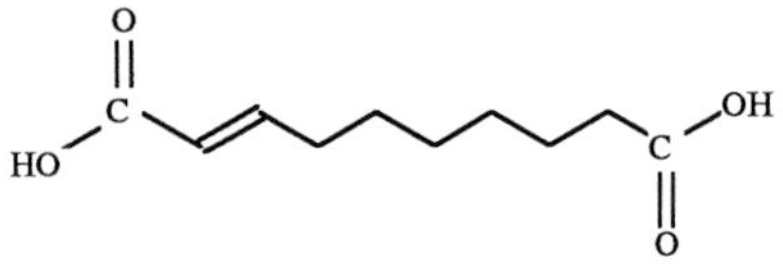

O C HO C OH O

trans-2-Decenoic acid (105)

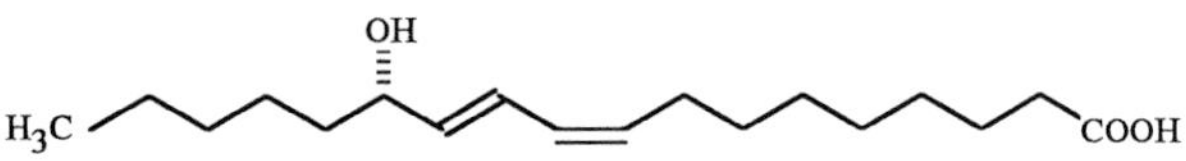

OH H_3C COOH

S-Coriolic acid (106)

H_3C COOH

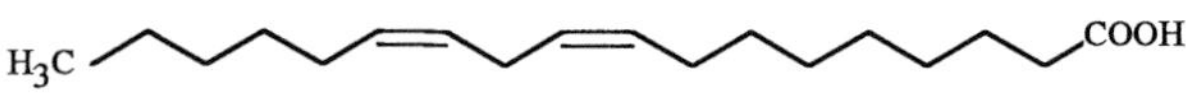

Linoleic acid (107)

Fig. 14. Insecticidal and nematicidal metabolites.

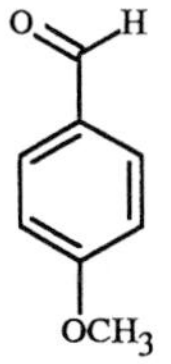

p-Anisaldeyde (108)

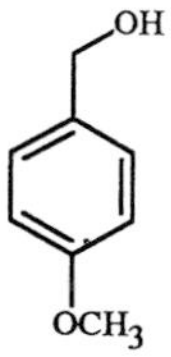

p-Anisyl alcohol (109)

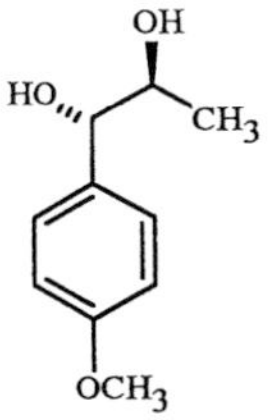

1-(4-Methoxyphenyl)-1,2-propanediol (110)

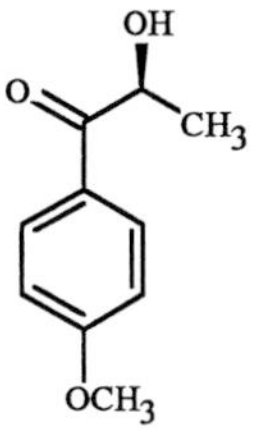

2-Hydroxy-(4´-methoxy)-propriophenone (111)

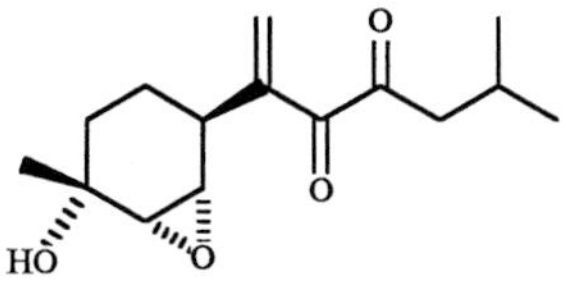

Cheimonophyllon A (112)

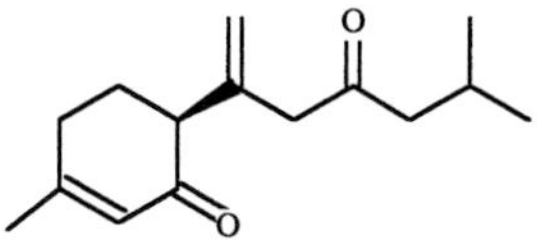

Cheimonophyllon B (113)

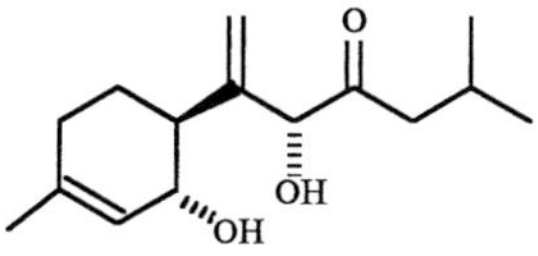

Cheimonophyllon C (114)

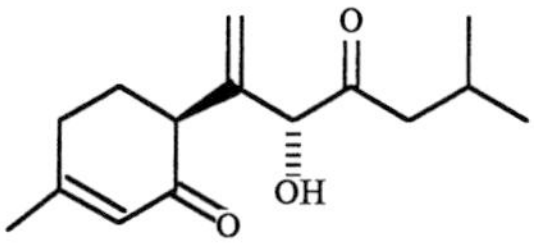

Cheimonophyllon D (115)

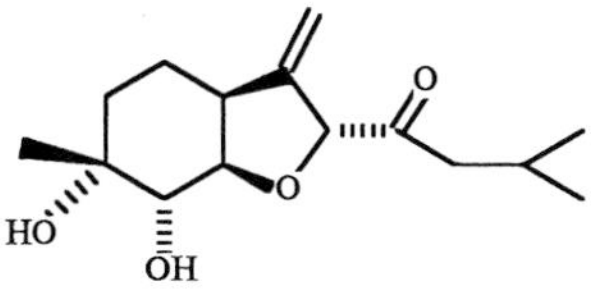

Cheimonophyllon E (116)

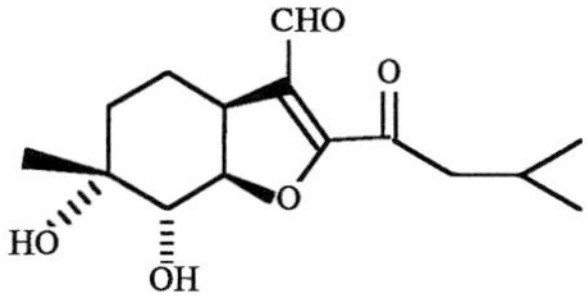

Cheimonophyllal (117)

Fig. 14

activity of different fatty acids depends on the chain length and the number of double bonds in the molecule. Furthermore, different species of nematodes varied in their sensitivity to various fatty acids.

Six different bisabolane sesquiterpenes, the cheimonophyllons A–E (112–116, Fig. 14), and cheimonophyllal (117, Fig. 14) from cultures of *Cheimonophyllum candidissimum* have recently been described (STADLER et al., 1994). Cheimonophyllons A, B, D, and cheimonophyllal exhibit nematicidal, cytotoxic, and antimicrobial activity, whereas cheimonophyllons C and E show weaker effects. The LD_{50} for *Caenorhabditis elegans* was 10 $\mu g\ mL^{-1}$ for cheimonophyllon A and D and 25 $\mu g\ mL^{-1}$ for cheimonophyllon B and cheimonophyllal.

5.9 Inhibitors of Cholesterol Biosynthesis

Drugs interfering with the biosynthesis of cholesterol are of potential value in the treatment of hypercholesterolemia which is one of the primary causes of arteriosclerosis and coronary heart disease. Mevinolin (monacolin K), a specific inhibitor of eukaryotic 3-hydroxy-3-methylglutaryl coenzyme A (HMG CoA) reductase isolated from cultures of *Monascus ruber* (ENDO, 1979) has been introduced into clinical practice.

Search for inhibitors of cholesterol biosynthesis in HeLa S3 cells resulted in the isolation of dihydroxerulin, xerulin, and xerulinic acid (118–120, Fig. 15), from surface cultures of *Xerula melanotricha* (KUHNT et al., 1990). Dihydroxerulin strongly interferes with the incorporation of ^{14}C acetate into cholesterol, while the incorporation of ^{14}C mevalonate is hardly affected. It was shown that the inhibitory effect on cholesterol biosynthesis is due to a strong inhibition of HMG CoA synthase at concentrations starting from 0.1 $\mu g\ mL^{-1}$ of dihydroxerulin. Similar results have been obtained with xerulin.

5.10 Inhibitors of Aminopeptidases

Among the enzymes bound to the outer surface of mammalian cells aminopeptidases have been reported to be potential targets for immunomodulating drugs. A prominent example is bestatin (ubenimex), a strong inhibitor of aminopeptidase B and leucine aminopeptidase, which was isolated from cultures of *Streptomyces olivoreticuli* (UMEZAWA et al., 1976). Clinical studies of this drug were published by MATHÉ (1987).

Tyromycin A (121, Fig. 16) from fermentations of *Tyromyces lacteus* is the first known naturally occurring citraconic anhydride derivative with two 3-methyl maleic anhydride units in its molecule. Tyromycin A strongly inhibits both leucine aminopeptidase and cysteine aminopeptidase bound to the outer surface of HeLa S3 cells. The K_i values were determined to be $4 \cdot 10^{-5}$ M for leucine aminopeptidase and $1.3 \cdot 10^{-5}$ M for cysteine aminopeptidase. Tyromycin A also inhibits cytosolic and microsomal leucine aminopeptidase of porcine kidney and carboxypeptidase of bovine kidney at concentrations of 25–60 $\mu g\ mL^{-1}$. The inhibitory activity of tyromycin A is due to the two maleic acid anhydride moieties. Tyromycin amide (122, Fig. 16) is devoid of any inhibitory activity on the cell-bound aminopeptidase of HeLa cells (WEBER et al., 1992).

5.11 Inhibitors of Phospholipases C and A_2

Caloporoside (123, Fig. 17), an inhibitor of phospholipase C, has been isolated from fermentations of *Caloporus dichrous* (WEBER et al., 1994). Caloporoside is a new glycosylated salicylic acid derivative which exhibits weak antibacterial and antifungal activity. Caloporoside shows a strong, selective inhibitory activity on phospholipase C from pig brain with a K_i value of 12.3 μM. Phospholipases C from *Clostridium welchii* and *Bacillus cereus,* which act on other substrates are inhibited to a much lesser extent. Phospholipases A_2 and D,

O
O
CH3
O
O
CH3
O
O
COOH

Dihydroxerulin (118) Xerulin (119) Xerulinic acid (120)

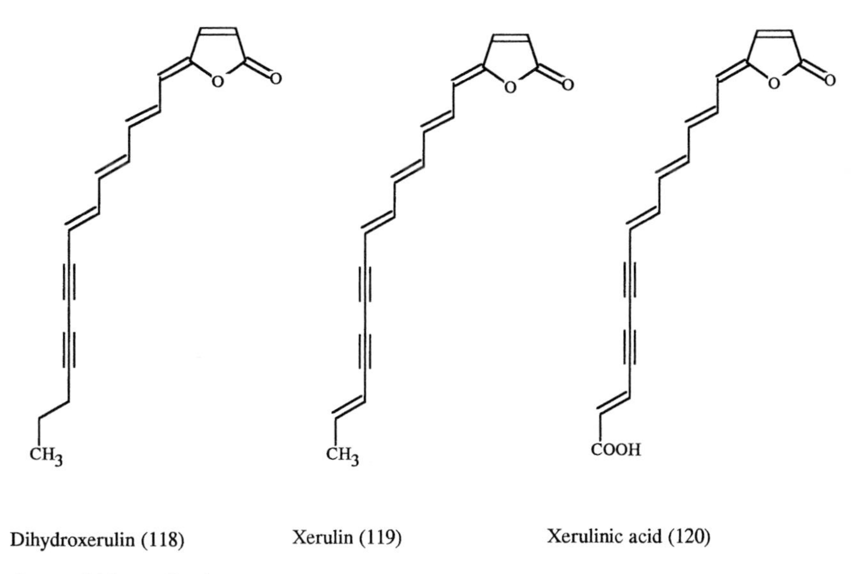

Fig. 15. Inhibitors of cholesterol biosynthesis.

triglyceride lipases, and acetylcholin esterase are not affected.

Several closely related derivatives of the aglycone part of caloporoside were isolated from fruiting bodies of *Merulius tremellosus* and *Phlebia radiata.* The merulinic acids A, B, and C (124–126, Fig. 17) exhibit antimicrobial and hemolytic activity which may be due to their lytic action on the cytoplasma membrane (GIANNETTI et al., 1978). Merulinic acids closely resemble the skin irritants of Anacardiaceae and of *Ginkgo biloba,* anacardic acid III, pelandjauic acid, and ginkgolic acid. Interestingly, mycelial cultures of *M. tremellosus* do not produce merulinic acids but the antibiotic sesquiterpenoid merulidial.

5-Hydroxy-3-vinyl-2(5H)-furanone (127, Fig. 17) has been isolated from cultures of *Calyptella* sp. (LORENZEN et al., 1995). With respect to the hydroxy-butenolide ring it resembles to the manoalides and luffarielloides which were isolated from *Luffariella variabilis* and related sponges (POTTS and FAULKNER, 1992). 5-Hydroxy-3-vinyl-2(5H)-furanone specifically inhibits the human synovial phospholipase A_2 with IC_{50} values of 100 nM. The compound also inhibits the aggregation of human and bovine platelets stimulated with different inducers.

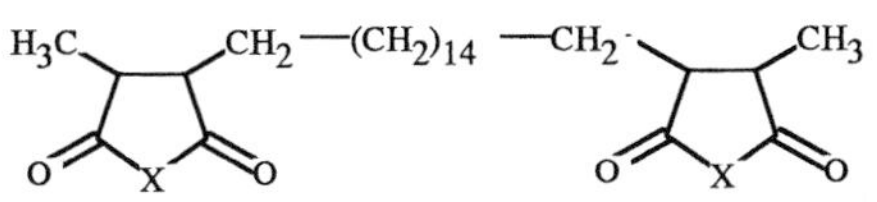

H3C
CH2—(CH2)14—CH2
CH3
O
X
O
O
X
O

Tyromycin A (121) X = O

Tyromycin amide (122) X = NH

Fig. 16. Inhibitors of aminopeptidases.

5.12 Inhibitors of $(Na^+–K^+)$-ATPases

The tricyclic sesquiterpenoids coriolin and coriolin B (128 and 129, Fig. 18) were isolated from cultures of *Coriolus consors* (TAKEUCHI et al., 1969). While coriolin B shows neither antitumor nor antimicrobial activity, its oxidation product diketocoriolin B (130, Fig. 18) does. Studies on the mode of action of diketocoriolin B revealed that the antitumor activity is due to the inhibition of $(Na^+–K^+)$-ATPase localized in the cell membrane of tumor cells which causes a cessation of growth (KUNIMOTO et al., 1973). By chemical modification of coriolin NISHIMURA et al. (1977) showed that the keto group at C-5 and the two epoxy

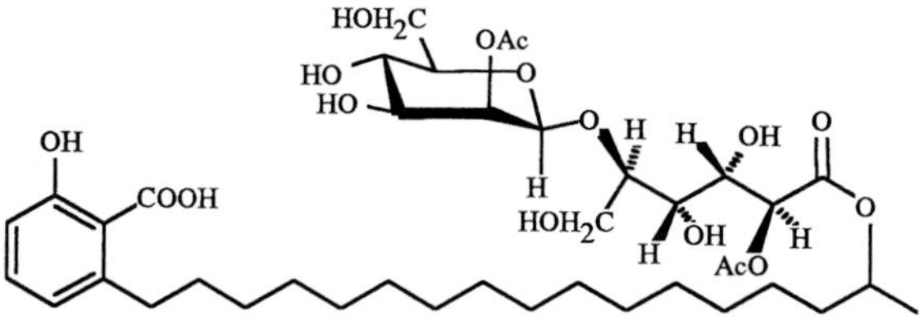

Caloporoside (123)

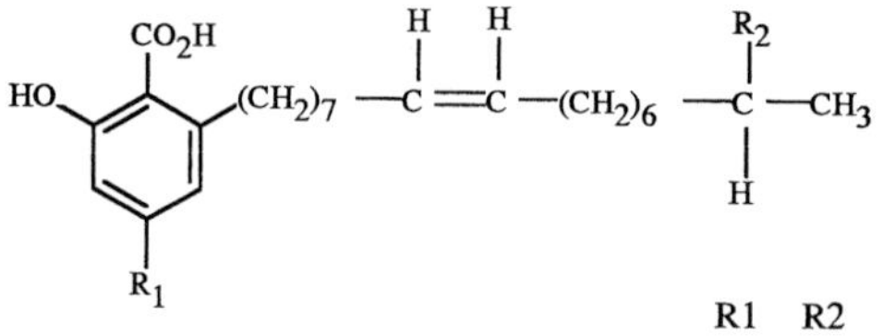

	R1	R2
Merulinic acid A (124)	OH	H
Merulinic acid B (125)	H	OH
Merulinic acid C (126)	H	H

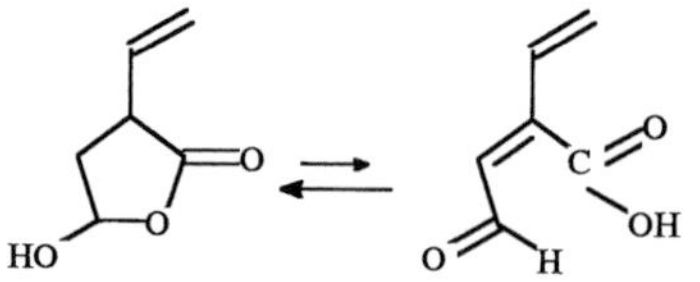

5-Hydroxy-3-vinyl-2(5H)-furanone (127)

Fig. 17. Inhibitors of phospholipases C and A_2.

groups greatly contribute to the antitumor and antibacterial activity. Diketocoriolin B augments antibody formation against sheep red blood cells (SRBC) *in vivo* at a concentration of 0.1 μg per mouse or in *in vitro* using spleen cell cultures at 0.01 ng per culture (ISHIZUKA et al., 1981).

5.13 Addendum

This section describes compounds isolated from 1995 until November 1996. They are presented under the appropriate section number of the main text.

Section 5.5: A screening for inhibitors of multiplication of vesicular stomatitis virus (VSV) in baby hamster kidney cells (BHK-21) resulted in the isolation of collybial (131, Fig. 19) from fermentations of *Collybia confluens* (SIMON et al., 1995). The propagation of VSV in BHK-21 cells was reduced by a factor of 10^3 at 21.5 μM of collybial with cytotoxic effects at 5-fold higher concentrations. Incorporation of labeled precursors into DNA, RNA, and proteins revealed that the antiviral effects of collybial are probably due to an interference with the macromolecular syntheses of the host. In addition, antibacterial activities against gram-positive bacteria were observed.

Section 5.8: In addition to the recently described bisabolanes cheimonophyllons A–E (112–116, Fig. 14) and cheimonophyllal (117,

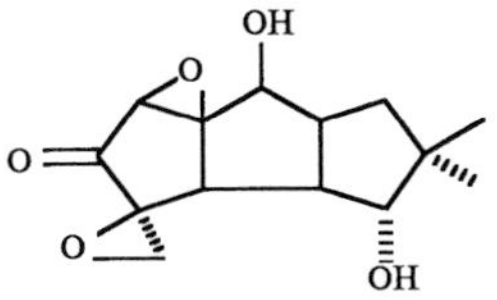

Coriolin (128)

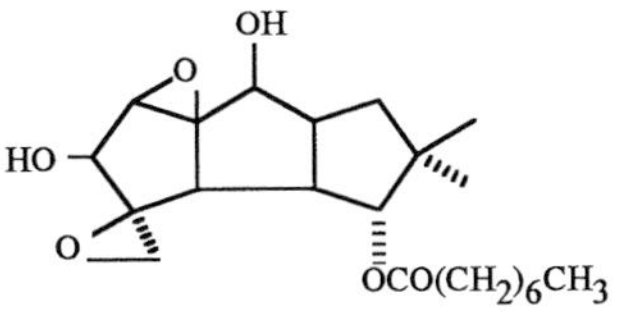

Coriolin B (129)

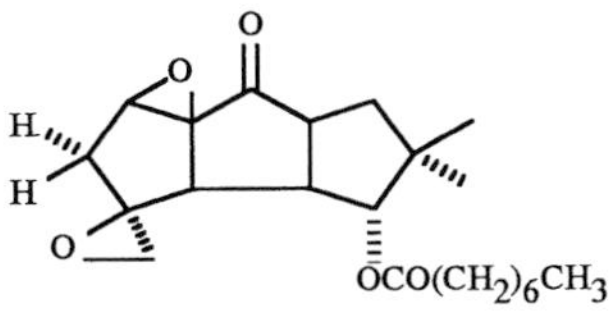

Diketocoriolin B (130)

Fig. 18. Inhibitors of (Na^+–K^+)-ATPases.

Fig. 14), the *p*-menthane 1,2-dihydroxyminthlactone (132, Fig. 19) was isolated as a minor nematicidal component from fermentations of *Cheimonophyllum candidissimum* (STADLER et al., 1995). The LD_{50} against the nematode *Caenorrhabditis elegans* was determined to 25 μg mL^{-1} without any additional antimicrobial and cytotoxic activity. The structurally related minthlactone has been previously reported as a constituent of peppermint (*Mentha piperita*) oil (TAKAHASHI et al., 1980).

Omphalotin (133, Fig. 19), a new cyclic dodecapeptide possessing strong and selective nematicidal activity against the plant pathogenic nematode *Meloidogyne incognita* was isolated from mycelial cultures of the basidiomycete *Omphalotus olearius* (MAYER et al., in press; STERNER et al., in press). The LD_{90} against *M. incognita* was determined to be 0.76 μg mL^{-1}, whereas the saprophytic nematode *C. elegans* was approximately 50 times less sensitive. Omphalotin exhibits no phytotoxic, antibacterial, or antifungal activities and is only weakly cytotoxic at high concentrations (100 μg mL^{-1}).

5.13.1 Inhibitors of Leukotriene Biosynthesis

Leukotrienes are potent biological mediators derived from arachidonic acid metabolism and are generated via the 5-lipoxygenase pathway. Leukotriene B_4, a dihydroxy derivative, causes adhesion and chemotactic movement of leukocytes, enzyme release, and generation of superoxide in neutrophiles. The sulfopeptide leukotrienes C_4, D_4, and E_4, are known as "slow reacting substances of anaphylaxis" and induce bronchoconstriction, stimulate mucus production, and increase vascular permeability (SAMUELSSON et al., 1987). Due to these effects the leukotrienes have been implicated as important mediators of inflammation and hypersensitivity reactions (FORD-HUTCHINSON et al., 1994; SALMON and GARLAND, 1991).

A screening for inhibitors of leukotriene C_4 biosynthesis resulted in the isolation of (+)-10α-hydroxy-4-muurolen-3-one (134, Fig. 20) from fermentations of an Ethiopian *Favolaschia* species (ZAPF et al., 1996). The IC_{50} value for the inhibition of leukotriene C_4 biosynthesis in rat basophilic leukemia (RBL-

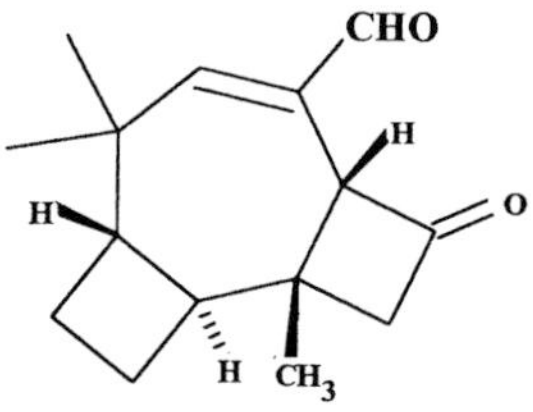

Collybial (131)

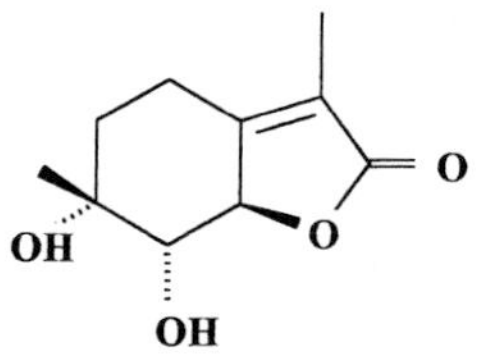

1,2-Dihydroxyminthlactone (132)

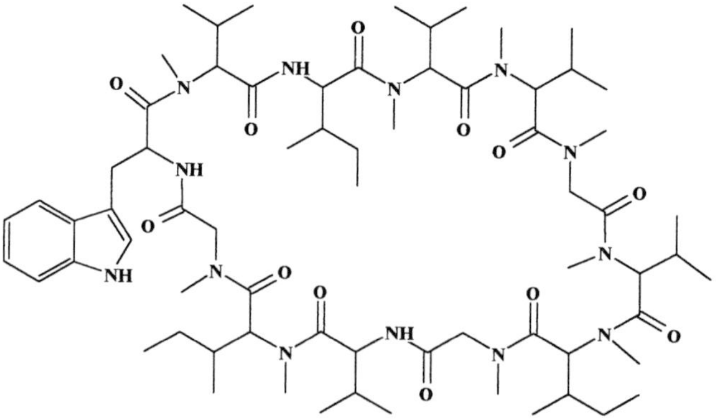

Omphalotin (133)

Fig. 19. Antiviral and nematicidal compounds.

1) cells was determined to be 5–10 μg mL^{-1}. Related cadinane sesquiterpenes, like (+)-T-cadinol and (−)-3-oxo-T-cadinol, have been reported before (CLAERSON et al., 1991) as constituents of scented myrrh (the resin of the plant *Commiphora guidotti*), and similar effects on leukotriene C_4 biosynthesis have been observed (ZAPF et al., 1996).

Three known lacterane type sesquiterpenoids, blennin A (135, Fig. 20), blennin C (136, Fig. 20), and deoxylactarorufin A (137, Fig. 20) were obtained from fermentations of *Lentinellus cochleatus* (WUNDER et al., 1996) together with the new metabolites (*Z*)-2-chloro-3-(4-methoxyphenyl)-2-propen-1-ol (138, Fig. 20) and lentinellone (139, Fig. 20), a protoilludane derivative. Blennin A (VIDARI et al., 1976), blennin C (DE BERNARDI et al., 1976), and deoxylactarorufin A (DANIEWSKI et al., 1977; DANIEWSKI and KROL, 1981) are strong inhibitors of leukotriene C_4 biosynthesis in RBL-1 cells with IC_{50} values of 5 μg mL^{-1} for blennin A, 4 μg mL^{-1} for blennin C, and 2 μg mL^{-1} for deoxylactarorufin A. (*Z*)-2-chloro-3-(4-methoxyphenyl)-2-propen-1-ol inhibited the leukotriene C_4 biosynthesis with an IC_{50} of 15 μg mL^{-1}, whereas lentinellone was inactive.

5.13.2 Inducers of Differentiation of Promyelocytic Leukemia Cells and Inhibitors of Signal Transduction in Tumor Cells

Among the phenotypic abnormalities in acute leukemia is a lack of granulocytes, macrophages, and platelets caused by the inability of the neoplastic leucocytes to undergo terminal differentiation and eventually apoptosis. The human HL-60 leukemia cell line is an excellent model for a study of functional and morphological differentiation *in vitro*, because the cells can be induced to differentiate into granulocytes or monocytes/macrophages. Differentiation may be followed by apoptosis, a process of active DNA fragmentation. The induction of differentiation and apoptosis are regulated by a network of signal transduction pathways and transcription factors which are possible targets for a rational antitumor therapy (HASS, 1992; LEVITZKI, 1994; THOMPSON, 1995; MANNING, 1996).

Pinicoloform (140, Fig. 21) an antibiotic from *Resinicium pinicola* was detected in a screening for metabolites inducing the differentiation of HL-60 cells to monocytes and macrophages (BECKER et al., 1994) at concentrations between 0.5–1 μg mL^{-1} (1.8–3.5 μM). Cytotoxic as well as antimicrobial activities have also been described at concentrations between 10–66 μM.

The diterpenes lepistal (141, Fig. 21) and lepistol (142, Fig. 21) were isolated from cultures of *Lepista sordida* (MAZUR et al., 1996). At a concentration of 0.2 μg mL^{-1} lepistal induces the differentiation of 20% of the HL-60 cells into granulocyte/monocyte-like cells and of 18% of the human histiocytic lymphoma (U-937) cells into monocyte-like cells. The related alcohol lepistol is 50–100 times less active. Cytotoxic activities of lepistal and lepis-

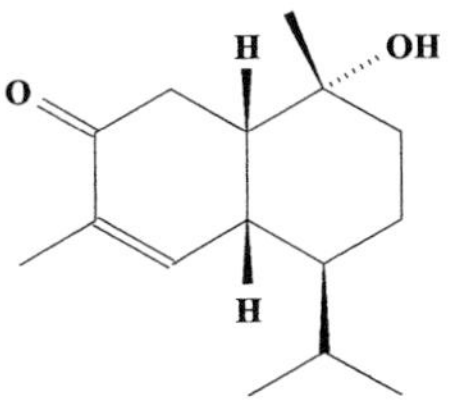

(+)-10α-hydroxy-4-muurolen-3-one (134)

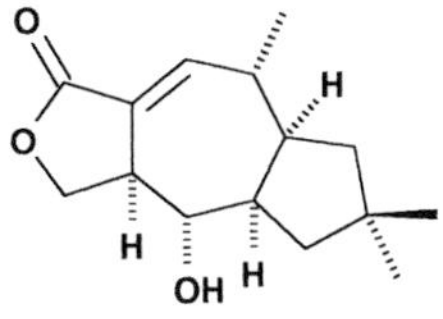

Blennin A (135)

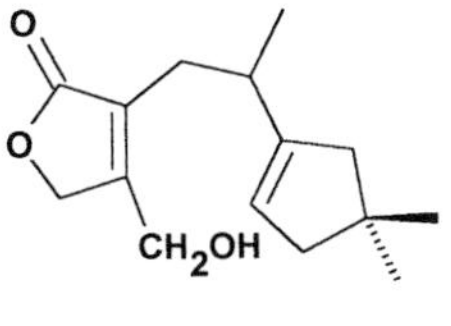

Blennin C (136)

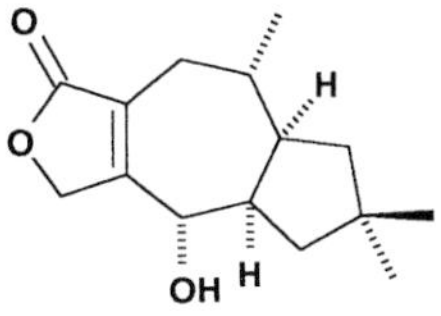

Deoxylactarorufin A (137)

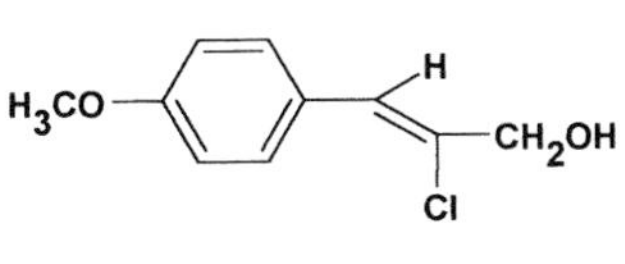

(*Z*)-2-chloro-3-(4-methoxyphenyl)-2-propen-1-ol (138)

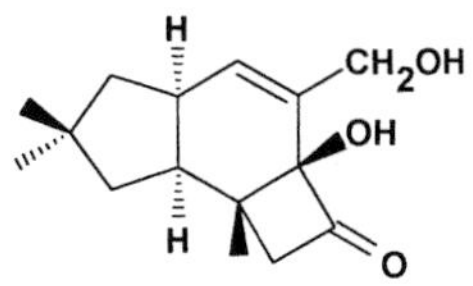

Lentinellone (139)

Fig. 20. Inhibitors of leukotriene biosynthesis.

tol were observed at 1 μg mL^{-1} and 50 μg mL^{-1}, respectively. Lepistal exhibits pronounced antimicrobial activity whereas lepistol shows no antibacterial and only weak antifungal activities. Therefore, the aldehyde function is considered to substantially contribute to the biological activity of lepistal.

Recently, two new bisabolane sesquiterpenes, nidulal (143, Fig. 21) and niduloic acid (144, Fig. 21), have been isolated from fermentations of *Nidula candida* (ERKEL et al., in press). Nidulal and niduloic acid induce differentiation of 15–25% of HL-60 cells at concentrations of 72 μM and 36 μM, respectively. It has been shown that induction of differentiation of HL-60 cells by nidulal is followed by apoptosis. In COS-7 cells (African green monkey) nidulal selectively activates the AP-1 dependent signal transduction pathways in a manner similar to the phorbol ester TPA, an activator of protein kinase C. In gel shift assays with extracts of nidulal-treated HL-60 cells a change of binding activities of the AP-1 transcription factor is observed, which may be the result of an altered composition of the AP-1 protein complex.

The novel norilludane puraquinonic acid (145, Fig. 21) was isolated from mycelial cultures of *Mycena pura* (BECKER et al., in press) as an inducer of morphological and

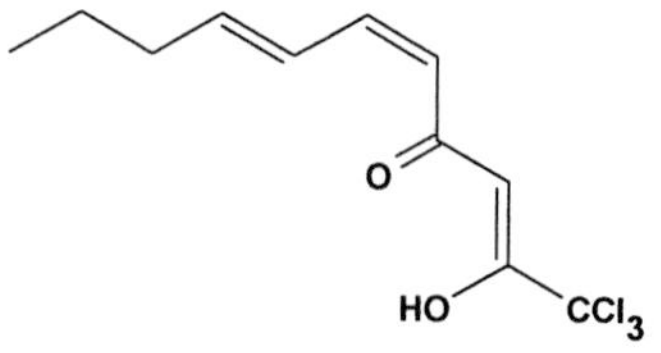

Pinicoloform (140)

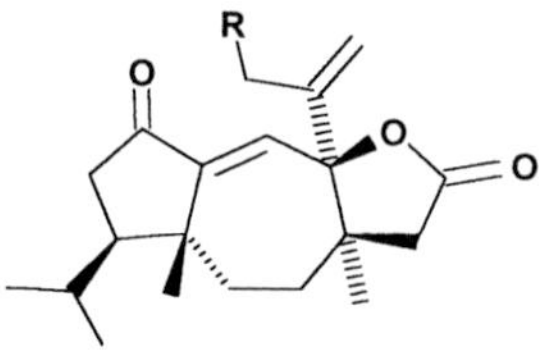

Lepistal (141) R= =O

Lepistol (142) R= —OH

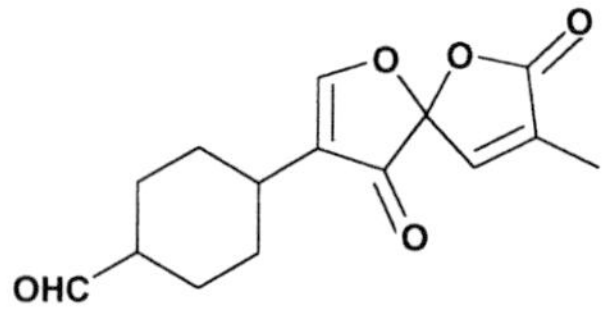

Nidulal (143)

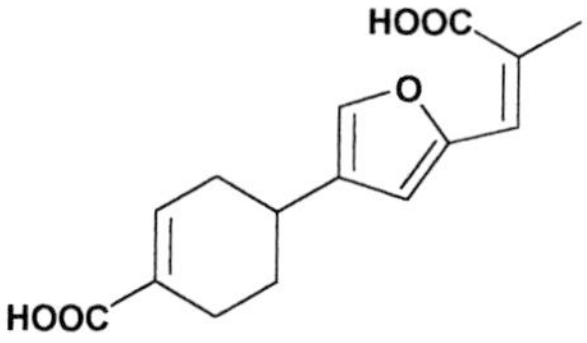

Niduloic acid (144)

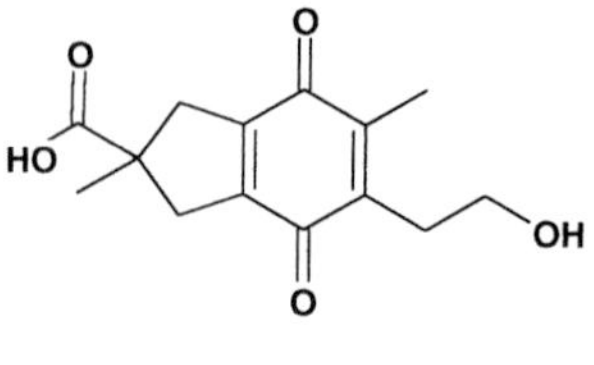

Puraquinonic acid (145)

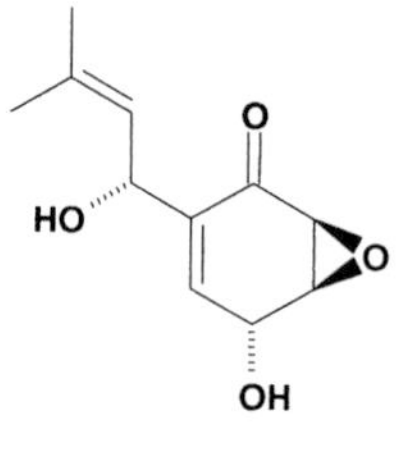

Panepoxydone (146)

Fig. 21. Inducers of differentiation and inhibitors of signal transduction in tumor cells.

physiological differentiation of mammalian cells. It induces differentiation of 30–40% of HL-60 cells into granulocyte- or monocyte/macrophage-like cells at 380 μM. At the same concentration U-937 cells, which are blocked at a later stage of development are affected to a much lesser extent.

NF-κB is an inducible, ubiquitous transcription factor, which regulates the expression of various cellular genes involved in immune response, inflammation, acute phase response, and several viral genes and inhibitors of NF-κB activation may, therefore, find broad application as novel therapeutics (BAEUERLE and HENKEL, 1994; BALDWIN, 1996; MANNING and ANDERSON, 1994).

In a search for new inhibitors of NF-κB-mediated signal transduction in COS-7 cells using the secreted alkaline phosphatase (SEAP) as reporter gene, panepoxydone (146, Fig. 21) was isolated from fermentations of the basidiomycete *Lentinus crinitus* (ERKEL et al., 1996). Panepoxydone and several related derivatives have been previously reported as secondary metabolites from *Panus rudis*, *P. conchatus* (KIS et al., 1970), and *Penicillium urticae* (SEGIGUCHI and GAUCHER, 1979). Panepoxydone inhibits the NF-κB activated expression of the SEAP with an IC_{50} of 1.5–2 μg mL^{-1} (7.15–9.52 μM). No inhibition of AP-1-mediated expression of the reporter gene could be observed at a concentration up to 5 μg mL^{-1} panepoxydone. Panepoxydone strongly reduces the TPA-, TNF-α-, and ocadaic acid-mediated binding of NF-κB to the high affinity consensus sequence in COS-7 and HeLa S3 cells as confirmed by electrophoretic mobility shift assays. Panepoxydone inhibits the phosphorylation of the inhibitory protein IκB and, therefore, sequesters the NF-κB complex in an inactive form. Recently, the related cycloepoxydon, an inhibitor of AP-1- and NF-κB-mediated gene expression, has been isolated from fermentations of a deuteromycete strain (GEHRT et al., in press).

6 Future Perspectives

It is evident that basidiomycetes provide a rich, yet quite untapped source of compounds with novel structures and in some cases very interesting biological activity. Many of the compounds isolated so far seem to be produced exclusively by this class of fungi. From the number of estimated basidiomycetes species (>30000) it may be assumed that there will be many more exciting discoveries made in the future.

7 References

AMES, B. N., MCCANN, J., YAMASAKI, E. (1975), Methods for detecting carcinogens and mutagens with the *Salmonella*/mammalian mutagenicity test, *Mut. Res.* **31**, 347–364.

ANCHEL, M., HERVEY, A., ROBBINS, W. J. (1950), Antibiotic substances from basidiomycetes. VII. *Clitocybe illudens, Proc. Natl. Acad. Sci. USA* **36**, 300–305.

ANKE, T., DIEKMANN, H. (1972), Metabolic products of microorganisms. 112. Biosynthesis of sideramines in fungi. Rhodotorulic acid synthetase from extracts of *Rhodotorula glutinis, FEBS Lett.* **27**, 259–262.

ANKE, T., STEGLICH, W. (1981), Screening of basidiomycetes for the production of new antibiotics, in: *Advances in Biotechnology*, Vol. **1** (MOO-YOUNG, M., Ed.), pp 35–40. Pergamon Press, Canada.

ANKE, T., STEGLICH, W. (1989), β-Methoxyacrylate antibiotics: from biological activity to synthetic analogues, in: *Biologically Active Molecules – Identification, Characterization, and Synthesis* (SCHLUNEGGER, U. P., Ed.), pp. 9–25. Berlin-Heidelberg: Springer-Verlag.

ANKE, H., ANKE, T., DIEKMANN H. (1973), Biosynthesis of sideramines in fungi. Fusigen synthetase from extracts of *Fusarium cubense, FEBS Lett.* **36**, 323–325.

ANKE, T., OBERWINKLER, F., STEGLICH, W., HÖFLE, G. (1977a), The striatins – new antibiotics from the basidiomycete *Cyathus striatus* (Huds. ex Pers.) Willd., *J. Antibiot.* **30**, 221–225.

ANKE, T., OBERWINKLER, F., STEGLICH, W., SCHRAMM, G. (1977b), The strobilurins – new antifungal antibiotics from the basidiomycete *Strobilurus tenacellus* (Pers. ex. Fr.) Sing., *J. Antibiot.* **30**, 806–810.

ANKE, T., HECHT, H. J., SCHRAMM, G., STEGLICH, W. (1979), Antibiotics from basidiomycetes. IX. Oudemansin, an antifungal antibiotic from *Oudemansiella mucida* (Schrader ex Fr.) Hoehnel (Agaricales), *J. Antibiot.* **32**, 1112–1117.

ANKE, T., KUPKA, J., SCHRAMM, G., STEGLICH, W. (1980), Antibiotics from basidiomycetes. X. Scorodonin, a new antibacterial and antifungal metabolite from *Marasmius scorodonius* (Fr.) Fr., *J. Antibiot.* **33**, 463–467.

ANKE, T., GIANETTI, B.M., STEGLICH W. (1982), Antibiotika aus Basidiomyceten. XV. 1-Hydroxy-2-nonin-4-on, ein antifungischer und cytotoxischer Metabolit aus *Ischnoderma benzoinum* (Wahl.) Karst., *Z. Naturforsch.* **37c,** 1–4.

ANKE, T., BESL, H., MOCEK, U., STEGLICH, W. (1983), Antibiotics from basidiomycetes. XVIII. Strobilurin C and oudemansin B, two new antifungal metabolites from *Xerula* species (Agaricales), *J. Antibiot.* **36**, 661–666.

ANKE, T., SCHRAMM, G. SCHWALGE, B., STEFFAN, B., STEGLICH, W. (1984), Antibiotika aus Basidiomyceten. XX. Synthese von Strobilurin A und Revision der Stereochemie der natürlichen Strobilurine, *Liebigs Ann. Chem.* 1984 1616–1625.

ANKE, T., BACKENS, S., STEGLICH, W. (1985a), A new antibiotic from *Paneolus* and *Psathyrella* species. *Abstract Annu. Meeting ASM.* Washington, D.C.: ASM Press.

ANKE, T., HEIM, J., KNOCH, F., MOCEK, U., STEFFAN, B., STEGLICH, W. (1985b), Crinipelline, die ersten Naturstoffe mit einem Terquinan-Gerüst, *Angew. Chem.* **27**, 714–716.

ANKE, T., STEGLICH, W., POMMER, E. H. (1986), The antifungal activities of striatals and striatins. *Abstracts 6th Int. Congr. Pesticide Chem.* (IUPAC), Ottawa.

ANKE, T., SCHRAMM, G., STEGLICH, W., VON JAGOW, G. (1988), Structure–activity relationships of natural and synthetic *E*-β-methoxyacrylates of the strobilurin and oudemansin series, in: *The Roots of Modern Biochemistry* (KLEINKAUF, H., VON DÖHREN, H., JÄNICKE, L., Eds.), pp. 657–662. Berlin: de Gruyter.

ANKE, H., HILLEN-MASKE, E., STEGLICH, W. (1988), 9-β-hydroxymarasmic acid and other sesquiterpenoids from submerged cultures of a basidiomycete, *Z. Naturforsch.* **44c**, 1–6.

ANKE, H., STERNER, O., STEGLICH, W. (1989), Structure–activity relationship for unsaturated dialdehydes. 3. Mutagenic, antimicrobial, cytotoxic and phytotoxic activities of merulidial derivatives, *J. Antibiot.* **42**, 738–744.

ANKE, T., WERLE, A., BROSS, M., STEGLICH, W. (1990), Antibiotics from basidiomycetes. XXXIII. Oudemansin X, a new antifungal *E*-β-methoxyacrylate from *Oudemansiella radicata* (Relhan ex Fr.) Sing., *J. Antibiot.* **43**, 1010–1011.

ANKE, T., ERKEL, G., KOKSCH, G., KUSCHEL, A., GIMENEZ, A., VELTEN, R., STEGLICH, W. (1993), Inhibitoren der HIV-1 Reversen Transcriptase und Induktoren der Differenzierung menschlicher Zellinien aus Pilzen, in: Wege zu neuen Produkten und Verfahren in der Biotechnologie (ANKE, T., ONKEN, U., Eds.). *DECHEMA Monographien* **129**, 15–25.

ARIGONI, D. (1962), La struttura di un terpene di nuovo genere, *Gazz. Chim. Itl.* **22**, 884–901.

AYER, W. A., BROWNE, L. M. (1981), Terpenoid metabolites of mushrooms and related basidiomycetes, *Tetrahedron Lett.* **17**, 2199–2248.

AYER, W. A., LEE, S. P., NAKASHIMA, T. T. (1979), Metabolites of bird's nest fungi. Part 12. Studies on the biosynthesis of cyathins, *Can. J. Chem.* **17**, 3338–3343.

BACKENS, S., STEFFAN, B., STEGLICH, W., ZECHLIN, L., ANKE, T. (1984), Antibiotika aus Basidiomyceten. XIX. Naematolin und Naematolon, zwei Caryophyllan-Derivate aus Kulturen von *Hypholoma*-Arten (Agaricales), *Liebigs Ann. Chem.* 1984, 1332–1342.

BACKENS, S., STEGLICH, W., BÄUERLE, J., ANKE, T. (1988), Antibiotika aus Basidimyceten, 28. Hydroxystrobilurin D, ein antifungisches Antibiotikum aus Kulturen von *Mycena sanguinolenta* (Agaricales), *Liebigs Ann. Chem.* 1988, 405–409.

BALDWIN, A. S. (1996), The NF-κB and IκB proteins: new discoveries and insights, *Annu. Rev. Immunol.* **14**, 649–681.

BARRON, G. L. (1977), The nematode-destroying fungi, in: *Topics in Mycobiology,* Vol. 1. Guelph, Ontario: Canadian Biological Publications.

BASSETT, C., SHERWOOD, R. T., KEPLER, J. A., HARMILTON, P. B. (1967), Production and biological activity of famannosin, a toxic sesquiterpene metabolite of *Fomes annosum, Phytopathology* **57**, 1046–1052.

BÄUERLE, J. (1981), Antibiotika aus Basidiomyceten der Gattungen *Clitopilus, Hohenbuehelia, Hemimycena,* und *Mycena* (Agaricales). *Thesis,* Department of Biology, University of Tübingen, FRG.

BÄUERLE, J., ANKE, T. (1980), Antibiotics from the genus *Mycena* and *Hydropus scabripes, Planta Med.* **39**, 195–196.

BÄUERLE, P. A., HENKEL, T. (1994), Function and activation of NF-κB in the immune system, *Annu. Rev. Immunol.* **12**, 141–179.

BÄUERLE, J., ANKE, T., JENTE, R., BOSOLD, F. (1982), Antibiotics from basidiomycetes. XVI. Antimicrobial and cytotoxic polyines from *Mycena viridimarginata, Arch. Microbiol.* **132**, 194–196.

Bäuerle, J., Anke, T., Hillen-Maske, E., Steglich, W. (1986), Hemimycin, a new antibiotic from two *Hemimycena* species (Basidiomycetes), *Planta Med.* **5**, 418.

Beale, M. H. (1990), The biosynthesis of C_5–C_{20} terpenoid compounds, *Nat. Prod. Rep.* **7**, 25–39, 387–407.

Becker, U., Anke, T., Sterner, O. (1994), A novel halogenated compound possessing antibiotic and cytotoxic activities isolated from the fungus *Resinicium pinicola* (J. Erikss.) Erikss. & Hjortst., *Z. Naturforsch.* **49c**, 772–774.

Becker, U., Erkel, G., Anke, T., Sterner, O. (in press), Puraquinonic acid, a novel inducer of differentiation of human HL-60 promyelocytic leukemia cells from *Mycena pura* (Pers. ex Fr.), *Nat. Prod. Lett.*

Birch, A. J., Cameron, D. W., Holzapfel, C. W., Rickards, R. W. (1963), The diterpenoid nature of pleuromutilin, *Chem Ind.* (London) 374–375.

Birch, A. J., Holzapfel, C. W., Rickards, R. W. (1966), The structure and some aspects of the biosynthesis of pleuromutilin, *Tetrahedron* (Suppl.) **8** Part II, 359–387.

Brandt, U., Schlägger, H., von Jagow, G. (1988), Characterisation of binding of the methoxyacrylate inhibitors to mitochondrial cytochrome c reductase, *Eur. J. Biochem.* **173**, 499–506.

Brandt, U., Haase, U., Schlägger, H., von Jagow, G. (1993), Speziesspezifität und Wirkmechansimus der Strobilurine, in: Wege zu neuen Produkten und Verfahren in der Biotechnologie (Anke, T., Onken, U., Eds). *DECHEMA Monographien* **129**, 27–38.

Brauer, H., Korn, A. (1993), Untersuchung der Umsatzleistung im Hubstrahl-Bioreaktor bei der biotechnologischen Produktion mit Pilzen, in: Wege zu neuen Produkten und Verfahren in der Biotechnologie (Anke, T., Onken, U., Eds.). *DECHEMA Monographien* **129**, 147–157.

Bresinsky, A., Besl, H. (1985), *Giftpilze.* Stuttgart: Wissenschaftliche Verlagsgesellschaft.

Cane, D. E., Nachbar, R. B. (1978), Stereochemical studies of isoprenoid biosynthesis. Biosynthesis of fomannosin from [1,2-$^{13}C_2$]acetate, *J. Am. Chem. Soc.* **100**, 3208–3212.

Chenina, R., Dochnahl, A., Huff, T., Kuball, H. G., Lange, B., Anke, H., Anke, T. (1993), Fluoreszenzspektrometer zur On-line-Verfolgung von Fermentationsprozessen, in: Wege zu neuen Produkten und Verfahren in der Biotechnologie (Anke, T., Onken, U., Eds.). *DECHEMA Monographien* **129**, 171–182.

Chibata, I., Okumura, K., Takeyama, S., Kotera, K. (1969), A new hypercholesterolemic substance in *Lentinus edodes, Experientia* **25**, 1237–1238.

Chihara, G., Hamuro, J., Madea, Y. Y., Arai, Y., Fukuoka, F. (1970), Fractionation and purification of the polysaccharides with marked antitumor activity, especially lentinan from *Lentinus edodes* (Berk.) Sing. (an edible mushroom), *Cancer Res.* **30**, 2776–2781.

Claerson, P., Andersson, R, Samuelsson, G. (1991), T-cadinol: a pharmacologically active constituent of scented myrrh: introductory pharmacological characterization and high field ^1H and ^{13}C NMR data, *Planta Med.* **57**, 352–356.

Clough, J. M. (1993), The strobilurins, oudemansins, and myxothiazols, fungicidal derivatives of β-methoxyacrylic acid, *Nat. Prod. Rep.* **10**, 565–574.

Connolly, J. D., Hill, R. A., Ngadjui, B. T. (1994), Triterpenoids, *Nat. Prod. Rep.* **11**, 91–117.

Daniewski, W. M., Krol, J. (1981), Constituents of higher fungi. Part XII, *Pol. J. Chem.* (*Rocz. Chem.*) **55**, 1247–1252.

Daniewski, W. M., Kocor, M., Krol, J. (1977), Constituents of higher fungi, *Pol. J. Chem.* (*Rocz. Chem.*) **51**, 1395–1398.

Davies, J., von Ahsen, U., Wank, H., Schroeder, R. (1992), Evolution of secondary metabolite production: potential role for antibiotics as prebiotic effectors of catalytic RNA reactions, in: Secondary Metabolites: Their Function and Evolution (Cadwick, D. J., Whelan, J., Eds.), pp. 24–44. *Ciba Foundation Symposium* 171.

De Bernardi, M., Fronza, G., Vidari, G., Vita-Finzi, P. (1976), Fungal metabolites II: new sesquiterpenes from *Lactarius scobiculatus* Scop. (Russulaceae), *Chim. Ind.* **58**, 177–178.

De Clercq, E. (1990), Perspectives for chemotherapy of the HIV infection: An introduction, *Pharmacochem. Libr.* **14** (Des. Anti-AIDS Drugs), 1–24.

Deml, G., Anke, T., Oberwinckler, F., Giannetti, B. M., Steglich, W. (1980), Schizonellin A and B, new glycolipids from *Schizonella melanogramma* (Ustilaginales), *Phytochemistry* **19**, 83–87.

Deol, B. S., Ridley D. D., Singh, P. (1978), Isolation of cyclodepsipetides from plant pathogenic fungi, *Aust. J. Chem.* **31**, 1397–1399.

Drews, J., Georgopoulos, A., Laber, G., Schütze, E., Unger, J. (1975), Antimicrobial activities of 81.723 hfu, a new pleuromutilin derivative, *Antimicrob. Agents Chemother.* **7**, 507–516.

Dugan, J. J., DeMayo, P., Nisbet, M., Robbinson, J. R., Anchel, M. (1966), Terpenoids XIV. The constitution and biogenesis of marasmic acid, *J. Am. Chem Soc.* **88**, 2838–2844.

EGGER, H., REINSHAGEN, H. (1976a), New pleuromutilin derivatives with enhanced antimicrobial activity. I. Synthesis, *J. Antibiot.* **29**, 915–922.

EGGER, H., REINSHAGEN, H. (1976b), New pleuromutilin derivatives with enhanced antimicrobial activity. II. Structure–activity correlations, *J. Antibiot.* **29**, 923–927.

ENDO, A. (1979), Monacolin K, a new hypercholesterolemic agent produced by a *Monascus* species, *J. Antibiot.* **32**, 852–854.

ERKEL, G., ANKE, T., VELTEN, R., STEGLICH, W. (1991), Podoscyphic acid, a new inhibitor of Avian myeloblastosis virus and Moloney murine leukemia virus reverse transcriptase from a *Podoscypha* species, *Z. Naturforsch.* **46c**, 442–450.

ERKEL, G., ANKE, T., GIMENENZ, A., STEGLICH, W. (1992), Antibiotics from basidiomycetes. XLI. Clavicornic acid, a novel inhibitor of reverse transcriptases from *Clavicorona pyxidata* (Pers. ex Fr.) Doty, *J. Antibiot.* **45**, 29–37.

ERKEL, G., ANKE, T., VELTEN, R., GIMENEZ, A., STEGLICH, W. (1994), Hyphodontal, a new antifungal inhibitor of reverse transcriptases from *Hyphodontia* sp. (Corticaceae, Basidiomycetes), *Z. Naturforsch.* **49c**, 561–570.

ERKEL, G., LORENZEN, K., ANKE, T., VELTEN, R., GIMENEZ, A., STEGLICH, W. (1995), Kuehneromycins A and B, two new biological active compounds from a tasmanian *Kuehneromyces* sp. (Strophariaceae, Basidiomycetes), *Z. Naturforsch.* **50c**, 1–10.

ERKEL, G., ANKE, T., STERNER, O. (1996), Inhibition of NF-κB activation by panepoxydone, *Biochem. Biophys. Res. Commun.* **226**, 214–221.

ERKEL, G., BECKER, U., ANKE, T., STERNER, O. (in press), Nidulal, a novel inducer of differentiation of human promyelocytic leukemia cells from *Nidula candida, J. Antibiot.*

FLOREY, H. W., CHAIN, W., HEATLEY, A., JENNINGS, N. G., ABRAHAM, E. P., FLOREY, M. E. (1949), *Antibiotics.* London: Oxford University Press.

FORD-HUTCHINSON, A. W., GRESSER, M., YOUNG, R. N. (1994), 5-Lipoxygenase, *Annu. Rev. Biochem.* **63**, 383–417.

FRAGA, B. M. (1990), Natural sesquiterpenoids, *Nat. Prod. Rep.* **7**, 61–84.

FREDENHAGEN, A., KUHN, A., PETER, H. H. (1990a), Strobilurins F, G, and H, three new antifungal metabolites from *Bolinea lutea.* I. Fermentation, isolation and biological activity, *J. Antibiot.* **43**, 655–660.

FREDENHAGEN, A., HUG, P., PETER, H. H. (1990b), Strobilurins F, G and H, three new antifungal metabolites from *Bolinea lutea.* II. Structure determination, *J. Antibiot.* **43**, 661–667.

FUJI, T., MAEDA, H., SUZUKI, F., ISHIDA, N. (1978), Isolation and characterization of a new antitumor polysaccharide, KS-2, extracted from culture mycelia of *Lentinus edodes, J. Antibiot.* **31**, 1079–1090.

FUJIMOTO, H., NAKAYAMA, Y., YAMAZAKI, M. (1993), Identification of immunosuppressive components of a mushroom, *Lactarius flavidulus, Chem. Pharm. Bull.* **41**, 654–658.

GEHRT, A., ERKEL, G., ANKE, H., ANKE, T., STERNER, O. (in press), New hexaketide inhibitors of eukaryotic signal transduction, *Nat. Prod. Lett.*

GERBER, N. N., SHAW, S. A., LECHEVALIER, H. (1980), Structures and antimicrobial activities of peniophorin A and B, two polyacetylenic antibiotics from *Peniophora affinins* Burt., *Antimicrob. Agents Chemother.* **17**, 636–641.

GERMERDONK, R., BECKER, P., GEHRIG, I. (1993), Verfahrenstechnische Methoden zur Verbesserung des Produktionsverhaltens bei der Fermentation des Pilzes *Cyathus striatus,* in: Wege zu neuen Produkten und Verfahren in der Biotechnologie (ANKE, T., ONKEN, U., Eds.). *DECHEMA Monographien* **129**, 159–169.

GIANNETTI, B. M., STEGLICH, W., QUACK, W., ANKE, T., OBERWINCKLER, F. (1978), Merulinsäuren A, B und C, neue Antibiotika aus *Merulius tremellosus* Fr. und *Phlebia radiata* Fr., *Z. Naturforsch.* **33c**, 807–816.

GILL, M. (1994), Pigments of fungi (Macromycetes), *Nat. Prod. Rep.* **11**, 67–90.

GILL, M., STEGLICH, W. (1987), Pigments of fungi (Macromycetes), in: *Progress in the Chemistry of Organic Natural Products,* Vol. 51 (ZECHMEISTER, L., Ed.), pp. 1–317. Wien, New York: Springer-Verlag.

GREENLEE, W. J., WOODWARD, R. B. (1976), Total synthesis of marasmic acid, *J. Am. Chem. Soc.* **98**, 6075–6076.

HANSON, J. R., MARTEN, T. (1973), Incorporation of [2-^{3}H2]- and [4(R)-4-^{3}H]-mevalonoid hydrogen atoms into the sesquiterpenoid illudin M., *J. Chem Soc. Chem. Comm.* 171–172.

HARA, M., YOSHIDA, M., MORIMOTO, M., NAKANO, H. (1987), 6-deoxyilludin M, a new antitumor antibiotic: Fermentation, isolation and structural identification, *J. Antibiot.* **40**, 1673–1646.

HARTTIG, U., ANKE, T., SCHERER, A., STEGLICH, W. (1990), Leianafulvene, a sesquiterpenoid fulvene derivative from cultures of *Mycena leaiana, Phytochemistry* **29**, 3942–3944.

HASS, R. (1992), Retrodifferentiation – an alternative biological pathway in human leukemia cells, *Eur. J. Cell Biol.* **58**, 1–11.

HASS, V. C., MUNACK, A. (1993), Modellierung und Regelung der Kultivierung von *Cyathius striatus*, in: Wege zu neuen Produkten und Verfahren in der Biotechnologie (ANKE, T., ON-

KEN, U., Eds.). *DECHEMA Monographien* **129**, 135–145.

HAUTZEL, R., ANKE, H., SHELDRICK, W. S. (1990), Mycenon, a new metabolite from a *Mycena* species TA 87202 (Basidiomycetes) as an inhibitor of isocitrate lyase, *J. Antibiot.* **43**, 1240–1244.

HAYASHI, M., WADA, K., MUNAKATA, K. (1981), New nematicidal metabolites from a fungus, *Irpex lacteus, Agric. Biol. Chem.* **45**, 1527–1529.

HECHT, H. J., HÖFLE, G., STEGLICH, W., ANKE, T., OBERWINKLER, F. (1978), Striatin A, B, and C: Novel diterpenoid antibiotics from *Cyathus striatus;* X-Ray crystal structure of striatin A, *J. Chem. Soc. Chem. Commun.* 665–666.

HEIM, J., ANKE, T., MOECK, U., STEFFAN, B., STEGLICH, W. (1988), Antibiotics from basidiomycetes. XXIX. Pilatin, a new biologically active marasmane derivative from cultures of *Flagelloscypha pilatii* Agerer, *J. Antibiot.* **41**, 1752–1757.

HERBERT, R. B. (1989), *The Biosynthesis of Secondary Metabolites.* London, New York: Chapman and Hall.

HIROTANI, M., O'REILLY, J., DONELLY, D. M. X. (1977), Fomannoxin – A toxic metabolite from *Fomes annosum, Tertahedron Lett.* **7**, 651–652.

HÖGENAUER, G. (1979), Tiamulin and pleuromutilin, in: *Antibiotics,* Vol. V-1 (HAHN, F. E., Ed.), pp. 340–360. Berlin, Heidelberg, New York: Springer-Verlag.

HOUCHINS, J., HIND, G. (1983), Flash spectroscopic characterization of photosynthetic electron transport in isolated heterocysts, *Arch. Biochem. Biophys.* **224**, 272–282.

HUNECK, S., CAMERON, A. F., CONOLLY, J. D., MCLAREN, M., RYCROFT, D. S. (1982), Hercynolactone, a new carotene sesquiterpenoid from the liverworts *Barbilophozia lycopodioides* and *B. hatcheri.* Crystal structure analysis, *Tetrahedron Lett.* **23**, 3959–3962.

IKEDA, M., NIWA, G., TOHAYAMA, K., SASSA, T., MIURA, Y. (1977), Structures of fasciculol C and its depsipeptides, new biologically active substances from *Nematoloma fasciculare, Agric. Biol. Chem.* **41**, 1803–1805.

ISAAC, B. G., AYER, S. W., LETENDRE, L. J., STONARD, R. J. (1991), Herbicidal nucleosides from microbial sources, *J. Antibiot.* **44**, 729–732.

ISHIBASHI, K., NOSE, K., SHINDO, T., ARAI, M., MISHIMA, H. (1968), Siccayne: a novel acetylenic metabolite of *Helminthosporium siccans, Ann. Sankyo Res. Lab.* **20**, 76–79.

ISHIZUKA, M., TAKEUCHI, T., UMEZAWA, H. (1981), Studies on the mechanism of action of deketocoriolin B to enhance the antibody formation, *J. Antibiot.* **34**, 95–102.

ISONO, K. (1988), Nucleoside antibiotics: structure, biological activity, and biosynthesis, *J. Antibiot.* **41**, 1711–1739.

JANSEN, B. J. M., DE GROOT, A. (1991), The occurrence and biological activity of drimane sesquiterpenoids, *Nat. Prod. Rep.* **8**, 309–318.

JELLISON, J., GOODELL, B., FEKETE, F., CHANDHOKE, V. (1990), Fungal siderophores and their role in wood biodegradation. The International Research Group On Wood Preservation. *Paper* prepared for the 21st Annual Meeting, Rotorua, New Zealand, 13–18 May, 1990.

JONES, R. H., THALLER, V. (1973), Microbial polyynes, in: *Handbook of Microbiology,* Vol. III, Microbial Products (LASKIN, a. I., LECHEVALIER, H. A., Eds.), pp. 63–74. Cleveland, OH: CRC Press.

KANEDA, M., TAKAHASHI, R., IITAKA, Y., SHIBATA, S. (1972), Retigeranic acid, a novel sesterpene isolated from the lichens of *Lobaria retigera* group, *Tetrahedron Lett.* **1972**, 4609–4611.

KAVANAGH, F., HERVEY, A., ROBBINS, W. J. (1949), Antibiotic substances from basidiomycetes. IV. *Marasmius conigenus, Proc. Natl. Acad. Sci. USA* **35**, 343–349.

KAVANAGH, F., HERVEY, A., ROBBINS, W. J. (1951), Antibiotic substances from basidiomycetes. VIII. *Pleurotus mutilus* (Fr.) Sacc. and *Pleurotus passeckerianus* Pilat, *Proc. Natl. Acad. Sci. USA* **37**, 570–574.

KIDA, T., SHIBAI, H., SETO, H. (1986), Structure of new antibiotics, pereniporins A and B, from a basidiomycete, *J. Antibiot.* **39**, 613–615.

KIS, Z., GLOSSE, A., SIGG, H. P., HRUBAN, L., SNATZKE, G. (1970), Die Struktur von Panepoxydon und verwandten Pilzmetaboliten, *Helv. Chem. Acta* **53**, 185–186.

KLEIN, K., ANKE, T., SHELDRICK, W. S., BROSS, M., STEFFAN, B., STEGLICH, W. (1990), Fulvoferruginin, a carotane antibiotic from *Marasmius fulvoferrugineus* Gilliam., *Z. Naturforsch.* **45c**, 845–850.

KLEINKAUF, H., VON DÖHREN, H. (1987), Biosynthesis of peptide antibiotics, *Annu. Rev. Microbiol.* **41**, 259–289.

KNAUSEDER, F., BRANDL, E. (1976), Pleuromutilins: Fermentation, structure and biosynthesis, *J. Antibiot.* **29**, 125–131.

KOBAYASHI H., MATSUNAGA, K., FUJII, M. (1993), PSK as chemopreventive agent, *Cancer Epidemiol. Biomarkers Prev.* **2**, 271–276.

KOMATSU, N., OKUBO, S., KIKUMOTO, S., KIMURA, K., SAITO, G., SAKAI, S. (1969), Host mediated antitumor action of schizophyllan, a glucan produced by *Schizophyllum commune, Gann* **60**, 137–144.

KRAICZY, P., HAASE, U., GENCIC, S., FLINDT, S., ANKE, T., BRANDT, U., VON JAGOW, G. (1996),

The molecular basis for the natural resistance of the cytochrome bc_1 complex from strobilurin producing basidiomycetes to center Q_P inhibitors, *Eur. J. Biochem.* **235**, 54–63.
KUBO, I., KIM, M., WOOD, W. F., NAOKI, H. (1986), Clitocine, a new insecticidal nucleoside from the mushroom *Clitocybe inversa, Tetrahedron Lett.* **27**, 4277–4280.
KUHNT, D., ANKE, T., BESL, H., BROSS, M., HERRMANN, R., MOCEK, U., STEFFAN, B., STEGLICH, W. (1990), Antibiotics from basidiomycetes. XXXVII. New inhibitors of cholesterol biosynthesis from cultures of *Xerula melanotricha* Dörfelt, *J. Antibiot.* **43**, 1413–1420.
KUNIMOTO, T., HORI, M., UMEZAWA, H. (1973), Mechanism of action of diketocoriolin B, *Biochim. Biophys.* **298**, 513–525.
KUPKA, J., ANKE, T., OBERWINCKLER, F., SCHRAMM, G., STEGLICH, W. (1979), Antibiotics from basidiomycetes. VII. Crinipellin, a new antibiotic from the basidiomyceteous fungus *Crinipellis stipitaria* (Fr.) Pat., *J. Antibiot.* **32**, 130–135.
KUPKA, J., ANKE, T., LIPMANN, F. (1980), Comparison of the cytotoxic effect of three antibiotics from basidiomycetes in different cell systems, p. 17. *Abstract Annu. Meeting ASM.* Washington, D.C.: ASM Press.
KUPKA, J., ANKE, T., GIANETTI, B.M., STEGLICH, W. (1981a), Antibiotics from basidiomycetes. XIV. Isolation and biological characterization of hypnophilin, pleurotellol, and pleurotellic acid from *Pleurotus hypnophilus* (Berk.) Sacc., *Arch. Microbiol.* **130**, 223–227.
KUPKA, J., ANKE, T., STEGLICH, W., ZECHLIN, L. (1981b), Antibiotics from basidiomycetes XI. The biological activity of siccayne, isolated from the marine fungus *Halocyphina villosa, J. Antibiot.* **34**, 298–304.
KUPKA, J., ANKE, T., MIZUMOTO, K., GIANETTI, B. M., STEGLICH, W. (1983), Antibiotics from basidiomycetes. XVII. The effect of marasmic acid on nucleic acid metabolism, *J. Antibiot.* **36**, 155–160.
KUSCHEL, A., ANKE, T., VELTEN, R., KLOSTERMEYER, D., STEGLICH, W. (1994), The mniopetals, new inhibitors of reverse transcriptases from a *Mniopetalum* species (Basidiomycetes), *J. Antibiot.* **47**, 733–739.
KWOK, O. C. H., PLATTNER, R., WEISLEDER, D., WICKLOW, D. T. (1992), A nematicidal toxin from *Pleurotus ostreatus* NRRL 3526, *J. Chem. Ecol.* **18**, 127–136.
LAUER, U., ANKE, T., SHELDRICK, W. S., SCHERER, A., STEGLICH, W. (1989), Antibiotics from basidiomyctes. XXXI. Aleurodiscal: an antifungal sesterpenoid from *Aleurodiscus mirabilis* (Berk. & Curt.) Höhn, *J. Antibiot.* **42**, 875–882.
LAUER, U., ANKE, T., HANSSKE, F. (1991), Antibiotics from basidiomycetes. XXXVIII. 2-methoxy-5-methyl-1,4-benzochinone, a thromboxane A_2 receptor antagonist from *Lentinus adherens, J. Antibiot.* **44**, 59–65.
LEE, S. G., ANKE, T. (1979), Die Wirkung von Striatin A, B, und C auf den Glucose-Transport in normalen und Rous-Sarkom-Virus-transformierten Hühner-Fibroblasten, *Hoppe-Seyler's Z. Physiol. Chem.* **360**, 1170.
LEMIEUX, R. U., THORN, J. A., BRICE, C., HASKINS, R. H. (1951), Biochemistry of the Ustilaginales. II. Isolation and partial characterization of ustilagic acid, *Can. J. Chem.* **29**, 409–414.
LEONHARDT, K., ANKE, T., HILLEN-MASKE, E., STEGLICH, W. (1987), 6-Methylpurine, 6-methyl-9-β-D-ribofuranosylpurine, and 6-hydroxymethyl-9-β-D-ribofuranosylpurineas antiviral metabolites of *Collybia maculata* (Basidiomycetes), *Z. Naturforsch.* **42c**, 420–424.
LEVITZKI, A. (1994), Signal-transduction therapy. A novel approach to disease managment, *Eur. J. Biochem.* **226**, 1–13.
LÖFGREN, N., LÜNING, B., HEDSTRÖM, H. (1954), The isolation of nebularin and the determination of its structure, *Acta Chem. Scand.* **8**, 670–680.
LORENZEN, K., ANKE, T., ANDERS, U., HINDERMAYR, H., HANSSKE, F. (1993), Two inhibitors of platelet aggregation from a *Panus* species (Basidiomycetes), *Z. Naturforsch.* **49c**, 132–138.
LORENZEN, K., ANKE, T., KONETSCHNY-RAPP, S., SCHEUER, W. (1995), 5-Hydroxy-3-vinyl-2(5H)-furanone – a new inhibitor of human synovial phospholipase A_2 and platelet aggregation from fermentations of a *Calyptella* species (Basidiomycetes), *Z. Naturforsch.* **50c**, 403–409.
MANNING, A. M. (1996), Transcription factors: a new frontier for drug discovery, *Drug Discovery Today* **1**, 151–160.
MANNING, A. M., ANDERSON, D. C. (1994), Transcription factor NF-κB: an emerging regulator of inflammation, in: *Annu. Reports Med. Chem.* **29** (BRISTOL, J. A., Ed.), 235–244.
MATHÉ, G. (1987), Bestatin compared to other pharmacologic immunoregulator or modulating agents, in: *Horizons on Antibiotic Research* (DAVIS, B. D., ICHIKAWA, T., MAEDA, K., MITSCHER, L. E., Eds.), pp. 44–65. Tokyo: Japan Antibiotics Research Association.
MAYER, A., ANKE, H., STERNER, O. (in press), Omphalotin, a new cyclic peptide with potent nematicidal activity from *Omphalotus olearius.* I. Fermentation and biological activity, *Nat. Prod. Lett.*
MAZUR, X., BECKER, U., ANKE, T., STERNER, O. (1996), Two new bioactive diterpenes from *Lepista sordida, Phytochemistry* **43**, 405–407.

McMorris, T. C., Anchel, M. (1965), Fungal metabolites. The structures of novel sesquiterpenoids Illudin S and M, *J. Am. Chem. Soc.* **87**, 1594–1600.

Morisaki, N., Furukawa, J., Nozoe, S., Itai, A., Itaka, Y. (1980), Synthetic studies on marasmane and isomarasmane derivatives, *Chem. Pharm. Bull.* **28**, 500–507.

Müller, E., Loeffler, W. (1982), *Mykologie.* Stuttgart: Thieme Verlag.

Musilek, V. (1969), *Czech. Patent* CS 136492, *Chem. Abstr.* 1969, 70, 18900y.

Nair, M. S. R., Anchel, M. (1977), Frustulosinol, an antibiotic metabolite of *Stereum frustulosum:* revised structure of frustulosin, *Phytochemistry* **16**, 390–392.

Nerud, F., Sedmera, P., Zouchova, Z., Musilek, V., Vondracek, M. (1982), Biosynthesis of mucidin, an antifungal antibiotic from basidiomycete *Oudemansiella mucida.* ^{2}H-, ^{13}C- and ^{14}C-labelling study, *Coll. Czech. Chem. Commun.* **47**, 1020–1025.

Nishimura, Y., Koyama, Y., Umezawa, S., Takeuchi, T., Ishizuka, M., Umezawa, H. (1977), Chemical modification of coriolin B, *J. Antibiot.* **30**, 59–65.

Obuchi, T., Kondoh, H., Watanabe, N., Tamai, M., Ōmura, S., Jun-Shan, Y., Xiao-Tian, L. (1990), Armillaric acid, a new antibiotic produced by *Armillaria mellea, Planta Med.* **56**, 198–201.

Potts, B. C. M., Faulkner, D. J. (1992), Phospholipase A_2 inhibitors from marine organisms, *J. Nat. Prod.* **55**, 1701–1717.

Price, M., Heinstein, P. (1978), Cell-free biosynthesis of illudins, *Lloydia* **41**, 574–577.

Quack, W., Anke, T., Oberwinkler, F., Giannetti, B. M., Steglich, W. (1978), Antibiotics from basidiomycetes. V. Merulidial, a new antibiotic from the basidiomycete *Merulius tremellosus* Fr., *J. Antibiot.* **31**, 737–741.

Rabe, U. (1989), Fermentation von *Cyathus striatus* (Huds. ex Pers.) Willd (Basidiomycetes) und Biosynthese der Striatine. *Dissertation,* University of Kaiserslautern, FRG.

Riedl, K. (1976), Studies on pleuromutilin and some of its derivatives, *J. Antibiot.* **29**, 132–139.

Roberts, D. W. (1981), Toxins of entomopathogenic fungi. In: *Microbial Control of Pests and Plant Diseases 1970–1980* (Burges, H. D., Ed.), pp. 441–464. London, New York: Academic Press.

Salmon, J. A., Garland, L. G. (1991), Leukotrienes, antagonists and inhibitors of leukotriene biosynthesis as potential therapeutic agents, in: *Progress in Drug Research* (Jucker, E., Ed.), pp. 9–90. Basel: Birkhäuser.

Samuelsson, B., Dahlen, S.-E., Lindgren, J. A., Rouzer, C., Serhan, C. N. (1987), Leukotrienes and lipoxins: structure, biosynthesis, and biological effects, *Science* **237**, 1171–1176.

Sasaki, T., Takasuka, N. (1976), Further study of the structure of lentinan, an antitumor polysaccharide from *Lentinus edodes, Carbohydr. Res.* **47**, 99–104.

Sauter, H., Ammermann, E., Benoit, R., Brand, S., Gold, R. E., Grammenos, W., Köhle, H., Lorenz, G., Müller, B., Röhl, F., Schirmer, U., Speakman, J. B., Wenderoth, B., Wingert, H. (1995), Mitochondrial respiration as a target for antifungals: lessons from research on strobilurins, in: *Antifungal Agents; Discovery and Mode of Action* (Dixon, G. K., Copping, L. G., Hollomon, D. W., Eds.), pp. 173–191. Oxford: BIOS Scientific Publishers.

Sayre, R. M., Walter, D. E. (1991), Factors affecting the efficacy of natural enemies of nematodes, *Annu. Rev. Phytopathol.* **29**, 149–166.

Schneider, H., Moser, A. (1987), Process kinetic analysis of pleuromutilin fermentation, *Bioproc. Eng.* **2**, 129–135.

Schramm, G. (1980), Neue Antibiotika aus Höheren Pilzen (Basidiomyceten). *PhD Thesis,* Department of Chemistry, University of Bonn, FRG.

Schramm, G., Steglich, W., Anke, T., Oberwinkler, F. (1978), Antibiotika aus Basidiomyceten, III. Strobilurin A und B, antifungische Stoffwechselprodukte aus *Strobilurus tenacellus, Chem. Ber.* **111**, 2779–2784.

Schramm, G. Steglich, W., Anke, T. (1982), Structure–activity relationship of strobilurins, oudemansin, and synthetic analogues, Abstract 457, in: *Abstracts 13th Int. Congr. Microbiol.,* August 8–13, Boston.

Schultes, R. E., Hofmann, A. (1980), *Pflanzen der Götter.* Bern, Stuttgart: Hallwag.

Schwitzgebel, K. (1992), Untersuchungen an holzabbauenden und holzverfärbenden Pilzen, *M. Sc. Thesis,* Department of Biology, University of Kaiserslautern, FRG.

Sedmera, P., Musilek, V., Nerud, F. (1981), Mucidin: Its identity with strobilurin A, *J. Antibiot.* **34**, 1069.

Segiguchi, J., Gaucher, M. (1979), Isoepoxydon, a new metabolite of the patulin pathway in *Penicillium urticae, Biochem. J.* **182**, 445–453.

Shinozawa, S., Tsutsui, K., Oda, T. (1979), Enhancement of the antitumor effect of illudin S by including it into liposomes, *Experientia* **35**, 1102–1103.

Simon, B. (1994), Antivirale und cytotoxische Wirkstoffe aus Basidiomyceten. *Ph.D. Thesis,* University of Kaiserslautern, FRG.

SIMON, B., ANKE, T., ANDERS, U., NEUHAUS, M., HANSSKE, F. (1995), Collybial, a new antibiotic sesquiterpenoid from *Collybia confluens* (Basidiomycetes), *Z. Naturforsch.* **50c**, 173–180.

STADLER, M., ANKE, T., DASENBROCK, J., STEGLICH, W. (1993a), Phellodonic acid, a new biologically active hirsutane derivative from *Phellodon melaleucus* (Thelephoraceae, Basidiomycetes), *Z. Naturforsch.* **48c**, 545–549.

STADLER, M., MAYER, A., ANKE, H., STERNER, O. (1993b), Fatty acids and other compounds with nematicidal activity from cultures of basidiomyocetes, *Planta Med.* **60**, 128–132.

STADLER, M., ANKE H., STERNER, O. (1994), New nematicidal and antimicrobial compounds from the basidiomycete *Cheimonophyllum candidissimum, J. Antibiot.* **47**, 1284–1289.

STADLER, M., FOURON, J.-F., STERNER, O., ANKE, H. (1995), 1,2-dihydroxyminthlactone, a new nematicidal monoterpene isolated from the basidiomycete *Cheimonophyllum candidissimum* (Berk & Curt.) Sing., *Z. Naturforsch.* **50c**, 473–475.

STÄRK, A., ANKE, T., MOECK, U., STEGLICH, W., KIRFEL, A., WILL, G. (1988), Lentinellic acid, a biologically active protoilludane derivative from *Lentinellus* species (Basidiomycetes), *Z. Naturforsch.* **43c**, 177–183.

STÄRK, A., ANKE, T., MOCEK, U., STEGLICH, W. (1991), Omphalone, an antibiotically active benzoquinone derivative from fermentations of *Lentinellus omphalodes, Z. Naturforsch.* **46c**, 989–992.

STERNER, O., BERGMAN, R., KIHLBERG, J., WICKBERG, B. (1985), The sesquiterpens of *Laccarius vellereus* and their role in a proposed chemical defense system, *J. Nat. Prod.* **48**, 279–288.

STERNER, O., CATRE, R., NILSON, L. (1987a), Structure–activity relationship for unsaturated dialdehydes. 1. The mutagenic activity of 18 compounds in the *Salmonella*/microsome assay, *Mutat. Res.* **188**, 169–174.

STERNER, O., STEFFAN, B., STEGLICH, W. (1987b), Novel azepine derivatives from the pungent mushroom *Chalciporus piperatus, Tetrahedron.* **43**, 1075–1082.

STERNER, O., ETZEL, W., MAYER, A., ANKE, H. (in press), Omphalotin, a new cyclic peptide with potent nematicidal activity from *Omphalotus olearius.* II. Isolation and structure determination, *Nat. Prod. Lett.*

STEWART, K. R. (1986), A method for generating protoplasts from *Clitopilus pinsitus, J. Antibiot.* **39**, 1486–1487.

SUGA, T., SHIO, T., MAEDA, Y. Y., CHIHARA, G. (1984), Antitumor activity of lentinan in murine syngeneic and autochthonous hosts and its suppressive effect on 3-methylcholantrene-induced carcinogenesis, *Cancer Res.* **44**, 5132–5137.

TAKAHASHI, K., SOMEYA, T., MURAKI, S., YOSHIDA, T. (1980), A new keto alkohol, (−)-minthlactone, (+)-isominthlactone and minor components in peppermint oil, *Agaric. Biol. Chem.* **44**, 1535–1543.

TAKAHASHI, A., KUSANO, G., OHTA, T., OHIZUMI, Y., NOZOE, S. (1989), Fasciculic acids A, B and C as calmodulin antagonists from the mushroom *Naematoloma fasciculare, Chem. Pharm. Bull.* **37**, 3247–3250.

TAKEUCHI, T., IINUMA, H., IWANAGA, J., TAKAHASHI, S., TAKITA, T., UMEZAWA, H. (1969), Coriolin, a new basidiomycete antibiotic, *J. Antibiot.* **22**, 215–217.

THOMPSON, C. B. (1995), Apoptosis in the pathogenesis and treatment of disease, *Science* **267**, 1456–1462.

TOMODA, H., HUANG, X. H., COA, J., NISHIDA, H., NAGAO, R., OKUDA, S., TANAKA, H., ŌMURA, S., ARAI, H., INOUE, K. (1992), Inhibition of acyl-CoA:Cholesterol acyltransferase activity by cyclodepsipeptide antibiotics, *J. Antibiot.* **45**, 1626–1632.

TSUGAGOSHI, S., HASHIMOTO, Y., FUJI, G., KOBAYASHI, H., NOMOTO, K., ORITA, K. (1984), Krestin (PSK), *Cancer Treat. Rev.* **11**, 131–155.

TURNER, W. B. (1971), *Fungal Metabolites.* London: Academic Press.

TURNER, W. B., ALDRIDGE, D. C. (1983), *Fungal Metabolites II.* London: Academic Press.

UMEZAWA, H., AOYAGI, T., SUDA, H., HAMADA, M., TAKEUCHI, T. (1976), Bestatin, an inhibitor of aminopeptidase B, produced by actinomycetes, *J. Antibiot.* **29**, 97–99.

VIDARI, G., DE BERNARDI, M., VITA-FINZI, P., FRONZA, G. (1976), Sesquiterpenes from *Lactarius blennius, Phytochemistry* **15**, 1953–1955.

VINING, L. C. (1990), Function of secondary metabolites, *Annu. Rev. Microbiol.* **44**, 395–427.

VON JAGOW, G., GRIBBLE, G. W., TRUMPOWER, B. L. (1986), Mucidin and strobilurin A are identical and inhibit electron transfer in the cytochrome bc_1 complex of the mitochondrial respiratory chain at the same site as myxothiazol, *Biochemistry* **25**, 775–780.

WEBER, W., ANKE, T., STEFFAN, B., STEGLICH, W. (1990a), Antibiotics from basidiomycetes. XXXII. Strobilurin E: a new cytostatic and antifungal (*E*)-β-methoxyacrylate antibiotic from *Crepidotus fulvotomentosus* Peck, *J. Antibiot.* **43**, 207–212.

WEBER, W., ANKE, T., BROSS, M., STEGLICH, W. (1990b), Strobilurin D and Strobilurin F: two new cytostatic and antifungal (*E*)-β-methoxyacrylate antibiotics from *Cyphellopsis anomala, Planta Med.* **56**, 446–450.

WEBER, W., SEMAR, M., ANKE, T., BROSS, M., STEGLICH, W. (1992), Tyromycin A: a novel inhibitor of leucine and cysteine aminopeptidase from *Tyromycetes lacteus, Planta. Med.* **58**, 56–59.

WEBER, W., SCHU, P., ANKE, T., VELTEN, R., STEGLICH, W. (1994), Caloporosid, a new inhibitor of phospholipases C from *Caloporus dichrous* (Fr.) Ryv., *J. Antibiot.* **47**, 1188–1194.

WINKELMANN, G. (1986), Iron complex products, in: *Biotechnology* 1st Edn., Vol. 4 (REHM, H. J., REED, G., Eds.), pp. 216–243. Weinheim: VCH.

WUNDER, A., ANKE, T., KLOSTERMEYER, D., STEGLICH, W. (1996), Lactarane type sesquiterpenoids as inhibitors of leukotriene biosynthesis and other new metabolites from submerged cultures of *Lentinellus cochleatus* (pers. ex Fr.) Karst., *Z. Naturforsch.* **51c**, 493–499.

YANG, J. S., SU, Y. L., WANG, Y. L., FENG, X. Z., YU, D. Q., LIANG, X. T. (1991), Two novel protoilludane norsesquiterpenoid esters, armillasin and armillatin, from *Armillaria mellea, Planta Med.* **57**, 478–480.

ZÄHNER, H. (1982), Mikrobieller Sekundärstoffwechsel, in: *Handbuch der Biotechnologie* (PRÄVE, P., FAUST, U., SITTIG, W., SUKATSCH, D. A., Eds.). Wiesbaden: Akademsiche Verlagsgesellschaft.

ZAPF, S. (1994), Neue antifungische Antibiotika aus Pilzen. *Ph.D. Thesis,* University of Kaiserslautern, FRG.

ZAPF, S., WERLE, A., ANKE, T., KLOSTERMEYER, D., STEFFAN, B., STEGLICH, W. (1995), 9-Methoxystrobilurine – Bindeglieder zwischen Strobilurinen und Oudemansinen, *Angew. Chem.* **107**, 255–257.

ZAPF, S., WUNDER, A., ANKE, T., KLOSTERMEYER, D., STEGLICH, W., SHAN, R., STERNER, O., SCHEUER, W. (1996), (+)-10α-hydroxy-4-muurolen-3-one, a new inhibitor of leukotriene biosynthesis from a *Favolaschia* species. Comparison with other sesquiterpenes, *Z. Naturforsch.* **51c**, 487–492.

12 Cyclosporins: Recent Developments in Biosynthesis, Pharmacology and Biology, and Clinical Applications

JÖRG KALLEN
VINCENT MIKOL
VALÉRIE F. J. QUESNIAUX
MALCOLM D. WALKINSHAW

Basel, Switzerland

ELISABETH SCHNEIDER-SCHERZER
KURT SCHÖRGENDORFER
GERHARD WEBER

Kufstein-Schaftenau, Austria

HANS G. FLIRI

Vitry-sur-Seine, France

1 Introduction

Cyclosporins are a family of hydrophobic cyclic undecapeptides with a remarkable spectrum of diverse biological activities. The first member of this class to be discovered was named cyclosporin A (CsA; structure shown in Fig. 1). To this date, some 30 members of this family have been isolated from natural sources. In addition, a close analog named SDZ 214-103, incorporating a lactone function in place of a peptide bond and exhibiting a similar biological profile was also discovered from natural sources (Fig. 2). The discovery and structure elucidation of cyclosporins have been amply reviewed (WENGER 1986; WENGER et al., 1986; VON WARTBURG and TRABER, 1988; FLIRI and WENGER, 1990).

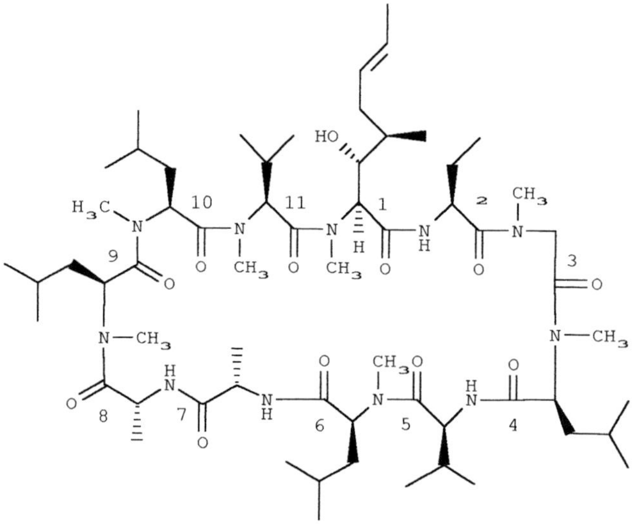

Fig. 1. Structure of cyclosporin A (Sandimmun®), including the numbering system.

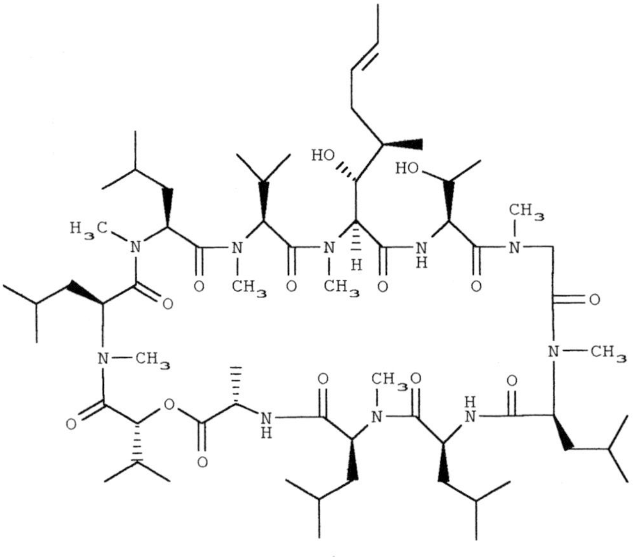

Fig. 2. Structure of SDZ 214-103.

In this chapter, emphasis is given to some more recent aspects of chemistry, biosynthesis, and biological activity.

2 Clinical Applications of Cyclosporins

2.1 Introduction

Cyclosporin A was initially isolated as an antifungal antibiotic. It was later shown to possess immunosuppressive properties of high therapeutic value. Since 1983, cyclosporin A, under the trade name Sandimmun®, has been in clinical use worldwide to prevent rejection of organ transplants. It has subsequently been approved for the therapy of certain autoimmune diseases. Since the time of market introduction of Sandimmun®, many additional biological activities of cyclosporins have been discovered, some of which may lead to novel clinical applications of cyclosporin A or of non-immunosuppressive analogs. At the time of market introduction of Sandimmun®, the mechanism by which this drug mediates immunosuppression was not understood at the molecular level nor was a receptor known. Since then, not only was a whole family of receptors discovered (i.e., the cyclophilins), but a possible role of these proteins for protein folding and cellular protein traffic has emerged. Much of what is known today about cyclosporins, cyclophilins and their biochemistry was greatly aided by the discovery of FK506, an immunosuppressive macrolide. This compound elicited much interest because, like cyclosporin A, it was a T cell selective immunosuppressant, but much more potent. This activity was soon shown to be based on a mechanism identical to that of cyclosporin. Search for FK506 receptors led to the discovery of the FK506 binding proteins (FKBPs), a novel protein family with no homologies to cyclophilins, yet with many properties in common. Like the cyclophilins, FKBPs appear to have a functional role in protein folding. Unveiling the mechanism of immunosuppressive activity of Sandimmun® and FK506 was greatly facilitated by the availability of a third compound, rapamycin, which binds to the same receptors as FK506, yet exhibits a different spectrum of biological activities. A detailed account of these aspects is given in Sects. 3.2. and 3.3. In this section present clinical applications of Sandimmun® are discussed. They include the following indications:

- allograft rejection,
- Behset's uveitis,
- rheumatoid arthritis,
- aplastic anemia (NDA's pending),
- nephrotic syndrome,
- atopic dermatitis (NDA's pending),
- psoriasis vulgaris.

2.2 Transplantation

Sandimmun® is a reversible inhibitor of the transcription of interleukin 2 (IL-2) and several other lymphokines, most notably in helper T lymphocytes (see Sect. 3.2). As a consequence, it suppresses the activation and/or maturation of various cell types, in particular those involved in cell-mediated immunity. Because of these properties, Sandimmun® has become the first-line immunosuppressant for prophylaxis and therapy of transplant rejection. In fact, the modern era of transplantation surgery was only possible after the availability of cyclosporin. The first patient to receive a kidney graft under CsA treatment was reported in 1978 (Calne et al., 1978). Soon thereafter, transplantations of liver, heart, and combined lung–heart commenced (Ernst, 1991; Barry, 1992; Kahan, 1992; Tsang et al., 1992). In 1991, only in Germany 450 liver transplantations were performed (Hopf et al., 1992). Organ availability has become a major limiting factor and numerous patients die while awaiting a donor organ. Therefore, organ preservation techniques have become an important aspect in the area of transplantation surgery. Histocompatibility matching, besides immunosuppression, is the key factor contributing to long-term graft survival. Currently, expected 10-year first graft survival rates for kidneys from HLA-identical siblings, 1-haplotype-matched relative, and cadaver donors are 74, 51, and 40%, respectively (BARRY, 1992). The survival probability after lung transplantation is approximately 65% after 1 year and 50% after 3 years (ERNST, 1991). Major problems encountered in transplantation surgery are technical difficulties during operation, serious infections, and acute rejection episodes during the first postoperative period, and chronic (long-term) rejection. Side effects can be classified into those associated to immunosuppression (lymphoproliferative disorders, infectious diseases caused by bacterial and fungal pathogens as well as viruses), and other adverse effects which are specific for the immunosuppressive drugs used. For Sandimmun® these include primarily impairment of renal function, hypertension, hirsutism, and gingival hyperplasia (MASON, 1989). Neurological and gastrointestinal effects are also common in Sandimmun® recipients but are usually mild to moderate and resolve on dosage reduction.

2.3 Autoimmune Diseases

Since Sandimmun® not only suppresses cell-mediated immunity but also humoral immune responses and inhibits chronic inflammatory reactions it appeared very promising in the treatment of autoimmune diseases. Prospective controlled trials performed in patients with autoimmune diseases have recently been reviewed (FREY, 1990; FAULDS et al., 1993). Efficacy could be proven for the following diseases: Endogenous uveitis, rheumatoid arthritis, Sjogren's syndrome, myasthenia gravis, psoriasis, atopic dermatitis, Crohn's disease. The drug is considered as a first-line therapy in patients with moderate or severe aplastic anemia who are not eligible for bone marrow transplantation. It may also be of benefit in patients with primary biliary cirrhosis and intractable pyoderma gangrenosum. Sandimmun® does not appear to be effective in patients with allergic contact dermatitis, multiple sclerosis, or amyotropic lateral sclerosis. Successful application in insulin-dependent diabetes will depend on the development of diagnostic tools indicating early disease onset before beta cell destruction has progressed too far and clinically overt diabetes is present. The most significant advantage in a number of indications is the steroid-sparing effect of Sandimmun®. To avoid relapse after control of active disease, patients should continue receiving Sandimmun® maintenance therapy at the lowest effective dose.

2.4 Activity Against Tumor Multidrug Resistance

Cellular resistance to cytotoxic drugs is often the cause of inefficient treatment of cancer with potent antitumor drugs. While many mechanisms of resistance occur, the mechanism of "multidrug resistance" (MDR) has received particular attention (ENDICOTT and LING, 1989). Most often, this type of resistance extends to several anticancer drugs of unrelated structural classes and mechanisms of action. A common feature of MDR is overexpression of a particular class of transmembrane glycoproteins called P-glycoproteins (Pgp) which serve as transport proteins rapidly effluxing antitumor drugs out of the tumor cells as soon as they have entered through the membrane. As a consequence, Pgp transporters decrease intracellular drug concentrations below their active threshold. Numerous *in vitro* studies have described agents which can restore the sensitivity of MDR tumor cells, including cyclosporin A at clinically achievable concentrations (TWENTYMAN, 1992). Moreover, this effect can be dissociated from immunosuppression, as non-immunosuppressive analogs have been shown to retain resistance modifier activity and some are even more potent than cyclosporin A. One such analog from Sandoz Pharma AG called SDZ PSC 833 is approximately tenfold more potent than Sandimmun® as a resistance modifier and is currently undergoing clinical trials (BOESCH et al., 1991). The structure of SDZ PSC 833 is shown in Fig. 3.

2.5 Anti-HIV Activity

A possible beneficial effect of Sandimmun® in HIV disease has been proposed as early as 1986 (ANDRIEU et al., 1986). The rationale is that activation of $CD4^+$ cells which is required for HIV replication (ZACK et al., 1990;

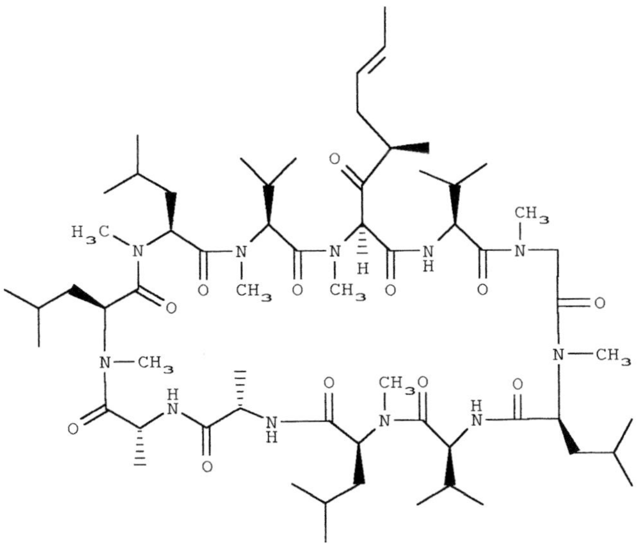

Fig. 3. Structure of SDZ PSC 833.

STEVENSON et al., 1990) is inhibited by CsA. In addition, CsA would inhibit the initiation of an autoimmune process involving killing of HIV infected lymphocytes by cytotoxic cells and may also counteract HIV-induced apoptotic cell death of $CD4^+$ cells (HABESHAW et al., 1990).

A thorough investigation of a series of immunosuppressive and non-immunosuppressive cyclosporins was performed in the preclinical research laboratories at the Sandoz Research Institute of Sandoz Pharma AG in Vienna, Austria. Compounds were evaluated for antiviral, cytotoxic, and for immunosuppressive activity *in vitro*. It was found that some non-immunosuppressive analogs of Sandimmun® were equal or even superior in their antiviral activity without being cytotoxic. One such analog, SDZ NIM 811 (Fig. 4), is comparable to Sandimmun® re-

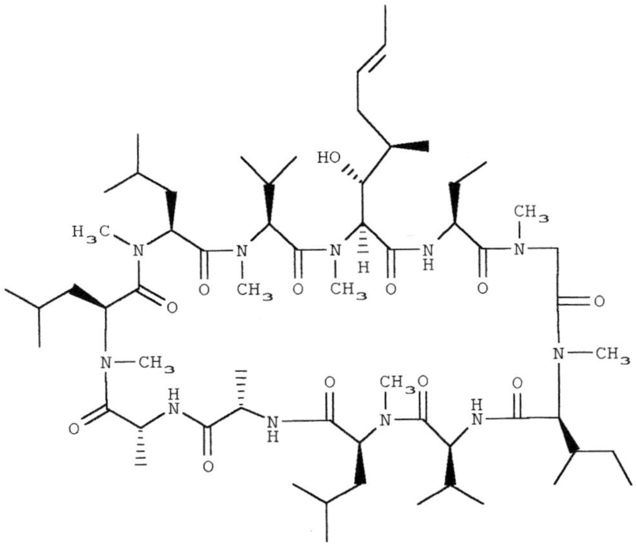

Fig. 4. Structure of SDZ NIM 811.

garding oral bioavailability and pharmacokinetics in animals and appears to be of lower nephrotoxicity (ROSENWIRTH et al., 1994).

3 Mode of Immunosuppressive Action

In addition to its clinical use as an immunosuppressant CsA has also been widely employed as an experimental tool for basic research. It has helped to understand the biochemical events needed to translate a signal from the T cell surface to the nucleus and the pathophysiological processes involving lymphocyte activation in a variety of diseases.

Early immunological studies revealed that CsA exerts specific effects on T cell lymphokine transcription (KRONKE et al., 1984). Because T cells are prominent in the cellular immune response, studies on the mechanism of immunosuppression by CsA have mainly focused on its role in regulating gene expression in T lymphocytes. To place CsA activity in perspective a summary on T cell activation is first given below.

3.1 Introduction to T Cell Activation

A schematic representation of the cellular immune response with emphasis on the central role of the activated T lymphocytes is shown in Fig. 5. For T cell activation, the antigen receptor on the T cell surface interacts with the processed antigen exposed in the proper histocompatibility context on the surface of the antigen presenting cell (APC; cf. Fig. 5). In the presence of additional accessory interactions between the T cell and the antigen presenting cell, antigen recognition leads to biochemical events which finally result in proliferation, differentiation, and maturation of the T cell to T effector cells with specific immunological function, e.g., helper and cytotoxic suppressor functions. The multiple steps involved in this process, namely signal recognition, signal transduction to the nucleus, resulting in gene activation, expression of growth factor receptors, growth factor synthesis and cell proliferation, are briefly reviewed (RYFFEL, 1989).

3.1.1 Signal Recognition

3.1.1.1 First Signal

After uptake and limited proteolysis, the antigen processed by the antigen presenting cell is recognized in the context of the major histocompatibility complex (MHC) by the antigen receptor on T lymphocytes (Fig. 5). The binding of antigen to the T cell receptor is an absolute requirement for T cell activation under physiological conditions. The T cell receptor is a multicomponent structure consisting of the clonotypic α and β (or γ and δ) chains and the invariant CD3 subunits γ, δ, ε, ζ, and η. The complete assembly of all components is required for cell surface expression, and thus for antigen receptor function. The ζ and η subunits of CD3 most likely transduce to the cytoplasm the activating signals originating from antigen recognition by the T cell receptor. Antibodies against the CD3 complex can induce T cell functional responses that are identical to antigen-induced responses, regardless of antigen specificity. In addition, the two transmembrane proteins CD4 and CD8 expressed on helper and cytotoxic T cells participate in the interaction between the T cell and the antigen presenting cell by binding to MHC class II and I molecules, respectively. Originally, they were called coreceptors because their association with an intracellular enzyme facilitates signaling during T cell activation (JANEWAY et al., 1989).

3.1.1.2 Costimulatory Signals

In addition to the molecular interactions between the T cell receptor, CD3, CD4, or CD8 and the antigen presenting cell, costimu-

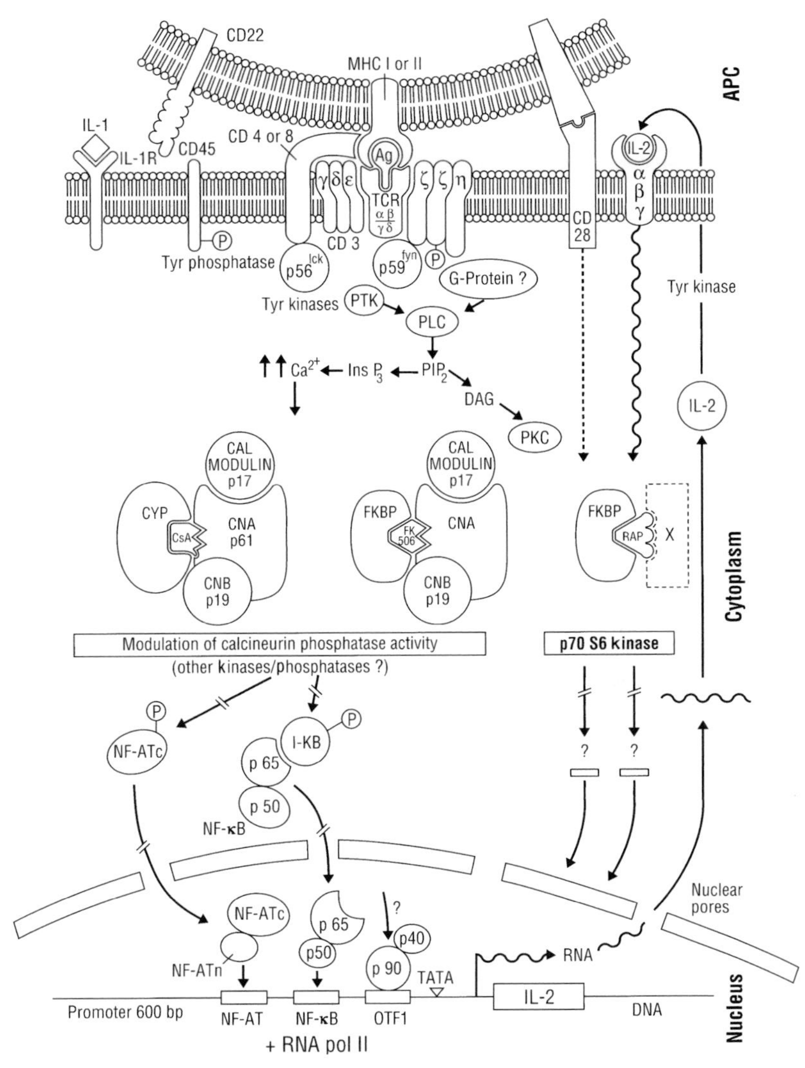

Fig. 5. CsA and FK506 both interfere, by binding to their respective immunophilins, with the function of intracellular molecules that transmit calcium-associated signals between the T-cell receptor (TCR) and the activation of lymphokine genes (IL-2) in the nucleus. Transcriptional regulation of IL-2 gene expression is modulated by the combination of transcription factors (e.g., NF-AT, NFκB, OTF-1) interacting with their corresponding recognition sites at the IL-2 promoter. These DNA/protein complexes, together with RNA polymerase II (RNA pol II), result in the antigen-inducible transcription of IL-2. Potential intervention sites for the pentameric complex (calcineurin A (p61), B (p19), calmodulin (p17), immunophilin, drug), involving, e.g., modification and translocation of antigen-inducible transcription factors (NF-AT; NFκB (p50, p65)), are indicated (II). CsA and FK506 interfere with the Go to G1 transition of the cell cycle, whereas rapamycin interferes with the G1 to S transition (for details, see text) (adapted from BAUMANN and BOREL, 1992).

latory signals contribute to the T cell receptor-driven proliferative response. These costimulatory signals are thought to be delivered by antigen presenting cells but not by tissue cells and might play an important role in the self/non-self discrimination by influencing the consequences of the T cell receptor engagement (LIU and LINSLEY, 1992). Conversely, *in vitro* stimulation of T cells in the absence of costimulation might lead to functional inactivation of the T cells, i.e., clonal anergy, or to activation-induced cell death (apoptosis).

The molecular basis for T cell costimulation is not yet fully understood. The accessory molecules are membrane proteins including the lymphoid adhesion molecules. They are invariant and bind to ligands expressed on the surface of other cells such as antigen presenting cells or target cells, thereby increasing the strength of adhesion between these cells and the T cells. In addition, accessory molecules may transduce biological signals to the T cell cytosol. CD2, e.g., which interacts with the leukocyte function-associated antigen-3 (LFA-3, CD58) augments specific signaling through the T cell receptor. In contrast, CD28 binding to the B7-1 and B7-2 molecules on antigen presenting cells initiate a signaling pathway which is distinct from the pathway emanating from the T cell receptor.

3.1.2 Signal Transduction

The antigen binding to the T cell receptor/CD3 complex in combination with costimulatory signals triggers a highly complicated signal transduction pathway which finally results in the pleiotropic T cell activation program. Although the molecular details of this program are still largely unknown, an attempt to summarize the current state of our knowledge with a focus on the mechanism of action of known immunosuppressants is schematically represented in Fig. 5 (IZQUIERDO and CANTRELL, 1992; ABRAHAM et al., 1992).

One of the earliest events after antigen binding is the activation of protein tyrosine kinases which initiates a cascade of downstream biochemical events (SEFTON and CAMPBELL, 1991). Since none of the subunits of the T cell receptor/CD3 complex reveals intrinsic catalytic domains, the protein tyrosine phosphorylation of CD3 is likely to be mediated by intracellular protein tyrosine kinases. Two members of the src family of protein tyrosine kinases have been identified in this context: $p59^{fyn}$ which is physically associated with the T cell receptor/CD3 complex, and $p56^{lck}$ which is associated with CD4 or CD8 and becomes activated upon coligation of CD4 or CD8 with the T cell receptor. How this signal is being transmitted from the T cell receptor-proximal protein tyrosine kinase activities to the cytoplasmic protein serine/threonine kinase and phosphatase cascades is less clear. In addition, the phosphorylation of tyrosine residues in the cytoplasmic domains of receptors such as CD3 ζ is required for the recruitment of several cellular enzymes like phospholipase C (PLC) to the cytoplasmic domain of the receptors. The γ1 and 2 isoforms of the phosphatidyl inositol-specific PLC depend on tyrosine phosphorylation for their activation. Once activated, PLC increases the hydrolysis of inositol phospholipids and produces two second messenger molecules. One, inositol triphosphate (IP3) binds to intracellular vesicles that store calcium ions (Ca^{2+}) causing them to release calcium into the cytoplasm. Ca^{2+} ions then bind to a small protein, calmodulin, which acts as a regulatory subunit for other enzymes essential for T cell activation such as the serine/threonine phosphatase calcineurin. Another second messenger molecule is diacylglycerol (DAG), a lipid molecule that remains in the membrane where it activates protein kinase C. Once activated, protein kinase C is translocated from the cytosol to the plasma membrane where it phosphorylates membrane-bound proteins. This event subsequently directs the modification of a set of other signal transmitting proteins. Finally, phosphorylation and dephosphorylation of certain transcription factors such as NFκB (nuclear factor for κ-light chain expression in B cells) and NF-AT (nuclear factor of activated T cells) induce or repress the transcription of their target genes such as the IL-2 gene. The effect of the modification is either direct by increasing or decreasing the affinity of the transcription factors for their specific binding site on the DNA or indirect via additional protein–protein interactions (HUNTER and KARIN, 1992).

3.1.3 Gene Activation

The metabolic events resulting from exposure of T cells to antigen culminate in blast transformation and progression through the cell cycle. Activated T cells express novel functions such as lymphokine secretion and lytic capability which are related to their effector role in the immune response. They increase in volume and undergo rapid increases in phosphatidyl inositol metabolism, in cytoplasmic pH, intracellular free calcium concentration, and in the serine/threonine and tyrosine phosphorylation pattern of various cellular proteins. Progression through the cell cycle is associated with a general increase in protein, lipid and RNA synthesis and with an increase of the mRNA level for a variety of activation-related genes. Many of these genes encode lymphokines and surface receptors necessary for the expansion and the immune functions of the activated T cells. About 70–100 genes are activated in T cells during the differentiation program which is taking place at times ranging from 15 min to 14 d following stimulation (CRABTREE, 1989).

By analogy to viral systems the T cell activated genes can be divided into three groups: the immediate, the early, and the late genes. Immediate genes (e.g., *c-fos, c-myc*) are transcribed after activation with no need for protein synthesis. Their products appear very soon after stimulation, usually within 10–30 min. Transcription of the immediate genes takes place during the transition from the quiescent state (Go) into the G1 phase of the cell cycle. Immediate genes are usually identified by their ability to be transcribed in the presence of a protein synthesis inhibitor such as cycloheximide or anisomycin. Early genes, which include the lymphokines (e.g., IL-2, IL-3, IL-4, interferon-γ, GM-CSF, and IL-2 receptor α) are transcribed after the immediate genes and require postactivation protein synthesis. The early genes are involved in the mid to late G1 phase of the cell cycle and include genes coding for products required for DNA synthesis and cell division. Late genes require both DNA synthesis and cell division for their expression. They are synthesized during the G2 phase of the cell cycle.

The manyfold mechanisms used by T cells to regulate their immediate, early, and late activation genes include alteration of the transcriptional rate, of initiation and termination of transcription, and of mRNA stability. It is the group of early genes that is most sensitive to immunosuppressants like CsA and FK506. The expression of this group of genes is mainly regulated at the level of transcription initiation. Transcription is initiated once the transcription factors such as NFκB, jun, fos, or oct-1, -2 that exist as precursors in the cytoplasm have been translocated into the nucleus and are activated by phosphorylation or dephosphorylation. This enables them to bind to their specific regulatory elements and participate in a functional transcription complex with RNA polymerase II (HUNTER and KARIN, 1992).

3.1.4 Cytokine Receptor Expression

T cells express a series of cytokine receptors on their membrane which allow them to respond to various cytokines including IL-1, IL-2, IL-4, IL-7, IL-9, IL-10, and IL-12. Some of these receptors seem to be upregulated during T cell activation. The receptor for IL-2, a major growth factor for T cells, is composed of three subunits (α, β, and γ) in its high-affinity form (TAKESHITA et al., 1992). The β and γ chains associate to form receptors of intermediate activity which are expressed on resting T cells. Upon T cell activation by antigen or by IL-2 itself, the expression of the α chain is upregulated and contributes to form high-affinity receptors. The IL-4 receptor 130 kDa protein is also upregulated during T cell activation by mitogenic ligands or by IL-4 (ARMITAGE et al., 1990).

3.1.5 Release of Cytokines

Several cytokines are produced by T cells upon activation, namely IFN-γ, IL-2, IL-3,

IL-4, IL-5, IL-6, IL-9, IL-10, TNFα, and GM-CSF. On the other hand, antigen presenting cells release IL-1, IL-6, IL-12, IL-13, TNFα and β and other cytokines. These cytokines have multiple regulatory effects such as the autocrine stimulation of helper T cells and the differentiation of cytotoxic and suppressor T lymphocytes, natural killer cells, and B lymphocytes. A prominent cytokine for further T cell activation is IL-2 since T cell activation can be induced by IL-2 itself, even in the absence of antigen, *in vitro*.

3.2 Current Knowledge of the Mode of Immunosuppressive Action of CsA

In the following section, our current knowledge as to how CsA interferes with the immune response at the molecular level is reviewed with some reference to two other immunosuppressive drugs of microbial origin, FK506 and rapamycin. Like CsA, FK506 and rapamycin were initially discovered and characterized as antifungal antibiotics. The structures of FK506 and rapamycin, together with that of cyclosporin A, are shown in Fig. 6. FK506 is a neutral macrolactone produced by *Streptomyces tsukubaensis*, a soil microorganism collected in the Tsukuba area of northern Japan (KINO et al., 1987). FK506 prolongs the survival of skin and organ allografts in experimental animal models and is active at approximately one tenth of the CsA dose required for the same effects. The initial clinical trials with FK506 showed a remarkable effect in liver transplantation and in rescuing drug-resistant rejection in organ transplantation. FK506 has recently been approved in the US for the indication of transplantation (trade name Prograf®). Like CsA, FK506 interferes with the process of T cell activation by specifically inhibiting the transcription of lymphokines (TOCCI et al., 1989). Rapamycin, a macrolide isolated from cultures of the soil microorganism *Streptomyces hygroscopicus* originates from Easter Island. Although the immunosuppressive properties of rapamycin were recognized early (MARTEL et al., 1977) they have been intensively investigated only recently, primarily after the discovery of FK506

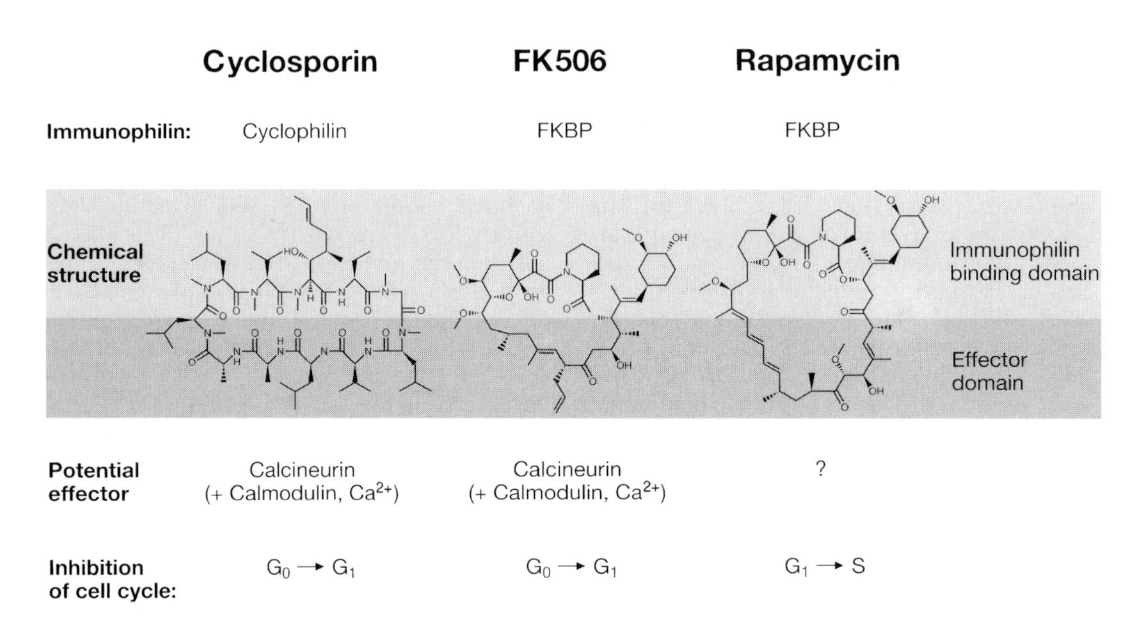

Fig. 6. Dual domain concept for the immunosuppressants of microbial origin CsA, FK506, and rapamycin (adapted from BAUMANN and BOREL, 1992).

and because of the striking structural similarity of the two compounds. Rapamycin is effective in many experimental transplantation models. A single dose of rapamycin leads to almost indefinite survival of cardiac allografts in the rat. It also showed activity in several murine autoimmune disease models (MORRIS, 1993). Rapamycin inhibits T cell activation at concentrations comparable to those of FK506 but with a completely different mechanism (DUMONT et al., 1990a, b).

3.2.1 Cyclosporin Receptors

A specific, saturable and reversible binding of CsA on mononuclear blood leukocytes was shown early (RYFFEL et al., 1982). Although membrane binding specific for CsA was recently reported (CACALANO et al., 1992) no membrane-bound receptors have been isolated to this point. On the other hand, accumulation of CsA within the cell suggested the existence of an intracellular receptor. Indeed, CsA, FK506, and rapamycin have been shown to bind to cytosolic proteins termed immunophilins (SCHREIBER, 1991). CsA predominantly binds to cyclophilins (CYP), a family of highly conserved proteins comprising both ubiquitous and tissue-specific proteins (SCHREIBER and CRABTREE, 1992). Cyclophilin A, the first protein identified, is a ubiquitous cytosolic 18 kDa protein (HANDSCHUMACHER et al., 1984). Cyclophilins B, C, and D, all with a molecular mass of about 22 kDa were reported later (HASEL et al., 1991; CARONI et al., 1991; PRICE et al., 1991; BERGSMA et al., 1991). Cyclophilin B contains an endoplasmic reticulum retention signal and is located in calcium-containing intracellular vesicles. Cyclophilin C seems to have a more restricted tissue distribution with low expression in lymphoid tissue. Murine cyclophilin C was reported to be highly expressed in kidney and was thus suggested to be involved in CsA nephrotoxicity (FRIEDMAN and WEISSMAN, 1991). However, this was not observed with human cyclophilin C (SCHNEIDER et al., 1994). In addition, a 40 kDa protein binding to CsA was also shown to share homology with cyclophilin A (KIEFER et al., 1992). The relative abundance and high conservation of the different cyclophilins across species suggests an important role for the normal cell function. All the members of the cyclophilin family show enzyme activity as peptidyl–prolyl *cis-trans* isomerases ("rotamases"), a property which enables them to accelerate the *cis-trans* isomerization of peptide bonds involving a prolyl residue. Interconversion of *cis* and *trans* conformers at peptide bonds to proline has been recognized as a rate determining step of protein folding *in vitro*. Therefore, it is not unlikely that the rotamase activity of cyclophilins might also facilitate protein folding *in vivo*. This activity of cyclophilins is potently inhibited by CsA (FISCHER et al., 1989; KERN et al., 1993) and some of its analogs, depending on the respective binding affinity of the compounds. The 40 kDa cyclophilin-like protein has lower affinity to CsA and its rotamase activity is also less sensitive to inhibition by this drug. FK506 binds to a separate group of ubiquitous immunophilins termed FK-binding proteins, FKBPs (SIEKIERKA et al., 1989b). Both a cytosolic form, FKBP12, and a membrane-associated form, FKBP13, sharing the same binding and enzymatic sites were reported (HAYANO et al., 1991; JIN et al., 1991). A higher molecular mass protein, FKBP59, was shown to be related to and associated with heat shock proteins and with the corticosteroid receptor (TAI et al., 1992). Strikingly, the FKBPs also exhibit peptidyl–prolyl *cis-trans* isomerase activity which is blocked when FK506 is bound (SIEKIERKA et al., 1989a). Neither CsA nor FK506 appear to cross-react for the binding to their respective immunophilins. Since both drugs block the induction of lymphokine gene transcription at the early stage of antigen-induced helper T cell activation, it has been postulated that T cell activation requires the separate activity of both immunophilins. In addition, the rotamase activity and the resulting protein folding was thought to be required for nuclear transport of some factors involved in signal transduction. However, there seems to be no correlation between inhibition of rotamase activity and inhibition of T cell activation, as demonstrated with the antagonists of FK506 and CsA, rapamycin and MeVal^4-Cs or SDZ NIM 811 (ZENKE et al., 1993).

It has been shown by competition experiments that FK506 and rapamycin bind to a common intracellular receptor since the two drugs act as reciprocal antagonists (DUMONT et al., 1990a). Indeed, rapamycin binds to all the FKBPs described above. In contrast, FKBP25 which shares homology with FKBP12 in its C-terminal region but differs in the N-terminal part selectively binds rapamycin (GALAT et al., 1992). Rapamycin inhibits the rotamase activity of the FKBPs but does not inhibit lymphokine transcription (DUMONT et al., 1990b). This clearly demonstrates that the inhibition of the rotamase activity of FKBP is insufficient *per se* to mediate the biological effect of FK506.

The relevance of the different isoforms of cyclophilins or FKBPs for inhibiting T cell activation was addressed in two different ways. (1) Overexpression of CYP-A or CYP-B but not of CYP-C increased T cell sensitivity to CsA. Similarly, overexpression of FKBP12 but not of FKBP13 or FKBP25 increased T cell sensitivity to FK506 (BRAM et al., 1993). Subcellular localization of CYP-B and CYP-C was also essential for their activity (BRAM et al., 1993). (2) By studying the binding of human cyclophilins A, B, and C to a series of cyclosporin derivatives it could be determined that the three cyclophilins recognize very similar sites on the CsA molecule. However, CYP-C can accommodate larger residues at position 2 of the cyclosporin molecule than CYP-A (SCHNEIDER et al., 1994). Therefore, a series of cyclosporin derivatives modified at position 2 could be selected that bind CYP-C up to tenfold better than CYP-A in comparison with CsA. Within this series of derivatives, the immunosuppressive activity measured by inhibition of IL-2 gene transcription correlated better with the binding to CYP-A than to CYP-C (QUESNIAUX et al., unpublished observations). Both approaches suggest that CYP-C might be less essential than CYP-A for T cell sensitivity to CsA.

3.2.2 Active Sites on the Drugs

The current model of the mode of action of immunosuppressant macrolides suggests that FK506 and rapamycin act via two regulatory domains: an immunophilin(FKBP)-binding domain which is shared between FK506 and rapamycin and an effector domain which is specific for each drug (BIERER et al., 1991) and accounts for their different activities. The same dual domain concept has also been demonstrated for CsA (Fig. 6). The binding site for cyclophilin was first mapped by immunochemical methods (QUESNIAUX et al., 1987) and later confirmed by NMR analysis (WEBER et al., 1991) and X-ray crystallography (PFLUEGL et al., 1993; MIKOL et al., 1993; see also Sect. 5.5). Cyclophilin A was shown to bind to residues 1, 2, 9, 10, and 11 of CsA, residues which were recognized early on to be essential for CsA immunosuppressive activity. The binding of CsA analogs to cyclophilin A also showed some correlation with their immunosuppressive activity (HANDSCHUMACHER et al., 1984; QUESNIAUX et al., 1988; DURETTE et al., 1988). To define the effector site of the cyclosporin A molecule several cell-permeable, cyclophilin-binding, but non-immunosuppressive cyclosporin derivatives were selected. These analogs that are modified in the "effector site" of CsA, like $MeVal^4$-Cs or SDZ NIM 811, antagonize the immunosuppressive effect of CsA on T cells (ZENKE et al., 1993). They effectively inhibit the *cis-trans* isomerase activity of cyclophilin providing again compelling evidence that inhibition of rotamase activity is insufficient for explaining CsA immunosuppression (SIGAL et al., 1991; BAUMANN et al., 1992). However, these results do not suggest that rotamase activity in general is irrelevant for the cell. Although the biological role of the two highly conserved families of rotamases, cyclophilins and FKBPS, is poorly understood at this time, there are a few examples for their role in protein transport. For example, the product of the *Drosophila* Nina A gene, a membrane-bound isomerase homologous to CYP-A, selectively transports rhodopsins in photoreceptor cells (STAMMES et al., 1992). Similarly, preliminary evidence suggests that cellular cyclophilin appears to be required for nuclear transport of DNA transcripts of HIV-1 (ROSENWIRTH et al., 1994; see Sect. 3.3.2).

3.2.3 Target Proteins of the Drug–Immunophilin Complexes

The structure–activity relationship of a large number of CsA or FK506 derivatives showed that the binding of CsA or FK506 to their respective immunophilins is necessary but *per se* not sufficient to inhibit T cell activation and thus raised the question of the molecular target of the drug–immunophilin complexes, CsA/CYP and FK506/FKBP. Indeed, CsA and FK506, when attached to their respective immunophilins could be part of a macromolecular complex and act on different primary targets or on different sites of a common primary target. Such a target would most probably be a component of the signal transduction pathway the activation of which ultimately leads to lymphokine transcription. Recent biochemical experiments support this model. The complexes of FK506/FKBP12 or CsA/CYP-A but not of rapamycin/FKBP12 were found to interact with calcineurin, a calcium/calmodulin-dependent protein phosphatase (LIU et al., 1991). Calcineurin is composed of two subunits, the catalytic, calmodulin binding 61 kDa A subunit and the regulatory, calcium binding 19 kDa B subunit. The drug–immunophilin complexes bind to and modulate the serine/threonine phosphatase activity of calcineurin *in vitro*. However, calcineurin does not bind uncomplexed CsA, FK506, cyclophilin, or FKBP. In T cell lysates containing natural immunophilins both CsA and FK506 inhibit calcineurin activity at concentrations which effectively block IL-2 production in activated T cells (FRUMAN et al., 1992a). Rapamycin has no effect on calcineurin activity. Furthermore, excess concentrations of rapamycin compete the effects of FK506, apparently by displacing FK506 from FKBP. Cyclophilins A, B, C and CYP-40, when complexed with CsA, bind calcineurin. However, the cyclophilin B complex is 2–5-fold more potent than that of cyclophilin A to inhibit calcineurin phosphatase activity (SWANSON et al., 1992). Of the FKBP complexes only FK506/FKBP12 seems to inhibit calcineurin. Interestingly, CsA/CYP-A and FK506/FKBP12 compete for binding to calcineurin despite absence of obvious structural similarities.

The pharmacological relevance of the inhibition of calcineurin phosphatase activity by CsA or FK506 for immunosuppression has been substantiated by the good correlation between calcineurin activity and IL-2 production in T cells which is dose-dependently inhibited by CsA or FK506 (FRUMAN et al., 1992b). In addition, overexpression of the calcineurin catalytic subunit in the Jurkat T cell line rendered the transfected cells more resistant to the immunosuppressive effects of CsA and FK506 (O'KEEFE et al., 1992; CLIPSTONE and CRABTREE, 1992). These results clearly showed that calcineurin is a target for drug–immunophilin complexes *in vivo* and suggested a physiological role for calcineurin in T cell activation. Since calcineurin is present in small amounts in T cells, it might be a rate determining enzyme in the T cell activation pathway. Such a situation would provide a rationale for the profound effects of CsA and FK506 on T cell activation.

A 77 kDa cyclophilin C-associated protein which binds CYP-C with high affinity in the absence but not in the presence of CsA was described (FRIEDMAN et al., 1993). The role of this protein is still unknown. A cellular function for FKBP12 has been reported recently, following the observation that FKBP12 co-purifies with the ryanodine receptor (BRILLANTES et al., 1994). Four ryanodine receptors associate to form intracellular Ca^{2+} release channels of the sarcoplasmic and endoplasmic reticula. FKBP12 was shown to stabilize channel gating by improving conductance increasing mean open time. This effect was reversed by both FK506 and rapamycin suggesting that blocking the *cis-trans* isomerase catalytic site is sufficient for inhibition by the drugs.

3.2.4 Downstream Effects in the Nucleus

Both CsA and FK506 act specifically on activation pathways mediated by the T cell receptor that induce an increase in intracellular Ca^{2+} concentration. The actual molecular target of the Ca^{2+}-dependent serine/threonine phosphatase calcineurin is not known.

Highly attractive candidates include cytosolic components of the transcription complex that need to be assembled in the nucleus and/or activated for expression of early T cell activation genes. Calcineurin may modify the phosphorylation of downstream components such as the antigen-inducible transcription factors NFκB (BAUMANN et al., 1991) and NF-AT (EMMEL et al., 1989) which are essential for IL-2 transcription. Both NFκB and NF-AT are present as precursors in the cytoplasm. They require protein modification such as phosphorylation or dephosphorylation to be translocated into the nucleus where they bind to the DNA and participate in the formation of a functional transcription complex leading to the transcription of the IL-2 gene. It has been shown recently that CsA and FK506 inhibit dephosphorylation of NF-AT (MCCAFFREY et al., 1993). CsA and FK506 bound to their respective immunophilin most probably interfere at this level by inhibiting the phosphatase activity of calcineurin. This might directly or indirectly alter the protein modification of the transcription factors and consequently block lymphokine transcription (SCHREIBER and CRABTREE, 1992). Rapamycin is not active on this pathway.

As implicated by the mutually antagonistic activity of rapamycin and FK506 the mechanism of rapamycin action in lymphoid cells is likely to depend on immunophilin/drug complex formation. Rapamycin does not affect the transcription of genes involved in early activation of T cells, such as IL-2, but appears instead to block later events leading to T cell activation, such as the signal transduction pathway driven by the interaction of IL-2 with the high-affinity IL-2 receptor. More generally, rapamycin was shown to inhibit growth factor receptor-mediated activation signals in a number of different cells as well as *in vivo* after myelodepression (QUESNIAUX et al., 1994). The rapamycin/FKBP complex and the putative associated effector molecules interfere with the activation of the p70 S6 protein kinase in response to growth factor stimulation (CHUNG et al., 1992). This is a key regulatory step in the cell cycle progression from G1 to S phase. The direct target of rapamycin appears to be a crucial element linking growth factor receptors to subsequent intracellular processes that regulate proliferation (PRICE et al., 1992) and has been identified recently by SCHREIBER and collaborators (BROWN et al., 1994). Previously, it had been suspected that it might be a proximate upstream activator of the p70 S6 protein kinase such as an activating p70 S6 kinase-kinase or a regulator of such an enzyme.

3.3 Molecular Evidence for other Biological Activities of Cyclosporins

3.3.1 Antiinflammatory Effects

Crossreactivity of antiserum against human IL-8 with cyclophilins led to the finding that IL-8 binds CsA but not some of its non-immunosuppressive analogs. Although IL-8 bears some sequence similarities to cyclophilin, it has no rotamase activity (BANG et al., 1993) and the relevance of these findings still needs further clarification.

Cyclophilin A shows proinflammatory activity and leukocyte chemotactic activity which can be inhibited by CsA but not by a non-immunosuppressive analog (XU et al., 1992; SHERRY et al., 1992). CYP-A is secreted by macrophages in response to endotoxin and it was proposed that CYP-A may function as a cytokine (SHERRY et al., 1992). FKBP was also shown to display some leukocyte chemotactic activity which is inhibited by FK506 (LEIVA and LYTTLE, 1992). Cyclophilin 40 was shown to share homology with P59, a member of the steroid receptor complex (KIEFFER et al., 1993).

3.3.2 Cyclosporins as Anti-HIV Agents

The clinical potential of the anti-HIV activity of cyclosporins, in particular of SDZ NIM 811, was briefly discussed in Sect. 2.4 without addressing possible mechanisms of this effect which is clearly different from that of all other anti-HIV agents described to date. In cell-free assay SDZ NIM 811 does not inhibit reverse transcriptase, protease, or integrase. There is some evidence that the cyclosporin-sensitive step in HIV replication is an event after virus penetration, possibly nuclear translocation of viral DNA (ROSENWIRTH et al., 1994). It has been reported recently that some cyclophilins bind to the HIV-1 protein $p55^{gag}$ as well as the capsid protein p24 (LUBAN et al., 1993). In a cell-free system, this interaction is disrupted by both cyclosporin A and SDZ NIM 811. Furthermore, inhibition of the gag–cyclophilin complex formation by cyclosporins correlates with the cyclophilin binding capacity of the compounds but not with their immunosuppressive potential. The same correlation was found for the anti-HIV activity of these derivatives (ROSENWIRTH et al., 1994). Whether inhibition of gag–cyclophilin complex formation by cyclosporins explains the antiviral effect remains to be elucidated.

3.3.3 Cyclosporins as Drug Resistance Modifiers

The clinical potential of the non-immunosuppressive cyclosporin SDZ PSC 833 to sensitize multidrug resistant tumor cells to chemotherapeutic agents was described in Sect. 2.3. This resistance phenomenon (MDR) is thought to be associated with the p-glycoprotein transporter. Direct binding of CsA to p-glycoprotein could be demonstrated by using a photoaffinity-labeled cyclosporin derivative (FOXWELL et al., 1989). Interestingly, while the immunosuppressive activity of cyclosporins is mediated by complex formation with cyclophilin (see Sect. 3.2.3) SDZ PSC 833 has no measurable affinity to cyclophilin. Similarly, other cyclosporins not binding to cyclophilins (e.g., cyclosporin H) also bind to p-glycoprotein. Resistance modifier activity of cyclosporins appears thus to be a cyclophilin-independent effect.

3.3.4 Antimalarial Activity of Cyclosporins

Activity of cyclosporin A against *Plasmodium* spp. *in vitro* and in rodent models *in vivo* was reported as early as 1981 by THOMMEN-SCOTT. Interestingly, treatment was most effective when started at a time when parasitemia was already established suggesting a direct toxic effect on the parasite. Recently, with the availability of cyclosporin analogs binding to cyclophilins but devoid of immunosuppressive activity (such as SDZ NIM 811) as well as of cyclosporins without any measurable affinity to cyclophilins, it has become possible to discern between cyclosporin effects mediated by inhibition of calcineurin by the cyclosporin–cyclophilin complex (e.g., immunosuppression), mediated by cyclophilin binding only (such as, e.g., the anti-HIV activity of SDZ NIM 811), from effects that are independent of cyclophilin (such as, e.g., the MDR activity of SDZ PSC 833 which does not bind to cyclophilin). With these reagents, a re-investigation of antimalarial activities of cyclosporins was performed (BELL et al., 1994). While CsA-sensitive peptidyl–prolyl *cis-trans* isomerase activity could be detected in extracts of *P. falciparum* the highest activity against *P. falciparum* was exhibited by SDZ PSC 833 suggesting that inhibition of rotamase activity may not be the lethal target of cyclosporins in *P. falciparum*.

4 Chemistry

4.1 Structural Aspects

Cyclosporins are composed of 11 aliphatic lipophilic amino acids of which four are leucines and three do not occur in mammalian proteins. In CsA these are: (4R)-4-[(E)-2-butenyl]-4-methyl-L-threonine (Bmt) in position 1, (L)-α-amino butyric acid in position 2, and (D)-alanine in position 8. Of the 11 peptide bonds 7 are N-methylated. This feature has several important implications: Firstly, the N-methylated peptide bonds and the cyclic structure of the molecule render cyclosporins stable towards mammalian digestive and systemic proteases (Sandimmun® is metabolized extensively in animals and man, however, exclusively by cytochrome P450-mediated oxidative transformations). Therefore, cyclosporins are not only well absorbed when given by the oral route, high and long lasting plasma levels are commonly obtained. These properties are essential prerequisites for a successful drug. A second consequence of the N-methylation pattern is a certain conformational rigidity in non-polar environment characterized by intramolecular hydrogen bonds and the lipophilic side chains being oriented towards the (hydrophobic) environment.

The high number of leucines in the cyclosporins is striking, especially in view of the well recognized role of leucines in protein–protein interactions. Their role for biological activity will be discussed below.

The structures of cyclosporins in apolar as well as polar solution have been determined by NMR spectrospic methods as well as by X-ray crystallography. This work is extensively discussed in Sect. 5. In non-polar solvents, the conformation of cyclosporins is very similar to that found in the crystalline state. Cyclosporin A in polar environment exhibits a different conformation from that found in apolar systems (KO and DALVIT, 1992). Cyclosporin A is not soluble enough in aqueous solution to allow a full 3D-structure determination by NMR but the ^{1}H-NMR spectra in water indicate the presence of a family of conformations. A single predominant conformation of cyclosporin is observed only in complex with certain metal ions (KOECK et al., 1992) or in complex with cyclophilin. There exist derivatives, however, which are more water-soluble than CSA and some modifications (e.g., a methyl group in position 3) of cyclosporin seem to stabilize one conformation in water. Recently, the 3D-structure of such a derivative, namely of [D-MeSer^{3}-D-Ser-(O-Gly)^{8}]-cyclosporin, in DMSO and in water has been determined by NMR (WENGER et al., 1994). Fig. 7 shows the conformation of this derivative in DMSO which is identical to the one in water. Strikingly, this structure is very similar to the one found for CsA bound to cyclophilin (Fig. 8) thus giving evidence for the first time that CsA adopts the cyclophilin-bound conformation (among many other conformations) in aqueous solution. An "induced-fit" hypothesis for the mode of binding of cyclosporins to cyclophilin would thus not seem necessary.

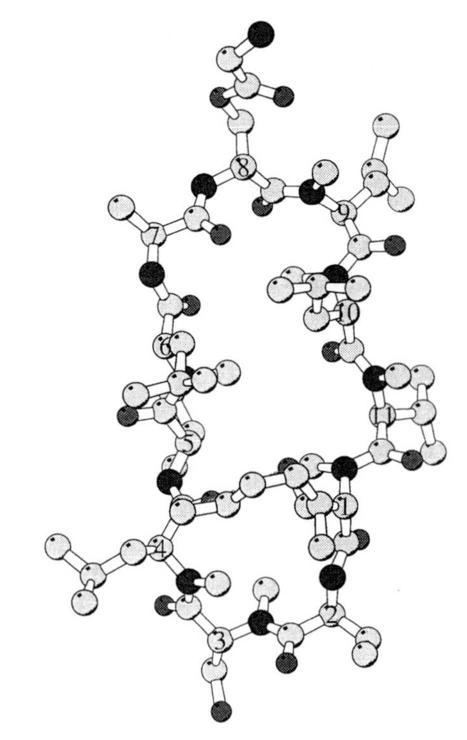

Fig. 7. NMR-structure of [D-MeSer^{3}-D-Ser-(O-Gly)^{8}]-cyclosporin, not bound to a receptor protein, in DMSO (identical to the one in water).

To date, some 30 cyclosporins have been isolated from natural sources. Their structures are shown in Tab. 1.

4.2 Chemical Synthesis and Production

The method of choice for the production of cyclosporins on a large scale is fermentation. Invariably, complex mixtures containing the desired cyclosporin along with a plethora of minor metabolites are obtained. To achieve sufficient purity of the final product repeated chromatography operations are necessary. For the production of cyclosporins on a large scale (tons), substantial investments in large-scale chromatography facilities are therefore

Tab. 1. Cyclosporins Isolated from Fermentation Broths[a]

Name	Amino Acid in Position										
	1	2	3	4	5	6	7	8	9	10	11
CsA	Me Bmt	Abu	Sar	Me Leu	Val	Me Leu	Ala	(D)-Ala	Me Leu	Me Leu	Me Val
CsB		Ala									
Csc		Thr									
CsD		Val									
CsE											Val
CsF	Deoxy-MeBmt Abu^{2}										
CsG		Nva									
CsH											(D)-Me-Val
CsI		Val								Leu	
CsK	Deoxy-MeBmt Val^{2}										
CsL	Bmt										
CsM		Nva			Nva						
CsN		Nva								Leu	
CsO	Me Leu	Nva									
CsP	Bmt	Thr									
CsQ				Val							
CsR						Leu				Leu	
CsS		Thr		Val							
CsT										Leu	
CsU						Leu					
CsV							Abu				
CsW		Thr									Val
CsX		Nva							Leu		
CsY		Nva				Leu					
CsZ	Me Aoc[b]										
Cs26		Nva			Leu						
Cs27	Bmt	Val									
Cs28	Me Leu										
Cs29				Me Ile							
Cs30	Me Leu	Val									
Cs31				Ile							
Cs32			Gly					(D)-Ser			
FR 901 495		Thr			Leu					Leu	

[a] Only residues different from those in Cyclosporin A are given.
[b] MeAoc = N-Methyl-(L)-amino octanoic acid.

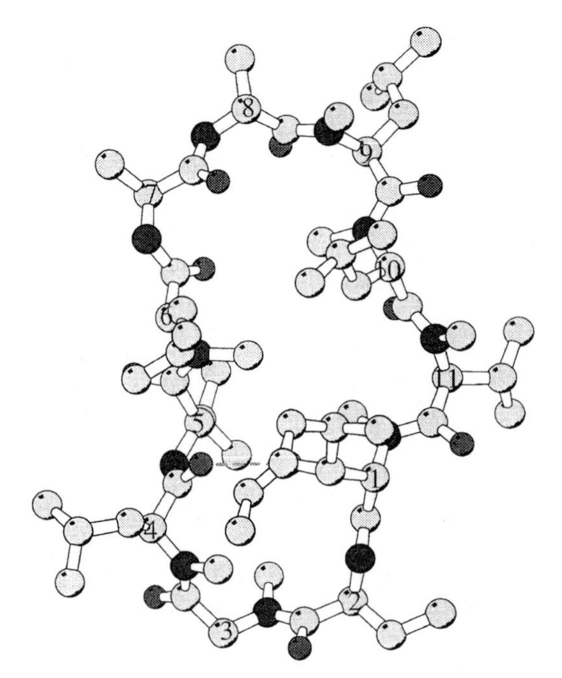

Fig. 8. X-ray structure of CsA when bound to CYP-A, crystallized from an aqueous solution. There is only one intramolecular hydrogen bond, namely between the hydroxyl group of MeBmt[1] and Leu[4](CO). All amide bonds are in the *trans* conformation.

necessary. For structure–activity studies, a large number of analogs has been prepared by total synthesis according to the method developed by Wenger (WENGER, 1983) or by semisynthetic transformations of natural cyclosporins. In addition, a solid phase-based methodology has been developed in the preclinical research laboratories of Sandoz Pharma AG allowing the preparation of almost any analog on a small scale in a relatively short time (BOBE, unpublished work).

Semisynthesis of cyclosporins involves several main approaches: Modification of the MeBmt aliphatic side chain, chemical transformations of the cyclic peptide backbone, and modifications based on selective ring opening reactions.

4.2.1 MeBmt Transformations and Variations

The olefinic double bond of MeBmt offers a number of possibilities of selective chemical transformations. In the preclinical research laboratories of Sandoz Pharma AG many analogs have been obtained by ozonolysis

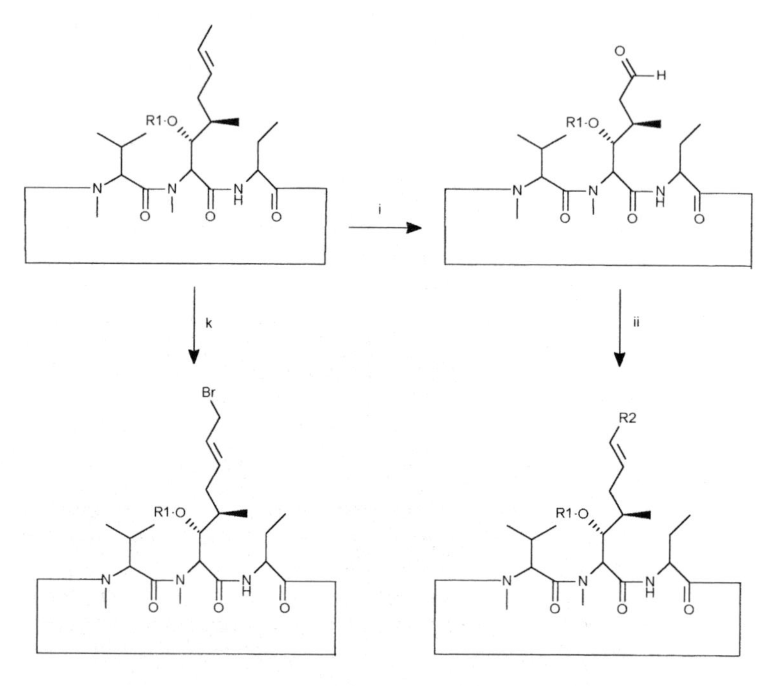

Fig. 9. Some chemical transformations of the aliphatic side chain of MeBmt. R1 = protective group (most often acetyl); reagents: i = ozone, ii = R2—CH=P(phenyl)$_3$; k = N-bromo-succinimide. For references, see text.

and subsequent Wittig reactions of the resulting aldehydes; some of these transformations have also been reported by other groups (PARK and MEIER, 1989). An overview of this chemistry is given in Fig. 9. Selective bromination of the terminal methyl group could be achieved by N-bromosuccinimid (EBERLE and NUNINGER, 1992). This reaction was used to prepare one of the principal Sandimmun® metabolites, OL-17 (EBERLE and NUNINGER, 1992). An analog of MeBmt carrying an additional methyl group at position 4 of the aliphatic carbon chain was synthesized by the group of RICH (AEBI et al., 1990). The corresponding cyclosporin derivative has been reported to exhibit very low affinity to cyclophilin, yet to be a surprisingly strong immunosuppressant (SIGAL et al., 1991). A possible explanation of this unexpected result is given in Sect. 5.6.1. Cyclosporin analogs carrying MeBmt modifications at position 1 have also been reported by the research laboratories of Merck, Sharpe & Dohme (SIGAL et al., 1991; WITZEL, 1989).

4.2.2 Chemical Transformations of the Cyclic Peptide Backbone

It was first shown by SEEBACH that cyclosporins can be converted into polyanions by treatment with an excess of a strong base and that such polyanions could be reacted with a variety of electrophiles to give derivatives of a N-methyl-D-amino acid in position 3 (SEEBACH et al., 1993). Many of these derivatives retain potent immunosuppressive properties. The chemistry is illustrated in Fig. 10.

Treatment of cyclosporins with phosphorous pentasulfide or its more modern equivalents (e.g., Lawesson's reagent) yields, depending on conditions, product mixtures from which mono- or dithioamide derivatives can be isolated (SEEBACH et al., 1991; EBERLE and NUNINGER, 1993). While most of these derivatives are devoid of immunosuppressive activity, due to the presence of the sulfur atom, they offer selective ring cleavage possibilities (see below).

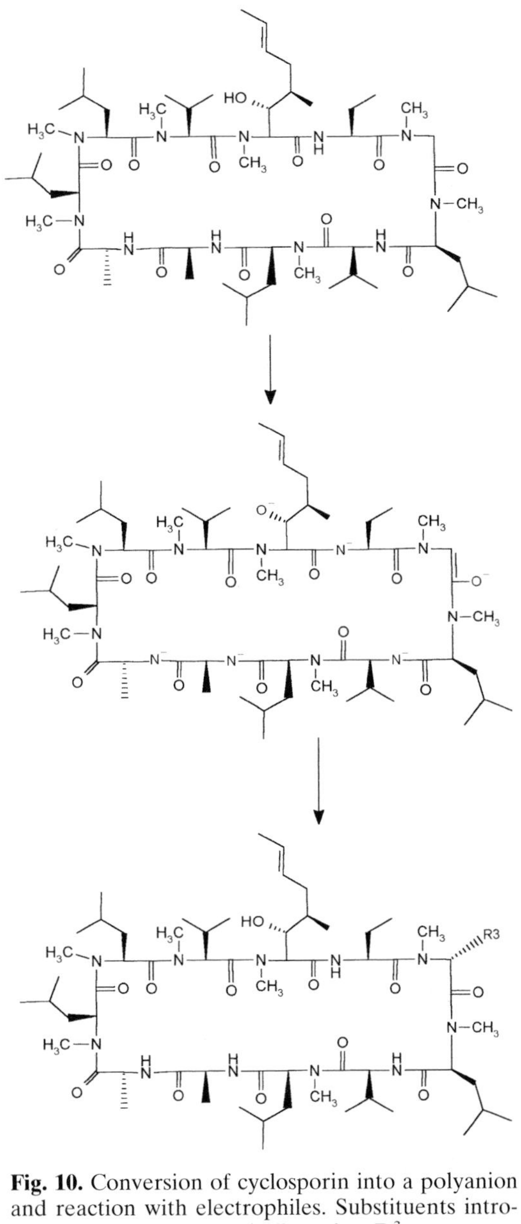

Fig. 10. Conversion of cyclosporin into a polyanion and reaction with electrophiles. Substituents introduced in position 3 are indicated as R^3.

4.2.3 Selective Ring Opening Reactions

A number of methods have been developed to selectively cleave the cyclosporin macrocycle at defined positions. From the linear

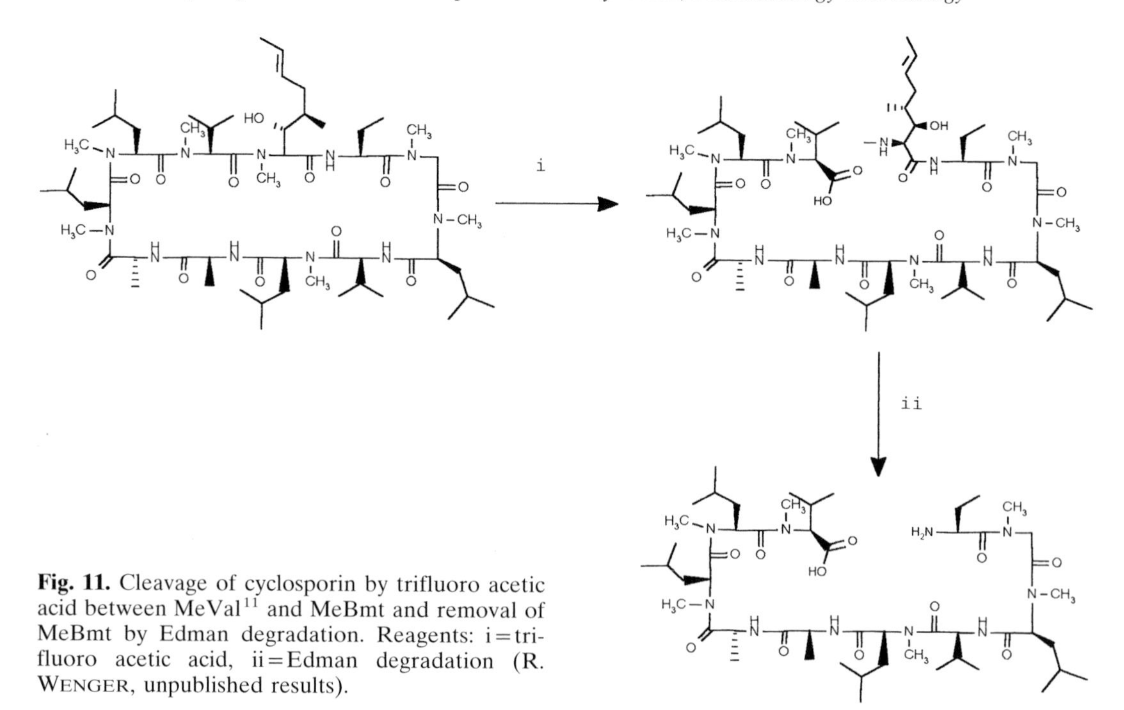

Fig. 11. Cleavage of cyclosporin by trifluoro acetic acid between MeVal^{11} and MeBmt and removal of MeBmt by Edman degradation. Reagents: i = trifluoro acetic acid, ii = Edman degradation (R. WENGER, unpublished results).

undecapeptide products thus obtained one or more amino acids can be removed by degradation methods (e.g., Edman methodology) and replaced by other amino acids or peptide fragments. Cyclosporin analogs are finally obtained by cyclization. One such example is the cleavage of cyclosporin by trifluoro acetic acid between MeVal^{11} and MeBmt, forming a linear undecapeptide with MeBmt at the N-terminus (WENGER, R., unpublished observations). Removal of MeBmt by Edman degradation forms a decapeptide lacking MeBmt; this compound serves as a starting material for cyclosporins modified at position 1 and this approach is complementary to that described in Sect. 4.2.1. The chemistry is outlined in Fig. 11.

A general method to cleave peptides at bonds to serine or threonine is based on activation of these bonds by transformation into an oxazolidinone with N,N′-carbonyl diimidazole (WENGER, R., unpublished observations). In such molecules the exocyclic amide bond is preferentially susceptible to attack by nucleophiles; reaction with, e.g., sodium methoxide in methanol forms a linear methyl ester which can be further modified by obvious methods. This approach is illustrated in Fig. 12.

An example of the use of thioamide derivatives of cyclosporins for selective ring opening is shown in Fig. 13.

4.2.4 Cyclosporins Incorporating other Non-Proteinogenic Amino Acids

Early attempts (i.e., prior to the structure determination of the CsA–cyclophilin complex) to stabilize the β-turn adopted by CsA in non-polar solution led to the synthesis of analogs incorporating lactam-bridged dipeptide units as depicted in Fig. 14 (LEE et al., 1990; WENGER, 1989). Not surprisingly in retrospect, these analogs lacked immunosuppressive activity. Recently, on the basis of the X-ray crystal structure of the CsA–cyclophilin complex, the dipeptide shown in Fig. 15 has been designed and synthesized by the group of S. L. SCHREIBER (ALBERG and SCHREI-

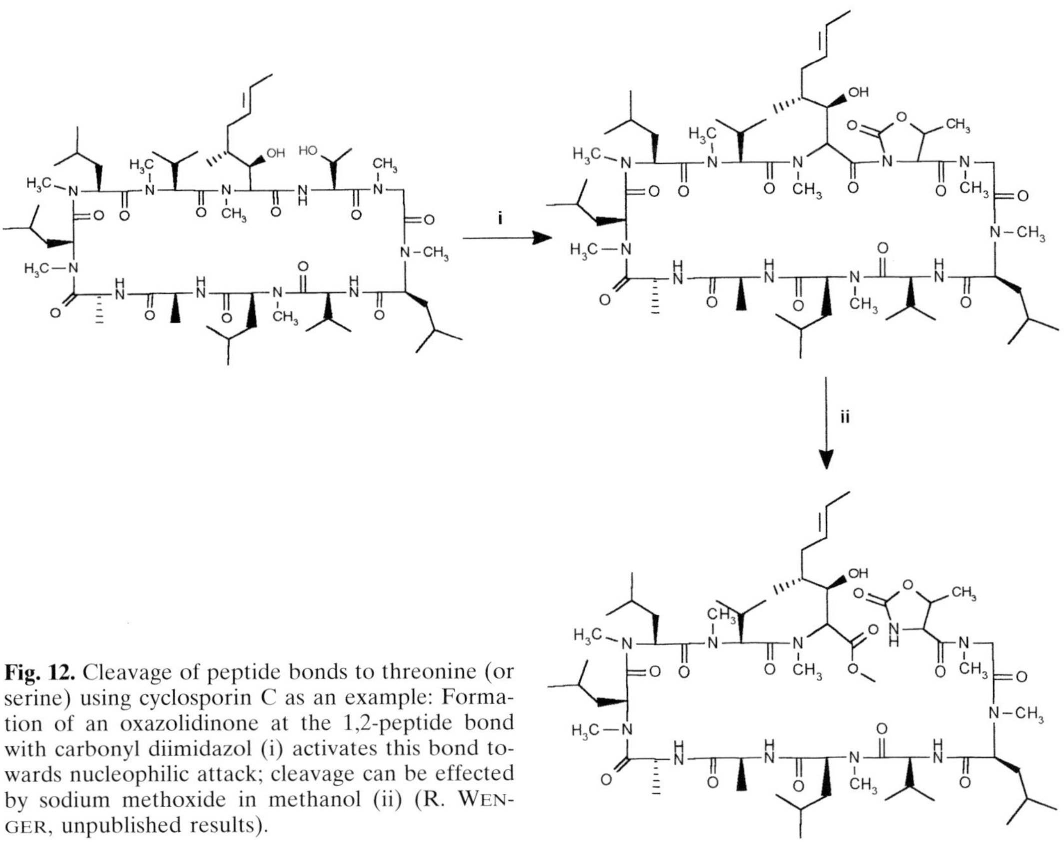

Fig. 12. Cleavage of peptide bonds to threonine (or serine) using cyclosporin C as an example: Formation of an oxazolidinone at the 1,2-peptide bond with carbonyl diimidazol (i) activates this bond towards nucleophilic attack; cleavage can be effected by sodium methoxide in methanol (ii) (R. WENGER, unpublished results).

BER, 1993). The cyclosporin derivative incorporating this compound in place of amino acids 7 and 8 indeed shows improved immunosuppressive activity *in vitro* when compared with CsA. Of all the cyclosporins published it is the most potent immunosuppressant.

Several cyclosporins incorporating fluorinated amino acids were reported by the research laboratories of Merck, Sharpe & Dohme. β-Fluoro-(D)-alanine[8]-cyclosporin was obtained by microbiological methods. This compound exhibits immunosuppressive potency comparable with that of CsA (HENSENS et al., 1992; PATCHETT et al., 1992). By elimination of HF this compound was converted into the dehydro alanine derivative from which by Michael addition reactions derivatives of D-cysteine[8] were obtained, some of them exhibiting potent immunosuppressive activity such as the thioether shown in Fig. 16.

4.3 Structure–Activity Relationships

A first impression of the structural features required for immunosuppressive activity of cyclosporins can be derived from the naturally occurring analogs. Furthermore, chemical derivation of MeBmt reveals that both the hydroxyl and the methyl group are indispensable and that the stereochemistry of their attachment to the carbon chain is equally critical (RICH et al., 1989). Hydrogenation of the double bond causes a slight loss of activity. Introduction of polar groups at the terminal carbon of the side chain also results in a decrease of immunosuppressive activity.

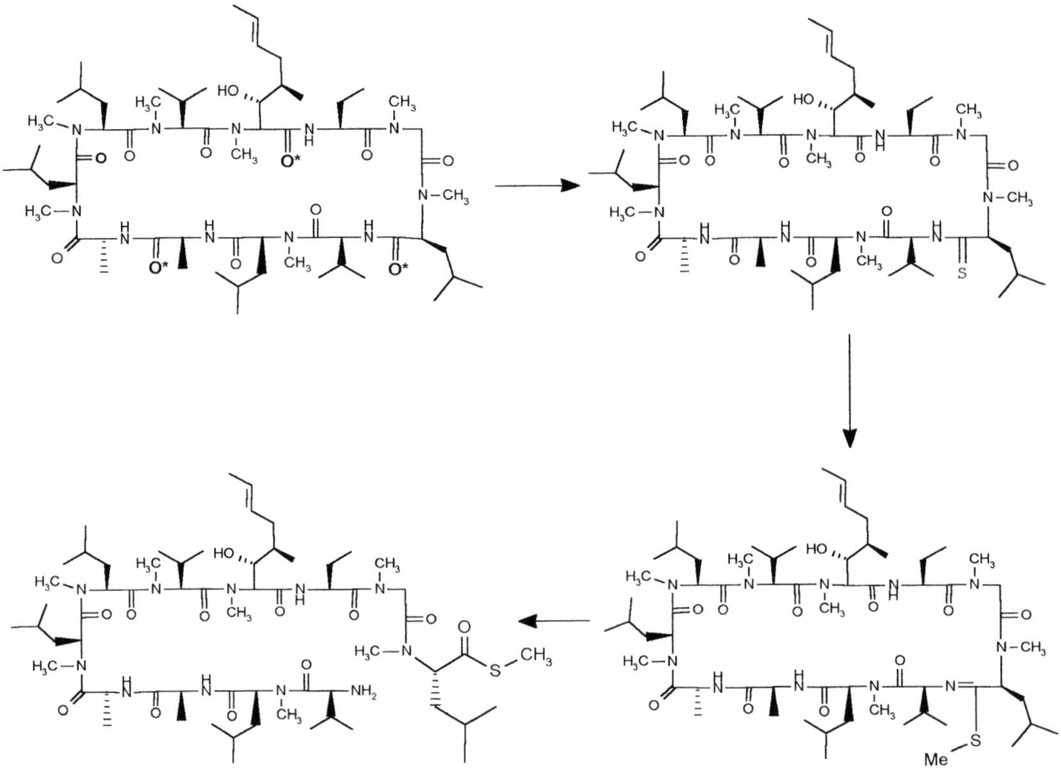

Fig. 13. Treatment of cyclosporin with phosphorous pentasulfide or Lawesson's reagent forms mixtures of mono- and dithio amides. The oxygen atoms replaced by sulfur atoms during these reactions are indicated with an asterisk. Cleavage of the thioamides can be achieved by alkylation at the sulfur atom, followed by acid hydrolysis.

As has been outlined in Sect. 3.2, the immunosuppressive activity of cyclosporins is based on an unusually complex mechanism which in a first step requires binding of the drug to a receptor of the family of cyclophilins and in a second step binding of this complex to protein phosphatase 2B (calcineurin), thereby inhibiting its catalytic activity. This activity is crucially dependent on cyclosporin; cyclophilin without any bound drug does not interact with calcineurin. There is also evidence that binding of the cyclosporin–cyclophilin complex to calcineurin involves both residues of cyclosporin as well as part of the cyclophilin molecule. This means that the molecule causing immunosuppression is the entire cyclosporin–cyclophilin complex. Consequently, cyclosporin structure–activity relationships must be analyzed both in terms of cyclophilin binding and in terms of mediating the calcineurin interaction of the complex. Much insight into the nature of this interaction was gained from analyzing the crystal structures of many cyclosporin–cyclophilin complexes (Sects. 5.5–5.8). As indicated above, both CsA and SDZ 214-103 contain a strikingly high number of leucines. In the cyclophilin complexes of these molecules the side chains of the leucines in position 4, 6, and 10 are exposed next to each other on the surface; they form a "leucine cluster" on the protein surface. It has long been recognized that leucine side chains in a specific three-dimensional disposition on the surface of a protein play a crucial role in mediating protein–protein interactions, a prominent example being the dimerization of transcription factors through "leucine zipper" domains (LAND-

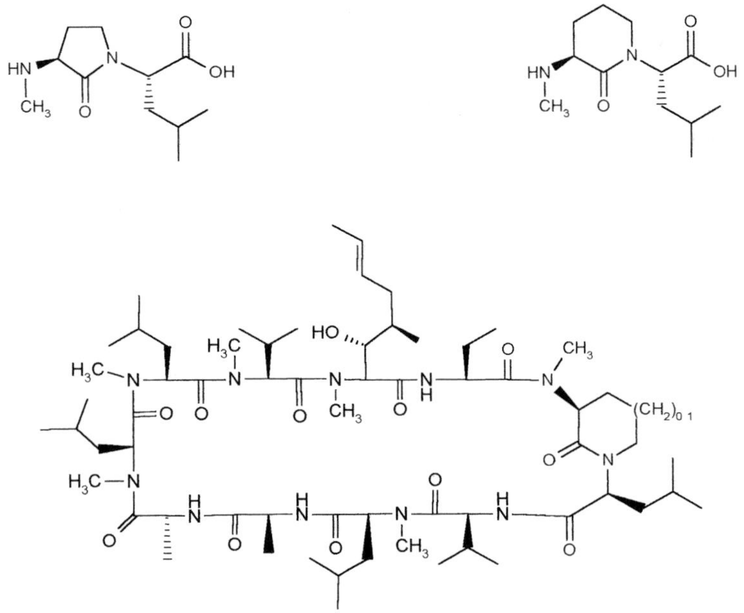

Fig. 14. Cyclosporins incorporating rigid dipeptide fragments at the the β-turn region in the absence of cyclophilin. For references, see text.

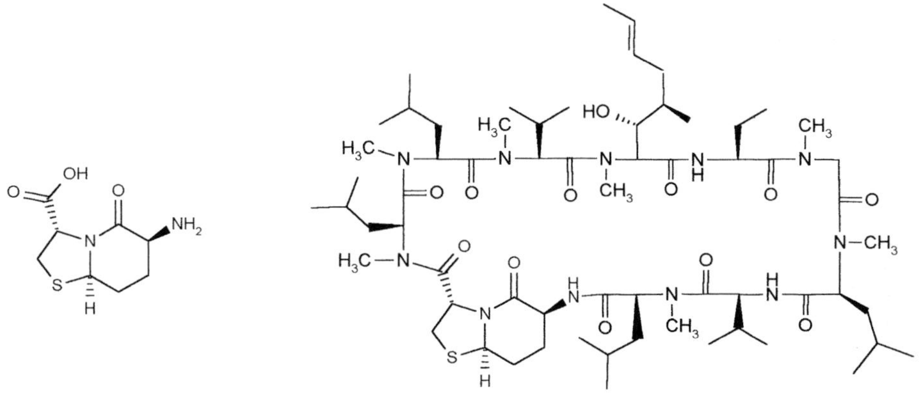

Fig. 15. Design and synthesis of a "calcineurin-bridging" ligand (ALBERG and SCHREIBER, 1993). The bicyclic amino acid is incorporated into the cyclosporin molecule in place of alanine[7] and (D)-alanine[8].

SCHULTZ et al., 1989). The presence of a leucine cluster on the surface of the cyclosporin–cyclophilin complex prompted an analysis of the role of these leucines by substituting each of them for valine or alanine. A shortened version of the outcome of this analysis is shown in Tab. 2.

From these results it is evident that in positions 4 and 6 the cyclosporin leucine side chains have crucial functions for the calcineurin interaction; they can be substituted for valine or alanine without much affecting the affinity of the compounds for cyclophilin. However, the affinity of their respective complexes to calcineurin has been lost as evidenced by the lack of immunosuppressive properties of these derivatives. In line with this interpretation none of these compounds

Tab. 2. Effects of Valine for Leucine Substitutions in Cyclosporin A on Cyclophilin Binding and Immunosuppressive Activity

Cyclosporin Analog	Cyclophilin Binding	Immunosuppressive Activity (relat. IC_{50})
CsA	1	1
MeVal[4]-Cs	0.5	>2,500
MeVal[6]-CS	3	46
MeVal[10]-Cs	120	190

inhibited the phosphatase activity of calcineurin *in vitro* when complexed to cyclophilin (FLIRI et al., 1993). In position 10 substituting leucine for valine or alanine largely abolishes the affinity of these derivatives for cyclophilin. The function of CsA as an immunosuppressant can therefore be viewed as that of a "molecular adapter": By docking into cyclophilin a leucine cluster is introduced into the surface of the protein. This leucine cluster in turn allows the complex to bind to calcineurin and to inhibit its phosphatase activity.

5 Structural Investigations of Cyclosporins, Cyclophilins, and their Complexes

5.1 Introduction

An introduction to the immunophilins (cyclophilins and FK506 binding proteins) is given in Sect. 3.2.1. The most likely target of CsA in human T cells mediating immunosuppression is cyclophilin A (BRAM et al., 1993).

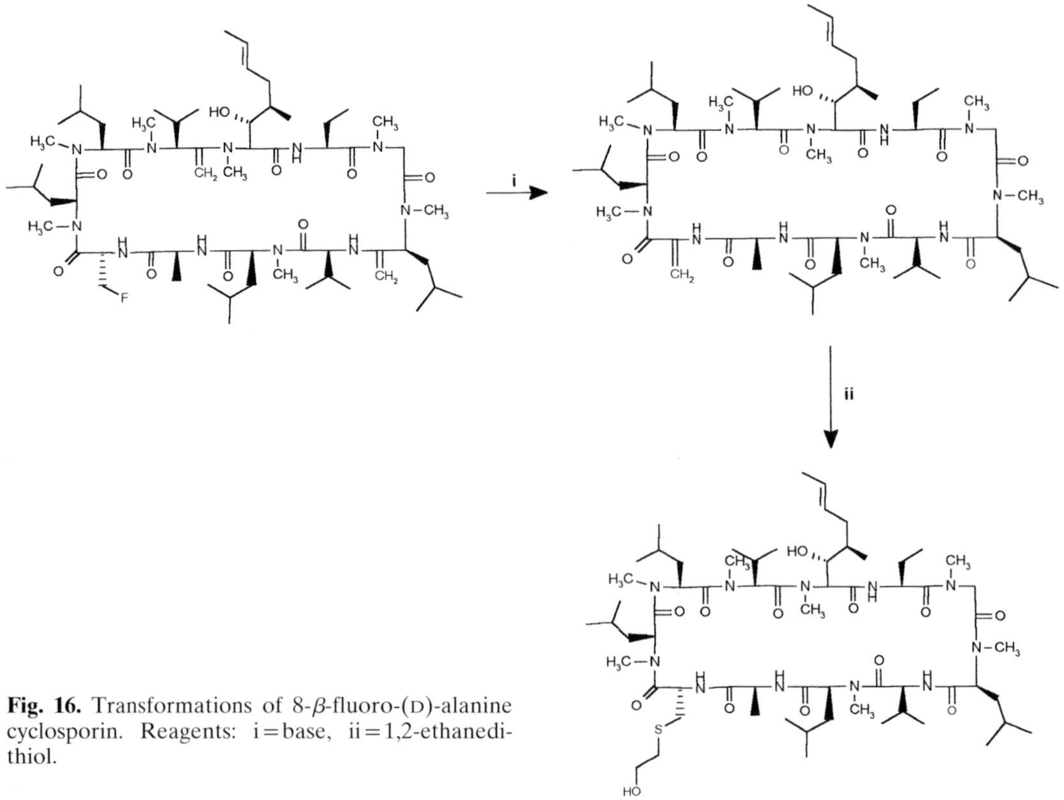

Fig. 16. Transformations of 8-β-fluoro-(D)-alanine cyclosporin. Reagents: i = base, ii = 1,2-ethanedithiol.

This protein has 165 amino acids and is found in the cytoplasm. Cyclophilin B has 208 residues and is found in the endoplasmic reticulum. Cyclophilin C has 194 residues and is thought to have some tissue specificity for kidney, at least in the mouse (FRIEDMAN and WEISSMAN, 1991). There is no sequence homology between cyclophilins and FKBPs and no obvious three-dimensional structural similarity.

Despite the chemical dissimilarity of the ligands and the structural differences between the proteins there is an intriguing overlap of biochemical and biological activity between the two immunophilin families (see Sect. 3.2). The common target of their complexes with the cognate drugs, calcineurin, is a heterodimer composed of subunit A (61 kDa) and subunit B (19 kDa) which has been shown to have affinity only for the immunophilin–drug complexes but not for the drug alone or the immunophilin alone (LIU et al., 1991). Furthermore, the binding of the FKBP12–FK506 complex to calcineurin competes with the binding of the CsA–CYP-A complex. Natural ligands for cyclophilins or FKBPs which could regulate phosphatase activity have not yet been discovered. Another puzzling coincidence between the two immunophilin families is their shared peptidyl–prolyl isomerase (PPIase) activity which led to the re-discovery of cyclophilin (TAKAHASHI et al., 1989; FISCHER et al., 1989) as an enzyme catalyzing protein folding *in vitro*. For proteins to adopt the correctly folded conformation the X-Pro amide bonds must be in the correct *cis* or *trans* conformation. Using model peptide substrates it has been found that immunophilins lower the energy of activation for isomerization of this amide bond by over 6 $kcal \cdot mol^{-1}$ from about 20 $kcal \cdot mol^{-1}$ down to 14 $kcal \cdot mol^{-1}$ (PARK et al., 1992). Both FKBP and cyclophilin have also been shown to accelerate the refolding of a number of proteins *in vitro*, presumably by catalyzing the rate determining step of proline isomerization (SCHOENBRUNNER et al., 1991; FRANSSON et al., 1992). Specific cellular targets of the PPIases are not yet known but it has been suggested that they play a role in the folding of newly synthesized proteins (GETHING and SAMBROOK, 1992). Blocking the *cis-trans* isomerase activity is, however, insufficient to induce immunosuppression (SIGAL et al., 1991).

There is a growing body of available 3D-structural information on immunophilins and their ligands. This work is directed at trying to understand both the biological and enzymatic activity. Protein X-ray crystal structures of cyclophilin A complexed with a tetrapeptide substrate (KALLEN et al., 1991), a dipeptide substrate (KE et al., 1993a), without substrate (KE et al., 1991), with CsA (PFLUEGL et al., 1993; MIKOL et al., 1993), and with a cyclosporin derivative modified in position 1 (KE et al., 1994; MIKOL et al., 1994b) have also been published as well as the X-ray structure of a Fab–CsA complex (ALTSCHUH et al., 1992). The 3D-structure of the CsA–CYP-A complex has also been determined by NMR techniques (THERIAULT et al., 1993), as well as the 3D-structure of a water soluble cyclosporin derivative (WENGER et al., 1994).

5.2 Structural Investigations of Uncomplexed Cyclosporins

5.2.1 The 3D-Structure of Uncomplexed CsA in Apolar Environment

Prior to the studies of the conformation of CsA bound to CYP-A the 3D-structure of free CsA had been determined. Because of the very low solubility of CsA in aqueous solution, crystallization and NMR studies were done using apolar solvents. The NMR structure in chloroform at 20 °C (Fig. 17) is virtually identical to the X-ray crystal structure (LOOSLI et al., 1985). The backbone forms a twisted β-sheet that involves residues 1, 2, 5, 6, 7, and 11 and a type II′ turn at Sar^3 and $MeLeu^4$. There is a *cis*-peptide bond between $MeLeu^9$ and $MeLeu^{10}$. All the four amide protons are involved in hydrogen bonds (three transannular ones):
$Abu^2(NH)-Val^5(CO)$,
$Val^5(NH)-Abu^2(CO)$, and

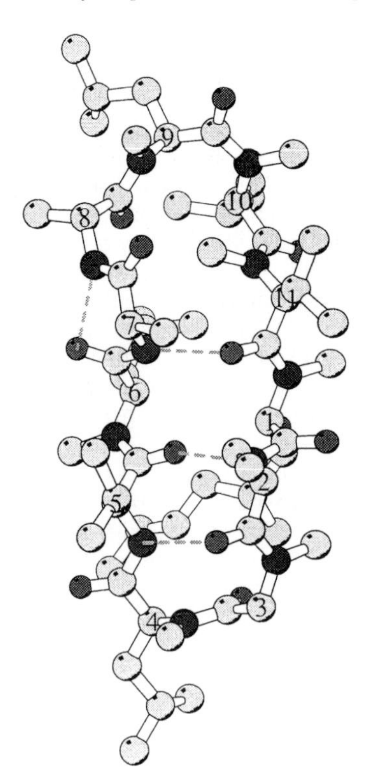

Fig. 17. Conformation of cyclosporin A in chloroform.

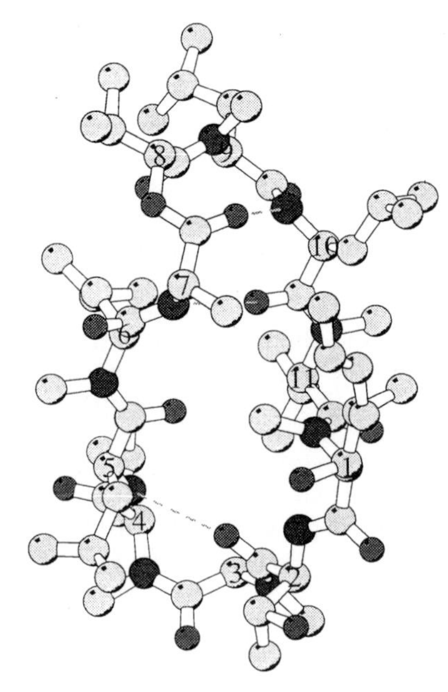

Fig. 18. X-ray structure of SDZ 214-103 (not bound to cyclophilin), crystallized from diethyl ether.

Ala^7(NH)—MeVal^{11}(CO), and one additional hydrogen bond:
D-Ala^8(NH)—MeLeu^6(CO).

5.2.2 The 3D-Structure of the Uncomplexed Peptolide SDZ 214-103 in Apolar Environment

The peptolide SDZ 214-103 differs chemically from CsA in the following way: Thr^2 instead of Abu^2, Leu^5 instead of Val^5, D-Hiv^8 instead of D-Ala^8, and Leu^{10} instead of MeLeu^{10}. The X-ray structure (Fig. 18) of the uncomplexed peptolide (crystallized from an organic solvent) is very different from the one of uncomplexed CsA (Fig. 30) or CsA bound to CYP-A (Fig. 8): There is a *cis* amide bond between residues 3 and 4 and a β-bend involving residues 7, 8, 9, 10. The following intramolecular hydrogen bonds are formed:
Thr^2(CO)—Leu^5(NH),
Ala^7(NH)—Leu^{10}(CO), and
Ala^7(CO)—Leu^{10}(NH).
The hydroxyl group of MeBmt^1 is not involved in an intramolecular hydrogen bond, but makes a packing interaction in the crystal with Sar^3(CO) of a neighboring molecule.

5.2.3 The 3D-Structure of Uncomplexed CsA in Polar Environment

CsA is not soluble enough in aqueous solution to allow a full 3D-structure determination by NMR but the ^{1}H-NMR spectra of CsA in water indicate the presence of a family of conformations (CsA adopts a single, predominant conformation only if complexed with certain metal cations (KOECK et al., 1992) or when bound to cyclophilin). There exist cyclosporin derivatives, however, that are more water soluble than CsA. Furthermore, introduction of a substituent in position 3 seems to stabilize one conformation in water. Recently, the 3D-structure of such a derivative, namely of D-MeSer^3-D-Ser-(O-Gly)^8-cyclosporin in DMSO and in water, has been determined by NMR (WENGER et al., 1994). The derivatization in position 8 makes the derivative more water-soluble and the substitution of D-MeSer for Sar in position 3 stabilizes a single conformation. Fig. 7 shows the conformation of this derivative in DMSO which is identical to the one in water. Strikingly, this structure is very similar to the one found for cyclosporins bound to cyclophilin (Fig. 8) thus giving evidence for the first time that cyclosporins preadopt the cyclophilin-bound conformation (among many other conformations) in aqueous solution. An "induced-fit" hypothesis for the mode of binding of cyclosporins to cyclophilin is thus not required.

5.3 The 3D-Structure of Cyclophilin A Complexed with a Tetrapeptide

At Sandoz Pharma AG the first three-dimensional X-ray structure of CYP-A with a model substrate (N-acetyl-Ala-Ala-Pro-Ala-amidomethyl coumarin) bound to its active site (two complexes per asymmetric unit) was elucidated (KALLEN et al., 1991; KALLEN and WALKINSHAW, 1992). The X-ray structure of unliganded CYP-A (1 molecule per asymmetric unit) was determined by KE (KE et al., 1991; KE, 1992). The same group also determined the structure of a CYP-A complex with the dipeptide substrate Ala-Pro (KE et al., 1993a).

The structure of CYP-A was determined in a collaborative effort using NMR and X-ray methods: The secondary structure of CYP-A was determined by NMR methods while simultaneously the tertiary structure was determined by X-ray methods (KALLEN et al., 1991; KE et al., 1991). Via chemical shift changes of certain residues in CYP-A upon complexation of the protein with CsA the NMR technique was also able to identify residues probably interacting with cyclosporin (KALLEN et al., 1991).

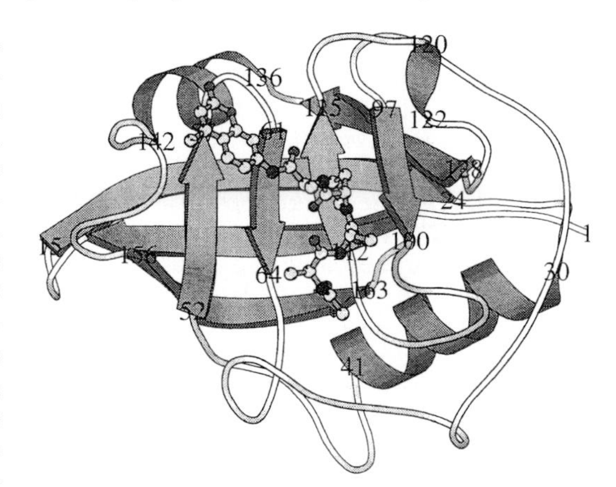

Fig. 19. The architecture of CYP-A consists of an eight-stranded antiparallel β-barrel the ends of which are closed off by two α-helices. The tetrapeptide (a substrate for the PPIase activity of CYP-A) binds to the outside of the β-barrel. Selected Cα-positions of CYP-A are indicated by their sequence numbers.

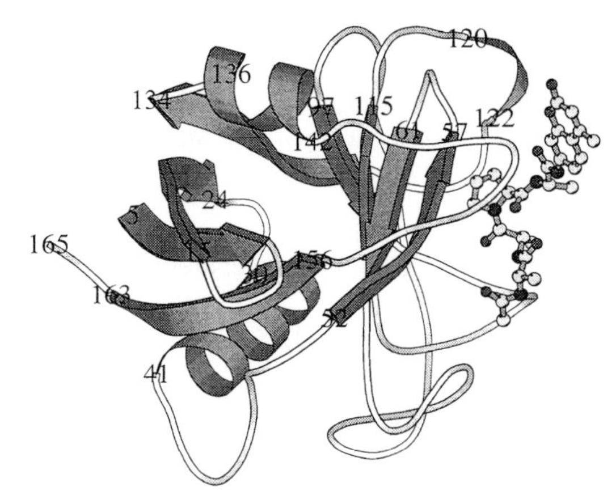

Fig. 20. The complex CYP-A-tetrapeptide rotated by 90° with respect to Fig. 19 around a vertical axis in the picture plane.

Cyclophilin is an approximately spherical molecule with a radius of ca. 17 Å (Figs. 19, 20). The main structural feature is the eight-stranded antiparallel β-barrel consisting of two roughly perpendicular four-stranded β-sheets with a +3, −1, −2, +1, −2, −3 topology. Both ends of the barrel are closed off with an α-helix. The cyclophilin β-barrel is similar in some respects to the superfamily of proteins involved in ligand transport, including retinol binding protein (RBP), bilin binding protein, β-lactoglobulin, fatty acid binding protein, and P2 myelin (COWAN et al., 1990). Most of these molecules encapsulate their ligand in the β-barrel core. In contrast, the barrel core in CYP-A is tightly packed with hydrophobic residues and the ligand binding site is on the outside of the barrel. The topology of CYP-A also differs from the simple [+1]n-up-and-down fold found in the RBP class of proteins or the [−3,+1,+1]-Greek key topology that is most frequently found in β-barrel proteins.

5.4 The Peptidyl–Prolyl Isomerase Active Site of Cyclophilin A

Two X-ray structures of CYP-A–substrate complexes have been published: a complex of CYP-A with the tetrapeptide N-acetyl-Ala-Ala-Pro-Ala-amidomethylcoumarin (Nac-AAPA-amc) (KALLEN et al., 1991) and a complex of CYP-A with the dipeptide Ala-Pro (KE et al., 1993a). In both cases the Ala-Pro amide bond adopts a *cis* conformation.

In the following, details of the CYP-A tetrapeptide structure will be given (Fig. 21). CYP-A residues which have one atom within 4 Å of any atom of the active site-bound Nac-AAPA-amc are: Arg^{55}, Ile^{57}, Phe^{60}, Gln^{63}, Ala^{101}, Asn^{102}, Gln^{111}, Phe^{113}, Leu^{122}, His^{126}, and Arg^{148}. The enzyme active site is a channel sitting on top of two antiparallel β-strands (with contacts from residues Phe^{60}, Gln^{63}, Gln^{111}, and Phe^{113}). Two loops protrude out from the surface of the barrel and provide a distinctive grooved protein surface. One loop from residues 101 to 110 contains the contact residues Ala^{101} and Asn^{102}. A narrow pass separates this loop from the second which comprises residues 69 to 74. Another important topological feature of the binding site is the wall composed of residues 118–126 in a close to helical conformation. Leu^{122} and His^{126} are in contact with the peptide substrate. The *cis*-proline of Nac-AAPA-amc sits in a rather deep pocket made principally by Leu^{122}, His^{126}, Phe^{113}, Phe^{60}, and Met^{61}.

There are three hydrogen bonds formed between the peptide and CYP-A (Fig. 21).

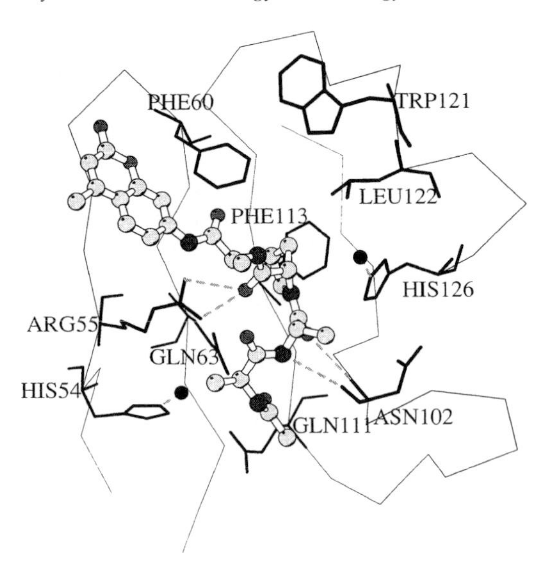

Fig. 21. Details of the interactions of the tetrapeptide with the enzyme active site of CYP-A. Intermolecular hydrogen bonds are formed between the main chain N and O atoms of Ala^{2} (from the peptide) and the main chain N and O atoms of Asn^{102} (from CYP-A) as well as between the carbonyl oxygen of Pro^{3} (from the peptide) and NH1 and NH2 of Arg^{55} (from CYP-A).

The guanidinium group of Arg^{55} forms hydrogen bonds to carbonyl oxygen of Pro^{3} in the peptide. It also makes a hydrogen bond to Gln^{63} which in turn is hydrogen bonded to the side chain of Gln^{111}. The other two direct substrate–protein hydrogen bonds are in the form of a short stretch of antiparallel β-sheet between the main chain N and O atoms of Ala^{2} and the main chain N and O atoms of Asn^{102}. The formation of a short stretch of antiparallel sheet is a common feature in many enzyme inhibitor complexes including aspartate proteases and serine proteases (BLUNDELL et al., 1987). The side chain of His^{126} is close to the alanyl–prolyl *cis* amide bond, however, the X-ray refinement suggests a conformation in which the histidine side chain preferentially hydrogen bonds to solvent water and protein main chain rather than to the substrate. The mechanism of isomerase action is not yet clear, however, the Ala-Pro *cis* amide bond is significantly twisted out of plane. Until now, four independent determinations of the ω angle have been performed that was found to vary between 20° and 45°. This is consistent with a mechanism of catalysis by distortion (HARRISON and STEIN, 1990; PARK et al., 1992; LIU et al., 1990; ROSEN et al., 1990) in which the immunophilin would bind the X-Pro amide bond of the substrate with a twisted, high-energy conformation. The transition state could also be stabilized by hydrogen bonding to the proline amide nitrogen (KOFRON et al., 1991) possibly via a water molecule or His[126] or Arg[55].

Based on his X-ray structure of the CYP-A-AP complex (not showing a significantly distorted amide bond) KE proposed a mechanism in which the transition state is stabilized via a hydrogen bond of a water molecule to the carbonyl oxygen of the amide bond (KE et al., 1993a).

A comparison of the active site of CYP-A when complexed with a substrate (cf. Fig. 21) and in the uncomplexed form (Fig. 22) shows that there is practically no structural change for CYP-A. The biggest movement occurs for the side chain of Arg[55] and shows a movement of its guanidyl group by about 2 Å. The only well-ordered water molecule (with a crystallographic B-factor <40 Å^2) that has to

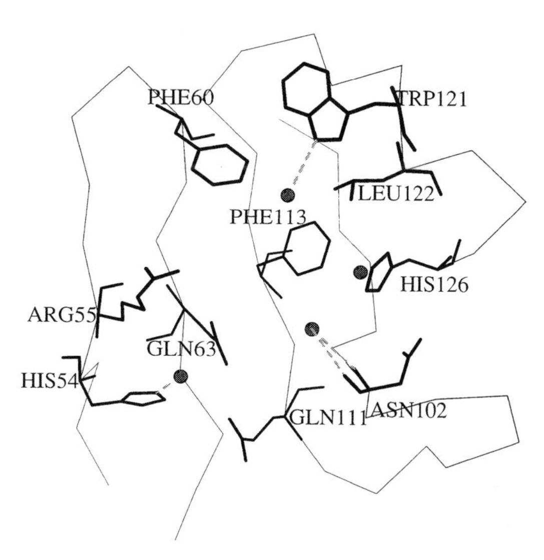

Fig. 22. Details of the enzyme active site of CYP-A in the unliganded form. Only selected water molecules with B-factors <40 Å^2 are indicated (spheres) as well as their hydrogen bonds to atoms from CYP-A.

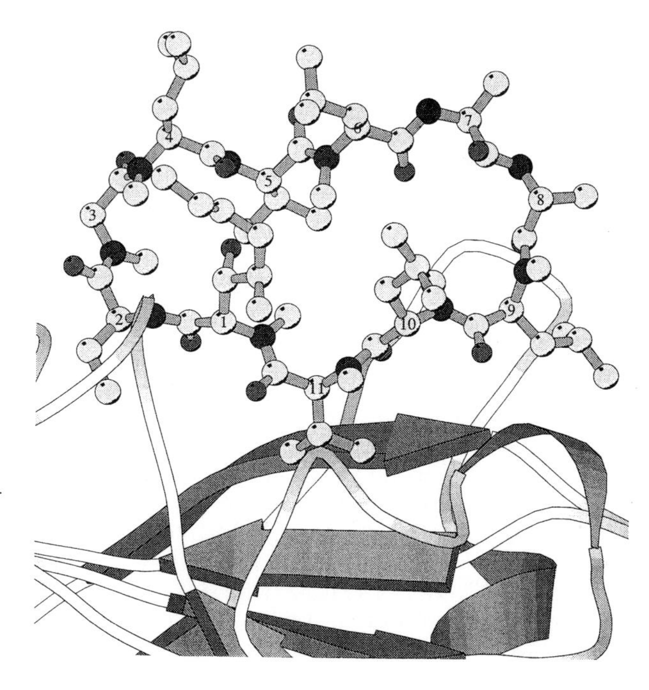

Fig. 23. A close-up view of the CsA–CYP-A complex which shows the distinction between residues of CsA making mostly contacts with CYP-A ("the CYP-A-binding region") and residues available for interactions with calcineurin ("the effector region").

be displaced upon binding of the Ala-Pro moiety of the tetrapeptide is the one hydrogen-bonded to the main chain O and N of Asn^{102}. These hydrogen bonds are replaced, in the complex, by hydrogen bonds between Ala^{2} of the tetrapeptide and Asn^{102} (cf. Fig. 21). Interestingly, the water molecules hydrogen bonded to His^{54} and His^{126} are conserved in the two structures.

5.5 The 3D-Structure of Cyclophilin A Complexed with CsA

The NMR structure of CsA when bound to CYP-A had been determined (Weber et al., 1991; Neri et al., 1991) before the 3D-structure of CYP-A was known and was found to be substantially different from the conformation of uncomplexed CsA as determined by NMR in chloroform or by X-ray crystallography. The main features of the CYP-A-bound conformation as determined by NMR are that all amide bonds are in the *trans* conformation and at first no intramolecular hydrogen bonds were found (subsequently, the X-ray structure of the CsA–CYP-A complex has shown that there is an intramolecular hydrogen bond between the hydroxyl group of $MeBmt^{1}$ and the carbonyl oxygen of $MeLeu^{4}$). This contrasts with the structure of uncomplexed CsA in the crystal (cf. Fig. 30) or in chloroform where the amide bond between $MeLeu^{9}$ and $MeLeu^{10}$ is *cis* and the four free NH groups are all involved in intramolecular hydrogen bonds (Loosli et al., 1985).

Using this NMR structure of CsA and 32 intermolecular NOEs (nuclear Overhauser effects) between CsA and CYP-A as distance constraints CsA was subsequently docked into CYP-A (Spitzfaden et al., 1992). The refined docked structure of the complex showed that 13 residues of cyclophilin are in contact with CsA. This docked structure agreed well with the structure–activity hypothesis based on the binding of a series of CsA derivatives and highlighted the importance of residues 10, 11, 1, 2, 3 (Fig. 23) for CYP-A binding (Quesniaux et al., 1987). The direction of the chain in the docked CsA is in the opposite direction to that observed in the tetrapeptide–cyclophilin complex. Arg^{55} is also involved as a hydrogen bond donor in both structures, to the carbonyl oxygens of $MeLeu^{10}$ in CsA and Pro^{3} in the peptide. An important result of the collaborative X-ray, NMR, and modelling work on the CsA–CYP-A complex was to show conclusively that the cyclosporin binding site was identical to the peptidyl–prolyl isomerase active site (cf. Figs. 19 and 24).

The NMR structure of the full CsA–CYP-A complex (Theriault et al., 1993) and the X-ray structure of a decameric CsA–CYP-A complex have now been solved (Pfluegl et al., 1993). Both structures are very similar and broadly confirm the results of the previous docking studies with the following corrections: There is actually an intramolecular hydrogen bond between the hydroxyl group of MeBmt and the carbonyl oxygen of $MeLeu^{4}$ and an intermolecular hydrogen bond between Abu^{2}(NH) of CsA and Asn^{102}(CO) of CYP-A, both not found in the docking model. The biological relevance of the decameric form (with a pentamer of CsA–CYP-A complexes per asymmetric unit; Fig. 25) is unclear. Direct interactions between the effector domain of cyclosporin and calcineurin would practically be impossible since about 80% of the CsA surface are buried in the decameric complex.

Subsequently, it was possible to grow crystals and solve the X-ray structure of a monomeric CsA–CYP-A complex at 2.1 Å resolution (Mikol et al., 1993). In contrast to the decameric structure there are no intermolecular contacts between CsA and neighboring CYP-A molecules in the crystal. The monomeric CsA–CYP-A complex shows the following features (cf. Fig. 26): The binding pocket for CsA is a mainly hydrophobic crevice formed by the following 13 residues of CYP-A which are within 4.0 Å of CsA: Arg^{55}, Phe^{60}, Met^{61}, Gln^{63}, Gly^{72}, Ala^{101}, Asn^{102}, Ala^{103}, Glu^{111}, Phe^{113}, Trp^{121}, Leu^{122}, and His^{126}. The binding site rests on three of the antiparallel strands of the eight-stranded β-barrel involving Phe^{60}, Met^{61}, Gln^{63}, Phe^{113}, Gln^{111}, and Arg^{55}. Other clusters of residues (Trp^{121}, Leu^{122}, His^{126}), (Ala^{101}, Asn^{102}, Ala^{103}), and (Gly^{72}) are located on three sep-

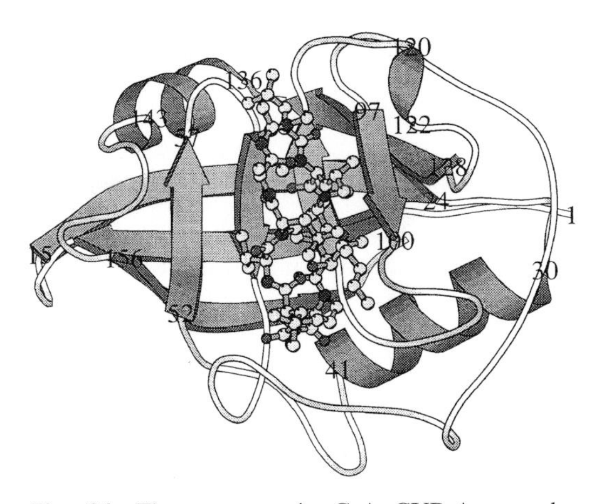

Fig. 24. The monomeric CsA–CYP-A complex viewed in the same orientation as the CYP-A–tetrapeptide complex in Fig. 19 showing that the cyclosporin binding site is identical to the peptidyl–prolyl active site (but CsA and the tetrapeptide bind with opposite polarities from N- to C-terminus).

arate loop regions which protrude some 10 Å–15 Å from the surface of the barrel forming a deeply grooved surface (cf. Fig. 24). CsA docks into CYP-A like a coin going part way into a slot machine; only one rim of the circular cyclosporin (formed by residues 9, 10, 11, 1, 2, 3) is in contact with CYP-A. The complementarity of fit is particularly good for MeVal^{11} which lies in the center of this binding rim. The valine side chain fits snugly into the deep proline-binding pocket formed by Phe^{60}, Met^{61}, Phe^{113}, and Leu^{122} (Fig. 26).

There are five direct hydrogen bonds between CsA and CYP-A: MeBmt^{1}(CO)-Gln^{63}(NH) (not unambiguously identified in the NMR structure of the complex), Abu^{2}(NH)-Asn^{102}(CO), MeLeu^{9}(CO)-Trp^{121}(NE), MeLeu^{10}(CO)-Arg^{55}(NH1), and MeLeu^{10}(CO)-Arg^{55}(NH2) (not reported in the NMR structure of the complex). There are five clearly identified water molecules mediating interactions between CsA and CYP-A: MeBmt^{1}(CO)-Wat^{18}-His^{54}(NE^{2}), MeBmt^{1}(CO) - Wat^{18} - Wat^{19} - Asn^{71}(CO), Abu^{2}(CO)-Wat^{31}-Thr^{73}(CO), and MeLeu^{6}-(CO)-Wat^{87}-Wat^{107}-Arg^{55}(NH2).

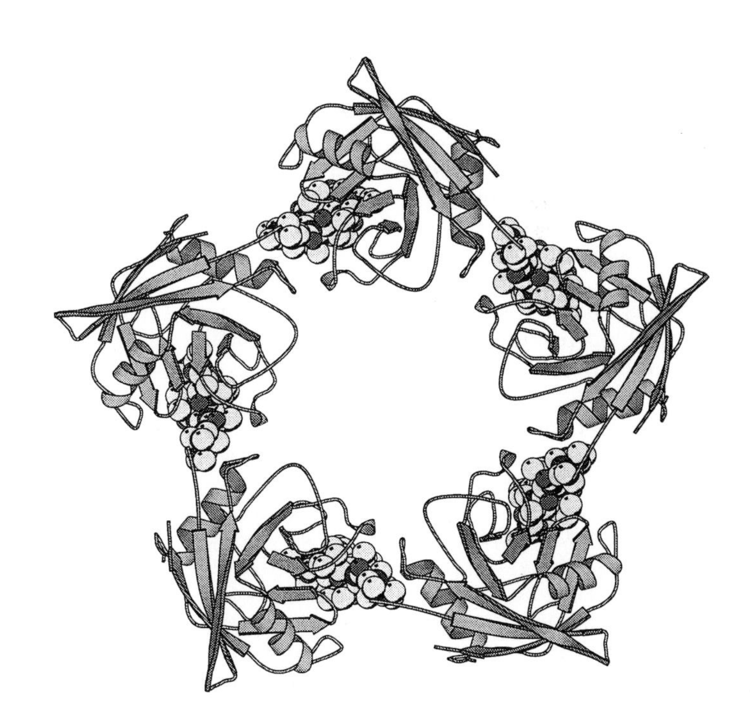

Fig. 25. The pentamer half of the decameric CsA–CYP-A complex.

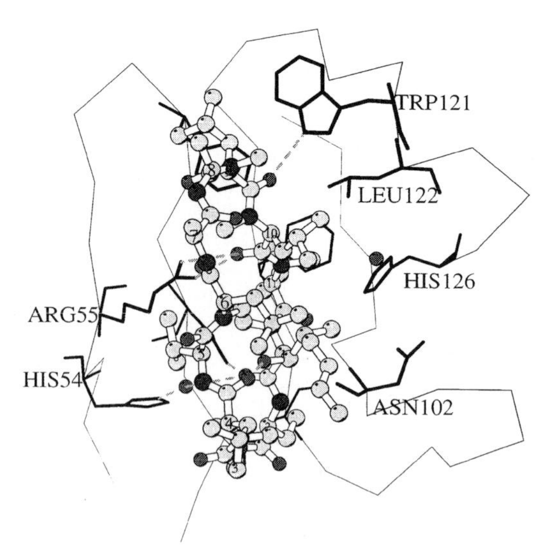

Fig. 26. Details of the interactions between CsA and CYP-A in the monomeric complex.

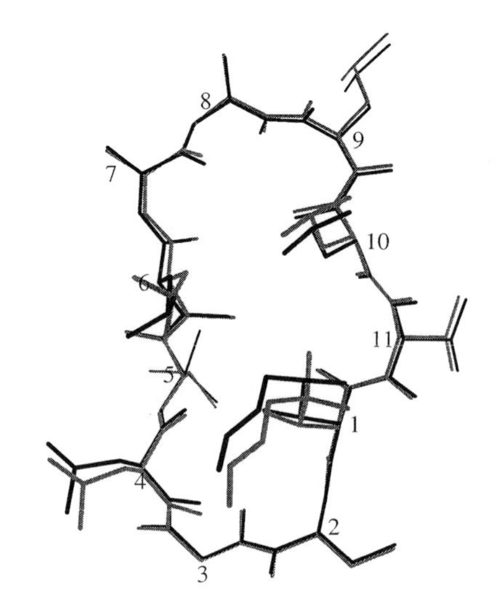

Fig. 27. A superposition (using C_α of CYP-A) of CsA (black) and [$MeBm_2t$]-cyclosporin (grey).

5.6 The X-Ray Structures of Cyclophilin A Complexed with Cyclosporin Derivatives

Using the crystal form with one CsA–CYP-A complex per asymmetric unit and cross-seeding techniques it was possible to obtain crystals and to solve the structures of 15 complexes of cyclophilin A and different cyclosporin derivatives. For all derivatives studied (variations in positions 1, 2, 3, 4, 8) the backbone conformation of the Cs-ring is practically unchanged (as well as the interactions with CYP-A). There are also practically no changes in the CYP-A-structure. In the following, the results from two complexes will be summarized.

5.6.1 [$MeBm_2t$]-Cyclosporin

[$MeBm_2t$]-Cs differs from CsA in having two (instead of one) methyl groups on the C_γ of residue 1 (AEBI et al., 1990). Its affinity for CYP-A is about 1% of that of CsA whereas its immunosuppressive activity *in vitro* is only reduced to 30% of that of CsA (SIGAL et al., 1991). The X-ray structure of CYP-A/([$MeBm_2t$]-Cs) (MIKOL et al., 1994b) shows that a change with respect to the structure of CsA/CYP-A has occurred for the side chains of residues 1 and 10: The C_γ of $MeLeu^{10}$ and of $MeBm_2t$ move apart by about 0.9 Å to accommodate the presence of the additional methyl group on $MeBm_2t$ (cf. Fig. 27). This small structural change must be beneficial for the interaction with calcineurin A/B. On the other hand, the affinity for CYP-A has decreased (as compared to CsA), although all the structurally mediated interactions to CYP-A are conserved. This finding suggests that for [$MeBm_2t$]-Cs the equilibrium (attained in aqueous solution for the uncomplexed ligand) between "nonbinding" and "binding" conformation is shifted to the "nonbinding" side, indicating a higher kinetic barrier for complex formation.

5.6.2 $MeIle^4$-Cs (SDZ NIM 811)

$MeIle^4$-Cs has the same affinity for CYP-A as CsA but no immunosuppressive activity. The X-ray structure (J. KALLEN et al., to be submitted) shows that the only change with respect to the structure of CsA/CYP-A has

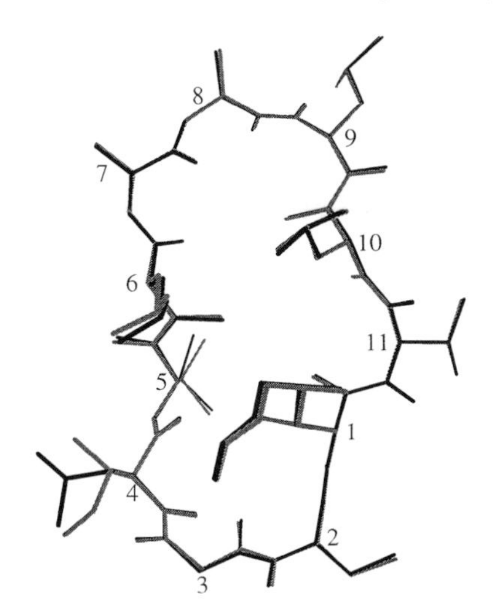

Fig. 28. A superposition (using C_α of CYP-A) of CsA (black) and MeIle[4]-cyclosporin (SDZ NIM 811) (grey). The substitution of Ile for Leu in position 4, without any other structural change, leads to a <1000-fold reduced immunosuppressive activity.

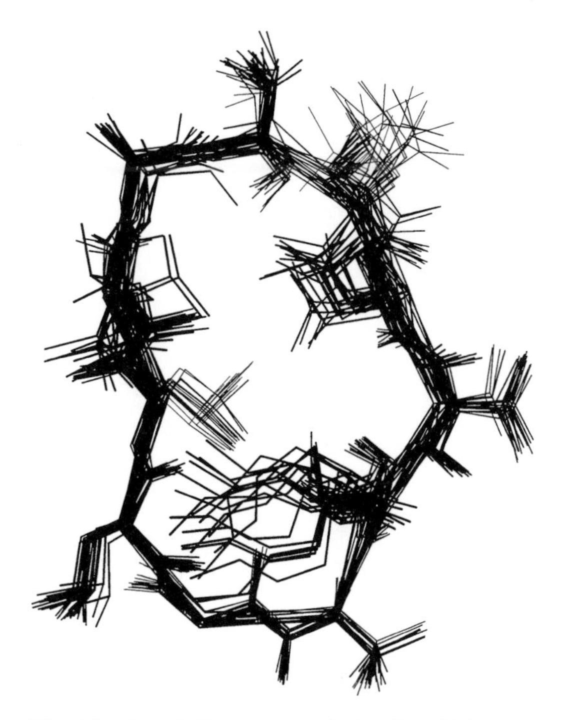

Fig. 29. The NMR-structure (a family of 20 structures with low NOE violations of maximally 0.29 Å) of SDZ 214-103 bound to CYP-A.

occurred for the side chain of residue 4 in that Leu is replaced by Ile (Fig. 28). The mere presence of a branched (vs. unbranched) C_β for residue 4 is thus sufficient to drastically impair binding to calcineurin A/B (which must, therefore, have a "tight-binding" pocket for this region (PAPAGEORGIOU et al., 1994); cf. the discussion in Sect. 4.3.

5.6.3 The 3D-Structure of SDZ 214-103 Bound to Cyclophilin A

The 3D-structure of CYP-A bound SDZ 214-103 has been determined by NMR methods (WIDMER et al., unpublished results). 95 structurally relevant intramolecular NOEs were used in structure calculations with the program DIANA. The non-hydrogen atoms of a family of 20 structures (with low residual NOE-violations of maximally 0.29 Å) are depicted in Fig. 29. A comparison with the X-ray structure of CsA when bound to CYP-A (Fig. 30) shows that the backbone conformations are very similar (slight differences might be due to measurement errors of the two techniques). On the other hand, the X-ray structures of free CsA and free SDZ 214-103 are quite different (cf. Figs. 18 and 30).

5.7 The X-Ray Structure of Cyclophilin B Complexed with a Cyclosporin Derivative

The affinity of cyclophilin B for CsA is approximately 10-fold higher as that of cyclophilin A (SCHNEIDER et al., 1994). Similarly, the complex CypB–CsA is known to be 5- to 10-fold more effective than CsA–CYP-A in inhibiting the Ser/Thr phosphatase activity of calcineurin (SWANSON et al., 1992). At Sandoz Pharma AG the X-ray structure of a complex of CypB with a cyclosporin derivative modified in position 8 ([O-(cholinylester)-D-Ser[8]]-cyclosporin) was determined in order to elucidate possible structural origins responsible for this difference (MIKOL et al., 1994a).

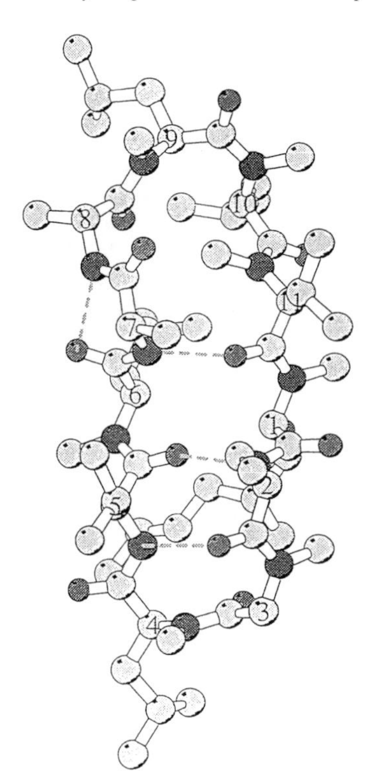

Fig. 30. The X-ray structure of CsA (not bound to cyclophilin) crystallized from acetone. C_α atoms are indicated by the residue numbers (1–11) and hydrogen bonds are indicated by dashed lines. All the four amide protons are involved in hydrogen bonds (three transannular ones: Abu^2(NH)—Val^5(CO), Val^5(NH)—Abu^2(CO), and Ala^7(NH)—MeVal^{11}-(CO), and one additional hydrogen bond: D-Ala^8(NH)—MeLeu^6(CO)).

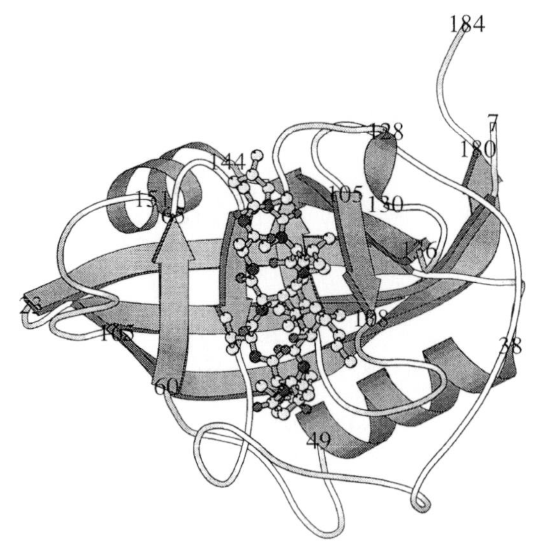

Fig. 31. The architecture of CypB (in complex with cyclosporin), viewed in the same orientation as CYP-A in Fig. 8. The major structural differences with respect to CYP-A occur at the N- and C-termini and in the loops 19–24 and 152–164 (residue numbering of CypB).

The overall structures of CypB and CYP-A are relatively similar (cf. Figs. 31, 32, 19, and 20). However, significant differences occur in two loops (residues 19–24 and 152–164 of CypB) and at the N- and C-termini. The active site regions of CypB and CYP-A are very similar. Indeed, there are practically no differences for any residues of the cyclosporin-binding pocket within 5.0 Å of cyclosporin (rmsd for all these non-hydrogen atoms is less than 0.15 Å). The cholinylester derivatization in position 8 could not be seen in the electron density map, probably because of too high conformational mobility. The binding and conformation of the cyclosporin residues are

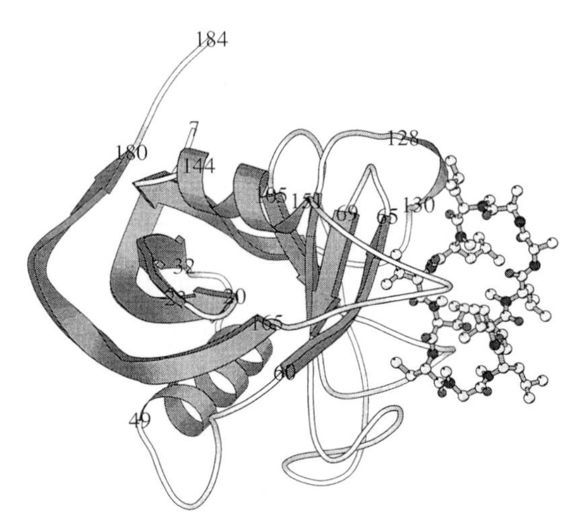

Fig. 32. The complex CypB–CsA rotated by 90° with respect to Fig. 30 around a vertical axis in the picture plane, to be compared with the view of CYP-A depicted in Fig. 20.

practically identical for the two complexes. Candidates for an explanation of the increased potency of CypB-Cs complexes are the following residues of CypB: Arg^{90}, Lys^{113}, Ala^{128}, and the loop containing Arg^{158}.

5.8 The X-Ray Structure of Cyclophilin C Complexed with CsA

The structure of CypC (KE et al., 1993b) shows that it is similar to CYP-A, with the exception of the loops Asp7-Lys9, Met70-Ile76, and Gln79-Thr189. The cyclosporin binding pockets are practically identical as well as the conformation of cyclosporin itself for the complexes CypC–CsA and CsA–CYP-A.

5.9 Conclusions

Cyclosporins can adopt a variety of conformations depending on the molecular environment. If the environment is hydrophobic, the number of intramolecular hydrogen bonds is maximized by adopting a conformation as shown in Fig. 30 (all four amide protons are involved in intramolecular hydrogen bonds). In a polar environment, on the other hand, the seven N-methyl groups are shielded from the solvent by adopting a conformation as shown in Fig. 8 (there is just one intramolecular hydrogen bond between the hydroxyl group of $MeBmt^1$ and the carbonyl oxygen of $MeLeu^4$). The transition between these two conformations resembles the "inversion of a glove". The NMR structure of a water-soluble cyclosporin derivative in a polar solvent in the absence of cyclophilin shows that cyclosporins can preadopt the cyclophilin-bound conformation in the absence of the protein and there is no need for an "induced-fit" hypothesis to explain the binding of cyclosporins to cyclophilins. The CYP-A-bound backbone conformation seems to be an energetically favorable one. It is "robust" against perturbations of cyclosporin side chains: In all cyclosporin derivatives analyzed structurally until now the backbone conformation is practically identical in all cases; (see, e.g., Fig. 29 for an "extreme" derivatization of CsA). The cyclosporin structure is also preserved in complexes with different receptors (as seen in the complex structures with CYP-A, CypB, CypC). The division of the cyclosporin molecule into a "binding region" (making binding interactions with cyclophilin) and an "effector region" (important for the immunosuppressive effect by mediating interaction of the cyclophilin complex with calcineurin) is visualized in Fig. 23. In the effector region there are some sensitive positions, i.e., residues in very close contact with calcineurin such as, e.g., residue 4: The mere change from $MeLeu^4$ to $MeIle^4$ without any other structural change is sufficient to completely abolish immunosuppressive activity (cf. Fig. 28).

CsA alone cannot exert an inhibitory effect on calcineurin, it needs to be complexed with cyclophilin. Hence, there must be crucial direct interactions between cyclophilin and calcineurin residues also. Evidence for this hypothesis is based on the different inhibitory potencies of cyclophilins A, B, and C when complexed with CsA (FLIRI et al., 1993). Extensive site-directed mutagenesis experiments on CYP-A have been done (ETZKORN et al., 1994) in order to find such crucial residues (e.g., the CYP-A mutant Arg148Glu complexed with CsA shows a 20-fold improved inhibition of calcineurin). The X-ray structure of CypB (the complex CypB-CsA is at least 10-fold more effective than CsA–CYP-A in inhibiting calcineurin) suggests also that the loop containing Arg^{158}, which corresponds to the loop containing Arg^{148} in CYP-A, might be important for modulating the interactions with calcineurin (apart from the possible candidates Arg^{90}, Lys^{113}, and Ala^{128} of CypB). A final answer to these questions will be given, of course, by the experimentally determined 3D-structure of the CsA–CYP-A-calcineurin complex. The additional availability of a 3D-structure of the FKBP12–FK506–calcineurin complex would also answer the question of how the obviously structurally dissimilar CsA–CYP-A and FKBP12–FK506 complexes can both compete for interaction with calcineurin.

6 Biosynthesis of Cyclosporins

6.1 Introduction

Some structural features of cyclosporin A are suggestive of a non-ribosomal biosynthetic origin: the unusual amino acids (4R)-4-[(E)-2-butenyl]-4-methyl-L-threonine (Bmt, position 1), α-amino butyric acid (position 2), and D-alanine (position 8) as well as the cyclic structure and the N-methylation of seven peptide bonds. Indeed, cyclosporins are biosynthesized by an extraordinary large multienzyme called cyclosporin synthetase. However, efficient cyclosporin biosynthesis not only needs a high activity of this central enzyme but also of pathways providing the unusual components as well as a sufficiently high internal pool of the methyl group donor S-adenosyl-methionine (Tab. 3). Whereas α-amino butyric acid is most likely derived from the common amino acid pool (SENN et al., 1991) Bmt and D-alanine have to be supplied by separate pathways. Hence, general interest in cyclosporin biosynthesis has been focused on three enzymes or enzymatic pathways, respectively: (1) cyclosporin synthetase, (2) alanine racemase, (3) Bmt-synthesizing enzymes.

6.2 Cyclosporin Synthetase

6.2.1 Enzymatic Activities of Cyclosporin Synthetase

Enzymatic activity in protein fractions of *Tolypocladium niveum* that could be ascribed to cyclosporin synthetase was first detected by ZOCHER et al. (1986). Initial *in vitro* studies led to the enzymatic synthesis of cyclo-(D-alanyl-N-methylleucyl) diketopiperazine (D-DKP) rather than to the synthesis of cyclosporin A. The reason for this is not known, but considering the finding that mutants of *T. niveum* expressing non-functional cyclosporin synthetase and, therefore, lacking the competence to produce cyclosporin A produce D-DKP instead (DITTMANN et al., 1990) it can be concluded that D-DKP is a bypass-product of cyclosporin synthetase under non-optimal conditions. First *in vitro* enzymatic synthesis

Tab. 3. Requirements for Cyclosporin Biosynthesis

Source	Enzyme(s)	Building Units and Cofactors for Cs-Biosynthesis
Common amino acid pool		Alanine, α-amino butyric acid, glycine, leucine, valine
(L)-Alanine	Alanine racemase	(D)-Alanine
Acetate, SAM, NADPH O_2, N_2	Bmt-polyketide synthase Transformation enzymes	Bmt
C_1-Pool	SAM Synthetase	SAM
Glucose		ATP ($MgCl_2$)
Pantothenate		4′-Phosphopantetheine covalently attached to cyclosporin synthetase
	Cyclosporin Synthetase ↓	
		Cyclosporins

of the complete cyclosporin molecule from the constituent amino acids, ATP, $MgCl_2$, and S-adenosyl-methionine as a methyl group donor was successful in 1987 (BILLICH and ZOCHER, 1987).

Whereas enzymes synthesizing other D-amino acid-containing peptides, e.g., gramicidin S and tyrocidine (KLEINKAUF and VON DÖHREN, 1990), harbor an integral epimerase function which epimerizes the respective L-amino acid into the D-form following its activation, cyclosporin synthetase rather incorporates D-alanine in position 8 only if already supplied in its D-form. Hence, epimerization of L-alanine has to be carried out by a distinct enzyme (see Tab. 3). In contrast, the methyltransferase activity for the N-methylation of the peptide bonds is an integral part of the purified enzyme (LAWEN and ZOCHER, 1990). Proteolysis experiments indicated the existence of several methyltransferase domains in one enzyme molecule, probably one for each of the seven methylation steps, which has been confirmed by cloning and sequencing the gene (see below). Alltogether, cyclosporin synthetase catalyzes at least 40 partial reaction steps: 11 aminoadenylation reactions, 11 transthiolation reactions, 7 N-methylation reactions, 10 elongation reactions, and the final cyclization reaction.

6.2.2 Characterization of the Enzyme

Purification and analysis by SDS-PAGE clearly demonstrated that the complete reaction sequence of cyclosporin biosynthesis is coded by a single multienzyme polypeptide (LAWEN and ZOCHER, 1990) but lacking adequate molecular weight markers it was a difficult task to determine the molecular mass of this giant protein. Analytical ultracentrifugation indicated that the enzyme most likely has a discus-like structure with a diameter of about 33 nm, a thickness of 4.6 nm, and a molecular mass of about 1.4 MDa (SCHMIDT et al., 1992). From cloning and sequencing studies of other peptide synthetases responsible for biosynthesis of, e.g., δ-(α-L-aminoadipyl)-L-cysteinyl-D-valine (SMITH et al., 1990; DIEZ et al., 1990; MACCABE et al., 1991; GUTIERREZ et al., 1991), enniatins (HAESE et al., 1993), gramicidin S (TURGAY et al., 1992; HORI et al., 1989; KRAETZSCHMAR et al., 1989), tyrocidine A (WECKERMANN et al., 1988), HC-toxin (SCOTT-CRAIG et al., 1992), and surfactin (COSMINA et al., 1993) it became clear that such enzymes are composed of domains. Each of these domains is responsible for the recognition and activation of one amino acid and for the peptidation reaction with the amino acid activated by the neighboring domain. In the case of methylated enniatins the corresponding domain harbors an additional module responsible for the methylation step (HAESE et al., 1993). Taking the molecular masses of all these peptide synthetases into account a molecular mass for cyclosporin synthetase of at least 1.6 MDa can be extrapolated.

6.2.3 Isolation and Characterization of the Cyclosporin Synthetase Gene

The gene coding for cyclosporin synthetase has recently been cloned and sequenced (LEITNER et al., 1994; WEBER et al., 1994). In order to obtain partial amino acid sequences to derive specific oligonucleotide probes the cyclosporin synthetase was purified from mycelia of *T. niveum* and partially digested with endoproteinases. 18 fragments were isolated, purified, and used for sequence determination. One of these fragments was identified by photoaffinity labeling with S-adenosyl-methionine and a second fragment by its capacity to activate L-alanine (LAWEN and ZOCHER, 1990; LEITNER et al., 1994; WEBER et al., 1994).

The cyclosporin synthetase gene (*sim*A) was cloned using an oligonucleotide probe derived from one of these amino acid sequences. A genomic library of *T. niveum* ATCC 34921 in λEMBL3 was screened, allowing determination of a nucleotide sequence of 46889 bp with an ORF of 45823 bp. The gene product (CYSYN) corresponds to 15281 amino acids with a predicted molecular mass of 1.7 MDa. A data bank search showed characteristic similarities to known peptide synthetases. 11 regions of CYSYN show high similarity. Two different types are found within the 11 domains of CYSYN. The first type (type I) is about 1000 amino acids in size and very similar to the one already detected in other multifunctional peptide synthetases. This type I is shown in Fig. 34 (non-filled rectangle). The second type (type II) is larger than the first type and includes an approximately 447 amino acid polypeptide (Fig. 33 and 34). This polypeptide sequence has probably N-methyltransferase activity as demonstrated by several lines of evidence: (1) An experimentally derived amino acid sequence of a 45 kDa fragment of cyclosporin synthetase having methyltransferase activity was found in the deduced amino acid sequence at a position corresponding to one of these additional sequences (Fig. 33). (2) This additional sequence is very similar to a corresponding sequence found in the enniatin synthetase (HAESE et al., 1993). (3) All seven sequences show similarity to a postulated consensus sequence for S-adenosyl-methionine binding and all seven CYSYN sequences are highly conserved in this region (LEITNER et al., 1994; WEBER et al., 1994).

There are seven type II and four type I domains. All 11 domains contain in the same relative position the putative amino acid binding and phosphopantetheine acid attachment sites found in other peptide synthetases. The 130000 Da cyclosporin synthetase fragment could be characterized by its capacity to activate L-alanine. This permits an assignment of its N-terminal sequence to CYSYN (Fig. 33). The fragment corresponds well to the 11th domain (LEITNER et al., 1994; WEBER et al., 1994). L-Alanine is known as the last amino acid added to the growing peptide chain (DITTMANN et al., 1994).

The order of domains – with or without a putative methyltransferase activity – corresponds to the biosynthetic order of methylated and non-methylated amino acids (DITTMANN et al., 1994). The authors, therefore, concluded that there is a correspondence of the order of the constitutive amino acids of cyclosporin A and the order of the 11 domains as shown in Fig. 33 (LEITNER et al., 1994; WEBER et al., 1994). A similar colinearity has been shown for other peptide synthetases (LAWEN et al., 1992).

The C-terminal end of domain 11 is not the C-terminus of CYSYN. There are approximately 500 amino acid residues of non-domain character. As the biosynthesis of cyclosporins also includes ring closure of the peptide one could speculate that this non-domain (i.e., showing no significant homology to domains 1–11) peptide sequence harbors this function (WEBER et al., 1994).

6.2.4 Manipulation of the Cloned Cyclosporin Synthetase Gene by Integrative Transformation

The description of the *sim*A gene and the correlation of the order of protein domains and constituent amino acids of cyclosporin will enable the construction of new fungal strains by exchange of domain-specific parts of the *sim*A gene by gene replacement (Fig. 35). Necessary for such experiments is a transformation system which allows the re-introduction of *in vitro* manipulated DNA by homologous recombination.

Plasmids composed of *A. nidulans* promoters fused to a bacterial hygromycin phosphotransferase gene have been used successfully in a number of fungal species (FINCHAM, 1989). Such plasmids can also be used for the transformation of *T. niveum* but there is a high proportion of multiple tandem integrations of the plasmid DNA.

Transformation systems for *T. niveum* were described by LEITNER and WEBER (1994). The authors isolated the cyclophilin gene as a source of a homologous promoter element and called the isolated gene *cpt*A. The entire gene consists of 890 bp, including the three introns of 220 bp, 57 bp, and 60 bp respectively. The gene codes for a protein (CPT) with a molecular mass of 19569 Da. The cDNA corresponding to *cpt*A is similar to the cyclophilin cDNA of *Neurospora crassa* (80%) which was used as a probe to isolate the *T. niveum* gene. At the amino acid level the similarity is also 80%. The promoter of the *T. niveum* gene was used for plasmid constructions in which this promoter is fused to a bacterial hy-

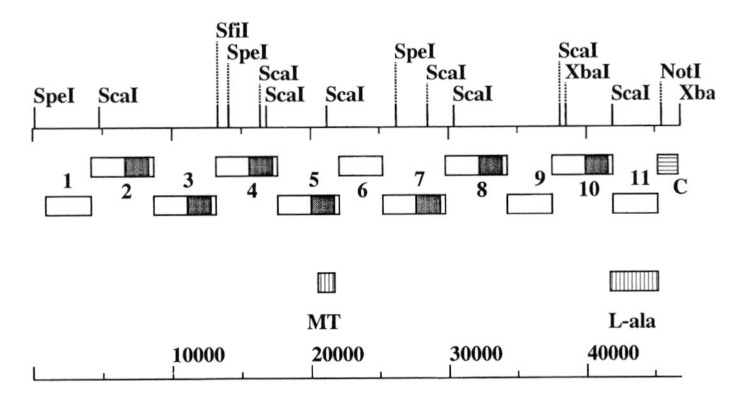

Fig. 33. Structure of the cyclosporin synthetase gene (*sim*A) and the derived polypeptide (CYSYN). A partial restriction map of the 47 kb gene region is shown. The structure of the derived translation product is illustrated by horizontal boxes. There are two different types of domains which are described in more detail in Fig. 34. The grey parts identify the N-methyltransferase subdomains. The box labeled with a C (horizontal lines) indicates the about 500 amino acid long C-terminal part of the translation product. The two boxes with vertical lines indicate the postulated positions of the N-methyltransferase fragment (MT) and the L-alanine activating fragment (L-ala) based on the N-terminal sequences and the observed molecular masses (WEBER et al., 1994).

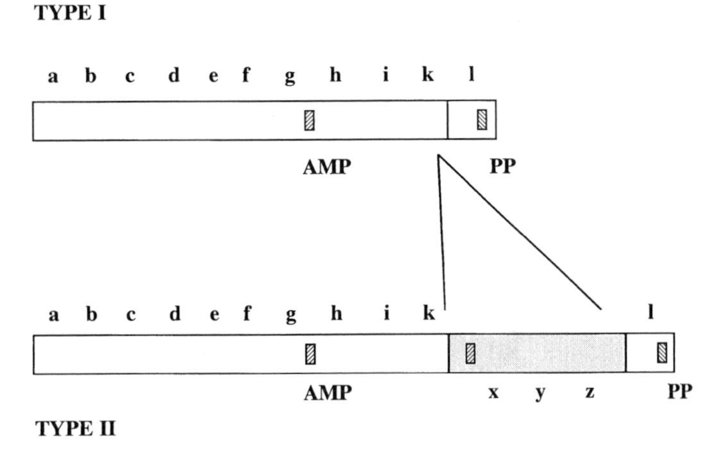

Fig. 34. Relationship of the two domain types. The two types of domains are aligned in order to illustrate their relationships. "a" to "l" illustrate similar amino acid sequences. The grey part of the lower box stands for the N-methyltransferase fragment. The small box within this region symbolizes putative S-adenosyl methionine binding sites. Further putative binding sites for AMP and phosphopantethein cofactor are also indicated (AMP, PP) (WEBER et al., 1994).

gromycin phosphotransferase gene. A fragment containing 2.1 kb upstream of the putative start codon of the *cpt*A gene was amplified by PCR and ligated with a 1.76 kb *Cla*I-*Xba*I fragment from pCSN44. This 1.76 kb pCSN44 fragment covers a bacterial hygromycin phosphotransferase gene and the fragment with the transcriptional terminator of the *A. nidulans trp*C gene (STABEN et al., 1989). The resulting plasmid was called pSIM10. Protoplasts of *T. niveum* could be transformed with purified plasmid pSIM10 DNA in the presence of PEG and hygromycin-resistant colonies were obtained (LEITNER and WEBER, 1994; WEBER et al., 1994). As it was intended to use the transformation system to manipulate the gene for the cyclosporin synthetase of *T. niveum*, derivatives of pSIM10 were constructed containing internal fragments of this gene. The first example, pSIM11, contains a 3.6 kb *Xho*I restriction fragment of the *sim*A gene. A crossover be-

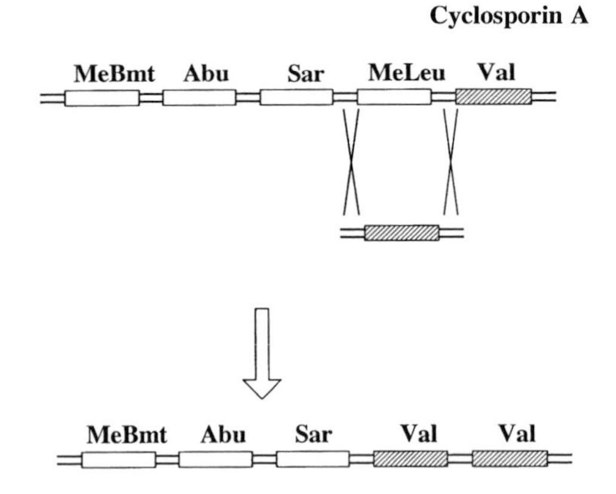

Fig. 35. Theoretical reprogramming of cyclosporin synthetase. The partial structure of the cyclosporin synthetase gene is shown schematically. The gene regions coding for domains are shown as boxes and labeled by the names of the corresponding amino acids. A fragment of the gene coding for a domain which activates valine (hatched box) is modified *in vitro* by adding the flanking regions of another part of the gene coding for a domain which activates leucine. Due to this sequences the DNA coding for the valine-specific domain can replace the DNA coding for the leucine-specific domain following recombination. The recombined gene codes for a cyclosporin synthetase that preferentially produces cyclosporin Q instead of cyclosporin A. This is a hypothetical scheme.

tween the cloned and the genomic version of DNA should lead to insertion of the plasmid DNA and to a partial duplication of the target DNA. As the cloned DNA does not contain the 5′ or 3′ end of the gene the insertion inactivates the gene. *T. niveum* protoplasts were transformed with pSIM11 following linearization of the plasmid DNA. Out of 81 pSIM11 transformants 50 (62%) were found that do not produce cyclosporin. Several of the transformants which lost the ability to produce cyclosporin were verified to contain the expected fragment sizes by Southern hybridizations (LEITNER and WEBER, 1994; WEBER et al., 1994).

The second example, pSIM13, includes a 2.1 kb *Eco*RI fragment of the cyclosporin synthetase gene cloned into pUC18. A 3.7 kb *Xho*I fragment including the promoter of the cyclophilin gene, the hygromycin phosphotransferase gene, and a *A. nidulans trp*C transcription terminator fragment as inserted into the central *Xho*I site of the cloned *sim*A fragment. Following transformation, this construct can be inserted into the genomic DNA by a double crossover (Fig. 36). pSIM13 was used for transformation of *T. niveum* ATCC 34921 protoplasts and the transformants were analyzed for cyclosporin production. DNA from pSIM13 transformants was analyzed by Southern hybridization. For the three different restriction enzymes used to digest the DNA the sites expected for the DNA of a mutant generated by gene replacement were identified (Fig. 37) (WEBER et al., 1994). The high frequency of transformants which do not produce cyclosporin indicates that only one copy of the *sim*A gene is present in the genome of *T. niveum*. A cyclosporin non-producing mutant of *T. inflatum* Cyb156 accumulating the cyclosporin precursor amino acid Bmt has been described by SANGLIER et al. (1990). This mutant showed reduced sporulation and reverted to cyclosporin formation at high frequency. In contrast, the pSIM11-transformants show normal morphology and growth characteristics. Transformants were also analyzed for accumulation of Bmt but only small amounts could be detected even if high-yielding *T. niveum* strains were used for transformation (LEITNER and WEBER, 1994; WEBER et al., 1994).

The transformation system described proved to be a powerful tool for gene disruption of the cyclosporin synthetase (*sim*A) gene of *T. niveum*. It is intended to use these methods for gene replacement experiments in which parts of the *sim*A gene coding for amino acid specific domains are exchanged. The mutated genes will direct the synthesis of new cyclosporins or of cyclosporins which are up to now only by-products of the biosynthesis.

6.2.5 Mechanistic Aspects of Biosynthesis

Detailed studies of the mechanism of cyclosporin biosynthesis provided strong evi-

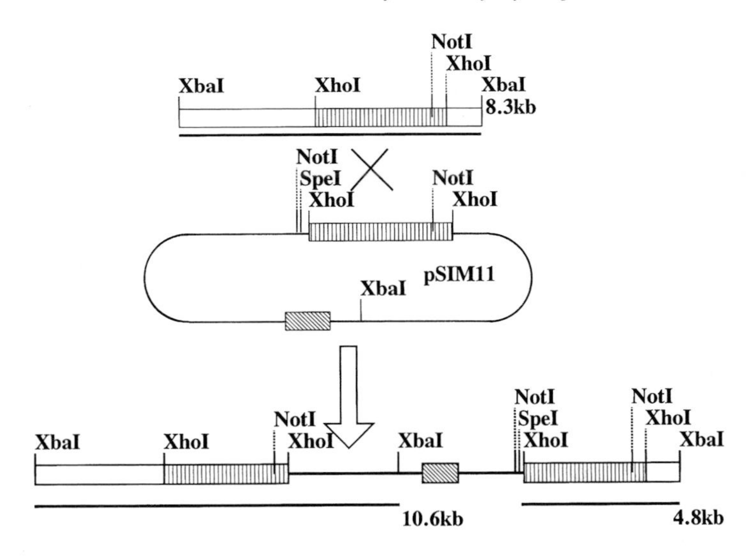

Fig. 36. Gene disruption with pSIM11. The upper part of the figure shows the *Xba*I fragment of the *Tolypocladium niveum sim*A gene. The *Xho*I fragment cloned in pSIM11 is indicated by vertical lines. The middle part of the figure shows pSIM11, the hph coding region is indicated by a box with diagonal lines. The lower part shows the predicted structure of the DNA following a single recombination event in the region cloned in pSIM11. The restriction fragments hybridizing with the labeled *Xho*I fragment are indicated (WEBER et al., 1994).

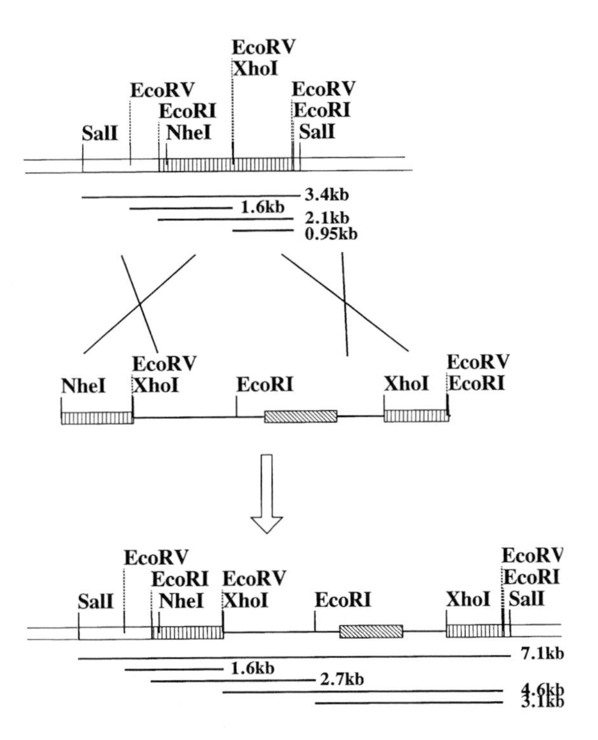

Fig. 37. Gene disruption with the NheI *Eco*RI fragment of pSIM13. The upper part of the figure shows a region of the *Tolypocladium niveum sim*A gene. The *Eco*RI fragment cloned and disrupted in pSIM13 is indicated by a box with vertical lines. The middle part of the figure shows the *Nhe*I *Eco*RI fragment of pSIM13 used for transformation; the hph coding region is indicated by a box with diagonal lines. The lower part shows the predicted structure of the DNA following a double recombination event in the region cloned in pSIM13. The restriction fragments hybridizing with the labeled *Eco*RI fragment if the DNA is digested with *Eco*RI or *Eco*RV or *Spe*I are indicated. For *Sal*I, e.g., this is a 3.4 kb fragment for the nondisrupted gene and a 7.1 kb fragment following gene disruption (WEBER et al., 1994).

dence that binding of D-alanine is the initial reaction of biosynthesis, followed by the stepwise synthesis of a single linear undecapeptide precursor which is finally cyclized to cyclosporin A (DITTMANN et al., 1994); (cf. Fig. 38). However, the details of the whole biosynthetic cycle are still unclear. As early as 1973, a thiotemplate mechanism had been suggested for the nonribosomal biosynthesis of peptides (LALAND and ZIMMER, 1973). Peptide synthetases activate their respective substrate amino acids in two steps, involving aminoacyl adenylates and thioesters as reactive intermediates, as shown in the following equation:

$$E_a + aa + MgATP^{2-} \rightarrow E(aaAMP) \rightarrow E-S-aa$$

E_a enzyme, adenylation site
$E-S-$ enzyme, amino acid binding site
aa amino acid

The assembly of the peptide chain is achieved by repeated transpeptidation and transthiolation reactions facilitated by 4′-phosphopantetheine as a carrier which interacts with the SH groups of the peripheral amino acid activation centers (KLEINKAUF and VON DÖHREN, 1990). The 4′-phosphopantetheine molecule has a length of 2 nm and has been postulated to reach up to six of the active centers of a synthetase (KLEINKAUF and VON DÖHREN, 1987). Although 4′-phosphopantetheine is also an essential component of cyclosporin synthetase (LAWEN and ZOCHER, 1990), in view of the size of this enzyme the model of a central swinging phosphopantetheine arm appears not applicable. Recent results with gramicidin S synthetase as well as with surfactin synthetase strongly indicate that the crucial amino acid for binding of the substrate amino acid is serine (SCHLUMBOHM et al., 1991; VOLLENBROICH et al., 1993; D'SOUZA et al., 1993), leading the authors to suggest a modified version of the thiotemplate mechanism in which each substrate amino acid is bound by the SH group of a phosphopantetheine arm which in turn is attached to the crucial serine residue of the respective activation center. This concept is further supported by the finding of the consensus amino acid sequence for attachment of phosphopantetheine in each of the 11 domains of cyclosporin synthetase (see Sect. 6.2.3). A model for cyclosporin biosynthesis, including the concept of a modified thiotemplate mechanism, is shown in Fig. 38.

6.2.6 Biosynthesis of Cyclosporin Variants

As common in nonribosomal peptide synthesis, the amino acid specificity of cyclosporin synthetase is rather low. This is reflected by the finding of more than 30 naturally occurring variants of cyclosporin A synthesized by *T. niveum* (TRABER et al., 1987); (see Tab. 1), the relative amount of the cyclosporin variants being considerably dependent on the fermentation conditions (KOBEL and TRABER, 1982). Strikingly, only some positions, especially position 2, show variability in their amino acid incorporation whereas other positions, especially positions 3 and 8, are relatively invariant. Detailed *in vitro* studies showed a somewhat divergent behavior of the isolated enzyme: Position 1, e.g., was found to show substantial flexibility and position 8 exhibits low substrate specificity as well (LAWEN et al., 1989, 1992; LAWEN and TRABER, 1993). Even incorporation of β-alanine in position 8 (and 7, respectively) was shown to be possible, causing formation of a 34-membered ring in contrast to the 33-membered ring of cyclosporins (LAWEN et al., 1994). However, only few of these *in vitro* accessible compounds can be synthesized *in vivo* by precursor directed biosynthesis (TRABER et al., 1988, 1994; HENSENS et al., 1992) perhaps due to unsuitable physiological conditions and/or metabolic utilization of the added precursors.

Fig. 38. Model for cyclosporin biosynthesis by a ▶ modified thiotemplate mechanism. Cyclosporin synthetase activates its respective substrate amino acids in two steps involving adenylation (indicated as a1, a2, etc.) followed by thio-esterification, most probably by a phosphopantetheine arm attached to a serine symbolized by —O— (SCHLUMBOHM et al., 1991). 7 out of 11 amino acids are methylated at this stage by integral methyl transfer units. Repeated transpeptidation and transthiolation reactions, (D)-alanine being the N-terminal amino acid (DITTMAN et al., 1994), lead to the assembly of the linear undecapeptide precursor which is finally cyclized to cyclosporin. The numbers 1–11 do not correlate with the numbering system of the chemical nomenclature but rather with the sequence of reactions of the biosynthetic pathway.

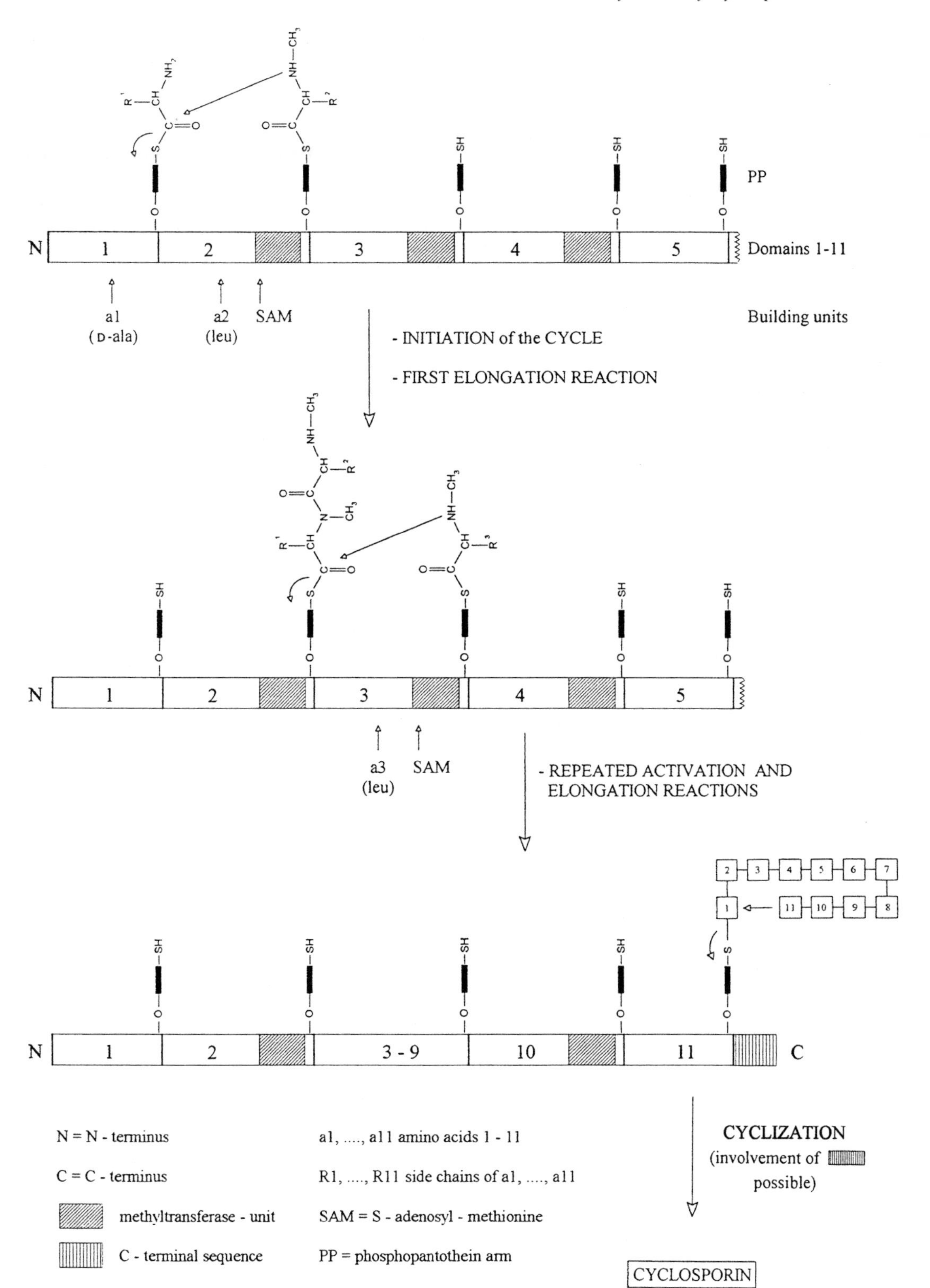
PP
N
Domains 1-11
a1
(D-ala)
a2
(leu)
SAM
Building units
- INITIATION of the CYCLE
- FIRST ELONGATION REACTION
a3
(leu)
SAM
- REPEATED ACTIVATION AND ELONGATION REACTIONS
C
CYCLIZATION
(involvement of possible)
CYCLOSPORIN
N = N - terminus
C = C - terminus
methyltransferase - unit
C - terminal sequence
a1, ..., a11 amino acids 1 - 11
R1, ..., R11 side chains of a1, ..., a11
SAM = S - adenosyl - methionine
PP = phosphopantothein arm

6.2.7 Related Enzymes: Peptolide SDZ 214-103 Synthetase

In addition to the multitude of cyclosporin variants which are synthesized by cyclosporin synthetase *in vitro* and/or *in vivo*, the existence of many related compounds with more pronounced structural deviations, synthesized by homologous multienzymes, can be expected. One example is SDZ 214-103 (Fig. 2), the cyclosporin-related peptolide produced by the fungus *Cylindrotrichum oligospermum* (Corda) Bonorden (HOFFMANN et al., 1994). The multienzyme responsible for biosynthesis of this compound was isolated from *C. oligospermum*, found to have a similar molecular weight as cyclosporin synthetase and to show cross reactivity with cyclosporin synthetase-specific antibodies (LAWEN et al., 1991, 1992). Despite the obvious similarity of the enzymes as well as of their respective products, cyclosporin synthetase and SDZ 214-103 synthetase can not substitute enzymatically for each other. Furthermore, detailed *in vitro* incorporation studies point to a more restricted substrate specificity of SDZ 214-103 synthetase (LAWEN and TRABER, 1993).

6.3 Alanine Racemase

6.3.1 Alanine Racemizing Activity

As mentioned above, cyclosporin synthetase is the first example of a peptide synthetase which lacks an integral activity epimerizing L-alanine to D-alanine. Hence, the existence of a distinct alanine racemase had to be postulated. This was somewhat unexpected since the existence of alanine racemases had previously been restricted to prokaryotic organisms where they are primarily involved in cell wall biosynthesis (ADAMS, 1972; WALSH et al., 1985; SODA et al., 1986). Enzymatic analysis of crude extracts of *T. niveum* indeed led to the detection of an alanine racemizing activity and allowed purification and characterization of the respective protein (HOFFMANN et al., 1994). Interestingly, no corresponding activity could be detected in strains of *C. oligospermum* (Corda) Bonorden (HOFFMANN et al., 1994), the producer of the SDZ 214-103 which contains an ester linked D-2-hydroxyisovaleric acid instead of D-alanine (cf. Fig. 2). In view of this result, it seems very likely that the exclusive and indispensable role of the detectable alanine racemase is supplying D-alanine for cyclosporin biosynthesis. In line with this assumption, inhibition of alanine racemase activity *in vivo* destroys the competence of *T. niveum* for biosynthesis of cyclosporin A, shown by feeding experiments with 3-fluoro-(D)-alanine, a well known inhibitor of prokaryotic alanine racemases (HENSENS et al., 1992). Since activation of D-alanine is the first step in cyclosporin biosynthesis (DITTMANN et al., 1994) and the affinity of cyclosporin synthetase for D-alanine is remarkably high (HOFFMANN et al., 1994) the role of this enzyme seems not to be restricted only to supply one of the structural components of cyclosporin A but it appears to be rather a rate-limiting pacemaker of the whole biosynthetic cycle. This view is supported by the absence of free D-alanine in extracts of cyclosporin producing strains of *T. niveum*.

6.3.2 Characterization of the Enzyme

Like many of the prokaryotic homologs the *T. niveum* enzyme consists of several subunits of a molecular mass of about 40000 Da (estimated by SDS-gel electrophoresis) and depends on pyridoxal phosphate as the exclusive cofactor (HOFFMANN et al., 1994). Consequently, the enzyme is susceptible to the classical racemase suicide inhibitors *in vivo* (HENSENS et al., 1992) as well as *in vitro*, as shown for D- and L-(1-aminoethyl)-phosphonate (HOFFMANN et al., 1994). *In vitro* studies with a purified enzyme fraction further revealed that the specificity of the racemase for L- and D-alanine is high but not exclusive. Compared to the reaction velocity of the epimerization of L-alanine, L-serine, 2-L-aminobutyric acid, and L-leucine, e.g., are racemized with relative reaction rates of 23%, 15%, and 13%, respectively (HOFFMANN et al.,

1994). Whether the enzyme is related to the prokaryotic homologs in an evolutionary sense will be elucidated by cloning and sequencing the corresponding gene which is in progress (K. SCHÖRGENDORFER et al., unpublished results).

6.4 Bmt-Synthesizing Enzymes

6.4.1 Polyketide Origin of Bmt

Bmt with its long aliphatic chain is the most unusual amino acid of cyclosporins. The structure suggested an acetate-derived biosynthetic pathway, confirmed by KOBEL et al. (1983) who were able to show by feeding experiments that four ^{13}C-labeled acetate units are coupled in a head-to-tail fashion and processed to finally yield Bmt. Methionine was identified as the source of the methyl group in position 4, whereas the origin of the amino group remained unknown. These results were later confirmed by feeding experiments using selectively ^{13}C-labeled glucose (SENN et al., 1991). Hence, Bmt can be classified as a polyketide.

These findings led to a model for the biosynthetic pathway of Bmt which, in accordance with the current concepts of polyketide biosynthesis, most likely takes place in at least two distinct phases (for recent reviews, see KATZ and DONADIO, 1993; JORDAN and SPENCER, 1993; O'HAGAN, 1991; ROBINSON, 1991; HOPWOOD and SHERMAN, 1990; SIMPSON, 1989): (1) a basic assembly process involving the coupling of four acetate units, the respective reduction and dehydration steps as well as the methylation reaction; and (2) a transformation process introducing the amino group (OFFENZELLER et al., 1993). Based on this concept it was obvious that identification of the basic assembly product would provide the first clues for elucidation of the biosynthetic route.

6.4.2 Identification of the Basic Assembly Product and Characterization of Bmt-Polyketide Synthase

NMR analysis of cyclosporin A produced by feeding experiments with [1-^{13}C,$^{18}O_2$]acetate demonstrated retention of the acetate-derived ^{18}O-isotope during biosynthesis (measured as an upfield shift of the ^{13}C-signal in position 3 of Bmt; OFFENZELLER et al., 1993). Assuming that all condensation, reduction and dehydration steps as well as the methylation reaction occur in the course of the basic assembly, this result was a first conclusive proof of 3(R)-hydroxy-4(R)-methyl-6(E)-octenoic acid to be the biosynthetic key intermediate (Fig. 39). Subsequently, detailed enzymatic studies led to the identification of this compound as a product of cell free enzyme fractions from *T. niveum* (OFFENZELLER et al., 1993). The respective Bmt-polyketide synthase has been characterized and its isolation, which is in progress, will allow its thorough characterization and provide the basis for cloning the corresponding gene. One interesting feature of the enzyme is that it releases its product as a coenzyme A thioester (OFFENZELLER et al., 1993), a process also known from fatty acid synthases of several fungi (LYNEN, 1980). Details of the basic assembly process, especially the stage of the methylation reaction (see possible routes in Fig. 39), remain to be elucidated and are under investigation.

6.4.3 The Transformation Process

According to the current understanding of Bmt-biosynthesis transformation most likely takes place by hydroxylation of the basic assembly product at the C_2-atom with subsequent oxidation and transamination to Bmt (cf. Fig. 39). It seems unlikely that these transformation reactions are performed by Bmt polyketide synthase itself suggesting the involvement of additional enzymes in Bmt biosynthesis, namely of a hydroxylase, an oxidase, and a transaminase. This model is supported by *in vivo* incorporation of the am-

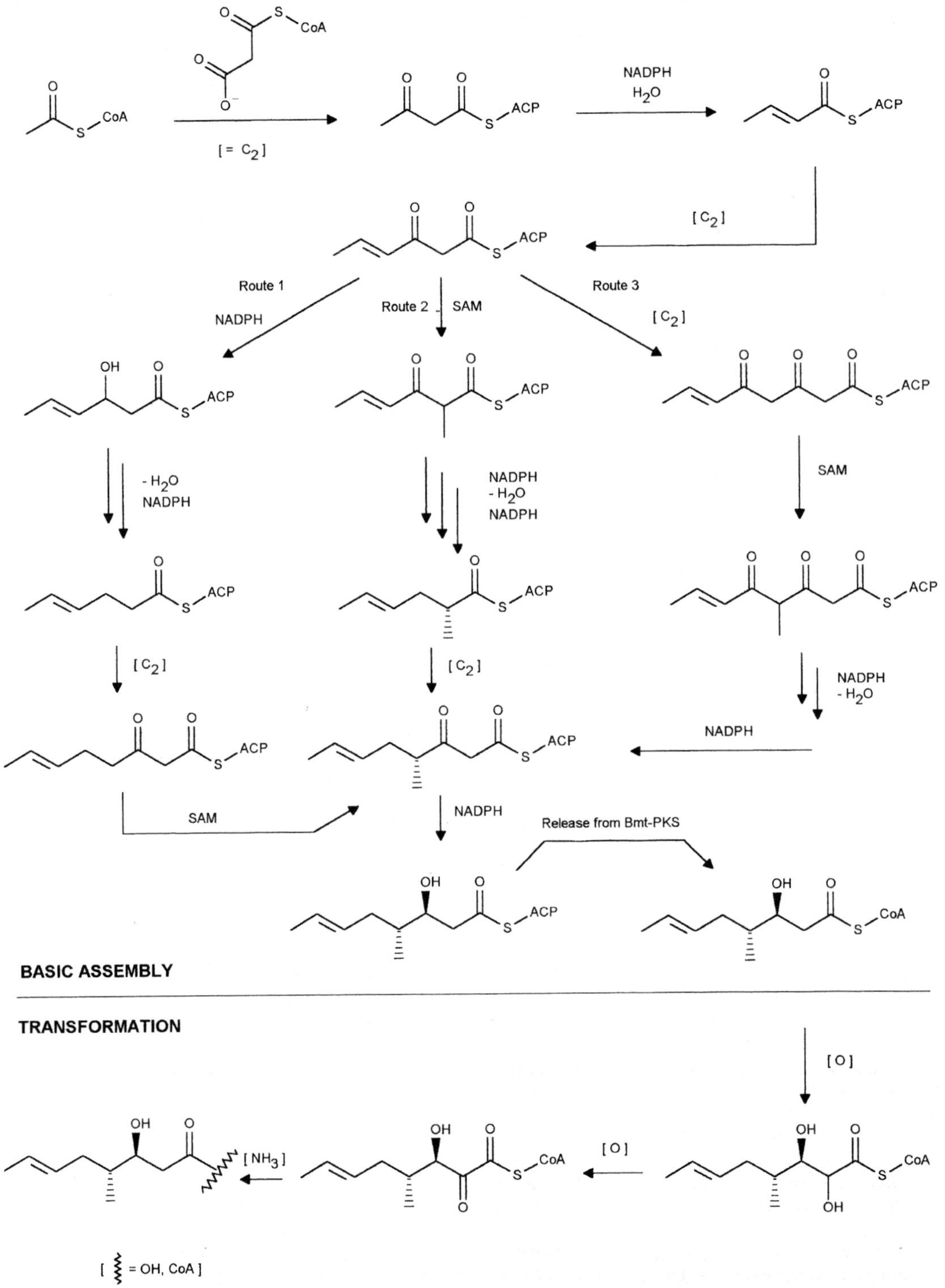

BASIC ASSEMBLY

TRANSFORMATION

Fig. 39. Biosynthetic pathway to Bmt. Abbreviations: S-adenosyl methionine, SAM; acyl carrier protein, ACP.

ino group of ^{15}N-labeled glutamate into Bmt which indicates participation of transaminases in the biosynthetic route (R. TRABER and K. MEMMERT, personal communication).

Note added in proof

Enzyme-bound 3-oxo-4-hexenoic acid could be identified as the direct methylation precursor (OFFENZELLER, M., SAUTER, G., TOTSCHING, K., SU, Z., MOSER, H., TRABER, R., SCHNEIDER-SCHERZER, E. (1996). Biosynthesis of the unusual amino acid (4*R*)-4-[(E)-2-butenyl]-4-methyl-L-threonine of cyclosporin A: enzymatic analysis of the reaction sequence including identification of the methylation precursor in a polyketide pathway (*Biochemistry* **35**, 8401–8412).

7 References

ABRAHAM, R. T., KARNITZ, L. M., SECRIST, J. P., LEIBSON, P. J. (1992), Signal transduction through the T-cell antigen receptor, *Trends Biochem. Sci.* **17**, 434–438.

ADAMS, E. (1972), Amino acid racemases and epimerases, in: *The Enzymes* (BOYER, B. D., Ed.), 3rd Edn., pp. 479–507. New York: Academic Press.

AEBI, J. D., DEYO, D. T., SUN, C. Q., GUILLAUME, D., DUNLAP, B., RICH, D. H. (1990), Synthesis, conformation, and immunosuppressive activities of three analogues of cyclosporin A modified in the 1-position, *J. Med. Chem.* **33**, 999–1009.

ALBERG, D. G., SCHREIBER, S. L. (1993), Structure-based design of a cyclophilin–calcineurin bridging ligand, *Science* **262**, 248–250.

ALTSCHUH, D., VIX, O., REES, B., THIERRY, J. C. (1992), A conformation of cyclosporin A in aqueous environment revealed by the X-ray structure of a cyclosporin–Fab complex, *Science* **256**, 92–94.

ANDRIEU, J. M., EVEN, P., VENET, A. (1986), AIDS and related syndromes as a viral-induced autoimmune disease of the immune system: an anti-MHC II disorder. Therapeutic implications, *AIDS Res.* **2**, 163–174.

ARMITAGE, R. J., BECKMANN, M. P., IDZERDA, R. L., ALPERT, A., FANSLOW, W. C. (1990), Regulation of interleukin 4 receptors on human T cells, *Int. Immunol.* **2**, 1039–1045.

BANG, H., BRUNE, K., NAGER, C., FEIGE, U. (1993), Interleukin-8 is a cyclosporin A binding protein, *Experientia* **49**, 533–538.

BARRY, J. M. (1992), Immunosuppressive drugs in renal transplantation. A review of the regimens, *Drugs* **44**, 554–566.

BAUMANN, G., BOREL, J. F. (1992), Mécanismes moléculaires de l'action des agents immunosuppresseurs, *médecine/sciences* **8**, 366–371.

BAUMANN, G., GEISSE, S., SULLIVAN, M. (1991), Cyclosporin A and FK-506 both affect DNA binding of regulatory nuclear proteins to the human interleukin-2 promoter, *New Biologist* **3**, 270–278.

BAUMANN, G., ZENKE, G., WENGER, R. M., HIESTAND, P. C., QUESNIAUX, V. F. J., ANDERSEN, E., SCHREIER, M. H. (1992), Molecular mechanisms of immunosuppression, *J. Autoimmun.* **5** (Suppl. A), 67–72.

BELL, A., WERNLI, B., FRANKLIN, R. M. (1994), Roles of peptidyl-prolyl *cis-trans* isomerase and calcineurin in the mechanisms of antimalarial action of cyclosporin A, FK506, and rapamycin, *Biochem. Pharmacol.* **48**, 495–503.

BERGSMA, D. J., EDER, C., GROSS, M., KERSTEN, H., SYLVESTER, D., APPELBAUM, E., CUSIMANO, D., LIVI, G. P., MCLAUGHLIN, M. M., KASYAN, K. ET AL. (1991), The cyclophilin multigene family of peptidyl-prolyl isomerases. Characterization of three separate human isoforms, *J. Biol. Chem.* **266**, 23204–23214.

BIERER, B. E., JIN, Y. J., FRUMAN, D. A., CALVO, V., BURAKOFF, S. J. (1991), FK 506 and rapamycin: molecular probes of T-lymphocyte activation, *Transplant. Proc.* **23**, 2850–2855.

BILLICH, A., ZOCHER, R. (1987), Enzymatic synthesis of cyclosporin A, *J. Biol. Chem.* **262**, 17258–17259.

BLUNDELL, T. L., COOPER, J., FOUNDLING, S. I., JONES, D. M., ATRASH, B., SZELKE, M. (1987), On the rational design of renin inhibitors: X-ray studies of aspartic proteinases complexed with transition state analogues, *Biochemistry* **26**, 5585–5595.

BOESCH, D., GAVERIAUX, C., JACHEZ, B., POURTIER MANZANEDO, A., BOLLINGER, P., LOOR, F. (1991), *In vivo* circumvention of P-glycoprotein-mediated multidrug resistance of tumor cells with SDZ PSC 833, *Cancer Res.* **51**, 4226–4233.

BRAM, R. J., HUNG, D. T., MARTIN, P. K., SCHREIBER, S. L., CRABTREE, G. R. (1993), Identification of the immunophilins capable of mediating inhibition of signal transduction by cyclosporin A and FK506: roles of calcineurin binding and cellular location, *Mol. Cell Biol.* **13**, 4760–4769.

BRILLANTES, A.-M. B., ONDRIAS, K., SCOTT, A., KOBRINSKY, E., ONDRIASOVA, E., MOSCHELLA, M. C., JAYARAMAN, T., LANDERS, M., EHRLICH, B. E., MARKS, A. R. (1994), Stabilization of calcium release channel (ryanodine receptor) function by FK506-binding protein, *Cell* **77**, 513–523.

BROWN, E. J., ALBERS, M. W., SHIN, T. B., ICHIKAWA, K., CURTIS, K. T., LANE, W. S., SCHREIBER, S. L. (1994), A mammalian protein targeted by G1-arresting rapamycin-receptor complex, *Nature* **369**, 756–758.

CACALANO, N. A., CHEN, B. X., CLEVELAND, W. L., ERLANGER, B. F. (1992), Evidence for a functional receptor for cyclosporin A on the surface of lymphocytes, *Proc. Natl. Acad. Sci. USA* **89**, 4353–4357.

CALNE, R. Y., WHITE, D. J., THIRU, S., EVANS, D. B., MCMASTER, P., DUNN, D. C., CRADDOCK, G. N., PENTLOW, B. D., ROLLES, K. (1978), Cyclosporin A in patients receiving renal allografts from cadaver donors, *Lancet* **2** (8104–5), 1223–1227.

CARONI, P., ROTHENFLUH, A., MCGLYNN, E., SCHNEIDER, C. (1991), S-cyclophilin. New member of the cyclophilin family associated with the secretory pathway, *J. Biol. Chem.* **266**, 10739–10742.

CHUNG, J., KUO, C. J., CRABTREE, G. R., BLENIS, J. (1992), Rapamycin-FKBP specifically blocks growth-dependent activation of and signaling by the 70 kD S6 protein kinases, *Cell* **69**, 1227–1336.

CLIPSTONE, N. A., CRABTREE, G. R. (1992), Identification of calcineurin as a key signalling enzyme in T-lymphocyte activation, *Nature* **357**, 695–697.

COSMINA, P., RODRIGUEZ, F., DE FERRA, F., GRANDI, G., PEREGO, M., VENEMA, G., VAN SINDEREN, D. (1993), Sequence and analysis of the genetic locus responsible for surfactin synthesis in *Bacillus subtilis*, *Mol. Microbiol.* **8**, 821–831.

COWAN, S. W., NEWCOMER, M. E., JONES, T. A. (1990), Crystallographic refinement of human serum retinol binding protein at 2 Å resolution, *Proteins* **8**, 44–61.

CRABTREE, G. R. (1989), Contingent genetic regulatory events in T lymphocyte activation, *Science* **243**, 355–361.

D'SOUZA, C., NAKANO, M. M., CORBELL, N., ZUBER, P. (1993), Amino-acylation site mutations in amino acid-activating domains of surfactin synthetase: Effects on surfactin production and competence development in *Bacillus subtilis*, *J. Bacteriol.* **175**, 3502–3510.

DIEZ, B., GUTIERREZ, S., BARREDO, J. L., VON SOLINGEN, P., VAN DER VOORT, L. H. M., MARTIN, J. F. (1990), The cluster of Penicillin biosynthetic genes, *J. Biol. Chem.* **265**, 16358–16365.

DITTMANN, J., LAWEN, A., ZOCHER, R., KLEINKAUF, H. (1990), Isolation and partial characterization of cyclosporin synthetase from a cyclosporin non-producing mutant of *Beauveria nivea*, *Biol. Chem. Hoppe-Seyler* **371**, 829–834.

DITTMANN, J., WENGER, R. M., KLEINKAUF, H., LAWEN, A. (1994), Mechanism of cyclosporin A biosynthesis. Evidence for synthesis via a single linear undecapeptide precursor, *J. Biol. Chem.* **269**, 2841–2846.

DUMONT, F. J., MELINO, M. R., STARUCH, M. J., KOPRAK, S. L., FISCHER, P. A., SIGAL, N. H. (1990a), The immunosuppressive macrolides FK-506 and rapamycin act as reciprocal antagonists in murine T cells, *J. Immunol.* **144**, 1418–1424.

DUMONT, F. J., STARUCH, M. J., KOPRAK, S. L., MELINO, M. R., SIGAL, N. H. (1990b), Distinct mechanisms of suppression of murine T cell activation by the related macrolides FK-506 and rapamycin, *J. Immunol.* **144**, 251–258.

DURETTE, P. L., BOGER, J., DUMONT, F. J., FIRESTONE, R., FRANKSHUN, R. A., LIN, C. S., MELINO, M. R., PESSOLANO, A. A. (1988), A study of the correlation between cyclophilin binding and *in vitro* immunosuppressive activity of cyclosporin A and analogues, *Transplant. Proc.* **20**, 51–57.

EBERLE, M. K., NUNINGER, F. (1992), Synthesis of the main metabolite (OL-17) of cyclosporin A, *J. Org. Chem.* **57**, 2689–2691.

EBERLE, M. K., NUNINGER, F. (1993), Preparation of (D-cysteine)8-cyclosporin via intramolecular sulfur transfer reaction, *J. Org. Chem.* **58**, 673–677.

EMMEL, E. A., VERWEIJ, C. L., DURAND, D. B., HIGGINS, K. M., LACY, E., CRABTREE, G. R. (1989), Cyclosporin A specifically inhibits function of nuclear proteins involved in T cell activation, *Science* **246**, 1617–1620.

ENDICOTT, J. A., LING, V. (1989), The biochemistry of P-glycoprotein mediated multidrug resistance, *Annu. Rev. Biochem.* **58**, 137–171.

ERNST, P. (1991), Lung transplantation: The current state of knowledge, *Union Med. Can.* **120**, 64–66.

ETZKORN, F., CHANG, Z., STOLZ, L., WALSH, C. T. (1994), Cyclophilin residues that affect noncompetitive inhibition of the protein serine phosphatase activity of calcineurin by the cyclophilin–cyclosporin A complex, *Biochemistry* **33**, 2380–2388.

FAULDS, D., GO, A. S., BENFIELD, P. (1993), Cyclosporin. A review of its pharmacodynamic and pharmacokinetic properties and therapeutic use in immunoregulatory disorders, *Drugs* **45**, 953–1040.

FINCHAM (1989), Transformation in fungi, *Microb. Rev.* **53**, 148–170.

FISCHER, G., WITTMANN-LIEBOLD, B., LANG, K., KIEFHABER, T., SCHMID, F. X. (1989), Cyclophilin and peptidyl-prolyl *cis-trans* isomerase are probably identical proteins, *Nature* **337**, 476–478.

FLIRI, H. G., WENGER, R. M. (1990), Cyclosporins: synthetic studies, structure–activity relationships, biosynthesis, and mode of action, in: *Biochemistry of Peptide Antibiotics* (KLEINKAUF, H., VON DÖHREN, H., Eds.), 1st Edn., pp. 246–287. Berlin, New York: Walter de Gruyter.

FLIRI, H. G., BAUMANN, G., ENZ, A., KALLEN, J., LUYTEN, M., MIKOL, V., MOVVA, R., QUESNIAUX, V. F. J., SCHREIER, M. H., WALKINSHAW, M. D., WENGER, R. M., ZENKE, G., ZURINI, M. G. (1993), Cyclosporins. Structure–activity relationships, *Ann. N. Y. Acad. Sci.* **696**, 47–53.

FOXWELL, B. M. J., LING, V., RYFFEL, B. (1989), Identification of the multidrug resistance related p-glycoprotein as a cyclosporin binding protein, *Mol. Pharmacol.* **36**, 543–546.

FRANSSON, C., FRESKGARD, P. O., HERBERTSSON, H., JOHANSSON, A., JONASSON, P., MARTENSSON, L. G., SVENSSON, M., JONSSON, B. H., CARLSSON, U. (1992), *Cis-trans* isomerization is rate-determining in the reactivation of denatured human carbonic anhydrase II as evidenced by proline isomerase, *FEBS Lett.* **296**, 90–94.

FREY, F. J. (1990), Cyclosporin in autoimmune diseases, *Schweiz. Med. Wochenschr.* **120**, 772–786.

FRIEDMAN, J., WEISSMAN, I. (1991), Two cytoplasmic candidates for immunophilin action are revealed by affinity for a new cyclophilin: one in the presence and one in the absence of CsA, *Cell* **66**, 799–806.

FRIEDMAN, J., TRAHEY, M., WEISSMAN, I. (1993), Cloning and characterization of cyclophilin C-associated protein: a candidate natural cellular ligand for cyclophilin C, *Proc. Natl. Acad. Sci. USA* **90**, 6815–6819.

FRUMAN, D. A., KLEE, C. B., BIERER, B. E., BURAKOFF, S. J. (1992a), Calcineurin phosphatase activity in T lymphocytes is inhibited by FK 506 and cyclosporin A, *Proc. Natl. Acad. Sci. USA* **89**, 3686–3690.

FRUMAN, D. A., MATHER, P. E., BURAKOFF, S. J., BIERER, B. E. (1992b), Correlation of calcineurin phosphatase activity and programmed cell death in murine T cell hybridomas, *Eur. J. Immunol.* **22**, 2513–2517.

GALAT, A., LANE, W. S., STANDAERT, R. F., SCHREIBER, S. L. (1992), A rapamycin-selective 25-kDa immunophilin, *Biochemistry* **31**, 2427–2434.

GETHING, M. J., SAMBROOK, J. (1992), Protein folding in the cell, *Nature* **355**, 33–45.

GUTIERREZ, S., DIEZ, B., MONTENEGRO, E., MARTIN, J. F. (1991), Characterization of the *Cephalosporium acremonium* pcbAB gene encoding α-aminoadipyl-cysteinyl-valine synthetase, a large multidomain peptide synthetase: Linkage to the pcbC gene as a cluster of early cephalosporin biosynthetic genes and evidence of multiple functional domains, *J. Bacteriol.* **173**, 2354–2365.

HABESHAW, J. A., DALGLEISH, A. G., BOUNTIFF, L., NEWELL, A. L., WILKS, D., WALKER, L. C., MANCA, F. (1990), AIDS pathogenesis: HIV envelope and its interaction with cell proteins, *Immunol. Today* **11**, 418–425.

HAESE, A., SCHUBERT, M., HERRMANN, A., ZOCHER, R. (1993), Molecular characterization of the enniatin synthetase gene encoding a multifunctional enzyme catalyzing N-methyl-depsipeptide formation in *Fusarium scirpi*, *Mol. Microbiol.* **7**, 905–914.

HANDSCHUMACHER, R. E., HARDING, M. W., RICE, J., DRUGGE, R. J., SPEICHER, D. W. (1984), Cyclophilin: A specific cytosolic binding protein for cyclosporin A, *Science* **226**, 544–547.

HARRISON, R. K., STEIN, R. L. (1990), Mechanistic studies of peptidyl prolyl *cis-trans* isomerase: evidence for catalysis by distortion, *Biochemistry* **29**, 1684–1689.

HASEL, K. W., GLASS, J. R., GODBOUT, M., SUTCLIFFE, J. G. (1991), An endoplasmic reticulum-specific cyclophilin, *Mol. Cell Biol.* **11**, 3484–3491.

HAYANO, T., TAKAHASHI, N., KATO, S., MAKI, N., SUZUKI, M. (1991), Two distinct forms of peptidylprolyl-*cis-trans*-isomerase are expressed separately in periplasmic and cytoplasmic compartments of *Escherichia coli* cells, *Biochemistry* **30**, 3041–3048.

HENSENS, O. D., WHITE, R. F., GOEGELMAN, R. T., INAMINE, E. S., PATCHETT, A. A. (1992), The preparation of [2-deutero-3-fluoro-D-ala^8]cyclosporin A by directed biosynthesis, *J. Antibiot.* **45**, 133–135.

HOFFMANN, K., SCHNEIDER-SCHERZER E., KLEINKAUF, H., ZOCHER, R. (1994), Purification and characterization of an eukaryotic alanine racemase, acting as a key enzyme in cyclosporin biosynthesis. *J. Biol. Chem.* **269**, 12710–12714.

HOPF, U., NEUHAUS, P., KONIG, V., BAUDITZ, J., KUTHER, S., SCHMIDT, C. A., STEFFEN, R., BLUMHARDT, G., BECHSTEIN, W. O., NEUHAUS, R. ET AL. (1992), Orthotopic liver transplantation in hepatic cirrhosis: on the problem of infection of the transplant with persistent hepatitis viruses, *Z. Gastroenterol.* **30**, 576–582.

HOPWOOD, D. A., SHERMAN, D. H. (1990), Molecular genetics of polyketides and its comparison to fatty acid biosynthesis, *Annu. Rev. Genet.* **24**, 37–66.

HORI, K., YAMAMOTO, Y., MINETOKI, T., KUROTSU, T., KANDA, M., MIURA, S., OKAMURA, K., FURUYAMA, J., SAITO, Y. (1989), Molecular cloning and nucleotide sequence of the Gramicidin S synthetase 1 gene, *J. Biochem.* **196**, 639–645.

HUNTER, T., KARIN, M. (1992), The regulation of transcription by phosphorylation, *Cell* **70**, 777–789.

IZQUIERDO, M., CANTRELL, D. A. (1992), T-cell activation, *Trends Cell Biol.* **2** (9), 268–271.

JANEWAY, C. A., ROJO, J., SAIZAWA, K., DIANZANI, U., PORTOLES, P., TITE, J., HAQUE, S., JONES, B. (1989), The Co-receptor function of murine CD4, *Immunol. Rev.* **109**, 77–92.

JIN, Y., ALBERS, M. W., LANE, W. S., BIERER, B. E., SCHREIBER, S. L., BURAKOFF, S. J. (1991), Molecular cloning of a membrane-associated human FK506- and rapamycin-binding protein, FKBP-13, *Proc. Natl. Acad. Sci. USA* **88**, 6677–6681.

JORDAN, P. M., SPENCER, J. B. (1993), The biosynthesis of tetraketides: Enzymology, mechanism, and molecular programming, *Biochem. Soc. Trans.* **21**, 222–228.

KAHAN, B. D. (1992), Immunosuppressive therapy, *Curr. Opin. Immunol.* **4**, 553–560.

KALLEN, J., WALKINSHAW, M. D. (1992), The X-ray structure of a tetrapeptide bound to the active site of human cyclophilin A, *FEBS Lett.* **300**, 286–290.

KALLEN, J., SPITZFADEN, C., ZURINI, M. G., WIDER, G., WIDMER, H., WUTHRICH, K., WALKINSHAW, M. D. (1991), Structure of human cyclophilin and its binding site for cyclosporin A determined by X-ray crystallography and NMR spectroscopy, *Nature* **353**, 276–279.

KATZ, L., DONADIO, S. (1993), Polyketide synthesis – prospects for hybrid antibiotics, *Annu. Rev. Microbiol.* **47**, 875–912.

KE, H. M. (1992), Similarities and differences between human cyclophilin A and other beta-barrel structures. Structural refinement at 1.63 Å resolution, *J. Mol. Biol.* **228**, 539–550.

KE, H. M., ZYDOWSKY, L. D., LIU, J. WALSH, C. T. (1991), Crystal structure of recombinant human T-cell cyclophilin A at 2.5 Å resolution, *Proc. Natl. Acad. Sci. USA* **88**, 9483–9487.

KE, H. M., MAYROSE, D., CAO, W. (1993a), Crystal structure of cyclophilin A complexed with substrate Ala–Pro suggests a solvent-assisted mechanism of *cis-trans* isomerization, *Proc. Natl. Acad. Sci. USA* **90**, 3324–3328.

KE, H. M., ZHAO, Y., LUO, F., WEISSMAN, I., FRIEDMAN, J. (1993b), Crystal structure of murine cyclophilin C complexed with immunosuppressive drug cyclosporin A, *Proc. Natl. Acad. Sci. USA* **90**, 11850–11854.

KE, H. M., MAYROSE, D., BELSHWA, P. J., ALBERG, D. G., SCHREIBER, S. L., CHANG, Z., ETZKORN, F., HO, S., WALSH, C. T. (1994), Crystal structures of cyclophilin A complexed with cyclosporin A and N-methyl-[(E)-2-butenyl]-4,4-dimethylthreonine cyclosporin A, *Structure* **2**, 33–44.

KERN, D., DRAKENBERG, T., WIKSTROM, M., FORSEN, S., BANG, H., FISCHER, G. (1993), The *cis/trans* interconversion of the calcium regulating hormone calcitonin is catalyzed by cyclophilin, *FEBS Lett.* **323**, 198–202.

KIEFFER, L. J., THALHAMMER, T. HANDSCHUMACHER, R. E. (1992), Isolation and characterization of a 40-kDa cyclophilin-related protein, *J. Biol. Chem.* **267**, 5503–5507.

KIEFFER, L. J., SENG, T. W., LI, W., OSTERMAN, D. G., HANDSCHUMACHER, R. E., BAYNEY, R. M. (1993), Cyclophilin-40, a protein with homology to the P59 component of the steroid receptor complex. Cloning of the cDNA and further characterization, *J. Biol. Chem.* **268**, 12303–12310.

KINO, T., HATANAKA, H., HASHIMOTO, M., NISHIYAMA, M., GOTO, T., OKUHARA, M., KOHSAKA, M., AOKI, H., IMANAKA, H. (1987), FK-506, A novel immunosuppressant isolated from a *Streptomyces*. I. Fermentation, isolation, and physico-chemical and biological characteristics, *J. Antibiot.* **40**, 1249–1255.

KLEINKAUF, H., VON DÖHREN, H. (1987), Biosynthesis of peptide antibiotics, *Annu. Rev. Microbiol.* **41**, 259–289.

KLEINKAUF, H., VON DÖHREN, H. (1990), Nonribosomal biosynthesis of peptide antibiotics, *Eur. J. Biochem.* **192**, 1–15.

KO, S. Y., DALVIT, C. (1992), Conformation of cyclosporin A in polar solvents, *Int. J. Pept. Protein Res.* **40**, 380–382.

KOBEL, H., TRABER, R. (1982), Directed biosynthesis of cyclosporins, *Eur. J. Appl. Microbiol. Biotechnol.* **14**, 237–240.

KOBEL, H., LOOSLI, H. R., VOGES, R. (1983), Contribution to knowledge of the biosynthesis of cyclosporin A, *Experientia* **39**, 873–876.

KOECK, M., KESSLER, H., SEEBACH, D., THALER, A. (1992), Novel backbone conformation of cyclosporin A: The complex with lithium chloride, *J. Am. Chem. Soc.* **114**, 2676–2686.

KOFRON, J. L., KUZMIC, P., KISHORE, V., COLON BONILLA, E., RICH, D. H. (1991), Determination of kinetic constants for peptidyl prolyl *cis-trans* isomerases by an improved spectrophotometric assay [published erratum appears in *Biochemistry* 1991 Nov 5; **30** (44): 10818], *Biochemistry* **30**, 6127–6134.

KRAETZSCHMAR, J., KRAUSE, M., MARAHIEL, M. A. (1989), Gramicidin S biosynthesis operon containing the structural genes *grsA* and *grsB* has an open reading frame encoding a protein homologous to fatty acid thioesterase, *J. Bacteriol.* **171**, 5422–5429.

KRONKE, M., LEONARD, W. J., DEPPER, J. M., ARYA, F., WONG-STAAL, F., GALLO, R. C., WALDMANN, T. A., GREENE, W. C. (1984), Cyclosporin A inhibits T-cell growth factor gene expression at the level of mRNA transcription, *Proc. Natl. Acad. Sci. USA* **81**, 5214–5218.

LALAND, S. G., ZIMMER, T.-L. (1973), The protein thiotemplate mechanism of synthesis for the peptide antibiotics produced by *Bacillus brevis*, *Essays Biochem.* **9**, 31–42.

LANDSCHULTZ, W. H., JOHNSON, P. F., MCKNIGHT, S. L. (1989), The DNA binding domain of the rat liver nuclear protein C/EBP is bipartite, *Science* **243**, 1681–1688.

LAWEN, A., TRABER, R. (1993), Substrate specificities of cyclosporin synthetase and peptolide SDZ 214-103 synthetase. Comparison of the substrate specificities of the related multifunctional polypeptides, *J. Biol. Chem.* **268**, 20452–20465.

LAWEN, A., ZOCHER, R. (1990), Cyclosporin synthetase. The most complex peptide synthesizing multienzyme polypeptide so far described, *J. Antibiot.* **42**, 1283–1289.

LAWEN, A., TRABER, R., GEYL, D., ZOCHER, R., KLEINKAUF, H. (1989), Cell-free biosynthesis of new cyclosporins, *J. Antibiot.* **42**, 1283–1289.

LAWEN, A., TRABER, R., GEYL, D. (1991), *In vitro* biosynthesis of [Thr^2,Leu^5,D-Hiv^8,Leu^{10}]-cyclosporin A, a cyclosporin-related peptolide, with immunosuppressive activity by a multienzyme polypeptide, *J. Biol. Chem.* **266**, 15567–15570.

LAWEN, A., DITTMANN, J., SCHMIDT, B., RIESNER, D., KLEINKAUF, H. (1992), Enzymatic biosynthesis of cyclosporin A and analogues, *Biochimie* **74**, 511–516.

LAWEN, A., TRABER, R., REUILLE, R., PONELLE, M. (1994), *In vitro* biosynthesis of ring-extended cyclosporins, *Biochem. J.* **300**, 395–399.

LEE, J. P., DUNLAP, B. RICH, D. H. (1990), Synthesis and immunosuppressive activities of conformationally restricted cyclosporin lactam analogues, *Int. J. Pept. Protein Res.* **35**, 481–494.

LEITNER, E., WEBER, G. (1994), Disruption of the cyclosporin synthetase gene of *Tolypocladium niveum*, *Curr. Genet.* **26**, 461–467.

LEITNER, E., SCHNEIDER-SCHERZER, E., SCHOERGENDORFER, K., WEBER, G. (1994), Cyclosporin synthetase, *European Patent Application* 578616.

LEIVA, M. C., LYTTLE, C. R. (1992), Leukocyte chemotactic activity of FKBP and inhibition by FK506, *Biochem. Biophys. Res. Commun.* **186**, 1178–1183.

LIU, Y., LINSLEY, P. S. (1992), Costimulation of T-cell growth, *Curr. Opin. Immunol.* **4**, 265–270.

LIU, J., ALBERS, M. W., CHEN, C., SCHREIBER, S. L., WALSH, C. T. (1990), Cloning, expression, and purification of human cyclophilin in *Escherichia coli* and assessment of the catalytic role of cysteines by site-directed mutagenesis, *Proc. Natl. Acad. Sci. USA* **87**, 2304–2308.

LIU, J., FARMER, J. D. JR., LANE, W. S., FRIEDMAN, J., WEISSMAN, I., SCHREIBER, S. L. (1991), Calcineurin is a common target of cyclophilin–cyclosporin A and FKBP–FK506 complexes, *Cell* **66**, 807–815.

LOOSLI, H. R., KESSLER, H., OSCHKINAT, H., WEBER, H. P., PETCHER, T. (1985), The conformation of cyclosporin A in the crystal and in solution, *Helv. Chim. Acta* **68**, 682–704.

LUBAN, J., BOSSOLT, K. L., FRANKE, E. K., KALPANA, G. V., GOFF, S. P. (1993), Human immunodeficiency virus type 1 Gag protein binds to cyclophilins A and B, *Cell* **73**, 1067–1078.

LYNEN, F. (1980), On the structure of fatty acid synthase of yeast, *Eur. J. Biochem.* **112**, 431–442.

MACCABE, A. P., VAN LIEMPT, H., PALISSA, H., RIACH, M. B. R., PFEIFER, E., VON DÖHREN, H., KINGHORN, J. R. (1991), δ-(L-α-aminoadipyl)-L-cysteinyl-D-valine synthetase from *Aspergillus nidulans*, *J. Biol. Chem.* **266**, 12646–12954.

MARTEL, R. R., KLICIUS, J., GALET, S. (1977), Inhibition of the immune response by rapamycin, a new antifungal antibiotic, *Can. J. Physiol. Pharmacol.* **55**, 48–51.

MASON, J. (1989), Pharmacology of cyclosporine (Sandimmune) VII. Pathophysiology and toxicology of cyclosporine in humans and animals, *Pharmacol. Rev.* **42**, 423–434.

MCCAFFREY, P. G., PERRINO, B. A., SODERLING, T. R., RAO, A. (1993), NF-ATp, a T lymphocyte DNA-binding protein that is a target for calcineurin and immunosuppressive drugs, *J. Biol. Chem.* **268**, 3747–3752.

MIKOL, V., KALLEN, J., PFLUEGL, G., WALKINSHAW, M. D. (1993), X-ray structure of a monomeric cyclophilin A–cyclosporin A crystal complex at 2.1 Å resolution, *J. Mol. Biol.* **234**, 1119–1130.

MIKOL, V., KALLEN, J., WALKINSHAW, M. D. (1994A), X-ray structure of a cyclophilin B/cyclosporin complex: comparison with cyclophilin A and delineation of its calcineurin binding domain. *Proc. Natl. Acad. Sci. USA* **91**, 5183–5186.

MIKOL, V., KALLEN, J., WALKINSHAW, M. D. (1994b), The X-ray structure of $(MeBm_2t)^1$-cyclosporin complexed with cyclophilin A provides an explanation for its anomalously high immunosuppressive activity, *Protein Eng.* **7** (5), 597–603.

MORRIS, R. E. (1993), New small molecule immunosuppressants for transplantation: review of essential concepts, *J. Heart Lung Transplant.* **12**, S275–S286.

NERI, P., GEMMECKER, G., ZYDOWSKY, L. D., WALSH, C. T., FESIK, S. W. (1991), NMR studies of [U-(13)C]cyclosporin A bound to human cyclophilin B, *FEBS Lett.* **290**, 195–199.

O'HAGAN, D. (1991), *The Polyketide Metabolites.* (MELLOR, J., Ed.). Chichester: Ellis Horwood Series in Organic Chemistry.

O'KEEFE, S. J., TAMURA, J., KINCAID, R. L., TOCCI, M. J., O'NEILL, E. A. (1992), FK-506- and CsA-sensitive activation of the interleukin-2 promoter by calcineurin, *Nature* **357**, 692–694.

OFFENZELLER, M., SU, Z., SANTER, G., MOSER, H., TRABER, R., MEMMERT, K., SCHNEIDER-SCHERZER, E. (1993), Biosynthesis of the unusual amino acid (4R)-4-[(E)-2-butenyl]-4-methyl-L-threonine of cyclosporin A. Identification of 3(R)-hydroxy-4(R)-methyl-6(E)-octenoic acid as a key intermediate by enzymatic *in vitro* synthesis and by *in vivo* labeling techniques, *J. Biol. Chem.* **268**, 26127–26134.

PAPAGEORGIOU, C., FLORINETH, A., FRENCH, R. (1994), Calcineurin has a very tight-binding pocket for the side chain of residue 4 of cyclosporin, *Bioorg. Med. Chem. Lett.* **4**, 267–272.

PARK, S. B., MEIER, G. P. (1989), A semisynthetic approach to olefinic analogues of amino acid one (MeBmt) in cyclosporin A, *Tetrahedron Lett.* **30**, 4215–4218.

PARK, S. T., ALDAPE, R. A., FUTER, O., DECENZO, M. T., LIVINGSTON, D. J. (1992), PPIase catalysis by human FK506-binding protein proceeds through a conformational twist mechanism, *J. Biol. Chem.* **267**, 3316–3324.

PATCHETT, A. A., TAUB, D., HENSENS, O. D., GOEGELMAN, R. T., YANG, L. H., DUMONT, F. J., PETERSON, L., SIGAL, N. H. (1992), Analogues of cyclosporin A modified at the D-Ala8 position, *J. Antibiot.* **45**, 94–102.

PFLUEGL, G., KALLEN, J., SCHIRMER, T., JANSONIUS, J. N., ZURINI, M. G., WALKINSHAW, M. D. (1993), X-ray structure of a decameric cyclophilin–cyclosporin crystal complex, *Nature* **361**, 91–94.

PRICE, E. R., ZYDOWSKY, L. D., JIN, M. J., BAKER, C. H., MCKEON, F. D., WALSH, C. T. (1991), Human cyclophilin B: A second cyclophilin gene encodes a peptidyl-prolyl isomerase with a signal sequence, *Proc. Natl. Acad. Sci. USA* **88**, 1903–1907.

PRICE, D. J., GROVE, J. R., CALVO, V., AVRUCH, J., BIERER, B. E. (1992), Rapamycin-induced inhibition of the 70-kilodalton S6 proteinkinase, *Science* **257**, 973–977.

QUESNIAUX, V. F. J., SCHREIER, M. H., WENGER, R. M., HIESTAND, P. C., HARDING, M. W., VAN REGENMORTEL, M. H. V. (1987), Cyclophilin binds to the region of cyclosporine involved in its immunosuppressive activity, *Eur. J. Immunol.* **17**, 1359–1365.

QUESNIAUX, V. F. J., SCHREIER, M. H., WENGER, R. M., HIESTAND, P. C., HARDING, M. W., VAN REGENMORTEL, M. H. V. (1988), Molecular characteristics of cyclophilin–cyclosporine interaction, *Transplantation* **46**, 23–28.

QUESNIAUX, V. F. J., WEHRLI, S., WIOLAND, C., SCHULER, W., SCHREIER, M. H. (1994), Effects of rapamycin on hematopoiesis, *Transplant. Proc.* **26**, 3135–3140.

RICH, D. H., SUN, C. Q., GUILLAUME, D., DUNLAP, B., EVANS, D. A., WEBER, A. E. (1989), Synthesis, biological activity, and conformational analysis of (2S, 3R, 4S)-MeBmt[1]-cyclosporin, a novel 1-position epimer of cyclosporin A, *J. Med. Chem.* **32**, 1982–1987.

ROBINSON, A. (1991), Polyketide synthase complexes: Their structure and function in antibiotic biosynthesis, *Philos. Trans. R. Soc. London B* **332**, 107–114.

ROSEN, M. K., STANDAERT, R. F., GALAT, A., NAKATSUKA, M., SCHREIBER, S. L. (1990), Inhibition of FKBP rotamase activity by immunosuppressant FK506: twisted amide surrogate, *Science* **248**, 863–866.

ROSENWIRTH, B., BILLICH, A., DATEMA, R., DONATSCH, P., HAMMERSCHMID, F., HARRISON, R. K., HIESTAND, P. C., JAKSCHE, H., MAYER, P., PEICHL, P., QUESNIAUX, V. F. J., SCHATZ, F., WENGER, R. M., WOLFF, B., ZENKE, G., ZURINI, M. G. (1994), Inhibition of HIV-1 replication by SDZ NIM 811, a non-immunosuppressive cyclosporin A analogue. *Antimicrob. Agents Chemother.* **38**, 1763–1772.

RYFFEL, B. (1989), Pharmacology of cyclosporine. VI. Cellular activation: Regulation of intracellular events by cyclosporine, *Pharmacol. Rev.* **41**, 407–423.

RYFFEL, B., GOETZ, U., HEUBERGER, B. (1982), Cyclosporin receptors on human lymphocytes, *J. Immunol.* **129**, 1978–1982.

SANGLIER, J. J., TRABER, R., BUCK, R. H., HOFMANN, H., KOBEL, H. (1990), Isolation of (4R)-4-[(E)-2-butenyl]-4-methyl-L-threonine, the characteristic structure element of cyclosporins, from a blocked mutant of *Tolypocladium inflatum*, *J. Antibiot.* **43**, 707–714.

SCHLUMBOHM, W., STEIN, T., ULLRICH, C., VATER, J., MARAHIEL, M. A., KRUFT, V., WITTMANN-LIEBOLD, B. (1991), An active serine is involved in covalent substrate amino acid binding at each reaction center of gramicidin S synthetase, *J. Biol. Chem.* **266**, 23135–23141.

SCHMIDT, B., RIESNER, D., LAWEN, A., KLEINKAUF, H. (1992), Cyclosporin synthetase is a 1.4 MDa multienzyme polypeptide. Re-evaluation of the molecular mass of various peptide synthetases, *FEBS Lett.* **307**, 355–360.

SCHNEIDER, H., CHARARA, N., SCHMITZ, R., WEHRLI, S., MIKOL, V., ZURINI, M. G., QUESNIAUX, V. F. J., MOVVA, N. R. (1994), Human cyclophilin C: Primary structure, tissue distribution, and determination of binding specificity for cyclosporins. *Biochemistry* **33**, 8218–8224.

SCHOENBRUNNER, E. R., MAYER, S., TROPSCHUG, M., FISCHER, G., TAKAHASHI, N., SCHMID, F. X. (1991), Catalysis of protein folding by cyclophilins from different species, *J. Biol. Chem.* **266**, 3630–3635.

SCHREIBER, S. L. (1991), Chemistry and biology of the immunophilins and their immunosuppressive ligands, *Science* **251**, 283–287.

SCHREIBER, S. L., CRABTREE, G. R. (1992), The mechanism of action of cyclosporin A and FK506, *Immunol. Today* **13**, 136–142.

SCOTT-CRAIG, J. S., PANACCIONE, D. G., POCARD, J.-A., WALTON, J. D. (1992), A large, multifunctional cyclic peptide synthetase catalyzing HC-toxin production in the plant pathogenic fungus *Cochliobolus carbonum* encoded by a 15.7-kb open reading frame, *J. Biol. Chem.* **267**, 26044–26049.

SEEBACH, D., KO, S. Y., KESSLER, H., KOECK, M., REGGELIN, M., SCHMIEDER, P., WALKINSHAW, M. D., BOELSTERLI, H. J., BEVEC, D. (1991), Thiocyclosporins: Preparation, solution and crystal structure, and immunosuppressive activity, *Helv. Chim. Acta* **74**, 1953–1990.

SEEBACH, D., BECK, A. K., BOSSLER, H. G., GERBER, C., KO, S. Y., MURTIASHAW, C. W., NAEF, R., SHODA, S., THALER, A., WENGER, R. M. (1993), Modification of cyclosporine-A (CS) – Generation of an enolate at the sarcosine residue and reactions with electrophiles, *Helv. Chim. Acta* **76**, 1564–1590.

SEFTON, B. M., CAMPBELL, M. A. (1991), The role of tyrosine protein phosphorylation in lymphocyte activation, *Annu. Rev. Cell Biol.* **7**, 257–274.

SENN, H., WEBER, C., KOBEL, H., TRABER, R. (1991), Selective ^{13}C-labelling of cyclosporin A, *Eur. J. Biochem.* **199**, 653–658.

SHERRY, B., YARLETT, N., STRUPP, A., CERAMI, A. (1992), Identification of cyclophilin as a proinflammatory secretory product of lipopolysaccharide-activated macrophages, *Proc. Natl. Acad. Sci. USA* **89**, 3511–3515.

SIEKIERKA, J. J., HUNG, S. H., POE, M., LIN, C. S., SIGAL, N. H. (1989a), A cytosolic binding protein for the immunosuppressant FK506 has peptidyl-prolyl isomerase activity but is distinct from cyclophilin, *Nature* **341**, 755–757.

SIEKIERKA, J. J., STARUCH, M. J., HUNG, S. H. Y., SIGAL, N. H. (1989b), FK-506, a potent novel immunosuppressive agent, binds to a cytosolic protein which is distinct from the cyclosporin A-binding protein, cyclophilin, *J. Immunol.* **143**, 1580–1583.

SIGAL, N. H., DUMONT, F. J., DURETTE, P. L., SIEKIERKA, J. J., PETERSON, L., RICH, D. H., DUNLAP, B. E., STARUCH, M. J., MELINO, M. R., KOPRAK, S. L. ET AL. (1991), Is cyclophilin involved in the immunosuppressive and nephrotoxic mechanism of action of cyclosporin A? *J. Exp. Med.* **173**, 619–628.

SIMPSON, T. J. (1989), Biological chemistry. Part (II) biosynthesis, *Annu. Rep. Prog. Chem. Sect. B* **85**, 321–351.

SMITH, D. J., EARL, A. J., TURNER, G. (1990), The multifunctional peptide synthetase performing the first step of penicillin biosynthesis in *Penicillium chrysogenum* is a 421073 Dalton protein similar to *Bacillus brevis* peptide antibiotic synthetases, *EMBO J.* **9**, 2743–2750.

SODA, K., TANAKA, H., TANIZAWA, K. (1986), Pyridoxyl phosphate enzymes catalyzing racemization, in: *Vitamin B6, Pyridoxyl Phosphate: Chemical, Biochemical and Medical Aspects. Coenzymes and Cofactors* (DOLPHIN, D., POULSON, R., AVRAMOVIC, O., Eds.), pp. 223–251. New York: Wiley & Sons.

SPITZFADEN, C., WEBER, H. P., BRAUN, W., KALLEN, J., WIDER, G., WIDMER, H., WALKINSHAW, M. D., WUTHRICH, K. (1992), Cyclosporin A–cyclophilin complex formation. A model based on X-ray and NMR data, *FEBS Lett.* **300**, 291–300.

STABEN, C., JENSEN, P. B., SINGER, M., POLLOCK, J., SCHECHTMAN, M., KINSEY, J., SELKER, E. (1989), Use of a bacterial hygromycin resistance gene as a dominant selectable marker in *Neurospora crassa* transformation, *Fung. Genet. Newslett.* **36**, 79–81.

STAMNES, M. A., RUTHERFORD, S. L., ZUKER, C. S. (1992), Cyclophilins: A new family of proteins involved in intracellular folding, *Trends Cell Biol.* **2**, 272–276.

STEVENSON, M., STANWICK, T. L., DEMPSEY, M. P., LAMONICA, C. A. (1990), HIV-1 replication is controlled at the level of T cell activation and proviral integration, *EMBO J.* **9**, 1551–1560.

SWANSON, S. K., BORN, T., ZYDOWSKY, L. D., CHO, H., CHANG, H. Y., WALSH, C. T. RUSNAK, F. (1992), Cyclosporin-mediated inhibition of bovine calcineurin by cyclophilins A and B, *Proc. Natl. Acad. Sci. USA* **89**, 3741–3745.

TAI, P., ALBERS, M. W., CHANG, H., FABER, L. E., SCHREIBER, S. L. (1992), Association of a 59-kilodalton immunophilin with the glucocorticoid receptor complex, *Science* **256**, 1315–1318.

TAKAHASHI, N., HAYANO, T., SUZUKI, M. (1989), Peptidyl-prolyl *cis-trans* isomerase is the cyclosporin A-binding protein cyclophilin, *Nature* **337**, 473–475.

TAKESHITA, T., ASAO, H., OHTANI, K., ISHII, N., KUMAKI, S., TANAKA, N., MUNAKATA, H., NAKAMURA, M., SUGAMURA, K. (1992), Cloning of the gamma chain of the human IL-2 receptor, *Science* **257**, 379–382.

THERIAULT, Y., LOGAN, T. M., MEADOWS, R., YU, L., OLEJNICZAK, E. T., HOLZMAN, T. F., SIMMER, R. L., FESIK, S. W. (1993), Solution structure of the cyclosporin A/cyclophilin complex by NMR, *Nature* **361**, 88–91.

THOMMEN-SCOTT, K. (1981), Antimalarial activity of cyclosporin A, *Agents Actions* **11**, 770–773.

TOCCI, M. J., MATKOVICH, D. A., COLLIER, K. A., KWOK, P., DUMONT, F. J., LIN, S., DEGUDICIBUS, S., SIEKIERKA, J. J., CHIN, J., HUTCHINSON, N. I. (1989), The immunosuppressant FK506 selectively inhibits expression of early T cell activation genes, *J. Immunol.* **143**, 718–726.

TRABER, R., HOFMANN, H., LOOSLI, H. R., PONELLE, M., VON WARTBURG, A. (1987), Neue Cyclosporine aus *Tolypocladium inflatum*. Die Cyclosporine K–Z, *Helv. Chim. Acta* **70**, 13–36.

TRABER, R., HOFMANN, H., KOBEL, H. (1988), Cyclosporins – new analogues by precursor directed biosynthesis, *J. Antibiot.* **42**, 591–597.

TRABER, R., KOBEL, H., LOOSLI, H. R., SENN, H., ROSENWIRTH, B., LAWEN, A. (1994), [MeIle[4]]cyclosporin, a novel natural cyclosporin with anti-HIV activity: Structural elucidation, biosynthesis, and biological properties, *Antiviral Chem. Chemother.* **5**, 331–339.

TSANG, V., HODSON, M. E., YACOUB, M. H. (1992), Lung transplantation for cystic fibrosis, *Br. Med. Bull.* **48**, 949–971.

TURGAY, K., KRAUSE, M., MARAHIEL, M. A. (1992), Four homologous domains in the primary structure of GrsB are related to domains in a superfamily of adenylate-forming enzymes, *Mol. Microbiol.* **6**, 529–546.

TWENTYMAN, P. R. (1992), Cyclosporins as drug resistance modifiers, *Biochem. Pharmacol.* **43**, 109–117.

VOLLENBROICH, D., KLUGE, B., D'SOUZA, C., ZUBER, P., VATER, J. (1993), Analysis of a mutant amino acid-activating domain of surfactin synthetase bearing a serine-to-alanine substitution at the site of carboxyl thioester formation, *FEBS Lett.* **325**, 220–224.

VON WARTBURG, A., TRABER, R. (1988), Cyclosporins. Fungal metabolites with immunosuppressive activities, *Progr. Med. Chem.* **25** (ELLIS, G. P., WEST, G. B., Eds.), pp. 1–33. Amsterdam: Elsevier Science Publishers, B.V., Biomedical Division.

WALSH, C. T., BADET, R., DAUB, E., ESAKI, N., GALAKATOS, N. (1985), Bacterial alanine racemases: Targets for antibacterial agents, *Spec. Publ. – R. Soc. Chem.* **55** (SCI-RSC Med. Chem. Symp. 3rd), 193–209.

WEBER, C., WIDER, G., VON FREYBERG, B., TRABER, R., BRAUN, W., WIDMER, H., WUTHRICH, K. (1991), The NMR structure of cyclosporin A bound to cyclophilin in aqueous solution, *Biochemistry* **30**, 6563–6574.

WEBER, G., SCHOERGENDORFER, K., SCHNEIDER-SCHERZER, E., LEITNER, E. (1994), The peptide synthetase catalyzing cyclosporin production in *Tolypocladium niveum* is encoded by a giant 45.8-kilobase open reading frame, *Curr. Genet.* **26**, 120–125.

WECKERMANN, R., FUERBASS, R., MARAHIEL, M. A. (1988), Complete nucleotide sequence of the *tycA* gene encoding the tyrocidine synthetase 1 from *Bacillus brevis*, *Nucleic Acids Res.* **16**, 11841.

WENGER, R. M. (1983), Synthesis of cyclosporine, *Helv. Chim. Acta* **67**, 502–525.

WENGER, R. M. (1989), Pharmacology of cyclosporin (Sandimmune). II. Chemistry, *Pharmacol. Rev.* **41**, 243–247.

WENGER, R. M. (1986), Cyclosporine and analogues: Structural requirements for immunosuppressive activity, *Transplant. Proc.* **18**, 213–218.

WENGER, R. M., PAYNE, T. G., SCHREIER, M. H. (1986), Cyclosporine: Chemistry, structure–activity relationships and mode of action, *Progr. Clin. Biochem. Med.* **3**, 157–191.

WENGER, R. M., FRANCE, J., BOVERMANN, G., WALLISER, L., WIDMER, A., WIDMER, H. (1994), The 3D structure of a cyclosporin analogue in water is nearly identical to the cyclophilin-bound conformation, *FEBS Lett.* **340**, 255–259.

WITZEL, B. (1989), New cyclosporin analogues with modified "C-9 amino acids", *US Patent* no. 4798823.

XU, Q., LEIVA, M. C., FISCHKOFF, S. A., HANDSCHUMACHER, R. E., LYTTLE, C. R. (1992), Leukocyte chemotactic activity of cyclophilin, *J. Biol. Chem.* **267**, 11968–11971.

ZACK, J. A., ARRIGO, S. J., WEITSMAN, S. R., GO, A. S., HAISLIP, A., CHEN, I. S. Y. (1990), HIV-entry into quiescent primary lymphocytes: molecular analysis reveals a labile, latent viral structure, *Cell* **61**, 213–222.

ZENKE, G., BAUMANN, G., WENGER, R. M., HIESTAND, P. C., QUESNIAUX, V. F. J., ANDERSEN, E., SCHREIER, M. H. (1993), Molecular mechanisms of immunosuppression by cyclosporins, *Ann. N. Y. Acad. Sci.* **685**, 330–335.

ZOCHER, R., NIHIRA, T., PAUL, E., MADRY, N., PEETERS, H., KLEINKAUF, H. (1986), Biosynthesis of cyclosporin A: Partial purification and properties of a multifunctional enzyme from *Tolypocladium inflatum*, *Biochemistry* **25**, 550–553.

13 Secondary Products from Plant Cell Cultures

JOCHEN BERLIN

Braunschweig, Federal Republic of Germany

1 Introduction

Higher plants produce a great variety of secondary metabolites, some of which are an indispensable source of commercially important compounds. These include pharmaceuticals (such as steroids, alkaloids, and glucosides), natural flavors, fragrances, dyes, and gums (such as natural rubber). As only a small portion of plants have as yet been analyzed for secondary products, it is not surprising that isolation and characterization of new compounds continue unabatedly. The array of assays for detecting biological activities of natural compounds has been enlarged and simplified, and has been paralleled by an increase in assay sensitivity. Thus, the screening of plant extracts for new biologically active compounds still seems to be a promising approach today (BALANDRIN et al., 1985).

However, it should also be mentioned that the screening of some 40000 plants by the NCI (National Cancer Institute, USA) yielded only three really novel anticancer compounds (vincristine, vinblastine, and taxol). It is doubtful whether this result helps to attract more industrial investments to this area of plant research. Five *Rauwolfia* alkaloids with tranquilizing and antihypertensive activities and the three anticancer substances were the only plant-derived substances with therapeutic efficacy and utitility, proven beyond doubt (TYLER, 1988), which have been introduced to the American market during the last 40 years. All other plant-derived compounds used as drugs today have been known for a much longer time and their production is well established. Thus, a general concern of pharmacognosists is how to make plant analyses more efficient with respect to drug development (TYLER, 1988).

If the plant product under consideration cannot be synthesized chemically it has to be isolated from wild or field-grown plants. Consequently, plants synthesizing products in high demand are usually grown in large-scale plantations. Since the establishment of cell cultures of any plant species is nowadays routine in most cases, plant cell biomass can also be produced in huge fermentors. Since in addition, reasonable amounts of secondary metabolites have been produced from some plant cell cultures it seems obvious to suggest that commercially interesting compounds might be produced in a factory type production in large bioreactors, such as those used in the production of microbial drugs. The advocates of the new technology have greatly overestimated the problems with agriculturally grown plant material while the biological, technical, legal, and financial problems of tissue culture technology have been generally underestimated. Interestingly, advocates of the tissue culture technology have come mainly from universities and public research centers, while in industry (perhaps with the exception of some Japanese industries) the interest was in general very meager worldwide.

Thus, one can name immediately two reasons why the tissue culture technology was in a difficult position from the beginning: Firstly, the lack of new plant compounds with very special, extraordinary pharmacological characteristics for which it would be worth-while developing novel tissue culture processes; secondly, the difficulty of replacing an established and legally approved production process by a new technique. It is indeed difficult to replace a conventional, approved production process based on the extraction of field-grown plants by a plant tissue culture production process. This could only happen if the calculation of costs showed an extreme advantage for tissue cultures, and if the procedure of obtaining a new legal approval required for a product from a new source is not too time-consuming and expensive. If a new product is under consideration, the superiority of the new drug must be clear and evident to justify the long and expensive procedures needed nowadays for introducing a new pharmaceutical to the market. In view of these two obstacles we have to leave it to industry and biotechnological companies to decide whether they consider production by tissue cultures as an attractive alternative. For each product under consideration the decision may be quite different from company to company and from country to country.

Updates of almost all the products found in plant tissue cultures were published by ELLIS (1988), CONSTABEL and VASIL (1988), and

BANTHORPE (1994). However, for many of the compounds mentioned in these reviews it is not readily clear whether they are of any biotechnological importance. Therefore, I will concentrate on some groups of compounds about which sufficient information is available in order to evaluate the biotechnological impact of plant cell cultures as producers. The main objective of my previous review (BERLIN, 1986) was to describe the experimental methods by which high yielding cultures were obtained and to demonstrate that the same techniques that could be applied to some pathways failed for others. Thus the aim was to provide the reader with a better understanding of what is possible today and what might become possible tomorrow.

In this updated review I will expand on this aspect, explaining why certain compounds are easily produced in cultured cells while other metabolites have proved recalcitrant to all known techniques for improved production. As I will describe mainly the same groups of compounds as in 1986 in the first edition of "*Biotechnology*", the extent of progress with respect to production levels during the last 10 years will become apparent. Some compounds have been replaced by others which are presently of more biotechnological interest; the new developments are included in this review. The possibility of altering production characteristics of pathways by genetic engineering opened a new area. This will be analyzed critically to whether it will improve the situation of tissue cultures as producers of commercially interesting compounds.

2 Some General Conclusions and Suggestions Based on the Present Biotechnological Impact of Plant Cell Cultures as Producers

PETERSEN and ALFERMANN (1993) have already described special features of plant cell cultures and the various culture types and techniques in great detail in Vol. 1 of this multi-volume comprehensive treatise. Due to their recent review and those of other (PARR, 1989; CHARLWOOD and RHODES, 1990; PAYNE et al., 1991; BUITELAAR et al., 1992; BANTHORPE, 1994; MISAWA, 1994) it is not necessary to repeat here the general aspects and specific characteristics of plant cell cultures and plant secondary metabolism outlined in my previous article (BERLIN, 1986). However, before reviewing individual culture systems and documenting the state of the art, I would like to make some general statements which may be regarded as suggestions for future research programs and which gain support from the analyses of the individual systems.

As indicated in Sect. 1, plant tissue cultures are presently not well accepted as a source of commercially interesting products. Indeed, the impact of tissue culture technology has not increased, but rather decreased despite all efforts and scientific progress. Some think that this technique is a futile approach; others believe that the scientific breakthrough has not yet been achieved for a true evaluation of the potential of plant cell cultures as producers. This prevents a larger engagement of companies. Thus, a critical review should analyze in which areas research efforts must be intensified and identify which approaches can be reduced today. The analysis of the field after 20 years of rather intensive research allows indeed some clear conclusions to be made about promising directions and about what should no longer be tried.

Today it is possible to analyze the scientific usefulness of systems for improving the general standing of plant tissue cultures as tools of biotechnology. The term "scientific usefulness" means that biotechnologically relevant studies must not be restricted to cultures of plant species synthesizing commercially important compounds. Meaningful studies on model systems include studies on the regulation and expression of metabolic pathways and the possibilities of their manipulation. These are presently at least as important for the future of this field as working on so-called commercially attractive pathways. The seemingly most attractive pathways yielding commercially useful products can presently not be expressed very well in morphologically undifferentiated cells, and thus such cultures are not very suitable for biochemical or molecular studies.

In the past the efforts of most tissue culture groups were aimed at establishing highly productive plant cell cultures by conventional techniques. The product level was the most important goal. These studies included the initiation of many individual cultures from different explants of one or more plants, of one or more species or varieties. The cultures were grown on media with different phytohormone compositions. Screening or selection, as well as trials for enhancing productivity by media variation or by the use of inducer compounds have been part of a good program for optimizing a culture system. The application of the same techniques led in some cases to product levels in the range of $g\,L^{-1}$ within a few days, while in other cases amounts from zero to a few $\mu g\,L^{-1}$ were produced within a month. Some products were generally found at high levels while others were always found at low levels or even lacking in a culture. These findings were confirmed by laboratories from all over the world. Indeed, most tissue cultures of one plant species, after full adaptation to the culture conditions, have very similar production characteristics, independent of the laboratory where they were established. The extent of expression of a pathway is usually not restricted to one plant species. If, for example, a pathway is highly repressed in the cultures of one plant species, it is most likely that cultures of other plant species containing the same or related pathways will also be poor producers. If there are clear indications that a certain pathway is poorly expressed in cultured cells, it is unlikely that one can change this through the analyses of many more individually established cultures from various explants, by the application of screening and selection, by media variation, or by the use of elicitors (BERLIN, 1988). Thus, it does not make sense to waste time with such poorly producing cultures, unless new approaches can be applied. There are some reports in the literature claiming production characteristics which seem to contradict all other experiences. If this is a biotechnologically relevant finding, e.g., a true variant line, similar lines can easily be isolated by other researchers by employing a corresponding screening, selection, or an extended culture initiation program (BERLIN, 1988). If seemingly unique lines with surprisingly good or different production characteristics cannot be detected by other groups despite all efforts it is certainly a transient trait, not useful for any biotechnological purposes. Important and stable production improvements are usually confirmed by independent laboratories.

As pointed out, some pathways are spontaneously well expressed in cultured cells allowing product formation rates in the range of $g\,L^{-1}$ in a rather short culture period. A prerequisite for such biotechnologically relevant levels is that product formation occurs in rapidly growing cells, parallels growth, or is easily induced in a production medium. If a well expressed product is of commercial interest (shikonin, berberines), industry can substantially improve the production rates in shake flasks or small fermentors by optimizing all process parameters. If industry has taken over a process, it does not make much sense for research groups to continue with that culture system in shake flask systems or small bioreactors if the only aim is to demonstrate that another medium or elicitor, or an altered fermentation protocol improves productivity. Unfortunately, most products found at biotechnologically relevant levels in cultured cells are not of great commercial importance. Thus, it is even more questionable to spend too much time on further productivity improvements for such compounds.

Nowadays, the main question should be whether a previously established product expression level, or its manipulation by changing the culture conditions, allows meaningful biochemical and molecular studies. Without knowledge of the enzymes involved in the pathways, without recognizing the rate-limiting steps within complex and branched pathways, and without identifying the factors controlling the often organ-specific expression of pathways, it is unlikely that plant tissue culture based production processes (especially in view of the general obstacles discussed in Sect. 1) will more often be considered as an attractive alternative. From this point of view, all research which aims to elucidate specific or general features of pathway regulation with the ultimate goal of manipulating pathway expression is important in keeping the tissue culture technology alive. Thus, it is hoped that more and more biotechnologically orientated groups use their established tissue culture systems for such purposes during the next decade.

3 Secondary Product Formation in Suspension Cultures

Well-growing cell suspension cultures have been regarded for a long time as the only relevant system for biotechnological production processes. Most recently, rapidly growing hairy root cultures with their root-specific production characteristics have also been grown in larger bioreactors (CURTIS, 1993). There are also a few groups who believe that immobilized plant cells are better suited for plant cell production processes than freely suspended cells (TANAKA, 1994). However, only cell suspension cultures have been scaled up to volumes of industrial relevance. For example, suspension cultures of *Echinacea purpurea* have successfully been grown in bioreactors of up to 75 m^3 (WESTPHAL, 1990). Thus, without a doubt they remain the most attractive system from a technological point of view. The disadvantage of rapidly growing, morphologically undifferentiated callus cultures and cell suspensions is that in such a cell state only a small part of the biosynthetic potential is expressed. Nevertheless, some products are found in suspension cultures at levels which exceed those of the differentiated plant. The few pathways which are extremely well expressed in suspension cultures will be described first, followed by a few examples of pathways which yield reasonable product levels. For the latter group I have chosen compounds of at least some biotechnological interest and about which sufficient information has been gathered for an evaluation. Subsequently, I will give an overview of the seemingly commercially most important pathways which have remained recalcitrant to improvements despite all efforts. For information about pathways not mentioned here the reader is referred to 34 special reviews in the book of CONSTABEL and VASIL (1988) which covers nearly all groups of secondary compounds found in plant tissue cultures.

3.1 Products Accumulating at High or Good Levels in Suspension Cultures

3.1.1 Cinnamic Acid Derivatives

Esters and amides of cinnamic acids, mainly of caffeic acid, form a major, widespread group of phenylpropanoid metabolites in plants. This group of compounds accumulates spontaneously, often at high levels, in cell suspension cultures. Two groups independently reported the accumulation of high levels of rosmarinic acid in *Coleus blumei* (Fig. 1) (RAZZAQUE and ELLIS, 1977; ZENK et al., 1977a). The cultures were initiated on the widely used B5 medium with 2,4-D and kinetin as phytohormones. These cultures were maintained in suspension for several years without loosing their capacity for synthesizing and accumulating rosmarinic acid. RAZZAQUE and ELLIS (1977) reported the accumulation of 8–11% rosmarinic acid (corresponding to 1.2–1.5 $g\,L^{-1}$) on the B5 growth medium. The synthesis of rosmarinic acid can be stimulated by increasing the sucrose levels to 5–8%. Under these conditions, the content of rosmarinic acid increased up to 15% (ZENK et al., 1977a). The specific values exceed those of the various organs of the plants by a factor of 5. From their experiments in a volume of 25 mL ZENK's group calculated a yield of 3.6 $g\,L^{-1}$ rosmarinic acid within 13 days (0.3 $g\,L^{-1}\,d^{-1}$). However, their first attempts at scaling up rosmarinic acid production in a 30 L airlift bioreactor showed that shake flask experiments are not necessarily transferable to larger-scale fermentations

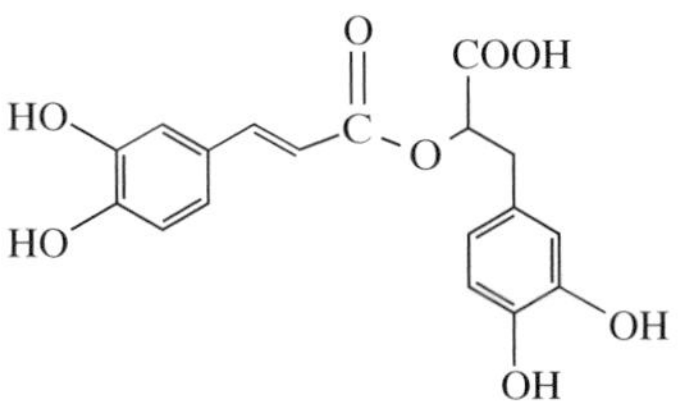

Rosmarinic acid
Coleus blumei
Anchusa officinalis

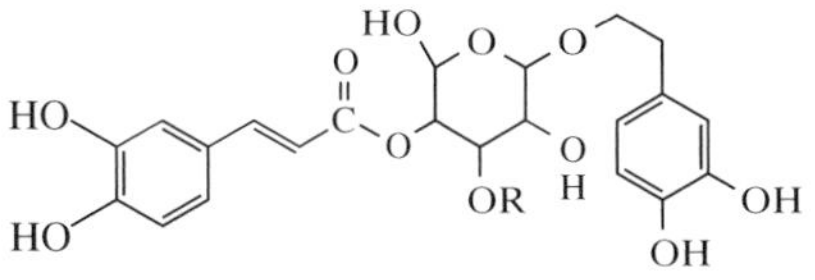

Verbascoside R = Rhamnose
Syringa vulgaris

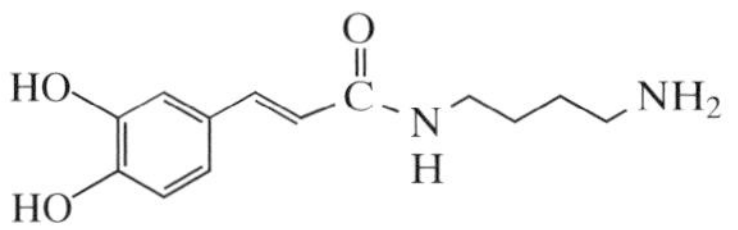

Caffeoyl putrescine
Nicotiana tabacum

Fig. 1. Caffeic acid derivatives accumulating at very high levels in cultured cells (see text).

since productivity dropped to 15% of the shake flask experiments. As rosmarinic acid has recently been shown to have good antiphlogistic activity, the Nattermann Company, Cologne, investigated this culture further. They established new lines of *C. blumei* which spontaneously synthesized rosmarinic acid with yields of 200 mg L^{-1} on the growth medium. This level could be enhanced 20-fold by transferring the cells to a simple production medium with an optimized concentration of sucrose as the main stimulator for maximum production (ULBRICH et al., 1985). As expected from the experience with microbial fermentations, these yields were further increased in 30 L airlift or stirred reactors up to 21% or 5.6 g L^{-1}. The output of rosmarinic acid was thus raised to 0.93 g L^{-1} d^{-1} (ULBRICH et al., 1985). This was at that time the highest amount of a defined secondary metabolite ever produced with plant cell cultures. This production level justifies a serious consideration of establishing a culture process as an alternative to harvesting from lower producing plants. However, the pharmacological efficacy of rosmarinic acid was evidently not high enough to encourage the Nattermann Company to further continue work on this compound. Indeed, the company has given up all previous activities in plant tissue cultures, and presently there seems to be no commercial interest in rosmarinic acid.

Returning to the discussion in Sect. 2, which of the points raised are applicable to rosmarinic acid? (1) Rosmarinic acid belongs to the group of compounds which is spontaneously accumulated at reasonably high to high levels in typical plant cell culture media. It is not only produced at high levels by *C. blumei* but also by several other plant species which are able to biosynthesize rosmarinic acid (WHITAKER et al., 1984). Cultures of *Anchusa officinalis* spontaneously accumulated 6% rosmarinic acid (DE-EKNAMKUL and ELLIS, 1984). (2) The production of rosmarinic acid can be greatly improved by very simple media variation (DE-EKNAMKUL and ELLIS, 1985a, b). (3) As with most spontaneous high producer lines production is rather stable or can easily be re-established by controlled culture conditions.

The fact that high levels of rosmarinic acid can be easily produced by suitable tissue cultures was well established in 1985. The hope for a commercial process was abandoned two or three years later. Thus, even higher production levels than those reported by ULBRICH et al. (1985) will not change the situation. A recent report that an optimized medium resulted in product levels of 6.4 g L^{-1} in a culture of *Salvia officinalis* (HIPPOLYTE et al., 1992) is thus only a strong confirmation of previous findings. In terms of biotechnological relevance, it is clear that the importance of rosmarinic acid for the field is not so much a question of product levels. More important is how this system is used to determine at the molecular level why this pathway is so well expressed and why, for example, sucrose sometimes acts as a strong inducer. Media variation experiments aimed elucidating the physiological or biochemical basis of observed production increases are of course also still important. Thus, it has recently been noted that the effect of sucrose depends on the carbohydrate level in the medium at the time when phosphate limitation occurs (GERTLOWSKI and PETERSEN, 1993). This observation might explain why not all rosmarinic acid cell cultures react to the same extent to increased sucrose supply. At least two groups have used cultures of *A. officinalis* and *C. blumei* for identifying all enzymes involved in rosmarinic acid biosynthesis (MIZUKAMI and ELLIS, 1991; PETERSEN et al., 1993) and some of these enzymes have been purified so that cloning of the corresponding genes could now proceed. Culture systems expressing a biosynthetic pathway quite well under all culture conditions may not be very suitable for the identification of regulatory factors controlling expression. Recently, it has been shown that rosmarinic acid and the enzymes involved in its biosynthesis can be induced in *Lithospermum erythrorhizon* and *Orthosiphon aristatus* by elicitors such as yeast and methyl jasmonate from nearly zero up to ca. 1.5% of dry mass (MIZUKAMI et al., 1993; SUMARYONO et al., 1991). Identification of the factors which allow spontaneous overproduction of rosmarinic acid in cultures of some plant species, and clarification of why other plant cells require inducer compounds will hopefully provide some useful hints for manipulating pathway controls.

Another caffeoyl derivative is verbascoside (acteoside) (Fig. 1). ELLIS (1983) established various lines of *Syringa vulgaris* on B5 medium, all of which are produced spontaneously high levels of up to 1.4 g L^{-1} verbascoside. Rapidly growing suspension cultures of *S. vulgaris* were found to contain a higher specific content of verbascoside (15%) than callus cultures (5–8%) (ELLIS 1983, 1985). Due to reports that verbascoside is a biologically active compound with antibacterial, antiviral, antihypertensive, and immunosuppressive properties, the interest in verbascoside-producing tissue cultures has increased, especially because plants generally contain only low amounts of this compound (see literature cited by INAGAKI et al., 1991). The initial observation of ELLIS that verbascoside belongs to the group of compounds whose production is favored under cell culture conditions has been confirmed. Cultures of *Hydrophila erecta* (HENRY et al., 1987) or *Leucosceptrum japonicum* (INAGAKI et al., 1991) yielded product levels in the range of 2 g L^{-1} without optimization. The analyses of cultures of several other plant species at the callus level indicate that many more highly effective systems for the production of verbascoside can be established (DELL et al., 1989; INAGAKI et al., 1991). However, there is currently no commercial interest in the production of these compounds by tissue cultures. Though the pathway is well expressed in cultured cells it has not yet been used for biochemical and regulatory studies.

The presence of various hydroxycinnamoyl putrescines (Fig. 1) in cultured cells of *Nicotiana tabacum* was first reported by MIZUSAKI et al. (1971). That these compounds have received some more attention during the last years is due to accidental finding. When selecting the *p*-fluorophenylalanine resistant cell line TX4 from a widely distributed XD (TX1) line (*N. tabacum* cv. Xanthi) PALMER and WIDHOLM (1975) noted that the levels of phenolics were increased manifold in TX4 cells. The phenolics were identified as hydroxycinnamoyl putrescines with caffeoyl putrescine as main component (Fig. 1) (BERLIN et al., 1982). TX1 cells accumulated between 0.6–1% hydroxycinnamoyl putrescines on the growth medium, while TX4 cells contained up to 10% of these compounds on a dry mass basis. This system has been used in two directions: optimization of product formation and comparison of the biochemical differences which lead to the different productivities. Growth limiting conditions (e.g., phosphate limitation) stimulated product formation (SCHIEL et al., 1984a). It is a frequently observed phenomenon that growth and secondary metabolism are countercurrent processes in cultured cells. Thus, even the formation of products accumulating at high levels on the growth medium can often be strongly enhanced by growth limiting conditions. The negative effect of accumulated phosphate on the synthesis of hydroxycinnamoyl putrescines suggested that the employment of fedbatch fermentation would give highest yields. The productivity of the high producing variant TX4 was indeed increased to 1.5 g L^{-1} by a

70 L fed-batch fermentation with phosphate as a limiting nutrient (SCHIEL et al., 1984b). Shake flask and batch fermentation of TX4 cells usually yielded 0.8–1.2 g L^{-1} (BERLIN et al., 1982). The yield of TX1 cells was increased by the fedbatch techniques from 160–200 mg L^{-1} to 300–400 mg L^{-1}. This shows the importance of using the best possible line for product optimizations. TX4 cells were the first biochemically selected variant line which showed overproduction of secondary metabolites and were thus interesting for biotechnological studies. It is also noteworthy that this highly productive variant line has maintained its production potential for more than 15 years (MEURER-GRIMES et al., 1989). Since the last review (BERLIN, 1986) no further studies regarding production improvements of hydroxycinnamoyl putrescines have been performed.

Though hydroxycinnamoyl putrescines are of no commercial interest the results obtained with TX4 cells during product optimization stimulated the analysis of other culture systems. A biochemical comparison of the low and high producing tobacco lines showed that enhanced hydroxycinnamoyl putrescine formation was due to distinctly enhanced activities of biosynthetic enzymes providing the cinnamoyl and amine moieties in TX4 cells (BERLIN et al., 1982) while the activity of the conjugating enzyme was similar in TX1 and TX4 cells (MEURER-GRIMES et al., 1989). Though the biochemical comparison of parent and variant lines provided some clues as to the biochemical requirements of overproducing lines, we did not analyze the system at the molecular level. One reason for this was the finding that not only the hydroxycinnamoyl putrescine pathway but also other enzyme activities such as tyrosine decarboxylase (WALKER et al., 1986) were altered in the *p*-fluorophenylalanine resistant cell line. Another reason was the fact that not only one but several enzyme activities should be "co-enhanced" in order to improve hydroxycinnamoyl putrescine production by genetic techniques. It has recently been shown that phenylalanine ammonia lyase is the rate-determining step in phenyl propanoid biosynthesis (BATE et al., 1994). Overexpression of this enzyme in transgenic tobacco callus, however, did not enhance the content of the major compound, the flavonoid rutin, but instead increased the levels of many other phenolics which were only present in trace amounts in wild-type callus (BATE et al., 1994).

3.1.2 Naphthoquinones and Anthraquinones

Quinones comprise a large group of secondary metabolites that are widely distributed in the plant kingdom. Since they are colored and, therefore, visible compounds, they have been a favored target of tissue culture research. Naphthoquinones and anthraquinones sometimes accumulate in cultured cells at levels far exceeding the amounts found in the intact plant. Some of these structures represent the active components of drugs. They are also important as natural dyes.

The red shikonin pigments of the cork layer of the roots of *Lithospermum erythrorhizon* are derivatives of 1,4-naphthoquinones (Fig. 2). These compounds have been used medicinally in Japan for the treatment of burns and skin disease and are now mainly used as a dye for lipsticks and for staining

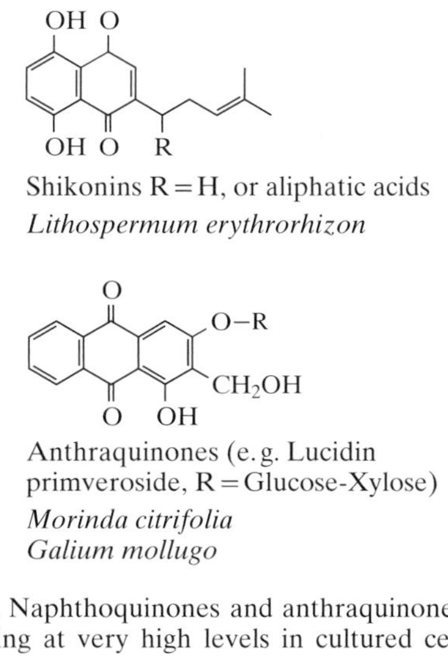

OH O

OH O R

Shikonins R = H, or aliphatic acids
Lithospermum erythrorhizon

O
O–R
CH_2OH
O OH

Anthraquinones (e.g. Lucidin primveroside, R = Glucose-Xylose)
Morinda citrifolia
Galium mollugo

Fig. 2. Naphthoquinones and anthraquinones accumulating at very high levels in cultured cells (see text).

silk. The plants have to be grown for 3–4 years before a yield of 1–2% shikonin is achieved in the roots (FUJITA, 1988). The total amount of *Lithospermum* roots used each year in Japan is 10000 kg. From this an annual demand of 150 kg shikonins can be calculated. As the plant cannot be grown in commercial quantities in Japan it has to be imported from Korea and China. It is often argued by Japanese scientists in industry (KOMAMINE et al., 1991) that due to the geographic situation the indigenous supply of plant material is not sufficient, and that plant tissue culture technology is, therefore, perhaps a more attractive alternative in Japan than elsewhere. This would explain the commercial production of shikonins from *L. erythrorhizon* cell suspension cultures by the Mitsui Company. Callus cultures of *L. erythrorhizon* were found to accumulate shikonin derivatives (TABATA et al., 1974). By repeated analytical screening over a period of two years two highly productive strains containing 20-fold increased levels of 1 mg g^{-1} fresh mass (ca. 10 mg g^{-1} dry mass) were isolated (MIZUKAMI et al., 1978). FUJITA et al. (1981a, b) investigated the effects of all media constituents on growth and production. They developed a production medium yielding 1.4 g L^{-1} shikonin derivatives within 23 d or 12% on a dry mass basis. Combining the two media in a two-stage process (1st stage 200 L growth medium, 2nd stage 750 L production medium) the yield was increased to 3.7 g g^{-1} dry mass inoculum within 23 d (FUJITA et al., 1982). By screening protoplast derived clones they isolated lines with an improved growth and higher productivity. The best line had a specific content of 23.2% and yielded 6.45 g shikonin per g inoculum (FUJITA et al., 1985). Production characteristics during the fermentation process were optimized by high density cultivation, fedbatch technique, controlled oxygen supply, and by the development of a rotating cylindrical bioreactor (FUJITA and HARA, 1985; FUJITA, 1988; TAKAHASHI and FUJITA, 1991). Since 1983 the Mitsui Company produces shikonins by this technology for cosmetics and dyes (TAKAHASHI and FUJITA, 1991). Comparison of the composition of shikonins extracted from cell cultures with that of various roots showed that the ratio of the individual shikonins was different (FUJITA, 1988). The altered ratio of the various shikonin derivatives in the cell cultures is no problem if the mixture of compounds is to be used for cosmetic purposes or as a dye. However, if one wants to replace a known approved plant drug by a tissue culture extract extensive evaluation of the pharmacological properties and equivalence studies are required. Such studies have recently been initiated for *L. erythrorhizon* cultures by the Mitsui Company (OZAKI et al., 1990; SUZUKI et al., 1991).

There are scattered reports in the literature of increased shikonin production in *L. erythrorhizon* cultures by *in situ* extraction, elicitation, or media variation. Though the observed effects are well demonstrated their final levels remain far below those reported by FUJITA et al. Such studies would deserve more attention if the production levels at the Mitsui Company could additionally be improved by the newly recommended techniques. This remark seems to be especially valid in view of the report of the Mitsui Company that two-phase cultures did not improve productivity of their high yielding line (DENO et al., 1987). In general, naphthoquinones and benzoquinones seem to belong to the groups of compounds which might readily be formed in cultured cells of various plant species (FUKUI et al., 1983; INOUE et al., 1984). It is evident that the highly expressed naphthoquinone biosynthetic pathway would be a good system for biochemical and molecular studies. The first results on the regulation of shikonin biosynthesis have been presented by HEIDE et al. (1989), showing that the ratio of *p*-hydroxybenzoic acid geranyltransferase and *p*-hydroxybenzoic acid glucosyltransferase activities is one of the regulatory controls of shikonin biosynthesis.

Anthraquinones in higher plants are formed either via the acetate polymalonate pathway or via the *o*-succinylbenzoic acid pathway. Their production in cell cultures has been reviewed in great detail by KOBLITZ (1988). Highly productive suspension cultures of plant species (e.g., of *Cassia* spp. or *Rhamnus* spp.) producing anthraquinones via the acetate polymalonate pathway have never been reported (VAN DEN BERG et al., 1988). However, it has been documented in numerous publications that the *o*-succinylbenzoic acid derived anthraquinones are well expressed in rapidly growing cell suspension cultures and accumulate sometimes at extraordinary levels. A good example are cell cultures of *Morinda citrifolia.* ZENK et al. (1975) tested a large variety of nutritional factors for their effect on growth and anthraquinone production. They established a production medium yielding 2.5 g L^{-1} anthraquinones corresponding to more than 10% of dry mass which exceeds the concentration of the root by a factor of 10. Important for high production levels were the replacement of the phytohormone 2,4-D by NAA and an increase of the sucrose level to 7%. The anthraquinones of cell cultures of *M. citrifolia* consist of a mixture of at least 12 aglyca and glucosides (LEISTNER, 1975; INOUE et al., 1981). The main components are lucidin derivatives (Fig. 2). Some of them have not yet been found in the intact plant (INOUE et al., 1981). The spontaneous high production of these anthraquinones were maintained in various bioreactors (WAGNER and VOGELMANN, 1977). The yields of anthraquinones in an airlift reactor were 30% higher than in experiments with shake flasks. An interesting example of manipulating the expression of pathways in cultured cell was reported for cell suspension cultures of *Morinda lucida.* In photoautotrophic cell cultures (chlorophyllous, no sugar in the medium) lipoquinones were the main components while anthraquinones were not found. When these cultures were transferred into the dark and sugar was added to the medium, anthraquinone biosynthesis was induced and lipoquinone formation was repressed (IGBAVBOA et al., 1985).

The results achieved with *M. citrifolia* cultures can also be obtained with cultures of *Galium mollugo* (BAUCH and LEISTNER, 1978). A B5-NAA medium with 7% sucrose gave the highest yields with ca. 2 g L^{-1} within 14 days. Lucidin primveroside (Fig. 2) was the main component. While the pathway of anthraquinones was readily expressed in cultured cells, the biosynthesis of iridoids remained repressed under all culture conditions. SCHULTE et al. (1984) optimized cell suspension culture media of 19 different Rubiaceae species for optimal yields of *o*-succinylbenzoic acid derived anthraquinones. 17 of the cultures yielded anthraquinone levels higher than those found in the corresponding plants. It was shown that nutritional and hormonal requirements of the various anthraquinone producing cultures, even those of the one family, may be quite different. Cultures of *Rubia cordifolia* maintained their high productivity when scaled up to 75 L (SUZUKI and MATSUMOTO, 1988), and cultures of *Rubia tinctorum* have been studied at San-Ei Chemical Industries (ODAKE et al., 1991) for the development of a commercial production process of anthraquinone pigments (e.g., of alizarin, purpurin). It is clear that these culture systems are also suitable for biochemical and regulatory studies. The first enzymes involved in the biosynthesis of *o*-succinylbenzoic acid (SIMANTIRAS and LEISTNER, 1989) and its further metabolism (SIEWEKE and LEISTNER, 1992) have been studied in cell cultures of *Galium* spp. The cultures of *G. mollugo* have also been used to overproduce shikimic acid. The addition of glyphosate inhibits the formation of *o*-succinylbenzoic acid and anthraquinones in these cultures thus causing the accumulation of the biosynthetic precursor, shikimic acid (10% of dry mass, 1.2 g L^{-1}) (STEINRÜCKEN and AMRHEIN, 1980).

3.1.3 Protoberberines and Benzophenanthridine Alkaloids

Isoquinoline alkaloids represent one of the largest groups of alkaloids in the plant kingdom. Common to all these alkaloids is that they are derived from tyrosine via (*S*)-norcoclaurine as a central intermediate (RUEFFER and ZENK, 1987) and not, as initially thought, via (*S*)-norlaudanosoline synthase (RUEFFER et al., 1981). Since the pathway is highly branched, a great variety of very different structures results from this central intermediate. Some of the branches are readily expressed in cultured cells, while others such as the morphinan alkaloids remain mostly repressed.

There are many reports showing that protoberberine alkaloids (Fig. 3) spontaneously

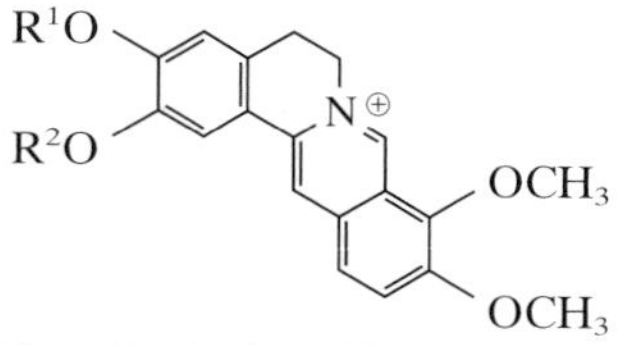

Protoberberine Alkaloids
$R^1 + R^2 = CH_2$ Berberine
$R^1 = CH_3$, $R^2 = H$ Columbamine
$R^1 = H$, $R^2 = CH_3$ Jatrorhizine
e.g., *Coptis japonica*
Berberis species
Thalictrum rugosum

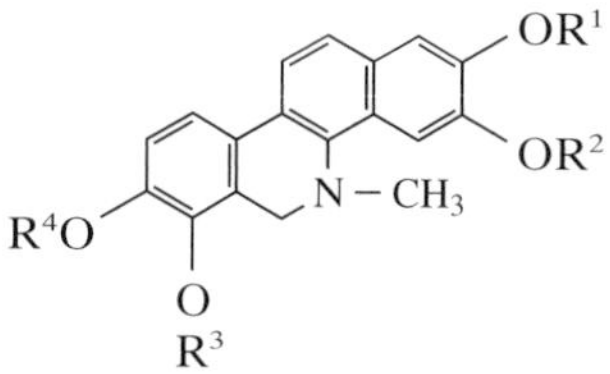

Benzophenanthridine alkaloids
$R^1 + R^2 = R^3 + R^4 = CH_2$ Dihydro-sanguinarine
$R^1 + R^2 = CH_2$, $R^3 + R^4 = CH_3$ Dihydro-chelerythrine
e.g., *Papaver somniferum*
Eschscholtzia californica

Fig. 3. Structures of protoberberine and benzophenanthridine alkaloids.

accumulate at high levels in cultured cells. Various cultures of *Berberis* spp. accumulated between 0.2 and 1.7 g protoberberine alkaloids, mainly jatrorrhizine, on a growth medium with 3.5% sucrose (HINZ and ZENK, 1981). BREULING et al. (1985) optimized the production to 3 g L^{-1} in a 20 L airlift bioreactor. Independently, two Japanese groups reported the accumulation of high levels of protoberberines in cultures of *Coptis japonica* (FUKUI et al., 1982; YAMADA and SATO, 1982). FUKUI et al. established a line accumulating berberine, jatrorrhizine, palmatine, and coptisine at a ratio of 50:22:22:6. Interestingly, the content of alkaloids increased gradually during subculturing from 8–15% which was paralled by increased growth (FUKUI et al., 1982). This shows that protoberberine formation is indeed a favored pathway in these rapidly growing cell cultures. Thus, alkaloid levels of up to 1.7 g L^{-1} were readily achieved in normal growth medium. The *Coptis* line described by YAMADA and SATO (1982) contained mainly berberine and only trace amounts of other alkaloids. By screening small cell aggregates for high berberine (yellow) producing clones cell lines were established which produced more than 1 g berberine L^{-1} or 10% on a dry weight basis (SATO and YAMADA, 1984).

Another important species spontaneously accumulating high levels of protoberberines is *Thalictrum*. While cell cultures of *Thalictrum minus* secrete most berberine into the medium (NAGAKAWA et al., 1984) those of *T. flavum*, *T. dipterocarpum* or *T. rugosum* accumulate the alkaloids within the cells (SUZUKI et al., 1988; PIEHL et al., 1988). Interestingly, when cells of *T. rugosum* were transferred to fresh medium lacking phosphate, protoberberines were released into the medium (BERLIN et al., 1988a). There are other reports indicating that cultures of plant species with the capability to biosynthesize protoberberines will spontaneously yield cultures accumulating these colored alkaloids in high amounts (0.2–1 g L^{-1}). The levels of protoberberines are so high that the Mitsui Company has decided to develop a production process for berberine with *Coptis japonica* cell cultures (MATSUBARA and FUJITA, 1991). As in the case of shikonins the medium was first optimized. For example, a 10-fold increase of Cu^{2+} increased the berberine content by 40% (MORIMOTO et al., 1988). The addition of very low levels of gibberelic acid increased the content by 30% (HARA et al., 1988). Since cell aggregate screening was regarded as very time-consuming and not very efficient, protoplasts with high berberine content were isolated with a cell sorter (HARA et al., 1989). All selected lines contained higher levels than the original culture. The next step was to develop culture conditions for a high density culture (MATSUBARA et al., 1989) so that 70 g dry mass per L (corresponding to ca. 700 g fresh weight) were obtained. This required increasing the nutrition and oxygen supply by fedbatch-perfusion cultivation (MATSUBARA and FUJITA, 1991). Today, the Mitsui Company produces berberines by a high density continuous culture method over a period of several months with a yield of 0.65 g L^{-1} d^{-1}. This is 8 times more than the value obtained in batch cultures (MATSUBARA and FUJITA, 1991). This demonstration of the potential of plant cell culture process optimization by the bioengineers suggests that further attempts to optimize protoberberine production should start from initial product levels as high as those published by the Mitsui Company. It was speculated that cultures which release their compounds into the medium might be favored by industry because this would diminish the problem of sacrificing the slowly growing plant cells for product extraction. In the case of protoberberines, an alternative production process seemed to be possible with the highly productive *T. minus* line. *T. minus* cells were probably tested by the Mitsui Company in an attempt to develop a new bioassay-based screening method for high berberine-producing cell colonies (SUZUKI et al., 1987). The real advantage of *T. minus* cells can only be exploited if cells are immobilized for berberine production. KOBAYASHI et al. (1988) obtained a production rate of 50 mg L^{-1} d^{-1} in batch and semicontinuous bioreactors over a period of at least 60 days of cultivation. This was undoubtedly an exciting result. However, in view of the improvements obtained with the *C. japonica* culture, it is clear that *T. minus* cultures are not yet competitive for use in a commercial process. Berberine chloride is used in Japan as a medicine for intestinal disorders and treatments of abnormal zymosis and has previously been obtained by extraction from the roots of *C. japonica* and the cortex of *Phellodendendron amurens* (MATSUBARA and FUJITA, 1991). Berberine produced with tissue culture has not yet been approved as a drug in Japan. However, pharmacological studies towards this goal are underway (SUZUKI et al., 1993a, b).

Another group of isoquinoline alkaloids which accumulate spontaneously in cultured cells are benzophenanthridine (Fig. 3) and the related protopine alkaloids. Cell cultures of most Papaveraceae contain alkaloids of this group. FURUYA et al. (1972) were the first to describe the occurrence of sanguinarines, protopines, and the aporphine magnoflorine in callus cultures of *Papaver somniferum*. When they analyzed the alkaloid pattern of callus cultures of 11 other species of Papaveraceae, almost identical alkaloid spectra were found (IKUTA et al., 1974). Since then, the occurrence of these alkaloids has been confirmed in numerous reports. Cell culture conditions seem to favor the synthesis of benzophenanthridine alkaloids which are found in the corresponding intact plants often at rather low levels (WILLIAMS and ELLIS, 1993). However, there is a significant difference between the production of benzophenanthridine and protoberberine alkaloids in cultured cells. While in *C. japonica*, e.g., protoberberine production increased with increasing growth rate (FUKUI et al., 1982), benzophenanthridine synthesis decreased with better growth (BERLIN et al., 1985). A suspension culture of *P. somniferum* with a growth cycle of 28 d contained nearly 6% sanguinarines (360 mg L^{-1}) after the 5th subcultivation. During further subcultivation the yield decreased to 200 mg L^{-1} after 20 subcultures and to 20 mg L^{-1} after 35 subcultures. Biomass production of this culture increased from 6 g in 28 d to 11 g in 10 d (BERLIN et al., 1985). The "biotechnological" interest in benzophenanthridine alkaloids increased only when it was noted that these alkaloids can be induced greatly by stress factors and elicitors. Thus, the yields of benzophenanthridine alkaloids of suspension cultures of *Eschscholtzia californica* were increased 10-fold to 150 mg L^{-1} by increasing the sucrose concentration in the medium to 8% (BERLIN et al., 1983). The observation of EILERT et al. (1985) that sanguinarine levels of *P. somniferum* were enhanced from 0.01% to 2.9% by treatment with fungal elicitors received even more interest. The advantage of the elicitor treatment in comparison with the altered production media is that the cells respond more rapidly, and the induced alkaloids in *Papaver* cell cultures are released into the medium (EILERT et al., 1985; CLINE and COSCIA, 1988). In the case of *E. californica*, the alkaloids, induced by various fungal and yeast preparations evidently remained within the cells (SCHUMACHER et al., 1987). These authors reported product levels in the same range (160 mg L^{-1}) as were found in the sucrose induced cells (BERLIN et al., 1983).

There was, however, one significant difference: the elicited cells contained the quarternary alkaloids while the sucrose-induced cells contained the corresponding dihydro forms. It is interesting to note that ZENK's group did not consider elicitor technology to be helpful for increasing productivity for commercial purposes (SCHUMACHER et al., 1987). This was based on the finding that lines containing high levels of the alkaloids in the unelicited state did not produce higher levels after elicitation than low yielding lines. This technology has nevertheless been used in efforts of developing a commercial production process for sanguinarine. This compound has an antibiotic activity against oral microorganisms causing periodontal disease (SOUTHARD et al., 1984). The chances of plant cell cultures to be used as a source were improved by the fact that cultured cells contained much higher levels of sanguinarine than tissues of intact plants and that an established production process did not exist.

Consequently, research has concentrated on the optimization and scale-up of sanguinarine production. It was shown that the mildly elicited *P. somniferum* could be re-elicited after a period of regrowth suggesting that a semicontinuous production process with re-elicitation could be established (TYLER et al., 1988). Due to the fact that a great portion of the elicited alkaloids were released, the suitability of surface-immobilized cells with adsorption of the released alkaloid to a resin was tested (KURZ et al., 1990). The total yields, however, were reduced. The highest yield published up to now is 300 mg L^{-1} in a 300 L airlift reactor (PARK et al., 1992). The cultures were grown from ca. 20 to 180 g fresh mass per L and were elicited after 7 d or when ca. 2 g L^{-1} glucose were left in the medium, and were harvested two days later. The authors believed that further improvements might be possible by optimizing the amount of dissolved oxygen in the medium. If, in addition, conditions for elicitation of a high density culture could be established, a commercial production would be possible.

There have also been efforts to select cell lines with higher production potential for sanguinarine (SONGSTAD et al., 1990). The stability of these cell lines during production and other characteristics important for the suitability of a selected line for scale up purposes remain to be shown. Although a commercial enterprise (Vipont) has been involved in the development of this process, it is unclear whether an industrial production will result. Further attempts to optimize sanguinarine production with cultures of *P. somniferum*, *E. californica*, or *Sanguinaria canadensis* should be left to industry if product level improvements are the only goal. They know how much the productivity has to be enhanced for a tissue culture process to be superior to conventional extraction of field grown plant material.

It is evident that the culture systems producing protoberberines and benzophenanthridines are useful systems for studying the enzymology of these pathways. All 13 enzymes required for the synthesis of berberine have been identified. A summary of the present knowledge, derived mainly from studies of ZENK's group at the University of Munich and from YAMADA's group at the University of Kyoto, has been presented by HASHIMOTO and YAMADA (1994). The enzymology and molecular biology of benzophenanthridine alkaloid biosynthesis has been reviewed by KUTCHAN and ZENK (1993). Thus, the biotechnological value of the cultures producing these alkaloids does not only lie in their impressive productivities but also in their potential to provide a deep insight into the regulation of the expression of these complex and branched pathways.

3.1.4 Monoterpene Indole Alkaloids

About 1200 alkaloids derived from tryptophan have been isolated from higher plants, which corresponds to about one quarter of all alkaloids (GRÖGER, 1980). Several of these alkaloids from *Rauwolfia*, *Catharanthus*, and *Cinchona* are used medicinally. Therefore, it was of great interest to see whether these alkaloids also accumulate in cultured cells. The dimeric monoterpene indole alkaloids, vinblastine and vincristine, used as antileukemia agents, are only present in trace amounts in the whole *Catharanthus roseus* plant. Indeed, these two compounds were initially the main reason why cultures of *C. roseus* received so much attention. These two alkaloids as well as the medicinally used alkaloids of *Cinchona* spp. will be discussed later in Sect. 3.2 in connection with poorly expressed compounds.

More than 50 monoterpene indole alkaloids have been isolated from various Apocynaceae, e.g., *Catharanthus* and *Rauwolfia* spp. The chemical identification of their complex alkaloid mixtures was achieved mainly by three groups (STÖCKIGT and SOLL, 1980; KOHL et al., 1982; KUTNEY et al., 1983). A list of almost all monoterpene indole alkaloids found in plant tissue cultures of various plant species is given by ELLIS (1988). Most of the alkaloids accumulate at low levels in the cultures. Sufficiently confirmed quantitative data allowing an evaluation of the biotechnological relevance are most often not available. Most studies regarding product levels have concentrated on the optimization of ajmalicine, serpentine (oxidized form of ajmalicine) and catharanthine (Fig. 4). There has been one report that a cell suspension culture of *Rauwolfia serpentina* accumulated 1.6 g raucaffricine per liter of medium (SCHÜBEL et al., 1989). However, this level dramatically exceeds the levels of all other monoterpene indole alkaloids ever measured in cultured cells and has not yet been confirmed in further publications by this or any other group working with *R. serpentina* cell cultures.

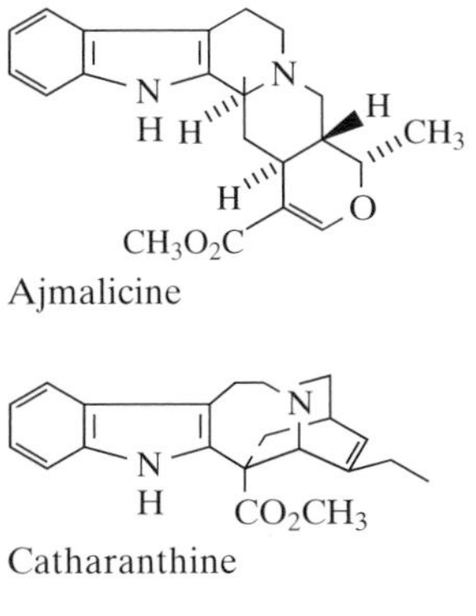

Ajmalicine

Catharanthine

Fig. 4. Monoterpene indole alkaloids of *Catharanthus roseus*.

Therefore, this claim cannot be regarded as being biotechnologically relevant. Cell cultures of *R. serpentina* were scaled up to 75 m^3 by the Diversa Company and no alkaloids were found in the rapidly growing cell cultures (WESTPHAL, 1990). Indeed, it is sometimes very difficult to evaluate the impact of product level claims, and it is futile to discuss such claims here with respect to their biotechnological impact. Information regarding this issue can be found in several special reviews covering the monoterpene alkaloids in cultured cells (BALSEVICH, 1988; DELUCA and KURZ, 1988; VAN DEN HEIJDEN et al., 1989; MORENO et al., 1995) or in the article of ELLIS (1988). The biotechnological progress on the improvement of product levels of ajmalicine/serpentine and catharanthine, about which reports from many different laboratories are available, is discussed below.

Callus cultures of *C. roseus* were found to contain low levels of monoterpene indole alkaloids (CAREW, 1975). However, suspension cultures accumulated no or only trace amounts of these compounds when grown on a growth medium with 2,4-D as phytohormone. ZENK's group was the first to develop a production medium (ZENK et al., 1977b). When the cells were transferred from the growth to a production medium, alkaloid formation was resumed after 3–5 d. This medium has successfully been used by several other laboratories. Of 458 independently established cell lines, 312 (approx. 75%) produced alkaloids when transferred to ZENK's production medium (KURZ et al., 1980). This production medium, however, had no special composition; the sole transfer of the cells into a 2,4-D-free medium allows the accumulation of reasonable levels of indole alkaloids (ZENK et al., 1977b; KNOBLOCH and BERLIN, 1980; PAREILLEUX and VINAS, 1984). The increase of sucrose to 5–8% had an additional beneficial effect. The phytohormone composition (indole acetic acid and benzylaminopurine) of ZENK's medium may have an additional stimulatory effect. The stimulatory effect of the production medium probably depends on the physiological state of the cells at the time of transfer from the growth to the production medium. The level of phosphate accumulated in the cells seems to play an important role in the inducibility of alkaloid formation in a phosphate-free production medium (KNOBLOCH and BERLIN, 1983; SCHIEL et al., 1987).

Suspension cultures of *C. roseus* grew rapidly, were easy to scale up in fermentors and yields of 30–70 mg L^{-1} serpentine were usually obtained within 20–30 days (ZENK et al., 1977b; WAGNER and VOGELMANN, 1977; SMART et al., 1982). There have been several efforts to isolate higher producing cell lines using screening methods. The group at the NRC in Saskatoon established more than 2000 individual lines from three *C. roseus* cultivars and screened them for alkaloid patterns and levels (KURZ et al., 1985). They found a substantial variation among the lines, and one variety seemed to be more suitable for catharanthine formation. However, this tremendous effort did not evidently result in the isolation of an exeptionally high producing line. ZENK's group has screened established lines for high yielding clones using specific radioimmunoassays and the fluorescence microscope (ZENK et al., 1977b; DEUS and ZENK, 1982). Indeed, they found clones having accumulated high levels of serpentine or ajmalicine of which theoretical levels of up to 400 mg L^{-1} were calculated. The isolated clones were, however, unstable, and alkaloid levels rapidly dropped to the values of the unscreened culture (DEUS-NEUMANN and ZENK, 1984). The behavior of the clones indicates that the they were not true variants as claimed but instead were cell colonies in a transient physiological state. Reasons for the failure to select true variants in this system by the applied screening method have been discussed in detail (BERLIN and SASSE, 1985; BERLIN, 1986).

For many years cultures of *C. roseus* were grown on media with 2,4-D. Alkaloid formation thus required transfer to a production medium. MORRIS (1986) found that cultures of *C. roseus* can not only be established and maintained on a medium with NAA/kinetin lacking 2,4-D but that they also produce alkaloids on this medium. These findings have also improved the chances of selecting clones with higher production. As predicted at a meeting in 1988 (BERLIN, 1990), high yielding clones were isolated by consecutive repeated analytical screening of cultures growing on NAA/kinetin medium. Again, researchers of the Mitsui Company (FUJITA et al., 1990) obtained this success. Repeated analytical screening improved the specific catharanthine content from 0.1% to 0.7%, and these lines have maintained their superior productivity for at least 10 subcultures. Additionally, a biochemical selection for 5-methyl-tryptophan-tolerant cell lines was applied. (FUJITA et al., 1990). Although the specific catharanthine content was further increased, it is not clear why these cell lines form higher levels of alkaloids. The productivity of the isolated lines was further optimized by optimizing the medium composition, inoculum density, and fermentation parameters. Finally, the 5-MT-tolerant line produced 230 mg L^{-1} catharanthine in a 1.7 L bioreactor within one week (specific content 1.1% of dry mass).

There have been tremendous efforts to stimulate the production of ajmalicine and serpentine in cultures of *C. roseus*. Many approaches have been used such as the addition of various biotic and abiotic elicitors, the use of high and low density culture, feeding and elicitation, alteration of media composition, immobilization, etc. (for reviews, see VAN DEN HEIJDEN et al., 1989; MORENO et al., 1995). Despite the many reports claiming to have demonstrated increased alkaloid production, one has to realize that the product levels have not been improved since the first findings 15–20 years ago. Thus, the product levels of bioreactor studies remain at 10–50 mg L^{-1} ajmalicine or serpentine (corresponding to 0.1 to 0.5% of dry mass) (SCHIEL and BERLIN, 1987; SCRAGG et al., 1989; SCRAGG et al., 1990; SCHLATMANN et al., 1994). One reason for this undoubtedly poor progress could be the fact that most often only a single factor was analyzed at any one time. Furthermore, analysis of the various factors did not take advantage of the previously improved culture levels. A comparison of the many studies done in universities and research centers in an attempt to optimize monoterpene alkaloid production in *C. roseus* with the industrial approach, e.g., of the Mitsui Company reveals that the former group of scientists is interested in detecting factors influencing alkaloid levels, but not really in improving product levels. Despite nearly 20 years of bioreactor studies with *C. roseus* it has been reported that the inconsistencies and the often contradictory results are likely due to insufficient characterization of the inoculum material (VAN GULIK et al., 1994). The authors suggested to use identical inoculum material when analyzing the effects of factors on growth and alkaloid production. For industrial scientists, productivity improvement is a step-by-step enhancement of product levels per volume and time and this requires that each step is reproducible. Thus, it seems that the maximum potential of *C. roseus* cultures for serpentine and ajmalicine production may only be found when a commercial enterprise takes over. A first step would be a screening of cultures growing on NAA kinetin medium for true variants overproducing ajmalicine or serpentine. However, a comparison of the production level improvements for berberine, shikonins, and catharanthine by the Mitsui Company shows that the maximum yield possible is mainly determined by the biology of the cell. The rather low levels of the monoterpene indole alkaloids found to date suggest that at best levels in the range of 200–500 mg L^{-1} within 1–2 weeks might be produced by an optimized culture. According to an economic assessment of the production of ajmalicine by *C. roseus* cultures, the present levels have to be increased at least by a factor of 40 to make the process competitive (DRAPEAU et al., 1987).

In contrast to the poor progress at the product level cell cultures of *C. roseus* have been used quite well for studies on the regulation of alkaloid formation. The correlation of enzyme activity pattern and product accumulation curves and the conclusions drawn by various groups are not, however, always consistent. Therefore, only facts which are generally accepted are mentioned here. It has repeatedly been shown that tryptophan decarboxylase is induced when alkaloids are produced. However, it has also been shown that induction of this enzyme does not necessarily lead to enhanced alkaloid formation (for a review see VAN DEN HEIJDEN et al., 1989). Nevertheless, since tryptophan decarboxylase is readily inducible two groups have cloned the cDNA of this enzyme from *C. roseus* (DELUCA et al., 1989; GODDIJN, 1992). The negative effect of phytohormones on alkaloid formation can perhaps now be explained by the finding that auxins down-regulate transcription levels of the tryptophan decarboxylase gene (GODDIJN et al., 1992). Geraniol-10-hydroxylase is regarded as a regulatory enzyme for the monoterpene moiety (secologanin) of the indole alkaloids. Indeed, this enzyme is also induced during initiation of alkaloid biosynthesis (SCHIEL et al., 1987). It has already been purified and its cloning is in progress (MEIJER et al., 1993a). The enzyme and the gene coding for strictosidine synthase which connects tryptamine and secologanin have been isolated (KUTCHAN, 1993). in some lines the enzyme was present in non-producing cells, in others it was induced when alkaloid formation was elicited. Molecular studies suggest a coordinated regulation of the genes coding for tryptophan decarboxylase and strictosidine synthase (PASQUALI et al., 1992). Detailed overviews of the enzymology and regulation of monoterpene indole alkaloids have been given recently (DELUCA, 1993; MEIJER et al., 1993b). The results of molecular studies should show whether genetic improvements of the biosynthesis of certain monoterpene alkaloids are an attainable goal.

3.1.5 Anthocyanins and Betalains

Anthocyanins have been found in many cultured cells of many plants (*Daucus*, *Haplopappus*, *Catharanthus*, *Petunia*, *Mathiola*, *Euphorbia*, *Perilla*, *Vitis*, *Aralia*, and many others). For some cultures the individual anthocyanin components have been identified, for others only the total amount of the pigments has been given (ELLIS, 1988; SEITZ and HINDERER, 1988; YAMAMOTO, 1991). In a few culture systems anthocyanins are formed in dark grown cells, but in most cases light is required for optimal production (SAKAMOTO et al., 1994). Due to their low toxicity, anthocyanins are widely used in food additives and also as a dye for silk (YAMAMOTO, 1991). While anthocyanin extracts are rather inexpensive the prices for individual anthocyanins seem to be very high which would justify the use of tissue culture based processes. Again, Japanese scientists and industries are leaders in the development of biotechnologically relevant culture systems.

Since anthocyanin producing cells are easily detected in callus cultures they have become an early target for visual screening (KINNERSLEY and DOUGALL, 1980). The most impressive success was reported by YAMAMOTO et al. (1982) when they established high yielding cell lines of *Euphorbia milli* by consecutive repeated screening. Only after 24 clonal selections were stable lines found; this required permanent screening over a period of many months. NOZUE et al. (1987) were also successful in selecting high anthocyanin yielding cell lines from sweet potato by consecutive repeated screening. This shows that even for compounds formed in the growth medium, stable clones are only obtained by repeated screenings. This result casts some doubt to claims of the instability of high producing variant lines if permanent screening was not applied over a very long period (BERLIN, 1990). As for many other secondary metabolites, the culture medium can also be optimized for anthocyanin production. The optimization of *E. millii* cultures by the Nippon Paint Corporation improved the productivity 4.5-fold to 32 mg L^{-1} d^{-1} (YAMAMOTO et al., 1989). The main component of the *E. millii* cell culture is cyanidin-3-arabinoside (Fig. 5) (YAMAMOTO, 1991). A great disadvantage of the *Euphorbia* system in commercial production is the requirement of light for pigment production. The same holds true for a highly productive culture of *Perrilla frutescens* (ZHONG and YOSHIDA, 1995) which produces up to 5.8 g L^{-1} anthocyanins in Iod under optimized culture conditions and light. Therefore, the best line for the commercial production of anthocyanins may be cultures of *Aralia cordata* which produce high levels in the dark. As with the cultures of *Euphorbia* and *Ipomea*, continuous cell aggregate cloning led to a highly productive and fast growing line (SAKAMOTO et al., 1994). 90% of all cells of the cloned line produced anthocyanins, a result which is only possible if product formation and growth are closely interrelated. Optimization of the basic inorganic salt concentration, the carbohydrate concentration, the ratio NO_3^-/NH_4^+, and the total nitrogen levels improved the anthocyanin content to 10.3% of dry mass. This culture was scaled up to 300 L in a 500 L jar fermentor (KOBAYASHI et al., 1993). After 16 d of cultivation 69 kg fresh mass with 545 g anthocyanin (corresponding to 17.2% of dry mass) were obtained. The further improvement of productivity during scale-up was evidently due to the controlled supply of CO_2 to the cultures. Efforts to identify the anthocyanins present in this high yielding culture system are underway. As is typical in Japan the project is being done with close cooperation between researchers from university and industry (Kyowa Hakko Kogyo Co. and Tonen Co.).

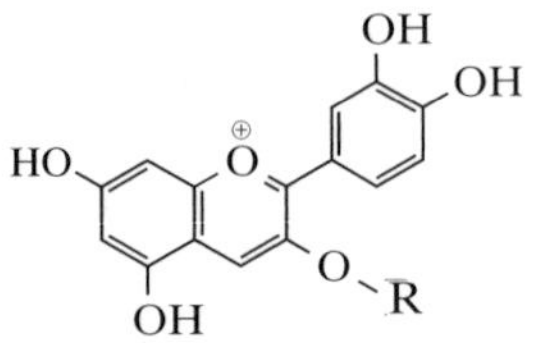

R = arabinoside
Cyanidin-3-arabinoside *(Euphorbia millii)*

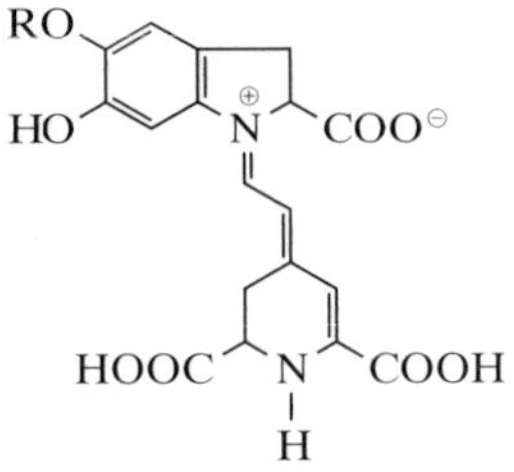

Betanidin, R = H
Betanin, R = Glucose
(various Centrospermae)

Fig. 5. Structures of anthocyanin and betacyanin.

Tissue culture systems may only become competitive for the production of individual anthocyanins and not for mixtures of anthocyanins. Therefore, it might be useful to look not only at total yields during selection and media optimization but also at whether such methods affect the percentage of any one anthocyanin in a complex mixture (DO and CORMIER, 1991).

As shown in the previous sections optimized culture systems have often successfully been applied for biochemical and regulatory studies by those interested in their biotechnological application. Although this has not yet been the case, our knowledge about their biosynthesis and the regulation of this pathway far exceeds that of all other secondary pathways (for reviews see DIXON and LAMB, 1990; FORKMANN, 1993; DOONER et al., 1991; KOES et al., 1994). Thus several genetic manipulations of these pathways have been made with the aim of altering pigmentation of flowers (MEYER et al., 1987; VAN DER KROL, 1990).

Other pigments often found in cell cultures of Centrospermae are the betalains which include red betacyanins and yellow betaxanthins (BÖHM and RINK, 1988). Cultures of *Chenopodium rubrum* (BERLIN et al., 1986), *Phytolaca americana* (SAKUTA et al., 1987), *Amaranthus tricolor* (BIANCO-COLOMAS and HUGUES, 1990), *Beta vulgaris* (GIROD and ZRYD, 1991), and *Portulaca* (KISHIMA et al., 1991) produce rather high levels of betacyanins under light conditions. Betacyanin formation is more or less growth-related, which explains the spontaneous accumulation of these pigments in cell cultures. By visual screening higher productive lines were isolated from the parent callus line and media compositions were optimized. Often, quantification was based only on the absorption at 535 nm. In the case of *Chenopodium rubrum* we identified amaranthin, celosianin, and betanin as the main betacyanins (Fig. 5) (BERLIN et al., 1986). Under optimized cultured conditions 35–45 mg L^{-1} betacyanins were produced in a modified Murashige & Skoog growth medium supplemented with tyrosine (BERLIN et al., 1986). From this line a superior subline was selected by screening. This line produced not only 3–4 times more betacyanins but also had an altered pigmentation due to an altered ratio of betacyanins (with celosianin as the main compound) (BERLIN, unpublished results). This line has not yet been optimized for highest product levels but was successfully used for the detection of enzymes involved in celosianin biosynthesis (BOKERN et al., 1991). The question of whether further optimizations are justified arises if the only aim is to make betacyanin tissue culture production more competitive. Although cell cultures of *Beta vulgaris* contain higher levels of betacyanins than all plant tissues tested, it is unlikely that cell cultures will be used for the production of betacyanins since the marked price for red beet extracts is extremely low (LEATHERS et al., 1992).

3.1.6 Steroidal Compounds

There are numerous reports on the occurrence of sterols, steroids, sapogenins, saponins, and steroidal alkaloids (ELLIS, 1988). In my previous review (BERLIN, 1986) the efforts to optimize diosgenin production in cultures of *Dioscorea deltoidea* were described in some detail. As no significant progress towards a commercial production has been reported since then, I will concentrate here on the ginseng saponins, which were only briefly mentioned in the last review.

The roots of *Panax ginseng* C. A. Meyer have widely been used as a tonic and precious medicine in oriental countries since ancient times. It is regarded as an adjuvant to prevent health disorders and is considered to be a "miraculous" drug for preserving health and promoting longevity (MISAWA, 1994). The active component with proven pharmacological effects are saponins, e.g., ginsenoside-Rb and -Rg (Fig. 6). Due to the worldwide increased demand for ginseng extracts and the fact that a cultivation period of 5–6 years is necessary before harvest, production of ginseng extracts or saponins using tissue cultures was regarded as an alternative source of supply. FURUYA's group analyzed a number of callus cultures of *Panax ginseng* with different phythohormone requirements (FURUYA, 1988), optimized

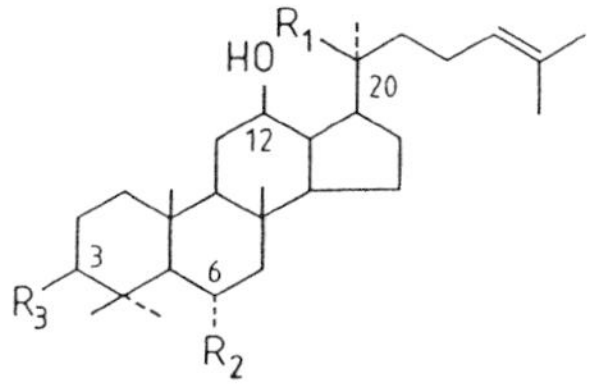

Fig. 6. Saponin aglycone of *Panax ginseng*. R_1 and R_3 are often diglycosides in R_b-ginsenosides. In Rg-ginsenoides R_2 is glucosylated.

their growth conditions, and compared their saponin patterns with those of *P. ginseng* roots. The patterns were quite comparable and levels of total saponins reached up to 0.7% of dry mass or 50 mg L^{-1} within 4 weeks in a 30 L jar fermenter. It was shown that some morphological differentiation increased saponin formation, and thus a rapidly growing embryo-like cell line was selected for scale-up to 20 m^3 (USHIYAMA, 1991). A biomass production of 700 mg L^{-1} d^{-1} was obtained.

Since 1988 the use of the tissue culture material has been approved for the Nitto Denko Co. by the Japan Ministry of Welfare and Health. The extracts from the tissue culture material have been added to wines, tonic drinks, soups, herbal liqueurs, and other food preparations since 1989 (USHIYAMA, 1991). For the approval, a comparison of the constituents of plant extracts and plant tissue culture derived material was necessary. In addition, some biological assays (Ames test, acute virulence test) and dietary tests with livestock feeding containing high amounts of the tissue culture material had to be performed. It is a little astonishing that the chemical comparison of plant and marketed tissue culture extracts does not contain details about the saponins and the ratio of Rb:Rg saponins (USHIYAMA, 1991). In view of the large-scale production of ginseng extract using an embryo-like suspension culture originally established by FURUYA's group (USHIYAMA, 1991), it is also somewhat surprising that one of the latest reports from FURUYA's group described the growth of an embryo culture in various 3 L bioreactors (ASAKA et al., 1993). This culture produced 0.1 mg L^{-1} d^{-1} saponins.

3.1.7 Immunologically Active Polysaccharides

There are a few reports that plant cell cultures sometimes release considerable amounts of polysaccharides into the culture medium. These polysaccharides however, have rarely been analyzed for biological activity. During investigations on the identification of the immunologically active components of *Echinacea* drugs, two polysaccharides were isolated and characterized (PROKSCH and WAGNER, 1987). The large amounts of these polysaccharides needed for further *in vivo* experiments were difficult to obtain from the plant material. Therefore, cell cultures of *E. purpurea* were established. The cultures also produced two major active polysaccharides and released them into the culture medium; however, they were not identical with the polysaccharides of the intact plant (WAGNER et al., 1989). These cultures were optimized on a laboratory scale and then scaled up to 60 m^3 in a cascade of bioreactors (WESTPHAL, 1990). Especially noteworthy was that the specific yield was improved 2–3-fold and that the cultivation periods in various fermentors were greatly reduced during the optimization by the process engineers (WAGNER et al., 1989; WESTPHAL, 1990). The cooperation between researchers of the University of Munich, the pharmaceutical company Lomapharm, Emmenthal, and the bioengineering company DIVERSA, Hamburg, led to the production of commercially relevant levels of a tissue culture specific drug which entered clinical trials. However, DIVERSA has recently been taken over by Phyton, mainly known for its engagement in developing a production process for taxol® (palitaxel) and thus it is not clear whether production of the immunologically active polysaccharides will continue.

3.2 Products of Commercial Interest Accumulating in Traces or not at all in Suspension Cultures

Large efforts have been made in the past to express compounds which are widely used as pharmaceuticals in cell cultures. These include morphinan alkaloids, tropane alkaloids, quinoline alkaloids, and cardiac glycosides. In addition, substantial efforts have been made to develop cell culture systems producing antitumor compounds. In this section the progress in increasing the product levels of these compounds will be analyzed. In contrast to the previous review (BERLIN, 1986) a special section on monoterpenoids will not be given here. Production of these compounds has remained low with callus and suspension cultures (for a review see MULDER-KRIEGER et al., 1988; BANTHORPE, 1994). So far, tissue culture systems have not contributed very much to the analysis of lower terpenoid biosynthesis and its regulation, despite the fact that even a low producing tissue culture can be a good source for related enzymes (BANTHORPE, 1994). Organ cultures are usually much better producers of lower terpenoids (CHARLWOOD et al., 1990; BANTHORPE, 1994). Nevertheless, the development of a biotechnologically viable process for these compounds is unlikely.

3.2.1 Morphinan Alkaloids

A number of reports have claimed that morphinan alkaloids are present in callus and cell suspension cultures of *Papaver somniferum* and *P. bracteatum* (for a review see KAMO and MAHLBERG, 1988; KAMIMURA, 1991). However, there are also several reports where no morphinan alkaloids were detected (BERLIN, 1986; KAMO and MAHLBERG, 1988). The elicitation of poppy cell cultures was not initially planned for the production of sanguinarines but instead for the induction of morphinan alkaloids. EILERT et al. (1985) clearly stated that morphinan alkaloids could not be induced by any of the elicitors tested. Claims in the literature of substantial morphinan alkaloid levels of up to 0.15% dry mass are of no biotechnological relevance since such data were not confirmed in further investigations. Evidently, some morphological differentiation is required for expression of the morphinan alkaloid pathway. Thus, freshly initiated cultures having retained the capacity to redifferentiate when stressed by elicitors or altered phytohormone composition, are able to synthesize and accumulate low levels of these alkaloids (KAMO and MAHLBERG, 1988). Such cultures usually exhibit poor growth and are not very stable, and thus no efforts have been made to produce these alkaloids in larger volumes. The most recent report on morphinan alkaloids in cell cultures claims a production of 2.5 mg morphine and 3.0 mg codeine per g dry mass (SIAH and DORAN, 1991). However, considering that these very high levels correspond only to 10–14 mg L^{-1} after a period of 56 d using phytohormone-free medium which must be replaced several times, it is clear that such findings are of no biotechnological relevance. They only show that under extreme stress conditions the morphinan branch is weakly expressed. Progress in the production of morphinan alkaloids by undifferentiated, rapidly growing cell suspension cultures will, therefore, be made only if factors are identified which prevent expression of the pathway in better growing cultured cells. A breakthrough could indeed result from the recent detection of the enzyme channelling (R)-reticuline via salutaridine into the morphinan alkaloid branch (GERARDY and ZENK, 1993). This membrane-bound microsomal enzyme is evidently only expressed in tissues and cells able to synthesize morphinan alkaloids. This might, however, be changed in the near future by genetic engineering.

3.2.2 Tropane Alkaloids

Hyoscyamine and scopolamine are the most important medicinal tropane alkaloids. Hence, several groups have worked on the production of these compounds in cultured cells. However, the alkaloid levels were found to be extremely low in callus and suspension cultures of *Hyoscyamus*, *Atropa*, *Datura*, *Duboisia*, and *Scopolia*. By using a squash technique in screening for alkaloid producing clones, YAMADA and HASHIMOTO (1982) selected a cell line of *Hyoscyamus niger* containing just 0.01–0.02% hyoscyamine per g dry weight in suspension cultures. Scopolamine levels were 10-fold lower. Also media variation experiments were not successful in increasing tropane alkaloid levels. Evidently, some root formation is required to express the tropane alkaloid pathway to a reasonable extent (see Sect. 4).

There has only been one report (BALLICA and RYU, 1994) describing a selected cell suspension culture derived from stem cells of *Datura stramonium* which seems to produce up to 80 mg L^{-1} total tropane alkaloids within 25 days. According to the authors, an integrated approach (selection, media optimization, elicitation, precursor feeding, optimization of process parameters) is necessary for enhancing productivity in cell cultures. However, it was not clearly shown that a combination of all factors positively affecting productivity was indeed necessary to obtain the above yield. In any case it does not make much sense to continue studies on the optimization of tropane alkaloid formation in suspension cultures for production purposes. For studying the biosynthesis and regulation of this pathway, highly productive hairy root cultures of all these species are available (see Sect. 4). Using hairy root cultures, the initial steps of tropane/nicotine alkaloid formation have been identified, enzymes have been purified, and gene cloning is in progress (HASHIMOTO and YAMADA, 1994).

3.2.3 Quinoline Alkaloids

Cinchona trees have been grown in plantations since more than 130 years for the production of bark containing the antimalarial, antifever compound quinine. The related compund quinidine is used as a drug against cardiac arrhythmia. Though the compounds have repeatedly been found in cell suspension cultures of *Cinchona succirubra*, *C. pubescens*, and *C. ledgeriana* at levels of up to 0,9% (KOBLITZ et al., 1983), suspension cultures of *Cinchona* could not be further developed for biotechnological purposes. Attempts to increase the usually rather low levels by screening and media variation were not very successful (HARKES et al., 1985). In addition, growth of cultures of this species is generally slow (WJNSMA and VERPOORTE, 1988). Highest alkaloid production was obtained with cultures showing some degree of differentiation, e.g., roots, shoots (WIJNSMA and VERPOORTE, 1988), and compact globular structures (HOEKSTRA et al., 1990). Due to the difficulties of establishing reasonably productive culture systems, the progress in the elucidation of the biosynthetic pathway leading to quinoline alkaloids has remained slow (WIJNSMA and VERPOORTE, 1988).

3.2.4 Antitumor Compounds

The search for antitumor compounds in tissue cultures has been a favored goal of many researchers. The great interest in cell cultures of *Catharanthus roseus* resulted from the fact that this plant species synthesizes in very low amounts (0.0005% of dry mass) the expensive dimeric monoterpene indole alkaloids vinblastine (Fig. 7) and vincristine which are the most active agents used in the treatment of certain forms of cancer (MISAWA and ENDO, 1988). HIRATA et al. (1989) determined a vinblastine content of 1.4 mg g^{-1} of dry mass in leaves of *C. roseus*, a level which would reduce the need for a tissue culture derived process. It is now generally accepted that cell suspension cultures of *C. roseus* are not able to synthesize the dimeric alkaloids; vindoline, one of the monomeric precursors, is not formed in cultured cells but in leaves (DECAROLIS and DELUCA, 1993). Research has gone into two directions – development of shoot culture systems and semi-syntheses from the two monomers. However, as with many shoot culture systems, productivity is presently 100-fold lower than in the leaves of intact plants (HIRATA et al., 1989). More efficient is the approach of coupling vindoline and catharanthine (derived from plant tissue culture) by chemical or enzymatic means to form 3′,4′-anhydrovinblastine and vinblastine (GOODBODY et al., 1988). The coupling of catharanthine and vinblastine with peroxidases or ferric ions and subsequent reduction yielded 50–70% anhydrovinblastine and 12–28% vinblastine (DICOSMO, 1990). The results obtained by GOODBODY and coworkers at Allelix looked promising from a commercial point of view. Nevertheless, the project was discontinued. Evidently, there was no need for developing a tissue-culture-based commercial process. High levels of catharanthine and vindoline for chemical coupling are readily isolated as by-products from *C. roseus* plants. Since vindoline is not found in cell suspension cultures, tissue cultures are not needed for efficient synthesis of vinblastine.

Presently, there is a tremendous interest in developing a tissue culture process for the production of taxol® (taxol is a registered trademark of Bristol Myers-Squibb for pacli-

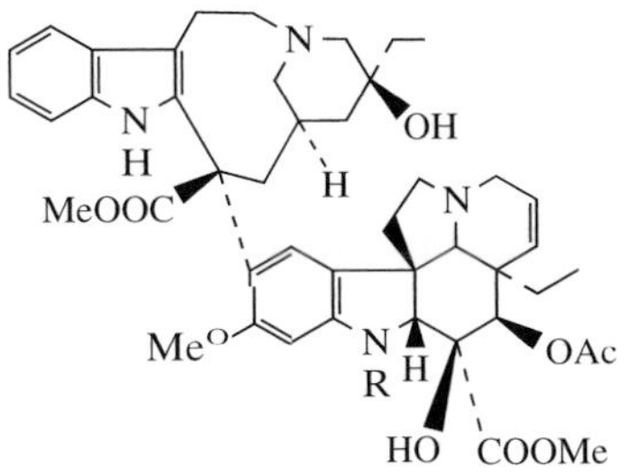

Vinblastine *(Catharanthus roseus)*

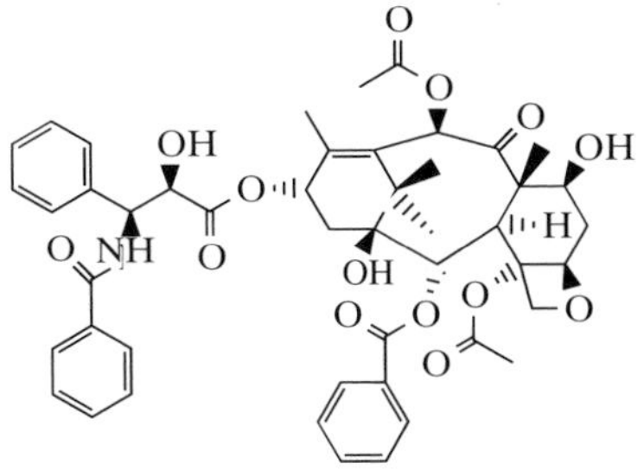

Taxol (*Taxus* spp.)

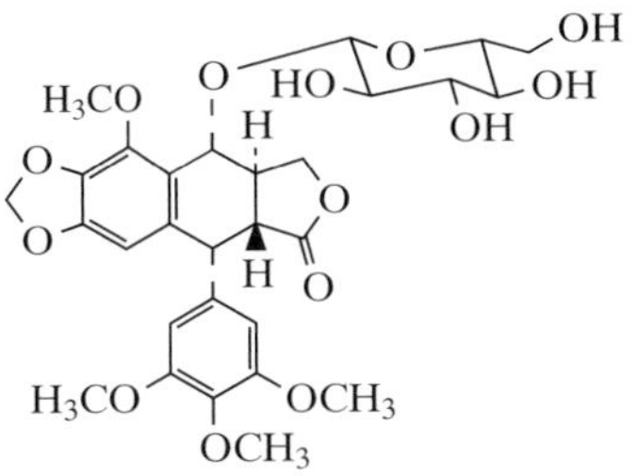

5-methoxypodopyllotoxin-4β-D-glucoside *(Linum flavum)*

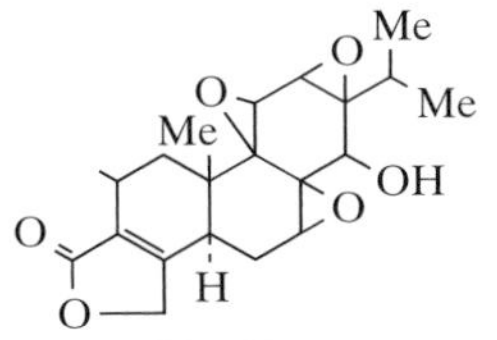

Triptolide *(Tripterygium wilfordii)*

Fig. 7. Antitumor compounds of some higher plants.

taxel) (Fig. 7). Detected during the screening program of the National Cancer Institute (NCI), it was considered the most interesting compound among 110000 tested structures (NICOLAOU et al., 1994). Taxol has been proven to be a very effective drug against several forms of cancer. It was approved by the US Food and Drug Administration (FDA) for the treatment of ovarian cancer in 1992. The chemistry and the möde of action of taxol has been reviewed recently (NICOLAOU et al., 1994; KINGSTON, 1994). A significant problem is a shortage in taxol supply (CRAGG et al., 1993). To isolate 1 kg taxol 10000 kg of dried bark from 3000 *Taxus brevifolia* trees have to be extracted. As nearly 2 g taxol are needed for the treatment of one patient it is evident that research has been initiated to improve taxol production in trees, to look for alternative sources of taxol, and to develop a total or semi-synthetic chemical synthesis (NICOLAOU et al., 1994; KINGSTON, 1994). Presently, taxol is produced commercially by semi-synthesis from 10-deacteyl baccatin III from needles of young *Taxus* trees grown in plantations. One alternative is to produce taxol by using tissue cultures and, indeed, as pointed out by EDGINGTON (1991) "if ever there was a plain target for plant tissue culture, this is it". In 1991, two American companies, Phyton Inc. and ESCAgenetics, announced that the development of their tissue culture-based processes was to be completed within 2–5 years. It was impossible to estimate how realistic these claims were, because both companies did not release scientifically sound information about their processes. Thus, conclusions could only be drawn from the results published by researchers from universities and from the improvements of systems with similar productivity obtained when industry was involved. In the meantime, ESCAgenetics dropped out of the race, while Phyton licensed their tissue culture process to Bristol-Myers Squibb. It has to be seen whether the huge fermentation facilities of Phyton in Germany will be used for commercial taxol production. In contrast to the American companies, the Mitsui Company has published some of their research results on taxol production in *Taxus* cultures (YUKIMUNE et al., 1996). According to the published data, the Japanese company again is ahead of all others.

Early publications on taxol production in tissue cultures of various *Taxus* species looked not very promising and did not support the taxol levels (153 mg L^{-1} after 6 weeks) claimed in the patent of BRINGI and KADKADE (1993). WICKREMESINHE and ARTECA (1993) induced callus cultures of various *Taxus* spp. and screened them for taxol production. The levels ranged from 0.0001–0.0131% of dry mass, and older browning callus produced more taxol than young pale callus. When they established cell suspension cultures from taxol producing callus, the specific taxol content could not be determined because the taxol peak was overlaid by other compounds (WICKREMESINHE and ARTECA, 1994a). From 28 g dry mass 120 μg pure taxol (4.3×10^{-5}%) were extracted. Thus quantitative data based only on HPLC determinations should be regarded with caution. The group of DICOSMO at Toronto University published several articles regarding the optimization of taxol production in *Taxus* callus and suspension cultures. They reported initial yields of 0.02% for a culture of *Taxus cuspidata* calli (FETT-NETO et al., 1992). Subsequently, it was shown that the medium composition affects taxol production and growth. As only percent of controls were given, it is not clear whether the experiments resulted in really higher productivity (FETT-NETO et al., 1993). When the cells were cultivated in shake flasks productivity went down by a factor of 10 in rapidly growing cultures (FETT-NETO et al., 1993), and in a more productive suspension culture levels of 0.15 mg L^{-1} were detected (FETT-NETO et al., 1994a). Taxol production could only be improved a little by feeding of potential precursors and thus, levels of 0.01% dry mass were found (FETT-NETO et al., 1994b). Substantially higher levels were reported by two groups from Cornell University, Ithaca (the city where Phyton is located); during culture media optimization experiments they reached levels of up to 15 mg L^{-1} taxol using lines of *T. baccata, T. canadensis,* and *T. cuspidata* (HIRASUNA et al., 1996; KETCHUM and GIBSON, in press). However, substantial variations of yields in individual experiments were observed making interpretations quite difficult. To overcome the oscillations of taxol production might be a severe problem in developing a large-scale process. In a bioreactor experiment with a working volume of 600 mL levels of up to 22 mg L^{-1} or 1.1 mg $L^{-1}d^{-1}$ taxol were found of which 90% were extracellular (PESTCHANKER et al., 1996). A breakthrough maybe was the finding that methyl jasmonate is an inducer of taxol biosynthesis in cultures of various *Taxus* species (MIRJALILI and LINDEN, 1996). Addition of 10μM methyl jasmonate increased taxol productivity 19-fold, from 0.2 mg L^{-1} to 3.4 mg L^{-1}. YUKIMUNE and coworkers (1996) added higher concentrations of methyl jasmonate (100 μM), used a cell line of *T. media* with a higher capacity for taxol biosynthesis, and observed taxol accumulation after induction for a longer period. Although taxol levels increased only 5-fold compared with controls, a productivity of 110 mg L^{-1} within 14 d was obtained. The specific content of taxol obtained was 0,6% and thus exceeded by far the specific contents found in any tissue of the intact plant.

Recently, transformed phytohormone-independent callus cultures were found to contain taxol levels up to 16 μg g^{-1} dry mass (HAN et al., 1994). Embryo culture has also been tested for taxol production (FLORES et al., 1993). However, taxol and taxane yields were not very different from those reported initially for callus and suspension cultures. A recent publication suggests roots of hydroponically grown *Taxus* plants as a source for taxol and related taxanes (WICKREMESINHE and ARTECA, 1994b), which would indicate that transformed root cultures should also be a suitable source for these compounds. Indeed, a recent patent application of Celex-Laboratories (1994) describes the production of taxol via hairy root cultures. Studying the published literature on taxol production in cell culture systems (for a review, see JAZIRI et al., in press), one might conclude that most cell lines of all *Taxus* species contain low levels of taxol, that some more productive lines can be obtained by screening, that methyl jasmonate is presently the only unambigously proven enhancer of taxol synthesis, and that production levels of higher producing lines often oscillate to an unacceptable extent. Nevertheless, if the published production rates can reproducibly be obtained in large volumes of several thousand liters and if they are optimized further by biochemical engineers, culture systems may compete with semi-synthesis. More knowledge about the biochemistry, regulation and limiting steps of the taxol pathway may eventually help to engineer lines with improved productivity (SRINIVASAN et al., 1996). They provide some evidence that taxol biosynthesis is located in plastids. They also conclude that the conversion of phenylalanine to phenylisoserine is a rate-limiting step of taxol biosynthesis. It might be important to stimulate the flux of the primary precursors into the taxane skeleton. Previously, it was assumed that mevalonate is the precursor of the isoprenoids. However, the findings of EISENREICH et al. (1996) suggest that the isoprenoid moieties of the taxane molecule are derived from a yet unknown precursor which is likely to be derived from a novel pathway of isoprenoid biosynthesis (using triose phosphate-type compounds and activated acetaldehyde), recently detected by ROHMER et al. (1993). CROTEAU's group demonstrated that taxa-4(5),11(12)-diene is the first intermediate in taxane biosynthesis, and recently they have purified the corresponding enzyme taxadiene synthase cyclizing the universal diterpene precursor geranylgeranyl pyrophosphate to taxa-4(5),11(12)-diene in a single step (HEZARI et al., 1995). Due to the importance of taxol it is predicted that further progress in the elucidation of taxol biosynthesis and its regulation will soon be made and that genetically engineered lines might be created in the near future.

Another important, commercially used antitumor compound is etoposide, a semisynthetic podophyllotoxin. Production of podophyllotoxin by tissue cultures of *Podophyllum peltatum* was first attempted by KADKADE (1982). However, the levels remained disappointingly low. The interest in podophyllotoxins was revived when root cultures of *Linum flavum* were found to contain high levels (1% of dry mass) of 5-methoxypodophyllotoxin (Fig. 7) and its glucoside (BERLIN et al., 1988b). Two Dutch groups confirmed this result and showed that suspension cultures of *L. flavum* also contained some podophyllotoxins (VAN UDEN et al., 1990; WICHERS et al., 1991). However, reasonable productivity required some morphological differentiation, e.g., root formation (VAN UDEN et al., 1991). Other podophyllotoxin producing cell cultures are those of *L. album* (SMOLLNY et al., 1993) and *P. hexandrum* (WOERDENBAG et al., 1990). The levels of podophyllotoxins vary from 0.05 to 0.3% of dry mass depending on the culture conditions and the tendency to differentiate. Thus root cultures (producing at least 1% podophyllotoxins in dry weight) seem to be most suitable for future research efforts, e.g., for biochemical studies (OOSTHAM et al., 1993).

A recent review by MISAWA and ENDO (1988) describes the productivity of cell culture systems for other potential antitumor compounds produced by plants and plant cell cultures. These include camptothecine, homoharringtonine, and maytansine. It must be concluded that the levels found in the culture systems are far too low to be attractive from a biotechnological point of view. The same holds true for tripdiolide and triptolide (Fig. 7) produced at 0.01% of dry mass by *Tripterygium* species (TAKAYAMA, 1994).

3.2.5 Cardiac Glycosides

The formation of cardiac glycosides in undifferentiated cell cultures of various *Digitalis* species was found to be very low or even lacking in rapidly growing suspension cultures. In slowly growing green callus cultures or in embryogenic cultures some cardiac glycoside accumulation was observed (for review see LUCKNER and DIETTRICH, 1988). However, low- or non-producing plant cell suspension cultures of *Digitalis lanata* are effective in the biotransformation of cardenolides and for this reason they are still rather attractive from a biotechnological point of view (see Sect. 6.2).

3.2.6 Vanillin and Vanilla Aroma

Vanilla is probably the most widely used flavor in food industries with a worldwide annual consumption of 1200–1500 t in 1993 (HAVKIN-FRENKEL, 1994). The prices for one kg natural vanillin from cured beans of *Vanilla planifolia* amount to $ 3000–4000 while the price for synthetic vanillin is less than $ 20 per kg. The flavor of natural vanilla extracts from beans of different origin varies. Though vanillin is the most important constituent of the aroma other compounds seem to affect substantially the taste of this flavor. Due to the high price of natural vanilla flavor plant tissue cultures, mainly of *V. planifolia*, have been regarded as an alternative source for the production of vanilla aroma and vanillin. The first publication on phenylpropanoid metabolism in *V. planifolia* showed that vanillin was not produced under normal culture conditions (FUNK and BRODELIUS, 1990). Formation of vanillic acid, however, was observed when 3,4-(methylenedioxy)-cinnamic acid, an inhibitor of *p*-coumarate CoA ligase, was added to the suspension cultures. This led to a shift from lignin to benzoate biosynthesis. After feeding potential cinnamic acid precursors there were indications that the cell culture must contain enzymes involved in benzoate biosynthesis. Finally, it was found that kinetin is an efficient elicitor of vanillic acid formation (FUNK and BRODELIUS, 1992). Vanillic acid levels of up to 0.1% of dry mass were detected. Thus, according to the literature, one would assume that cultures of *V. planifolia* are of no commercial interest. It was, therefore, somewhat surprising that ESCAgenetics (KNUTH and SAHAI, 1991) filed a patent in 1988 (issued in 1991) for the production of vanilla aroma (PhytoVanilla^{TM}) and biosynthesized vanillin (Phytovanillin^{TM}) by suspension cultures of *V. fragrans*. The aroma compounds are secreted into the medium and bind to absorbents. The patent states that 16–18 mg vanillin per L of medium were produced within ca. 45 days. A comparison of the vanilla flavor profiles of beans with that of tissue cultures as presented in the patents reveals that vanillin is the main component in both extracts, but that otherwise the composition of the two extracts is quite different. Thus, one can conclude that the suspension cultures produce a new vanilla aroma for which a market may or may not exist. ESCAgenetics announced that they would be able to produce tissue culture flavors such as vanilla at an incredibly low price of \$ 50–100 kg^{-1} (MOSHY et al., 1989). STAHLHUT (1993) reported that the vanilla flavor from suspension cultures of *V. fragrans* "exited the lab and entered commercial development". Unfortunately, the company was not willing to support their announcements by providing any scientific data or comments of their "marketed" product for this review. Thus, some doubts remained as to whether "PhytoVanilla^{TM} is considerably more economic to produce than natural vanilla extract" (GOLDSTEIN, ESCAgenetics). It has to be seen whether another company will continue the process of vanilla aroma production after the shutting down of ESCAgenetics.

Based on the literature and the patent it is now clear that vanillin production can be induced and optimized in cultured cells of *Vanilla* spp. by media composition, selection, elicitation, and feeding of suitable precursors. The highest specific yields (0.16% vanillin on a dry mass basis) were reported for an embryo culture of *V. planifolia* grown in small bioreactors (KNORR et al., 1993; HAVKIN-FRENKEL, 1994). A comparison of an HPLC extract of the embryo culture and a Bourbons vanilla bean extract exhibited much more similarities of the components than shown in the patents of ESCAgenetics. HAVKIN-FRENKEL (1994) also compared the production costs. If the cultures produce 2% vanillin in large vessels, the estimated costs for the production of 1 kg vanillin would be between \$ 500–1 000 (investment for bioreactors not included). Although the embryo culture system contains the highest published content of vanillin reported up to now it is commercially not yet competitive. In this context it should be mentioned that alternative biotechnological approaches for the production of vanillin have been investigated (CHEETHAM, 1993). For example, root tissue of *V. planifolia* converts ferulic acid to vanillin at production rates of 400 mg kg^{-1} d^{-1} root tissue and concentrations of 7 g kg^{-1} roots were obtained (WESTCOTT et al., 1994). This suggests that hairy root cultures should be tested for the production of vanillin by biotransformation. The production of vanillin and related compounds may also be improved in the near future on the basis of present studies on the enzymology of the formation of benzoic acids from cinnamic acids (LÖSCHER and HEIDE, 1994).

4 Secondary Product Formation in Hairy Root Cultures

The attempts at enhancing product formation in rapidly growing suspension cultures showed that some pathways are spontaneously well expressed under such conditions. Furthermore, their product formation can easily be optimized by rather simple methods. Poorly expressed pathways, however, do not respond well to manipulations aimed at obtaining productivity improvements. There are unfortunately many more compounds, not mentioned in the previous sections, which are only formed at low levels in undifferentiated cell cultures. After induction of some morphological differentiation, however, enhanced biosynthesis of many of these compounds was observed. Organ cultures of plant cells, e.g., root and shoot cultures, are known for quite some time. However, from a biotechnological point of view these cultures were of low interest because they are difficult to establish, they often grow slowly and are sometimes very unstable. An overview of the nevertheless substantial studies on the accumulation of secondary compounds by organized shoot and root plant cultures has been given by CHARLWOOD et al. (1990). The image of organ cultures has greatly changed since the finding that for most dicotyledonous plants rapidly growing hairy root cultures can be initiated by transformation with *Agrobacterium rhizogenes*. Besides the progress in the biochemistry and molecular biology of plant secondary pathways the broad application of hairy root cultures has been the most important stimulus for the field. The special characteristics of transformed hairy root cultures in comparison with normal root cultures have been reviewed by RHODES et al. (1990). They are phytohormone-independent, exhibit rapid growth and a high degree of genetic and biochemical stability. Their biosynthetic capacity is equivalent to the corresponding plant root, and they can be grown in bioreactors. The most important biotechnological value of these cultures is perhaps that many pathways are accessible to biochemical and molecular studies. Indeed, they are also amenable to genetic manipulations.

The first report on the initiation of hairy root cultures for production purposes was in 1985 when FLORES and FILNER showed that *Hyoscyamus* spp. hairy root cultures produced 0.5% tropane alkaloids on a dry mass basis and those of tobacco contained more than 3% nicotine. This was followed by a huge number of reports on the initiation of metabolite producing hairy root cultures. Thus, this review cannot cover all the productive hairy root culture systems which have been described in the recent years (for a review or literature see SIGNS and FLORES, 1990; RHODES et al., 1990; PAYNE et al., 1991; SAITO et al., 1992; TOIVONEN, 1993; BANTHORPE, 1994). Naturally, plant species which do not produce the desired commercially important compounds in suspension cultures were the main target. The greatest biotechnological efforts have been put into hairy root cultures of tropane alkaloid producing Solanaceae. According to the literature several hundred individually transformed hairy root cultures have been established and analyzed with respect to their ability to form tropane alkaloid (KNOPP et al., 1988; PARR et al., 1990) and hairy root cultures of all known plants (e.g., *Atropa*, *Datura*, *Hyoscyamus*, *Brugmansia*, *Duboisia*, *Scopolia*) producing these alkaloids have been established. In general, it was found that hairy root cultures contained roughly the same amount of alkaloids as that found in the roots of the corresponding intact plant species. Root cultures may show some variation in growth rate, alkaloid content, and productivity (MANO et al., 1989). Thus, screening for a high producing line seems to be worthwhile. However, the productivity differences between 60 different lines of one species were only 2–3-fold on average. The medium composition is also important for growth and alkaloid production (MANO et al., 1989; BERLIN et al., 1990). In particular, the amount of nitrogen and the ratio NO_3^-/NH_4^+ seem to affect branching of roots and product formation. With the addition of phytohormones to the growth medium of root cultures, more callus-like root cultures are formed which produce much lower levels of alkaloids. In contrast to suspension cultures where improved production is often observed under growth limiting conditions hairy root cultures produce best when root formation and growth is optimal. In agreement with this elicitors (stress factors) had no great effect on product levels although it induced phytoalexin production (FURZE et al., 1991). In general, it was found that the specific production of secondary metabolites cannot be manipulated as readily as in productive suspension cultures (BERLIN et al., 1990; TOIVONEN, 1993).

In theory, all dicot plants should be susceptible to transformation by *A. rhizogenes* and should yield good growing hairy root cultures. However, some plant species have been found to be rather recalcitrant. For example, transformation of *Papaver somniferum* with *A. rhizogenes* resulted in phytohormone-independent cell suspension cultures rather than root cultures (WILLIAMS and ELLIS, 1993). Since normal root cultures of *Linum flavum* produce high levels of 5-methoxypodophylotoxin (BERLIN et al., 1988b) it was logical to initiate hairy root cultures of this species. A successful transformation has been reported only once (OOSTHAM et al., 1993). However, this culture exhibited a lower growth rate than normal root cultures. Our own efforts to establish hairy root cultures of *L. flavum* have not yet been successful (BERLIN and KUZOVKINA, unpublished results).

In addition to the usefulness of hairy root cultures in biochemical and molecular studies, they might also be useful for the production of desired metabolites in large-scale volumes due to their good productivity. A recent review on the cultivation of root cultures in bioreactors (CURTIS, 1993) shows that the process engineers are just at the beginning of developing suitable reactors and operation schemes. Several hairy root cultures have been grown in bioreactors with capacities of 1–20 L. For example, hairy root cultures of *Coleus forskohlii* grown in 20 L glass jar fermentors produced forskolin (a novel heart active and blood pressure-lowering compound) with a yield of 14 mg L^{-1} after 3 weeks (KROMBHOLZ et al., 1992). In suspension cultures the forskolin content was extremely low. Many different reactor configurations have been used for growing hairy root cultures. Good results were obtained with mechanically agitated fermentors (KONDO et al., 1989; BUITELAAR et al., 1991). High growth rates were also achieved in reactors in which the roots were fixed to supports of stainless steel or polyurethane foam and sprayed with nutrient (WILSON et al., 1990; DI IORIO et al., 1992). A technical problem is the scale-up of hairy root cultures from small to very large bioreactors. Hairy roots form usually large aggregates which cannot be pumped through pipes from one reactor to another. We suggested (BERLIN et al., 1990) to grow the phytohormone-independent hairy root cultures initially in the presence of phytohormones. The resulting callus-root culture with substantially shorter roots would be used for scale up. In the last bioreactor full root formation could then be reinduced by omission of the phytohormone. TAKAYAMA et al. (1994) demonstrated the feasibility of this approach for a hairy root culture of *Hyoscyamus niger* which was scaled up in the presence of 0.3 mg L^{-1} NAA in an agitated bioreactor. They obtained a root suspension which could be transferred through pipes (13 mm internal diameter) from one reactor to the next. Tropane alkaloid production (3% of dry mass) reached the original levels after transfer to the phytohormone-free medium.

Sometimes the products of hairy root cultures are spontaneously released into the medium, from which they can easily be recovered. A scopolamine secreting line of *Duboisia leichhardtii* was cultivated in a bioreactor with continuous exchange of the medium. The product was recovered by using a XAD-2 column. Under optimized conditions (stainless steel mesh as a support, turbine-blade reactor, two-stage culture for growth and product release) a total of 1.3 g L^{-1} scopolamine was recovered from the XAD-column during 11 weeks of continuous operation (MURANAKA et al., 1993). High levels of thiophenes are only produced in root cultures of *Tagetes* ssp. BUITELAAR et al. (1991) grew hairy roots of *T. patula* with 1.6% thiophenes on a dry mass basis in various bioreactors and continuously harvested 70% of the lipophilic compounds in a two-liquid-phase system. Interestingly, product formation and secretion of secondary products into the medium was often enhanced when two-phase culture or absorption to resins was used (TOIVONEN, 1993).

5 Plant Tissue Cultures as a Source of New Chemicals?

Plant tissue cultures are not usually used as producers for commercially interesting compounds since alternative production methods (field grown plants, chemical synthesis) have been established by industry. The observation that under cell culture conditions new metabolites sometimes accumulate which had not previously been detected in the intact plant suggested that plant cell cultures may be used as a source of new chemicals. In addition, components which are present in the intact plant at low levels were found to be present at high levels in cultured cells in some cases, for example, sanguinarine in poppy plants (WILLIAMS and ELLIS, 1993). Dedifferentiation of the cultured plant cell leads to a regulatory state which is not present in the developing plant, and this may lead to the expression of novel product patterns (ARENS et al., 1982). The idea of screening plant cell cultures for novel pharmacologically active compounds was mainly followed by the Nattermann Company in Cologne and researchers of the Pharmaceutical Institute of the University of Munich. Indeed, some new biologically active compounds were detected by such screenings (ARENS et al., 1986). However, from a biotechnological point of view this approach would only be successful if the new compound had unique (superior) pharmacological characteristics and if product levels were high enough to encourage further development. Taking into account the number of extracts which had to be screened by the NCI program before a unique compound such as taxol was detected, and that product patterns and formation are much lower in cultured cells than in intact plants (BERLIN, 1986), the chances of a successful outcome of this approach were questionable from the beginning. In the meantime, the Nattermann Company has given up this approach. Although the number of "new" compounds found exclusively or originally in tissue cultures has steadily increased (RUYTER and STÖCKIGT, 1989), the importance of cell cultures as a source of novel compounds has not increased.

6 Biotransformations with Cultured Plant Cells

The previous sections dealt mainly with the *de novo* synthesis of secondary metabolites by cell suspension and hairy root cultures. The production of valuable plant products from cheap precursors by biotransformation is another possibility of using plant tissue cultures. Indeed, the 12-β-hydroxylation of β-methyldigitoxin to β-methyldigoxin by cell suspension cultures of *Digitalis lanata* was expected to become the first commercial tissue culture process (REINHARD and ALFERMANN, 1980). This example showed that even cultures unable to synthesize a particular compound *de novo* can carry out specific reactions of the corresponding pathway. Many other examples demonstrate the potential of plant cell cultures to biotransform exogenous substrates (for a review see SUGA and HIRATA, 1990). However, the main question remains as to whether plant cell cultures can compete with the well-known capabilities of microorganisms in the field of biotransformations. Thus, one should look for plant-specific biotransformation reactions which are not performed by microorganisms. Glycosylation of foreign products, for example, is a reaction which is not performed by bacteria. Plant cells should also be superior to microorganisms if stereospecific enzymatic reactions of plant specific pathways are required. The two examples given below demonstrate how plant cell cultures can be used for biotechnologically relevant biotransformations.

6.1 Arbutin

Arbutin has been used as a urethral disinfectant for many years and has been shown to be a potent suppressor of melanin synthesis in human skin. It is thus used as a skin depigmentation agent. Although arbutin is found at rather high levels (5–8%) in several members of the Ericaceae, it is presently produced by chemical synthesis. The Japanese Shiseido Company is trying to develop an alternative production process by the glucosylation of hydroquinone added to cell suspension cultures of *Catharanthus roseus*, although arbutin is not a natural constituent of this plant (YOKOYAMA and YANAGI, 1991). First the culture stage during which highest glucosylation occurs was determined. Then the medium composition was optimized, and the biotransformation efficiency of various strains and cultures were compared. By fed-batch cultivation and continuous feeding of the high density cultures with substrate (14 d old, 5 L impeller driven bioreactor) arbutin formation was improved to nearly 10 g L^{-1} within 4 days. A specific content of 45% arbutin per g dry mass was tolerated by the cells which is the highest specific content of a secondary metabolite reported for cultured plant cells. It was stated that the production costs for arbutin by chemical synthesis or biotransformation in tissue cultures are very similar (YOKOYAMA and YANAGI, 1991). They are rather optimistic that the biotechnological production of useful glucosides will become increasingly inexpensive relative to chemical synthesis. Recently, it was shown that cell suspension cultures of *Rauwolfia serpentina* are also able to form efficiently arbutin from hydroquinone (18 g L^{-1} in 7 d) (LUTTERBACH and STÖCKIGT, 1992).

6.2 Biotransformation of Cardiac Glycosides

Digitalis lanata plants contain two main cardiac glycosides which upon hydrolysis yield digoxin and digitoxin, drugs which are widely used in the treatment of heart diseases. Digitoxin is used to a much lesser extent in therapy, and therefore conversion of digitoxin to digoxin by 12-β-hydroxylation was regarded as a biotechnologically useful biotransformation reaction. Cell cultures of *D. lanata,* although unable to produce cardiac glycosides *de novo*, were found to contain high 12-β-hydroxylation activity (REINHARD and ALFERMANN, 1980). However, β-methyldigitoxin was initially found to be the only substrate which was exclusively converted into a digoxin derivative. With all other digitoxins, side reactions were observed. The use of β-methyldigitoxin as a substrate seemed to be feasible since it was already being produced by Boehringer Mannheim. This company initially was interested in applying cell cultures of *D. lanata* in the biotransformation of β-methyldigitoxin. However, the idea of using this technology for production was dropped in the early 1980s. Nevertheless, research on this system has continued. REINHARD and coworkers at the University of Tübingen optimized the biotransformation up to a 210 L working volume in an airlift reactor (REINHARD et al., 1989). During a semicontinuous process (6 repeated batch cultivations) 513 g β-methyldigoxin was produced from 641 g β-methyldigitoxin within 3 months. Since the product is released into the medium further improvements in shortening the production period are expected. Optimization of the biotransformation process did not only include studies on optimal substrate supply and medium composition but also selection of an improved strain for highest hydroxylation capacity by repeated cell aggregate cloning (REINHARD et al., 1989).

When digitoxin was added to cell cultures of *D. lanata* two reactions occurred: 12-β-hydroxylation and 16′-O-glucosylation yielding deacetyllanatoside C as the main product. KREIS and REINHARD (1990) developed a two-stage semicontinuous process by transferring a selected cell line into a production medium (8% glucose) for the biotransformation reaction. On average 400 mg L^{-1} deacetyllanatoside C were produced from 600 mg digitoxin in a 20 L airlift fermentor within 7 d. To achieve a high biotransformation of digitoxin into one major cardenolide, it was important to change from a growth to a production medium. In contrast to β-methyldigoxin, deacetyllanatoside C is mainly found within the cells. The main goal, however, was to develop a biotransformation production process from digitoxin to digoxin by preventing the 16′-O-glucosylation. Since the optimum temperature of 12-β-hydroxylase is around 20 °C while that of glucosyltransferase is 37 °C, lower temperatures (e.g., 19 °C) would therefore favor hydroxylation (KREIS and REINHARD, 1992). Deacetyllanatoside is mainly stored in the vacuoles of the cells while digoxin is released into the medium. The lower the cell density was, the lower was the storage capacity for deacetyllanatoside and the higher was the percentage of digoxin formation. Based on this knowledge a semicontinuous production process in a 300 L airlift reactor was developed with 8% glucose as a production medium and incubation times of 40–60 h. From 0.8 mmol digitoxin, ca. 0.6 mmol digoxin were obtained, 80% of which was found in the medium (KREIS and REINHARD, 1992). The latter result demonstrates in particular the importance of understanding the physiology and biochemistry of a pathway if natural products are used in biotransformation processes (KREIS et al., 1993).

7 Metabolic Engineering of Secondary Pathways in Cultured Cells

Many reviews have pointed out that more knowledge of the regulation of secondary pathways at the enzyme and gene level is needed before the true potential of cultured plant cells for metabolite production can be recognized and exploited. The above analysis of the field clearly supports this view. Despite all the progress, the fact remains that the biology of the cell is the limiting factor since only those pathways which are spontaneously well expressed under culture conditions or easily inducible by simple modifications are accessible for the development of biotechnological processes. This can only be changed if we succeed in altering the expression of pathways by genetic engineering techniques. Indeed, several groups worldwide have used their culture systems (as indicated in the previous sections) not only for optimizing product levels but also for biochemical and molecular studies. Thus, our knowledge of the enzymology of several pathways has greatly increased during the last few years. The genes coding for several of these enzymes have been cloned. For the phenylpropanoid/flavonoid pathway, which has been analyzed the most extensively, the first regulatory elements have been detected (FORKMANN, 1993; KOES et al., 1994). Thus, we seem to be at the beginning of an era when aimed manipulations of secondary pathways might become possible. This might also tremendously increase the production potential of plant cell cultures and thus their biotechnological impact. However, progress in this direction is likely to be slow since even in microbial systems the success of metabolic engineering is not yet overwhelming (CAMERON and TONG, 1993; STEPHANOPOLOUS and SINSKY, 1993). In order to devise rational approaches for metabolic engineering a knowledge of the limiting steps of the pathways is required (STEPHANOPOULOS and SINSKY, 1993). Locating such critical steps is often rather difficult. In the case of non-expressed pathways in plant cells, a one-step manipulation will usually not be sufficient. Nevertheless, the first genetic manipulations of secondary pathways in plant cells have been performed. However, only a few of them have been directed towards the production of secondary metabolites. In most cases, the establishment of transgenic plants rather than cultures with altered characteristics has been attempted. As the expression of secondary pathways is quite different in intact plants and in cultured cells, the metabolic effects of a genetic transformation in these systems will also be different. This review focusses on the manipulation of product formation in cultured cells which might later be grown in bioreactors. Thus, only the results of transgenic cell cultures which were created for the overproduction of desired compounds by various biosynthetic pathways will be described in some detail. Nevertheless, it should be mentioned that genetic engineering of transgenic plants with desired characteristics is in full progress. For example, Calgene and Suntory have announced the development of blue roses (HOLTON and TANAKA, 1994). A general overview of metabolic engineering of commercially useful biosynthetic pathways in transgenic plants and plant cells were given by KISHORE and SOMERVILLE (1993) and NESSLER (1994).

7.1 Serotonin Biosynthesis in *Peganum harmala*

Cell cultures of *Peganum harmala* loose their capability of synthesizing serotonin (Fig. 8) during prolonged subcultivation. It was found that this was due to the lack of tryptophan decarboxylase (TDC) activity in fully undifferentiated cell suspension cultures. The second and final enzymatic step of serotonin biosynthesis, the 5-hydroxylation of tryptamine, however, remains highly expressed in all *P. harmala* cell cultures (SASSE et al., 1987; COURTOIS et al., 1988). Thus, it was a rational approach to introduce a constitutively expressed *tdc* gene into *P. harmala* cell suspension cultures to restore or maintain serotonin formation. A corresponding cDNA of *Catharanthus roseus* (GODDIJN, 1992) was introduced into *P. harmala* cells under the control of the 35S promoter of the Cauliflower Mosaic Virus via *Agrobacterium tumefaciens* (BERLIN et al., 1993). Several cell suspension cultures were obtained with constitutively expressed TDC activities of around 30 pkat/mg protein and serotonin levels of 1–2% of dry mass; the serotonin levels of the controls were below 0.1%. It was also shown that in these transgenic cultures tryptophan supply is the next rate-limiting factor (BERLIN et al., 1993) which has to be overcome to obtain even higher levels of serotonin. Although serotonin biosynthesis is not representative of a typical secondary pathway because of its simplicity, this example shows that removal of a rate-limiting step (clearly identified by the feeding experiments with tryptamine) can help to increase the production rate.

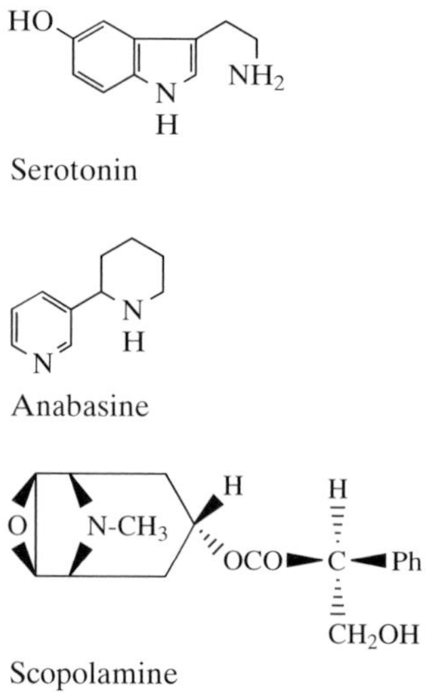

The same *tdc* gene was also overexpressed in *C. roseus*, the plant from which it was derived. However, as expected from literature data, the constitutive expression of TDC only led to a large accumulation of tryptamine and did not affect alkaloid levels (GODDIJN, 1992). Feeding of tryptamine does not enhance alkaloid formation in *Catharanthus* cell suspension cultures. The overproduction of tryptamine by the constitutive expression of the *tdc* gene had evidently no negative effect on tryptophan supply for protein synthesis. Anthranilate synthase levels (the proposed regulatory enzyme of tryptophan biosynthesis) were unchanged in transgenic tobacco plants (POULSEN et al., 1994). An interesting, yet inexplicable observation was made by SONGSTAD et al. (1991). They found that *tdc*-transgenic tobacco plants not only overproduce tryptamine but also produce high levels of tyramine, although the engineered enzyme does not accept tyrosine as substrate.

Fig. 8. Secondary metabolites whose levels were enhanced in cultured cells by metabolic engineering.

7.2 Affecting Nicotine Alkaloid Biosynthesis in Tobacco

The regulation of nicotine biosynthesis is relatively well understood (for a review see HASHIMOTO and YAMADA, 1994). The activity levels of putrescine N-methyltransferase (PMT) determine whether reasonable levels of nicotine can be formed in tobacco suspension cultures. As this enzyme is also a key enzyme in tropane alkaloid biosynthesis, great efforts have been made to clone this gene, even by researchers from industry. Philip Morris (NAKATANI and MALIK, 1992) filed a patent on the production of transgenic tobacco plants overexpressing PMT. However, reports on such transgenic plants overproducing PMT have not yet been published.

The first attempts to affect nicotine biosynthesis genetically have used the gene of a yeast ornithine decarboxylase (*odc*) (HAMILL et al., 1990). This enzyme activity, however, is not generally regarded as a rate-limiting step of the pathway since feeding putrescine does not affect nicotine levels. Transgenic root cultures of *Nicotiana rustica* constitutively expressing the yeast gene under the control of the CaMV35S promoter contained distinctly enhanced ODC activity. However, the high enzyme activity did not result in corresponding high levels of nicotine as product levels were only increased 2-fold. In view of the fact that putrescine is not a limiting intermediate in nicotine biosynthesis, the outcome was not so surprising. It is not yet clear whether the 2-fold increase is really due to the increased enzyme activity. Feeding of the cultures with ornithine could provide more insight into the influence of the engineered ODC on the nicotine levels.

Anabasine (Fig. 8) is an analog of nicotine in which the N-methyl pyrrolinium ring is replaced by a Δ^1-piperideinium ring. This ring is derived from lysine/cadaverine, and it has clearly been shown that cadaverine is the rate-limiting intermediate in anabasine biosynthesis (WALTON and BELSHAW, 1988). Therefore, a bacterial lysine decarboxylase (*ldc*) gene was introduced into tobacco root cultures (BERLIN et al., 1994, HERMINGHAUS et al., 1996). When the gene was placed under the control of the CaMV35S promoter and fused to the coding sequence of the transit peptide of the small subunit of ribulose diphosphate carboxylase, many root cultures with LDC activity were obtained. This enzyme activity was not detected in any of the control cultures. It was shown that the enzyme had been targeted to leucoplasts, the site of lysine biosynthesis in root cells. High levels of cadaverine (0.5% of dry mass; control cultures, ca. 0.01%) and anabasine (0.5% of dry mass; controls, 0–0.02%) were found in the most productive root culture (HERMINGHAUS et al., 1996). Feeding of lysine enhanced both cadaverine and anabasine levels to more than 1% of dry mass while metabolite levels of control cultures were hardly affected (HERMINGHAUS et al., 1996). Nicotine was still the main alkaloid in these cultures. However, anabasine became the second most abundant alkaloid. Improving the internal supply of lysine by altering the feedback control of lysine biosynthesis (SHAUL and GALILI, 1991) might lead to even higher levels of anabasine. Lysine decarboxylase expressed in the leaves resulted in the production of cadaverine only (HERMINGHAUS et al., 1996). This shows that genetic manipulation at one biosynthetic node will lead to significant improvements of product levels only when several prerequisites are fulfilled. Firstly, the target enzyme must be the rate-limiting step; secondly, the engineered enzyme must be sufficiently supplied with substrate; the reaction product must be transported to the site from where it is channelled into the related pathway. The latter means that the other enzymes of the pathway should be expressed quite well. Therefore, genetic manipulations in undifferentiated cells, in organ cultures, or in whole plants might yield completely different results depending on the overall expression of a biosynthetic pathway in various tissues.

7.3 Enhancing Scopolamine Production in *Atropa belladonna*

Hyoscyamine 6β-hydroxylase catalyzes the oxidative reactions in the biosynthetic pathway leading from hyoscyamine to scopolamine (Fig. 8). A cDNA coding for this enzyme with epoxidating and hydroxylating activities was cloned using RNA from a root culture of *Hyoscyamus niger* (reviewed by HASHIMOTO and YAMADA, 1994). The cDNA was cloned into a binary vector for plant transformation under the control of the CaMV35S promoter and integrated into *Agrobacterium rhizogenes* 15834 (HASHIMOTO et al., 1993). Hairy root cultures were obtained by infecting *Atropa belladonna* leaf disks. Hyoscyamine 6β-hydroxylase activities were 3- to 5-fold increased in the transgenic root cultures compared to the controls. Scopolamine levels of the controls were ca. 0.05% of dry mass, hyoscyamine being the main alkaloid. In the best transgenic lines scopolamine levels were up to 5-fold higher while the hyoscyamine content was greatly reduced (HASHIMOTO et al., 1993). However, there is evidence that the expression of the engineered enzyme is not yet optimal in root tissues. When the vector was introduced into an *Agrobacterium tumefaciens* strain transgenic *A. belladonna* plants which accumulated high levels of scopolamine in leaves and stems were obtained; the amount of scopolamine formation in the roots of the same plants was not very high (YUN et al., 1992).

7.4 Does Genetic Engineering Improve the Potential of Plant Tissue Cultures as Producers of Interesting Metabolites?

The examples given above show that the levels of end products of biosynthetic pathways can be modified by genetic engineering methods. Best results will be obtained when the overproduced enzyme activity is clearly the rate-limiting step of the pathway. However, usually a pathway has a series of rate-limiting steps. Thus, if one step has been removed by genetic engineering, the next (e.g., substrate supply) may soon become active and limit the extent of overproduction. The systems we have been working on (serotonin and anabasine biosynthesis) could be used to check whether by a second transformation the second rate-limiting step might also be overcome. The genes controlling lysine and tryptophan biosynthesis have been cloned. On the other hand, it is clear that the identification of rate-limiting steps of pathways is usually not as easy as in our systems where simple feeding experiments provided the information. Thus, one can understand and share the sceptical view of HASHIMOTO and YAMADA (1994) that single site manipulations of secondary pathways will usually not be sufficient for high increases of pathway end product levels. On the other hand, one should realize that we are just at the beginning of a new era. Therefore, conclusions should not be drawn before the first double transformants have been obtained. In addition, the product levels which are not overwhelming so far may increase further when expression and targeting of the engineered protein is optimized.

The greatest potential of metabolic engineering for improving the standing of plant cell cultures will, however, only become apparent if we succeed in identifying regulatory genes which control the expression of whole pathways in cultured cells. It is possible that pathways which are not usually in operation might become active in undifferentiated cells. We are still far away from the identification of elements which regulate the tissue- and organ-specific initiation of a whole secondary pathway. However, a promoter analysis and the search for transduction signals and transcription factors involved in the regulation of secondary pathways have already been started (DIXON and LAMB, 1990; KOES et al., 1994). Thus, it may be possible to express more pathways in cultured cells in the not too distant future.

In some situations it may be desirable to turn off or fortify one branch of a pathway so that more or less carbon enters that branch. Overexpression of a tryptophan decarboxylase gene in *Brassica napus* led to the accumulation of high levels of tryptamine, while the levels of unwanted indole glucosinolates in seeds of the transgenic plants were only 3% of that found in seeds of untransformed plants (CHAVADEJ et al., 1994). Reduction of metabolite levels has also been achieved by expressing antisense RNA which is complementary to the mRNA encoding a pathway enzyme. Antisense expression of a chalcone synthase gene resulted in distinctly decreased flower pigmentation (VAN DER KROL et al., 1988).

Overexpression of an engineered enzyme may lead to an accumulation of intermediates of a pathway to high levels. Examples are the overexpression of the amino acid decarboxylases leading to high accumulation levels of the corresponding amines (BERLIN et al., 1994). In addition new products may be formed in cultured cells. For example, formation of resveratrol in tobacco occurred only if stilbene synthase genes from *Arachis hypogaea* or *Vitis vinifera* were expressed (HAIN et al., 1990; 1993). Expression of a coriander desaturase results in the formation of petroselinic acid, a novel product of tobacco (CAHOON et al., 1992). However, engineering of novel enzyme reactions in a plant cell will only be useful if a corresponding substrate is naturally formed in the transgenic cells. Expression of a strictosidine synthase gene, coding for a key enzyme of monoterpene alkaloid biosynthesis, in tobacco, led to an active enzyme without function because its substrates tryptamine and secologanin are not formed in tobacco (MCKNIGHT et al., 1991).

The enzymes of secondary metabolism in plant cell cultures have been regarded as a "pot of gold" (ZENK, 1991). There is no doubt that certain plant cell culture systems have been most important for recent progress in the enzymology of secondary pathways. It remains to be seen whether these enzymes are also a "pot of gold" for biotechnology purposes. However, it is now possible to create plant cell cultures which overproduce certain enzymes of secondary metabolism by genetic engineering. When the enzyme itself is the target compound for overproduction, plant cell cultures are not considered to be competitive (KUTCHAN et al., 1994). These authors recommended the heterologous expression of plant genes in insect cultures as a more convenient approach, because of their higher productivity and secretion of the enzymes into the medium from which they are easily purified. Using a cell culture of *Spodoptera frugiperda* (*Sf9*) up to 4 mg L^{-1} strictosidine synthase of berberine bridge enzyme was produced (KUTCHAN et al., 1994). The real potential of transgenic plant cell cultures as enzyme producers, however, has not yet been exploited. Using targeting signals and suitable promoters it may be possible to produce comparable levels of a desired protein in plant cells.

8 Conclusions and Outlook

Substantial progress has been made in the biochemistry and molecular biology of secondary pathways of higher plants since the first edition of *"Biotechnology"*. Many novel enzyme reactions have been detected, and a number of genes encoding components of secondary metabolism have been cloned. Successful alterations of secondary pathways by genetic transformation have been achieved. With the availability of stable and rapidly growing hairy root cultures, the number of products found at high levels in cultured cells has increased greatly. The technological feasibility of large-scale fermentations of suspension cultures has been clearly demonstrated. The technical problems of growing hairy root cultures in huge bioreactors are surmountable. However, despite all scientific progress, the enthusiasm of industry for plant cell cultures as producers has not yet correspondingly increased. Indeed, there is no reason to believe that industries will change their reserved attitude in the foreseeable future. One can only hope that this conclusion is wrong, and that the following statement made at the International Tissue Culture Congress is correct: "Scientists from universities and research centers have not yet realized what is really going on industry with respect to the tissue culture technology" (SMITH, 1995). Although this optimistic view is shared by some other American colleagues (TATICEK et al., 1994) the future does not look so promising. As a consequence of the overall low interest of the industrial sector in plant secondary metabolites and plant tissue culture technology, the governmental support for this field of biotechnology was reduced. Several leading groups of biotechnologically orientated governmental research centers, e.g., in England and Canada, were recently forced to leave the field. Thus, this field will remain an exciting area of basic research but not of biotechnology. It remains to be seen whether the future progress in understanding and manipulating regulatory controls of secondary pathways will eventually enhance the biotechnological importance of plant cell cultures as producers. The decisive question, which can only be answered by each individual company, is whether the commercial importance of any plant-tissue-culture-derived product is high enough to justify the undoubtedly high investments needed for large-scale production.

Acknowledgement

I would like to thank Dr. HARA, Mitsui Co., Dr. SMITH, Urbana, and Dr. TAKAYAMA, Shizuoka, for providing me with new and unpublished information. Special thanks are due to Dr. HAVKIN-FRENKEL, David Michael & Co,, for her helpful advice regarding flavor production in tissue cultures.

References

ARENS, H., BORBE, H. O., ULBRICH, B., STÖCKIGT, J. (1982), Detection of pericine, a new CNS-active indole alkaloid from *Picralima nitida* cell suspension culture by opiate receptor binding studies, *Planta Med.* **40**, 218–223.

ARENS, H., ULBRICH, B., FISCHER, H., PARNHAM, M. J., RÖMER, A. (1986), Novel antiinflammatory flavonoids from *Podophyllum verspille* cell culture, *Planta Med.* **52**, 468–473.

ASAKA, I., II, I., HIROTANI, M., ASADA, Y., FURUYA, T. (1993), Production of ginsenoside saponins by culturing ginseng (*Panax ginseng*) embryogenic tissues in bioreactors, *Biotechnol. Lett.* **15**, 1259–1264.

BALANDRIN, M. F., KLOCKE, J. A., WURTELE, E. S., BOLLINGER, W. H. (1985), Natural plant chemicals: Sources of industrial and medicinal materials, *Science* **228**, 1154–1160.

BALLICA, R., RYU, D. D. Y. (1994), Tropane alkaloid production from *Datura stramonium*: An integrated approach to bioprocess optimization of plant cultivation, in: *Advances in Plant Biotechnology. Studies in Plant Sciences*, Vol. 4 (RYU, D. D. Y., FURASAKI, S., Eds.), pp. 221–254. Amsterdam: Elsevier.

BALSEVICH, J. (1988), Monoterpene indole alkaloids from Apocynaceae other than *Catharanthus roseus*, in: *Cell Culture and Somatic Cell Genetics*, Vol. 5: *Phytochemicals in Plant Cell Cultures* (CONSTABEL, F., VASIL, I. K., Eds.), pp. 371–384. San Diego: Academic Press.

BANTHORPE, D. V. (1994), Secondary metabolism in plant tissue culture: scope and limitations, *Nat. Prod. Rep.* **11**, 303–328.

BATE, N. J., ORR, J., NI, W., MEROMI, A., NADLER-HASSAR, T., DOERNER, P. W., DIXON, R. A., LAMB, C. J, ELKIND, Y. (1994), Quantitative relationship between phenylalanine ammonia-lyase levels and phenylpropanoid accumulation in transgenic tobacco identifies rate-determining step in natural product synthesis, *Proc. Natl. Acad. Sci. USA* **91**, 7608–7612.

BAUCH, H. J., LEISTNER, E. (1978), Aromatic metabolites in cell suspension cultures of *Galium mollugo*, *Planta Med.* **33**, 105–127.

BERLIN, J. (1986), Secondary products from plant cell cultures, in: *Biotechnology*, Vol. 4, 1st Edn. (REHM, H.-J., REED, G., Eds.), pp. 630–658. Weinheim: VCH.

BERLIN, J. (1988), Formation of secondary metabolites in cultured plant cells and its impact on pharmacy, in: *Biotechnology in Agriculture and Forestry*, Vol. 4 (BAJAJ, Y. P. S., Ed.), pp. 37–59. Berlin: Springer-Verlag.

BERLIN, J. (1990), Screening and selection for variant cell lines with increased levels of secondary metabolites, in: *Secondary Products from Plant Tissue Culture* (CHARLWOOD, B. V., RHODES, M. J. C., Eds.), pp. 119–137. Oxford: Clarendon.

BERLIN, J., SASSE, F. (1985), Selection and screening techniques for plant cell cultures, *Adv. Biochem. Eng.* **31**, 99–132.

BERLIN, J., KNOBLOCH, K. H., HÖFLE, G., WITTE, L. (1982), Biochemical characterization of two tobacco cell lines with high and low yields of cinnamoyl putrescines, *J. Nat. Prod.* **45**, 83–87.

BERLIN, J., FORCHE, E., WRAY, V., HAMMER, J., HÖSEL, W. (1983), Formation of benzophenanthridine alkaloids by suspension cultures of *Eschscholtzia californica Z. Naturforsch.* **38c**, 346–352.

BERLIN, J., BEIER, H., FECKER, L., FORCHE, E., NOÉ, W., SASSE, F., SCHIEL, O., WRAY, V. (1985), Conventional and new approaches to increase alkaloid production of plant cell cultures, in: *Primary and Secondary Metabolism of Plant Cell Cultures* (NEUMANN, K. H., BARZ, W., REINHARD, E., Eds.), pp. 272–280. Berlin: Springer-Verlag.

BERLIN, J., SIEG, S., STRACK, D., BOKERN, M., HARMS, H. (1986), Production of betalains by suspension cultures of *Chenopodium rubrum*, *Plant Cell Tissue Organ Cult.* **5**, 163–174.

BERLIN, J., MOLLENSCHOTT, C., WRAY, V. (1988a), Triggered efflux of protoberberine alkaloids from cell suspension cultures of *Thalictrum rugosum*, *Biotechnol. Lett.* **10**, 193–198.

BERLIN, J., BEDORF, N., MOLLENSCHOTT, C., WRAY, V., SASSE, F., HÖFLE, G. (1988b), On the podophyllotoxins of root cultures of *Linum flavum*, *Planta Med.* **54**, 204–206.

BERLIN, J., MOLLENSCHOTT, C., GREIDZIAK, N., ERDOGAN, S., KUZOVKINA, I. (1990), Affecting secondary product formation in suspension and hairy root cultures – a comparison, in: *Progress in Plant Cellular and Molecular Biology* (NIJKAMP, H. J. J., VAN DER PLAS, L. H. W., VAN AARTRIJK, J., Eds.), pp. 763–768. Dordrecht: Kluwer.

BERLIN, J., RÜGENHAGEN, C., DIETZE, P., FECKER, L. F., GODDIJN, O. J. M., HOGE, J. H. C. (1993), Increased production of serotonin by suspension and root cultures of *Peganum harmala* transformed with a tryptophan decarboxylase cDNA from *Catharanthus roseus*, *Transgen. Res.* **2**, 336–344.

BERLIN, J., FECKER, S., HERMINGHAUS, C., RÜGENHAGEN (1994), Genetic modification of plant secondary metabolism: alteration of product levels by overexpression of amino acid decarboxylases, in: *Advances in Plant Biotechnology. Studies in Plant Science*, Vol. 4 (RYU, D. D. Y., FURASAKI, S., Eds.), pp. 57–81. Amsterdam: Elsevier.

BIANCO-COLOMAS, J., HUGUES, M. (1990), Establishment and characterization of a betacyanin producing cell line of *Amaranthus tricolor*: Inductive effects of light and cytokinins, *J. Plant Physiol.* **136**, 734–739.

BÖHM, H., RINK, E. (1988), Betalains, in: *Cell Culture and Somatic Cell Genetics of Plants*, Vol. 5: *Phytochemicals in Plant Cell Cultures* (CONSTABEL, F., VASIL, I. K., Eds.), pp. 449–463. San Diego: Academic Press.

BOKERN, M., WRAY, V., STRACK, D. (1991), Accumulation of phenolic conjugates and betacyanins, and changes in the activities of enzymes involved in feruloylglucose metabolism in cell suspension cultures of *Chenopodium rubrum*, *Planta* **184**, 261–270.

BREULING, M., ALFERMANN, A. W., REINHARD, E. (1985), Cultivation of cell cultures of *Berberis wilsonae* in 20-L airlift bioreactors, *Plant Cell Rep.* **4**, 220–223.

BRINGI, V., KADKADE, P. (1993), Enhanced production of taxol and taxanes by cell cultures of *Taxus* species, WO 93/17121.

BUITELAAR, R. M., LANGENHOFF, A. A. M., HEIDSTRA, R., TRAMPER, J. (1991), Growth and thiophene production by hairy root cultures of *Tagetes patula* in various two-liquid-phase bioreactors, *Enzyme Microb. Technol.* **13**, 487–494.

BUITELAAR, R. M., TRAMPER, J. (1992), Strategies to improve the production of secondary metabolites with plant cell cultures: a literature review, *J. Biotechnol.* **23**, 111–141.

CAHOON, E. B., SHANKLIN, J., OHLROGGE, J. B. (1992), Expression of a coriander saturase results in petroselinic acid production in transgenic tobacco, *Proc. Natl. Acad. Sci. USA* **89**, 11184–11188.

CAMERON, D. D., TONG, I. T. (1993), Cellular and metabolic engineering – an overview, *Appl. Biochem. Biotechnol.* **38**, 105–140.

CAREW, D. P. (1975), Tissue culture studies of *Catharanthus roseus*, in: *The Catharanthus Alkaloids* (TAYLOR, W. I., FARNSWORTH, N. R., Eds.), pp. 193–208. New York: Marcel Dekker.

Celex-Laboratories (1994), Production of *Taxus* sp. hairy root culture – taxol, taxane and diterpene production by *Agrobacterium rhizogenes* transformed yew culture, WO 94/20606.

CHARLWOOD, B. V., RHODES, M. J. C. (1990), Secondary products from plant tissue culture, *Proc. Phytochem. Soc.,* Vol. 30, Oxford: Clarendon.

CHARLWOOD, B. V., CHARLWOOD, K. A., MOLINA-TORRES, J. (1990), Accumulation of secondary compounds by organized plant cultures, in: *Secondary Products from Plant Tissue Culture* (CHARLWOOD, B. V., RHODES, M. J. C., Eds.), pp. 167–200. Oxford: Clarendon.

CHAVADEJ, S., BRISSON, N., MCNEIL, J. N., DELUCA, V. (1994), Re-direction of tryptophan leads to production of low indole glucosinolate canola, *Proc. Natl. Aad. Sci. USA* **91**, 2166–2170.

CHEETHAM, P. S. J. (1993), The use of biotransformations for the production of flavours and frangrances, *TIBTECH* **11**, 478–488.

CLINE, S. D., COSCIA, C. J. (1988), Stimulation of sanguinarine production by combined fungal elicitation and hormonal deprivation in cell suspension cultures of *Papaver bracteatum*, *Plant Physiol.* **86**, 161–165.

CONSTABEL, F., VASIL, I. K. (Eds.) (1988), *Cell Culture and Somatic Cell Genetics of Plants*, Vol. 5: *Phytochemicals in Plant Cell Cultures*. San Diego: Academic Press.

COURTOIS, D., YVERNEL, D., FLORIN, B., PETIARD, V. (1988), Conversion of tryptamine to serotonin by cell supension cultures of *Peganum harmala*, *Phytochemistry* **27**, 3137–3141.

CRAGG, G. M., SCHEPARTZ, S. A., SUFFNESS, M., GREVER, M. R. (1993), The taxol supply crisis. New NCI policies for handling the large-scale production of novel anticancer and anti-HIV agents, *J. Nat. Prod.* **56**, 1657–1668.

CURTIS, W. R. (1993), Cultivation of roots in bioreactors, *Curr. Opin. Biotechnol.* **4**, 205–210.

DE CAROLIS, E., DE LUCA, V. (1993), Purification, characterization, and kinetic analysis of a 2-oxoglutarate-dependent dioxygenase involved in vindoline biosynthesis from *Catharanthus roseus*, *J. Biol. Chem.* **268**, 5504–5511.

DE-EKNAMKUL, W., ELLIS, B. E. (1984), Rosmarinic acid production and growth characteristics of *Anchusa officinalis* cell suspension cultures, *Planta Med.* **50**, 346–350.

DE-EKNAMKUL, W., ELLIS, B. E. (1985a), Effects of auxins and cytokinins on growth and rosmarinic acid formation in cell suspension cultures of *Anchusa offcinalis*, *Plant Cell Rep.* **4**, 50–53.

DE-EKNAMKUL, W., ELLIS, B. E. (1985b), Effects of macronutrients on growth and rosmarinic acid formation in cell suspension cultures of *Anchusa officinalis*, *Plant Cell Rep.* **4**, 46–49.

DELL, B., ELSEGOOD, C. L., GHISALBERTI, E. L. (1989), Production of verbascoside in callus tissue of *Eremophila* spp., *Phytochemistry* **28**, 1871–1872.

DELUCA, V. (1993), Enzymology of indole alkaloids, in: *Methods of Plant Biochemistry*, Vol. 9 (DEY, P. M., HARBORNE, J. B., Eds.), pp. 345–368. London: Academic Press.

DELUCA, V., KURZ, W. G. W. (1988), Monoterpene indole alkaloids (*Catharanthus* alkaloids), in: *Cell Culture and Somatic Cell Genetics of Plants*, Vol. 5: *Phytochemicals in Plant Cell Cultures* (CONSTABEL, F., VASIL, I. K., Eds.), pp. 385–402. San Diego: Academic Press.

DELUCA, V., MARINEAU, C., BRISSON, N. (1989), Molecular cloning and analysis of a cDNA encoding plant tryptophan decarboxylase: comparison with animal dopa decarboxylase, *Proc. Natl. Acad. Sci. USA* **86**, 9969–9973.

DENO, H., SUGA, C., MORIMOTO, T., FUJITA, Y. (1987), Production of shikonin derivatives by cell suspension cultures of *Lithospermum erythrorhizon*. VI. Production of shikonin derivatives by a two layer culture containing an organic solvent, *Plant Cell Rep.* **6**, 197–199.

DEUS, B., ZENK, M. H. (1982), Exploitation of plant cells for the production of natural compounds, *Biotechnol. Bioeng.* **24**, 1965–1974.

DEUS-NEUMANN, B., ZENK, M. H. (1984), Instability of alkaloid production in *Catharanthus roseus* cell suspension cultures, *Planta Med.* **50**, 427–431.

DICOSMO, F. (1990), Strategies to improve yields of secondary metabolites to industrially interesting levels, in: *Progress in Plant Cellular and Molecular Biology* (NIJKAMP, H. J. J., VAN DER PLAS, L. H. W., VAN AARTRIJK, J., Eds.), pp. 717–724. Dordrecht: Kluwer.

DI IORIO, A. A., CHEETHAM, R. D., WEATHERS, P. J. (1992), Growth of transformed roots in a nutrient mist bioreactor: Reactor performance and evaluation, *Appl. Microbiol. Biotechnol.* **37**, 457–462.

DIXON, R. A., LAMB, C. J. (1990), Regulation of secondary metabolism at the biochemical and genetic levels, in: *Secondary Products from Plant Tissue Culture* (CHARLWOOD, B. V., RHODES, M. J. C., Eds.), pp. 103–118. Oxford: Clarendon.

DO, C. B., CORMIER, F. (1991), Effects of high ammonium concentrations on growth and anthocyanin formation in grape (*Vitis vinifera* L.) cell suspension cultured in a production medium, *Plant Cell Tissue Organ Cult.* **27**, 169–174.

DOONER, H. K., ROBBINS, T. P., JORGENSEN, R. A. (1991), Genetic and developmental control of anthocyanin biosynthesis, *Annu. Rev. Genet.* **25**, 173–199.

DRAPEAU, D., BLANCH, H. W., WILKE, C. R. (1987), Economic assessment of plant cell culture for the production of ajmalicine, *Biotechnol. Bioeng.* **30**, 946–953.

EDGINGTON, S. M. (1991), Taxol – out of the woods, *Bio/Technology* **9**, 933–938.

EISENREICH, W., MENHARD, B., HYLANDS, P. J., ZENK, M. H., BACHER, A. (1996), Studies on the biosynthesis of taxol: the taxane carbon skeleton is not of mevalonoid origin, *Proc. Natl. Acad. Sci. USA* **93**, 6431–6436.

EILERT, U., KURZ, W. G. W., CONSTABEL, F. (1985), Stimulation of sanguinarine accumulation in *Papaver somniferum* cells by fungal homogenates, *J. Plant Physiol.* **119**, 65–76.

ELLIS, B. E. (1983), Production of hydroxyphenylethanol glycosides in suspension cultures of *Syringa vulgaris*, *Phytochemistry* **22**, 1941–1943.

ELLIS, B. E. (1985), Metabolism of caffeoyl derivatives in plant cell cultures, in: *Primary and Secondary Metabolism of Plant Cell Cultures* (NEUMANN, K. H., BARZ, W., REINHARD, E., Eds.), pp. 164–173, Berlin: Springer-Verlag.

ELLIS, B. E. (1988), Natural products from plant tissue culture, *Nat. Prod. Rep.* **5**, 581–612.

FETT-NETO, A. G., DICOSMO, F., REYNOLDS, W. F., SAKATA, K. (1992), Cell culture of *Taxus* as a source of the antineoplastic drug taxol and related taxanes, *Bio/Technology* **10**, 1572–1575.

FETT-NETO, A. G., MELANSON, S. J., SAKATA, K., DICOSMO, F. (1993), Improved growth and taxol yield in developing calli of *Taxus cuspidata* by medium composition modification, *Bio/Technology* **11**, 731–734.

FETT–NETO, A. G., ZHANG, W. Y., DICOSMO, F. (1994a), Kinetics of taxol production, growth and nutrient uptake in cell suspensions of *Taxus cuspidata*, *Biotechnol. Bioeng.* **44**, 205–210.

FETT-NETO, A. G., MELANSON, S. J., NICHOLSON, S. A., PENNINGTON, J. J., DICOSMO, F. (1994b),

Improved taxol yield by aromatic and amino acid feeding to cell cultures of *Taxus cuspidata*, *Biotechnol. Bioeng.* **44**, 967–971.

FLORES, H. E., FILNER, P. (1985), Metabolic relationships of putrescine, GABA and alkaloids in cell and root cultures of Solanaceae, in: *Primary and Secondary Metabolism of Plant Cell Cultures* (NEUMANN, K. H., BARZ, W., REINHARD, E., Eds.), pp. 174–185. Berlin: Springer-Verlag.

FLORES, T., WAGNER, L. J., FLORES, H. E. (1993), Embryo culture and taxane production in *Taxus* spp., *In Vitro Cell. Dev. Biol.* **29P**, 160–165.

FORKMANN, G. (1993), Control of pigmentation in natural and transgenic plants, *Curr. Opin. Biotechnol.* **4**, 159–165.

FUJITA, Y. (1988), Shikonin: Production by (*Lithospermum erythrorhizon*) cell cultures, in: *Biotechnology in Agriculture and Forestry*, Vol. 4 (BAJAJ, Y. P. S., Ed.), pp. 225–236. Berlin: Springer-Verlag.

FUJITA, Y., HARA, Y. (1985), The effective production of shikonin by cultures with an increased cell population, *Agric. Biol. Chem.* **49**, 2071–2075.

FUJITA, Y., HARA, Y., OGINO, T., SUGA, C. (1981a), Production of shikonin derivatives by cell suspension cultures of *Lithospermum erythrorhizon.* I. Effects of nitrogen sources on the production of shikonin derivatives, *Plant Cell Rep.* **1**, 59–60.

FUJITA, Y., HARA, Y., SUGA, C., MORIMOTO, T. (1981b), Production of shikonin derivatives by cell suspension cultures of *Lithospermum erythrorhizon*. II. New medium for the production of shikonin derivatives, *Plant Cell Rep.* **1**, 61–63.

FUJITA, Y., TABATA, M., NISHI, A., YAMADA, Y. (1982), New medium and production of secondary compounds with the two-stage culture methods, in: *Plant Tissue Culture 1982* (FUJIWARA, A., Ed.), pp. 399–400. Tokyo: Maruzen Press.

FUJITA, Y., TAKAHASHI, S., YAMADA, Y. (1985), Selection of cell lines with high productivity of shikonin derivatives by protoplast culture of *Lithospermum erythrorhizon* cells, *Agric. Biol. Chem.* **49**, 1755–1759.

FUJITA, Y., HARA, Y., MORIMOTO, T., MISAWA, M. (1990), Semisynthetic production of vinblastine involving cell cultures of *Catharanthus roseus*, in: *Progress in Plant Cellular and Molecular Biology* (NIJKAMP, H. J. J., VAN DER PLAS, L. H. W., VAN AARTRIJK, J., Eds.), pp. 763–768. Dordrecht: Kluwer.

FUKUI, H., NAGAKAWA, K., TSUDA, S., TABATA, M. (1982), Production of isoquinoline alkaloids by cell suspension cultures of *Coptis japonica*, in: *Plant Cell Culture 1982* (FUJIWARA, A., Ed.), pp. 313–314. Tokyo: Maruzen Press.

FUKUI, H., TSUKADA, M., MIZUKAMI, H., TABATA, M. (1983), Formation of stereoisomeric mixtures of naphthoquinone derivatives in *Echium lycopsis* callus cultures, *Phytochemistry* **22**, 453–456.

FUNK, C., BRODELIUS, P. (1990), Influence of growth regulators and elicitor on phenylpropanoid metabolism in suspension cultures of *Vanilla planifolia*, *Phytochemistry* **29**, 845–848.

FUNK, C., BRODELIUS, P. (1992), Phenylpropanoid metabolism in suspension cultures of *Vanilla planifolia*. IV. Induction of vanillic acid formation, *Plant Physiol.* **99**, 256–262.

FURUYA, T. (1988), Saponins (Ginseng saponins), in: *Cell Culture and Somatic Cell Genetics of Plants*, Vol. 5: *Phytochemicals in Plant Cell Cultures* (CONSTABEL, F., VASIL, I. K., Eds.), pp. 213–236. San Diego: Academic Press.

FURUYA, T., IKUTA, A., SYONO, K. (1972), Alkaloids from callus tissue of *Papaver somniferum*, *Phytochemistry* **11**, 3041–3044.

FURZE, J. M., RHODES, M. J. C., PARR, A. J., ROBINS, R. J., WHITEHEAD, I. M., THRELFALL, D. R. (1991), Abiotic factors elicit sesquiterpenoid phytoalexin production but not alkaloid production in transformed root cultures of *Datura stramonium*, *Plant Cell Rep.* **10**, 111–114.

GERARDY, R., ZENK, M. H. (1993), Formation of salutaridine from (*R*)-reticuline by a membrane bound cytochrome-P-450 enzyme from *Papaver somniferum*, *Phytochemistry* **32**, 79–86.

GERTLOWSKI, C., PETERSEN, M. (1993), Influence of the carbon source on growth and rosmarinic acid production in suspension cultures of *Coleus blumei*, *Plant Cell Tissue Organ Cult.* **34**, 183–190.

GIROD, P. A., ZRYD J. P. (1991), Secondary metabolism in cultured red beet (*Beta vulgaris* L.) cells. Differential regulation of betaxanthins and betacyanin biosynthesis, *Plant Cell Tissue Organ Cult.* **25**, 1–12.

GODDIJN, O. J. M. (1992), Regulation of terpenoid indole alkaloid biosynthesis in *Catharanthus roseus*: The tryptophan decarboxylase gene. *PhD Thesis*, Rijksuniversiteit Leiden.

GODDIJN, O. J. M., DE KAM, R. J., ZANETTI, A., SCHILPEROORT, R. A., HOGE, J. H. C. (1992), Auxin rapidly down-regulates transcription of the tryptophan decarboxylase gene from *Catharanthus roseus Plant Mol. Biol.* **18**, 1113–1120.

GOODBODY, A. E., ENDO, T., VUKOVIC, J., KUTNEY, J. P., CHOI, L. S. L., MISAWA, M. (1988), Enzymic coupling of catharanthine and vindoline to form 3′,4′-anhydrovinblastine by horseradish peroxidase, *Planta Med.* **54**, 136–140.

GRÖGER, D. (1980), Alkaloids derived from tryptophan and anthranilic acid, in: *Secondary Plant Products. Encyclopedia of Plant Physiology,* Vol. 8 (BELL, E. A., CHARLWOOD, B. V., Eds.), pp. 128–159. Berlin: Springer-Verlag.

HAIN, R., BIESELER, B., KINDL, H., SCHRÖDER, G., STÖCKER, R. (1990), Expression of a stilbene synthase gene in *Nicotiana tabacum* results in synthesis of the phytoalexin resveratrol, *Plant Mol. Biol.* **15**, 325–335.

HAIN, R., REIF, H. J., KRAUSE, E., LANGEBARTELS, R., KINDL, H., VORNAM, B., WIESE, W., SCHMELZER, E., SCHREIER, P. H., STÖCKER, R. H., STENZEL, K. (1993), Disease resistance results from foreign phytoalexin expression in a novel plant, *Nature* **361**, 153–156.

HAMILL, J. D., ROBINS, R. J., PARR, A. J., EVANS, D. M., FURZE, J. M., RHODES, M. J. C. (1990), Overexpressing a yeast ornithine decarboxylase gene in transgenic roots of *Nicotiana rustica* can lead to enhanced nicotine accumulation, *Plant Mol. Biol.* **15**, 27–38.

HAN, K. H., FLEMING, P., WALKER, K., LOPER, M., CHILTON, W. S., MOCEK, U., GORDON, M. P., FLOSS, H. G. (1994), Genetic transformation of mature *Taxus*: an approach to genetically control the *in vitro* production of the anticancer drug taxol, *Plant Sci.* **95**, 187–196.

HARA, Y., YOSHIOKA, T., MORIMOTO, T., FUJITA, Y., YAMADA, Y. (1988), Enhancement of berberine production in suspension cultures of *Coptis japonica* by gibberellic acid treatment, *J. Plant Physiol.* **133**, 12–15.

HARA, Y., YAMAGATA, H., MORIMOTO, T., HIRATSUKA, J., YOSHIOKA, T., FUJITA, Y., YAMADA, Y. (1989), Flow cytometric analysis of cellular berberine contents in high- and low-producing cell lines of *Coptis japonica*, *Planta Med.* **55**, 151–154.

HARKES, P. A. A., KRIJBOLDER, L., LIBBENGA, K. R., WJNSMA, R., NSENGIYAREMGE, T., VERPOORTE, R. (1985), Influence of various media constituents on the growth of *Cinchona ledgeriana* tissue cultures and the production of alkaloids and anthraquinones therein, *Plant Cell Tissue Organ Cult.* **4**, 199–214.

HASHIMOTO, T., YAMADA, Y. (1994), Alkaloid biogenesis: Molecular aspects, *Annu. Rev. Plant Physiol. Plant Mol. Biol.* **45**, 257–285.

HASHIMOTO, T., YUN, D. J., YAMADA, Y. (1993), Production of tropane alkaloids in genetically engineered root cultures, *Phytochemistry* **32**, 713–718.

HAVKIN-FRENKEL, D. (1994), Vanilla flavor production: Tissue culture and the whole plant. *VIII. IAPTC Congress*, Firenze (Abstract), S18–56.

HEIDE, L., NISHIOKA, N., FUKUI, H., TABATA, M. (1989), Enzymatic regulation of shikonin biosynthesis in *Lithospermum erythrorhizon* cell cultures, *Phytochemistry* **28**, 1873–1877.

HENRY, M., ROUSSEL, J. L., ANDARY, C. (1987), Verbascoside production in callus and suspension cultures of *Hygrophila erecta*, *Phytochemistry* **26**, 1961–1963.

HERMINGHAUS, S., THOLL, D., RÜGENHAGEN, C., FECKER, L. F., LEUSCHNER, C., BERLIN, J. (1996), Improved metabolic action of a bacterial lysine decarboxylase gene in tobacco hairy root cultures by its fusion to a *rbcS* transit peptide coding sequence, *Transgen. Res.* **5**, 193–201.

HEZARI, M., LEWIS, N. G., CROTEAU, R. (1995), Purification and characterization of taxa-4(5),11(12)-diene synthase from Pacific yew (*Taxus brevifolia*) that catalyzes the first committed step of taxol biosynthesis, *Arch. Biochem. Biophys.* **322**, 437–444.

HINZ, H., ZENK, M. H. (1981), Production of protoberberine alkaloids by cell suspension cultures of *Berberis* species, *Naturwissenschaften* **68**, 620–621.

HIPPOLYTE, I., MARIN, B., BACCOU, J. C., JONARD, R. (1992), Growth and rosmarinic acid production in cell suspension cultures of *Salvia officinalis*, *Plant Cell Rep.* **11**, 109–112.

HIRASUNA, T. J., PESTCHANKER, L. J., SRINIVASAN, V., SHULER, M. L. (1996), Taxol production in suspension cultures of *Taxus baccata, Plant Cell Tiss. Org. Cult.* **44**, 95–102.

HIRATA, K., KOBAYASHI, M., MIYAMOTO, K., HASHI, T., OKAZAKI, M., MIURA, Y. (1989), Quantitative determination of vinblastine in tissue cultures of *Catharanthus roseus* by radioimmunoassay, *Planta Med.* **55**, 262–264.

HOEKSTRA, S. S., HARKES, P. A. A., VERPOORTE, R., LIBBENGA, K. R. (1990), Effect of auxin on cytodifferentiation and production of quinoline alkaloid in compact globular structures of *Cinchona ledgeriana*, *Plant Cell Rep.* **8**, 571–574.

HOLTON, T. A., TANAKA, Y. (1994), Blue roses – a pigment of our imagination? *TIBTECH* **12**, 40–41.

IGBAVBOA, U., SIEWEKE, H. J., LEISTNER, E., RÖWER, I., HÜSEMANN, W., BARZ, W. (1985), Alternative formation of anthraquinones and lipoquinones in heterotrophic and photoautotrophic cell suspension cultures of *Morinda lucida*, *Planta* **166**, 537–544.

IKUTA, A., SYONO, K., FURUYA, T. (1974), Alkaloids of callus tissues and redifferentiated plantlets in the Papaveraceae, *Phytochemistry* **13**, 2175–2179.

INAGAKI, N., NISHIMURA, H., OKADA, M., MITSU-

HASHI, H. (1991), Verbascoside production by plant cell cultures, *Plant Cell Rep.* **9**, 484–487.

INOUE, K., NAYESHIRO, H., INOUYE, H., ZENK, M. H. (1981), Anthraquinones in cell suspension cultures of *Morinda citrifolia*, *Phytochemistry* **20**, 1693–1700.

INOUE, K., UEDA, S. NAYESHIRO, H., MORITOME, N., INOUYE, H. (1984), Biosynthesis of naphthoquinones and anthraquinones in *Streptocarpus dunnii* cell cultures, *Phytochemistry* **23**, 313–318.

JAZIRI, M., ZHIRI, A., GUO, Y. W., DUPONT, J. P., SHIMOMURA, K., HAMADA, H., VANHAELEN, M., HOMÉS, J. (in press), *Taxus* sp. cell, tissue and organ cultures as alternative sources for taxoid production: a literature survey, *Plant Cell Tiss. Org. Cult.*

KADKADE, P. G. (1982), Growth and podopyllotoxin production in callus tissues of *Podophyllum peltatum*, *Plant Sci. Lett.* **25**, 107–115.

KAMIMURA, S. (1991), Production of morphinan alkaloids, in: *Plant Cell Culture in Japan. Progress in Production of Useful Plant Metabolites by Japanese Enterprises Using Plant Cell Culture Technology* (KOMAMINE, A., MISAWA, M., DICOSMO, F., Eds.), pp. 27–38. Tokyo: CMC Co.

KAMO, K. K., MAHLBERG, P. G. (1988), Morphinan alkaloids: Biosynthesis in plant (*Papaver* spp.) tissue cultures, in: *Biotechnology in Agriculture and Forestry*, Vol. 4 (BAJAJ, Y. P. S., Ed.), pp. 251–263. Berlin: Springer-Verlag.

KETCHUM, R. E. B., GIBSON, D. M. (in press) Paclitaxel production in suspension cultures of *Taxus, Plant Cell Tiss. Org. Cult.*

KINGSTON, D. G. I. (1994), Taxol: the chemistry and structure relationships of a novel anticancer agent, *TIBTECH* **12**, 222–227.

KINNERSLEY, A. M., DOUGALL, D. K. (1980), Increase in anthocyanin yield from wild carrot cell cultures by a selection based on cell-aggregate size, *Planta* **149**, 200–204.

KISHIMA, Y., NOZALI, K., AKASHI, R., ADACHI, T. (1991), Light-inducible pigmentation in *Portulaca* callus; selection of a betalain producing cell line, *Plant Cell Rep.* **10**, 304–307.

KISHORE, G. M., SOMERVILLE, C. R. (1993), Genetic engineering of commercially useful biosynthetic pathways in transgenic plants, *Curr. Opin. Biotechnol.* **4**, 152–158.

KNOBLOCH, K. H., BERLIN, J. (1980), Influence of the medium composition on the formation of secondary compounds in cell suspension cultures of *Catharanthus roseus*, *Z. Naturforsch.* **35c**, 551–556.

KNOBLOCH, K. H., BERLIN, J. (1983), Influence of phosphate on the formation of the indole alkaloids and phenolic compounds in cell suspension cultures of *Catharanthus roseus*. I. Comparison of enzyme activities and product accumulation, *Plant Cell Tissue Organ Cult.* **2**, 333–340.

KNOPP, E., STRAUSS, A., WEHRLI, W. (1988), Root induction on several Solanaceae species by *Agrobacterium rhizogenes* and determination of root tropane alkaloid content, *Plant Cell Rep.* **7**, 590–593.

KNORR, D., CASTER, C., DÖRNENBURG, H., GRÄF, S., HAVKIN-FRENKEL, D., PODSTOLSKI, A., WERRMANN, U. (1993), Biosynthesis and yield improvement of food ingredients from plant cell and tissue cultures, *Food Technol.* **47** (12), 57–63.

KNUTH, M. E., SAHAI, O. P. (1991), Flavor composition and method, *US Patents* No. 5,057,424 and 5,068,184.

KOBAYASHI, Y., FUKUI, H., TABATA, M. (1988), Berberine production by batch and semi-continuous cultures of immobilized *Thalictrum* cells in an improved bioreactor, *Plant Cell Rep.* **7**, 249–252.

KOBAYASHI, Y., AKITA, M., SAKAMOTO, K., LIU, H., SHIGEOKA, T., KOYANA, T., KAWAMURA, M., FURUYA, T. (1993), Large-scale production of anthocyanin by *Aralia cordata* cell suspension cultures, *Appl. Microbiol. Biotechnol.* **40**, 215–218.

KOBLITZ, H. (1988), Anthraquinones, in: *Cell Culture and Somatic Cell Genetics of Plants*, Vol. 5: *Production of Phytochemicals in Plant Cell Cultures* (CONSTABEL, F., VASIL, I. K., Eds.), pp. 113–142. San Diego: Academic Press.

KOBLITZ, H., KOBLITZ, D., SCHMAUDER, H. P., GRÖGER, D. (1983), Studies on tissue cultures of genus *Cinchona* L.: Alkaloid production in cell suspension cultures, *Plant Cell Rep.* **2**, 122–125.

KOES, R. E., QUATTROCCHIO, F., MOL, J. N. M. (1994), The flavonoid biosynthetic pathway in plants: function and evolution, *BioEssays* **16**, 123–132.

KOHL, W., WITTE, B., HÖFLE, G. (1982), Alkaloids from *Catharanthus roseus* tissue cultures III, *Z. Naturforsch.* **37b**, 1346–1351.

KOMAMIME, A., MISAWA, M., DICOSMO, F. (Eds.) (1991), *Plant Cell Culture in Japan. Progress in Production of Useful Plant Metabolites by Japanese Enterprises Using Plant Cell Culture Technology.* Tokyo: CMC Co.

KONDO, O., HONDA, H., TAYA, M., KOBAYASHI, T. (1989), Comparison of growth properties of carrot hairy root in various bioreactors, *Appl. Microbiol. Biotechnol.* **32**, 291–294.

KREIS, W., REINHARD, E. (1990), Two-stage cultivation of *Digitalis lanata* cells: semicontinuous production of desacetyllanatoside C in 20-litre airlift bioreactors, *J. Biotechnol.* **16**, 123–126.

KREIS, W., REINHARD, E. (1992), 12β-Hydroxylation of digitoxin by suspension-cultured *Digitalis lanata* cells: Production of digoxin in 20-litre and 300-litre air-lift bioreactors, *J. Biotechnol.* **26**, 257–273.

KREIS, W., HOELZ, H., SUTOR, R., REINHARD, E. (1993), Cellular organization of cardenolide biotransformation in *Digitalis grandiflora*, *Planta* **191**, 246–251.

KROMBHOLZ, R., MERSINGER, R., KREIS, W., REINHARD, E. (1992), Production of forskolin by axenic *Coleus forskohlii* roots cultivated in shake flasks and 20-L glass jar bioreactors, *Planta Med.* **58**, 328–333.

KURZ, W. G. W., CHATSON, K. B., CONSTABEL, F., KUTNEY, J. P., CHOI, L. S. L., KOLODZIEJCZYK, P., SLEIGH, S. K., STUART, K. L., WORTH, B. R. (1980), Alkaloid production in *Catharanthus roseus* cell cultures: initial studies on cell lines and their alkaloid content, *Phytochemistry* **19**, 2583–2587.

KURZ, W. G. W., CHATSON, R. B., CONSTABEL, F. (1985), Biosynthesis and accumulation of indole alkaloids in *Catharanthus roseus* cultivars, in: *Primary and Secondary Metabolism in Plant Cell Cultures* (NEUMANN, K. H., BARZ, W., REINHARD, E., Eds.), pp. 143–153. Berlin: Springer-Verlag.

KURZ, W. G. W., PAIVA, N. L., TYLER, R. T. (1990), Biosynthesis of sanguinarine by elicitation of surface immobilized cells of *Papaver somniferum*, in: *Progress in Plant Cellular and Molecular Biology* (NIJKAMP, H. J. J., VAN DER PLAS, L. H. W., VAN AARTRIJK, J., Eds.), pp. 682–688. Dordrecht: Kluwer.

KUTCHAN, T. M. (1993), Strictosidine: From alkaloid to enzyme to gene, *Phytochemistry* **32**, 493–506.

KUTCHAN, T. M., ZENK, M. H. (1993), Enzymology and molecular biology of benzophenanthridine biosynthesis, *J. Plant Res.* (Spec. Issue) **3**, 165–173.

KUTCHAN, T. M., BOCK, A., DIETTRICH, H. (1994), Heterologous expression of the plant proteins strictosidine synthase and berberine bridge enzyme in insect culture, *Phytochemistry* **35**, 353–360.

KUTNEY, J. P., AWERYN, B., CHOI, L. S. L., HONDA, T., KOLDZIEJZCYK, P., LEWIS, N. G., SATO, T., SLEIGH, S. K., STUART, K. L., WORTH, B. R., KURZ, W. G. W., CHATSON, K. B., CONSTABEL, F. (1983), Studies in plant tissue culture: The synthesis and biosynthesis of indole alkaloids, *Tetrahedron* **39**, 3781–3795.

LEATHERS, R. R., DAVIN, C., ZRYD, J. P. (1992), Betalain producing cell cultures of *Beta vulgaris* L., var. bikores monogerm (red beet). *In Vitro Dev. Biol. Plant.* **28**, 39–45.

LEISTNER, E. (1975), Isolierung, Identifizierung und Biosynthese von Anthrachinonen in Zellsuspensionskulturen von *Morinda citrifolia*, *Planta Med.* (Suppl.) 214–224.

LÖSCHER, R., HEIDE, L. (1994), Biosynthesis of *p*-hydroxybenzoate from *p*-coumarate and *p*-coumarate-coenzyme A in cell-free extracts of *Lithospermum erythrorhizon* cell cultures, *Plant Physiol.* **106**, 271–279.

LUCKNER, M., DIETTRICH, B. (1988), Cardenolides, in: *Cell Culture and Somatic Cell Genetics of Plants*, Vol. 5: *Phytochemicals in Plant Cell Cultures* (CONSTABEL, F., VASIL, I. K., Eds.), pp. 193–212. San Diego: Academic Press.

LUTTERBACH, R., STÖCKIGT, J. (1992), High-yield formation of arbutin from hydroquinone by cell suspension cultures of *Rauwolfia serpentina*, *Helv. Chim. Acta* **75**, 2009–2011.

MANO, Y., OHKAWA, H., YAMADA, Y. (1989), Production of tropane alkaloids by hairy root cultures of *Duboisia leichhardtii* transformed by *Agrobacterium rhizogenes*, *Plant Sci.* **59**, 191–201.

MATSUBARA, K., FUJITA, Y. (1991), Production of berberine, in: *Plant Cell Culture in Japan. Progress in Production of Useful Plant Metabolites by Japanese Enterprises Using Plant Cell Culture Technology* (KOMANINE, A., MISAWA, M., DICOSMO, F., Eds.), pp. 39–44. Tokyo: CMC Co.

MATSUBARA, K., KITANI, S., YOSHIOKA, Y., MORIMOTO, T., FUJITA, Y., YAMADA, Y. (1989), High density culture of *Coptis japonica* cells increases berberine production, *J. Chem. Technol. Biotechnol.* **46**, 61–69.

MCKNIGHT, T. D., BERGEY, D. R., BURNETTE, R. J., NESSLER, C. L. (1991), Expression of an enzymatically-active and correctly targeted strictosidine synthase in transgenic tobacco plants, *Planta* **185**, 148–152.

MEIJER, A. H., LOPES CARDOSO, M. I., VOSKUILEN, J. T., DE WAAL, A., VERPOORTE, R., HOGE, J. H. C. (1993a), Isolation and characterization of a cDNA clone from *Catharanthus roseus* encoding NADPH:cytochrome P-450 mono-oxygenase in plants, *Plant J.* **4**, 47–60.

MEIJER, A. H., VERPOORTE, R., HOGE, J. H. C. (1993b), Regulation of enzymes and genes involved in monoterpene indole alkaloid biosynthesis in *Catharanthus roseus*, *J. Plant Res.* (Spec. Issue) **3**, 145–164.

MEURER-GRIMES, B., BERLIN, J., STRACK, D. (1989), Hydroxycinnamoyl-CoA:putrescine hydroxycinnamoyltransferase in tobacco cell cultures with high and low levels of caffeoylputrescine, *Plant Physiol.* **89**, 488–492.

MEYER, P., HEIDMANN, I., FORKMANN, G., SAEDLER, H. (1987), A new petunia flower colour generated by transformation of a mutant with a maize gene, *Nature* **330**, 677–678.

MIRJALILI, N., LINDEN, J. C. (1996) Methyl jasmonate induced production of taxol in suspension cultures of *Taxus cuspidata:* ethylene interaction and induction models, *Biotechnol. Prog.* **12**, 110–118.

MISAWA, M. (1994), Plant tissue culture: an alternative for production of useful metabolites, *FAO Agric. Serv. Bull.* **108**. Rome: FAO.

MISAWA, M., ENDO, T. (1988), Antitumor compounds, in: *Cell Culture and Somatic Cell Genetics*, Vol. 5: *Phytochemicals in Plant Cell Cultures* (CONSTABEL, F., VASIL, I. K., Eds.), pp. 553–568. San Diego: Academic Press.

MIZUKAMI, H., ELLIS, B. E. (1991), Rosmarinic acid formation and differential expression of tyrosine aminotransferase isoforms in *Anchusa officinalis* cell suspension cultures, *Plant Cell Rep.* **10**, 321–324.

MIZUKAMI, H., KONOSHIMA, M., TABATA, M. (1978), Variation in pigment production in *Lithospermum erythrorhizon* callus cultures, *Phytochemistry* **17**, 95–97.

MIZUKAMI, H., TABIRA, Y., ELLIS, B. E. (1993), Methyl jasmonate-induced rosmarinic acid biosynthesis in *Lithospermum erythrorhizon* cell suspension cultures, *Plant Cell Rep.* **12**, 706–709.

MIZUSAKI, S., TANABE, Y., NOGUCHI, M., TAMAKI, E. (1971), *p*-Coumaroyl putrescine, caffeoyl putrescine and feruloyl putrescine from callus tissue culture of *Nicotiana tabacum*, *Phytochemistry* **10**, 1347–1350.

MORENO, P. R. H., VAN DER HEIJDEN, R., VERPOORTE, R. (1995), Cell and tissue cultures of *Catharanthus roseus:* a literature survey. II. Updating from 1988–1993, *Plant Cell Tiss. Org. Cult.* **42**, 1–25.

MORIMOTO, M., HARA, Y., KATO, Y., HIRATSUKA, J., YOSHIOKA, T., FUJITA., Y., YAMADA, Y. (1988), Berberine production by cultured *Coptis japonica* cells in a one stage culture using medium with a high copper concentration, *Agric. Biol. Chem.* **52**, 1835–1836.

MORRIS, P. (1986), Regulation of product synthesis in cell cultures of *Catharanthus roseus*. II. Comparison of production media, *Planta Med.* **52**, 121–126.

MOSHY, R. J., NIEDER, M. H., SAHAI, O. P. (1989), Biotechnology in the flavor and food industry of the USA, in: *Biotechnology. Challenges for the Flavor and Food Industry* (LINDSEY, R. L., WILLIS, B. J., Eds.), pp. 145–163. London: Elsevier.

MULDER-KRIEGER, T., VERPOORTE, R., BAERHEIM-SVENDSEN, A., SCHEFFER, J. J. C. (1988), Production of essential oils and flavours in plant cell and tissue cultures. A review, *Plant Cell Tissue Organ Cult.* **13**, 85–154.

MURANAKA, T., OHKAWA, H., YAMADA, Y. (1993), Continuous production of scopolamine by a culture of *Duboisia leichardtii* hairy root clone in a bioreactor system, *Appl. Microbiol. Biotechnol.* **40**, 219–223.

NAGAKAWA, K., KONAGAI, A., FUKUI, H., TABATA, M. (1984), Release and crystallization of berberine in liquid medium of *Thalictrum minus* cell suspension cultures, *Plant Cell Rep.* **3**, 254–257.

NAKATANI, H., MALIK, V. S. (1992), Putrescine N-methyltransferase, recombinant DNA molecules encoding putrescine N-methyltransferase, and transgenic tobacco plants with altered nicotine content, *Eur. Patent Application* 9113283.6, Publication Nr. 0486214 A2.

NESSLER, C. R. (1994), Metabolic engineering of plant secondary products, *Transgen. Res.* **3**, 109–115.

NICOLAOU, K. C., DAI, W. M., GUY, R. K. (1994), Chemie und Biologie von Taxol, *Angew. Chem.* **106**, 38–69.

NOZUE, M., KAWAI, J., YOSHITAMA, K. (1987), Selection of a high anthocyanin-producing cell line of sweet potato cell cultures and identification of the pigment, *J. Plant Physiol.* **129**, 81–88.

ODAKE, K., ICHI, T., KUSUHARA, K. (1991), Production of madder colorants, in: *Plant Cell Culture in Japan. Progress in the Production of Useful Plant Metabolites by Japanese Enterprises Using Plant Tissue Culture Technology* (KOMAMINE, A., MISAWA, M., DICOSMO, F., Eds.), pp. 138–146. Tokyo: CMC Co.

OOSTHAM, A., MOL, J. N. M., VAN DER PLAS, L. H. W. (1993), Establishment of hairy root cultures of *Linum flavum* producing the lignan 5-methoxypodophyllotoxin, *Plant Cell Rep.* **12**, 474–477.

OZAKI, Y., SUGA, C., YOSHIOKA, T., MORIMOTO, T., HARADA, M. (1990), Evaluation of equivalence on pharmacological properties between natural crude drugs and their cultured cells based on their components. Accelerative effect of *Lithospermi radix* and inhibitory effect of *Coptidis rhizoma* on proliferation and granulation tissue, *Yakugaku Zasshi* **110**, 268–272.

PALMER, J. E., WIDHOLM, J. M. (1975), Characterization of carrot and tobacco cell cultures resistant to *p*-fluorophenylalanine, *Plant Physiol.* **56**, 233–238.

PAREILLEUX, A., VINAS, R. (1984), A study on the alkaloid production of resting cell suspensions of

Catharanthus roseus in a continuous flow reactor, *Appl. Microbiol. Biotechnol.* **19**, 316–320.

PARK, J. M., YOON S. Y., GILES, K. L., SONGSTAD, D., EPPSTEIN, D., NOVAKOVSKI, D., FRIESEN, L., ROEWER, I. (1992), Production of sanguinarine by suspension cultures of *Papaver somniferum* in bioreactors, *J. Ferment. Bioeng.* **74**, 292–296.

PARR, A. J. (1989), The production of secondary metabolites by plant cell cultures, *J. Biotechnol.* **10**, 1–26.

PARR, A. J., PAYNE, J., EAGLES, J., CHAPMAN, B. T., ROBINS, R. J., RHODES, M. J. C. (1990), Variation in tropane alkaloid accumulation within the Solanaceae and strategies for its exploitation, *Phytochemistry* **29**, 2545–2550.

PASQUALI, G., GODDIJN, O. J. M., DE WALL, A., VERPOORTE, R., SCHILPEROORT, R. A., HOGE, J. H. C., MEMELINK, J. (1992), Coordinated regulation of two indole alkaloid biosynthetic genes from *Catharanthus roseus* by auxin and elicitors, *Plant Mol. Biol.* **18**, 1121–1131.

PAYNE, G. F., BRINGI, V., PRINCE, C., SHULER, M. L. (1991), *Plant Cell and Tissue Culture in Liquid Systems.* Munich: Hanser.

PESTCHANKER, L. J., ROBERTS, S. C., SHULER, M. L. (1996), Kinetics of taxol production and nutrient use in suspension cultures of *Taxus cuspidata* in shake flasks and a Wilson-type reactor, *Enzyme Microb. Technol.* **19**, 256–260.

PETERSEN, M., ALFERMANN, A. W. (1993), Plant cell cultures, in: *Biotechnology*, Vol. 1, 2nd Edn. (REHM, H.-J., REED, G., Eds.), pp. 578–614. Weinheim: VCH.

PETERSEN, M., HÄUSLER, E., KARWATZKI, B., MEINHARD, J. (1993), Proposed biosynthetic pathway for rosmarinic acid in cell suspension cultures of *Coleus blumei*, *Planta* **189**, 10–14.

PIEHL, G. W., BERLIN, J., MOLLENSCHOTT, C., LEHMANN, J. (1988), Growth and alkaloid production of a cell suspension culture of *Thalictrum rugosum* in shake flasks and membrane-stirrer reactors with bubble free aeration, *Appl. Microbiol. Biotechnol.* **29**, 456–461.

POULSEN, C., GODDIJN, O. J. M., HOGE, J. H. C., VERPOORTE, R. (1994), Anthranilate synthase and chorismate mutase activities in transgenic tobacco plants overexpressing tryptophan decarboxylase from *Catharanthus roseus*, *Transgen. Res.* **3**, 43–49.

PROKSCH, A., WAGNER, H. (1987), Structural analysis of a 4-*O*-methylglucuronoarabinoxylan with immuno-stimulating activity from *Echinacea purpurea*, *Phytochemistry* **26**, 1989–1993.

RAZZAQUE, A., ELLIS, B. E. (1977), Rosmarinic acid production in *Coleus blumei*, *Planta* **137**, 287–291.

REINHARD, E., ALFERMANN, A. W. (1980), Biotransformation of plant cell cultures, *Adv. Biochem. Eng.* **16**, 49–83.

REINHARD, E., KREIS, W., BARTHLEN, U., HELMBOLD, U. (1989), Semicontinuous cultivation of *Digitalis lanata* cells: Production of β-methyldigoxin in a 300-L airlift bioreactor, *Biotechnol. Bioeng.* **34**, 502–508.

RHODES, M. J. C., ROBINS, R. J., HAMILL, J. D., PARR, A. J., HILTON, M. G., WALTON, N. J. (1990), Properties of transformed roots, in: *Secondary Products of Plant Tissue Culture* (CHARLWOOD, B. V., RHODES, M. J. C., Eds.), pp. 201–225. Oxford: Clarendon.

ROHMER, M., KNANI, M., SIMONIN, P., SUTTER, B., SAHM, H. (1993), Isoprenoid biosynthesis in bacteria: a novel pathway for the early steps leading to isopentenyl diphosphate, *Biochem. J.* **295**, 517–524.

RUEFFER, M., ZENK, M. H. (1987), Distant precursors of benzylisoquinoline alkaloids and their enzymatic formation, *Z. Naturforsch.* **42c**, 319–332.

RUEFFER, M., EL-SHAGI, H., NAGAKURA, N., ZENK, M. H (1981), *S*-Norlaudanosoline synthase: the first enzyme in the benzylisoquinoline biosynthetic pathway, *FEBS Lett.* **129**, 5–9.

RUYTER, C. M., STÖCKIGT, J. (1989), Neue Naturstoffe aus pflanzlichen Zellkulturen, *GIT Fachz. Lab.* **33**, 283–293.

SAITO, K., YAMAZAKI, M., MURAKOSHI, I. (1992), Transgenic medicinal plants: *Agrobacterium*-mediated foreign gene transfer and production of secondary metabolites, *J. Nat. Prod.* **55**, 149–162.

SAKAMOTO, K., IIDA, K., SAWAMURA, K., HAJIRO, K., ASADA, Y., YOSHIKAWA, T., FURUYA, T. (1994), Anthocyanin production in cultured cells of *Aralia cordata*, *Plant Cell Tissue Organ Cult.* **36**, 21–26.

SAKUTA, M., TAKAGI, T., KOMAMINE, A. (1987), Effects of nitrogen source on betacyanin accumulation and growth in suspension cultures of *Phytolaca americana*, *Physiol. Plant.* **71**, 459–463.

SASSE, F., WITTE, L., BERLIN, J. (1987), Biotransformation of tryptamine to serotonin by cell suspension cultures of *Peganum harmala*, *Planta Med.* **53**, 354–359.

SATO, F., YAMADA, Y. (1984), High berberine-producing cultures of *Coptis japonica* cells, *Phytochemistry* **23**, 281–285.

SCHIEL, O., BERLIN, J. (1987), Large-scale fermentation and alkaloid production of cell suspension cultures of *Catharanthus roseus*, *Plant Cell Tissue Organ Cult.* **8**, 153–161.

SCHIEL, O., JARCHOW-REDECKER, K., PIEHL, G. W., LEHMANN, J., BERLIN, J. (1984a), Increased formation of cinnamoyl putrescines by fed-batch fermentation of cell suspension cultures of *Nicotiana tabacum*, *Plant Cell Rep.* **3**, 18–20.

SCHIEL, O., MARTIN, B., PIEHL, G. W., NOWAK, J., HAMMER, J., SASSE, F., SCHAER, W., LEHMANN, J., BERLIN, J. (1984b), Some technological aspects on the production of cinnamoyl putrescines by cell suspension cultures of *Nicotiana tabacum*, in: *3rd Eur. Congr. Biotechnol.*, Vol. 1, pp. 167–172. Weinheim: VCH.

SCHIEL, O., WITTE, L., BERLIN, J. (1987), Geraniol-10-hydroxylase activity and its relation to monoterpene indole alkaloid accumulation in cell suspension cultures of *Catharanthus roseus*, *Z. Naturforsch.* **42c**, 1075–1081.

SCHLATMANN, J. E., MORENO, P. R. H., VINKE, J. L., TEN HOOPEN, H. J. G., VERPOORTE, R., HEIJNEN, J. J. (1994), Effect of oxygen and nutrient limitation on ajmalicine production and related enzyme activities in high density cultures of *Catharanthus roseus*, *Biotechnol. Bioeng.* **44**, 461–468.

SCHÜBEL, H., RUYTER, C. M., STÖCKIGT, J. (1989), Improved production of raucaffricine by cultivated *Rauwolfia* cells, *Phytochemistry* **28**, 491–494.

SCHULTE, U., EL-SHAGI, H., ZENK, M. H. (1984), Optimization of 19 Rubiaceae species in cell culture for the production of anthraquinones, *Plant Cell Rep.* **3**, 51–54.

SCHUMACHER, H. M., GUNDLACH, H., FIEDLER, F., ZENK, M. H. (1987), Elicitation of benzophenanthridine alkaloid biosynthesis in *Eschscholtzia californica* cell cultures, *Plant Cell Rep.* **6**, 410–413.

SCRAGG, A. H., CRESSWELL, R. C., ASHTON, S., YORK, A., BOND, P. A., FOWLER, M. W. (1989), Growth and alkaloid production in bioreactors by a selected *Catharanthus roseus* cell line, *Enzyme Microb. Technol.* **11**, 329–333.

SCRAGG, A. H., ASHTON, S., YORK, A., STEPAN-SARKISSIAN, G., GREY, D. (1990), Growth of *Catharanthus roseus* suspensions for maximum biomass and alkaloid accumulation, *Enzyme Microb. Technol.* **12**, 292–298.

SEITZ, H. U., HINDERER, W. (1988), Anthocyanins, in: *Cell Culture and Somatic Cell Genetics of Plants*, Vol. 5: *Phytochemicals in Plant Cell Cultures* (CONSTABEL, F., VASIL, I. K., Eds.), pp. 49–76. San Diego: Academic Press.

SHAUL, O., GALILI, G. (1991), Increased lysine synthesis in transgenic tobacco plants expressing a bacterial dihydrodipicolinate synthase in their chloroplasts, *Plant J.* **2**, 203–209.

SIAH, C. L., DORAN, P. M. (1991), Enhanced codeine and morphine production in suspended *Papaver somniferum* cultures after removal of exogenous hormones, *Plant Cell Rep.* **10**, 349–353.

SIEWEKE, H. J., LEISTNER, E. (1992), *o*-Succinylbenzoate:CoA ligase from anthraquinone producing cell cultures of *Galium mollugo*, *Phytochemistry* **31**, 2329–2335.

SIGNS, M. W., FLORES, H. E. (1990), The biosynthetic potential of plant roots, *BioEssays* **12**, 7–13.

SIMANTIRAS, M., LEISTNER, E. (1989), Formation of *o*-succinylbenzoic acid from isochorismic acid in protein from anthraquinone-producing plant cell suspension cultures, *Phytochemistry* **28**, 1381–1382.

SMART, N. J., MORRIS, P., FOWLER, M. W. (1982), Alkaloid production by cells of *Catharanthus roseus* grown in airlift fermenter systems, in: *Plant Tissue Culture 1982* (FUJIWARA A., Ed.), pp. 397–398. Tokyo: Maruzen Press.

SMITH, M. A. L. (1995), Large-scale production of secondary metabolites, in: *Current Issues in Plant Molecular and Cellular Biology* (TERZI, M., CELLA, R., FALAVIGNA, A., Eds.), pp. 669–674. Dordrecht: Kluwer.

SMOLLNY, T., WICHERS, H., DE RIJK, T., VAN ZWAM, A., SHASAVARI, A., ALFERMANN, A. W. (1993), Formation of lignans in suspension cultures of *Linum album. 3rd Workshop on "Primary and Secondary Metabolism of Plants and Plant Cell Cultures"*. Abstract P13, Leiden University.

SONGSTAD, D. D., GILES, K. L., PARK, J. M., NOVAKOVSKI, D., EPPSTEIN, D., FRIESEN, L. (1990), Use of nurse cultures to select for *Papaver somniferum* cell lines capable of enhanced sanguinarine production, *J. Plant Physiol.* **136**, 236–239.

SONGSTAD, D. D., KURZ, W. G. W., NESSLER, C. L. (1991), Tyramine accumulation in *Nicotiana tabacum* transformed with a chimeric tryptophan decarboxylase gene, *Phytochemistry* **30**, 3245–3246.

SOUTHARD, G. L., GROZNIK, W. J., BOULWARE, R. T., THORNE, E. E., WALBORN, D. E., YANKELL, S. L. (1984), Sanguinarine, a new antiplaque agent: retentention and plaque specificity, *J. Am. Dent. Assoc.* **108**, 339–341.

SRINIVASAN, V., CIDDI, V., BRINGI, V., SHULER, M. (1996), Metabolic inhibitors, elicitors, and precursors as tools for probing yield limitation in taxane production by *Taxus chinensis* cell cultures, *Biotechnol. Prog.* **12**, 457–466.

STAHLHUT, R. (1993), ESCAgenetics Corporation, *IAPTC Newslett.* **73**, 12–15.

STEINRÜCKEN, H. C., AMRHEIN, N. (1980), The herbicide glyphosate is a potent inhibitor of 5-enolpyruvylshikimic acid 3-phosphate synthase,

Biochem. Biophys. Res. Commun. **94**, 1207–1212.

STEPHANOPOULOS, G., SINSKY, A. J. (1993), Metabolic engineering – methodologies and future prospects, *TIBTECH* **11**, 392–396.

STÖCKIGT, J., SOLL, H. J. (1980), Indole alkaloids from cell suspension cultures of *Catharanthus roseus* and *C. ovalis*, *Planta Med.* **40**, 22–30.

SUGA, T., HIRATA, T. (1990), Biotransformation of exogenous substrates by plant cell cultures, *Phytochemistry* **29**, 2393–2406.

SUMARYONO, W., PROKSCH, P., HARTMANN, T., NIMTZ, M., WRAY, V. (1991), Induction of rosmarinic acid accumulation in cell suspension cultures of *Orthosiphon aristatus* after treatment with yeast extract, *Phytochemistry* **30**, 3267–3271.

SUZUKI, H., MATSUMOTO, T. (1988), Anthraquinone: Production by plant cell culture, in: *Biotechnology in Agriculture and Forestry*, Vol. 4 (BAJAJ, Y. P. S., Ed.), pp. 237–250. Berlin: Springer-Verlag.

SUZUKI, T., YOSHIOKA, T., HARA, Y., TABATA, M., FUJITA, Y. (1987), A new bioassay system for screening high berberine-producing cell colonies of *Thalictrum minus*, *Plant Cell Rep.* **6**, 194–196.

SUZUKI, M., NAKAGAWA, K., FUKUI, H., TABATA, M. (1988), Alkaloid production in cell suspension cultures of *Thalictrum flavum* and *T. dipterocarpum*, *Plant Cell Rep.* **7**, 26–29.

SUZUKI, H., SUGA, C., MORIMOTO, T., HARADA, M. (1991), Quantitative analysis of plant hormones, auxins, in biotechnologically cultured products of medicinal plants, *Shoyakugaku Zasshi* **45**, 137–141.

SUZUKI, H., OZAKI, Y., SUGA, C., MORIMOTO, T., SATAKE, M., HARADA, M. (1993a), Dissolution tests of *Coptis* rhizome and cultured cells of *Coptis japonica*, *Shoyakugaku Zasshi* **47**, 311–315.

SUZUKI, H., OZAKI, Y., SATAKE, M. (1993b), Changes of components – dissolution from prescription containing *Coptis* rhizome or cultured cells of *Coptis japonica* into the detection, *Shoyakugaku Zasshi* **47**, 396–401.

TABATA, M., MIZUKAMI, H., HIRAOKA, N., KONOSHIMA, M. (1974), Pigment formation in callus cultures of *Lithospermum erythrorhizon*, *Phytochemistry* **13**, 927–932.

TAKAHASHI, S., FUJITA, Y. (1991), Production of shikonin, in: *Plant Cell Culture in Japan. Progress in the Production of Useful Plant Metabolites by Japanese Enterprises Using Plant Cell Culture Technology* (KOMANINE, A., MISAWA, M., DICOSMO, F., Eds.), pp. 72–78. Tokyo: CMC Co.

TAKAYAMA, S. (1994), *Tripterygium wilfordii*: *In vitro* culture and the production of anticancer compounds tripdiolide and tripdolide, in: *Biotechnology in Agriculture and Forestry*, Vol. 28 (BAJAJ, Y. P. S., Ed.), pp. 457–468. Berlin: Springer-Verlag.

TAKAYAMA, S., TAKIZAWA, N., KUROYANAGI, M. (1994), The method and system for large-scale culture of plant roots using aeration-agitation bioreactor. Strategies for scale-up. *VIII. IAPTC Congr.*, Firenze (Abstract) S20–22.

TANAKA, A. (1994), Immobilization of plant cells, in: *Advances in Plant Biotechnology. Studies in Plant Science*, Vol. 4 (RYU, D. D. Y., FURASAKI, S., Eds.), pp. 209–220. Amsterdam: Elsevier.

TATICEK, R. A., LEE, C. W. T., SHULER, M. L. (1994), Large-scale insect and plant cell culture, *Curr. Opin. Biotechnol.* **5**, 165–174.

TOIVONEN, L. (1993), Utilization of hairy root cultures for production of secondary metabolites, *Biotechnol. Prog.* **9**, 12–20.

TYLER, V. E. (1988), Medicinal Plant Research: 1953–1987, *Planta Med.* **54**, 95–100.

TYLER, R. T., EILERT, U., RIJNDERS, C. O. M., ROEWER, I. A., KURZ, W. G. W. (1988), Semicontinuous production of sanguinarine and dihydrosanguinarine by *Papaver somniferum* L. cell suspension cultures treated with fungal homogenate, *Plant Cell Rep.* **7**, 410–413.

ULBRICH, B., WIESNER, W., ARENS, H. (1985), Large-scale production of rosmarinic acid from plant cell cultures of *Coleus blumei*, in: *Primary and Secondary Metabolism of Plant Cell Cultures* (NEUMANN, K. H., BARZ, W., REINHARD, E., Eds.), pp. 293–303. Berlin: Springer-Verlag.

USHIYAMA, K. (1991), Large-scale culture of ginseng, in: *Plant Cell Culture in Japan. Progress in the Production of Useful Plant Metabolites by Japanese Enterprises Using Plant Tissue Culture Technology* (KOMAMINE, A., MISAWA, M., DICOSMO, F., Eds.), pp. 92–98. Tokyo: CMC Co.

VAN DEN BERG, A. J. J., RADEMA, M. H., LABADIE, R. P. (1988), Effects of light on anthraquinone production in *Rhamnus purshiana* suspension cultures, *Phytochemistry* **27**, 415–417.

VAN DEN HEIJDEN, R., VERPOORTE, R., TEN HOOPEN, H. J. G. (1989), Cell and tissue cultures of *Catharanthus roseus*, *Plant Cell Tissue Organ Cult.* **18**, 231–280.

VAN DER KROL, A. R., LENTING, P. E., VEENSTRA, J., VAN DER MEER, I. M., KOES, R. E., GERATS, A. G. M., MOL, J. N. M., STUITJE, A. R. (1988), An antisense chalcone synthase gene in transgenic plants inhibits flower pigmentation, *Nature* **333**, 866–869.

VAN DER KROL, A. R., MUR, L. A., BELD, M., MOL, J. N. M., STUITJE, A. R. (1990), Flavonoid genes in *Petunia*: addition of a limited number of gene copies may lead to a suppression of gene expression, *Plant Cell* **2**, 218–221.

VAN GULIK, W. M., NUUTILA, A. M., VINKE, K. L., TEN HOOPEN, H. J. G., HEIJNEN, J. J. (1994), Effects of carbon dioxide, air flow rate, and inoculation density on batch growth of *Catharanthus roseus* cell suspensions in stirred fermenters, *Biotechnol. Prog.* **10**, 335–339.

VAN UDEN, W., PRAS, N., VOSSEBELD, E. M., MOL, J. N. M., MALINGRÉ, T. M. (1990), Production of 5-methoxypodophyllotoxin in cell suspension cultures of *Linum flavum*, *Plant Cell Tissue Organ Cult.* **20**, 81–87.

VAN UDEN, W., PRAS, N., HOMAN, B., MALINGRÉ, T. M. (1991), Improvement of the production of 5-methoxypodophyllotoxin using a new selected root culture of *Linum flavum*, *Plant Cell Tissue Organ Cult.* **27**, 115–121.

WAGNER, F., VOGELMANN, H. (1977), Cultivation of plant cell cultures in bioreactors and formation of secondary metabolites, in: *Plant Tissue Culture and its Biotechnological Application* (BARZ, W., REINHARD, E., ZENK, M. H., Eds.), pp. 245–252. Heidelberg: Springer-Verlag.

WAGNER, H., STUPPNER, H., PUHLMANN, J., BRÜMMER, B., DEPPE, K., ZENK, M. H. (1989), Gewinnung von immunologisch aktiven Polysacchariden aus *Echinacea*-Drogen und -Gewebekulturen, *Z. Phytother.* **10**, 35–38.

WALKER, M. A., ELLIS, B. E., DUMBROFF, E. B., DOWNER, R. G., MARTIN, R. J. (1986), Changes in amines and biosynthetic enzyme activities in *p*-fluorophenylalanine resistant and wild type tobacco cell cultures, *Plant Physiol.* **80**, 825–828.

WALTON, N. J., BELSHAW, N. J. (1988), The effect of cadaverine on the formation of anabasine from lysine in hairy root cultures of *Nicotiana hesperis*, *Plant Cell Rep.* **7**, 115–118.

WESTCOTT, R. J., CHEETHAM, P. J. S., BARRACLOUGH, R. J. (1994), Use of organized viable plant vanilla aerial roots for the production of natural vanilla, *Phytochemistry* **35**, 135–138.

WESTPHAL, K. (1990), Large-scale production of new biologically active compounds in plant cell cultures, in: *Progress in Plant Cellular and Molecular Biology* (NIJKAMP, H. J. J., VAN DER PLAS, L. H. W., VAN AARTRIJK, J., Eds.), pp. 601–608. Dordrecht: Kluwer.

WHITAKER, R. J., HASHIMOTO, T., EVANS, D. A. (1984), Production of secondary metabolite rosmarinic acid by plant cell suspension cultures, *Ann. N. Y. Acad. Sci.* **435**, 364–366.

WICHERS, H. J., VERSLUIS-DE HAAN, G. G., MARSMAN, J. W., HARKES, M. P. (1991), Podophyllotoxins in plants and cell cultures of *Linum flavum*, *Phytochemistry* **30**, 3601–3604.

WICKREMESINHE, E. R. M., ARTECA, R. N. (1993), *Taxus* callus cultures: Initiation, growth, optimization, characterization and taxol production, *Plant Cell Tissue Organ Cult.* **35**, 181–193.

WICKREMESINHE, E. R. M., ARTECA, R. N. (1994a), *Taxus* cell suspension cultures: optimization of growth and production of taxol, *J. Plant Physiol.* **144**, 183–188.

WICKREMESINHE, E. R. M., ARTECA, R. N. (1994b), Roots of hydroponically plants as a source of taxol and related taxanes, *Plant Sci.* **101**, 125–135.

WJNSMA, R., VERPOORTE, R. (1988), Quinoline alkaloids of *Cinchona*, in: *In Cell Culture and Somatic Cell Genetics of Plants*, Vol. 5: *Phytochemicals in Plant Cell Cultures* (CONSTABEL, F., VASIL, I. K., Eds.), pp. 337–356. San Diego: Academic Press.

WILLIAMS, R. D., ELLIS, B. E. (1993), Alkaloids from *Agrobacterium rhizogenes*-transformed *Papaver somniferum* cultures, *Phytochemistry* **32**, 719–723.

WILSON, P. D. G., HILTON, M. G., MEEHAN, P. T. H., WASPE, C. R., RHODES, M. J. C. (1990), The cultivation of transformed roots from laboratory to pilot plant, in: *Progress in Plant Cellular and Molecular Biology* (NIJKAMP, H. J. J., VAN DER PLAS, L. H. W., VAN AARTRIJK, J., Eds.), pp. 700–705. Dordrecht: Kluwer.

WOERDENBAG, H. J., VAN UDEN, W., FRIJLINK, H. W., LERK, C. F., PRAS, N., MALINGRÉ, T. M. (1990), Increased podophyllotoxin production in *Podophyllum hexandrum* cell suspension cultures after feeding coniferyl alcohol as a β-cyclodextrin complex, *Plant Cell Rep.* **9**, 97–100.

YAMADA, Y., HASHIMOTO, T. (1982), Production of tropane alkaloids in cultured cells of *Hyoscyamus niger*, *Plant Cell Rep.* **1**, 101–103.

YAMADA, Y., SATO, F. (1982), Production of berberine in cultured cells of *Coptis japonica*, *Phytochemistry* **20**, 545–547.

YAMAMOTO, Y. (1991), Anthocyanin production in plant cell cultures, in: *Plant Cell Culture in Japan. Progress in the Production of Useful Plant Metabolites by Japanese Enterprises Using Plant Cell Culture Technology* (KOMAMINE, A., MISAWA, M., DICOSMO, F., Eds.), pp. 114–126. Tokyo: CMC Co.

YAMAMOTO, Y., MIZUGUCHI, R., YAMADA, Y. (1982), Selection of a high and stable-pigment producing strain in cultured *Euphorbia millii* cells, *Theor. Appl. Genet.* **61**, 113–116.

YAMAMOTO, Y., KINOSHITA, Y. WATANABE, S., YAMADA, Y. (1989), Anthocyanin production in suspension cultures of high producing cells of *Euphorbia millii*, *Agric. Biol. Chem.* **53**, 417–423.

YOKOYAMA, M., YANAGI, M. (1991), High-level production of arbutin by biotransformation, in: *Plant Cell Culture in Japan. Progress in the Production of Useful Plant Metabolites by Japanese Enterprises Using Plant Cell Culture Technology* (KOMAMINE, A., MISAWA, M., DICOSMO, F., Eds.), pp. 78–91. Tokyo: CMC Co.

YUKIMUNE, Y., TABATA, H., HIGASHI, Y., HARA, Y. (1996), Methyl jasmonate-induced overproduction of paclitaxel and baccatin III in *Taxus* cell suspension cultures, *Nat. Biotechnol.* **14**, 1129–1132.

YUN, D. J., HASHIMOTO, T., YAMADA, Y. (1992), Metabolic engineering of medicinal plants: Transgenic *Atropa belladonna* with an improved alkaloid composition, *Proc. Natl. Acad. Sci. USA* **89**, 11799–11803.

ZENK, M. H. (1991), Chasing enzymes of secondary metabolism – Plant cell culture a pot of gold, *Phytochemistry* **30**, 3861–3863.

ZENK, M. H., EL-SHAGI, H., SCHULTE, U. (1975), Anthraquinone production by cell suspension cultures of *Morinda citrifolia*, *Planta Med.* (Suppl.), 79–101.

ZENK, M. H., EL-SHAGI, H., ULBRICH, B. (1977a), Production of rosmarinic acid by cell suspension cultures of *Coleus blumei*, *Naturwissenschaften* **64**, 585–586.

ZENK, M. H., EL-SHAGI, H., ARENS, H., STÖCKIGT, J., WEILER, E. W., DEUS, B. (1977b), Formation of indole alkaloids serpentine and ajmalicine in cell suspension cultures of *Catharanthus roseus*, in: *Plant Tissue Culture and its Biotechnological Application* (BARZ, W., REINHARD, E., ZENK, M. H., Eds.), pp. 27–43. Heidelberg: Springer-Verlag.

ZHONG, J. J., YOSHIDA, T. (1995), High density cultivation of *Perilla frutescens* cell suspensions for anthocyanin production: Effects of Sucrose concentration and inoculum size, *Biotechnol. Bioeng.* **38**, 653–658.

14 Biotechnical Drugs as Antitumor Agents

Udo Gräfe, Klausjürgen Dornberger,
Hans-Peter Saluz

Jena, Germany

1 Introduction

Malignant tumors are a major cause of mortality, ranking only second to cardiovascular diseases in industrialized countries. The frequency of cancer increases with age, and the general increase in life span in this century renders the prevention and treatment of cancer a serious problem.

The aim of this chapter is to review the structures and modes of action of drugs from microorganisms, such as doxorubicin, bleomycin, mitomycin C, and others which have been used in the past as anticancer chemotherapeutics. The treatment of cancer by microbial metabolites was proposed already in 1955 when the cytotoxic effect of actinomycin D on tumor cells was first described. Subsequently, a plethora of other cytotoxic metabolites of microbial and plant origin was discovered (e.g., vincristine, camptothecin, maytansin, podophyllotoxins, taxol, and taxoids); however, only a few of these have found application (Tab. 1).

The high toxicity of most of these drugs and the non-specific interaction with normal and tumor cells were the major hurdles for their therapeutic application. Today, a small spectrum of natural products (antibiotics, plant alkaloids, and terpenes), synthetic alkylating agents (nitrosoureas, busulfan), heavy metal complexes (*cis*-diammine-dichloro-platinum, etc.), and antifolates (5-fluorouracil, methothexate) are used in the treatment of cancer (GALE et al., 1981; WILMAN, 1990). In general, the use of cytotoxic drugs in chemotherapy is accompanied by severe side effects such as gastrointestinal disorders, cardiovascular toxicity, nausea, and vomiting. Many tumors are not susceptible (while slowly dividing) or even become resistant to cytostatic

Tab. 1. Biotechnical and Plant-Derived Drugs as Therapeutic Antitumor Agents

Name	Structural Type	Mode of Activity
Actinomycin D (Dactinomycin)	Chromopeptide	Intercalation into the DNA, inhibition of RNA polymerase
Daunorubicin Doxorubicin Carminomycin Nogalamycin	Anthracycline	Intercalation into the DNA, formation of free radicals, induction of DNA strand breaks, inhibition of topoisomerases
Bleomycin (Peplomycin, Phleomycin)	Glycopeptide	Formation of free radicals, DNA strand breaking agent
Anthramycins	Benzodiazepine	Formation of free radicals and induction of DNA strand breaks
Mitomycin C, porfiromycin	Quinone	Formation of free radicals and induction of DNA strand breaks
Endynamicin	Enediyne	Formation of free radicals and induction of DNA strand breaks
Taxol, taxoids	Diterpene	Inhibition of tubulin depolymerization
Vincistine Vinblastin Vindesin	Alkaloids	Inhibition of mitosis by interference with tubulin polymerization
Camptothecin	Terpenoid plant toxins	Topoisomerase II inhibitor
Etoposid, teniposid	Podophyllum toxins	Topoisomerase I inhibitor
Maytanosides, ansamitocin, geldanamycin	Ansamacrolide	Inhibition of mitosis

drugs under therapeutic conditions (ASZALOS, 1988). In the 1970s it became evident that cytotoxic drugs were not the only answer to the cancer problem. New efforts were needed to uncover the molecular causes of malignant cellular growth and to develop more specifically acting anticancer agents. The similarity of the metabolic pattern of normal and cancer cells, the high diversity and different sensitivity of human tumors to anticancer agents made the search for an anticancer "wonder" drug a rather hopeless enterprise. This is illustrated by the moderate effects of chemical derivatives on the improvement of the activity of common anticancer drugs and the reduction of their toxic side effects. More recent efforts towards a more specific delivery of cytoxic drugs, such as the development of immunotoxins or drug encapsulations into liposomes, did not improve this general picture. A major aim of modern anticancer chemotherapy is to reduce the side effects, e.g., nausea and vomiting.

Major approaches to both the elucidation of the causes of malignant diseases and the search for new anticancer drugs have been promoted by recent advances of molecular biology and biomedical pharmacy. Although it is not within the scope of this survey to discuss these in detail a short reference is given below.

Dividing mammalian cells undergo a cell cycle (mitosis). Somatic cells in G_0 phase enter the G_1 phase and subsequently the S phase during which chromosome replication occurs. Cell division is accomplished during the subsequent G_2 and M phases. Subsequently, the two cell copies either enter a new G_0 phase or differentiate to a mature non-dividing cell. Cancer cells are unable to differentiate and hence they undergo a new cell cycle and divide repeatedly. Tumor cells can regulate their growth autonomously by secreting hormone-like proteins such as bombesin. Bombesin is a tetradecapeptide produced by most of the small-cell lung carcinomas. It is a potent mitogen which stimulates growth of small-cell lung cancer (SCLC) cells in serum cultures (MÜLLER, 1986). Several of the "classical" antitumor agents were shown to affect individual phases of the cell cycle in a specific manner, e.g., daunorubicin and *cis*-platinum act on G_1 phase cells; arabinosyl cytosine, thioguanine, and doxorubicin act on S-phase cells; bleomycin, and *cis*-diammine-dichloro-platinum act on G_2 phase cells; vincristine, doxorubicin, colchicine, etoposid, and bleomycin act on M phase cells. This suggested that they could change specific signaling pathways regulating cell division. The recent discovery of the cycline family of eukaryotic regulatory proteins unraveled specific signaling factors which trigger the coordinated events of cell division in yeast (SCHWOB and NASMYTH, 1993).

Identical copies of the parental cell are formed in mammals in response to outer signals such as growth hormones (Fig. 1) (KAHN and GRAF, 1986; BURCK et al., 1988). These exogenous signals are recognized by mem-

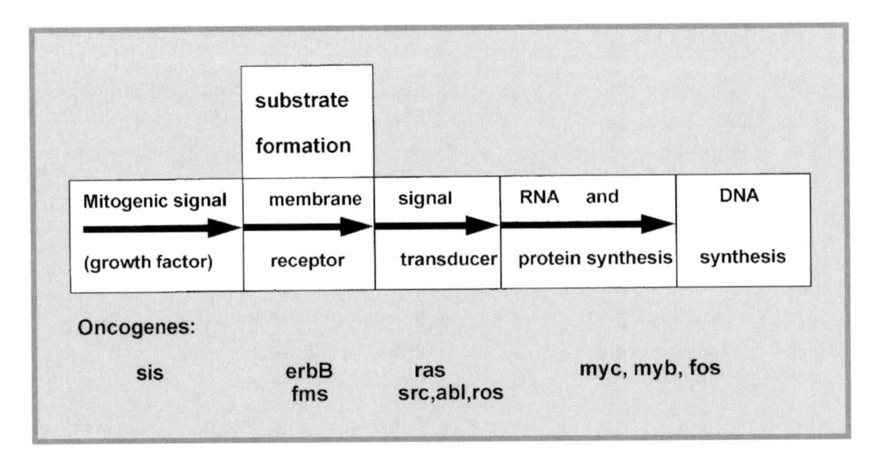

Fig. 1. Signal transduction pathway in cellular growth regulation (KAHN and GRAF, 1986).

brane receptors which transmit the message via GTP-dependent membrane proteins (G proteins) to proteins located at the inner side of the cytoplasmic membrane. In response, the latter become activated and modify (e.g., phosphorylate, farnesylate, etc.) proteins which by switching on subsequent events of the cellular growth-regulating pathway, act as a kind of an operational amplifier (YARDIN and ULLRICH, 1988). A normal cell only divides in response to a given signal, but a cancer cell has lost all growth control and repeatedly enters a new cell cycle. Hence, a deregulated expression of potential cancer genes (proto-oncogenes) encoding growth factors, membrane receptors, transcriptional factors, and regulatory enzymes has been suggested. In turn, autonomous secretion of growth factors, deregulated responses to growth factors, and continuous cell division will occur. The existence of tumor suppressor genes has been proven, the products of which prevent the outbreak of malignant growth.

Moreover, it became evident that carcinogenesis can be caused by chemical DNA strand-breaking and/or alkylating agents, and transforming oncogenic viruses (BRADSHAW and PRENTIS, 1987; PIMENTEL, 1987). Insertion of the latter into the DNA causes particular genes (proto-oncogenes) involved in growth control to become cancer genes ("oncogenes"). Usually, oncogene-transformed cells can be distinguished from the ancestral type by

- the overproduction of growth factors (autocrine secretion of growth hormones),
- altered structures of membrane receptors rendering them permanently activated even in the absence of exogenous growth factors,
- altered structures of growth factors giving rise to permanent (irreversible) receptor activation, and
- altered production rates, structures, and functions of intracellular signaling proteins such as protein kinases, phospholipases, inositol kinase.

Different types of oncogene-expressing (transformed) cells are now available as tools in the search for specific inhibitors of the pertinent, deregulated cellular growth function. A series of receptor proteins and growth-regulating enzymes has been well characterized and *in vitro* screening assays have been developed (Tab. 2). Since the 1980s, many new structures which interfere with cellular signal transduction have been discovered. These are regarded as "soft" anticancer drugs. Their particular mode of action renders them invaluable "biochemical tools" in detailed investigations of malignant growth. The first part of this article surveys microbial products which are used as "classical", cytotoxic antitumor agents. They interfere with DNA replication, transcription, and mitosis. In the second part, "non-classical" approaches with new and low-toxicity anticancer agents are discussed (UMEZAWA, 1989; WILMAN, 1990). Some of these new drugs amplify the cytotoxicity of classical anticancer agents, even in resistant cell lines, and inhibit metastasis and angiogenesis of tumors. In addition, inhibitors of growth regulatory enzymes such as protein kinases, inositol kinases, and phospholipases will be reviewed.

2 Prescreening for Antineoplastic Agents

In order to screen microbial cultures for antitumor drugs, predictive prescreens are required because animal antitumor assays have poor sensitivity. For DNA-damaging and cytotoxic agents a series of specific high capacity assays has been developed (FOX, 1991). Promising results have been obtained with microorganisms which express special genes (such as *recA*) in response to DNA damage. Moreover, enhanced activity against repair-deficient microbial mutants is another indication for cytotoxic drugs interfering with DNA. Studies on DNA-replication in a cell-free system have led to the detection of covalently modifying agents (GREENSTEIN and MAIESE, 1984).

Neoplastic cell lines have been widely used in many laboratories to screen microbial cul-

Tab. 2. Inhibitors of Enzymes of the Cellular Growth-Regulating Signal Transmission Pathway and Drugs Exerting a Non-Classical Action on Tumors

Enzyme/Screening Feature	Inhibitor
Protein kinase C	Calphostin, staurosporin, RK-286c, K-252a, UCN-01, BE-13793c, Sch45752, balanol, MS-282a
Protein tyrosin kinase	BE-23372M, emodin, lavendustin, erbstatin, genistein, epiderstatin
Protein phosphatase	Tautomycin, okadaic acid, dephostatin, microcystin, calyculin
Calmodulin and cAMP-Phosphodiesterase	KS-505a
Phospholipase D	Sch49210, Sch53514, Sch53517
Inositol-specific phospholipase C	Hispidospermidin, psi-tectorigen, Q12713
Inositol phosphate receptor agonist	Adenophostins
Inositol kinase	Herbimycin, isoflavones, echiguanin, inostamycin, piericidin B1-N-oxide, echigramins, piericidin, benzaldehyde
Diacylglycerol kinase	Cochlioquinone, temphone
Inositol monophosphatase	L-671-776
DNA topoisomerase I	Dotriacolide, camptothecins (from plants), rebeccamycin
DNA topoisomerase II	UCT4B, saintopin, BE-10988, streptonigrin, anthracyclines, BE-22179, cyclothialidin
Poly(ADP-ribose)polymerase	2-Methylquinazolines, benzamides (synth.)
Glyoxylase I	Glyo I and II
Induction of differentiation of cancer cell lines	Hygrolidine, cycloheximide, trichostatins, reveromycin, differanisol, differenol A, microcystilide A, herbimycin, indolecarbazoles (staurosporin), lavanducyanin, oxanosin, kasuzamycin
Metastasis inhibitors and inhibitors of angiogenesis	TAN-1120, TAN-1323, U-77863, U-77864, matlystatins, siastatin B, WF-16775, erbstatin, herbimycin, lactoquinomycin
Glutathion S-transferase	TA-3037A, benastatins, rishirilid, cysfluoretin, bequinostatin
Enhancer of the cytotoxicity of antitumor agents	Piperafizins, bisucaberin, verapamil (synth.), BE-13793c, cyclosporin A, resorthiomycin, rubiginon, polyethers (laidlomycin), hatomarubigin
Reduction of side effects of highly toxic anticancer agents	Conagenin
Ras-farnesyltransferase	Pepticinnamine, streptonigrin, indolocarbazoles
Immunostimulators	Forphenicinol, FK-156, bestatin
Activity against tumor cells, resistant to other antitumor drugs	Kazusamycin, lactoquinomycin
Immunosuppressants	Spergualin, 15-deoxyspergualin
Adenosine deaminase	Adecypenol, pentostatin
5′-Nucleotidase	Nucleoticidin
Inhibitors of the eukaryotic cell cycle	Reveromycins
Melanogenesis	OH-3981

ture or plant extracts; however, better purified fractions and components are required for results which are more reliable. Thus, the test program of the National Cancer Institute of the USA operates *in vitro* disease-oriented screens which are directed specifically to cell type specific agents. New compounds are tested against a panel of approximately 60 cell lines derived from human solid tumors such as colon tumors, melanoma, kidney tumors, ovarian tumors, brain tumors, and leukemia (GREVER et al., 1992; KAO and COLLINS, 1989). A more recent approach is the usage of cell cultures in the screening for inducers of tumor cell apoptosis (YAMAZAKI et al., 1995).

Increasing use is been made of transformed cells (such as those transformed by retroviruses, e.g., the Rous sarcoma virus) to search for agents which display a selective toxicity or which alter tumor cell morphology. In addition, these cell lines are useful as tools in screening for inhibitors of special oncogene-encoded cellular functions (UMEZAWA, 1989). Another more recent, promising approach concerns the protein targets of drugs interfering with the regulation of the cell cycle. High-throughput screening assays have been developed for a series of receptors and enzymes of the cellular signaling cascade which are crucial in the development of malignancy.

Despite these promising *in vitro* approaches trials with animals will finally be required to reduce the number of prescreening-positive compounds to an amount which justifies continuing investigations, including clinical trials. In this context, the use of human tumor xenograft models represents a new approach to the evaluation of putative anticancer agents.

An effective screening program for new anticancer agents will need new sources of drugs such as rare microorganisms, plants, and animals. During the last ten years, marine organisms such as tunicates, molluscs, dinoflagellates, and bryozoa have become the subject of an expanding field of research in the search for new antitumor agents (see, e.g., dolastatins, spongistatins, bryostatins, okadaic acid, and other cytotoxic metabolites) (JENSEN and FENICAL, 1994).

3 Classical Anticancer Drugs

3.1 Nucleoside Antibiotics and Analogs of Nucleobases

The structural class of the naturally occurring nucleoside antbiotics comprises approximately 150 representatives (ISONO, 1988; ASZALOS and BERDY, 1978). Numerous analogs have been obtained by chemical synthesis. They exert their antimicrobial and cytotoxic activity as "pseudo" nucleobases or nucleosides which are activated to "pseudo" nucleotides. Thereafter, they act via a negative feedback mechanism on nucleotide-forming biosynthetic pathways or inhibit DNA- and RNA-dependent polymerases. However, the incorporation of "pseudo" nucleotides may also diminish the template function of DNA. Antagonists of folic acid, purines, and pyrimidines have been used in anticancer chemotherapy.

Nucleobase analogs such as 5-fluorouracil (synthetic), 6-mercaptopurine (synthetic), thioguanine, thiouridine (NISHIKORI et al., 1992), 7-hydroxy-guanine (*Streptomyces* sp.) (KITAHARA et al., 1985), and "pseudo" nucleosides such as arabinosyl cytosine (synthetic), cadeguomycin (TSUCHIYA et al., 1992), oxanosin (*Streptomyces hygroscopicus*) (YUAN et al., 1985), neplanocin (*Ampullariella regulans*) (YAGINUMA et al., 1981), spicamycin, its derivative SPM VIII (KAMISHOHARA et al., 1994) and others (see, e.g., ISONO, 1988) display antitumor and antiviral activities against various cells and viruses. The general therapeutic problem associated with these agents is their poor selectivity and relatively high toxicity. However, several examples of a more selective action of nucleosides against viruses are known; synthetic drugs such as acyclovir and its derivatives are powerful inhibitors of herpes virus thymidylate kinase, and hence invaluable therapeutic agents (Fig. 2, see also the therapeutic effect of 2,3-dideoxynucleosides on HIV virus infections).

A more specific antitumor activity has been demonstrated for inhibitors of enzymes asso-

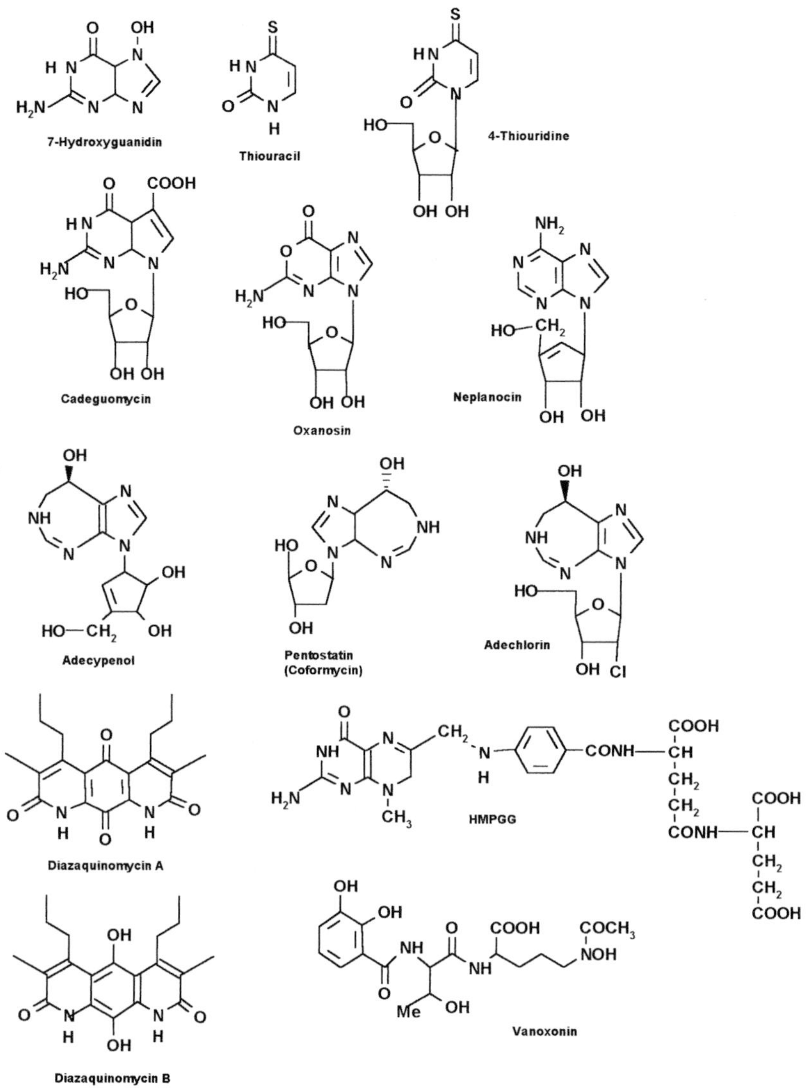

7-Hydroxyguanidin Thiouracil 4-Thiouridine Cadeguomycin Oxanosin Neplanocin Adecypenol Pentostatin (Coformycin) Adechlorin Diazaquinomycin A HMPGG Vanoxonin Diazaquinomycin B

Fig. 2. Structures of neoplasm inhibitory nucleobase analogs, nucleosides, and antifolates.

ciated with special kinds of malignancies, such as adenosine deaminase.

The importance of purine metabolism for the immune system was recognized due to the observable associaton of deficiencies in purine salvaging enzymes, such as adenosine deaminase and purine nucleoside phosphorylase, with heritable immunodeficiency. Coadministration of adenosine deaminase inhibitors in the treatment of leukemia by arabinosyl adenosine caused syndromes similar to those observed in immunodeficient patients. Inhibitors of this enzyme such as pentostatin (2-deoxy-coformycin; *Streptomyces antibioticus*) (HOLLIS-SHOEWALTER et al., 1992), and adechlorin (*Actinomadura* sp. OMR-37) (ŌMURA et al., 1985) (Fig. 2) were aimed to prevent the degradation of arabinosyl adenosine by this enzyme (AGARWAL et al., 1983; ŌMURA et al., 1986).

In eukaryotic cells, 5′-nucleotidase is necessary for the formation of the 5′cap structure of mRNA. Hence, inhibitors of this enzyme such as the polysaccharide nucleoticidin

(*Pseudomonas* sp.) (OGAWARA et al., 1985) inhibit malignant cell growth. Moreover, folate metabolism was suggested as an additional promising target for antitumor drugs (see, e.g., the role of methotrexate as an inhibitor of tumor cell dihydrofolate reductase; ROSOWSKI et al., 1992) (Fig. 2). More recently, thymidylate synthase (VAN DER WILT et al., 1994; KALMAN, 1989) and serine hydroxymethyltransferase (RAO, 1991) have been suggested as targets for new anticancer drug development.

The search for inhibitors of folic acid metabolism from microbial cultures (e.g., by the use of *Entercoccus facium* as an assay organism) led to the discovery of analogs and inhibitors of thymidylate synthase as "antifolates". Thus, 7-hydro-8-methylpteroylglutaminylglutamic acid (HMPGG) was isolated as a folic acid analog from an *Actinomyces* culture (MURATA, et al., 1987). Diazaquinomycin A (MURATA et al., 1985) and vanoxonin (KAMAI et al., 1985) are microbial inhibitors of thymidylate synthase which have cytotoxic effects on cell cultures (Fig. 2).

Moreover, a series of nucleotide analogs has been isolated from microorganisms, which are active as antitumor and antimetastatic agents due to their interaction with sugar nucleotide glycosyltransferase (see below) (KHAN and MALTA, 1992).

3.2 Drugs Binding Non-Covalently to the DNA

Netropsin and distamycin (pyrrolamidine antbiotics from *Streptomyces netropsis* and *S. diastallus,* respectively) are invaluable biochemical tools in recent DNA research

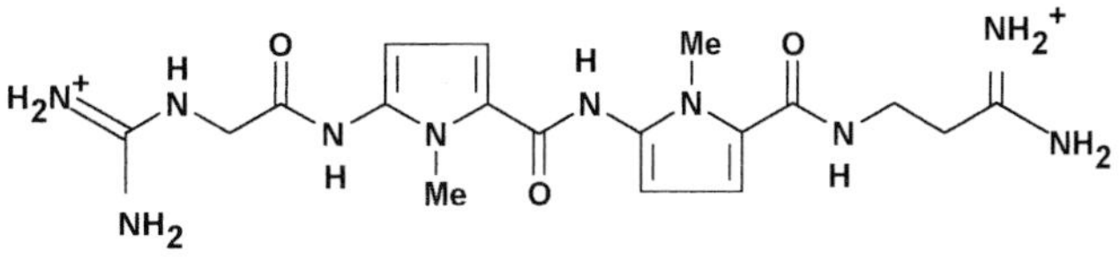

Netropsin

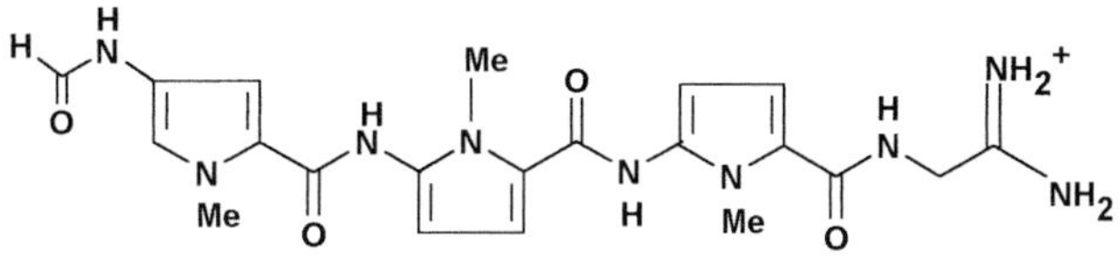

Distamycin

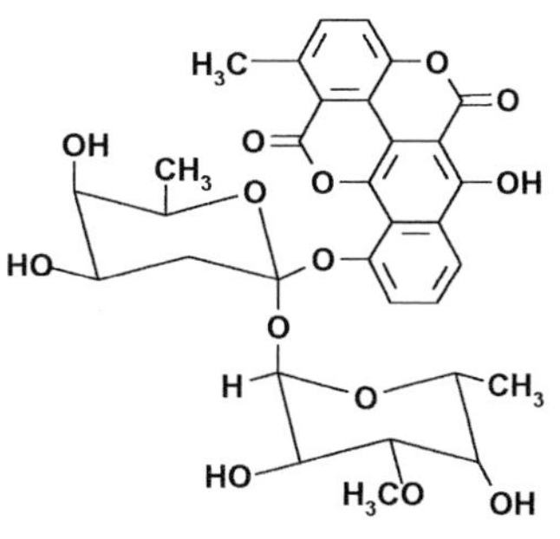

Chartreusin

Fig. 3. Structures of antibiotics binding non-covalently to the DNA.

(Fig. 3). Both antibiotics bind to the minor groove preferably at regions rich in adenosine and thymidine. The complex formation of netropsin and the duplex DNA (d(5′CGCGAATTCGCG3′)) suggested that at the 5′AATT sequence in the center the three NH groups of netropsin form hydrogen bonds with O-2 of thymine and N-3 of adenine (KOPKA et al., 1985) (see Fig. 3). Distamycin which is a closely related antibiotic forms van der Waals bonds between its 5 NH-groups and O-2 of thymidine and the N-3 of adenine in the DNA minor groove (COLL et al., 1987; DERVAN et al., 1987) (see Fig. 3).

As a consequence of binding, the unwinding of DNA during replication and transcription as well as the access of the pertinent enzymes and proteins to specific DNA sites are prevented (ZIMMER et al., 1990). Chemical modifications of distamycin were done to improve the antiviral properties of its structures (ARCAMONE, 1994).

Non-covalent binding to DNA at different sequences was also shown for a series of polycyclic aromatic antbiotics such as chartreusin (KRÜGER et al., 1986) (Fig. 3), chromomycin (KOAMURA et al., 1988; UCHIDA et al., 1985), angucyclins (saquayamycin), and saframycin (see Fig. 8a). However, none of these compounds are used in anticancer therapy.

Both chromomycin A3 and olivomycin inhibit the synthesis of DNA and RNA *in vitro* and the function of RNA polymerase and DNA polymerase I *in vivo.* Binding of chromomycin to DNA is magnesium-dependent and occurs preferably at regions with a high G/C content. The complex formed is stable to nuclease digestion. Similar to chromomycin A3, an antibiotic from *Streptomyces plicatus,* mithramycin binds preferably to d(5′ATGCAT3′)2 regions (BEAUVILLE et al., 1990). Olivomycin interferes with the elongation step of RNA polymerase and inhibits DNA polymerase I by binding to guanine- and cytosine-containing DNA regions.

3.3 Intercalating Antibiotics

Intercalating drugs have been important both as tools of biochemical research and as anticancer agents. Their molecular interactions with DNA have been studied extensively by physicochemical methods (WANG et al., 1987; WANG, 1992).

The term intercalation refers to the insertion of flat, planar, aromatic molecules (quinolines, acridines, phenanthridines, phenoxazines, anthraquinones, fluorenes) into the minor groove of DNA, between two adjacent base pairs (WANG, 1992; WARING, 1975). Polar substituents such as amino acids or sugars (see the structures of actinomycin D and doxorubicin, Fig. 4) stabilize the intercalation complex by the formation of hydrogen bonds with the phosphodiester groups of the deoxyribose moieties of DNA (MANGER, 1980; KNUGH et al., 1980). Intercalation results in the distortion of the DNA and steric stress whereby neighboring bonds are weakened. As a consequence, the template function of the DNA is reduced, and single and double strand breaks as well as frameshift mutations may occur. In addition to some synthetic and plant-derived drugs (flavines, acridine dyes, ellipticin), the actinomycins (KNUGH et al., 1980; ADAMSON et al., 1979) and the anthracyclines (EL KHADEM, 1982; CASSIDY and DOUROS, 1988; LOWN, 1988) are well known intercalators of microbial origin. The latter type of antbiotics differs from the intercalating drugs by its radical-forming properties and its inhibitory effect on DNA topoisomerase II (see below).

3.3.1 Ellipticin

Ellipticin and its 9-methoxy derivative were isolated from plants such as *Excavatia coccinea* and *Chrosia moorie.* They inhibit DNA synthesis by binding to the minor grove of double-stranded DNA.

3.3.2 Actinomycins

Already in 1955 the antitumor activity of actinomycin D (Fig. 4) (from *Streptomyces antibioticus*) was observed (TSCHAGOSHI et al., 1986). Representatives of the actinomycin chromopeptide family are extremely potent inhibitors of RNA polymerase, and they have widely been used as a biochemical tool to

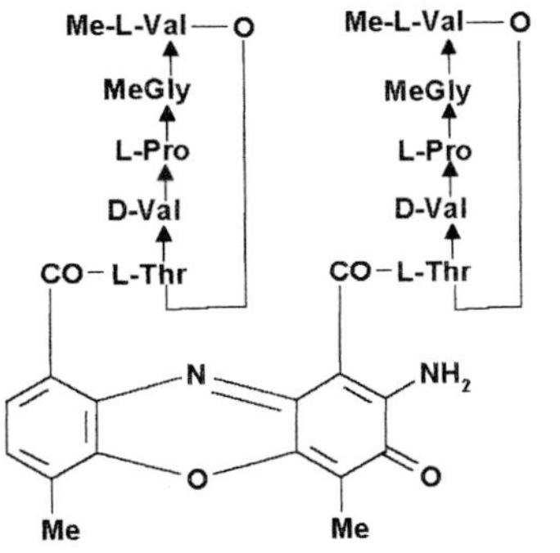

Actinomycin D

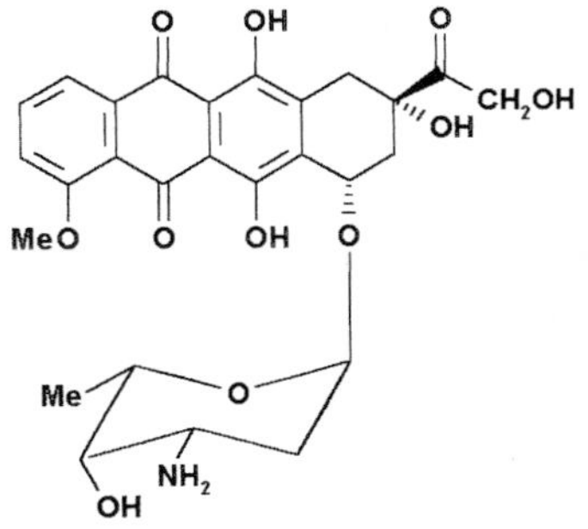

Doxorubicin

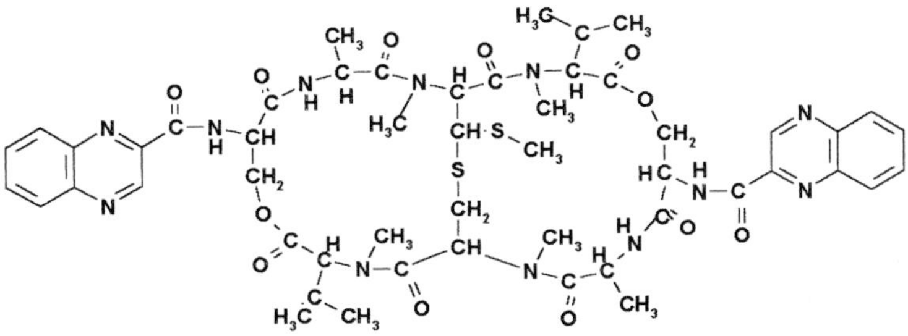

Echinomycin

UK 63.052

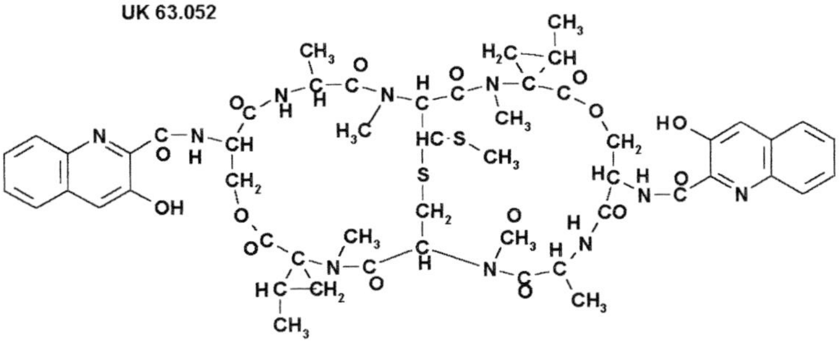

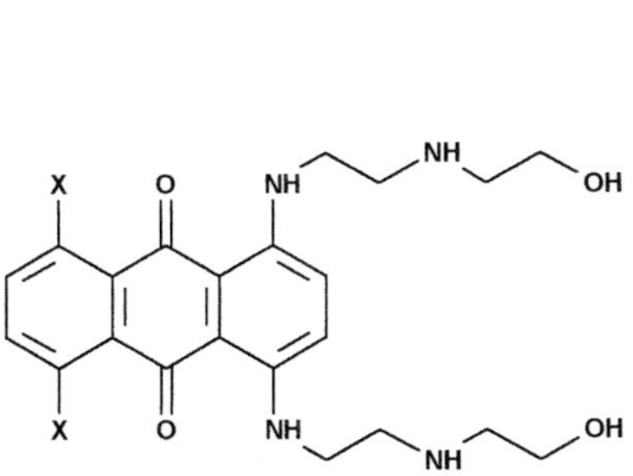

Mitoxantron (X=OH)

Ametantron (X=H)

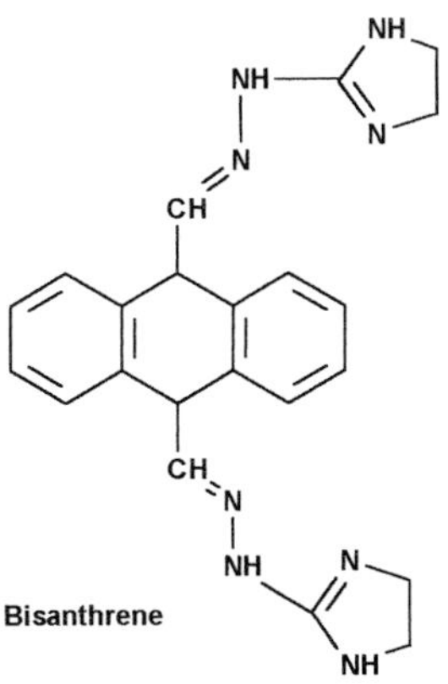

Bisanthrene

Fig. 4. Structures of intercalating drugs.

study RNA synthesis (GALE et al., 1981; FORNICA, 1977). Actinomycins are composed of a phenoxazinone chromophore, which is the intercalating unit, and two linked pentapeptidolactone moieties (MAYER and KATZ, 1978; KATAGIRI, 1975). The homologs are distinguishable by their amino acid composition. The compositon of the actinomycin complex depends on the available amino acid precursors (FORNICA, 1977). Intercalation of the actinomycins requires the 2-NH_2 group of the phenoxazinone chromophore and the quinoid 2-C O_2; DNA binding occurs at $(d(CATGAT))_2$ sequences (ZHOU et al., 1989). Intercalation of actinomycin D and its analogs in DNA was extensively studied by spectroscopic methods (KNUGH et al., 1980). Computational techniques have been employed to calculate the substituent effect on free energy of binding to DNA (LYBRAND, 1988). In the past, actinomycin D (dactinomycin, Fig. 4) was used in clinical anticancer trials but due to its high toxicity its therapeutic use in Wilms tumor, choriocarcinoma, testis carcinoma, neuroblastoma, and sarcoma is disputed. In order to reduce toxic side effects and to improve the efficacy against certain tumor cells, a series of analogs of actinomycin D was obtained by semisynthesis, but none of them appeared to be more promising than the parent compound (ADAMSON et al., 1979; FORNICA, 1977).

3.3.3 Quinoxaline Antibiotics

The anticancer activity of echinomycin (Fig. 4) was rediscovered due to investigations of its mode of action at the molecular level. It was the first DNA bis-intercalator to be identified (WARING, 1992).

Bis-intercalation involves the binding of a drug to DNA via the quinoxaline rings, with the peptide moieties binding to the DNA minor groove (ADDES et al., 1992). Representatives of this group of antibiotics display different sequence specificity (ADDESS et al., 1992), whereby CpG sequences are preferred. Triostin and luzopeptin are related antibiotics which display similar bis-intercalating properties (WANG, 1992) as detected by X-ray diffraction and NMR spectroscopy. Recently, sandramycin, a related 3-hydroxy-quinoxalinic acid cyclopeptide antibiotic was isolated, which lacks a sulphide bridge (MATSON et al., 1993). The compound UK-63.052 (*Streptomyces braegensis*) (Fig. 4) is a new representative of the quinomycin group of antibiotics (RANCE et al., 1989).

In every case, the alanine residue of the octapeptide ring of echinomycin type antibiotics is a critical determinant of the observed specificity for CpG dinucleotide sequences (WARING 1979, 1990). This is due to hydrogen binding interactions which involve NH and CO groups of the alanines together with the 2-amino group and N-3 of guanine in the minor groove of the helix. The binding of echinomycin type antibiotics to DNA is very tight and remains stable even during gel electrophoresis of DNA fragments in footprinting analysis (FOX, 1990).

3.3.4 Anthracyclines and Related *p*-Quinones

Both, daunomycin and doxorubicin (Fig. 4) (a semisynthetic derivative of daunomycin) bind to DNA sites which contain either only (G:C) or (A:T) base pairs (ASZALOS and BERDY, 1978; SALUZ and WIEBAUER, 1995; NEIDLY and WARING, 1983; WILMAN, 1990). In general, a three base pair binding site is required for the intercalation. The aminosugar at C-7-O of the benzo[a]naphthacene quinoid aglycone is needed to replace water from the minor groove and to anchor the molecule to the phosphodiester linkages between the adjacent deoxyribose moieties (WANG et al., 1987). Due to these properties some representatives of the anthracyclines induce a small degree of helical unwinding.

Doxorubicin and less frequently daunorubicin are used in the treatment of acute lymphatic leukemia, acute myeloid leukemia, lymphoma, sarcoma, Wilms tumor, and cancer of the breast, lung, bladder, thyroid, and prostate. After recognizing the anthracyclines as lead structures a series of intercalating anthraquinone type anticancer agents such as mitoxantron, ametantron (Fig. 4), anthropyrazol, and bisanthrene (Fig. 4) was developed as anticancer drugs and used therapeutically, e.g., in the treatment of breast cancer (ASZALOS, 1988; HOLLIS-SHOEWALTER, 1988).

3.4 Inhibitors of Enzymes of DNA Replication and Transcription: DNA Topoisomerases

Dramatic advances have been achieved in the field of DNA topoisomerases. These include the cloning of new topoisomerase genes involved in DNA recombination and segregation of replicated DNA (WATT and HICKSON, 1994; POOT and HOEHN, 1993; GIACCONE, 1994). Hence, topoisomerases have become promising targets of modern anticancer agents (PAOLETTI, 1993). Antitumor agents such as epipodopyllotoxins (etoposid, teniposid), acridine dyes, anthracyclines, and ellipticins were found to be specific inhibitors of topoisomerase II, and the plant alkaloid camptothecin was recognized as a specific inhibitor of mammalian topoisomerase I (Fig. 5a) (FOSTEL and SHEN, 1994).

Topoisomerases as targets of antibiotics and antitumor drugs, their relation to the topological stage of intracellular DNA, and their function in replication have been discussed in a series of reviews (ZIMMER et al., 1990; FOSTEL and SHEN, 1994; CHEN and LEROY, 1994). The enzymes catalyze the concerted breakage and rejoining of the DNA backbone, and hence they are indispensable for DNA replication, transcription, chromosome segregation, and recombination. Mammalian type I topoisomerases mediate the relaxation of negatively or positively supercoiled DNA by transiently breaking and releasing one DNA strand in such a manner that the linking number changes by steps of one. Type II topoisomerases break and religate phosphodiester bonds in double-stranded DNA passing another duplex region through the break, thereby altering DNA topology. In addition, they are structural proteins involved in the spatial organization of chromatin (ZUNINO and CAPRANICO, 1990). One group of agents such as coumarin derivatives (novobiocin, coumermycin) and the synthetic quinolone antibacterial agents act at the level of the enzyme subunits (FOSTEL and SHEN, 1994; CHAKRABORTY et al., 1994; CHEN and LEROY, 1994; HSIEH, 1992), either by interfering with ATP hydrolysis or with the DNA cleaving subunit A.

Drugs introducing lesions in the DNA trap the transient topoisomerase II–DNA intermediate which is held together by two covalent bonds. Anthracyclines (see Fig. 7), ellipticin, and epipodophyllotoxins (etoposid and tenoposid) (Fig. 5b) are the major antitumor drugs that form cleavable complexes in eukaryotes.

In a search for new inhibitors of topoisomerase II (HECHT et al., 1992) a benzoanthraquinone (UCE 1022) has recently been isolated from *Paecilomyces* sp. (FIJII et al., 1994). Other inhibitors of microbial origin are dotriacolide (FIJII et al., 1994) (inhibitor of topoisomerase I and DNAse), TAN-1496A (a diketopiperazine, *Streptomyces* sp.; Fig. 5a) (FUNABASHI et al., 1994), and UCE6 (a non-glycosylated anthraquinone, *Streptomyces* sp.; Fig. 5a) (FIJII et al., 1993). Anthracyclines were also shown to inhibit toposomerase I in addition to topoisomerase II (CROW and CROTHERS, 1994).

Recent screening approaches in the search for microbial inhibitors of type II topoisomerases (PAOLETTI, 1993) revealed new inhibitors, such as BE-22179 (a peptide from *Streptomyces*) (OKADA et al., 1994), saintopin (a benzo[a]anthraquinone from *Paecilomyces* sp.; Fig. 5b) (YAMASHITA et al., 1990a), BE-10988 (*Streptomyces xanthocidicus*; Fig. 5b) (OKA et al., 1991), streptonigrin (TOLSTIKOV et al., 1992; YAMASHITA et al., 1990b) and its derivatives, and UCT4B (terpentecine type, *Streptomyces* sp.; Fig. 6b) (UOSAKI et al., 1993; KAWADA et al., 1992a). The majority of these inhibitors exhibit potent antitumor activity. Some also affect bacterial topoisomerase II (gyrase) which is the target of several important antibacterial agents. A cytotoxic quinoid system, popolahuanone E (Fig. 5b), was isolated from a Pohnpei sponge. It is selectively toxic for the A549 non-small cell human lung cancer cell line (CARNEY and SCHEUER, 1993). In a series of comparative studies it was shown that the interaction with mammalian topoisomerase II is one of the major causes of cytotoxicity of the anthracycline type antibiotics (CAPRANICO and ZUNI-

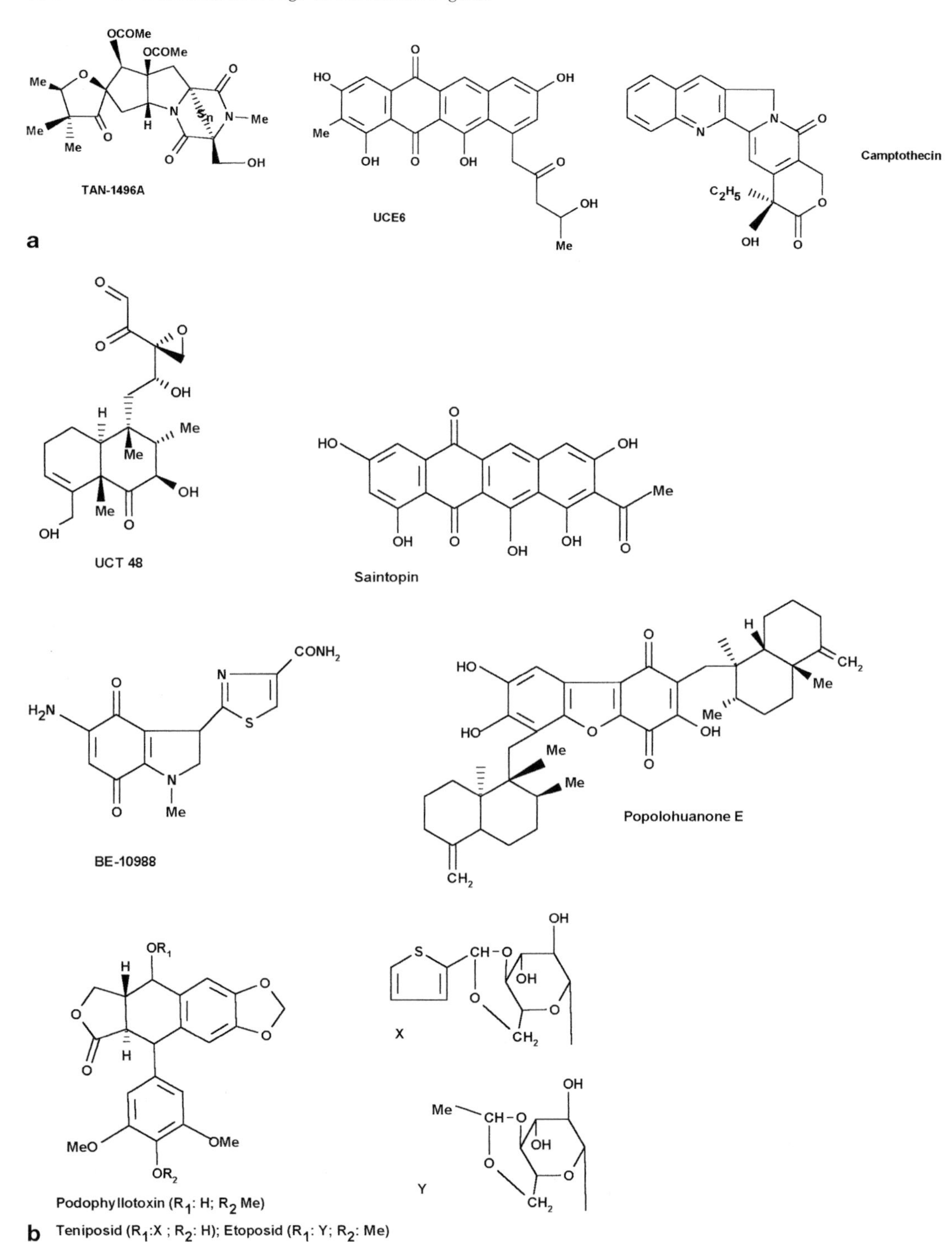

Fig. 5. Structures of topoisomerase inhibitors. **a** Inhibitors of topoisomerase I, **b** inhibitors of topoisomerase II.

NO, 1990; ZUNINO and CAPRANICO, 1990; GREVER et al., 1992). Although interference with the functions of DNA topoisomerases I and II holds great promise for the treatment of cancer, these drugs may themselves be dangerous due to their mutagenic and carcinogenic potential (ANDERSON and BERGER, 1994). DNA helicase has recently been proposed as a target involved in DNA replication, repair, recombination and transcription by unwinding of double-stranded DNA. Heliquinomycin as a new spirocyclic aromatic compound from *Streptomyces* sp. effectively inhibits the enzyme (CHINO et al., 1996).

Podophyllotoxins as inhibitors of topoisomerase II:

The antitumor drugs podophyllotoxin (Fig. 5b), peltatins, and other lignane type compounds have been extracted from *Podophyllum peltatum* L. These compounds interact non-covalently with tubulin and inhibit mitosis during metaphase (see Sect. 3.6). Due to their high toxicity for mammalian organisms they cannot be used for therapeutic purposes. Semisynthetic derivatives, such as tenoposid, etoposid (Fig. 5b), and mitopodosid appear to be more promising. They poorly interfere with tubulin and mitosis but instead, inhibit the transport of nucleosides in cells and the incorporation of thymine and uridine into DNA and RNA due to interaction with topoisomerase I (FRANZ, 1990).

Camptothecin type compounds as inhibitors of topoisomerase II:

Camptothecins are plant dimeric indole alkaloids which were discovered as selective growth regulators for several species of mono- and dicotyledonous plants (BUTA and KALINSKI, 1988; WALL and WANI, 1993; SLICHENMYER et al., 1993; CURRAU, 1993) (Fig. 5a). Chemical studies on inhibitors of DNA topoisomerase I which lead to the production of semisynthetic derivatives, such as CPT-11 and topotecan suggested a therapeutical potential for these types of compounds (ANDOH et al., 1993). Topotecan and CPT-11 are active against a broad spectrum of human tumors (SLICHENMYER et al., 1993). They slow down the religation step of topoisomerase I and stabilize the covalent adduct between the enzyme and DNA. In S phase cells double-stranded breaks occur at the position of the topoisomerase I–DNA adducts which is probably the reason for the high cytotoxicity of these agents. Both drugs are now under clinical investigation (SLICHENMYER et al., 1993; CURRAU et al., 1993).

3.5 Agents Forming Covalent Bonds with DNA

Covalent modification of DNA by interaction of reactive molecules with bases or sugars blocks its template function and induces the onset of repair processes. As an initial step, radical species are generated by the drug molecule due to its interaction with metabolic enzymes such as cytochrome-dependent and other oxidoreductases. The result is interstrand crosslinking and single- or double-stranded DNA cleavages. Some of the prominent representatives of this type of commercially available agent will be discussed (PINEDO, 1980).

3.5.1 Mitomycin C

Mitomycin C (*Streptomyces caespitosus*) (Fig. 6) is used in anticancer chemotherapy by virtue of its inhibitory effect on DNA synthesis and its ability to induce strand cleavages (TOMASZ, 1994). It is known as one of the most potent antitumor antibiotics. The molecular mechanism of action of this highly toxic antibiotic involves the reduction of the quinone and subsequent activation of C-1 (PINEDO, 1992; TOMASZ, 1994; SARTORELLI et al., 1993). Mitomycin C and its analogs, such as porfiromycin and mitiromycin (KASAI et al., 1991; ARAI et al., 1994) are used as an alternative to radiation during the treatment of hypoxic solid tumors which have acquired enhanced sensitivity to this bioreductive alkylating agent. It has been suggested that in hypoxic cells, one-electron transfer enzymes, such as diaphorase, control bioreductive alkylation. At least six different enzymes are capable of activating mitomycin C and other similar drugs. The nature of the activated molecular species and the resultant biological lesions can vary with the activating enzyme. This, in turn, causes variations in toxicity for different cell types. The process of activation is accompanied by the opening of the conformationally constrained aziridine ring to yield free radicals as intermediates. The double-bonded structure formed interacts with guanine bases in opposite DNA strands, thereby crosslinking them in a bifunctional manner (RAUTH and RAYMOND, 1993; TOMASZ, 1994; ROCKWELL et al., 1993; WARDMAN, 1990; BUTLER and HAEY, 1987). Frameshift mutations appear to be another consequence of DNA repair and cause of potent mutagenic and teratogenic activity of the mitomycins. In order to obtain less toxic and more potent mitomycin C analogs, both congeners of mitomycin fermentations (see, e.g., albomitomycin KW-2149) and synthetic derivatives have been isolated (KONO et al., 1991, 1995; KASAI et al., 1991). DNA damage appears to be critical for the cytotoxicity of mitomycins. Both interstrand and intrastrand crosslinking occur in mitomycin-treated cells.

3.5.2 Anthramycins

Anthramycins (pyrrolo(1,4)-benzodiazepines) bind tightly via their C-11 atom to the 2-amino group of guanine bases with the elimination of water (Fig. 6). The amino groups of these antibiotics play an important role in their molecular mechanism of action (BARKLEY et al., 1986; MORRIS et al., 1990). Representatives of these agents include anthramycin (Fig. 6), sibiromycin, tomaymycin, neothramycin, and porothramycin (TSUNAKAWA et al., 1988). Due to the high risk of hazardous side effects, the anthramycins offer no advantage in cancer treatment compared to other cytotoxic antitumor drugs. Some anthramycins were evaluated in clinical trials but they have not come to general use.

3.5.3 Cyclopropane, Aziridines, and Epoxide Compounds

The compound CC-1065 (*Streptomyces zelensis*; Fig. 6) (MARTIN et al., 1980) is a heterocyclic antibiotic which forms a DNA adduct via the initial interaction of the DNA guanine residues with a cyclopropane ring attached to a cryptic *p*-quinoid structure. This is followed by molecular rearrangements and cleavage of the phosphodiester bond from deoxyribose. The compound CC-1065 displays sequence specificity in that it binds to regions rich in guanine (SCAHILL et al., 1990).

A series of microbial products containing aziridine and epoxide structures, such as azinomycins (*Streptomyces griseofuscus*), WF-3405 (*Amauroascus aureus*), and duocarmycins A and SA (*S. zelensis*) (Fig. 6; ICHIMURA et al., 1991) was suggested to form covalent bonds with DNA. The exhibit extremely potent cytotoxic activity ($IC_{50}=10^{-12}$ M–10^{-9} M for Hela S3 cells). The constrained three-membered ring structure in addition to particular steric and electronic conditions permits a nucleophilic attack on DNA bases.

Duocarmycin SA (ICHIMURA et al., 1990) was found to be the most stable and most potent cytotoxin of these agents. They alkylate DNA in mechanistically similar manner to CC-1065 (ICHIMURA et al., 1990, 1991). DNA alkylation occurs via addition of adenine N3 to the least substituted carbon of the activated cyclopropane within AT-rich minor groove sites (BOGER and JOHNSON, 1996). In this context, scirpene type mycotoxins such as trichothecin (Fig. 6), trichodermole, anguidine, nivalenol, crotocin, and their analogs were tentatively used as antitumor agents. They are frequently occurring products of Fungi imperfecti (*Trichothecium, Fusarium, Myrothecium, Trichoderma*) and inhibit DNA and protein synthesis.

3.5.4 Streptonigrin

Streptonigrin (bruneomycin) (Fig. 6) and its natural analogs (LIN et al., 1992) are cytotoxic radical-forming anticancer drugs which modify DNA covalently and cause strand

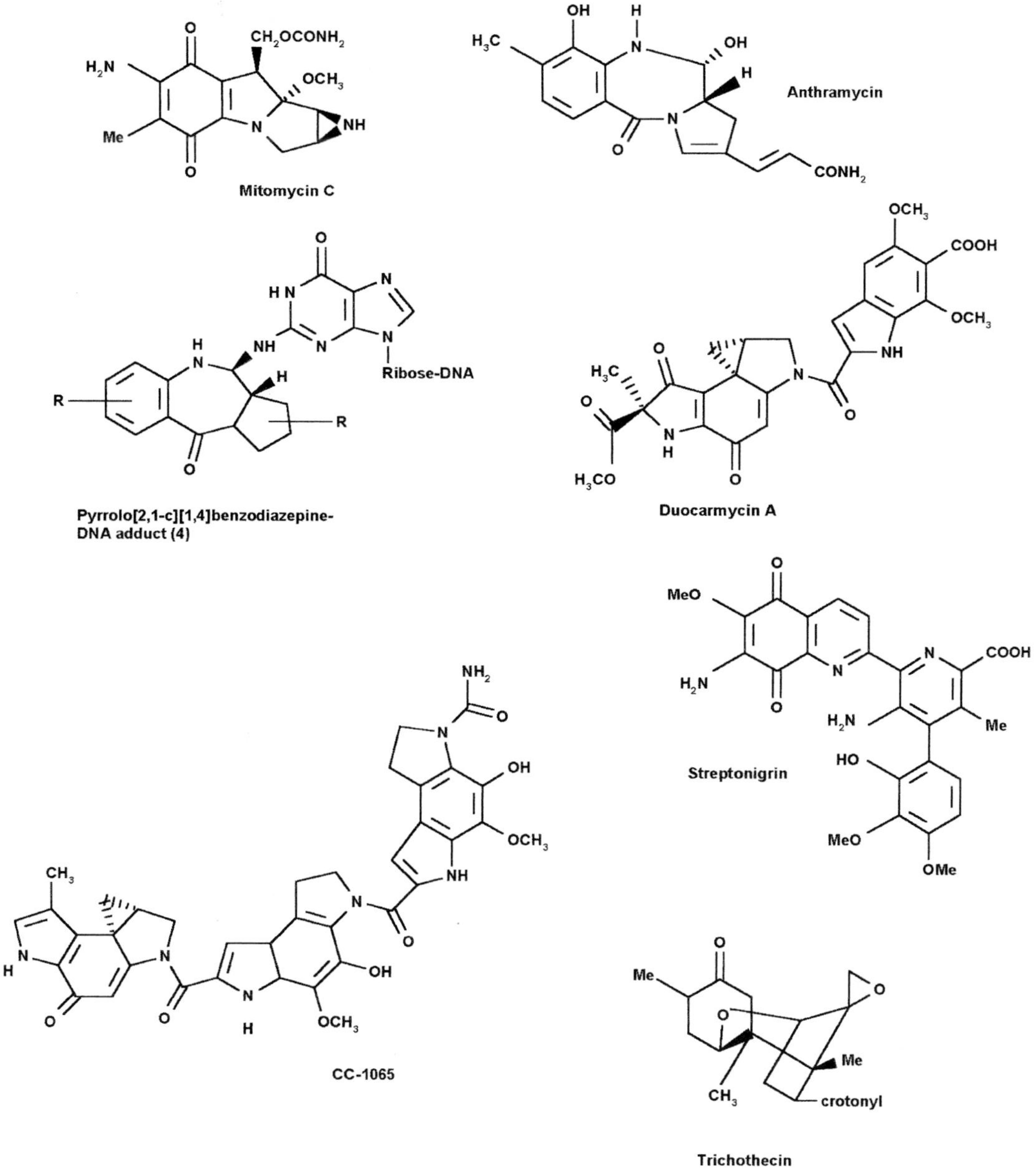

Fig. 6. Structures of drugs binding covalently to the DNA.

breaks. Interactions with topoisomerase II appear to be involved in this activity (YAMASHITA et al., 1990b). Streptonigrin and a series of chemical derivatives were shown to inhibit HIV reverse transcriptase (TAKE et al., 1989).

3.5.5 Anthracyclines

The anthracyclines (Fig. 7a) belong to a class of antitumor drugs which exhibit activity against a spectrum of human cancers. Only a few tumors, such as colon cancer, melanoma,

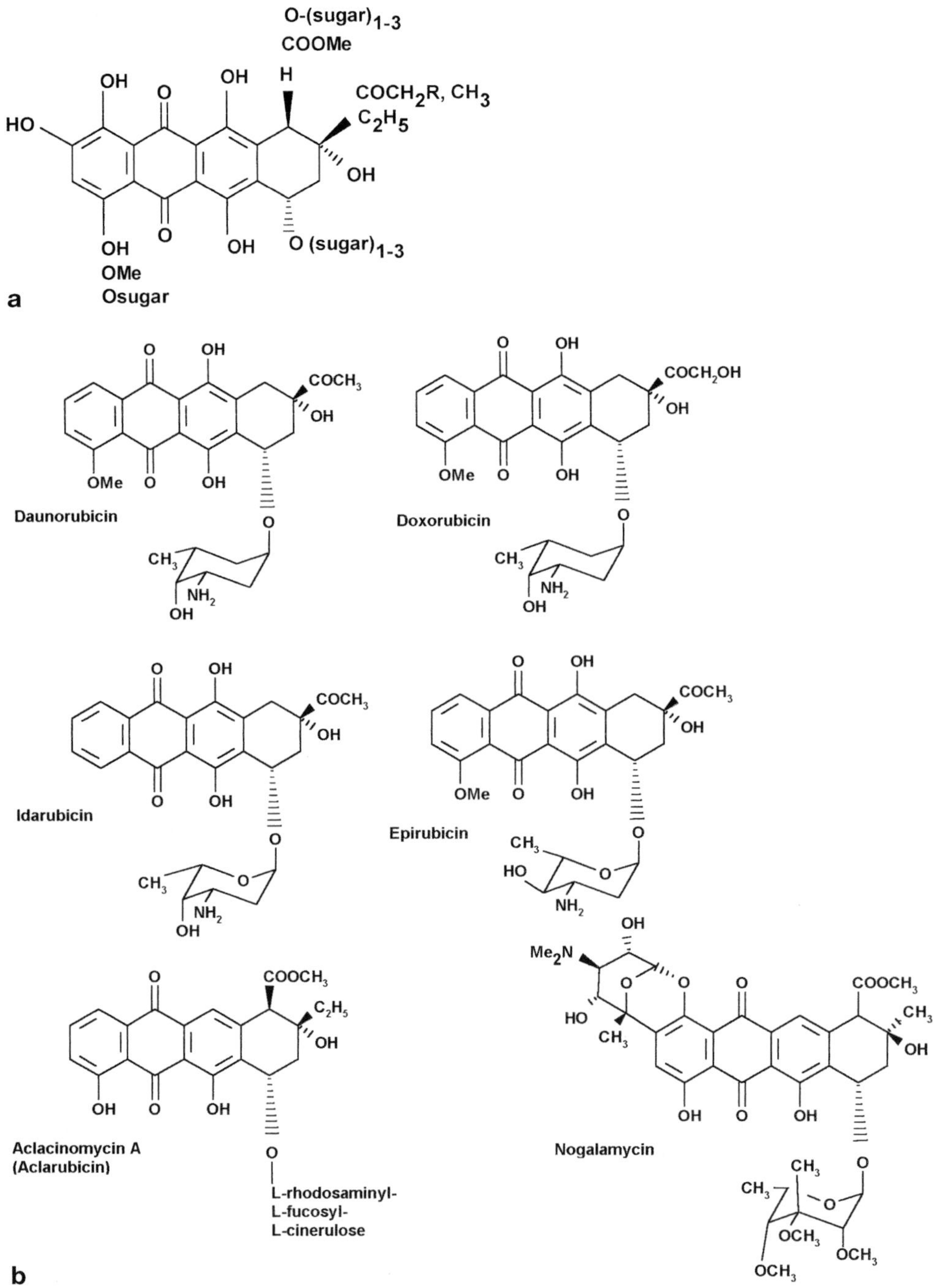

Fig. 7. Structures of anthracycline type antibiotics.
a General substitution pattern of anthracyclinone aglycones of anthracyclines, **b** anthracycline structures, **c** recently discovered new anthracycline structures.

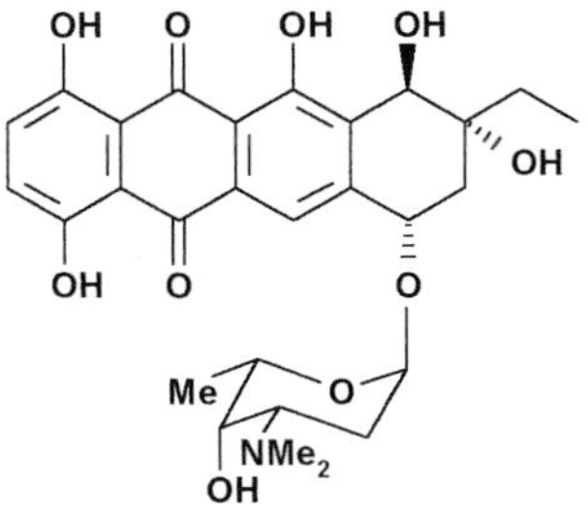

Alldimycin

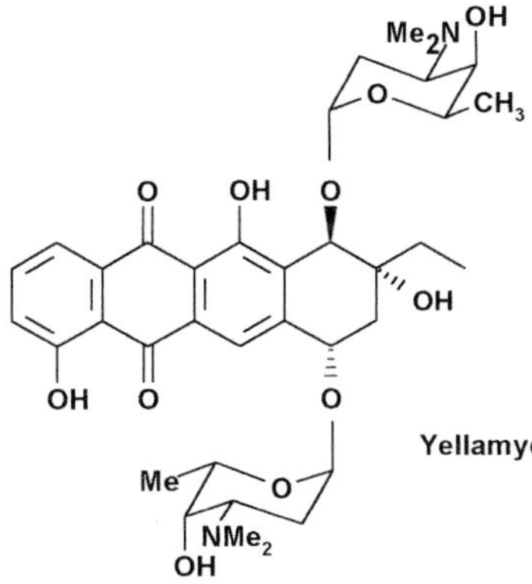

Yellamycin B

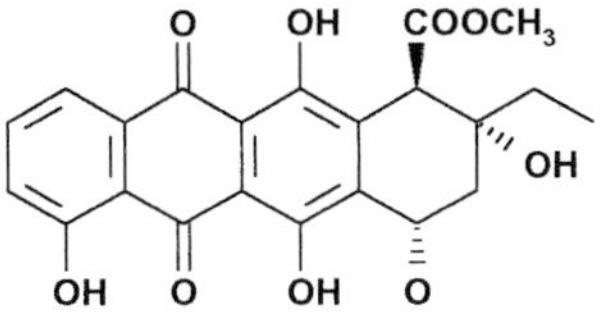

Epelmycin A

L-rhodosamin-
L-rhodosaminyl-
L-cinerulose A

Fig. 7c

chronic leukemias, and renal cancer are unresponsive to them. The first clinically effective anthracycline, daunorubicin (DNR), was discovered independently in 1963 at Farmitalia (daunomycin) and Rhône-Poulonc (rubidomycin). Later on, adriamycin (doxorubicin, DOX) was discovered as the product of a mutant strain and was also obtained semisynthetically.

In 1969, it was demonstrated that DOX was less toxic and more active against a much broader spectrum of tumors than DNR. The former rapidly replaced the older drug in clinical applications. In general, cumulative cardiotoxicity limits treatment with DOX to approximately nine months at usual dosages, and most cancers develop resistance to this agent.

DOX differs from DNR only by an additional hydroxyl group. This fact has encouraged researchers worldwide to search for analogs of DOX that display lower acute toxicity or cardiomyopathy, that can be administered orally, and that have different or greater antitumor efficacy. The aim of this research was three-fold:

(1) Synthesis of new anthracycline analogs:
This process has continued from the 1960s up to the present and might also continue in the future (LOWN, 1993). No one has kept a count of the anthracycline analogs synthesized over the past 25 years, but their number probably exceeds 2000, and more are being reported every month.

(2) Coadministration of other agents with DOX:
Researchers have attempted worldwide to administer doxorubicin in conjunction with other substances that will mitigate cardiotoxicity or overcome drug resistance of the cancer cells (STEINHERZ and STEINHERZ, 1991; VAN KALKEN et al., 1991).
Moreover, immunoconjugates of DOX and different anthracyclines incorporated into liposomes are being evaluated in clinical trials to minimize heart exposure to the drug while maintaining antitumor efficacy (PEREZ-SOLER et al., 1995).

(3) Screening for new anthracycline compounds of microbial origin:

DNR and DOX as well as their biosynthetic congeners are all derived from actinomycetes, particularly from the genus *Streptomyces.* They are obtained by using the well known fermentation and recovery techniques generally used in antibiotic technology. Many of the more than 400 anthracyclines isolated and characterized to date have been isolated from blocked mutants of a variety of strains. The main anthracyclines selected for clinical evaluation are shown in Fig. 7b. Alldimycins (JOHDO et al., 1991a), yellamycins (JOHDO et al., 1991b), epelmycins (Fig. 7c) (JOHDO et al., 1991c), respinomycin (UBUKATA et al., 1993), cororubicin (ISHIGAMI et al., 1994), mutactimycin (MAEDA et al., 1992), betaclamycin B (YOSHIMOTO et al., 1992), cinerubin R (NAKATA et al., 1992), and rubomycins F and H (FOMICHOWA et al., 1992) are examples of the previously discovered new anthracycline type structures.

Anthracyclines consist of a tetracyclic aglycone (anthracyclinone, tetrahydro-naphthacene-quinone) which is linked with up to ten sugars usually attached at positions C-7 or C-10; C-4-glycosylated derivatives have also been reported. The general substitution pattern of the anthracyclinones is shown in Fig. 7a. The nogalamycin family is distinguishable by the unusual substitution pattern of the left-side aromatic ring by a bis-oxa-bicyclo[3.3.1.]nonane ring system (DUMITRIU, 1996).

Fig. 7c shows a series of anthracycline structures representing various building patterns. The aglycones differ in the hydroxylation of the aromatic nucleus and in the substituents at positions 7 and 10. Modifications of the side chain (e.g., by oxidations, reductions) and/or methylations are a consequence of numerous enzymatic and non-enzymatic reactions occurring during biosynthesis. The anthracyclinone part of the anthracyclines is produced via a polyketide synthase system by a series of reactions similar to that of fatty acid biosynthesis (VANEK et al., 1977). The intermediate undergoes hydroxylations, methylations, glycosylations, and other modifications. Aklanonic acid and aklaviketone are the earliest detectable intermediates of anthracycline biosynthesis (WAGNER et al., 1991; ECKARDT and WAGNER, 1988). More recently, the biosynthesis of anthracycline aglycones was studied in detail by a genomic analysis (STROHL et al., 1989; BARTEL et al., 1990). The biosynthetic origin of many of the known anthracycline aglycones, such as ε- and β-pyrromycinones, ε-, β-, α-, and α_2-rhodomycinones, ε-*iso*-rhodomycinones, daunomycinone, and adriamycinone could be ascribed to aklanonic acid as a single intermediate.

The pigmented antibiotics are formed in the mycelium of microorganisms grown in shake flasks or in stirred aerated fermenters on media containing the usual organic nutrients and inorganic salts. For the biotechnical production of daunorubicin, the fermenters are operated usually for up to 10 days. Because anthracyclines are hazardous chemicals, due to their cardiotoxicity and/or mutagenicity, special care is needed when isolating and purifying these agents from the fermentation broth (UMEZAWA et al., 1987). Recovery and purification of the anthracyclines from large-scale fermentations is described elsewhere (WHITE and STROSHANE, 1984).

New anthracycline analogs: the next generation:

In the 1970s and the early 1980s the main aims were to expand the antitumor spectrum of DOX, to reduce its cardiotoxicity, and to develop an orally administrable compound. The outcome of this research was a few analogs which display moderate clinical advantages (WEISS, 1992). In the 1980s developmental efforts in this field also turned to the problem of multi-drug resistance of cancer cells. For instance, F860191, AD-198, FCE23762, ME2303, and MX-2 were reported to be active against DOX resistant and multi-drug resistant cell lines and even to exceed DOX activity sometimes by a factor of up to ten (for a survey, see WEISS, 1992).

Formation of oxygen radicals by anthracyclines:

As a general feature, *p*-quinoid structures such as mitomycin C, streptonigrin, tetracenomycin type antibiotics, angucyclines, and the representatives of the anthracycline family (BUTLER and HAEY, 1987; KONO et al., 1991; KASAI et al., 1991; BARKLEY et al., 1986; FEIG and LIPPARD, 1994) can be reduced enzymatically by cytochrome-dependent enzymes to yield semiquinone radicals (Fig. 8). These will combine with molecular oxygen to form superoxide anion and hydroxyl ion radicals. Thereafter, the free radical species react preferably with the deoxyribose moieties of DNA splitting single bonds between neighboring sugars. Although the anthracyclines could be considered as intercalating agents and inhibitors of topoisomerase II (see above) they are also potent producers of free radicals. The anthracycline semiquinones are subsequently stabilized by deglycosylation of the C-7-O-linked sugar. As a consequence of free radical formation, frameshift mutations are induced. Thus the anthracyclines, bleomycin (see below), mitomycin C, and similar anticancer agents are hazardous chemicals (UMEZAWA et al., 1987).

3.5.6 Bleomycin

Bleomycin is a glycopeptide antbiotic produced by *Streptomyces verticillatus* as a complex of not less than sixteen related antitumor antibiotics (TAKITA, 1984; MURADA, 1988) (Fig. 9). The main component bleomycin A_2 was originally used clinically in the treatment of lung carcinoma, plate epithelium carcinoma, Hodgkins lymphoma, glioma, skin carcinoma, and testis carcinoma (GIRI and WANG, 1989; SOLAIMAN, 1988; POVIRK and AUSTIN, 1991; TAKITA et al., 1989). Pulmonary toxicity is dose-limiting. More than 200 semisynthetic derivatives have been obtained in order to improve activity and reduce toxic side effects (MURAOKA et al., 1988; TAKAHASHI et al., 1987a, b; TAKITA and OGINO, 1987). The naturally occurring bleomycins are distinguishable by the amino side chain. Directed biosynthesis of individual components by the feeding of amino acids has been reported (TAKITA, 1984). Peplomycin and liblomycin (A, B, C) (KURAMOCHI et al., 1988; KURAMOCHI-MOTEGI et al., 1991; TAKAHASHI et al., 1987c) are semisynthetic derivatives of bleomycins which contain a more space-filling side chain. The aim of their synthesis was to improve the stability against hydrolysis and to increase lipophilicity (OTSUKA et al., 1988; SEBTI and LAZO, 1988; NISHIMURA et al., 1987; OHNO, 1989). A general characteristic of these agents is that they form heavy metal complexes with divalent cations such as Cu^{2+}, Co^{2+}, Zn^{2+}, Fe^{2+}, and can even be isolated as heavy metal (Cu^{2+}) chelates from the aqueous solution (Fig. 9).

The mode of antitumor action of bleomycin has been subject to detailed investigations (MATSHURA, 1988; STUBBE and KOZARICH, 1987; STREKOWSKI, 1992). The antibiotic and its derivatves are commonly thought to exert their biological effects as metal–drug complexes which bind to the DNA. Thereafter, the adduct causes oxidative strand cleavage, and hence may be regarded as a low-molecular weight "DNAse" (HECHT, 1994). The disaccharide moiety appears to be necessary to enable penetration of the antibiotic through the cytoplasmic membrane barrier, while the thiazolopeptide side chain is essential for the contact to the DNA (POVIRK and AUSTIN, 1991). The cytotoxicity of bleomycin for some kind of tumor cells results from the reductive activation of dioxygen by metallobleomycins. Iron-II ions are able to transfer electrons to molecular oxygen to form reactive and damaging intermediates of oxygen, such as superoxide anion radical. Radical generation ($O_2^{-}\bullet$) is initiated by the binding of iron-II ions to the β-aminoalanine-pyrimidine core of bleomycins. Interaction of Fe-II with molecular oxygen generates the superoxide anion radical to form Fe(III)-bound bleomycin. The formation of radicals probably occurs in close proximity to the DNA, since there is intercalative binding of the bisthiazole moiety of bleomycin to special minor groove DNA sequences. It was also shown that the bleomycin–Fe(II)–O_2 complex splits double-stranded DNA specifically at the GC(5′-3′) and GT(5′-3′) sequences. It has been proposed that upon cleavage of the DNA the bleomy-

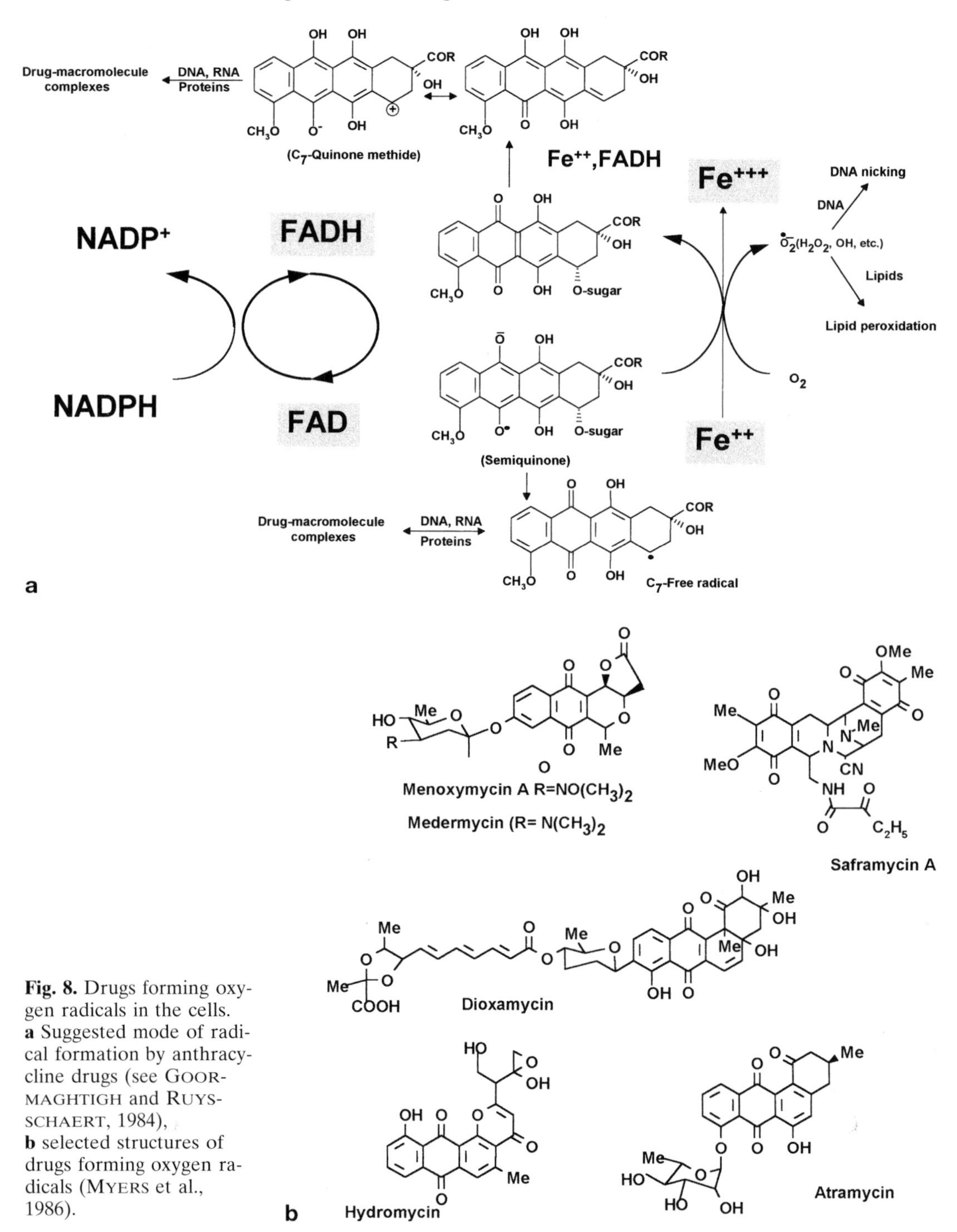

Fig. 8. Drugs forming oxygen radicals in the cells. **a** Suggested mode of radical formation by anthracycline drugs (see GOORMAGHTIGH and RUYSSCHAERT, 1984), **b** selected structures of drugs forming oxygen radicals (MYERS et al., 1986).

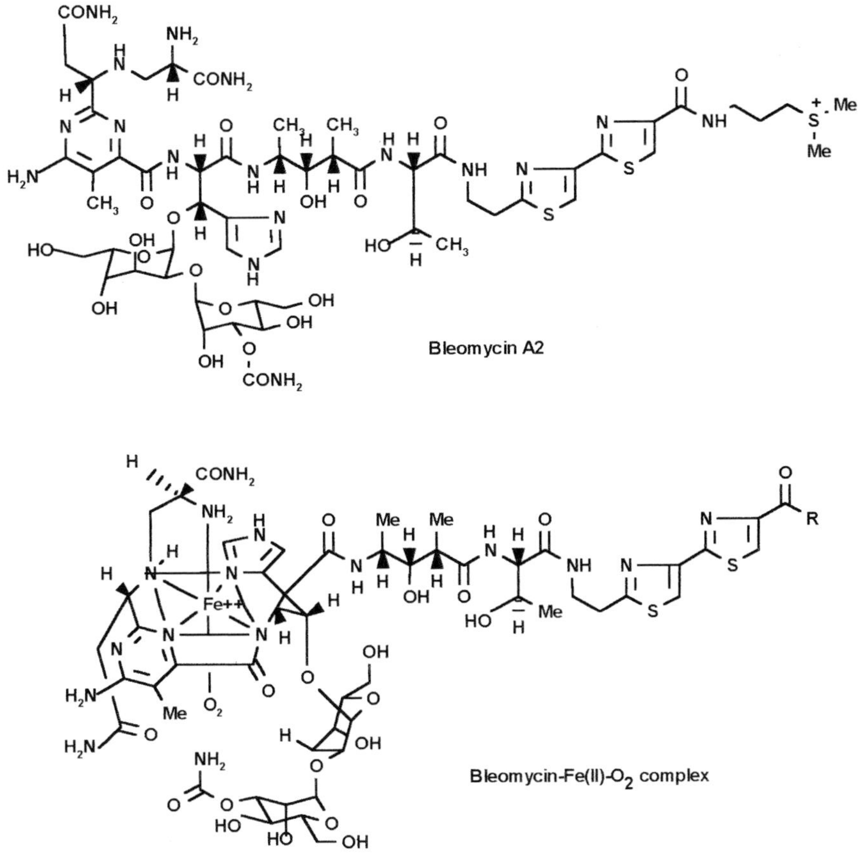

Fig. 9. Structure of bleomycin and the bleomycin–Fe(II)–O_2 complex (STUBBE and KOZARICH, 1987; MATSUHARA, 1988).

cin–Fe(III) complex dissociates from the target and is reduced to the (Fe-II) complex. Enzymatic modification of the antibiotic occurs by serum proteases which hydrolyze the amide structure (SEBTI and LAZO, 1988). Hence, inhibitors of proteases can be used to improve the efficacy of bleomycin. Moreover, bleomycin–Fe(II)–O_2 and bleomycin–Fe(III)–O_2 complexes catalyze lipid peroxidation concomitantly with singlet oxygen evolution. Thus, oxidative damage of the cellular membranes may be one of the mechanisms determining the cytotoxic activity of this antibiotic (KIKUCHI and TETSUKA, 1992).

3.5.7 Macromolecular Antitumor Antibiotics and the Enedyine Family of Cytotoxic Drugs

The most recent discovery in the field of radical-forming, cytotoxic antitumor antibiotics concerned the extremely potent structural group of the enedyines. In general, these antibiotics are highly toxic for tumor cells, such as P388 leukemia and B16 melanoma. Enedyine chromophores are frequent constituents of macromolecular antitumor antbiotics, such as neocarzinostatin and C-1027 (HOFSTEAD et al., 1992; ISHIDA et al., 1965; OTANI, 1993). The enedyine subunits display antitumor activity even in the absence of the pertinent apoprotein. Other members of this protein group are auromomycin (YAMASHITA et al., 1979), macromomycin (CHIMURA et al., 1968), kedarcidin (LEET, 1992; HOFSTEAD et al., 1992), largomycin (YAMAGUCHI et al., 1970; MURAMATSU et al., 1991), maduropeptin (HANADA et al., 1991a), and actinoxanthin (KHOKHLOV et al., 1969), but the chemical nature of their chromophores has not always been explored in detail. Similar to neocarzinostatin, madurapeptin is composed of a large peptide backbone and an aromatic chromophore which is responsible for the biological activity (HANADA et al., 1991a, b).

The agent C-1027 from *Streptomyces globisporus* is the most recently discovered representative of the high molecular-weight antitumor antibiotics. It is composed of a 110 amino acid backbone and an unstable enedyine chromophore (OTANI, 1993) (Fig. 10a). A synthetic route to C-1027 was recently elaborated (IIDA et al., 1993). The apoprotein of C-1027 displays some amino peptidase activity which is inhibited by amastatin and bestatin (SAKATA et al., 1992).

More recent research in the field of high molecular weight antitumor antibiotics has focussed on the nucleotide sequences of apoprotein genes (SAKATA et al., 1989, 1992). Advances in physicochemical analysis permitted the structural elucidation of the enedyine chromophore of neocarzinostatin from *Streptomyces carzinostaticus* (EDO et al., 1988). Several non-protein-bound representatives of the unique enedyine structure such as esperamicins (MAGNUS and BENNETT, 1989), calicheamicin (LEE, 1992), and dynemicins (KONISH et al., 1991) have been isolated from *Actinomyces* strains (Fig. 10a).

Mode of action of the enedyines:

Neocarzinostatin (NCS) was the first natural enedyine described which directly attacks the sugar moiety of DNA residues (GOLDBERG, 1986; GOLDBERG et al., 1989). In contrast to ionizing radiation or bleomycin which damage deoxyribose through reactive oxygen species, NCS produces DNA sugar lesions in a sequence-specific manner at closely corresponding sites of the two DNA strands. The mechanism of neocarzinostatin action was shown to involve the formation of bisradicals and subsequent damage at C-5′ of deoxyribose of thymidylate in the DNA. The neocarzinostatin chromophore binds to DNA in a two-step process. The first is an external binding and the second involves the intercalation of the naphthoate moiety between adjacent DNA base pairs in the minor groove and the electrostatic interaction of the amino sugar with the charged phosphate backbone of the DNA. Upon addition of thiol agents (R-SH) free radicals are produced which attack the DNA to cause double-stranded breaks (GOLDBERG, 1991; HENSENS and GOLDBERG, 1989; DUMITRIU, 1996).

The esperamicins (LAM et al., 1993) (Fig. 10a) as the main representatives of the non-protein group are composed of a bicyclic core (an enedyine, an allylic trisulfide, and an enone) which is bound to a trisaccharide and a substituted 2-deoxy-L-fucose. Esperamicin A_1 (*Actinomadura verrucosospora*) was produced on a large scale for clinical trials in cancer treatment (GOLDBERG et al., 1989). Biosynthetic studies on esperamicin A_1 (BENTLER et al., 1994) (Fig. 10b) investigated the incorporation of the single and double ^{13}C-labeled acetates L-(methyl-^{13}C) methionine and $Na^{34}SO_4$. The C_{15} bicyclic enedyine core is derived by head-to-tail condensation of seven acetate units. The S-methyl groups of the trisulfide, the thiosugar, and O-methyl sugars are incorporated from S-adenosyl-methionine. A minor congener of esperamicin A_1, esperamicin P, was isolated from the fermentation broth of *Actinomadura verrucospora* (GOLDBERG et al., 1989). It differs from esperamicin A_1 in that it contains a methyl tetrasulfide moiety instead of a methyl trisulfide.

The dynemicins (*Micromonospora chersina*) lack the apoprotein constituent of the enedyine chromophore. Dynemycin A contains the bicyclo[7.3.1]-1,5-diyn-3-ene and 1,4,6-trihydroxy-anthraquinone functions (Fig. 10a). It exhibits a potent antibacterial and antitumor activity against a wide range of bacteria and cells, respectively. Satellite compounds, such as L, M, N, O, P, and Q have recently been isolated as shunt metabolites in fermentations of *Micromonospora chersina* and from mutants of this strain (KONISHI et al., 1991; KAMEI et al., 1991; MIYOSHI-SAITOH et al., 1991).

The calicheamicins (Fig. 10a) belong to a family of seven glycosylated compounds from *Micromonospora echinaspora* ssp. *calichensis* which demonstrate potent activity *in vivo* against the murine tumors P388 and B16 (LEE et al., 1989, 1992). Enedyines, such as esperimicin have previously been synthesized via several routes (LAM et al., 1993; LEE, 1992).

3.5.8 Other Agents Forming Active Oxygen Radicals in the Cells

All normal cells have various defense systems against active oxygen species. However, several tumor cell types have lost a part of these defense mechanisms. Hence, substances generating active oxygen radicals such as quinoid structures have been suggested to exhibit selective cytotoxicity against such tumor cells (CURRAU et al., 1993). In screening for such antitumor agents *Streptomyces* sp. KB10 was found to produce the new menoxymycins A and B (related to medermycin) which actively generate superoxide radicals in N18-RE-105 cell lysates (HAYAKAWA et al., 1994) (Fig. 8b). Numerous *p*-quinoid structures such as dioxamycin (SAWA et al., 1991), hydramycin (HANADA et al., 1991b), the glycosylated saptomycins (ABE et al., 1993a), tetracenomycins, and angucycline type aromatic polycycles, such as saframycins (KANEDA et al., 1987) and atramycins (ABE et al., 1991) possibly owe their cytotoxic and cytostatic effects to the formation of active oxygen radicals during aerobic metabolism (Fig. 8b). Saframycin A was also reported to bind covalently to duplex DNA (KANEDA et al., 1987). In addition, various polycyclic aromatic compounds with quinone structures or an epoxy side chain display cytotoxic activities. This feature was also ascribed to the formation of free radicals or other reactive structures. Representatives of this kind of agents are the sapurimycins (*Streptomyces* sp.) (UOSAKI et al., 1991).

3.6 Inhibitors of Mitosis and the Microtubular System

The early discovery of the antimitotic activity of plant products such as colchicine (Fig. 11) led to some clinical trials for cancer treatment, but due to their high toxicity and relatively poor efficacy they have not found therapeutic application.

The complex indole type vinca alkaloids (*Catharanthus roseus*) (GUNDA et al., 1994; HEINSTEIN and CHANG, 1994; KINGSTON, 1994), such as vincristine and its derivatives (POITIER et al., 1994; JOEL, 1994), vinblastine, and the taxol type agents (from *Taxus brevifolia;* ATTA-UR-RACHMAN et al., 1994; NOBLE, 1990) are plant-derived antitumor drugs which prevent mitotic cell division (Fig. 11). While the vinca alkaloids inhibit tubulin polymerization, the latter (taxoids) appear as particularly promising due to their opposite effect of promoting the polymerization of tubulin.

In folk medicine, leaf extracts of the subtropical plant *Catharanthus roseus* were known to be effective in the treatment of diabetes. Attempts to isolate the active hypoglycemic principle led, instead, to the discovery and isolation of cytotoxic vinblastine and vincristine. These complex indole alkaloids are now used in the clinical treatment of a variety of cancers (NOBLE, 1990) and in therapeutic tumor cell synchronization (CAMPLEJOHN, 1980). A major problem in cancer treatment by vincristine is the development of resistance due to P glycoprotein-mediated drug efflux (HILL, 1986). 500 kg of *Catharantus roseus* (*Vinca rosea*) have to be extracted and the extract fractionated to obtain 1 g of vincristine. Hence, partial synthesis of vincristine, vinblastine and vindesin is done using vindolin. Taxol (paclitaxel) (Fig. 11) was discovered in 1964 during a large screening for antitumor drugs at the National Cancer Institute of the USA. It is an exciting new anticancer drug, exhibiting clinical activity in the treatment of ovary and breast cancer (KINGSTON, 1994; MARTY et al., 1994; ROTHENBERG, 1993). Taxol was first isolated from the pacific yew tree (*Taxus brevifolia*). To overcome the problems of raw material supply,

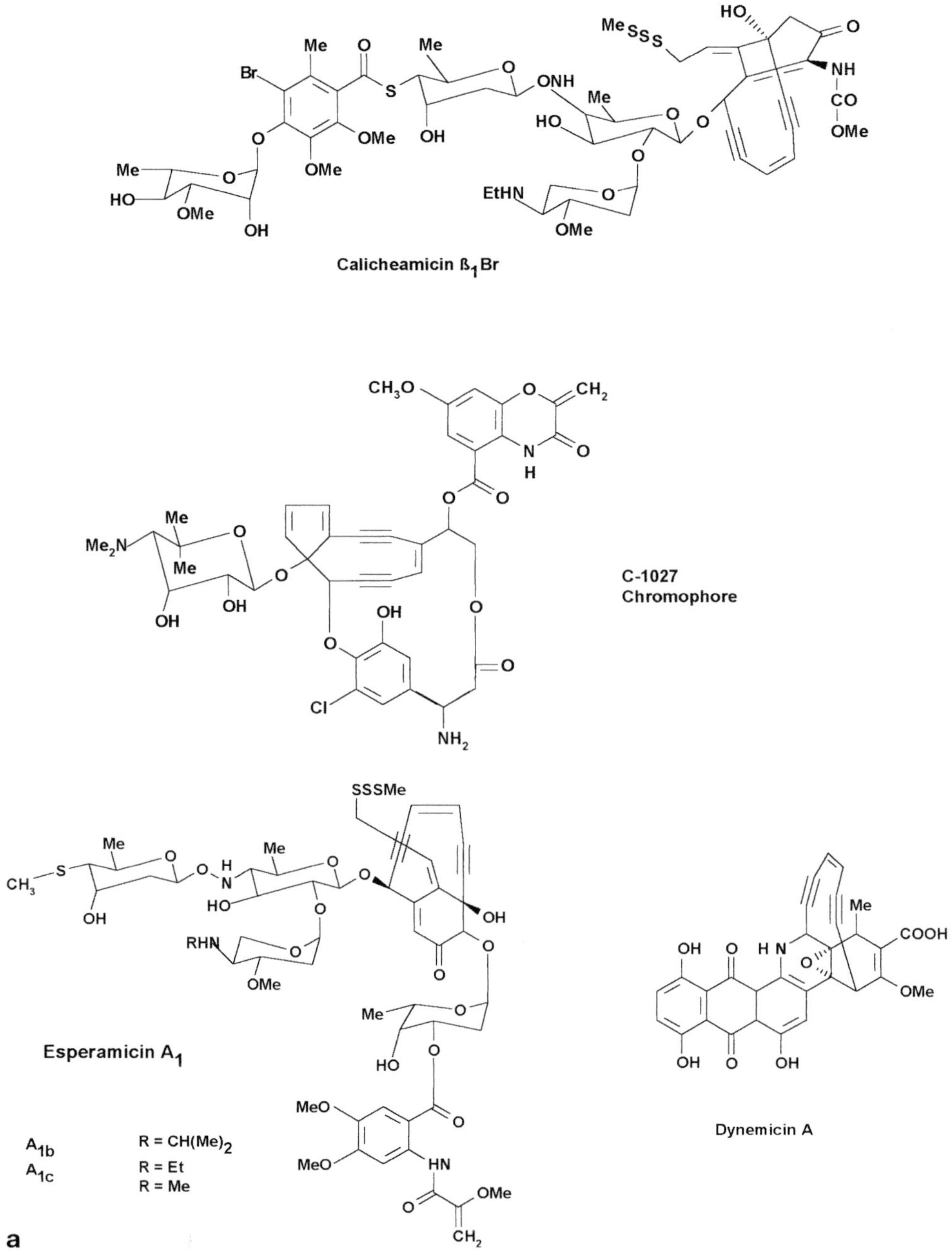

Calicheamicin ß₁Br

C-1027 Chromophore

Esperamicin A_1

A_{1b} R = CH(Me)$_2$
A_{1c} R = Et
R = Me

Dynemicin A

a

Fig. 10. Structures of enedyine drugs.
a Chromophore structures of esperamicin, calicheamicin, dynemicin, and C-1027,
b enrichment pattern of esperamicin A1 supplemented with ^{13}C-labeled single and double-labeled acetate. ^{13}C labeling is indicated by ●, ■, and ■●.

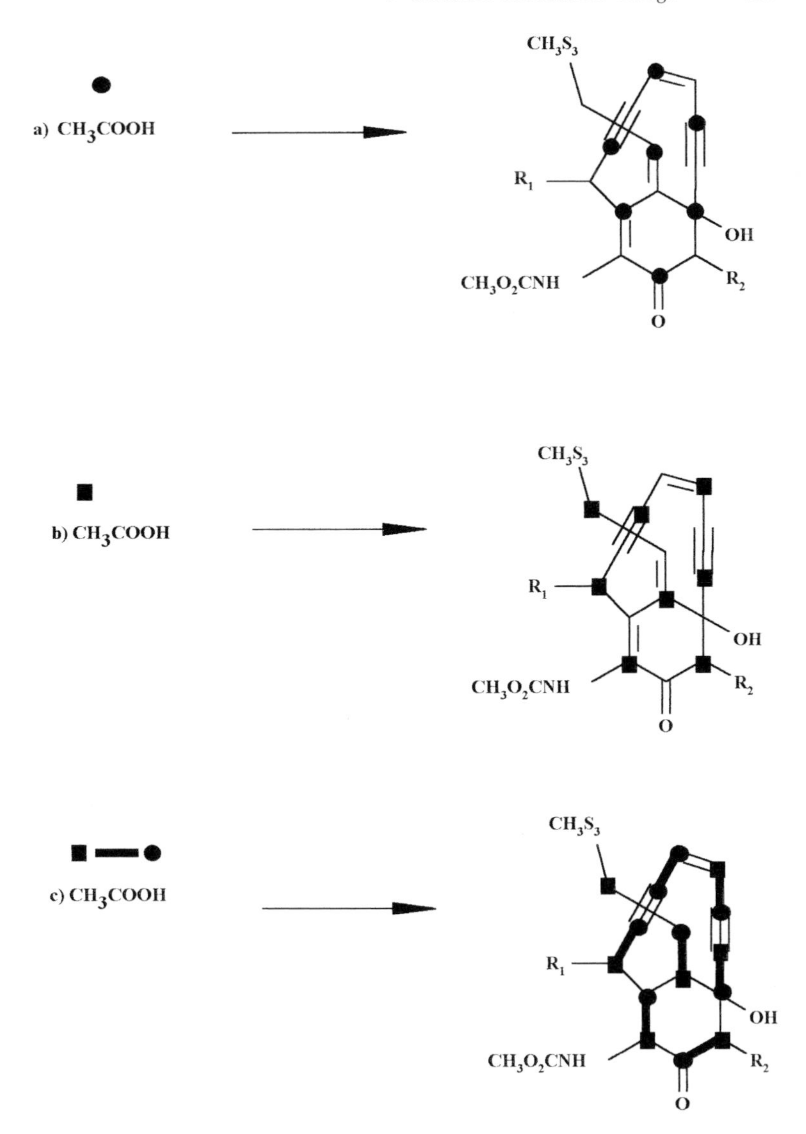

Fig. 10b

paclitaxel and docetaxel (taxotere) (Fig. 11) were developed (POITIER et al., 1994; MARTY et al., 1994). In particular, the semisynthesis from the related structures of the baccatins appear to be a feasible way of taxane production (HEINSTEIN and CHANG, 1994; KINGSTON, 1994). Baccatins are produced by the leaves of the European yew and can be chemically transformed to taxotere (docetaxel) (POITIER et al., 1994; JOEL, 1994). Recently, an endophytic hyphomyces fungus of *Taxus brevifolia* was reported to form low amounts of taxol, but so far a biotechnical procedure for the fermentative production of taxol has still not been developed (STIERLE et al., 1994). Taxoids are particularly promising antitumor agents due to their novel cancerostatic mechanism of action (ROTHENBERG, 1993). These compounds interfere with essential cellular processes such as mitosis, cellular motility, transport, and maintenance of cellular structures. The mechanism of action of

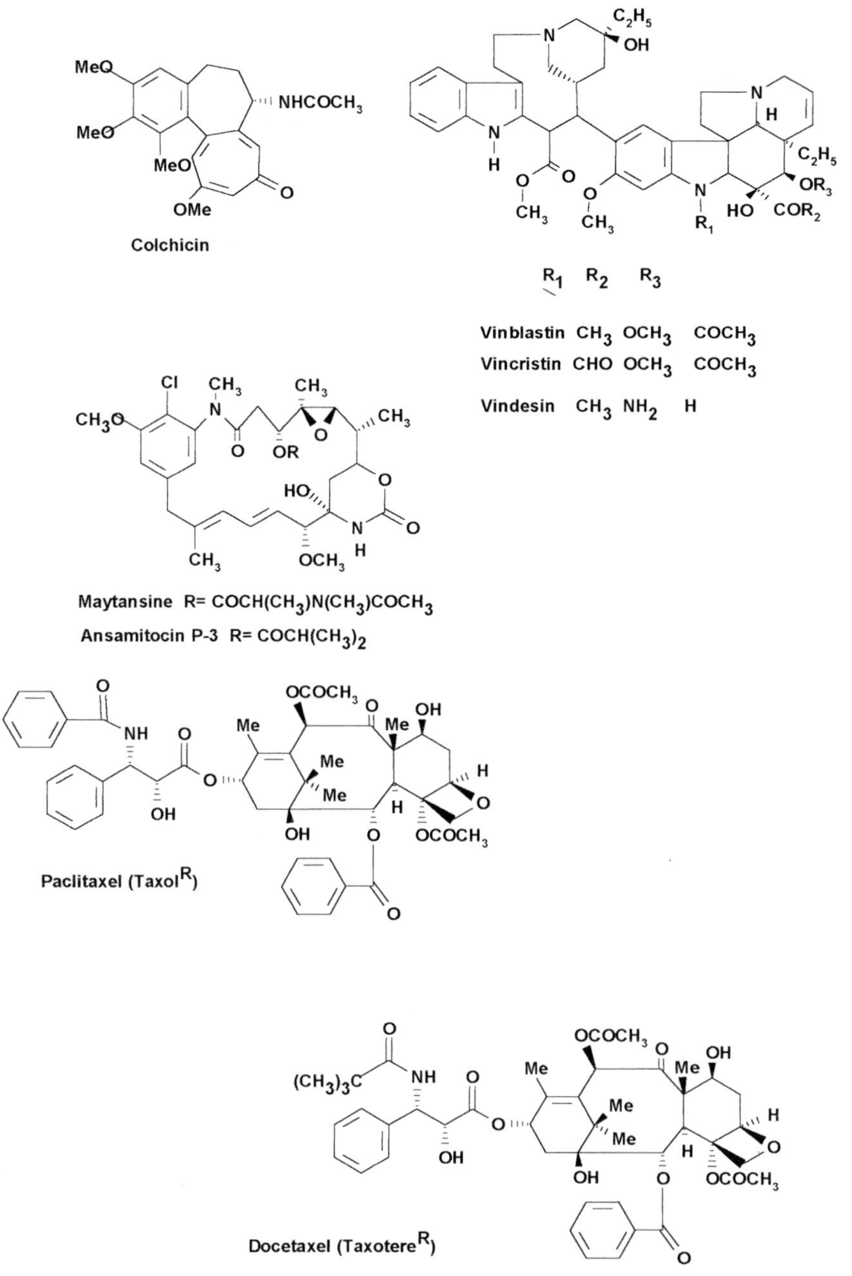

Colchicin

	R_1	R_2	R_3
Vinblastin	CH_3	OCH_3	$COCH_3$
Vincristin	CHO	OCH_3	$COCH_3$
Vindesin	CH_3	NH_2	H

Maytansine R= $COCH(CH_3)N(CH_3)COCH_3$

Ansamitocin P-3 R= $COCH(CH_3)_2$

Paclitaxel (Taxol[R])

Docetaxel (Taxotere[R])

Fig. 11. Inhibitors of the microtubular system.

taxanes is to promote the formaton of extremely stable non-functional microtubular aggregates. The cells are thus prevented from entering the G_2 or M phase of the cell cycle. Both paclitaxel and docetaxel are more potent against proliferating cells than non-proliferating cells *in vitro* (ROTHENBERG, 1993). Myelosuppression is a dose-limiting side effect. Current research is centered now around improving drug extraction, the semisynthetic production of the parent taxane structures (PAQUETTE, 1993; GUERITTE-VOEGELEIN et al., 1994), the enhancement of water solubility, and the identification of taxane analogs which are active against resistant cancer cells (ROTHENBERG, 1993; VYAS, 1993). Due to the impressive clinical activity of paclitaxel and docetaxel, the taxanes will be subject to preclinical and clinical development in the future (GUERITTE-VOEGELEIN et al., 1994). Moreover, structure–activity relationships (SAR) of drug–tubulin interactions are under investigation (GUERITTE-VOEGELEIN et al., 1994). Chemical modifications such as C-2 deoxygenation result in a total loss of tubulin function. A relatively small number of microbial drugs inhibits mitosis in tumor cells. Examples are rhizoxin (*Rhizopus chinensis*), a macrolide antibiotic and tubulin inhibitor of *Aspergillus nidulans* (KATO et al., 1991), ansamitocin from *Actinomyces* sp. (similar to maytansine from plant *Maytanus* sp.) (Fig. 11), and curacin A from the marine cyanobacterium *Lyngbya majuscula* (PIRRUNG and NAUHANS, 1994). Rhizoxin exhibits potent antimitotic activity by binding to β-tubulin, and is active against most eukaryotic cells. It shares the same binding site with maytansine and ansamitocin P-3 (Fig. 11) on porcine, brain, and fungal tubulin which is different from that of colchicine and vinblastine (IWASAKI, 1989).

3.7 Reduction of the Side Effects of Highly Toxic Anticancer Agents

Highly effective cytotoxic anticancer agents, such as the anthracyclines, vincristine, and bleomycin, usually have severe side effects such as cardiotoxicity, myelosuppression (FURK et al., 1989), release of interleukin 2, and others which are dosage limiting. Preinduction of metallothein synthesis, e.g., by bismuth salt was reported to exert a protective effect against various anticancer agents (e.g., bleomycin, *cis*-diammine-dichloro-platinum) (NAGAMURA et al., 1993).

Iron chelators and antioxidants have also been studied with regard to brain protection against neoplasm inhibitors (GUTTERIDGE, 1992). To overcome myelosuppression as one of the limiting factors in cancer chemotherapy by anthracyclines, a search was initiated for immunomodulatory stimulators of leukocyte and platelet formation (KAWATSU et al., 1993, 1994). As a result of this screening, conagenin was detected in cultures of *Streptomyces roseosporus.* It binds exclusively to T cells activated by concanavalin A or cytokines and enhances both T cell proliferation and lymphokine formation. Due to the induction of cytokines, the production of leukocytes and platelets is subsequently stimulated.

3.8 Cytotoxic Compounds with Poorly Characterized Mode of Activity

There are continuously references in the literature to new cytotoxic structures (see, e.g., reports on structures such as russuphelins (TAKAHASHI, 1993), heptelidic acid (KIWASHIMA et al., 1994), and stubomycin (KOMIYAMA et al., 1983). Cytotoxicity appears to be a very frequent property of new natural products. In a personal literature compilation, out of 4,000 microbial drugs, 600 were indicated to suppress cell growth. For the majority of these agents the mechanism of action has not been investigated. In the past, much interest has been focussed on new metabolites which might inhibit tumor cell lines that are intrinsically resistant to the known set of anticancer drugs or that have acquired multi-drug resistance. In the course of screening using adriamycin-resistant HL-60 cells, homooligomycins A and B (*Streptomyces bottropensis*) were detected as a result of their particular effect on colon-26 carcinoma (YAMAZAKI et al., 1992).

A further focus of interest have been the numerous neoplasm inhibitors which are selectively toxic against a special type of tumor cell. For instance, the homooligomycins A and B (*Streptomyces bottropensis*) were detected due to their particular effect against colon-26 carcinoma (MAGAE et al., 1993). In a screening program for immunomodulators, the melastins (*Streptomyces* sp.) (MAGAE et al., 1993) were similarly found to selectively inhibit the growth of leukemia cells and the lipopolysaccharide-induced blastogenesis of T cells.

4 Non-Classical Approaches to Antitumor Drugs

Much interest has been centered on new anticancer agents which exert their antitumor activity via a hitherto unknown interaction with cellular regulation and cellular transduction pathways. Among these new agents, inhibitors of tumor cell resistance to anticancer agents, tumor metastasis, and angiogenesis of tumors appear promising.

Otherwise, the discovery of the oncogenes and their role in malignant cell growth has opened new horizons for target-directed screening of new anticancer agents. This development was initated by the exploration of particular oncogene-encoded functions such as those of erb, sis, and ras in tumor cells. Oncogenes are modified forms of normal genes (proto-oncogenes) which have been altered either by chemical carcinogenesis or retroviruses in such a manner that they escape the normal pattern of control (FERDINAND, 1989). These alterations may result in the overexpression of the genes, the loss of down-regulation of the genes, and the formation of altered proteins. Proteins encoded by these genes, such as the epidermal growth factor receptor (EGF receptor) and membrane-associated *Ras*-proteins are abnormal derivatives of the normal proteins involved in cellular signaling and cell cycle regulation. The presently available information on cellular regulation appears to be the "tip of the iceberg". Recent additions to our knowledge were provided by the discovery of the cyclins, the tumor suppressor genes and their proteins. These findings will doubtlessly promote the discovery of new agents which interfere with these targets and module their activity. The following survey summarizes the results of recent screening approaches to these non-classical anticancer drugs.

4.1 Potentiators of Cytotoxic Antitumor Agents

DNA repair mechanisms serve as a useful target for modulating the cytotoxic and chemotherapeutic effects of those agents. Their mechanism of action causes the induction of DNA damage. Poly(ADP-ribose) polymerase responds to DNA breaks by cleaving the substrate NAD^+ and using the resultant ADP-ribose moieties to synthesize homopolymers of ADP-ribose. Inhibitors of this enzyme such as benzamide derivatives and benadrostin (Fig. 12) (YOSHIDA et al., 1988) prevent DNA repair and potentiate the tumoricidal effects of DNA strand-breaking agents such as bleomycin (BERGER et al., 1987; YOSHIDA et al., 1988; GAAL and PEARSON, 1986; SUZUKI et al., 1990). The simple compound, 2-methyl-4[3H]-quinazoline, was discovered in *Bacillus cereus* cultures and found to be an inhibitor of poly(ADP-ribose) synthetase from calf thymus (YOSHIDA et al., 1988). In addition, calcium channel antagonists such as verapamil and fendiline are known to potentiate the activity and toxicity of mitomycin C in mammalian and bacterial cells (SCHEID et al., 1991). The mechanism of this effect has not yet been explored in detail; however, it seems reasonable to suggest that P glycoprotein-mediated efflux of the drug is prevented. Similarly, a series of other agents were reported to increase the cytotoxic effect of neoplasm inhibitors. A major obstacle in cancer chemotherapy is the appearance of resistant tumor cells which are non-treatable by cytostatic drugs such as vincristine and doxorubicin. Several reasons for this are known (HILL, 1986; BIEDLER et al., 1993). Overexpression of metallothein in cancer cell lines, e.g., confers resistance to anticancer drugs such as platinum complexes (KELLEY et al., 1988). Detoxification of cytotoxic drugs may also occur by glucuronidation (BURCHELL et al., 1991). Other cancer cell lines become resistant to a series of anticancer drugs such as taxol, colchicine, doxorubicin, actinomycin, and vincristine (HILL, 1986). This so-called multidrug resistance (MDR) is frequently associated with increased drug efflux and decreased accumulation within the cells. The

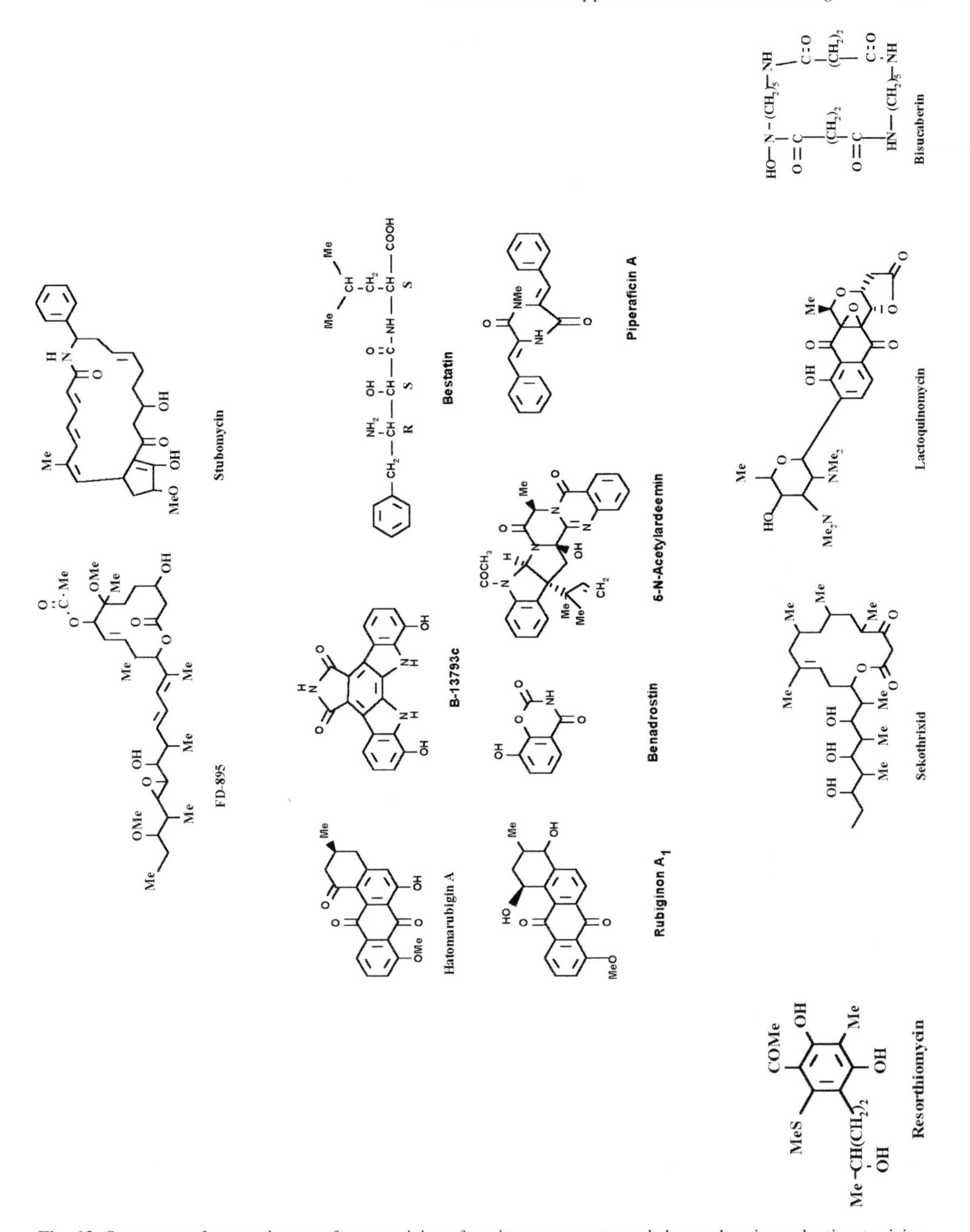

Fig. 12. Structure of potentiators of cytotoxicity of antitumor agents and drugs showing selective toxicity.

phenomenon is due to membrane proteins, such as P 170 glycoprotein, which is capable of conveying drugs out of the cells. This glycoprotein was detected in the mouse macrophage-like cell line J 774.2 and in other cells, and its role as mediator of the active efflux of cytotoxic drugs was studied (GUNICKE and HOFMANN, 1992; BORREL et al., 1994; REICHLE et al., 1991). Mobile ionophoric carriers, such as valinomycin, lasalocid, monensin, calcimycin, and others inhibit the efflux of anthracyclines, whereas channel-forming ionophores (e.g., gramicidin A) do not. Cyclosporine, a calcium chelator, also interfered with β-glycoprotein-mediated efflux of these drugs (BORREL et al., 1994; DIETEL, 1991).

Directed screening for drugs which inhibit the growth of multiresistant tumor cells (MDR) or which could restore and potentiate the antitumor activity of known agents thus appeared to yield promising substances. Examples of these agents (Fig. 12) are the hatomarubigins (*Streptomyces* sp., inhibitors of multi-drug resistant tumor cell lines) (HAYAKAWA et al., 1991), BE 137932c (*Streptomyces* sp., inhibits the growth of doxorubicin-resistant cancer cells) (KEFRI et al., 1991), the piperaficins (*Streptoverticillium aspergilloides,* potentiate vincristine activity against resistant murine P388 cell lines) (KAMEI et al., 1990), bestatin (*Streptoverticillium olivoreticuli*, an immunostimulator, increases the *in vivo* vincristine cytotoxicity against colorectal K562 carcinoma cells) (UMEZAWA, 1989), sekothrixid (an antitumor drug) (KIM et al., 1991), rubiginon (*Streptomyces griseorubigenosus,* restores colchicine sensitvity to resistant tumor cells) (OKA et al., 1990, 1991), resorthiomycin (potentiates the antitumor activity of vincristine) (TAKARA et al., 1990), laidlomycin (a polyether antibiotic from *Streptomyces* sp., potentiates anticancer drug activity against MDR carcinoma KB-4 cells) (KAWADA et al., 1992b), bisucaberin (*Alteromonas haloplanctis,* sensitizes tumor cells to cytolysis mediated by macrophages) (TAKAHASHI et al., 1987a), kazusamycin (see Fig. 18) (*Streptomyces* sp.) (YOSHIDA et al., 1987), lactoquinomycin (*Streptomyces tanashiriensis,* active against resistant L 51178y lymphoblastoma) (TANAKA et al., 1989), and semisynthetic derivatives of staurosporin (reversion of the multi-drug resistance of cancer cells) (WAKUSAWA et al., 1993), 5-N-actylardeemin (Fig. 12) (*Aspergillus fischeri,* reversion of multi-drug resistance in tumor cells) (HOCHLOWSKI et al., 1993) and BE-12406 A and B (*Streptomyces* sp., inhibition of vincristine- and doxorubicin-resistant P388 murine leukemia cells) (KOGIRI et al., 1991). A particular feature of the potentation of antitumor drug activity is the potentiating effect of inhibitors of bleomycin hydrolases (leupeptin, pepstatin) on bleomycin activity (NISHIMURA et al., 1987). Some tumor cells resistant to inhibitors of topoisomerase II display a type of MDR that differs from P glycoprotein-associated MDR and is restricted to drugs that inhibit topoisomerase II by stabilizing cleavable DNA–protein complexes (BECK et al., 1993).

4.2 Inhibitors of Glutathione S Transferase as Enhancers of Antitumor Activity of Drugs

Glutathione conjugate formation may constitute an important drug detoxification and bioactivation mechanism for several classes of mutagenic and carcinogenic compounds (SATO, 1989). For instance, the metabolism of nephrotoxic and nephrocarcinogenic haloalkenes involves glutathione S conjugate formation. Glutathione was suggested also to play a role in the toxicity or action of bleomycin, cyclophosphamide, and neocarcinostatin (ANDERS, 1991; SATO, 1989). An increased glutathione S transferase activity may create resistance to some anticancer drugs such as alkylating agents, doxorubicin, and *cis*-diammine-dichloro-platinum. Consequently, inhibitors of glutathione S transferase may improve the activity of antitumor drugs. Moreover, they could also act as antiinflammatory and antiallergic agents due to their effect on the prostaglandin and eicosanoid metabolism. In a worldwide screening approach several new inhibitors were discovered, such as benastatins C and D (*Streptomyces* sp.) (AOYAMA et al., 1993a), cysfluoretins (*Streptomyces* sp.) (AOYAMA et al., 1993b), TA-3037 A (KOMAGATA et al., 1992a), rishirilide B (KOmagata et al., 1992b), and bequinostatins C and D (YAMAZAKI et al., 1993) (Fig. 13).

4.3 Antimetastasis Drugs and Inhibitors of Angiogenesis

Prevention of metastasis of cancer cells and angiogenesis is a major aim of cancer chemotherapy. Angiogenesis is a tissue differentiation process which connects a growing tumor with the blood vessels. Recently, phospholipase D was proposed as a target for tumor invasion inhibitors. Drugs, such as Sch 49210, Sch 53514 and Sch 53517 from the fungus *Nattrasia mangifera* displayed potent activity in the antitumor invasion chamber assay (CHU, 1994). Among the metalloproteinases involved in tumor invasion, angiogenesis, and rheumatoid arthritis, the type "N" enzymes may play a particular role in the degradation of basement membranes. Searching for inhibitors of this enzyme, such as the matlystatins from *Actinomadura atramentaria,* has been proposed as a promising approach towards new antitumor drugs (HARUYAMA et al., 1994). Antimetastasis activity was reported for U-77863 (TROLL et al., 1987) and U-77864 (methylphenyl-propenyl-carboxamides) from *Streptomyces griseoluteus* (HARPER and WELCH, 1992). These compounds were shown to inhibit the growth of tumor xenograft models (Fig. 14). Nitrogen-containing pseudo-sugars such as siastatin B, nagstatin, nojirimycins and their chemically derived derivatives have recently been reported to prevent metastasis by the inhibition of β-glucuronidase, heparanase, and β-D-mannosidase of tumor cells (SATOH et al., 1996; TSURUOKA et al., 1996; KAWASE et al., 1996; TATSUDA et al., 1996).

Using the chicken embryo chorioallantoic membrane assay system, a series of angiogenesis inhibitors were disclosed, such as herbimycin A (*Streptomyces* sp.) (YAMASHITA et al., 1989; SAKAI et al., 1989), irsogladin (SATO et al., 1993), erbstatin (*Streptomyces* sp.) (OIKAWA et al., 1993), TAN-1120 (an anthracycline, *Streptomyces* sp.) (NOZAKI et al., 1993), TAN-1323 (*Streptomyces* sp. S-45628) (MUROI, 1990), analogs of siastatin B, WF-16775 A_1, A_2 (derivatives of pyridazine from *Chaetosbolisia erysiophoides*) (OTSUKA et al., 1992), 15-deoxyspergualin (NISHIKAWA et al., 1991a, b), and staurosporin (OIKAWA et al., 1992). These drugs are expected to prevent neovascularization and thus seem to be promising not only in the prevention of metastasis but also in the treatment of retinopathy and rheumatoid arthritis (OTSUKA et al., 1992).

4.4 Antitumor Effects of Immunomodulators

In the healthy human, the immune system recognizes both microbes and malignant cells and destroys them. The outbreak of cancer

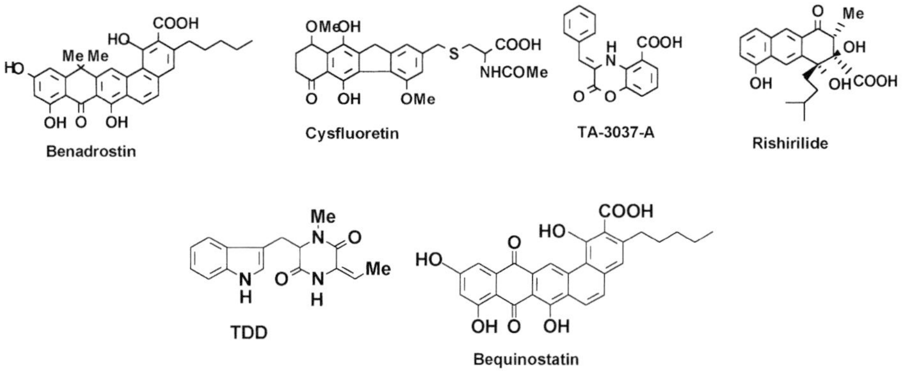

Benadrostin Cysfluoretin TA-3037-A Rishirilide TDD Bequinostatin

Fig. 13. Inhibitors of glutathione S transferase.

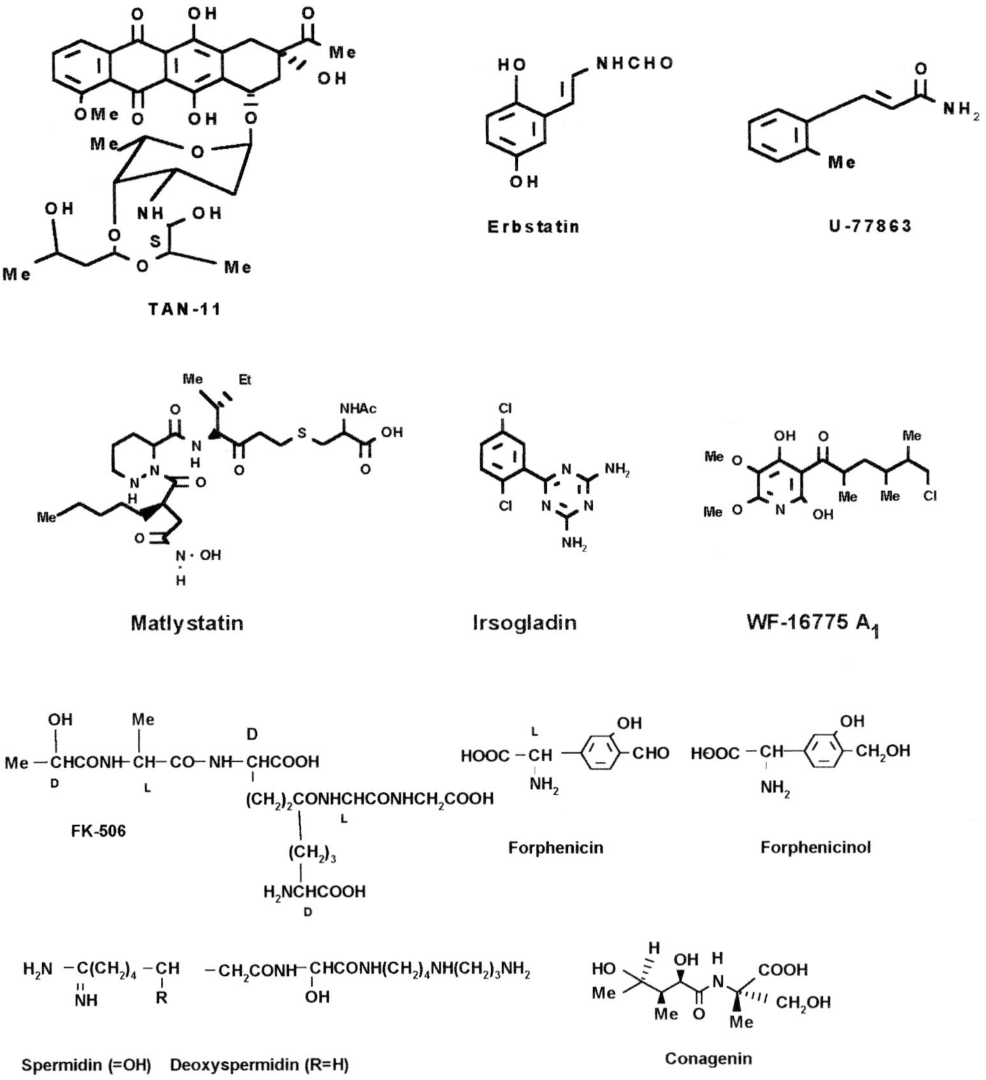

TAN-11

Erbstatin

U-77863

Matlystatin

Irsogladin

WF-16775 A_1

FK-506

Forphenicin

Forphenicinol

Spermidin (=OH) Deoxyspermidin (R=H)

Conagenin

Fig. 14. Selected structures of antimetastatic inhibitors and anti-angiogenetic and immunomodulatory drugs.

may be promoted by insufficient immunological defense and poor recognition of abnormal cells. Hence, immunostimulating and immunoregulatory agents have been considered as potential anticancer therapeutic agents. The immunostimulatory and antineoplastic properties of bacterial wall preparations, especially from *Mycobacterium* spp. were discovered already in the 1970s. High molecular weight glucans secreted into the medium, such as shizophyllan and lentinan, belong to a group of immunostimulatory compounds which exhibit antitumor activity. They are derived from higher fungi and mushrooms. Their antitumor activity has been attributed to the stimulatory effect on the immune sysem (LAATSCH, 1992). Subsequently, the chemical structures of other naturally occurring immunomodula-

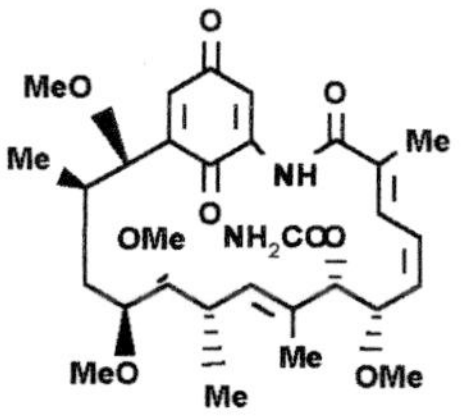

Herbimycin A

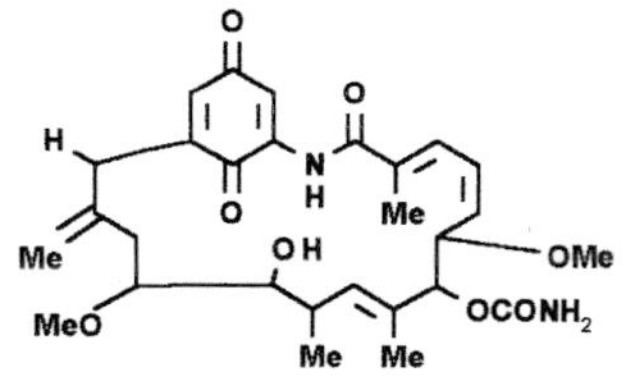

Herbimycin B

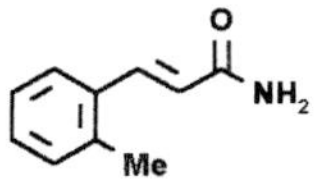

U-7786

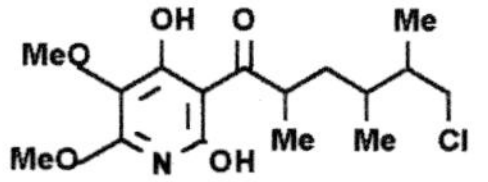

WF-16775

Fig. 14

tors were elucidated. These compounds may provide the pharmaceutical scientist with an armamentarium that can be used to mount a rational treatment of immunopathologies including cancer. For several immunomodulators of microbial origin, protective effects against experimental tumors have been established in mice (KLEGEMANN, 1993) (Fig. 14).

Bestatin (*Streptoverticillium olivoreticuli*) which was discovered by H. UMEZAWA's group in Japan (ABE et al., 1985) enhances the activity of antitumor drugs and is used in clinical applications. The drug was initially found to be an inhibitor of leucine aminopeptidase. Immunostimulatory and antitumor effects have been reported also for some other microbial protease inhibitors (BILLINGS, 1993). The compound FK-156 (*Streptomyces olivaceogriseus*) (IZUMI et al., 1983) and conagenin (*Streptomyces roseosporus*) (YAMASHITA et al., 1993; KAWATSU et al., 1993) are additional examples of microbial metabolites which may exert anticancer activity due to their stimulatory effect on cytokine production by T cells and macrophage activity. Forphenicin (*Streptomyces* sp.) and its biotransformation product forphenicinol inhibit chicken alkaline phosphatase. Forphenicinol induces γ-interferon production in mice which were sensitized by the BCG vaccine, and enhances macrophage activity. It also displays an antitumor effect on MetA fibrosarcoma and adenocarcinoma (OKURA et al., 1986). For a series of immunosuppressing agents such as spergualin (*Bacillus laterosporus*) and its semisynthetic 15-deoxy derivative, antitumor activities were demonstrated, e.g., against the murine leukemia P 388 (NISHIKAWA et al., 1991a, b). Retardation of malignant cell growth was ascribed to the cytotoxic and cytostatic effects of these drugs, but the mode of action of these and other immunosuppressants (see, e.g., cyclosporine A and FK-506) has not been explored in detail.

4.5 Inhibitors of the Cellular Mitogenic Signal Transduction Pathway

Fig. 15 shows the role of protein kinases, phosphoprotein phosphatases (NEER and CHAPHAM, 1988), phospholipases, and farnesyltransferases in the transduction of an extracellular signal, such as a growth hormone, into an intracellular response of DNA replication and mitogenesis. In oncogene-transformed cells these enzymes are often involved in neoplasmic cell growth. Much effort in the search for new anticancer agents is now focussed on specifically acting inhibitors of the cell cycle such as acetophthalidin from a marine *Penicillium* strain which arrests the mammalian cell cycle in the G2/M phase (CUI et al., 1996a, b).

So far, more than 40 distinguishable oncogenes have been identified. They can be classified into four main groups according to their function:

(1) oncogenes encoding tyrosine protein kinases such as *src* (similar to insulin receptor and catalytic chain of mammalian cAMP-dependent protein kinase from Rous sarcoma virus), *erbB, fgr,* and *fes* (see also Fig. 1),
(2) oncogenes encoding proteins involved in the metabolic regulation of GTP-binding proteins as essential parts of G protein-dependent membrane receptors (K-*Ras,* H-*Ras,* N-*Ras*),
(3) oncogenes encoding proteins which act at the level of gene regulation as transcription factors such as *Myc* and *Myb,* and
(4) oncogenes encoding growth factor-like proteins such as *Sis* (similar to platelet-derived growth factor).

To escape the normal pattern of control "proto-oncogenes" have to be transformed into oncogenes either by structural alterations, due to chemical carcinogenesis (GUENGRICH, 1988) or by integration into retroviruses and reintegration into the host genome (HART and TURTURRO, 1988; BERTRAM, 1990; WEINBERG, 1985).

4.5.1 Inhibitors of Protein Kinases and Protein Phosphatases

Protein kinases are ubiquitous regulatory proteins in microbial, plant, and animal cells which exert their activity by the ATP-dependent phosphorylation of serine, threonine, and tyrosine residues (Fig. 16). The phosphorylated proteins display an altered function (activated or inactivated) (BASU and LAZO, 1994; LEVITZKI, 1994). Phosphorylation and dephosphorylation of proteins can result in the intracellular amplification of a mitotic signal, generated by interaction of a growth hormone with a membrane receptor. The protein kinases share at least some common features in their secondary and tertiary structures (TAYLOR, 1989; TOWBRIDGE, 1991). Myosin light chain kinase, a cytoplasmic enzyme, has a Ca^{2+}/calmodulin-binding domain carboxy terminal to the catalytic core, and binding of ligands activates the kinase. Membrane-associated protein kinase C is activated by Ca^{2+}, diacylglycerol or phorbol ester type compounds, and phospholipids, and the recognition sites for these ligands lie amino terminal to the catalytic core. Tyrosine–protein kinase activity was first detected as a characteristic

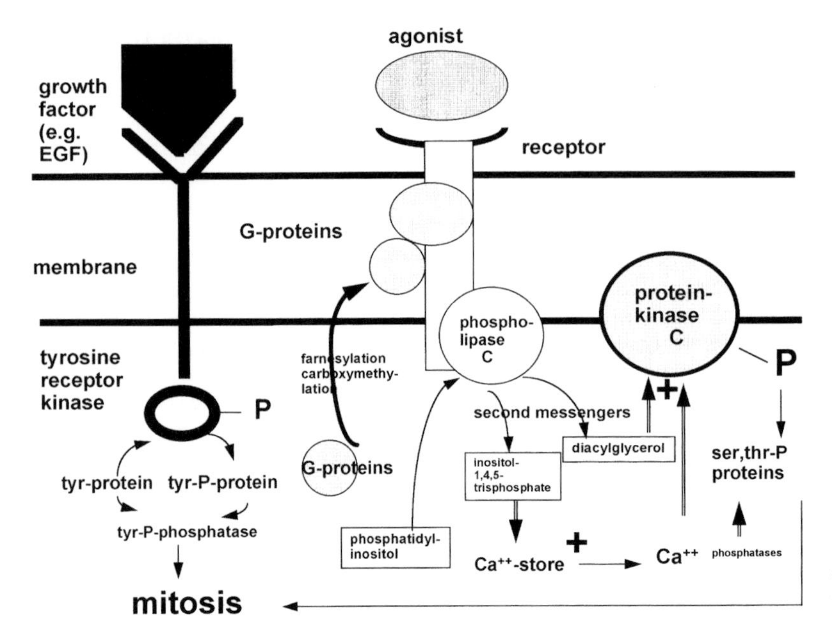

Fig. 15. Key steps in the cellular mitotic signal transduction pathway.

of the transforming protein from Rous sarcoma virus, pp60$^{v\text{-}src}$. Removal of a phosphorylation site of this protein converts the "protooncoprotein" into a transforming protein. Following myristylation (farnesylation) pp60$^{v\text{-}src}$ moves to the plasma membrane. Receptors of growth factors, such as the epidermal growth factor (EGF) receptor, span the membrane via a single membrane-spanning element. Binding of EGF to the outer domain activates the cytoplasmic tyrosine protein kinase activity. Another prototype of a tyrosine-protein kinase transmembrane protein is CD45 (T200 or leukocyte common antigen) (TROWBRIDGE, 1991).

Another type of protein kinase is activated by cyclic 3,5-adenosine monophosphate (cAMP). The activating ligand (cAMP) binds to a distinct regulatory subunit thereby inducing conformational changes that provoke dissociation of the holoenzyme (TAYLOR, 1989).

Reversible phosphorylation/dephosphorylation of proteins is involved in many cellular activities (KAHN and GRAF, 1986). For example, the M phase promoting factor (MPF) in mammalian cells is composed of a catalytical p34^{cdc2} protein which has protein kinase activity and a regulatory protein (cyclin B). The activity of MPF is regulated by reversible phosphorylation/dephosphorylation. Recently, tautomycin (*Streptomyces* sp.) was found to inhibit dephosphorylation of MPF thereby preventing mitosis (MAGAE et al., 1988, 1992).

In the search for inhibitors of protein kinase C, the morphogenic effect of phorbol esters on mammalian cells provided a useful screening feature (TAKAHASHI et al., 1987, 1989). Phorbol esters and indolactams (teleocidin and blastmycetin from *Streptoverticilium* sp.) (HAGIWARA et al., 1988) are known as tumor promoters. They stimulate growth of tumors but are not carcinogenic. The discovery of tumor promoters from Euphorbiaceae, such as phorbol esters contributed to a better understanding of the cellular mechanism of growth regulation. In addition to tumor promotion, other processes such as inflammation, mitogenesis of lymphocytes, platelet activation, etc. were affected by structural analogs of the second messenger diacylglycerol. The phorbol esters and similar structures (Fig. 16) activate protein kinase C. This effect has been visualized as bleb formation on the cell surface of K562 chronic myeloic leukemia cells which is induced in the presence of phorbol dibutyrate. Bleb induction can be prevented by specific inhibitors of protein kinase C and thus a practicable high-throughput screening assay was developed (OSADA et al., 1988). The same bleb formation assay has also been used to search for inhibitors of protein phosphatases. These enzymes provoke the same type of stimulaton of bleb formation which is a characteristic of the phorbol esters (MAGAE et al., 1988, 1992; BLUMBERG et al., 1989). On the other hand, differences in the effect on tetrazolium blue reduction of HL-60 cells of phorbol esters and tautomycin suggested that the latter has a different mode of action (MAGAE et al., 1992). For the sake of completeness, the bryostatins (from marine Bryozoa) should be mentioned since they are similar tumor promotors (Fig. 16). They are antitumor and antileukemic compounds isolated from marine animal organisms as a type of cyclic polyether which also mimics the structure of 1,2-diacylglycerol (RAMSDELL and PETTIT, 1986; WOLF and BAGGIOLINI, 1988; TAKAHASHI et al., 1987b). Similarly, novel inhibitor structures of protein kinase C and phosphoprotein phosphatases were discovered, such as staurosporin, UCN-1, UCN-2 (TAKASHI et al., 1989), tautomycin (*Streptomyces spiroverticillatus*) (MAGAE et al., 1992), calyculin, microcystin (Fig. 16), and okadaic acid (from a marine organism) (SUGANUMA et al., 1988). Calyculin A affects phosphoprotein phosphatase 1 (PP1) 30 to 250fold more than okadaic acid, and also inhibits protein phosphatase 2A (PP2A). Tautomycin inhibits several protein phosphatases, even those of smooth muscles. No activity was found on myosin light chain kinase or protein kinase C. Okadaic acid, a polyether-like shellfish toxin, inhibits serine-threonine specific protein phosphatases, particularly protein phosphatase 2A. Moreover, isopalimurin, a mild protein phosphatase inhibitor, has recently been isolated from a sponge (MURRAY et al., 1993) (Fig. 16).

Presently, much interest is focussed on phosphotyrosyl protein phosphatases which

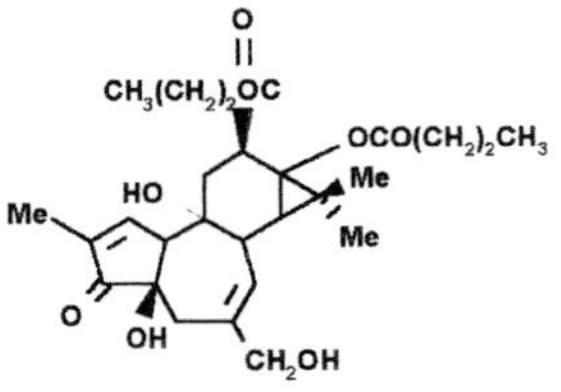

Phorbol 12,13-dibutyrat

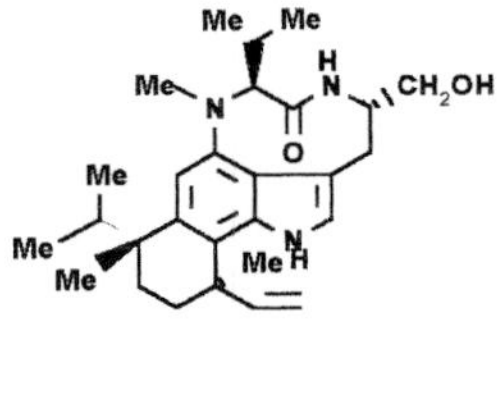

Teleocidin B-4

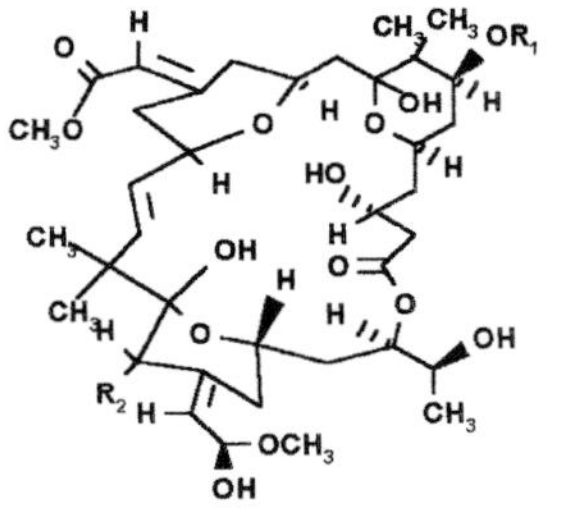

BRYOSTATIN 1

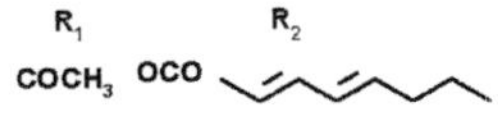

R1	R2
$COCH_3$	OCO

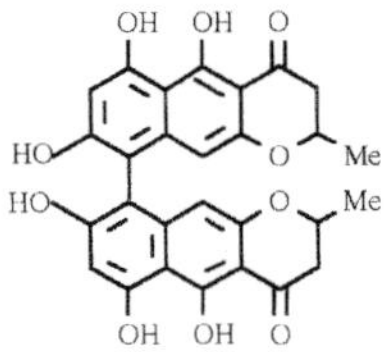

Sch 45752

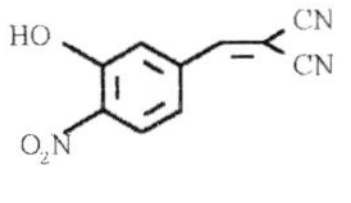

Tyrphostin

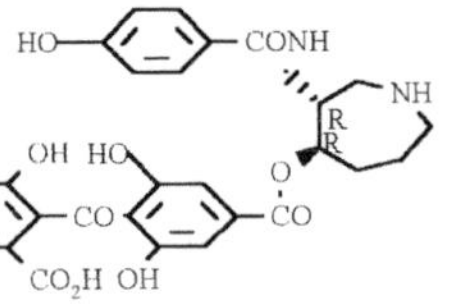

Azepinostatin

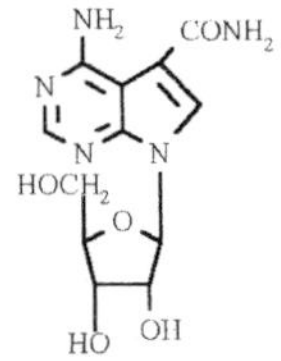

Sangivamycin

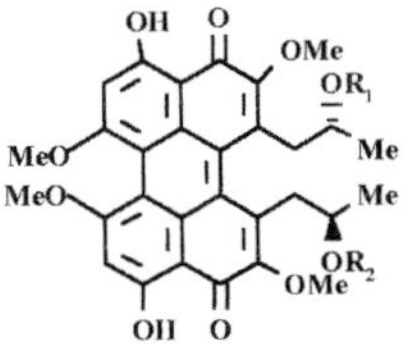

Calphostin

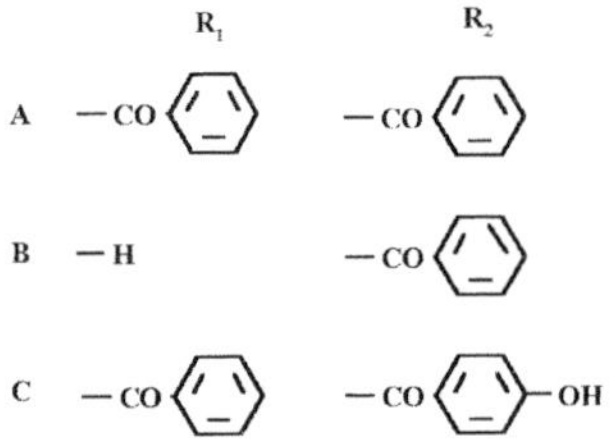

	R1	R2
A	—CO	—CO
B	—H	—CO
C	—CO	—CO ... OH

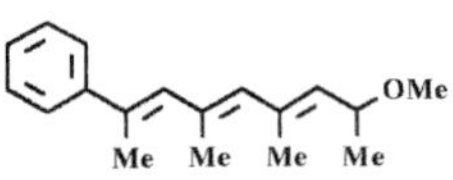

BE-23372M

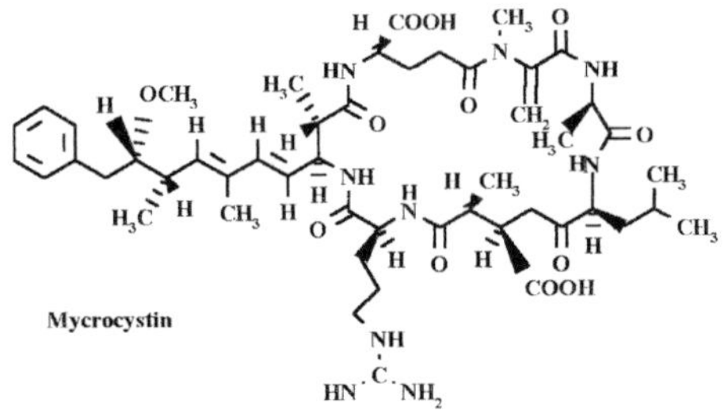

Mycrocystin

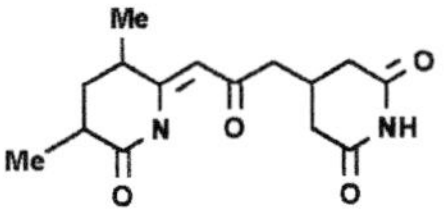

Epiderstatin

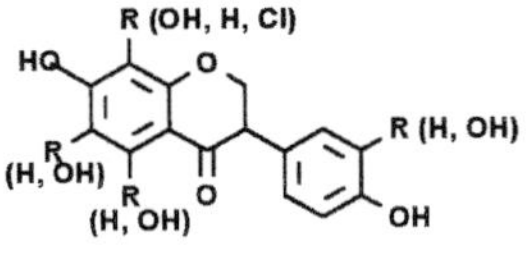

Isoflavonoids

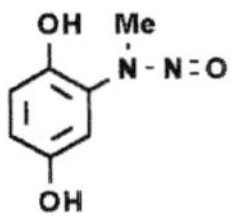

Dephostatin

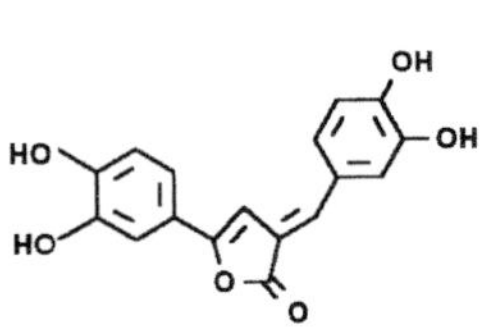

BE-23372M

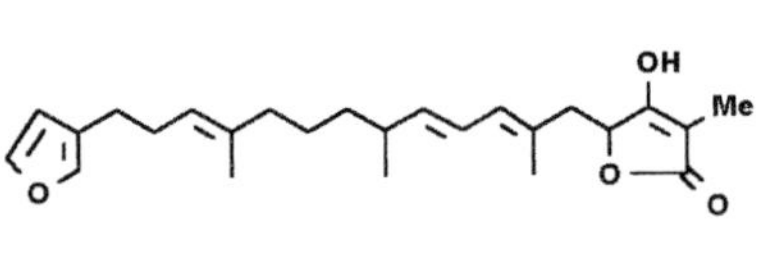

Isopalimurin

◀ **Fig. 16.** Selected structures of inhibitors of protein-kinase C, tyrosine protein kinase, and protein phosphatases.

are a particular family of transmembrane enzymes (LAN et al., 1989; KRUEGER et al., 1990; FISCHER et al., 1991).

4.5.1.1 Inhibitors of Protein Kinase C

Members of the indolocarbazole family of antibiotics such as staurosporin (*Streptomyces* sp.), UCN-01, UCN-02 (*Streptomyces* sp.) (AKINAGA et al., 1993), K-252a (*Nocardiopsis* sp.) (NAKANISHI et al., 1986), RK-286C (OSADA et al., 1990), KT 6006 and CGP41251, are potent inhibitors of protein kinase C (Fig. 16). These unique structures are produced by actinomycetes; however, staurosporin-like compounds have recently been discovered in extracts from marine tunicates. Derivatives of staurosporin such as MLR-52, display immunosuppressive activity. This effect is comparable to that of other immunosuppressors, such as cyclosporine A and FK-506, which affect protein phosphorylations and phosphoprotein dephosphorylations (MCALPINE et al., 1994).

Moreover, indolocarbazoles display remarkable antimicrobial activities against bacteria and fungi including *Candida albicans* and *Botrytis cinerea,* but no correlation was found between protein kinase inhibitory potencies and inhibition of *Streptomyces* sporulation or growth inhibition (SANCELME et al., 1994). The indolocarbazole compounds display antitumor activity against human and murine tumor cell lines both *in vitro* and *in vivo* (AKINAGA et al., 1993; TAKAHASHI et al., 1987b; WOLF et al., 1988). This type of structure (SANCELME et al., 1994) inhibits either protein kinase C (see above) or topoisomerase I (rebeccamycin, AT 2433, and derivatives) (OSADA et al., 1990; KANEKO et al., 1990). AT 2433 has an inhibitory effect on protein kinase C. Staurosporin and its analogs also inhibit the activity of the Rous sarcoma virus-transformed protein p60 (*scr* oncogene product) (NAKANO et al., 1987; CAI et al., 1996).

The antitumor antibiotic calphostin (UCN-1028) (Fig. 16) consists of five components (A, B, C, D, I) (IIDA et al., 1989). It is produced by the fungus *Cladosporium cladosporoides* and strongly inhibits protein kinase C. Another inhibitory structure is balanol (azepinostatin) (Fig. 16) isolated from *Verticillium balanoides* and *Fusarium merismoides* (BOROS et al., 1994; OHSHIMA et al., 1994). In addition nucleoside antibiotics, such as sangivamycin (Fig. 16), and membrane-active compounds, such as polymyxin B, were reported to inhibit protein kinases including protein kinase C (LOOMIS and BELL, 1988; ALLGAIER et al., 1986).

4.5.1.2 Inhibitors of Other Protein Kinases

The pamamycin-type agents MS-282a and MS-282b, are macrodiolide-type inhibitors of calmodulin-activated myosin light chain kinase from *Streptomyces tauricus* ATCC 27470 (NAKANISHI, 1994). Other protein kinases are not inhibited by these agents. The interaction of MS-282 compounds with the enzyme occurs at the calmodulin-binding site. Calmodulin-dependent nucleotide phosphodiesterase is also affected. The compound Sch45752 (Fig. 16) from the fungus SCF-125 is an inhibitor of various protein kinases (IIDA et al., 1989); however, it also inhibits calmodulin-sensitive cyclic nucleotide phosphodiesterase.

4.5.1.3 Inhibitors of Tyrosine Protein Kinases

Erbstatin (Fig. 14), which was isolated from the culture broth of *Streptomyces* sp. MH 435-UF3, is a potent inhibitor of the EGF associated protein tyrosine kinase (IMOTO et al., 1987; SONODA et al., 1989). It specifically inhibits the autophosphorylation of EGF receptor but does not inhibit either cyclic AMP-dependent protein kinases or protein kinase C (IMOTO et al., 1987; UMEZAWA et al., 1986). Also, synthetic peptide sequences were found to be good substrates for the EGF tyrosine kinase and have been used in kinetic studies. According to these investigations, erbstatin competes with the peptide substrate but not with ATP (UMEZAWA et al., 1986). 4′,7,8-trihydroxyisoflavone, 3′,4′,7-trihydroxyisoflavone, 8-chloro-3′,4′,5,7-tetrahydroxyisoflavone, and orobol were isolated from the culture broth of *Streptomyces* sp. OH-1049 are isoflavones (Fig. 17) which also inhibit EGF receptor tyrosine-protein kinase. This effect coincided with their increasing effect on the life span of tumor-bearing mice (S180, P388) (KOMIYAMA et al., 1989; OGAWARA et al., 1989a). This compounds do compete with ATP. The interaction of flavonoids with mammalian protein kinases was later found to be non-specific (END, 1987). Although they occur in microbial cultures, isoflavone glycosides such as genistein and daidzin appear to stem from the plant-derived nutrients. Microbial biotransformation results in the formation of substituted or even glycosylated products (ANGANWUTAKU et al., 1992).

More recently discovered structures with inhibitory activity against EGF receptor mediated protein-tyrosine kinase are, e.g., dephostatin (*Streptomyces* sp.) (KAKEYA et al., 1993), BE-13793C (TANAKA et al., 1992), BE-23372M (*Rhizoctonia* sp., TANAKA et al., 1994), tyrphostins (synthetic) (Fig. 16), and *p*-quinoid structures like paeciloquinones A–F (FREDENHAGEN et al., 1995). Epiderstatin (a substituted 2,6-piperidine-dione) (Fig. 16) was isolated from *Streptomyces pulveraceus* ssp. *epiderstagenes* (SONODA et al., 1989). The compound inhibits incorporation of [^{3}H]thymidine into quiescent cells stimulated by EGF and also reverts the morphology of ^{ts}*src*-transformed cells to normal.

Contact inhibition of growth by neighboring cells is an essential characteristic of normal cells. However, tumor cells lack this behavior. Herbimycin A (Fig. 14) (*Streptomyces* sp.) is a benzoquinoid ansamacrolide which was shown to prevent the interaction of v-*src* oncogene-expressed cells due to the inhibition of pp60$^{v\text{-}src}$ tyrosyl protein kinase and phosphatidylinositol kinase (IWAI et al., 1980; SUZUKAKE-TSUCHIYA et al., 1989). This antibiotic was originally screened for its herbicidal activity.

4.5.2 Inhibitors Affecting the Metabolism of Phosphoinositols

Phosphatidylinositol turnover appears to be correlated with cell transformation by some types of oncogenes (*ras*) and also with cellular responses to growth factors (EGF, PDGF) (FLEISCHMAN et al., 1986; JACHOWSKI et al., 1986; BERRIDGE and IRVINE, 1984, 1989). The signal generation induced by the enzymatic breakdown of phosphatidylinositol-4,5-diphosphate involves phospholipase C. This key enzyme is capable of forming two second messenger molecules, inositol-1,4,5-triphosphate and 1,2-diacylglycerol. Inositol-1,4,5-triphosphate mobilizes the intracellular calcium pool and 1,2-diacylglycerol activates protein kinase C (Fig. 15). Consequently, phospholipase C, the phosphatidylinositol turnover, the phosphatidylinositol receptor, protein kinase C (see above), and diacyglycerol kinase are potential targets in the search of cellular growth regulators. Q 12713 (*Actinomadura* sp.) (OGAWARA et al., 1992), (hispidospermidins (*Chaetomia* sp.) (YANAGISAWA, 1994a, b), and caloporoside (*Caloporus dichrous*) (WEBER et al., 1994; TATSUDA and YASUDA, 1996) which are inhibitors of phospholipase C have recently been discovered (Fig. 17). The search for inhibitors of the inositol-triphosphate turnover provided a series of different chemical structures, such as psitectorigen (an isoflavone from *Actinomyces* cultures and plants, Fig. 17) (IMOTO et al., 1988, 1991), inostamycin A (a polyether from a *Streptomyces* strain) (IMOTO et al., 1990), echiguanins (TAKEUCHI, 1992), and piericidins B1, B5 and their N′-oxides (NISHIOKA et al., 1991). Recently, automated screens have been developed for inhibitors of myo-inositol monophosphatase such as ATCC 20928 A–C (diterpene types; STEFANELLI et al., 1996). Piericidin B1 N-oxide (Fig. 17) was isolated from a *Streptomyces* strain as an inhibitor of phosphatdylinositol phosphate turnover (NISHIOKA, 1994). The drug did not inhibit the synthesis of DNA, RNA, or proteins. However, it reversibly reduced growth of A431 cells and Ehrlich carcinoma (NISHIOKA, 1994). Echiguanin (TAKEUCHI, 1992) was shown to inhibit phosphatidylinositol kinase. In con-

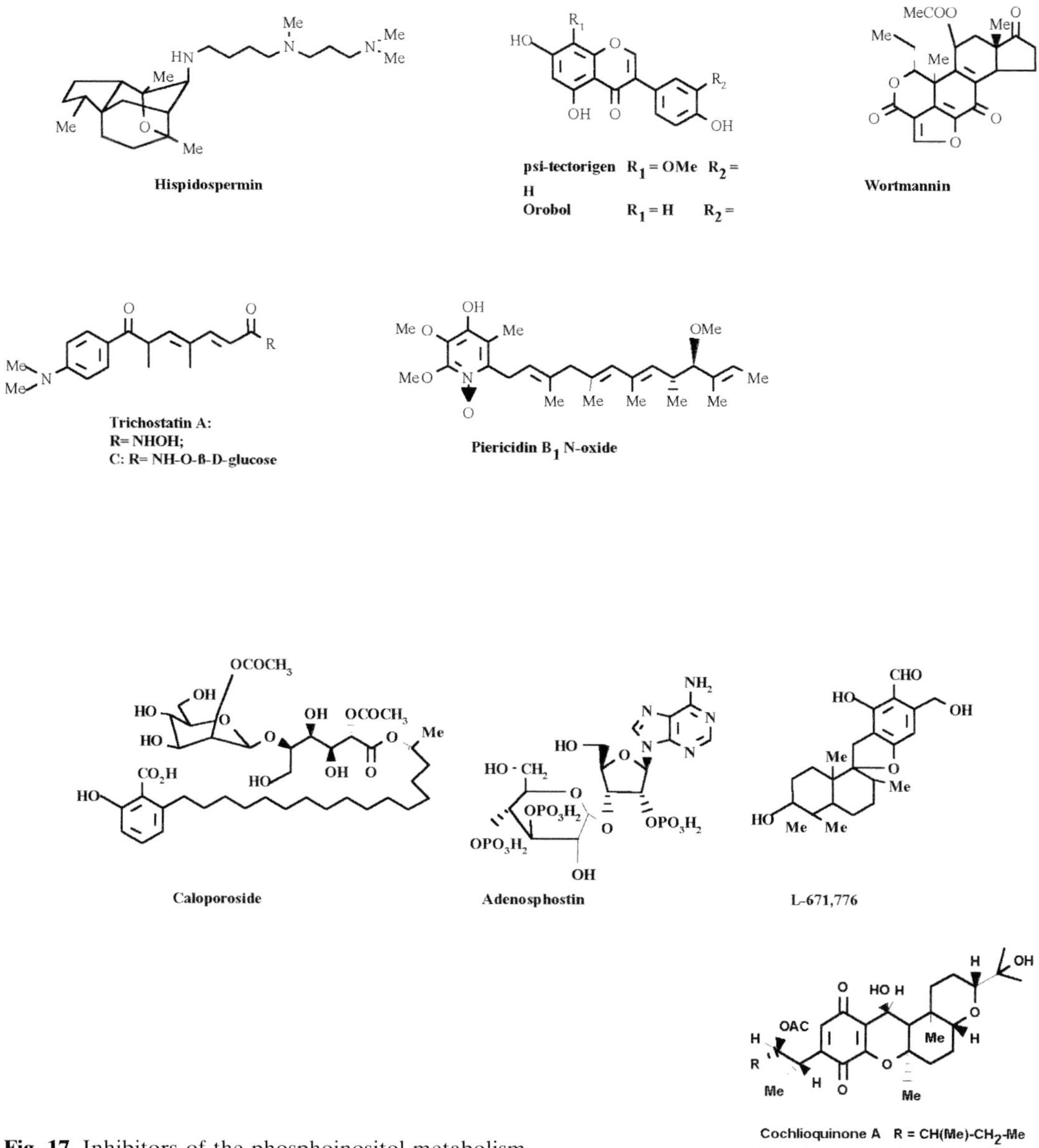

Hispidospermin

psi-tectorigen R_1 = OMe R_2 = H
Orobol R_1 = H R_2 =

Wortmannin

Trichostatin A:
R= NHOH;
C: R= NH-O-ß-D-glucose

Piericidin B_1 N-oxide

Caloporoside

Adenosphostin

L-671,776

Cochlioquinone A R = CH(Me)-CH_2-Me
Stemphone: R = -C(Me)=CHMe

Fig. 17. Inhibitors of the phosphoinositol metabolism.

trast to inostamycin, the inostamycins B and C displayed no inhibitory effect on phosphatidylinositol turnover despite their structural similarity (ODAI et al., 1994; IMOTO et al., 1990).

Herbimycin A (Fig. 14), a benzoquinoid antibiotic (IWAI et al., 1980; SUZUKAKE-TSUCHIYA et al., 1990) was reisolated as a substance altering the transformed morphology of v-*src*-expressed cells to normal morphology. These changes were accompanied by alterations in cytoskeletal organization, synthesis of fibronectin, and colony-forming ability on distinct media. The compound was shown to inhibit phosphoinositol kinase (SUZUKAKE-TSUCHIYA et al., 1989). The trichostatins (*Streptomyces toyocaensis*) (Fig. 17) are antifungal antibiotics but also potent inducers of erythroid differentiaton in mouse Friend leukemia cells. In low concentration trichostatin A arrested the cell cycle of normal fibroblast cells in both G_1 and G_2, and induced the formation of proliferative tetraploid cells after release from the G_2 arrest. Inhibition of phosphoinositol kinase was demonstrated for this agent (YOSHIDA et al., 1990).

The opposite effect of inositol phosphokinase inhibitors, i.e., an increase in cellular phosphatidylinositol concentration can be achieved by inhibitors of inositol phosphate phosphatases. This idea led to the discovery of L-671,776 from a hyphomycete (*Memnoniella echinata*). It was shown to inhibit both the myo-inositol-1,4-disphosphate phosphatase and the 1,4,5-triphosphate 5-phosphatase (LAM et al., 1992). The adenophostins A and B (Fig. 17) occur as phosphorylated adenosine analogs in cultures of *Penicillium brevicompactum* (TAKAHASHI et al., 1994). They are potent agonists of the inositol-1,4,5-triphosphate receptor and mimic second messenger activity.

Diacylglycerol kinase phosphorylates diacylglycerol to form phosphatidic acid. This enzyme is affected by extracellular stimulators and is involved in the regulation of protein kinase C by decreasing the intracellular concentration of diacylglycerol. In a search for inhibitors of this enzyme cochlioquinone and stemphone from *Drechslera sacchari* have recently been discovered in fungal strains (OGAWARA et al., 1994) (Fig. 17).

4.5.3 Substances Changing the Morphology of Oncogene-Transformed Cells or Showing Selective Toxicity

Oncogene-transformed (*ras, myc, fos, erb,* etc.) cell lines are now available. They are used to select for specific inhibitors of signal transduction pathways. The selection of these inhibitors is based on their capacity to restore normal cell morphology or selectively kill oncogene-expressing cells (SUZUKAKE-TSUCHIYA et al., 1990; UMEZAWA, 1989). For instance, Sch52900 and Sch52901 are new dimeric diketopiperazine structures from *Gliocladium* sp. which inhibit c-*fos* protooncogene induction as an early event in the transition of cells from the quiescent to the growing stage (CHU et al., 1995). Other fungal products, tryprostatins A and B (diketopiperazins from *Aspergillus fumigatus*), inhibit the cell cycle of murine tsFT210 cells by inhibition of Cdc-2-kinase (CUI et al., 1996a, b). Moreover, indolocarbazoles inhibited the cell cycle progression of *ras*-transformed rat fibroblasts (AKINAGA et al., 1993).

Ras-related proteins control a wide variety of cellular processes. They are associated with the cell membrane, and their function and activity is modified by prenylation, proteolysis, and carboxymethylation of the carboxyl terminus (KHOSRAVI-FAR et al., 1992). The dorrigocins A and B (*Streptomyces platensis*) and depudecins (Fig. 18) have recently been reported as new antifungal antibiotics that change the morphology of *ras*-transformed N/H/3T3 cells to that of normal cells by inhibiting protein carboxymethylation (KARWOWSKI et al., 1994; MATSUMOTO et al. 1992). Differanisol A (Fig. 18), from a *Chaetomium* strain induces the differentiation of Friend leukemic cells in mice. Interestingly, the chemical structure of this substance is very similar to that of the stalk cell differentiation inducing factor of the cellular slime mold *Dictyostelium discoideum* (KUBOHARA et al., 1993; MORI, 1989). The reveromycins A–D (Fig. 18) interrupt eukaryotic cell growth due to their antagonistic effect on the mitogenic activity of the growth hormone, EGF (KOSHINO et al., 1992). They cause morphological reversion of *src*$^{+s}$-NRK cells and exhibit antiproliferative activity for human tumor cell lines. Rhizopodin (Fig. 18) is a cytostatic compound from *Myxococcus stipitatus* which inhibits growth of various animal cultures without killing the cells. Fibroblast cells become larger and form long branches in an irreversible manner. Protein phosphorylation has been suggested to be the initial effect of this compound (SASSE et al., 1993).

Redifferentiation of cancer cells was also reported for the anthracycline type cosmomycins (*Streptomyces cosmosus,* redifferentiaton inducer of Friend leukemia cells) (KHOS-

RAVI-FAR et al., 1992), latosillan (*Alcaligenes latus,* inducer of differentiation of mouse myeloic leukemia cells; high-molecular weight polysaccharide) (HAYAKAWA et al., 1982), spicamycin (*Streptomyces alanosinicus,* differentiation inducer of myeloic leukemia cells; Fig. 18) (HAYAKAWA et al., 1983), citrinin (*Monosporascus carpoanus,* differentiation inducer of myeloic leukemia cells; Fig. 18) (KAWASHIMA et al., 1983), canacelunin (*Streptomyces* sp., inhibitor of cancer cell agglutination, inducer of lymphoblastoid transformation of mouse brain cells) (IKEKAWA et al., 1980), microcystilide A (*Microcystis seruginosa*) (TSUKAMOTO, 1993), hygrolidin (SUZUKAKE-TSUCHIYA et al., 1991), differenol A (*Streotomyces* sp., redifferentiation inducer of murine leukemia cells) (ASAKI et al., 1981), lavanducyanin (*Streptomyces alriufer,* stimulation of Hela cell proliferation; Fig. 18) (MATSUMOTO and SETO, 1991), kasuzamycin (*Streptomyces* sp., antitumor antibiotic, affects L1210 cell cycle; Fig. 18) (TAKAMIYA et al., 1988), trichostatins (*Streptomyces* sp., similar to leptomycin, inducer of erythroid differentiation in murine Friend leukemia cells, inhibition of the cell phase transfer G1 to G2; Fig. 18) (YOSHIDA et al., 1990), and hygrolidin (SUZUKAKE-TSUCHIYA et al., 1991).

Modulation of the actin filament network and/or protein phosphorylation is involved in tautomycin-induced bleb formation. Protein phosphorylation is also involved in recycling cell surface receptors for transferrin and EGF (KURISAKI et al., 1992).

Redifferentiation of *ras*-transformed cells to normal morphology is also a characteristic of flavones and some glutarimide antibiotics such as acetoxycycloheximide (Fig. 18) and cycloheximide (OGAWARA et al., 1989a, b). Herbimycin (Fig. 14) and sparoxomycins reverse the transformed morphology of temperature-sensitve Rous sarcoma virus-infected rat cells (ts/NRK) to the normal morphology concomitant with a drastic reduction in intracellular $p60^{src}$ kinase activity. Semisynthetic derivatives of herbimycin prolonged the life span of tumor-bearing mice (UEHARA et al., 1988; UBUKATA et al., 1996). Some macrolide antibiotics such as FD-891 (Fig. 18) and FD-892, which possess polyhydroxy-alkane-substituted side chains, induce morphological changes of HL-60 cells at low concentrations and display cytocidal activity at higher dosages (SEKI-ASANO et al., 1994). Antitumor activities are also a characteristic of macrolactams, such as leinamycin and hitachimycin (Fig. 18) (HARA et al., 1990; SHIBATA et al., 1988). The antibiotic FR 901228 from *Chromobacterium violaceum* (Fig. 18) also reverted the transformed morphology of a *ras* transformant cell line to normal and exhibited antitumor activity against murine and human tumors (UEDA et al., 1994). Melastin (mol. wt. 5000 ± 3000) suppresses lipopolysaccharide-induced blastogenesis of B cells more effectively than concanavalin A and selectively inhibits growth of several leukemia cells (MAEDA, 1993). Leptomycin B and leptolstatin from *Streptomyces* strains (Fig. 18) have recently been found to inhibit proliferation of cultured animal cells by blocking progression of the cell cycle in G1 and G2 phases (ABE et al., 1993b). Phosmidosin C nucleosides from *Streptomyces* sp. display morphology reversion activity against *src* transformed NRD cells (MATSUURA et al., 1996). Last but not least, reductioleptomycin A (*Streptomyces* sp.) (Fig. 18) was shown to flatten the morphology of v-ras^{ts} NRK cells (HOSAKAWA et al., 1993).

4.5.4 Inhibitors of *ras*-Farnesyltransferase

The *ras* genes are mutated in 50% of colon and 90% of pancreatic carcinoma. Farnesylprotein transferase catalyzes farnesylation of a protein encoded by *ras* (Ras or p21). This is essential for the associaton of Ras proteins with the cell membrane and *ras*-mediated transformation. Inhibition of the prenylating enzyme, *ras*-farnesyltransferase, was therefore proposed as a specific target of "soft" anticancer agents. In screening for representatives of such drugs, streptonigrin (*Streptomyces albus*) and its 10′-desmethyl derivative (VAN DER PYL et al., 1992), representatives of the indolcarbazole type inhibitors of protein kinase C (CGP 412511, staurosporin, UCN-01, K-252a), the pepticinnamins (lipopeptide, *Streptomyces* sp.) (ŌMURA et al.,

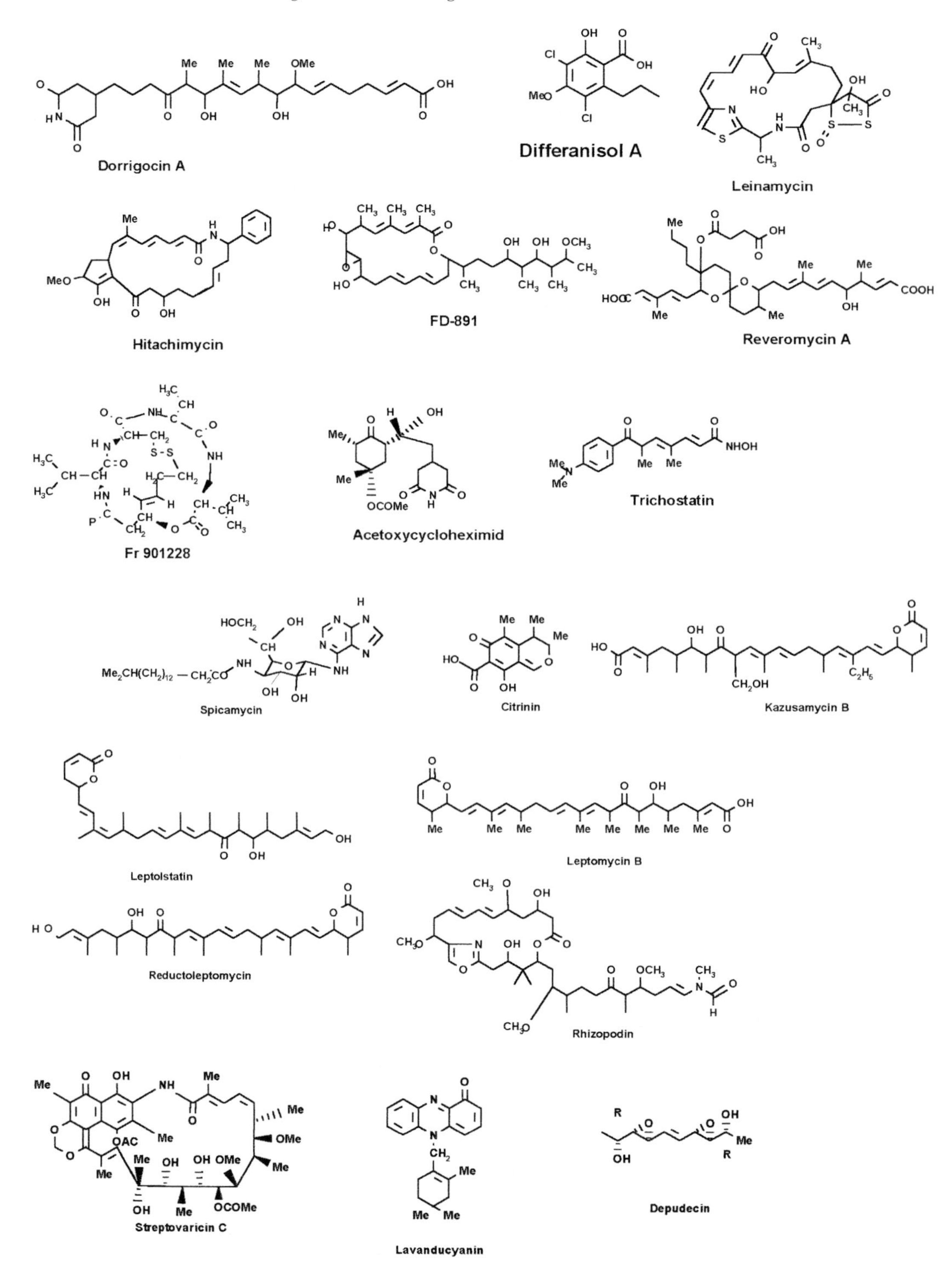

Fig. 18. Examples of drugs changing the morphology of oncogene-transformed cells or showing selective toxicity against oncogene-expressed cells.

1993; SHIOMI et al., 1993), saquayamicins (SEKIZAWA et al., 1996), and andrastins (*Penicillium* sp., ŌMURA et al., 1996) were found to be new inhibitors of this enzyme and blockers of the cell cycle progression of *ras*-transformed cells (Fig. 19). More recently, a series of other structures from *Actinoplanes* sp. such as actinoplanic acid A and its derivatives have been reported to inhibit *ras*-farnesyltransferase (SINGH et al., 1994a, b, c). In this context, fungal structures such as fusidienol (*Fusidium griseum*) (SINGH et al., 1994a): kurasoins A and B (*Paecilomyces* sp. FO-3648, UCHIDA et al., 1996), preussomerins, deoxypreussomerins, Sch 49210, Sch 53514-53517 (*Preusseria isomera, Harmonema demativides*) (SINGH et al., 1994c), chaetomellic acid (*Chaetomella acuseta*) (SINGH et al., 1993) (Fig. 19), gliotoxin, and acetylgliotoxin (VAN DER PYL et al., 1992) provided non-conventional leads in the search for inhibitors of this enzyme.

4.5.5 Inhibitors of Sexual Hormone Production and Hormone–Receptor Interactions

Suppression of the production of sexual hormones (estrogens and androgens) and of receptor interaction of these hormones were proposed as a therapeutical strategy in the treatment of hormone-dependent tumors. The therapeutic efficacy of synthetic aromatase inhibitors such as aminoglutethimides, imidazoles, and triazoles (VANDEN BOSCHE et al., 1994; BRAND et al., 1988) for breast cancer and tetrazoles motivated a search for nonsteroidal microbial compounds which could reduce the effects of estrogen by antagonizing its receptor in a manner similar to the synthetic drug tamoxifen. The compound R1128, an estrogen receptor antagonist, is an alkylated trihydroxyanthraquinone from *Streptomyces* sp. 1128 (HORI et al., 1993a, b). Related polycyclic phenolic structures are the napyradiomycins A and B (HORI et al., 1993c) which have also recently been isolated from a *Streptomyces* culture. A promising approach to the treatment of prostata carcinoma is the use of inhibitors of testosterone-5α-reductase. Non-steroidal inhibitor structures similar to pyocyanin, WS-9659A and B, were also obtained from *Streptomyces* cultures (NAKAYAMA et al., 1989) (Fig. 20).

4.5.6 Miscellaneous Drugs with Potential Antitumor Activity

The discovery of new cytotoxic compounds will continue. New structures such as duacins (HIDA et al., 1994), leptosins (TAKAHASHI et al., 1994), cochleamycin (SHINDO et al., 1996), and himastatin (LEET et al., 1996) may be suitable for studying structure–activity relationships and modes of action. A therapeutic potential was suggested for inhibitors of melanin synthesis such as OH-3984K1 and -K2 (*Streptoymces* sp.), members of the albocycline family of antibiotics (KOMIYAMA et al., 1993) (Fig. 21) and melanoxazal from the hemolymphe of the silkworm *Bombyx mori* (TAKAHASHI et al., 1996) as well as for inhibitors of the cAMP-phosphodiesterase (HEDGE et al., 1993), in addition to glyoxylase inhibitors such as glyo I and II (ALLEN et al., 1993; TORNALLAY, 1990).

Marine organisms have contributed many new and promising antitumor agents (Fig. 21). The spongistatins from the Eastern Indian Ocean sponge species are exceptionally potent growth inhibitors for human cancer cells. They comprise a series of homologous compounds distinguishable by chlorine substitutions (PETTIT et al., 1993a, b). Similarly, *Dolabella auricularia* from the Indian Ocean contains the cytotoxic dolastatins (see the dolastatin D) and its congeners (SAE et al., 1993). Other recent examples of cytotoxic compounds from marine organisms are phloeodictines A1–A7 and C1–C2 which are guanidine alkaloids from the New Caledonian sponge *Phloeodictyon* sp. (KOURANY-LEFOLL et al., 1994), phakellistatins from a Comoros marine sponge (PETTIT, 1994), agelasphins (sponge *Agelas mauritianus*) (NATORI et al., 1994; HARADA et al., 1993), amphidinolide F (KOBAYASHI et al., 1993) and bryostatins 16–18 from the marine bryozoan *Bugula neritina* (PETTIT et al., 1996). Cyanobacteria also supply a rich source of cytotoxic metabolites. However, due to their general toxicity against human cells a therapeutic application

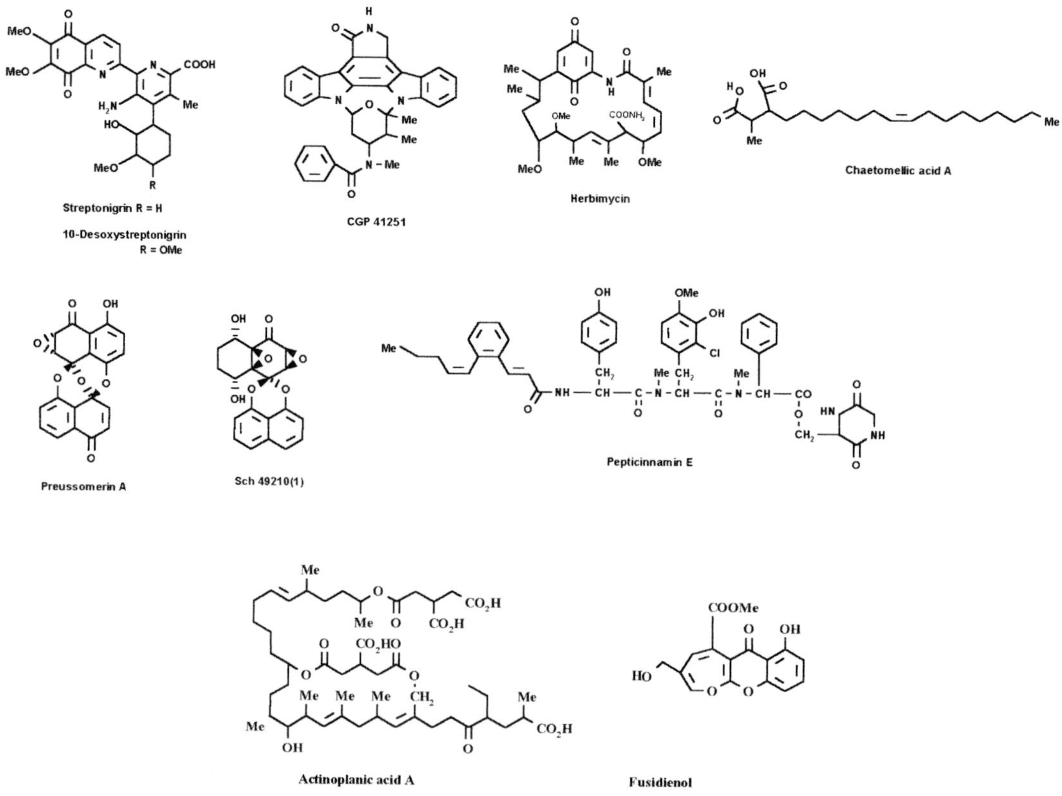

Streptonigrin R = H
10-Desoxystreptonigrin R = OMe
CGP 41251
Herbimycin
Chaetomellic acid A
Preussomerin A
Sch 49210(1)
Pepticinnamin E
Actinoplanic acid A
Fusidienol

Fig. 19. Examples of inhibitors of *ras*-farnesyltransferase.

R 1129 (R= Prop,But,Pent)
Napyradiomycin A
WS-9859 A

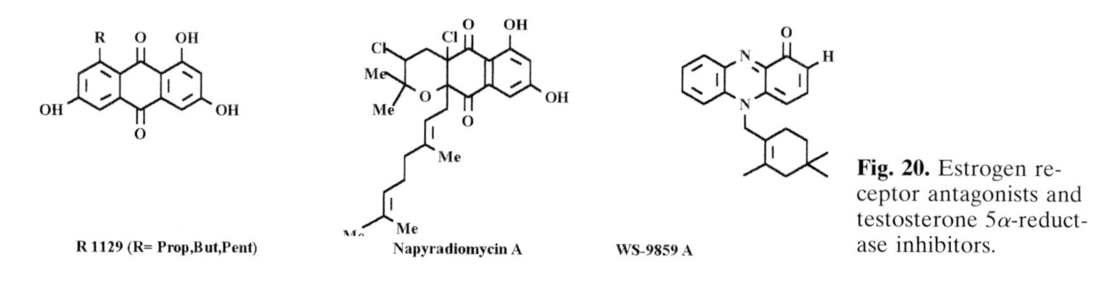

Fig. 20. Estrogen receptor antagonists and testosterone 5α-reductase inhibitors.

of these compounds not is foreseeable at present.

5 Closing Remarks

During the past ten years many new drug structures have been discovered from new sources, such as new microbes, plants, and marine animals, by using new target-oriented screening assays. Both in the field of classical cytotoxic drugs and in the recognition of specific inhibitors of the mammalian cell cycle, major advances have been made. Investigations with enedyines and taxol have led to new types of neoplasm inhibitors which are now being introduced to clinical therapy. Various novel effectors of the cell cycle and the signaling cascade are now used as invaluable biochemical tools which may help to gain

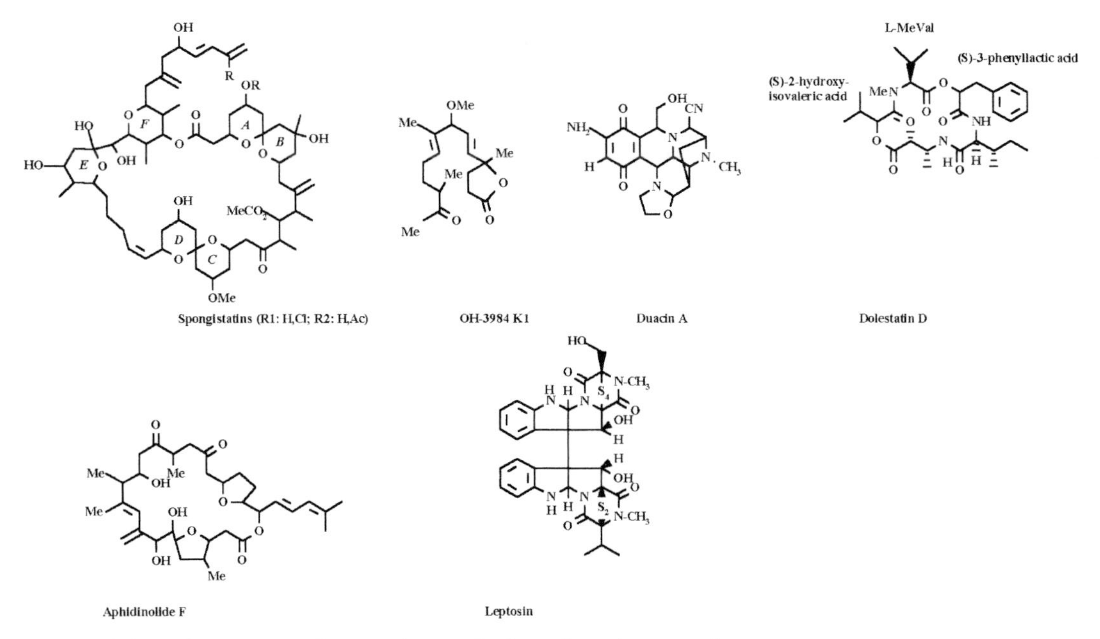

Fig. 21. Selected structures of cytotoxic metabolites of marine organisms.

a deeper insight into the mechanism of malignant cell growth. Moreover, their therapeutic potential is presently under investigation. It is likely that new antitumor drugs will be discovered in the future when new screening assays and natural sources become available.

Acknowledgement

We are deeply indebted to Mrs. HILTRUD KLOSE for typing and preparing the manuscript.

6 References

ABE, F., SHIBUYA, K., ASHIZAWA, Y., TAKAHASHI, K., HORINISHI, H., MATSUDA, A., ISHIZUKA, M., TAKEUCHI, T., UMEZAWA, H. (1985), Enhancement of antitumor effect of cytotoxic agents by bestatin, *J. Antibiot.* **38**, 411–414.

ABE, N., ENOKI, N., NAKAKITA, Y., UCHIDA, H., NAKAMURA, T. (1991), Isolation and characterization of atramycin B, new isotetracenone type antitumor antbiotic, *J. Antibiot.* **44**, 1025–1027.

ABE, N., ENOKI, N., NAKAKITA, Y., UCHIDA, H., NAKAMURA, T., MUREKATA, N. (1993a), Novel antitumor antibiotics septomycins. II. Isolation, physico-chemical properties and structure elucidation, *J. Antibiot.* **46**, 1536–1549.

ABE, K., YOSHIDA, M., NAOKI, H., HORINOUCHI, S., BEPPU, T. (1993b), Leptolstatin from *Streptomyces* sp. SAM 1595, a new GAP phase-specific inhibitor of the mammalian cell cycle. II. Physico-chemical properties and structure, *J. Antibiot.* **46**, 735–740.

ADAMSON, R. H., SIEBER, S. M., DOUROS, J. D. (1979), Development of actinomycin analogs. *Adv. Med. Oncol. Res. Educ. Proc. Int. Cancer Congr. 12th* (1979), Meeting Date 1978, Vol. 5, pp. 93–99. Oxford: Edn. B.W. Fox, Pergamon.

ADDESS, K. J., GILBERT, D. E., FEIGOU, J. (1992), Sequence specificity of quinoxaline antibiotics: a comparison of the binding of [N-MeCyS3] [N-MeCys7] tandem and echinomycin to DNA studied by proton NMR spectroscopy. *Struct. Funct. Proc. Conversation Discip. Biomol. Stereodyn. 7th* (1992), Meeting Date 1991, Vol. 1, pp. 147–164. (SARMA, R. H., SARMA, M. H., Eds.). New York: Adenine Press.

AGARWAL, R. P., BELL, R., LILLQUIST, A., MCCAFFEY, R. (1983), Purine metabolism in leukemia, *Ann. Natl. Acad. Sci.* **2**, 160–168.

AKINAGA, S., NOMURA, K., GOMI, K., OKABE, M. (1993), Diverse effects of indolecarbazole compounds on the cell cycle progression of *ras*-transformed rat fibroblast cells, *J. Antibiot.* **46**, 1767–1770.

ALLEN, R. E., LO, T. W. C., THOMALLY, P. J. (1993), Inhibitors of glyoxylase. I. Design, synthesis, inhibitory characteristcs and biological evaluation, *Biochem. Soc. Trans.* **21**, 535–540.

ALLGAIER, C., HERTTING, G. (1986), Polymyxin B, a selective inhibitor of protein kinase C, diminishes the release of noradrenaline and the enhancement of release caused by phorbol 12,13-dibutyrate, *Naunyn-Schmiedeberg's Arch. Pharmacol.* **333**, 218–221.

ANDERS, M. W. (1991), Glutathione-dependent bioactivation of xenobiotics: implications for mutagenicity and carcinogenicity, *Proc. Int. Symp. Princess Takamatsu Cancer Res. Fund* (1991) Vol. Date 190, 21 (Xenobiot. Cancer) 89–99.

ANDERSON, R. D., BERGER, N. A. (1994), Mutagenicity and carcinogenicity of topoisomerase interactive agents, *Mutat. Res.* **309**, 109–142.

ANDOH, T., KJELDSEN, E., BOUVEN, B. J., WESTERGAARD, O., ISHII, K., OKADA, K. (1993), Camptothecin-resistant DNA topoisomerase I, in: *DNA Topoisomerase in Cancer* (POTMESIL, M., KOHN, K. W., Eds.), pp. 249–295. New York: Oxford Univ. Press.

ANGANWUTAKU, I. O., ZIRBES, E., ROSAZZA, J. P. N. (1992), Isoflavonoids from streptomycetes: origins of genistein, 8-chlorogenistein, and 6,8-dichlorogenistein, *J. Nat. Prod.* **55**, 1498–1504.

AOYAMA, T., KOMIYAMA, F., YAMAZAKI, T., TATEE, T., ABE, F., MURAOKA, Y., NAGANAWA, H., AOYAGI, T., TAKEUCHI, T. (1993a), Benastatins C and D, new inhibitors of glutathione S-transferase, produced by *Streptomyces* sp. MI384-DF12: production, isolation, structure determination and biological activities, *J. Antibiot.* **46**, 712–718.

AOYAMA, T., ZHAO, W., KOJIMA, F., MURAOKA, Y., NAGANAWA, H., TAKEUCHI, T., AOYAGI, T. (1993b), Cysfluoretin, a new inhibitor of glutathione S-transferase produced by *Streptomyces* sp. MI384-DF-12, *J. Antibiot.* **46**, 1471–1474.

ARAI, H., KANDA, Y., ASHIZAWA, T., MORIMOTO, M., GOMI, K., KONO, M., KASAI, M. (1994), Mitomycin derivatives having unique condensed-ring structures, *J. Antibiot.* **47**, 1312–1321.

ARCAMONE, F. (1994), Design and synthesis of anthracycline and distamycin derivatives as new, sequence-specific DNA-binding pharmacological agents, *Gene* **149**, 57–61.

ASAKI, K., ONO, I., KUSUKABE, H., NAKAMURA, G., ISONO, K. (1981), Studies on differentiation-inducing substances of animal cells. I. Differenol A, a differentiation inducing substance against mouse leukemia cells, *J. Antibiot.* **34**, 919–920.

ASZALOS, A. (Ed.) (1988), Antitumor compounds of natural origin: chemistry and biology, Vol. II. Boca Raton, FL: CRC Press.

ATTA-UR-RACHMAN, A., IQBAL, Z. NASIR, H. (1994), Synthetic approaches to vinblastine and vincristine – anticancer alkaloids of *Catharanthus roseus. Stud. Nat. Prod. Chem.* **14** (Stereoselective Synthesis) (Pt. I.), 805–877.

BARKLEY, M. D., CHEATHAM, S., THURSTON, D. E., HURLEY, L. H. (1986), Pyrrolo[1,4]benzodiazepine antitumor antbiotics: evidence for two forms of tomaymycin bound to DNA, *Biochemistry* **25**, 3021–3031.

BARTEL, P. L., ZHU, C.-B., CAMPEL, Z. S., DOSCH, D. C., CONNORS, N. C., STROHL, W. R., BEALE, J. M., FLOSS, H. G. (1990), Biosynthesis of anthraquinones by interspecies cloning of actinorhodin biosynthesis genes in streptomycetes: clarification of actinorhodin gene functions, *J. Bact.* **172**, 4816–4826.

BASU, A., LAZO, J. S. (1994), Protein kinase C, in: *New Molecular Targets in Cancer Chemotherapy.* (KERR, D. J., WORKMAN, P., Eds.), pp. 121–14. Boca Raton, FL: CRC Press.

BEAUVILLE, D. L., KENIRY, M. A., SHAFER, R. H. (1990), NMR investigation of mithramycin A binding to d(ATGCAT)$_2$: a comparative study with chromomycin A$_3$, *Biochemistry* **29**, 9294–9304.

BECK, W. T., DANKS, M. K., WOLVERTON, J. S., KIM, R., CHEN, M. (1993) Drug resistance associated with altered DNA topoisomerase II, *Adv. Enzyme Regul.* **33**, 113–127.

BENTLER, J. A., CLARK, P., ALVAREDO, A. B. (1994), Esperamicin P, the tetrasulfide analog of esperamicin A1, *J. Nat. Prod.* **57**, 629–633.

BERGER, N. A., BERGER, S. J., GERSON, S. L. (1987), DNA repair, ADP-ribosylation and pyridine nucleotide metabolism as targets for cancer chemotherapy, *Anti-Cancer Drug Res.* **2**, 203–210.

BERRIDGE, M. J., IRVINE, R. F. (1984), Inositol trisphosphate, a novel second messenger in cellular signal transduction, *Nature* **312**, 315–321.

BERRIDGE, M. J., IRVINE, R. F. (1989), Inositol phosphates and cell signalling, *Nature* **341** (6239), 197–205.

BERTRAM, J. S. (1990), The chemoprevention of human cancer: an overview, in: *Effects of Therapy on Biology and Kinetics of the Residual Tumor.* Part A: *Preclinical Aspects,* pp. 345–360. New York, London: Wiley-Alan Liss.

BIEDLER, J. L., MYERS, M. B. (1993), Multidrug resistance (vinca alkaloids, actinomycin D, and anthracycline antibiotics), in: *Drug Resistance of*

Mammalian Cells, Vol. 2 (GUPTA, R. S., Ed.), pp. 57–88. Boca Raton, FL: CRC Press.

BILLINGS, C. P. (1993), Approaches to studying the target enzyme of anticarcinogenic protease inhibitors, in: *Protease Inhibitors as Cancer Chemotherapeutic Agents* (TROLL, W., KENNEDY, A. R., Eds.), pp. 191–198. New York: Plenum Press.

BLUMBERG, P. M., PETTIT, G. R., WARREN, B. S., SZALLASI, A., SCHUMAN, L. D., SHARKEY, N. A., NAKAMURA, H., DELL'AQUILA, M. L., DE ORIES, D. J. (1989), The protein kinase C pathway in tumor promotion, in: *Brain Carcinogenesis: Mechanisms and Human Relevance,* pp. 201–212. New York: Alan Liss.

BOGER, D. L., JOHNSON, D. S. (1996), CC-1065 and the duocarmycins: understanding their biological function through mechanistic studies, *Angew. Chem.* (Int. Edn. Engl.) **35**, 1438–1474.

BOROS, C., HAMILTON, S. M., KATZ, B., KULANTHAIVEL, P. (1994), Comparison of balanol from *Verticillium balanoides* and ophidiocordin from *Cordyceps ophioglassoides, J. Antibiot.* **47**, 1010–1016.

BORREL, M. N., PEREIRA, E., FIALLO, M. (1994), Analysis of multidrug transporters in living cells, *Met.-Based Drugs* **1**, 175–182.

BRADSHAW, R. A., PRENTIS, S. J. (1987), *Oncogenes and Growth Factors.* Amsterdam: Elsevier.

BRAND, M. E., PUETT, D., GAROLA, R., FENOLL, K., COVEY, D. F., ZIMNISKI, D. J. (1988), Aromatase and aromatase inhibitors: from enzymology to selective chemotherapy, in: *Hormones, Cell Biology and Cancer: Perspectives and Potentials,* pp. 65–84. New York: Alan Liss.

BURCHELL, B., BAIRD, S., COUGHTRIE, M. W. H. (1991), The role of xenobiotic glucuronidating enzymes in drug resistance of tumor cells and tissues, *Proc. Int. Symp. Princess Takamatsu Cancer Res. Found.,* Volume Date 1990, 21 (Xenobiot. Cancer) 263–275.

BURCK, K. B., LIN, E. T., LARRICK, J. W. (1988), *Oncogenes.* New York, Heidelberg: Springer-Verlag.

BUTA, J. G., KALINSKI, A. (1988), Camptothecin and other plant growth regulators in higher plants with antitumor activity, *ACS Symp. Ser.* **380**, 294–304.

BUTLER, J., HAEY, B. M. (1987), Are reduced quinones necessarily involved in the antitumor activity of quinone drugs? *Br. J. Cancer* (Suppl.) **55**, 53–59.

CAI, Y., FREDENHAGEN, A., HUG, P., MEYER, T., PETER, H. H. (1996), Further minor metabolites of staurosporin produced by a *Streptomyces longisporoflavus* strain, *J. Antibiot.* **49**, 519–526.

CAMPLEJOHN, R. S. (1980), A critical review of the use of vincristin, *Cell Tissue Kinet.* **13**(3), 327–335.

CAPRANICO, G., ZUNINO, F. (1990), Structural requirements for DNA topoisomerase II inhibition by anthracyclines, *Jerusalem Symp. Quantum Chem. Biochem.* **23** (Mol. Basis. Specif. Nucleic Acid Drug Interact.), 167–176.

CARNEY, J. C., SCHEUER, P. J. (1993), Polohuanone E, a topoisomerase-II inhibitor with selective lung tumor cytotoxicity from the Pohnpei sponge *Dysidea* sp., *Tetrahedron Lett.* **34**, 3727–3730.

CASSIDY, J. M., DOUROS, J. D. (1988), *Anticancer Agents Based on Natural Product Modes.* New York: Academic Press.

CHAKRABORTY, A. K., MAJUMDER, HEMATA, K., HODGSON, C. P. (1994), DNA topoisomerase inhibitors: potential uses in molecular medicine, *Expert Opin. Ther. Pat.* **4**, 655–668.

CHEN, A. Y., LEROY, L. F. (1994), DNA topoisomerases: essential enzymes and lethal targets, *Ann. Rev. Pharmacol. Toxicol.* **34** ,191–218.

CHIMURA, H., ISHIZUKA, M., HAMADA, H., HORI, S., KIMURA, K., IWANAGA, J., TAKEUCHI, T., UMEZAWA, H. (1968), A new antibiotic, macromomycin, exhibiting antitumor and antimicrobial activity, *J. Antibiot.* **21**, 44–49.

CHINO, M., NISHIKAWA, K., UMEKITA, M., HAYASHI, C., YAMAZAKI, T., TSUCHIDA, T., SAWA, T., HAMADA, M., TAKEUCHI, T. (1996), Heliquinomycin, a new inhibitor of DNA helicase, produced by *Streptomyces* sp. MJ929–SF2. I. Taxonomy, production, isolation, physico-chemical properties and biological activities, *J. Antibiot.* **49**, 752–757.

CHU, M. (1994), A novel class of antitumor metabolites from the fungus *Nattrasia mangifera, Tetrahedron Lett.* **35**, 1343–1346.

CHU, M., TRUUMEIS, I., ROTHOFSKY, M. L., PATEL, M. G., GENTILE, F., DAS, R. P., PUAR, S. M., UN, S. L. (1995), Inhibition of c-*fos* protooncogene induction by Sch 52900 and Sch 52901, novel diketopiperazines produced by *Gliocladium* sp., *J. Antibiot.* **48**, 1440–1445.

COLL, M., FREDERICH, C. A., WONG, A., RICH, A. (1987), A bifurcated hydrogen-bonded conformation in the d(A,T) base pairs of the DNA dodecamer d ($C6CA_3T_3GCG$) and its complex with distamycin, *Proc. Natl. Acad. Sci. USA* **84**, 8385–8389.

CROW, R. T., CROTHERS, D. M. (1994), Inhibition of topoisomerase I by anthracycline antbiotics. Evidence for general inhibition of topoisomerase I by DNA-binding agents, *J. Med. Chem.* **37**, 3191–3194.

CUI, C., UBUKATA, M., KAKEYA, H., ONOSE, R., OKADA, G., TAKAHASHI, I., ISONO, K., OSADA, H. (1996a), Acetophtalidin, a novel inhibitor of mammalian cell cycle, produced by a fungus isolated from a sea sediment, *J. Antibiot.* **49**, 216–219.

CUI, C., KAKEYA, H., OSATA, H. (1996b), Novel mammalian cell cycle inhibitors, tryprostatins A, B, and other diketopiperazines produced by *Aspergillus fumigatus.* II. Physico-chemical properties and structures, *J. Antibiot.* **49**, 534–540.

CURRAU, D. P. (1993), The camptothecins: a reborn family of antitumor agents, *J. Clin. Chem. Soc.* (Tapei) **40**(1), 1–6.

CURRAU, D. P., SISKO, J., YESKE, P. E., LIN, H. (1993), Recent applications of radical reactions in natural product synthesis, *Pure Appl. Chem.* **65**(6), 1153–1159.

DERVAN, P. D., SCOTT YOUNGQUIST, R., SLUKA, J. P. (1987), in: *Stereochemistry of Organic and Bioorganic Transformations* (BARTHMANN, W., SHARPLESS, K. B., Eds.), pp. 221–234. Weinheim: VCH.

DIETEL, M. (1991), What's new in cytostatic drug resistance and pathology, *Pathol. Res. Pract.* **187**, 892–905.

DUMITRIU, S. (1996), *Polysaccharides in Medicinal Applications.* New York, Basel, Hongkong: Marcel Dekker.

ECKARDT, K., WAGNER, C. (1988), Biosynthesis of anthracyclinones, *J. Basic Microbiol.* **28**, 137–144.

EDO, K., SAITO, K., AKIYAMA-MURAI, Y., MIZUGAKI, M., KOIDE, Y., ISHIDA, N. (1988), An antitumor polypeptide antibiotic neocarzinostatin: the mode of apo-protein – chromophore interaction, *J. Antibiot.* **41**, 554–562.

EL KHADEM, H. S. (Ed.) (1982), *Anthracycline Antibiotics.* New York: Academic Press.

END, D. W. (1987), Non-selective inhibition of mammalian protein kinases by flavinoids *in vitro, Res. Commun. Chem. Pathol. Pharmacol.* **56**, 76–86.

FEIG, A. L., LIPPARD, S. J. (1994), Reactions of non-hemic iron (II) centers with dioxygen in biology and chemistry, *Chem. Rev.* (Washington D.D.) **94**, 759–805.

FERDINAND, F. J. (1989), Mechanism of tumor induction by human retroviruses, *Arzneim. Forsch.* **39**(6), 735–740.

FIJII, N., YAMASHITA, Y., CHIBA, S., UOSAKI, Y., SAITOH, Y., TUJI, Y., NAKANO, H. (1993), UCE 6, a new antitumor antibiotic with topoisomerase I medaited DNA cleavage activity from Actinomycetes, *J. Antibiot.* **46**, 1173–1174.

FIJII, F., YAMASHITA, Y., CHIBATA, S., SAITOH, Y., NAKANO, H. (1994), UCE 1022, a new antitumor antibiotic with topoisomerase I mediated DNA cleavage activity from *Paecilomyces, J. Antibiot.* **47**, 949–951.

FISCHER, E. H., CHARBONNEAN, H., TACHS, N. K. (1991), Protein tyrosine phosphatases: a diverse family of intracellular and transmembrane enzymes, *Science* **253**, 401–406.

FLEISCHMANN, F. F., CHAHWALA, S. B., COUNTLEY, L. (1986), Ras transformed cells: altered levels of phosphatidylinositol 4,5-bisphosphate and catabolites, *Science* **232**, 407–410.

FOMICHOVA, E. V. (1992), Design of enzyme inhibitors as drugs, *J. Antibiot.* **45**, 1185–1186.

FORNICA, J. (1977), The biogenesis of actinomycins, *Actinomycetes Relat. Org.* **12**(4), 12–20.

FOSTEL, J., SHEN, L. L. (1994), *DNA Topoisomerases, Design of Enzyme Inhibitors as Drugs,* Vol. 2 (SANDLER, M., SMITH, H. J., Eds.), pp. 564–624. Oxford: Oxford University Press.

FOX, K. R. (1990), Footprinting studies of the interaction of quinomycin antibotic UK 63052 with DNA: comparison with echinomycin, *J. Antibiot.* **43**, 1307–1315.

FOX, B. W. (1991), Natural products in cancer treatment from bench to the clinic, *Trans. R. Soc. Trop. Med. Hyg.* **85**, 22–25.

FRANZ, G. (1990), Biogene Cytostatica, *PharmuZ* **6**, 257–262.

FREDENHAGEN, A., HUG, P., SAUTER, H., PETER, H. H. (1995), Paeciloquinones A, B, C, D, E and F: New potent inhibitors of protein tyrosine kinase produced by *Paecilomyces carneus.* II. Characterization and structure determination, *J. Antibiot.* **48**, 199–204.

FUNABASHI, Y., HORIGUCHI, T., IINUMA, S., TANIDA, S., HORADA, S. (1994), TAN-1496A, C and E, diketopiperazine antibiotics with inhibitory activity against mammalian DNA topoisomerase I, *J. Antibiot.* **47**, 1202–1218.

FURK, J. L., ABDUL HAMIED, T. A., PARKER, D. (1989), Potentiation of interleukin-2 release by anticancer drugs, *Agents Actions* **26**(1–2), 156–157.

GAAL, J. C., PEARSON, C. K. (1986), Covalent modification of proteins by ADP-ribosylation, *TIBS* **11**, 171–175.

GALE, E. F., CUNDLIFFE, E., REYNOLDS, P. E., RICHMOND, M. H., WARING, M. J. (1981), *The Molecular Basis of Antibiotic Action.* New York, Sydney, Toronto: Wiley.

GIACCONE, G. (1994), DNA topoisomerases and topoisomerase inhibitors, *Pathol. Biol.* **42**, 346–352.

GIRI, S., WANG, Q. (1989), Mechanism of bleomycin-induced lung injury, *Comments Toxicol.* **3**, 145–179.

GOLDBERG, H. I. (1986), Molecular mechanisms of DNA sugar damage by antitumor antibiotics, *Pontif. Acad. Sci. Scr. Varia* **70**, 425–462.

GOLDBERG, H. I. (1991), Mechanism of neocarzinostatin action: role of DNA microstructure in determinaton of chemistry of bistranded oxidative damage, *Acc. Chem. Res.* **24**, 192–196.

GOLDBERG, H. I., KAPPA, L. S., CHIN, D. H., LEE, S. H. (1989), Bistranded oxidative DNA sugar damage by targeted antibiotic diradicals, *New Methods Drug Res.* **3**, 27–42.

GOORMAGHTIGH, E., RUYSSCHAERT, J. M. (1984), Anthracycline glycoside-membrane interactions, *Biochim. Biophys. Acta* **779**, 271–288.

GREENSTEIN, M., MAIESE, W. M. (1984), Prescreens for novel antineoplastic agents, *Dev. Ind. Microbiol.* **25**, 267–275.

GREVER, M. R., SCHEPARTZ, S. A., CHAPNER, B. A. (1992), The national cancer institute: cancer drug development and development program, *Semin. Oncol.* **19**, 622–638.

GUENGRICH, F. P. (1988), Roles of cytochrome P-450 enzymes in chemical carcinogenesis and cancer chemotherapy, *Cancer Res.* **48**, 2946–2954.

GUERITTE-VOEGELEIN, F., GUENARD, D., DUBOIS, J., WAHL, A., POITIER, P. (1994), Chemical and biological studies in the taxol (pacitaxel) and taxotere (docetaxel) series of new antitumor agents, *J. Pharm. Belg.* **49**, 193–205.

GUNDA, G. I., ALI, S. M., ZYGMUNT, J., JAGASINGHE, L. R. (1994), Taxol: a novel antitumor agent, *Expert Opin. Ther. Pat.* **4**, 109–120.

GUNICKE, H., HOFMANN, J. (1992), Cytotoxic and cytostatic effects of antitumor agents induced at the plasma membrane level, *Pharmacol. Ther.* **55**, 1–30.

GUTTERIDGE, J. M. C. (1992), Iron and oxygen radicals in brain, *Ann. Neurol.* **32** (Suppl.), 516–521.

HAGIWARA, N., IRIE, K., FUNAKI, A., HAYASHI, H., ARAI, M., KOSHIMIZU, K. (1988), Structure and tumor-promoting activity of new teleocidin-related metabolites (blastmycetins) from *Streptoverticillium blastmyceticum, Agric. Biol. Chem.* **52**, 641–648.

HANADA, M., OKKUMA, H., YONEMOTO, T., TONITA, T, OHBAYASHI, M., KAMEI, H., HIYAKI, T., KONISHI, M., KAWAGUCHI, H. (1991a), Maduropeptin, a complex of new macromolecular antitumor antibiotic, *J. Antibiot.* **44**, 403–414.

HANADA, M., KANEDA, K., NISHIYAMA, Y., HOSHINO, Y., KONISHI, M., OKI, T. (1991b), Hydramycin, a new antitumor antbiotic. Taxonomy, isolation, physico-chemical properties, structure and biological activity, *J. Antibiot.* **44**, 824–830.

HARA, M., ASANO, K., KAWAMOTO, I., TAKIGUCHI, T., KATSUMATA, S., TAKAHASHI, K., NAKANO, H. (1990), Leinamycin, a new antitumor antibiotic from *Streptomyces.* Producing organism, fermentation and isolation, *J. Antibiot.* **42**, 1768–1774.

HARADA, K., MAGUMI, T., SHIMADA, T., SUZUKI, M. (1993), Occurrence of four depsipeptides aerugicopeptins, together with microcystins from toxic cyanobacteria, *Tetrahedron Lett.* **34**, 6091–6094.

HARPER, D. E., WELCH, D. R. (1992), Isolation, purification, synthetics and antiinvasive/antimetastatic activity of U-77863 and U-77864 from *Streptomyces griseoluteus* strain WS6724, *J. Antibiot.* **45**, 1827–1836.

HART, R. W., TURTURRO, A. (1988), Current views of the biology of cancer, in: *Carcinogen Risk Assessment* (TRAVIS, C. C., Ed.), New York: Plenum Publishing Corp.

HARUYAMA, H., OHKUMA, Y., NAGAKI, H., OGITA, T., TANAKI, K., KINOSHITA, T. (1994), Matlystatins, new inhibitors of type IV collagenases from *Actinomadura atramentaria.* III. Structure elucidation of matlystatins A to F, *J. Antibiot.* **47**, 1473–1480.

HAYAKAWA, Y., NAKAGAWA, M., ANDO, T., SHIMAZU, A., SETO, H., OTAKE, N. (1982), Studies on the differentiation inducers of myceloid leukemic cells. I. Latosillan, a new inducer of the differentiation of M1 cells, *J. Antibiot.* **35**, 1252–1254.

HAYAKAWA, Y., NAKAGAWA, M., KAWAI, H., TANABE, K., NAKAYAMA, H., SHIMAZU, A., SETO, H., OSAKE, N. (1983), Studies on the differentiation of inducers of myeloid leukemic cells. III. Spicamycin, a new inducer of differentiation of HL-60 human promyeolocytic leukemia cells, *J. Antibiot.* **36**, 934–937.

HAYAKAWA, Y., HA, S., KIM, Y. J., FURIKATA, K., SETO, H. (1991), Hamatorubicins A, B, C and D, new isotetracenone antibiotics effective against multidrug-resistant tumor cells, *J. Antibiot.* **44**, 1179–1186.

HAYAKAWA, Y., ISHIGAMI, K., SHIN-YA, K., SETO, H. (1994), Menoxymycins A and B, antitumor antibiotics generating active oxygen in tumor cells, *J. Antibiot.* **47**, 1344–1347.

HECHT, S. M. (1994), RNA degradation by bleomycin, a naturally occurring bioconjugate, *Bioconjugate Chem.* **5**, 513–526.

HECHT, S. M., BERRY, D. E., MACKENZIE, L. J., DUSOY, K. W., NASUTI, C. A. (1992), A strategy for identifying novel mechanistically unique inhibitors of topoisomerase I, *J. Nat. Prod.* **55**, 401–413.

HEDGE, V. R., MILLER, J. R., PATEL, M. G., KING, A. H., PUAR, M. S., HORAN, A., HART, R., YARBOROUGH, R., GULLO, V. (1993), Sch-45752 – an inhibitor of calmodulin-sensitive cyclic nucleotide phosphodiesterase activity, *J. Antibiot.* **46**, 207–213.

HEINSTEIN, P. F., CHANG. C. J. (1994), *Annu. Rev. Plant. Physiol. Plant. Mol.-Biol.* **45**, 663–674.

HENSENS, O. D., GOLDBERG, I. H. (1989), Mechanism of activation of the antitumor antibiotic neocarzinostatin by mercaptan and sodium borohydride, *J. Antibiot.* **42**, 761–767.

HIDA, T., MERROI, M., TANIDA, S., HARADA, S. (1994), Structures of duacin A_1 and B_1 new naphthyridinomycin-type antitumor antibiotics, *J. Antibiot.* **47**, 917–921.

HILL, B. T. (1986), Resistance of mammalians tumor cells to anticancer drugs: mechanisms and concepts relating specificity to methotrexate and vincristine, *J. Antimicrob. Chemother.* **18** (Suppl. B), 61–73.

HOCHLOWSKI, J. E., MULLALLY, M. M., SPANTON, S. G., WHITTERN, D. N., HILL, P., MCALPINE, J. B. (1993), Reversion of multi-drug resistance to antitumor antibiotics was also reported for 5-*N*-acetylardeemin from *Aspergillus fischeri, J. Antibiot.* **46**, 380–386.

HOFSTEAD, S. J., MATSON, J. A., MALACKO, A. R., MARQUARDT, H. (1992), Kedarcidin, a new chromoprotein antitumor antibiotic. II. Isolation, purification and physico-chemical properties, *J. Antibiot.* **45**, 1250–1292.

HOLLIS-SHOEWALTER, H. D. (1988), Design, tumor biology, and biochemical pharmacology of anthrapyrazoles, in: *Anthracycline and Anthracenedione-Based Anticancer Agents* (LOWN, J. W., Ed.). Amsterdam, New York: Elsevier.

HOLLIS-SHOEWALTER, H. D., BUNGE, R. H., FRENCH, J. C., HURLEY, T. R., LEEDS, R. L., LEJA, B., MCDONNELL, P. D., EDMUNDS, C. R. (1992), Improved production of pentostatin and identification of fermentation metabolites, *J. Antibiot.* **45**, 1914–1918.

HORI, Y., ABE, Y., EZAKI, M., GOTO, T., OKUNAMA, M., KOHSAKA, M. (1993a), R1128 substances, novel non-steroidal estrogen-receptor antagonists produced by a *Streptomyces.* I. Taxonomy, fermentation, isolation and biological properties, *J. Antibiot.* **46**, 1055–1062.

HORI, Y., ABE, Y., NISHIMURA, M., GOTO, T., OKUHARA, M., KOHSAKA, M. (1993b), R1128 substances, novel non-steroidal estrogen-receptor antagonists produced by a *Streptomyces.* III. Pharmacological properties and antitumor activities, *J. Antibiot.* **46**, 1069–1075.

HORI, Y., ABE, Y., SHIGEMATSU, N., GOTO, T., OKUNARA, M., KOHSAKA, M. (1993c), Napyradiomycins A and B: non-steroidal estrogen receptor antagonists produced by a *Streptomyces, J. Antibiot.* **46**, 1890–1893.

HOSOKAWA, N., IIMURA, H., NAGANAWA, H., HAMADA, M., TAKEUCHI, T. (1993), A new antibiotic, structurally related to leptomycin A, flattens the morphology of r-ras^{ts} NRK cells, *J. Antibiot.* **46**, 676–679.

HSIEH, T. S. (1992), DNA topoisomerases, *Curr. Opin. Cell Biol.* **4**, 396–400.

ICHIMURA, M., OGAWA, T., TAKAHASHI, K., KOBAYASHI, E., KAWAMOTA, I., YASUZAWA, T., TAKAHASHI, I., NAKANO, H. (1990), Duocarmycin SA, a new antitumor antibiotic from *Streptomyces* sp., *J. Antibiot.* **43**, 1037–1038.

ICHIMURA, M., OGAWA, T., KATSUMATA, S., TAKAHASHI, K., TAKAHASHI, I., NAKANO, H. (1991), Duocarmycin S, new antitumor antibiotics produced by streptomycetes: producing organism and improved production, *J. Antibiot.* **44**, 1045–1053.

IIDA, T., KOBAYASHI, E., YOSHIDA, M., SANO, H. (1989), Calphostins, novel and specific inhibitors of protein kinase. II. Chemical structures, *J. Antibiot.* **42**, 1475–1481.

IIDA, K., ISHII, T., HIRAMA, M., OTANI, T., MINAMI, Y., YOSHIDA, K. (1993), Synthesis and absolute stereochemistry of the aminosugar moiety of antibiotic C-1027 chromophore, *Tetrahedron Lett.* **34**, 4079–4082.

IKEKAWA, T., ASARI, T., MANABE, T., UMEJI, M., YANOMA, S. (1980), Canacelunin, a cancer cell agglutinin from *Streptomyces* spec., *J. Antibiot.* **33**, 776–777.

IMOTO, M., UMEZAWA, K., ISSHIKI, K., KUNIMOTO, S., SAWA, T., TACKEUCHI, T., UMEZAWA, H. (1987), Kinetic studies of tyrosine kinase inhibition by erbstatin, *J. Antibiot.* **40**, 1471–1473.

IMOTO, M., YAMASHITA, T., SAWA, T., KURASAWA, S., NAGANAWA, H., TAKEUCHI, T., BAOQUAN, Z., UMEZAWA, K. (1988), Inhibition of cellular phosphatidylinositol turnover by *psi*-tectorigen, *FEBS Lett.* **230**, 43–46.

IMOTO, M., UMEZAWA, K., TAKAHASHI, Y., NAGANAWA, H., JITAKA, Y., NAKAMURA, H., KAZUMI, Y., SASAKI, Y., HAMADA, M., SAWA, T., TAKEUCHI, T. (1990), Isolation and structure determination of inostamycin, a novel inhibitor of phosphatidylinositol turnover, *J. Nat. Prod.* **53**, 825–829.

IMOTO, M., SHIMURA, N., UMEZAWA, K. (1991), Inhibition of epidermal growth factor-induced activation of phospholipase C by *psi*-tectorigen, *J. Antibiot.* **44**, 915–917.

ISHIDA, N., MIYAZAKI, K., KUMAGAI, K., RIKIMARU, K. (1965), Neocarzinostatin, an antitumor antibiotic of high molecular weight. Isolation, physicochemical properties and biological activities, *J. Antibiot. Ser. A* **18**, 68–76.

ISHIGAMI, K., HAYAKAWA, Y., SETO, H. (1994), Cororubicin, a new anthracycline antibiotic generating active oxygen in tumor cells, *J. Antibiot.* **47**, 1219–1225.

ISONO, K. (1988), Nucleoside antibiotics: Structure, biological activity and biosynthesis, *J. Antibiot.* **41**, 1711–1752.

IWAI, Y., NAKAGAWA, A., SADAKANE, N., ŌMURA, S. (1980), Herbimycin B, a new benzoquinoid ansamycin with anti-TMV and herbicidal activities, *J. Antibiot.* **38**, 1114–1119.

IWASAKI, S. (1989), *Rhizoxin, an Inhibitor of Tubulin. Novel Microbial Products for Medicine and Agriculture* (DEMAIN, A. L., SAMKUTI, G. A., HUNTER-CEVERA, J. C., ROSSMORE, H. W., Eds.). Washington, D. C.: Society for Industrial Microbiology.

IZUMI, S., NAKAHARA, K., GOTOH, T., HASHIMOTO, S., KINO, T., OKUHARA, M., ADRI, V., IMANAKA, H. (1983), Antitumor effects of novel immunoactive peptides, FK-156, and its synthetic derivatives, *J. Antibiot.* **36**, 566–574.

JACHOWSKI, S., RETTENMEIR, C. W., SHERR, C. J., ROCK, C. O. (1986), A guanine nucleotide-dependent phosphatidylinositol 4,5-diphosphate phospholipase C in cells transformed by the V-fms and V-fes oncogenes, *J. Biol. Chem.* **261**, 4978–4985.

JENSEN, P. R., FENICAL, W. (1994), Strategies for the discovery of secondary metabolites from marine bacteria: ecological perspectives, *Annu. Rev. Microbiol.* **48**, 559–584.

JOEL, P. S. (1994), Taxol and taxotere: from yew tree to tumor cell, *Chem. Ind.* (London) (**5**), 172–175.

JOHDO, O., TONE, H., OKAMOTO, R., YOSHIMOTO, A., NAKAGAWA, H., SAWA, T., TAKEUCHI, T. (1991a), Anthracycline metabolites from *Streptomyces violaceus* A262. V. New anthracycline alldimycin A: a minor component isolated from obelmycin beer, *J. Antibiot.* **44**, 1160–1164.

JOHDO, O., WATANABE, Y., ISHIKURA, T., YOSHIMOTO, A., NAGANAWA, H., SAWA, T., TAKEUCHI, T. (1991b), Anthracycline metabolites from *Streptomyces violaceus* A262. II. New anthracycline epelmycins produced by a blocked mutant strain SU2-730, *J. Antibiot.* **44**, 1121–11.

JOHDO, O., TONE, H., OKAMOTO, R., YOSHIMOTO, A., NAGANAWA, H., SAWA, T., TAKEUCHI, T. (1991c), Anthracycline metabolites from *Streptomyces violaceus* A262. IV. New anthracycline yellamycins produced by a variant strain SC-7, *J. Antibiot.* **44**, 1155–1159.

KAHN, P., GRAF, T. (1986), *Oncogenes and Growth Control.* Berlin, Heidelberg: Springer-Verlag.

KAKEYA, H., IMOTO, M., TAKAHASHI, Y., NAGANAWA, H., TAKEUCHI, T., UMEZAWA, K. (1993), Dephostatin, a novel protein tyrosine phosphatase inhibitor produced by *Streptomyces.* II. Structure determination, *J. Antibiot.* **46**, 1716–1719.

KALKEN VAN, C. K., HOEVEN VAN DER, J. J. M., JONG DE, J. et al. (1991), Bepridil in combination with anthracyclines to reverse anthracycline resistance in cancer patients, *Eur. J. Cancer* **27**, 739–744.

KALMAN, T. I. (1989), Inhibition of thymidylate synthase cycle of leukemia cells by antifolates: the usefulness of cellular tritium assay, in: *Chem. Biol. Pteridines Proc. Int. Symp. Pteridines Folic Acid Deriv. 9th (1980) Meeting* Date 1989 (CURTINS, H., GHISLA, S., BLAN, N., Eds.), pp. 1192–1197. Berlin: DeGruyter.

KAMAI, F., ISSHIKI, K., UMEZAWA, Y., MORISHIMA, H., NAGANAWA, H., TAKITA, T., TAKEUCHI, T., UMEZAWA, H. (1985), Vanoxonin, a new inhibitor of thymidylate synthetase. II. Structure determination and total synthesis, *J. Antibiot.* **38**, 31–38.

KAMEI, H., OKA, M., HAMAGISHI, Y., TOMITA, K., KONISHI, M., OKI, T. (1990), Piperafizines A and B, potentiators of cytotoxocity of vincristine, *J. Antibiot.* **43**, 1918–1920.

KAMEI, H., NISHIYAMA, Y., TAKAHASHI, A., OBI, Y., OKI, T. (1991), Dynemicins, new antibiotics with the 1,5-diyn-3-ene and anthraquinone subunit. II. Antitumor activity of dynemicin A and its triacetyl derivative, *J. Antibiot.* **44**, 1306–1311.

KAMISHOHARA, M., KAWAI, H., ODAGAWA, A., ISOE, T., MOCHIZUKI, J.-I., UCHIDA, T., HAYAKAWA, Y., SETO, H., TSURUO, T., OTAKE, N. (1994), Antitumor activity of SPM VIII, a derivative of the nucleoside antibiotic spicamycin, against human tumor xenografts, *J. Antibiot.* **47**, 1305–1312.

KANEDA, S., YOUNG, C., YAZAWA, K., TAKAHASHI, K., MIKAMI, Y., ARAI, T. (1987), Biological activities of newly prepared saframycins, *J. Antibiot.* **40**, 1640–1642.

KANEKO, T., WONG, H., UTRIG, J., SCHURIG, J., DOYLE, T. W. (1990), Water soluble derivatives of rebeccamycin, *J. Antibiot.* **43**, 125–127.

KAO, J. W. Y., COLLINS, J. L. (1989), A rapid *in vitro* screening system for the identification and evaluation of anticancer drugs, *Cancer Invest.* **7**, 303–311.

KARWOWSKI, J. P., JACKSON, M., SUNGA, G., SHELDON, P., PODDIG, J. B., KOHL, W. L., KADAM, S. (1994), Dorrigocins: novel antifungal antibiotics that change the morphology of ras-transformed N/H/3T3 cells to that of nomal cells. I. Taxonomy of the producing organism, *J. Antibiot.* **45**, 862–689.

KASAI, M., KONO, M., SHIRAHATA, K. (1991), The derivation of a novel mitomycin skeleton: 3α-alkoxymitomycin, *J. Antibiot.* **44**, 301–308.

KATAGIRI, K. (1975), Quinoxaline, in: *Antibiotics,* Vol. 3, *Mechanism of Action of Antimicrobial and Antitumor Agents* (CORCORAN, J. W., HAHN, F. E., Eds.). Berlin, Heidelberg, New York: Springer-Verlag.

KATO, Y., OSAWA, Y., IMADA, T., IWASAKI, S., SHIMAZAKI, N., KOBAYASHI, T., KOMAI, T. (1991), Studies on macrocyclic lactone antibiotics. XIII. Anti-tubulin activity and cytotoxicity of rhizoxin derivates: synthesis of a photoaffinity derivate, *J. Antibiot.* **44**, 66–75.

KAWADA, S., YAMASHIDA, Y., UOSAKI, Y., GOMI, K., IWASAKI, T., TAKIGUCHI, T., NAKANO, H. (1992a), UCT4B, a new antitumor antibiotic with topoisomerase II metiated DNA cleavage activity, from *Streptomyces* sp., *J. Antibiot.* **45**, 1182–1184.

KAWADA, M., SUMI, S., UMEZAWA, K., INONYE, S., SAWA, T., SETO, H. (1992b), Circumvention of multidrug resistance in human carcinoma KB cells by polyether antibiotics, *J. Antibiot.* **45**, 556–562.

KAWASE, Y., TAKAHASHI, M., TAKATSU, T., ARAI, M., NAKAJIMA, M., TANZAWA, K. (1996), A-72363 A-1, A-2, and C, novel heparanase inhibitors from *Streptomyces nobilis* SANK 60192. II. Biological activities, *J. Antibiot.* **49**, 61–64.

KAWASHIMA, A., NAKAGAWA, M., HAYAKAWA, Y., KAWAI, H., SETO, H., OTAKE, N. (1983), Studies on differentiation inducers of myeloid leukemia cells. II. Citrinin, a new inducer of the differentiation of M1 cells, *J. Antibiot.* **36**, 173–174.

KAWATSU, M., YAMASHITA, T., OSONO, M., ISHIZUKA, M., TAKEUCHI, T. (1993), T cell activation by conagenin in mice, *J. Antibiot.* **46**, 1687–1691.

KAWATSU, M., YAMASHITA, T., ISHIZUKA, M., TAKEUCHI, T. (1994), Effect of conagenin of thrombocytopenia induced by antitumor agents in mice, *J. Antibiot.* **47**, 1123–1129.

KEFRI, K., KONDO, H., YOSHINORI, T., ARAKAWA, H., NAKAGAMA, S., SATOH, F., KAWAMURA, K., OKURA, A., SUDA, H., OKANISHI, M. (1991), A new antitumor substance, BE-13793c produced by a streptomycete. Taxonomy, fermentation, isolation structure determination and biological activity, *J. Antibiot.* **44**, 723–728.

KELLEY, S. L., BASU, A., TEICHER, B. A., HACKER, M. P., HAMER, P. H., LAZO, J. S. (1988), Overexpression of metallothionein confers resistance to anticancer drugs, *Science* **241**(4874), 1813–1815.

KHAN, S. H., MALTA, K. L. (1992), *Recent Advances in the Development of Potential Inhibitors of Glycosyltransferase. Glycoconjugates* (ALLEN, H. J., KISAILUS, E. C., Eds.), pp. 361–378. New York: Marcel Dekker.

KHOKHLOV, A. S., CHERCHES, B. Z., RESHETOV, P. D., SMIRNOVA, G. M., SOROKINA, I. B., PROKOPTZEWA, T. A., KOLODITSKAYA, T. A., SMIRNOV, V. V., NAVASHIN, S. M., FOMINA, I. P. (1969), Physico-chemical and biological studies on actinoxanthin, an antibiotic from *Actinomyces globisporus* 1131, *J. Antibiot.* **22**, 541–544.

KHOSRAVI-FAR, R., LOX, A. D., KATO, K., DER, C. J. (1992), Protein prenylation: key to ras function and cancer intervention, *Cell Growth Differ.* **3**, 461–469.

KIKUCHI, H., TETSUKA, T. (1992), On the mechanism of lipoxygenase-like action of bleomycin–iron complexes, *J. Antibiot.* **45**, 548–555.

KIM, Y. J., FURIKATA, K., SHIMAZU, A., FURIKATA, K., SETO, H. (1991), Isolation and structural elucidation of sekothrixide, a new macrolide effective to overcome drug-resistance of cancer cells, *J. Antibiot.* **44**, 1280–1282.

KINGSTON, D. G. I. (1994), Taxol: the chemistry and structure–activity relationship of a novel anticancer agent, *Trends Biotechnol.* **12**, 222–227.

KITAHARA, M., ISHII, K., KUMADA, Y., SHIRAISHI, T., FURUTA, T., MIWA, T., KAWAHANADA, H., WATANABE, K. (1985), 7-hydroxyguanine, a novel antimetabolite from a strain of *Streptomyces purpurasceus.* I. Taxonomy of the producing organism, fermentation, isolation and biological activity, *J. Antibiot.* **38**, 972–978.

KIWASHIMA, J., ITO, F., KATO, T., NIWANO, M., KOSHINO, H., URAMOTO, M. (1994), Antitumor activity of heptelidic acid chlorohydrin, *J. Antibiot.* **47**, 1562–1563.

KLEGEMANN, M. E. (1993), Use of bacterial immunomodulators for the treatment of cancer, in: *Recent Adv. Pharm. Ind. Biotechnol. Minutes Int. Pharm. Technol. Symp., 6th* (1993), Meeting Date 1992 (HINCAL, A. A., KAS, S. H., Eds.), pp. 179–195. Paris: Santé.

KNUGH, T. R., HOOK, J. W. III., BALAKRISHNAN, M. S., CHEN. F. M. (1980), Spectroscopic studies of actinomycin and ethidium-complexes with deoxyribonucleic acids, in: *Nucleic Acids Geometry and Dynamics* (SARMA, R., Ed.), pp. 351–366. Elmsford, NY: Pergamon Press.

KOAMURA, M., YOSHIMURA, Y., MATSUMOTO, K., TERUI, Y. (1988), New amedic acid antibiotics. II. Structure elucidation, *J. Antibiot.* **41**, 68–72.

KOBAYASHI, J., TSUDA, M., ISHIBASHI, M., SHIGEMORI, H., YAMASHI, T., HIROTA, H., SAZAKI, T. (1993), Amphidinolide F, a new cytotoxic macrolide from the marine dinoflagellate *Amphidinum* spec., *J. Org. Chem.* **58**, 2645–2646.

KOGIRI, K., ARAKAWA, H., SATOH, F., KAWAMURA, K., OKURA, A., SUDA, H., OKANISHI, M. (1991), New antitumor substances, BE-12406 A and BE-12406 B produced by a streptomycete. I. Taxonomy, fermentation, isolation, physico-chemical and biological properties, *J. Antibiot.* **44**, 1054–1060.

KOMAGATA, D., SAWA, T., MURAOKA, Y., IMADA, C., OKAMI, Y., TAKEUCHI, T. (1992a), TA-3037A, a new inhibitor of glutathione S-transferase, produced by actinomycetes. I. Production, isolation, physio-chemical properties and biological activities, *J. Antibiot.* **45**, 1117–1122.

KOMAGATA, D., SAWA, R., KINOSHITA, N., IMADA, C., SAWA, T., NAGANAWA, H., HAMADA, M., OKAMI, Y., TAKEUCHI, T. (1992b), Isolation of glutathione transferase inhibitors, *J. Antibiot.* **45**, 1681–1683.

KOMIYAMA, K., EDANAMI, K., TANOH, A., YAMAMOTO, H., UMEZAWA, I. (1983), Studies on the biological activity of stubomycin, *J. Antibiot.* **36**, 301–311.

KOMIYAMA, K., FUNAYAMA, S., ANRAKU, Y., MITO, A., TAKAHASHI, Y., ŌMURA, S., SHIMASAKI, H. (1989), Isolation of isoflavonoids possessing antioxidant activity from the fermentation broth of *Streptoymces* sp., *J. Antibiot.* **42**, 1344–1349, 1350–1355.

KOMIYAMA, K., TAKAMATSU, S., TAKAHASHI, Y., SHINOSE, M., HAYASHI, M., TANAKA, H., IWAI, Y., ŌMURA, S. (1993), New inhibitors of melanogenesis OH-3984 K1 and K2. I. Taxonomy, fermentation, isolation, and biological activities. II. Physico-chemical properties and structural elucidation, *J. Antibiot.* **46**, 1520–1525, 1526–1529.

KONISHI, M., OHKUMA, H., MATSUMOTO, K., SAITOH, K., MIGAKI, T., OKI, T., KAWAGUCHI, H. (1991), Dynemicins, new antibiotics with the 1,5-diyn-3-ene and anthraquinone subunit. I. Production, isolation and physico-chemical properties, *J. Antibiot.* **44**, 1300–1305.

KONO, M., KASAI, M., SHIRAHATA, K., HIRAYAMA, N. (1991), The configuration of mitiromycin and its derivation from mitomycin B, *J. Antibiot.* **44**, 309–315.

KONO, M., KASAI, M., SHIRAHATA, K., HIRAYAMA, N. (1995), Isolation of albomitomycins from a solution from a solution of 7-amino substituted mitomycins; mitomycin C and KW-2149, *J. Antibiot.* **48**, 179–181.

KOPKA, M. L., YOON, C., GOODSELL, D., PIMA, P., DICKERSON, R. E. (1985), The molecular origin of DNA-drug specificity in netropsin and distamycin, *Proc. Natl. Acad. Sci. USA* **82**, 1376–1380.

KOSHINO, H., TAKAHASHI, H., OSADA, H., ISONO, K. (1992), Reveromycins, new inhibitors of eukaryotic cell growth. III. Structures of reveromycins A, B, C and D. Inhibition of the mitogenic activity of epidermal growth factor, *J. Antibiot.* **45**, 1420–1427.

KOURANY-LEFOLL, E., CAPREROTE, O., SÉVENET, T., MONTAGNAC, A., PAIS, M. (1994), Pulveodictines A1–A7, and C1–C2, antibiotic and cytotoxic guanidine alkaloids from the new caledonian sponge *Sphleodictyon* sp., *Tetrahedron* **11**, 3415–3426.

KRUEGER, N. X., STREULI, M., SAITO, H. (1990), Structural diversity and evolution of human receptor-like protein tyrosine phosphatases, *EMBO J.* **9**, 3241–3252.

KRÜGER, W. C., PSCHIGODA, L. M., MOSCOWITZ, A. (1986), The binding of the antitumor antibiotic chartreusin to poly(dA-dT)-poly(dA-dT), poly(dG-dC)-poly(dG-dC) calf thymus DNA, transfer RNA and ribosomal RNA, *J. Antibiot.* **39**, 1298–1303.

KUBOHARA, Y., OKAMOTO, K., TANAKA, Y., ASAHI, K., SAKURAI, A., TAKAHASHI, N. (1993), Differanisole A, an inducer of the differentiation of Friend leukemic cells, induces stalk cell differentiation in *Dictyostelium discoideum, FEBS Lett.* **322**, 73–75.

KURAMOCHI, H., MOTEGI, A., TAKAHASHI, K., TAKEUCHI, T. (1988), DNA cleavage activity of ablomycin (NK313), a novel analog of bleomycin, *J. Antibiot.* **41**, 1846–1853.

KURAMOCHI-MOTEGI, A., KURAMOCHI, H., TAKAHASHI, K., TAKEUCHI, T. (1991), Cell killing mode of liblomycin (NK313), a novel dose-survival relationship different from bleomycins, *J. Antibiot.* **44**, 429–434.

KURISAKI, T., MAGAE, J., ISONO, K., NAGAI, K., YANASAKI, M. (1992), Effects of tautomycin, a protein phosphatase inhibitor, on recycling of mammalian cell surface molecules, *J. Antibiot.* **45**, 252–257.

LAATSCH, H. (1992), A fungal polysaccharide exhibiting antitumor activity, *PharmuZ* **21**, 159–166.

LAM, Y. K. T., WICHMANN, C. F., HEINZ, M. S., GUARIGLIA, L., GIACOBBE, R. A., MOCHALES, S., KONG, L., HONEYCUTT, S. S., ZINK, D., BILLS, G. F., HUANG, L., BURG, R. W., MONAGHAN, R. L., JACKSON, R., REID, G., MAC-

GUIRE, J. J., McKNIGHT, A. T., RAGAN, C. I. (1992), A novel phosphoinositol monophosphatase inhibitor from *Memnoniella echinata, J. Antibiot.* **45**, 1397–1404.

LAM, K. S., VEITCH, J. A., GOLIK, J., KRISHNAN, B., KLOHR, S. E., VOLK, K. J., FORENZA, S., DOYLE, T. W. (1993), Biosynthesis of esperamicin A_1, an enedyine antitumor antibiotic, *J. Am. Chem. Soc.* **115**, 12340–12345.

LAN, K. H. W., FARLEY, J. R., BAYLINK, D. J. (1989), Phosphotyrosyl protein phosphatase, *Biochem. J.* **257**, 23–36.

LEE, M. D. (1992), Calicheamicins, a novel family of antitumor antbiotics. 4. Structure elucidation of calicheamicins β1Br, γ1Br, *J. Am. Chem. Soc.* **114**, 985–997.

LEE, M. D., MANNING, J. K., WILLIAMS, D. R., KUCK, N. A., TESTA, R. T., BORDERS, D. B. (1989), Calicheamicins, a novel family of antitumor antibiotics. 3. Isolation, purification, and characterization of calicheamicins, *J. Antibiot.* **42**, 1070–1087.

LEET, J. E. (1992), Kedarcidin, a new chromoprotein antibiotic: structural elucidation of the kedarcidin chromophore, *J. Am. Chem. Soc.* **114**, 7946–7948.

LEET, J. E., SCHROEDER, D. R., GOLIK, J., MATSON, J. A., DOYLE, T. W., LAM, K. S., HILL, S. E., LEE, M. S., WHITNEY, J. L., KRISHNAN, B. S. (1996), Himastatin, a new antitumor antibiotic from *Streptomyces hygroscopicus.* III. Structural elucidation, *J. Antibiot.* **49**, 299–311.

LEVITZKI, A. (1994), Protein tyrosine kinase inhibitors, in: *New Molecular Targets in Cancer Chemotherapy* (KERR, D. J., WORKMAN, P., Eds.), pp. 67–79. Boca Raton, FL: CRC Press.

LIU, W. C., BARBACID, M., BULGAR, M., CLARK, J. M., ROSSWELL, A. R., DEAN, L., DOYLE, T. W., FERNANDES, P. B., HUANG, S., MANNE, V., PIRNITZ, D. M., WELLS, J. S., MYERS, E. (1992), 10′-Desmethoxystreptonigrin, a novel analog of streptonigrin, *J. Antibiot.* **45**, 454–457.

LOOMIS, C. R., BELL, R. M. (1988), Sangivamycin, a nucleoside analogue, is a potent inhibitor of protein kinase C, *J. Biol. Chem.* **263**, 1682–1692.

LOWN, J. W. (Ed.) (1988), *Anthracyclines and Anthracene-Dione-Based Anticancer Agents.* Amsterdam, New York: Elsevier.

LOWN, J. W. (1993), Anthracycline and anthraquinone anticancer agents: current status and recent developments, *Pharmacol Ther.* **60**, 185–214.

LYBRAND, T. P. (1988), Interaction of peptide antibiotics with DNA. Peptides. Chemistry and biology, *Proc. Amer. Pept. Symp. 10th,* meeting date 1987 (MARSHALL, G. R., Ed.), pp. 416–419. Leiden: ESCOM Sci. Publ.

MAEDA, J. (1993), Melastin, a novel product of *Streptomyces* that selectively inhibits leukemia cell growth, *Biosci. Biotech. Biochem.* **57**, 969–972.

MAEDA, J., YAZAWA, K., MIKAMI, Y., ISHIBASHI, M., KOBAYASHI, J. (1992), The producer and biological activities of SO-075R1, a new mutactimycin group antibiotic, *J. Antibiot.* **45**, 1448–1852.

MAGAE, J., WATANABE, C., OSADA, H., CHENG, X., ISONO, K. (1988), Induction of morphological changes of human myeloid leukemia and activation of protein kinase by a novel antibiotic, tautomycin, *J. Antibiot.* **41**, 932–937.

MAGAE, J., HINO, A., ISONO, K., NAGAI, K. (1992), Respiratory burst induced by phorbol ester in the presence of tautomycin, a novel inhibitor of protein phosphatases, *J. Antibiot.* **45**, 246–251.

MAGAE, J., TSUJI, R. F., WANG, Z., KATAOKA, T., LEE, M.-H., HANADA, T., KURISAKI, T., URAMOTO, M., YAMASAKI, M., NAGAI, K. (1993), Melastin, a novel product of *Streptomyces* that selectively inhibits leukemia cell growth, *Biosci. Biotech. Biochem.* **57**, 969–972.

MAGNUS, P., BENNETT, F. (1989), Synthetic studies on the esperamicin/calicheamicin antitumor antibiotics, *Tetrahedron Lett.* **30**, 3637–3641.

MANGER, A. B. (1980), The actinomycins, *Top. Antibiot. Chem.* **5**, 223–306.

MARTIN, D. G., CHICHESTER, C. G., DUCHAMP, D. J., MISZAK, S. A. (1980), Structure of CC-1065 (NSC-298223), a new antitumor antibiotic, *J. Antibiot.* **33**, 902–903.

MARTY, M., EXTRA, J. M., GIACCHETTI, S., CUVIER, C., ESPIE, M. (1994), Taxoids: a new class of cytotoxic agents, *Nouv. Rev. Fr. Hematol.* **36** (Suppl. 1), 525–528.

MATSHURA, T. (1988), Photochemical and oxidative strand cleavage of DNA, *Stud. Org. Chem. Amsterdam* **88**, 353–366.

MATSON, J. A., COLSON, K. L., BELOFSKY, G. N., BLUMBERG, B. B. (1993), Sandramycin, a novel antitumor antibiotic produced by a *Nocardioides* sp. II. Structure determination, *J. Antibiot.* **46**, 162–166.

MATSUMOTO, M., SETO, H. (1991), Stimulation of mammalian cell proliferation by lavanducyanin, *J. Antibiot.* **44**, 1471–1473.

MATSUMOTO, M., MATSUTANI, S., SUGITA, K., YOSHIDA, H., HAYASHI, F., TERNI, Y., NAKAI, H., NOTANI, N., KAWAMURA, Y., MATSUMOTO, K., SHOJI, J., YOSHIDA, T. (1992), Depudecin: a novel inhibitor of NIH3T3 cells doubly transformed by ras- and src-oncogene, produced by *Alternaria brassicola, J. Antibiot.* **45**, 879–885.

MATSUURA, N., ONOSE, R., OSADA, H. (1996), Morphology reversion activity of phosmidosine and phosmidoxine B, a newly isolated derivative, on src-transformed NRK cells, *J. Antibiot.* **49**, 361–365.

MAYER, A. B., KATZ, E. (1978), Actinomycins, *J. Chromatogr. Libr.* **15** (Antibiotics: isolation, separation, purification), 1–33.

MCALPINE, J. B., KARWOWSKI, J. P., JACKSON, M., MULLALY, M. M., HOCHLOWSKI, J. E., PREMACHANDRAN, U., BURRES, N. S. (1994), MLR-52, (4′-demethylamino-4′,5′-dihydroxy-staurosporine), a new inhibitor of protein kinase C with immunosuppressive activity, *J. Antibiot.* **47**, 281–286.

MIYOSHI-SAITOH, M., MORISAKI, N., TOKIWA, Y., IWASAKI, S., KONISHI, M., SAITOH, K., OKI, T. (1991), Dynemicins O, P and Q: novel antibiotics related to dynemicin A. Isolation, characterization and biological activity, *J. Antibiot.* **44**, 1037–1044.

MORI, K. (1989), Synthesis of differanisol A, an inducer of differentiation, *Liebigs Ann. Chem.* **3**, 303–305.

MORRIS, S. J., THURSTON, D. E., NEVELL, T. G. (1990), Evaluation of the electrophilicity of DNA-binding pyrrolo[2,1-c][1,4]benzodiozepines by HPLC, *J. Antibiot.* **43**, 1286–1294.

MÜLLER, R. (1986), Proto-oncogenes and differentiaton, *TIBS* **11**, 129–132.

MURADA, Y. (1988), Semisynthetic bleomycins, in: *Horizons on Antibiotic Research* (DAVIES, B. D., ISHIKAWA, T., MAEDA, K., MITSCHER, L. A., Eds.), pp. 88–101. Toyko: Japan Antibiotics Association.

MURAMATSU, R., ABE, S., HAYASHI, H., YAMAGUCHI, K., JINDA, K., SAKANO, K., NAKAMURA, S. (1991), Complete amino acid sequence of phenomycin, an antitumor polypeptide antibiotic, *J. Antibiot.* **44**, 1222–1227.

MURAOKA, Y., SAITO, S., NOGAMI, T., UMEZAWA, K., TAKITA, T., TAKEUCHI, T., UMEZAWA, H., SAKATA, N., HORI, M., TAKAHASHI, K., EKIMOTO, H., MINAMIDES, S., NISHIKAWA, K., KURAMOCHI, H., MOTEGI, A., FUKUOKA, T., NAKATANI, T., FUJII, A., MATSUDA, A. (1988), Bleomycin: structure–activity relationship in semisynthetic bleomycin, in: *Horizons on Antibiotic Research* (DAVIES, B. D., ICHIKAWA, T., MAEDA, K., MITSCHER, L. A., Eds.), pp. 88–101. Tokyo: Japan Antibiotics Research Association.

MURATA, M., MIYASAKA, T., TANAKA, H., ŌMURA, S. (1985), Diazoquinomycin A, a new antifolate antibiotic inhibits thymidylate synthase, *J. Antibiot.* **38**, 1025–1033.

MURATA, M., TANAKA, H., ŌMURA, S. (1987), 7-hydro-8-methyl-pteroylglutaminyl-glutamic acid, a new antifolate from an actinomycete, *J. Antibiot.* **40**, 251–257.

MUROI, M. (1990), Angiogenesis inhibitors TAN-1323 C and D manufacture with *Streptomyces*, *Jpn. Patent* JP 03290193.

MURRAY, L., SIM, A. T. R., POSTAS, J. A. P., CAPON, R. J. (1993), Isopalimurin: a mild protein phosphatase inhibitor from a Southern Australian marine sponge *Dysidea* sp., *Aust. J. Chem.* **46**, 1291–1294.

NAGAMURA, A., SATOH, M., IMURA, N. (1993), Metallothionein III (*International Conference of Metallothionein, 3rd*) Meeting Date 1992 (SUZUKI, K. T., IMURA, N., KIMURA, M., Eds.). Basel: Birkhäuser.

NAKANISHI, S. (1994), MS-282a and MS-282b, new inhibitors of calmodulin-activated myosin light chain kinase from *Streptomyces tauricus* ATCC 27470, *J. Antibiot.* **47**, 855–861.

NAKANISHI, S., MATSUDA, Y., IWAHASHI, K. KOSE, H. (1986), K-252b, c and d, potent inhibitors of protein kinase C from microbial origin, *J. Antibiot.* **39**, 1067–1071.

NAKANO, H., KOBAYASHI, E., TAKAHASHI, I., TAMAOKI, T., KUZUU, Y., IBA, H. (1987), Staurosporine inhibits tyrosine-specific protein kinase activity of Rous sarcoma virus transforming protein p60, *J. Antibiot.* **40**, 706–708.

NAKATA, M., SAITO, M., INOUYE, Y., NAKAMURA, S., HAYAKAWA, Y., SETO, H. (1992), A new anthracycline antibiotic, cinerubin R. Taxonomy, structural elucidation and biological activity, *J. Antibiot.* **45**, 1599–1608.

NAKAYAMA, O., SHIGEMATSU, N., KATAYAMA, A., TAKASE, S., KIYOTO, S., HASHIMOTO, M., KOHSAKA, M. (1989), WS-9659A and B, novel testosteron 5α reductase inhibitors isolated from a *Streptomyces.* II. Structural elucidation of WS-9659 A and B, *J. Antibiot.* **42**, 1230–1234.

NATORI, T., MORITA, M., AKIMOTO, K., KOEZUKA, Y. (1994), Agelasphins, novel antitumor and immunostimulatory cerebrosides from the marine sponge *Angelas induritianus, Tetrahedron* **50**, 2771–2784.

NEER, E. J., CHAPHAM, D. E. (1988), Roles of G protein subunits in transmembrane signalling, *Nature* **333**, 129–134.

NEIDLY, S., WARING, M. J. (Eds.) (1983), *Molecular Aspects of Anticancer Drug Action.* Weinheim: Verlag Chemie.

NISHIKAWA, K., SHIBASAKI, C., HIRATSUKA, M., ARAKAWA, M., TAKAHASHI, K., TAKEUCHI, T. (1991a), Antitumor spectrum of deoxyspergualin and its lack of cross-resistance to other antitumor antibiotics, *J. Antibiot.* **44**, 1201–1209.

NISHIKAWA, K., SHIBASAKI, C., UCHIDA, T., TAKAHASHI, K., TAKEUCHI, T. (1991b), The nature of *in vivo* cell-killing of deoxyspergualin and its replication in combination with other antitumor agents, *J. Antibiot.* **44**, 1237–1246.

NISHIKORI, T., HIRUMA, S., KUROHAWA, T., SAITO, S., SHIMADA, N. (1992), Microbial production of 4-thymidine, *J. Antibiot.* **45**, 1376–1377.

NISHIMURA, C., NISHIMURA, T., TANAKA, N., SUZUKI, H. (1987), Potentiation of the cytotoxicity of peplomycin against Ehrlich Ascites Carcinoma by bleomycin hydrolase inhibitor, *J. Antibiot.* **40**, 1794–1795.

NISHIOKA, H. (1994), Antitumor effect of piericidin B_1 N-oxide inhibitors of phosphatidylinositolphosphate turnover, *J. Antibiot.* **47**, 447–452.

NISHIOKA, H., SAWA, T., ISSHIKI, K., TAKAHASHI, Y., NAGANAWA, H., MATSUDA, N., HATTORI, S., HAMADA, M., TAKEUCHI, T., UMEZAWA, H. (1991), Isolation and structure determinaton of a novel phosphatidylinositol turnover inhibitor, piericidin B_1 N-oxide, *J. Antibiot.* **44**, 1283–1285.

NOBLE, R. L. (1990), The discovery of the vinca alkaloids – chemotherapeutic agents against cancer, *Biochem. Cell. Biol.* **68**, 1344–1351.

NOZAKI, Y., HIDA, T., IINUMA, S., ISHII, T., SUDO, K., MUROI, M., KANAMARU, T. (1993), TAN-1120, a new anthracycline with potent angiostatic activity, *J. Antibiot.* **46**, 569–579.

ODAI, H., SHINDO, K., ODAGAWA, A., MOCHIZUKI, J., HAMADA, M., TAKEUCHI, T. (1994), Inostamycins B and C, new polyether antibiotics, *J. Antibiot.* **47**, 939–941.

OGAWARA, H., UCHINO, K., AKIYAMA, T., WATANABE, S. (1985), A new 5′-nucleotidase inhibitor, nucleoticidin. I. Taxonomy, fermentation, isolation and biological properties, *J. Antibiot.* **38**, 153–156.

OGAWARA, H., AKIYAMA, T., WATANABE, S., ISO, N., KOBORI, M., SEODA, Y. (1989a), Inhibition of tyrosine protein kinase activity by synthetic isoflavones and flavones, *J. Antibiot.* **42**, 340–343.

OGAWARA, H., HASUMI, Y., HIGASHI, K., ISHII, Y., SAITO, T., WATANABE, S., SUZUKI, K., KOBORI, M., TANAKA, K. (1989b), Acetoxycycloheximide and cycloheximide convert transformed morphology of ras-transformed cells to normal morphology, *J. Antibiot.* **42**, 1530–1533.

OGAWARA, H., HIGASHI, K., MANITA, S., TANAKA, K., SHIMIZU, Y., SHUFANG, L. (1992), An inhibitor for inositol-specific phospholipase C from *Actinomadura* sp., *J. Antibiot.* **45**, 1365–1367.

OGAWARA, H., HIGASHI, K., MACHIDA, T., TAKASHIMA, J., CHIBA, N., MIKAWA, T. (1994), Inhibitors of diacylglycerol kinase from *Drechslera sacchari*, *J. Antibiot.* **47**, 499–501.

OHNO, M. (1989), From natural bleomycins to man-designed bleomycin, *Pure Appl. Chem.* **61**, 581–584.

OHSHIMA, S., YANAGISAWA, M., KATOH, A., FUJII, T., SANO, T., MATSUKAMA, S., FURUMAI, T., FUJIU, M., WATANABE, K., YAHOSE, K., ARISAWA, M., OKUDA, T. (1994), *Fusarium merismoides* Corola Nr. 6356, the source of the protein kinase C inhibitor, azepinostatin. Taxonomy, yield improvement, fermentation and biological activity, *J. Antibiot.* **47**, 639–647.

OIKAWA, T., SHIMAMURA, M., ASHINO, H., NAKAMURA, O. (1992), Inhibition of angiogenesis by staurosporine, a potent protein kinase inhibitor, *J. Antibiot.* **45**, 1155–1160.

OIKAWA, T., ASHINO, H., SHIMAMURA, M., HASEGAWA, M., MORITA, I., MUROTA, S., ISHIZUNKA, M., TAKEUCHI, T. (1993), Inhibition of angiogenesis by erbstatin, an inhibitor of tyrosine kinase, *J. Antibiot.* **46**, 785–790.

OKA, M., KAMEI, H., HAMAGISHI, Y., TANITA, K., MIGAKI, T., KONISHI, M., OKI, T. (1990), Chemical and biological properties of rubiginone, a complex of new antibiotics with vincristine-cytotoxicity potentiating activity, *J. Antibiot.* **43**, 967–976.

OKA, H., YOSHINARI, T., MURAI, T., KAWAMURA, K., SATOH, F., FUNAISHI, K., OKURA, A., SUDA, H., OKANISHI, M., SHIZURI, Y. (1991), A new topoisomerase-II inhibitor, BE-10988, produced by a streptomycete. I. Taxonomy, fermentation, isolation and characterization, *J. Antibiot.* **44**, 486–491.

OKADA, H., SUZUKI, H., YOSHINARI, T., ARAKAWA, H., OKURA, A., SUDA, H., YAMADA, A., UEMURA, D. (1994), A new topoisomerase II inhibitor, BE 22179, produced by a streptomycete I producing strain, fermentation, isolation and biological activity, *J. Antibiot.* **47**, 129–135.

OKURA, A., NAKADAIRA, M., NAITO, K., ISHIZUKA, M., TAKEUCHI, T., UMEZAWA, H. (1986), Effect of forphenicinol on γ-interferon production in mice sentisized with BCG, *J. Antibiot.* **39**, 569–580.

ŌMURA, S., IMAMURA, N., KUGA, H., ISHIKAWA, H., YAMAZAKI, Y., OKANO, K., KIMURA, K., TAKAHASHI, Y., TANAKA, H. (1985), Adechlorin, a new adenosine deaminase inhibitor containing chlorine. Production, isolation, properties, *J. Antibiot.* **38**, 1008–1015.

ŌMURA, S., TANAKA, H., KUGA, H., IMAMURA, N. (1986), Adecypenol, a unique adenosine deaminase inhibitor containing homopurine and cyclopentene rings, *J. Antibiot.* **39**, 309–310, 1219–1224.

ŌMURA, S., VAN DER PYL, D., INOKASHI, J., TAKAKASHI, Y., TAKESHIMA, H. (1993), Pepticinnamins, new farnesyl-protein transferase inhibitors produced by an actinomycete. I. Producing strain, fermentation, isolation and biological activity, *J. Antibiot.* **46**, 222–228.

ŌMURA, S., INOKOSHI, J., UCHIDA, R., SHIOMI, K., MASUMA, R., KAWAKUBO, T., TANAKA, H., IWAI, Y., KOSEMURA, S., YAMAMURA, S. (1996), Andrastins A–C, new protein farnesyltransferase inhibitors produced by *Penicillium* sp. FO-3929. I. Producing strain, fermentation, isolation, and biological activities, *J. Antibiot.* **49**, 414–417.

OSADA, H., MAGAE, J., WATANABE, C., ISONO, K. (1988), Rapid screening method for inhibitors of protein kinase C, *J. Antibiot.* **41**, 925–931.

OSADA, H., TAKAHASHI, H., TSUNODA, K., KUSAKABE, H., ISONO, K. (1990), A new inhibitor of protein kinase C, RK-286C (4′-Demethylamino-4′-hydroxystaurosporin). I. Screening, taxonomy, fermentation and biological activity, *J. Antibiot.* **43**, 163–167.

OTANI, T. (1993), Conformation studies and assessment by spectral analysis of the protein-chromophore interaction of the macromolecular antitumor antibiotic C-1027, *J. Antibiot.* **46**, 791–802.

OTSUKA, T., TAKASE, S., TERANO, H., OKUHARA, M. (1992), New angiogenesis inhibitors WF-16775 A_1 and A_2, *J. Antibiot.* **45**, 1970–1973.

PAOLETTI, C. (1993), The localization of topoisomerase II cleavage sites on DNA in the presence of antitumor drugs, *Pharmacol. Ther.* **60**, 381–387.

PAQUETTE, L. A. (1993), Synthetic progress toward powerfully cytotoxic taxol and related taxane diterpenes, in: *Organic Chemistry: Its State of Art* (KISAHUEREK, M. V., Ed.), pp. 103–115. Basel: Helvetica Chimica Acta Publishers.

PEREZ-SOLER, R., SUGARMAN, S., ZOU, Y., PRIEBE, W. (1995), Use of drug carries to ameliorate the therapeutic index of anthracycline antibiotics, *ACS Symp. Ser.* **574** (Anthracycline Antibiotics), 300–319.

PETTIT, G. R., CHICHACZ, Z. A., GAO, F., HERALD, C. L., BOYD, M. R. (1993a), Isolation and structure of the human cancer cell growth inhibitors. Spongistatins 2 and 3 from an Eastern Ocean *Spongia* sp., *J. Chem. Soc. Chem. Commun.* **1993**, 1166–1193.

PETTIT, G. R., HERALD, C. L., CHICHACZ, Z. A., GAO, F., SCHMIDT, Y. M., BOYD, M. R. (1993b), Isolation and structure of the powerful human cancer cell growth inhibitors spongistatins 4 and 5 from an African *Spirastrella spinispinilifera* (Porifera), *J. Chem. Soc. Chem. Commun.* **24**, 1805–1807.

PETTIT, G. R., TAN, R., HERALD, D. L., CERNY, R. C., WILLIAMS, M. D. (1994), Antineoplastic agents 277. Isolation and structure of phakellistatin 3 and isophakellistatin 3 from a Republic of Comoras marine sponge, *J. Org. Chem.* **59**, 1593–1595.

PETTIT, G. R., GAO, F., BLUMBERG, P. M., HERALD, C. L., COLL, J. O., KAMANO, Y., LEWIN, N. E., SCHMIDT, J. M., CAAPIUS, J. C. (1996), Antineoplastic agents 340. Isolation and structural elucidation of bryostatins 16–18, *J. Nat. Prod.* **59**, 286–289.

PIMENTEL, E. (1987), *Oncogenes.* Boca Raton, FL: CRC Press.

PINEDO, H. M. (Ed.) (1980), *Cancer Chemotherapy: The EORT Cancer Chemotherapy.* Amsterdam, Oxford: Annual Excerpta Medica.

PIRRUNG, M. C., NAUHANS, S. K. (1994), Structure of curacin A, a novel antimitotic, antiproliferative, and bine shrimp toxic natural product from the marine cyanobacterium *Lyngbya majuscula, Chemtracts. Org. Chem.* **7**, 128–129.

POITIER, C. P., GUERITTE-VOEGELEIN, F., GUENARD, D. (1994), Taxoids, a new class of antitumor agents of plant origin: recent results, *Nouv. Rev. Fr. Hematol.* **36** (Suppl. 1), S21–S23.

POOT, M., HOEHN, H. (1993), DNA topoisomerases and the DNA lesion in human genetic instability, *Toxicol. Lett.* **61**, 297–308.

POVIRK, C., AUSTIN, M. J. F. (1991), Genotoxicity of bleomycin, *Mutat. Res.* **257**(2), 127–143.

RAMSDELL, J. S., PETTIT, R. G. (1986), Three activators of protein kinase C, bryostatins, calyculin and phorbol esters show differing specificities of action of CH_4 pituitary cells, *J. Biol. Chem.* **261**, 17073–17080.

RANCE, M. J., RUDDOCK, J. C., PACEY, M. S., CULLEN, W. P., HUANG, L. H., JEFFERSON, M. T., MAEDA, H., TONE, J. (1989), UK-63052 complex, new quinomycin antibiotics from *Streptomyces braegensis* subsp. *taponicus.* Taxonomy, fermentation, isolation, characterization and antimicrobial activity, *J. Antibiot.* **42**, 206–217.

RAO, A. N. (1991), Serine hydroxymethyl transferase: a target for cancer chemotherapy, *New Trends Biol. Chem.* (OZAWA, T., Ed.), pp. 333–340. Tokyo: Japan Sci. Soc. Press.

RAUTH, M. A., RAYMOND, R. S. (1993), Cellular approaches to bioreductive drug mechanisms, *Cancer Metastasis Rev.* **12**, 153–164.

REICHLE, H., DIDDENS, H., RASTATTER, J., BERDEL, W. E. (1991), Resistenzmechanismen maligner Zellen gegenüber Zytostatika, *Dtsch. Med. Wochenschr.* **116**, 186–191.

ROCKWELL, S., SARTOLINI, A. C., TOMASZ, M., KENNEDY, K. A. (1993), Cellular pharmacology of quinone bioreductive alkylating agents, *Cancer Metastasis Rev.* **12**, 165–176.

ROSOWSKY, A., FORSCH, R. A., REICH, V. E., FREISHEIM, J. H., MORAN, R. G. (1992), Side chain modified 5-deazafolate and 5-deazafolate analogues as mammalian folylpolyglutamate synthetase and glycinamide ribonucleotide formyltransferase inhibitors. Synthesis and *in vitro* biological evaluation, *J. Med. Chem.* **35**, 1578–1588.

ROTHENBERG, M. L. (1993), Taxol, taxotere and other new taxanes, *Curr. Opin. Invest. Drugs* **2**, 1269–1277.

SAE, H., NEMOTO, T., ISHIWATA, H., OJIKA, M., YAMADA, K. (1993), Isolation structure and synthesis of dolastatin D, a cytotoxic cyclic depsipeptide from the sea hore *Dolabella auricularia, Tetrahedron Lett.* **34**, 8449–8452.

SAKAI, M., KAWAI, Y., AONO, M., TAKAHASHI, K. (1989), A new activity of herbimycin A: inhibition of angiogenesis, *J. Antibiot.* **42**, 1015–1017.

SAKATA, N., KARUBE, T., TANAKA, M., HAYASHI, H., HORI, M., HOTTA, K., HAMADA, M. (1989), Nucleotide sequence of the macromomycin apoprotein gene and its expression in *Streptomyces macromomyceticus, J. Antibiot.* **42**, 1704–1712.

SAKATA, N., SUZUKAKE-TSUCHIYA, K., MORIYA, Y., HAYASHI, H., HORI, M., OTANI, T., NAKAI, M., AOYAGI, T. (1992), Aminopeptidase activity of an antitumor antibiotic, C-1027, *J. Antibiot.* **45**, 113–117.

SALUZ, H. P., WIEBAUER, K. (Eds.) (1995), Natural products for studying the structure of nucleic acids and DNA biomolecular complexes, Chapter 3 in: *DNA and Nucleoprotein Structure in vivo. Investigations on Protein/DNA Interactions and Nucleic Acid Structures.* Austin, TX: Landes Company.

SANCELME, M., FABRE, S., PRUDHOMME, M. (1994), Antimicrobial activities of indolocarbazole and bis-indole protein kinase C inhibitors, *J. Antibiot.* **47**, 792–798.

SARTORELLI, A. C., TOMASZ, M., ROCKWELL, S. (1993), Studies on the mechanism of the cytotoxic action of the mitomycin antibiotics in hypoxic and oxygenated EMT6 cells, *Adv. Enzyme Regul.* **33**, 3–17.

SASSE, F., STEINMETZ, H., HÖLLE, G., REICHENBACH, H. (1993), Rhizopodin, a new compound from *Myxococcus stipitatus* (Myxobacteria) causes formation of rhizopodia-like structures in animal cell cultures, *J. Antibiot.* **46**, 741–748.

SATO, K. (1989), Glutathione transferases as markers of preneoplasia and neoplasia, *Adv. Cancer Res.* **52**, 205–255.

SATO, Y., MORIMOTO, A., KINE, A., OKAMURA, K., HAMANAKA, R., KOHNO, K., KUWANO, M., SAKATA, T. (1993), Irsogladine is a potent inhibitor of angiogenesis, *FEBS Lett.* **322**, 155–158.

SATOH, T., NISHIMURA, Y., KONDO, S., TAKEUCHI, T. (1996), Synthesis and antimetastatic activity of 6-trichloroacetamido and 6-guanidino analogues of siastatin B (1996), *J. Antibiot.* **49**, 321–325.

SAWA, R., MATSUDA, N., UCHIDA, T., IKEDA, T., SAWA, T., NAGANAWA, H., HAMADA, M., TAKEUCHI, T. (1991), Dioxamycin, a new benz[a]anthraquinone antibiotic, *J. Antibiot.* **44**, 396–402.

SCAHILL, T. A., JENSEN, R. M., SWENSON, D. H., HAKENBUHLER, N. T., PEKOLD, G., WIERENGA, W., BRAHME, N. D. (1990), An NMR study of the covalent and noncovalent interactions of CC-1065 and DNA, *Biochemistry* **29**, 2852–2860.

SCHEID, W., WEBER, J., ROETTGERS, U., TRAUT, H. (1991), Enhancement of the mutagenicity of anticancer drugs by the calcium agonists verapamil and fendilihe, *Arzneim. Forsch.* **41**, 901–904.

SCHWOB, E., NASMYTH, K. (1993), CLB5 and CLB6, a new pair of B cyclins involved in DNA replication in *Saccharomyces cerevisiae, Genes Dev.* **7**, 1160–1175.

SEBTI, S. M., LAZO, J. S. (1988), Metabolic inactivation of bleomycin analogs by bleomycin hydrolase, *Pharmacol. Ther.* **38**, 321–329.

SEKI-ASANO, M., OKAZAKI, T., YAMAGICHI, M., SAKAI, N., HAMADA, K., MIZORIE, K. (1994), Isolation and characterization of new 18-membered macrolides FD-891 and FD-892, *J. Antibiot.* **47**, 1226–1233.

SEKIZAWA, R., IINUMA, H., NAGANAWA, H., HAMADA, M., TAKEUCHI, T., YAMAZUMI, J., UMEZAWA, K. (1996), Isolation of novel saquayamicins as inhibitors of farnesyl-protein transferase, *J. Antibiot.* **49**, 487–490.

SHIBATA, K., SATSUMABAYASHI, S., SANO, H., KOMIYAMA, K., NAKAGAWA, A., ŌMURA, S. (1988), Chemical modification of hitachimycin. Synthesis, antibacterial, cytocidal, and *in vivo* antitumor activities of hitachimycin derivatives, *J. Antibiot.* **41**, 614–623.

SHINDO, K., MATSUOKA, M., KAWAI, H. (1996a), Studies on cochleamycins, novel antitumor antibiotics. I. Taxonomy, production, isolation and biological activities, *J. Antibiot.* **49**, 241–244.

SHIOMI, K., YANG, Y., INOKOSHI, J., VAN DER PYL, D., NAKAGAWA, A., TAKESHIMA, H., ŌMURA, S. (1993), Pepticinnamins, new farnesyl-protein transferase inhibitors produced by an actinomycete. II. Structural elucidation of pepticinnamin E, *J. Antibiot.* **46**, 229–234.

SINGH, S. B., ZINK, D. L., LIESCH, J. M., GOETZ, M. A., JENKINS, R. J., NALLIN-OMSTEAD, M., SILVERMAN, K. C., BILLS, G. F., MOSLEY, R. T., GIBBS, J. B., ALBERS-SCHÖNBERG, G., LINGHAM, R. B. (1993), Isolation and structure of chaetomellic acids A and B from *Chaetomella acutiseta:* farnesyl pyrophosphate mimic inhibitors of ras farnesyl-protein transferase, *Tetrahedron* **49**, 5917–5926.

SINGH, S. B., JONES, E. T, GOETZ, M. A., BILES, G. F., NALLIN-OUSTEAD, M., JENKINS, R. G., LINGHAM, R. B., SILVERMAN, K. C., GIBBS, J. B. (1994a), Fusidienol: a new inhibitor of ras farnesyl-protein transferase from *Fusidium griseum, Tetrahedron. Lett.* **35**, 4693–4696.

SINGH, S. B., NESCH, J. M., LINGHAM, R. B., GOETZ, M. A., GIBBS, J. B. (1994b), Actinoplanic acid A: a macrocyclic polycarboxylic acid which is a potent inhibitor of ras farnesyl-protein transferase, *J. Am. Chem. Soc.* **116**, 11606–11607.

SINGH, S. B., ZINK, D. C., NESCH, J. M., BALL, R. G., GOETZ, M. A., BOLESSA, E. A., GIACOBBE, R. A., SILVERMAN, K. C., BILLS, G. F., PELAEZ, F., CASCALES, C., GIBBS, J. B., LINGHAM, R. B. (1994c), Preussomerins and deoxypreussomerins: novel inhibitors of ras farnesyl-protein transferase, *J. Org. Chem.* **59**, 6296–6302.

SLICHENMYER, W. J., ROWINSKY, E. K., DONEKOWER, R. C., KAUFMANN, S. H. (1993), The current status of camptothecin analogs as antitumor agents, *J. Nat. Cancer Inst.* **85**(4), 271–291.

SOLAIMAN, D. (1988), *Bleomycin and its Metal Complexes. Metal-Based Anti-Tumour Drugs* (GIELEN, M. F., Ed.), pp. 235–256. London: Publisher Freund.

SONODA, T., OSADA, H., URAMOTO, M., UZAWA, J., ISONO, K. (1989), Epiderstatin, a new inhibitor of the mitogenic activity induced by epidermal growth factor. II. Structure elucidation, *J. Antibiot.* **42**, 1607–1609.

STEFANELLI, S., SPONGA, F., FERRARI, P., SOTTANI, C., CORTI, E., BRUNATI, C., ISLAM, K. (1996), Inhibition of myo-inositol monophosphatase ATCC20928 factors A and C. Isolation, physico-chemical characterization and biological properties, *J. Antibiot.* **49**, 611–616.

STEINHERZ, L., STEINHERZ, P. (1991), Delayed cardiac toxicity from anthracycline therapy, *Pediatrician* **18**, 49–52.

STIERLE, A., STIERLE, D., STROBEL, G., BIGNAMI, G., GROTHANS, P. (1994), Endophytic fungi of Pacific yew (*Taxus brevifolia*) as a source of taxol, taxanes and other pharmacophores, *ACS Symp. Ser.* **557** (Bioregulators for Crop Protection and Pest Control), 64–77.

STREKOWSKI, L. (1992), Molecular basis for the enhancement and inhibition of bleomycin-mediated degradation of DNA and DNA-binding compounds, *Adv. Detailed React. Mech.* **2** (Mech. Biol. Importance), 61–109.

STROHL, W. R., BARTEL, P. L., CONNORS, N. C., ZHU, C.-B., DOSCH, D. C., BEALE, J. M., JR., FLOSS, H. G., STUTZMAN-ENGWALL, K., OTTEN, S. L., HUTCHINSON, C. R. (1989), Biosynthesis of natural and hybrid polyketides by anthracycline producing Streptomycetes, in: *Genetics and Molecular Biology of Industrial Microorganisms* (HERSHBERGER, C. L., QUEENER, S. W., HEGEMAN, G., Eds.), pp. 68–84. Washington: American Society for Microbiology.

STUBBE, J., KOZARICH, J. W. (1987), Mechanism of bleomycin-induced DNA degradation, *Chem. Rev.* **87**, 1107–1136.

SUGANUMA, M., YOSHIZAWA, S., HIROTA, M., NAKOYASU, M., OJIKA, M., FUJUKI, H., SUGORI, H., WAKAMATSU, K., YAMADA, K., SIGUMURA, T. (1988), Okadaic acid: an additional non-phorbol-12-tetradecanoiete-13-acetate-type tumor pormoter, *Proc. Natl. Acad. Sci. USA* **85**, 1768–1771.

SUZUKAKE-TSUCHIYA, K., MORIGA, Y., HORI, M. (1989), Induction by herbimycin A of contact inhibition in v-src-expressed cells, *J. Antibiot.* **42**, 1831–1834.

SUZUKAKE-TSUCHIYA, K., MORIYA, Y., YAMAZAKI, K., HORI, M., HOSOKAWA, N., SAWA, T., IINUMA, H., NAGANAWA, H., IMADA, C., HAMADA, M. (1990), Screening for antibiotics preferentially active against ras oncogene-expressed cells, *J. Antibiot.* **43**, 1489–1495.

SUZUKAKE-TSUCHIYA, K., MORIYA, Y., KAWAI, H., HORI, M., UEKARA, Y., IINUMA, H., NAGANAWA, H., TAKEUCHI, T. (1991), Inhibition of pinocytosis by hygrolidin family antibiotics: Possible correlation with their selective effects on oncogene-expressed cells, *J. Antibiot.* **44**, 344–348.

SUZUKI, H., MENEGAZZI, M., CARCERERI, D., PRATI, A., GEROSA, F., TAMMASI, M., SCARPA, A., SORRENTINO, S., DE PRISCO, R., LIBONATI, M. (1990), Studies on potential markers in human prostatic carcinoma and on a possible role of poly(ADP-ribose) polymerase in carcinogenesis, in: *Pathology of Gene Expression* (FRATI, L., AARONSON, S. A., Eds.). New York: Raven Press.

TAKAHASHI, A. (1993), Russuphelins B, C, D. E and F, new cytotoxic substances from the mushroom *Russula subniglicans* HONGO, *Chem. Pharm. Bull.* **41**, 1726–1729.

TAKAHASHI, M. (1994), Adenophostins A and B: potent agonists of inositol 1,4,5-trisphosphate receptor produced by *Penicillium brevicompactum.* Taxonomy, fermentation, isolation, physico-chemical and biological properties, *J. Antibiot.* **47**, 1643–1647.

TAKAHASHI, A., NAKAMURA, H., KAMEYAMA, T., KURASAWA, S., NAGANAWA, H., OKAMI, Y., TAKEUCHI, T., UMEZAWA, H. (1987a), Bisucaberin, a new siderophore sensitizing tumor cells to macrophage-mediated cytolysis. II. Physicochemical properties and structure elucidation, *J. Antibiot.* **40**, 1671–1676.

TAKAHASHI, I., KOBAYASHI, E., ASANO, K., YOSHIDA, M., NAKANO, H. (1987b), UCN-01, a selective inhibitor of protein kinase C from *Streptomyces, J. Antibiot.* **40**, 1782–1784.

TAKAHASHI, K., KATSUTOSHI, E., HISAO, M., SEIKI, N. (1987c), Liblomycin, a new analogue of bleomycin, *Cancer Treat. Rev.* **14**, 169–177.

TAKAHASHI, I., ASONO, K., KAWAMOTO, I., TAMAOKI, T., NAKANO, H. (1989), UCN-01 and UCN-02, new selective inhibitors of protein kinase C. I. Screening, producing organism and fermentation, *J. Antibiot.* **42**, 564–570.

TAKAHASHI, C., NUMATA, A., MATSUMURA, E., MJNOURA, K., ETO, H., SHINGU, T., ITO, T., HASEGAWA, T. (1994), Leptosins I and J, cytotoxic substances produced by a *Leptosphaeria* sp. Physico-chemical properties and structures, *J. Antibiot.* **47**, 1242–1249.

TAKAHASHI, S., HASHIMOTO, P., HAMANO, K., SUZUKI, T., NAKAGAWA, A. (1996), Melanoxazal, new melanin biosynthesis inhibitors discovered by using the larval haemolymph of the silkworm *Bombyx mori.* Production, isolation, structural elucidation and biological properties, *J. Antibiot.* **49**, 513–518.

TAKAMIYA, K., YOSHIDA, E., TAKAHASHI, T., OKURA, A., OKANISHI, M., KOMIYAMA, K., UMEZAWA, I. (1988), The effect of kazusamycin B on the cell cycle and morphology of cultured L1210 cells, *J. Antibiot.* **41**, 1854–1861.

TAKARA, M., TOMIDA, A., NISHIMORA, T., YAMAGUCHI, H., SUZUKI, H. (1990), Resorthiomycin, a novel antitumor antibiotic. III. Potentiation of antitumor drugs and its mechanism of action, *J. Antibiot.* **43**, 138–142.

TAKE, Y., KUBO, T., TAKEMORI, E., INOYE, Y., NAKAMURA, S., NISHIMURA, T., SUZUKI, H., YAMASUCHI, H. (1989), Biological properties of streptonigrin derivatives. III. *In vitro* and *in vivo* antitumor activities, *J. Antibiot.* **42**, 968–1976.

TAKEUCHI, T. (1992), Novel physiologically active substance, *Jpn. Patent* JP 04217681.

TAKITA, T. (1984), The bleomycins: properties, biosynthesis and fermentation, in: *Biotechnology of Industrial Antibiotics* (VANDAMME, E. J., Ed.), pp. 595–621. New York: Marcel Dekker.

TAKITA, T., MURAOKA, Y. (1990), *Biosynthesis and Chemical Synthesis of Bleomycin. Biochemistry of Peptide Antibiotics* (KLEINKAUF, H., VON DOEHREN, H., Eds.), pp. 289–309. Berlin: DeGruyter.

TAKITA, T., OGINO, T. (1987), Peplomycin and liblomycin, new analogs of bleomycin, *Biomed. Pharmacother.* **41**, 219–226.

TAKITA, T., MURAOKA, Y., TAKAHASHI, K. (1989), Bleomycins: basic research, *Gann Monogr. Cancer Res.* **36** (Antitumor. Nat. Prod.), 59–70.

TANAKA, N., OKABE, T., ISONO, F., KASHIWAGI, M., NOMOTO, K., TAKAHASHI, M., SHIMAZU, A., NISHIMURA, T. (1989), Lactoquinomycin, a novel anticancer antibiotic. I. Taxonomy, isolation and biological activity, *J. Antibiot.* **38**, 1327–1332.

TANAKA, S., OHKUBO, M., KOJIRI, K., SUDA, H. (1992), A new indolocarbazole antitumor substance ED-110, a derivative of BE-13793c, *J. Antibiot.* **45**, 1797–1798.

TANAKA, S., OKABE, T., NAKAJIMA, S., YOSHIDA, E., SUDA, H. (1994), BE-23372M, a novel protein tyrosine kinase inhibitor. I. Producing organism, fermentation, isolation, and biological activities, *J. Antibiot.* **47**, 289–293.

TATSUDA, K., YASUDA, S. (1996), Synthesis and biological evaluation of caloporoside analogs, *J. Antibiot.* **49**, 713–715.

TATSUDA, K., IKEDA, Y., MIURA, S. (1996), Synthesis and glycosidase inhibitory activities of nagstatin triazole analogs, *J. Antibiot.* **49**, 836–837.

TAYLOR, G. S. (1989), cAMP-dependent protein kinases, *J. Biol. Chem.* **264**, 8443–8446.

THORNALLEY, P. J. (1990), The glyoxylase system: new developments towards functional characterization of a metabolic pathway fundamental to biological life, *Biochem. J.* **269**, 1-11.

TOLSTIKOV, V. V., HOLPNE KOZLOVA, N. Y., ORESHKINA, T. D., OSIPOVA, T. V., PREOBRAZHENSKAYA, M. N., SZTARICSKAI, F., BALZARINI, J., DE CLERCQ, E. (1992), Amides of antibiotic streptonigrin and amino dicarboxylic acids or amino-sugar. Synthesis and biological evaluation, *J. Antibiot.* **45**, 1020–1025.

TOMASZ, M. (1992), Mitomycin C: DNA sequence specificity of a natural DNA cross-linking agent, *Adv. DNA sequence specific Agents* **1**, 247–261.

TOMASZ, M. (1994), DNA-adducts of mitomycin. *IARC Sci. Publ.* **125** (DNA-Adducts: Identification and Biological Significance), 349–357.

TROLL, W., WIESNER, R., FRENKEL, K. (1987), Anticarcinogenic action of protease inhibitors, *Adv. Cancer Res.* **49**, 265–283.

TROWBRIDGE, I. S. (1991), CD45, *J. Biol. Chem.* **266**, 23517–23520.

TSCHAGOSHI, S., TAKENDO, T. UMEZAWA, H. (1986), Antitumor substances, in: *Biotechnology,* Vol. 4 (REHM, H.-J., REED, G., Eds.), pp. 509–530. Weinheim: VCH.

TSUCHIYA, K. S., MORIYA, Y., HON, M., ITOH, O., TAKEUCHI, T., EKIMOTO, H., HIRATSUKA, M. (1992), Synergism between 5-fluorouracil and oxanosine in inhibition of growth of ras-expressed cells *in vitro* and *in vivo, J. Antibiot.* **45**, 283–285.

TSUKAMOTO, S. (1993), Microcystilide A. A novel cell-differentiation-promoting depsipeptide from

Microcystis seruginosa NO-15-1840, *J. Amer. Chem. Soc.* **115**, 11046.

TSUNAKAWA, M., KAMEI, H., KONISHI, M., MIGAKI, T., OKI, T., KAWAGUCHI, H. (1988), Porothramycin, a new antibiotic of the anthramycin group: production, isolation, structure and biological activity, *J. Antibiot.* **41**, 1366–1373.

TSURUOKA, T., FUKUYASHI, H., ISHI, M., USUI, T., SHIBAHARA, S., INOUYE, S. (1996), Inhibition of mouse tumor metastasis with nojirimycin-related compounds, *J. Antibiot.* **49**, 155–161.

UBUKATA, M., UZAWA, J., OSADA, H., ISONO, K. (1993), Respinomycins A1, A2, B, C, and D, a novel group of anthracycline antibiotics. II. Physicochemical properties and structure elucidation, *J. Antibiot.* **46**, 942–951.

UBUKATA, M., MORITA, T., URAMOTO, M., OSADA, H. (1996), Sparoxomycins A1 and A2, new inducers of the flat reversion of NRK cells transformed by temperature sensitive Rons sarkoma virus. II. Isolation, physicochemical properties and structure elucidation, *J. Antibiot.* **49**, 65–70.

UCHIDA, T., IMOTO, M., WATANABE, Y., MIURA, K., DOBASHI, T., MATSUDA, N., SAWA, T., NAGANAWA, H., HAMADA, M., TAKEUCHI, T., UMEZAWA, H. (1985), Saquayamycins, new aquayamycin-group antibiotics, *J. Antibiot.* **38**, 1171–1181.

UCHIDA, R., SHIOMI, K., INOKOSHI, J., MASUMA, R., KAWAKUBO, T., TANAKA, H., IWAI, Y., ŌMURA, S. (1996), Kurasoins A and B, new protein farnesyltransferase inhibitors produced by *Paecilomyces* sp. FO-3684, *J. Antibiot.* **49**, 932–934.

UEDA, H., NAKAJAMA, H., HORI, Y., FUJITA, T., NISHIMURA, M., GOTO, T., OKUHARA, M. (1994), FR 901228, a novel antitumor bicyclic depsipeptide produced by *Chromobacterium violaceum,* No. 968, I. Taxonomy, fermentation, isolation, physico-chemical and biological properties, and antitumor activities, *J. Antibiot.* **47**, 310.

UEHARA, Y., MURAKAMI, Y., SUZUKAKE-TSUCHIYA, K., MORIYA, Y., SANO, H., SHIKATA, K., ŌMURA, S. (1988), Effects of herbimycin derivatives on src oncogene function in relation to antitumor activity, *J. Antibiot.* **41**, 831–834.

UMEZAWA, K. (1989), Screening of oncogene function inhibitors, possible antitumor agents. Anticancer Drugs, *Colloques INSERM,* Vol. 191, (TAPIERC, H., Ed.), pp. 13–21.

UMEZAWA, H., IMOTO, M., SAWA, T., ISSHIKI, K., MATSUDA, N., IINUMA, H., HAMADA, M., TAKEUCHI, T. (1986), Studies on a new epidermal growth factor-receptor kinase inhibitor, erbstatin, produced by MH435-hF3, *J. Antibiot.* **39**, 170–173.

UMEZAWA, K., HARESAKU, M., MURAMATSU, M., MATSUSHIMA, T. (1987), Mutagenicity of anthracycline glycosides and bleomycins in a *Salmonella* assay system, *Biomed. Pharmacother.* **41**, 214–218.

UOSAKI, Y., YASUZAWA, T., HARA, M., SAITOH, Y., SANO, H. (1991), Sapurimycin, new antitumor antibiotic produced by streptomyces structure determinaton, *J. Antibiot.* **44**, 40–44.

UOSAKI, Y., KAWADA, S., NAKANO, H., SAITOH, Y., SANO, H. (1993), UCT4B, a new antitumor antibiotic with topoisomerase II mediated DNA cleavage activity. Structure determination, *J. Antibiot.* **46**, 235–240.

VAN DER PYL, D., INOKOSHI, J., SHIOMI, K., YANG, H., TAKESHIMA, H., ŌMURA, S. (1992), Inhibition of farnesyl protein transferase by gliotoxin and acetylglyotoxin, *J. Antibiot.* **45**, 1803–1805.

VAN DER WILT, C. L., PETERS, G. J. (1994), New targets for pyrimidine antimetabolites in the treatment of solid tumors. I. Thymidylate synthase, *Pharm. World Sci.* **16**, 84–103.

VANDEN BOSCHE, H., MOLREELS, H., KOYMANS, L. M. H. (1994), Aromatase inhibitors – mechanisms for nonsteroidal inhibitors, *Breast Cancer Res. Treat.* **30**, 43–55.

VANEK, Z., TAX, J., KOMERSOVA, I., SEDMERA, P., VOKOUN, J. (1977), Anthracyclines, *Folia Microbiol.* **22**, 139–159.

VYAS, D. M. (1993), Recent advances in medical chemistry of taxol, *Pharmacochem. Libr.* **20** (Trends in Drug Research), 261–270.

WAGNER, C., ECKARDT, K., IHN, W., SCHUMANN, G., STENGEL, C., FLECK, W. F., TRESSELT, D. (1991), Biosynthese der Anthracycline: eine Neuinterpretation der Ergebnisse zur Daunomycin-Biosynthese, *J. Basic Microbiol.* **31**, 223–240.

WAKUSAWA, S., INOKO, K., MIGAMOTO, K. (1993), Staurosporine derivatives reverse multi drug resistance without correlation with their protein kinase inhibitory activities, *J. Antibiot.* **46**, 353–355.

WALL, M. E., WANI, M. C. (1993), Camptothecin and analogs: synthesis, biological *in vitro* and *in vivo* activities, and clinical possibilities, *ACS Symp. Ser.* **534** (Human Medical Agents from Plants), 146–169.

WANG, H. H. J. (1992), Intercalative drug binding to DNA, *Curr. Opin. Struct. Biol.* **2**, 361–368.

WANG, A. H. J., LEGHETTO, G., QUIGLEY, G. J., RICH, A. (1987), Interactions between an anthracycline antibiotic and DNA: molecular structure of daunomycin complexed to d(CGTACG) at 1,2A-resolution, *Biochemistry* **26**, 1152–1163.

WARDMAN, P. (1990), Bioreductive activation of quinones: redox properties and thiol reactivity, *Free Radicals Res. Commun.* **8,** 219–229.

WARING, M. E. (1975), Ethidium and propidium, in: *Antibiotics,* Vol. 3, *Mechanisms of Action of Antimicrobial and Antitumor Agents* (CORCORAN, J. W., HAHN, F. E., Eds.), pp. 141–166. Berlin, Heidelberg, New York: Springer-Verlag.

WARING, M. J. (1979), Echinomycin, triostin and related antibiotics, in: *Antibiotics,* Vol. 5/2, *Mechanisms of Action of Antibacterial Agents* (HAHN, F. E., Ed.), pp. 173–194. Berlin, Heidelberg, New York: Springer-Verlag.

WARING, M. J. (1990), The molecular basis of specific recognition between echinomycin and DNA. *Jerusalem Symp. Quantum Chem. Biol.* **23** (1990) (Molecular Basis of Specific Nucleic Acid-Drug Interactions) 225–24.

WARING, M. J. (1992), Echinomycin, *Pathol. Biol.* **40**, 1022–1024.

WATT, P. M., HICKSON, I. D. (1994), Structure and function of type II DNA topoisomerases, *Biochem. J.* **303**, 681–695.

WEBER, H., SCHU, P., ANKE, T., VELTEN, R., STEGLICH, W. (1994), Caloporoside, a new inhibitor of phospholipases C from *Caloporus dichrous* (Fr.) Ryr., *J. Antibiot.* **47**, 1188–1194.

WEINBERG, R. A. (1985), The action of oncogenes in the cytoplasm and nucleus, *Science* **230**, 770–776.

WEISS, R. B. (1992), The anthracyclines will ever find a better doxorubicin? *Semin Oncol.* **19**, 670–686.

WHITE, J. R., STROSHANE, R. M. (1984), Daunorubicin and adriamycin: properties, biosynthesis, and fermentation, Chapter 19, in: *Biotechnology of Industrial Antbiotics* (VANDAME, E. J., Ed.), pp. 569–594. New York, Basel: Marcel Dekker.

WILMAN, D. E. V. (Ed.) (1990), The chemistry of antitumor agents. New York: Chapman and Hall.

WOLF, M., BAGGIOLINI, M. (1988), The protein kinase inhibitor staurosporin, like phorbol esters, induces the association of protein kinase C with membranes, *Biochem. Biophys. Res. Commun.* **154**, 1273–1279.

YAGINUMA, S., MOTO, N., TSUJINO, M., SUDATE, Y., HAGASHI, M., OTANI, M. (1981), Neplanocin A, a new antitumor antibiotic. I. Producing organism, isolation and characterization, *J. Antibiot.* **34**, 359–366.

YAMAGUCHI, T., FURUMAI, T., SATO, M., OKUDA, T., ISHIDA, N. (1970), Studies on a new antitumor antibiotic, largomycin. I. Taxonomy of the largomycin-producing strain and production of the antbiotic, *J. Antibiot.* **23**, 369–372.

YAMASHITA, T., NAOI, N., HIDAKA, T., HORI, S., KIMURA, K., TAKEUCHI, T., UMEZAWA, H. (1979), Studies on auromomycin, *J. Antibiot.* **32**, 330–339.

YAMASHITA, T., SAKAI, M., KAWAI, Y., AONO, M., TAKAHASHI, K. (1989), A new activity of herbimycin A: inhibition of angiogenesis, *J. Antibiot.* **42**, 1015–1017.

YAMASHITA, Y., SAITOH, Y., ANDO, K., TAKAHASHI, K., OHNO, H., NAKANO, H. (1990a), Saintopin, a new antitumor antibiotic with topoisomerase II dependent DNA cleavage activity, from *Paecilomyces, J. Antibiot.* **43**, 1344–1346.

YAMASHITA, Y., KAWADA, S., FUJII, N., NAKANO, H. (1990b), Induction of mammalian DNA topoisomerase II dependent DNA cleavage by antitumor antibiotic streptonigrin, *Cancer Res.* **50**, 5841–5844.

YAMASHITA, T., OSONO, M., MASUDA, T., ISHIZUKA, M., TAKEUCHI, T., KAWATSU, M. (1993), Effect of conagenin in tumor bearing mice. Antitumor activity, generation of effector cells and cytokine production, *J. Antibiot.* **46**, 1692–1698.

YAMAZAKI, M., YAMASHITA, T., HARADA, T., NISHIKORI, T., SAITO, S., SHIMADA, N., FUGII, A. (1992), 44-Homooligomycins A and B, new antitumor antibiotics from *Streptomyces bottropensis, J. Antibiot.* **45**, 171–179.

YAMAZAKI, T., TATEE, T., HOYAMA, T., KOJINA, F., TAKEUCHI, T. (1993), Bequinostatins C and D, new inhibitors of glutathione S-transferase produced by a *Streptomyes* spec. MB84-DF12, *J. Antibiot.* **46**, 1309–1311.

YAMAZAKI, K., AMEMIAY, M., ISHIZUKA, M., TAKEUCHI, T. (1995), Screening for apoptosis inducers in microbial products and induction of apoptosis by cytostatin, *J. Antibiot.* **48**, 1138–1140.

YANAGISAWA, M. (1994a), Hispidospermidin, a novel phospholipase C inhibitor produced by *Chaetomia hispidulum* (Cda) Moesz NR 7127. I. Screening, taxonomy, and fermentation, *J. Antibiot.* **47**, 1–5; II. Isolation, characterization and structural elucidation, *J. Antibiot.* **47**, 6–17.

YANAGISAWA, M. (1994b), Hispidospermidin, a novel phospholipase C inhibitor produced by *Chaetomia hispidulum* (Cda) Moesz NR 7127. II. Isolation, characterization and structural elucidation, *J. Antibiot.* **47**, 6–17.

YARDIN, Y., ULLRICH, A. (1988), Molecular targets of signal transduction by growth factors, *Biochemistry* **27**, 3113–3119.

YOSHIDA, E., KOMIYAMA, K., NORITO, K., WATANABE, Y., TAKAMIYA, K., OKURA, A., FUNAISHI, K., KAWAMURA, K., FUNAYAMA, S., UMEZAWA, I. (1987), Antitumor effect of kazusamycin B on experimental tumors, *J. Antibiot.* **40**, 1596–1604.

YOSHIDA, S., NAGAWARA, H., AOYAGI, T., TAKEUCHI, T., UMEZAWA, A. (1988), Benadrostin, new inhibitor of poly(ADP-ribose) synthetase, produced by actinomycetes. II. Structure elucidation, *J. Antibiot.* **41**, 1015–1018.

YOSHIDA, M., HOSHIKAWA, Y., KOSEKI, K., HORI, K., BEPPU, T. (1990), Structural specificity for biological activity of trichostatin A, a specific inhibitor of mammalian cell cycle with potent differentiation-inducing activity in Friend leukemia cells, *J. Antibiot.* **43**, 1101–1106.

YOSHIMOTO, A., JOHDO, O., WATANABE, Y., NISHIDA, H., OKAMOTO, R. (1992), Production of a new anthracycline antibiotic betaclamycin B by microbial conversion with a specific aclacinomycin-negative mutant, *J. antibiot.* **45**, 1005–1007.

YUAN, B. D., WU, R. T., SATO, I., OKABE, T., SUZUKI, H., NISHIMURA, T., TANAKA, N. (1985), Biological activity of cadeguomycin. Inhibition of tumor growth and metastasis. Immunostimulation, and potentiation of 1-β-D-arabino-furanosylcytosine, *J. Antibiot.* **38**, 642–648.

ZHOU, N., JAMES, T. L., SHAFER, R. H. (1989), Binding of actinomycin D to $(d(ATCGAT))_2$: NMR evidence of multiple complexes, *Biochemistry* **28**, 5231–5239.

ZIMMER, C., STÖRL, K., STÖRL, J. (1990), Microbial DNA topoisomerases and their inhibition by antibiotics, *J. Basic Microbiol.* **30**, 209–224.

ZUNINO, F., CAPRANICO, G. (1990), DNA topoisomerase II as the primary target of antitumor anthracyclines, *Anti-Cancer Drug Res.* **5**, 307–317.

Index

C

G

H

I

K

T